原子光谱分析前沿

侯贤灯　王秋泉　史建波　吕　弋　江桂斌　等　编著

科学出版社
北　京

内 容 简 介

原子光谱分析是痕量元素及其形态分析的最常用和最准确的方法，几十年来其商品化仪器更是蓬勃发展，在各行各业获得广泛应用。本书邀请涉及原子光谱分析相关研究的课题组参与撰写了相关研究进展，主要包括原子吸收光谱分析、原子发射光谱分析、原子荧光光谱分析、（多接收）电感耦合等离子体质谱分析、激光电离质谱分析等。本书共三部分：方法基础与仪器装置、进样与联用技术、分析应用。内容涵盖从样品前处理到进样技术、仪器部件和仪器、机理研究、组学分析、同位素分析、环境分析、生物分析、材料分析、食品分析和地质样品分析，以及相关标准与标准物质等。

本书适合从事原子光谱分析和无机质谱分析相关研究与应用的科技工作者，特别是相关专业师生、研究人员、仪器研发和分析测试技术员等阅读与参考。

图书在版编目（CIP）数据

原子光谱分析前沿 / 侯贤灯等编著. —北京：科学出版社，2022.9
ISBN 978-7-03-072928-6

Ⅰ. ①原⋯ Ⅱ. ①侯⋯ Ⅲ. ①原子光谱-光谱分析-研究 Ⅳ. ①O657.31

中国版本图书馆 CIP 数据核字（2022）第 152620 号

责任编辑：杨新改／责任校对：杜子昂
责任印制：吴兆东／封面设计：东方人华

科 学 出 版 社 出版
北京东黄城根北街 16 号
邮政编码：100717
http://www.sciencep.com

北京建宏印刷有限公司 印刷
科学出版社发行　各地新华书店经销
*

2022 年 9 月第 一 版　开本：889×1194　1/16
2023 年 1 月第二次印刷　印张：47
字数：1 300 000

定价：298.00 元

（如有印装质量问题，我社负责调换）

前　言

原子光谱分析法包括原子吸收光谱分析法、原子发射光谱分析法、原子荧光光谱分析法和无机质谱分析法，至今仍然是痕量元素分析最常用和最准确的方法，几十年来其商品仪器更是蓬勃发展，在各行各业获得广泛应用。

20世纪50年代出现的原子吸收光谱分析法特别是其后的石墨炉原子化（现在统称为电热原子化）技术，使原子吸收光谱定量分析研究活跃了近40年；60年代，电感耦合等离子体原子发射光谱分析法的出现，特别是80年代电感耦合等离子体质谱分析法的提出，由于多元素同时检测与高灵敏度的显著优势，原子吸收光谱分析的应用领域逐渐被电感耦合等离子体技术占领；但同时期，与原子吸收光谱分析法相关联的氢化物发生原子荧光光谱分析法以其独特的性能和廉价的优势在国内获得长足发展；虽然可氢化物发生的元素数目有限，但这些元素是地质、环境和食品分析中经常需要测定的元素，氢化物发生原子荧光光谱分析法分析这些元素具有独特的优势，因此得以迅猛发展，至今生机勃勃。如今，各种原子光谱分析技术在元素定性定量分析、形态分析、成像分析等方面都获得了广泛应用，特别是激光光源的应用、同位素标记和溯源等技术使原子光谱分析法在食品分析、环境监测、生物分析、组学研究等应用领域得到不断拓宽。

半个世纪以来，我国原子光谱分析工作者不断努力，不仅使原子光谱分析和原子质谱技术在国内取得巨大成功，助力我国工农业建设和经济发展，还将继续为改善环境、确保食品安全发挥不可替代的作用。在原子光谱分析科研工作方面，我国科学家在国际舞台发挥着越来越重要的作用。

为系统总结我国原子光谱分析相关领域近几年来的研究进展，也便于国内同行交流，经"第六届全国原子光谱及相关技术学术会议"组委会讨论，认为编辑出版《原子光谱分析前沿》有利于促进和推动我国原子光谱分析的发展。本书集结了众多对原子光谱分析研究和应用最有热情的杰出同行参与撰写，内容涵盖从样品前处理到进样技术、仪器部件和仪器、机理研究、组学分析、同位素分析、环境分析、生物分析、食品分析和地质样品分析等。激光、纳米材料、微流控芯片、元素标记、微等离子体等技术的引入，激发了原子光谱分析研究新动力。

感谢参与撰写的各位同行专家、老师和研究生，也感谢科学出版社提供的出版机会以及编辑付出的辛勤劳动。

值此全国原子光谱及相关技术学术会议十周年之际，出版此书具有特别的意义，在此祝愿原子光谱分析科学工作者再接再厉，再立新功！我坚信，几年之后，将有更多的科学家参与撰写本书的续篇。

2020年12月

目　　录

前言
第 0 章　导论 ·· 1
　　参考文献 ·· 5

第一部分　方法基础与仪器装置

第 1 章　化学改进剂与原子化机理研究进展 ·· 11
　1.1　化学改进剂 ··· 12
　1.2　氢化物原位富集 ··· 15
　1.3　石墨管的持久化学改进技术 ··· 16
　1.4　电热原子化过程的动力学研究 ··· 17
　1.5　结束语 ··· 27
　　参考文献 ·· 27

第 2 章　激光电离质谱法用于固体样品元素直接分析 ···································· 34
　2.1　激光微探针飞行时间质谱 ··· 36
　2.2　激光电离四极杆质谱 ··· 39
　2.3　激光电离双聚焦质谱 ··· 40
　2.4　激光微探针傅里叶变换质谱 ··· 42
　2.5　激光电离飞行时间质谱 ··· 44
　2.6　总结与展望 ··· 54
　　参考文献 ·· 54

第 3 章　MC-ICP-MS 同位素比值的测定 ··· 61
　3.1　引言 ··· 62
　3.2　MC-ICP-MS 的非质量分馏 ··· 63
　3.3　同位素分馏校正模型 ··· 67
　3.4　总结与展望 ··· 73
　　参考文献 ·· 73

第 4 章　元素介导的电感耦合等离子体质谱定量生物分析 ···························· 85
　4.1　引言 ··· 86

4.2 内源性报告元素	87
4.3 外源性报告元素	90
4.4 基于 ICP-MS 的单细胞分析技术	105
4.5 展望	106
参考文献	107

第5章 基于介质阻挡放电微等离子体的原子光谱分析 · 121

5.1 DBD 低温原子化器	123
5.2 DBD 等离子体激发源	127
5.3 DBD 等离子体蒸气发生	134
5.4 其他 DBD 元素分析技术	142
5.5 总结与展望	143
参考文献	143

第6章 常压辉光放电等离子体发射光谱分析 · 150

6.1 常压辉光放电等离子体简介	151
6.2 常压辉光放电应用于发射光谱辐射源	153
6.3 液体电极-金属电极辉光放电诱导的蒸气发生技术	164
6.4 常压辉光放电等离子体的机理研究进程	165
参考文献	168

第7章 现场环境元素分析 · 176

7.1 概述	177
7.2 现场环境元素分析的样品预处理	178
7.3 现场环境元素分析的进样方法	179
7.4 现场环境分析方法与装置	185
7.5 展望	186
参考文献	187

第二部分 进样与联用技术

第8章 化学蒸气发生法 · 197

8.1 化学蒸气发生法概述	198
8.2 氢化物发生	198
8.3 光化学蒸气发生	201
8.4 其他化学蒸气发生法	206
8.5 展望	208
参考文献	209

第9章 微等离子体原子光谱仪进样技术 ... 218
- 9.1 概述 ... 219
- 9.2 原子发射光谱中的微等离子体激发源 ... 220
- 9.3 原子吸收/原子荧光光谱中的微等离子体原子化器与捕集器 ... 232
- 9.4 质谱中的微等离子体离子源 ... 234
- 9.5 微等离子体诱导化学反应 ... 236
- 9.6 总结与展望 ... 239
- 参考文献 ... 239

第10章 单纳米颗粒ICP-MS分析技术 ... 251
- 10.1 单纳米颗粒与ICP-MS分析 ... 252
- 10.2 单纳米颗粒ICP-MS在环境分析中的应用 ... 255
- 10.3 单纳米颗粒ICP-MS在生物分析中的应用 ... 257
- 10.4 单纳米颗粒ICP-MS分析技术的前景与展望 ... 261
- 10.5 总结 ... 264
- 参考文献 ... 265

第11章 微流控芯片–等离子体质谱联用技术用于细胞中痕量元素及其形态分析 ... 270
- 11.1 引言 ... 271
- 11.2 基于微流控芯片的微萃取技术 ... 272
- 11.3 基于等离子体质谱的单细胞分析 ... 282
- 11.4 总结与展望 ... 294
- 参考文献 ... 294

第12章 激光剥蚀等离子体质谱分析 ... 301
- 12.1 引言 ... 302
- 12.2 激光剥蚀元素分馏机理及其在元素分析中的应用进展 ... 303
- 12.3 激光在副矿物U-Th-Pb定年中的应用 ... 304
- 12.4 激光在放射性同位素分析中的应用 ... 305
- 12.5 激光在稳定同位素分析中的应用 ... 309
- 12.6 激光在单个流体包裹体分析中的应用 ... 311
- 12.7 激光分析样品前处理技术 ... 313
- 12.8 激光剥蚀液体样品进样技术 ... 316
- 12.9 LA-ICP-MS专业数据处理软件 ... 317
- 12.10 展望 ... 319
- 参考文献 ... 319

第13章 激光剥蚀等离子体质谱法深度剖析技术 ... 335
- 13.1 引言 ... 336
- 13.2 深度剖析技术的优势与局限 ... 336

13.3　LA-ICP-MS 深度剖析技术原理 337
13.4　深度剖析性能的影响因素 340
13.5　LA-ICP-MS 深度剖析的应用 351
13.6　展望 354
参考文献 354

第 14 章　色谱–原子光谱/质谱联用技术 358
14.1　引言 359
14.2　气相色谱–原子光谱/质谱联用技术 359
14.3　液相色谱–原子光谱/质谱联用技术 362
14.4　毛细管电泳–电感耦合等离子体质谱联用技术 372
14.5　展望 379
参考文献 379

第 15 章　原子光谱分析中样品制备技术 402
15.1　固体分析法 404
15.2　试液制备法 406
15.3　相分离法 409
15.4　场辅助样品制备法 419
15.5　样品制备/原子光谱分析在线联用技术 426
15.6　总结与展望 430
参考文献 431

第三部分　分析应用

第 16 章　金属组学发展现状 467
16.1　金属组学研究内容及技术 468
16.2　金属蛋白质组学研究 475
16.3　金属组学在金属药物研究中的应用 479
16.4　展望 482
参考文献 482

第 17 章　金属组学：高通量分析技术 491
17.1　金属组学简介 492
17.2　金属组学高通量分析方法 494
17.3　总结与展望 499
参考文献 499

第 18 章　砷的环境行为、代谢机理、健康效应 507
18.1　砷形态分析方法简介 509

18.2 砷形态代谢研究：以洛克沙胂为例 ………………………………………………… 511
18.3 砷的主要环境暴露 ……………………………………………………………… 517
18.4 砷的健康效应研究：以砷与蛋白质的结合为例 ………………………………… 522
18.5 总结与展望 ……………………………………………………………………… 529
参考文献 …………………………………………………………………………………… 530

第19章 原子光谱在汞研究中的应用 …………………………………………………… 544
19.1 原子光谱法是汞分析的重要手段 ……………………………………………… 545
19.2 基于原子光谱的汞分析方法 …………………………………………………… 545
19.3 原子光谱在汞研究中的应用进展 ……………………………………………… 551
参考文献 …………………………………………………………………………………… 557

第20章 MC-ICP-MS 在生物和环境同位素分析中的应用 ……………………………… 568
20.1 引言 …………………………………………………………………………… 569
20.2 MC-ICP-MS 的应用原理 ……………………………………………………… 569
20.3 MC-ICP-MS 分析方法最新进展 ……………………………………………… 571
20.4 MC-ICP-MS 在环境同位素分析中的应用 …………………………………… 574
20.5 MC-ICP-MS 在生物同位素分析中的应用 …………………………………… 578
20.6 展望 …………………………………………………………………………… 579
参考文献 …………………………………………………………………………………… 580

第21章 LA-ICP-MS 的生物元素成像应用 ……………………………………………… 588
21.1 概述 …………………………………………………………………………… 589
21.2 重金属原位成像 ………………………………………………………………… 589
21.3 金属类药物原位成像 …………………………………………………………… 592
21.4 疾病诊断 ………………………………………………………………………… 596
21.5 蛋白识别 ………………………………………………………………………… 600
21.6 细胞检测 ………………………………………………………………………… 605
21.7 纳米材料原位成像 ……………………………………………………………… 607
21.8 展望 …………………………………………………………………………… 610
参考文献 …………………………………………………………………………………… 610

第22章 原子光谱技术在纳米材料分析中的应用 ……………………………………… 621
22.1 不同介质中纳米材料的识别和定量 …………………………………………… 622
22.2 尺寸表征 ………………………………………………………………………… 628
22.3 表面性质分析 …………………………………………………………………… 632
22.4 生物体内成像 …………………………………………………………………… 633
22.5 形态分析 ………………………………………………………………………… 634
22.6 展望 …………………………………………………………………………… 638
参考文献 …………………………………………………………………………………… 639

第 23 章　食品分析、卫生检验与临床检验 · 644
23.1　食品分析研究进展 · 645
23.2　临床样品分析研究进展 · 660
23.3　展望 · 669
参考文献 · 669

第 24 章　地质样品中微量及痕量元素分析 · 679
24.1　微量元素分析 · 680
24.2　稀土元素分析 · 683
24.3　稀有金属元素分析 · 684
24.4　稀散金属元素分析 · 684
24.5　贵金属元素分析 · 685
24.6　卤族元素分析 · 686
24.7　展望 · 688
参考文献 · 688

第 25 章　标准与标准物质 · 697
25.1　概述 · 698
25.2　标准物质技术发展 · 702
25.3　标准物质在典型领域的应用 · 731
25.4　展望 · 736
参考文献 · 737

第 0 章 导 论

（侯贤灯[1]，江桂斌[2]）

---

[1] 四川大学，成都；[2] 中国科学院环境生态研究中心，北京

传统原子光谱分析法包括原子吸收光谱分析法、原子发射光谱分析法、原子荧光光谱分析法，至今仍然是痕量元素分析最常用和最准确的方法，在众多领域获得广泛应用。如今，基于电感耦合等离子体的质谱分析法以及基于激光和微等离子体的各种原子光谱分析的仪器和方法研究相当活跃，各种联用技术、新型样品处理和进样方法层出不穷。我国科技工作者在原子光谱分析和无机质谱分析技术开发、仪器（部件）研制、分析方法和应用研究等方面都做了重要的贡献，取得了显著的科研成果，已有若干文章专门介绍了中国原子光谱分析研究的情况[1-4]。我国改革开放后，始于1985年的北京分析测试学术报告会暨展览会（Beijing Conference and Exhibition on Instrumental Analysis，BCEIA），一开始就是分析仪器与仪器分析的国际学术会议，让中国光谱分析和其他分析科学与技术一起走向世界舞台[5]；进入新世纪后，2007年在厦门召开了第35届国际光谱会议（The 35th Colloquium Spectroscopicum Internationale，CSI）[6]，2010年在成都召开了第四届亚太地区等离子体光谱化学冬季会议（The 4th Asia-Pacific Winter Conference on Plasma Spectrochemistry，2010 APWC）[7]，同期在成都召开了第一届全国原子光谱及相关技术学术会议，中国的原子光谱分析研究更加受到国际同行关注。图0-1和图0-2是我们统计的近十年来中国在分析化学综合性期刊（以美国化学会的 *Analytical Chemistry* 为例）和原子光谱专业性期刊（以英国皇家化学会的 *Journal of Analytical Atomic Spectrometry* 为例）发表原子光谱分析和无机质谱分析论文的情况。可以看出，过去十年来，中国科技工作者发表的原子光谱分析和无机质谱分析的论文在综合性期刊中，从最初的四分之一达到了最近两年的半壁江山（图0-1）；在原子光谱专业性期刊中，也呈现了从不到五分之一逐步迈向三分之一的趋势。过去十年的汇总数据表明，在 *Analytical Chemistry* 发表的原子光谱/质谱分析论文中，最热门的等离子体质谱[特别是电感耦合等离子体质谱分析法（inductively coupled plasma mass spectrometry，ICP-MS）]占三分之二，其中中国贡献了等离子体质谱分析类论文的三分之一；而原子荧光分析，仅占不到十分之一，但其中中国贡献了原子荧光分析论文的80%，这主要得益于中国有自主知识产权的氢化物发生-原子荧光光谱仪（hydride generation-atomic fluorescence spectrometer，HG-AFS）的蓬勃发展及其在地矿、环境和食品分析中的广泛应用；此外，值得一提的是，中国在原子发射光谱分析领域所占份额也超过40%。在 *Journal of Analytical Atomic Spectrometry* 发表的论文中（图0-2），等离子体质谱占60%，其中中国贡献了20%；原子荧光分析仅占4%，但其中中国差不多贡献了一半。以上分析表明，中国在原子光谱/无机质谱分析研究领域的发展势头看好，在各个方面都做出了卓越的贡献。

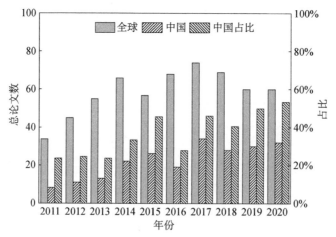

图0-1　在2011～2020年期间，中国在美国化学会期刊 *Analytical Chemistry* 上发表论文情况

数据来源：Web of Science

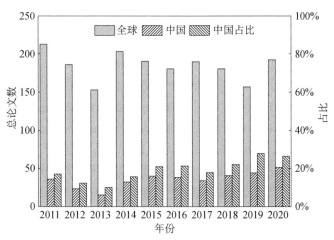

图 0-2 在 2011～2020 年期间，中国在英国皇家化学会期刊
Journal of Analytical Atomic Spectrometry 上发表论文情况

数据来源：Web of Science

除了等离子体质谱分析外，原子发射光谱的研究仍然相当活跃，特别是电感耦合等离子体发射光谱分析法（inductively coupled plasma optical emission spectrometry，ICP-OES）以及基于新型微等离子体的原子发射光谱分析法。而原子吸收光谱分析的研究发展相对平稳，占 10%左右，主要原因是 ICP-MS 和 ICP-OES 的高灵敏多元素分析能力的冲击以及原子吸收光谱分析技术相对成熟度更高。连续光源原子吸收光谱仪的出现实现了多元素同时测定，还可以测定没有相应空心阴极灯的元素，但并没有得到普遍应用。应该看到，原子吸收光谱技术依然是单一元素测定及开展各种研究稳定、准确、可靠的技术。激光激发原子荧光光谱分析法虽然灵敏度高，但近年来少有相关报道，可能的原因包括可波长调谐的激光器昂贵以及 HG-AFS 的成功。在现代原子光谱分析研究中，激光更多地被应用于激光蒸发或称之为激光剥蚀（laser vaporization 或 laser ablation，LA）的进样技术中；此外，很大程度上得益于许多现场分析对固体直接进样与仪器小型化的需求，激光诱导击穿光谱分析法（laser-induced breakdown spectroscopy，LIBS）的研究近年来有所发展，LIBS 仪器也在众多领域逐渐获得应用[8-10]。

原子光谱分析在元素定性定量分析、形态分析、成像分析等方面都获得了广泛应用，特别是激光光源的应用、各种组学的发展、同位素标记和溯源等技术使原子光谱与质谱分析在食品分析、环境监测、地质样品分析、生物分析、卫生检验与临床分析等应用领域得到不断拓宽。激光、纳米材料、微流控芯片、元素标记、各种微等离子体等技术的引入，激发了原子光谱分析研究新动力。除了无机元素分析领域，近年来，其在有机分析特别是生物分析方面的应用已显示出一些独特的优越性[11,12]。

本书分为三大部分：方法基础与仪器装置；进样与联用技术；分析应用。

在方法基础与仪器装置研究方面，我国学者首创钯化学改进剂（即原来的基体改进剂，现统称化学改进剂）[13]，为消除电热原子吸收光谱分析中的基体效应提供了通用的化学改进剂并建立了相应的原子化模型，获得国际同行的高度认可，在实践中得到了广泛应用；激光电离质谱法可直接分析固体样品，实现无标样定量分析[14]；在减少和消除多接收电感耦合等离子体质谱分析（multi-collector inductively coupled plasma mass spectrometry，MC-ICP-MS）中质量分馏对绝对同位素比值测量的影响方面，近年来发展的质量偏倚校正模型，提高了绝对同位素比值测量的精准度[15]，并已有测量结果获得 IUPAC 认可；利用生物分子的内源性元素和外源性元素的选择性标记技术，以元素为桥梁并实现信号放大，将 ICP-MS 用于定量生物分析方面，获得重要成果[16]；近年来，微等离子体在原子光谱/原子质谱分析中的应用方兴未艾，特别是介质阻挡放电（dielectric barrier discharge，

DBD）[17]等离子体和辉光放电（glow discharge，GD）[18]等离子体在原子光谱分析中的应用；常压微等离子体和电热钨丝原子化器的应用，结合基于微型电荷耦合器件的微型光谱仪，促进了小型化甚至便携式原子光谱仪的研制，特别适合于现场环境元素分析应用[19,20]。

联用技术研究方面的发展体现了解决复杂科学问题的需求。各种原子光谱/质谱检测技术可以与气相色谱（gas chromatography，GC）、高效液相色谱（high performance liquid chromatography，HPLC）、离子色谱（ion chromatography，IC）和毛细管电泳（capillary electrophoresis，CE）联用，金属组学的发展正是得益于原子光谱/质谱的高灵敏（通常也具有高分辨）检测与色谱的高分辨分离的联姻。在色谱-原子光谱/质谱联用技术接口的发展和多维色谱联用技术及其应用方面，CE 和 ICP-MS 联用的接口、雾室和雾化器的改进方面有了新的发展[21]；微流控芯片与 ICP-MS 的联用颇具特色，微流控芯片在微萃取及其在（单）细胞中痕量元素/形态分析中具有新的应用[22]；基于单纳米粒子的 ICP-MS 技术，提高了检测灵敏度，拓展了 ICP-MS 在生物、环境和临床分析中的应用，并有可能在单细胞和单分子分析中获得应用[23,24]；进样技术被称之为原子光谱分析的阿喀琉斯之踵（Achilles heel），当然也包括样品制备技术[25]，化学蒸气发生法（chemical vapor generation，CVG；包括氢化物发生、光化学蒸气发生及其他化学蒸气发生法），目前研究的热门是光化学蒸气发生，既适用于传统原子光谱/质谱仪的进样，也适用于微等离子体原子光谱仪的进样[26,27]。激光剥蚀最常用于 ICP-MS 的进样，广泛地应用于地质样品的同位素分析中[28]，以及多种材料深度分析[29]；在实际样品分析中，样品前处理往往是分析程序中的第一步，其组分的有效性和转移的准确定量性至关重要，本书中也对该专题进行了全面的介绍[30]。

原子光谱/质谱在生物分析研究中的应用最为活跃，最有代表性的体现是其奠定了金属组学的技术基础并推动了该学科的发展。本书综合讨论了高通量分析技术和金属组学发展现状[31-33]。金属组学的研究技术被国内外团队应用于砷的环境行为、代谢机理和健康效应的研究，取得了新的认识，对减少人体暴露于有毒的含砷化合物、促进和保护公共健康，具有重要意义[34-37]；关于汞及其形态和同位素分析，研究对象包括大气系统、水生生态系统（含稻田），从生物地球化学循环的角度审视和探讨含汞物种的迁移、积累与演化[38-40]；利用 MC-ICP-MS 进行了生物和环境同位素分析，发展了重金属示踪、大气细颗粒物示踪和纳米颗粒物示踪新技术[41,42]；结合 LA 采样，利用 ICP-MS 生物元素成像技术实现了生物组织和细胞内元素的原位成像分析，该技术可望用于元素代谢分布特征研究和生物分子作用机理的解析[43,44]；纳米科学的相关研究工作一直是热点，本书也专题总结了原子光谱技术在各种纳米材料分析中的应用，新的技术包括纳米材料尺寸的原子光谱法表征等[45-47]；食品分析、卫生检验和临床分析是直接与人民健康密切相关的大领域，也是原子光谱/质谱分析的大舞台，通过本书可以了解国内外相关应用进展[48,49]；20 世纪中叶，国内原子光谱分析技术的先行发展及其国产仪器的开发在很大程度上受益于地质找矿和冶金分析的需求驱动，如今，原子光谱分析特别是 ICP-MS 分析技术在地质样品分析中仍然是主力军，在地质样品的微量及痕量元素（同位素）分析中发挥了不可替代作用[50]。众所周知，在使用各种分析技术分析实际样品时，一般都需要标准和标准物质以确保分析结果的可靠性，原子光谱分析也不例外，因此，标准和标准物质研制不可或缺[51]。

原子光谱分析及相关技术是科学研究的重要工具之一，同时也是环境、食品、国家安全等方面依赖的重要技术手段。撰写本书的主要目的，是希望涵盖与总结这一技术发展的历史与现状，从基础、仪器到应用进行其全景描述。但限于作者水平，所述内容方面难以避免遗漏乃至不恰当的描述，敬请读者批评指正。我国已经跨入原子光谱和质谱分析研究与应用的国际前列，期待通过不断的总结凝练，活跃学术交流，持续推动其方法、技术和仪器（部件）方面的创新，以满足不断涌现的新型应用需求。这本《原子光谱分析前沿》只是开篇，让我们一起努力，期待续篇。

参 考 文 献

[1] Hou X D. Analytical atomic spectrometry: An active research area in China. Journal of Analytical Atomic Spectrometry, 2010, 25: 447-452.

[2] Gao S. The rise of atomic spectrometry in China over the past 25 years. Journal of Analytical Atomic Spectrometry, 2010, 25: 1803-1807.

[3] Hang W. Booming atomic spectrometry in China. Journal of Analytical Atomic Spectrometry, 2015, 30: 850-851.

[4] 杭乐, 徐周毅, 杭纬, 黄本立. 中国原子光谱技术及应用发展近况. 光谱学与光谱分析, 2019, 39: 1329-1339.

[5] 马锡冠. 前进中的北京分析测试学术报告会与展览会 (BCEIA). 现代科学仪器, 1997, 3: 3-4.

[6] 黄本立. 我国申办国际光谱会议成功——第 35 届国际光谱会议将于 2007 年在厦门召开. 光谱学与光谱分析, 2006, 26: 974.

[7] Hou X D. The Fourth Asia-Pacific Winter Conference on Plasma Spectrochemistry. Journal of Analytical Atomic Spectrometry, 2011, 26: 1113-1114.

[8] Zhang Y, Zhang T L, Li H. Application of laser-induced breakdown spectroscopy (LIBS) in environmental monitoring. Spectrochimica Acta Part B: Atomic Spectroscopy, 2021, 181: 106218.

[9] Senesi G S, Harmon R S, Hark R R. Field-portable and handheld laser-induced breakdown spectroscopy: Historical review, current status and future prospects. Spectrochimica Acta Part B: Atomic Spectroscopy, 2021, 175: 106013.

[10] Wang Q Q, Xiangli W T, Teng G, Cui X T, Wei K. A brief review of laser-induced breakdown spectroscopy for human and animal soft tissues: Pathological diagnosis and physiological detection. Applied Spectroscopy Reviews, 2021, 56: 221-241.

[11] Hu J, Yang P, Hou X D. Atomic spectrometry and atomic mass spectrometry in bioanalytical chemistry. Applied Spectroscopy Reviews, 2019, 54: 180-203.

[12] Chen B W, Hu L G, He B, Luan T G, Jiang G B. Environmetallomics: Systematically investigating metals in environmentally relevant media. Trends in Analytical Chemistry, 2020, 126: 115875.

[13] 单孝全, 倪哲明. 基体改进效应应用于石墨炉原子吸收测定汞. 化学学报, 1979, 37: 261-266.

[14] Yu Q, Huang R, Li L, Lin L, Hang W, He J, Huang B L. Applicability of standardless semiquantitative analysis of solids by high-irradiance laser ionization orthogonal time-of-fight mass spectrometry. Analytical Chemistry, 2009, 81(11): 4343-4348.

[15] Yang L, Tong S, Zhou L, Hu Z, Mester Z, Meija J. A critical review on isotopic fractionation correction methods for accurate isotope amount ratio measurements by MC-ICP-MS. Journal of Analytical Atomic Spectrometry, 2018, 33(11): 1849-1861.

[16] Liang Y, Yang L M, Wang Q Q. An ongoing path of element-labeling/tagging strategies toward quantitative bioanalysis using ICP-MS. Applied Spectroscopy Reviews, 2016, 51:117-128.

[17] Zhu Z L, Chan G C Y, Ray S J, Zhang X R, Hieftje G M. Microplasma source based on a dielectric barrier discharge for the determination of mercury by atomic emission spectrometry. Analytical Chemistry, 2008, 80(22): 8622-8627.

[18] Peng X X, Wang Z. Ultrasensitive determination of selenium and arsenic by modified helium atmospheric pressure glow discharge optical emission spectrometry coupled with hydride generation. Analytical Chemistry, 2019, 91:10073-10080.

[19] Jiang X M, Chen Y, Zheng C B, Hou X D. Electrothermal vaporization for universal liquid sample introduction to

dielectric barrier discharge microplasma for portable atomic emission spectrometry. Analytical Chemistry, 2014, 86: 5220-5224.

[20] 吴鹏, 温晓东, 吕弋, 侯贤灯. 钨丝在原子吸收光谱分析中的应用. 分析化学, 2006, 34: 278-282.

[21] Liu L, Yun Z, He B, Jiang G B. Efficient interface for online coupling of capillary electrophoresis with inductively coupled plasma-mass spectrometry and its application in simultaneous speciation analysis of arsenic and selenium. Analytical Chemistry, 2014, 86(16):8167-8175.

[22] He M, Chen B B, Wang H, Hu B. Microfluidic chip-inductively coupled plasma mass spectrometry for trace elements and their species analysis in cells. Applied Spectroscopy Reviews, 2019, 54 (3): 250-263.

[23] Hu J Y, Deng D Y, Liu R, Lv Y. Single nanoparticle analysis by ICPMS: A potential tool for bioassay. Journal of Analytical Atomic Spectrometry, 2018, 33: 57-67.

[24] Hu S H, Liu R, Zhang S C, Huang Z, Xing Z, Zhang X R. A new strategy for highly sensitive immunoassay based on single-particle mode detection by inductively coupled plasma mass spectrometry. Journal of the American Society for Mass Spectrometry, 2009, 20: 1096-1103.

[25] Cai Y, Li S H, Dou S, Yu Y L, Wang J H. Metal carbonyl vapor generation coupled with dielectric barrier discharge to avoid plasma quench for optical emission spectrometry. Analytical Chemistry, 2015, 87(2): 1366-1372 .

[26] Zheng C B, Li Y, He Y H, Ma Q, Hou X D. Photo-induced chemical vapor generation with formic acid for ultrasensitive atomic fluorescence spectrometric determination of mercury: Potential application to mercury speciation in water. Journal of Analytical Atomic Spectrometry, 2005, 20 (8): 746-750.

[27] Hu J, Chen H J, Hou X D, Jiang X M. Cobalt and copper ions synergistically enhanced photochemical vapor generation of molybdenum: Mechanism study and analysis of water samples. Analytical Chemistry, 2019, 91 (9): 5938-5944.

[28] Hu Z C, Zhang W, Liu Y S, Gao S, Li M, Zong K Q, Chen H H, Hu S H. "Wave" signal-smoothing and mercury-removing device for laser ablation quadrupole and multiple collector ICPMS analysis: Application to lead isotope analysis. Analytical Chemistry, 2015, 87(2): 1152-1157.

[29] Luo T, Ni Q, Hu Z C, Zhang W, Shi Q H, Gunther D, Liu Y S, Zong K Q, Hu S H. Comparison of signal intensities and elemental fractionation in 257 nm femtosecond LA-ICP-MS using He and Ar as carrier gases. Journal of Analytical Atomic Spectrometry, 2017, 32(11): 2217-2225.

[30] Xia L, Yang J, Su R, Zhou W, Zhang Y, Zhong Y, Huang S, Chen Y, Li G. Recent progress in fast sample preparation techniques. Analytical Chemistry, 2020, 92 (1): 34-48.

[31] Chen C, Chai Z, Gao Y. Nuclear Analytical Techniques for Metallomics and Metalloproteomics. Cambridge: RSC Publishing, 2010.

[32] Chen C, Li Y-F, Qu Y, Chai Z, Zhao Y. Advanced nuclear analytical and related techniques for the growing challenges in nanotoxicology. Chemical Society Reviews, 2013, 42(21): 8266-8303.

[33] Li H, Wang R, Sun H. Systems approaches for unveiling the mechanism of action of bismuth drugs: New medicinal applications beyond *Helicobacter pylori* infection. Accounts of Chemical Research, 2019, 52:216-227.

[34] Liu Q Q, Lu X F, Peng H Y, Popowich A, Tao J, Uppal J S, Yan X W, Boe D, Le X C. Speciation of arsenic: A review of phenylarsenicals and related arsenic metabolites. TrAC Trends in Analytical Chemistry, 2018, 104: 171-182.

[35] Peng H Y, Hu B, Liu Q Q, Li J H, Li X F, Zhang H Q, Le X C. Methylated phenylarsenical metabolites discovered in chicken liver. Angewandte Chemie International Edition, 2017, 56: 6773-6777.

[36] Shen S, Li X F, Cullen W R, Weinfeld M, and Le X C. Arsenic binding to proteins. Chemical Reviews, 2013, 113: 7769-7792.

[37] Cui J L, Shi J B, Jiang G B, Jing C Y. Arsenic levels and speciation from ingestion exposures to biomarkers in ShanXi,

China: Implications for human health. Environmental Science & Technology, 2013, 47: 5419-5424.

[38] Li P, Du B Y, Maurice L, Laffont L, Lagane C, Point D, Sonke J E, Yin R S, Lin C J, Feng X B. Mercury isotope signatures of methylmercury in rice samples from the Wanshan mercury mining area, China: Environmental implications. Environmental Science & Technology, 2017, 51(21): 12321-12328.

[39] Li L, Wang F, Meng B, Lemes M, Feng X, Jiang G B. Speciation of methylmercury in rice grown from a mercury mining area. Environmental Pollution, 2010, 158(10): 3103-3107.

[40] Gao E, Jiang G, He B, Yin Y, Shi J. Speciation of mercury in coal using HPLC-CV-AFS system: Comparison of different extraction methods. Journal of Analytical Atomic Spectrometry, 2008, 23(10): 1397-1400.

[41] Lu D, Liu Q, Zhang T, Cai Y, Yin Y, Jiang G. Stable silver isotope fractionation in the natural transformation process of silver nanoparticles. Nature Nanotechnology, 2016, 11(8): 682-686.

[42] Lu D, Luo Q, Chen R, Zhuansun Y, Jiang J, Wang W, Yang X, Zhang L, Liu X, Li F, Liu Q, Jiang G B. Chemical multi-fingerprinting of exogenous ultrafine particles in human serum and pleural effusion. Nature Communications, 2020, 11(1): 2567.

[43] Xu M, Bijoux H, Gonzalez P, Mounicou S. Investigating the response of cuproproteins from oysters (*Crassostrea gigas*) after waterborne copper exposure by metallomic and proteomic approaches. Metallomics, 2014, 6(2):338-346.

[44] Xu M, Frelon S, Simon O, Lobinski R, Mounicou S. Development of a non-denaturing 2D gel electrophoresis protocol for screening *in vivo* uranium-protein targets in *Procambarus clarkii* with laser ablation ICP MS followed by protein identification by HPLC-Orbitrap MS. Talanta, 2014, 128: 187-195.

[45] Liu J F, Chao J B, Liu R, Tan Z Q, Yin Y G, Wu Y, Jiang G B. Cloud point extraction as an advantageous preconcentration approach for analysis of trace silver nanoparticles in environmental waters. Analytical Chemistry, 2009, 81(15): 6496-6502.

[46] Tan Z Q, Liu J F, Guo X R, Yin Y G, Byeon S K, Moon M H, Jiang G B. Toward full spectrum speciation of silver nanoparticles and ionic silver by on-line coupling of hollow fiber flow field-flow fractionation and minicolumn concentration with multiple detectors. Analytical Chemistry, 2015, 87 (16): 8441-8447.

[47] Tan Z Q, Yin Y G, Guo X R, Amde M, Moon M H, Liu J F, Jiang G B. Tracking the transformation of nanoparticulate and ionic silver at environmentally relevant concentration levels by hollow fiber flow field-flow fractionation coupled to ICPMS. Environmental Science & Technology, 2017, 51 (21): 12369-12376.

[48] Yang G, Zheng J, Chen L, Lin Q, Zhao Y, Wu Y, Fu F. Speciation analysis and characterisation of arsenic in lavers collected from coastal waters of Fujian, south-eastern China. Food Chemistry, 2012, 132: 1480-1485.

[49] Liu L, Yun Z, He B, Jiang G B. Efficient interface for online coupling of capillary electrophoresis with inductively coupled plasma-mass spectrometry and its application in simultaneous speciation analysis of arsenic and selenium. Analytical Chemistry, 2014, 86: 8167-8175.

[50] Qi L, Gao J, Huang X, Hu J, Zhou M F, Zhong H. An improved digestion technique for determination of platinum group elements in geological samples. Journal of Analytical Atomic Spectrometry, 2011, 26(9): 1900-1904.

[51] Josephs R D, Daireaux A, Li M, Choteau T, Martos G, Westwood S, Wielgosz R I, Li H, Wang S, Feng L, Huang T, Pan M, Zhang T, Gonzalez-Rojano N, Balderas-Escamilla M, Perez-Castorena A, Perez-Urquiza M, Ün I, Bilsel M. Pilot study on peptide purity-synthetic oxytocin. Metrologia, 2020, 57: Number 1A.

第一部分　方法基础与仪器装置

第1章　化学改进剂与原子化机理研究进展

(吴　鹏[①]，严秀平[②*])

- ▶ 1.1　化学改进剂 / 12
- ▶ 1.2　氢化物原位富集 / 15
- ▶ 1.3　石墨管的持久化学改进技术 / 16
- ▶ 1.4　电热原子化过程的动力学研究 / 17
- ▶ 1.5　结束语 / 27

①四川大学，成都；②江南大学，无锡。*通讯作者联系方式：xpyan@jiangnan.edu.cn

本章导读

- 我国学者首创钯化学改进剂，为消除电热原子吸收光谱分析中的基体干扰提供了通用的方法。
- 我国学者提出测定电热原子化过程动力参数的通用方法，为从正常分析条件下单一原子吸收信号获取原子化过程的动力学参数提供了定量的方法。
- 我国学者提出的钯化学改进剂和原子化模型，得到了国际同行的公认和广泛应用。

石墨炉原子吸收（也称电热原子吸收）光谱技术具有试样消耗量低（μL）和灵敏度高等突出优势，是痕量和超痕量元素分析的首选方法之一。但由于石墨管体系比较封闭、体积小且温度不均匀，容易导致严重的背景信号和基体干扰。因此，石墨炉原子吸收技术中的基体干扰消除及相关机理的研究，一直是原子光谱学界研究的重点。本章从化学改进剂和原子化机理研究两个方面，介绍这些领域的研究进展。主要通过介绍倪哲明和林铁铮先生等领导的实验室在这些领域的研究工作，以便读者了解我国学者在这些领域所做的主要贡献。

1.1 化学改进剂

化学改进效应（chemical modification effect）是指在试样中加入某种化学试剂，使样品基体转化为易挥发的化学形态，或将被测元素转化为更加稳定的化学形态，以便在灰化热解阶段使用更高的温度驱尽基体而又不损失待测元素的现象[1]。所加入的化学试剂称为化学改进剂（chemical modifier）[1]。自1972年Ediger等[2]在原子吸收光谱分析中使用硝酸铵作为化学改进剂消除氯化钠对铜和镉测定的干扰以来，化学改进技术已获得非常广泛的应用。化学改进剂最初是以"基体改进剂"（matrix modifier）一词问世的，随着化学改进技术的发展，化学改进剂显示了对原子化过程中所有组分（分析物、基体、原子化器表面和气相、原子化过程等）多方面的效应，其含义不断得到丰富。国际纯粹与应用化学联合会建议将化学改进剂定义为：为了以所希望的方式影响在原子化器内发生的过程而加入的试剂[3]。

1979年，倪哲明先生及其同事在用冷原子吸收法测定汞时，发现仅有1 μg钯等贵金属就严重抑制了汞蒸气的逸出，因而提出了痕量钯作为汞在石墨炉中的稳定剂以避免汞的损失，使石墨炉原子吸收光谱直接测定挥发性汞成为可能。他们使用钯为化学改进剂稳定汞，直接测定了废水中的汞，允许灰化温度高达500℃（图1-1），检测限达到0.2 ng，回收率80%~110%[4]。这是首次在国际上提出使用钯作为化学改进剂，比挪威奥斯陆大学Weibust等[5]的报道还要早两年。钯化学改进剂的出现，为一直被认为难以用石墨炉原子吸收光谱法测定的一些易挥发元素如砷、硒、汞、铅和铊等提供了可靠的测定途径，已成为应用广泛的通用化学改进剂。表1-1主要概括了倪哲明和单孝全先生等的相关工作。

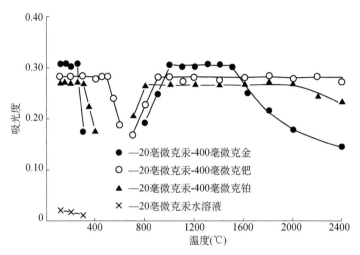

图 1-1　汞的灰化和原子化温度曲线[4]

表 1-1　钯化学改进剂应用于挥发性元素的石墨炉原子吸收测定

改进剂	测定元素	样品基体	灰化温度（℃）	文献
Pd	Hg	废水	500	[4]
Pd	Pb	海水	1200	[6]
Pd	Bi	废水、海水、尿样	1200	[7]
Pd 或 Ir	Pb	血样	1200/900	[8]
Pd、Pt、Ir	Te	水样	1300	[9]
Pd、Pt、Ir	Te	土壤、粮食、蔬菜	1300	[10]
Pd	Pb	尿样	1200	[11]
Pd	As	土壤、煤灰、生物样品	1300	[12]
Pd	Tl	废水	1000	[13]
Pd	In	矿样、底泥、煤灰	1200/1000	[14]

化学改进剂由于减少/消除了基体干扰、背景吸收和分析物灰化损失，因此在提高分析灵敏度和准确度等方面起了重要作用[15]。从本质上讲，化学改进技术是一种在线化学处理技术，既保留了常规化学处理消除干扰和提高测定灵敏度的优点，又避免了常规化学处理费时费力、操作烦琐、易沾污和损失的缺点，已广泛应用于石墨炉原子吸收光谱分析中。在倪哲明先生提出的钯化学改进剂的基础上，科学家又发展了许多其他化学改进剂，但大多以铂族金属为基础[16,17]。尤其是 Perkin-Elmer 公司的 Welz 等发展的 "$Pd(NO_3)_2 + Mg(NO_3)_2$" 化学改进剂[18-22]，已列入诸多商品化石墨炉原子吸收仪器的操作手册中。Bermejo-Barrera 等[23-25]发现在悬浮进样石墨炉原子吸收测定中，"$Pd(NO_3)_2+Mg(NO_3)_2$" 化学改进剂亦有很好的背景消除效果。此外，在石墨炉原子吸收中使用的一些化学改进剂，也同样在基于钨丝的电热原子吸收中具有广泛的应用，显著提高了一些易挥发元素（镉、铅和铜等）的测定效果[26-29]。

1.1.1　化学改进剂的种类和作用

根据化学分类、作用性质和作用对象三者综合考虑，可将现有化学改进剂分为如下几类。

1. 金属盐类化合物

这类化学改进剂主要是无机金属盐类，属于热稳定型化学改进剂，既可以与基体形成热稳定化合物，使之更好地与易挥发分析元素分离以减少干扰，又可以与分析元素形成热稳定化合物或合金以减少灰化损失，因此可通过提高灰化温度分离易挥发基体。同时，有些难熔、易形成热稳定碳化物的金属改进剂，通过形成碳化物石墨管涂层，改善了分析元素原子化条件，提高了分析灵敏度。钙、镁、镍、铜、钼、钨、锆和钽等盐类都得到了广泛应用。尤其是铂族金属系列，在倪哲明先生等倡导下，无论是单独的金属钯和铂还是其复合物等，作为多功能通用化学改进剂均成功地应用于许多领域，得到了国际同行的公认[16,17]。

2. 铵盐和无机酸

铵盐和无机酸化学改进剂的主要作用是使基体转化为易挥发铵盐或酸类，使之在灰化阶段除尽而与分析元素分离，从而达到减少/消除基体干扰目的。这类化学改进剂常用的有硝酸铵、磷酸铵、硝酸和盐酸等。有些盐或酸类的阴离子对分析元素有热稳定作用，如硫酸、硫化铵和磷酸等。

3. 有机酸

有机酸的化学改进作用在于其络合作用和热解产物的还原性，使分析元素形成易解离和具有挥发性的物质，以减少/消除基体的干扰。通常在无机试样中加入有机物既能改变试样表面物化性质，也能减少基体对分析元素的包藏。常用的有机酸包括抗坏血酸、草酸、柠檬酸、酒石酸和乙二胺四乙酸等。

4. 纳米材料

在铂族元素系列改进剂的基础上，国际同行进一步将钯/铂纳米粒子应用于化学改进剂。与传统的钯盐溶液相比，钯纳米粒子的改进效果更佳，易挥发元素的信号更强[30,31]。

5. 其他化学改进剂

氩气是石墨炉中常用的惰性保护气体。如果在氩气中混入一定比例的氢气，可增加还原气氛；混入氧气则有助于基体完全灰化。已有报道表明，活性气体介入灰化和原子化过程，并影响石墨炉中试样的原子化过程，如氧的存在会使一些元素出峰时间向高温区位移[32,33]，以减少基体干扰。

1.1.2 化学改进剂的作用机理

化学改进机理大致可分为化学机理、物理机理和电化学机理。在许多场合，化学机理与物理机理是同时存在的。例如，铂族金属化学改进剂在低温时主要是通过化学吸附使挥发性分析物变得稳定，而在较高温度的热解阶段主要是催化石墨还原分析物或催化分析物热分解生成分析元素单质，再与铂族金属形成相应的固溶体或化合物，然后在原子化阶段完全分解。其中既有物理作用又有化学作用，不能将物理作用与化学作用完全分开。

1. 化学机理

化学机理是指化学改进剂与基体或分析元素发生化学反应，改变基体或分析元素的化学形态，扩大基体与分析元素之间的差异，从而达到消除基体干扰的目的。如海水样品中加入硝酸铵，可使高熔点和高沸点的氯化钠转化为易分解挥发的硝酸钠和氯化铵，使之在分析元素原子化之前蒸发或挥发除去，从而消除氯化钠对海水中痕量元素测定时产生的严重背景吸收干扰[2]。产生的反应如下：

$$NaCl+NH_4NO_3 \longrightarrow NaNO_3+NH_4Cl$$

硝酸钙能显著提高铝、硼、铋、镉、锗和锡的灰化温度，降低原子化温度，增强抗干扰能力，提高灵敏度。其化学改进效应是在石墨管表面发生炭化反应，生成的 $CaC_{2(s)}$ 在固相还原 $Al_2O_{3(s)}$ 为 $Al_{(g)}$；$Ca_{(g)}$ 在气相中分别还原 BeO 和 GeO_2 为 $Be_{(g)}$ 和 $Ge_{(g)}$，钙在固相对锗没有增感作用；对于 SnO_2，既有在石墨管表面炭化反应生成的 $CaC_{2(s)}$ 在固相还原 SnO_2，又有 $Ca_{(g)}$ 在气相还原 SnO_2，导致原子化效率提高[34]。

由于分析物化学吸附在钯、铂、铑、钌和铱化学改进剂或石墨管表面，使挥发性分析物在低温条件下变得稳定。在较高温度的热解阶段，铂族金属改进剂催化石墨还原分析物或催化分析物热分解，单质态分析物与铂族金属形成相应的固溶体或化学计量化合物。在原子化阶段，分析物-铂族元素化合物完全分解。因此，从减少分析物损失的观点考虑，在热解阶段形成这些化合物是关键。然而，现有的理论还不能解释钯、铂、铑、钌和铱化学改进剂在效率上的显著差别[16]。

2. 物理机理

物理机理是指化学改进剂与基体或分析元素发生物理作用，形成固溶体或金属间化合物，降低熔点或沸点等，促使基体或分析元素提前或延迟蒸发或挥发。

杨芃原等[35]用扫描电子显微镜（SEM）、X 射线光电子能谱（XPS）和 X 射线衍射（XRD）等技术证实，在灰化阶段钯与分析物铅、锌和砷形成了金属间固溶物，分析物包含在钯的晶格内，直到石墨炉温度升到足以使晶格破裂再将分析物释放出来。钯在石墨炉内的存在状态对化学改进的效果有重要影响。经抗坏血酸处理后产生的还原态钯的粒度细且表面积大，增强了吸附和催化作用或改变了原子化历程，因而化学改进的效果更好，而氧化钯则没有增感效果[36]。梁春穗等[37]用分步加样法研究了钯的化学改进机理。钯和锗在灰化阶段在石墨管表面发生下列固相反应：

$$PdO_{(s)} + C_{(s)} = Pd_{(s)} + CO_{(g)}$$

$$Ge_xO_y + yCO_{(g)} + xPd_{(s)} = xGe\text{-}Pd_{(s,1)} + yCO_{2(g)}$$

这些固相反应生成热稳定物质，提高了灰化温度。这种稳定作用持续到原子化开始阶段。

有机酸或硝酸镁单独存在都不能阻止磷的损失，钯化学改进剂的效果取决于钯的量、钯的预处理及混合化学改进剂中第二组分的特性等因素。若钯在平台上的分布是不均匀的，沉积表面的有些部位没有钯改进剂，则不能对分析物起稳定作用，结果在不同的原子化阶段仍有磷损失。硝酸镁能促使钯分布在整个表面，分析物与钯改进剂能更有效地相互作用，因此钯-硝酸镁混合改进剂对磷的稳定效果最好[38]。

在进样之前先引入钯化学改进剂并进行热处理，再将 0.02～0.05 ng 样品单独或与钯一起蒸发，微量（0.1 μg）钯明显地改变了原子蒸气的蒸发模式和停留时间。由于蒸气被管壁吸附和在蒸气传输中分子扩散，钙、锰、铅和锡原子蒸气转移速度比根据扩散消失机理预期的速度慢 1.5～1.8 倍。将钯的量增加到 2.5 μg 时，由于扩散速度增大，导致停留时间和峰面积减少，伴随着温度升高。用双线法测温表明，蒸发 0.5～4 μg 钯，温度提高了 75～200 K，证实了钯蒸气与石墨相互作用是放热反应[39]。

3. 电化学机理

吴华等[40]用标准电极电位的观点解释了镁、镍和钯对测定铜的化学改进效应的差异。Mg^{2+}/Mg、Ni^{2+}/Ni、Cu^{2+}/Cu 和 Pd^{2+}/Pd 的标准电极电位 $E^{\ominus}$ 分别为 -2.37 V、-0.23 V、0.340 V 和 0.951 V。在高温灰化时，较高 $E^{\ominus}$ 的钯首先得到电子而形成亚分子层，随后是铜、镍和镁。铜容易被钯亚分子层吸附包埋，形成稳定的 Pd—Cu 键，故钯的化学改进效应最好，镍和镁次之[40]。

1.2 氢化物原位富集

氢化物原位富集（*in situ* concentration）也称原位捕集（*in situ* trapping），是将形成的氢化物在线富集在修饰过的石墨炉或石墨平台表面，然后快速升温原子化，从而获得更高的原子吸收信号强度，并消除氢化物发生动力学对信号的影响，显著地改善了测定的灵敏度和精密度。Lee[41]于 1982 年首先将石墨炉氢化物原位富集技术用于环境样品中铋的测定。产生的 BiH_3 由氩气流载带，通过预热到 300℃的修饰过的碳棒原子化器 90 s，BiH_3 被富集，随后升温到 1850℃原子化 2 s，测定天然水、海藻和沉积物中的铋。倪哲明等[42,43]在钯化学改进剂的基础上，提出了在涂钯石墨管内原位捕集氢化物的方法，进一步提高了分析灵敏度。这是现有氢化物形成元素如砷、硒、锑、铋、碲、锗和铅等最灵敏的分析方法之一，可直接测定试样中痕量硒和砷等元素，为生物、医药和环境等领域提供了测定痕量元素的灵敏方法，被国家自然科学基金委员会列入 1995 年重要成果。

在倪哲明先生等的工作之前，氢化物大部分都直接捕集于石墨管表面[44-49]。但这种捕集方法所需温度较高（>600℃），因而效率较低（如 Lee 最初的工作，捕集效率仅为 24%）。1989 年，倪哲明等[42,43]、Sturgeon 等[50]和 Rettberg 等[51]各自报道了涂钯石墨管可在较低温度下（200~300℃）高效捕集氢化物。在氢化物发生之前，先在石墨管内注入约 50 μL 的钯溶液，然后升温蒸发水分，实现钯的管内涂覆。然后，启动氢化物发生装置并将石墨炉温度设置为 200~300℃，将氢化物导出管口直接对准石墨管的进样口以实现氢化物在线原位捕集。待捕集完成，再将石墨管温度升高至原子化温度，实现捕集的氢化物原子化。与直接石墨管内捕集相比，涂钯石墨管的捕集温度显著降低，捕集效率也得以大幅提高，进而显著提高了灵敏度。部分使用涂钯石墨管捕集氢化物的文献总结于表 1-2 中。

表 1-2 涂钯石墨管用于氢化物在线捕集及测定

测定元素	捕集温度（℃）	特征质量（pg）	样品基体	文献
Bi	200	28.3		
Te	200	22.0	生物材料与天然水	[43]
Ge	600	25.9		
As	300	10.0		
Sb	300	13.1	尿与水样	[42]
Se	300	14.7		
Hg	250	114	海水与废水	[52]
As	600	10.6	海水与河水	[53]
Sn	300	7	钢材、底泥、树叶、牛肝	[54]
In	800	630	—	[55]
Tl	500	920	—	[56]
Se	200	40	高温镍基合金	[57]
Te	200	35	高温镍基合金	[58]
Au	400		矿石	[59]

1.3　石墨管的持久化学改进技术

从石墨炉内涂钯可增加氢化物的捕集效率出发，科学家进而提出了石墨管的持久改性技术。对石墨管进行一次改性处理，可使改性后的石墨管使用多次甚至上千次。该技术首次由 Shuttler 等提出，注入 50 μg 钯和 50 μg 铱到横向加热石墨原子化器的利沃夫平台上，经加热程序处理形成持久改进剂，用氢化物发生-石墨炉原子吸收光谱测定砷、铋和硒，平台使用寿命达 300 次[66]。持久化学改进剂的显著特点是使用寿命长，同时也节省了贵金属化学改进剂的用量与加快了分析速度。Lima 等提出的

"W+Rh"持久性改进剂，不仅可大幅延长石墨管的使用寿命，而且相比于传统的化学改进技术降低了贵金属的使用量，更可用于悬浮进样[66-70]。

可用作石墨管持久改性的化学改进剂的元素包括高熔点贵金属铱、钯、铂、铑和钌，生成难熔化合物的金属铪、钼、铌、铼、钽、钛、钒、钨和锆及生成"共价"碳化物的元素硼和硅等。中等挥发性的贵金属银、金和钯不宜于单独用作持久化学改进剂，只有与其他低挥发金属形成化（混）合物如 Pd-Ir、Pd-Rh 和 Au-Rh，或在碳化物涂层表面结合形成 Pd-Zr 和 Pd-W，提高其热稳定性之后才能用作持久化学改进剂。在这两种情况下，银、金和钯都会逐渐蒸发并在原子化器内重新分布，引起灵敏度漂移、涂层寿命缩短和记忆效应。生成"似盐"碳化物的元素钪、钇和镧系元素，由于容易水解使石墨表面受到腐蚀，热解石墨层脱落，导致灵敏度漂移和精密度降低等，不适于用作持久化学改进剂[71]。

制备持久涂层的方法有热解还原沉积法、电沉积法和阴极溅射法。热解还原沉积法是直接将贵金属溶液注入热解或碳化物涂层石墨管内，再按一定的升温程序处理，热解还原生成贵金属沉积层[72]。电沉积法是先将石墨管在适当温度下进行净化，再移取一定体积贵金属溶液注入石墨管，以铂或铱为阳极插入贵金属溶液中，石墨管本身为阴极，在一定电压和电流下电沉积贵金属。Bulska 等[73]详细研究了电沉积铱和钯等的制备方法。阴极溅射法制备持久化学改进剂，因需要专门的实验设备，在一般实验室应用不多。

石墨管持久改进的效果除了受涂层制备方法影响之外，还取决于试样基体的类型和质量，热解、原子化、净化温度和时间以及通保护气的情况。显而易见的是，石墨管持久改进技术有着明显的优越性和发展潜力。但在实际工作中，也显示出某些缺点和限制，如管与管之间重复性差，为避免和减少化学改进剂的损失，石墨管所使用的热解、原子化和净化温度较低等[74]。

钨丝是另外一种有潜力的电热原子化器，其持久化学改进也颇受关注。侯贤灯等[75]用沉积在钨丝原子化器上的铱为化学改进剂测定硒，原子出峰时间延迟约 0.2 s。由于铱在钨丝原子化阶段损失，因此延迟的程度随钨丝使用时间而减少。铱持久化学改进剂的使用寿命可达 300～400 次，钨丝寿命可延长到 1600 次。所以，每当钨丝点火 300～400 次后要重新在钨丝上镀铱。同样，用钨丝电热原子吸收测定血和尿生物体液中的铅时，通过加热钨丝和铑溶液使钨丝表面生成铑薄层作为永久性化学改进剂，不仅能在灰化阶段稳定铅，而且在净化阶段使残留物易于除去[76,77]。此外，钨丝上涂覆的铂族金属，也能用于氢化物的在线捕集，实现灵敏度的大幅提升[78]。

1.4 电热原子化过程的动力学研究

1.4.1 概述

电热原子吸收光谱分析由于基体干扰严重，因此阻碍了其应用于复杂基体中痕量元素的直接测定。为了从理论上预测理想的分析条件以及了解基体干扰的信息，从而达到控制和消除干扰的目的，研究电热原子化器中气态原子的形成过程尤为重要。20 世纪 70 年代，石墨炉原子吸收创始人利沃夫（L'vov）[79]提出了信号轮廓随时间变化的概念，即逐渐升温条件下的 L'vov 模型。Fuller[80]使原子化过程在基本恒定温度下进行，在最佳的动力学条件下测量石墨炉中原子浓度的变化，即 Fuller 模型。这两种方法奠定了研究石墨炉原子化过程动力学的理论基础。此后，许多学者致力于该方面的理论和实践研究，比较典型的包括 Sturgeon 模型[81]和 Smets 模型[82]。

自 20 世纪 80 年代以来，我国学者应用现代物理技术以及动力学和热力学相结合的方法，对原子化机理进行了深入系统的研究。如邓勃先生领导的实验室将石墨炉探针技术用于原子化机理研究[83-88]。由于探针可以直接进行 XRD、SEM 和 XPS 分析，因此这是一种简便的研究原子化机理的方法。此外，我国学者在已有原子化模型的基础上，改进和丰富了相关理论，并创新提出了通用性的理论模型和见解[89-101]。如严秀平在林铁铮和倪哲明先生指导下，从改进已有原子化动力学模型开始，到提出通用的原子化模型和方法，直至应用于原子化机理和干扰机理的研究，为原子化机理、化学改进机理和干扰机理研究提供了新的方法学[91,93,96,97,99]。所提出的正常分析条件下从单一原子吸收信号获取升温原子化动力学参数的方法[96]，成为电热原子化动力学研究的常用方法，被国际同行称为"严方法"或"严图"（Yan method 或 Yan plot）[102-111]。该法不仅避免了 Smets 模型的主要缺陷如稳定态近似、原子化为一级动力学等不合理假定，以及不能用于正常分析条件下原子化机理的研究和不能（定量）测定原子化动力学级数等缺点，而且为分数级数和多重机理的原子化动力学的研究提供了定量方法。所提出的等温原子化模型被"分析化学丛书"以"等温原子化严秀平模式"为标题编入《原子吸收及原子荧光光谱分析》一书中[112]。

1.4.2 等温原子化模型

在常规管壁蒸发原子化方式中，由于测量原子吸收信号时原子化器内的气相温度未能达到预定设置的温度，容易引起气相干扰。为了解决这一问题，通常采用"石墨平台"或"探针"原子化技术，使分析物在原子化器的气相温度达到预定设置的温度后进行原子化，即"等温原子化"。因此，研究等温原子化机理对于进一步提出消除气相干扰的有效措施具有指导意义。

严秀平等[91-93]将原子形成的速率方程表示为

$$-\frac{dN}{dt} = k_1 N^x \tag{1-1}$$

式中，N 是未原子化的被测元素的量，k_1 和 x 分别是原子形成过程的速率常数和反应级数。为了从原子吸收信号求得 N 值，假定：①被测元素的总量 N_0 正比于峰高吸收 A_{max}；②时刻 t 已原子化的被测元素的量 $N_0 - N$ 正比于吸光度 A，即可得到时刻 t 未原子化的被测元素的量为

$$N = P(A_{max} - A) \tag{1-2}$$

式中，P 为比例常数。在恒温条件下，从式（1-1）和式（1-2）得到

$$Y = -k_1 P^{x-1} t + 常数 \tag{1-3}$$

其中，

$$Y = \begin{cases} \ln(A_{max} - A) & (x = 1) \\ (A_{max} - A)^{1-x} / (1-x) & (x \neq 1) \end{cases}$$

在等温原子化条件下，由于 k_1 不随 t 变化，因此从 Y 对 t 图的线性情况可得反应级数 x，而从 Y 对 t 图的直线斜率可得 $k_1 P^{x-1}$。考尚铭等[92]利用此法研究了一些元素在钨丝探针上的等温原子化机理。

为了从原子吸收信号求得 N 值，严秀平等[93]也将时刻 t 未原子化的被测元素的量考虑为

$$N = P\left(k_2 \int_t^\infty A dt - A\right) \tag{1-4}$$

式中，k_2 为原子消失速率常数，从而得到

$$\ln\left(k_2 \int_t^\infty A dt - A\right) = -k_1 t + 常数 \quad (x = 1) \tag{1-5}$$

$$\left(k_2\int_t^\infty A\mathrm{d}t - A\right)^{1-x} = -k_1(1-x)P^{x-1}t + 常数 \quad (x \neq 1) \quad (1\text{-}6)$$

并研究了铜、铁、铝、钴和钼的原子化机理[93]。

梁彦忠等[94]将石墨炉内气态原子浓度 n 随时间的变化率考虑为

$$\frac{\mathrm{d}n}{\mathrm{d}t} = k_1 N^x - k_2 n \quad (1\text{-}7)$$

从式（1-4）和式（1-7）可以得到

$$\ln\left(\frac{\mathrm{d}A}{\mathrm{d}t} + k_2 A\right) = x\ln\left(k_2\int_t^\infty A\mathrm{d}t - A\right) + \ln[k_1(1-x)P^{x-1}] \quad (1\text{-}8)$$

在等温原子化条件下，从式（1-8）即可得到原子化过程的动力学级数 x 和原子形成速率常数 k_1。

1.4.3 升温原子化模型

常规电热原子吸收光谱分析通常是在升温原子化方式下进行的。在常规管壁蒸发原子化方式中，待测元素的原子吸收信号是在原子化器温度上升的过程中测量的（图 1-2）。因此，升温原子化过程要比等温原子化过程复杂得多。

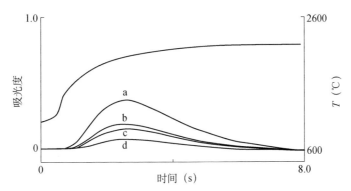

图 1-2 不同质量铜的电热原子吸收信号轮廓[96]

(a) 2.5 ng；(b) 1.25 ng；(c) 1 ng；(d) 0.5 ng。干燥温度 120℃（升温 5 s，保持 30 s），灰化温度 1200℃（升温 5 s，保持 30 s），原子化温度 2300℃（升温 1 s，保持 10 s）

Smets[82]在原子形成为一级动力学的假设下，采用稳定态近似得到

$$k_1 = A / \int_t^\infty A\mathrm{d}t \quad (1\text{-}9)$$

与等温原子化不同，升温原子化条件下的原子形成速率常数 k_1 和原子化器温度 T 均随时间 t 的变化而变化。Smets 模型[式（1-9）]是当时测定电热原子化过程活化能 E_a 最常用的方法。但是，由于 Smets 模型得到的 Arrhenius 图在高温区常常弯向 $1/T$ 轴，因此只能应用于原子化初试阶段即低温区。此外，Smets 模型采用了原子形成为一级动力学和稳定态近似等不合理假定，因此不能应用于原子形成过程的动力学参数的测定。

Chung[95]以式（1-2）表示时刻 t 未原子化的被测元素的量 N，在原子形成为一级动力学和稳定态近似的假定下，得到了计算原子形成速率常数公式：

$$k_1 = k_2 A / (A_{\max} - A) \quad (1\text{-}10)$$

Chung[95]的结果表明利用式（1-10）得到的 Arrhenius 图在原子吸收信号出现到极大值的温度范围内

具有良好的线性。然而，后来的研究表明 Chung 型 Arrhenius 图在高温区明显弯离 $1/T$ 轴[96,113]，这可能是由于在式（1-10）中低估了未原子化的被测元素的量之故[114,115]。

为了消除 Smets 模型得到的 Arrhenius 图在高温区弯曲的现象，严秀平等[97]将时刻 t 未原子化的被测元素的量 N 修正为以式（1-4）表示，从而将 Smets 公式改进为

$$k_1 = k_2 A / \left(k_2 \int_t^\infty A \mathrm{d}t - A \right) \quad (1\text{-}11)$$

式中，k_2 为原子消失速率常数。由式（1-9）和式（1-11）得到的 Arrhenius 图在低温区基本重合，但在高温区前者明显地弯向 $1/T$ 轴而后者仍然保持良好的线性（图 1-3）[97]。严秀平等[98]根据式（1-11）研究了银和钠的电热原子化机理。

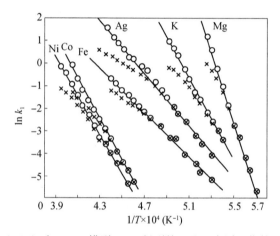

图 1-3　由式（1-11）（○）和 Smets 模型（×）得到的一些元素原子化的 Arrhenius 图比较[97]

尽管式（1-11）消除了 Arrhenius 图在高温区弯曲的现象，但仍采用了原子形成为一级动力学和稳定态近似的假定。因此，严秀平等[99]再次改进了 Smets 模型，避免了稳定态近似和原子形成为一级动力学的假定，将原子形成速率表示为

$$-\frac{\mathrm{d}N}{\mathrm{d}t} = k_1 N^x = k_0 N^x \exp(-E_\mathrm{a}/RT) \quad (1\text{-}12)$$

由式（1-4）和式（1-12）得到

$$\ln Z = -\frac{E_\mathrm{a}}{RT} + \ln\left[k_0 P^{x-1} R \left(1 - \frac{2RT}{E_\mathrm{a}}\right) / (\alpha E_\mathrm{a}) \right] \quad (1\text{-}13)$$

其中，

$$Z = \begin{cases} T^{-2} \ln\left[k_2 \int_0^\infty A \mathrm{d}t / \left(k_2 \int_t^\infty A \mathrm{d}t - A \right) \right] & (x=1) \\ \left[\left(k_2 \int_t^\infty A \mathrm{d}t - A \right)^{1-x} - \left(k_2 \int_0^\infty A \mathrm{d}t \right)^{1-x} \right] / [T^2 (x-1)] & (x \neq 1) \end{cases}$$

式中，α 为加热速率。对于大多数反应的 E_a 值及反应所发生的温度范围，式（1-13）右边的第二项可近似为常数[116]。这样，使 $\ln Z$ 对 $1/T$ 图成线性的 x 值即为反应级数，而从 $\ln Z$ 对 $1/T$ 图的直线斜率可得 E_a 值。严秀平等[99]利用式（1-13）测定了镍和锂电热原子化的反应级数，并发现由式（1-13）得到的 Arrhenius 图具有较宽的线性范围。

严秀平等[91,100]又将式（1-2）代入式（1-12），这样式（1-13）中的 Z 应具有以下形式：

$$Z = \begin{cases} T^{-2} \ln[A_{\max} / (A_{\max} - A)] & (x = 1) \\ \left[(A_{\max} - A)^{1-x} - A_{\max}^{1-x} \right] / \left[T^2 (x-1) \right] & (x \neq 1) \end{cases}$$

并据此研究了铁的原子化机理[100]。

Chakrabarti 等[113]假定：①在通气状态下原子化时，原子消失速率比原子形成速率快；②被测元素的气态原子在气相中的扩散过程没有活化能垒；③在恒定加热速率下，原子形成速率常数可用 Arrhenius 定理近似；④样品在恒定、低加热速率下原子化，从而导出了下列两个方程：

$$\ln\left\{\left[-\ln\left(\frac{\int_t^\infty A \mathrm{d}t}{\int_0^\infty A \mathrm{d}t}\right)\right] / T^2\right\} = -\frac{E_a}{RT} + \ln k_0 \quad (1\text{-}14)$$

$$\ln\left(\int_0^t A \mathrm{d}t / \int_0^\infty A \mathrm{d}t\right) = -\Delta E_a + \ln(k_0 / k_0') \quad (1\text{-}15)$$

式中，ΔE_a 是被测元素的原子从原子化器表面释放过程的活化能与其再沉积过程的活化能之差，k_0 和 k_0' 分别为上述两个过程的频率因子。式（1-14）和式（1-15）均是在原子形成为一级动力学过程的假定下推导出的。前者适用于无原子蒸气再沉积情况，而后者考虑了原子蒸气在管表面的再沉积过程。据报道，上述两式得到的 Arrhenius 图具有较宽的线性范围[113]。

Cathum 等[117]在原子消失速率大于原子形成速率以及原子形成为一级动力学的条件下，推导出下列两个方程：

$$\ln(A_m T_m^2) = \ln\left(\int_{t_m}^\infty A \mathrm{d}t\right) + \ln(\alpha E_a / R) \quad (1\text{-}16)$$

$$n_m = \left[\alpha E_{-a} / (RT_m^2)\right]\left(\int_{t_m}^\infty A \mathrm{d}t / A_m\right) \quad (1\text{-}17)$$

式中，A_m、t_m、T_m 和 n_m 分别为最大吸光度及其对应时间、温度和反应级数，α 为石墨炉升温速率。因此，对于不同样品量的原子吸收信号，若将 $\ln(A_m T_m^2)$ 对 $\ln\left(\int_{t_m}^\infty A \mathrm{d}t\right)$ 作图得一直线，则可证明原子形成为一级动力学的假定正确，且从直线的截距可得 E_a。式（1-16）仅适用于测定吸收信号最大值时的动力学级数，且计算时所需的活化能常需由其他方法预先测得以及石墨炉升温速率必须为恒值。

Fonseca 等[111]对 Cathum 法进行了评论，认为根据式（1-16）利用 $\ln(A_m T_m^2)$ 对 $\ln\left(\int_{t_m}^\infty A \mathrm{d}t\right)$ 图是否为直线来验证原子形成为一级动力学的假定是不可靠的，因为即使对具有分数级数动力学的金和银的原子化，$\ln(A_m T_m^2)$ 对 $\ln\left(\int_{t_m}^\infty A \mathrm{d}t\right)$ 图也具有较好的线性。此外，若将式（1-16）和式（1-17）应用于银的原子化，得到的反应级数在 2.03~2.56 之间；对于金的原子化，用式（1-16）和式（1-17）求得的反应级数为 2.2±0.2。这说明由式（1-16）和式（1-17）计算得到的反应级数比实际值高得多。然而，由于在 Fonseca 等[111]的实验中需要采用停气及较高升温速率实现原子化，因此不能满足式（1-16）和式（1-17）的先决条件，故仅从 Fonseca 等[111]的实验结果也难以对 Cathum 法做出否定的结论。

Rojas 等[118]用数学方法将复杂的升温原子化进行"恒温"化处理，并利用吸收信号及吸光度变化率信号的一些特征值，推导了计算升温原子化过程的活化能和频率因子的方法。设吸收信号的下降部分位于恒温区 T_p，若以 t_p 和 t_m 分别表示炉温达到恒定温度 T_p 所对应的时间和吸光度达到最大值所对应的时间，并且 $t_m > t_p$，而吸光度变化率达最大值和最小值所对应的时间分别为 t_x 和 t_y，则有 $t_x < t_m$ 和 $t_y > t_m$。在原子形成及原子消失均为一级动力学过程以及恒速升温的条件下，可得下列方程：

$$A(t_y) / \int_{t_m}^{\infty} A \mathrm{d}t = k_p \exp\left[-k_p\left(t_y - t_m\right)\right] \tag{1-18}$$

式中，k_p 为温度 T_p 时的原子形成速率常数。因此，只要从不同 T_p 温度时的吸收信号求得特征值 t_y、$A(t_y)$、t_m 及 $\int_{t_m}^{\infty} A \mathrm{d}t$，就可根据式（1-18）计算出不同 T_p 时的原子形成速率常数 k_p，然后由 Arrhenius 定理求得原子形成过程的活化能和频率因子。此方法仍未摆脱原子形成为一级动力学的假定，并且需要一定的实验条件使得 $t_m > t_p$，以及需要不同原子化温度下的吸收信号轮廓，以便求得原子形成过程的活化能和频率因子。

McNally 和 Holcombe[119]以式（1-12）为原子供给函数，结合简单的扩散消失函数，用计算机模拟技术研究了不同反应级数时样品浓度对吸收信号位置的影响，从而提出了从不同样品浓度所对应的吸收信号的位置变化情况估计反应级数的方法。此方法称为"浓度研究"（concentration study）。例如，对于零级过程，随样品量的增加，峰值向高温方向移动，且信号的上升部分几乎重叠；对于一级动力学过程，随样品量的增加，峰值位置几乎不变，但吸收信号的出现温度向低温位移；对于原子形成为分数级动力学过程，吸收信号的峰位置随样品量增加向高温移动而出现温度略向低温移动；对于动力学级数大于 1 的原子化过程，吸收信号的峰位置随样品量的增加反而向低温区移动。用"浓度研究"法判断原子形成过程的动力学级数较为直观，但难以确定准确的反应级数。此外，此法需要不同浓度时的吸收信号并且只有在原子化机理不受样品量影响的条件下才能应用。

以上方法仍然难以应用于正常分析条件下电热原子化过程动力学参数的定量测定，特别是对于具有分数级数和多重机理的原子化过程。因此，严秀平等[96]将正常分析条件下原子吸光度随时间的变化率用式（1-19）描述：

$$\frac{\mathrm{d}A}{\mathrm{d}t} = P^{x-1} k_0 \left(\int_t^{\infty} k_2 A \mathrm{d}t - A\right)^x \exp\left(-\frac{E_a}{RT}\right) - k_2 A \tag{1-19}$$

从而导出了式（1-20）：

$$\ln\left[\frac{\dfrac{\mathrm{d}A}{\mathrm{d}t} + k_2 A}{\left(\int_t^{\infty} k_2 A \mathrm{d}t - A\right)^x}\right] = -\frac{E_a}{RT} + \ln\left(k_0 P^{x-1}\right) \tag{1-20}$$

Rojas 等[120]也推导了式（1-20）。由式（1-20）知，对于由单一动力学机理控制的原子化过程，在假定的 x 值下，若将 $\ln\left[\dfrac{\dfrac{\mathrm{d}A}{\mathrm{d}t} + k_2 A}{\left(\int_t^{\infty} k_2 A \mathrm{d}t - A\right)^x}\right]$ 对 $1/T$ 作图是一直线，则所假定的 x 值正确，并从直线的斜率可得 E_a，从直线的截距可得 k_0 值。但是，这种用尝试法难以适用于具有分数级数和多重机理的原子化过程。

为了克服式（1-20）的上述缺陷，严秀平等[96]进一步利用差分法处理式（1-20）得到

$$\frac{\Delta\ln\left(\dfrac{\mathrm{d}A}{\mathrm{d}t}+k_2A\right)}{\Delta\ln\left(\int_t^\infty k_2A\mathrm{d}t-A\right)}=-\frac{\left(\dfrac{E_\mathrm{a}}{R}\right)\Delta\left(\dfrac{1}{T}\right)}{\Delta\ln\left(\int_t^\infty k_2A\mathrm{d}t-A\right)}+x \quad (1\text{-}21)$$

式（1-21）表明：若原子化过程由单一动力学机理控制，则 $\dfrac{\Delta\ln\left(\dfrac{\mathrm{d}A}{\mathrm{d}t}+k_2A\right)}{\Delta\ln\left(\int_t^\infty k_2A\mathrm{d}t-A\right)}$ 对 $\dfrac{\Delta\left(\dfrac{1}{T}\right)}{\Delta\ln\left(\int_t^\infty k_2A\mathrm{d}t-A\right)}$ 作图应为一直线，从直线的斜率可得 E_a 值，截距即为 x（图1-4）。若原子化过程由多重原子化机理控制，则 $\dfrac{\Delta\ln\left(\dfrac{\mathrm{d}A}{\mathrm{d}t}+k_2A\right)}{\Delta\ln\left(\int_t^\infty k_2A\mathrm{d}t-A\right)}$ 对 $\dfrac{\Delta\left(\dfrac{1}{T}\right)}{\Delta\ln\left(\int_t^\infty k_2A\mathrm{d}t-A\right)}$ 作图应在不同温度范围内表现为多条不同的直线，从各直线的斜率和截距可得相应原子化过程的 E_a 和 x 值。计算机模拟研究及电热原子吸收实验证明了式（1-20）和式（1-21）的有效性[96]。式（1-20）和式（1-21）为从正常分析条件下单一原子吸收信号同时、准确地求取原子形成过程的反应级数、活化能和频率因子提供了可靠的方法，尤其对于具有分数级数以及多重机理的原子化过程的研究更为优越[96,122]。

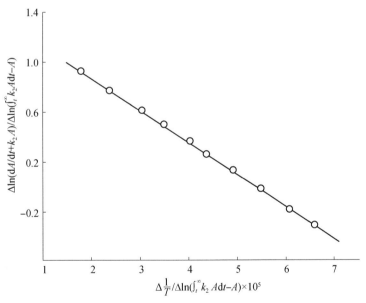

图1-4 从图1-2（b）原子吸收信号轮廓得到的 $\dfrac{\Delta\ln\left(\dfrac{\mathrm{d}A}{\mathrm{d}t}+k_2A\right)}{\Delta\ln\left(\int_t^\infty k_2A\mathrm{d}t-A\right)}$ 对 $\dfrac{\Delta\left(\dfrac{1}{T}\right)}{\Delta\ln\left(\int_t^\infty k_2A\mathrm{d}t-A\right)}$ 图[96]

严秀平等[96]提出的测定升温原子形成过程的反应级数和活化能的方法，被国际同行认为是现有方法中假设最少的、更准确的电热原子化动力学参数测定方法，已成为国际同行研究电热原子动力学的常用方法之一，被称为"严方法"或"严图"（Yan method 或 Yan plot）[103-111,121,122]。对于由

单一机理控制的原子化过程，由"严方法"[96]得到的 Arrhenius 图在整个原子化温度范围内具有很好的线性，而由 Sturgoen 模型[81]、Smets 模型[82]和 Chung 模型[95]得到的 Arrhenius 图在高温区明显弯曲（图 1-5）。国际著名光谱学家 Holcombe 研究组[122]将"严方法"[96]与 Rojas 方法[120]和 McNally 方法[119]作了详细的比较研究。结果表明 McNally 方法只能得到半定量的动力学级数，Rojas 方法由于需要用尝试法确定原子化过程的动力学级数和活化能，因此具有相当大的主观性，而"严方法"不仅能方便地得到定量的原子化过程的动力学级数和活化能数据，而且没有上述提到的 McNally 方法和 Rojas 方法的缺点。

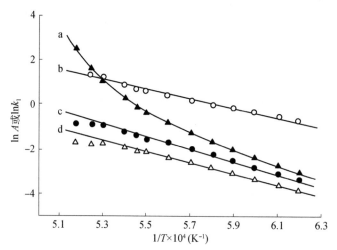

图 1-5 利用不同模型/方法从图 1-2（b）原子吸收信号轮廓得到的铜原子化 Arrhenius 图
(a) Chung 模型[95]；(b) 严方法[96]；(c) Smets 模型[82]；(d) Sturgeon 模型[81]

1.4.4 电热原子化动力学模型的应用

我国学者发展的电热原子化模型和方法被国际同行应用于原子化机理、干扰机理和化学改进机理的研究[103-111,121,122]。所改进的 Smets 模型[99]被 Mazzucotelli 和 Grotti[102]应用于研究电热原子吸收测定硒时的干扰机理和化学改进机理。所提出的升温原子化模型[96]被 Thomaidis 和 Piperaki 应用于研究化学改进剂存在时，金的电热原子化动力学[103]以及铬的电热原子化过程[104]，被 Fischer 和 Rademeyer 应用于无化学改进剂时硒的电热原子化动力学[105]以及钯化学改进剂存在时硒的电热原子化动力学[106]的研究，被 Eleni 等[107]应用于研究铂在水溶液和血清样品中的电热原子化机理。Alvariz 等利用"严方法"研究了不同原子化器表面和化学改进剂对铜[109]的电热原子化动力学的影响。Fischer[110]还采用"严方法"研究了钯化学改进剂消除磷酸盐对硒电热原子化过程干扰的机理。表 1-3 和表 1-4 总结了应用我国学者提出的原子化模型研究原子化机理、干扰机理和化学改进机理的主要研究工作。

第1章 化学改进剂与原子化机理研究进展

表 1-3 一些元素在有无钯、铑或锆化学改进剂时的升温原子化动力学参数

元素	动力学级数				活化能 E_a (kJ/mol)				使用模型	文献
	热解石墨管（PGT）	PGT+Pd	PGT+Rh	涂锆 PGT	PGT	PGT+Pd	PGT+Rh	涂锆 PGT		
In	0.98±0.02（低温） 0.37±0.02（高温）	1.06±0.03	—	0.64±0.05	138±8（低温） 51±7（高温）	421±27	—	186±13	式（1-21）	[123]
Sn	1.10±0.04（低温） 0.47±0.06（高温）	1.04±0.07	—	0.98±0.05	450±30（低温） 211±24（高温）	792±27	—	436±10	式（1-21）	[124]
Pb	0.95±0.06	0.94±0.07	—	0.28±0.05	312±17	519±50	—	188±21	式（1-21）	[125]
Au	0.31±0.03	—	—	—	293±8	—	—	—	式（1-21）	[126]
Se	1.57±0.21	1.29±0.15	1.08~1.11	—	239±29（高温）	203±20	—	—	式（1-21）	[105, 106]
Au	0.32~0.71	0.96~1.05	—	—	298~352	244~258	149~160	—	式（1-21）	[103]
Pt	1.52±0.06	—	—	—	242±13	—	—	—	式（1-21）	[107]
Cr	1.00~1.07	—	1.09~1.12	—	228~254	—	313~338	—	式（1-21）	[104]
Cu	1.00±0.08	0.40±0.3	—	—	83±3	198±11	—	—	式（1-21）	[109]

表 1-4 一些元素等温原子化动力学参数和原子化机理

元素/基体	温度范围（℃）	原子化表面	动力学级数	活化能 E_a（kJ/mol）	可能的决速步骤	使用模型	文献
Cu/HNO$_3$	900~2250	PGT	0	167	Cu$_{(g)}$→Cu$_{(g)}$	式(1-5)和(1-6)	[93]
Fe/HCl	1800~2100	PGT	0	184	FeO$_{(s)}$+C$_{(s)}$→Fe$_{(l)}$+CO$_{(g)}$	式(1-5)和(1-6)	[93]
Co/HCl	2100~2400	PGT	0	180	CO$_{2(g)}$→Co$_{(g)}$	式(1-5)和(1-6)	[93]
Al/HNO$_3$	2300~2400	PGT	0	301	Al$_{(l)}$→Al$_{(g)}$	式(1-5)和(1-6)	[93]
Mo/NH$_3$·H$_2$O	2400~2650	PGT	0	615	Mo$_{(s)}$→Mo$_{(g)}$	式(1-5)和(1-6)	[93]
Al/HNO$_3$	2150~2350	PGT 中钨丝探针		904	Al$_2$O$_{3(s)}$→Al$_{(g)}$+O$_{2(g)}$	式(1-3)	[92]
Ca/HNO$_3$	1850~2150	PGT 中钨丝探针		459	CaO$_{(g)}$→Ca$_{(g)}$+O$_{(g)}$	式(1-3)	[92]
Cr/HNO$_3$	1950~2250	PGT 中钨丝探针		429	CrO$_{(g)}$→Cr$_{(g)}$+O$_{(g)}$	式(1-3)	[92]
Cu/HNO$_3$	1800~2200	PGT 中钨丝探针		206	Cu$_{(g)}$→Cu$_{(g)}$	式(1-3)	[92]
Pb/HNO$_3$	1700~2000	PGT 中钨丝探针		174	Pb$_{(l)}$→Pb$_{(g)}$	式(1-3)	[92]
Zn/HNO$_3$	1500~1873	PGT 中钨丝探针		140	Zn$_{(l)}$→Zn$_{(g)}$	式(1-3)	[92]
Bi/HNO$_3$	1957~2357	PGT 中钨丝探针	0.67±0.02	302±8	从氧化 Bi 表面释放	式(1-8)	[94]
Ge/HNO$_3$	2077~2327	PGT 中钨丝探针	1.01±0.11	109±2	GeO$_{(g)}$→Ge$_{(g)}$+O$_{(g)}$	式(1-8)	[94]
Pb/HNO$_3$	1777~2227	PGT 中钨丝探针	0.46±0.03	159±2	从 Pb-W 表面释放	式(1-8)	[94]
Mn/HNO$_3$	1847~2227	PGT 中钨丝探针	0.97±0.09	372±5	MnO$_{(g)}$→Mn$_{(g)}$+O$_{(g)}$	式(1-8)	[94]
Cu/HNO$_3$	1797~2227	PGT 中钨丝探针	0.61±0.02	193±24	从探针表面解吸附	式(1-8)	[127]
Ag/HNO$_3$	1427~1977	PGT 中钨丝探针	0.45±0.09	248±5	从二维小岛上释放	式(1-8)	[127]
In/HNO$_3$	1857~2357	PGT 中钨丝探针	0.46±0.04	123±8	从二维小岛上释放	式(1-8)	[127]
Ni/HNO$_3$	2157~2357	PGT 中钨丝探针	0.99±0.11	297±6	从探针表面释放	式(1-8)	[127]
Mn/HNO$_3$	1100~1700	普通石墨管	0	396	MnO$_{(g)}$→Mn$_{(g)}$+O$_{2(g)}$	式(1-3)	[128]

1.5 结　束　语

本章介绍了电热原子吸收光谱分析中化学改进剂和原子化机理的研究进展，回顾了我国原子光谱分析研究人员对这些领域的主要贡献。然而，遗憾的是还有许多其他优秀工作因篇幅有限而未能一一介绍。在过去的四十多年里，我国学者在钯化学改进剂领域的开拓性和系统性研究以及在原子化机理领域的创新研究，极大地推动了电热原子吸收光谱分析的发展。

参 考 文 献

[1] Schlemmer G, Welz B. Palladium and magnesium nitrates, a more universal modifier for graphite furnace atomic absorption spectrometry. Spectrochimica Acta Part B, 1986, 41: 1157-1165.

[2] Ediger R D, Peterson G E, Kerber J D. Application of the graphite furnace to saline water analysis. Atomic Absorption Newsletters, 1974, 13: 61-64.

[3] 邓勃. 持久化学改进技术——一种新的化学改进技术. 现代仪器, 2003, 9: 12-14.

[4] 单孝全, 倪哲明. 基体改进效应用于石墨炉原子吸收测定汞. 化学学报, 1979, 37: 261-266.

[5] Weibust G, Langmyhr F J, Thomassen Y. Thermal stabilization of inorganic and organically-bound tellurium for electrothermal atomic absorption spectrometry. Analytica Chimica Acta, 1981, 128: 23-29.

[6] 单孝全, 倪哲明. 基体改进效应用于石墨炉原子吸收测定海水中铅. 环境科学, 1980, 5: 24-28.

[7] Jin L Z, Ni Z M. Matrix modification for the determination of trace amounts of bismuth in wastewater, sea-water and urine by graphite-furnace atomic-absorption spectrometry. Canadian Journal of Spectroscopy, 1981, 26: 219-223.

[8] 孙汉文, 单孝全, 倪哲明. 基体改进效应用于石墨炉原子吸收测定血中铅. 原子光谱分析, 1981, 1: 10-13.

[9] 单孝全, 倪哲明. 基体改进效应用于石墨炉原子吸收测定水中四价和六价碲. 环境科学学报, 1981, 1: 74-80.

[10] 单孝全, 孙汉文, 倪哲明. 石墨炉原子吸收测定土壤、粮食和蔬菜中痕量碲. 环境科学丛刊, 1981, 2: 5-8.

[11] Shan X Q, Ni Z M. Matrix modification for the determination of lead in urine by graphite-furnace atomic-absorption spectrometry. Canadian Journal of Spectroscopy, 1982, 27: 75-81.

[12] Shan X Q, Ni Z M, Zhang L. Determination of arsenic in soil, coal fly ash and biological samples by electrothermal atomic absorption spectrometry with matrix modification. Analytica Chimica Acta, 1983, 151: 179-185.

[13] Shan X Q, Ni Z M, Zhang L. Application of matrix-modification in determination of thallium in wastewater by graphite-furnace atomic-absorption spectrometry. Talanta, 1984, 31: 150-152.

[14] Shan X Q, Ni Z M, Yuan Z N. Determination of indium in minerals, river sediments and coal fly ash by electrothermal atomic absorption spectrometry with palladium as a matrix modifier. Analytica Chimica Acta, 1985, 171: 269-277.

[15] 单孝全, 倪哲明. 基体改进效应石墨炉原子吸收测定易挥发性元素砷、硒、银和铋. 化学学报, 1981, 39: 575-578.

[16] Volynsky A B. Mechanisms of action of platinum group modifiers in electrothermal atomic absorption spectrometry. Spectrochimica Acta Part B, 2000, 55: 103-150.

[17] Volynskii A B. Chemical modifiers based on platinum-group metal compounds in electrothermal atomic absorption spectrometry. Journal of Analytical Chemistry, 2004, 59: 502-520.

[18] Welz B, Schlemmer G, Mudakavi J R. Palladium nitrate-magnesium nitrate modifier for graphite furnace atomic absorption spectrometry. Part 1. Determination of arsenic, antimony, selenium and thallium in airborne particulate matter. Journal of Analytical Atomic Spectrometry, 1988, 3: 93-97.

[19] Welz B, Schlemmer G, Mudakavi J R. Palladium nitrate-magnesium nitrate modifier for graphite furnace atomic absorption spectrometry. Part 2. Determination of arsenic, cadmium, copper, manganese, lead, antimony, selenium and thallium in water. Journal of Analytical Atomic Spectrometry, 1988, 3: 695-701.

[20] Welz B, Schlemmer G, Mudakavi J R. Palladium nitrate-magnesium nitrate modifier for electrothermal atomic absorption spectrometry. Part 3. Determination of mercury in environmental standard reference materials. Journal of Analytical Atomic Spectrometry, 1992, 7: 499-503.

[21] Welz B, Bozsai G, Sperling M, Radziuk B. Palladium nitrate-magnesium nitrate modifier for electrothermal atomic absorption spectrometry. Part 4. Interference of sulfate in the determination of selenium. Journal of Analytical Atomic Spectrometry, 1992, 7: 505-509.

[22] Welz B, Schlemmer G, Mudakavi J R. Palladium nitrate-magnesium nitrate modifier for electrothermal atomic absorption spectrometry. Part 5. Performance for the determination of 21 elements. Journal of Analytical Atomic Spectrometry, 1992, 7: 1257-1271.

[23] Bermejo-Barrera P, Barciela-Alonso C, Aboal-Somoza M, Bermejo-Barrera A. Slurry sampling for the determination of lead in marine sediments by electrothermal atomic absorption spectrometry using palladium-magnesium nitrate as a chemical modifier. Journal of Analytical Atomic Spectrometry, 1994, 9: 469-475.

[24] Bermejo-Barrera P, Moreda-Pineiro J, Moreda-Pineiro A, Bermejo-Barrera A. Palladium as a chemical modifier for the determination of mercury in marine sediment slurries by electrothermal atomization atomic absorption spectrometry. Analytica Chimica Acta, 1994, 296: 181-193.

[25] Bermejo-Barrera P, Barciela-Alonso M C, Moreda-Pineiro J, Gonzalez-Sixto C, Bermejo-Barrera A. Determination of trace metals (As, Cd, Hg, Pb and Sn) in marine sediment slurry samples by electrothermal atomic absorption spectrometry using palladium as a chemical modifier. Spectrochimica Acta Part B, 1996, 51: 1235-1244.

[26] Bruhn C G, Neira J Y, Valenzuela G D, Nobrega J A. Chemical modifiers in a tungsten coil electrothermal atomizer. Part 1. Determination of lead in hair and blood. Journal of Analytical Atomic Spectrometry, 1998, 13: 29-35.

[27] Bruhn C G, San Francisco N A, Neira J Y, Nobrega J A. Determination of cadmium and lead in mussels by tungsten coil electrothermal atomic absorption spectrometry. Talanta 1999, 50: 967-975.

[28] Silva S G, Nobrega J A, Jones B T, Donati G L. Magnesium nitrate as a chemical modifier to improve sensitivity in manganese determination in plant materials by tungsten coil atomic emission spectrometry. Journal of Analytical Atomic Spectrometry, 2014, 29: 1499-1503.

[29] Bruhn C G, Huerta V N, Neira J Y. Chemical modifiers in arsenic determination in biological materials by tungsten coil electrothermal atomic absorption spectrometry. Analytical and Bioanalytical Chemistry, 2004, 378: 447-455.

[30] Resano M, Florez M R. Direct determination of sulfur in solid samples by means of high-resolution continuum source graphite furnace molecular absorption spectrometry using palladium nanoparticles as chemical modifier. Journal of Analytical Atomic Spectrometry, 2012, 27: 401-412.

[31] Yi Y Z, Jiang S J, Sahayam A C. Palladium nanoparticles as the modifier for the determination of Zn, As, Cd, Sb, Hg and Pb in biological samples by ultrasonic slurry sampling electrothermal vaporization inductively coupled plasma mass spectrometry. Journal of Analytical Atomic Spectrometry, 2012, 27: 426-431.

[32] Salmon S G, Holcombe J A. Alteration of metal release mechanisms in graphite furnace atomizers by chemisorbed oxygen. Analytical Chemistry, 1982, 54: 630-634.

[33] Salmon S G, Davis R H, Holcombe J A. Time shifts and double peaks for lead caused by chemisorbed oxygen in electrothermally heated graphite atomizers. Analytical Chemistry, 1981, 53: 324-330.

[34] 姚金玉, 蒋永清. 石墨炉原子吸收法中硝酸钙作为基体改进剂的作用. 光谱学与光谱分析, 1991, 11: 45-50.

[35] Yang P Y, Ni Z M, Zhuang Z X, Xu F C, Jiang A B. Study of palladium-analyte binary system in the graphite furnace by surface analytical techniques. Journal of Analytical Atomic Spectrometry, 1992, 7: 515-519.

[36] 渠荣遴, 何琲. 石墨炉原子吸收分析中抗坏血酸预还原钯基体改进剂测定微量锰的研究. 光谱学与光谱分析, 1997, 17: 61-65.

[37] 梁春穗, 黄妙英. 电热原子吸收光谱测定锗时钯的基体改进作用及其在食品分析中的应用. 分析测试通报, 1992, 11: 65-68.

[38] Caraballo E A H, Alvarado J, Arenas F. Study of the electrothermal atomization of phosphorus in transversely-heated graphite atomizers. Spectrochimica Acta Part B, 2000, 55: 1451-1464.

[39] Sadagov Y M, Katskov D A. Effect of palladium modifier on the analyte vapor transport in a graphite furnace atomizer. Spectrochimica Acta Part B, 2001, 56: 1397-1405.

[40] 吴华, 吴福全, 李绍南. 赛曼石墨炉原子吸收中用钯消除氯化物干扰的机理. 福建分析测试, 1998, 7: 892-895.

[41] Lee D S. Determination of bismuth in environmental samples by flameless atomic absorption spectrometry with hydride generation. Analytical Chemistry, 1982, 54: 1682-1686.

[42] Zhang L, Ni Z M, Shan X Q. *In situ* concentration of metallic hydrides in a graphite furnace coated with palladium. Spectrochimica Acta Part B, 1989, 44: 339-346.

[43] Zhang L, Ni Z M, Shan X Q. *In situ* concentration of metallic hydride in a graphite furnace coated with palladium-determination of bismuth, germanium and tellurium. Spectrochimica Acta Part B, 1989, 44: 751-758.

[44] Andreae M O. Determination of inorganic tellurium species in natural waters. Analytical Chemistry, 1984, 56: 2064-2066.

[45] Sturgeon R E, Willie S N, Berman S S. Hydride generation-atomic absorption determination of antimony in seawater with *in situ* concentration in a graphite furnace. Analytical Chemistry, 1985, 57: 2311-2314.

[46] Sturgeon R E, Willie S N, Berman S S. Preconcentration of selenium and antimony from seawater for determination by graphite furnace atomic absorption spectrometry. Analytical Chemistry, 1985, 57: 6-9.

[47] Sturgeon R E, Willie S N, Berman S S. Hydride generation-graphite furnace atomic absorption spectrometry: New prospects. Fresenius' Journal of Analytical Chemistry, 1986, 323: 788-792.

[48] Willie S N, Sturgeon R E, Berman S S. Hydride generation atomic absorption determination of selenium in marine sediments, tissues and seawater with *in-situ* concentration in a graphite furnace. Analytical Chemistry, 1986, 58: 1140-1143.

[49] Sturgeon R E, Willie S N, Sproule G I, Berman S S. Sorption and atomisation of metallic hydrides in a graphite furnace. Journal of Analytical Atomic Spectrometry, 1987, 2: 719-722.

[50] Sturgeon R E, Willie S N, Sproule G I, Robinson P T, Berman S S. Sequestration of volatile element hydrides by platinum group elements for graphite furnace atomic absorption. Spectrochimica Acta Part B, 1989, 44: 667-682.

[51] Doidge P S, Sturman B T, Rettberg T M. Hydride generation atomic absorption spectrometry with *in situ* pre-concentration in a graphite furnace in the presence of palladium. Journal of Analytical Atomic Spectrometry, 1989, 4: 251-255.

[52] Yan X P, Ni Z M, Guo Q L. *In situ* concentration of mercury vapour in a palladium-coated graphite tube: Determination of mercury by atomic absorption spectrometry. Analytica Chimica Acta, 1993, 272: 105-114.

[53] Sturgeon R E, Willie S N, Sproule G I, Robinson P T, Berman S S. Sequestration of volatile element hydrides by

platinum group elements for graphite furnace atomic absorption. Spectrochimica Acta Part B, 1989, 44: 667-682.

[54] Zhang L, McIntosh S, Carnrick G R, Slavin W. Hydride generation flow injection using graphite furnace detection: Emphasis on determination of tin. Spectrochimica Acta Part B, 1992, 47: 701-709.

[55] Liao Y P, Li A M. Indium hydride generation atomic absorption spectrometry with *in situ* preconcentration in a graphite furnace coated with palladium. Journal of Analytical Atomic Spectrometry, 1993, 8: 633-636.

[56] Liao Y P, Chen G, Yan D, Li A M, Ni Z M. Investigation of thallium hydride generation using *in situ* trapping in graphite tube by atomic absorption spectrometry. Analytica Chimica Acta, 1998, 360: 209-214.

[57] 彭兰乔, 姚金玉. 氢化物石墨炉原子吸收光谱法直接测定高温镍基合金中的硒. 分析化学, 1994, 22: 1135-1137.

[58] 彭兰乔, 姚金玉, 谢文兵. 氢化物发生石墨炉原位富集直接测定镍基合金中碲. 理化检验 (化学分册), 1995, 31: 75-76.

[59] Ma H B, Fan X F, Zhou H Y, Xu S K. Preliminary studies on flow-injection *in situ* trapping of volatile species of gold in graphite furnace and atomic absorption spectrometric determination. Spectrochimica Acta Part B, 2003, 58: 33-41.

[60] Ni Z M, Zhang D Q. Influence of sample deposition and coating with Zr and Pd on the atomization kinetics of germanium in graphite furnace atomic absorption spectrometry. Spectrochimica Acta Part B, 1995, 50: 1779-1786.

[61] Garboś S, Walcerz M, Bulska E, Hulanicki A. Simultaneous determination of Se and As by hydride generation atomic absorption spectrometry with analyte concentration in a graphite furnace coated with zirconium. Spectrochimica Acta Part B, 1995, 50: 1669-1677.

[62] Ni Z M, Hang H B, Li A, He B, Xu F Z. Determination of tributyltin and inorganic tin in sea-water by solvent extraction and hydride generation electrothermal atomic absorption spectrometry. Journal of Analytical Atomic Spectrometry, 1991, 6: 385-387.

[63] Yan X P, Ni Z M. Determination of lead by hydride generation atomic absorption spectrometry with *in situ*, concentration in a zirconium coated graphite tube. Journal of Analytical Atomic Spectrometry, 1991, 6: 483-486.

[64] 马玉平, 古丽克孜, 方新红. 流动注射氢化物石墨炉原子吸收法测定环境与生物样品中的痕量汞. 光谱学与光谱分析, 1993, 13: 107-110.

[65] Shuttler I L, Feuerstein M, Schlemmer G. Long-term stability of a mixed palladium-iridium trapping reagent for *in situ* hydride trapping within a graphite electrothermal atomizer. Journal of Analytical Atomic Spectrometry, 1992, 7: 1299-1301.

[66] Lima E C, Krug F J, Jackson K W. Evaluation of tungsten-rhodium coating on an integrated platform as a permanent chemical modifier for cadmium, lead, and selenium determination by electrothermal atomic absorption spectrometry. Spectrochimica Acta Part B, 1998, 53: 1791-1804.

[67] Lima E C, Krug F J, Ferreira A T, Barbosa F. Tungsten-rhodium permanent chemical modifier for cadmium determination in fish slurries by electrothermal atomic absorption spectrometry. Journal of Analytical Atomic Spectrometry, 1999, 14: 269-274.

[68] Lima E C, Barbosa F, Krug F J, Guaita U. Tungsten-rhodium permanent chemical modifier for lead determination in digests of biological materials and sediments by electrothermal atomic absorption spectrometry. Journal of Analytical Atomic Spectrometry, 1999, 14: 1601-1605.

[69] Lima E C, Barbosa F, Krug F J. Tungsten-rhodium permanent chemical modifier for lead determination in sediment slurries by electrothermal atomic absorption spectrometry. Journal of Analytical Atomic Spectrometry, 1999, 14: 1913-1918.

[70] Lima E C, Barbosa F, Krug F J. The use of tungsten-rhodium permanent chemical modifier for cadmium determination in decomposed samples of biological materials and sediments by electrothermal atomic absorption spectrometry.

Analytica Chimica Acta, 2000, 409: 267-274.

[71] Tsalev D L, Slaveykova V I, Lampugnani L, D'Ulivo A, Georgieva R. Permanent modification in electrothermal atomic absorption spectrometry-advances, anticipations and reality. Spectrochimica Acta Part B, 2000, 55: 473-490.

[72] da Silva, A F, Welz B, Curtius A J. Noble metals as permanent chemical modifiers for the determination of mercury in environmental reference materials using solid sampling graphite furnace atomic absorption spectrometry and calibration against aqueous standards. Spectrochimica Acta Part B, 2002, 57: 2031-2045.

[73] Bulska E, Liebert-Ilkowska K, Hulanicki A. Optimization of electrochemical deposition of noble metals for permanent modification in graphite furnace atomic absorption spectrometry. Spectrochimica Acta Part B, 1998, 53: 1057-1062.

[74] Volynsky A B, Wennrich R. Mechanisms of the action of platinum metal modifiers in electrothermal atomic absorption spectrometry: Aims and existing approaches. Spectrochimica Acta Part B, 2002, 57: 1301-1316.

[75] Hou X D, Yang Z, Jones B T. Determination of selenium by tungsten coil atomic absorption spectrometry using iridium as a permanent chemical modifier. Spectrochimica Acta Part B, 2001, 56: 203-214.

[76] Zhou Y, Parsons P J, Aldous K M, Brockman P, Slavin W. Atomization of lead from whole blood using novel tungsten filaments in electrothermal atomic absorption spectrometry. Journal of Analytical Atomic Spectrometry, 2001, 16: 82-89.

[77] Zhou Y, Parsons P J, Aldous K M, Brockman P, Slavin W. Rhodium as permanent modifier for atomization of lead from biological fluids using tungsten filament electrothermal atomic absorption spectrometry. Spectrochimica Acta Part B, 2002, 57: 727-740.

[78] Liu R, Wu P, Xu K L, Lv Y, Hou X D. Highly sensitive and interference-free determination of bismuth in environmental samples by electrothermal vaporization atomic fluorescence spectrometry after hydride trapping on iridium-coated tungsten coil. Spectrochimica Acta Part B, 2008, 63: 704-709.

[79] L'vov B V. Atomic Absorption Spectrochemical Analysis. London: Adam Hilger, 1970.

[80] Fuller C W. A kinetic theory of atomisation for non-flame atomic-absorption spectrometry with a graphite furnace. The kinetics and mechanism of atomisation for copper. Analyst, 1974, 99: 739-744.

[81] Sturgeon R E, Chakrabarti C L, Langford C H. Studies on the mechanism of atom formation in graphite furnace atomic absorption spectrometry. Analytical Chemistry, 1976, 48: 1792-1807.

[82] Smets B. Atom formation and dissipation in electrothermal atomization. Spectrochimica Acta Part B, 1980, 35: 33-42.

[83] 邓勃, 罗燕飞. 元素在石墨炉内石墨探针表面上的原子化机理研究Ⅰ. 镉与铝的原子化机理. 高等学校化学学报, 1990, 11: 780-782.

[84] 邓勃, 王建平. 石墨炉内石墨探针表面上的原子化机理研究Ⅲ.铬的原子化机理. 化学学报, 1991, 49: 1124-1128.

[85] 王建平, 邓勃. 石墨炉内石墨探针表面上的原子化机理研究Ⅻ.钴的原子化机理. 分析化学, 1991, 19: 1358-1362.

[86] 王建平, 邓勃. 石墨炉内石墨探针表面上的原子化机理研究Ⅴ. 锰的原子化机理, 高等学校化学学报, 1991, 12: 1323-1325.

[87] 王建平, 邓勃. 石墨炉内石墨探针表面上的原子化机理研究Ⅳ. 钒的原子化机理. 光谱学与光谱分析, 1992, 12: 83-88.

[88] 王建平, 邓勃. 石墨炉内石墨探针表面上的原子化机理研究Ⅷ. 锡的原子化机理. 分析测试通报, 1992, 11: 7-12.

[89] Yan X P, Ni Z M. Kinetic studies on the mechanism of atomization in electrothermal atomic absorption spectrometry with and without chemical modifiers. Fresenius' Journal of Analytical Chemistry, 2001, 370: 1052-1060.

[90] 严秀平, 林铁铮. 电热原子吸收光谱法中原子化过程的动力学研究. 分析试验室, 1989, 8: 30-37.

[91] 严秀平, 林铁铮. 测定电热原子吸收光谱法中原子形成过程动力学参数的新方法. 光谱学与光谱分析, 1989, 9: 45-47.

[92] 考尚铭, 严秀平, 林铁铮. 一些元素在钨丝探针上的等温原子化机理. 分析化学, 1989, 17: 481-484.

[93] 严秀平, 林铁铮. 电热原子吸收光谱法中原子形成过程的探讨. 化学学报, 1989, 47: 1139-1145.

[94] Liang Y Z, Ni Z M, Yan X P. Determination of kinetic parameters for atom formation at constant temperature in graphite furnace atomic absorption spectroscopy. Spectrochimica Acta Part B, 1995, 50: 725-737.

[95] Chung C H. Atomization mechanism with Arrhenius plots taking the dissipation function into account in graphite furnace atomic absorption spectrometry. Analytical Chemistry, 1984, 56: 2714-2720.

[96] Yan X P, Ni Z M, Yang X T, Hong G Q. An approach to the determination of the kinetic parameters for atom formation in electrothermal atomic absorption spectrometry. Spectrochimica Acta Part B, 1993, 48: 605-624.

[97] 严秀平, 刘志军, 林铁铮. 电热原子吸收光谱法中原子形成过程的研究Ⅰ. Smets 模型的改进. 分析化学, 1987, 15: 1072-1075.

[98] 严秀平, 林铁铮. 电热原子吸收光谱法中原子形成过程的研究Ⅱ. 银和钠原子化机理探讨. 分析化学, 1990, 18: 174-176.

[99] Yan X P, Lin T Z, Liu Z J. Improvement of the Smets method in electrothermal atomic-absorption spectrometry. Talanta, 1990, 37: 167-171.

[100] 严秀平, 林铁铮. 电热原子吸收光谱法中原子形成过程动力学参数的测定及其应用. 分析化学, 1990, 18 (3): 271-274.

[101] 徐腾, 柳志龙, 张国莹, 沈石年. 电热原子化机理研究Ⅰ. 测定原子化过程活化能的新方法. 光谱学与光谱分析, 1989, 9: 52-56.

[102] Mazzucotelli A, Grotti M. Effects of interfering elements and chemical modifier on the activation energy of electrothermal atomization of selenium. Spectrochimica Acta Part B, 1995, 50: 1897-1904.

[103] Thomaidis N S, Piperaki E A. Determination of the kinetic parameters for the electrothermal atomization of gold with and without chemical modifiers. Spectrochimica Acta Part B, 1999, 54: 1303-1320.

[104] Thomaidis N S, Piperaki E A. Effect of chemical modifiers on the kinetic parameters characterizing the electrothermal atomization of chromium. Spectrochimica Acta Part B, 2000, 55: 611-627.

[105] Fischer J L, Rademeyer C J. Kinetics of selenium atomization in electrothermal atomization atomic absorption spectrometry (ETA-AAS). Part 1: Selenium without modifiers. Spectrochimica Acta Part B, 1998, 53: 537-548.

[106] Fischer J L, Rademeyer C J. Kinetics of selenium atomization in electrothermal atomization atomic absorption spectrometry (ETA-AAS). Part 2: Selenium with palladium modifiers. Spectrochimica Acta Part B, 1998, 53: 549-567.

[107] Eleni P S, Thomaidis N S, Piperaki E A. Investigation of the mechanism of the electrothermal atomization of platinum in a graphite furnace from aqueous solutions and serum samples. Journal of Analytical Atomic Spectrometry, 2005, 20: 111-117.

[108] Alvarez M A, Carrion N, Gutierrez H. Effects of atomization surfaces and modifiers on the electrothermal atomization of cadmium. Spectrochimica Acta Part B, 1995, 50: 1581-1594.

[109] Alvarez M A, Carrion N, Gutierrez H. Effects of atomization surfaces and modifiers on the kinetics of copper atomization in electrothermal atomic absorption spectrometry. Spectrochimica Acta Part B, 1996, 51: 1121-1132.

[110] Fischer J L. Electrothermal atomization of palladium stabilized selenium in the presence of phosphate. Spectrochimica Acta Part B, 2002, 57: 525-533.

[111] Fonseca R W, Guell O A, Holcombe J A. Comments on the determination of the order of release during electrothermal atomization. Spectrochimica Acta Part B, 1992, 47: 573-576.

[112] 李安模, 魏继中. 原子吸收及原子荧光光谱分析. 北京: 科学出版社, 2000: 164.

[113] Chakrabarti C L, Cathum S J. Arrhenius plots for activation energy of atomization in graphite-furnace atomic-absorption spectrometry. Talanta, 1991, 38: 157-166.

[114] Akman S, Bektas S, Genc O. A novel approach to the interpretation of graphite furnace atomic absorption signals. Spectrochimica Acta Part B, 1988, 43: 763-772.

[115] Akman S, Genc O, Bektas S, Investigation of the atomization mechanisms of copper, platinum, iridium and manganese in graphite-furnace atomic-absorption spectrometry. Spectrochimica Acta Part B, 1991, 46: 1829-1839.

[116] Coats A W, Redfern J P. Kinetic parameters from thermogravimetric data. Nature, 1964, 201: 68-69.

[117] Cathum S J, Chakrabarti C L, Hutton J C. Investigation of the order of copper atomization at the absorbance maximum in graphite furnace atomic absorption spectrometry. Spectrochimica Acta Part B, 1991, 46: 35-44.

[118] Rojas D, Olivares W. Simple mathematical treatment for non-isothermal atomization. Journal of Analytical Atomic Spectrometry, 1989, 4: 613-617.

[119] McNally J, Holcombe J A. Existence of microdroplets and dispersed atoms on the graphite surface in electrothermal atomizers. Analytical Chemistry, 1987, 59: 1105-1112.

[120] Rojas D, Olivares W. A method for the determination of the kinetic order and energy of the atom formation process in electrothermal atomization atomic absorption spectrometry (ETA-AAS). Spectrochimica Acta Part B, 1992, 47: 387-397.

[121] Styris D L, Redfield D A. Perspectives on mechanisms of electrothermal atomization. Specrochimica Acta Reviews, 1993, 15: 71-123.

[122] Fonseca R W, Pfefferkorn L L, Holcombe J A. Comparisons of selected methods for the determination of kinetic parameters from electrothermal atomic absorption data. Spectrochimica Acta Part B, 1994, 49: 1595-1608.

[123] Yan X P, Ni Z M, Yang X T, Hong G Q. Kinetics of indium atomization from different atomizer surfaces in electrothermal atomic absorption spectrometry (ETAAS). Talanta, 1993, 40: 1839-1846.

[124] Quan Z, Ni Z M, Yan X P. Influence of atomizer surface on the kinetics of Tin atomization in electrothermal atomic-absorption spectrometry. Canadian Journal of Spectroscopy, 1994, 39: 54-59.

[125] Yan X P, Ni Z M. Electrothermal atomization of lead from different atomizer surfaces. Spectrochimica Acta Part B, 1993, 48: 1315-1323.

[126] 严秀平, 倪哲明. 金从电热原子化器表面原子化的动力学. 光谱学与光谱分析, 1994, 14: 59-62.

[127] Liang Y Z, Ni Z M, Yan X P. Atomization of Cu, Ag, Ni and In at constant temperature from tungsten probe surface. Chemical Analysis (Warsaw), 1996, 41: 601-613.

[128] 祖莉莉, 李安模. 石墨炉原子吸收分析中锰的原子化过程探讨. 分析化学, 1993, 21: 467-469.

第 2 章 激光电离质谱法用于固体样品元素直接分析

(毕梦香[①],陆 桥[①],徐周毅[①],杭 乐[①],王煜兵[①],张振建[①],杭 纬[①]*,黄本立[①])

▶ 2.1 激光微探针飞行时间质谱 / 36

▶ 2.2 激光电离四极杆质谱 / 39

▶ 2.3 激光电离双聚焦质谱 / 40

▶ 2.4 激光微探针傅里叶变换质谱 / 42

▶ 2.5 激光电离飞行时间质谱 / 44

▶ 2.6 总结与展望 / 54

①厦门大学,厦门。*通讯作者联系方式:weihang@xmu.edu.cn

第2章 激光电离质谱法用于固体样品元素直接分析

本章导读

- 质谱法在固体样品直接分析方面因具有较高的灵敏度（ppm～ppb），且能够定性定量地检测未知样品，而被广泛应用。
- 固体样品直接分析质谱法已经有了长足的发展，但在定量分析方面仍然受到诸多限制。半定量分析为在缺乏标准样品的情况下，提供了一种获得固体样品元素组成的方法。
- 激光电离质谱法在进行固体样品直接分析时对元素没有选择性，定量分析时无需标准样品的校准。激光离子源可以结合不同的质量分析器对固体样品进行直接分析。
- 飞行时间质谱（time-of-flight mass spectrometry，TOFMS）扫描速度快、可进行全谱分析，并且能够与脉冲式激光电离源完美结合。

在现代仪器测定样品中痕量和超痕量元素领域，固体样品元素分析至关重要[1]。与其他常规分析方法相比，固体样品直接分析技术具有以下优点，如很少或无需样品预处理，较少的样品消耗以及分析速度快等。尤其是在分析难以消解、有害或前处理过程烦琐耗时的固体样品时，固体样品直接分析技术优势显著[2]。

固体样品直接分析方法以光谱法和质谱法为主。在光谱法中，代表性技术是火花发射光谱法（spark optical emission spectrometry，Spark-OES）[3-6]、X射线荧光光谱法（X-ray fluorescence spectrometry，XRF）[7]和激光诱导击穿光谱法（laser-induced breakdown spectrometry，LIBS）[8]。但是这些基于原子发射光谱法的分析技术得到的谱图存在大量干扰谱线，背景噪声大且检出限有限[9]。在质谱法中，已有众多类型的质谱技术用于固体样品的元素分析，例如二次离子质谱（secondary ion mass spectrometry，SIMS）[10-12]、辉光放电质谱（glow discharge mass spectrometry，GDMS）[13,14]和电感耦合等离子体质谱法（inductively coupled plasma mass spectrometry，ICPMS）[15-18]。其中，SIMS易受到基体效应的影响，仪器价格昂贵，并且维护成本较高。尽管碰撞反应池可以在一定程度上缓解干扰，但ICPMS和GDMS都易受到碰撞气体和多原子物质的干扰[19]。此外，上述用于固体样品直接分析技术通常都需要固体标准品进行校准以完成定量分析。为了降低元素分馏效应和基体效应，这些固体标准品既要与基体匹配，又要包含必要的分析物元素，且需要适用于校准程序的适当浓度。但是缺乏合适的固体标准样品，而适当的固体标准样品又通常难以制造或价格昂贵，这便限制了固体样品直接分析的应用发展[20]。因此，无标样固体样品直接分析越来越引起人们的关注，它为固体样品分析提供了新的解决途径。

当前，有许多用于固体样品直接分析的经典离子源，例如火花源、辉光放电源和激光剥蚀电感耦合等离子体离子源。作为最悠久的技术之一，火花源在多种类型材料的多元素分析中发挥了重要作用，尤其是对高纯度样品的分析[21]。其原理是将高压施加到由分析材料制成的电极上，产生火花等离子体。由于所生成离子的动能分散较大，通常使用昂贵的双聚焦质谱仪才能获得足够的质量分辨率。考虑到火花等离子体的重现性差以及需要制成导电样品电极的要求[22]，火花离子源随后被辉光放电（glow discharge，GD）源所取代。GD通常在低压条件下工作，产生低能量的离子。当与质谱仪结合使用时，具有比火花离子源更高的检测灵敏度。但是，常规的直流辉光放电质谱法（GDMS）难以分析非导电材料。另外，由于商品化射频辉光放电质谱仪价格昂贵，因此GDMS的应用在一定程度上受到限制[23]。近年来，激光剥蚀-电感耦合等离子体质谱法（laser ablation inductively coupled plasma mass spectrometry，LA-ICPMS）已被广泛使用[24]，其中剥蚀和电离过程分为两个步骤，使得可以对两个过程进行单独优化，减少了基体效应的干扰。激光剥蚀与ICPMS的结合为固体样品的表面微区分析提供了灵敏而精准的工具。

自Honig和Woolston首次将激光技术用于元素直接分析以来，激光电离质谱技术就被认为是一种强大的分析工具[25-27]。激光具备方向性好、单色性好、相干性好、亮度高四大特性，且对作用的样品没有导电率的要求等优势[28]。激光照射样品之后，样品表面物质瞬间被加热熔化、蒸发、激发和电离。已有大量研究阐述了激光与物质的相互作用机理和作用过程产生的等离子体的性质[29]。Amoruso等[30]总结了在高、低激光功率密度时激光和物质相互作用的不同过程。他们指出，由脉冲激光的前沿产生的羽流是一种稀薄的介质，并且激光束通过由低激光功率密度产生的羽流过程中，其能量几乎没有衰减。然而在高激光功率密度的照射下，气态样品的温度高到足以引起原子激发和电离，该羽流又继续吸收入射激光的能量，从而导致样品分子解离和等离子体形成。许多研究表明，当激光功率密度在 $10^9 \sim 10^{10}$ W/cm^2 的范围内时，各个元素的相对灵敏度因子（relative sensitivity factor，RSF），也即是相对灵敏度系数（relative sensitivity coefficient，RSC）都非常接近，趋近于1，有利于元素分析[31]。值得说明的是，RSC表征的是待测元素相对于内标元素的灵敏度，一般认为RSC越接近于1，各元素的响应越均一，分析结果越准确。基于以上优势，激光电离质谱（laser ionization mass spectrometry，LIMS）技术为固体样品的多元素直接分析提供了可行性方案，最近几十年来吸引了众多原子光谱学家从事该领域的研究。

与LA-ICPMS和GDMS相比，LIMS的谱图干扰更小，对样品平整度和几何形状的要求也较GDMS宽松[32]；与LA-ICPMS和SIMS相比，LIMS需要的仪器更简单，离子的动能分散更少，并且相对于SIMS，基体效应更少[33]。重要的是，LIMS可用于非金属元素的定量分析[34]，这是大多数无机质谱仪的主要难点。

本章的讨论集中在激光电离质谱分析方法上，这种类型的仪器已经成为固体样品直接元素分析的多功能工具。我们将主要介绍与激光电离源结合的分析仪器及其代表性应用，包括激光微探针飞行时间质谱（laser microprobe time-of-flight mass spectrometry，LAMMA）、激光电离四极杆质谱（laser ionization quadropole mass spectrometry，LI-QMS）、激光电离双聚焦质谱（laser ionization double focusing mass spectrometry，LI-DFMS）、激光电离傅里叶变换质谱（laser ionization Fourier transform mass spectrometry，LI-FTMS），重点介绍在发展激光电离飞行时间质谱仪进程中涌现的代表性仪器及其应用，包括典型的激光电离飞行时间质谱仪（typical laser ionization time-of-flight mass spectrometer）、带有轴对称静电扇区质谱仪的激光电离飞行时间质谱仪（laser ionization time-of-flight with axially-symmetric electrostatic sector mass spectrometer）、带离子引导冷却池的激光电离飞行时间质谱仪（laser ionization time-of-flight mass spectrometry with ion guide cooling cell，LI-TOFMS-CC）、气体辅助激光电离飞行时间质谱仪（buffer-gas-assisted laser ionization time-of-flight mass spectrometry，BGA-LI-TOF），并以实验数据证实了BGA-LI-TOF对固体金属样品（NIST标准样品SRM663）的半定量分析能力。最后，介绍了一些具有代表性微型LI-TOFMS及其在航空航天工作中的应用，凸显其在太空探索方面的巨大潜力。

2.1　激光微探针飞行时间质谱

LAMMA由Fenner和Daly于1966年首次报道[35]。在实验中，使用钇铝石榴石激光器（Nd:YAG）对待测样品进行电离，并使用带有静电扇区的TOF作为质量分析器，但质量分辨率较低，检出限较高。此后，在接下来的几十年中，该仪器的灵敏度和分辨率都获得了显著提升[36]。

LAMMA 将激光束（激光功率密度在 $10^7 \sim 10^9$ W/cm^2 的范围内可调）聚焦在样品表面，用来解吸和电离微米级的固体样品。使用中等的激光功率密度，可以将剥蚀后的样品分解并电离成不同的离子，例如单原子离子、分子离子和同位素簇离子。由于单脉冲激光只会解吸出极微量的样品，因此通常需要具有高传输效率和超高灵敏度的质量分析器来进行分析。飞行时间质谱较适合担任这项工作，除此之外，它还具备响应时间短、可以同时进行多离子检测、机械结构简单及理论上没有质量上限等优点[37]。更重要的是，TOFMS 的脉冲离子提取特性完美匹配脉冲激光离子源。激光电离和 TOFMS 结合使用的可行性已经通过商品化的 LAMMA 仪器得以验证，该技术仅通过改变加速系统电极电位和电极的电场方向即可用于正离子和负离子的检测。该仪器是可用于生命科学、气溶胶、冶金、半导体、有机化合物和单细胞研究的灵敏微量分析工具。

2.1.1 仪器与方法

LAMMA 最具有代表性的三款商业仪器是 LAMMA 500（Leybold Heraeus）[38]、LAMMA 1000（Leybold Heraeus）[39]和 LIMA-2A（Cambrige Mass Spectrometry Ltd）[40]。自 1975 年开发出第一台商用仪器以来，LAMMA 在分析领域已经进行了许多出色的研究并得到广泛应用。

LAMMA 500 仪器最初是为分析薄层样品（例如组织切片）而开发的。LAMMA 500 仪器示意图如图 2-1 所示。离子源系统配有两个激光器，其中波长为 613 nm 的 He-Ne 激光器用于解吸样品，激光束通过透紫外浸没物镜（32×和 100×）被聚焦到质谱分析区域。而单脉冲调 Q 激光（波长为 266 nm，脉宽为 15 ns）用于电离样品，激光功率密度可以通过电动调节激光能量衰减控制。将样品放置在真空腔体中（气压为 10^{-4} Pa）一个二维样品台上，可以通过薄石英真空密封观察窗查看。脉冲激光发出的光束聚焦到显微镜下的样品上，瞬态记录仪接收到激光脉冲同步的光电二极管发出的触发信号。该仪器以透射模式工作，激光束穿透样品，离子将从样品的另一侧飞出并通过一组离子透镜进行聚焦，再通过 1.8 m 长的无场飞行区后，进入反射区校正离子的初始能量发散，提高质量分辨率[41]。最后，自由飞行管的末端通过开放式二次电子倍增器检测到离子，输出信号存储在具有 8 位精度和 10 ns 时间分辨率的瞬态记录仪中。输出信号可以显示在示波器上，也可以传输到计算机中进行数据处理。

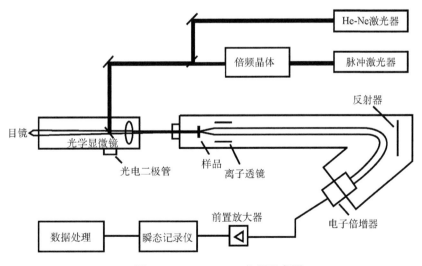

图 2-1　LAMMA 500 仪器示意图

透射模式的优点是可以将光学透镜放置在非常靠近样品的位置，从而使激光光斑的尺寸接近衍射极限[42]。但是，透射模式会降低离子产率，因此根据传输模式的特性，样品通常需要足够薄（薄片通常为0.1～2 μm，颗粒通常为0.5～5 μm）以使激光束能够透过。结果发现，不同几何形状的样品可能导致元素定量误差[43]。值得说明的是，在实际中难以制备可支撑在透射电子显微镜（TEM）栅网上的超薄样品。对于薄组织切片，最常用的方法是在切片之前将样品嵌入在环氧树脂中。另外，将样品安装在真空窗口后面的正确位置更是一个耗时的过程。

LAMMA 500的检出限和横向分辨率主要取决于激光功率密度。通过调节滤光片衰减器可以将激光束衰减到1%，从而产生较低的激光功率密度和较小的激光光斑直径。但是，若要获得更高的横向分辨率，将会降低仪器的检测灵敏度。另外，当激光束从样品的一侧穿透到另一侧时，会对测试的样品造成一定的破坏性。因此，通过该技术无法对待测样品进行深度剖面分析。

LAMMA 500仪器由于其配置限制，难以对待测样品进行整体分析，所以该仪器主要用于薄层样品的分析。因此，开发了一种"两步分析"法以扩展LAMMA 500仪器在对样品进行整体分析的应用[44]。第一步，样品在没有加速电压的情况下被入射激光束蒸发并重新沉积到石英真空密封垫上。第二步，用与第一步相同的激光剥蚀电离法对沉积样品进行分析。尽管这种"两步分析"法略有成效，但它的作用过程比较复杂，且仅适用于样品成分通过等离子体重组而不发生改变的特殊样品。

LAMMA 1000在LAMMA 500的基础上进行了改进，克服了LAMMA 500受限于样品形状的不足，可分析的样品类别也得到很大程度的扩展，例如金属、地质材料、不规则几何形状的半导体等。该仪器的原理与LAMMA 500相似，主要区别表现在离子采样方式的不同。在LAMMA 1000中，激光束由45°方向照射到样品表面，离子沿法线方向离开样品表面，然后加速到TOF分析器中。此外，LAMMA 1000仪器将样品安装在三维样品平台上，而不是LAMMA 500中的二维平台上（见图2-2）。

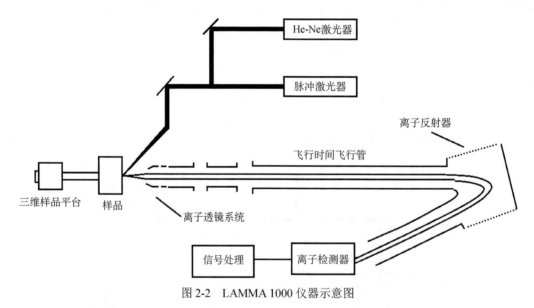

图2-2 LAMMA 1000仪器示意图

LIMA-2A仪器同时具备LAMMA 500和LAMMA 1000的采样方式。对于薄片样品，该仪器以透射模式运行，而对于块状样品，则以反射模式运行。该仪器采用TOF作为质量分析器，配备了可容纳8个样品的可旋转式样品台。随着性能的增强，LIMA-2A被广泛用于电介质中杂质的鉴别、微区分析以及较厚样品深度剖析等领域。

2.1.2 元素分析应用

LAMMA的卓越功能使我们能够分析多种类型的样品材料，尤其是随着LAMMA 1000的发展，对多种样品进行整体分析已经成为人们关注的热点。但是，由于缺乏离子形成过程的详细理论描述，使得 LAMMA 的定量分析可信度大打折扣。Hercules 等[45]建立了解吸/电离模型，以供进一步研究。结果表明，在激光解吸/电离的过程中，至少有五个过程，这是基于直接电离区域在充分接近热平衡的条件下进行的。整个过程包括激光将固体直接电离、在激光作用周围的区域中固体的解吸和电离、表面电离、气相反应和中性粒子的发射与离子形成。但是，目前还是无法提供作为待测样品和激光束参数的函数理论模型，仍然需要使用经验程序来对样品元素浓度进行量化。对 Na 和 K 元素的定量分析结果显示，与原子吸收光谱法（atomic absorption spectroscopy，AAS）和仪器中子活化分析（instrumental neutron activation analysis，INAA）所得的结果相比，差异约为 4 倍[46]。不仅如此，制备具有均匀元素分布的标准样品和寻找合适的解决方案相当复杂且困难。同时，每个脉冲激光能量的变化以及各元素之间相对灵敏度的巨大差异也限制了其定量分析应用。

2.1.3 激光微探针质谱的特点

如上所述，LAMMA 技术允许在高激光功率密度辐照下进行不同样品的元素分析。以下部分总结了 LAMMA 在元素分析中的潜力和局限性。

潜力：

（1）空间分辨率高。可以将激光微探针聚焦成微米级乃至亚微米级的光斑，这对样品的微区分析是相当有利的。

（2）可进行正负离子分析。通过更改相关的电极电位，LAMMA 仪器可在正负两种模式下运行。

（3）可获得元素和分子信息。根据激光功率密度，可以获得有关单原子离子、分子离子和其他各种碎片离子信息。

（4）检出限低。元素的相对检出限在 ppm 量级，金属中某些元素的绝对检出限可以达到 10^{-20} g[47]。

局限性：

（1）质量分辨率差。由于用于解吸和电离的激光功率密度高，并且缺乏有效的方法来冷却具有较大动能分散的离子，因此质量分辨率差是 LAMMA 技术不可避免的缺点。

（2）精度低。由于分析过程中探测器离子电流的不稳定，造成准确度相对较低，并且实验结果的精度与实验人员的操作经验关联很大。

（3）严重的光谱干扰。高激光功率密度下的多电荷离子和低激光功率密度的分子离子是光谱干扰的主要来源。

（4）难以进行定量分析。重现性差会导致相对灵敏度发生较大变化，而且标准样品的制备相当困难。

（5）两个激光系统导致仪器复杂且昂贵。

总体来说，该激光微探针质谱仪作为一种强大的定性分析工具，可提供固体样品微米尺度的化学成分信息。对样品局部分析的特点使 LAMMA 技术在科学技术中应用广泛。例如，生物医学需要激光微探针来研究亚细胞水平的有害元素、药物和代谢产物[48]。然而，由于其局限性和其他先进技术的出现，激光微探针飞行时间质谱仪已逐渐被其他新颖技术（如 LA-ICPMS）取代。

2.2 激光电离四极杆质谱

在当前常规分析中，激光电离四极杆质谱仪因具有低成本、小尺寸、机械结构简单等优异特性而被广泛使用。许多商业仪器已采用 QMS 作为优选的质量分析器，如气相色谱-质谱（GCMS）和电感耦合等离子体质谱（ICPMS）。此外，除被用作质量分析仪外，四极杆还可被用作碰撞诱导解离的碰撞池[49]、冷却池和离子传输装置等[50]。

基于四极杆质谱上述诸多优点，其被广泛用于激光电离质谱。例如，Hardin 及合作者[51]开发了一种采用常规四极杆系统的激光电离质谱仪，用于分析非挥发性样品，但是其重现性较差。Kuzuya 等[52]开发了另一种基于四极杆的仪器，被认为是商用四极杆质量分析器和激光微探针系统的组合。该仪器使用调 Q 脉冲固体激光器（波长为 1064 nm，脉宽为 10 ns）作为电离源，使用脉冲发生器将脉冲激光与四极杆质量检测同步。对包括纯金属、合金和陶瓷在内的不同固体样品进行定量分析，结果显示出较好的重现性[相对标准偏差（relative standard deviation，RSD）约为 1%]，并且同位素丰度比与自然值之间具有良好的一致性。QMS 中的元素检测通常以单扫描或多扫描模式进行，但是在单扫描模式下，对每次激光照射，只能重复分析选定质量的离子。即使可以在多重扫描模式下获得完整的质谱图，也因质谱扫描速度低而造成整个分析非常耗时。此外，在两种模式获得的质谱图中均表现出较低的质量分辨率和高背景干扰。

Beccaglia 等[53]将激光电离四极杆质谱仪用于定量分析土壤样品中的砷元素和钙元素。将激光（1064 nm，10 ns，50 mJ）聚焦在目标样品表面上，光斑直径为 480 μm。实验中研究了激光重复频率、能量等参数，同时也对基体效应进行了评估。结果表明，不同元素在锌基体中的基体效应较铝基体更小；在定量分析时，使用信号强度不同的元素组成进行校准，获得 As 的检出限为 70 ppm，但是 As(Ⅲ) 和 As(Ⅴ) 之间没有形态上的区别；Ca 元素测量结果与通过 X 射线发射获得的值非常吻合。

作为激光电离质谱仪的质量分析器，四极杆保持了上述固有优势。同时，质量分辨率低、质量歧视和具有较低的质量上限等也在一定程度上限制了其应用。但除了元素分析外，四极杆还与激光技术结合进行其他课题研究。例如，由 Torrisi 等[54]开发的 LIQMS 仪器结合了静电离子偏转技术，用于研究激光诱导等离子体的特性。该静电装置用于监测从等离子体中喷出的粒子，粒子质量范围为 1~300 amu，由中性和带电粒子组成，能量分布为 1 eV 至 1 keV。采用具有 10^9 W/cm^2 激光功率密度的二倍频固体激光器（532 nm，3 ns）处理不同金属（包括铝、锌、钽和铅）。Sage 等[55]设计并验证了带有激光电离源的四极杆质谱仪，可用于测量真空中固体样品（包括 Ni、Al 和 ZnO）离子的动能。实验中使用高激光功率密度的固体激光器产生大量带电离子，而后选取了 Ni（$n \leqslant 3$）的多重电荷离子进行了详细研究，结果表明不同种类离子的速度依次为 $v(Ni^+) < v(Ni^{2+}) < v(Ni^{3+})$。通过麦克斯韦位移函数（shifted Maxwellian function）较好地描述了速度分布。

2.3　激光电离双聚焦质谱

扇形磁质量分析器结构也十分简单，但其在可接受离子能量范围和质量分辨率方面性能较差。为寻求仪器的高分辨率，通常将扇形磁场与扇形电场结合，即在扇形磁场之前加一个扇形电场，组成双聚焦质量分析器。双聚焦质量分析器根据两个扇形场区之间的不同角度，可分为 Mattauch-Herzog、Nier-Johnson、Baubrige-Jordan、Hintenberger-Koning、Takeshita 和 Matsuda 等结构，其中最常用的是前两种结构[56-58]。最初将双聚焦质量分析器与火花源离子源结合用于固体样品分析，但是火花源在离子化方面的适用性低于激光离子源[59]。由于激光离子源和双聚焦质量分析器的组合具有低噪声背景和高质量分辨率等诱人的性能，所以激光离子源也逐渐取代了火花源。因此，在本节中，将主要讨论在当前分析中应用广泛的 Mattauch-Herzog 结构的激光电离双聚焦质谱仪。

图 2-3 列出了 Mattauch-Herzog 结构的激光电离质谱仪的示意图，该几何结构是由 31.82°的扇形电场区和与离子飞行方向具有相反曲率的 90°扇形磁场区组成。这两个扇形场区可用于离子飞行方向和速度的聚焦。扇形电场区部分由两个同心弯曲的板组成，在这些板上施加一定的电压，可控制离子束在通过分析器时严格弯曲。此外，在扇形电场区的平面上放置了一个能量狭缝，该狭缝可传输较窄能量范围的离子。一种特殊的感光光板被采用作为该质谱仪的检测器。值得说明的是，扇形电场区中离子轨迹的半径取决于离子的动能，因此扇形电场区充当飞入离子的能量过滤器。随后的扇形磁场区根据动量和电荷将磁场中的离子分离。在整个分析过程中，可以通过同时扫描加速电位和磁场来获得完整的质谱图。

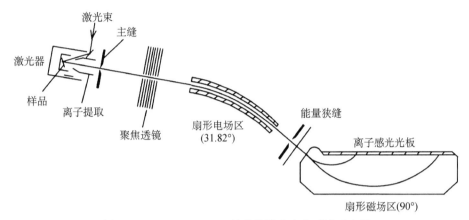

图 2-3 Mattauch-Herzog 结构的激光电离质谱示意图

由 Honig 和 Woolsen 研制的激光电离质谱耦合了一个双聚焦的 Mattauch-Herzog 质谱分析仪 (AEI-MS7)，利用微秒脉冲激光聚焦到待测样品上进行离子化，每一个激光脉冲产生大约 1010 个离子。利用该装置，他们在真空中分析了一些具有代表性的金属、半导体和绝缘体样品，将其中样品表面杂质的检出限降到了几个 ppm 的水平。

Jochum 等[60]还采用了配有激光电离源的双聚焦质谱仪，用于同时检测岩石样品中约 30 种微量元素。在他们的研究中使用了具有 Mattauch-Herzog 结构的 AEI-MS 7 商品化仪器。将波长为 1064 nm 的脉冲激光束聚焦在样品表面，可以产生直径 40 μm、深度约 3 μm 的弹坑。总离子量由位于磁区之前的检测器测量，使用感光片进行多元素检测。他们还研究了激光能量、聚焦位置等实验参数对离子产率的影响。结果表明离子产率在焦点位置处显示出最大值，在高激光能量（1～2 mJ）下，总离子产量随 M 型曲线变化而增加，这意味着有两个最佳聚焦位置可获得最大离子产量。而当激光能量大于 2 mJ 时，离子产率几乎与激光能量甚至样品材料无关。此外，他们还使用内标元素和相对灵敏度因子校准了未知地质样品的离子强度。地质基质样品的整体分析结果表明，除部分碱性元素电离电位低外，其余元素的 RSF 值均接近于 1，通过激光电离质谱法测量的浓度与文献中表明的约四个数量级的结果非常吻合，微量元素的检出限约为 0.01～0.1 ppm。

AEI-MS 7 仪器的另一个应用是对地质玻璃（geological glasses）[61]的微量元素分析。它使用激光功率密度约 10^{10} W/cm^2 的激光束产生直径约 20 μm、深度约 3 μm 的弹坑。同时作者还开发了一种特殊程序，用于使用内标元素和相对灵敏度因子定量测定元素浓度，从而提高了分析精度和准确性，并分别校正了不同元素在离子形成、传输和检测过程中的误差。值得说明的是，RSF 值与元素电离能、原子半径和沸点有关，根据元素电离能、原子半径和沸点的经验公式计算的结果，实验的 RSF 值小于 10%。RSF 实验值与根据经验公式计算的值相符。在 0.1～10 000 ppm 的浓度范围内，大多数元素的结果与实际值非常吻合。同时，通过与其他技术[例如 LA-ICPMS、INAA、同步加速 X 射线荧光光谱法（synchrotron X-ray fluorescence，SY-XRF）和热电离质谱（thermal ionization mass spectrometry，TIMS）]获得的结果进行比较，验证了该技术分析的准确性。但由于光电板离子检测器的特性，所以 AEI-MS 7 仪器分析费时且灵敏度和精度均有限。

在双聚焦质谱分析中，离子信号的采集效率是一个重要的参数。Conzemius 及其合作者[62]使用了一种特殊的激光器，该激光器以高重复频率运行产生离子，与其他类似系统相比，该信号采集所花费的时间缩短了约 20 倍。施加到样品表面的激光功率密度约为 10^9 W/cm^2，并且离子检测器能够同时测量从 M 到 $36M$ 的质量范围（M 可变，如果 $M=7$，则可检测元素范围覆盖 Li 到 U）。通过使用标准样品，获得了相对均匀的离子强度，用于钢样品中痕量元素的测量。然而，该系统的主要缺点是在质谱图中会出现大量的多电荷离子。

激光电离双聚焦质谱仪显示出常规元素分析的多项优势。它以其高灵敏度、较好的重现性和高质量分辨率以及所有质谱仪中最佳的定量能力而闻名。但是，与其他使用动态分离系统的仪器（如 TOFMS 和 QMS）相比，双聚焦质量分析仪结构更加复杂、成本也更高。另外，由于双聚焦扇区的分辨能力由狭缝宽度确定，通过减小狭缝宽度虽然可以获得更高的质量分辨率，但是却会降低仪器的检测灵敏度。

2.4　激光微探针傅里叶变换质谱

目前，使用 TOF 分析器的激光微探针质谱分析已经得到充分的研究，并为多种材料的定性分析提供了有力手段。但是，由于质量分辨率低、准确性低及定量分析可信度较低，导致使用 TOF 分析的 LAMMA 技术并未被广泛接受。TOF 分析器未能充分发挥出激光在实际识别方面的潜力，因此开发了激光微探针傅里叶变换质谱。激光离子源与 FTMS 的结合为待测样品成分的分析提供了强大的识别能力。因为它是唯一提供微区分析以及高质量分辨率（通常超过 100 000）和高质量精度（通常优于 1 ppm）的仪器。

2.4.1　仪器和原理

将 FTMS 引入激光微探针质谱的首次尝试是在梅茨大学开始的[63]。据报道，该仪器所测元素离子的质量分辨率超过 100 000，质量精度优于 10 ppm。该系统通常由磁体、分析器单元、超高真空系统和复杂的数据处理系统四个部分组成。根据分析器系统的复杂框架，FTMS 通常与外部激光离子源耦合。需要特别指出的是，在这些仪器中，可以通过引入第二束激光进行后电离来增强电离效率。图 2-4 是由 Van Vaeck 等开发的代表性激光微探针 FTMS 的仪器示意图，固体激光器（λ=266 nm，τ=10 ns，E=10 mJ/脉冲）以 45°的入射角作用在样品表面[64]。与 LAMMA 500 仪器相同，使用第二个 He-Ne 激光器照射分析区域，一般可以获得优于 5 μm 的横向分辨率。经激光作用产生的离子经离子传输区被引入至圆柱形开放式分析器单元，该单元分别由捕获电极、激发电极和检测电极的六个电极组成。对该单元进行定向，以使圆柱体的主轴与磁场对齐。该装置中的捕获电极是电极末端的两个圆柱体，中心圆柱体被分为四个电极，分别用作激发板和检测板。施加到捕获电极的电压极性决定了电极中是否保留了正离子或负离子。在分析器单元中，离子受到垂直于离子速度方向和超导体磁场（通常为 3～5 T）的洛伦兹力的约束，该力使离子沿着以回旋加速器频率为特征的圆形轨道行进，该回旋加速器频率仅与磁场保持恒定时的质荷比相对应。对于 ^{28}Si$^+$ 离子，质量分辨率高达 2 200 200。

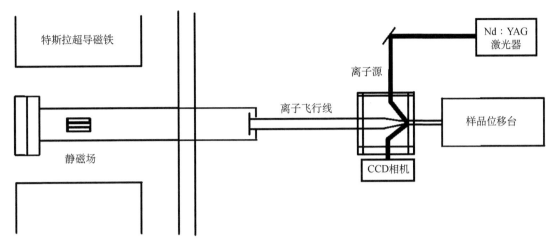

图 2-4 外部激光离子源耦合的激光微探针傅里叶变换质谱仪示意图

2.4.2 元素分析的功能和应用

带有 FTMS 的激光微探针兼具微探针和 FTMS 结合的显著优点，这使其成为分析领域的重要工具。首先，通过快速扫描可以同时检测具有不同 m/z 的离子，而不需要像其他质量分析器（四极杆、飞行时间质量分析器等）将离子彼此分离，因此所有的离子都能被充分地利用以获得高灵敏度。其次，激光微探针 FTMS 的质量分辨率比任何其他激光微探针质谱仪都要高得多，通常在 10 万以上，即使在高质量范围也不会影响其灵敏度和精度。此外，分辨能力仅与回旋加速器频率和瞬态信号的持续时间相对应[65]，从而可以使用长瞬态信号获得高分辨率。第三，在高稳定的超导磁场中，可以准确地检测回旋频率，具有较高的精度。因此，如果仅保持分析器的高真空度并适当控制其中的离子数，即使没有内标，也可以在 ppm 范围内准确测定质量。值得说明的是，FTMS 最吸引人的特点是广泛的离子捕获和操纵功能，这为详细探测离子结构提供了便利。由于激光微探针 FTMS 在元素分析方面具有显著的优势，因此涌现出了大量的相关研究。此外，它还广泛用于分子分析[66]和无机化合物的直接形态测定。

Bakker 等[67]通过激光微探针 FTMS 研究了离体大鼠心脏的局部钙浓度，该离体大鼠无论是否有随后的再灌注，都经历了短暂的局部缺血。将该方法与形态电子显微镜（EM）技术进行了比较，结果表明：对连续分析样品进行局部浓度比较评估是可行的。Kockx 等[68]进行了另一项与移植的人体静脉中钙沉积有关的研究，在该研究中他们使用激光微探针 FTMS 来确定钙沉积的性质，获得的正负离子质谱图可以识别样品中的离子是自由还是成键的。Erel 等[69]使用激光微探针 FTMS 来区分天然和人造蛋白石宝石。在确定不同蛋白石的指纹谱之前，他们还研究了激光波长、激光功率密度和样品制备对实验结果的影响。实验使用波长为 355 nm 的倍频固体激光器（脉宽为 4.3 ns，脉冲能量 0.80 mJ）和另一种波长为 221.67 nm 的 TDL 90 染料激光器（脉宽为 4.3 ns，脉冲能量 0.25 mJ）作为外部离子源。在两种离子检测模式下检测特定物种，可确保获得重要的区分信息。在正离子模式下，人工蛋白石的特征是含有铅和大量的锆原子，在负离子模式下发现氧化态，而在天然蛋白石中发现铝、钛、铁和铷。另外，两种蛋白石中的特殊离子也可为鉴别提供帮助。值得说明的是，与其他用于分析这些珍贵宝石的质谱技术相比，激光微探针 FTMS 是低损的。

激光微探针 FTMS 的优势非常明显，它可以在一秒钟内提供具有超高分辨率的完整质谱图。而且由于高分辨率，可以简单地避免质量峰的某些重叠。除此之外，在生物分子的结构分析中，该技术可以使用串级质谱来对分子结构进行分析。但是由于整个系统的高成本以及烦琐的日常维护，限制了激光微探针 FTMS 仪器的应用范围。除了价格高昂之外，FTMS 系统的复杂性几乎消除了内部建造的可行性，即只有商业仪器可用。此外，激光微探针 FTMS 需要熟练的操作人员，而激光微探针 TOFMS 更适合多用户操作[70]。和 LAMMA 受限的原因类似，激光微探针傅里叶变换质谱也很难进行定量分析。

2.5 激光电离飞行时间质谱

LIMS 是以激光直接电离样品作为离子源的质谱技术的总称，在 Honig 和 Woolston 于 1963 年首次应用后，引起了业界的广泛关注[71-74]。LIMS 技术的优势是：可以对任何具有吸光特性的固体样品进行分析，包括导体、半导体、电介质和粉末。不仅如此，在 10^9~10^{10} W/cm² 的激光照射下，各元素的相对灵敏度趋于均匀[22,31]，所以只需要使用某一元素的峰面积除以谱图中所有元素的峰面积，即可得到该元素在待测固体样品中的组分。激光离子源可以结合不同的质量分析器对固体样品进行直接分析，但由于飞行时间分析器（TOF）理论上无质量上限、扫描速度快、可进行全谱分析，并且能够与脉冲式电离源完美结合，故其发展较快。此外，激光电离飞行时间质谱在离子源内使用缓冲气体可以大大降低多电荷离子引起的谱图干扰，提高仪器的分辨率，即使在缺乏标准样品的情况下，也可实现固体样品的半定量分析。

本节将介绍激光电离飞行时间质谱仪在不断改进和发展中涌现的代表性仪器及其应用，包括典型的激光电离飞行时间质谱仪（typical laser ionization time-of-flight mass spectrometer）、带有轴对称静电扇区质谱仪的激光电离飞行时间质谱仪（laser ionization time-of-flight with axially-symmetric electrostatic sector mass spectrometer）、带离子引导冷却池的激光电离飞行时间质谱仪（laser ionization time-of-flight mass spectrometry with ion guide cooling cell，LI-TOFMS-CC）、气体辅助激光电离飞行时间质谱仪（buffer-gas-assisted laser ionization time-of-flight mass spectrometry，BGA-LI-TOF），结合笔者课题组自行搭建的 BGA-LI-TOF 并以实验数据证实了其对固体金属样品（NIST 标准样品 SRM663）半定量分析能力。最后，介绍了具有代表性的微型 LI-TOFM 及其在航空航天工作中的应用，凸显其在太空探索方面的巨大潜力。

2.5.1 典型的激光电离飞行时间质谱仪

与 LAMMA 技术不同，当前激光电离飞行时间质谱仪通常使用 10^9~10^{10} W/cm² 的激光功率密度。在如此高的激光功率密度下，分子和多原子离子的含量将大大降低，但却无法避免离子动能分散和多电荷离子的产生。如果没有适当的装置来降低离子的动能，其质量分辨率将被限制在 LAMMA 技术水平。但是随着计时电路和高速数据采集系统的发展，激光电离飞行时间质谱仪的分辨率能力已经有了很大的提高。

图 2-5 是 Joseph 等开发的激光电离飞行时间质谱仪的示意图[75]，使用基频波长为 1064 nm、脉宽为 8 ns 的固体激光器剥蚀样品。建立了 Mamyrin 型反射式飞行时间质谱仪，其中检测器由两个微通道板（micro channel plate，MCP）组成，信号通过前置放大器进行放大，并由数字示波器进行记录，该仪器的质量分辨率约为 1000。该质谱仪用于研究掺杂有某些轻质量稀土元素（La、Ce、Sm、Nd）的 UO_2 的固体和液体样品，其中液体样品干燥以形成薄片，而固体样品直接粘贴到铝板上即可，处理完成的待测样品置于 $1×10^{-6}$ Torr 压力下的腔室中进行分析。后来该仪器被用于测量硼元素在硼酸

溶液中的同位素比[76]。实验中还尝试了使用二倍频波长，但是获得的激光功率密度相对较低。在硼酸样品中，所测量的 ^{10}B 同位素组成的精度在±1%范围内。同时，将结果与通过热电离质谱法测定的值进行了比较，结果表明两者之间具有较好的一致性。

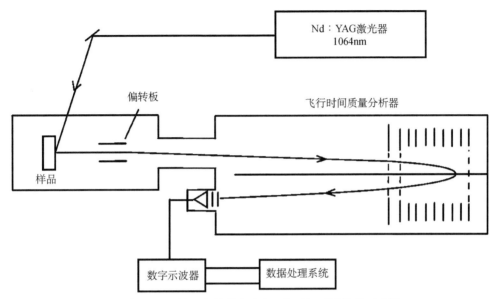

图 2-5 Joseph 等开发的激光电离飞行时间质谱仪的示意图

Mróz 等[77]开发了一种激光电离反射式飞行时间质谱仪，用于定量测量 Al-Li 样品中轻元素（H、C、O）的化学成分。使用脉冲激光器（1064 nm，5 ns）作为电离源，采用大约 $3×10^9$ W/cm^2 的激光功率密度来减少多电荷离子的数量。他们使用伽利略微通道板模拟粒子增益的测量方法对结果进行定量处理，但所得结果中锂与铝之比的测量值远高于真实值。

在常规 LIMS 中，常用低激光功率密度来进行电离，这不适用于获取样品的化学计量组成，所以难以获得令人满意的定量分析结果。但是，在高激光功率密度电离的情况下，会形成大量的多电荷离子。另外，由于被高能激光剥蚀后的离子的动能分布较大，因此不可避免造成质量分辨率不足，难以消除同质量数元素干扰。

2.5.2 带有轴对称静电扇区质谱仪的激光电离飞行时间质谱

Eubank 等在 1963 年研发了首个具有轴对称静电扇区的 TOFMS 系统，该系统仅由飞行管和圆柱形冷凝器组成[78]。Sysoev 等[79]开发了一种结构紧凑的 TOFMS 仪器（LAMAS-10M），用激光对固体样品进行照射时具有较好的重现性，尤其是该分析仪在静电部分和飞行时间质量分析器之间具有两步聚焦过程。该仪器可用于地质、环境监测、无机和有机化学等多个领域研究。

LAMAS-10M 结构设计紧凑，其原理图如图 2-6 所示。真空、机械和电子设备的所有组件均安装在尺寸为 55×55×80 cm^3 的外壳中。离子源使用激光电离源，其发射波长为 1064 nm，脉宽为 7 ns，重复频率为 50 Hz。激光入射角为 45°时，聚焦光斑大小约为 50～70 μm，激光功率密度为 $3×10^9$～$6×10^9$ W/cm^2。值得注意的是，该离子源通过一个可调衰减器调节激光功率密度。该仪器安装了自动扫描系统，可扫描表面积约 20 mm^2 的样品，并在真空腔体与换样腔体二者之间装有一个插板阀便于快速更换样品。为了观察聚焦点和激光等离子体的产生，采用了 70×的光学显微镜。激光剥蚀样品产生的离子在萃取电压作用下传输到飞行时间质量分析器。静电分析仪以圆柱形离子冷却池的形式制成，离子冷却池中的离子偏转角为509°。TOF 区具有两个轴向对称的无场飞行区，它们分别位于静电场区之前和之后。在离子冷却池出口和检测器之间，使用电位管降低离子飞行速度，从而可以有效降低离子的空间分散。使用二次电子倍增管作为检测器，并通过扫描静电分析仪的幅值获得样品质谱图。

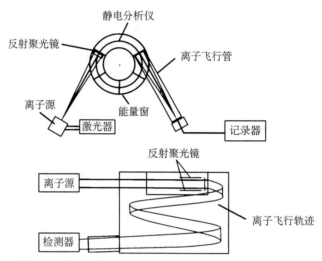

图 2-6　激光电离飞行时间质谱仪 LAMAS-10M 示意图

LAMAS-10M 被用于金属和地质样品中的元素分析[79]。对钨进行杂质分析发现 $^{43}Ca^+$ 的信噪比约为 10，检出限约为 3~4 ppm。对富硼的样品分析发现，测得的同位素丰度与标准比值吻合得很好，相对标准偏差小于 2%。此外，该仪器还用于定量分析块状和粉末状样品。例如，测定了块状标准样品青铜 663，结果发现主要元素 Cu 的 RSD 在 0.4%~0.7% 的范围内，而次要元素的 RSD 较高。在这个测量中，元素的浓度通过公式（2-1）计算得到，在计算中忽略了少数双电荷离子。公式（2-1）中 I_i 是质谱图中存在的单电荷离子的强度，单电荷离子的 RSC 由经验方程式[公式（2-2）]计算得出，其中 m_i 为原子质量，φ_{li} 为第一电离势，$\sigma(\varphi_{li})$ 为单电荷离子的电离截面，k 为玻尔兹曼常数，T 为绝对温度。

$$C_i = I_i \mathrm{RSC}_i / \sum I_i \mathrm{RSC}_i \tag{2-1}$$

$$\mathrm{RSC}_i = \frac{\sqrt{m_i}}{\sigma(\varphi_{li})} \exp\left(\frac{\varphi_{li}}{kT}\right) \tag{2-2}$$

将两个硫化物矿石粉末参考样品（Zh928-76 和 RAS-4）压制成片并进行分析。在测定 Zh928-76 中的贵金属时，考虑了来自同质量数元素和双电荷离子的峰干扰。Rh 和 Pd 的结果与标准值相符，Pt 和 Ir 的结果相对较差。这些贵重元素的检出限为 20~50 ppb。Zh928-76 样品在不同实验中的信噪比在 5 倍左右变化。对于 RAS-4 样品，所测元素的组成与大多数元素的标准值相一致。由于该样品的结构非常复杂，因此将另一种方法用于元素分析。以 Fe 为内标，按公式（2-3）计算所有杂质浓度：

$$C_i = C_{im} \frac{C_{\mathrm{Fe,s}}}{C_{\mathrm{Fe,m}}} \times \frac{\mathrm{RSC}_i}{\mathrm{RSC}_{\mathrm{Fe}}} \times \frac{M_i}{M_{\mathrm{Fe}}} \tag{2-3}$$

其中，C_i 为确定的杂质浓度，C_{im} 为从相对质谱信号强度得出的杂质浓度，$C_{\mathrm{Fe,s}}$ 和 $C_{\mathrm{Fe,m}}$ 为已知浓度和从相对质谱信号强度得出的内标浓度，RSC_i 和 $\mathrm{RSC}_{\mathrm{Fe}}$ 为杂质与铁内标的相对灵敏度系数，M_i 和 M_{Fe} 为杂质和铁的平均同位素质量。

与常规的反射式飞行时间质谱不同，LAMAS-10M 的轴对称扇形分析仪中没有栅网，从而避免了直接撞击栅格线或散布在栅格中而造成的离子损失[80]。但是由于该仪器设有能量窗口，离子的传输效率将会受到一定的限制。尽管由于高激光功率密度未观察到簇峰的干扰，但是仍然存在一些双电荷离子和同质量数元素引起质量干扰。在分析能力方面，该仪器的质量分辨率通常为 1000，最佳检测限可以低至 ppb 级，并且具有良好的重现性。

2.5.3 带有离子引导冷却池的激光电离飞行时间质谱

为了解决上述常见的 LI-TOFMS 的局限性，厦门大学杭纬课题组[81-83]将缓冲气体引入离子源，由于氦气质量轻且化学性质稳定，因此被用作缓冲气体。随后，将飞出的具有高动能的离子通过射频多极杆装置进行冷却，并导入反射式飞行时间质量分析器中。与真空中的激光剥蚀过程相比，激光剥蚀羽流与背景气体的相互作用是一个更为复杂的气体动力学过程，涉及剥蚀物质的减速、衰减和热电离，以及冲击波的形成[84]。在适当的离子源气压下加入缓冲气体，多电荷离子可通过三体碰撞复合（three-body collisional recombination）以及电子和缓冲气体中性分子重新结合的过程转化为单电荷离子。同时，使用缓冲气体还可以减小离子的动能分散[85,86]。综上，在采用高激光功率密度的情况下，使用缓冲气体可以大大降低多电荷离子引起的光谱干扰。

射频多极杆装置已被广泛用作不同系统的碰撞池[19,87,88]。电离后进入电场的离子被限制在靠近轴的位置，而阻力则是通过与缓冲气体的碰撞引起的。在多极杆装置中，离子信号受射频电压和工作频率的影响，并且离子在飞行过程中存在质量歧视。研究发现，较轻的离子在较高的频率和较低的射频电压下传输效率更高，而较重的离子在较低频率和较高幅度下传输效率更高。在当前的研究中，大多使用适当的频率和电压应用于射频六极杆导向碰撞池，以获得满意的信号强度和不同元素的均匀传输效率。飞行时间质谱法中的质量分辨率主要受到推斥区中离子的能量分散和空间分散的限制[89]。因此，为了提高分辨率，杭纬课题组采用了由 Dawson 和 Guilhaus[90]所描述的垂直引入式飞行时间质量分析器。

图 2-7 是带有离子导向冷却室的内置激光电离垂直引入式飞行时间质谱仪的示意图。脉冲激光（532 nm，4.4 ns）用于激光剥蚀和电离。将待测样品安装在置于氦气低压环境中的直接进样杆（DIP）上，然后调制激光入射角为 45° 聚焦到样品表面上。第一个六极杆用于传输离子并减少离子的分散，而第二个六极杆用于进一步冷却聚焦离子。锥孔体 2 之后的直流四极杆可以将圆形的束流截面变成压缩的椭圆形，从而让更多的离子通过狭缝。经过离子透镜聚焦之后，离子束到达推斥区并被正交加速。MCP 输出的信号被放大后经过一个时间数值转换器（time-to-digital converter，TDC）以脉冲计数模式来记录离子[77]。

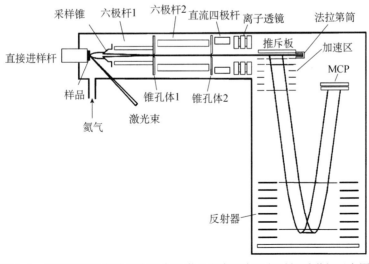

图 2-7 带有离子导向冷却室的内置激光电离正交飞行时间质谱仪示意图

激光电离飞行时间质谱仪通常在 $10^9 \sim 10^{10}$ W/cm² 甚至更高的激光功率密度下工作，以此来减少激光电离过程中的元素分馏效应。由于高强度激光等离子体产生的离子具有较大的能量分布，此时离子源内气压值会对该系统中离子信号和相对灵敏因子影响显著。例如，在高压力下生成的离子可能被过度冷却，从而导致灵敏度降低。在低压情况下，离子与缓冲气体之间的碰撞不足会导致质量分辨率降低，并且多电荷离子会产生严重干扰。此外，激光功率密度、样品位置、采样锥的电势和射频频率等也是影响实验结果的重要参数。

该技术已应用于包括金属、地质和生物样品在内的多类样品的半定量分析。钢样品的质谱图中未发现质谱干扰和多电荷离子峰，同位素比与自然丰度也非常吻合，质量分辨率约为 7000，检测限达到 ppm 级别。此外，无需标准样品即可获得不同元素几乎相同的相对灵敏因子。大量的实验表明，在离子源和六极冷却池中使用缓冲气体可以有效地减少多离子峰干扰。总之，通过这种技术可以实现理想的半定量分析[82]。激光电离飞行时间质谱仪与离子引导冷却技术相结合是一种用于元素定量分析的新技术。在适当的实验条件下，无需基质匹配的标准样品即可完成半定量分析。该仪器的检测极限为亚 ppm 级，但是目前对轻离子的质量分辨率较差。

2.5.4 氦气辅助激光电离飞行时间质谱

由于 TOF 具有脉冲式离子采集的特性，因此 TOFMS 特别适合用于测量来自脉冲激光源的离子。但是，在高激光功率密度下产生的离子具有数百 eV 的动能分散[91]。而 TOFMS 即使在反射模式下，也只能分辨动能分散小于 20 eV 的离子。因此，如何减少离子的动能分散便成为提高激光电离飞行时间质谱仪分辨率的难点。

前面已经简述杭纬课题组创造性地把低压氦气引入离子源系统中，通过缓冲气体与高速运动的离子之间频繁的碰撞来减弱其动能，此时控制合适的气压便能使离子冷却，达到减少离子动能分散的目的。另外，通过对惰性气体辅助激光离子源的理论研究和实验结果发现，高激光功率密度下生成的大量多价离子由于和缓冲气体发生了三体碰撞复合反应，均被降价为单价离子，从而降低了谱图的干扰。下面将详细介绍低压氦气氛围下的激光电离过程和三体复合理论。

与高真空下的激光电离过程不同，氦气氛围中的离子会频繁发生碰撞和复合重组，从而影响离子动能和离子组成。高压等离子体在形成后从样品表面快速膨胀，然后通过与惰性气体分子的碰撞而减慢其膨胀速度。从理论上讲，等离子体在膨胀过程中受到缓冲气体的阻碍，并且等离子体密度长时间保持在数百微秒的高水平，同时伴随着电子温度的迅速下降。从等离子体中逸出的电子被周围的气体俘获[92]，并且这些电子作为第三体与高价离子发生三体复合重组[93,94]。结果显示，电子和离子之间的碰撞导致离子的电荷减少。

$$M^{n+} + e + e \longrightarrow M^{(n-1)+} + e \tag{2-4}$$

三体复合的速率系数（$k_{\text{3b-recom}}$）为

$$k_{\text{3b-recom}} = a\rho T_e^{-5} \text{ cm}^6/\text{s} \qquad (T_e < 3100 \text{ K}) \tag{2-5}$$

$$k_{\text{3b-recom}} = b\rho T_e^{-5.8} \text{ cm}^6/\text{s} \qquad (T_e > 3100 \text{ K}) \tag{2-6}$$

式中，a 和 b 是与离子种类有关的常数，T_e 是电子温度，而 ρ 是库仑对数。

$$\rho = i^3 \ln\left[\sqrt{(i^2+1)}\right] \tag{2-7}$$

这些公式清楚地表明，电子在三体复合中优先被高价离子吸引，例如双电荷离子的复合速率系数比单电荷离子的复合速率系数高约 20 倍。因此，在三体复合过程中，高价离子优先降价，在实验中的最佳条件下，双电荷离子的数量可以减少至少两个数量级[95]。因此，在氦气辅助激光电离飞行时间质谱中，低压氦气的引入不仅可以实现和高能离子碰撞降低动能分散的效果，也可以通过三体碰撞复合减少多价离子的干扰。

如前所述，当前的直接固体分析技术通常需要一系列固体标准品用于元素定量分析。即使对于所谓的"无标准"技术（例如XRF），也必须根据标准的分析结果来开发复杂的数据处理软件，并且只能应用于具有特定矩阵的样品。对于质谱分析，一种无标定量分析的简便方法是使用代表特定元素的离子流除以总离子流，以直接获得元素浓度。但是，这种方法的前提是要实现对不同元素的统一相对灵敏度和可忽略的干扰。

为了实现不同元素的统一相对灵敏度，可通过增大激光功率密度来减少元素响应差异。将氦气引入激光离子源后，在高激光功率密度和合适的缓冲气体压力下，来自多电荷离子和多原子离子的干扰都可以忽略不计，并且不同元素的RSF稳定且接近统一。所以，氦气的引入使LI-TOFMS满足固体样品的直接无标样分析的条件。综上，使用氦气辅助激光电离飞行时间质谱可以实现固体样品的无标半定量分析，这是元素分析中的一项重大进展[96-98]。

2.5.4.1 仪器和原理

低压氦气引入离子源后，大大提高了仪器的分辨率。杭纬课题组利用此技术开发了具有垂直几何结构的高激光功率密度电离飞行时间质谱仪，有效减少了离子动能分散对谱图的影响[81]。激光电离正交飞行时间质谱（laser ionization orthogonal time-of-flight mass spectrometry，LI-O-TOFMS）最初使用射频六极杆传输离子，但它导致了对轻离子（$m/z<40$）的质量歧视[82]。之后，该课题组采用一组圆柱Einzel透镜代替LI-O-TOFMS离子光学系统中的射频多极杆，并将采样锥垂直于激光羽流放置，以便采取具有较小动能分散的离子[99]。通过直接从等离子体中采样离子，LI-O-TOFMS还可用于诊断和表征激光等离子体[100]，从而发展了与缓冲气体辅助离子源相关的等离子体理论[101]。

根据对LIMS的研究，该课题组对高辐射激光电离正交飞行时间质谱进行了改进，用来解决质谱质量分辨率差和严重的多价离子干扰[99]等问题，其示意图如图2-8所示。离子源包括离子源室、自制的进样杆、固定在DIP头部的二维移动平台、调Q脉冲Nd∶YAG激光和激光聚焦透镜。实验时，先将固体样品置于二维移动平台上，再将超高纯氦气引入样品腔，并使用流量计控制样品腔压力。采样锥垂直于激光束安装，以获得碰撞冷却后动能小的离子。一组圆柱形Einzel透镜位于采样锥后部，用来将离子束空间聚焦到狭缝中。Einzel透镜由每侧三个共六个长方体电极组成，这样设置是为了将原始的圆形横截面离子束成形为椭圆形横截面束，所以稍微调整Einzel透镜中两个中间电极的电势，就可以精确地引导离子束穿过狭缝。

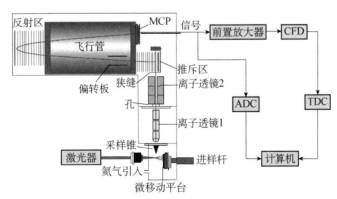

图2-8　具有两种不同数据采集模式的LI-O-TOFMS示意图

质量分析器是自主搭建的小型TOF分析仪，其中推斥板和反射板之间的距离仅约为50 cm。但是，该质量分析器为元素分析提供了足够的质量分辨能力（>2000）。为了保障进入推斥区的离子能够顺利到达检测器，特使用偏转板来调整离子的飞行轨迹，确保离子飞行时间在 X 方向和 Y 方向保持同步。LI-O-TOFMS 的更多细节和主要参数可在参考文献中找到（见参考文献[102]、[103]）。

LI-O-TOFMS 中使用了两种不同的数据采集系统。在第一个系统中，来自 MCP 的信号依次经过前置放大器、峰值鉴别器（constant-fractional discriminator，CFD）、时间-数字转换器（time-to-digital converter，TDC），最后计算机对信号进行数据采集和处理。TDC 采集系统通常可提供具有高信噪比的超清晰质谱图，在第二个系统中，信号直接由示波器显示和记录，然后由计算机对采集信号进行处理。示波器实际上起着模拟-数字转换器（analog-to-digital converter，ADC）的作用。

LI-O-TOFMS 的元素分析能力表现为：一个钢铁标准样品，使用 ADC 数据采集系统，只需 10 s 就可以通过 100 次激光获得清晰的谱图。从谱图结果来看，信号干扰非常小，它们对总离子流的影响或对大多数元素峰的干扰都可以忽略不计。

为了充分发掘 LI-O-TOFMS 在固体样品分析方面的潜力，杭纬课题组对激光电离飞行时间质谱进行了不断地改进和发展，并深入研究了低压氦气辅助激光电离的理论机理。为了优化仪器，还研究了激光功率密度、激光波长、缓冲气体压力和基体效应的相关参数对仪器灵敏度及分辨率的影响[83,102,103]，并对 LI-O-TOFMS 仪器进行了改进，自主搭建了最新的高分辨气体辅助激光电离飞行时间质谱仪（BGA-LI-TOF），该仪器采用了以下技术，大大提高了仪器的分辨率。

（1）采用反射式飞行时间质量分析器。相比于直线式飞行时间质谱，在离子飞行时间上，反射式飞行时间质谱的离子飞行时间是其两倍，飞行距离的增加有助于提高仪器分辨率。另外，离子经过反射区时，初始动能大的离子较初始动能小的离子进入反射区的距离更长，这样离子在返回路程中的一定位置聚焦，这便将离子的初始动能分散便转换为了空间分散，从而实现了离子的空间聚焦，再次提高了仪器的分辨能力。

（2）离子垂直引入。较高的激光功率密度造成了离子的初始速度分散[104]，但是离子反射式飞行并不能解决回头时间造成的初始速度分散，因此采用垂直式将离子引入分析器时，垂直引入的离子在 Y 方向上的动能几乎为零，可以有效减少离子在轴向方向上的能量分散。离子的垂直式引入结合反射式飞行，很大程度上提高了激光电离飞行时间质谱仪的分辨率。

（3）引入低压氦气辅助。将低压氦气引入离子源，高价离子通过三体碰撞复合变成低价离子，多原子离子通过碰撞解离得到分解[93,105-108]。因此，在 BGA-LI-TOF 中，低压氦气的引入对离子的碰撞冷却解离及降价起到了关键作用。

（4）三维移动样品台的使用。可编程控制的压电式三维移动样品台不仅保证了采样过程中离子信号的稳定性，同时实现了实验操作中对样品的自动化操作，不仅分析效率得到了提升，还可以实现对样品进行扫描式三维成像。

（5）采用脉冲串推斥技术。采用脉冲串推斥可以减少多原子离子和多电荷离子的干扰[95,98]。例如，如果 Nd∶YAG 激光器的最大重复频率为 10 Hz，则每个激光脉冲的离子检测时间为 100 ms。从离子的时间分布来看，在激光脉冲之后的第一毫秒对排斥区域中的离子信号进行采样，离子包中分布了多电荷离子、单电荷离子和多原子离子。因此，可以在 TOFMS 检测中使用特殊的排斥方案，在新型脉冲序列模式下，仅在指定的时间段内执行推斥，并且可以调整每列的脉冲数以及激光脉冲和排斥脉冲列之间的延迟时间，选择性地收集单电荷离子。该方案不仅可以有效地从离子包中得到信号较强的单价离子，消除了多原子离子干扰，还可以通过减少无用的推斥脉冲来降低背景噪声，提高信噪比。

2.5.4.2 元素分析的功能和应用

为了体现高分辨氦气辅助激光电离质谱对固体样品的定性和半定量分析能力，我们用上述

BGA-LI-TOF 技术对 SRM663 标准样品进行了分析,并获取其元素含量信息。实验条件如表 2-1 所示,其中激光作用位点与采样锥锥口的轴向距离和垂直距离均为 10 mm。样品腔气压 160 Pa。

表 2-1 BGA-LI-TOF 的运行参数

部件	电压（V）	部件	电压（V）
Skimmer	60	Slit	−100
Extract 1	−190	MCP	800
Extract 2	0	Grid	0
Len 1	−350	Back	575
Len 2	0	ACCE	−2000
Len 3	−350	Steer X	−2550
Len 4	−350	Steer Y	−2000

图 2-9 为 SRM663 的质谱图,从图中可以明显看到,谱图明显较为干净,几乎不存在干扰峰,这是因为在高激光功率密度的作用下,所有元素的 RSC 值趋于一致。虽然剥蚀出的离子具有较强的动能及很大的动能分散,但是由于低压氦气的引入,使其与产生的离子发生频繁的三体碰撞复合反应,生成的高价离子被降价为单价离子,多原子离子也被分解为单原子离子,从而降低了谱图的干扰。BGA-LI-TOF 谱图中所呈现的离子强烈依赖于激光强度和环境条件,所以 SRM663 谱图中存在的 C、N、O 元素峰主要是受环境的影响。谱图中存在的 NbO 和 ZrO,主要是因为元素 Nb 和 Zr 与样品腔及样品表面的元素 O 结合,由于结合后形成的氧化物具有极大的键能,即使使用较高的激光功率密度照射也不能使其断裂,所以在谱图中存在氧化物峰。虽然 BGA-LI-TOF 谱图中也存在着部分无法识别的干扰峰,但其占总离子流的比例很少,对总离子流的影响或对大多数元素峰的干扰都可以忽略不计。

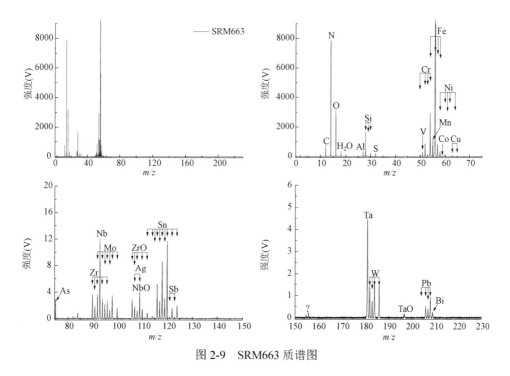

图 2-9 SRM663 质谱图

SRM663 样品的定量分析结果见表 2-2,从表中可以看到金属元素的实测浓度非常接近标准浓度,大多数金属元素的偏差都在一个数量级之内。但是由于辅助气体和样品腔中残留空气的干扰,以及非金属元素较高的电离能导致的低离子产率，导致大多数非金属元素的实测浓度偏离标准浓度。总的来说，除了干扰严重的非金属元素(C、N 和 O)，BGA-LI-TOF 能够满足对 SRM663 样品的半定量分析。

表 2-2　SRM663 样品中元素的平均测量浓度结果

元素	标准浓度（ppm）	实测浓度（ppm）	元素	标准浓度（ppm）	实测浓度（ppm）
B	9	1.0	Ni	3 200	4 146.1
C	6 260	114.1	Cu	980	1 423.6
N	41	16 454.3	Zn	4	46.5
O	7	7 502.1	As	100	63.7
Mg	5	25.6	Zr	500	274.6
Al	2 400	2 185.6	Nb	490	471.5
Si	7 400	6 194.1	Mo	300	100.3
P	290	119.0	Ag	37	144.2
S	57	241.0	Sn	1 040	1 626.0
Ti	500	454.0	Sb	20	18.7
V	3 100	4 785.1	Ta	530	435.0
Cr	13 100	19 344.6	W	460	453.0
Mn	15 000	23 013.3	Pb	22	170.7
Fe	943 627	882 364.7	Bi	8	22.6
Co	480	469.0			

2.5.5　微型激光电离飞行时间质谱仪

质谱仪是分析化学、生物和其他类型材料的最强大的实验室工具之一。但是常规实验室质谱仪尺寸大、质量较大，且具有高功率要求，导致仪器成本高并且限制了其商业应用。因此，微型质谱仪的发展在分析领域，尤其是在太空研究中，具有发展前景。在 20 世纪 80~90 年代，已经设计了几种微型质谱仪，例如用于火卫一任务的 LIMA-D 和用于火星勘探的 LASMA 仪器[109]。

最近，研究者研制了几种新颖的微型激光电离飞行时间质谱仪，它们具有较小的尺寸（体积为 84 cm^3），更轻的重量和可移动性能。例如 Brinckerhoff 等[26,109-111]开发了一种微型仪器，该仪器是基于先前在火卫一上进行的 LIMA-D 实验搭建的。它主要用于将来的太空原位分析，并可以从微量的固体样品中获得主要、微量及痕量成分。该仪器的示意图如图 2-10 所示。

由于采用全同轴设计,因此该仪器结构非常紧凑。整个系统的质量约为 2 kg，体积小于 2×10^3 cm^3。使用调 Q 固体激光器（1064 nm，8 ns）在真空中将样品电离，产生的激光功率密度约为 10^9 W/cm^2，可以通过可变衰减器对激光功率密度进行调节。激光通过聚焦透镜直接聚焦到样品上，激光剥蚀光斑直径约为 30~100 μm。在当前设置中，样品与分析器间的可调距离约为 2~5 cm。只需将仪器降低到着陆器或流动站的适当高度，就可以不受干扰地分析风化层样品，而深层样品采集则应使用取芯钻或其他设备。激光剥蚀出的离子不进行预加速，就以自然动能分布被提取，在反射器中实现能量聚焦，该反射器由定义推斥区的很多环组成。同时设计有能量窗口（VA、V2）以防止未聚焦的离子到达检测器。

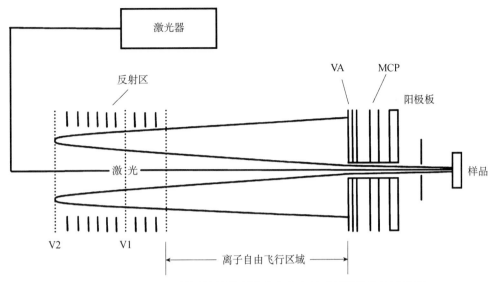

图 2-10　Brinkerhoff 等设计的微型激光电离飞行时间质谱仪的示意图

该仪器已用于分析金属、岩石、土壤、气溶胶和其他样品。对 NIST 标准材料钢的研究表明，如果选择适当的能量窗口，则可以对特定元素进行半定量分析[110]。据报道，该仪器的剥蚀效率与传输效率之间存在 8 个数量级差异。这种传输衰减是由三个因素引起的，包括 MCP 孔的接收角度、栅网的穿透度以及能量窗口的可过滤性。此外，对钢样品和地质材料的分析结果表明，该仪器可以检测到金属和非金属元素，检出限约为几个 ppm，并且在相对较宽的质量范围内，质量分辨率约为 250。

根据欧洲航天局 BepiColombo 前往水星的任务计划，Rohner 等[112]设计了一个小型实验室激光电离飞行时间质谱仪，旨在将其部署在无气行星表面上，用以确定岩石和土壤的元素和同位素组成。该仪器的体积为 650 cm^3，质量为 500 g（包括所有电子设备），整个系统可以在非常低的功率（3 W）下运行。采用二极管泵浦的调 Q 激光器 Laser 99[113,114]，其波长为 1064 nm，脉宽为 0.66 ns，可用于极其紧凑的设计，最大提供 12 μJ 的激光能量。激光束斜向聚焦在目标上，产生直径为 15 μm 的椭圆形凹坑。产生的正离子从样品表面上方的激光等离子体通过探测器中心的孔进行加速和聚焦，然后被传输到轴向对称的无栅网反射质量分析器中。将整个质量分析器放置在一个大型超高真空室中，将其抽至 10^{-8} mbar 的压力。对于陨石样品，该仪器在 FWHM 下的质量分辨率为 600，动态检测范围为 4 个数量级。钢中同位素比值的测量值在可接受值的 4% 以内。但是，由于低的激光功率密度（<10^9 W/cm^2），此技术出现了严重的元素分馏效应和质量干扰。

基于上述微型质谱仪，Rohner 等[115]设计了另一种用于行星漫游车的高度小型化的激光电离飞行时间质谱仪。该仪器的质量为 280 g，包括激光和所有电子设备，体积为 84 cm^3。在相同的脉冲宽度为 0.66 ns 的情况下，分别使用 532 nm（最大输出能量为 4 μJ）和 1064 nm（最大输出能量为 12 μJ）激光。产生的离子由加速板加速，在离子透镜的作用下聚焦在静电分析仪的入口上（参见图 2-11），该静电分析仪不仅用于离子聚焦，而且还用作选择离子的能量窗口。该仪器质量分辨率超过 180，动态范围为 5 个数量级，但是信噪比相对较低，并且在该仪器中观察到与以前的原型相同的元素分馏和质量干扰问题。

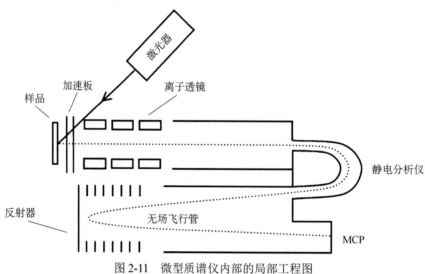

图 2-11　微型质谱仪内部的局部工程图

由于体积小且性能出色，微型 LI-TOFMS 在空间分析中非常重要。对于约 10^{-5} Torr 气压的空间条件（火星空间条件），则不需要样品采集和真空泵系统。实际上，这些仪器目前仅在实验室中进行了一些相关研究。但是，这些仪器的质量分辨率仅能分离低质量（即 $M<56$）的离子，而在较高质量范围内离子受到严重干扰。

2.6　总结与展望

本章概述了激光电离源结合不同的质量分析器形成的质谱技术，如激光微探针飞行时间质谱分析仪、激光电离四极杆质谱、激光电离双聚焦质谱、激光微探针傅里叶变换质谱以及激光电离飞行时间质谱仪。主要介绍了仪器的基本原理、特征、元素分析应用以及最新发展。这些直接的固体采样技术基于聚焦激光束对样品材料的剥蚀和离子化，已被证明是对固体样品直接分析的强大工具。需要特别指出的是，飞行时间质量分析器理论上无质量上限、扫描速度快、可进行全谱分析，并且能够与脉冲式电离源完美结合，故其发展较快。另外，低压氦气的引入和脉冲串推斥技术的应用，大大提高了激光电离飞行时间质谱仪的分辨率。

参 考 文 献

[1] Adams F, Vertes A. Inorganic mass spectrometry of solid samples. Fresenius' Journal of Analytical Chemistry, 1990, 337(6): 638-647.

[2] Russo R E, Mao X, Liu H, Gonzalez J, Mao S S. Laser ablation in analytical chemistry: A review. Talanta, 2002, 57(3): 425-451.

[3] Fagundes L, Dieguez A C. Optimization of nitrogen analysis by spark optical emission spectrometry at Arcelor Mittal Tubarao. Metallurgical Analysis, 2012, 32(7): 25-31.

[4] Ure A M, Bacon J R. Comprehensive analysis of soils and rocks by spark-source mass spectrometry. Analyst, 1978,

103(1229): 807-822.

[5] Bacon J R, Ure A M. Spark-source mass spectrometry: Recent developments and applications. Analyst, 1984, 109(10): 1229-1254.

[6] Taylor S R, Gorton M P. Geochemical application of spark source mass spectrography. Geochimica et Cosmochimica Acta, 1977, 41(9): 1375-1380.

[7] Margui E, Queralt I, Hidalgo M. Application of X-ray fluorescence spectrometry to determination and quantitation of metals in vegetal material. Trends in Analytical Chemistry, 2009, 28(3): 362-372.

[8] Winefordner J D, Gornushkin I B, Correll T, Gibb E, Smith B W, Omenetto N. Comparing several atomic spectrometric methods to the super stars: Special emphasis on laser induced breakdown spectrometry, LIBS, a future super star. Journal of Analytical Atomic Spectrometry, 2004, 19(9): 1061-1083.

[9] Khuder A, Bakir M A, Karjou J, Sawan M K. XRF and TXRF techniques for multi-element determination of trace elements in whole blood and human hair samples. Journal of Radioanalytical and Nuclear Chemistry, 2007, 273(2): 435-442.

[10] Van Ham R, Adriaens A, Van Vaeck L, Adams F. The use of time-of-flight static secondary ion mass spectrometry imaging for the molecular characterization of single aerosol surfaces. Analytica Chimica Acta, 2006, 558(1-2): 115-124.

[11] Benninghoven A. Surface investigation of solids by the statical method of secondary ion mass spectroscopy (SIMS). Surface Science, 1973, 35(2): 427-457.

[12] Metson J B, Bancroft G M, Nesbitt H W, Jonasson R G. Analysis for rare earth elements in accessory minerals by specimen isolated secondary ion mass spectrometry. Nature, 1984, 307(5949): 347-349.

[13] Hang W, Walden W O, Harrison W W. Microsecond pulsed glow discharge as an analytical spectroscopic source. Analytical Chemistry, 1996, 68(7): 1148-1152.

[14] Coburn J W, Eckstein E W, Kay E. Elemental composition profiling in thin films by glow-discharge mass spectrometry. Journal of Applied Physics, 1975, 46(7): 2828-2830.

[15] Liang Q, Jing H, Gregoire D C. Determination of trace elements in granites by inductively coupled plasma mass spectrometry. Talanta, 2000, 51(3): 507-513.

[16] Gray A L. Solid sample introduction by laser ablation for inductively coupled plasma source-mass spectrometry. Analyst, 1985, 110(5): 551-556.

[17] Wanner B, Moor C, Richner P, Bronnimann R, Magyar B. Laser ablation inductively coupled plasma mass spectrometry (LA-ICP-MS) for spatially resolved trace element determination of solids using an autofocus system. Spectrochimica Acta Part B: Atomic Spectroscopy, 1999, 54(2): 289-298.

[18] Heinrich C, Pettke T, Halter W, Aigner-Torres M, Audetat A, Gunther D, Hattendorf B, Bleiner D, Guillong M, Horn I. Quantitative multi-element analysis of minerals, fluid and melt inclusions by laser-ablation inductively-coupled-plasma mass-spectrometry. Geochimica et Cosmochimica Acta, 2003, 67(18): 3473-3497.

[19] Mason P R D, Kraan W J. Attenuation of spectral interferences during laser ablation inductively coupled plasma mass spectrometry (LA-ICP-MS) using an rf only collision and reaction cell. Journal of Analytical Atomic Spectrometry, 2002, 17(8): 858-867.

[20] Ishida T, Akiyoshi T, Sakashita A, Kinoshiro S, Fujimoto K, Chino A. A new laser ablation system for quantitative analysis of solid samples with ICP-MS. Analytical Sciences, 2008, 24(5): 563-569.

[21] Liu X D, Verlinden J, Adams F, Adriaenssens E. Analysis of high purity arsenic by spark-source mass spectrometry. Analytica Chimica Acta, 1986, 180: 341-348.

[22] Matus L, Seufert M, Jochum K P. Ion yield of a laser plasma mass spectrometer. International Journal of Mass

Spectrometry and Ion Processes, 1988, 84(1-2): 101-111.

[23] Stuewer D. Glow discharge mass spectrometry: A versatile tool for elemental analysis. Fresenius' Journal of Analytical Chemistry, 1990, 337(7): 737-742.

[24] Liu Y, Hu Z, Gao S, Guenther D, Xu J, Gao C, Chen H. *In situ* analysis of major and trace elements of anhydrous minerals by LA-ICP-MS without applying an internal standard. Chemical Geology, 2008, 257(1-2): 34-43.

[25] Becker J S, Dietze H J. Laser ionization mass-spectrometry in inorganic trace analysis. Fresenius' Journal of Analytical Chemistry, 1992, 344(3): 69-86.

[26] Brinckerhoff W B, Cornish T J, McEntire R W, Cheng A F, Benson R C. Miniature time-of-flight mass spectrometers for *in situ* composition studies. Acta Astronautica, 2003, 52(2-6): 397-404.

[27] Rohner U, Whitby J A, Wurz P. A miniature laser ablation time-of-flight mass spectrometer for *in situ* planetary exploration. Measurement Science and Technology, 2003, 14(12): 2159-2164.

[28] Honig R E, Woolston J R. Laser-induced emission of electrons, ions, and neutral atoms from solid surfaces. Applied Physics Letters, 1963, 2: 138-139.

[29] Pesme D, Hueller S, Myatt J, Riconda C, Maximov A, Tikhonchuk V T, Labaune C, Fuchs J, Depierreux S, Baldis H A. Laser-plasma interaction studies in the context of megajoule lasers for inertial fusion. Plasma Physics and Controlled Fusion, 2002, 44(12B): B53-B67.

[30] Amoruso S, Bruzzese R, Spinelli N, Velotta R. Characterization of laser-ablation plasmas. Journal of Physics B: Atomic, Molecular and Optical Physics, 1999, 32(14): R131-R172.

[31] Becker J S, Dietze H J. Laser ionization mass spectrometry in inorganic trace analysis. Fresenius' Journal of Analytical Chemistry, 1992, 344(3): 69-86.

[32] Resano M, Garcia-Ruiz E, Belarra M A, Vanhaecke F, McIntosh K S. Solid sampling in the determination of precious metals at ultratrace levels. Trends in Analytical Chemistry, 2007, 26(5): 385-395.

[33] Boxer S G, Kraft M L, Weber P K. Advances in imaging secondary ion mass spectrometry for biological samples. Annual Review of Biophysics, 2009, 38: 53-74.

[34] Li L, Zhang B, Huang R, Hang W, He J, Huang B. Laser ionization orthogonal time-of-flight mass spectrometry for simultaneous determination of nonmetallic elements in solids. Analytical Chemistry, 2010, 82(5): 1949-1953.

[35] Denoyer E, Van Grieken R, Adams F, Natusch D F S. Laser microprobe mass spectrometry. 1. Basic principles and performance characteristics. Analytical Chemistry, 1982, 54(1): 26.

[36] Beusen J M, Surkyn P, Gijbels R, Adams F. Quantitative analysis of silicate minerals by secondary ion mass spectrometry and laser microprobe mass analysis: A comparative study. Spectrochimica Acta, Part B: Atomic Spectroscopy, 1983, 38B(5-6): 843-851.

[37] Hang W, Lewis C, Majidi V. Practical considerations when using radio frequency-only quadrupole ion guide for atmospheric pressure ionization sources with time-of-flight mass spectrometry. Analyst, 2003, 128(3): 273-280.

[38] Vogt H, Heinen H J, Meier S, Wechsung R. LAMMA 500 principle and technical description of the instrument. Fresenius Zeitschrift Fur Analytical Chemistry, 1981, 308(3): 195-200.

[39] Heinen H J, Meier S, Vogt H, Wechsung R. LAMMA 1000, a new laser microprobe mass analyzer for bulk samples. International Journal of Mass Spectrometry and Ion Processes, 1983, 47: 19-22.

[40] Southon M J, Witt M C, Harris A, Wallach E R, Myatt J. Laser-microprobe mass-analysis of surface layers and bulk solids. Vacuum, 1984, 34(10-11): 903-909.

[41] Hercules D M, Novak F P, Viswanadham S K, Wilk Z A. Applications of laser microprobe mass spectrometry in organic analysis. Analytica Chimica Acta, 1987, 195: 61-71.

[42] Simons D S. Laser microprobe mass spectrometry: Description and selected applications. Applied Surface Science, 1988, 31(1): 103-117.

[43] Musselman I H, Simons D S, Linton R W. Effects of sample geometry on interelement quantitation in laser microprobe mass spectrometry. International Journal of Mass Spectrometry and Ion Processes, 1992, 112(1): 19-43.

[44] Kaufmann R, Wieser P, Wurster R. Application of the laser microprobe mass analyzer LAMMA in aerosol research. Scanning Electron Microscopy, 1980, (2): 607-622.

[45] Hercules D M, Day R J, Balasanmugam K, Dang T A, Li C P. Laser microprobe mass spectrometry. 2. Applications to structural analysis. Analytical Chemistry, 1982, 54(2): 280.

[46] Seydel U, Lindner B. Application of the laser microprobe mass analyzer (LAMMA) to qualitative and quantitative single cell analysis. International Journal of Quantum Chemistry, 1981, 20(2): 505-512.

[47] Guest W H. Recent developments of laser microprobe mass analyzers, LAMMA 500 and 1000. International Journal of Mass Spectrometry and Ion Processes, 1984, 60: 189-199.

[48] Vaeck L V, Struyf H, Roy W V, Adams F. Organic and inorganic analysis with laser microprobe mass spectrometry. Part II: Applications. Mass Spectrometry Reviews, 1994, 13(3): 209-232.

[49] Mansoori B A, Dyer E W, Lock C M, Bateman K, Boyd R K. Analytical performance of a high-pressure radio frequency-only quadrupole collision cell with an axial field applied by using conical rods. Journal of the American Society for Mass Spectrometry, 1998, 9(8): 775-788.

[50] Lock C M, Dyer E. Characterization of high pressure quadrupole collision cells possessing direct current axial fields. Rapid Communications in Mass Spectrometry, 1999, 13(5): 432-448.

[51] Hardin E D, Vestal M L. Laser ionization mass spectrometry of nonvolatile samples. Analytical Chemistry, 1981, 53(9): 1492-1497.

[52] Kuzuya M, Ohoka Y, Katoh H, Sakanashi H. Application of a quadrupole mass filter to laser ionization mass spectrometry: Synchronization between the laser pulse and the mass scan. Spectrochimica Acta, Part B: Atomic Spectroscopy, 1998, 53B(1): 123-129.

[53] Beccaglia A M, Rinaldi C A, Ferrero J C. Analysis of arsenic and calcium in soil samples by laser ablation mass spectrometry. Analytica Chimica Acta, 2006, 579(1): 11-16.

[54] Torrisi L, Lorusso A, Nassisi V, Picciotto A. Characterization of laser ablation of polymethylmethacrylate at different laser wavelengths. Radiation Effects and Defects in Solids, 2008, 163(3): 179-187.

[55] Sage R S, Cappel U B, Ashfold M N R, Walker N R. Quadrupole mass spectrometry and time-of-flight analysis of ions resulting from 532 nm pulsed laser ablation of Ni, Al, and ZnO targets. Journal of Applied Physics, 2008, 103(9), 2903604.

[56] Solyom D A, Burgoyne T W, Hieftje G M. Plasma-source sector mass spectrometry with array detection. Journal of Analytical Atomic Spectrometry, 1999, 14(8): 1101-1110.

[57] Hilmer R M, Taylor J W. Instrumental and numerical considerations for on-line interpretation of high resolution mass spectral data. Analytical Chemistry, 1973, 45(7): 1031-1045.

[58] Belshaw N S, Freedman P A, O'Nions R K, Frank M, Guo Y. A new variable dispersion double-focusing plasma mass spectrometer with performance illustrated for Pb isotopes. International Journal of Mass Spectrometry, 1998, 181: 51-58.

[59] Bingham R A, Salter P L. Analysis of solid materials by laser probe mass spectrometry. Analytical Chemistry, 1976, 48(12): 1735-1740.

[60] Matus L, Seufert H M, Jochum K P. Microanalysis of geological samples by laser plasma ionization mass spectrometry (LIMS). Fresenius' Journal of Analytical Chemistry, 1994, 350(4-5): 330-337.

[61] Seufert H M, Jochum K P. Trace element analysis of geological glasses by laser plasma ionization mass spectrometry

(LIMS): A comparison with other multielement and microanalytical methods. Fresenius' Journal of Analytical Chemistry, 1997, 359(4-5): 454-457.

[62] Conzemius R J, Svec H J. Scanning laser mass spectrometer milliprobe. Analytical Chemistry, 1978, 50(13): 1854-1860.

[63] Pelletier M, Krier G, Muller J F, Weil D, Johnston M. Laser microprobe Fourier transform mass spectrometry. Preliminary results for feasibility and evaluation. Rapid Communications in Mass Spectrometry, 1988, 2(7): 146-150.

[64] Aubriet F. Laser-induced Fourier transform ion cyclotron resonance mass spectrometry of organic and inorganic compounds: Methodologies and applications. Analytical and Bioanalytical Chemistry, 2007, 389(5): 1381-1396.

[65] Dorrestein P C, Kelleher N L. Dissecting non-ribosomal and polyketide biosynthetic machineries using electrospray ionization Fourier-transform mass spectrometry. Natural Product Reports, 2006, 23(6): 893-918.

[66] Van Roy W, Mathey A, Van Vaeck L. *In-situ* analysis of lichen pigments by Fourier transform laser microprobe mass spectrometry with external ion source. Rapid Communications in Mass Spectrometry, 1996, 10(5): 562-572.

[67] Bakker A, De Nollin S, Van Vaeck L, Slezak J, Ravingerova T, Jacob W, Ruigrok T J C. Lidocaine does not prevent the calcium paradox in rat hearts: A laser microprobe mass analysis (LAMMA) study. Life Sciences, 1995, 56(19): 1601-1611.

[68] Kockx M M, Cambier B A, Bortier H E, De M G R, Declercq S C, van C P A, Bultinck J. Foam cell replication and smooth muscle cell apoptosis in human saphenous vein grafts. Histopathology, 1994, 25(4): 365-371.

[69] Erel E, Aubriet F, Finqueneisel G, Muller J-F. Capabilities of laser ablation mass spectrometry in the differentiation of natural and artificial opal gemstones. Analytical Chemistry, 2003, 75(23): 6422-6429.

[70] Bakker A, Van Vaeck L, Jacob W. Applications of laser microprobe mass spectrometry in biology and medicine. Scanning Microscopy, 1996, 10(3): 753-775.

[71] Dimov S S, Chryssoulis S L, Lipson R H. Quantitative elemental analysis for rhodium and palladium in minerals by time-of-flight resonance ionization mass spectrometry. Analytical Chemistry, 2003, 75(23): 6723-6727.

[72] Seufert H M, Jochum K P. Trace element analysis of geological glasses by laser plasma ionization mass spectrometry (LIMS). A comparison with other multielement and microanalytical methods. Fresenius' Journal of Analytical Chemistry, 1997, 359(4-5): 454-457.

[73] Sysoev A A, Sysoev A A. Can laser-ionization time-of-flight mass spectrometry be a promising alternative to laser ablation/inductively-coupled plasma mass spectrometry and glow discharge mass spectrometry for the elemental analysis of solids? European Journal of Mass Spectrometry, 2002, 8(3): 213-232.

[74] Vadillo J M, Garcia C C, Alcantara J F, Laserna J J. Thermal-to-plasma transitions and energy thresholds in laser ablated metals monitored by atomic emission/mass spectrometry coincidence analysis. Spectrochimica Acta, Part B: Atomic Spectroscopy, 2005, 60B(7-8): 948-954.

[75] Joseph M, Manoravi P, Sivakumar N, Balasubramanian R. Laser mass spectrometric studies on rare earth doped UO_2. International Journal of Mass Spectrometry, 2006, 253(1-2): 98-103.

[76] Manoravi P, Joseph M, Sivakumar N, Balasubramanian R. Determination of isotopic ratio of boron in boric acid using laser mass spectrometry. Analytical Sciences, 2005, 21(12): 1453-1455.

[77] Mróz W, Prokopiuk A, Kozlov B, Czujko T, Jozwiak S, Krzywinski J, Stockli M P, Fehrenbach C. Quantitative measurements of the chemical composition using a reflectron mass analyzer with a microchannel plate detector. Review of Scientific Instruments, 2000, 71(3): 1425-1428.

[78] Eubank H P, Wilkerson T D. Ion-energy analyzer for plasma measurements. Review of Scientific Instruments, 1963, 34: 12-18.

[79] Sysoev A A, Sysoev A A, Poteshin S S, Pyatakhin V I, Shchekina I V, Trofimov A S. Direct sampling time-of-flight mass

spectrometers for technological analysis. Fresenius' Journal of Analytical Chemistry, 1998, 361(3): 261-266.

[80] Sysoev A A. Time-of-flight analysers with sector fields: Advances and prospects. European Journal of Mass Spectrometry, 2000, 6(6): 501-513.

[81] Hang W. Laser ionization time-of-flight mass spectrometer with an ion guide collision cell for elemental analysis of solids. Journal of Analytical Atomic Spectrometry, 2005, 20(4): 301-307.

[82] He J, Zhong W, Mahan C, Hang W. Laser ablation and ionization time-of-flight mass spectrometer with orthogonal sample introduction and axial field rf-only quadrupole cooling. Spectrochimica Acta, Part B: Atomic Spectroscopy, 2006, 61B(2): 220-224.

[83] Peng D, He J, Yu Q, Chen L, Hang W, Huang B. Parametric evaluation of laser ablation and ionization time-of-flight mass spectrometry with ion guide cooling cell. Spectrochimica Acta, Part B: Atomic Spectroscopy, 2008, 63B(8): 868-874.

[84] Harilal S S, O'Shay B, Tao Y, Tillack M S. Ambient gas effects on the dynamics of laser-produced tin plume expansion. Journal of Applied Physics, 2006, 99(8): 083303/083301-083303/083310.

[85] Kellerbauer A, Kim T, Moore R B, Varfalvy P. Buffer gas cooling of ion beams. Nuclear Instruments & Methods in Physics Research, Section A: Accelerators, Spectrometers, Detectors, and Associated Equipment, 2001, 469(2): 276-285.

[86] Koslovsky V, Fuhrer K, Tolmachev A, Dodonov A, Raznikov V, Wollnik H. Cooling of direct current beams of low mass ions. International Journal of Mass Spectrometry, 1998, 181: 27-30.

[87] Krutchinsky A N, Chernushevich I V, Spicer V L, Ens W, Standing K G. Collisional damping interface for an electrospray ionization time-of-flight mass spectrometer. Journal of the American Society for Mass Spectrometry, 1998, 9(6): 569-579.

[88] Boulyga S F, Becker J S. ICP-MS with hexapole collision cell for isotope ratio measurements of Ca, Fe, and Se. Fresenius' Journal of Analytical Chemistry, 2001, 370(5): 618-623.

[89] Drewnick F, Wieser P H. A laser ablation electron impact ionization time-of-flight mass spectrometer for analysis of condensed materials. Review of Scientific Instruments, 2002, 73(8): 3003-3006.

[90] Guilhaus M. Essential elements of time-of-flight mass spectrometry in combination with the inductively coupled plasma ion source. Spectrochimica Acta, Part B: Atomic Spectroscopy, 2000, 55B(10): 1511-1525.

[91] Tyrrell G C, Coccia L G, York T H, Boyd I W. Energy-dispersive mass spectrometry of high energy ions generated during KrF excimer and frequency-doubled Nd：YAG laser ablation of metals. Applied Surface Science, 1996, 96-98: 227-232.

[92] Ramendik G I, Kryuchkova O I, Tyurin D A, McHedlidze T R, Kaviladze M S. Factors affecting the relative sensitivity coefficients in spark and laser plasma source mass spectrometry. International Journal of Mass Spectrometry and Ion Processes, 1985, 63(1): 1-15.

[93] Harilal S S, Bindhu C V, Tillack M S, Najmabadi F, Gaeris A C. Internal structure and expansion dynamics of laser ablation plumes into ambient gases. Journal of Applied Physics, 2003, 93(5): 2380-2388.

[94] Namba S, Nozu R, Takiyama K, Oda T. Spectroscopic study of ablation and recombination processes in a laser-produced ZnO plasma. Journal of Applied Physics, 2006, 99(7): 073302/073301-073302/073309.

[95] Yu Q, Huang R, Li L, Lin L, Hang W, He J, Huang B. Applicability of standardless semiquantitative analysis of solids by high-irradiance laser ionization orthogonal time-of-fight mass spectrometry. Analytical Chemistry, 2009, 81(11): 4343-4348.

[96] Chen L, Lin L, Yu Q, Yan X, Hang W, He J, Huang B. Semiquantitative multielemental analysis of biological samples by a laser ionization orthogonal time-of-flight mass spectrometer. Journal of the American Society for Mass Spectrometry, 2009, 20(7): 1355-1358.

[97] Tong Q, Yu Q, Jin X, He J, Hang W, Huang B. Semi-quantitative analysis of geological samples using laser plasma time-of-flight mass spectrometry. Journal of Analytical Atomic Spectrometry, 2009, 24(2): 228-231.

[98] Yu Q, Huang R, Li L, Lin L, Hang W, He J, Huang B. Applicability of standardless semiquantitative analysis of solids by high-irradiance laser ionization orthogonal time-of-fight mass spectrometry. Analytical Chemistry, 2009, 81(11): 4343-4348.

[99] He J, Huang R, Yu Q, Lin Y, Hang W, Huang B. A small high-irradiance laser ionization time-of-flight mass spectrometer. Journal of Mass Spectrometry, 2009, 44(5): 780-785.

[100] Douglas D J, French J B. Collisional focusing effects in radio frequency quadrupoles. Journal of the American Society for Mass Spectrometry, 1992, 3(4): 398-408.

[101] Lin Y, Yu Q, Huang R, Hang W, He J, Huang B. Characterization of laser ablation and ionization in helium and argon: A comparative study by time-of-flight mass spectrometry. Spectrochimica Acta, Part B: Atomic Spectroscopy, 2009, 64B(11-12): 1204-1211.

[102] Huang R, Yu Q, Tong Q, Hang W, He J, Huang B. Influence of wavelength, irradiance, and the buffer gas pressure on high irradiance laser ablation and ionization source coupled with an orthogonal time of flight mass spectrometer. Spectrochimica Acta, Part B: Atomic Spectroscopy, 2009, 64B(3): 255-261.

[103] Yu Q, Li L, Zhu E, Hang W, He J, Huang B. Analysis of solids with different matrices by buffer-gas-assisted laser ionization orthogonal time-of-flight mass spectrometry. Journal of Analytical Atomic Spectrometry, 2010, 25(7): 1155-1158.

[104] Coles J, Guilhaus M. Orthogonal acceleration: A new direction for time-of-flight mass spectrometry: Fast, sensitive mass analysis for continuous ion sources. Trends in Analytical Chemistry, 1993, 12(5): 203-213.

[105] Huang R, Yu Q, Li L, Lin Y, Hang W, He J, Huang B. High irradiance laser ionization orthogonal time-of-flight mass spectrometry: A versatile tool for solid analysis. Mass Spectrometry Reviews, 2011, 30(6): 1256-1268.

[106] Harilal S S. Expansion dynamics of laser ablated carbon plasma plume in helium ambient. Applied Surface Science, 2001, 172(1-2): 103-109.

[107] Harilal S S, Bindhu C V, Tillack M S, Najmabadi F, Gaeris A C. Plume splitting and sharpening in laser-produced aluminium plasma. Journal of Physics D: Applied Physics, 2002, 35(22): 2935-2938.

[108] Laska L, Jungwirth K, Kralikova B, Krasa J, Pfeifer M, Rohlena K, Skala J, Ullschmied J, Badziak J, Parys P, Wolowski J, Woryna E, Gammino S, Torrisi L, Boody F P, Hora H. Generation of multiply charged ions at low and high laser-power densities. Plasma Physics and Controlled Fusion, 2003, 45(5): 585-599.

[109] Brinckerhoff W B. Pulsed laser ablation TOF-MS analysis of planets and small bodies. Applied Physics A: Materials Science & Processing, 2004, 79(4-6): 953-956.

[110] Brinckerhoff W B, Managadze G G, McEntire R W, Cheng A F, Green W J. Laser time-of-flight mass spectrometry for space. Review of Scientific Instruments, 2000, 71(2): 536-545.

[111] Brinckerhoff W B. On the possible *in situ* elemental analysis of small bodies with laser ablation TOF-MS. Planetary and Space Science, 2005, 53(8): 817-838.

[112] Rohner U, Whitby J A, Wurz P. A miniature laser ablation time-of-flight mass spectrometer for *in situ* planetary exploration. Measurement Science & Technology, 2003, 14(12): 2159-2164.

[113] Zayhowski J J. Q-switched operation of microchip lasers. Optics Letters, 1991, 16(8): 575-577.

[114] Zayhowski J J. Passively Q-switched Nd：YAG microchip lasers and applications. Journal of Alloys and Compounds, 2000, 303-304: 393-400.

[115] Rohner U, Whitby J A, Wurz P, Barabash S. Highly miniaturized laser ablation time-of-flight mass spectrometer for a planetary rover. Review of Scientific Instruments, 2004, 75(5): 1314-1322.

第3章 MC-ICP-MS 同位素比值的测定

(Lu Yang[①]*,何 娟[②],侯贤灯[②],Juris Meija[①],Zoltán Mester[①])

▶ 3.1 引言 / 62

▶ 3.2 MC-ICP-MS 的非质量分馏 / 63

▶ 3.3 同位素分馏校正模型 / 67

▶ 3.4 总结与展望 / 73

[①]National Research Council Canada,加拿大;[②]四川大学,中国。*通讯作者联系方式:lu.yang@nrc-cnrc.gc.ca

本章导读

- 综述了 MC-ICP-MS 同位素分馏/质量偏倚的类型。
- 讨论了质量分馏对绝对同位素比值测量的影响。
- 详细论述了常用质量偏倚校正模型的优缺点。
- 对 MC-ICP-MS 同位素比值的测定进行了展望。

3.1 引 言

多接收器电感耦合等离子体质谱（MC-ICP-MS）是高精度测定同位素比值最常用的两种技术之一，广泛应用于考古学、物源研究、医学、核科学、法医学、地球科学和环境科学等多个学科[1-9]。与传统的热电离质谱（TIMS）相比，MC-ICP-MS 的样品引入方式简单、灵敏度高，并具有测量高电离势元素的能力，但它会产生更大的同位素分馏/质量偏倚[1,7]。用 MC-ICP-MS 测得锂的同位素比值与真实值之间的偏差高达 25%[10]（而 TIMS 中锂的偏差仅为 1%~2%[11]）。

ICP-MS 中出现同位素分馏现象的原因至今还不完全清楚，但很可能是由于离子通过取样锥时的超音速气流膨胀，以及在截取锥区域的空间电荷效应引起的。这两个过程都有利于将较重的同位素传输到质谱仪中，导致整个质量区间产生不均匀的响应（灵敏度）[12-14]。质量偏倚的大小也随操作条件而变化，例如样品气流速、采样锥和等离子体炬管末端之间的距离以及样品深度等[15]。此外同位素分馏也随时间而漂移，并受到样品基体的影响。因此合适的质量偏倚校正是 MC-ICP-MS 获得精准同位素比值的关键。

自 20 世纪 90 年代初 MC-ICP-MS 上市以来，传统的同位素分馏校正模型（如指数定律、Russell 定律和双稀释剂法）已被广泛应用于 MC-ICP-MS[1,6,7]。其中 Russell 定律[16]是最广泛使用的模型之一：

$$R_{i/j} = r_{i/j} \left(\frac{m_i}{m_j} \right)^f \tag{3-1}$$

式中，r 和 R 分别表示同位素比值的测量值和经质量偏倚校正后的真实值，m_i 和 m_j 分别为两种同位素的核素质量，f 为质量偏倚校正因子。通常认为，从单个同位素比值中获得的同位素分馏因子 f 可用于其他同位素比值测量值的校正。这种校正方法的准确性取决于公式本身的有效性，因为公式(3-1)毕竟只是经验模型。值得注意的是，传统的同位素分馏校正模型是为了校正同位素分馏随核素质量变化而变化的质量分馏（MDF）。此外，该校正模型假设内标物（标准物）和分析物具有相同的质量分馏因子。一方面，内标物（标准物）可以是与分析物相同的元素的另一对同位素，例如，$R^{Sr}_{88/86}$ 同位素比值被广泛应用于 $R^{Sr}_{87/86}$ 的质量分馏校正[17-29]；$R^{Gd}_{160/156}$ 和 $R^{Sm}_{147/149}$ 分别被用于 Gd 和 Sm 的其他同位素对的质量分馏校正[30]。另一方面，公式（3-1）更常用于另一种元素同位素比值的校正，例如，Tl 被用作 Pb[31-36]和 Hg[37,38]同位素比值的内标校正物，Zn 和 Cu 相互校正[39]、Ni 校正 Fe[40]、Zr 或 Ru 校正 Mo[41]、Zr 校正 Rb[42]等。然而近年来，随着对同位素分馏行为更充分的认识，我们发现分馏因子 f 值在不同元素之间存在显著差异[43-51]。众所周知，锂同位素在 MC-ICP-MS 中的同位素分馏远比铀同位素明显。当 $R^{Hg}_{202/201}$ 和 $R^{Os}_{187/188}$ 的比值分别通过 $R^{Tl}_{205/203}$ 和 $R^{Os}_{192/188}$ 使用公式（3-1）进行校正时，其误差高达 0.5%[51,52]。此外，使用公式（3-1）得到的"校正"同位素比值可能还依赖于浓度与基质[51]，这进一步质疑了该校正模型对于准确测量同位素比值的有效性。

如果同位素分馏因元素而异，便没有理由相信同一元素的不同同位素之间具有相同的质量分馏。事实上最近的研究表明，在 MC-ICP-MS 中同一元素的不同同位素对，例如 Nd[53-56]、Ce[54]、W[57-59]、Sr[60]、Ge[61]、Pb[61]、Hg[61]、Si[62]、Hf[63,64]、Ba[65]、Os[52]和 Gd[66]等元素的同位素，在同位素分馏方面确实存在差异。这种同位素分馏被称为非质量分馏（MIF）。重要的是，MIF 的出现对同位素分馏校正模型的选择有着重要的影响。用 MDF 校正模型校正 MC-ICP-MS 中具有 MIF 的同位素，很可能会得到有偏差的结果[52,53,59,61-66]。例如 ^{187}Os，采用 Russell 定律校正同位素比值 $R^{Os}_{187/188}$，可观察到接近 0.5%的偏差[52]，因此同位素分馏校正仍然是该领域的巨大挑战。目前针对同位素分馏校正的研究也取得了一定进展[67]。此外，最频繁使用的校正模型[1,7]，如双稀释剂同位素分馏校正模型（DS）、标准-样品交叉同位素分馏校正模型（SSB）、标准-样品交叉结合内标同位素分馏校正模型（C-SSBIN）、线性回归模型，都依赖于同位素基准物质的使用，而这些基准物通常是有限的。

本章概述了利用 MC-ICP-MS 精准测量同位素比值的最新进展，重点介绍了 MC-ICP-MS 本身 MIF 的研究现状，及其对几种广泛使用的同位素分馏校正模型[如优化的线性回归模型（ORM）和全同位素质量混合校正模型（FGIM）]的影响。还详细讨论了每个模型的核心概念、假设、利弊以及同位素比值测量的应用实例。

3.2　MC-ICP-MS 的非质量分馏

由于 MIF 在各种生物、地质和化学过程研究中的重要应用，关于自然界中 MIF 现象的研究一直是同位素测量科学领域的热点[9,68-73]。MIF 是指其同位素分馏大小与核素质量没有严格关系的同位素分馏类型。自然界中 MIF 现象最典型的例子便是氧[74,75]和硫的同位素[76]。例如，对流层臭氧中的氧相对空气来说稍重一些。根据经典的质量分馏模型，由于臭氧中 ^{18}O/^{16}O 同位素之间的质量差是 ^{17}O/^{16}O 的两倍，因此 ^{18}O/^{16}O 同位素比值（相对于空气）将比后者增加约两倍。然而，^{17}O 和 ^{18}O 在臭氧中的相对富集程度几乎相同[75]。与在自然界中观察到的 MIF 类似，MC-ICP-MS 本身在某种程度上也存在类似于自然界中同一元素的不同同位素的非质量分馏。事实证明，在 2002 年首次发现 Nd[53]同位素的非质量分馏后，更多的同位素在 MC-ICP-MS 中的 MIF 被实验证实（例如 Nd[53-56]、Ce[54]、W[57-59]、Sr[60]、Ge[61]、Pb[61]、Hg[61]、Si[62]、Hf[63,64]、Ba[65]、Os[52]和 Gd[66]等元素的同位素）。不仅如此，研究发现 MC-ICP-MS 中的 MIF 会随着等离子体条件的变化而变化[52,59,62,64,66,77-79]，这给正确校正质量偏倚以获得准确同位素比值带来了更大的挑战。可以判定将来还会在 MC-ICP-MS 中发现更多显示 MIF 的元素，例如，Schmitt 等[80]在两年的研究期间并未报道 MC-ICP-MS 中 Cd 同位素的 MIF 现象，但是作者无论是使用 Cd 的另一对同位素结合指数定律还是采用双稀释剂校正模型，都无法实现重复和准确的 Cd 同位素比值测量。这种现象很可能是由 MC-ICP-MS 中 Cd 同位素的 MIF 效应造成的。为了避免使用指数同位素分馏模型（质量相关模型），并通过 MC-ICP-MS 获得准确的 Cd 同位素比值，Pritzkow 等用镉的七种同位素稀释剂，采用基准方法——全同位素质量混合校正模型（FGIM）进行了 Cd 的同位素分馏校正[81]。显然，MC-ICP-MS 自身的 MIF 现象并不局限于等离子体模式或高灵敏度截取锥[54]；相反，在常规的液体样品引入模式中，许多同位素体系也都产生了这一现象，表明 MIF 是 MC-ICP-MS 中普遍存在的。所以在用 MC-ICP-MS 进行同位素比值测量之前，研究 MC-ICP-MS 中是否存在 MIF，以及选择相应的同位素分馏校正模型都是十分必要的。需要注意的是，本章中所有关于 MIF 的后续讨论都是针对发生在 MC-ICP-MS 仪器内部的 MIF，而非自然样品中的 MIF。

3.2.1 MC-ICP-MS 同位素分馏种类的评估方法

评估同位素分馏目前已有多种方法,其中最常用于区分 MIF 和 MDF 的方法为三同位素图[57,82,83],该方法依赖于同一元素两对同位素比值之间的协同变化。由于三同位素曲线图中只有斜率是相关的,研究表明通过两对测量同位素比值的自然对数值 $\ln r_{i/j}$ vs. $\ln r_{k/j}$ 的三同位素曲线图,便可以确定 MC-ICP-MS 中同位素分馏的种类[52,59,61,62,64,66]。这种方法不需要元素的真实同位素组成就可以判定 MIF。根据分馏过程的潜在机制,仅从核素质量就可以预测该图的斜率。例如,动力学控制的单向化学反应的三同位素图的斜率为

$$\beta_{\text{kinetic}} = \frac{\ln m_j - \ln m_i}{\ln m_j - \ln m_k} \tag{3-2}$$

热力学控制的平衡过程的斜率为

$$\beta_{\text{equilibrium}} = \frac{\dfrac{1}{m_i} - \dfrac{1}{m_j}}{\dfrac{1}{m_k} - \dfrac{1}{m_j}} \tag{3-3}$$

以上两种理想化的过程可以统一使用以下的幂校正方程来表示:

$$\beta_{\text{generalized power}} = \frac{m_i^n - m_j^n}{m_k^n - m_j^n} \tag{3-4}$$

式中,n 是歧视指数,$n=-1$ 对应于热力学平衡过程,$n\to 0$ 对应于动力学过程,Russell 定律[公式(3-1)]则属于动力学模型。

如图 3-1 所示,上述方法产生了三种可能的情况:①$\ln r_{i/j}$ vs. $\ln r_{k/j}$ 曲线的测量结果显示出与理论质量分馏模型(如 Russell 定律)一致的斜率,证实为 MDF[图 3-1(a)];②同位素比值的斜率与任何理论模型明显不同,从而证实了 MIF[图 3-1(b)];③不相关的数据在三同位素图[图 3-1(c)]中非系统地分散,这种情况可能是由仪器测量误差引起的。正如最近钨同位素[59]的研究所示,$\ln r_{180/184}$ 与 $\ln r_{186/184}$ 的实际斜率不仅显示出与两种理论模型(动力学质量相关定律和平衡质量相关定律)的显著偏离,而且还显示出不同测量阶段的显著变化,如图 3-2 所示。

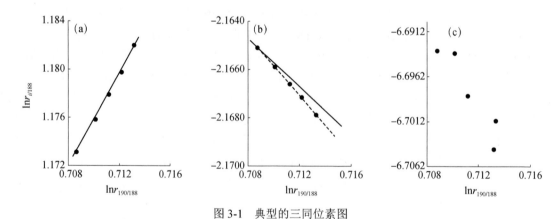

图 3-1 典型的三同位素图

实线表示通过计算方程式理论预测的质量分馏的斜率 β_{Russell};(a)质量分馏,$r_{192/188}$ vs. $r_{190/188}$;(b)非质量分馏,$r_{187/188}$ vs. $r_{190/188}$;(c)使用标准的 10^11 电阻器,$^{184}\text{Os}^+$ 信号过低 $r_{184/188}$ vs. $r_{190/188}$ 的仪器测量误差。数据来自于文献[52],数据图改编自文献[7]并获得了英国皇家化学学会的版权许可

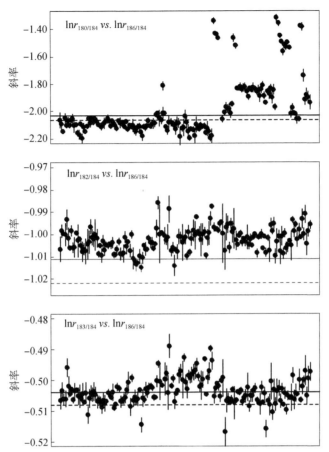

图 3-2 MC-ICP-MS 测量 WOLF-1 所得三同位素图的斜率（2018 年 2～8 月的 159 组数据）的随机变化

该斜率既偏离了 Russell 定律（即动力学控制的质量分馏，水平实线）也不同于热力学控制的平衡过程的质量分馏（水平虚线）。数据图来自 Zhang 等的实验结果[59]并获得了 Elsevier 版权许可

我们建议使用该方法对 MIF 进行预判，用 MC-ICP-MS 在 4～5 个递增的射频功率下，短时间内（约 10 min）对单元素标准溶液进行测量[52,59,64,66,77,78]，便能在最终同位素比值测量之前就确定此元素是否发生了 MIF。

3.2.2 MC-ICP-MS 中 MIF 现象的可能原因

MC-ICP-MS 中产生 MIF 的确切原因仍未有定论[54,55,57,61]。核场效应（NFS），也称核体积效应，和同位素磁效应（或核自旋效应）被认为是最可能导致 MIF 的原因[54,70,84,85]。文献报道了对这两种效应的详细介绍[85,86]。简而言之，NFS 效应来自原子的大小和形状不同产生的同位素效应。NFS 效应的大小通过原子核直径 r_a^2 的偏差来预测，r_a^2 是核素质量 m_a 的函数。通常以 Δr_a^2 vs. $\Delta m_a/(m_{a,1}m_{a,2})$ 的形式绘制标准曲线，若偏离线性曲线则被识别为 NFS 效应。核素的核电荷均方根半径和核素质量均有报道[87,88]。如图 3-3 所示，同位素 ^{199}Hg、^{204}Hg、^{73}Ge、^{76}Ge、^{189}Os、^{174}Hf、^{174}Hf、^{183}W 和 ^{186}W 等均偏离其预期的线性关系。因此，这些同位素产生 MIF 很可能是归因于 NFS 效应。

最近的研究[52,59,61,64]证实了一些具有 NFS 效应的同位素表现出显著的 MIF 现象，如 ^{73}Ge、^{76}Ge、^{199}Hg、^{201}Hg、^{174}Hf、^{179}Hf 和 ^{183}W。然而，也有同位素（^{189}Os）表现出明显的 NFS 效应，但在 MC-ICP-MS 中却没有显著的 MIF；反之，也有一些没有 NFS 效应但有显著 MIF 的同位素（^{200}Hg、^{180}W 和 ^{182}W）。因此 NFS 并不能准确预测 MC-ICP-MS 的 MIF 现象。

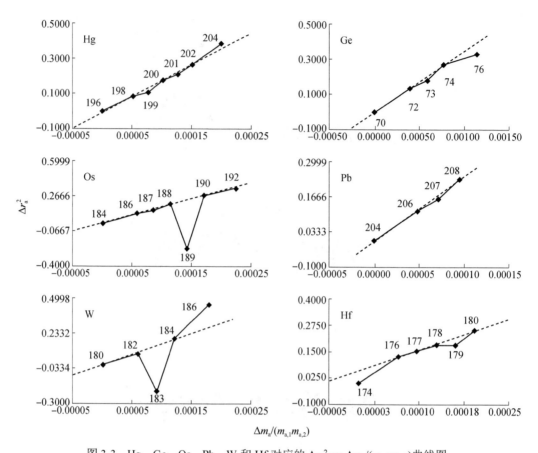

图 3-3 Hg、Ge、Os、Pb、W 和 Hf 对应的 Δr_a^2 vs. $\Delta m_a/(m_{a,1}m_{a,2})$ 曲线图

Δr_a^2 分别使用 ^{196}Hg、^{70}Ge、^{184}Os、^{204}Pb、^{180}W 和 ^{174}Hf 计算得到。数据图部分改编自文献[7]并获得了皇家化学学会的版权许可

磁同位素效应来自具有非零核自旋耦合常数的原子核在磁场中的自旋-自旋耦合相互作用,例如 ^{199}Hg（自旋量子数 $I=-1/2$）、^{201}Hg（$I=-3/2$）、^{187}Os（$I=-1/2$）、^{189}Os（$I=-3/2$）、^{179}Hf（$I=+9/2$）以及 ^{183}W（$I=-1/2$）[71,84,85,89]。磁同位素效应被认为是在涉及自由基和重金属（如汞）[71,84,85]的化学反应中观察到 MIF 的主要原因。尽管在 MC-ICP-MS 中确实观察到许多磁同位素的 MIF 现象,如 ^{73}Ge、^{201}Hg、^{207}Pb、^{187}Os、^{179}Hf 和 ^{183}W 等,但对于核自旋数为零的同位素也观察到了 MIF（例如 ^{70}Ge、^{200}Hg、^{204}Hg、^{204}Pb、^{192}Os、^{186}W、^{180}Hf、^{184}Hf）。此外,显示出最大质量偏倚的同位素不一定是具有非零核自旋的同位素。例如,根据 Russell 定律,汞同位素中的 ^{201}Hg 和 ^{204}Hg 相对于 ^{202}Hg/^{198}Hg 具有同样程度的质量偏倚（通常相当于 0.1%）[61]。因此,同位素的分馏行为不能简单地区分为非磁性原子核与磁性原子核两类。

最近的研究表明[54,55],观察到的钕同位素的 MIF 可能是由于在 Nu MC-ICP-MS 仪器（Nu Plasma HR 和 Nu Plasma 1700, Nu Instruments Ltd, Wrexham, UK）上使用的高灵敏度 X 型截取锥（Nu Instruments Ltd, Wrexham, UK）使氧化物形成的水平增加所致。因为 MIF 程度确与 NdO$^+$/Nd$^+$ 比值的增加有关。另外,与标准的高性能 H 型锥相比,使用 X 型截取锥得到的 Nd 同位素的 MIF 程度更大（H 型锥具有圆柱形入口和喇叭形出口,而 X 型截取锥则完全为喇叭形）。与之相反,在另一项研究中[57],使用 Neptune MC-ICP-MS（Thermo Fisher Scientific, Bremen, Germany）上的 X 型截取锥（Thermo Fisher Scientific, Bremen, Germany）观察到的 W 同位素 MIF 比使用 H 型锥要小得多,仅为后者的一半。研究者认为在 W 同位素中观察到的 MIF 归因于其较高的第一电离能（$E_{i,1}$）,并且该作者认为 $E_{i,1} \geqslant 8$ eV 的其他元素也很有可能在 MC-ICP-MS 中表现出 MIF[57]。然而,进一步的研究[62]又报道了在 Neptune MC-ICP-MS 中使用 X 型截取锥与使用标准 H 型截取锥观察到的 ^{29}Si 的 MIF 大小相似。此外，与先前研究[61]报道的 Neptune MC-ICP-MS 中 Ge、Hg 和 Pb 的 MIF 相似，使用这两种截取锥观察到的 MIF 大小均表现出随时间的显著变化。据 Andrén 等的报道[15]，同位素分馏程度也受样品气体流速和采样深度的影响。上述结果说明，在 MC-ICP-MS 中观察到 MIF 程度的不同可能是由于等离子体条件不同引起的。最近通过 Neptune Plus MC-ICP-MS（Thermo Fisher Scientific，Bremen，Germany）对 Os[52]、W[59]与 Gd[66]同位素的研究中也发现了在不同等离子体条件下观察到的 MIF 大小的变化。此外，在正常的 MC-ICP-MS 操作条件下，无论同位素 $E_{i,1}$ 值是高（如 Hg、Si 和 Os）还是低（如 Nd 和 Sr），都有可能观察到 MIF 现象。这些实验结果证实了先前使用不同截取锥观测到的 Nd 和 W 同位素[54,55,57]的 MIF 程度不一致很可能是由于等离子体条件的不同，而非 X 型截取锥导致的氧化程度增强或高电离能[54,55,57]。

综上所述，NFS 效应、磁同位素效应、氧化物水平和高电离电位都不能完全解释 MC-ICP-MS 中观察到的 MIF 现象。尽管 MIF 很可能与等离子体条件有关，但具体原因仍未可知。因此需要在这方面进行更多的基础研究，以促进我们对 MC-ICP-MS 中同位素分馏的更深理解。

3.2.3 MIF 对 MC-ICP-MS 中同位素校正模型选择的影响

近年来，人们对 MC-ICP-MS 中 MIF 的研究兴趣主要集中在产生该现象的可能原因上（核场效应、磁矩、氧化水平或高电离势）[54,55,57]，而忽略了 MIF 对 MC-ICP-MS 精确测量同位素比值的潜在影响。如果使用质量分馏的校正模型去校正具有非质量分馏的同位素，会导致结果的不准确。因此，MC-ICP-MS 中的 MIF 现象使得选择正确的同位素分馏校正模型至关重要。直到最近，双稀释剂同位素分馏校正模型（DS）被认为是除全同位素质量混合法之外最精确的同位素校正模型之一[1,90]，并被广泛应用于 MC-ICP-MS 同位素比值测定中[1,7,91-107]。然而，多同位素元素 MIF 的最新发现表明 DS 模型在 MC-ICP-MS 质量分馏校正中的应用还有待商榷。因为 DS 校正模型是基于质量分馏定律（如线性定律、指数定律或 Russell 定律）衍生而来，该模型依赖于同一元素的不同同位素具有相同质量偏倚校正因子这一假设，而实验证明这一假设并不成立[52,59,61,64,108]。

文献中同位素比值大多使用 δ 值[公式（3-5）]来表示，这使得微小的同位素差异也可以明确地表示出来，而不需要对常用标准物质的绝对同位素比值有确切的了解。值得注意的是，MIF 对选择同位素分馏校正模型以测定同位素比值 δ 值的影响并不大，因为 δ 值是一个相对比值，分析物和标准物的测量误差得以相互抵消。因此，所有现有的同位素分馏校正模型都可以使用。然而，为了准确测定不同基体样品中的同位素 δ 值（或绝对同位素比值），通常需要从样品基质中提纯分析物[1]。

$$\delta = \left(\frac{R_{i/j}^{\text{sample}}}{R_{i/j}^{\text{std}}} - 1 \right) \tag{3-5}$$

本章主要讨论 MIF 对几种常用同位素分馏模型的影响，并对这些模型的优缺点进行了详细讨论。

3.3 同位素分馏校正模型

3.3.1 双稀释剂同位素分馏校正模型

传统的双稀释剂同位素分馏校正模型（DS）包含对两种样品中同一元素四种同位素的测量，一种是天然样品，另一种是在天然样品中加入已知同位素比值的两种富集同位素即"稀释剂"形成的混合样品，再根据"稀释剂"已知的同位素比值计算得到样品中同位素的比值。DS 方法最初由 Dietz 等提出并用于 TIMS 测定铀的同位素组成中[109]。后来，Dodson[110,111]、Russell[112]和 Cumming[113]提出了数学模型以及更深入合理的不确定度传递的处理方法，以优化稀释剂中两个富集同位素的比值。天然样品（N）与稀释剂（S）混合得到混合物（M），三者的同位素比值（相对于共同的参考同位素 4）之间的关系可用公式（3-6）表示：

$$R_{N,i} - R_{M,i} = q \cdot (R_{M,i} - R_{S,i}) \tag{3-6}$$

式中，q 是参考同位素 4 在稀释剂与它在混合样品中之比，R 分别是天然样品、稀释剂和混合物中的同位素真值；下标 $i=1、2、3$ 表示元素的其他三种同位素与参考同位素 4 之间的比值。与传统的同位素分馏校正模型一样，DS 方法依赖于同位素分馏模型的选择。最初采用的是线性同位素分馏定律[1,6,109-113]，后来为了获得更精确的结果而采用了 Russell 定律。根据 Russell 定律，天然样品和混合样品中同位素比的测量值与真实值具有如下关系：

$$R_{N,i} = r_{N,i} \cdot \left(\frac{m_i}{m_4}\right)^{f_N} \tag{3-7}$$

$$R_{M,i} = r_{M,i} \cdot \left(\frac{m_i}{m_4}\right)^{f_M} \tag{3-8}$$

式中，f_N 和 f_M 分别是天然样品和混合样品的同位素分馏因子；m_i（$i=1, 2, 3$）表示同位素的核素质量。结合方程式（3-8）得出：

$$r_{N,i} \cdot \left(\frac{m_i}{m_4}\right)^{f_N} - r_{M,i} \cdot \left(\frac{m_i}{m_4}\right)^{f_M} - r_{M,i} \cdot \left(\frac{m_i}{m_4}\right)^{f_M} \cdot q + q \cdot R_{S,i} = 0 \tag{3-9}$$

对于三个未知变量（f_N、f_M 和 q），可以通过求解上述三个方程（i 的每个值对应一组方程）得到。由于 DS 法可以直接测定每个样品的同位素分馏因子，因此不需要两个样品的基质相匹配。得益于此优势，在过去的二十年中，DS 方法已广泛地应用于 MC-ICP-MS 对许多元素同位素的测定中[1,7,91-107,114-118]，而该模型在实践方面存在的一些缺点被忽略了。比如，必须确保富集双稀释剂的有效性、必须对稀释剂同位素组成进行准确标定、必须避免样品之间可能存在的交叉污染，另外还必须满足分析物至少存在四种无干扰同位素的基本要求。近年来，通过对天然样品和天然样品与双稀释剂形成的混合物的测量，DS 被改进为更适用于三同位素体系的模型（Mg，Si）[119,120]。详细的数学计算过程可见其他报道[119,120]。

考虑到 DS 模型依赖于如 Russell 定律等同位素分馏模型的选择才能从同位素测量值中计算得出同位素真实值[公式（3-7）和公式（3-8）]，因而该模型是建立在质量分馏模式的假设上，如在体系存在 MIF 的情况下则无法得到准确的同位素比值[1,7]。因此，对于显示 MIF 的同位素，使用 MC-ICP-MS 与 DS 模型得出的同位素比值可能是错误的，例如 Cd 元素[80]。不过，DS 方法仍然可以与 MC-ICP-MS 联用测定仅显示 MDF 的同位素体系。故而在用 MC-ICP-MS 进行同位素组成测量之前，对目标分析物的同位素分馏性质进行评估非常重要。

值得注意的是，用 DS 方法结合标准-样品交叉同位素分馏校正模型（SSB），可以同时对 MIF 和 MDF 两种分馏模式进行准确校正[65,101]。最近的一项研究报道[101]首先使用 ^{98}Ru-^{101}Ru 的 DS 模型校正 $R^{Ru}_{102/99}$ 的同位素分馏，然后使用标准-样品交叉测定法对 DS 校正过的 $R^{Ru}_{102/99}$ 值进行进一步校正，最终得到了准确的 $R^{Ru}_{102/99}$ 值。

3.3.2 标准-样品交叉同位素分馏校正模型

在 MC-ICP-MS 使用的众多同位素分馏校正模型中[1,7]，只要有已知同位素组成的同一元素的标准物质作为基准，便能通过标准-样品交叉同位素分馏校正模型（SSB），得到准确的同位素比值[36,57,121-125]。由于每个同位素比值都是单独校正的，MIF 对模型的精准度并没有影响。该模型首先根据公式（3-10）使用相邻的标准样品的实验值与真实值计算，得到同位素比的校正系数（$K_{i/j}$），再根据公式（3-11）将两个 $K_{\text{std},i/j}$ 值的平均值用于计算样品中分析物的同位素比的真实值。

$$K_{\text{std},i/j} = \frac{R_{\text{std},i/j}}{r_{\text{std},i/j}} \tag{3-10}$$

$$R_{\text{sample},i/j} = \overline{K_{\text{std},i/j}} \cdot r_{\text{sample},i/j} \tag{3-11}$$

由于 SSB 方法是按照"标准-样品-标准"的顺序进行连续测试（外部校准），必须确保待测样品和标准物质具有一致的同位素分馏行为。需要保证样品与标准溶液的基质与浓度完全匹配，样品基体的有效分离也至关重要。此外，需要较长的预热时间来保证仪器的稳定性，以及较短的测量时间以减少测试过程中可能存在的信号漂移。只要满足这些条件，就可以利用 SSB 进行同位素分馏校正，获得准确的同位素比值。然而在实际应用中，当基体分离不完全或是在 MC-ICP-MS 中同位素分馏不稳定时，会导致准确度和精密度的下降。正如 Pietruszka 和 Reznik 最近报道的[121]，在柱分离过程中引入的基质导致了 SSB 技术无法精确地获得钼的同位素比值。另外，在样品和标准品中添加相同的基质以减少由于基体不完全分离而产生的基质效应，可能是获得更加精确的同位素分馏校正的有效途径[126]。

3.3.3 标准-样品交叉结合内标同位素分馏校正模型（C-SSBIN）

SSB 模型不能完全校正测量过程中同位素信号随时间漂移带来的质量偏倚[1,7,127,128]，这一缺点可以通过引入另一种元素作为内标来解决。根据 Russell 定律，C-SSBIN 利用标准样品中待测元素已知的同位素比来标定内标元素的同位素比，然后反过来使用该值校准样品中待测元素的同位素比[32,40,45,49,50,60,129-132]。该方法已应用于多种同位素体系，如 Cu 校正 Zn、Tl 校正 Pb、Zr 校正 Sr、W 校正 Re、Pd 校正 Ag、Gd 校正 Eu、Ga 校正 Cu 等。这种组合式同位素分馏校正方法的主要优点在于，在测量过程中仪器造成的同位素分馏的随机变化也可以得到校正。如 Yang 等[49]所示，与单纯的 SSB 技术相比，以 Zr 为内标测得的 $R_{87/86}^{\text{Sr}}$ 和 $R_{88/86}^{\text{Sr}}$ 比值的精度大约提高了 3 倍。同样，最近的一项研究中，使用 Ga 为内标的 C-SSBIN 法比 SSB 法测得的铜同位素比值的精度提高了 4 倍[132]。虽然基质诱导的同位素分馏可以在一定程度上得到校正[1,7,133]，但这种 SSB 和内标同位素分馏校正方法（C-SSBIN）无法完全校正同位素比值对样品浓度依赖产生的偏差[49]。因此，通常需要对标准品和样品中的分析物和内标物进行浓度匹配，以获得准确的同位素比值。该方法中，内标元素是标准物质与样品间进行准确同位素分馏校正的桥梁。尽管受到所采用的同位素分馏校正模型的局限（例如，$f_{\text{Sr}}=f_{\text{Zr}}$），Zr 同位素比值本身可能会出现偏差，但在 SSB 校准的第二步中（Zr→Sr），这种偏差基本上可被抵消掉。只有当基质完全分离以及浓度完全匹配时，才会使这种误差得以完全消除。

值得注意的是，除了 C-SSBIN 校正模型外，同位素分馏还可以通过首先应用 Russell 定律，然后使用 SSB 进行第二次校正从而实现同位素比值的精确校正[134-137]。

3.3.4 线性回归同位素分馏校正模型

如前文所述，由于质量分馏模型必须建立在待测元素与内标元素具有一致的同位素分馏因子的假设上，因此 MC-ICP-MS 中 MIF 的存在使传统的质量分馏模型无法获得准确的同位素比值。通过使用 Maréchal 等 30 年前开创的方法[138]，利用 MC-ICP-MS 中同位素信号随时间的漂移建立起的线性回归关系，可以巧妙地规避这些缺点。线性回归方法依赖于已知同位素比值的标准物质实现对待测同位素的校正。校准物质可以是样品待测元素的一对同位素或是加入到样品中的另一种元素的一对同位素。根据先前的研究所报道的[31,41,44,48,52,59,64,66,77,79,138-144]，使用后者的情况更为普遍。

Yang 团队在原始的线性回归模型上做了进一步简化与改进[7,52,59,66,77-79,143,144]，该模型建立于待测元素和内标元素同位素比值之间的时间相关漂移，其模型如下：

$$\ln r_{i/j} = a + b \cdot \ln r_{\text{ref}} \tag{3-12}$$

式中，系数 a 和 b 是相应线性回归曲线的截距和斜率，通过使用最小二乘法拟合 $\ln r_{i/j}$ 与 $\ln r_{\text{ref}}$ 获得。同位素比值的真实值（R）与测量值（r）通过校正因子 K 直接联系起来：$R_{\text{ref}} = K_{\text{ref}} \times r_{\text{ref}}$ 以及 $R_{i/j} = K_{i/j} \times r_{i/j}$。联立上述两个方程最终可得到校正同位素比值，且该校正方法不依赖于内标元素与待测元素具有相同的质量分馏因子这一假设：

$$R_{i/j} = e^a \cdot R_{\text{ref}}^b \tag{3-13}$$

基于该原理，可以有效地根据校准物质已知的同位素比值来校准目标元素的同位素比值，而无需假设两者必须具有相同的质量分馏（事实上两者并不相同）。值得注意的是，公式(3-13)的线性回归模型并不是由指数定律或 Russell 定律等传统的同位素校正方法衍生而来[138]。重要的是，该模型能够同时校正 MC-ICP-MS 中发生的质量相关以及非质量相关分馏[7,77,143]。与早期只使用截距的公式不同[145,146]，该回归模型属于双参数体系（截距和斜率）。此外，分析物和校准物质都是在同一溶液中同时测量的，因此消除了样品的基质效应。

早期线性回归模型的最大缺点[140,143,144]是两对同位素比值发生明显漂移并形成线性对数关系所需的测量时间较长。线性回归模型依赖于观测到的同位素比值随时间的不稳定性（即漂移），而 MC-ICP-MS 仪器（预热后）的高稳定性与之相悖。为建立 $\ln r_{i/j}$ 和 $\ln r_{\text{ref}}$ 之间的线性关系，线性回归模型通常需要很长的测量时间，单次测量时长可达 15 小时。这反过来又需要大量的样本。据报道，通过添加基体元素[46,147]或对等离子体射频功率施加微小的增量变化[52,59,64,66,77-79,141]，可以在较短的时间尺度上诱导同位素分馏使比值产生明显漂移。其中后一种方法可将测量时间从每次数小时缩短至仅需 10~30 min，使优化后的线性回归模型（ORM）适用于样本量较小的体系。最重要的是，该优化线性回归模型的准确性已通过不同方式得到了验证：首先，分别使用两种独立的内标元素（Re 和 Tl）[77]对铱同位素比值进行校正，得到了一致的铱同位素比值，其差别在 10^{-4} 数量级。其次，用全同位素质量混合法（基准方法）对钼[141]、铅[79]同位素比值的线性回归测量结果进行了交叉验证，两者的结果在 10^{-4} 数量级范围内具有良好的一致性。

3.3.5 （全）同位素质量混合校正模型

以上讨论的 DS、SSB、C-SSBIN 等同位素分馏校正模型和优化的线性回归模型（ORM）均依赖于初始同位素标准物质的准确性。如何获取同位素组成的初始值仍然是一个巨大的挑战，同时对新的校准模型的需求也逐渐上升，有时甚至以牺牲测量精度为代价。近几年来，利用已知化学纯度的富集同位素通过重量法配制混合物来获得同位素比值的校正方法受到了越来越广泛的关注[79,81,148-161]，该方法由 Nier 在 1950 年首次提出并用于氩同位素的测量[162]。在本章中，这种类型的同位素分馏校正模型被称为同位素质量混合校正模型（GIM）。这里我们以 Li、Ag 和 Tl 等双同位素体系为例来阐述该模型的基本原理：根据两种单独的富集同位素（A 和 B）以及这两种同位素标准品按质量配制的混合物（AB）的所有测量值，它们的同位素比值真实值（同位素分馏校正后的值）可以分别使用以下方程式表示（假设三次测量中同位素分馏保持相同）：

$$R_{A,i/j} = K_{i/j} \cdot r_{A,i/j} \tag{3-14}$$

$$R_{B,i/j} = K_{i/j} \cdot r_{B,i/j} \tag{3-15}$$

$$R_{AB,i/j} = K_{i/j} \cdot r_{AB,i/j} \tag{3-16}$$

式中，R 和 r 分别是同位素比值的真实值（未知的）和测量值；K 是这些测量结果中的同位素比值校正因子（也未知）。必须注意的是，混合物 AB 是 A 和 B 按质量配制的，因此 $R_{AB,i/j}$ 的值可根据 A 和 B 的同位素组成（尚未可知）以及混合物中使用的纯 A 和 B 的质量计算得出：

$$R_{AB,i/j} = \frac{n_{AB,i}}{n_{AB,j}} = \frac{x_{A,i} \cdot \left(\dfrac{m_A}{A_{r,A}}\right) + x_{B,i} \cdot \left(\dfrac{m_B}{A_{r,B}}\right)}{x_{A,j} \cdot \left(\dfrac{m_A}{A_{r,A}}\right) + x_{B,j} \cdot \left(\dfrac{m_B}{A_{r,B}}\right)} \tag{3-17}$$

式中，x 和 A_r 分别是材料 A 和 B 的同位素丰度和原子量，m_A 和 m_B 分别是用于制备混合物 AB 的富集同位素 A 和 B 的质量。同位素丰度可以用同位素比值表示，原子量则由同位素丰度计算。比如：

$$x_{A,i} = \frac{R_{A,i/j}}{1 + R_{A,i/j}} \tag{3-18}$$

$$A_{r,A} = m_i \cdot x_{A,i} + m_j \cdot x_{A,j} \tag{3-19}$$

式中，m_i 和 m_j 分别是元素两种同位素的核素质量。将等式（3-17）与式（3-18）代入式（3-16），得出 $R_{AB,i/j}$ 的表达式：

$$R_{AB,i/j} = \frac{R_{A,i/j} \cdot m_A \cdot (m_i \cdot R_{B,i/j} + m_j) + R_{B,i/j} \cdot m_B \cdot (m_i \cdot R_{A,i/j} + m_j)}{m_A \cdot (m_i \cdot R_{B,i/j} + m_j) + m_B \cdot (m_i \cdot R_{A,i/j} + m_j)} \tag{3-20}$$

得到的三个方程组[式（3-13）、式（3-14）和式（3-19）]可以求解出三个未知 $K_{i/j}$、$R_{A,i/j}$ 和 $R_{B,i/j}$。比如：

$$K_{i/j} = \frac{m_j}{m_i} \cdot \frac{m_A \cdot (r_{A,i/j} - r_{AB,i/j}) + m_B \cdot (r_{B,i/j} - r_{AB,i/j})}{m_A \cdot r_{B,i/j} \cdot (r_{AB,i/j} - r_{A,i/j}) + m_B \cdot r_{A,i/j} \cdot (r_{AB,i/j} - r_{B,i/j})} \tag{3-21}$$

在成功应用 MC-ICP-MS 结合经典的"单一混合"策略测定 Li 同位素比值之后[148]，同位素质量混合校正模型的应用范围在一定程度上已扩展到铁[149]和钕[151]等多同位素元素的测量。如研究报道[148,151]，仅使用两种富集同位素和两者的混合物，例如铁的 ^{56}Fe 和 ^{54}Fe、Nd 的 ^{142}Nd 和 ^{144}Nd，便可推导出 $K_{54/56}$（或 $K_{144/142}$）的同位素比校正因子。假设同一元素的不同同位素具有相同的同位素分馏，再根据 $K_{54/56}$ 结合 Russell 定律[公式（3-1）]计算得到其他同位素对（例如 $K_{57/56}$ 和 $K_{58/56}$）的校正因子。这种方法主要是为了获得一种元素单一同位素对的校正结果。在最近测定 Zn[150]、Se[159] 和 Yb[160] 同位素比值时，并未使用所有稳定同位素的高纯富集同位素，而只使用了其中三种富集同位素。尽管这类简化的双同位素和三同位素混合法避免了使用所有富集同位素，但它们需要利用质量分馏模型来推导出其他同位素对的 K 值，这反过来可能会对在 MC-ICP-MS 中显示 MIF 的同位素的测量结果带来偏差[7,61]。

为了获得具有 N 种稳定同位素元素的准确同位素比值（它具有 $N-1$ 个同位素比值），需要 N 个富集同位素试剂。该系统包含 (N^2-1) 个未知变量：$(N-1)$ 个同位素分馏校正因子和 N 个独立富集样品的 $N(N-1)$ 组同位素比值。该体系的解析需要测量 $(N-1)$ 组独立的二元混合物的同位素比值。对于一个三同位素系统，至少需要测量五种独立同位素样品。举个例子，早期对硅的绝对同位素比值的测量就是通过测量五种样品完成的：包括三种纯的富集同位素样品和两种混合物（高纯的硅-28、硅-29、硅-30、硅-28 和硅-29 的 1∶1 混合物，以及硅-28 和硅-30 的 1∶1 混合物）[152]。我们将这种同位素分馏校正模型称为全同位素质量混合校正模型（FGIM）。如上所述，对于双同位素体系，同位素分馏因子的解析并不难，但对于三同位素体系（硅）而言，其分析过程变得愈加复杂了[154]。随着元素同位素数量的增加，K_s 解析的相应数学运算的复杂性会急剧上升。最近，Stoll-Werian 等已经报道了高达 12 种同位素体系的 K_s 值的解析方法[163]。

另一种方法是采用数值迭代方法来推导同位素比校正因子[1,7,159,160]。例如，对于双同位素体系，在首次迭代中，K 的值被设置为 1，$R_{A,i/j}$ 和 $R_{B,i/j}$ 的值 [由 $r_{A,i/j}$、$r_{B,i/j}$ 通过式（3-14）、式（3-15）计算得到] 结合公式（3-20）来计算 $R_{AB,i/j}$。在第二次迭代中，从式（3-16）反推 K 值，然后循环重复此过程，直到获得稳定不变的 K 值。虽然迭代法缺乏数学上的严谨性，并且难以用于最终结果不确定度传递的计算中，但该方法使用起来很简单。最近的研究表明，这种情况下的不确定度计算可以通过蒙特卡罗模拟法实现[1,7,159,160]。

由于 FGIM 是基于富集同位素及其混合物的连续测量，因此需要对分析物的基体以及元素浓度进行匹配。此外，需要较长的仪器预热时间以确保仪器的稳定性，而较短的测量时间对于减少同位素分馏的时间漂移至关重要。FGIM 同位素分馏校正模型的主要优点之一是适用于样本量小的体系。更重要的是，与 SSB、C-SSBIN 和 ORM 不同，FGIM 是一种基准校正方法[164]，并不依赖任何其他已知同位素组成的标准物质。换言之，不需要事先知道富集同位素的真实同位素组成，只需知道其化学纯度以定量制备混合物。Henrion 引入了同位素比值 1∶1 匹配的概念[165]，以减少同位素稀释结果的潜在误差。在更复杂的高阶同位素稀释模型中也可以使用这一理念[166]。这些观察结果得到了合乎逻辑的结论，表明当在混合物中使用最佳的 1∶1 同位素比例匹配时，FGIM 显示出更稳定的结果。这一点也在最近铅的同位素比值的研究中得到了证实[79]。

然而随着元素稳定同位素数量的增加，所需富集同位素的成本、对化学纯度表征的难度以及数学计算的复杂性都急剧上升。例如，与双同位素体系相比，硅的三同位素体系就是一个典型例子。在上述几点问题中，对富集同位素的化学纯度进行表征是至关重要的。除了传统的库仑法、滴定法和重量法外，辉光放电质谱法（GDMS）是一种可以提供高纯金属纯度且可溯源到国际单位制 SI 的替代技术[167-169]。因此，只要金属纯度在 0.999 g/g 至 0.999 999 9 g/g 的区间内，GDMS 均可用于这些富集同位素的纯度测量。

值得注意的是，如最近一篇文章所述，这种从同位素混合物的分析中得到同位素比值的基准校正方法可以通过不同的方式进行[164]。例如，校准双同位素体系仅需分析一种混合物样品，也可以调整为分析多种混合物，从而避免分析所有的富集同位素样品。具体而言，在双同位素体系中，如果由两种富集同位素配制成三种不同组成的二元混合物，则不需要直接测量单个富集同位素样品。实验设计中这种巧妙的组合为绕过高富集同位素样品的测量提供了可能（或便利）[152,153]。而这种方法是以提高了校准系数的不确定度为代价的。

近年来，全蒸发（TE）TIMS 作为一种成熟的同位素绝对比值测量方法，获得了广泛的关注和成功[170-185]。这种技术的主要优点是它提供了绝对同位素比值，而不需要富集同位素或其他同位素标准物质进行质量偏倚校正，因为同位素的信号强度是在整个蒸发过程中收集的。与通常的 MC-ICP-MS 测量结果相比，该方法测量的同位素比值的不确定度可能更大。

3.4 总结与展望

最近，多同位素元素的非质量分馏研究表明了 MIF 是 MC-ICP-MS 中确切存在的常见现象。迄今为止，MC-ICP-MS 中出现 MIF 的真正原因仍未可知，现有的假设，如核场（核体积）效应、磁同位素（核自旋）效应、X 型截取锥造成的高氧化物水平或元素具有较高的第一电离能等都不能系统地解释观察到的 MIF 现象。本章概述的 MC-ICP-MS 中 MIF 对准确测量同位素比值的影响表明，使用传统质量相关校正模型，如 Russell 定律和一度被认为最精确的 DS 模型，以获得校正同位素比值的方法并不适合于 MC-ICP-MS 中显示出 MIF 的同位素。同理，能够同时校正 MDF 和 MIF 的同位素分馏校正模型，如优化的线性回归模型和全同位素质量混合校正模型等，更加具有吸引力。为了更合理地选择与应用这些校正模型，我们必须了解它们的利弊和与之相关的假设。

目前，大多数发表的成果仅报道了同位素结果的标准偏差，而充分考虑测量过程中产生的所有可能的不确定度来源，对于得到准确的、可溯源至国际单位制 SI 的结果至关重要[1,186]。与方法相关的综合标准不确定度建议根据 JCGM 100：2008"测量数据评估-测量不确定度表达指南"[186]进行报道，以便对不同研究小组获得的同位素比值结果进行有意义的比较。

当今 MC-ICP-MS 已成为高准确度和高精度同位素比值测量最常用的两种技术之一，它有望在未来更广泛地应用于其他学科。希望看到对 MC-ICP-MS 中发生的非质量相关的分馏机理的深入研究，从而有助于我们理解并促进同位素分馏校正模型的进一步发展。

参 考 文 献

[1] Yang L. Accurate and precise determination of isotopic ratios by MC-ICP-MS: A review. Mass Spectrometry Reviews, 2009, 28(6): 990-1011.

[2] Hoefs J. Stable Isotope Geochemistry. Vol. 285. Berlin: Springer, 2009.

[3] Heumann K G, Vanhaecke F. Isotope ratios in analytical chemistry. Analytical and Bioanalytical Chemistry, 2008, 390: 433-435.

[4] Balcaen L, Moens L, Vanhaecke F. Determination of isotope ratios of metals (and metalloids) by means of inductively coupled plasma-mass spectrometry for provenancing purposes: A review. Spectrochimica Acta, Part B: Atomic Spectroscopy, 2010, 65: 769-786.

[5] Laeter J D. Mass spectrometry and geochronology. Mass Spectrometry Reviews, 1998, 17: 97-125.

[6] Vanhaecke F, Degryse P. MC-ICP-MS in isotopic analysis: Fundamentals and applications using ICP-MS. John Wiley & Sons, 2012.

[7] Yang L, Tong S, Zhou L, Hu Z, Mester Z, Meija J. A critical review on isotopic fractionation correction methods for accurate isotope amount ratio measurements by MC-ICP-MS. Journal of Analytical Atomic Spectrometry, 2018, 33(11): 1849-1861.

[8] Desaulty A M, Petelet-Giraud E. Zinc isotope composition as a tool for tracing sources and fate of metal contaminants in rivers. Science of the Total Environment, 2020, 728: 138599.

[9] Tsui M T K, Blum J D, Kwon S Y. Review of stable mercury isotopes in ecology and biogeochemistry. Science of the Total Environment, 2020, 716: 135386.

[10] Millot R, Guerrot C, Vigier N. Accurate and high-precision measurement of lithium isotopes in two reference materials by MC-ICP-MS. Geostandards and Geoanalytical Research, 2004, 28: 153-159.

[11] Kasemann S A, Jeffcoate A B, Elliott T. Lithium isotope composition of basalt glass reference material. Analytical Chemistry, 2005, 77(16): 5251-5257.

[12] Platzner T I, Habfast K, Walder A J, Goetz A. Modern Isotope Ratio Mass Spectrometry. Vol. 145. New York: Wiley, 1997.

[13] Heumann K G, Gallus S M, Radlinger G, Vogl J. Precision and accuracy in isotope ratio measurements by plasma source mass spectrometry. Journal of Analytical Atomic Spectrometry, 1998, 13(9): 1001-1008.

[14] Gillson G R, Douglas D J, Fulford J E, Halligan K W, Tanner S D. Nonspectroscopic interelement interferences in inductively coupled plasma mass spectrometry. Analytical Chemistry, 1988, 60(14): 1472-1474.

[15] Andrén H, Rodushkin I, Stenberg A, Malinovs.ky D, Baxter D C. Sources of mass bias and isotope ratio variation in multi-collector ICP-MS: Optimization of instrumental parameters based on experimental observations. Journal of Analytical Atomic Spectrometry, 2004, 19(9): 1217-1224.

[16] Russell W A, Papanastassiou D A, Tombrello T A. Ca isotope fractionation on the earth and other solar system materials. Geochimica et Cosmochimica Acta, 1978, 42(8): 1075-1090.

[17] McArthur J M, Rio D, Massari F, Castradori D, Bailey T R, Thirlwall M, Houghton S. A revised pliocene record for marine-$^{87}Sr/^{86}Sr$ used to date an interglacial event recorded in the Cockburn Island Formation, Antarctic Peninsula. Palaeogeography, Palaeoclimatology, Palaeoecology, 2006, 242(1-2): 126-136.

[18] Smellie J L, McArthur J M, McIntosh W C, Esser R. Late Neogene interglacial events in the James Ross Island region, northern Antarctic Peninsula, dated by Ar/Ar and Sr-isotope stratigraphy. Palaeogeography, Palaeoclimatology, Palaeoecology, 2006, 242(3-4): 169-187.

[19] Fortunato G, Mumic K, Wunderli S, Pillonel L, Bosset J O, Gremaud G. Application of strontium isotope abundance ratios measured by MC-ICP-MS for food authentication. Journal of Analytical Atomic Spectrometry, 2004, 19(2): 227.

[20] Galler P, Limbeck A, Boulyga S F, Stingeder G, Hirata T, Prohaska T. Development of an on-line flow injection Sr/matrix separation method for accurate, high-throughput determination of Sr isotope ratios by multiple collector-inductively coupled plasma-mass spectrometry. Analytical Chemistry, 2007, 79(13): 5023-5029.

[21] Balcaen L, Schrijver I D, Moens L, Vanhaecke F. Determination of the $^{87}Sr/^{86}Sr$ isotope ratio in USGS silicate reference materials by multi-collector ICP-mass spectrometry. International Journal of Mass Spectrometry, 2005, 242(2-3): 251-255.

[22] Barbaste M, Robinson K, Guilfoyle S, Medina B, Lobinski R. Precise determination of the strontium isotope ratios in wine by inductively coupled plasma sector field multicollector mass spectrometry (ICP-SF-MC-MS). Journal of Analytical Atomic Spectrometry, 2002, 17(2): 135-137.

[23] Rodushkin I, Bergman T, Douglas G, Engstrom E, Sorlin D, Baxter D C. Authentication of Kalix (N.E. Sweden) vendace caviar using inductively coupled plasma-based analytical techniques: Evaluation of different approaches. Analytica Chimica Acta, 2007, 583(2): 310-318.

[24] Garcia-Ruiz S, Moldovan M, Fortunato G, Wunderli S, Garcia Alonso J I. Evaluation of strontium isotope abundance ratios in combination with multi-elemental analysis as a possible tool to study the geographical origin of ciders. Analytica Chimica Acta, 2007, 590(1): 55-66.

[25] Goswami V, Singh S K, Bhushan R, Rai V K. Temporal variations in $^{87}Sr/^{86}Sr$ and $^{\varepsilon}Nd$ in sediments of the southeastern

Arabian Sea: Impact of monsoon and surface water circulation. Geochemistry, Geophysics, Geosystems, 2012, 13(1). doi.org/10.1029/2011GC003802.

[26] Laffoon J E, Davies G R, Hoogland M L P, Hofman C L. Spatial variation of biologically available strontium isotopes ($^{87}Sr/^{86}Sr$) in an archipelagic setting: A case study from the Caribbean. Journal of Archaeological Science, 2012, 39(7): 2371-2384.

[27] Tappe S, Simonetti A. Combined U-Pb geochronology and Sr-Nd isotope analysis of the Ice River perovskite standard, with implications for kimberlite and alkaline rock petrogenesis. Chemical Geology, 2012, 304-305: 10-17.

[28] Yang Z, Fryer B J, Longerich H P, Gagnon J E, Samson I M. 785 nm femtosecond laser ablation for improved precision and reduction of interferences in Sr isotope analyses using MC-ICP-MS. Journal of Analytical Atomic Spectrometry, 2011, 26(2): 341-351.

[29] Hobbs J A, Lewis L S, Ikemiyagi N, Sommer T, Baxter R D. The use of otolith strontium isotopes ($^{87}Sr/^{86}Sr$) to identify nursery habitat for a threatened estuarine fish. Environmental Biology of Fishes, 2010, 89(3-4): 557-569.

[30] Isnard H, Brennetot R, Caussignac C, Caussignac N, Chartier F. Investigations for determination of Gd and Sm isotopic compositions in spent nuclear fuels samples by MC-ICP-MS. International Journal of Mass Spectrometry, 2005, 246(1-3): 66-73.

[31] Collerson K D, Kamber B S, Schoenberg R. Applications of accurate, high-precision Pb isotope ratio measurement by multi-collector ICP-MS. Chemical Geology, 2002, 188: 65-83.

[32] Weiss D J, Kober B, Dolgopolova A, Gallagher K, Spiro B, Le Roux G, Mason T F D, Kylander M, Coles B J. Accurate and precise Pb isotope ratio measurements in environmental samples by MC-ICP-MS. International Journal of Mass Spectrometry, 2004, 232(3): 205-215.

[33] Chernyshev I V, Chugaev A V, Shatagin K N. High-precision Pb isotope analysis by multicollector-ICP-mass-spectrometry using $^{205}Tl/^{203}Tl$ normalization: Optimization and calibration of the method for the studies of Pb isotope variations. Geochemistry International, 2007, 45(11): 1065-1076.

[34] Gallon C, Aggarwal J, Flegal A R. Comparison of mass discrimination correction methods and sample introduction systems for the determination of lead isotopic composition using a multicollector inductively coupled plasma mass spectrometer. Analytical Chemistry, 2008, 80(22): 8355-8363.

[35] Woodhead J, Hellstrom J, Maas R, Drysdale R, Zanchetta G, Devine P, Taylor E. U-Pb geochronology of speleothems by MC-ICP-MS. Quaternary Geochronology, 2006, 1(3): 208-221.

[36] Chew D M, Sylvester P J, Tubrett M N. U-Pb and Th-Pb dating of apatite by LA-ICPMS. Chemical Geology, 2011, 280(1-2): 200-216.

[37] Kritee K, Blum J D, Johnson M W, Bergquist B A, Barkay T. Mercury stable isotope fractionation during reduction of Hg(II) to Hg(0) by mercury resistant microorganisms. Environmental Science & Technology, 2007, 41(6): 1889-1895.

[38] Zheng W, Foucher D, Hintelmann H. Mercury isotope fractionation during volatilization of Hg(0) from solution into the gas phase. Journal of Analytical Atomic Spectrometry, 2007, 22(9): 1097.

[39] Mason T F D, Weiss D J, Horstwood M, Parrish R R, Russell S S, Mullane E, Coles B J. High-precision Cu and Zn isotope analysis by plasma source mass spectrometry. Journal of Analytical Atomic Spectrometry, 2004, 19(2): 218.

[40] Baxter D C, Rodushkin I, Engström E, Malinovsky D. Revised exponential model for mass bias correction using an internal standard for isotope abundance ratio measurements by multi-collector inductively coupled plasma mass spectrometry. Journal of Analytical Atomic Spectrometry, 2006, 21(4): 427.

[41] Anbar A D, Knab K A, Barling J. Precise determination of mass-dependent variations in the isotopic composition of molybdenum using MC-ICPMS. Analytical Chemistry, 2001, 73: 1425-1431.

[42] Waight T, Baker J, Willigers B. Rb isotope dilution analyses by MC-ICPMS using Zr to correct for mass fractionation: Towards improved Rb-Sr geochronology? Chemical Geology, 2002, 186(1-2): 99-116.

[43] Hirata T. Lead isotopic analyses of NIST standard reference materials using multiple collector inductively coupled plasma mass spectrometry coupled with a modified external correction method for mass discrimination effect. Analyst, 1996, 121(10): 1407-1411.

[44] White W M, Albarede F, Telouk P. High-precision analysis of Pb isotope ratios by multi-collector ICP-MS. Chemical Geology, 2000, 167(3-4): 257-270.

[45] Luo Y, Dabek-Zlotorzynska E, Celo V, Muir D C G, Yang L. Accurate and precise determination of silver isotope fractionation in environmental samples by multicollector-ICPMS. Analytical Chemistry, 2010, 82(9): 3922-3928.

[46] Woodhead J. A simple method for obtaining highly accurate Pb isotope data by MC-ICP-MS. Journal of Analytical Atomic Spectrometry, 2002, 17(10): 1381-1385.

[47] Thirlwall M F. Multicollector ICP-MS analysis of Pb isotopes using a ^{207}Pb-^{204}Pb double spike demonstrates up to 400 ppm/amu systematic errors in Tl-normalization. Chemical Geology, 2002, 184: 255-279.

[48] Xie Q, Lie S, Evans D, Dillon P, Hintelmann H. High precision Hg isotope analysis of environmental samples using gold Trap-MC-ICP-MS. Journal of Analytical Atomic Spectrometry, 2005, 20(6): 515-522.

[49] Yang L, Peter C, Panne U, Sturgeon R E. Use of Zr for mass bias correction in strontium isotope ratio determinations using MC-ICP-MS. Journal of Analytical Atomic Spectrometry, 2008, 23(9): 1269.

[50] Yang L, Dabek-Zlotorzynska E, Celo V. High precision determination of silver isotope ratios in commercial products by MC-ICP-MS. Journal of Analytical Atomic Spectrometry, 2009, 24(11): 1564-1569.

[51] Yang L, Sturgeon, Ralph E. Isotopic fractionation of mercury induced by reduction and ethylation. Analytical and Bioanalytical Chemistry, 2009, 393(1): 377-385.

[52] Zhu Z, Meija J, Tong S, Zheng A, Zhou L, Yang L. Determination of the isotopic composition of osmium using MC-ICP-MS. Analytical Chemistry, 2018, 90(15): 9281-9288.

[53] Vance D, Thirlwall M. An assessment of mass discrimination in MC-ICP-MS using Nd isotopes. Chemical Geology, 2002, 185(3-4): 227-240.

[54] Newman K, Freedman P A, Williams J, Belshaw N S, Halliday A N. High sensitivity skimmers and non-linear mass dependent fractionation in ICP-MS. Journal of Analytical Atomic Spectrometry, 2009, 24(6): 742-751.

[55] Newman K. Effects of the sampling interface in MC-ICP-MS: Relative elemental sensitivities and non-linear mass dependent fractionation of Nd isotopes. Journal of Analytical Atomic Spectrometry, 2012, 27(1): 63-70.

[56] Xu L, Hu Z, Zhang W, Yang L, Liu Y, Gao S, Luo T, Hu S. *In situ* Nd isotope analyses in geological materials with signal enhancement and non-linear mass dependent fractionation reduction using laser ablation MC-ICP-MS. Journal of Analytical Atomic Spectrometry, 2015, 30(1): 232-244.

[57] Shirai N, Humayun M. Mass independent bias in W isotopes in MC-ICP-MS instruments. Journal of Analytical Atomic Spectrometry, 2011, 26(7): 1414-1420.

[58] Mei Q F, Yang J H, Yang Y H. An improved extraction chromatographic purification of tungsten from a silicate matrix for high precision isotopic measurements using MC-ICP-MS. Journal of Analytical Atomic Spectrometry, 2018, 33(4): 569-577.

[59] Zhang R, Meija J, Huang Y, Pei X, Mester Z, Yang L. Determination of the isotopic composition of tungsten using MC-ICP-MS. Analytica Chimica Acta, 2019, 1089: 19-24.

[60] Irrgeher J, Prohaska T, Sturgeon R E, Mester Z, Yang L. Determination of strontium isotope amount ratios in biological tissues using MC-ICP-MS. Analytical Methods, 2013, 5(7): 1687-1694.

[61] Yang L, Mester Z, Zhou L, Gao S, Sturgeon R E, Meija J. Observations of large mass-independent fractionation occurring in MC-ICP-MS: Implications for determination of accurate isotope amount ratios. Analytical Chemistry, 2011, 83(23): 8999-9004.

[62] Yang L, Zhou L, Hu Z, Gao S. Direct determination of Si isotope ratios in natural waters and commercial Si standards by ion exclusion chromatography multicollector inductively coupled plasma mass spectrometry. Analytical Chemistry, 2014, 86(18): 9301-9308.

[63] Hu Z, Liu Y, Gao S, Liu W, Zhang W, Tong X, Lin L, Zong K, Li M, Chen H, Zhou L, Yang L. Improved in situ Hf isotope ratio analysis of zircon using newly designed X skimmer cone and jet sample cone in combination with the addition of nitrogen by laser ablation multiple collector ICP-MS. Journal of Analytical Atomic Spectrometry, 2012, 27(9): 1391-1399.

[64] Tong S Y, Meija J, Zhou L, Mester Z, Yang L. Determination of the isotopic composition of hafnium using MC-ICP-MS. Metrologia, 2019, 56(4): 044008.

[65] Miyazaki T, Kimura J I, Chang Q. Analysis of stable isotope ratios of Ba by double-spike standard-sample bracketing using multiple-collector inductively coupled plasma mass spectrometry. Journal of Analytical Atomic Spectrometry, 2014, 29(3): 483.

[66] He J, Yang L, Hou X, Mester Z, Meija J. Determination of the isotopic composition of gadolinium using MC-ICP-MS. Analytical Chemistry, 2020, 92(8): 6103-6110.

[67] Walczyk T. Tims versus multicollector-ICP-MS: Coexistence or struggle for survival? Analytical and Bioanalytical Chemistry, 2004, 378(2): 229-231.

[68] Weyer S, Anbar A D, Gerdes A, Gordon G W, Algeo T J, Boyle E A. Natural fractionation of $^{238}U/^{235}U$. Geochimica et Cosmochimica Acta, 2008, 72(2): 345-359.

[69] Sherman L, Blum J, Johnson K, Keeler G, Barres J, Douglas T. Use of mercury isotopes to understand mercury cycling between arctic snow and atmosphere. Nature Geoscience, 2010, 3: 173-177.

[70] Fujii T, Moynier F, Albarède F. The nuclear field shift effect in chemical exchange reactions. Chemical Geology, 2009, 267(3): 139-156.

[71] Moynier F, Fujii T, Telouk P. Mass-independent isotopic fractionation of tin in chemical exchange reaction using a crown ether. Analytica Chimica Acta, 2009, 632(2): 234-239.

[72] Moynier F, Fujii T, Brennecka G A, Nielsen S G. Nuclear field shift in natural environments. Comptes Rendus Geoscience, 2013, 345(3): 150-159.

[73] Malinovs.ky D, Vanhaecke F. Mercury isotope fractionation during abiotic transmethylation reactions. International Journal of Mass Spectrometry, 2011, 307(1-3): 214-224.

[74] Clayton R N, Grossman L, Mayeda T K. A component of primitive nuclear composition in carbonaceous meteorites. Science, 1973, 182(4111): 485.

[75] Heidenreich J E, Thiemens M H. A non-mass-dependent oxygen isotope effect in the production of ozone from molecular oxygen: The role of molecular symmetry in isotope chemistry. The Journal of Chemical Physics, 1983, 78: 892-895.

[76] Farquhar J, Bao H, Thiemens M. Atmospheric influence of earth's earliest sulfur cycle. Science, 2000, 289(5480): 756.

[77] Zhu Z, Meija J, Zheng A, Mester Z, Yang L. Determination of the isotopic composition of iridium using multicollector-ICP-MS. Analytical Chemistry, 2017, 89(17): 9375-9382.

[78] He J, Meija J, Hou X, Zheng C, Mester Z, Yang L. Determination of the isotopic composition of lutetium using MC-ICP-MS. Analytical and Bioanalytical Chemistry, 2020, 412: 6257-6263.

[79] Tong S, Meija J, Zhou L, Methven B, Mester Z, Yang L. High-precision measurements of the isotopic composition of common lead using MC-ICP-MS: Comparison of calibration strategies based on full gravimetric isotope mixture and regression models. Analytical Chemistry, 2019, 91(6): 4164-4171.

[80] Schmitt A D, Galer S J G, Abouchami W. High-precision cadmium stable isotope measurements by double spike thermal ionisation mass spectrometry. Journal of Analytical Atomic Spectrometry, 2009, 24(8): 1079-1088.

[81] Pritzkow W, Wunderli S, Vogl J, Fortunato G. The isotope abundances and the atomic weight of cadmium by a metrological approach. International Journal of Mass Spectrometry, 2007, 261(1): 74-85.

[82] Young E D, Galy A, Nagahara H. Kinetic and equilibrium mass-dependent isotope fractionation laws in nature and their geochemical and cosmochemical significance. Geochimica et Cosmochimica Acta, 2002, 66(6): 1095-1104.

[83] Galy A, Belshaw N S, Halicz L, O'Nions R K. High-precision measurement of magnesium isotopes by multiple-collector inductively coupled plasma mass spectrometry. International Journal of Mass Spectrometry, 2001, 208(1): 89-98.

[84] Malinovsky D, Vanhaecke F. Mass-independent isotope fractionation of heavy elements measured by MC-ICP-MS: A unique probe in environmental sciences. Analytical and Bioanalytical Chemistry, 2011, 400(6): 1619-1624.

[85] Bigeleisen J. Nuclear size and shape effects in chemical reactions. Isotope chemistry of the heavy elements. Journal of the American Chemical Society, 1996, 118(15): 3676-3680.

[86] Buchachenko A L. Magnetic isotope effect: Nuclear spin control of chemical reactions. The Journal of Physical Chemistry A, 2001, 105(44): 9995-10011.

[87] Wang M, Audi G, Kondev F G, Huang W J, Naimi S, Xu X. The AME2016 atomic mass evaluation (Ⅱ). Tables, graphs and references. Chinese Physics C, 2017, 41(3): 030003.

[88] Angeli I. A consistent set of nuclear rms charge radii: Properties of the radius surface $R(N, Z)$. Atomic Data and Nuclear Data Tables, 2004, 87(2): 185-206.

[89] JAEA. Tables of Nuclear Data. http: //wwwndc.Jaea.Go.Jp/nuc/index. modified at 2016/06/01.

[90] Meija J. An ode to the atomic weights. Nature Chemistry, 2014, 6(9): 749-750.

[91] Gault-Ringold M, Stirling C H. Anomalous isotopic shifts associated with organic resin residues during cadmium isotopic analysis by double spike MC-ICP-MS. Journal of Analytical Atomic Spectrometry, 2012, 27(3): 449-459.

[92] Millet M A, Baker J A, Payne C E. Ultra-precise stable Fe isotope measurements by high resolution multiple-collector inductively coupled plasma mass spectrometry with a ^{57}Fe-^{58}Fe double spike. Chemical Geology, 2012, 304-305: 18-25.

[93] Bonnand P, Parkinson I J, James R H, Karjalainen A-M, Fehr M A. Accurate and precise determination of stable Cr isotope compositions in carbonates by double spike MC-ICP-MS. Journal of Analytical Atomic Spectrometry, 2011, 26(3): 528.

[94] Lacan F, Radic A, Labatut M, Jeandel C, Poitrasson F, Sarthou G, Pradoux C, Chmeleff J, Freydier R. High-precision determination of the isotopic composition of dissolved iron in iron depleted seawater by double spike multicollector-ICPMS. Analytical Chemistry, 2010, 82(17): 7103-7111.

[95] Li J, Zhu X K, Tang S H. Effects of acid and concentration on Cu and Zn isotope measurements by multicollector-inductively coupled plasma mass spectrometry. Chinese Journal of Analytical Chemistry, 2008, 36(9): 1196-1200.

[96] Fletcher I R. Using the common-Pb standards SRM-981 and SRM-982 as double spikes. International Journal of Mass Spectrometry, 2007, 261(2-3): 234-238.

[97] Baker J, Peate D, Waight T, Meyzen C. Pb isotopic analysis of standards and samples using a ^{207}Pb-^{204}Pb double spike and thallium to correct for mass bias with a double-focusing MC-ICP-MS. Chemical Geology, 2004, 211(3-4): 275-303.

[98] Gopalan K, Macdougall D, Macisaac C. Evaluation of a ^{42}Ca-^{43}Ca double-spike for high precision Ca isotope analysis. International Journal of Mass Spectrometry, 2006, 248(1-2): 9-16.

[99] Schoenberg R, Zink S, Staubwasser M, von Blanckenburg F. The stable Cr isotope inventory of solid earth reservoirs determined by double spike MC-ICP-MS. Chemical Geology, 2008, 249(3-4): 294-306.

[100] Millet M A, Dauphas N. Ultra-precise titanium stable isotope measurements by double-spike high resolution MC-ICP-MS. Journal of Analytical Atomic Spectrometry, 2014, 29(8): 1444-1458.

[101] Hopp T, Fischer-Gödde M, Kleine T. Ruthenium stable isotope measurements by double spike MC-ICP-MS. Journal of Analytical Atomic Spectrometry, 2016, 31(7): 1515-1526.

[102] Creech J B, Moynier F, Badullovich N. Tin stable isotope analysis of geological materials by double-spike MC-ICP-MS. Chemical Geology, 2017, 457: 61-67.

[103] Creech J B, Baker J A, Handler M R, Bizzarro M. Platinum stable isotope analysis of geological standard reference materials by double-spike MC-ICP-MS. Chemical Geology, 2014, 363: 293-300.

[104] Klaver M, Coath C D. Obtaining accurate isotopic compositions with the double spike technique: Practical considerations. Geostandards and Geoanalytical Research, 2019, 43(1): 5-22.

[105] Wu G, Zhu J M, Wang X, Johnson T M, Han G. High-sensitivity measurement of Cr isotopes by double spike MC-ICP-MS at the 10 ng level. Analytical Chemistry, 2019, 92(1): 1463-1469.

[106] Inglis E C, Creech J B, Deng Z, Moynier F. High-precision zirconium stable isotope measurements of geological reference materials as measured by double-spike MC-ICP-MS. Chemical Geology, 2018, 493: 544-552.

[107] Pons M L, Millet M A, Nowell G N, Misra S, Williams H M. Precise measurement of selenium isotopes by HG-MC-ICP-MS using a 76-78 double-spike. Journal of Analytical Atomic Spectrometry, 2020, 35(2): 320-330.

[108] Meija J, Yang L, Sturgeon R, Mester Z. Mass bias fractionation laws for multi-collector ICPMS: Assumptions and their experimental verificatio. Analytical Chemistry, 2009, 81: 6774-6778.

[109] Dietz L A, Paghugki C F, Land G A. Internal standard technique for precise isotopic abundance measurments in thermal ionization mass spectrometry. Analytical Chemistry, 1962, 34(6): 709.

[110] Dodson M H. A theoretical study of the use of internal standards for precise isotopic analysis by the surface ionization technique: Part I—General first-order algebraic solutions. Journal of Scientific Instruments, 1963, 40(6): 289.

[111] Dodson M H. Simplified equations for double-spiked isotopic analyses. Geochimica et Cosmochimica Acta, 1970, 34(11): 1241-1244.

[112] Russell R D. The systematics of double spiking. Journal of Geophysical Research, 1971, 76(20): 4949-4955.

[113] Cumming G L. Propagation of experimental errors in lead isotope ratio measurements using the double spike method. Chemical Geology, 1973, 11(3): 157-165.

[114] Wasserman N L, Johnson T M. Measurements of mass-dependent Te isotopic variation by hydride generation MC-ICP-MS. Journal of Analytical Atomic Spectrometry, 2020, 35(2): 307-319.

[115] Gaspers N, Magna T, Ackerman L. Molybdenum mass fractions and stable isotope compositions of sedimentary carbonate and silicate reference materials. Geostandards and Geoanalytical Research, 2020, 44(2): 363-374.

[116] Arnold T, Schönbächler M, Rehkämper M, Dong S, Zhao F J, Kirk G J D, Coles B J, Weiss D J. Measurement of zinc stable isotope ratios in biogeochemical matrices by double-spike MC-ICP-MS and determination of the isotope ratio pool available for plants from soil. Analytical and Bioanalytical Chemistry, 2010, 398(7-8): 3115-3125.

[117] Inglis E C, Moynier F, Creech J, Deng Z, Day J M D, Teng F Z, Bizzarro M, Jackson M, Savage P. Isotopic fractionation of zirconium during magmatic differentiation and the stable isotope composition of the silicate earth. Geochimica et Cosmochimica Acta, 2019, 250: 311-323.

[118] Feng L, Zhou L, Hu W, Zhang W, Li B, Liu Y, Hu Z, Yang L. A simple single-stage extraction method for Mo separation from geological samples for isotopic analysis by MC-ICP-MS. Journal of Analytical Atomic Spectrometry,

2020, 35(1): 145-154.

[119] Coath C D, Elliott T, Hin R C. Double-spike inversion for three-isotope systems. Chemical Geology, 2017, 451: 78-89.

[120] Chew G, Walczyk T. Measurement of isotope abundance variations in nature by gravimetric spiking isotope dilution analysis (GS-IDA). Analytical Chemistry, 2013, 85(7): 3667-3673.

[121] Pietruszka A J, Reznik A D. Identification of a matrix effect in the MC-ICP-MS due to sample purification using ion exchange resin: An isotopic case study of molybdenum. International Journal of Mass Spectrometry, 2008, 270(1-2): 23-30.

[122] Petit J C J, Jeroen D J, Lei C, Nadine M. Development of Cu and Zn isotope MC-ICP-MS measurements: Application to suspended particulate matter and sediments from the scheldt estuary. Geostandards and Geoanalytical Research, 2010, 32(2): 149-166.

[123] Bizzarro M, Paton C, Larsen K, Schiller M, Trinquier A, Ulfbeck D. High-precision Mg-isotope measurements of terrestrial and extraterrestrial material by HR-MC-ICP-MS-implications for the relative and absolute Mg-isotope composition of the bulk silicate earth. Journal of Analytical Atomic Spectrometry, 2011, 26(3): 565-577.

[124] Amrani A, Sessions A L, Adkins J F. Compound-specific $\delta^{34}S$ analysis of volatile organics by coupled GC/multicollector-ICPMS. Analytical Chemistry, 2009, 81(5-6): 523-528.

[125] Kasyanova A V, Streletskaya M V, Chervyakovs.kaya M V, Kiseleva D V. A method for $^{87}Sr/^{86}Sr$ isotope ratio determination in biogenic apatite by MC-ICP-MS using the SSB technique. AIP Conference Proceedings. AIP Publishing LLC, 2019, 2174(1): 020028.

[126] Peel K, Weiss D, Chapman J, Arnold T, Coles B. A simple combined sample-standard bracketing and inter-element correction procedure for accurate mass bias correction and precise Zn and Cu isotope ratio measurements. Journal of Analytical Atomic Spectrometry, 2008, 23(1): 103-110.

[127] Fietzke J, Eisenhauer A. Determination of temperature-dependent stable strontium isotope ($^{88}Sr/^{86}Sr$) fractionation via bracketing standard MC-ICP-MS. Geochemistry, Geophysics, Geosystems, 2006. DOI: 10.1029/2006GC001243.

[128] Halicz L, Segal I, Fruchter N, Stein M, Lazar B. Strontium stable isotopes fractionate in the soil environments? Earth and Planetary Science Letters, 2008, 272(1-2): 406-411.

[129] Miller C A, Peucker-Ehrenbrink B, Ball L. Precise determination of rhenium isotope composition by multi-collector inductively-coupled plasma mass spectrometry. Journal of Analytical Atomic Spectrometry, 2009, 24(8): 1069.

[130] Zhang T, Zhou L, Yang L, Wang Q, Feng L P, Liu Y S. High precision measurements of gallium isotopic compositions in geological materials by MC-ICP-MS. Journal of Analytical Atomic Spectrometry, 2016, 31(8): 1673-1679.

[131] Arantes de Carvalho G G, Oliveira P V, Yang L. Determination of europium isotope ratios in natural waters by MC-ICP-MS. Journal of Analytical Atomic Spectrometry, 2017, 32(5): 987-995.

[132] Sullivan K, Layton-Matthews D, Leybourne M, Kidder J, Mester Z, Yang L. Copper isotopic analysis in geological and biological reference materials by MC-ICP-MS. Geostandards and Geoanalytical Research, 2020, 44(2): 349-362.

[133] Anoshkina Y, Costas-Rodríguez M, Vanhaecke F. Iron isotopic analysis of finger-prick and venous blood by multi-collector inductively coupled plasma-mass spectrometry after volumetric absorptive microsampling. Journal of Analytical Atomic Spectrometry, 2017, 32(2): 314-321.

[134] Rua-Ibarz A, Bolea-Fernandez E, Vanhaecke F. An in-depth evaluation of accuracy and precision in Hg isotopic analysis via pneumatic nebulization and cold vapor generation multi-collector ICP-mass spectrometry. Analytical and Bioanalytical Chemistry, 2016, 408(2): 417-429.

[135] Devulder V, Lobo L, Van Hoecke K, Degryse P, Vanhaecke F. Common analyte internal standardization as a tool for correction for mass discrimination in multi-collector inductively coupled plasma-mass spectrometry. Spectrochimica

Acta, Part B: Atomic Spectroscopy, 2013, 89: 20-29.

[136] Yuan H, Cheng C, Chen K, Bao Z. Standard-sample bracketing calibration method combined with Mg as an internal standard for silicon isotopic compositions using multi-collector inductively coupled plasma mass spectrometry. Acta Geochimica, 2016, 35(4): 421-427.

[137] Chen K Y, Yuan H L, Liang P, Bao Z A, Chen L. Improved nickel-corrected isotopic analysis of iron using high-resolution multi-collector inductively coupled plasma mass spectrometry. International Journal of Mass Spectrometry, 2017, 421: 196-203.

[138] Maréchal C N, P.Telouk, Albarede F. Precise analysis of copper and zinc isotopic compositions by plasma-source mass spectrometry. Chemical Geology, 1999, 156: 251-273.

[139] Albarède F, Telouk P, Blichert-Toft J, Boyet M, Agranier A, Nelson B. Precise and accurate isotopic measurements using multiple-collector ICPMS. Geochimica et Cosmochimica Acta, 2004, 68(12): 2725-2744.

[140] Meija J, Yang L, Sturgeon R E, Mester Z. Certification of natural isotopic abundance inorganic mercury reference material NIMS-1 for absolute isotopic composition and atomic weight. Journal of Analytical Atomic Spectrometry, 2010, 25(3): 384.

[141] Malinovs.ky D, Dunn P J H, Goenaga-Infante H. Calibration of Mo isotope amount ratio measurements by MC-ICPMS using normalisation to an internal standard and improved experimental design. Journal of Analytical Atomic Spectrometry, 2016, 31(10): 1978-1988.

[142] Poitrasson F, Freydier R. Heavy iron isotope composition of granites determined by high resolution MC-ICP-MS. Chemical Geology, 2005, 222(1-2): 132-147.

[143] Yang L, Meija J. Resolving the germanium atomic weight disparity using multicollector ICPMS. Analytical Chemistry, 2010, 82: 4188-4193.

[144] Yang L, Sturgeon R E, Mester Z, Meija J. Metrological triangle for measurements of isotope amount ratios of silver, indium, and antimony using multicollector-inductively coupled plasma mass spectrometry: The 21st century harvard method. Analytical Chemistry, 2010, 82(21): 8978-8982.

[145] Baxter D C, Rodushkin I, Engström E, Malinovsky D. Revised exponential model for mass bias correction using an internal standard for isotope abundance ratio measurements by multi-collector inductively coupled plasma mass spectrometry. Journal of Analytical Atomic Spectrometry, 2006, 21(4): 427-430.

[146] Baxter D C, Rodushkin I, Engström E. Isotope abundance ratio measurements by inductively coupled plasma-sector field mass spectrometry. Journal of Analytical Atomic Spectrometry, 2012, 27(9): 1355-1381.

[147] Archer C, Vance D. Mass discrimination correction in multiple-collector plasma source mass spectrometry: An example using Cu and Zn isotopes. Journal of Analytical Atomic Spectrometry, 2004, 19(5): 656.

[148] Qi H P, Berglund M, Taylor P D P, Hendrickx F, Verbruggen A, Bièvre P D. Preparation and characterisation of synthetic mixtures of lithium isotopes. Fresenius' Journal of Analytical Chemistry, 1998, 361(8): 767-773.

[149] Zhou T, Zhao M, Wang J, Lu H. Absolute measurements and certified reference material for iron isotopes using multiple-collector inductively coupled mass spectrometry. Rapid Communications in Mass Spectrometry, 2008, 22(5): 717-720.

[150] Ponzevera E, Quétel C R, Berglund M, Taylor P D P, Evans P, Loss R D, Fortunato G. Mass discrimination during MC-ICPMS isotopic ratio measurements: Investigation by means of synthetic isotopic mixtures (IRMM-007 series) and application to the calibration of natural-like zinc materials (including IRMM-3702 and IRMM-651). Journal of the American Society for Mass Spectrometry, 2006, 17(10): 1413-1428.

[151] Zhao M, Zhou T, Wang J, Lu H, Xiang F. Absolute measurements of neodymium isotopic abundances and atomic

weight by MC-ICPMS. International Journal of Mass Spectrometry, 2005, 245(1-3): 36-40.

[152] Rienitz O, Pramann A, Schiel D. Novel concept for the mass spectrometric determination of absolute isotopic abundances with improved measurement uncertainty: Part 1—Theoretical derivation and feasibility study. International Journal of Mass Spectrometry, 2010, 289(1): 47-53.

[153] Pramann A, Rienitz O, Schiel D, Güttler B, Valkiers S. Novel concept for the mass spectrometric determination of absolute isotopic abundances with improved measurement uncertainty: Part 3—Molar mass of silicon highly enriched in ^{28}Si. International Journal of Mass Spectrometry, 2011, 305(1): 58-68.

[154] Yang L, Mester Z, Sturgeon R E, Meija J. Determination of the atomic weight of ^{28}Si-enriched silicon for a revised estimate of the avogadro constant. Analytical Chemistry, 2012, 84(5): 2321-2327.

[155] Malinovsky D, Dunn P J H, Goenaga-Infante H. Determination of absolute ^{13}C/^{12}C isotope amount ratios by MC-ICPMS using calibration with synthetic isotope mixtures. Journal of Analytical Atomic Spectrometry, 2013, 28(11): 1760-1771.

[156] Dunn P J H, Malinovs.ky D, Goenaga-Infante H. Calibration strategies for the determination of stable carbon absolute isotope ratios in a glycine candidate reference material by elemental analyser-isotope ratio mass spectrometry. Analytical and Bioanalytical Chemistry, 2015, 407(11): 3169-3181.

[157] Vogl J, Yim Y H, Lee K S, Goenaga-Infante H, Malinowskiy D, Ren T, Wang J, Vocke R D, Murphy K, Nonose N, Rienitz O, Noordmann J, Näykki T, Sara-Aho T, Ari B, Cankur O. Final report of the key comparison CCQM-K98: Pb isotope amount ratios in bronze. Metrologia, 51(1A), 08017.

[158] Narukawa T, Hioki A, Kuramoto N, Fujii K. Molar-mass measurement of a ^{28}Si-enriched silicon crystal for determination of the avogadro constant. Metrologia, 2014, 51(3): 161-168.

[159] Wang J, Ren T, Lu H, Zhou T, Zhao M. Absolute isotopic composition and atomic weight of selenium using multi-collector inductively coupled plasma mass spectrometry. International Journal of Mass Spectrometry, 2011, 308(1): 65-70.

[160] Wang J, Ren T, Lu H, Zhou T, Zhou Y. The absolute isotopic composition and atomic weight of ytterbium using multi-collector inductively coupled plasma mass spectrometry and development of an Si-traceable ytterbium isotopic certified reference material. Journal of Analytical Atomic Spectrometry, 2015, 30(6): 1377-1385.

[161] Vogl J, Brandt B, Noordmann J, Rienitz O, Malinovskiy D. Characterization of a series of absolute isotope reference materials for magnesium: *Ab initio* calibration of the mass spectrometers, and determination of isotopic compositions and relative atomic weights. Journal of Analytical Atomic Spectrometry, 2016, 31(7): 1440-1458.

[162] Nier A O. A redetermination of the relative abundances of the isotopes of neon, krypton, rubidium, xenon, and mercury. Physical Review, 1950, 79(3): 450-454.

[163] Stoll-Werian A, Flierl L, Rienitz O, Noordmann J, Kessel R, Pramann A. Absolute isotope ratios-analytical solution for the determination of calibration factors for any number of isotopes and isotopologues. Spectrochimica Acta, Part B: Atomic Spectroscopy, 2019, 157: 76-83.

[164] Meija J. Calibration of isotope amount ratios by analysis of isotope mixtures. Analytical and Bioanalytical Chemistry, 2012, 403(8): 2071-2076.

[165] Henrion A. Reduction of systematic errors in quantitative analysis by isotope dilution mass spectrometry (IDMS): An iterative method. Fresenius' Journal of Analytical Chemistry, 1994, 350(12): 657-658.

[166] Pagliano E, Mester Z, Meija J. Reduction of measurement uncertainty by experimental design in high-order (double, triple, and quadruple) isotope dilution mass spectrometry: Application to GC-MS measurement of bromide. Analytical and Bioanalytical Chemistry, 2013, 405(9): 2879-2887.

[167] Meija J, Methven B, Sturgeon R E. Uncertainty of relative sensitivity factors in glow discharge mass spectrometry. Metrologia, 2017, 54(5): 796-804.

[168] Sturgeon R E, Methven B, Willie S N, Grinberg P. Assignment of purity to primary metal calibrants using pin-cell VG 9000 glow discharge mass spectrometry: A primary method with direct traceability to the SI international system of units? Metrologia, 2014, 51(5): 410-422.

[169] Vogl J, Kipphardt H, Richter S, Bremser W, Del Rocio Arvizu Torres M, Lara Manzano J V, Buzoianu M, Hill S, Petrov P, Goenaga-Infante H, Sargent M, Fisicaro P, Labarraque G, Zhou T, Turk G C, Winchester M, Miura T, Methven B, Sturgeon R, Jahrling R, Rienitz O, Mariassy M, Hankova Z, Sobina E, Krylov A I, Kustikov Y A, Smirnov V V. Establishing comparability and compatibility in the purity assessment of high purity zinc as demonstrated by the CCQM-P149 intercomparison. Metrologia, 2018, 55(2): 211-221.

[170] Fiedler R. Total evaporation measurements: Experience with multi-collector instruments and a thermal ionization quadrupole mass spectrometer. International Journal of Mass Spectrometry and Ion Processes, 1995, 146-147(C): 91-97.

[171] Fujii T, Suzuki D, Watanabe K, Yamana H. Application of the total evaporation technique to chromium isotope ratio measurement by thermal ionization mass spectrometry. Talanta, 2006, 69(1): 32-36.

[172] Mialle S, Gourgiotis A, Aubert M, Stadelmann G, Gautier C, Isnard H, Chartier F. Improvement in thermal-ionization mass spectrometry (TIMS) using total flash evaporation (TFE) method for lanthanides isotope ratio measurements in transmutation targets. 2011 2nd International Conference on Advancements in Nuclear Instrumentation, Measurement Methods and their Applications. IEEE, 2011: 1-5.

[173] Richter S, Kühn H, Aregbe Y, Hedberg M, Horta-Domenech J, Mayer K, Zuleger E, Bürger S, Boulyga S, Köpf A, Poths J, Mathew K. Improvements in routine uranium isotope ratio measurements using the modified total evaporation method for multi-collector thermal ionization mass spectrometry. Journal of Analytical Atomic Spectrometry, 2011, 26(3): 550-564.

[174] Jakopič R, Sturm M, Kraiem M, Richter S, Aregbe Y. Certified reference materials and reference methods for nuclear safeguards andsecurity. Journal of Environmental Radioactivity, 2013, 125: 17-22.

[175] Kraiem M, Essex R M, Mathew K J, Orlowicz G J, Soriano M D. Re-certification of the CRM 125-A UO_2 fuel pellet standard for uranium isotopic composition. International Journal of Mass Spectrometry, 2013, 352: 37-43.

[176] Wegener M R, Mathew K J, Hasozbek A. The direct total evaporation (DET) method for TIMS analysis. Journal of Radioanalytical and Nuclear Chemistry, 2013, 296(1): 441-445.

[177] Mialle S, Richter S, Hennessy C, Truyens J, Jacobsson U, Aregbe Y. Certification of uranium hexafluoride reference materials for isotopic composition. Journal of Radioanalytical and Nuclear Chemistry, 2015, 305(1): 255-266.

[178] Mathew K J, Haoszbek A. Comparison of mass spectrometric methods (TE, MTE and conventional) for uranium isotope ratio measurements. Journal of Radioanalytical and Nuclear Chemistry, 2016, 307(3): 1681-1687.

[179] Fukami Y, Tobita M, Yokoyama T, Usui T, Moriwaki R. Precise isotope analysis of sub-nanogram lead by total evaporation thermal ionization mass spectrometry (TE-TIMS) coupled with a ^{204}Pb-^{207}Pb double spike method. Journal of Analytical Atomic Spectrometry, 2017, 32(4): 848-857.

[180] Quemet A, Ruas A, Dalier V, Rivier C. Americium isotope analysis by thermal ionization mass spectrometry using the total evaporation method. International Journal of Mass Spectrometry, 2018, 431: 8-14.

[181] Wakaki S, Ishikawa T. Isotope analysis of nanogram to sub-nanogram sized Nd samples by total evaporation normalization thermal ionization mass spectrometry. International Journal of Mass Spectrometry, 2018, 424: 40-48.

[182] Wang J, Lu H, Ren T, Zhou T, Song P, Wang S. Development of isotopic certified reference materials by metrological

approach. Journal of Physics: Conference Series, 2018, 1065(24): 242003.

[183] Quemet A, Ruas A, Dalier V, Rivier C. Development and comparison of high accuracy thermal ionization mass spectrometry methods for uranium isotope ratios determination in nuclear fuel. International Journal of Mass Spectrometry, 2019, 438: 166-174.

[184] Song P, Wang J, Zhang Y, Lu H, Ren T. Total evaporation technique for high-accuracy isotopic analysis of isotopically enriched molybdenum by negative thermal ionization mass spectrometry. Metrologia, 2019, 56(2): 024005.

[185] Wang S, Wang J, Lu H, Fang X, Song P, Ren T. Investigation of the crucial factors affecting accurate measurement of strontium isotope ratios by total evaporation thermal ionization mass spectrometry. Rapid Communications in Mass Spectrometry, 2019, 33(9): 857-866.

[186] JCGM 100: 2008. Evaluation of measurement data: Guide to the expression of uncertainty in measurement. International Bureau of Weights and Measures, Sèvres, France, 2008.

第4章 元素介导的电感耦合等离子体质谱定量生物分析

(严晓文[①]，杨利民[①]，王秋泉[①*])

▶ 4.1 引言 / 86

▶ 4.2 内源性报告元素 / 87

▶ 4.3 外源性报告元素 / 90

▶ 4.4 基于 ICP-MS 的单细胞分析技术 / 105

▶ 4.5 展望 / 106

---

①厦门大学，厦门。*通讯作者联系方式：qqwang@xmu.edu.cn

本章导读

- 获取生物分子信息是理解生命运行机制、发现生物标志物和药物靶点并应用于疾病诊断和药物开发的基础；身处极其复杂生物体系中的种类繁多、性质差异大的生物分子，在缺少标准品的情况下难以准确定量；ICP-MS 因其在元素/同位素定量分析上的独特优势，在生物分子定量分析领域崭露头角。
- 基于生物分子的内源性元素的 ICP-MS 定量生物分析。
- 外源性元素选择性标记技术可实现蛋白质、核酸和细胞（细菌、病毒）更为准确和灵敏的 ICP-MS 定量生物分析。
- ICP-MS 生物分析中的信号放大和倍增策略。
- 除了更深入的基础研究还需进一步加强外，ICP-MS 生物分析的实际应用研究也应同步展开。针对典型目标生物标志物分子或细胞（细菌和病毒）的元素化学和生物标记方法需要进一步标准化，相应的元素标签和信号放大/倍增试剂"工具箱"要同时配备，使基于 ICP-MS 的生物分析技术和方法，既开花也结出丰硕的果实，造福人类。

4.1 引 言

以细胞为基本单元，生命体是由核酸、蛋白质等物质组成的复杂分子体系，具有对外部环境产生反应以及不断代谢并繁殖后代的能力。核酸是生命遗传信息的储存载体，其序列和含量相对稳定；而蛋白质是生命功能的具体执行者，发挥着生命活动所必需的各种生物功能。在生命过程中新蛋白质不断表达的同时伴随着旧蛋白质的降解，蛋白质的种类、含量、翻译后修饰状态高度动态变化。获取生命分子，特别是蛋白质分子的种类、含量、翻译后修饰等信息对于我们理解生命运行机制，以及发现疾病生物标志物和药物靶点用于疾病诊断和药物开发都至关重要[1]。由于生物背景极其复杂、分子种类繁多且性质差异大，在缺少对应的标准品的情况下，生物分子的定量分析十分困难[2,3]。由不同原子经化学共价链接构成的生物分子或通过物理或化学相互作用组装成各种各样的超分子集合体在生命过程中发挥着不可或缺的特殊功能。作为碳基生物，地球上的生命体主要由碳、氢、氧、氮、磷和硫六种基本元素组成，期间各种微量元素作为辅助因子发挥着奇妙的作用。是否可以通过直接测定这些基本元素获取生命分子的信息？理论上通过元素分析技术可以测定纯化后的生物分子的元素/原子组成，但是尚无法得到其结构、含量等更高维度的信息。例如，蛋白质的含氮量大约占其总质量的 16%，通过传统的凯氏定氮法测定生物制品中的含氮量便可估算出其总蛋白质的含量[4]；但是当一些含氮量高的非蛋白质分子（如三聚氰胺）混入共存时会导致错误的蛋白质检测结果。另一方面，虽然晶体衍射、核磁共振、冷冻电镜等技术可以在原子级别分辨率水平上分析这些基本元素的空间分布，因而可用于生命分子的结构解析，但前提是需要制备高纯度、高浓度的样品以弥补其抗干扰能力弱、灵敏度不足等问题，用于复杂生物样品的直接定量分析还是一个尚待完成的艰巨任务。在这种情况下，各种类型的标记技术应运而生，以获得灵敏且具有选择性的信号来解决所面临的问题。标记技术的基本原理是通过共价反应将信号报告分子标记到目标分子上或通过非共价相互作用改变信号分子的信号强度，通过对信号分子的测定得到目标生命分子的种类、含量等信息[5-7]。

在众多的标记技术中，通过抗体将放射元素标记到目标生物分子上的放射免疫分析技术最早得到应用[8]。这种技术具有专一性强和灵敏度高等优点，但也面临放射元素获取困难以及需要专业化实验室降低放射性射线对人体造成健康损伤等问题。因而，非放射性光学标记技术（如荧光、化学发光）逐渐得到发展和广泛应用[9-12]。值得注意的是，由于光学信号自身的半峰宽普遍很大且稳定性与信号来源分子的结构和能级相关，目前应用的光学信号分子大多受限于谱峰重叠干扰和光漂白等不足，难以用于多目标分子的同时准确定量[7]。目前，基于软电离技术的分子质谱是最为主流的生物分子鉴定和定量技术[13-17]。作为一种非标记分析技术，分子质谱通过检测分子离子和碎片离子，结合软件技术，可对生物分子进行鉴定和非标记相对定量分析。与稳定同位素标记技术相结合可实现更高灵敏度和准确性的蛋白质相对定量分析[18-24]；通过合成稳定同位素标记的标准品，还可以进一步实现对目标生物分子的绝对定量分析。但其主要问题是生命分子种类繁多，合成与目标分子对应的同位素标准品十分困难；即使可以制备，也面临高昂的成本问题[25-27]。

21 世纪初，通过选择性免疫反应对生物分子进行元素标记，本以检测元素/同位素为特征的电感耦合等离子体质谱分析法（ICP-MS）被率先应用到生物分析中[28-30]。此后，基于 ICP-MS 的生物分析经历近 20 年的发展逐渐凸显出其在目标生物分子和细胞（包括细菌和病毒）绝对定量分析领域的独特优势[31-61]。除了具有同位素信号灵敏、稳定、线性范围宽等特点外，与其他软电离源质谱技术（如 ESI-和 MALDI-MS）相比，ICP-MS 最为突出的优势是元素同位素的信号响应与元素存在的化学/物理形态无关，即生物分子（细胞）本身含有的内源性元素或标记在生物分子（细胞）上的外源性元素与其游离状态的离子在高温硬电离源-电感耦合等离子体（ICP）中的信号响应几乎相同。因此，当与同位素稀释技术相结合应用时，只需使用一种简单的元素同位素内标即可实现对多种不同生物分子的绝对定量分析；此外，元素同位素质谱信号轮廓相对于光谱信号要窄得多，克服了光谱分析在检测多种组分时遇到的光谱重叠的问题，选择性/分辨率都有极大的提升。

基于 ICP-MS 的定量生物分析方法成功与否之关键在于元素同位素的质谱信号强度或同位素比值是否客观地反映了生物分子（细胞）的含量（数量）。不论是以生物分子（细胞）本身含有的内源性元素（如蛋白质和核酸分子骨架上的硫和磷以及一些与蛋白质结合的类金属或金属元素等），还是通过化学选择性或生物特异性反应标记在生物分子（细胞）上的外源性元素（如灵敏度高、干扰小、生物背景低的镧系元素）对生物分子（细胞）进行定量（计数）时，所测定的元素/同位素应与目标生物分子（细胞）间有稳定准确的化学计量关系。为此，本章将围绕元素介导的 ICP-MS 生物分析这一主题，回顾近年来相关方法/策略的发展，并展望未来的发展趋势。

4.2 内源性报告元素

4.2.1 硫

硫（S）是生命体中广泛存在的一种非金属元素。由于较低的氧化还原电位，且可以与碳和氧原子形成共价键并与软金属离子形成准共价键，S 主要存在于含 S 无机阴离子、金属-S 原子簇合物、谷胱甘肽、糖、酯等小分子化合物，并以半胱氨酸或蛋氨酸的形式存在于蛋白质等大分子的一级序列中，参与金属配位、酶催化、氧化还原平衡调节等生命过程。虽然 S 在蛋白质中的总含量大约只有 2%，但是分布较为均匀。在人类蛋白质组的胰蛋白酶酶解肽段中，26.6%的肽段中至少含有一个半胱氨酸，25.5%的肽段中至少含有一个蛋氨酸，这些肽段分别存在于 96.1%和 98.9%的蛋白质中。因此，理论上可以通过对 S 的测定实现对绝大部分蛋白质的定量[62]，是一种普适性的可行定量方法。

这种通过 S 的含量进行蛋白质定量的方法适用于纯化后的单一蛋白质定量分析，否则只能给出蛋白质的总量信息。例如，采用硝酸将牛血清白蛋白（BSA）上的半胱氨酸和蛋氨酸消解为无机形态的硫酸盐，可以利用流动注射进样 ICP-AES 对 S 的测定，从而定量分析 BSA[63]。以硫胺为内标，采用 μLC-ICP-MS 联用技术，可以实现对胰岛素和在 E. coli 中过量表达的 cheAH(3—137) 和 cheA-C(257—513) 两个蛋白质片段的定量分析[64]。由于 S 含有多个稳定同位素 ^{32}S(95.02%)、^{33}S(0.75%)、^{34}S(4.21%) 和 ^{36}S(0.02%)，因此可以利用同位素稀释技术对蛋白质进行更精确的绝对定量[65]。但是，使用柱后同位素稀释 μLC-ICP-MS 联用技术对人类载脂蛋白 A-I（human apolipoprotein A-I）和甲胎蛋白（α-fetoprotein）两种标准蛋白质定量时，蛋白质的定量准确性受到色谱柱回收率的影响（人类载脂蛋白 A-I 的回收率达到 100%，而甲胎蛋白的回收率只有 60%）[66]。在色谱流动相中加入 ^{34}SO$_4^{2-}$ 盐，应用柱前同位素稀释法不但可以回避色谱回收率问题的影响，而且消除了因仪器漂移产生的偏差，进而可以对人血清白蛋白和酵母蛋白酶切后的含硫肽段进行 nanoLC-ICP-MS 的精确定量，检测限达到 pg 水平[67]。利用 ICP-MS 的多同位素同时检测能力，通过柱后同位素稀释技术对 ^{32}S/^{34}S 和 ^{54}Fe/^{56}Fe 四种同位素同时监测，完成了对人血清中转铁蛋白（transferrin）和白蛋白（albumin）的定量[68]。利用高分辨的多接收（multicollector）ICP-MS 还可对经 ^{34}SO$_4^{2-}$ 培养的酵母的蛋白质生物合成进行追踪研究[69,70]。

虽然通过 S 进行蛋白质定量具有众多优点，但其较高的第一电离能（10.357 eV）使其在 Ar-ICP 中的电离效率较低（10%～14%），而且存在严重的多原子离子干扰（例如，^{16}O$_2^+$ 和 ^{16}O^{18}O$^+$），使只有 0.5～1 原子质量单位（amu）分辨率（$m/\Delta m$）的单四极杆 ICP-MS 显得力不从心。例如，丰度最高的 ^{32}S（31.972 07 amu）与 ^{16}O$_2^+$（31.989 83 amu）的质量差别只有不到 0.02 amu，因此只有分辨率大于 1800 的质谱仪才能将它们区分开。在普通的四极杆 ICP-MS 配备"动态反应池"（dynamic reaction cell, DRC）将 ^{32}S$^+$ 转变为 ^{32}S^{16}O$^+$ 可以避开 ^{16}O$_2^+$ 的干扰[71] 或使用串联三重四极杆 ICP-QQQ-MS[72] 以克服多原子离子的干扰极大地提高了 S 的检测准确度，为通过 S 的检测实现多肽和蛋白质定量开辟了一条有效的途径[73]。

除了蛋白质，LC-ICP-MS 也应用于含硫氨基酸[74]、环状多肽[75] 和其他含硫化合物的直接定量分析[57]。

4.2.2 磷

磷（P）是一种重要的生物生命体必需的非金属元素。P 是核酸骨架和细胞膜磷脂的主要成分，并参与了蛋白质磷酸化后修饰过程。蛋白质的磷酸化是一个动态过程，由蛋白质激酶和磷酸脂酶分别催化蛋白质的磷酸化和去磷酸化过程[76]，真核细胞内有 30% 以上的蛋白质同时进行着这一过程[77]。因此，对磷酸化蛋白质进行快速定量分析具有重要意义。常规的凝胶电泳等间接的蛋白质磷酸化研究方法难以对磷酸化过程进行快速分析，ICP-MS 也因此成为一种通过直接检测 P 来快速分析磷酸化蛋白质的技术。值得注意是，由于 P 元素的第一电离能较高（10.484 eV），致使其在 Ar-ICP（Ar, 15.76 eV）中的离子化效率相对较低（约 33%）；同时在四极杆 ICP-MS 中，^{31}P 受到多原子离子（如 ^{14}N^{16}O^{1}H$^+$）的严重干扰。近年来，随着高分辨质谱（HR-MS）和 DRC/碰撞池（collision cell, CC）技术的出现，背景干扰问题得到了一定程度的解决，使得通过 P 对磷酸化蛋白质进行定量分析成为可能。利用小体积雾化器或直接注入式高效雾化器（direct injection high efficiency nebulizer, DIHEN）偶联微口径色谱与 HR-MS（μLC-ICP-HRMS），通过测定 ^{31}P 实现了对不同的磷酸化多肽和 β-酪蛋白（β-casein）的酶解产物进行定性与定量分析[78,79]。同时定量 ^{31}P 和 ^{32}S 的含量，还可以实现对磷酸化蛋白质和多肽上磷酸化程度的定量[80-84]。使用配备 DRC 单元的四极杆 ICP-MS，通过向 DRC 中加入

O_2 反应气将 ^{31}P 和 ^{32}S 分别转化为 $^{31}P^{16}O$ 和 $^{32}S^{16}O$，可以避免多原子离子干扰，实现生物样品中 P 和 P/S 比例的测定以及磷酸化蛋白质（如 β-酪蛋白）的检测[71,82]。

梯度洗脱反相色谱具有良好的分离效果，被广泛用于多肽和蛋白质的分离。值得注意的是，使用 ICP-MS 作为反相色谱的元素检测器面临的一个问题是 ICP-MS 的相应信号因流动相中有机溶剂的含量的变化发生改变。测定拥有不同质量同位素的元素时，柱后同位素稀释技术可以消除信号漂移；但是，P 只有一个稳定同位素 ^{31}P，无法使用同位素稀释技术消除这一影响。为了消除梯度洗脱造成的 ICP-MS 信号漂移，在 μHPLC 柱后加入含高浓度乙腈的鞘流来缓冲流动相的组成，从而使 ^{31}P 的信号维持稳定；利用易于得到的含磷标准品双(对硝基苯基)磷酸酯[bis(4-nitrophenyl) phosphate, BNPP]为内标，μHPLC-ICP 结合 ESI-MS 实现了 β-酪蛋白酶解产物中的磷酸化多肽进行绝对定量与鉴定，检出限达到 110 fmol[83]。基于类似的策略，在反相色谱柱后加入与梯度洗脱相反的鞘液来抵消流动相中有机溶剂含量的变化，同样实现了对磷酸化多肽的绝对定量[84]。

4.2.3 硒

硒（Se）作为人体的必需微量元素得到了人们广泛的关注。近年来，不仅 Se 形态学得到了研究[85]，Se 蛋白和含硒蛋白质的定量研究也逐渐成为研究热点[86,87]。Se 元素主要通过两种途径进入蛋白质肽链：①硒代半胱氨酸（SeCys）通过基因编码的方式被 UGA 密码子导入到蛋白质上；②硒代蛋氨酸（SeMet）以随机取代蛋氨酸（Met）的方式进入蛋白质肽链中[88]。到目前为止，哺乳动物中已经发现了 22 种含硒蛋白质[89]，科学家推测人类基因组中至少有 25 种硒蛋白[90]。由于 Se 原子的第一电离能为 9.75 eV，比磷和硫的第一电离能略低，因此在 ICP 中的离子化效率相对较高；但是在 Ar-ICP 中含有大量的 $^{40}Ar_2^+$ 离子会干扰丰度最高的 $^{80}Se^+$ 的测定。借助 DRC 技术，使用 CH_4 为反应气可消除 $^{40}Ar_2^+$ 对 $^{80}Se^+$ 的干扰。柱后同位素稀释阴离子交换色谱-ICP-DRC-MS 技术被用于研究人体血浆中的硒蛋白 P、谷胱甘肽过氧化物酶、血清白蛋白以及一些小分子的 Se 化合物，实现了对血浆中 Se 不同存在形态的定量分析，发现硒蛋白 P 可以作为评价人体 Se 含量水平的标志物[91]。尺寸排阻色谱与柱后同位素稀释-ICP-MS 联用也可定量血红细胞中的谷胱甘肽过氧化物酶，发现这个蛋白质可以作为人体氧化还原水平的标志物[92]。此外，由于一些多肽药物含 Se，因此也可以通过 LC-ICP-MS 进行定量分析和质量控制[93]。由于硒蛋白的数量有限，因此通过 Se 元素对硒蛋白组进行定量相对容易，但是 Se 不像 S 那样广泛存在于蛋白质中，难以推广到其他不含 Se 的蛋白质的直接定量分析，仅局限于含 Se 的蛋白质的定量研究。

4.2.4 金属元素

生物体的每三个蛋白质中，就有一个蛋白质需要与一种或几种金属离子结合来完成催化、信号调节或稳定结构的功能，这些蛋白质被称为"金属蛋白"（metalloprotein）[94]。目前已经在自然界中发现大约有 20 种金属元素，如铜、铁、锌、钴或钼等，是生命的必需元素。这些金属蛋白在生命体中发挥了各种各样的功能。例如，金属硫蛋白（metallothionein）和铁蛋白（ferritin）对于细胞内的金属动态平衡和解毒过程起到重要作用[95,96]，金属伴侣蛋白（metallochaperone）保护并且协助金属酶（metalloenzyme）行使正常功能[97]。

ICP-MS 对金属元素的定量有很大优势，人们很自然地想到通过对金属离子的测定来对金属蛋白进行定量。金属离子通常与蛋白质的精氨酸、天冬氨酸、谷氨酸和半胱氨酸等氨基酸残基的 N、O 和 S 等原子进行非共价配位结合，蛋白质的三维构象对于维持这种弱相互作用十分重要。一些变性试剂，例如有机溶剂、尿素、表面活性剂或者极端酸性都会导致金属离子从蛋白质上"脱落"。因此对金属蛋白进行定量分析时要尽量维持在生理条件下进行，以保持金属蛋白的稳定性。尺寸排阻色谱和离子交换色谱通常也被用于分离金属蛋白。在使用阴离子交换色谱对金属蛋白进行分离时，为了维持蛋白质的构象和其中金属离子的稳定性，生理 pH 的 NH_4AC 或 Tris-HCl 缓冲溶液体系（pH 7.4）常作为流动相，利用 ^{57}Fe-、^{65}Cu-和 ^{67}Zn-柱后同位素稀释-ICP-MS 对正常人和正在进行血液透析患者的血浆中含 Fe、Cu 和 Zn 的蛋白质进行分离和定量，表明该方法可以用于疾病诊断[98]。虽然富集金属同位素蛋白标准不易制备或获得，但是柱前同位素稀释技术能校正蛋白质在样品处理和色谱分离过程中的损失，使定量结果更为准确[99]。通过结合的金属对蛋白质进行定量分析不仅可以得到蛋白质的含量信息，还能够帮助人们了解蛋白质组中金属元素的形态和不同形态的含量分布，这对于揭示金属元素在生命体中所发挥的作用有重要意义。科学家们相继提出了金属组（metallome）的概念[100]，它的目的是系统地研究细胞或者组织中所有的金属离子的分布、形态、平衡或者在某个组织、细胞和器官中的金属元素的组成。对于金属组的研究，以及金属组与基因组、转录组、蛋白组和代谢组的关系研究被称为"金属组学（metallomics）"[101-106]。金属组学对于揭示金属元素在生命、医学、疾病、环境中发挥的作用有重要意义，相关工作已经得到系统的总结[107-112]。孙红哲教授在 2013 年组织中国学者于 *Metallomics* 杂志上出版了一个专刊，加之本书中有两个章节对金属组学的研究方法和技术进行阐述，在此不再赘述。

4.3 外源性报告元素

上述的"内源报告元素"中，S、P 和 Se 都存在一个共同的特点，就是它们的第一电离能相对较高，限制了它们在 Ar-ICP 中的电离效率的进一步提升；虽然 DRC、CC 和 HR-MS 可以消除多原子离子的干扰提高信噪比，但绝对灵敏度因电离效率所限并没有实质性的提高。另一方面，以配位键或电荷相互作用结合在金属蛋白中的金属离子稳定性和来自样品基体中相同金属离子/形态的干扰问题限制了以内源元素的检测来定量蛋白的实际应用。幸运的是，在组成蛋白分子的氨基酸和蛋白质翻译后修饰结构中有很多可化学修饰/反应的"位点"，这些位点为实现蛋白分子"外源报告元素"的标记提供了可能性（图 4-1）[41]。实现化学选择性标记的同时，我们可以灵活地筛选生物背景低、在 Ar-ICP 中离子化效率高的外源元素，为基于 ICP-MS 生物分析灵敏度和准确度的进一步提高提供了一个契机。

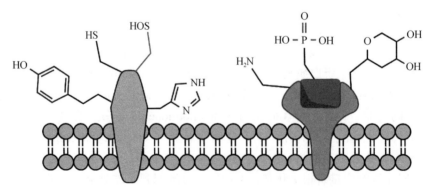

图 4-1 生物分子上用于外源性元素标记的化学基团

4.3.1 选择性化学标记

4.3.1.1 针对巯基的外源元素标记

汞（Hg）致毒原因之一就是其与蛋白质分子中还原性巯基间（Hg—S）的选择性准共价相互作用。利用还原型巯基容易与 Hg 选择性反应的特性，近年来，一系列通过巯基进行金属标记策略发展起来，将以 S 进行的蛋白质定量分析转换成通过所标记的低生物背景且无多原子离子干扰的 Hg 来实现蛋白质的定量检测。基于此策略，甲基汞、乙基汞和对氯汞基苯甲酸被用作 Hg-标签用于蛋白质巯基的标记。通过先后标记原有的还原性自由巯基和被三(2-羧乙基)膦(TCEP)还原二硫键释放的巯基，利用 ESI-MS 测定汞标记前后的质量差异和同位素分布，实现了谷胱甘肽（glutathione）、植物螯合肽（phytochelatins）、溶菌酶（lysozyme）和乳球白蛋白（lactoglobulin）几种多肽和蛋白质分子中自由巯基和二硫键的精确测定[113]。在此基础上，使用 HPLC-ICP-MS 结合 CH_3HgCl 探针对巯基的标记对牛胰腺核糖核酸酶 A、溶菌酶和胰岛素 B 链几种蛋白质进行了定量，检出限达到 pmol 水平，相比于硫的直接测定有显著提高[114]。几乎是同一时期，四羟基汞苯甲酸（p-hydroxymercuribenzoic acid，pHMB）也被用于标记巯基，结合 MALDI-/ESI-MS、HPLC-ICP-MS、GE-LA-ICP-MS 和同位素稀释等技术，实现了卵清白蛋白（ovalbumin）和胰岛素（insulin）等模型蛋白质的定量检测[115-117]。这些小分子 Hg 化合物易于进入蛋白质内部与巯基反应，对巯基有较高的标记效率。例如，相对于蛋白质 10 倍过量的 pHMB 可以在一个小时内标记卵清白蛋白和胰岛素上 95% 和 90%～110% 的巯基。但是，由于这些有机汞化合物有很强的毒性，限制了它们的实际应用。面对这个问题，动态 Hg 标记技术被发展起来以降低 Hg-标签的毒性[118]。硫柳汞（C_2H_5Hg-THI）正常的生理条件下非常稳定，不容易发生解离，被广泛用作疫苗、化妆品等的抗菌剂[118]。当遇到含有还原型巯基的分子时，C_2H_5Hg-THI 解离释放出 $C_2H_5Hg^+$ 离子与巯基准共价结合；所合成的 CH_3Hg-THI 和 $CH_3^{204}Hg$-THI 两种标签，相较于 C_2H_5Hg-THI，CH_3Hg-THI 上的 CH_3Hg^+ 比 $C_2H_5Hg^+$ 有更强的巯基结合能力，更有利于蛋白质的快速动态标记；$CH_3^{204}Hg$-THI 标签可用于制备 ^{204}Hg 标记蛋白质作为柱前同位素稀释定量标准品，成功地实现了对卵清白蛋白、乳球白蛋白和谷胱甘肽进行动态标记和绝对定量[119]（图 4-2）。这一方法大大减少了应用 Hg 试剂时的毒性，而且能够对蛋白质进行快速标记，有较大的使用价值。进一步利用不同尺寸的 Hg-标签结合 ICP-MS 定量技术探测了蛋白质巯基所处的微环境，拓展了这一策略的应用领域[120]。

不限于 ICP-MS，Hg 标记策略也与 CVG-AFS 结合用于巯基小分子和蛋白质的定量分析。利用 pHMB 标记巯基，在 RPLC-CVG-AFS 联用技术上使用 UV/HCOOH 作为还原体系，实现对半胱氨酸、半胱氨酰甘氨酸、高半胱氨酸、γ-谷氨酰半胱氨酸、谷胱甘肽、N-乙酰半胱氨酸一系列小分子巯基氨基酸的标记和定量[121]。此外，使用微波辅助光化学反应器，在无需加入氧化剂的条件下可将 89.0% 的 pHMB 氧化为 Hg(Ⅱ)，进一步与 CVG-AFS 在线联用实现了对血清白蛋白酶解多肽的定量分析[122]。在对二硫键的 Hg 标记条件研究过程中发现在碱性 NaOH 溶液中（pH=13.5），无需使用 TCEP 等试剂预还原，即可对二硫键进行 pHMB 标记，进一步使用 SEC-CVG-AFS 技术实现了对溶菌酶、α-乳白蛋白、β-乳球蛋白 A、抑蛋白酶多肽、人血清清蛋白、细胞色素 C 的定量分析[123]。虽然目前已经发展了多种类型的 Hg 标记策略和原子光谱/质谱技术用于模型蛋白质、多肽的定量研究，但是由于人们对 Hg 的惧怕心理，限制了其更为广泛的应用。值得指出的是，虽然 Hg 不受多原子离子的干扰，生物背景值低，而且标记步骤简单，在 Ar-ICP-MS 中的离子化效率也只有大约 40%，相对于 S（10%～14%）的离子化效率虽有所提升，但是非常有限。

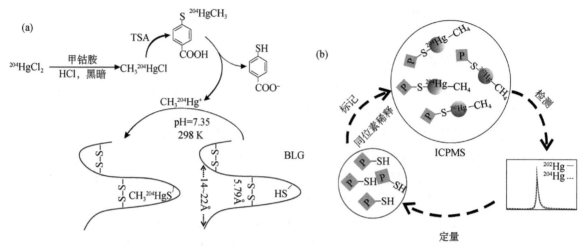

图 4-2 蛋白质巯基的（a）CH₃HgCl 探针动态标记和（b）ICP-MS 同位素稀释定量策略[119]

稀土元素包含镧（La）、铈（Ce）、镨（Pr）、钕（Nd）、钷（Pm）、钐（Sm）、铕（Eu）、钆（Gd）、铽（Tb）、镝（Dy）、钬（Ho）、铒（Er）、铥（Tm）、镱（Yb）和镥（Lu）15 个镧系元素以及与它们的性质接近的钪（Sc）和钇（Y）共 17 种元素。它们的第一电离能在 5.5~6.2 eV 之间，远低于 ICP 形成气体氩的第一电离能 15.7 eV，在 Ar-ICP 中的离子化效率近乎达到 100%，检出限可以达到 pg/L 的水平；因原子质量超过 100 Da，多原子离子的干扰极少（轻稀土的氧化物离子对其后的重稀土离子存在干扰，在选择监测同位素时应注意避开！）；稀土元素的生物背景也极低，是理想的"外源报告元素"。镧系元素处于元素周期表中的同一个位置，使用含有镧系元素的金属编码标签（metal-coded tag，MeCT）可取代价格昂贵的稳定同位素标签来针对蛋白质巯基进行标记和相对定量[124]。虽然当时他们并未提出利用 ICP-MS 对稀土标记的蛋白质进行绝对定量，但不失为是生物分子的镧系元素（Ln）化学标记和 ICP-MS 定量分析策略的序曲之一。通过使用马来酰亚胺功能化 Ln-DOTA 配合物（MMA-DOTA-Ln）对蛋白质巯基进行标记，结合 HPLC-ESI-MS 对二硫键还原和标记条件进行系统优化，实现了对溶菌酶、胰岛素和核糖核酸酶 A 三种模型蛋白质巯基的 Ln 完全标记；进一步利用 HPLC-ICP-MS 结合柱后 ¹⁵³Eu 同位素稀释技术对酶解多肽和完整蛋白质定量分析进行比较研究，发现受到酶解不完全和色谱分离能力有限等问题的限制，难以通过酶解多肽对蛋白质进行准确的定量；相比之下，针对完整蛋白质进行定量避免了以上问题，取得了准确的定量结果[125]（图 4-3）。这一工作论证了基于 ICP-MS 的蛋白质定量技术路线，表明对完整蛋白质进行金属标记与定量可以取得更为准确的定量结果。在此工作基础上，使用类似的马来酰亚胺功能化 Ln-配合物标记模型蛋白质，色谱纯化标记了 Ln 的酶解多肽，以纯化后的 Ln 标记多肽作为定量标准品，可以降低蛋白质酶解不完全和多肽在色谱柱上的丢失问题对蛋白质定量准确性的影响[126]。将凝胶电泳与 LA-ICP-MS 结合，完成了对人血清[127]和 E. coli[128]样品中蛋白质的巯基 Ln 标记与定量分析。

细胞内的蛋白质巯基受到 H_2O_2、NO、ONOO⁻等 ROS/RNS 物质的氧化会转变为氧化态的次磺酸（sulfenic acid，R-SOH）、亚磺酸（sulfinic acid，R-SO₂H）和磺酸（sulfonic acid，R-SO₃H）等形式。其中，次磺酸的形成是可逆的，可以被一些小分子巯基化合物或氧化还原酶[谷氧还蛋白（glutaredoxins）和硫氧还蛋白（thioredoxins）]还原成巯基。巯基-次磺酸间的平衡反映了细胞的氧化还原状态，也是一种重要的调控蛋白质功能的翻译后修饰过程。为此，1,3-环己二酮修饰的 Ln 络合物（Ln-DOTA-Dimedone）被设计合成并用于次磺酸的 Ln 标记[129]。在 Ln-DOTA-Dimedone 对次磺酸有良好专一性的基础上，含有碘乙酰胺基团的 Ln-MeCAT-IA 被同时用于巯基的标记，结合 HPLC-ESI-MS 和 SDS-PAGE-LA-ICP-MS 等技术，测定了 β-乳球蛋白在不同 H_2O_2 氧化条件下的巯基/次磺酸的含量变化；也可利用 CuAAC 点击反应偶联炔基-β-酮酯(KE)和叠氮-DOTA，Ln-DOTA-KE 标

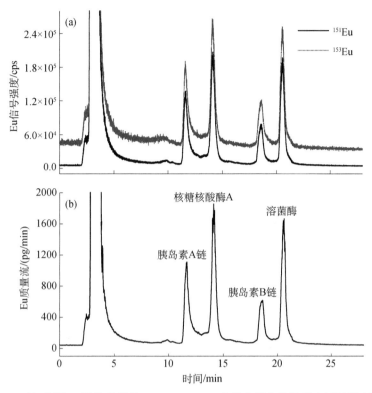

图 4-3　柱后 ^{153}Eu 同位素稀释 HPLC-ICP-MS 定量分析完整的稀土标记蛋白[125]

记次磺酸，用于 BSA 在不同 H_2O_2 氧化条件下的巯基/次磺酸含量的 SEC-ICP-MS 测定[130]。这一过程可以分两步进行：先用炔基-β-酮酯把端炔修饰到次磺酸上，接着通过 CuAAC 点击反应把叠氮-DOTA-Ln 标记到炔基上，应用亲和纯化和同位素稀释技术用于人血清的白蛋白的次磺酸的定量[131]。

4.3.1.2　针对氨基的外源元素标记

蛋白质的赖氨酸侧链和末端氨基是除了巯基以外最常用的标记位点。氨基属于硬碱，难以与金属共价结合；通常使用含有氨基反应基团的双功能金属配体与其反应，达到金属标记的目的。目前已报道的针对氨基进行金属标记的反应基团主要包括酸酐（anhydride）、异硫氰酸酯（isothiocyanate，SCN）和 N-羟基琥珀酰亚胺酯（NHS 酯）。

使用 SCN 修饰的螯合配体 SCN-DOTA 标记蛋白质氨基时，即使使用 40 倍浓度过量的 SCN-DOTA-Eu 与蛋白质进行反应，也未能实现完全标记（每个 BSA 分子平均标记了 2.4 个 Eu 元素）[132]。相比于 SCN，带有酸酐基团的双功能金属螯合配体，如二亚乙基三胺五乙酸二酐（DTPAA），更为普遍地用于蛋白质氨基的金属标记。使用 Y 和 Tb 编码的 DTPAA 标签对标准蛋白质酶解混合物进行标记，结合 HPLC-ESI-MS 技术，对金属-二亚乙基三胺五乙酸(DTPA)-多肽复合物在色谱分离过程中的稳定性和在质谱中的离子化效率、CID 效率和信号相应的线性动态范围进行考察，实现了对脱铁转铁蛋白、牛血清白蛋白、肌红蛋白、β-乳球蛋白、α-乳白蛋白、溶菌酶六种标准蛋白质的鉴定和相对定量[133]；采用 DTPAA 将 Eu^{3+} 标记到多肽氨基上，利用同位素稀释 HPLC-ICP-MS 可定量多肽[134]。通过金属标记进行蛋白质定量的关键在于金属标记的特异性和效率，如果标记选择性不好或反应不完全，就难以确定金属与蛋白质之间的化学计量比，金属定量结果也就无法转化为蛋白质的准确含量。为此，采用螯合了 Lu 的 DTPAA（Lu-DTPAA）对肽段进行标记，通过对标记反应条件进行优化，提高了蛋白质和肽段的稀土标记效率和专一性，发展了基于 ^{176}Lu 柱后同位素稀释的蛋白质高灵敏定量检测方法[135]，回收率达到 100%，精密度为 4.9%，Lu-DTPAA 标记肽段的检测限达到 179 amol，灵敏度相比于对 S、P 进行直接定量提高了 4 个数量级。利用 DTPAA 分别标记 Ce 和 Sm 两种稀土元素到蛋白质上，利用阳离子交换色谱与 ICP-MS 联用对两份蛋白质样品的 Ce 和 Sm 进行检测时，得到了准确的蛋白质（核糖核酸酶 A、细胞色素 C 和溶菌酶）相对定量结果[136]。除了 SDS-PAGE 和 HPLC，CE-ICP-MS 联用技术也被用于 Ln-DTPAA 标记多肽[137]和蛋白质[138]的高灵敏度定量分析，检出限达到几 amol 水平。

DTPAA 含有两个酸酐基团，因此一个 DTPAA 分子有可能与两个氨基反应，增加了定量分析难度；只含单个 NHS 基团的金属配合物则可以避免这个问题，其中 DOTA-NHS 酯最为常用。使用 DOTA-NHS 酯将 Eu 标记到模型多肽，接着用 ^{151}Eu 柱后同位素稀释 LC-ICP-MS 对多肽进行了定量分析[139]。使用同样的标签可制备稀土标记多肽作为标准品以替代同位素标签用于人体肝脏组织中细胞色素 P450 和 5'-二磷酸葡萄糖醛酸转移酶的定量分析[140]。相比于同位素标记多肽，稀土标记多肽标准品的使用不仅降低了试剂成本，也减少了分析时间。此外，将 Ln-DOTA-NHS 酯和 MeCAT-IA 结合起来可同时标记 BSA 和 HAS 酶解多肽的氨基和巯基。同时标记两个位点带来的优势包括：当使用两种稀土分别标记氨基和巯基时，可以测定氨基和巯基的化学计量比；当使用同一种稀土标记时，可以增加每条多肽上的稀土标记个数，提高灵敏度[141]。不限于稀土元素，使用 NHS 活化的联吡啶钌（Ru-NHS）也被用于标记肌红蛋白、转铁蛋白、甲状腺球蛋白三个模型蛋白质，并利用 ^{99}Ru 柱后同位素稀释 SEC-ICP-MS 对完整的模型蛋白进行定量分析。NHS 酯通常在中性偏碱溶液中与氨基反应，在这一条件下，NHS 酯也容易水解，这一竞争反应导致氨基标记效率降低。为此，通过微波辅助金属标记策略可显著提高 Ln-DOTA-NHS 酯与多肽氨基的反应效率和速度[142]。虽然目前已经发展了多种类型的氨基金属标记探针，由于酸胺、SCN 和 NHS 酯等都有在水溶液中容易水解的问题，对含有多个氨基的完整蛋白质进行完全标记仍然比较困难。

4.3.1.3 针对酪氨酸和组氨酸的碘标记

^{125}I 在放射性免疫分析中被广泛用于蛋白质标记和检测。Markwell 在 1982 年优化了 pH、温度、缓冲液、浓度和反应容器等条件，建立了使用 Na^{125}I 和表面修饰的 N-氯苯磺酰胺的聚苯乙烯微球（iodo-beads）来碘化蛋白质的方法。iodo-beads 起到将反应性的 I 转移到蛋白质上作用，从而减少氧化剂与蛋白质直接接触造成蛋白质的氧化降解[143]。使用 Markwell 的方法对牛血清白蛋白、溶菌酶和胃蛋白酶等几个模型蛋白质进行碘标记，ESI-MS/MS 对多肽的鉴定结果表明酪氨酸和组氨酸是主要的碘化位点；使用 SDS-PAGE 分离这些模型蛋白质并用 LA-ICP-MS 进行了胶上定量分析[144]。此方法虽然可以高效地碘化蛋白质，但是试剂较为昂贵、标记步骤较为复杂且时间较长，不可避免地会导致蛋白质降解，进一步发展的使用 KI 和碘单质混合体系（KI$_3$）来碘化蛋白质的方法相比于 NaI/iodo-beads 体系，KI$_3$ 可以保证蛋白质碘化的前提下，简化标记步骤、降低成本、减少蛋白质降解、提高 LA-ICP-MS 定量灵敏度[145]。虽然 NaI/iodo-beads 和 KI$_3$ 体系都可以实现蛋白质的碘标记，但是由于标记位点专一性不强，导致碘和蛋白质的化学计量比难以控制，效率尚待提高。为此，使用对酪氨酸有高度专一性的碘化试剂双(吡啶)四氟硼化碘（IPy$_2$BF$_4$）对多肽进行快速（2 min）、完全的碘标记具有一定的优势，并用于蛋白质的 HPLC-ICP-MS 绝对定量分析[146]。

除了对蛋白质进行共价碘标记外，利用碘芬酸（iophenoxic acid，IPA）与人血清白蛋白（HSA）非共价结合的特性，HSA 的 IPA 标记策略也被发展并用于 SEC-ICP-MS 定量分析。通过同时监测 I 和 Pt 的信号，研究了 HSA 与顺铂的相互作用[147]。除了用于定量分析，使用稳定同位素 ^{127}I 来标记促肾上腺皮质激素（ACTH）4—10 片段 ^{127}I-ACTH（4—10）上的苯丙氨酸，用 ICP-MS 分析了 ^{127}I-ACTH（4—10）的代谢过程和机理。研究证实了 I 标记对 ACTH（4—10）的代谢过程和酶解位点没有影响。相比于放射性同位素 ^{125}I，使用 ^{127}I 标记和 ICP-MS 检测解决了 ^{125}I 较短的半衰期（60 天）和放射性危害问题，是一种有潜力的研究多肽药物代谢动力学的分析技术[148]。

4.3.1.4 针对磷酸基团的金属螯合标记

虽然已有很多工作成功利用 P 对蛋白质进行定量，但如前所述由于 P 的离子化效率比较低，灵敏度还难以满足低丰度磷酸化多肽的定量要求；而且，生物体内的含磷分子种类繁多，磷的背景值比较大，只有借助高效的分离技术对磷酸化蛋白质进行基线分离才能进行准确定量；但是目前还没有这样的分离技术能够对整个蛋白质组中所有的蛋白质进行完全分离，广谱的富集技术结合生物特异性释放是目前选择性定量或磷酸化多肽/蛋白质的一条可行途径。为此，一种针对磷酸化酪氨酸（pY）的特异性镓（Ga）标记技术得到了发展。他们把配体与光裂解基团邻硝基苄醇偶联并修饰到磁珠上，构建了一个集磷酸化多肽的Ga标记、捕获、纯化、定量功能于一体的Ga标签。利用酪氨酸磷酸酶选择性释放pY，成功用于MCF-7细胞的pY的选择性定量[149]（图4-4）。

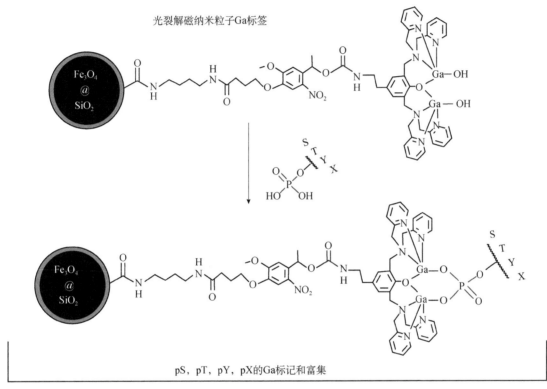

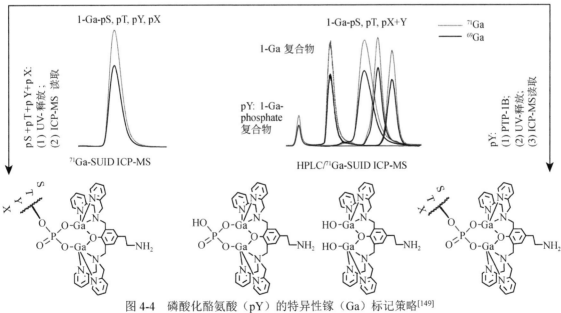

图4-4　磷酸化酪氨酸（pY）的特异性镓（Ga）标记策略[149]

pX表示其他含有磷酸根的分子/离子

4.3.2 生物亲和标记

4.3.2.1 元素免疫亲和标记

虽然原子光谱/质谱技术早已用于分析金属结合蛋白，但是直到近二十年才出现将元素标记策略与 ICP-MS 结合用于蛋白质定量研究工作。利用稀土配合物修饰的抗体用于人促甲状腺激素（thyroid stimulating hormone）的元素免疫标记和 ICP-MS 定量研究是 ICP-MS 生物分析的开山之作[28]。此后，不同类型的金属配合物、聚合物、纳米粒子修饰的抗体发展起来，并且结合各式各样的信号产生方式，用于蛋白质、核酸以及细胞的元素免疫标记与定量分析。目前已有综述对元素免疫标记与定量研究进行精彩总结、讨论和展望[150-155]，由于篇幅有限，本节就不做重复介绍。

4.3.2.2 元素核酸适配体亲和标记

除了抗体，使用其他亲和试剂替代抗体用于目标蛋白质的元素标记和定量的研究也在不断尝试。核酸适配体（aptamer）是一种经指数富集的配体系统进化技术（systematic evolution of ligands by exponential enrichment，SELEX），得到的结构化寡核苷酸序列（RNA 或 DNA）与相应的靶标分子有高度的识别能力和亲和力。通过选择凝血酶（α-Thrombin）作为模型蛋白，并且使用 Apt15 和 Apt29 两种与凝血酶不同位点结合的核酸适配体分别修饰到磁纳米粒子和金纳米粒子表面可进行三明治夹心法金属标记，ICP-MS 定量结果显示使用核酸适配体作为亲和标记试剂具有良好的专一性和线性范围，且检出限达到 0.5 fmol[156]。基于同样的原理，把金/银两种纳米粒子分别标记到细胞色素 C 和胰岛素对应的适配体上，可对这两种蛋白质进行 ICP-MS 同时定量分析[157]。纳米粒子信号放大技术极大地提高了 ICP-MS 检测灵敏度，可以在 1 mL 血清中检测 100 个癌细胞[158]。虽然核酸适配体具有灵敏度高、批次差异性小、低成本、稳定易保存等优点，且元素标记定量策略已经证明可行，但是由于 SELEX 筛选过程是在人为环境中完成的，人为环境与生理环境有所不同，因此得到的适配体的亲和性没有抗体的高且对环境敏感，可用的适配体明显少于抗体也限制了它的推广使用。

4.3.2.3 凝集素亲和元素标记

糖基化是一种重要的蛋白质翻译后修饰方式。蛋白质的糖基化程度和糖链构型/结构变化影响胚胎发育、细胞凋亡、病毒/细菌感染和人体免疫等生命过程，并与糖尿病、癌症等疾病紧密相关。因此，对糖基化修饰进行定量分析具有深远意义。凝集素（lectin）是从植物、动物中提取的一种可以与单糖或多糖结合的蛋白质。由于对某些糖构型具有一定的亲和力和专一性，因此被用于糖蛋白的分析检测。将五种不同稀土聚合物分别修饰到五种识别不同糖型的凝集素上，把蛋白质固定到载体上后，接着利用修饰了稀土的凝集素对糖蛋白进行标记与 ICP-MS 检测，完成了对 BSA、卵清白蛋白、猪胃黏蛋白、牛颌下腺黏蛋白、胎牛胎球蛋白等模型蛋白质糖型的同时分析[159]。此外，可将凝集素修饰到磁珠上用于模型糖蛋白的捕获，利用稀土配合物修饰的抗体对捕获的糖蛋白进行金属标记和 ICP-MS 检测[160]。基于凝集素与糖亲和的原理，将杂交链反应信号放大与 ICP-MS 检测结合在一起，可实现对 H9N2 病毒颗粒的高灵敏度检测[161]。

虽然凝集素已用于糖蛋白的捕获与分析，但是由于其体积较大且亲和力不高，在元素质谱生物分析中的应用还相对有限。糖是多羟基化合物，某些糖型的顺式邻二醇基团易与硼酸在碱性条件下形成可逆的共价键。利用这一性质将巯基苯硼酸修饰到磁珠上，使用磁珠捕获人血清中的糖蛋白，利用修饰了 Au 和 Ag 纳米粒子的抗体对甲胎蛋白 (AFP)和癌胚抗原 (CEA)两种糖蛋白进行特异性标记和 ICP-MS 检测，取得了与临床上使用的化学发光免疫分析法一致的结果[162]；利用苯硼酸在生理 pH 条件下易与唾液酸结合的性质，将苯硼酸修饰的生物素标记到细胞表面，使用亲和素修饰 AuNPs 与生物素结合实现了对细胞表面唾液酸的金属标记和 ICP-MS 定量[163]。

生物酶切/组装策略结合生物正交点击反应提供了另一种对蛋白质糖基化位点进行金属标记的策略。例如，使用 β-N-乙酰葡萄糖苷内切酶（endo S-glycosidase，endo S）剪切糖基化位点的多糖，留下 N-乙酰葡糖胺（N-acetylglucosamine，GlcNAc），接着通过半乳糖转移酶（β-GalT1）将叠氮修饰半乳糖胺（UDP-N-azidoacetylgalactosamine，UDP-GalNAz）连接到 GlcNAc 上实现对蛋白质糖基化位点的叠氮修饰；继而通过点击反应将稀土配体（DIBO-DOTA-Tb）修饰到叠氮基团上，完成对蛋白质糖基化位点的金属标记。虽然只使用 HPLC-ESI-MS 分析了抗体的糖基化多肽，但是这一工作提供了另外一种策略用于糖蛋白金属标记[164]。

4.3.2.4 细胞表面抗原的元素亲和标记

目前荧光成像或流式细胞技术是最为流行的以荧光染料为信号报告分子的细胞分析方法。由于荧光光谱重叠、光猝灭等问题的存在，这些技术难以用于多目标物的定量分析。在这种情况下，基于 ICP 质谱的细胞金属标记与定量分析方法逐渐发展起来以解决这些问题。其中，最为成功的是质谱流式技术（mass cytometry，CyTOF）[166]。CyTOF 使用稀土同位素编码的抗体标记对应的目标抗原，借助 ICP-TOF-MS 的多元素分析能力，可同时定量分析数十种细胞表面的目标抗原[165]。这方面的工作已有综述进行总结[154-155,167]，本书的其他章节也会对此内容做详细介绍，在此不再赘述。在这里我们主要介绍近年来发展的基于常规四极杆 ICP-MS 和荧光技术的金属/荧光双标记策略及其在细胞分析中的应用工作。

将荧光素和稀土配体同时修饰到 RGD 环肽上，制成一种可特异性靶向到癌细胞表面 $\alpha_v\beta_3$ 整联蛋白（$\alpha_v\beta_3$-integrin）的三功能标签，可对 $\alpha_v\beta_3$ 阳性细胞进行"观察和计数"荧光成像和 ICP-MS 定量分析，获得了不同癌细胞表面 $\alpha_v\beta_3$ 整联蛋白含量在 69.2～309.4 amol 水平的基础信息[168]（图 4-5）；在人红细胞白血病细胞（HEL）表面 $\alpha_{IIb}\beta_3$ 整联蛋白的识别多肽 KQAGDV 上修饰 $Au_{24}Peptide_8$ 金纳米簇作为荧光和 ICP-MS 报告分子，用于 HEL 细胞的成像和定量分析[169]。将荧光素天线分子、Nd-DOTA 和前列腺特异性膜抗原（PSMA）的多肽抑制剂分子 DUPAaFFC（2-[3-(1,3-dicarboxypropyl)-ureido]pentanedioic acid (DUPA)-8-Aoc-Phe-Phe-Cys）化学链接组装，获得了具有近红外发光和质谱报告特性的 PSMA 探针，可用于靶向前列腺癌细胞表面 PSMA 的荧光/ICP-MS 二维定量分析。通过 PSMA 的定量结果不仅获得了区分前列腺癌细胞和良性前列腺肥大患者的 PSMA 阈值，而且可检测 PSMA-阳性循环肿瘤细胞，并与肿瘤的 Gleason 评分有很好的关联[170]。除了多肽，抗体和适配体也被用作亲和试剂用于细胞的荧光/金属同时标记。将 Cy3/UCNPs[171]和 QDs[172]修饰到抗体上，用于 HepG2 癌细胞的成像和 ICP-MS 定量分析，使用变构适配体也达到了同样的成像和定量分析结果[173]。金属配合高聚物或金属纳米粒子的使用提高了检测稀少癌细胞的灵敏度，但是这类材料尚难以精确控制金属原子标记的化学计量比。以 MS2 噬菌体外壳蛋白颗粒为模板，可控修饰准确数量的稀土离子和亲和配体，精确地控制金属原子标记的化学计量比，成功用于细胞的元素标记和信号倍增的绝对定量分析[174]。

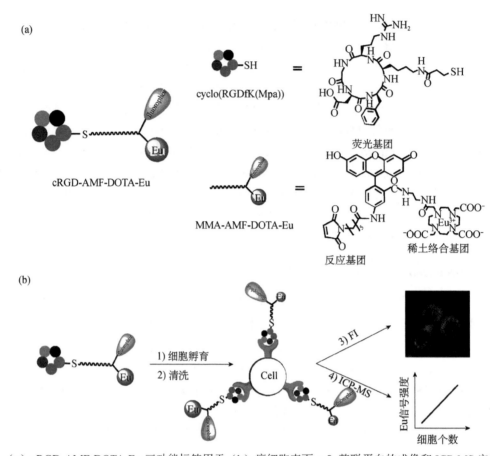

图 4-5 （a）cRGD-AMF-DOTA-Eu 三功能标签用于（b）癌细胞表面 $\alpha_v\beta_3$ 整联蛋白的成像和 ICP-MS 定量分析[168]

4.3.3 生物选择性反应介导的元素标记策略

以上所介绍的选择性金属标记策略与 ICP-MS 定量分析方法的分析对象主要为模型蛋白质，尚较少用于实际样品分析，主要原因是这些标记位点广泛分布于蛋白质组中。以含有巯基的半胱氨酸为例，虽然在蛋白质组中的比例较低，但是分布广泛，蛋白质组中 96.1%的蛋白质含有半胱氨酸。如果这些蛋白质全都被金属标记，现有的分离技术（如色谱）的分离能力还无法实现它们的基线分离，准确定量也就无从谈起。绝大部分蛋白质含有不同个数的可标记氨基酸残基，对每个蛋白质进行准确化学计量比的完全元素标记尚有难度。实际上，这也是目前不论是基于分子还是元素质谱蛋白质定量方法面临的主要瓶颈之一。此外，是否需要对蛋白质组中的每一个蛋白分子进行定量也有不同的观点，在不断的争论中。近年来，目标蛋白质组学，即只定量生命体中某一类功能蛋白质，逐渐成为定量蛋白质组学的热点研究方向。根据蛋白质性质的不同，发展选择性金属标记策略对蛋白质进行分类标记，以减少一个分析流程中蛋白质的数量，降低对色谱分离的压力，是实现实际生物样品定量分析较为可行的技术路线。

4.3.3.1 酶抑制剂介导的元素标记

传统的化学共价反应本身可以选择性地与某一官能团反应，但当众多蛋白质分子结构中都含有这一官能团（如半胱氨酸的巯基或赖氨酸的氨基）时，针对蛋白质的选择性将不复存在，难以选择性地标记某一类蛋白质。随着活性蛋白质谱（ABPP）概念的提出和相关技术的发展，利用酶活性选择性抑制原理来实现蛋白质的特异性标记是突破蛋白质高选择性标记瓶颈的一个机会。基于 ABPP 的稀土标记策略得到发展并应用于实际生物样品中活性丝氨酸蛋白酶（serine protease，SP）的 ICP-MS 绝对定量分析[175]（图 4-6）。丝氨酸活性位点上的羟基由于其周围氨基酸的活化作用，比非活性位点的丝氨酸羟基具有更高的反应活性。带有苯磺酰氟（SF）基团的 SP 抑制剂可以与 SPs 活性位点丝氨酸羟基选择性结合，保证每个 SP 只标记一个稀土元素/原子。利用 CuAAC 点击反应，将 4-(2-氨乙基)苯磺酰氟[4-(2-aminoethyl)benzenesulfonyl fluoride hydrochloride，AEBSF]与 Eu-DOTA 偶联，制备了对 SP 具有特异性的 Eu-标签（SF-Eu），成功地对多种蛋白质的混合样品中典型 SP、胰凝乳蛋白酶（chymotrypsin，CT）和弹性蛋白酶（elastase）进行了 ICP-MS 定量，并进一步应用于人体血浆和老鼠不同组织样品中 SP 定量测定，证明了蛋白质选择性稀土标记策略在实际样品定量分析中的应用价值。需要指出的是，ICP-MS 可定量 SF-Eu 标记的 SP，同时需要借助分子质谱（ESI-MS）对标记的蛋白质进行鉴定以确认金属与蛋白质的化学计量比。在此工作基础上，一个集活性蛋白质选择性稀土标记、分离、富集、纯化和定量功能于一体的活性蛋白质分析平台（multifunctional active protein quantification platform，MAPQP）被设计和构建起来，其具有普适性的应用前景。MAPQP 主要包括：①活性蛋白标签（ABP）用于活性蛋白质的特异标记；② DOTA-Ln 用于 ICP-MS 的定量分析，可根据不同的靶点选择不同的镧系元素（Ln）；③生物素用于被标记蛋白质的纯化和富集；④光致裂解基团用于 Ln 标记蛋白质的释放。MAPQP 是一个"化学枢纽"，集成了多个模块，可同时进行 ICP-MS 定量分析和 ESI-MS 结构鉴定，不仅可以实现已知蛋白分子的定量分析，更为重要的是有利于寻找蛋白酶家族中尚待发现的"新成员"[176]（图 4-7）。

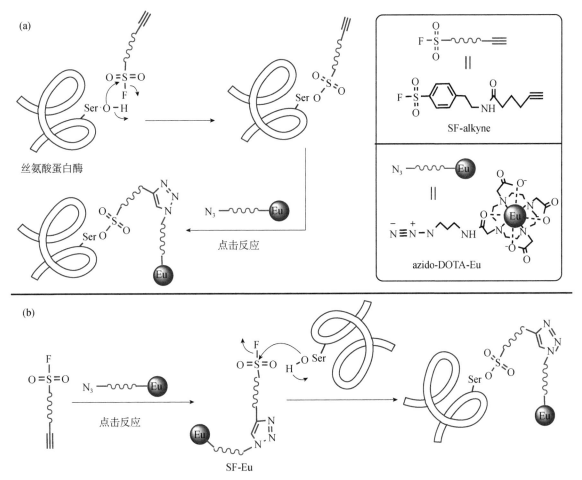

图 4-6　基于磺酰氟抑制剂的丝氨酸蛋白酶活性位点羟基选择性稀土标记策略[175]
(a) SF-alkyne 先标记丝氨酸蛋白酶活性位点羟基，接着通过点击反应标记 azido-DOTA-Eu；
(b) SF-alkyne 和 azido-DOTA-Eu 先通过点击反应等合成 SF-Eu，接着直接用于标记丝氨酸蛋白酶

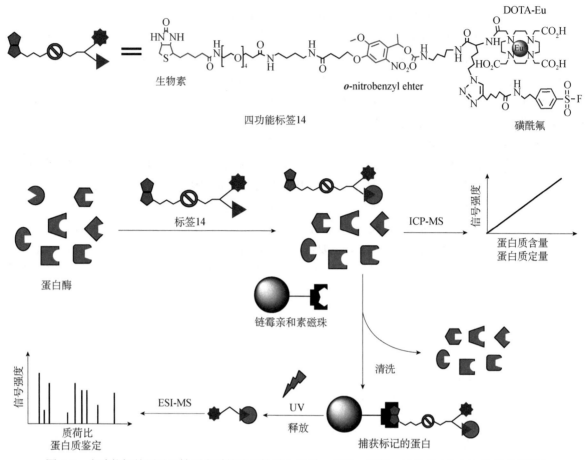

图 4-7 多功能标签用于活性蛋白质的选择性稀土标记、分离、富集、纯化和 ICP-MS 定量分析[176]

基于类似的原理，17α-乙炔基雌二醇（17α-ethynylestradiol）稀土标签被设计合成并用于人肝微粒体和血清中细胞色素 P450 3A4（CYP3A4）的选择性稀土标记和柱后同位素稀释 HPLC-ICP-MS 定量分析[177]；另一种二苯并环辛烯（dibenzylcyclooctyne，DBCO）修饰的氯乙酰胺标签（DBCO-ChAcA）也被设计出来，用于选择性标记谷胱甘肽 S-转移酶 ω1（glutathione S-transferase omega 1，GSTO1）活性位点的半胱氨酸，同时利用无铜催化点击反应，将稀土和荧光团修饰到 DBCO 基团上，可用于细胞中 GSTO1 的 ICP-MS 定量与荧光成像分析[178]。稀土修饰的多肽抑制剂也被用于肝细胞溶酶体中组织蛋白酶活性位点半胱氨酸的标记和定量分析，通过组织蛋白酶的活性评价可实现正常和肝癌细胞中的 pH 值的探测[179]（图 4-8）。

除了把金属元素修饰到抑制剂上用于酶的标记，一些药物本身也含有金属，可以用于酶的直接金属标记与定量分析。金诺芬（auranofin）是一种含金元素的口服抗风湿药，可与硫氧还蛋白还原酶 1（TrxR1）活性位点的硒代半胱氨酸结合。利用这一性质，使用亲和层析和离子交换色谱从血清样品中分离出 TrxR1，接着与金诺芬混合标记；通过 HPLC-ICP-MS 同时测定 Se 和 Au 元素的信号，实现了对 TrxR1 的定量和活性测定，发现血清中只有大约 9%的酶处于活性状态[180]。

4.3.3.2 特异性酶切底物介导的元素释放策略

通过化学或酶与抑制剂选择性反应将金属修饰到蛋白质上是较为常用的元素标记和 ICP-MS 信号读取策略，但是这些技术仍然依赖于蛋白质的基线分离，用于实际复杂生物样品分析多颇费周折。

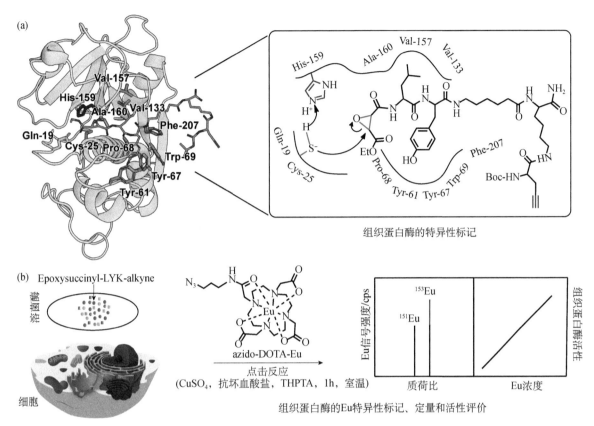

图 4-8　靶向组织蛋白酶活性位点半胱氨酸的稀土标记多肽抑制剂标签[179]

近年来,其他免色谱分离的元素信号获取方式也被发展起来用于生物分子的快速定量分析。利用蛋白酶具有特异性剪切多肽底物的特性,稀土编码多肽纳米标签成功地应用于蛋白酶的 ICP-MS 定量与活性评价[181](图 4-9)。将多肽底物的一端修饰在纳米粒子上,另一端标记上稀土元素;当样品中含有目标蛋白酶时,多肽底物被选择性切断从而释放出相应的稀土元素配合物,通过 ICP-MS 测定离心得到上清液中的稀土元素的含量,即可得到蛋白酶的含量与活性信息。在整个分析过程中无需对蛋白酶进行稀土标记和色谱分离,因此是一种免标记、免色谱分离的定量分析方法。利用 ICP-MS 的多元素分析能力,可在不同多肽底物上标记不同的稀土元素,实现了对多种蛋白酶的同时定量分析,这一工作从不同角度提供了一种新型元素信号产生和定量方式。基于类似的原理,将稀土标记多肽作为 MALDI-MS 的同位素标签用于蛋白酶活性测定[182]。稀土荧光配体也被标记到 caspase-3 的多肽底物上,并修饰到金纳米粒子表面,由于金纳米粒子猝灭荧光,因此完整的标签无荧光;当标签进入细胞被 caspase-3 剪切,释放出稀土荧光配体,荧光得到恢复,从而实现荧光成像和 ICP-MS 定量分析的双重信息[183]。使用金纳米团簇作为 ICP-MS 信号报告分子标记多肽底物末端,并修饰到负载了抗癌药物的金属有机框架(MOFs)纳米颗粒上用于荧光成像和 ICP-MS 定量分析。当纳米颗粒进入细胞后释放出抗癌药物,诱导细胞凋亡,产生了可作为细胞凋亡指示蛋白酶的 caspase-3 剪切多肽底物,释放出发荧光的金纳米团簇,实现了对抗癌药物的输送和细胞响应的监测[184]。

4.3.3.3　碱基互补配对介导的元素标记策略

随着 ICP-MS 技术在蛋白质定量分析研究领域的开展,研究人员也开始关注其他种类生物分子的

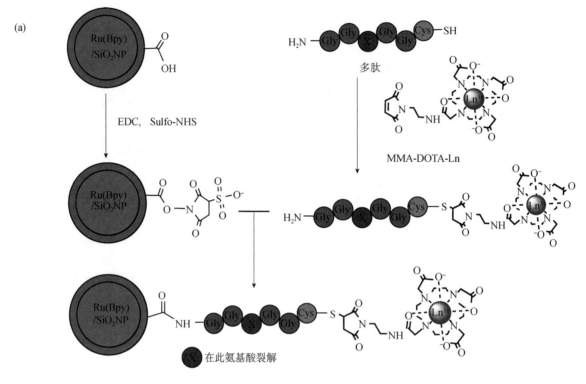

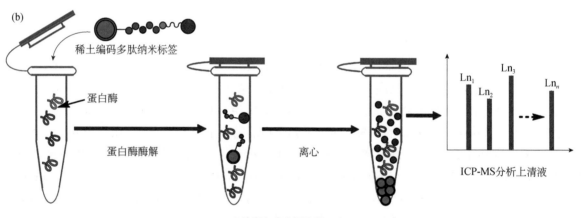

图4-9 （a）稀土编码多肽纳米标签的设计合成和（b）用于蛋白酶的 ICP-MS 免标记定量与活性评价[181]

定量研究，以进一步挖掘 ICP-MS 技术的定量优势，核酸就是其中之一。核酸与蛋白质的最大差别是核酸可以通过碱基互补配对精确地结合，这一性质为高选择性标记和检测特定序列的核酸提供了原理性的保障。类似于抗原抗体相互作用，夹心法是最常采用的方式。利用夹心法，将互补序列分别修饰在作为信号分子的金纳米粒子和用于磁分离的磁球上，样品中存在的目标 DNA 将金纳米粒子结合到磁球上，用于 ICP-MS 定量分析，检出限达到 80 zmol[185]。利用单颗粒 ICP-MS 区分单分散和交联金纳米粒子的性质，可实现对 DNA 进行免分离、一步法定量分析[186]。使用金、银、铂三种不同的金属纳米粒子标记不同 DNA，单颗粒 ICP-MS 还可以实现对多种 DNA 序列的同时检测[187]。基于单颗粒 ICP-MS 的夹心法还被用于测定 *E. coli* O157 16S rRNA 的含量[188]。

除了纳米粒子，稀土配合物也被用于核酸的夹心法标记与定量分析。使用不同稀土元素编码不同的信号DNA，同时将稀土同位素修饰的核酸结合到捕获DNA上作为同位素稀释标准品，实现了对15种DNA序列的绝对与相对定量分析[189]。将核酸稀土夹心标记法与滚环扩增（RCA）信号放大技术相结合，可实现对90 zmol HBV病毒核酸的高灵敏度检测[190]。稀土配体用作信号报告分子用于核酸的杂交过程研究[191]和单核苷酸多态性分析[192]也有报道。

前面介绍的核酸分析主要利用了碱基互补配对原理对目标分子进行元素标记。实际上，早期的核酸元素标记定量研究更多是利用核酸与固相载体的非特异性吸附的性质实现DNA的金属标记与ICP-MS定量[193]。首先将c-myc多肽修饰到DNA分子上，利用DNA可以吸附在硝酸纤维素膜并被UV光交联的性质将DNA固定到硝酸纤维素膜上，接着利用修饰了金纳米粒子的抗c-myc抗体对DNA进行金属标记与ICP-MS定量分析，检出限达到0.2 pmol。基于类似的原理，利用生物素与亲和素选择性结合的性质，将生物素修饰DNA与亲和素修饰的金纳米粒子进行孵育，DNA的含量与金纳米粒子的含量相关，最后可利用SEC-ICP-MS分离和检测结合了DNA和没有结合DNA的金纳米粒子，实现对DNA含量的测定[194]。以上两种方法虽然实现了DNA的ICP-MS定量分析，但是只是利用了DNA修饰官能团与固相载体吸附的性质，只能用于DNA的总量测定，不具有DNA序列特异性。

4.3.3.4 核酸T-Hg-T标记

离子可以选择性地插入到T：T碱基错配位点，形成T-Hg-T复合物。利用这一性质，可以实现对T：T错配DNA的特异性Hg标记。以T-Hg-T错配为茎干区的分子信标，当没有目标分子时，Hg与分子信标保持结合；如果待测DNA序列结合并打开分子信标的茎干区时，Hg被释放出来。使用CE-ICP-MS可以分离和检测游离的或与分子信标结合的汞离子，游离汞离子的含量即代表目标DNA的信号[195]。经典分子信标通过环状区DNA与目标DNA的互补配对成刚性双链DNA后将分子信标打开，若环状区替换为具有变构功能的核酸适配子，则形成可对蛋白质等非核酸靶点产生信号响应的分子信标，这种分子信标与T-Hg-T结合在一起，可用于蛋白质、DNA等分子的CVG-AFS/ICP-MS定量分析[196-198]。

4.3.3.5 基于核酸信号放大技术的元素标记核酸分析

核酸不仅具有"碱基互补配对"的选择性，而且可以利用PCR技术进行扩增。作为一种可编程的生物分子，可以设计成各种模块用于信号放大，已经发展了多种以荧光染料为信号分子的核酸信号放大技术。元素质谱与这些技术的结合，一系列基于ICP-MS的核酸特异性识别和可扩增特性的信号放大技术得到发展，主要有RCA[190,199-200]、杂交链式反应（hybridization chain reaction，HCR）[201-205]、连接介导扩增（ligation-mediated amplification）[206]、DNA纳米机器（DNA nano-machine）[207,208]、终点聚合酶链式扩增（end-point PCR amplification）[209]、双重特异核酸酶信号放大[duplex-specific nuclease (DSN) signal amplification][210]等元素信号放大技术。此外，通过在带负电荷的DNA上直接合成金属纳米粒子可以直接实现元素信号放大功能[211,212]。由于篇幅所限，本书也有专章介绍，这里就不做详细介绍。

4.3.3.6 细胞自身代谢机制和生物正交反应介导元素标记策略

除了使用多肽、抗体和适配子作为亲和试剂可以直接标记细胞表面的特异性抗原，近几年还发展了一种基于细胞自身代谢机制和生物正交反应的元素标记策略。以叠氮修饰的氮乙酰甘露糖胺作为唾液酸的从头生物合成底物，利用细胞的自身代谢通路将叠氮基团组装到细胞表面唾液酸糖基化位点的末端，接着通过点击反应将 DBCO-DOTA-Eu 或 DBCO-PEG4-BODIPY 标记到叠氮基团上，完成了对细胞的稀土/荧光双标记。唾液酸生物代谢和生物正交反应的专一性保证了 ICP-MS 准确定量细胞表面的唾液酸含量，并发现在 1 μmol/L 紫杉醇的孵育下，细胞表面的唾液酸含量降低 67%，预示着紫杉醇通过干预糖代谢组装治疗癌症的新机制（图 4-10）[213]。基于类似的代谢标记策略，进一步利用细菌自身的代谢通路将炔基修饰 D 型丙氨酸组装到细菌表面的肽聚糖中用于稀土编码标记；使用单颗粒 ICP-MS 实现对细菌的计数，研究了万古霉素和纳米银对细菌的不同作用机制（图 4-11）[214]。

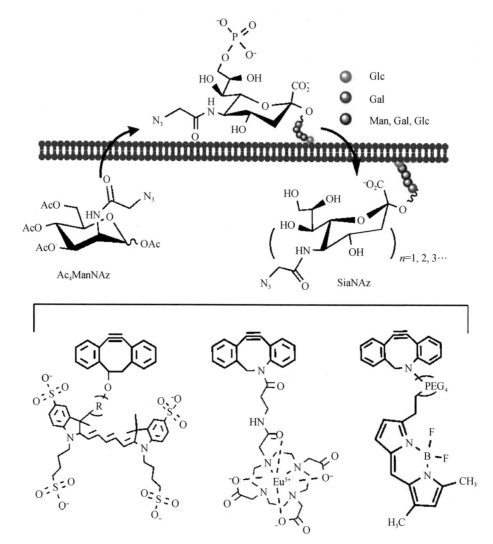

图 4-10　基于唾液酸代谢通路的细胞表面糖基化位点稀土标记策略[213]

第 4 章 元素介导的电感耦合等离子体质谱定量生物分析

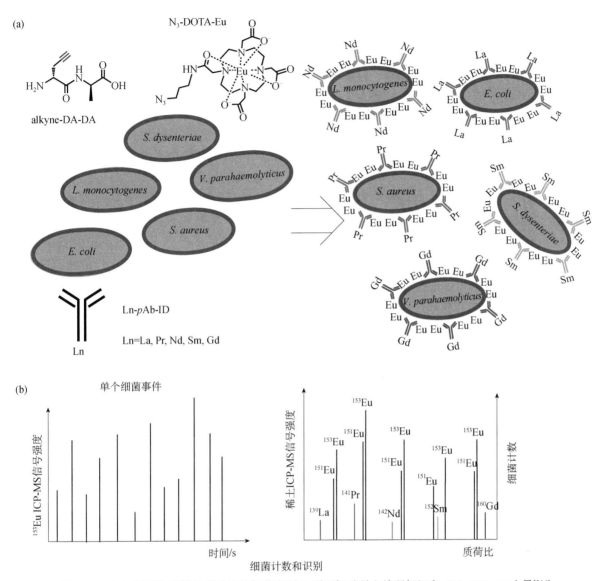

图 4-11 （a）利用肽聚糖代谢通路的细菌表面 D 型丙氨酸稀土编码标记和（b）ICP-MS 定量[214]

4.4 基于 ICP-MS 的单细胞分析技术

与传统的细胞分析法只能获得细胞群体的平均性质相比，单细胞分析提供了细胞个体的差异化信息，有利于研究细胞的基本化学生物学性质以及低丰度癌细胞的检测。单细胞分析的关键之一在于把细胞样品的细胞逐个依次地引入到 ICP-MS 中进行检测。除了 CyTOF 分选技术外，研究人员发展了各种单细胞进样技术以满足这一需求。通常情况下，细胞的直径约为 10 μm，体积约为 1 pL，单细胞以极快的速度通过，因此对 ICP-MS 的分析速度要求很高。时间分辨 ICP-MS（TR-ICP-MS）是最常使用的元素检测器。V 凹槽雾化器被用作 TR-ICP-MS 的进样器分析小球藻中 Mg、Mn、Cu 等金属含量和分布[215]。相似的雾化器也被用于分析铬金属形态在癌细胞中的形态分布[216]。微同心雾化器（microconcentric nebulizer，MCN）也作为单细胞进样器用于研究幽门螺杆菌对抗溃疡铋金属药物的摄取过程和含量，发现抗溃疡铋金属药物可能通过铁转运通路进入细胞[217]。传统雾化器的单细胞进样效率较低，大约只有1%，为此，高效同心雾化器和小体积雾室被结合起来用于单细胞进样，使进样效率提高到70%[218]。在同心雾化器和ICP炬管之间连接一个加热电阻丝缠绕的雾化室可以提高单细胞进样效率，并用于癌细胞中金属元素单细胞分析[219]。为了减少细胞团聚，提高单细胞分散性，在细胞样品注射器和雾化器之间连接细内径毛细管作为细胞流通管路，同时加入鞘液隔绝细胞可提高细胞分散性，使得血红细胞的进样效率大大提高[220]。最近发展的毛细管单细胞 SC-ICP-MS 在线分析系统也被用于分析四膜虫对汞的摄取和富集规律[221]。

微流控芯片技术可在几十到几百纳米的尺度上高通量操控细胞。利用这个优势将牛血细胞包裹在油包水的微液滴中，得到分散良好和信号稳定的 SC-ICP-MS 定量结果，分析过程只需 1 μL 样品[222]。微液滴 SC-ICP-MS 联用技术的进一步发展实现了细胞中的 Zn[223]、Au[224]、Ag[225] 等纳米粒子的定量分析。油包水通常使用的全氟己烷等有机溶剂的生物兼容性较差且有可能污染质谱，因此微液滴在进入质谱之前需要使用筒式加热器（cartridge heater）和膜脱溶器（membrane desolvator）或富氧试剂把有机溶剂在四极杆质量分析器之前去除。螺旋微通道惯性聚焦芯片无油分选单细胞技术的发展极大地克服了有机质油对后续 ICP-MS 分析的干扰[226]。不仅如此，现有的单细胞进样效率尚待进一步提高，考虑到细胞本身就是一种由磷脂双分子层细胞膜包裹的"液滴"，一种双聚焦单细胞间隔时间可控列队无油微流控-细胞直接注入进样系统被设计出来，用于 SC-ICP-MS 分析[227]（图 4-12）。不仅只使用了可热降解的 NH_4HCO_3 溶液进行细胞的分选列队，具有可控时间间隔的列队细胞直接注入 ICP-MS 中便于具有不同停留时间的 ICP-MS 仪器进行检测，细胞进样效率和检测灵敏度有了极大的提升，可以直接检测单细胞中的痕量元素。

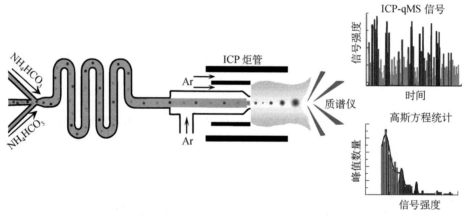

图 4-12　可热降解 NH_4HCO_3 双聚焦无油微流控-直接进样单细胞 ICP-MS 分析系统[227]

4.5　展　　望

纵观 ICP-MS 生物分析近 20 年的发展历史，解决生物问题是牵引力，仪器的进步和元素标记策略的发展是两个推动力。不论是生物分子或细胞本身含有的还是通过化学或生物途径标记到生物分子或细胞上的元素都为 ICP-MS 生物分析奠定了基础；为了突破 ICP-MS 的检测极限，针对超痕量关键标志性生物分子或作为生命活动基本单元集成了各种生物分子协同作用信息的单细胞分析,相应的信号放大和倍增策略也得到了发展,特别是在具有明确结构和修饰官能团的生物纳米颗粒上进行靶向和元素报告基团的可控编码为平衡含量或数目相差巨大的生物分子或细胞的同时定量分析提供了可能性[174](图4-13)。ICP-MS区别于其他分析工具的特性之一是超强的ICP离子源可把形态各异的分子甚至细胞中的元素几乎无差别地转变成"原子离子(atomic ion)",为进行更为简单的不依赖形态的(species-independent)同位素稀释绝对定量分析创造了机会。除了更深入的基础研究还需进一步加强以解决伙伴分析技术(方法)难以解决的问题外,ICP-MS生物分析的实际应用研究也应同步展开。针对典型目标生物标志物分子或细胞(细菌和病毒)的元素化学和生物标记(或介导)的方法需要进一步标准化,相应的元素标签和信号放大/倍增试剂"工具箱"要同时配备,这样才能使基于ICP-MS生物分析技术和方法能够被不具专业背景知识的普通工作人员使用,既开花也结出丰硕的果实,造福人类。

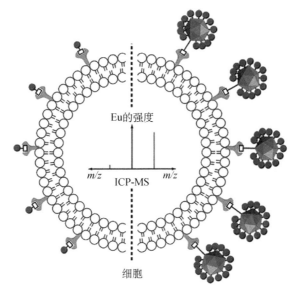

图4-13 以噬菌体衣壳蛋白为模板的稀土元素和靶向亲和配体可控修饰标签[174]

参 考 文 献

[1] Pandey A, Mann M. Proteomics to study genes and genomes. Nature, 2000, 405: 837-846.

[2] Wilkins M R. Progress with proteome projects: Why all proteins expressed by a genome should be identified and how to do it. Biotechnology & Genetic Engineering Reviews, 1996, 13: 19-50.

[3] Tyers M, Mann M. From genomics to proteomics. Nature, 2003, 422: 193-197.

[4] Kjeldahl J. New method for the determination of nitrogen in organic substances. Zeitschrift für Analytische Chemie, 1883, 22: 366-383.

[5] Dallongeville S, Garnier N, Rolando C, Tokarski C. Proteins in art, archaeology, and paleontology: From detection to identification. Chemical Reviews, 2016, 116: 2-79.

[6] Kingsmore S F. Multiplexed protein measurement: Technologies and applications of protein and antibody arrays. Nature

Reviews Drug Discovery, 2006, 5: 310-320.

[7] Chattopadhyay P K, Price D A, Harper T F, Betts M R, Yu J, Gostick E, Perfetto S P, Goepfert P, Koup R A, De Rosa S C, Bruchez M P, Roederer M. Quantum dot semiconductor nanocrystals for immunophenotyping by polychromatic flow cytometry. Nature Medicine, 2006, 12: 972-977.

[8] Yalow R S, Berson S A. Assay of plasma insulin in human subjects by immunological methods. Nature, 1959, 184: 1648-1649.

[9] Fazekas de St, Groth S, Webster R G, Datyner A. Two new staining procedures for quantitative estimation of proteins on electrophoretic strips. Biochimica et Biophysica Acta, 1963, 71: 377-391.

[10] Meyer T S, Lambert B L. Use of Coomassie brilliant blue R250 for the electrophoresis of microgram quantities of parotid saliva proteins on acrylamide-gel strips. Biochimica et Biophysica Acta, 1965, 107: 144-145.

[11] Berggren K. Background-free, high sensitivity staining of proteins in one-and two-dimensional sodium dodecyl sulfatepolyacrylamide gels using a luminescent ruthenium complex. Electrophoresis, 2000, 21: 2509-2521.

[12] Patton W F, Beechem J M. Rainbow's end: The quest for multiplexed fluorescence quantitative analysis in proteomics. Current Opinion in Chemical Biology, 2002, 6: 63-69.

[13] Fenn J B, Mann M, Meng C K, Wong S F, Whitehouse C M. Electrospray ionization for mass spectrometry of large biomolecules. Science, 1989, 246: 64-71.

[14] Tanaka K, Waki H, Ido Y, Akita S, Yoshida Y, Yoshida T. Protein and polymer analyses up to m/z 100 000 by laser ionization time-of-flight mass spectrometry. Rapid Communications in Mass Spectrometry, 1988, 2: 151-153.

[15] Ong S, Mann M. Mass spectrometry: Based proteomics turns quantitative. Nature Chemical Biology, 2005, 1: 252-262.

[16] Gygi S P, Aebersold R. Mass spectrometry and proteomics. Current Opinion in Chemical Biology, 2000, 4: 489-494.

[17] Zhang Y Y, Fonslow B R, Shan B, Baek M C, Yates J R. Protein analysis by shotgun/bottom-up proteomics. Chemical Reviews, 2013, 113: 2343-2394.

[18] Ong S E, Blagoev B, Kratchmarova I, Kristensen D B, Steen H, Pandey A, Mann M. Stable isotope labeling by amino acids in cell culture, SILAC, as a simple and accurate approach to expression proteomics. Molecular & Cellular Proteomics, 2002, 1: 376-386.

[19] Gygi S P, Rist B, Gerber S A, Turecek F, Gelb M H, Aebersold R. Quantitative analysis of complex protein mixtures using isotope-coded affinity tags. Nature Biotechnology, 1999, 17: 994-999.

[20] Ross P L, Huang Y N, Marchese J N, Williamson B, Parker K, Hattan S, Khainovski N, Pillai S, Dey S, Daniels S, Purkayastha S, Juhasz P, Martin S, Bartlet-Jones M, He F, Jacobson A, Pappin D J. Multiplexed protein quantitation in Saccharomyces cerevisiae using amine-reactive isobaric tagging reagents. Molecular & Cellular Proteomics, 2004, 3: 1154-1169.

[21] Liu Z, Cao J, He Y F, Qiao L, Xu C J, Lu H J, Yang P Y. Tandem ^{18}O stable isotope labeling for quantification of N-glycoproteome. Journal of Proteome Research, 2010, 9: 227-236.

[22] Wang W X, Zhon H, Lin H, Roy S, Shaler T A, Hill L R, Norton S, Kumar P, Anderle M, Becker C H. Quantification of proteins and metabolites by mass spectrometry without isotopic labeling or spiked standards. Analytical Chemistry, 2003, 75: 4818-4826.

[23] Nesvizhskii A I, Vitek O, Aebersold R. Analysis and validation of proteomic data generated by tandem mass spectrometry. Nature Methods, 2007, 4: 787-797.

[24] Huang T J, Armbruster M R, Coulton J B, Edwards J L. Chemical tagging in mass spectrometry for systems biology. Analytical Chemistry, 2019, 91: 109-125.

[25] Gerber S A, Rush J, Stemman O, Kirschner M W, Gygi S P. Absolute quantification of proteins and phosphoproteins

from cell lysates by tandem MS. Proceedings of the National Academy of Sciences of the United States of America, 2003, 100: 6940-4945.

[26] Beynon R J, Doherty M K, Pratt J M, Gaskell S J. Multiplexed absolute quantification in proteomics using artificial QCAT proteins of concatenated signature peptides. Nature Methods, 2005, 2: 587-589.

[27] Pan S, Aebersold R, Chen R, Rush J, Goodlett D R, McIntosh M W, Zhang J, Brentnall T A. Mass spectrometry based targeted protein quantification: Methods and applications. Journal of Proteome Research, 2009, 8: 787-797.

[28] Zhang C, Wu F, Zhang Y, Wang X, Zhang X. A novel combination of immunoreaction and ICP-MS as a hyphenated technique for the determination of thyroid-stimulating hormone (TSH) in human serum. Journal of Analytical Atomic Spectrometry, 2001, 16: 1393-1396.

[29] Zhang C, Zhang Z, Yu B, Shi J, Zhang X R. Application of the biological conjugate between antibody and colloid Au nanoparticles as analyte to inductively coupled plasma mass spectrometry. Analytical Chemistry, 2002, 74: 96-99.

[30] Baranov V I, Quinn Z, Bandura D R, Tanner S D. A Sensitive and quantitative element-tagged immunoassay with ICPMS detection. Analytical Chemistry, 2002, 74: 1629-1636.

[31] Lobinski R, Schaumloffel D, Szpunar J. Mass spectrometry in bioinorganic analytical chemistry. Mass Spectrometry Reviews, 2006, 25: 255-289.

[32] Sanz-Medel A, Montes-Bayon M, del Rosario Fernandez de la Campa M, Encinar J R, Bettmer J. Elemental mass spectrometry for quantitative proteomics. Analytical and Bioanalytical Chemistry, 2008, 390: 3-16.

[33] Prange A, Pröfrock D. Chemical labels and natural element tags for the quantitative analysis of bio-molecules. Journal of Analytical Atomic Spectrometry, 2008, 23: 432-459.

[34] Bettmer J, Montes-Bayon M, Ruiz-Encinar J, Fernández-Sánchez M L, del Rosario Fernández de la Campa M, Sanz-Medel A. The emerging role of ICP-MS in proteomic analysis. Journal of Proteomics, 2009, 72: 989-1005.

[35] Becker J S, Jakubowski N. The synergy of elemental and bimolecular mass spectrometry: New analytical strategies in life sciences. Chemical Society Reviews, 2009, 38: 1969-1983.

[36] Tholey A, Schaumloffel D. Metal labeling for quantitative protein and proteome analysis using ICPMS. TrAC Trends in Analytical Chemistry, 2010, 29: 399-408.

[37] Sanz-Medel A. ICP-MS for multiplex absolute determinations of proteins. Analytical and Bioanalytical Chemistry, 2010, 398: 1853-1859.

[38] Bomke S, Sperling M, Karst U. Organometallic derivatizing agents in bioanalysis. Analytical and Bioanalytical Chemistry, 2010, 397: 3483-3494.

[39] Bettmer J. Application of isotope dilution ICP-MS techniques to quantitative proteomics. Analytical and Bioanalytical Chemistry, 2010, 397: 3495-3502.

[40] Wang M, Feng W Y, Hao Y L, Chai Z F. ICP-MS-based strategies for protein quantification. Mass Spectrometry Reviews, 2010, 29: 326-348.

[41] 徐明, 严晓文, 杨利民, 王秋泉. 化学标记与蛋白质/多肽的识别检测. 中国科学: 化学, 2011, 41: 663-677.

[42] Sussulini A, Becker J S. Combination of PAGE and LA-ICP-MS as an analytical workflow in metallomics: State of the art, new quantification strategies, advantages and limitations. Metallomics, 2011, 3: 1271-1279.

[43] Sanz-Medel A, Montes-Bayon M, Bettmer J, Fernandez-Sanchez M L, Encinar J R. ICP-MS for absolute quantification of proteins for heteroatom-tagged, targeted proteomics. TrAC Trends in Analytical Chemistry, 2012, 40: 52-63.

[44] Kretschy D, Koellensperger G, Hann S. Elemental labelling combined with liquid chromatography inductively coupled plasma mass spectrometry for quantification of biomolecules: A review. Analytica Chimica Acta, 2012, 750: 98-110.

[45] Konz I, Fernández B, Fernández M L, Pereiro R, Sanz-Medel A. Laser ablation ICP-MS for quantitative biomedical

applications. Analytical and Bioanalytical Chemistry, 2012, 403: 2113-2125.

[46] Yan X W, Yang L M, Wang Q Q. Detection and quantification of proteins and cells by use of elemental mass spectrometry: Progress and challenges. Analytical and Bioanalytical Chemistry, 2013, 405: 5663-5670.

[47] Swart C. Metrology for metalloproteins: Where are we now, where are we heading. Analytical and Bioanalytical Chemistry, 2013, 405: 5697-5723.

[48] Schwarz G, Mueller L, Beck S, Linscheid M W. DOTA based metal labels for protein quantification: A review. Journal of Analytical Atomic Spectrometry, 2014, 29: 221-233.

[49] Liu R, Wu P, Yang L, Hou X D, Lv Y. Inductively coupled plasma mass spectrometry based immunoassay: A review. Mass Spectrometry Reviews, 2014, 33: 373-393.

[50] Campanella B, Bramanti E. Detection of proteins by hyphenated techniques with endogenous metal tags and metal chemical labelling. Analyst, 2014, 139: 4124-4153.

[51] de Bang T C, Husted S. Lanthanide elements as labels for multiplexed and targeted analysis of proteins, DNA and RNA using inductively-coupled plasma mass spectrometry. TrAC Trends in Analytical Chemistry, 2015, 72: 45-52.

[52] Liang Y, Yang L M, Wang Q Q. An ongoing path of element-labeling/tagging strategies toward quantitative bioanalysis using ICP-MS. Applied Spectroscopy Reviews, 2016, 51: 117-128.

[53] Liu R, Zhang S X, Wei C, Xing Z, Zhang S C, Zhang X R. Metal stable isotope tagging: Renaissance of radioimmunoassay for multiplex and absolute quantification of biomolecules. Accounts of Chemical Research, 2016, 49: 775-783.

[54] Maes E, Tirez K, Baggerman G, Valkenborg D, Schoofs L, Encinar J R, Mertens I. The use of elemental mass spectrometry in phosphoproteomic applications. Mass Spectrometry Reviews, 2016, 35: 350-360.

[55] Liu Z R, Li X T, Xiao G Y, Chen B B, He M, Hu B. Application of inductively coupled plasma mass spectrometry in the quantitative analysis of biomolecules with exogenous tags: A review. TrAC Trends in Analytical Chemistry, 2017, 93: 78-101.

[56] Bishop D P, Hare D J, Clases D, Doble P A. Applications of liquid chromatography-inductively coupled plasma-mass spectrometry in the biosciences: A tutorial review and recent developments. TrAC Trends in Analytical Chemistry, 2018, 104: 11-21.

[57] Calderon-Celis F, Encinar J R, Sanz-Medel A. Standardization approaches in absolute quantitative proteomics with mass spectrometry. Mass Spectrometry Reviews, 2018, 37: 715-737.

[58] Raab A, Feldmann J. Biological sulphur-containing compounds: Analytical challenges. Analytica Chimica Acta, 2019, 1079: 20-29.

[59] Linscheid M W. Molecules and elements for quantitative bioanalysis: The allure of using electrospray, MALDI, and ICP mass spectrometry side-by-side. Mass Spectrometry Reviews, 2019, 38: 169-186.

[60] Hu J, Yang P, Hou X D. Atomic spectrometry and atomic mass spectrometry in bioanalytical chemistry. Applied Spectroscopy Reviews, 2019, 54: 180-203.

[61] Cruz-Alonso M, Lores-Padín A, Valencia E, González-Iglesias H, Fernández B, Pereiro R. Quantitative mapping of specific proteins in biological tissues by laser ablation-ICP-MS using exogenous labels: Aspects to be considered. Analytical and Bioanalytical Chemistry, 2019, 411: 549-558.

[62] Zhang H, Yan W, Aebersold R. Chemical probes and tandem mass spectrometry: A strategy for the quantitative analysis of proteomes and subproteomes. Current Opinion in Chemical Biology, 2004, 8: 66-75.

[63] Svantesson E, Pettersson J, Markides K E. The use of inorganic elemental standards in the quantification of proteins and biomolecular compounds by inductively coupled plasma spectrometry. Journal of Analytical Atomic Spectrometry, 2002, 17: 491-496.

[64] Wind M, Wegener A, Eisenmenger A, Kellner R, Lehmann W D. Sulfur as the key element for quantitative protein analysis by capillary liquid chromatography coupled to element mss spectrometry. Angewandte Chemie International Edition, 2003, 42: 3425-3427.

[65] Rappel C, Schaumlöffel D. The role of sulfur and sulfur isotope dilution analysis in quantitative protein analysis. Analytical and Bioanalytical Chemistry, 2008, 390: 605-615.

[66] Zinn N, Krüger R, Leonhard P, Bettmer J. μLC coupled to ICP-SFMS with post-column isotope dilution analysis of sulfur for absolute protein quantification. Analytical and Bioanalytical Chemistry, 2008, 391: 537-543.

[67] Schaumlöffel D, Giusti P, Preud'Homme H, Szpunar J, Łobiński R. Precolumn isotope dilution analysis in nanoHPLC-ICPMS for absolute quantification of sulfur-containing peptides. Analytical Chemistry, 2007, 79: 2859-2868.

[68] Feng L, Zhang D, Wang J, Li H M. Simultaneous quantification of proteins in human serum via sulfur and iron using HPLC coupled to post-column isotope dilution mass spectrometry. Anal Methods, 2014, 6: 7655-7662.

[69] Martınez-Sierra J G, Sanz F M, Espilez P H, Marchante Gayo J M, Garcia Alonso J I. Biosynthesis of sulfur-34 labelled yeast and its characterisation by multicollector-ICP-MS. Journal of Analytical Atomic Spectrometry, 2007, 22: 1105-1112.

[70] Martinez-Sierra J G, Sanz F M, Espilez P H, Santamaria-Fernandez R, Marchante Gayon J M, Garcia Alonso J I. Evaluation of different analytical strategies for the quantification of sulfur-containing biomolecules by HPLC-ICP-MS: Application to the characterisation of ^{34}S-labelled yeast. Journal of Analytical Atomic Spectrometry, 2010, 25: 989-997.

[71] Bandura D R, Baranov V I, Tanner S D. Detection of ultratrace phosphorus and sulfur by quadrupole ICPMS with dynamic reaction cell. Analytical Chemistry, 2002, 74: 1497-1502.

[72] Fernandez S D, Sugishama N, Encinar J R, Sanz-Medel A. Triple quad ICPMS (ICPQQQ) as a new tool for absolute quantitative proteomics and phosphoproteomics. Analytical Chemistry, 2012, 84: 5851-5857.

[73] Wang M, Feng W Y, Lu W W, Li B, Wang B, Zhu M, Wang Y, Yuan H, Zhao Y L, Chai Z F. Quantitative analysis of proteins via sulfur determination by HPLC coupled to isotope dilution ICPMS with a hexapole collision cell. Analytical Chemistry, 2007, 79: 9128-9134.

[74] Rampler E, Dalik T, Stingeder G, Hanna S, Koellensperger G. Sulfur containing amino acids-challenge of accurate quantification. Journal of Analytical Atomic Spectrometry, 2012, 27: 1018-1023.

[75] Adaba R I, Mann G, Raab A, Houssen W E, McEwan A R, Thomas L, Tabudravu J, Naismith J H, Jaspars M. Accurate quantification of modified cyclic peptides without the need for authentic standards. Tetrahedron, 2016, 72: 8603-8609.

[76] Reinders J, Sickmann A. State-of-the-art in phosphoproteomics. Proteomics, 2005, 5: 4052-4061.

[77] Mann M, Ong S E, Gronborg M, Steen H, Jensen O N, Pandey A. Analysis of protein phosphorylation using mass spectrometry: Deciphering the phosphoproteome. Trends in Biotechnology, 2002, 20: 261-268.

[78] Wind M, Edler M, Jakubowski N, Linscheid M, Wesch H, Lehmann W D. Analysis of protein phosphorylation by capillary liquid chromatography coupled to element mass spectrometry with P-31 detection and to electrospray mass spectrometry. Analytical Chemistry, 2001, 73: 29-35.

[79] Wind M, Eisenmenger A, Lehmann W D. Modified direct injection high efficiency nebulizer with minimized dead volume for the analysis of biological samples by micro-and nano-LC-ICPMS. Journal of Analytical Atomic Spectrometry, 2002, 17: 21-26.

[80] Wind M, Wesch H, Lehmann W D. Protein phosphorylation degree: Determination by capillary liquid chromatography and inductively coupled plasma mass spectrometry. Analytical Chemistry, 2001, 73: 3006-3010.

[81] Krüger R, Kübler D, Pallissé R, Burkovski A, Lehmann W D. Protein and proteome phosphorylation stoichiometry analysis by element mass spectrometry. Analytical Chemistry, 2006, 78: 1987-1994.

[82] Bandura D R, Ornatsky O, Liao L. Characterization of phosphorus content of biological samples: Potential tool for cancer research. Journal of Analytical Atomic Spectrometry, 2004, 19: 96-100.

[83] Pereira Navaza A, Ruiz Encinar J, Sanz-Medel A. Absolute and accurate quantification of protein phosphorylation by using an elemental phosphorus standard and element mass spectrometry. Angewandte Chemie International Edition, 2007, 46: 569-571.

[84] Pröfrock D, Prange A. Compensation of gradient related effects when using capillary liquid chromatography and inductively coupled plasma mass spectrometry for the absolute quantification of phosphorylated peptides. Journal of Chromatography A, 2009, 1216: 6706-6715.

[85] Rayman M P. The importance of selenium to human health. Lancet, 2000, 356: 233-241.

[86] Ruiz Encinar J, Ouerdane L, Buchmann W, Tortajada J, Lobinski R, Szpunar J. Identification of water-soluble selenium-containing proteins in selenized yeast by size-exclusion-reversed-phase HPLC/ICPMS followed by MALDI-TOF and electrospray Q-TOF mass spectrometry. Analytical Chemistry, 2003, 75: 3765-3774.

[87] Ballihaut G, Mounicou S, Lobinski R. Multitechnique mass-spectrometric approach for the detection of bovine glutathione peroxidase selenoprotein: Focus on the selenopeptide. Analytical and Bioanalytical Chemistry, 2007, 388: 585-591.

[88] Stadtman T C. Selenocysteine. Annual Review of Biochemistry, 1996, 65: 83-100.

[89] Behne D, Kyriakopoulos A. Mammalian selenium-containing proteins. Annual Review of Nutrition, 2001, 21: 453-473.

[90] Kryukov G V, Castellano S, Novoselov S V, Lobanov A V, Zehtab O, Guigo R, Gladyshev V N. Characterization of mammalian selenoproteomes. Science, 2003, 300: 1439-1443.

[91] Xu M, Yang L M, Wang Q Q. Quantification of selenium-tagged proteins in human plasma using species-unspecific isotope dilution ICP-DRC-qMS coupled on-line with anion exchange chromatography. Journal of Analytical Atomic Spectrometry, 2008, 23: 1545-1549.

[92] Gomez-Espina J, Blanco-Gonzalez E, Montes-Bayon M, Sanz-Medel A. Elemental mass spectrometry for Se-dependent glutathione peroxidase determination in red blood cells as oxidative stress biomarker. Journal of Analytical Atomic Spectrometry, 2012, 27: 1949-1954.

[93] Møller L H, Gabel-Jensen C, Franzyk H, Bahnsen J S, Stürup S, Gammelgaard B. Quantification of pharmaceutical peptides using selenium as an elemental detection label. Metallomics, 2014, 6: 1639-1647.

[94] Tainer J A, Roberts V A, Getzoff E D. Metal-binding sites in proteins. Current Opinion in Biotechnology, 1991, 2: 582-591.

[95] Vašák M, Hasler D W. Metallothioneins: New functional and structural insights. Current Opinion in Chemical Biology, 2000, 4: 177-183.

[96] Chasteen N D, Ferritin. Uptake, storage, and release of iron. Metal Ions in Biological Systems, 1998, 35: 479-514.

[97] Liberek K, Lewandowska A, Ziętkiewicz S. Chaperones in control of protein disaggregation. EMBO Journal, 2008, 27: 328-335.

[98] Muñiz C S, Gayón J M M, Alonso J I G, Sanz-Medel A. Speciation of essential elements in human serum using anion exchange chromatography coupled to post-column isotope dilution analysis with double focusing ICP-MS. Journal of Analytical Atomic Spectrometry, 2001, 16: 587-592.

[99] Busto M E, Montes-Bayón M, Sanz-Medel A. Accurate determination of human serum transferrin isoforms: Exploring metal-specific isotope dilution analysis as a quantitative proteomic tool. Analytical Chemistry, 2006, 78: 8218-8226.

[100] Williams R J P. Chemical selection of elements by cells. Coordination Chemistry Reviews, 2001, 216: 583-595.

[101] Haraguchi H. Metallomics as integrated biometal science. Journal of Analytical Atomic Spectrometry, 2004, 19: 5-14.

[102] Szpunar J. Metallomics: A new frontier in analytical chemistry. Analytical and Bioanalytical Chemistry, 2004, 378: 54-56.

[103] 江桂斌, 何滨. 金属组学及其研究方法与前景. 中国科学基金, 2005, 3: 151-155.

[104] Mounicou S, Szpunar J, Łobiński R. Metallomics: The concept and methodology. Chemical Society Reviews, 2009, 38: 1119-1138.

[105] Łobiński R, Becker J S, Haraguchi H, Sarkar B. Metallomics: Guidelines for terminology and critical evaluation of analytical chemistry approaches (IUPAC Technical Report). Pure and Applied Chemistry, 2010, 82: 493-504.

[106] 王秋泉. 金属组学// "10000 个科学难题"化学编委会·10000 个科学难题: 化学卷. 北京: 科学出版社, 2011: 391-392.

[107] Li Y F, Chen C Y, Qu Y, Gao Y X, Li B, Zhao Y L, Chai Z F. Metallomics, elementomics, and analytical techniques. Pure and Applied Chemistry, 2008, 80: 2577-2594.

[108] Hu L G, He B, Wang Y C, Jiang G B, Sun H Z. Metallomics in environmental and health related research: Current status and perspectives. Chinese Science Bulletin, 2013, 58: 169-176.

[109] Vogiatzis C G, Zachariadis G A. Tandem mass spectrometry in metallomics and the involving role of ICP-MS detection: A review. Analytica Chimica Acta, 2014, 819: 1-14.

[110] Wang Y C, Wang H B, Li H Y, Sun H Z. Metallomic and metalloproteomic strategies in elucidating the molecular mechanisms of metallodrugs. Dalton Transactions, 2015, 44: 437-447.

[111] Haraguchi H. Metallomics: The history over the last decade and a future outlook. Metallomics, 2017, 9: 1001-1013.

[112] Montes-Bayón M, Shararb M, Corte-Rodriguez M. Trends on (elemental and molecular) mass spectrometry based strategies for speciation and metallomics. TrAC Trends in Analytical Chemistry, 2018, 104: 4-10.

[113] Guo Y, Chen L, Yang L, Wang Q Q. Counting sulfhydryls and disulfide bonds in peptides and proteins using mercurial ions as an MS-tag. Journal of the American Society for Mass Spectrometry, 2008, 19: 1108-1113.

[114] Guo Y, Xu M, Yang L, Wang Q Q. Strategy for absolute quantification of proteins: CH_3Hg^+ labeling integrated molecular and elemental mass spectrometry. Journal of Analytical Atomic Spectrometry, 2009, 24: 1184-1187.

[115] Kutscher D J, Busto M E C, Zinn N, Sanz-Medel A, Bettmer J. Protein labelling with mercury tags: Fundamental studies on ovalbumin derivatised with p-hydroxymercuribenzoic acid (pHMB). Journal of Analytical Atomic Spectrometry, 2008, 23: 1359-1364.

[116] Kutscher D J, Bettmer J. Absolute and relative protein quantification with the use of isotopically labeled p-hydroxymercuribenzoic acid and complementary MALDI-MS and ICPMS detection. Analytical Chemistry, 2009, 81: 9172-9177.

[117] Kutscher D J, Fricker M B, Hattendorf B, Bettmer J, Günther D. Systematic studies on the determination of Hg-labelled proteins using laser ablation-ICPMS and isotope dilution analysis. Analytical and Bioanalytical Chemistry, 2011, 401: 2691-2698.

[118] Sattler W, Yurkerwich K, Parkin G. Molecular structures of protonated and mercurated derivatives of thimerosal. Dalton Transactions, 2009, 22: 4327-4333.

[119] Xu M, Yan X W, Xie Q Q, Yang L M, Wang Q Q. Dynamic labeling strategy with ^{204}Hg-isotopic methylmercurithiosalicylate for absolute peptide and protein quantification. Analytical Chemistry, 2010, 82: 1616-1620.

[120] Xu M, Yang L M, Wang Q Q. A way to probe the microenvironment of free sulfhydryls in intact proteins with a series of monofunctional organic mercurials. Chemistry: A European Journal. 2012, 18: 13989-13993.

[121] Tang L, Chen F, Yang L M, Wang Q Q. The determination of low-molecular-mass thiols with 4-(hydroxymercuric) benzoic acid as a tag using HPLC coupled online with UV/HCOOH-induced cold vapor generation AFS. Journal of

Chromatography B, 2009, 877: 3428-3433.

[122] Campanella B, Rivera J G, Ferrari C, Biagi S, Onor M, D'Ulivo A, Bramanti E. Microwave photochemical reactor for the online oxidative decomposition of *p*-hydroxymercurybenzoate (*p*HMB)-tagged proteins and their determination by cold vapor generation-atomic fluorescence detection. Analytical Chemistry, 2013, 85: 12152-12157.

[123] Campanella B, Onor M, Ferrari C, D'Ulivo A, Bramanti E. Direct, simple derivatization of disulfide bonds in proteins with organic mercury in alkaline medium without any chemical pre-reducing agents. Analytica Chimica Acta, 2014, 843: 1-6.

[124] Whetstone P A, Butlin N G, Corneillie T M, Meares C F. Element-coded affinity tags for peptides and proteins. Bioconjugate Chemistry, 2004, 15: 3-6.

[125] Yan X W, Xu M, Yang L M, Wang Q Q. Absolute quantification of intact proteins via 1, 4, 7, 10-tetraazacyclododecane-1, 4, 7-trisacetic acid-10-maleimidoethylacetamide-europium labeling and HPLC coupled with species-unspecific isotope dilution ICPMS. Analytical Chemistry, 2010, 82: 1261-1269.

[126] Esteban-Fernández D, Scheler C. Linscheid M W. Absolute protein quantification by LC-ICP-MS using MeCAT peptide labeling. Analytical and Bioanalytical Chemistry, 2011, 401: 657-666.

[127] Esteban-Fernandez D, Bierkandt F S, Linscheid M W. MeCAT labeling for absolute quantification of intact proteins using label-specific isotope dilution ICP-MS. Journal of Analytical Atomic Spectrometry, 2012, 27: 1701-1708.

[128] Bergmann U, Ahrends R, Neumann B, Scheler C, Linscheid M W. Application of metal-coded affinity tags (MeCAT): Absolute protein quantification with top-down and bottom-up workflows by metal-coded tagging. Analytical Chemistry, 2012, 84: 5268-5275.

[129] El-Khatib A H, Esteban-Fernández D, Linscheid M W. Inductively coupled plasma mass spectrometry-based method for the specific quantification of sulfenic acid in peptides and proteins. Analytical Chemistry, 2008, 80: 4455-4486.

[130] Sharar M, Saied E M, Rodriguez M C, Arenz1 C, Montes-Bayón M, Linscheid M W. Elemental labelling and mass spectrometry for the specific detection of sulfenic acid groups in model peptides: A proof of concept. Analytical and Bioanalytical Chemistry, 2017, 409: 2015-2027.

[131] Sharar M, Rodríguez-Solla H, Linscheid M W, Montes-Bayón M. Detection of sulfenic acid in intact proteins by mass spectrometric techniques: Application to serum samples. RSC Advances, 2017, 7: 44162-44168.

[132] Jakubowski N, Waentig L, Hayen H, Venkatachalam A, Bohlen A, Roos P H, Manz A. Labelling of proteins with 2-(4-isothiocyanatobenzyl)-1, 4, 7, 10-tetraazacyclododecane-1, 4, 7-tetraacetic acid and lanthanides and detection by ICP-MS. Journal of Analytical Atomic Spectrometry, 2008, 23: 1497-1507.

[133] Liu H, Zhang Y, Wang J, Wang D, Zhou C, Cai Y, Qian X H. Method for quantitative proteomics research by using metal element chelated tags coupled with mass spectrometry. Analytical Chemistry, 2006, 78: 6614-6621.

[134] Patel P, Jones P, Handy R. Isotopic labelling of peptides and isotope ratio analysis using LC-ICP-MS: A preliminary study. Analytical and Bioanalytical Chemistry, 2008, 390: 61-65.

[135] Rappel C, Schaumloffel D. Absolute peptide quantification by lutetium labeling and nanoHPLC-ICPMS with isotope dilution analysis. Analytical Chemistry, 2009, 81: 385-393.

[136] Zheng L N, Wang M, Wang H J, Wang B, Li B, Li J J, Zhao Y L, Chai Z F, Feng W Y. Quantification of proteins using lanthanide labeling and HPLC/ICP-MS detection. Journal of Analytical Atomic Spectrometry, 2011, 26: 1233-1236.

[137] Yang M W, Wang Z W, Fang L, Zheng J P, Xu L J, Fu F F. Simultaneous and ultra-sensitive quantification of multiple peptides by using europium chelate labeling and capillary electrophoresis-inductively coupled plasma mass spectrometry. Journal of Analytical Atomic Spectrometry, 2012, 27: 946-951.

[138] Yang M W, Wu W H, Ruan Y J, Huang L M, Wu Z J, Cai Y, Fu F F. Ultra-sensitive quantification of lysozyme based on

element chelate labeling and capillary electrophoresis-inductively coupled plasma mass spectrometry. Analytica Chimica Acta, 2014, 812: 12-17.

[139] Liu R, Hou X D, Lv Y, McCooeye M, Yang L, Mester Z. Absolute quantification of peptides by isotope dilution liquid chromatography-inductively coupled plasma mass spectrometry and gas chromatography/mass spectrometry. Analytical Chemistry, 2013, 85: 4087-4093.

[140] Lv Y Y, Zhang H Q, Wang G B, Xia C S, Gao F Y, Zhang Y J, Qiao H L, Xie Y P, Qin W J, Qian X H. A novel mass spectrometry method for the absolute quantification of several cytochrome P450 and uridine 5′-diphospho-glucuronosyltransferase enzymes in the human liver. Analytical and Bioanalytical Chemistry, 2020, 412: 1729-1740.

[141] El-Khatib A H, Esteban-Fernández D, Linscheid M W. Dual labeling of biomolecules using MeCAT and DOTA derivatives: Application to quantitative proteomics. Analytical and Bioanalytical Chemistry, 2012, 403: 2255-2267.

[142] Christopher S J, Kilpatrick E L, Yu L L, Davis W C, Adair B M. Preliminary evaluation of a microwave-assisted metal-labeling strategy for quantification of peptides via RPLC-ICP-MS and the method of standard additions. Talanta, 2012, 88: 749-758.

[143] Markwell M A K. A new solid-state reagent to iodinate proteins. Analytical Biochemistry, 1982, 125: 427-432.

[144] Jakubowski N, Messerschmidt J, Anorbe M G, Waentig L, Hayen H, Roos P H. Labelling of proteins by use of iodination and detection by ICP-MS. Journal of Analytical Atomic Spectrometry, 2008, 23: 1487-1496.

[145] Waentig L, Jakubowski N, Hayen H, Roos P H. Iodination of proteins, proteomes and antibodies with potassium triodide for LA-ICP-MS based proteomic analyses. Journal of Analytical Atomic Spectrometry, 2011, 26: 1610-1618.

[146] Navaza A P, Encinar J R, Ballesteros A, Gonzalez J M, Sanz-Medel A. Capillary HPLC-ICPMS and tyrosine iodination for the absolute quantification of peptides using generic standards. Analytical Chemistry, 2009, 81: 5390-5399.

[147] Dersch J M, Nguyen T T T N, Østergaard J, Stürup S, Gammelgaard B. Selective analysis of human serum albumin based on SEC-ICP-MS after labelling with iophenoxic acid. Analytical and Bioanalytical Chemistry, 2015, 407: 2829-2836.

[148] Lim H, Cao Y, Qiu X, Silva J, Evans D C. A nonradioactive approach to investigate the metabolism of therapeutic peptides by tagging with ^{127}I and using inductively-coupled plasma mass spectrometry analysis. Drug Metabolism and Disposition, 2015, 43: 17-26.

[149] Tang N N, Li Z X, Yang L M, Wang Q Q. ICPMS-based specific quantification of phosphotyrosine: A gallium-tagging and tyrosine-phosphatase mediated strategy. Analytical Chemistry, 2016, 88: 9890-9896.

[150] Tanner S D, Bandura D R, Ornatsky O, Baranov V I, Nitz M, Winnik M A. Flow cytometer with mass spectrometer detection for massively multiplexed single-cell biomarker assay. Pure and Applied Chemistry, 2008, 80: 2627-2641.

[151] Careri M, Elviri L, Mangia A. Element-tagged immunoassay with inductively coupled plasma mass spectrometry for multianalyte detection. Analytical and Bioanalytical Chemistry, 2009, 393: 57-61.

[152] Giesen C, Waentig L, Panne U, Jakubowski N. History of inductively coupled plasma mass spectrometry-based immunoassays. Spectrochimica Acta Part B, 2012, 76: 27-39.

[153] Liu R, Wu P, Yang L, Hou X D, Lv Y, Mueller L, Traub H, Jakubowski N, Drescher D, Baranov V I, Kneipp J. Trends in single-cell analysis by use of ICP-MS. Analytical and Bioanalytical Chemistry 2014, 406: 6963-6977.

[154] Spitzer M H, Nolan G P. Mass cytometry: Single cells, many features. Cell, 2016, 165: 780-791.

[155] Zhang L W, Vertes A. Single-cell mass spectrometry approaches to explore cellular heterogeneity. Angewandte Chemie International Edition, 2018, 57: 4466-4477.

[156] Zhao Q, Lu X F, Yuan C G, Li X F, Le X C. Aptamer-linked assay for thrombin using gold nanoparticle amplification and inductively coupled plasma-mass spectrometry detection. Analytical Chemistry, 2009, 81: 7484-7489.

[157] Liu J M, Yan X P. Ultrasensitive, selective and simultaneous detection of cytochrome c and insulin based on immunoassay and aptamer-based bioassay in combination with Au/Ag nanoparticle tagging and ICP-MS detection. Journal of Analytical Atomic Spectrometry, 2011, 26: 1191-1197.

[158] Yang W J, Xi Z M, Zeng X X, Fang L, Jiang W J, Wu Y N, Xu L J, Fu F F. Magnetic bead-based AuNP labelling combined with inductively coupled plasma mass spectrometry for sensitively and specifically counting cancer cells. Journal of Analytical Atomic Spectrometry, 2016, 31: 679-685.

[159] Leipold M D, Herrera I, Ornatsky O, Baranov V, Nitz M. ICP-MS-based multiplex profiling of glycoproteins using lectins conjugated to lanthanide-chelating polymers. Journal of Proteome Research, 2009, 8: 443-449.

[160] Peng H Y, Jiao Y, Xiao X, Chen B B, He M, Liu Z R, Zhang X, Hu B. Magnetic quantitative analysis for multiplex glycoprotein with polymer-based elemental tags. Journal of Analytical Atomic Spectrometry, 2014, 29: 1112-1119.

[161] Zhang X, Xiao G Y, Chen B B, He M, Hu B. Lectin affinity based elemental labeling with hybridization chain reaction for the sensitive determination of avian influenza A (H9N2) virions. Talanta, 2018, 188: 442-447.

[162] Zhang X, Chen B B, He M, Zhang Y W, Xiao G Y, Hu B. Magnetic immunoassay coupled with inductively coupled plasma mass spectrometry for simultaneous quantification of alpha-fetoprotein and carcinoembryonic antigen in human serum. Spectrochimica Acta Part B, 2015, 106: 20-27.

[163] Zhang X, Chen B B, He M, Zhang Y, Peng L, Hu B. Boronic acid recognition based-gold nanoparticlelabeling strategy for the assay of sialic acid expression on cancer cell surface by inductively coupled plasma mass spectrometry. Analyst, 2016, 141: 1286-1293.

[164] Ruhe L, Ickert S, Beckl S, Linscheid M W. A new strategy for metal labeling of glycan structures in antibodies. Analytical and Bioanalytical Chemistry, 2018, 410: 21-25.

[165] Bendall S C, Simonds E F, Qiu P, Amir E D, Krutzik P O, Finck R, Bruggner R V, Melamed R, Trejo A, Ornatsky O I, Balderas R S, Plevritis S K, Sachs K, Pe'er D, Tanner S D, Nolan G P. Single-cell mass cytometry of differential immune and drug responses across a human hematopoietic continuum. Science, 2011, 332: 687-696.

[166] Bandura D R, Baranov V I, Ornatsky O I, Antonov A, Kinach R, Lou X D, Pavlov S, Vorobiev S, Dick J E, Tanner S D. Mass cytometry: Technique for real time single cell multitarget immunoassay based on inductively coupled plasma time-of-flight mass spectrometry. Analytical Chemistry, 2009, 81: 6813-6822.

[167] Han G, Spitzer M H, Bendall S C, Fantl W J, Nolan G P. Metal-isotope-tagged monoclonal antibodies for high-dimensional mass cytometry. Nature Protocols, 2018, 13: 2121-2148.

[168] Zhang Z B, Luo Q, Yan X W, Li Z X, Luo Y C, Yang L M, Zhang B, Chen H F, Wang Q Q. Integrin-targeted trifunctional probe for cancer cells: A "seeing and counting" approach. Analytical Chemistry, 2012, 84: 8946-8951.

[169] Zhai J, Wang Y L, Xu C, Zheng L N, Wang M, Feng W Y, Gao L, Zhao L N, Liu R, Gao F P, Zhao Y L, Chai Z F, Gao X Y. Facile approach to observe and quantify the $\alpha_{IIb}\beta_3$ integrin on a single-cell. Analytical Chemistry, 2015, 87: 2546-2549.

[170] Liu C L, Lu S, Yang L M, Chen P J, Bai P M, Wang Q Q. Near-infrared neodymium tag for quantifying targeted biomarker and counting its host circulating tumor cells. Analytical Chemistry, 2017, 89: 9239-9246.

[171] Yang B, Zhang Y, Chen B B, He M, Yin X, Wang H, Li X T, Hu B. A multifunctional probe for ICP-MS determination and imaging of cancer cells. Biosensors & Bioelectronics, 2017, 96: 77-83.

[172] Yang B, Chen B B, He M, Hu B. Quantum dots labeling strategy for "counting and visualization" of HepG2 cells. Analytical Chemistry, 2017, 89: 1879-1886.

[173] Yang B, Chen B B, He M, Yin X, Xu C, Hu B. Aptamer-based dual-functional probe for rapid and specific counting and imaging of MCF-7 cells. Analytical Chemistry, 2018, 90: 2355-2361.

[174] Yuan R, Ge F C, Liang Y, Zhou Y, Yang L M, Wang Q Q. Viruslike element-tagged nanoparticle inductively coupled

plasma mass spectrometry signal multiplier: Membrane biomarker mediated cell counting. Analytical Chemistry, 2019, 91: 4948-4952.

[175] Yan X W, Luo Y C, Zhang Z B, Li Z X, Luo Q, Yang L M, Zhang B, Chen H F, Bai P M, Wang Q Q. Europium-labeled activity-based probe through click chemistry: Absolute serine protease quantification using ^{153}Eu isotope dilution ICP/MS. Angewandte Chemie International Edition, 2012, 51: 3358-3363.

[176] Yan X W, Li Z X, Liang Y, Yang L M, Zhang B, Wang Q Q. A chemical "hub" for absolute quantification of a targeted protein: Orthogonal integration of elemental and molecular mass spectrometry. Chemical Communications, 2014, 50: 6578-6581.

[177] Liang Y, Yan X W, Li Z X, Yang L M, Zhang B, Wang Q Q. Click chemistry mediated Eu-tagging: Activity-based specific quantification and simultaneous activity evaluation of CYP3A4 using ^{153}Eu species-unspecific isotope dilution inductively coupled plasma mass spectrometry. Analytical Chemistry, 2014, 86: 3688-3692.

[178] Liang Y, Jiang X, Tang N N, Yang L M, Chen H F, Wang Q Q. Quantification and visualization of glutathione S-transferase omega 1 in cells using inductively coupled plasma mass spectrometry (ICP-MS) and fluorescence microscopy. Analytical and Bioanalytical Chemistry, 2015, 407: 2373-2381.

[179] Ji C X, Liang Y, Ge F C, Yang L M, Wang Q Q. Inhibitory covalent labeling and clickable-Eu-tagging-based ICPMS: Measurement of pH-dependent absolute activities of the cathepsins in hepatocyte lysosomes. Analytical Chemistry, 2019, 91: 7032-7038.

[180] Gomez-Espina J, Blanco-Gonzalez E, Montes-Bayon M, Sanz-Medel A. HPLC-ICP-MS for simultaneous quantification of the total and active form of the thioredoxin reductase enzyme in human serum using auranofin as an activity-based probe. Journal of Analytical Atomic Spectrometry, 2016, 31: 1895-1903.

[181] Yan X W, Yang L M, Wang Q Q. Lanthanide-coded protease-specific peptide-nanoparticle probes for label-free multiplex protease assay using element mass spectrometry: A proof-of-concept study. Angewandte Chemie International Edition, 2011, 50: 5130-5133.

[182] Gregorius B, Jakoby T, Schaumlöffel D, Tholey A. Monitoring of protease catalyzed reactions by quantitative MALDI-MS using metal labeling. Analytical Chemistry, 2013, 85: 5184-5190.

[183] Feng D, Tian F, Qin W J, Qian X H. A dual-functional lanthanide nanoprobe for both living cell imaging and ICP-MS quantification of active protease. Chemical Science, 2016, 7: 2246-2250.

[184] Yin X, Yang B, Chen B B, He M, Hu B. Multifunctional gold nanocluster decorated metal-organic framework for real-time monitoring of targeted drug delivery and quantitative evaluation of cellular therapeutic response. Analytical Chemistry, 2019, 91: 10596-10603.

[185] Hsu I H, Chen W H, Wu T K, Sun Y C. Gold nanoparticle-based inductively coupled plasma mass spectrometry amplification and magnetic separation for the sensitive detection of a virus-specific RNA sequence. Journal of Chromatography A, 2011, 1218: 1795-1801.

[186] Han G J, Xing Z, Dong Y H, Zhang S C, Zhang X R. One-step homogeneous DNA assay with single-nanoparticle detection. Angewandte Chemie International Edition, 2011, 50: 3462-3465.

[187] Zhang S X, Han G J, Xing Z, Zhang S C, Zhang X R. Multiplex DNA assay based on nanoparticle probes by single particle inductively coupled plasma mass spectrometry. Analytical Chemistry, 2014, 86: 3541-3547.

[188] Xu X M, Chen J Y, Li B R, Tang L J, Jiang J H. Single particle ICP-MS-based absolute and relative quantification of E. coli O157 16S rRNA using sandwich hybridization capture. Analyst, 2019, 144: 1725-1730.

[189] Han G J, Zhang S C, Xing Z, Zhang X R. Absolute and relative quantification of multiplex DNA assays based on an elemental labeling strategy. Angewandte Chemie International Edition, 2013, 52: 1466-1471.

[190] Luo Y C, Yan X W, Huang Y S, Wen R B, Li Z X, Yang L M, Yang C Y, Wang Q Q. ICP-MS-based multiplex and ultrasensitive assay of viruses with lanthanide-coded biospecific tagging and amplification strategies. Analytical Chemistry, 2013, 85: 9428-9432.

[191] Lopez-Fernandez L, Blanco-Gonzalez E, Bettmer J. Determination of specific DNA sequences and their hybridisation processes by elemental labelling followed by SEC-ICP-MS detection. Analyst, 2014, 139: 3423-3428.

[192] Lareo P L, Linscheid M W, Seitz O. Nucleic acid and SNP detection via template-directed native chemical ligation and inductively coupled plasma mass spectrometry. Journal of Mass Spectrometry, 2019, 54: 676-683.

[193] Merkoçii A, Aldavert M, Tarrason G, Eritja R, Alegret S. Toward an ICPMS-linked DNA assay based on gold nanoparticles immunoconnected through peptide sequences. Analytical Chemistry, 2005, 77: 6500-6503.

[194] Kerr S L, Sharp B. Nano-particle labelling of nucleic acids for enhanced detection by inductively-coupled plasma mass spectrometry (ICP-MS). Chemical Communications, 2007, 43: 4537-4539.

[195] Li Y, Sun S K, Yang J L, Jiang Y. Label-free DNA hybridization detection and single base-mismatch discrimination using CE-ICP-MS assay. Analyst, 2011, 136: 5038-5045.

[196] Chen P P, Wu P, Chen J B, Yang P, Zhang X F, Zheng C B, Hou X D. Label-free and separation-free atomic fluorescence spectrometry-based bioassay: Sensitive determination of single-strand DNA, protein and double-strand DNA. Analytical Chemistry, 2016, 88: 2065-2071.

[197] Chen P P, Wu P, Zhang Y X, Chen J B, Jiang X M, Zheng C B, Hou X D. Strand displacement-induced enzyme-free amplification for label-free and separation-free ultrasensitive atomic fluorescence spectrometric detection of nucleic acids and proteins. Analytical Chemistry, 2016, 88: 12386-12392.

[198] Chen P P, Huang K, Dai R, Sawyer E, Sun K, Ying B W, Wei X W, Geng J. Sensitive CVG-AFS/ICP-MS label-free nucleic acid and protein assays based on a selective cation exchange reaction and simple filtration separation. Analyst, 2019, 144: 2797-2802.

[199] He Y, Chen D L, Li M X, Fang L, Yang W J, Xu L J, Fu F F. Rolling circle amplification combined with gold nanoparticles-tag for ultrasensitive and specific quantification of DNA by inductively coupled plasma mass spectrometry. Biosensors & Bioelectronics, 2014, 58: 209-213.

[200] Li X M, Luo J, Zhang N B, Wei Q L. Nucleic acid quantification using nicking-displacement, rolling circle amplification and bio-bar-code mediated triple-amplification. Analytica Chimica Acta, 2015, 881: 117-123.

[201] Deng C, Zhang C H, Tang H, Jiang J H. ICP-MS DNA assay based on lanthanide labels and hybridization chain reaction amplification. Analytical Methods, 2015, 7: 5767-5771.

[202] Liu Y, Ding Y G, Gao Y, Liu R, Hu X R, Lv Y. Enzyme-free amplified DNA assay: Five orders of linearity provided by metal stable isotope detection. Chemical Communications, 2018, 54: 13782-13785.

[203] Liu X, Zhang S Q, Cheng Z H, Wei X, Yang T, Yu Y L, Chen M L, Wang J H. Highly sensitive detection of microRNA-21 with ICPMS via hybridization accumulation of upconversion nanoparticles. Analytical Chemistry, 2018, 90: 12116-12122.

[204] He Y, Chen S L, Huang L, Wang Z W, Wu Y N, Fu F F. Combination of magnetic-beads-based multiple metal nanoparticles labeling with hybridization chain reaction amplification for simultaneous detection of multiple cancer cells with inductively coupled plasma mass spectrometry. Analytical Chemistry, 2019, 91: 1171-1177.

[205] Li B R, Tang H, Yu R Q, Jiang J H. Single-nanoparticle ICPMS DNA assay based on hybridization-chain-reaction-mediated spherical nucleic acid assembly. Analytical Chemistry, 2020, 92: 2379-2382.

[206] Brückner K, Schwarz K, Beck S, Linscheid M W. DNA quantification via ICP-MS using lanthanide-labeled probes and ligation-mediated amplification. Analytical Chemistry, 2014, 86: 585-591.

[207] Li X T, Chen B B, He M, Hu B. A dual-functional probe for quantification and imaging of intracellular telomerase. Sensor and Actuators B: Chemical, 2018, 277: 164-171.

[208] Wang C Q, Liu R, Hu J Y, Lv Y. Ratiometric DNA walking machine for accurate and amplified bioassay. Chemistry: A European Journal, 2019, 25: 12270-12274.

[209] Iglesias T, Espina M, Sierra L M, Bettmer J, Gonzalez E B, Montes-Bayon M, Sanz-Medel A. Enhanced detection of DNA sequences using end-point PCR amplification and on-line gel electrophoresis (GE)-ICP-MS: Determination of gene copy number variations. Analytical Chemistry, 2014, 86: 11028-11032.

[210] Zhang S X, Liu R, Xing Z, Zhang S C, Zhang X R. Multiplex miRNA assay using lanthanide-tagged probes and duplex-specific nuclease amplification strategy. Chemical Communications, 2016, 52: 14310-14313.

[211] Liu R, Wang C Q, Xu Y M, Hu J Y, Deng D Y, Lv Y. Label-free DNA assay by metal stable isotope detection. Analytical Chemistry, 2017, 89: 13269-13274.

[212] Liu R, Hu J Y, Chen Y X, Jiang M, Lv Y. Label-free nuclease assay with long-term stability. Analytical Chemistry, 2019, 91: 8691-8696.

[213] Liang Y, Jiang X, Yuan R, Zhou Y, Ji C X, Yang L M, Chen H F, Wang Q Q. Metabolism-based click-mediated platform for specific imaging and quantification of cell surface sialic acids. Analytical Chemistry, 2017, 89: 538-543.

[214] Liang Y, Liu Q, Zhou Y, Chen S, Yang L M, Zhu M, Wang Q Q. Counting and recognizing single bacterial cells by a lanthanide-encoding inductively coupled plasma mass spectrometric approach. Analytical Chemistry, 2019, 91: 8341-8349.

[215] Ho K S, Chan W T. Time-resolved ICP-MS measurement for single-cell analysis and on-line cytometry. Journal of Analytical Atomic Spectrometry, 2010, 25: 1114-1122.

[216] Wei X, Hu L L, Chen M L, Yang T, Wang J H. Analysis of the distribution pattern of chromium species in single cells. Analytical Chemistry, 2016, 88: 12437-12444.

[217] Tsang C N, Ho K S, Sun H Z, Chan W T. Tracking bismuth antiulcer drug uptake in single helicobacter pylori cells. Journal of the American Chemical Society, 2011, 133: 7355-7357.

[218] Groombridge A S, Miyashita S I, Fujii S I, Nagasawa K, Okahashi T, Ohata M, Umemura T, Takatsu A, Inagaki K, Chiba K. High sensitive elemental analysis of single yeast cells (*Saccharomyces cerevisiae*) by time-resolved inductively-coupled plasma mass spectrometry using a high efficiency cell introduction system. Analytical Science, 2013, 29: 597-603.

[219] Wang H L, Wang M, Wang B, Zheng L N, Chen H Q, Chai Z F, Feng W Y. Interrogating the variation of element masses and distribution patterns in single cells using ICP-MS with a high efficiency cell introduction system. Analytical and Bioanalytical Chemistry, 2017, 409: 1415-1423.

[220] Cao Y P, Feng J S, Tang L F, Yu C H, Mo G C, Deng B Y. A highly efficient introduction system for single cell-ICP-MS and its application to detection of copper in single human red blood cells. Talanta, 2020, 206: 120174.

[221] Shi J B, Ji X M, Wu Q, Liu H W, Qu G B, Yin Y G, Hu L G, Jiang G B. Tracking mercury in individual tetrahymena using a capillary single-cell inductively coupled plasma mass spectrometry online system. Analytical Chemistry, 2020, 92: 622-627.

[222] Verboket P E, Borovinskaya O, Meyer N, Günther D, Dittrich P S. A new microfluidics-based droplet dispenser for ICP-MS. Analytical Chemistry, 2014, 86: 6012-6018.

[223] Wang H, Chen B B, He M, Hu B. A facile droplet-chip-time-resolved inductively coupled plasma mass spectrometry online system for determination of zinc in single cell. Analytical Chemistry, 2017, 89: 4931-4938.

[224] Wei X, Zheng D H, Cai Y, Jiang R, Chen M L, Yang T, Xu Z R, Yu Y L, Wang J H. High-throughput/high-precision

sampling of single cells into ICPMS for elucidating cellular nanoparticles. Analytical Chemistry, 2018, 90: 14543-14550.

[225] Yu X X, Chen B B, He M, Wang H, Hu B. 3D Droplet-based microfluidic device easily assembled from commercially available modules online coupled with ICPMS for determination of silver in single cell. Analytical Chemistry, 2019, 91: 2869-2875.

[226] Zhang X, Wei X, Men X, Jiang Z, Ye W Q, Chen M L, Yang T, Xu Z R, Wang J H. Inertial-force-assisted, high-throughput, droplet-free, single-cell sampling coupled with ICP-MS for real-time cell analysis. Analytical Chemistry, 2020, 92: 6604-6612.

[227] Zhou Y, Chen Z Q, Zeng J X, Zhang J X, Yu D X, Zhang B, Yan X W, Yang L M, Wang Q Q. Direct infusion ICP-qMS of lined-up single-cell using an oil-free passive microfluidic system. Analytical Chemistry, 2020, 92: 5286-5293.

第5章 基于介质阻挡放电微等离子体的原子光谱分析

(朱振利[1]*，刘 星[1]，杨 春[1])

▶ 5.1 DBD 低温原子化器 / 123

▶ 5.2 DBD 等离子体激发源 / 127

▶ 5.3 DBD 等离子体蒸气发生 / 134

▶ 5.4 其他 DBD 元素分析技术 / 142

▶ 5.5 总结与展望 / 143

---

[1]中国地质大学（武汉），武汉。*通讯作者联系方式：zlzhu@cug.edu.cn

本章导读

- 介绍了基于 DBD 技术的低温原子化器的开发及应用。
- 介绍了基于 DBD 技术的等离子体激发源的开发及应用。
- 介绍了基于 DBD 技术的样品引入策略的开发及应用。

等离子体技术的发展极大地推动了原子光谱分析方法以及仪器的进步，从原子吸收、原子荧光光谱的空心阴极灯（光源）到发射光谱、无机质谱的电感耦合等离子体（ICP）激发源都是等离子体技术在原子光谱分析中的应用。介质阻挡放电（dielectric barrier discharge，DBD）也称无声放电和静默放电，是目前研究热门的等离子体技术。DBD 等离子体结构简单，最早由 Lichtenberg 在 1778 年提出，是指在放电空间中至少有一层绝缘介质插入的一种气体放电，因其放电无明显声音被称之为静默（无声）放电。DBD 常见的放电结构主要包括平板式、线筒式、毛细管式和喷雾式等，图 5-1 给出了几种典型的放电结构示意图。

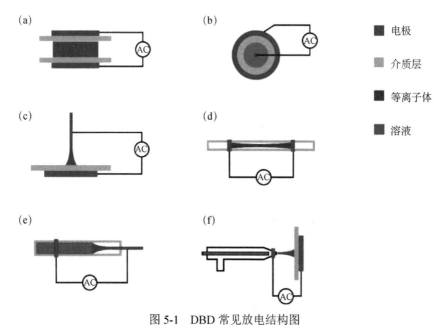

图 5-1 DBD 常见放电结构图
(a, c) 平板式；(b) 线筒式；(d, e) 毛细管式；(f) 喷雾式

DBD 是一种典型的低温等离子体，工作气体通常是氩气、氮气、氦气或者空气。此外，DBD 可在大气压下工作、温度一般不到 100℃无需额外冷却、功率低等特点让其在仪器小型化和原位检测方面体现巨大潜力，近年来引起了国内外学者广泛的研究兴趣，被广泛应用于低温原子化器、激发源、离子源以及等离子体蒸气发生引入技术的开发。本章对 DBD 技术在元素分析方面多年的发展做一简单的归纳和总结，如图 5-2 列出了 DBD 低温原子化器、微等离子体激发源以及 DBD 样品引入方法报道的分析元素，而针对有机物分析的 DBD 离子源并没有涉及。此外，需要指出的是，本章仅试图对 DBD 元素分析的主要文献进行总结，如有相关文献工作遗漏敬请谅解。

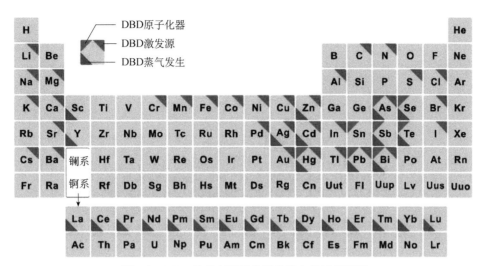

图 5-2　DBD 技术在原子光谱分析中的应用范围

5.1　DBD 低温原子化器

5.1.1　原子吸收光谱 DBD 原子化器

DBD 作为一种低温等离子体，可产生化学性质活跃的自由基、高能电子或离子，可有效地促进不同分子的分解，因此，可用来实现元素的原子化。传统的热原子化器（火焰型或电热型）的原子化温度太高（900℃），需要冷却装置等，难以实现仪器的小型化。相比之下，DBD 作为一种低温原子化器，具有体积小、能耗少等显著特点，在分析仪器的微型化和便携化研究中具有显著的优势。2000 年，DBD 首次被作为原子化器与二极管激光吸收光谱法联用，实现了卤代烃的裂解和氟氯原子的激发。自此之后 DBD 作为氢化物的原子化器也被国内外学者逐渐应用到原子吸收光谱（AAS）和原子荧光光谱（AFS）领域[1]。例如，将平板型 DBD 等离子体（70 mm×15 mm×5 mm）原子化器应用于原子吸收光谱（AAS），实现了无机砷氢化物和有机砷氢化物的低温原子化，在接近室温条件、低功耗下达到与电热石英管原子化器可媲美的检测效果[2]。在此基础上，与高效液相色谱（HPLC）联用（图 5-3），实现了砷的形态分析。该 HPLC-HG-DBD-AAS 系统对 As（Ⅲ）、As（Ⅴ）、甲基砷（MMA）和二甲基砷（DMA）的检出限分别为 1.0 μg/L、11.8 μg/L、2.0 μg/L、18.0 μg/L。该课题组还利用 DBD 实现了 Se、Sb、Sn 三种元素氢化物的原子化，利用 AAS 检测的检出限为 13.0 μg/L、0.6 μg/L、10.6 μg/L[3]。

除了砷的形态分析，研究者们还在该 DBD-AAS 方法的基础上建立了一种非色谱联用直接快速分析鱼肉中无机汞和甲基汞的方法[4]。先通过 $NaBH_4$ 将无机汞和甲基汞（CH_3Hg^+）分别还原成 Hg^0 和 CH_3HgH，DBD 原子化器开启时可有效实现 CH_3HgH 的原子化从而获得无机汞和甲基汞的总信号，因此利用 DBD 低温原子化器比较无机汞的吸光度（DBD 关）以及无机汞与甲基汞的总吸光度（DBD 开）就实现了两种不同形态汞的测定。此外，也有学者基于微顺序注射系统建立了一种检测汞及其形态的 DBD 微型化长光程原子吸收光谱系统，其中同样采用 DBD 低温原子化器来实现 CH_3HgH 氢化物的原子化，并选择长光程流通池作为光谱吸收的检测单元来提高吸收光谱的检测灵敏度（无机汞与甲基汞的检出限分别为 0.3 μg/L 和 0.4 μg/L）[5]。

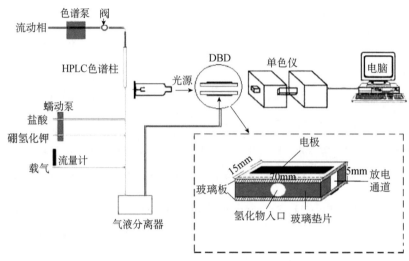

图 5-3　HPLC-HG-DBD-AAS 砷形态分析装置示意图[2]

国外学者后续也在探究 DBD 作为原子化器的机理和性能方面做了较为系统的研究。例如，将商品化原子吸收光谱仪中的石英管原子化器（QTA）替换成内表面有二甲基二氯硅烷（DMDCS）涂层的石英平板型 DBD 原子化器，可实现水中 Bi 的高灵敏检测[6]。研究结果表明 DMDCS 涂层的存在虽可将信号提高 2～4 倍，但使用 DBD 原子化器时的灵敏度仍低于 QTA 原子化器，不过研究发现 Ar 中有无 O_2 明显影响了 DBD 对 Bi 的捕获释放情况。此外，该研究者还比较了三种不同原子化器的 HG-AAS 系统对 Se 的分析效果，如图 5-4 所示，这三种原子化器分别是石英平板型 DBD 原子化器、多重火焰石英原子化器（MMQTA）和外加热石英平板原子化器（EHPA），结果发现 DBD 原子化器对 Se 的检出限（0.24 ng/mL）虽然略高于 MMQTA 原子化器（0.15 ng/mL），但 DBD 原子化器具有更好的稳定性[7]。

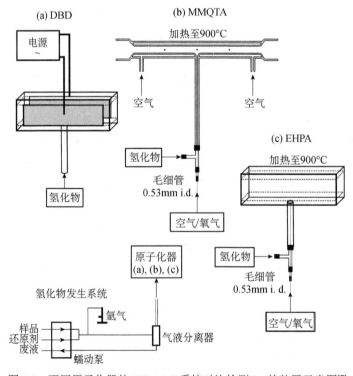

图 5-4　不同原子化器的 HG-AAS 系统对比检测 Se 的装置示意图[7]

石英 DBD 作为原子阱对 As[8]、Sb[9]、Se[10]元素展现出了良好的富集效率,可以大幅改善 HG-DBD-AAS 方法的检出限。研究结果表明 DBD 对于不同元素具有不同的富集效率,比如在 Ar/O$_2$ 气氛下对 As[8]、Sb[9]氢化物的富集效率可高达 100%,但对 Bi[6]、Se[11]氢化物只有 60%~70%。进一步的,研究者通过同位素示踪法研究了 DBD 对 Se 的捕获/释放效率,发现捕获效率大于 90%,但受放电电压限制,释放效率只有 70%[10]。除了使用预富集手段,还可通过改变 DBD 原子化器的操作条件来改善分析方法的灵敏度:有研究证实选择方波而非正弦波,使用电沉积法制作电极而非胶水黏合法,以及利用干燥剂减少进入 DBD 原子化器中气体的湿度都可以改善 Sn 的分析效果[12]。为进一步对比研究 DBD 原子化器和 QTA 原子化器,通过应用激光诱导荧光光谱法(LIF),学者们获得了 DBD 原子化器和 QTA 原子化器在原子化 PbH$_4$ 时两种原子化器内自由 Pb 原子的空间分布情况和原子化效率[13]。如图 5-5 所示,自由 Pb 原子主要在 DBD 中部且原子化效率约为 23%;自由 Pb 原子却均匀分布在整个 QTA 中,且原子化效率更高(88%)。

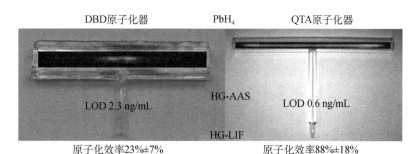

图 5-5 DBD 原子化器和 QTA 原子化器对 PbH$_4$ 的原子化效率[13]

5.1.2 原子荧光光谱的 DBD 原子化器

除了 AAS 的原子化器,DBD 也被应用于 AFS 仪器的原子化器研究。2008 年,有学者将平板型 DBD(长方体放电腔 3 mm×4 mm×50 mm)作为 AFS 系统的原子化器用于 As 的测定,并与氢化物发生-"阀上实验室"(HG-LOV)联用实现了系统的微型化[14]。然而与 AAS 系统不同,平板型 DBD 在非色散的 AFS 系统中会产生较高的背景噪声,该 DBD-AFS 系统对 As 的检出限为 0.03 μg/L,多次测量 2.0 μg/L As 的 RSD 为 2.8%(n=11)。在 AAS 低温 DBD 原子化器的基础上,有学者进一步对 DBD 原子化器的结构进行改进,发展了同心石英管结构的 DBD(图 5-6),该 DBD 的内管通 Ar 载气作为放电反应室,为屏蔽外部空气对等离子体的影响并保护还原的自由原子,外管中也通入 Ar 气作为屏蔽气,该 HG-DBD-AFS 系统对 As 的检出限为 0.04 μg/L,11 次测量 10 μg/L 和 50 μg/L As 的 RSD 分别为 1.1%和 1.6%[15]。此后,朱振利等[16]接着证实了该 DBD 原子化器还拓展到了其他元素,成功测定了样品中的 Se、Pb[16]、Bi[17]、As、Sb[18]、Te[19],进一步证实了 DBD 原子化器的潜力。

国内学者也利用 AFS 系统详细研究了 DBD 原子化器对不同元素的捕集释放效果。研究者将双同心石英管 DBD 与 HG-AFS 系统联用,发现 HG 生成的 As 的氢化物被可以 O$_2$/Ar 气氛 9.2 kV 放电的 DBD 有效捕获,而简单将 DBD 放电条件变为 H$_2$/Ar 气氛和 9.5 kV,就可将捕获的分析物有效释放,从而将 As 信号提高了 8 倍[20]。随后,使用三层同心石英管的优化设计,并且改进气路系统,研发出性能更好的新型原位 DBD 原子阱(绝对信号灵敏度提高了 4 倍)[21]。该三层同心石英管 DBD 的示意图见图 5-7(a),内管中的金属电极用作高压电极,中间管的外管壁上连接接地电极,而内管和中间管之间的石英层充当介质阻挡层。在中间管和外管之间通入屏蔽气,使等离子体与空气隔离,可

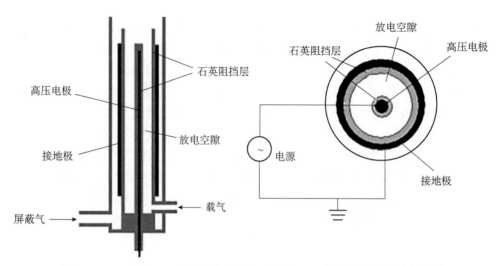

图 5-6　HG-DBD-AFS 系统中的双同心石英管 DBD 原子化器结构示意图[15]

（左）横向视图；（右）俯视图

避免荧光猝灭和空气氧化。DBD 原子阱可视为一种简单、高效、低耗的气相富集（GPE）技术，可以有效地消除基体干扰，对于基体简单的样品，DBD 原子阱的应用可实现其在无需样品前处理情况下就被快速灵敏检测。进一步的，有学者将三层同心石英管 DBD-AFS 系统与悬浮液氢化物发生进样（SLS-HG）联用，成功检测出微生物样品中的痕量 As。该 SLS-HG-原位 DBD-AFS 方法灵敏度高，且几乎不用样品前处理，可实现对生物样品中 As 的快速分析[22]。该作者还将三层同心石英管 DBD-AFS 系统与直接进样氢化物发生（DS-HG）联用，实现对血液中痕量 As 的直接检测[23]。基于 DBD 的高效富集优势，将紫外蒸气发生（UVG）技术与 DBD-AFS 系统联用还可实现 Se 的高灵敏测定［图 5-7（b）］。DBD 的富集使方法灵敏度提高了约 16 倍，方法的检测限（LOD）为 4 pg（0.004 μg/L）。文献中还使用原位光纤光谱仪在 DBD 出口检测光谱，XRF 检测 DBD 内管壁元素，以及 $Pb(CH_3COO)_2$ 试纸检测等方法，研究了硒在捕获、释放和运输中的机理：Se 溶液经 UVG 产生 SeCO 和 H_2Se 物质，SeCO 和 H_2Se 物质在 O_2-Ar 气氛下反应被氧化为 SeO_2 或亚硒酸盐后被捕获在石英表面，最后在 H_2-Ar 气氛下从石英表面释放，进入到 AFS 中检测[24]。

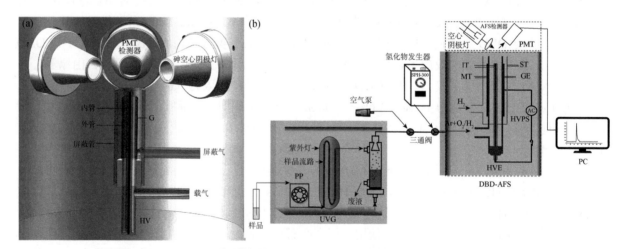

图 5-7　（a）DBD 原子阱[21] 和（b）UVG-DBD-AFS 系统实验装置示意图[24]

除了元素浓度分析，还有研究者结合色谱利用 DBD 原子化器实现了形态分析。例如，有学者在 GC-AFS 系统基础上借助 DBD 原子化器建立了吹扫预富集测定海水中的甲基汞的方法，海水经

NaBEt$_4$ 衍生化后，通入 N$_2$ 将衍生物吹扫捕集，接着在加热捕集阱的同时通入 Ar 载气将释放出的衍生物带入到 GC 中进行分离，分离后的气态甲基乙基汞被 DBD 原子化器还原成 Hg^0，最后用 AFS 检测。该 GC-DBD-AFS 方法对甲基汞的 LOD 为 0.0008 ng/L[25]。相较于高温的加热型原子化器，虽然 DBD 的激发能力和原子化效率有限，但仍可以满足多数常见元素（如 As、Se、Sb、Sn、Hg、Bi、Pb、Te）的原子化，且具有较好的稳定性。另外，DBD 作为原子化器向原子阱延伸的发展过程中，通过改变 DBD 电压、气氛等放电条件可实现其对 As、Se、Sb、Bi 等元素的捕集释放，提高了分析方法的灵敏度和抗干扰能力，简单、低耗、易控、易于小型化，具有发展成为一种新的高效气相预富集技术的潜力。DBD 原子化器在仪器微型化的发展中展现了巨大的应用潜力，但其性能尚有提升空间，须扩大分析样品的应用范围。若可通过简单的 DBD 操作直接实现对复杂基体样品中分析元素的原子化或预富集，以满足实际样品分析的需要，将在现场分析应用中发挥重要意义。

5.2 DBD 等离子体激发源

原子发射光谱分析是一种重要的元素分析技术，在生产生活中仍应用广泛。传统原子光谱分析激发源如 ICP 激发源具备多元素同时分析、高精度、高灵敏度等优点，但价格昂贵、维护复杂、功耗高、惰性气体消耗高等限制了其现场、高效、低成本的分析应用，而微等离子体作为激发光源具有功耗低、尺寸小、激发能力强、易于维护等特点，非常适合发展现场、高效、低成本分析检测技术。DBD 等离子体除了作为 AAS 或 AFS 的原子化器使用，也可以作为激发源实现发射光谱的检测，而且与 ICP 等源相比，DBD 产生的连续背景发射大幅降低进一步改善了信噪比，很多研究证实了 DBD 激发源在发射光谱分析上的潜力。基于 DBD 等离子体激发源的发射光谱检测系统，结构简单、易于搭建且分析性能优异，自 2008 年首次用于 Hg 元素检测以来成为元素分析仪器小型化的研究热点。DBD 可以直接分析气态物种，通过与多种进样方式如化学蒸气发生（CVG）、氢化物发生（HG）、紫外光发生（PVG）、电热蒸发（ETV）等联用大幅扩展了 DBD 作为激发源的应用范围和分析效果。此外，DBD 作为色谱检测器与色谱联用也备受关注，溶液样品 DBD 直接检测也有诸多报道。

5.2.1 气态物种进样

CVG 是最早被用于 DBD-OES 的样品引入技术，朱振利[26]和于永亮[27]等几乎同时开展了 DBD 激发源的研究工作。图 5-8（a）为基于 CVG 进样的大气压平板 DBD 激发源（尺寸：0.6×1×10 mm^3，功耗<1 W），通过发射光谱检测了 Hg[26]。通过 CVG 进样将溶液中的 Hg 离子转化为单质 Hg 并在自制气液分离器中完成分析物与废液分离，未除水的单质 Hg 由载气（Ar 或 He）携带进入 DBD 激发源激发检测。该装置在不同载气下 Hg 检出限分别为 14 ng/L（He-DBD）和 43 ng/L（Ar-DBD），该工作中采用了 0.5 m 焦距的光谱仪。图 5-8（b）则展示了另一种以 QE65000 微型光谱仪作为检测器的平板型 DBD-OES 微等离子体检测技术，实现了 Hg 元素的检测[27]。由注射泵和六通阀组成顺序注射的进样系统可节省样品用量，实验表明 5% NaCl 以及 5 mg/L 的 Cd^{2+}、Fe^{3+}、Cu^{2+}、Zn^{2+}、Cr^{3+}、Ni^{2+}、Pb^{2+}、Sb^{3+}、AsO$_2^-$ 和 SeO$_3^{2-}$ 不会对 Hg 的检测产生明显干扰，通过标准物质的检测也验证了该方法的准确和可靠性，Hg 的检测限为 0.2 μg/L。此外，也有学者使用显微镜玻璃片、有机玻璃和铝箔电极组成了易于微型化和制造 DBD 放电腔，使用 SnCl$_2$ 还原含 Hg 样品得到汞蒸气进入 DBD 检

测，检出限为 2.8 μg/L，并对等离子体特性进行表征，其电子温度>激发温度>振动温度，表明产生的为非热（non-thermal）等离子体[28]。

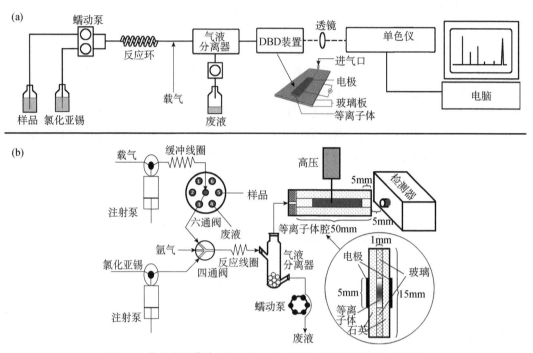

图 5-8 化学蒸气发生 DBD-OES 汞元素分析的装置示意图[26,27]

为了实现气体中 NH_3 的检测，有研究者设计了封闭式的平板 DBD，该装置能避免空气中组分尤其是含氮物种对检测 NH_3 时产生光谱干扰[29]。图 5-9 展示的是一种基于微空心阴极放电（MHCD）的介质阻挡-微空心阴极放电（DB-MHCD）[30]，其放电腔孔径为 200 μm，厚度为 300 μm，由脉冲电源供能维持等离子体，该装置能实现样品无损检测并连续工作几天，其检测 He 气中气相 Cl 的检出限为 27 μg/L，相比于 MHCD，DB-MHCD 放电更加稳定。类似的封闭式毛细管 DBD 结构还被用于气相中 H_2S 和液相中 S^{2-} 的检测，其检出限分别为 1.4 mg/m^3 和 11.2 mg/L[31]。

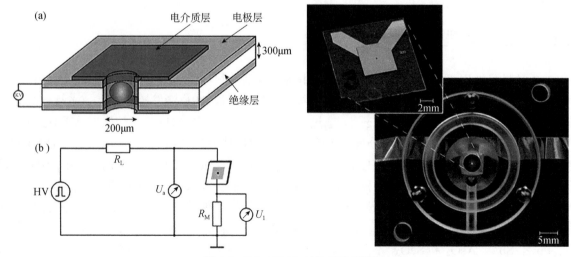

图 5-9 DB-MHCD 结构示意图[30]

基于 CVG 进样的毛细管 DBD 激发源还可用于卤素及其化合物的准确测定[32,33]。以 H_2O_2 与碘化物反应得到 I_2 蒸气由载气引入毛细管 DBD 激发源检测,该方法 I(905 nm)检出限为 30 μg/L 并成功用于 GBW10023 紫菜、食用盐及西地碘含片等实际样品检测。进一步地,DBD-OES 被应用于溴酸盐和溴离子的检测:需先将溶液中溴酸盐预还原成溴离子再使用强氧化剂氧化 Br^- 得到 Br_2 蒸气,结果表明 $KMnO_4/H_2SO_4$ 的氧化剂体系比 $K_2S_2O_8$ 和 H_2O_2 体系反应速率更快,可直接在线检测,不需预先反应收集再引入 DBD 检测。方法证实基体中 3000 mg/L 的 Na^+、K^+、Mg^{2+}、Al^{3+}、Fe^{3+}、Cu^{2+}、Zn^{2+}、NO_3^- 和 SO_4^{2-} 对 Br 的检测无明显干扰,因此该方法可以快速筛查环境水样中溴酸盐和溴离子的污染。基于 $KMnO_4$ CVG 进样与 DBD-OES 相结合的方法,可实现 Cl(837 nm)、Br(827 nm)、I(905 nm)的同时检测,并展现出了对于共存碱金属、碱土金属以及过渡金属元素离子较高的抗干扰能力,成功地对标准认证物(CRM)和海水实际样品进行准确的定量分析[34]。

此外,还有学者利用微波辅助氧化环境水样含碳有机物,并将产物 CO_2 引入 DBD-OES 获取 C 原子特征发射光谱从而实现原位、简单、快速水样总有机碳的检测,其检出限(以 C 计)为 0.01 mg/L,该装置实际水样总有机碳检测结果与商品化仪器检测一致,表明检测准确可靠[35]。基于 DBD-分子发射(MES)法在测定水体中总硫含量时展现出了潜力,实验中首先将水体中的 S 转化为 H_2S 引入到 DBD 中激发,通过对 H_2S 的分子发射峰(301.9 nm)的监控,间接获取水体中硫含量的信息,并且通过超声加热进一步提高了 H_2S 的气体产率,检出限可达 0.10 mg/L,可对溶液进行在线长时间持续的监控工作[36]。该 DBD 装置还实现了食物中总 SO_2 的分析,开发了一种快速、准确、无损的低成本食品中 SO_2 的分析方法,通过 SO_2 分子发射峰(301.9 nm),成功定量分析了葡萄酒、糖、水果蔬菜等食品中的总 SO_2 含量,检出限可达 0.01 mg/L[37]。

HG 进样属于 CVG 进样一种,但因其应用广泛而单独讨论。通常 HG 进样伴生的氢气会导致低功率 DBD 等离子体不稳定甚至熄灭,从而使 HG 和 DBD-OES 联用具有挑战性。朱振利等[38]将氢化物发生(HG)与 DBD-OES 联用首次实现 As 元素检测,如图 5-10(a)设计的线筒式 DBD(<30 W)内外电极相距约 5 mm,内电极为钨棒、外电极为铜线圈,该 DBD 结构可以很好地与 HG 耦合产生稳定的等离子体并且观测到 5 条 As 原子发射线(193.7 nm、197.2 nm、200.3 nm、228.8 nm 和 234.9 nm)。最优条件下 As 检出限为 4.8 μg/L,可以基本满足水样中 As 的检测。此外该 DBD 结构有望用于 Se、Sb、Cd 等氢化物发生元素检测。进一步地,为了提高 DBD 激发源的分析效果,对 As 在毛细管 DBD 中的发射信号进行了时空分辨研究[图 5-10(b)],通过增强电荷耦合检测器(ICCD)可获取某一时刻 As 最佳信噪比信号,As 的检出限为 93 ng/L[39]。之后,分别对两种不同放电气体(Ar 和 He)下 As 发射的时空分辨进行表征[40],间接提供了 As 在整个放电通道上被原子化激发的证据,解释了 DBD 在原子吸收光谱测量中的良好适用性。

PVG 是一种简单绿色的进样技术,也被作为一种进样方法用于 DBD 激发源元素分析。如图 5-11(a)所示,PVG-DBD-OES 技术可用于直接检测疫苗中硫柳汞,平板 DBD 功率≤18 W,待测样品检测前仅需加入甲酸,可以实现每小时 30 个样品的高通量检测[41]。考虑到疫苗基体的基质效应,该方法验证了一定范围内 NaCl、Na_2HPO_4、NaH_2PO_4、Al^{3+}、Co^{2+}、Ni^{2+} 等对 Hg 检测的干扰,结果表明无显著影响,其对 Hg^{2+} 和硫柳汞(Hg)检出限分别为 0.19 μg/L 和 0.17 μg/L。此外,还可利用石英材质的紫外灯反应器将待测溶液中的 Ni 转化成挥发性 $Ni(CO)_4$[图 5-11(b)],经由气液分离器后 $Ni(CO)_4$ 随载气进入毛细管 DBD 激发检测,其中 Ni 检出限为 1.3 μg/L,实验表明 PVG 金属羰基化进样不会产生氢气,不会引起 DBD 等离子体熄灭等问题,因此与 DBD 微等离子体源联用潜力巨大[42]。

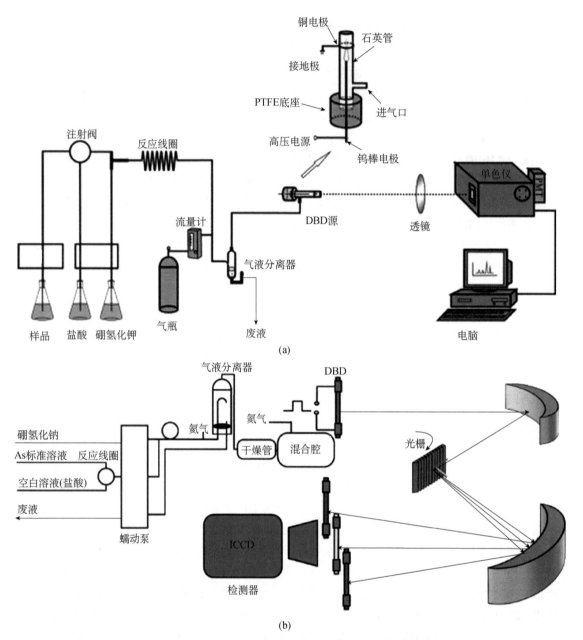

图 5-10 HG-DBD-OES 砷元素分析的装置示意图[38,39]

ETV 进样可以实现微量液体样品高效引入，避免液体样品直接进入 DBD 时负载大而导致灵敏度受限甚至熄灭，一般通过小型钨丝电热蒸发实现引入。如图 5-12（a）所示，将 10 μL 样品加到钨丝上，通过钨丝分步升温先去溶剂再原子化/蒸发待测元素，可实现基体与待测元素的分离从而减少基体干扰，随后分析物由 He 气带入 DBD 激发获取特征发射光谱成功对水样中的 Cd 和 Zn 进行了准确测定[43]。研究发现，在 DBD 外缠绕加热丝辅助加热可以降低水汽影响提升 DBD 激发能，而 Cd 和 Zn 检出限分别为 0.8 μg/L(0.008 ng)和 24 μg/L(0.24 ng)。此外，还有学者[44]通过钨丝电热蒸发将 20 μL 含 Pb 样品蒸发引入 DBD-OES 检测，干扰离子实验表明 50 mg/L 的 Ba^{2+}、Ca^{2+}、Cd^{2+}、Co^{2+}、Cr^{3+}、Cu^{2+}、Fe^{3+}、K^+、Mg^{2+}、Mn^{2+}、Na^+和 Zn^{2+}对 0.1 mg/L 的 Pb 检测无明显影响，Pb 的检出限为 7.7 μg/L，单个样品检测时间约 3 min，表明该方法可用于环境水体中 Pb 的快速、灵敏检测[图 5-12（b）]。

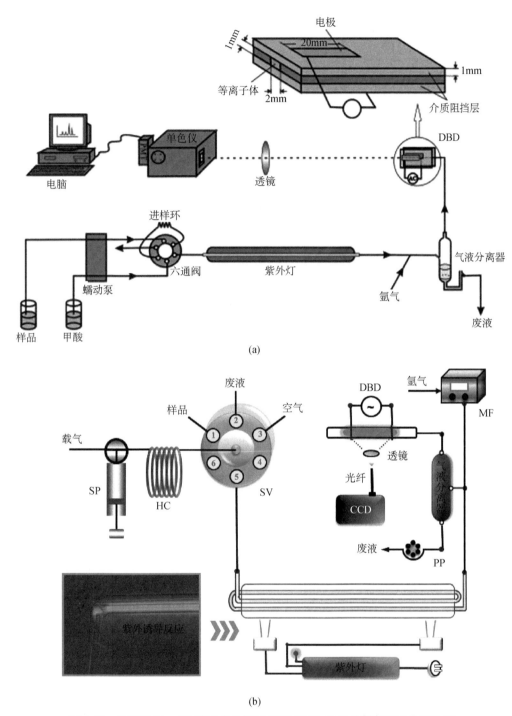

图 5-11 应用于 Hg、Ni 等元素分析的 PVG-DBD-OES 分析装置示意图[41,42]

ETV 耦合 DBD-MES 还可对液相和气相中的丙酮（分析线 519 nm）进行准确的定量分析，通过微量（7 μL）液体进样，在没有额外的色谱柱进行分离的情况下得到准确的分析结果，检出限可以达到 4.4 ng[45]。姜杰课题组[46]研制了一种 ETV-DBD-OES 便携式仪器，仪器总质量 4.5 kg，功耗 37 W，在较短的分析时间（3 min）和微量样品消耗（3 μL）下，可同时进行多元素分析，其中 Zn、Pb、Ag、Cd、Au、Cu、Mn、Fe、Cr 和 As 的检出限范围为 0.16～11.65 μg/L。随后，该仪器成功对大米中的 Cd 进行了快速定量分析，整个样品分析流程不到 11 min，其检出限满足欧盟检测标准[47]。此外，金汞齐富集电热释放也被应用于 DBD 发射光谱汞的灵敏检测。例如，通过涂覆金的钨丝捕获单元

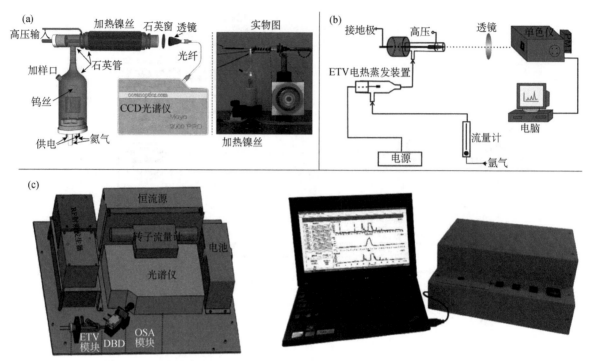

图 5-12 已报道的不同 ETV-DBD-OES 系统的装置示意图及样机照片[43,44,46]

可以预富集大气中的 Hg^0（预富集时间为 2 min）并排除水汽对 Hg 激发检测的影响从而大幅改善了 DBD-OES 检测 Hg 的灵敏度，检出限达 0.12 ng/L（Ar-DBD），最终搭建了一款自动化大气 Hg 分析仪[48]。此外，也有研究通过金汞齐预富集后热解析释放的方法对烟气中的 Hg^0 捕获后再利用 DBD-OES 检测 Hg，其毛细管 DBD 为圆柱陶瓷管（内径 1 mm，外径 2 mm），并与金属电极组成（两电极间距 23 mm）的封闭放电结构，该装置可以很好地避免空气及烟气中无机气态小分子对 Hg 检测产生的光谱干扰[49]。

DBD 作为色谱检测器用于有机物及金属元素形态分析也备受关注。有学者设计了线筒式 DBD 装置，由铜内电极（直径 1.7 mm）、铜线圈外电极和石英管组成，并在 DBD 前端连接了 GC 实现不同卤代烃的分离检测，该装置成功用于 CCl_4、$CHCl_3$、CH_2Cl_2、CH_3I、CH_3CH_2I 和 $(CH_3)_2CHI$ 检测，这表明 DBD 是很有前景的 GC 检测器，此外，也有学者将线筒式 DBD-MES 应用于苯、甲苯、二甲苯（BTX）的检测，并且发现 DBD 电极表面介质层涂覆纳米 MnO_2 对 BTX 的检测有显著的增敏效果，其 BTX 检出限范围为 0.2~0.5 μg/L[51]。在前期研究基础上，研究者们将线筒式 DBD-OES 与 GC 联用首次用于含碳有机物检测，观察到 C 193.0 nm 和 247.8 nm 的原子发射线，并且 DBD 加热（300℃）可以大幅提高 C 的特征发射，改善分析性能，DBD-OES 与 GC 联用可以广泛用于挥发性有机物检测，并且能检测 GC-FID 难以检测的 HCHO、CO 和 CO_2[52]。在环境领域，对于重金属的形态分析越来越引起人们的注意，如图 5-13 所示，将 DBD-OES 作为气相色谱（GC）的检测器与顶空固相微萃取结合可应用于对水稻中不同形态的 Hg 的准确定量分析[53]。DBD 的两个电极材料均为 Cu，总功率约为 12 W，Ar 载气流速维持在 200 mL/min，在低功耗、低气耗的条件下对溶液中的无机汞（Hg^{2+}）、甲基汞（CH_3Hg^+）和乙基汞（$CH_3CH_2Hg^+$）进行了准确的定量分析，检出限分别可以达到 0.5 μg/L、0.75 μg/L 和 1.0 μg/L。使用该 DBD 装置学者们还完成了对 Hg 的简单快速非色谱形态分析，通过 Ar 中的微量 N_2 与甲基汞中的 C 络合，生成 CN 分子发射峰（358.3 nm），通过对 CN 的定量分析，确定甲基汞的含量，采用此方法可同时完成对总汞、甲基汞和无机汞的定量分析[54]。利用 Hg^{2+} 衍生化对福美双进行标记，福美双中的二硫键（—S—S—）被还原为硫基（—SH），可实现瓜果中福美双的高灵敏、精准定量分析，这种新方法检出限比传统的 GC 和 LC-MS 低一个数量级，拓展了

DBD-OES 在大分子有机农药中的应用[55]。随后，DBD-OES 激发源还被用作液相色谱的检测器，对蔬菜水果中的二硫代氨基甲酸盐（DTCs）进行了准确的定量分析，与电子捕获检测器相比，检出限相对较高，但可以克服传统方法无法区分不同DTCs亚类分析物的缺陷[56]。

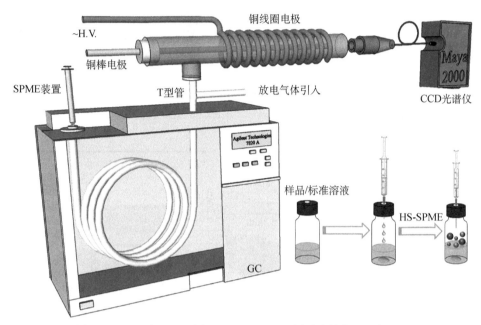

图 5-13　汞元素形态分析 GC-DBD-OES 分析系统的装置示意图[53]

5.2.2　气溶胶/液体进样

气溶胶/液体样品直接引入是光谱分析中最常用的进样技术，也有很多研究考察了气溶胶/液体直接进样下 DBD 激发源的效果。图 5-14（a）展示了一种液体电极介质阻挡放电（LE-DBD）微等离子体激发源，铜箔外电极包裹毛细管上，内电极为钨丝，LE-DBD 电极并不与溶液直接接触能很好避免电化学反应产生气泡的影响。LE-DBD-OES 成功用于溶液中 Sr、Pb 和 Hg 检测，检出限分别为 18 mg/L、40 mg/L 和 42 mg/L[57]。

此后，学者们进一步改进了 LE-DBD 及进样流速，进样流速为 20 μL/min 且低于溶液在等离子体区的消耗，溶液先是与钨电极间距为 200 μm 时自动点燃然后随着溶液消耗等离子体区拉长直至熄灭，这样获得了周期脉冲的放电循环，从而利于维持放电的长期稳定性[58]。随后，该课题组探讨了 LE-DBD 的工作机理，表明类似电喷射过程是仅次于热蒸发过程导致 LE-DBD 中溶液向等离子体传输的重要路径[59]。还有学者设计了一种介质阻挡放电射流（DBD-pencil）激发源，气动雾化产生的样品气溶胶由 Ar 携带进入 DBD-pencil 并在射流区采集发射光谱检测，其转动温度和激发温度分别为 800 K 和 4000 K，多种金属元素检出限分别为 27 μg/L(Ca)、49 μg/L(Cu)、58 μg/L(Mg)、40 μg/L(Li)、13 μg/L(Na)、180 μg/L(Zn)[60]。

液膜介质阻挡放电（LF-DBD）是一种静态式放电装置，避免了进样泵的使用，简化了装置结构及缩小装置体积[61]。如图 5-14（b）所示，LF-DBD 由钨丝、载玻片及铜箔组成，铜箔黏覆在载玻片下方，载玻片凹槽（直径 15 mm，深 5 mm）用于盛放样品，开始工作是钨丝距离液面 2 mm 自动点燃维持稳定等离子体，随着样品的逐渐消耗，放电间距拉大直至熄灭，一个放电流程一般在 60 s 以内。该装置成功用于水体中多种金属元素检测，其中 Na、K、Cu、Zn、Cd 的检测限分别为 7 μg/L、25 μg/L、74 μg/L、79 μg/L、38 μg/L。LF-DBD 功耗≤18 W，无需辅助气，样品消耗≤80 μL 且 1 min

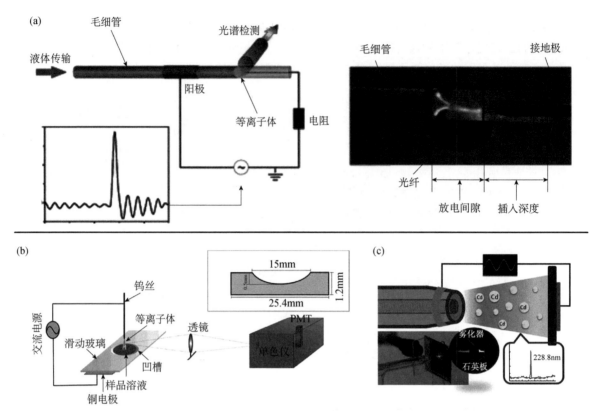

图 5-14 溶液/气溶胶 DBD-OES 分析系统的（a）LE-DBD、（b）LF-DBD 和（c）LS-DBD 装置示意图[57,61,62]

内可以实现多元素分析，用于发展小型化现场分析设备前景广阔。图 5-14（c）展示了一种将电极线圈直接套在气动雾化器出口附近设计了喷雾 DBD（LS-DBD）装置，出口右侧为贴有电极的石英板，利用其中产生的 DBD 放电实现了对液体喷雾中的 Cd 的高灵敏度直接分析，并通过添加小分子有机物有效地提升了发射信号的灵敏度，对某些元素获得了与 ICP-OES 相当的检出限[62]。该团队通过阀内微珠注射进样和柱后衍生化提供效果显著的富集作用，将 Cd 的检出限降低为 0.06 μg/L[63]。此后，又采用这种气动雾化 DBD 对液体样品中的 Zn、Cd、Hg 进行了直接分析，研究了基体匹配方法解决基体干扰的良好效果，并探究了乙醇浓度、溶液电导率和 pH 值对发射信号灵敏度的影响，进一步扩大了 LS-DBD-OES 的应用范围[64]。

5.3 DBD 等离子体蒸气发生

众所周知，进样过程在很大程度上决定了元素分析的灵敏度和准确性[65]。CVG 方法由于其较高的灵敏度和选择性而被广泛应用于原子光谱和质谱分析[66]。尽管常规 CVG 方法在元素和同位素分析中得到了广泛的应用，但还存在一些限制，包括易受到过渡金属干扰、需消耗大量昂贵且不稳定化学试剂，这些都限制了其进一步的应用[67]。

等离子体与液体接触过程可诱导发生等离子体化学反应生成大量活性物质，如自由基、离子和分子（O、O_3、$^1O_2$、•OH、•H、•NO、H_2O_2、NO_2^- 和 NO_3^- 等）[68,69]。这些活性物质溶解于液体中，可与目标反应物相互作用，引发一系列等离子体化学氧化还原反应。近年来，研究人员开发了许多基于等离子体技术的蒸气发生方法，其蒸气发生效率高、化学动力学速度快，且避免了使用化学还原/氧化试剂[67,70,71]。其中，DBD 技术由于器件结构简单、易于制造等优点，在等离子体蒸气发生方法开发研究领域也取得了较好的研究成果。迄今为止，DBD-CVG 技术已成功地应用于 As、Sb、Bi、Se、Te、Hg、Pb、Cd、Zn、Ag、Sc、Y、La、Ce、Pr、Nd、Sm、Eu、Gd、Tb、Dy、Ho、Er、Tm、Yb 和 Lu 等元素的蒸气发生（图 5-15）。下面主要总结了 DBD-CVG 技术的发展和应用，包括等离子体反应器的设计、反应条件、反应机理和产物、共存离子干扰、应用范围、优缺点和未来可能的发展方向。

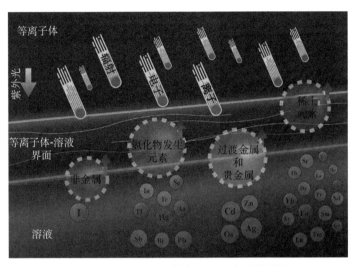

图 5-15　DBD-CVG 技术的元素应用范围[67]

5.3.1　DBD-CVG 等离子体反应器

研究表明，等离子体反应器的设计对蒸气发生具有非常显著的影响。自 DBD-CVG 报道以来，研究者相继发展了同轴式、液膜式、喷雾式等多种结构，有效改善了蒸气发生效果和分析元素的范围。2011 年，侯贤灯课题组[72]提出了一种基于同轴线筒式 DBD 装置的等离子体 CVG 技术，并实现了溶液中 Hg 的高效蒸气发生。该 DBD 装置由两个同轴石英管组成，作为双介质阻挡层，如图 5-16（a）所示。两根铜线分别插入石英内管及缠绕在石英外管外侧，作为放电电极。由于内电极直接插入石英内管中，既可以避免电极对反应溶液带来污染的潜在风险，同时还可以防止铜线被溶液腐蚀起到保护电极的作用。研究发现在含 Hg 溶液中加入不同反应介质，直接影响着 Hg 的分析灵敏度[图 5-16（b）]。其中，当样品中加入 10%甲酸时，与传统的 PN-ICP-OES 系统相比，采用 DBD-CVG 技术 Hg 的信号强度提高了 38 倍，Hg 的检出限可以低至 0.09 μg/L，为水样中痕量 Hg 的测定提供了一种灵敏的方法。接下来研究工作中，通过该 DBD 装置还实现了 Se(Ⅳ)蒸气的产生，信号强度比传统溶液雾化（PN）提高了 10 倍以上[73]。值得注意的是，这种方法不能实现 Se(Ⅵ)的蒸气生成。虽然所提出的 DBD-CVG 技术比传统的 PN 样品引入系统具有更高的引入效率，但应用范围仍然有限，仅被用于 Hg 和 Se(Ⅳ)的测定。

考虑到 DBD-CVG 技术是一种具有潜力的高效、绿色的进样技术，研究者们在 DBD-CVG 方法开发上做出了不同的尝试。非流动液体薄膜 DBD（TFDBD）蒸气发生技术摒弃了额外的样品引入装置，通过与 AFS 联用成功测定了 Hg^{2+}、MeHg、EtHg 和硫柳汞，检出限分别为 0.02 μg/L、0.02 μg/L、0.02 μg/L 和 0.06 μg/L[74,75]。与同轴线筒式 DBD 反应器相比[72]，二者在样品的引入方式上存在显著的差别。如图 5-17 所示，该装置的内石英管可拆卸，分析过程中，首先使用石英管蘸取样品溶液。

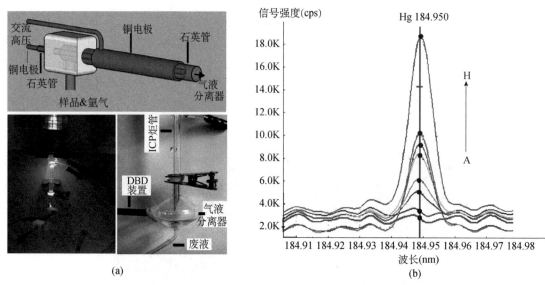

图 5-16 （a）同轴 DBD-CVG 装置示意图及实物照片；（b）不同模式下的 Hg 蒸气发生效率[72]

在 DBD 等离子体存在下，石英管表面样品中的 Hg 迅速反应生成 Hg 蒸气。由于样品以非流动液体薄膜的形式吸附在石英内管上，样品与等离子体的反应效率较高，适合于有机汞和无机汞的高灵敏度检测。这种非流动 TFDBD 装置只需要 6 μL 的样品溶液，就可用于微量样品中微量 Hg 的测定。

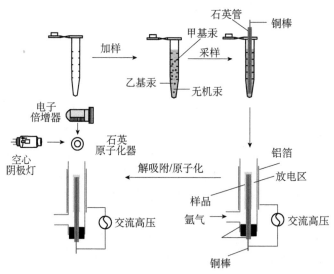

图 5-17 非流动进样 TFDBD-CVG 装置示意图[74]

朱振利等[76-79]进一步设计了一种采用流动进样的 TFDBD 反应器，并成功地实现了 Cd、Hg、Zn、MeHg 和 EtHg 的蒸气发生。图 5-18（a）为流动式进样的 TFDBD 装置的示意图。溶液从 DBD 等离子体反应器顶部引入时，可在内层玻璃介质阻挡层的外表面形成流动的液膜。如图 5-18（b）所示，无论石英原子化器处于关闭或者打开状态都能获得一定的 Cd 荧光信号，但是当原子化器处于开启状态时，Cd 荧光信号发生了显著的增加，这意味着 Cd 的挥发性物种主要是气态分子化合物[76,78]。该课题组的工作还证实 DBD-CVG 反应中生成的 Hg 和 Zn 的挥发性物种是原子（Hg^0 和 Zn^0）[77,78]。需要指出的是，该 TFDBD-CVG 中，需要添加氢气作为反应辅助气体，才能实现 Cd 和 Zn 的高效蒸气发生。随后的研究中，还发现，非离子表面活性剂的存在可以使 Cd 和 Hg 的信号强度分别提高 5.4 倍和 5.1 倍，可能的原因是非离子表面活性剂的存在改善了反应动力学过程以及挥发性物质从溶液进入气相的转移效率[78]。除了元素分析，TFDBD-CVG 还可作为 HPLC 和 AFS 的接口，并成功用于 Hg^{2+}、MeHg 和 EtHg 的形态分析[79]。由于 DBD-CVG 技术具有设备简单、功耗低的优点，还可将流动式进样的 TFDBD-CVG 与一种小型化的 DBD-OES 系统结合，用于水样中痕量 Hg 的测定[80]。

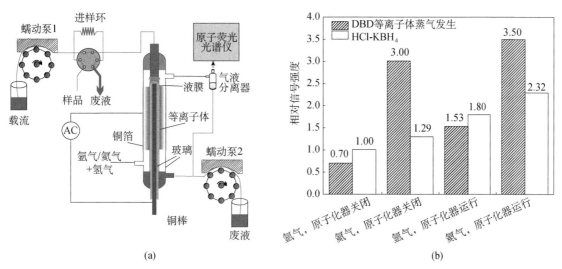

图 5-18 （a）流动进样 TFDBD-CVG 装置示意图；（b）不同条件下的 Cd 蒸气发生效率对比[76]

液膜式 DBD 等离子体条件下，利用氢气做反应气使得难以常压下发生的反应得以发生，甚至实现了活泼 Zn 的蒸气发生，拓展了蒸气发生的元素范围，但蒸气发生元素范围仍然有限，研究者们还发展了液体喷雾式 DBD 反应器。2014 年，清华大学的研究者们[81]对 DBD 反应装置进行了创新，该方法首先通过雾化器将样品溶液雾化成气溶胶，随后气溶胶与氢气混合进入同轴型 DBD 装置发生等离子体化学反应生成挥发性物质。如图 5-19 所示，显著区别于前期的 DBD-CVG 装置，在该系统中，氩不仅作为放电气体产生等离子体，同时还用于溶液的雾化，以增加样品与等离子体的反应效率。同时，与使用 TF-DBD 实现 Cd[76]和 Zn[77]的蒸气发生相似，实验中加入了氢气（H_2/Ar～37.5%）以辅助等离子体化学反应的进行，最终成功实现了 As、Te、Se 和 Sb 四种元素的蒸气发生[图 5-19（b）]。通过原子荧光光谱仪测得 As、Te、Se 和 Sb 的检出限分别为 30 μg/L、50 μg/L、70 μg/L 和 60 μg/L。此外，该方法可与高效液相色谱法（HPLC）联用，用于测定 As(Ⅲ)、As(Ⅴ)、一甲基砷酸（MMA）和二甲基亚砷酸（DMA）。该方法具有样品消耗少的潜在优势，但其灵敏度比传统的 CVG 方法低 10～100 倍，这可能是因为雾化器的进样效率通常较低，而该装置中样品的雾化过程与等离子体反应过程相对独立，这可能导致大多数样品并不能进入等离子体区发生反应。

得益于这种气溶胶进样模式下溶液与等离子体之间的有效接触，液体喷雾 DBD 还被成功应用于稀土元素的检测[82]，包括 Sc、Y、La、Ce、Pr、Nd、Sm、Eu、Gd、Tb、Dy、Ho、Er、Tm、Yb 和 Lu 等，检测灵敏度提升 4～10 倍并有良好的测定精度。尽管 16 种稀土元素的样品引入效率在 22.1%±0.1%（La）和 30.9%±0.4%（Er）之间，但是考虑到稀土元素往往难以实现蒸气发生，雾化进样本就可以实现一定程度的样品引入，而气溶胶在等离子体环境中发生库仑爆炸同样可能进一步提高样品引入效率，因此对这 16 种稀土元素是否确定发生了等离子体化学反应形成了挥发性物质仍需要进一步的证据确定。

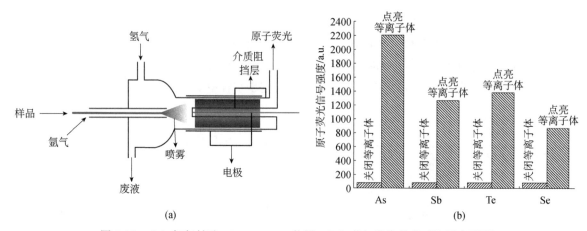

图 5-19 (a) 氢气辅助 LSDBD-CVG 装置;(b) 蒸气发生效率对比示意图[81]

考虑到 DBD 等离子体反应系统中,气溶胶进样模式下溶液与等离子体之间具有更高效接触,可能带来更高的蒸气发生效率。图 5-20 为一种新型液体喷雾 DBD(LSDBD)蒸气发生反应器的示意图[83]。该 LSDBD 反应器中,缠绕在同心雾化器喷嘴外围的铜箔用作内电极,黏附在圆柱形反应器底部玻璃平台外侧的铜箔充当外电极。两个电极之间的距离设定为 3 mm 左右。在雾化器引入样品后,在两个电极之间形成了锥形气溶胶区,接通交流高压电源即可在气溶胶带的中心区域产生稳定等离子体,首次实现了 Pb 的蒸气发生。区别于上述其他 LSDBD 方法[81,82],该方法中雾化器的喷嘴处即存在等离子体,因此并不受雾化效率的影响,几乎所有溶液在形成喷雾后都能与 DBD 等离子体发生有效接触。与传统的 PN-ICP-MS 相比,LSDBD-CVG 测试 Pb 的灵敏度提高了 12 倍,Pb 的检出限为 0.003 μg/L,与传统的 HG-ICP-MS[84]和 PVG-MC-ICP-MS[85]方法相当。值得注意的是,该 LSDBD-CVG 除了避免了氢气的使用,而且显著地提高了 DBD-CVG 技术对溶液中共存离子的耐受能力,更有利于实际样品的应用。基于 LSDBD-CVG 方法突出的共存离子耐受能力,刘星等[86]后续成功将其应用于水稻中的 Cd 的灵敏、可靠检测,满足国家标准对大米中 Cd 的检测需求。

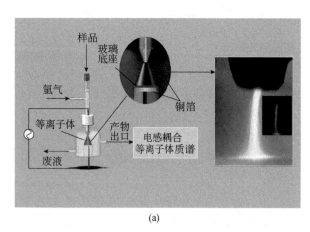
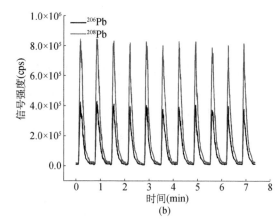

图 5-20 LSDBD-CVG 测定 Pb[83]

微样品中痕量、超痕量元素的精准分析仍存在较大的困难。CVG 方法具有较高的样品引入效率,有望在该领域得到应用。但是,实现不同元素的蒸气发生所需的化学反应条件不同,同时实现多种元素的高效氢化物发生仍存在较大的难度。为了解决这些问题,研究者在前期的研究基础上进一步提出了一种高效的 LSDBD-CVG 方法用于同时测定微量样品中 Se、Ag、Sb、Pb 和 Bi[87]。该工作将

DBD-CVG 拓展到了 Ag 和 Bi 的蒸气发生。该 LSDBD 发生器采用了类似于双通道雾室的设计，可以显著减少进入 ICP 中的大液滴数量，有利于保持 ICP 的稳定。此外，采用了样品提升量更低的雾化器（样品流速：0.3 mL/min）进样，以在减少样品消耗的同时仍维持较高的雾化效率。测得 Se、Ag、Sb、Pb 和 Bi 的 LOD 分别为 10 ng/L、2 ng/L、5 ng/L、4 ng/L 和 3 ng/L，其中 Se(Ⅵ)、Ag、Sb(Ⅴ)、Pb 和 Bi 的样品引入效率分别为 41%±5%、67%±6%、46%±4%、47%±5% 和 87%±4%。此外，使用该 LSDBD-CVG 方法无须对样品溶液进行预还原或预氧化，因而减少了试剂消耗，节省了分析时间，只需消耗 20 μL 样品即可实现对待测元素的快速高灵敏度分析。

随着 DBD 反应器设计的改进，DBD-CVG 方法的应用范围从 Hg、Se、MeHg、EtHg 和硫柳汞扩展到现在包括 Se、Cd、Zn、As、Sb、Te、MMA、DMA、Ag、Pb、Bi、Sc、Y、La、Ce、Pr、Nd、Sm、Eu、Gd、Tb、Dy、Ho、Er、Tm、Yb 和 Lu。从最初的同轴式反应器结构[72]到液膜式进样结构[76]，再到液体喷雾式进样结构[83]的发展，结果表明样品溶液与等离子体相互作用的方式直接影响了 DBD-CVG 方法的蒸气生成效率。通过改善样品与等离子体之间的有效接触面积，可能会进一步扩大 DBD-CVG 技术的应用范围和蒸气发生效果。

5.3.2　DBD-CVG 中的共存离子干扰

从表 5-1 可以看出，在 DBD-CVG 方法中 Hg 的信号强度易受到样品中氯离子的影响。当 DBD-CVG 中[Hg]/[Cl⁻]为 1∶1000 时，Hg 信号的回收率仅为 67%。但是，令人惊讶的是，当[Hg]/[Cl⁻]变为 1∶100 时，Hg 信号的回收率高达 191%。此外，与无机 Hg 相比，Cl⁻对有机汞（MeHg、EtHg 和硫柳汞）的影响则较小[74,75]。除 Cl⁻外，共存离子如 Au^{3+}、CO_3^{2-}、Fe^{3+}、Cu^{2+}、Cl^-、Br^-、I^-、$Cr_2O_7^{2-}$、MnO_4^-、Na^+ 和 NH_4^+ 也会抑制 Hg 的蒸气发生效率[72,74,78,80]。相比之下，Cl⁻对 Hg 的传统 CVG 过程影响较小甚至没有影响。在 TFDBD-CVG 中，当 Cd 与过渡金属[如 $NiCl_2$、$CoCl_2$、$Pb(NO_3)_2$、$Cu(NO_3)_2$ 和 $Zn(NO_3)_2$]的浓度比为 1∶2000 时，Cd 信号的回收率介于 41%~85% 之间[76,78]。在碱金属和碱土金属的存在下，对 Cd 的反应同样有抑制作用。然而，LSDBD-CVG 方法对样品中共存离子有较好的耐受性，当[Cd]∶[M]为 1∶5000 时，Cd 的回收率在 90%~110% 之间[86]。同样，当 LSDBD-CVG 用于 Se、Ag、Sb、Pb 和 Bi 的蒸气发生时，也显示出对共存离子（碱金属、过渡金属，甚至氢化物形成元素）的良好耐受能力[87]。由于其对共存离子的高耐受能力，LSDBD-CVG 方法已成功应用于测定如水、土壤、沉积物、水稻等环境样品中的待测元素。从表 5-1 可以看出，LSDBD-CVG 方法对共存离子的耐受能力具有一定的优势。但总体而言，DBD-CVG 方法对共存离子的耐受能力有待进一步提高，对共存离子造成影响的原因同样仍需要进一步阐释。

表 5-1　共存离子对 DBD-CVG 方法的影响

分析物	DBD 类型	共存离子	[T]∶[M]	回收率（%）	参考文献
Hg	DBD	$AuCl_3$、KCr_2O_7、$Fe(NO_3)_3$、HCl	1∶200、1∶200、1∶2 000、1∶70 000	6、83、85、39	[72]
Hg	TFDBD	CO_3^{2-}、Cl^-	1∶100、1∶100（1∶1000）	87、191（67）	[74]
MeHg、EtHg	TFDBD	Cl^-	都为 1∶1 000	86、89	[74]
硫柳汞	TFDBD	KNO_3、$NaNO_3$、$Ca(NO_3)_2$、Na_2HPO_4、NaCl、KH_2PO_4、$Cu(NO_3)_2$、$Al_2(SO_4)_3$	都为 1∶2 000	无显著影响	[75]
Hg	TFDBD	Fe^{3+}、Cu^{2+}、Cl^-、Br^-、I^-、$Cr_2O_7^{2-}$、MnO_4^-	1∶4 000、1∶4 000、1∶4 000、1∶4 000、1∶4 000、1∶400、1∶400	82、115、80、50、20、88、85	[80]

续表

分析物	DBD 类型	共存离子	[T]:[M]	回收率（%）	参考文献
Hg	TFDBD	$NaNO_3$、$Fe(NO_3)_2$、NH_4NO_3	都为 1:10 000	83、77、77	[78]
Cd	TFDBD	$NaNO_3$、KNO_3、$NiCl_2$、$CoCl_2$、$Pb(NO_3)_2$、$Cu(NO_3)_2$、$Zn(NO_3)_2$、NaCl	都为 1:2 000	64、72、41、65、78、55、85、71	[76]
Cd	LSDBD	Na^+、Ca^{2+}、Mg^{2+}、Mn^{2+}、Ni^{2+}、Co^{2+}、Cu^{2+}、Zn^{2+}、Hg^{2+}	都为 1:5 000	无显著影响	[86]
Cd	TFDBD	$Ca(NO_3)_2$、$Mg(NO_3)_2$	1:1 000、1:1 000	73、65	[78]
Zn	TFDBD	$NaNO_3$、$Ca(NO_3)_2$、$Fe(NO_3)_3$、$Mn(NO_3)_2$、$Cu(NO_3)_2$、$Pb(NO_3)_2$、$Cd(NO_3)_2$、$Co(NO_3)_2$、$Ni(NO_3)_2$、NH_4Cl	1:500	55、47、32、17、16、29、12、50、22、74	[77]
Se(IV)	DBD	Cu^{2+}	1:1	显著抑制	[73]
Pb	LSDBD	K^+、Na^+、Ca^{2+}、Mg^{2+}、Mn^{2+}、Ni^{2+}、Co^{2+}、Cu^{2+}、Zn^{2+}、Hg^{2+}、Fe^{3+}、As^{5+}、Se^{6+}、Cd^{2+}	1:1 000	无显著影响	[83]
Se、Ag、Sb、Bi	LSDBD	K、Ni、Mg、Co、Zn、Cu、Ca、Mn、Se、Ag、Sb、Bi	1:5 000 (K、Ni、Mg、Co、Zn、Cu、Ca、Mn)，1:500 (Se、Ag、Sb、Bi)	无显著影响	[87]
Ce、Nd、Sm、Gd、Tm	NFDBD	Na、Mg、Ba、Zn、Cu、Fe、Co、Ni、Pb、Ce、Nd、Sm、Gd、Tm	1:200 000 (Na、Mg)，1:2 000 (Ba、Zn、Cu、Fe、Co、Ni、Pb)，1:200 (Ce、Nd、Sm、Gd、Tm)	无显著影响	[82]

5.3.3 DBD-CVG 技术的应用

目前为止，DBD-CVG 已经作为 AFS[74-81,86]、ICP-OES[72,73]、DBD-OES[80]、ICP-MS[82,83,87]等仪器的进样方法，成功实现了 As、Sb、Bi、Se、Te、Hg、Pb、Cd、Zn、Ag、Sc、Y、La、Ce、Pr、Nd、Sm、Eu、Gd、Tb、Dy、Ho、Er、Tm、Yb 和 Lu 等元素的检测。表 5-2 显示，在 DBD-CVG 方法中有机汞和无机汞均具有良好蒸气发生效率（>89%）。使用 DBD-CVG 方法时，Cd、Zn 和 Bi 的蒸气发生效率相比传统 CVG 方法同样具有优势。尽管 As 的等离子体蒸气发生效率并未给出，但考虑到其检出限较高，为 30 μg/L，意味着其蒸气发生效率仍较低。从表中统计结果可以看出，等离子体诱导 CVG 系统中大多数元素的检出限可以满足痕量或超痕量级别元素的定量分析。同时，非流动 TFDBD-CVG 和 LSDBD-CVG 方法的样品消耗量分别低至 6 μL 和 20 μL，非常适于微量样品的分析[74,87]。

表 5-2 DBD-CVG 方法部分操作参数及分析性能

分析物	样品引入方式	DBD 类型	溶液组成	样品引入效率（%）	LOD（μg/L）	参考文献
Hg	CF, 2.0 mL/min	DBD	2%甲酸	NR	0.09	[72]
Hg、MeHg、EtHg	6 μL	TFDBD	1%甲酸	NR	0.02	[74]
Hg、硫柳汞	8 μL	TFDBD	去离子水	NR	0.03、0.06	[75]
Hg、MeHg、EtHg	FI, 1.2 mL/min	HPLC-TFDBD	去离子水	NR	1.6、0.42、0.75	[79]
Hg	CF, 3 mL/min	TFDBD	0.5%甲酸	NR	0.2	[80]
Cd	4.5 mL/min	TFDBD	去离子水	NR	0.03	[76]
Cd	CF, 2.0 mL/min	LSDBD	2%甲醇 0.01 mol/L KCl	68	0.01	[86]

续表

分析物	样品引入方式	DBD 类型	溶液组成	样品引入效率（%）	LOD（μg/L）	参考文献
Hg、Cd	CF，3.0 mL/min、3.7 mL/min	TFDBD	1 CMC Triton X-114	82、80	0.0045、0.0024	[78]
Zn	CF，2.5 mL/min	TFDBD	去离子水	27	0.2	[77]
Se^{4+}	CF，2.0 mL/min	DBD	15%甲酸	NR	6	[73]
As、Te、Sb、Se	10 μL/min	LSDBD	5% HCl	NR	30、50、70、60	[81]
Pb	CF，1.2 mL/min	LSDBD	5%甲酸 0.01 mol/L HCl	42	0.003	[83]
Se、Ag、Sb、Pb、Bi	CF，0.3 mL/min	LSDBD	7%甲酸	Se(Ⅵ) 41、67、Sb(Ⅴ) 46、47、87	0.01、0.002、0.005、0.004、0.003	[87]
16 种稀土元素	FI，0.9 mL/min	NFDBD	2% HNO_3	22～31	$2×10^{-6}$～$4.22×10^{-4}$	[82]

CF：连续进样系统；FI：流动注射进样系统；NR：未报道。

表 5-3 统计了 DBD-CVG 技术的主要应用范围，如河流沉积物、土壤、玄武岩、工业废水、鱼组织、冻干尿液、大米、疫苗、海带、紫菜、古菌细胞、牙形石等。此外，由于 DBD-CVG 具有较强的反应效率，可与高效液相色谱联用，以实现不同形态分析 Hg（Hg^{2+}、MeHg、EtHg）[79]和 As[As(Ⅲ)、As(Ⅴ)、MMA 和 DMA][81]的分析测定。遗憾的是，除 LSDBD-CVG 方法外，DBD-CVG 方法仍然容易受到基体干扰的影响，在实际样品分析中的应用有限，主要集中在标准样品的测量上。

表 5-3 DBD-CVG 技术的应用

分析物	样品性质	检测方法	校正方式	参考文献
Hg	矿泉水	DBD-ICP-OES	EC	[72]
Hg、MeHg、EtHg	冻干尿液*、模拟天然水*、金枪鱼*	TFDBD-AFS	EC 和 SA	[74]
Hg、硫柳汞	疫苗	TFDBD-AFS	EC 和 SA	[75]
Hg、MeHg、EtHg	金枪鱼*	HPLC-TFDBD-AFS	SA	[79]
Hg	水	TFDBD-DBD-OES	SA	[80]
Cd	模拟天然水*、大米*	TFDBD-AFS	EC 和 SA	[76]
Cd	大米	LSDBD-AFS	EC 和 SA	[86]
Hg、Cd	模拟天然水*	TFDBD-AFS	EC 和 SA	[78]
Zn	水	TFDBD-AFS	EC 和 SA	[77]
Se^{4+}	水	DBD-ICP-OES	NR	[73]
As^{3+}、As^{5+}、MMA、DMA、Te、Sb、Se	水系沉积物*、海带*、紫菜*	LSDBD-AFS	NR	[81]
Pb	河流沉积物*、土壤*、玄武岩*、模拟天然水*	LSDBD-ICP-MS	EC 和 SA	[83]
Se、Ag、Sb、Pb、Bi	模拟天然水*、牙形石、古菌	LSDBD-ICP-MS	EC	[87]
16 种稀土元素	水、沉积物	NFDBD-ICP-MS	EC	[82]

*标准物质。NR：未报道；EC：标准曲线法；SA：标准加入法。

近年来，基于 DBD-CVG 技术在痕量/超痕量元素测定和形态分析领域已经取得一定的进展。主要优势有：①避免在蒸气发生过程中使用不稳定的还原剂，有利于降低样品被污染的风险，同时降低了分析成本和简化了分析流程；②相较于传统 CVG 方法，除 As 外均具有更高或相当的蒸气发生效率；③DBD 装置结构简单，能耗低；④无需预还原或预氧化步骤，可与色谱法联用进行元素形态分析；⑤适用于微量样品中的微量元素分析。值得一提的是，尽管取得了一定进展，仍然需要进一步工作，未来的研究方向可能主要包括以下几个方面：①通过发展等离子体与样品反应效率更高的 DBD 装置，以进一步拓展其应用范围，以更好地应对基体复杂样品测试过程中存在的困难。②目前氢气辅助等离子体化学反应在 DBD-CVG 中已经展现出一定的潜力，那么其他可能在等离子体作用下生成活性物质的气体，包括甲烷、一氧化碳、乙烯，甚至碘甲烷等，是否也能对 DBD-CVG 方法的分析性能带来提升仍值得探究。③等离子体催化反应已得到广泛应用，那么在 DBD-CVG 方法中添加适当的催化剂以进一步提高其分析性能同样值得重视。④为了进一步研究 DBD-CVG 乃至等离子体蒸气发生方法的反应机理，同样需要对反应过程中的生成的活性物质以及最终挥发性产物进行准确定性。

5.4 其他 DBD 元素分析技术

相比于液体放电技术，DBD 技术在装置搭建、等离子体形成等方面具有较大的优势。液体放电技术中等离子体形成于液体与电极之间，而 DBD 还能在两电极外形成等离子体射流。因此，除了可以实现液体样品中待测元素的高效蒸气发生，DBD 射流也被尝试应用于非液体样品中待测元素的解离。当 DBD 等离子体射流与样品表面接触时，高能量的粒子流会溅射并剥蚀样品表面，从而实现待测元素在非液体样品中的转移，作为一种固体进样方式。

DBD 在原子光谱中作为解吸装置，集进样与解吸与一体，无需复杂的样品预处理，能够快速对样品表面待测元素进行分析检测。图 5-21 为一种基于 DBD 等离子体射流技术的解析源，它由一个 30 mm 长的石英毛细管和两个包裹在毛细管外的铜环组成。两个铜环的宽度为 1 mm，它们之间的间距为 12 mm[88]。离子源安置在剥蚀池的上盖上，与剥蚀池的上盖成 45°。接通高压电源后，在微等离子体射流的剥蚀下，经薄层色谱板分离后的 Hg 的化合物被解吸附/原子化，产生的 Hg^0 在氩气吹扫下进入原子荧光检测器检测，Hg^{2+}、MeHg 和 PhHg 的检测限分别为 0.51 pg、0.29 pg 和 0.34 pg。DBD 等离子体探针还可作为一种固体进样方法，其工作原理为通过 DBD 产生低温等离子体对塑料表面进行剥蚀，解吸出的 Hg 元素在载气的作用下引入到 AFS 中进行直接定量分析[87]。此外，该低温等离子体探针-原子荧光光谱法的分析元素范围被进一步拓展，实现了聚偏二氟乙烯膜上 Cd 含量直接检测。通过该方法，可以实现固体样品表面中 Cd 的剥蚀及原子化[90]。近年，研究者还通过将 H_2 引入 DBD 等离子体中，开发了一种由 H_2 辅助的 DBD 等离子体的直接固体采样技术：含 H_2 的氩等离子体与含硒的固体样品直接相互作用，在 DBD 等离子体作用下，固体样品表面上的 Se 元素有效地解吸并转化为挥发性分子物质，在载气吹扫下进入 AFS 进行检测分析。该方法硒的检出限为 19.7 pg[91]。

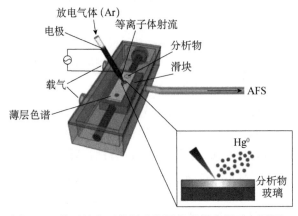

图 5-21 基于等离子体射流的固体进样装置示意图[88]

DBD具有装置简单、离解能力强和能耗低等优势，已成功用于废水和废气中有机污染物的快速降解。在原子光谱领域，常压DBD还可用于大米样品的消解。该DBD装置由两个同心石英管制成，其中铜棒作为内电极插入内管中，铜丝作为外电极均匀地紧紧缠绕在外管壁[92]。大米样品粉末和消解液均匀混合后被循环泵入两个石英管之间的环形空间，在等离子体的作用下样品溶液由浑浊逐渐变为澄清。与常规消解方法相比，该方法无需使用除H_2O之外的其他化学试剂便可实现元素的高灵敏检测，更环保、更节能、更安全。等离子体中产生的高反应性物质，例如羟基自由基（·OH）等强氧化性自由基和O_3，可以有效地氧化和降解大米样品中的有机化合物。DBD不仅可以消解液体样品中有机物，对挥发性有机化合物（VOC）也具有分解能力。双层石英管的同轴DBD反应器还被应用于取代商用AFS汞分析仪中的催化热解炉，促使水生食物样品中有机基质成分产生的VOC的分解和汞原子化，并成功用于固体样品中汞分析[93]。在线DBD反应器与金阱的结合能够完全消除基质干扰。该方法分析时间少于5 min并且最多可以分解12 mg的干燥水生食物粉末，LOD为0.5 μg/kg。

5.5 总结与展望

DBD等离子体装置结构简单、操作温度低，但富含电子等活性基团，具有优异的解离、激发能力，已被广泛应用于低温原子化器、微型等离子体激发源的开发，成功实现了Hg、As、Sb、Pb、Cd、Ni等金属元素以及卤素、C、S等非金属元素的分析检测，并可与色谱等装置联用以实现元素形态的分析，此外还可以作为色谱的检测器实现有机物的分析。与传统技术相比，DBD原子化器及激发源具有功耗低、体积小的优势，非常适合便携式仪器的研发，通过与CVG、ETV、PVG等高效进样方法的结合，可满足很多元素的检测需求。此外，DBD低温等离子体中富含的各种活性粒子可以诱导发生各类化学反应，从而实现元素的化学蒸气发生，氢化物的在线捕集与释放以及样品的消解，为元素分析提供了新的思路和策略。但需要指出的是，现有DBD等离子体元素分析技术仍存在一定不足，需要进一步拓展其元素分析的范围、效果以及实际样品的分析能力，这就迫切需要对DBD等离子体进行深入性能表征，探究原子化、激发以及蒸气发生的机理，从而提升DBD等离子体的分析性能。

参 考 文 献

[1] Miclea M, Kunze K, Musa G, Franzke J, Niemax K. The dielectric barrier discharge: A powerful microchip plasma for diode laser spectrometry. Spectrochimica Acta Part B: Atomic Spectroscopy, 2001, 56(1): 37-43.

[2] Zhu Z, Zhang S, Lv Y, Zhang X. Atomization of hydride with a low-temperature, atmospheric pressure dielectric barrier discharge and its application to arsenic speciation with atomic absorption spectrometry. Analytical Chemistry, 2006, 78(3): 865-872.

[3] Zhu Z, Zhang S, Xue J, Zhang X. Application of atmospheric pressure dielectric barrier discharge plasma for the determination of Se, Sb and Sn with atomic absorption spectrometry. Spectrochimica Acta Part B: Atomic Spectroscopy, 2006, 61(8): 916-921.

[4] Zhu Z, Liu Z, Zheng H, Hu S. Non-chromatographic determination of inorganic and total mercury by atomic absorption

spectrometry based on a dielectric barrier discharge atomizer. Journal of Analytical Atomic Spectrometry, 2010, 25(5): 697-703.

[5] Yu Y, Gao F, Chen M, Wang J. A miniaturized long-optical path atomic absorption spectrometer with dielectric barrier discharge as atomizer for mercury and methylmercury. Acta Chimica Sinica, 2013, 71(8): 1121-1124.

[6] Duben O, Boušek J, Dědina J, Kratzer J. Dielectric barrier discharge plasma atomizer for hydride generation atomic absorption spectrometry: Performance evaluation for selenium. Spectrochimica Acta Part B: Atomic Spectroscopy, 2015, 111: 57-63.

[7] Kratzer J, Bousek J, Sturgeon R E, Mester Z, Dedina J. Determination of bismuth by dielectric barrier discharge atomic absorption spectrometry coupled with hydride generation: Method optimization and evaluation of analytical performance. Analytical Chemistry, 2014, 86(19): 9620-9625.

[8] Novak P, Dedina J, Kratzer J. Preconcentration and atomization of arsane in a dielectric barrier discharge with detection by atomic absorption spectrometry. Analytical Chemistry, 2016, 88(11): 6064-6070.

[9] Zurynkova P, Dedina J, Kratzer J. Trace determination of antimony by hydride generation atomic absorption spectrometry with analyte preconcentration/atomization in a dielectric barrier discharge atomizer. Analytica Chimica Acta, 2018, 1010: 11-19.

[10] Kratzer J, Musil S, Dědina J. Feasibility of *in situ* trapping of selenium hydride in a DBD atomizer for ultrasensitive Se determination by atomic absorption spectrometry studied with a ^{75}Se radioactive indicator. Journal of Analytical Atomic Spectrometry, 2019, 34(1): 193-202.

[11] Kratzer J, Musil S, Marschner K, Svoboda M, Matousek T, Mester Z, Sturgeon R E, Dedina J. Behavior of selenium hydride in heated quartz tube and dielectric barrier discharge atomizers. Analytica Chimica Acta, 2018, 1028: 11-21.

[12] Juhászová L, Burhenn S, Sagapova L, Franzke J, Dědina J, Kratzer J. Hydride generation atomic absorption spectrometry with a dielectric barrier discharge atomizer: Method optimization and evaluation of analytical performance for tin. Spectrochimica Acta Part B: Atomic Spectroscopy, 2019, 158: 105630.

[13] Albrecht M, Mrkvičková M, Svoboda M, Hraníček J, Voráč J, Dvořák P, Dědina J, Kratzer J. Atomization of lead hydride in a dielectric barrier discharge atomizer: Optimized for atomic absorption spectrometry and studied by laser-induced fluorescence. Spectrochimica Acta Part B: Atomic Spectroscopy, 2020, 166: 105819.

[14] Yu Y L, Du Z, Chen M L, Wang J H. A miniature lab-on-valve atomic fluorescence spectrometer integrating a dielectric barrier discharge atomizer demonstrated for arsenic analysis. Journal of Analytical Atomic Spectrometry, 2008, 23(4): 493-499.

[15] Zhu Z, Liu J, Zhang S, Na X, Zhang X. Evaluation of a hydride generation-atomic fluorescence system for the determination of arsenic using a dielectric barrier discharge atomizer. Analytica Chimica Acta, 2007, 607(2): 136-141.

[16] Zhu Z, Liu J, Zhang S, Na X, Zhang X. Determination of Se, Pb, and Sb by atomic fluorescence spectrometry using a new flameless, dielectric barrier discharge atomizer. Spectrochimica Acta Part B: Atomic Spectroscopy, 2008, 63(3): 431-436.

[17] Xing Z, Wang J, Zhang S, Zhang X. Determination of bismuth in solid samples by hydride generation atomic fluorescence spectrometry with a dielectric barrier discharge atomizer. Talanta, 2009, 80(1): 139-142.

[18] Xing Z, Kuermaiti B, Wang J, Han G, Zhang S, Zhang X. Simultaneous determination of arsenic and antimony by hydride generation atomic fluorescence spectrometry with dielectric barrier discharge atomizer. Spectrochimica Acta Part B: Atomic Spectroscopy, 2010, 65(12): 1056-1060.

[19] 别克赛力克·库尔买提, 王娟, 韩国军, 邢志. 低温等离子体原子化器-原子荧光光谱法测定碲. 分析化学, 2010(3): 66-70.

[20] Mao X, Qi Y, Huang J, Liu J, Chen G, Na X, Wang M, Qian Y. Ambient-temperature trap/release of arsenic by dielectric barrier discharge and its application to ultratrace arsenic determination in surface water followed by atomic fluorescence spectrometry. Analytical Chemistry, 2016, 88(7): 4147-4152.

[21] Qi Y, Mao X, Liu J, Na X, Chen G, Liu M, Zheng C, Qian Y. In situ dielectric barrier discharge trap for ultrasensitive arsenic determination by atomic fluorescence spectrometry. Analytical Chemistry, 2018, 90(10): 6332-6338.

[22] Liu M, Liu T, Liu J, Mao X, Na X, Ding L, Chen G, Qian Y. Determination of arsenic in biological samples by slurry sampling hydride generation atomic fluorescence spectrometry using in situ dielectric barrier discharge trap. Journal of Analytical Atomic Spectrometry, 2019, 34(3): 526-534.

[23] Liu M, Liu T, Mao X, Liu J, Na X, Ding L, Qian Y. A novel gas liquid separator for direct sampling analysis of ultratrace arsenic in blood sample by hydride generation in-situ dielectric barrier discharge atomic fluorescence spectrometry. Talanta, 2019, 202: 178-185.

[24] Liu M, Liu J, Mao X, Na X, Ding L, Qian Y. High sensitivity analysis of selenium by ultraviolet vapor generation combined with microplasma gas phase enrichment and the mechanism study. Analytical Chemistry, 2020, 92(10): 7257-7264.

[25] He Q, Yu X, Li Y, He H, Zhang J. Dielectric barrier discharge induced atomization of gaseous methylethylmercury after NaBEt4 derivatization with purge and trap preconcentration for methylmercury determination in seawater by GC-AFS. Microchemical Journal, 2018, 141: 148-154.

[26] Zhu Z L, Chan G C Y, Ray S J, Zhang X R, Hieftje G M. Microplasma source based on a dielectric barrier discharge for the determination of mercury by atomic emission spectrometry. Analytical Chemistry, 2008, 80(22): 8622-8627.

[27] Yu Y L, Du Z, Chen M L, Wang J H. Atmospheric-pressure dielectric-barrier discharge as a radiation source for optical emission spectrometry. Angewandte Chemie-International Edition, 2008, 47(41): 7909-7912.

[28] Abdul-Majeed W S, Parada J H L, Zimmerman W B. Optimization of a miniaturized DBD plasma chip for mercury detection in water samples. Analytical and Bioanalytical Chemistry, 2011, 401(9): 2713-2722.

[29] Wu Z C, Chen M L, Li P, Zhu Q Q, Wang J H. Dielectric barrier discharge non-thermal micro-plasma for the excitation and emission spectrometric detection of ammonia. Analyst, 2011, 136(12): 2552-2557.

[30] Meyer C, Demecz D, Gurevich E L, Marggraf U, Jestel G, Franzke J. Development of a novel dielectric barrier microhollow cathode discharge for gaseous atomic emission spectroscopy. Journal of Analytical Atomic Spectrometry, 2012, 27(4): 677-681.

[31] Wu Z C, Jiang J, Li N. Cold excitation and determination of hydrogen sulfide by dielectric barrier discharge molecular emission spectrometry. Talanta, 2015, 144: 734-739.

[32] Yu Y L, Dou S, Chen M L, Wang J H. Iodine excitation in a dielectric barrier discharge micro-plasma and its determination by optical emission spectrometry. Analyst, 2013, 138(6): 1719-1725.

[33] Yu Y L, Cai Y, Chen M L, Wang J H. Development of a miniature dielectric barrier discharge-optical emission spectrometric system for bromide and bromate screening in environmental water samples. Analytica Chimica Acta, 2014, 809: 30-36.

[34] Zhang D J, Cai Y, Chen M L, Yu Y L, Wang J H. Dielectric barrier discharge-optical emission spectrometry for the simultaneous determination of halogens. Journal of Analytical Atomic Spectrometry, 2016, 31(2): 398-405.

[35] Han B J, Jiang X M, Hou X D, Zheng C B. Miniaturized dielectric barrier discharge carbon atomic emission spectrometry with online microwave-assisted oxidation for determination of total organic carbon. Analytical Chemistry, 2014, 86(13): 6214-6219.

[36] Qian B, Zhao J, Wu Q, He Y, Peng L X, Hian B J. Miniaturized heating/ultrasound assisted direct injection-dielectric

barrier discharge molecular emission spectrometry for determination of dissolved sulfide in environmental water. Microchemical Journal, 2020, 152. doi.org/10.1016/j.microc.2019.104442.

[37] Qian B, Zhao J, He Y, Peng L X, Ge H L, Han B J. Miniaturized dielectric barrier discharge-molecular emission spectrometer for determination of total sulfur dioxide in food. Food Chemistry, 2020, 317. doi.org/10.1016/j.foodchem. 2020.126437.

[38] Zhu Z L, He H Y, He D, Zheng H T, Zhang C X, Hu S H. Evaluation of a new dielectric barrier discharge excitation source for the determination of arsenic with atomic emission spectrometry. Talanta, 2014, 122: 234-239.

[39] Burhenn S, Kratzer J, Svoboda M, Klute F D, Michels A, Veza D, Franzke J. Spatially and temporally resolved detection of arsenic in a capillary dielectric barrier discharge by hydride generation high-resolved optical emission spectrometry. Analytical Chemistry, 2018, 90(5): 3424-3429.

[40] Burhenn S, Kratzer J, Klute F D, Dedina J, Franzke J. Atomization of arsenic hydride in a planar dielectric barrier discharge: Behavior of As atoms studied by temporally and spatially resolved optical emission spectrometry. Spectrochimica Acta Part B: Atomic Spectroscopy, 2019, 152: 68-73.

[41] He H Y, Zhu Z L, Zheng H T, Xiao Q, Jin L L, Hu S H. Dielectric barrier discharge micro-plasma emission source for the determination of thimerosal in vaccines by photochemical vapor generation. Microchemical Journal, 2012, 104: 7-11.

[42] Cai Y, Li S H, Dou S, Yu Y L, Wang J H. Metal carbonyl vapor generation coupled with dielectric barrier discharge to avoid plasma quench for optical emission spectrometry. Analytical Chemistry, 2015, 87(2): 1366-1372.

[43] Jiang X M, Chen Y, Zheng C B, Hou X D. Electrothermal vaporization for universal liquid sample introduction to dielectric barrier discharge micropasma for portable atomic emission spectrometry. Analytical Chemistry, 2014, 86(11): 5220-5224.

[44] Zheng H T, Ma J Z, Zhu Z L, Tang Z Y, Hu S H. Dielectric barrier discharge micro-plasma emission source for the determination of lead in water samples by tungsten coil electro-thermal vaporization. Talanta, 2015, 132: 106-111.

[45] Yang T, Gao D X, Yu Y L, Chen M L, Wang J H. Dielectric barrier discharge micro-plasma emission spectrometry for the detection of acetone in exhaled breath. Talanta, 2016, 146: 603-608.

[46] Li N, Wu Z C, Wang Y Y, Zhang J, Zhang X N, Zhang H N, Wu W H, Gao J, Jiang J. Portable dielectric barrier discharge-atomic emission spectrometer. Analytical Chemistry, 2017, 89(4): 2205-2210.

[47] Jiang J, Li Z J, Wang Y Y, Zhang X N, Yu K, Zhang H, Zhang J, Gao J, Liu X Y, Zhang H N, Wu W H, Li N. Rapid determination of cadmium in rice by portable dielectric barrier discharge-atomic emission spectrometer. Food Chemistry, 2020, 310. doi.org/10.1016/j.foodchem.2019.125824.

[48] Puanngam M, Ohira S I, Unob F, Wang J H, Dasgupta P K. A cold plasma dielectric barrier discharge atomic emission detector for atmospheric mercury. Talanta, 2010, 81(3): 1109-1115.

[49] Wu Z C, Chen M L, Tao L, Zhao D, Wang J H. Sequential monitoring of elemental mercury in stack gas by dielectric barrier discharge micro-plasma emission spectrometry. Journal of Analytical Atomic Spectrometry, 2012, 27(10): 1709-1714.

[50] Li W, Zheng C B, Fan G Y, Tang L, Xu K L, Lv Y, Hou X D. Dielectric barrier discharge molecular emission spectrometer as multichannel GC detector for halohydrocarbons. Analytical Chemistry, 2011, 83(13): 5050-5055.

[51] Jiang X, Li C H, Long Z, Hou X D. Selectively enhanced molecular emission spectra of benzene, toluene and xylene with nano-MnO_2 in atmospheric ambient temperature dielectric barrier discharge. Analytical Methods, 2015, 7(2): 400-404.

[52] Han B J, Jiang X M, Hou X D, Zheng C B. Dielectric barrier discharge carbon atomic emission spectrometer: Universal

GC detector for volatile carbon-containing compounds. Analytical Chemistry, 2014, 86(1): 936-942.

[53] Lin Y, Yang Y, Li Y X, Yang L, Hou X D, Feng X B, Zheng C B. Ultrasensitive speciation analysis of mercury in rice by headspace solid phase microextraction using porous carbons and gas chromatography-dielectric barrier discharge optical emission spectrometry. Environmental Science & Technology, 2016, 50(5): 2468-2476.

[54] 李成辉, 蒋小明, 侯贤灯. 氢化物发生-介质阻挡放电光谱分析法测定无机汞和甲基汞. 实验技术与管理, 2019, 36(2): 80-84.

[55] Han B J, Li Y, He Y, Lv D Z, Peng L X, Yu H M. Miniaturized dielectric barrier discharge-atomic emission spectrometer for pesticide: Sensitive determination of thiram after derivatization with mercurial ion. Microchemical Journal, 2018, 138: 457-464.

[56] Han B J, Li Y, Qian B, He Y, Peng L X, Yu H M. A novel liquid chromatography detector based on a dielectric barrier discharge molecular emission spectrometer with online microwave-assisted hydrolysis for determination of dithiocarbamates. Analyst, 2018, 143(12): 2790-2798.

[57] Tombrink S, Muller S, Heming R, Michels A, Lampen P, Franzke J. Liquid analysis dielectric capillary barrier discharge. Analytical and Bioanalytical Chemistry, 2010, 397(7): 2917-2922.

[58] Krahling T, Muller S, Meyer C, Stark A K, Franzke J. Liquid electrode dielectric barrier discharge for the analysis of solved metals. Journal of Analytical Atomic Spectrometry, 2011, 26(10): 1974-1978.

[59] Krahling T, Michels A, Geisler S, Florek S, Franzke J. Investigations into modeling and further estimation of detection limits of the liquid electrode dielectric barrier discharge. Analytical Chemistry, 2014, 86(12): 5822-5828.

[60] Novosad L, Hrdlicka A, Slavicek P, Otruba V, Kanicky V. Plasma pencil as an excitation source for atomic emission spectrometry. Journal of Analytical Atomic Spectrometry, 2012, 27(2): 305-309.

[61] He Q, Zhu Z L, Hu S H, Zheng H T, Jin L L. Elemental determination of microsamples by liquid film dielectric barrier discharge atomic emission spectrometry. Analytical Chemistry, 2012, 84(9): 4179-4184.

[62] Cai Y, Zhang Y J, Wu D F, Yu Y L, Wang J H. Nonthermal optical emission spectrometry: Direct atomization and excitation of cadmium for highly sensitive determination. Analytical Chemistry, 2016, 88(8): 4192-4195.

[63] Zhang Y J, Cai Y, Yu Y L, Wang J H. A miniature optical emission spectrometric system in a lab-on-valve for sensitive determination of cadmium. Analytica Chimica Acta, 2017, 976: 45-51.

[64] Cai Y, Gao X G, Ji Z N, Yu Y L, Wang J H. Nonthermal optical emission spectrometry for simultaneous and direct determination of zinc, cadmium and mercury in spray. Analyst, 2018, 143(4): 930-935.

[65] Browner R F, Boorn A W. Sample introduction: The Achilles' heel of atomic spectroscopy?. Analytical Chemistry, 1984, 56(7): 786A-798A.

[66] Gao Y, Liu R, Yang L. Application of chemical vapor generation in ICP-MS: A review. Chinese Science Bulletin, 2013, 58(17): 1980-1991.

[67] Liu X, Zhu Z, Xing P, Zheng H, Hu S. Plasma induced chemical vapor generation for atomic spectrometry: A review. Spectrochimica Acta Part B: Atomic Spectroscopy, 2020, 167: 105822.

[68] Rumbach P, Witzke M, Sankaran R M, Go D B. Decoupling interfacial reactions between plasmas and liquids: Charge transfer vs plasma neutral reactions. Journal of the American Chemical Society, 2013, 135(44): 16264-16267.

[69] Samukawa S, Hori M, Rauf S, Tachibana K, Bruggeman P, Kroesen G, Whitehead J C, Murphy A B, Gutsol A F, Starikovskaia S, Kortshagen U, Boeuf J P, Sommerer T J, Kushner M J, Czarnetzki U, Mason N. The 2012 plasma roadmap. Journal of Physics D: Applied Physics, 2012, 45(25): 253001-253048.

[70] He Q, Zhu Z, Hu S. Plasma-induced vapor generation technique for analytical atomic spectrometry. Reviews in Analytical Chemistry, 2014, 33(2): 111-121.

[71] Pohl P, Greda K, Dzimitrowicz A, Welna M, Szymczycha-Madeja A, Lesniewicz A, Jamroz P. Cold atmospheric plasma-induced chemical vapor generation in trace element analysis by spectrometric methods. TrAC Trends in Analytical Chemistry, 2019, 113: 234-245.

[72] Wu X, Yang W, Liu M, Hou X, Zheng C. Vapor generation in dielectric barrier discharge for sensitive detection of mercury by inductively coupled plasma optical emission spectrometry. Journal of Analytical Atomic Spectrometry, 2011, 26(6): 1204-1209.

[73] Yang W, Zhu X, Wu X. Dielectric barrier discharge induced chemical vapor generation for determination of selenium (Ⅳ) by ICP-OES. Chemical Research and Application, 2011, 23: 644-648.

[74] Liu Z, Zhu Z, Wu Q, Hu S, Zheng H. Dielectric barrier discharge-plasma induced vaporization and its application to the determination of mercury by atomic fluorescence spectrometry. Analyst, 2011, 136(21): 4539-4544.

[75] Wu Q, Zhu Z, Liu Z, Zheng H, Hu S, Li L. Dielectric barrier discharge-plasma induced vaporization for the determination of thiomersal in vaccines by atomic fluorescence spectrometry. Journal of Analytical Atomic Spectrometry, 2012, 27(3): 496-500.

[76] Zhu Z, Wu Q, Liu Z, Liu L, Zheng H, Hu S. Dielectric barrier discharge for high efficiency plasma-chemical vapor generation of cadmium. Analytical Chemistry, 2013, 85(8): 4150-4156.

[77] Zhu Z, Liu L, Li Y, Peng H, Liu Z, Guo W, Hu S. Cold vapor generation of Zn based on dielectric barrier discharge induced plasma chemical process for the determination of water samples by atomic fluorescence spectrometry. Analytical and Bioanalytical Chemistry, 2014, 406(29): 7523-7531.

[78] Li Y, Zhu Z, Zheng H, Jin L, Hu S. Significant signal enhancement of dielectric barrier discharge plasma induced vapor generation by using non-ionic surfactants for determination of mercury and cadmium by atomic fluorescence spectrometry. Journal of Analytical Atomic Spectrometry, 2016, 31(2): 383-389.

[79] Liu Z, Xing Z, Li Z, Zhu Z, Ke Y, Jin L, Hu S. The online coupling of high performance liquid chromatography with atomic fluorescence spectrometry based on dielectric barrier discharge induced chemical vapor generation for the speciation of mercury. Journal of Analytical Atomic Spectrometry, 2017, 32(3): 678-685.

[80] Leng A Q, Tian Y F, Wang M X, Wu L, Xu K L, Hou X D, Zheng C B. A sensitive and compact mercury analyzer by integrating dielectric barrier discharge induced cold vapor generation and optical emission spectrometry. Chinese Chemical Letters, 2017, 28(2): 189-196.

[81] Yang M, Xue J, Li M, Han G, Xing Z, Zhang S, Zhang X. Low temperature hydrogen plasma assisted chemical vapor generation for atomic fluorescence spectrometry. Talanta, 2014, 126: 1-7.

[82] He Q, Wang X, He H, Zhang J. A feasibility study of rare-earth element vapor generation by nebulized film dielectric barrier discharge and its application in environmental sample determination. Analytical Chemistry, 2020, 92(3): 2535-2542.

[83] Liu X, Zhu Z, Li H, He D, Li Y, Zheng H, Gan Y, Li Y, Belshaw N S, Hu S. Liquid spray dielectric barrier discharge induced plasma-chemical vapor generation for the determination of lead by ICPMS. Analytical Chemistry, 2017, 89(12): 6827-6833.

[84] Yilmaz V, Arslan Z, Rose L. Determination of lead by hydride generation inductively coupled plasma mass spectrometry (HG-ICP-MS): On-line generation of plumbane using potassium hexacyanomanganate (Ⅲ). Analytica Chimica Acta, 2013, 761: 18-26.

[85] Gao Y, Xu M, Sturgeon R E, Mester Z, Shi Z, Galea R, Saull P, Yang L. Metal ion-assisted photochemical vapor generation for the determination of lead in environmental samples by multicollector-ICPMS. Analytical Chemistry, 2015, 87(8): 4495-4502.

[86] Liu X, Zhu Z, Bao Z, Zheng H, Hu S. Determination of trace cadmium in rice by liquid spray dielctric barrier discharge induced plasma-chemical vapor generation coupled with atomic fluorescence spectrometry. Spectrochimica Acta Part B: Atomic Spectroscopy, 2018, 141: 15-21.

[87] Liu X, Zhu Z, Bao Z, He D, Zheng H, Liu Z, Hu S. Simultaneous sensitive determination of selenium, silver, antimony, lead, and bismuth in microsamples based on liquid spray dielectric barrier discharge plasma-induced vapor generation. Analytical Chemistry, 2019, 91(1): 928-934.

[88] Liu Z, Zhu Z, Zheng H, Hu S. Plasma jet desorption atomization-atomic fluorescence spectrometry and its application to mercury speciation by coupling with thin layer chromatography. Analytical Chemistry, 2012, 84(23): 10170-10174.

[89] 杨萌, 薛蛟, 李铭, 李佳, 黄秀, 邢志. 低温等离子体原子荧光光谱法直接测定固体样品中的汞. 分析化学, 2012(8): 1164-1168.

[90] 李铭, 李健, 陈帅, 杨萌, 黄秀, 冯璐, 范博文, 邢志. 低温等离子体探针原子荧光光谱法检测镉元素的方法研究. 分析仪器, 2017(2): 53-57.

[91] Li M, Xing Z, Sun G, Liang J, Huang X, Fan B, Wang, Q. A novel vapor generation method by hydrogen-containing plasma for Se direct solid sampling. Journal of Analytical Atomic Spectrometry, 35: 904-911.

[92] Luo Y J, Yang Y, Lin Y, Tian Y F, Wu L, Yang L, Hou X D, Zheng C B. Low-temperature and atmospheric pressure sample digestion using dielectric barrier discharge. Analytical Chemistry, 2018, 90(3): 1547-1553.

[93] Liu T, Liu M, Liu J, Mao X, Qian Y. On-line microplasma decomposition of gaseous phase interference for solid sampling mercury analysis in aquatic food samples. Analytica Chimica Acta, 2020, 1121: 42-49.

第6章　常压辉光放电等离子体发射光谱分析

（汪　正[①]*，彭晓旭[①]）

▶ 6.1　常压辉光放电等离子体简介 / 151

▶ 6.2　常压辉光放电应用于发射光谱辐射源 / 153

▶ 6.3　液体电极-金属电极辉光放电诱导的蒸气发生技术 / 164

▶ 6.4　常压辉光放电等离子体的机理研究进程 / 165

①中国科学院上海硅酸盐研究所，上海。*通讯作者联系方式：wangzheng@mail.sic.ac.cn

本章导读

- 系统总结了常压辉光放电等离子体发射光谱最近取得的研究进展。
- 概括了目前常压辉光放电等离子体应用于发射光谱辐射源的具体增敏方式。
- 分析了目前常压辉光放电等离子体的机理研究进展。

6.1 常压辉光放电等离子体简介

常压辉光放电微等离子体是一种通过在电极和对电极间施加高电压,在大气压环境中产生的尺寸为毫米量级的等离子体,是一种由中性原子和分子、自由基、激发态原子、离子和电子组成的物质的独特状态。根据产生等离子的电极和对电极的状态,可将常压辉光放电微等离子体分为液体电极-金属电极辉光放电(solution electrode glow discharge,SEGD)微等离子体以及金属电极-金属电极辉光放电(metal electrode glow discharge,MEGD)微等离子体。这里需要说明的是,上述几种常压辉光放电微等离子体可能在不同时期具有不同的命名方式,由于部分命名方式所涵盖的放电种类范围较广,使得上述几种微等离子体可能在部分研究中具有相似的命名方式。因此,这里以电极和对电极的形态对常压辉光放电等离子体进行了区分和命名。

常压辉光放电微等离子体的放电特性受多方面实验参数的影响,这包括常压辉光放电的电极结构、电极材料、放电参数(指放电过程中的放电电流、电极间距等)、等离子体气相组分以及液体电极组分。合理地优化影响常压辉光放电等离子体的参数能够显著提升其在特定领域的应用性能。

6.1.1 液体电极-金属电极辉光放电

液体电极-金属电极辉光放电(SEGD)可根据液体电极的极性分为液体阴极辉光放电(solution cathode glow discharge,SCGD)和液体阳极辉光放电(solution anode glow discharge,SAGD)。简单地改变液体电极极性即能产生两种具有不同放电行为的等离子体。两种等离子体对于放电条件(主要包括放电电流、电极间距、载酸种类及浓度)的要求以及等离子体的特性(包括其中元素或分子发射谱带的分布、I-V 特性曲线、激发温度等参数)是导致液体阴极辉光放电微等离子体和液体阳极辉光放电微等离子体在应用于发射光谱的辐射源时,所能够分析的元素种类相差较大的主要原因。

6.1.1.1 液体阴极-金属电极辉光放电

金属电极和电解质之间的放电现象最早是在 1887 年 Cubkin 进行化学实验的过程中发现的。随后,Cubkin 将放电称为辉光放电电解(glow discharge electrolysis,GDE)[1]。此后,在辉光放电电解所产生的等离子体中检测到了铜和铟(存在于阴极电解液中)的特征原子发射线。这为辉光放电微等离子体研究奠定了重要的基础。在 GDE 的基础上,Cserfalvi 等[2]提出了一种改进的辉光放电装置,称之为电解液阴极放电(electrolyte-cathode discharge,ELCAD)。为了进一步了解 ELCAD 的放电行为,一种能够改变 ELCAD 放电环境压力的装置随之出现[3]。

基于先前的相关基础研究,Yang 等[4]和 Kim 等[5]分别开发了封闭式 ELCAD[6]和敞开式 ELCAD[7]系统,并将上述两种具有不同放电结构的系统应用于水中痕量重金属的检测。Webb 通过进一步简化 ELCAD 的结构,提出了液体阴极辉光放电(SCGD)[8,9]。以上基于液体阴极-金属电极辉光放电的研究无疑对以后的SCGD应用和机理研究具有重要意义。

6.1.1.2　液体阳极-金属电极辉光放电

在1958年，首次观察到溶液阳极和金属阴极间同样可产生辉光放电，这一过程被称为液体阳极辉光放电（SAGD）[10]。在提出液体辉光放电的初期阶段，并未通过液体阳极辉光放电检测到阳极电解液中金属元素在等离子体中的发射谱线[2]，这导致常压辉光放电的研究重心主要集中在液体阴极辉光放电上，初期仅对液体阳极辉光放电的电学特性和光谱特性进行了研究并同液体阴极辉光放电进行了对比，未将液体阳极辉光放电应用于发射光谱的辐射源上[11-14]。

但是，液体阳极辉光放电在等离子体-液体界面处引发和控制电化学反应的能力为电化学打开了一个新的方向，装置结构如图6-1所示。这突显了液体阳极辉光放电等离子体气相-液相界面所存在的化学反应对于改变液体阳极电解质成分方面的关键作用[15-18]。在此研究的基础上，朱振利等提出一种半封闭式液体阳极辉光放电装置，并将其作为发射光谱法（OES）的辐射源用于高灵敏度地测定Cd和Zn[19]。此外，进一步的研究加深了这个话题，并扩展了带有冷却系统的SAGD的应用，例如用于Ag、Cd、Hg、Pb、Tl、In和Zn[20-22]等其他元素。Pohl等[23]描述了SAGD-OES的另一种结构设计，他通过将液体样品滴放在石墨盘阳极上，实现了基于单液滴液体样品的分析，显著减少了样品的消耗量。不局限于应用于OES的辐射源，SAGD的另一种应用是化学蒸气的产生，这比常规的电化学氢化物的蒸气产生效率高，且不需要其他还原剂[24]。

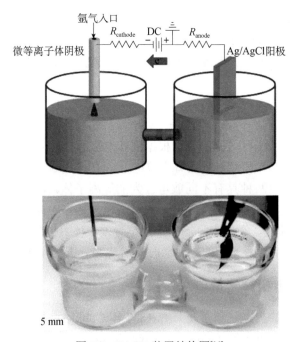

图6-1　SAGD装置结构图[16]

6.1.2　金属电极-金属电极辉光放电

金属电极-金属电极辉光放电（MEGD）微等离子体是一种在高压、直流、大气压环境下，两个固态电极间产生的尺寸在毫米量级的稳定等离子体。等离子体内部含有大量的高能粒子，能与进入其中的目标分析物发生碰撞，将其内部分子解离、激发和电离，因此可以作为原子发射光谱分析的辐射源。大气压辉光放电光谱技术具有稳定性好、元素选择性高、结构简单、运行成本低等优点。

1933 年，德国学者 Von Engel 等[25]在裸露的电极间通入空气或氢气，产生了直流辉光放电的现象，这种辉光放电稳定性很差，容易产生辉光-电弧转变，阴极需要持续冷却，放电必须在真空环境下开始，然后逐渐升至大气压。1988 年，日本学者 Okazaki 等[26]报道了一种在大气压惰性气体中产生的稳定辉光放电，自此世界各国的学者逐渐开始研究大气压辉光放电这一领域的课题。

6.2　常压辉光放电应用于发射光谱辐射源

6.2.1　液体电极-金属电极辉光放电应用于发射光谱辐射源

原子发射光谱（optical emission spectrometer，OES）技术因能够同时分析多元素、分析速度快、灵敏度高等优势，已被广泛应用于工业、环境、冶金等多个领域。其相较于电化学方法和比色法具备高通量和高灵敏的特点。简单、快速和低成本的重金属野外在线测试技术的关键在于仪器的小型化、便携化，而原子光谱仪器小型化的关键制约因素就是原子化器和激发源。常压辉光放电微等离子体作为光谱的辐射源时，产生的尺度为毫米量级甚至更低的等离子体在大气压环境下能够对多种元素进行有效的原子化及激发，同时无需借助特殊气体或只需要较低的气体消耗量，这对于实现元素的在线原位检测具有重要的研究意义。

液体电极-金属电极辉光放电微等离子体应用于原子光谱的辐射源时，无需特殊的样品引入系统（如用于对样品进行雾化的雾化室以及雾化器），溶液中的分析物即能够通过放电反应以一定方式进入气相等离子体中进行原子化和激发过程，最终能够提供具有较低的背景水平以及相对简单的原子发射光谱[4,27]。尤其最近提出的采用流动液体电极辉光放电微等离子体装置，其流动液体电极的设计能够保证样品引入及检测过程中不会受高浓度基质样品溶液的记忆效应或污染的影响。这使得液体电极-金属电极辉光放电微等离子体具有相当大的吸引力，可以应用于环境监测等研究领域中各种复杂基质样品中元素的检测（主要是金属）。现如今，液体电极-金属电极辉光放电微等离子体被认为是最有前途和替代性的激发源之一，与电感耦合等离子体光谱仪（inductive coupled plasma optical emission spectrometer，ICP-OES）相比，它可以提供相当或者更优的元素检出限[28-32]。

常压辉光放电等离子体发射光谱系统的基本构造主要分为四个部分：供电系统、进样系统、常压辉光放电辐射源系统、光谱信号检测系统。供电系统分别为进样系统、常压辉光放电辐射源系统、光谱信号检测系统，提供必要的电能。对于一些组成简单的液体分析物，可直接将该分析物溶于特定酸电解液中，通过进样系统的蠕动泵直接传输至 SEGD 辐射源系统中。当分析物的组成较为复杂时，可在进样系统中引入各种前端样品分离预富集技术，如氢化物发生（hydride generation，HG）、流动注射（flow injection，FI）、光化学蒸气发生（photochemical vapor generation，PVG）、固相萃取（solid phase extraction，SPE）等。分析物中的元素经等离子体-液体界面处的复杂反应过程或由载气吹扫进入放电产生的等离子体中，通过与等离子体中粒子相互作用进行激发并产生特征激发谱线。经光学元件对光辐射信号进行处理后传输至光谱检测器当中，检测器将接收到的光学信号转换成电子信号，处理并储存于计算机中，再以各种方式转换成为光谱图。综上，提升常压辉光放电等离子体发射光谱的性能可通过以下途径实现：优化分析物进样方式、优化常压辉光放电辐射源系统的激发能力、优化光谱信号检测系统的光辐射采集效率。

● 辉光放电等离子体发射光谱分析增敏方法

1. 等离子体装置结构优化增敏

前期研究主要在常压辉光放电的硬件上做了较多的改良，主要涉及常压辉光放电结构的改进，使得能够显著提升常压辉光放电辐射源系统对于进入其中元素的激发能力[14,19-23,31,33-46]。与液体阳极辉光放电相比，液体阴极辉光放电作为发射光谱辐射源受到了极大的关注。对于常压辉光放电而言，随放电电流的升高，放电的电流密度几乎保持恒定[36]。通过缩减常压辉光的放电体积，能够有效地提高常压辉光放电辐射源的电流密度[35,36,40]。基于这一设计理念，通过使用 J 形的玻璃管（内径 0.38 mm×外径 1.1 mm）作为电解液的传输通道，有效缩小了阴极电解液的表面积，这一设计允许液体阴极辉光放电的放电体积减小至接近 2 mm^3，相应得到更高放电电流密度，使得液体阴极辉光放电微等离子体对于一系列元素具有更佳的检测下限[36]。评估常压辉光放电发射光谱性能的另一项关键指标是检测的稳定性。蠕动泵作为液体阴极辉光放电进样系统中常用的流体传输装置，运作过程中会由于转轮的交替释放产生一个脉冲流，这对于液体阴极辉光放电产生持续稳定的等离子体是不利的。因此，脉冲阻尼的设计能够有效缓解由蠕动泵引起的脉冲流。结构最为简单且易于实施的脉冲阻尼系统可直接利用蠕动泵管实现。通过将连接蠕动泵和液体电极-金属电极辉光放电辐射源之间的蠕动泵管进行打结的方式构成一种简易的脉冲阻尼系统（又称菊花链），可以有效地降低由蠕动泵的脉冲所引起的液体阴极辉光放电等离子体的不稳定性[36,39-41]。在此基础上，通过借助若干玻璃球、硅胶管以及软管接头，提出了一种优化设计的脉冲阻尼器，进一步削弱了由蠕动泵脉冲流所引起的等离子体不稳定性，如图 6-2 所示[45]。

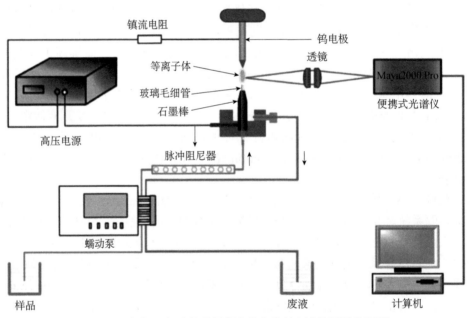

图 6-2　脉冲阻尼-液体阴极辉光放电发射光谱装置结构图[45]

此外，通过对液体阴极辉光放电系统的结构进行设计，同样能够提高等离子体的稳定性。从电解质毛细管顶部的中心到边缘切割一个宽度等于电解质毛细管内径的 V 形槽，能够保证电解液顺利向下流动，即使在较低的电解液流速下（0.8 mL/min）同样可以获得持续稳定的等离子体[31,37,38,43,46]。Wang 等[42]通过将液体阴极辉光放电辐射源系统中用于传输电解液的石英毛细管直接连接到石墨棒上，有效消除了溶液位差对于等离子体稳定性的影响，同时简化后的结构提高了常压辉光放电系统的便携性。此外，一些新设计的常压辉光放电原子光谱系统通过重新优化液体阴极辉光放电结构，有效降低了分析过程中所需消耗的样品量[33,44]，并进一步简化了液体阴极辉光放电辐射源的结构[33,42]，有效地推动了基于常压辉光放电的原子发射光谱向便携式仪器发展的进程。最近，通过将构成液体阴极辉光放电原子光谱装置的各功能系统，包括供电系统、进样系统、液体阴极辉光放电辐射源系统、光谱信号检测系统集成为一体，成功构建了基于 SCGD 的便携式高通量分析仪器，如图 6-3 所示。此外，通过借助锂电池为仪器进行供电，保证了该仪器能够在野外进行样品的在线分析。同时通过编写相应的仪器控制软件，仪器能够实现样品的自动检测和数据处理，大幅度提高了样品分析的通量[34]。

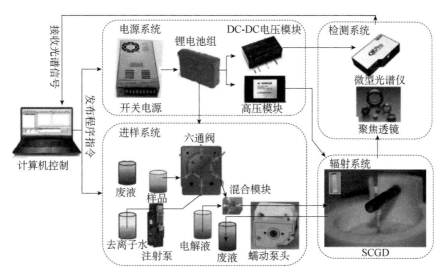

图 6-3　便携式高通量发射光谱分析仪器结构示意图[34]

以液体阳极辉光放电作为光谱辐射源的研究相对较晚。SAGD 的一个最新设计是，用装在与储水池相连的培养皿中的自来水作为液体阳极，以保持液面位置在良好的近似恒定的状态，用于测量 SAGD 等离子中 OH 密度的空间分辨及其等离子体的气体温度[14]。不仅限于 OH 发射谱带的检测，通过基于阴极池和阳极池组成的封闭 U 形管设计，SAGD 能够实现对于阳极电解液中镉、锌的高灵敏度测定。独立的放电电池阻止了氧从阳极迁移到阴极，并使等离子体保持在大气压氩气中[19]。不仅使用 SAGD-OES 对镉、锌进行了高度灵敏的测定，而且通过借助水冷模块对金属阴极进行冷却，避免了液体阳极辉光放电过程中金属阴极电极的过度烧蚀，有效地延长了电极的使用寿命并提高了 SAGD 的稳定性，同时扩大了可分析元素，包括 Ag、Hg、Pb 和 Tl 等[22]的范围。在此基础上，通过将水冷模块更换为黄铜散热片，在保证电极散热效果的基础上进一步简化了液体阳极辉光放电系统的结构，进一步消除了水冷模块的水消耗[20,21]。相比上述流动液体阳极的结构设计，通过使用单液滴电解液直接作为液体阳极辉光放电的液体阳极，同样能够对微量液体中的目标元素进行高灵敏度检测。同时，无需进样系统的特点保证了 SAGD 具有更为简易的结构[23]，如图 6-4 所示。

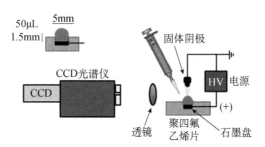

图 6-4　单液滴液体阳极辉光放电发射光谱结构示意图[23]

2. 改变液体电极组分

改变用作液体电极的电解质溶液的组成是提高液体电极-金属电极辉光放电性能的简单且有效的方法。具体的实施方法包括：①改变液体电极基体溶液成分，主要指液体电极的阳离子和阴离子种类及其含量。常规液体电极-金属电极辉光放电的液体电极为酸性电解质，不同酸性电解质显著影响液体电极-金属电极辉光放电等离子体中元素的发射强度。②在确定液体电极的基体电解质后，可通过向液体电极中添加低分子量有机化合物、离子或非离子表面活性剂改变液体电极的特性，进而改变放电过程中元素由液体电极至等离子体气相的效率或元素的激发效率，以此提高液体电极-金属电极辉光放电发射光谱对于元素的检测性能。

1）改变液体电极基体溶液成分

电解质的成分是影响 SEGD-OES 对于元素分析性能的另一项关键参数。对于 SCGD 而言，使用酸基质作为 SCGD 的阴极电解液比相同实验条件下使用盐溶液作为阴极电解液能够产生更强的元素发射，且相同酸浓度下，酸基质中不同的阴离子会对元素的发射产生不同程度的影响[47]。考虑到硝酸作为阴极电解液时所产生的较强元素发射以及较高的化学兼容性，在 SCGD 体系中通常使用 HNO_3 作为 SCGD 的阴极电解液[33,48]。

2）添加低分子量有机化合物

此外，向阴极电解液中添加低分子量有机化合物（low molecular weight organic compound，LMWOC），例如甲酸、甲醇、乙酸和乙醇，是进一步降低 SCGD-OES 对于电解质中部分元素检测限的一种简单且有效的方法[32,33,37,38,49]，特别是对于一些检测过程中存在严重基体干扰的样品，例如锆合金和高盐度盐水。通过向上述存在基质干扰的样品溶液中添加甲酸，无需对样品进行基质分离，即可通过 SCGD-OES 对其中痕量元素进行快速检测[32,37]。盐水样品中存在的干扰阴离子（如 SO_4^{2-}）同样可通过添加低分子量有机化合物（如甲酸、甘油和抗坏血酸）的方式来进行消除，保证了 SCGD-OES 对于盐水样品中痕量元素分析时的检测性能[32,37]。这里必须要注意的是，不同的低分子量有机化合物对元素的敏化作用也不同[29,32,33,37,38,49]。即使将相同种类以及浓度的低分子量有机化合物添加到 SCGD 阴极电解液中，不同 SCGD 放电条件下（酸浓度、流速等实验参数的不同）对相同元素的敏化效果也可能不同[29,38,49]。

为了明确低分子量有机化合物对于 SCGD 中元素光谱信号增敏的机制，通过测量和比较 SCGD 的阴极电解液的特性（电导率、pH 和密度）以及 SCGD 的电学和光谱学特性，以进一步阐明 SCGD 等离子体与低分子量有机化合物有关的过程。研究结果表明低分子量有机化合物在 SCGD 中的作用机制并不是单一简单的反应过程，而是可能同时存在多种有助于改善元素响应的过程。其中可能涉及低分子量有机化合物对于溶液蒸发速度的影响（对分析物信号的增强基本无贡献）、液体阴极表面液滴的形成（低分子量有机化合物可能改变液体阴极表面释放的含有分析物液滴的数量及其尺寸，进而影响分析物在等离子体中的发射强度）以及挥发性物质的产生及其从液滴中的挥发过程（分析物可能通过与有机化合物和/或其降解产物的反应转化为挥发性产物进入至等离子体中）[50]。低分子量有机化合物对于 SCGD 中元素光谱信号增敏的机制还有待进一步细化研究。

同样地，将上述低分子量有机化合物添加到 SAGD 的阳极电解液中同样能够改善 SAGD-OES 对于元素的检测性能，包括提高元素的发射信号强度和信背比以及降低背景水平及其波动[21,23,51]。然而，在 SAGD 体系中低分子量有机化合物对于元素的增敏效果受限于电解液的 pH 值。如最近研究发现，使用含有 1%甲醇的 pH=1 HNO_3 电解液作为 SAGD 系统的阳极电解液，Cd(I)在 228.8 nm 处的强度比没有甲醇时高 2.1 倍，而对于含有 1%甲醇的 pH=6 HNO_3 电解液，Cd(I)在 228.8 nm 处的发射强度会受到较为严重的抑制效应[23,51]。低分子量有机化合物在 SAGD 系统中对元素的增敏效应不仅受限于电解液的组分，同时对于不同元素存在显著的差异。相比银（Ag）、镉（Cd）、汞（Hg）、铅（Pb）、铊（Tl）和锌（Zn）几种元素，大多数LMWOC，特别是甲醇和乙醇，显著提高了SAGD-OES系统中铟（In）的发射强度[21]。尽管低分子量有机化合物在某些特定的实验条件下对于SAGD-OES系统中Cd、Pb和Ag的发射强度具有抑制效应，另一方面同时导致了Ag、Cd和Pb分析线附近背景的水平和波动相应降低。最终结果是Cd、Pb和Ag信背比提高了，因此可获得更佳的检出限[51]。

上述SAGD和SCGD对于低分子量有机化合物所具有的不同响应行为表明SAGD阳极电解液和SCGD阴极电极液中元素迁移至等离子体的过程是显著不同的。此外，在SAGD系统中，低分子量有机化合物的添加仅对In的发射强度具有显著的增敏效应。因此不能将低分子量有机化合物对于SAGD中元素发射的增敏机制简单地归因于阳极电解液的物理性质的变化，如阳极电解液表面张力、黏度或沸点的降低，更有可能的增敏机制是低分子量有机化合物的添加可能会引起SAGD阳极电解液中电化学反应过程的变化[21]。

3）添加表面活性剂

通过添加非离子性活性剂（Triton X-45、Triton X-100、Triton X-405、Triton X-114、Triton X-705）[31,52-54]或离子活性剂（十六烷基三甲基氯化铵）[30,55,56]至SCGD系统的阴极电解液中同样获得了类似的元素发射敏化效果。此外，相比添加低分子量有机化合物，添加上述表面活性剂至SCGD系统的阴极电解液中能够显著降低分子谱带的发射强度以及总体背景发射水平，同时降低了背景发射的波动幅度。上述抑制效应在添加低分子量有机化合物的过程中并不明显，这表明低分子量有机化合物和表面活性剂对于SCGD系统中元素的作用机理并不相同[49,52]。表面活性剂对于SCGD系统中元素的增敏效果取决于表面活性剂的浓度及其分子尺寸。相比分子量较低的非离子表面活性剂，分子量较大的非离子表面活性剂（如Triton X-405或Triton X-705）可能会更大幅度地增加电解质溶液的黏度，因此发射光谱中能够观察到更为明显的分子发射谱带的降低以及金属发射线的增强[53,54]。尽管分子量较重的非离子活性剂（Triton X-705）对SCGD系统中元素的发射具有更强的增敏作用，但是该非离子活性剂所具有的较高黏度导致其难以排出SCGD的阴极放电室，这就导致长期将添加有分子量较重的非离子表面活性剂的电解液作为液体阴极不便于SCGD-OES对于元素的检测[53]。通过将非离子表面活性剂Triton X-114和碘化钾的混合物添加至SCGD系统的阴极电解液中，同样能够显著提高SCGD-OES系统对于元素（Cd、Hg、Cr、Pb、Tl）测定的灵敏度。尤其是对于元素铊（Tl），添加Triton X-114和碘化钾的混合物的方式能够极大幅度地提高SCGD-OES系统对于Tl的检测灵敏度[54]。

将离子型表面活性剂，即十六烷基三甲基氯化铵（CTAC）添加到SCGD的液体阴极中，同样能够显著提升SCGD-OES对Cd、Hg、Pb和Cr的检测灵敏度，同时降低了背景发射强度以及上述元素原子发射谱线的波动[30]。但是，如果将目标元素的挥发性分析物由SCGD的金属阳极端引入，SCGD阴极电解液中非离子表面活性剂的存在会导致元素发射强度的降低。这表明SCGD系统中离子型表面活性剂对于阴极电解液中元素的增敏行为与离子型表面活性剂改善元素由液体阴极到等离子体的迁移效率有关[57]。

与SCGD系统相比，向液体阳极中添加非离子表面活性剂（0.5% m/v 的Triton X-405）的SAGD系统放电非常不稳定，这可能是由电解液阳极表面张力和黏度的突然变化所致。因此，添加表面活性剂至SAGD-OES系统中的相关研究并未被重视[23]。

3. 光辐射采集优化增敏

SCGD的物理结构包括负辉光（negative glow）（阴极表面上方约0.5 mm的强发射区域）、法拉第暗区（Faraday dark space）（负辉光上方约0.5 mm高的区域）、正柱（positive column）（法拉第暗区上方约3 mm高的区域）和阳极辉光（anode glow）（在阳极表面）[58]。通过检测在SCGD中原子和分子发射光谱的空间分布，证明了SCGD中元素和分子发射在空间上不是均匀的，而且在上述某一特定区域内具有较高的灵敏度以及较高的发射均匀性[8,9,58,59]。因此提高常压辉光放电发射光谱对于元素分析性能的另一项关键方法即是常压辉光放电辐射源中光辐射信号的采集优化。

通过分析元素发射在SCGD中的强度、信背比和信噪比的空间分布，确定目标元素的发射谱线最强以及背景发射（来自其他元素或分子源）较低的放电区域，收集该区域的发射光谱用于目标元素的检测，能够显著提升SCGD-OES对于目标元素的检测性能。因此，利用常压辉光放电等离子体中光谱发射的不均匀性可以显著提高SCGD对于目标分析的分析性能，同时能够降低甚至消除基体干扰[9,58]。考虑到SCGD中不同元素发射的空间分布差异，为保证SCGD-OES对于目标元素的分析性能，必须根据元素在SCGD中的最佳发射区域确定优化的空间窗口，以确保对于不同元素的分析性能。然而，该方法只能针对某个元素进行单独优化空间采光窗口，并且其应用局限于配备有二维探测器阵列（例如EMCCD、CCD等）的光谱仪或能够扫描穿过光谱仪入口狭缝的图像的仪器。利用一个单一的、折中的空间窗口，空间滤波可以有效地应用到没有二维探测器阵列的仪器上，并且可以为大多数元素提供相近的分析性能[58]。

4. 辉光放电等离子体发射光谱联用进样技术

一般情况下，大多数液体样品中元素的常规测定可通过SEGD-OES系统在连续流动进样模式下直接完成[20-23,30,31,34,36-38,41,42,45,46,48,49,51-56,60-70]。然而，面对精确测定复杂基体中元素的挑战，就需要对样品进行一个分离和预富集的过程，这就出现了一种将元素分离和预富集技术与SEGD耦合的采样技术。将流动注射（flow injection，FI）采样方法与SCGD系统耦合，以降低样品溶液中潜在干扰成分的放电负荷[8,28,32,38-41,49,50,65,71-73]，具体装置结构如图6-5所示。该流动注射取样系统需要通过标准六通阀和样品回路来实现，能够保证在较低的样品消耗量下完成元素的检测过程，同时样品检测的通量较高，特别适合于瞬态分析[32,38,40,41,49,50,65,73]。通过自动、在线生成校准曲线图，进一步提高了SCGD-OES的样品检测通量，这也最大限度地减少了离线样品制备所需的时间。上述自动、在线生成校准曲线图的具体实施方法：第一种方法使用梯度高效液相色谱泵对标准储备液、样品溶液和稀释剂进行在线混合和输送，以获得所需的溶液组成；另一种方法是使用一个由三个蠕动泵组成的简单系统来执行相同的在线溶液混合功能[39]。

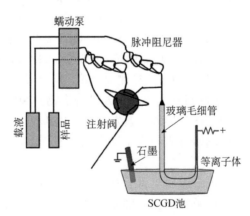

图6-5　FI-SCGD-OES装置结构图[40]

由于样品中不同元素的检测需要，建立了一种基于在线固相萃取（solid phase extraction，SPE）的SCGD-OES与流动注射耦合提高元素测定灵敏度的方法，并合成了固相微柱中填充的各种萃取材料，以实现对目标元素的高效分离和富集，比如L-半胱氨酸修饰介孔氧化硅[Hg(Ⅱ)][72]、赖氨酸修饰介孔二氧化硅[Cr(Ⅵ)][28]、介孔二氧化硅接枝氧化石墨烯[Pb(Ⅱ)][71]。为简化FI-SPE系统，将在线固相萃取系统的流出液直接引入SCGD的样品毛细管。

采用微流量气体射流喷嘴阳极为拓宽分析物向放电区域的传递途径和显著提高 SCGD-OES 系统的测定灵敏度提供了另一个机会。该设计允许挥发性分析物直接从阳极引入 SCGD 中，并为 SCGD 与蒸气发生器的耦合提供了可行性，例如，在冷蒸气发生（CVG）中形成 Hg、As、Se、Pb、Ge、Sn 和 Sb 的冷蒸气，并通过氩载体/喷射支持气体流从反应/分离系统中吹扫[74-78]出去。此外，光化学蒸气发生（photochemical vapor generation，PVG）作为一种有前途的挥发性物质生成技术，具有毒性小、操作简单、受过渡金属干扰小等优点，已成功地与 SCGD-OES 联用，用于高灵敏度的汞测定，装置结构如图 6-6 所示。不同 PVG 参数下 PVG 对元素形态的选择性产气特性为 PVG 应用于 Hg 等元素形态分析提供了重要的技术支持[79]。氢化物发生（HG）作为一种 CVG 技术，具有对元素价态（Sb、As、Se 等）进行选择性蒸气发生的特点，这为 SCGD 与 HG 体系耦合进行元素价态分析提供了可行性[75,76,80]。超声雾化（ultrasonic nebulization，USN）产生的含有干气溶胶的分析物也可以通过气体喷嘴射流阳极传输到 SCGD 系统中，使得 SCGD-OES 的灵敏度提高了一个数量级，证明了 SCGD 与 USN 进样技术相耦合的重要意义[57]。

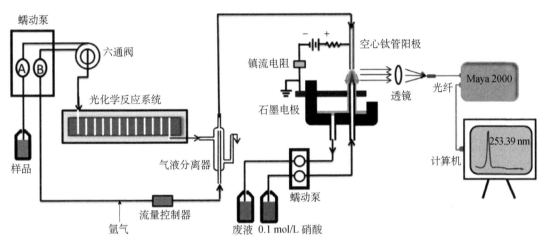

图 6-6　PVG-SCGD-OES 装置结构图[79]

考虑到部分阴离子对于人体或环境具有更为重要的影响，对于样品中部分阴离子的检测可能更为重要。将高效液相色谱仪（HPLC）、离子抑制器（IS）和微型置换柱（RC）耦合到 SCGD 系统中，建立了一种测定 F^-、Cl^-、Br^-、BrO_3^-、NO_3^-、CH_3COO^-、SO_4^{2-} 和 SO_3^{2-} 的新方法，装置结构如图 6-7 所示[81]。通过跟踪 SCGD 中的 Li 信号，可以间接地测定上述阴离子。最重要的是，由于 SCGD 对碱金属物种具有特殊的灵敏度[40]，上述系统有可能对离子色谱（IC）检测高度敏感[81]。

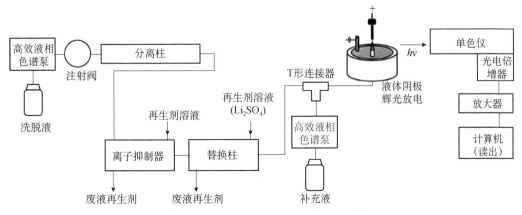

图 6-7　HPLC-SCGD-OES 装置结构图[81]

通过无界面模式构建了一种新型毛细管微等离子体分析系统（C-μPAS）。通过共享一个直流电，将毛细管电泳（capillary electrophoresis，CE）与SAGD放电结合在一起，可以测定样品中的阴离子（Cl^-、Br^-、CH_3COOCl^-、CH_3COOBr^-）。C-μPAS将样品引入系统、分析物分离以及分析信号检测系统集成到一个紧凑的单元中，这主要取决于通过SEGD放电过程能够引发CE过程的发现。此外，通过交换正极和负极形成的反向C-μPAS也可以测定金属阳离子[82]。

将SEGD与蒸气发生耦合的一个缺点是，它会降低原子化/激发源作为多元素检测器的能力。因为并非所有元素都适合蒸气发生，而且特定元素的蒸气的发生需要在特定实验条件下进行。

5. 等离子体气相组分优化增敏

另一种常见的SEGD-OES增敏方式即是改变放电等离子体气相组分，具体的实施方法包括以下两种：①可以通过使用空心管电极作为SEGD系统的放电电极，载气直接从空心阳极/阴极进入等离子体中[20,67,68,83,84]；②也可将SEGD放电体系设置于半密封腔室中，通过非电极管道向放电腔室通入特定载气来改变等离子体气相组分[15,19,47]。这两种气体引入方法都能保证载气能有效地参与等离子体气相反应，对改变放电稳定性、等离子体气相组分以及为SEGD提供更好的激发条件具有显著影响[20,47,68,83]。这里需要说明的是，SEGD中的支持气体对发射光谱和放电稳定性的影响受到多方面因素的限制，如载气成分、分析物导入至SEGD中的方法、液体电极的结构。

载气成分是指放电时气体气氛的组成，取决于放电室的结构和输入气体的种类。对于敞开式SEGD放电结构（等离子体暴露于大气环境中），即使向等离子体中输入特定的辅助气体（例如，CO_2[83]、Ar[15,68,83]、He[68,83]、Cl_2[47]、H_2-Ar[74-76]、H_2-He[78,80]、O_2[15]、N_2[15]），放电过程同样始终有空气参与排放。通过向敞开式SCGD放电系统中输入特定气体（Ar、He或CO_2），碱金属的原子发射线的灵敏度不会因为载气的输入或输入载气种类的差别而发生显著改变。对于在半封闭SCGD系统中，输入特定辅助气体（CO_2、Ar或He），可使分析物的原子发射强度下降数倍（从1.5~4倍），光谱干扰（尤其是NO分子带的干扰）同样显著降低。对比上述敞开式和半封闭式SCGD结构中载气组分改变而引起的原子发射线或分子发射谱带强度的变化，可得出结论：切断SCGD放电过程中空气的供应是产生上述发射光谱的主要原因[68,83]。这里需要说明的是，在气相等离子体反应过程中生成的产物（NO、NO_2等）能够溶解至液相并改变液相组成[15,85]。但是如果SCGD采用流动液体电极的电极结构，则与等离子体气相直接接触的液相部分是处于流动且不断更新的状态，避免了等离子体气相产物溶解至液体电极而对放电过程产生影响。另一方面，分析物进入至等离子体的方式同样直接关系到等离子体气相组分对于元素的影响效果。如果将分析物转化为挥发物直接从空心管电极导入至等离子体气相而不是经液体电极，则改变等离子体气相的组分直接会导致等离子体中的粒子发生改变，进而直接对元素产生不一样的效应[57,74,77-79]。

6.2.2 金属电极-金属电极辉光放电应用于发射光谱辐射源

6.2.2.1 金属电极-金属电极辉光放电结构

2006年，Andrade等[86]在铝合金腔室内，以锥形末端的铜棒为阳极、平面抛光末端钨棒为阴极，搭建了一台针-板式大气压辉光放电装置，该装置实物图及示意图如图6-8（a）所示。阳极呈锥形是为了提高尖端电子密度，阴极抛光是为了防止电荷分布不均引起电场密度畸变，两极间距1 cm。放电时，在腔室内充入氦气，接入直流电后形成了稳定的大气压辉光放电。文献中对辉光放电微等离子体的放电区域进行了研究并将其划分为负辉区（negative glow，NG）、法拉第暗区（Faraday dark space，FDS）和正柱区（positive column，PC），如图6-8（b）所示。阴极在离子流轰击下产生二次电子发射，辅助形成导电通道；负辉区是整个等离子体中最亮的区域，放电通道中大部分电压降落都发生在这一区域，大部分的放电功率也是在此处被耗散掉，电子在这一区域中被电场加速到能够产生电离和雪崩击穿的程度；法拉第暗区中的电子能量很低，由于粒子复合扩散的原因，此区域的电子密度也很低，净空间电荷很低；正柱区占据了绝大部分的放电区域，尺寸会随放电间距的改变而改变。Staack 等[87]对比了电极间距分别为 0.1 mm、0.5 mm、1 mm 和 3 mm 时的辉光放电图像，发现正柱区会随放电间距的增大而增大，在放电间距为 0.1 mm 时，正柱区消失不见，而负辉区的尺寸对电极间距的变化不敏感。

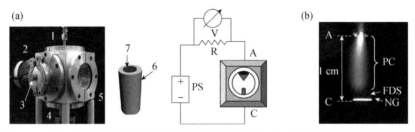

图 6-8 （a）针-板式大气压辉光放电装置示意图和（b）辉光放电分区示意图[86]

1. 氦气管路；2. 压力计；3. 阴极连接端；4. 真空管路；5. 观察窗口；6. 阴极氧化铝屏蔽壳；7. 阴极钨棒。A. 阳极；C. 阴极

除了上述极不对称的针-板式电极结构外，管-管式空心电极结构也得到了广泛的报道。2011 年，Gielniak 等[88]发现在 He 气氛中，两个空心电极之间可以实现非常稳定的大气压辉光放电。将放电区域收束在石英管中，可以更好地提高等离子体的稳定性以及原子发射强度。Wang 等[89]以外径为 1.6 mm 的不锈钢管作为阴极，外径为 3 mm 的钨管作为阳极，两者同轴放置在内径为 4 mm 的石英管内，也就是两极之间放电产生的等离子体被束缚在该石英管内，装置示意图如图 6-9 所示。氦气携带前端氢化物发生装置产生的气态分析物从阴极不锈钢管进入放电区域，光纤探头放置在钨管侧采光。束缚管的存在使得两极间的放电介质（惰性气体环境）更加稳定，因此产生的大气压辉光放电更加稳定。该装置对 Se 和 As 的检出限分别为 0.13 ng/mL 和 0.087 ng/mL。还有少数文献介绍过线-筒式、圆点-凹坑式以及针-柱式电极等，只要电极端部电子密度够高、两极间存在稳定的放电介质、两极间直流电压足够击穿放电介质，就能产生大气压辉光放电，电极样式可以根据需要灵活设计。

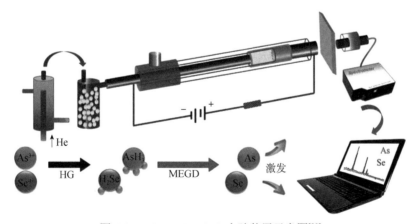

图 6-9 HG-MEGD-OES 实验装置示意图[89]

6.2.2.2 金属电极-金属电极辉光放电等离子体电学特性

等离子体放电的主要特性，如击穿电压、电压-电流特性和放电结构主要取决于电极形状、电极

间距和放电气体等。Staack 等[87]的研究发现，放电间距越小、放电电流越大，所需的放电电压越小；同一电极间距下，放电电流较大时，放电更稳定，不同电极间距的电压-电流特性图如图 6-10 所示。一般来说电极材料对辉光放电的电压-电流特性影响不大，但阴极需要承受较高的温度，当阴极较薄且材料不耐高温时，会出现严重的烧蚀现象。此外，虽然放电时会向两极之间通入惰性气体，但仍无法隔绝电极与空气的接触，放电时负辉光烧蚀过的地方会产生一层氧化涂层，当整个放电范围内的电极表面都被氧化后，放电将会无法进行，磨去电极表面的氧化层后，电极仍能继续使用。

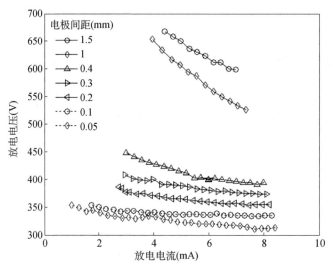

图 6-10　MEGD 不同电极间距的电压-电流特性图[87]

大气压辉光放电的击穿电压与电极间距的关系遵循 Paschen 定律，空气中最小击穿电压对应的理论最小电极间距为 7 μm，然而实验往往会受到灰尘、振动干扰以及定位不准的影响，可实现的最小电极间距为 20 μm。在氦气氛中放电时，最大电极间距可达 75 mm。目前广泛使用的大气压辉光放电电极间距通常在 1～10 mm 范围内，放电间距过大时，形成的辉光放电微等离子体会出现弯曲、漂移的现象，易受气流扰动而熄灭。在放电区域增加束缚管可以维持放电介质稳定、隔离环境气体扰动，使放电更加稳定。

6.2.2.3　金属电极-金属电极辉光放电等离子体气相组分的影响

大气压辉光放电装置中，往往需要向装置中通入惰性气体，如氦气[89]、氮气[90]、氖气、氩气[91]或某些混合气体。以空气为介质来实现大气压辉光放电比较困难，因为电子在放电电源停止时，会很快与正离子复合或者形成负离子，导致电子密度下降，而惰性气体特别是氦气具有能量较高的亚稳态，能够通过碰撞电离其他原子，减少负离子的形成，从而延长等离子体的寿命。这些气体最重要的作用是充当放电介质，其次是作为样品进入辐射源的载气，有时还可以作为辐射源的冷却气。不同的气氛中放电产生的等离子体颜色不同，这与气体分子受激后发射的特定波长谱线有关；氦气气氛中放电比氩气稳定，不容易出现辉光-电弧转变，这与气体的电离能有关；即使使用高纯气体，也无法完全排除其中的空气和水蒸气，放电光谱中会出现 NO、OH、N_2 的分子谱带，但使用氢-氦混合气体可以降低甚至消除部分分子谱带。

6.2.2.4　金属电极-金属电极辉光放电原子发射光谱的进样方式

以固态电极大气压辉光放电作为原子光谱分析的辐射源时，只能直接分析气态或气溶胶形式的样品，这就需要选用合适的进样技术把待测样品转化成这两种形态并送入辐射源。进样方法的选择是否合适，传输是否有效，传输效率如何，将直接影响 MEGD-AES 的分析性能和应用范围。常见的进样技术有气动雾化、超声雾化、化学蒸气发生、冷蒸气发生、氢化物法发生、电热蒸发、激光剥蚀等。分析时应根据样品的物理化学性能特点，选取合适的进样技术。

液体样品或已经消解为液体的固体样品可以通过雾化的方式进样，雾化方式可以细分为气动雾化、超声雾化、热喷雾法等。气动雾化（pneumatic nebulization，PN）是利用高速气流将毛细管中的试液吸出并击碎成细雾，再经雾化室去除大颗粒水滴。雾化进样的方式样品消耗量较大，雾化效率仅 2%～3%，绝大部分的溶液都成为废液被排出。Moβ 等[92]使用气动雾化与大气压辉光放电耦合的方式对水溶液中的 Cd、Cu、Mg、Mn 和 Na 元素进行检测，检出限分别为 190 μg/L、140 μg/L、110 μg/L、130 μg/L 和 10 μg/L。文献中还对比了干燥和湿润两种情况下，大气压辉光放电微等离子体的激发温度，发现湿润条件下转动温度（T_{rot}）比干燥时下降约 500 K，激发温度（T_{exc}）下降约 240～400 K。

超声波雾化是利用电子高频振荡将液态水分子间的分子键打散而产生自然扩散的水雾，用载气将产生的水雾带入辐射源即可。与气动雾化相比，超声雾化产生的雾珠量更大、更细、更均匀，单位时间内进入辐射源的气溶胶更多。这种方式同时会将大量的雾珠传输至大气压辉光放电等离子体中，这导致大气压辉光放电的稳定性快速下降，进而降低大气压辉光放电发射光谱的分析性能。因此超声雾化后的分析物须先通过干燥装置进行去溶剂化，之后干气溶胶进入大气压辉光放电辐射源进行原子化和激发，得到目标分析物的发射光谱图。

化学蒸气发生（chemical vapor generation，CVG）是通过化学反应的方法将待测元素转化成气态挥发性物质引入辐射源的分析方法，这种方法能够将待测元素从基质中分离出来，避免了基质光谱干扰，同时能够起到富集待测元素的作用，从而改善仪器的检出限和准确度。此外，CVG 对待测元素的存在形式和价态有一定的选择性。CVG 技术可以分为基于硼氢化物反应的氢化物发生、光诱导化学蒸气发生、介质阻挡蒸气发生、冷原子蒸气发生和电化学蒸气发生等。Zheng 等[91]用 1.5%(m/v) 的抗坏血酸和 1.0 mol/L 的盐酸将水溶液中的亚硝酸盐还原为挥发性一氧化氮，将产生的 NO 与水蒸气分离后，用氩载气吹扫至 MEGD 微等离子体中进行激发，用光谱仪记录 NO 谱线的发射情况。研究了操作参数对 NO 的产生和 MEGD 微等离子体源激发性能的影响，在优化条件下该方法对亚硝酸盐的检出限为 0.26 μg/mL，为饮用水中潜在亚硝酸盐的检测提供了新选择。

氢化物发生（hydride generation，HG）是通过化学反应将某些元素转化成具有挥发性的共价氢化物，再由载气将氢化物带入辐射源进行分析，是化学蒸气发生中的一种。然而，只有元素周期表第Ⅳ、Ⅴ、Ⅵ族中的一些元素如 As、Sb、Bi、Ge、Sn、Pb、Se、Te 能形成挥发性共价氢化物，因此氢化物发生这种进样方式的应用范围特别有限。Peng 等[89]通过耦合氢化物发生进样装置，将样品溶液中的 Se 和 As 离子还原为对应的氢化物形式，经气液分离和干燥之后，以 He 为载气将其引入大气压辉光放电辐射源进行光谱分析。该方法对 Se 和 As 的检出限分别为 0.13 ng/mL 和 0.087 ng/mL。

光化学蒸气发生（photochemical vapor generation，PVG）是将溶液样品泵入到紫外灯中，在紫外线的照射下溶液中会形成具有氧化性的羟基自由基将有机物氧化、分解，而还原性的氢自由基可以将金属离子还原并形成挥发性物质，经气液分离后可以进入辉光放电等离子体进行原子光谱分析。除传统的氢化物发生元素（As、Bi、Sb、Se、Sn、Pb、Cd、Te、Hg）之外，一些过渡族金属（Ni、Co、Cu、Fe）、贵金属（Ag、Au、Rh、Pd、Pt）和非金属（I、S）均可以在紫外光和小分子有机酸的存在下实现蒸气发生。Zhang 等[93]利用 PVG 耦合尖端放电光谱技术检测了液体样品中的 Hg、Fe、Ni 和 Co 元素含量，元素的检出限分别为 0.1 μg/L、10 μg/L、0.2 μg/L 和 4.5 μg/L。与传统的微等离子体分析相比，不仅扩宽了可检测的元素范围，还显著改善了 Hg 和 Ni 的检出限。

电热蒸发（electrothermal vaporization，ETV）技术采用电加热的方式，经过干燥、灰化和蒸发三个步骤将微升或毫克级试样转化为干气溶胶,再经载气引入辐射源进行检测。与雾化进样的方式相比,ETV 进样技术的传输效率可达 80%,进而可以将检出限改善 1~2 个数量级。研究发现使用适当的化学改性剂如 ETDA、抗坏血酸、HNO_3、H_2O_2、PTFE 等改变待测物质或基质的物理或化学性质,使用小体积的蒸发器和短距离传输管减少气溶胶在管路中的沉积,均能显著提高试样传输效率。Deng 等[94]用 ETV-MEGD 法检测了水稻中的 Cd 含量,毫克级的水稻粉末被放置在 Ni-Cr 线圈中脱水和热解,释放出 Cd 蒸气,经钨线圈捕获后再释放,将其与基质分离。使用含 10% H_2 的 Ar 作为 Cd 的载气和大气压辉光放电的介质,经参数优化后,该方法对 Cd 的检出限为 0.26 μg/kg。

激光剥蚀(laser ablation,LA)进样是将峰值功率很高的脉冲激光聚焦到待测样品表面,在极短时间内将样品表面升至高温,促使样品熔化和蒸发,再用载气将剥蚀出的气溶胶输送至辐射源即可检测。这种进样方式特别适合分析硬度高、熔点高等难以用常规酸解的样品,避免了烦琐的样品前处理过程,减少了污染的可能性,避免了溶液制备中的稀释效应,有利于改善仪器分析性能。同时由于激光剥蚀的尺寸可以控制在微米级,因此采用激光剥蚀技术进样可以实现对样品的选区、微区分析。Wang 等[95]用激光剥蚀与大气压辉光放电联用的方式分析了土壤样品中的 Zn、Pb 和 Cd 三种元素的含量,该方法对三种元素的检出限分别为 0.68 mg/kg、2.71 mg/kg 和 0.31 mg/kg,与激光诱导击穿光谱法相比,将 Zn 的检出限降低了一个数量级。

气相色谱法(gas chromatography,GC)是由惰性气体将气化后的试样带入含固定相的色谱柱中,由于不同的组分具有不同的物理和化学性质,在色谱柱固定相中流动的速度不同,到达色谱柱末端的时间也不同,因此可以被分离开来。根据固定相的形式不同,气相色谱法可以分为气固色谱和气液色谱两种。气相色谱法既可独自对样品进行定性和定量分析,也可以与质谱、电感耦合等离子体光谱以及串联质谱等联用分析。Eijkel 等[96]将气相色谱与芯片形式的大气压辉光放电室联用,对己烷的检出限为 10^{-12} g/s。

近年来,固态电极大气压辉光放电技术得到了国内外学者的广泛关注,由于其具有可在大气压下操作、结构简单、体积小、能耗小、元素选择性高等特点,加之可以耦合多种进样技术,使得 MEGD 在发展小型化分析仪器应用于微量元素分析方面具有良好的前景。

6.3 液体电极-金属电极辉光放电诱导的蒸气发生技术

常压辉光放电等离子体除应用于发射光谱的辐射源外,同样可应用于元素挥发性蒸气的产生,作为发射光谱分析技术领域的新型样品分离预富集系统对于提高发射光谱的分析性能具有重要的意义。目前,SEGD 系统在电感应耦合等离子体光学发射光谱(inductively coupled plasma optical emission spectrometer,ICP-OES)[97-100]、原子荧光光谱(atomic fluorescence spectroscopy,AFS)[24,101]和激光诱导击穿光谱法(laser-induced breakdown spectroscopy,LIBS)的蒸气产生技术领域取得了重要的研究进展[102]。

初期 SCGD 蒸气发生装置是采用 U 型管设计的半封闭式 SCGD 系统,利用该 SCGD 系统代替 ICP-OES 的雾化器单元并将其蒸气发生性能同常规气动雾化器进行比较,证明了 SCGD 作为蒸气发生技术的应用潜力[98]。为了进一步降低 SCGD 蒸气发生过程中的死体积和极限分散性,设计了一种更为紧凑的 SCGD 装置(腔体体积约 3 mL),该装置应用于溶液中汞、碘和铼元素的蒸气发生取得了较好的效果[97-101]。上述研究同时发现,SCGD 蒸气发生系统不仅能够将无机汞[Hg(Ⅱ)]快速转化为 Hg 蒸气，同样可以将有机汞（硫柳汞）以高的蒸气发生效率直接转化为挥发性的 Hg 蒸气，并且无需对有机汞进行预氧化。此外，与其他化学蒸气发生技术相比，SCGD 中元素蒸气发生过程中伴随存在的离子干扰较小[97]。同时，在 SCGD 诱导蒸气发生系统中，有机汞物种的分解和 Hg^{2+} 的还原是瞬时完成的，因此将 SCGD 作为在线耦合高效液相色谱（high-performance liquid chromatography，HPLC）和原子荧光光谱之间的接口，能够实现对于无机汞（Hg^{2+}）、甲基汞（MeHg）和乙基汞（EtHg）的形态分析[101]。对于碘元素而言，SCGD 同样能够将碘化物和碘酸盐同时以较高的蒸气发生效率直接转化为挥发性碘蒸气[99,100]。上述研究结果直接说明了 SCGD 作为发射光谱领域样品的蒸气发生技术具有重要的应用潜力。

为了进一步提升 SCGD 诱导蒸气发生系统对于部分元素（如 Hg）的蒸气发生效率，通过在样品溶液中添加低分子量有机酸（甲酸或乙酸）或醇（乙醇），能够有效提高溶液中 Hg 进入至等离子体中的效率，进而提高了 Hg 的蒸气发生效率[97]。除此之外，2-巯基乙醇在提高 SCGD 诱导蒸气发生系统对 Hg 的蒸气发生效率方面同样有效[101]。然而，对于碘元素而言，低分子量有机化合物（乙醇和乙酸）在增加 KIO_3 的蒸气产生效率的同时，KI 的蒸气产生效率被抑制，这可能与 KI 和 KIO_3 在 SCGD 诱导蒸气发生系统中蒸气的产生机理不同有关。还同时表明样品中所存在的低分子量有机化合物会对 SCGD 诱导蒸气发生系统应用于碘蒸气发生时造成一定的干扰[99]。

相比 SCGD 诱导蒸气发生技术，SAGD 诱导蒸气发生系统所能蒸气发生的元素种类同 SCGD 具有一定差异，例如 SAGD 能够快速将液体中的镉（Cd）和锌（Zn）转化为对应的挥发性蒸气，而且目前已证实由 SAGD 产生的 Cd 和 Zn 的蒸气物质均为分子形式。这足以表明：尽管 SCGD 和 SAGD 仅是液体电极和金属电极的极性对调，但是两者的放电机理存在本质区别，这也是导致 SCGD 和 SAGD 所能诱导蒸气发生的元素种类不同的原因[24]。

6.4 常压辉光放电等离子体的机理研究进程

尽管常压辉光放电微等离子体作为发射光谱辐射源已取得了重要的研究进展，但目前仍未有相关报道对于常压辉光放电微等离子体的放电机理提出明确且完整的阐释。进一步揭示常压辉光放电微等离子体中的放电行为（包括元素的原子化、激发以及电离过程）对于进一步深化对常压辉光放电的认知、提高常压辉光放电微等离子体在各领域中的性能以及拓宽常压辉光放电微等离子体的应用领域具有重要的研究意义。针对 SEGD 体系中元素原子化、激发、电离和分子解离或碎裂机理的现有研究表明，SEGD 中的放电行为是复杂的[15,16,50,58,98,103,104]，而且 SCGD 和 SAGD 中的放电行为存在显著区别。

6.4.1 液体阴极辉光放电机理研究进程

目前针对 SCGD 的机理研究所提出的模型主要包括阴极溅射模型[2,98,106]、类电喷雾模型[9,36,48,103-105]以及上述机理的组合[50,67,104]，但没有明确的实验结果直接表明 SCGD 的放电过程为上述某种机理模型或者一种全新的放电过程。这导致目前液体阴极辉光放电微等离子体应用于发射光谱辐射源时，放电过程中分析物的原子化、激发以及电离机制仍是一个有争议的问题[105]。

6.4.1.1 阴极溅射模型

液体阴极辉光放电的阴极溅射模型可根据其溅射粒子分为三种模型：①通过离子-离子碰撞所发生的离子溅射过程[2,98,106]；②由溶液溅射产生中性原子的阴极溅射过程[98]；③从溶液中溅射产生准中性化合物的阴极溅射过程[98]。

离子-离子碰撞溅射模型：对于直接通过离子-离子碰撞进行离子溅射的假设，来自等离子体的高能轰击离子直接撞击液体阴极溶液表面层中的金属离子。但考虑到阴极暗区（cathode dark space，CDS）层中存在电场强度约为 10^7 V/m 的电场，该电场的电场方向由放电阳极指向液体阴极，因此该模型中通过离子碰撞而溅射出溶液表面的带正电粒子会在阴极暗区受到强大的电场力而促使它们重新返回阴极，导致这些通过离子碰撞的正离子无法穿越阴极暗区进入常压辉光放电中的其他区域[98,106]。因此，在离子-离子碰撞的阴极溅射过程中存在一种结合机制，即三体碰撞过程（$M^+ + e + X \longrightarrow M + X$），其中正离子（$M^+$）与具有相当能量的电子（e）结合成为中性粒子，X（可能是中性原子或电子）作为第三体吸收正离子和电子结合过程中释放的能量，该过程中产生的中性粒子（M）呈电中性，可不受电场力的作用扩散到放电中的其他区域进行激发或电离[106]。然而，考虑到正离子尺寸范围，发生上述过程的理论碰撞截面非常低，因此阴极溅射过程中发生上述过程的可能性非常小[98]。综上，SCGD 的离子-离子碰撞溅射模型具有较低的发生可能性。

中性原子的阴极溅射过程：考虑到上述阴极暗区中存在的巨大电场力会对正离子的阳极运动过程产生巨大阻碍作用，因此提出了另一种假设，即从溶液中溅射产生的是中性原子。在此假设下，等离子体中的高能正离子通过与 H_2O 分子碰撞电离产生的电子与水分子结合，以 H_2O^-（水合电子）的形式存在。这类水合电子具有强还原性，能够将金属盐溶液中的金属离子还原成中性原子。考虑到水合电子所具有的上述将金属离子还原至金属原子的能力，在目标粒子从溶液表面溅射出来之前，金属离子通过捕获溶剂化电子而成为中性金属原子的可能性是相当大的。因此，基于上述模型所构建的放电过程中，中性原子可通过溅射进入至等离子体，之后通过简单的扩散（不受阴极暗区强电场的影响）进入辉光放电区域进行激发[98]。但是，研究表明通过溅射过程进入至等离子体中的金属通量与电子-金属离子结合的过程间的相关性很弱，这在很大程度说明上述中性原子溅射的模型并不能反映常压辉光放电等离子体的真实行为。

基于上述模型，另一种准中性化合物的溅射模型被提出，即在放电过程中，由溶液中溅射进入至等离子体的物种为准中性化合物溅射：假设在阴极溅射过程中由溶液表面溅射出的初级粒子是金属-水配合物或水团簇，从表面溅射（通过轰击正离子的碰撞作用）产生的水簇离子会由于放电过程中产生的热效应而去溶剂化；随后，生成的水合络合物离子通过在阴极暗区中的强电场中失去氢氧根离子而开始稳定。但是，如果通过溅射过程由溶液进入等离子体的粒子所具有的初动能低于一定水平，则其中带正电或后续成为带正电的粒子将返回至阴极溶液中。此外，通过羟基化合物的形成而成为中性电荷的粒子可能会由于其键的极化率而受到电场的进一步影响，即电场可能破坏上述化合物中的化学键，导致产生裸露的带正电粒子，在电场作用下该离子将返回阴极。只有具有较强 OH 键的化合物才能进入等离子体中的原子化区域，由于较高的气体温度，在此区域发生原子化产生分析物原子，产生的分析物原子进一步向上运动进入等离子体中的激发区，最后以气溶胶形式离开等离子体。通过质谱检测等离子体气相中的放电产物，证明了等离子体气相中存在溶剂化金属离子，这在一定程度上与上述机理模型中描述的放电过程具有较高的一致性[98]。SCGD 的中性化合物阴极溅射模型如图 6-11 所示。

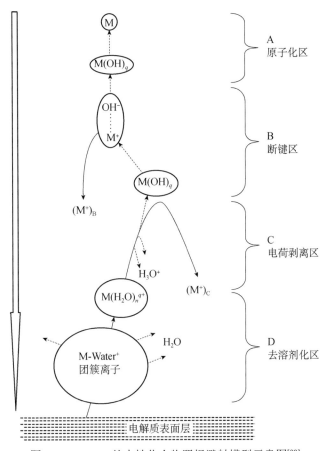

图 6-11　SCGD 的中性化合物阴极溅射模型示意图[98]

6.4.1.2　电喷雾样模型

SCGD 的多项研究得出这样的假设：在 SCGD 放电过程中，至少存在一部分阴极电解液以液滴的形式传输至放电等离子体中，这些液滴可能带电并以类似于电喷雾的方式从粗糙的溶液表面散射出来[9,36,48,103-105,107]。通过激光散射技术观测 SCGD 溶液-等离子体界面发现，SCGD 的溶液表面有大量的喷射液滴以及溶液羽流的形成，这一在 SCGD 中观察到的液滴喷射形状类似于电喷雾离子源中所观测到的泰勒锥结构，表明 SCGD 放电中可能存在类似于电喷雾机理的过程，导致 SCGD 的溶液表面产生喷射的液滴[105,108]。最重要的是，SCGD 液体阴极表面存在的约为 10^7 V/m[98]或 $6×10^6$ V/m[12,109]数量级的阴极暗区电场似乎支持 SCGD 液体阴极表面能够发生类电喷雾过程[105]。如果上述提到的类电喷雾过程不仅是 SCGD 液体阴极表面喷射液体形成的原因，同时负责将 SCGD 液体阴极中的分析物由溶液传输至 SCGD 等离子体中（分析物很可能是通过这些液滴一同进入至 SCGD 等离子体气相的），则 SCGD 放电过程中 SCGD 液体阴极表面的气溶胶和喷射液滴的形成随 SCGD 液体阴极中酸浓度降低的实验发现能够很好地用于解释为什么 SCGD 等离子体气相中分析物的发射强度随液体阴极酸度的降低而下降[105]。此外，当 SCGD 作为质谱离子源时，肾素底物 I 在经 SCGD 解离后产生了由单质子化和双质子化的分子离子为主的质谱图，这表明 SCGD 对于肾素底物 I 电离的质谱图类似于电喷雾电离质谱图。同时，双电荷分子离子的存在表明 SCGD 中正在发生类电喷雾的电离过程。然而，在目前的研究阶段，仅依据上述研究发现对于提出 SCGD 诱导分子碎片化的明确电离机制还为时过早[103]。

6.4.1.3 几种机制的结合

通过比较暴露于 SCGD 等离子体之前和之后溶液的性质以及测量有机添加剂如何影响这些性质而开展的研究工作更偏向于一种混合机理模型。该机理模型提出液体阴极辉光放电过程中，阴极电解液中的分析物由液体阴极传输至等离子体的过程是多种机理过程同时作用的结果，包括阴极表面溶液的蒸发、液滴的产生以及放电过程中分析物转化为挥发性物质[50]。

通过将 SCGD-MS 的质谱信息与碰撞诱导解离（collision-induced dissociation，CID）、电子俘获解离（electron-capture dissociation，ECD）、电子转移解离（electron-transfer dissociation，ETD）、光解离、热解离和化学裂解等方法得到的质谱进行比较，能够对 SCGD 中的解离机理有一个更为全面的认识。SCGD 中分析物（分子）的解离过程可能是多种机制的综合作用，包括分子与气相自由基的相互作用、等离子体中产生的紫外辐射或类电喷雾电离[104]。

6.4.2 SAGD 机理研究

相比较 SCGD，SAGD 具有显著不同的放电行为。首先，SAGD-OES 所能够检测的元素种类与 SCGD-OES 具有显著差异，SAGD 能够有效激发的元素多为一些易于转换为挥发性化合物的元素，如镉元素（Cd）和锌元素（Zn）。其次，向 SAGD 的液体阳极中添加低分子量有机试剂会产生不同于 SCGD 的响应结果。这表明 SAGD 和 SCGD 中，分析物由液体阳极和液体阴极传输至等离子体气相的过程是完全不同的[21,22]。此外，SCGD 和 SAGD 之间不同的电学特性以及发射光谱特性明确地表明了液体阳极和液体阴极具有显著不同的放电行为[12,14,109]。在 SAGD 放电过程中，溶液表面受到电子的轰击，因此，SCGD 中通过正离子轰击液体阴极表面所发生的溅射过程在 SAGD 中基本可以忽略[22]。分析物从 SAGD 液体阳极释放至等离子体气相的机理过程迄今无法解释，可能包括电荷转移反应导致其挥发性物质生成的过程，以及放电气相中电子的撞击导致金属离子的飞溅及其原子化过程[22]。

目前，已有研究表明，SAGD 等离子体气相中元素（Ag、Cd、Hg、In、Pb、Tl 和 Zn）的挥发性物质浓度比其在 SAGD 阳极液体中的浓度高几倍。上述研究结果表明 SAGD 放电系统中，诸如类电喷雾或溶液蒸发之类的过程并未对 SAGD 中元素由液体阳极进入至等离子体的过程发挥主要作用[20]。此外，目前已证明在 SAGD 系统的液相-等离子体气相界面处存在质子（H^+）还原为氢气以及铁氰化物还原为亚铁氰化物的过程[16]，这表明在 SAGD 体系的液相-等离子体气相交界处确实发生了电子转移反应。因此，有理由相信这种电子转移反应可能有助于形成上述元素的挥发物，进而元素能够进入至等离子体气相中[22]。即使确定 SAGD 系统的液体-等离子体气相界面处存在电子转移反应，但溶液中反应的空间分布仍不清楚，电子可能被溶剂化后在溶液中将元素离子还原为原子[17]。而且，最近一项研究通过使用全内反射几何结构直接测量了 SAGD 等离子体中产生的溶剂化电子，溶剂化电子的平均穿透深度估计为 (2.5 ± 1.0) nm，这直接证实了 SAGD 的阳极液体中确实存在溶剂化电子，并且明确了溶剂化电子的纵向空间分布[110]。但上述溶剂化电子是否在元素由液体阳极传输至等离子体气相的过程中发挥作用有待进一步研究。

参 考 文 献

[1] Gubkin D J. Electrolytische metallabscheidung an der freien oberflache einer salzlosung. Annalen Der Physik, 1887,

114-115.

[2] Cserfalvi T, Mezeit P, Apait P. Emission studies on a glow discharge in atmospheric pressure air using water as a cathode. Journal of Physics D: Applied Physics, 1993, 26: 2184-2188.

[3] Mezei P, Cserfalvi T, Jánossy M, Szöcs K, Kim H J. Similarity laws for glow discharges with cathodes of metal and an electrolyte. Journal of Physics D: Applied Physics, 1998, 31: 2818-2825.

[4] Yang S P, Ku S H, Hong H S, Piepmeier E H. Fundamental studies of electrolyte-as-cathode glow discharge-atomic emission spectrometry for the determination of trace metals in flowing water. Spectrochimica Acta Part B: Atomic Spectroscopy, 1998, 53: 1167-1179.

[5] Kim H J, Lee J H. Development of open-air type electrolyte-as-cathode glow discharge-atomic emission spectrometry for determination of trace metals in water. Spectrochimica Acta Part B: Atomic Spectroscopy, 2000, 55: 823-831.

[6] Hickling BY A, Ingram M D. Contact glow-discharge electrolysis. Homepage, 1964, 8: 65-81.

[7] Cserfalvi T, Mezei P. Direct solution analysis by glow discharge: Electrolyte-cathode discharge spectrometry. Journal of Analytical Atomic Spectrometry, 1994, 9: 345-349.

[8] Webb M R, Chan G C Y, Andrade F J, Gamez G, Hieftje G M. Spectroscopic characterization of ion and electron populations in a solution-cathode glow discharge. Journal of Analytical Atomic Spectrometry, 2006, 21: 525-530.

[9] Webb M R, Andrade F J, Gamez G, McCrindle R, Hieftje G M. Spectroscopic and electrical studies of a solution-cathode glow discharge. Journal of Analytical Atomic Spectrometry, 2005, 20: 1218-1225.

[10] Dwight E Couch, Brenner A. Glow discharge spectra of copper and indium above aqueous solutions. Journal of the Electrochemical Society, 1959, 106: 279-287.

[11] Gaisin A F. A vapor-air discharge between electrolytic anode and metal cathode at atmospheric pressure. High Temperature, 2005, 43: 680-687.

[12] Bruggeman P, Liu J, Degroote J, Kong M G, Vierendeels J, Leys C. Dc excited glow discharges in atmospheric pressure air in pin-to-water electrode systems. Journal of Physics D: Applied Physics, 2008, 41: 215201.

[13] Miao S Y, Ren C S, Wang D Z, Zhang Y T, Qi B, Wang Y N. Conical DC discharge in ambient air using water as an electrode. IEEE Transactions on Plasma Science, 2008, 36: 126-130.

[14] Xiong Q, Yang Z, Bruggeman P J. Absolute OH density measurements in an atmospheric pressure dc glow discharge in air with water electrode by broadband UV absorption spectroscopy. Journal of Physics D: Applied Physics, 2015, 48: 424008.

[15] Rumbach P, Witzke M, Sankaran R M, Go D B. Decoupling interfacial reactions between plasmas and liquids: Charge transfer vs plasma neutral reactions. Journal of the American Chemical Society, 2013, 135: 16264-16267.

[16] Richmonds C, Witzke M, Bartling B, Lee S W, Wainright J, Liu C C, Sankaran R M. Electron-transfer reactions at the plasma-liquid interface. Journal of the American Chemical Society, 2011, 133: 17582-17585.

[17] Witzke M, Rumbach P, Go D B, Sankaran R M. Evidence for the electrolysis of water by atmospheric-pressure plasmas formed at the surface of aqueous solutions. Journal of Physics D: Applied Physics, 2012, 45: 442001.

[18] Tochikubo F, Shimokawa Y, Shirai N, Uchida S. Chemical reactions in liquid induced by atmospheric-pressure DC glow discharge in contact with liquid. Japanese Journal of Applied Physics, 2014, 53: 126201.

[19] Liu X, Zhu Z, He D, Zheng H, Gan Y, Hu S, Wang Y. Highly sensitive elemental analysis of Cd and Zn by solution anode glow discharge atomic emission spectrometry. Journal of Analytical Atomic Spectrometry, 2016, 31: 1089-1096.

[20] Greda K, Gorska M, Welna M, Jamroz P, Pohl P. *In-situ* generation of Ag, Cd, Hg, In, Pb, Tl and Zn volatile species by flowing liquid anode atmospheric pressure glow discharge operated in gaseous jet mode: Evaluation of excitation processes and analytical performance. Talanta, 2019, 199: 107-115.

[21] Greda K, Burhenn S, Pohl P, Franzke J. Enhancement of emission from indium in flowing liquid anode atmospheric pressure glow discharge using organic media. Talanta, 2019, 204: 304-309.

[22] Greda K, Swiderski K, Jamroz P, Pohl P. Flowing liquid anode atmospheric pressure glow discharge as an excitation source for optical emission spectrometry with the improved detectability of Ag, Cd, Hg, Pb, Tl, and Zn. Analytical Chemistry, 2016, 88: 8812-8820.

[23] Jamroz P, Greda K, Dzimitrowicz A, Swiderski K, Pohl P. Sensitive determination of Cd in small-volume samples by miniaturized liquid drop anode atmospheric pressure glow discharge optical emission spectrometry. Analytical Chemistry, 2017, 89: 5729-5733.

[24] Liu X, Liu Z, Zhu Z, He D, Yao S, Zheng H, Hu S. Generation of volatile cadmium and zinc species based on solution anode glow discharge induced plasma electrochemical processes. Analytical Chemistry, 2017, 89: 3739-3746.

[25] Von Engel A, Seeliger R, Steenbeck M. Glow discharge at high pressures. Zeitschrift Fur Physik, 1933, 85: 144-160.

[26] Kanazawa S, Kogoma M, Moriwaki T, Okazaki S. Stable glow plasma at atmospheric-pressure. Journal of Physics D: Applied Physics, 1988, 21: 838-840.

[27] Jakubowski N, Dorka R, Steers E, Tempez A. Trends in glow discharge spectroscopy. Journal of Analytical Atomic Spectrometry, 2007, 22: 722-735.

[28] Ma J X, Wang Z, Li Q, Gai R, Li X. On-line separation and preconcentration of hexavalent chromium on a novel mesoporous silica adsorbent with determination by solution-cathode glow discharge-atomic emission spectrometry. Journal of Analytical Atomic Spectrometry, 2014, 29: 2315-2322.

[29] Yu J, Yang S, Sun D, Lu Q, Zheng J, Zhang X, Wang X. Simultaneously determination of multi metal elements in water samples by liquid cathode glow discharge-atomic emission spectrometry. Microchemical Journal, 2016, 128: 325-330.

[30] Zhang Z, Wang Z, Li Q, Zou H, Shi Y. Determination of trace heavy metals in environmental and biological samples by solution cathode glow discharge-atomic emission spectrometry and addition of ionic surfactants for improved sensitivity. Talanta, 2014, 119: 613-619.

[31] Shekhar R, Madhavi K, Meeravali N N, Kumar S J. Determination of thallium at trace levels by electrolyte cathode discharge atomic emission spectrometry with improved sensitivity. Analytical Methods, 2014, 6: 732-740.

[32] Yang C, Wang L, Zhu Z, Jin L, Zheng H, Belshaw N S, Hu S. Evaluation of flow injection-solution cathode glow discharge-atomic emission spectrometry for the determination of major elements in brines. Talanta, 2016, 155: 314-320.

[33] Świderski K, Pohl P, Jamróz P. A miniaturized atmospheric pressure glow microdischarge system generated in contact with a hanging drop electrode: A new approach to spectrochemical analysis of liquid microsamples. Journal of Analytical Atomic Spectrometry, 2019, 34: 1287-1293.

[34] Peng X X, Guo X H, Ge F, Wang Z. Battery-operated portable high-throughput solution cathode glow discharge optical emission spectrometry for environmental metal detection. Journal of Analytical Atomic Spectrometry, 2019, 34: 394-400.

[35] Jamróz P, Pohl P, Żyrnicki W. An analytical performance of atmospheric pressure glow discharge generated in contact with flowing small size liquid cathode. Journal of Analytical Atomic Spectrometry, 2012, 27: 1032-1037.

[36] Webb M R, Andrade F J, Hieftje G M. Compact glow discharge for the elemental analysis of aqueous samples. Analytical Chemistry, 2007, 79: 7899-7905.

[37] Manjusha R, Reddy M A, Shekhar R, Jaikumar S. Determination of major to trace level elements in zircaloys by electrolyte cathode discharge atomic emission spectrometry using formic acid. Journal of Analytical Atomic Spectrometry, 2013, 28: 1932-1939.

[38] Shekhar R. Improvement of sensitivity of electrolyte cathode discharge atomic emission spectrometry (ELCAD-AES)

for mercury using acetic acid medium. Talanta, 2012, 93: 32-36.

[39] Schwartz A J, Ray S J, Hieftje G M. Automatable on-line generation of calibration curves and standard additions in solution-cathode glow discharge optical emission spectrometry. Spectrochimica Acta Part B: Atomic Spectroscopy, 2015, 105: 77-83.

[40] Webb M R, Andrade F J, Hieftje G M. High-throughput elemental analysis of small aqueous samples by emission spectrometry with a compact, atmospheric-pressure solution-cathode glow discharge. Analytical Chemistry, 2007, 79: 7807-7812.

[41] Doroski T A, King A M, Fritz M P, Webb M R. Solution-cathode glow discharge-optical emission spectrometry of a new design and using a compact spectrograph. Journal of Analytical Atomic Spectrometry, 2013, 28: 1090-1095.

[42] Wang Z, Gai R, Zhou L, Zhang Z. Design modification of a solution-cathode glow discharge-atomic emission spectrometer for the determination of trace metals in titanium dioxide. Journal of Analytical Atomic Spectrometry, 2014, 29: 2042-2049.

[43] Manjusha R, Reddy M A, Shekhar R, Kumar S J. Determination of cadmium in zircaloys by electrolyte cathode discharge atomic emission spectrometry (ELCAD-AES). Analytical Methods, 2014, 6: 9850-9856.

[44] Zu W, Yang Y, Wang Y, Yang X, Liu C, Ren M. Rapid determination of indium in water samples using a portable solution cathode glow discharge-atomic emission spectrometer. Microchemical Journal, 2018, 137: 266-271.

[45] Wang J, Tang P, Zheng P, Zhai X. Analysis of metal elements by solution cathode glow discharge-atomic emission spectrometry with a modified pulsation damper. Journal of Analytical Atomic Spectrometry, 2017, 32: 1925-1931.

[46] Shekhar R, Karunasagar D, Ranjit M, Arunachalam J. Determination of elemental constituents in different matrix materials and flow injection studies by the electrolyte cathode glow discharge technique with a new design. Analytical Chemistry, 2009, 81: 8157-8166.

[47] Mezei P, Cserfalvi T, Kim H J, Mottaleb M A. The influence of chlorine on the intensity of metal atomic lines emitted by an electrolyte cathode atmospheric glow discharge. Analyst, 2001, 126: 712-714.

[48] Webb M R, Andrade F J, Hieftje G M. Use of electrolyte cathode glow discharge (ELCAD) for the analysis of complex mixtures. Journal of Analytical Atomic Spectrometry, 2007, 22: 766-774.

[49] Doroski T A, Webb M R. Signal enhancement in solution-cathode glow discharge: Optical emission spectrometry via low molecular weight organic compounds. Spectrochimica Acta Part B: Atomic Spectroscopy, 2013, 88: 40-45.

[50] Decker C G, Webb M R. Measurement of sample and plasma properties in solution-cathode glow discharge and effects of organic additives on these properties. Journal of Analytical Atomic Spectrometry, 2016, 31: 311-318.

[51] Swiderski K, Dzimitrowicz A, Jamroz P, Pohl P. Influence of pH and low-molecular weight organic compounds in solution on selected spectroscopic and analytical parameters of flowing liquid anode atmospheric pressure glow discharge (FLA-APGD) for the optical emission spectrometric (OES) determination of Ag, Cd, and Pb. Journal of Analytical Atomic Spectrometry, 2018, 33: 437-451.

[52] Greda K, Jamroz P, Dzimitrowicz A, Pohl P. Direct elemental analysis of honeys by atmospheric pressure glow discharge generated in contact with a flowing liquid cathode. Journal of Analytical Atomic Spectrometry, 2015, 30: 154-161.

[53] Greda K, Jamroz P, Pohl P. The improvement of the analytical performance of direct current atmospheric pressure glow discharge generated in contact with the small-sized liquid cathode after the addition of non-ionic surfactants to electrolyte solutions. Talanta, 2013, 108: 74-82.

[54] Greda K, Jamroz P, Pohl P. Effect of the addition of non-ionic surfactants on the emission characteristic of direct current atmospheric pressure glow discharge generated in contact with a flowing liquid cathode. Journal of Analytical Atomic

Spectrometry, 2013, 28: 134-141.

[55] Lu Q, Feng F, Yu J, Yang W. Determination of trace cadmium in zinc concentrate by liquid cathode glow discharge with a modified sampling system and addition of chemical modifiers for improved sensitivity. Microchemical Journal, 2020, 152: 104308.

[56] Lu Q, Yang S, Su M. Direct determination of Cu by liquid cathode glow discharge-atomic emission spectrometry. Spectrochimica Acta Part B: Atomic Spectroscopy, 2016, 125: 136-139.

[57] Greda K, Jamroz P, Pohl P. Ultrasonic nebulization atmospheric pressure glow discharge—Preliminary study. Spectrochimica Acta Part B: Atomic Spectroscopy, 2016, 121: 22-27.

[58] Schwartz A J, Hieftje G M. Spatially resolved measurements to improve analytical performance of solution-cathode glow discharge optical-emission spectrometry. Spectrochimica Acta Part B: Atomic Spectroscopy, 2016, 125: 168-176.

[59] Schwartz A J, Hieftje G M. Evaluation of interference filters for spectral discrimination in solution-cathode glow discharge optical emission spectrometry. Journal of Analytical Atomic Spectrometry, 2016, 31: 1278-1286.

[60] Yu J, Zhang X, Lu Q, Sun D, Zhang Z, Yang W. Evaluation of analytical performance for the simultaneous detection of trace Cu, Co and Ni by using liquid cathode glow discharge-atomic emission spectrometry. Spectrochimica Acta Part B: Atomic Spectroscopy, 2018, 145: 64-70.

[61] Yu J, Zhang X, Lu Q, Wang X, Sun D, Wang Y, Yang W. Determination of calcium and zinc in gluconates oral solution and blood samples by liquid cathode glow discharge-atomic emission spectrometry. Talanta, 2017, 175: 150-157.

[62] Zheng P C, Gong Y M, Wang J M, Zeng X B. Elemental analysis of mineral water by solution cathode glow discharge-atomic emission spectrometry. Analytical Letter, 2016, 50: 1512-1520.

[63] Yuan M, Peng X, Ge F, Wang Z. Simplified design for solution anode glow discharge atomic emission spectrometry device for highly sensitive detection of Ag, Bi, Cd, Hg, Pb, Tl, and Zn. Microchemical Journal, 2020, 155: 104785.

[64] Yu J, Yang S, Lu Q, Sun D, Zheng J, Zhang X, Wang X, Yang W. Evaluation of liquid cathode glow discharge-atomic emission spectrometry for determination of copper and lead in ores samples. Talanta, 2017, 164: 216-221.

[65] Wang Z, Schwartz A J, Hieftje G M. Determination of trace sodium, lithium, magnesium, and potassium impurities in colloidal silica by slurry introduction into an atmospheric-pressure solution-cathode glow discharge and atomic emission spectrometry. Journal of Analytical Atomic Spectrometry, 2013, 28: 234-240.

[66] Zu W, Wang Y, Yang X, Liu C. A portable solution cathode glow discharge-atomic emission spectrometer for the rapid determination of thallium in water samples. Talanta, 2017, 173: 88-93.

[67] Jamróz P, Gręda K, Pohl P, Żyrnicki W. Atmospheric pressure glow discharges generated in contact with flowing liquid cathode: Production of active species and application in wastewater purification processes. Plasma Chemistry and Plasma Processing, 2013, 34: 25-37.

[68] Greda K, Jamroz P, Pohl P. Comparison of the performance of direct current atmospheric pressure glow microdischarges operated between a small sized flowing liquid cathode and miniature argon or helium flow microjets. Journal of Analytical Atomic Spectrometry, 2013, 28: 1233.

[69] Webb M R, Hieftje G M. Spectrochemical analysis by using discharge devices with solution electrodes. Analytical Chemistry, 2009, 81: 862-867.

[70] He Q, Zhu Z, Hu S. Flowing and nonflowing liquid electrode discharge microplasma for metal ion detection by optical emission spectrometry. Applied Spectroscopy Reviews, 2013, 49: 249-269.

[71] Mo J M, Wang Z. On-line separation and pre-concentration on a mesoporous silica-grafted graphene oxide adsorbent coupled with solution cathode glow discharge-atomic emission spectrometry for the determination of lead. Microchemical Journal, 2017, 130: 353-359.

[72] Li Q, Zhang Z, Wang Z. Determination of Hg^{2+} by on-line separation and pre-concentration with atmospheric-pressure solution-cathode glow discharge atomic emission spectrometry. Analytica Chimica Acta, 2014, 845: 7-14.

[73] György K, Bencs L, Mezei P, Cserfalvi T. Novel application of the electrolyte cathode atmospheric glow discharge: Atomic absorption spectrometry studies. Spectrochimica Acta Part B: Atomic Spectroscopy, 2012, 77: 52-57.

[74] Cheng J Q, Li Q, Zhao M, Wang Z. Ultratrace Pb determination in seawater by solution-cathode glow discharge-atomic emission spectrometry coupled with hydride generation. Analytica Chimica Acta, 1077: 107-115.

[75] Guo X X, Peng X X, Li Q, Mo J M, Du Y, Wang Z. Ultra-sensitive determination of inorganic arsenic valence by solution cathode glow discharge-atomic emission spectrometry coupled with hydride generation. Journal of Analytical Atomic Spectrometry, 2017, 32: 2416-2422.

[76] Huang C C, Li Q, Mo J M, Wang Z. Ultratrace determination of tin, germanium, and selenium by hydride generation coupled with a novel solution-cathode glow discharge-atomic emission spectrometry method. Analytical Chemistry, 2016, 88: 11559-11567.

[77] Greda K, Jamroz P, Pohl P. Coupling of cold vapor generation with an atmospheric pressure glow microdischarge sustained between a miniature flow helium jet and a flowing liquid cathode for the determination of mercury by optical emission spectrometry. Journal of Analytical Atomic Spectrometry, 2014, 29: 893-902.

[78] Greda K, Jamroz P, Jedryczko D, Pohl P. On the coupling of hydride generation with atmospheric pressure glow discharge in contact with the flowing liquid cathode for the determination of arsenic, antimony and selenium with optical emission spectrometry. Talanta, 2014, 137: 11-17.

[79] Mo J M, Li Q, Guo X X, Wang Z. Flow injection photochemical vapor generation coupled with miniaturized solution-cathode glow discharge atomic emission spectrometry for determination and speciation analysis of mercury. Analytical Chemistry, 2017, 89: 10353-10360.

[80] Zhao M, Peng X, Yang B, Wang Z. Ultra-sensitive determination of antimony valence by solution cathode glow discharge optical emission spectrometry coupled with hydride generation. Journal of Analytical Atomic Spectrometry, 2020, 35: 1148-1155.

[81] Schwartz A J, Wang Z, Ray S J, Hieftje G M. Universal anion detection by replacement-ion chromatography with an atmospheric-pressure solution-cathode glow discharge photometric detector. Analytical Chemistry, 2013, 85: 129-137.

[82] Jiang X, Xu X, Hou X, Zheng C. A novel capillary microplasma analytical system: Interface-free coupling of glow discharge optical emission spectrometry to capillary electrophoresis. Journal of Analytical Atomic Spectrometry, 2016, 31: 1423-1429.

[83] Greda K, Swiderski K, Jamroz P, Pohl P. Reduction of spectral interferences in atmospheric pressure glow discharge optical emission spectrometry. Microchemical Journal, 2017, 130: 7-13.

[84] Shirai N, Nakazawa M, Ibuka S, Ishii S. Atmospheric DC glow microplasmas using miniature gas flow and electrolyte cathode. Japanese Journal of Applied Physics, 2009, 48: 036002.

[85] Zhou R, Zhou R, Prasad K, Fang Z, Speight R, Bazaka K, Ostrikov K. Cold atmospheric plasma activated water as a prospective disinfectant. The crucial role of peroxynitrite. Green Chemistry, 2018, 20: 5276-5284.

[86] Andrade F J, Wetzel W C, Chan G C Y, Webb M R, Gamez G, Ray S J, Hieftje G M. A new, versatile, direct-current helium atmospheric-pressure glow discharge. Journal of Analytical Atomic Spectrometry, 2006, 21: 1175.

[87] Staack D, Farouk B, Gutsol A, Fridman A. Characterization of a dc atmospheric pressure normal glow discharge. Plasma Sources Science and Technology, 2005, 14: 700-711.

[88] Gielniak B, Broekaert J A C. Study of a new direct current atmospheric pressure glow discharge in helium. Spectrochimica Acta Part B: Atomic Spectroscopy, 2011, 66: 21-27.

[89] Peng X X, Wang Z. Ultrasensitive determination of selenium and arsenic by modified helium atmospheric pressure glow discharge optical emission spectrometry coupled with hydride generation. Analytical Chemistry, 2019, 91: 10073-10080.

[90] Gherardi N, Gouda G, Gat E, Ricard A, Massines F. Transition from glow silent discharge to micro-discharges in nitrogen gas. Plasma Sources Science and Technology, 2000, 9: 340-346.

[91] Zheng H, Guan X, Mao X, Zhu Z, Yang C, Qiu H, Hu S. Determination of nitrite in water samples using atmospheric pressure glow discharge microplasma emission and chemical vapor generation of NO species. Analytica Chimica Acta, 2018, 1001: 100-105.

[92] Moβ K K, Broekaert J A C. Study of a direct current atmospheric pressure glow discharge in helium with wet aerosol sample introduction systems. Journal of Analytical Atomic Spectrometry, 2014, 29: 674.

[93] Zhang S, Luo H, Peng M, Tian Y, Hou X, Jiang X, Zheng C. Determination of Hg, Fe, Ni, and Co by miniaturized optical emission spectrometry integrated with flow injection photochemical vapor generation and point discharge. Analytical Chemistry, 2015, 87: 10712-10718.

[94] Deng Q, Yang C, Zheng H, Liu J, Zhu Z. Direct determination of cadmium in rice by solid sampling electrothermal vaporization atmospheric pressure glow discharge atomic emission spectrometry using a tungsten coil trap. Journal of Analytical Atomic Spectrometry, 2019, 34: 1786-1793.

[95] Ge F, Gao L, Peng X, Li Q, Zhu Y, Yu J, Wang Z. Atmospheric pressure glow discharge optical emission spectrometry coupled with laser ablation for direct solid quantitative determination of Zn, Pb, and Cd in soils. Talanta, 2020, 218: 121119.

[96] Eijkel J C T, Stoeri H, Manz A. A DC microplasma on a chip employed as an optical emission detector for gas chromatography. Analytical Chemistry, 2000, 72: 2547-2552.

[97] Zhu Z, Chan G C Y, Ray S J. Use of a solution cathode glow discharge for cold vapor generation of mercury with determination by ICP-atomic emission spectrometry. Analytical Chemistry, 2008, 80: 7043-7050.

[98] Cserfalvi T, Mezei P. Investigations on the element dependency of sputtering process in the electrolyte cathode atmospheric discharge. Journal of Analytical Atomic Spectrometry, 2005, 20: 939.

[99] Zhu Z, He Q, Shuai Q, Zheng H, Hu S. Solution cathode glow discharge induced vapor generation of iodine for determination by inductively coupled plasma optical emission spectrometry. Journal of Analytical Atomic Spectrometry, 2010, 25: 1390.

[100] Zhu Z, Huang C, Xiao Q, Liu Z, Zhang S, Hu S. On line vapor generation of osmium based on solution cathode glow discharge for the determination by ICP-OES. Talanta, 2013, 106: 133-136.

[101] He Q, Zhu Z, Hu S, Jin L. Solution cathode glow discharge induced vapor generation of mercury and its application to mercury speciation by high performance liquid chromatography-atomic fluorescence spectrometry. Journal of Chromatography A, 2011, 1218: 4462-4467.

[102] Zheng P, Liu H, Wang J, Shi M, Wang X, Zhang B, Yang R. Online mercury determination by laser-induced breakdown spectroscopy with the assistance of solution cathode glow discharge. Journal of Analytical Atomic Spectrometry, 2015, 30: 867-874.

[103] Schwartz A J, Williams K L, Hieftje G M, Shelley J T. Atmospheric-pressure solution-cathode glow discharge: A versatile ion source for atomic and molecular mass spectrometry. Analytica Chimica Acta, 2017, 950: 119-128.

[104] Schwartz A J, Shelley J T, Walton C L, Williams K L, Hieftje G M. Atmospheric-pressure ionization and fragmentation of peptides by solution-cathode glow discharge. Chemical Science, 2016, 7: 6440-6449.

[105] Schwartz A J, Ray S J, Elish E, Storey A P, Rubinshtein A A, Chan G C, Pfeuffer K P, Hieftje G M. Visual observations of an atmospheric-pressure solution-cathode glow discharge. Talanta, 2012, 102: 26-33.

[106] Mezei P, Cserfalvi T, Janossy M. Pressure dependence of the atmospheric electrolyte cathode glow discharge spectrum. Journal of Analytical Atomic Spectrometry, 1997, 12: 1203-1208.

[107] Pohl P, Jamroz P, Swiderski K, Dzimitrowicz A. Lesniewicz, critical evaluation of recent achievements in low power glow discharge generated at atmospheric pressure between a flowing liquid cathode and a metallic anode for element analysis by optical emission spectrometry. Trends in Analytical Chemistry, 2017, 88: 119-133.

[108] Moon D E, Webb M R. Imaging studies of emission and laser scattering from a solution-cathode glow discharge. Journal of Analytical Atomic Spectrometry, 2020, DOI: 10.1039/d0ja00134a.

[109] Bruggeman P, Guns P, Degroote J, Vierendeels J, Leys C. Influence of the water surface on the glow-to-spark transition in a metal-pin-to-water electrode system. Plasma Sources Science and Technology, 2008, 17: 045014.

[110] Rumbach P, Bartels D M, Sankaran R M, Go D B. The solvation of electrons by an atmospheric-pressure plasma. Nature Communications, 2015, 6: 7248.

第 7 章　现场环境元素分析

（郑成斌[①]*，邹志荣[②]，侯贤灯[①]）

▶ 7.1　概述 / 177

▶ 7.2　现场环境元素分析的样品预处理 / 178

▶ 7.3　现场环境元素分析的进样方法 / 179

▶ 7.4　现场环境分析方法与装置 / 185

▶ 7.5　展望 / 186

---

①四川大学，成都；②四川师范大学，成都。*通讯作者联系方式：abinscu@scu.edu.cn

本章导读

- 介绍了现场环境元素分析的重要性和必要性。
- 介绍了现场分析样品消解和现场分离富集金属的方法。
- 介绍了常用现场环境元素分析仪器装置的进样技术。
- 介绍了可用于现场环境元素分析的原子光谱分析方法与装置。
- 对现场环境元素分析进行了展望。

7.1 概 述

近年来，大量研究表明环境污染已成为影响人类健康的重要因素之一[1,2]。Landrigan 曾在《柳叶刀》发表调查报告：2015 年环境污染就造成了 900 万人死亡和 4.6 万亿美元的经济损失[3]。世界卫生组织（World Health Organization，WHO）也指出全球 70%的疾病和 40%的死亡人数与环境污染因素密切相关[4]。同时，国际癌症研究机构（International Agency of Research on Cancer，IARC）在 2018 年发布的《全球癌症报告》中估计 2018 年全球新增癌症病例约为 1810 万例，其中绝大多数新增病例是由于空气、饮用水和食品污染所致[5]。环境污染导致的严重后果引起世界各国的高度重视，我国也先后出台了《大气污染防治行动计划》《水污染防治行动计划》《土壤污染防治行动计划》等管理办法，简称为"水十条、土十条、气十条"。

重金属元素是一类典型的环境污染物，除天然源外，其主要来源于矿产开采、冶炼和加工以及农业生产和生活等活动。大多数有毒重金属易与蛋白质、生物酶以及其他生物分子发生作用，即使低浓度下也会导致生物体代谢紊乱，引起"致畸、致癌、致突变"等危害[6]。不同于有机化合物污染，重金属污染具有难降解、生物富集、难以通过人为处理和修复去除等特点，一旦释放到环境中就会参与环境系统循环，并通过食物链进入人体。近年来，有研究表明我国一些地区的大气、水体和土壤中重金属污染较为严重。如在 2000~2012 年期间，有研究对我国 30 个省、直辖市共 42 个城市的大气重金属污染物进行了调研，发现 18 个城市空气中的铅含量超过我国《环境空气质量标准》（GB 3095—2012）的年平均值（272 ng/m^3），对人们身体健康特别是儿童智力发育存在一定风险[7]；清华大学郝吉明院士团队基于多种方法估计 1978~2014 年我国大约向大气中排放了 13 294 吨汞，其中气态单质汞、气态氧化汞及颗粒态汞分别占 58.2%、37.1%和 4.7%[6]，并研究了我国各行业向大气中排放汞的情况（图 7-1）；2014 年 4 月环境保护部和国土资源部联合发布的《全国土壤污染状况调查公报》表明，全国土壤总的超标率达 16.1%，近 1/5 的耕地受到污染，其中轻微、轻度、中度和重度污染点位比例分别为 11.2%、2.3%、1.5%和 1.1%。为此，"十三五"规划将重金属污染综合防治作为环境污染治理的工作重点。

掌握重金属含量、迁移转化以及毒性毒理等信息是重金属污染防治的关键，而获取这些信息离不开灵敏和准确的环境分析方法开发和仪器研制。原子光谱分析法因其灵敏度高、选择性好、仪器稳定等优点，而成为最常用的重金属分析方法，是重金属准确测定的"金"标准。传统原子光谱分析仪体积较为庞大、能耗高，因而目前基于原子光谱的环境重金属分析通常只能在实验室内完成。然而，突发性环境污染事件时有发生，急需现场环境分析方法和装置，以便对污染物种类、浓度及扩散方向进行快速判断，以便制定正确的防控措施[8]。另一方面，当样品离开环境母体后，由于溶解氧、光照、

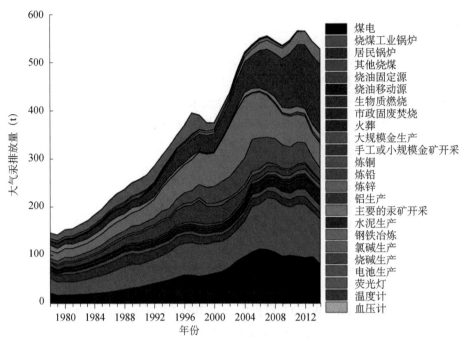

图 7-1　中国 1978~2014 年大气汞的排放趋势及其来源[6]

样品基质、压力等诸多因素发生了急剧变化，因而在样品采集、运输和储存过程不可避免地发生氧化还原、吸附、挥发等过程，导致待测分析物损失和污染等问题[9]。此外，现场环境分析技术可避免样品运输、保存等步骤，减轻了从事重金属环境污染物普查的工作者的负担。因此，研制可现场分析与监测的原子光谱分析仪具有非常重要的意义，已成为重金属污染物检测与防控的研究热点和难点。近年来，国内外科学工作者在现场环境重金属污染物分析方法和仪器装置方面取得了可喜进展。本章聚焦基于原子光谱技术的现场环境重金属分析方法、技术和仪器装置的研究进展，主要包括样品预处理、样品引入、分析装置等内容。

7.2　现场环境元素分析的样品预处理

样品处理技术既可高效富集分析物又能有效消除样品基体干扰，从而成为环境样品分析中至关重要的环节，其占据整个分析时间的 60% 以上，也是分析误差的主要来源。由于现场环境分析装置能耗低、仪器简单、抗干扰能力差，因此与传统实验室分析仪器相比，其对样品前处理技术的依赖程度更高。遗憾的是，样品前处理新方法、新技术及新装置的研究严重滞后，没有得到足够的重视，用于现场环境元素分析的样品前处理技术（样品消解、待测元素的分离富集等）虽然非常重要，但相关报道不多。

7.2.1　样品现场消解方法

虽然传统样品消解方法特别是微波消解方法可实现快速、样品损失小、无机酸和氧化剂消耗少的样品消解，但传统消解方法普遍存在能耗高、装置体积大，难以用于野外现场样品消解。基于芬顿或类芬顿反应的高级氧化法常用于有机污物降解，其主要通过 Fe^{2+}、Co^{2+}、Cu^{2+} 等过渡金属离子及其纳米材料催化 H_2O_2，产生羟基（·OH）等氧化性自由基。这些氧化性自由基可迅速矿化有机污染物，实现对其快速降解。基于此，我们将具有非均相类芬顿效应的纳米 Fe_3O_4、零价铁用于传统微波消解[10]、紫外光消解[11]。由于消解过程中产生大量的·OH，这类消解方法既显著降低了样品消解温度以及强酸、氧化剂的使用量，使得整个消解反应条件更加温和，有望实现环境样品野外现场消解。另一方面，磁性纳米 Fe_3O_4 对一些元素（如砷、镉等）具有极强的吸附能力，在加快样品消解的同时还能对待测元素离子进行有效的分离富集，从而可显著提高分析方法的灵敏度。以上方法虽然改善了传统消解方法的一些不足，但也存在纳米催化材料易团聚，导致催化效果不能有效利用的问题。含铁微马达材料同时兼具类芬顿催化和催化降解 H_2O_2 产生氧气泡的性能，可避免催化剂团聚，从而保持催化剂活性，实现了环境固态样品野外现场消解[12]。

微等离子体作用于氧气可产生臭氧、羟基自由基等氧化性物质，因而也常用于降解有机污染物或消解含有机物的样品。最近，我们构建了阵列式的介质阻挡放电（dielectric barrier discharge，DBD）样品消解新装置，通过微等离子体作用产生氧化性物质以氧化有机物，最终实现低功耗、高通量的样品消解[13]。该方法仅需空气，不再加入氧化剂，就可实现样品消解，具有绿色环保、现场消解等潜力。

7.2.2 待测金属离子现场分离富集

通常，环境样品中污染物浓度极低，同时一些共存离子或化合物也会对后续的检测造成干扰，因此需要发展现场分离富集技术。目前，用于元素现场分析的分离富集技术通常有液液萃取、固相萃取（solid phase extraction，SPE）以及固相微萃取（solid phase microextraction，SPME）等。例如，浊点萃取[14,15]、中空纤维支撑液膜萃取[16]、超声场辅助浊点萃取[17]、编结反应器[18]、液液分散萃取[19]、单液滴萃取[20]、离子液体辅助液相萃取[21]等样品分离富集技术均成功用于小型化钨丝电热原子吸收光谱分析中。这些技术的引入既消除了样品基质干扰，又改善了分析方法的灵敏度、抗干扰能力及检出限等各项指标。为了进一步提高小型化电热钨丝原子吸收光谱仪的灵敏度，侯贤灯等[22]发展了流动注射-氢化物发生-钨丝原子化器原位捕集技术，使钨丝电热原子吸收光谱测镉的灵敏度提高了 58 倍，检出限达 3 ng/L。液液萃取技术通常需要使用有机试剂。一方面，在现场环境分析中会产生二次污染物难以处理，存在污染环境的潜在风险；另一方面，有机试剂的引入严重影响原子光谱的原子化器、激发源（特别是微等离子体激发源）的原子化效率、激发性能和稳定性。为了避免有机试剂的干扰和提高原子光谱法的分析性能，科研工作者采用各种固相萃取技术对分析物进行分离富集[23-25]。Barua 等[26]利用 SPE 萃取酸性废水中的贵金属（Au、Pd 和 Pt），随后使用液体电极辉光放电原子发射光谱法对这些贵金属进行高灵敏分析。SPME 作为一种新颖、绿色的样品预处理技术，不但集样品分离、富集和进样于一体，还避免使用有机试剂，是理想的污染物现场分离富集技术。为此，顶空 SPME 技术被引入微等离子体原子发射光谱分析中，用于环境水样中铅和汞的测定。该方法在实现了 Hg 和 Pb 富集的同时，还消除了水汽对微等离子体的影响，使 Hg 和 Pb 的检出限分别达到 1 ng/L 和 3 ng/L[27,28]。SPME 由于受到萃取相的限制，难以进一步提高分析方法灵敏度，最近，在线吹扫捕集技术被引入微等离子体原子发射光谱分析，在汞的测定中获得了 0.08 ng/L 的检出限，实现了清洁环境水样中超痕量汞的分析测定[29]。

7.3 现场环境元素分析的进样方法

样品引入方法不但严重影响样品的进样效率，而且还决定基体与分析物能否有效分离，从而影响分析方法的灵敏度和抗干扰能力。由于小型原子光谱仪器装置简单、能耗低、抗干扰能力有限，因此对样品引入技术提出了更高的要求。为了满足现场环境分析的需求，科学工作者在传统直接液体进样的基础上，将化学蒸气发生（chemical vapor generation，CVG）、电热蒸发、SPME 等进样技术引入，显著提高了小型化原子光谱仪器的分析性能，其中 CVG 是最常用的样品引入方法，主要包括光化学蒸气发生（photochemical vapor generation，PVG）[30]、氢化物发生（hydride generation，HG）[31,32]、氧化物发生（oxidation vapor generation，OVG）[33]、微等离子体诱导化学蒸气发生（microplasma induced chemical vapor generation，MI-CVG）[34,35]等。为了进一步满足现场元素快速分析的需求，科研人员还进一步发展了多种小型化的进样装置/方法。

7.3.1 直接固体/液体样品分析与气动雾化进样

手持式 X 射线荧光光谱仪是最常用的元素现场分析仪器。该仪器可以直接用于岩矿、土壤、金属等样品中元素现场分析，但由于灵敏度不足以及缺乏合适的标准物质对照，该方法多用于定性和半定量分析。目前，尚未有将传统小型原子光谱仪直接用于固体样品分析的相关报道，通常需要将样品溶解成液体状态，随后采用各种进样方法引入至仪器中进行检测。液体样品直接分析或者气动雾化进样是原子光谱分析最常用的进样方法，该方法具有简单、快速和普适性强等优点。例如在小型钨丝电热原子吸收光谱分析中多采用直接注入液体样品进行分析。在早期的微等离子体原子发射光谱分析研究中也常将样品液体作为一个或两个电极，构建了液体电极辉光放电原子发射光谱仪，主要包括溶液阴极辉光放电（solution cathode glow discharge，SCGD）和溶液阳极辉光放电（solution anode glow discharge，SAGD）等类型。这类方法将等离子体直接作用于待测溶液，经过蒸发、熔融、原子化、激发等过程实现待测元素的原子发射光谱检测。

气动雾化进样是将液体样品雾化成为气溶胶，随后在载气的帮助下，将其吹入原子光谱仪中进行检测。该方法由于结构简单、易于操作等优点，目前仍然是市售原子光谱分析仪的最常用的进样方式。但是，由于气动雾化技术的进样效率低（约为 5%）、受基体干扰等缺点，该方法灵敏度的改善受到了一定的限制。此外，气动雾化进样还存在毛细管容易被悬浮小颗粒或析出的盐分堵塞等问题。对于微等离子体来说，水蒸气的存在还会降低微等离子体的稳定性和分析性能。因此，气动雾化进样在微等离子体中的应用尚不多见。王建华等巧妙地将气动雾化进样技术与 DBD 相结合，将气动雾化器喷嘴与钨片形成 DBD，将液体样品引入 DBD 中实现了待测元素的直接原子化和激发[36,37]。该方法随后还与阀上实验室（lab-on-valve，LOV）联用，用于环境水样品中痕量金属元素的分析[38]。此外，Greda 等[39]通过超声雾化进样装置产生气溶胶，随后将其引入大气压辉光放电-原子发射光谱仪（atmospheric pressure glow microdischarge-optical emission spectrometer，APGD-OES）中进行检测。

由于待测液体样品一般具有一定的导电性，因此，可将液体待测样品作为其中一个电极构建辉光放电体系，实现对液体样品中金属元素的检测，该方法称为电解液阴极辉光放电（electrolyte cathode discharge，ELCAD），其主要放电方式以溶液阴极辉光放电为主。目前，SCGD 的研究主要集中在提高对目标分析物的分析性能、装置小型化、减小样品进样量等方面。通过向电极溶液中添加低分子量有机酸[40,41]和表面活性剂[42]等方式可以提高 SCGD-OES 的灵敏度，改善分析体系的分析性能。为了构建便携式的 SCGD-OES 分析体系，汪正等[43]和 Webb 等[44]分别将微量注射管和毛细管直接与石墨管连接，并与钨棒构成一个新型的 SCGD 反应池，极大地简化了 SCGD-OES 装置。毛细管电泳（capillary electrophoresis，CE）具有快速、高效分离目标分析物的性能，因此被用于与液体电极辉光放电-原子发射光谱法（liquid-electrode glow discharge-optical emission spectrometry）联用，构建了一个小型化的微等离子体分析体系，不仅使分析装置变得更为紧凑，而且实现了微量样品进样和元素形态分析[45]。受该工作的启发，研究者们还建立了液滴阵列平台（图 7-2），通过毛细管虹吸效应实现无泵无阀的自动进样,并将该小型化的毛细管液体电极微等离子体-原子发射光谱仪用于血液样品和水样品中汞和镉的分析[46]。Jamroz 等[47]将钨电极与放在石墨片电极上的少量液体构建了液滴阳极-直流大气压辉光放电-原子发射光谱分析新装置,实现了对微量液体样品(50 μL)的分析。此外,一种样品消耗量少于 80 μL 的液膜-DBD-OES 分析系统也成功用于环境水样多元素的快速、同时分析[48]。

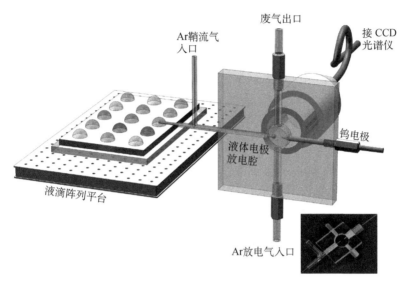

图 7-2　液滴阵列-毛细管液体电极微等离子体-原子发射光谱仪示意图[46]

7.3.2　电热蒸发进样

电热蒸发进样(electrothermal vaporization,ETV)是将微量样品置于石墨或者金属材料表面,通过一系列的程序升温将其转变为干燥气溶胶的进样方法[49]。该方法具有以下几个优点:①进样效率较高;②样品用量少,适用于微量样品的分析;③由于 ETV 是通过一定的程序实现升温,在待测物实现电热蒸发之前,部分其他杂质已经被去除,从而实现有效的基质分离,减少了来自基体的干扰;④对于复杂基质或者高盐度样品,该方法可以简化样品前处理步骤,减少其他污染物的引入;⑤有可能实现固体样品直接分析。石墨炉是商用原子光谱仪器中主要的 ETV 装置,此外,钨、钽、铂和钼等多种金属也被制成不同的装置用于原子光谱分析的 ETV 进样[50,51]。其中,钨丝由于其熔点高、体积小、导电性能好、价格低廉、能耗低等优点,是目前较为理想的石墨炉替代品,特别是在构建小型化/便携式分析原子光谱分析仪器的应用中[52,53]。

基于钨丝电热蒸发的优点,便携式、低能耗的钨丝 ETV 装置最近被成功用作 DBD-OES 的进样装置,构建了钨丝 ETV-DBD-OES 分析系统。该系统可先通过程序升温实现待测物与易挥发基质和水分的有效分离。随后,继续提高温度使待测物蒸发进入 DBD 中进行原子化/激发并检测,显著提高了方法抗干扰能力和分析灵敏度[54]。同时,钨丝 ETV 还被用于 PD-OES,进一步提高便携式微等离子体原子发射光谱灵敏度,实现多种元素的同时高灵敏度分析[55]。在此基础上,进一步将氢化物发生、钨丝在线捕集与 ETV-DBD-OES 联用,实现水中超痕量镉的高灵敏分析[56]。由于钨丝 ETV 装置具有体积小、蒸发速度快、装置简单等特点,其与微等离子体原子发射光谱联用报道后,受到了同行广泛关注,纷纷研制了各种 ETV-DBD-OES 系统,并用于水中铅、大米中镉的分析检测[57-59]。

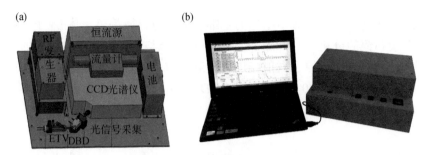

图 7-3　便携式 ETV-DBD-OES 仪器示意图（a）和实物图片（b）[58]

7.3.3　化学蒸气发生进样

对于原子光谱分析仪器来说，气态进样方式相对于传统的液体进样方式具有独特的优势，如进样效率高、基体干扰小等优点。化学蒸气发生特别是氢化物发生毫无疑问是原子光谱仪的最常见也是最为成熟的气态进样方式，通过 CVG 可以将溶液中的目标分析物（一般是离子状态）转变为其挥发性物质，从而有效地实现目标分析物的高效蒸气进样（甚至可能接近 100%）以及与复杂样品基质的有效分离，进而极大地改善目标分析物的灵敏度和检出限。至今已有多种 CVG 方法被报道作为原子光谱分析的气态进样方法，除了 HG，还有光化学蒸气发生、氧化物发生等。CVG 相关详细内容可参见本书第 8 章。这里主要简介 CVG 在微等离子体原子光谱分析中的应用。

7.3.3.1　氢化物发生

HG 是最常用的 CVG 方法之一，也是目前商品化原子光谱仪器中较为常用的进样手段之一。DBD 在原子光谱分析中一般作为原子化器[60,61]、激发源[62]和离子源[63]，其特点是低功率、小型化，但大量溶液雾化进样可能降低其激发能力，甚至熄灭等离子体。为了提高 DBD-OES 测定砷的性能[62]，可通过 HG 将砷转化为具有挥发性的 AsH_3，经气液分离后进入 DBD 中进行激发/原子化，最终实现对水样品中砷的灵敏分析。此外，DBD 还被用作商品化 AFS/AAS 的原子化器，结合 HG 实现了对 As、Sb、Pb、Se 和 Bi 等元素的检测[60,64,65]。为了提高 AFS 测砷的灵敏度，DBD 装置还被成功用于对砷氢化物在线高效捕集[66]。Franzke 等[67,68]通过 HG 将砷、碲等元素的氢化物引入毛细管 DBD 中，实现了对其空间和时间分辨的检测。

尖端放电（point discharge，PD）是通过在两个尖端电极上施加电压，使其形成稳定的放电形式。相比于 DBD 等其他微等离子体，PD 的能量更集中，因而具有更高的激发能力。其装置也更为简单，易于与各种进样技术联用。以 HG 为 PD-OES 进样方式，构建了小型化的 HG-PD-OES 仪器，并将其应用于痕量 As、Bi、Sb 和 Sn 的检测[69]。为了提高分析性能，可将该系统中的一根电极替换为空心管电极（hollow electrode，HE）。该电极既是放电电极也作为样品引入通道，即氢化物通过该空心通道直接进入微等离子体中，从而使进样效率和激发效率得到了较大的改善，提高了灵敏度和稳定性。3D 打印技术作为先进制造技术，近年在分析仪器制造方面展现出了无可比拟的优势。为此，3D 打印技术被用于成功构建并组装了小型化的 HG-HEPD-OES 分析装置[70]（图 7-4），为微型原子光谱仪研制提供了一条新路径。为了提高微等离子体体积，使得分析物尽可能被激发，可采用交叉双位点（四电极）尖端放电设计，相对于传统的尖端放电模式，其放电面积有效增大，提高了分析物激发概率，对 As、Hg 和 Pb 的分析性能增强了 3~4 倍[71]。传统的 HG 进样产生了较多的副产物氢气以及引入水蒸气，这些都会在一定程度上影响微等离子体的稳定性和分析性能，甚至可能会淬灭微等离子体。

为解决这一问题，可在将氢化物发生的产物引入 DBD 进行原子化/激发之前，将其通过浓硫酸进行干燥处理，减少了水蒸气的干扰[62,72]。

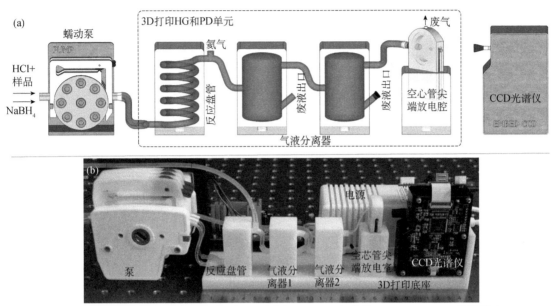

图 7-4　(a) HG-HEPD-OES 工作原理示意图；(b) 3D 打印的 HG-HEPD-OES 装置照片

SGD-OES 早期直接用于液体样品中金属元素分析，将其与 HG 进样技术联用，成功用于硒、锗、锡元素的检测[73]。这些元素的氢化物通过一个空心钛管引入 SCGD-OES 中，使 SCGD-OES 的分析性能得到了极大提升。随后，该装置还成功用于铅[74]和砷[75]的分析。此外，HG 技术还与 APGD-OES 联用以改善硒、砷和锑的分析灵敏度和检出限[76-78]。朱振利等[79]还发展了一个电池驱动的小型化 HG-SCGD-OES 分析仪器（图 7-5），并将其应用于地下水中砷的分析检测，该仪器具有功率低（<8 W）、耗气量小（100 mL/min）等优点，且其分析性能可与市售的 HG-AFS 仪器相媲美。

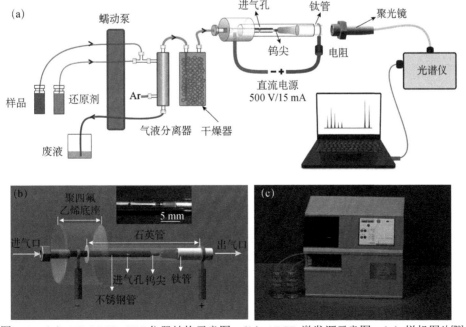

图 7-5　(a) HG-SCGD-OES 仪器结构示意图；(b) APGD 激发源示意图；(c) 样机图片[79]

7.3.3.2 光化学蒸气发生

光化学蒸气发生是近年来新发展的一种绿色、高效的气体进样方式。该方法由 Sturgeon 等于 2003 年首次提出，并将其应用于硒的检测分析[80]，随后该方法逐渐发展成为一种新的进样方式，在原子光谱分析领域得到了广泛应用[30,35,81-83]。近年来，随着各类高效光催化剂的发展和离子增敏剂的引入，使 PVG 可检测的元素范围得到了极大的拓展。目前该方法不仅适用于传统的氢化物发生元素，还可以应用于过渡金属元素和部分非金属元素的检测。作为一种理想的气态进样新方式，PVG 不仅保持了传统 CVG 的优点，还具有一些独特的优势，如绿色环保（仅使用低分子量有机酸）、受过渡金属干扰小、不产生或仅产生少量氢气的优点等[30,35,82,83]。PVG 产生氢气量少的优点相对于传统的 HG 来说，更适合与微等离子体技术联用。

PVG 与 DBD-OES 联用最先被用于镍的检测。首先通过 PVG 使 Ni 转化为易挥发的金属羰基化合物。由于整个 PVG 过程中，没有氢气的产生，成功地解决了 DBD 等离子体易被氢气淬灭的问题[84]。该体系还被用于疫苗中硫柳汞的检测[85]。鉴于 PD 具有更高的激发能力，PVG 还被用作 PD-OES 的进样方式，发展了 PVG-PD-OES 小型化分析系统（图 7-6），并成功用于 Hg、Fe、Ni 和 Co 同时高灵敏检测[86]。在此基础上，我们还发展出了光氧化蒸气发生-PD-OES 小型化分析体系。通过光氧化使水中有机碳氧化成二氧化碳，通过 PD-OES 测定碳原子发射，最终实现对环境水样中不可吹除有机碳的快速测定[87]。PVG 还被用于与液体阴极辉光放电联用，构建了小型化的 PVG-SCGDAES 分析系统用于汞的非色谱分离的形态分析[88]。

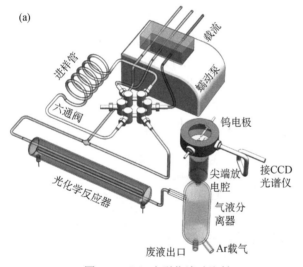

图 7-6 （a）小型化流动注射-PVG-PD-OES 分析体系示意图；（b）PD 激发源图片[86]

7.3.3.3 氧化物发生

金属元素一般可通过还原剂将其还原为挥发性气态物质，而对于非金属元素来说，一般则需要将其氧化为气态物质，随后将其引入微等离子体中进行分析，微波辅助过硫酸盐氧化物发生装置最先被用于氧化环境水样中有机物生成 CO_2[89]。随后，通过介质阻挡放电-碳原子发射光谱仪（dielectric barrier discharge-carbon atomic emission spectrometer，DBD-C-AES）解离激发生成的 CO_2，产生碳原子发射光谱并用 CCD 光谱仪加以检测，最终用于持续监测环境水样中总有机碳（total organic carbon，TOC）。随后，光氧化蒸气发生[87]、DBD 催化氧化蒸气发生[90]方法均被用作小型化 PD-OES 的进样方式，用于水中可溶性有机碳的连续监测。氧化化学蒸气发生不但可用于氧化有机物产生 CO_2，还可将 I⁻ 在线氧化物为碘蒸气，随后引入 DBD-OES 中加以检测，实现了紫菜、食盐和西地碘含片中碘的分析检测[91]。该方法还可以通过 KMnO₄ 等氧化剂将 Cl⁻、Br⁻ 和 I⁻ 氧化为挥发性卤素后，再通过 DBD-OES 对这些卤素离子进行分析[92,93]（图 7-7）。

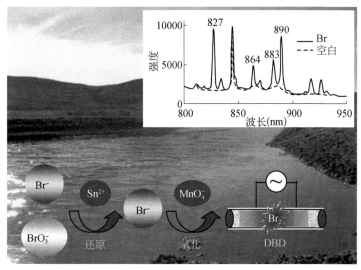

图 7-7　DBD-OES 检测溴离子和溴酸盐的示意图[92]

7.4　现场环境分析方法与装置

传统的原子光谱分析法，如原子吸收光谱法、原子荧光光谱法和电感耦合等离子体原子发射光谱法/质谱法常因体积大且笨重、能耗高等原因，使用范围一般仅限于实验室，这在一定程度上限制了其在环境分析中的应用，特别是现场环境分析中的应用。因此，科研人员们致力于发展小型化/便携式的原子光谱仪器，从进样装置、激发源/原子化器/离子源、检测器等多个方面进行研究整合，以期发展出可以实现现场环境分析的仪器[52]。钨丝电热原子吸收光谱法和微等离子体-原子光谱法是目前较为成熟的可以实现现场快速分析的两种原子光谱分析法。近年来，目视比色法因其装置简单、操作便捷、易于携带等优点，也在环境分析中得到了一定的应用。因此，本节简介这三类分析方法/装置在现场环境分析中的应用。

7.4.1　钨丝电热原子吸收光谱分析法

1972 年，Williams 等[94]最先将钨丝作为原子化器引入原子吸收光谱仪，随后，钨丝作为原子化器/电热蒸发器被广泛应用于原子光谱仪器中[95,96]。Jones 等[97,98]设计了钨丝电热原子吸收光谱仪，并将其应用于铅和镉的测定。侯贤灯等结合多年研究经验，研制了便携式钨丝电热原子吸收光谱仪，并与北京瑞利分析仪器公司共同研发了商品化 WFX-910 型便携式原子吸收光谱仪（图 7-8），该仪器采用钨丝电热原子化器，并通过 CCD 实现了光谱采集，仪器的质量和体积均得到了一定的缩减。整机具有工具箱式外壳，内置电池，整机质量为 18 kg，实现了真正意义上的便携。通过实验对 WFX-910 型便携式原子吸收光谱仪的分析性能进行验证，实验结果表明该仪器对部分元素（镉、铬、铜、锰和铅五种元素）的分析灵敏度略低于石墨炉原子吸收光谱法，但显著地高于火焰原子吸收光谱法[99]。

随后将该仪器与多种预富集技术（如浊点萃取[14,100]、固相萃取[101]、单液滴微萃取[20]等）联用，进一步提高了其分析性能。

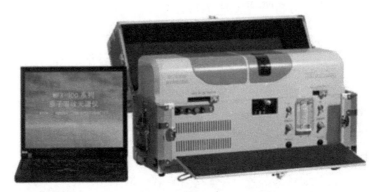

图 7-8　北京瑞利分析仪器公司生产的 WFX-910 型便携式原子吸收光谱仪

7.4.2　微等离子体-原子光谱分析法

微等离子体由于具有体积小、能耗低、温度低、结构简单、易于操作等优点，因此，特别适用于小型化的原子光谱分析装置。目前，已有基于各种微等离子体的小型化仪器/体系的报道，这些小型化的分析装置/仪器/通过与不同的进样方式联用，实现了对待测物的高灵敏检测。如 ETV-DBD-OES[58]、HG-SCGD-AES[79]、PVG-PD-OES[86]和 HG-HEPD-OES[70]等，这些小型化/便携式的分析仪器/装置/体系的成功构建，在很大程度上实现了对金属离子、有机污染物等分析物的现场实时快速检测分析。鉴于这些小型化/便携式装置/体系在前文已有相关介绍，在此不再赘述。

7.4.3　其他方法

目视比色法是一种快速简单的分析方法，它可以通过肉眼对溶液颜色、体积、长度等的变化来判定待测物的浓度。目视比色法一般需要的装置相对简单，特别适合于野外环境的快速分析。当然，目视比色法的灵敏度有待进一步优化提高，目前一般应用于半定量分析。

例如，通过 HG 将 Se(Ⅳ)还原为 H_2Se 气体，随后将该气体引入金纳米溶液中，使金纳米发生团聚。从而使溶液颜色由酒红色变为蓝色，实现了水样中 Se(Ⅳ)的目视比色分析[102]。PVG 也被与金纳米-目视比色法联用用于无机硒[Se(Ⅳ)和 Se(Ⅵ)]的非色谱形态分析[103]。此外，HG 和顶空固相微萃取-目视比色技术联用，可进一步提高目视比色法的分析灵敏度，从而实现了硒[104,105]、锌[106]、砷[12]等元素的快速灵敏分析。

7.5　展　　望

现代社会对先进的、快速的环境分析与监测方法和技术提出了新的要求。可以预见，基于原子光谱的现场环境分析方法及其仪器装置研究将会得到快速发展，其发展趋势和瓶颈主要有以下几个方面：

（1）先进制造技术和光电技术的发展，将有力地促进小型化原子光谱分析仪器的完善。随着 3D

打印技术、微电子集成技术等的引入及学科之间交叉的深入，必将进一步简化小型化原子光谱分析仪的研制，使小型化仪器装置的稳定性、灵敏度等性能得到显著改善。

（2）化学计量学的引入，可拓展现有小型原子光谱分析技术的应用空间。当前，大数据和人工智能方兴未艾，将其用于便携式原子光谱分析中，或许将有望解决环境样品基体干扰以及非色谱形态化合物分析难题，进一步拓展小型化原子光谱分析装置的应用范围。

（3）样品引入技术的发展，特别是一些特异性或者普适性强且能以"干"状态引入分析物的进样技术需要深入研究。最近我们发现一些环境污染物可以通过选择性的化学反应实现气态进样，如NO_2^-与甜蜜素选择性反应，生成环己烯气态化合物，故而可通过微等离子体-原子光谱仪测定环己烯来实现水样中亚硝酸盐氮的高灵敏、现场分析。

（4）拓展微等离子体-原子光谱在其他现场快速分析技术领域中的应用。原子光谱分析技术往往限于金属检测，因而如何拓展小型化原子光谱仪在金属形态化合物以及非金属污染物的检测是未来研究的又一重点。

（5）原子光谱分析装置技术转移和标准方法的制定。仪器检定规程和方法标准的制定，有利于推动创新仪器的产业化，未来对于新型小型化原子光谱仪需要加大力度重视这一方向的工作。

致谢：感谢国家自然科学基金（U21A20283，21622508）的资助。

参 考 文 献

[1] Li X D, Jin L, Kan H D. Air pollution: A global problem needs local fixes. Nature, 2019: 437-439.

[2] Zhang R Y, Wang G H, Guo S, Zamora M L, Ying Q, Lin Y, Wang W G, Hu M, Wang Y. Formation of urban fine particulate matter. Chemical Reviews, 2015, 115: 3803-3855.

[3] Landrigan P J, Fuller R, Acosta N J R, Adeyi O, Arnold R, Basu N, Baldé A B, Bertollini R, Bose-O'Reilly S, Boufford J I, Breysse P N, Chiles T, Mahidol C, Coll-Seck A M, Cropper M L, Fobil J, Fuster V, Greenstone M, Haines A, Hanrahan D, Hunter D, Khare M, Krupnick A, Lanphear B, Lohani B, Martin K, Mathiasen K V, McTeer M A, Murray C J L, Ndahimananjara J D, Perera F, Potočnik J, Preker A S, Ramesh J, Rockström J, Salinas C, Samson L D, Sandilya K, Sly P D, Smith K R, Steiner A, Stewart R B, Suk W A, van Schayck O C P, Yadama G N, Yumkella K, Zhong M. The Lancet Commission on pollution and health. The Lancet, 2018, 391: 462-512.

[4] Zhang J F, Mauzerall D L, Zhu T, Liang S, Ezzati M, Remais D V. Environmental health in China progress towards clean air and safe water. The Lancet, 2010, 375: 1110-1119.

[5] Bray F, Ferlay J, Soerjomataram I, Siegel R L, Torre L A, Jemal A. Global cancer statistics 2018: GLOBOCAN estimates of incidence and mortality worldwide for 36 cancers in 185 countries. CA: A Cancer Journal for Clinicians, 2018, 68: 394-424.

[6] Wu Q R, Wang S X, Li G L, Liang S, Lin C J, Wang Y F, Cai S Y, Liu K Y, Hao J M. Temporal trend and spatial distribution of speciated atmospheric mercury emissions in China during 1978—2014. Environmental Science & Technology, 2016, 50: 13428-13435.

[7] 邹天森, 张金良, 陈昱, 王慢想, 潘丽波, 王先良, 魏复盛. 中国部分城市空气环境铅含量及分布研究. 中国环境科学, 2015, 35: 23-32.

[8] Snyder D T, Pulliam C J, Ouyang Z, Cooks R G. Miniature and fieldable mass spectrometers: Recent advances.

Analytical Chemistry, 2016, 88: 2-29.

[9] Coyne R V, Collins J A. Loss of mercury from water during storage. Analytical Chemistry, 1972, 44: 1093-1096.

[10] Jia Y, Yu H M, Wu L, Hou X D, Yang L, Zheng C B. Three birds with one Fe_3O_4 nanoparticle: Integration of microwave digestion, solid phase extraction, and magnetic separation for sensitive determination of arsenic and antimony in fish. Analytical Chemistry, 2015, 87: 5866-5871.

[11] Yu H M, Ai X, Xu K L, Zheng C B, Hou X D. UV-assisted Fenton digestion of rice for the determination of trace cadmium by hydride generation atomic fluorescence spectrometry. Analyst, 2016, 141: 1512-1518.

[12] Luo Y J, Su Y B, Lin Y, He L B, Wu L, Hou X D, Zheng C B. $MnFe_2O_4$ micromotors enhanced field digestion and solid phase extraction for on-site determination of arsenic in rice and water. Analytica Chimica Acta, 2021, 1156: 338354.

[13] Luo Y J, Yang Y, Lin Y, Tian Y F, Wu L, Yang L, Hou X D, Zheng C B. Low-temperature and atmospheric pressure sample digestion using dielectric barrier discharge. Analytical Chemistry, 2018, 90: 1547-1553.

[14] Wen X D, Wu P, Chen L, Hou X D. Determination of cadmium in rice and water by tungsten coil electrothermal vaporization-atomic fluorescence spectrometry and tungsten coil electrothermal atomic absorption spectrometry after cloud point extraction. Analytica Chimica Acta, 2009, 650: 33-38.

[15] Wen X D, Zhang H Z, Deng Q W, Yang S C. Investigation of a portable tungsten coil electrothermal atomic absorption spectrometer for analysis of nickel after rapidly synergistic cloud point extraction. Analytical Methods, 2014, 6: 5047-5053.

[16] Zeng C J, Wen X D, Tan Z Q, Cai P Y, Hou X D. Hollow fiber supported liquid membrane extraction for ultrasensitive determination of trace lead by portable tungsten coil electrothermal atomic absorption spectrometry. Microchemical Journal, 2010, 96: 238-242.

[17] Chen L Q, Lei Z R, Yang S C, Wen X D. Application of portable tungsten coil electrothermal atomic absorption spectrometer for the determination of trace cobalt after ultrasound-assisted rapidly synergistic cloud point extraction. Microchemical Journal, 2017, 130: 452-457.

[18] Wen X D, Yang S C, Zhang H Z, Deng Q W. Combination of knotted reactor with portable tungsten coil electrothermal atomic absorption spectrometer for on-line determination of trace cadmium. Microchemical Journal, 2016, 124: 60-64.

[19] Wen X D, Yang S C, Zhang H Z, Wang J W. Determination of trace bismuth by using a portable spectrometer after ultrasound-assisted dispersive liquid-liquid microextraction. Analytical Methods, 2014, 6: 8773-8778.

[20] Wen X D, Deng Q W, Wang J W, Yang S C, Zhao X. A new coupling of ionic liquid based-single drop microextraction with tungsten coil electrothermal atomic absorption spectrometry. Spectrochimica Acta Part A: Molecular and Biomolecular Spectroscopy, 2013, 105: 320-325.

[21] Wen X D, Deng Q W, Wang J W, Zhao X. Application of portable spectrometer for ultra trace metal analysis after ionic liquid based microextraction. Analytical Methods, 2013, 5: 2978-2983.

[22] Chen P P, Deng Y J, Guo K C, Jiang X M, Zheng C B, Hou X D. Flow injection hydride generation for on-atomizer trapping: Highly sensitive determination of cadmium by tungsten coil atomic absorption spectrometry. Microchemical Journal, 2014, 112: 7-12.

[23] Van Khoai D, Kitano A, Yamamoto T, Ukita Y, Takamura Y. Development of high sensitive liquid electrode plasma: Atomic emission spectrometry (LEP-AES) integrated with solid phase pre-concentration. Microelectronic Engineering, 2013, 111: 343-347.

[24] Barua S, Rahman I M M, Miyaguchi M, Yunoshita K, Ruengpirasiri P, Takamura Y, Mashio A S, Hasegawa H. Speciation of inorganic selenium in wastewater using liquid electrode plasma-optical emission spectrometry combined with supramolecule-equipped solid-phase extraction system. Microchemical Journal, 2020, 159: 105490.

[25] Mo J M, Zhou L, Li X H, Li Q, Wang L J, Wang Z. On-line separation and pre-concentration on a mesoporous silica-grafted graphene oxide adsorbent coupled with solution cathode glow discharge-atomic emission spectrometry for the determination of lead. Microchemical Journal, 2017, 130: 353-359.

[26] Barua S, Rahman I M M, Miyaguchi M, Mashio A S, Maki T, Hasegawa H. On-site analysis of gold, palladium, or platinum in acidic aqueous matrix using liquid electrode plasma-optical emission spectrometry combined with ion-selective preconcentration. Sensors and Actuators B: Chemical, 2018, 272: 91-99.

[27] Zheng C B, Hu L G, Hou X D, He B, Jiang G B. Headspace solid-phase microextraction coupled to miniaturized microplasma optical emission spectrometry for detection of mercury and lead. Analytical Chemistry, 2018, 90: 3683-3691.

[28] He Z, Lin Y, Wang Y, He L B, Hou X D, Zheng C B. Growth of carbonaceous nanoparticles on steel fiber from candle flame for the long-term preservation of ultratrace mercury by solid-phase microextraction. Analytical Chemistry, 2020, 92: 9583-9590.

[29] Yuan X, Yang G, Ding Y, Li X M, Zhan X F, Zhao Z J, Duan Y X. An effective analytical system based on a pulsed direct current microplasma source for ultra-trace mercury determination using gold amalgamation cold vapor atomic emission spectrometry. Spectrochimica Acta Part B: Atomic Spectroscopy, 2014, 93: 1-7.

[30] Zou Z R, Hu J, Xu F J, Hou X D, Jiang X M. Nanomaterials for photochemical vapor generation-analytical atomic spectrometry. TrAC Trends in Analytical Chemistry, 2019, 114: 242-250.

[31] Long Z, Luo Y M, Zheng C B, Deng P C, Hou X D. Recent advance of hydride generation-analytical atomic spectrometry. Part I: Technique development. Applied Spectroscopy Reviews, 2012, 47: 382-413.

[32] Long Z, Chen C, Hou X D, Zheng C B. Recent advance of hydride generation-analytical atomic spectrometry. Part II: Analysis of real samples. Applied Spectroscopy Reviews, 2012, 47: 495-517.

[33] Wu P, He L, Zheng C B, Hou X D, Sturgeon R. Applications of chemical vapor generation in non-tetrahydroborate media to analytical atomic spectrometry. Journal of Analytical Atomic Spectrometry, 2010, 25: 1217.

[34] Liu X, Zhu Z L, Xing P J, Zheng H T, Hu S H. Plasma induced chemical vapor generation for atomic spectrometry: A review. Spectrochimica Acta Part B: Atomic Spectroscopy, 2020, 167: 105822.

[35] He Q, Zhu Z L, Hu S H. Plasma-induced vapor generation technique for analytical atomic spectrometry. Reviews in Analytical Chemistry, 2014, 33: 111-121.

[36] Cai Y, Gao X G, Ji Z N, Yu Y L, Wang J H. Nonthermal optical emission spectrometry for simultaneous and direct determination of zinc, cadmium and mercury in spray. Analyst, 2018, 143: 930-935.

[37] Cai Y, Zhang Y J, Wu D F, Yu Y L, Wang J H. Nonthermal optical emission spectrometry: Direct atomization and excitation of cadmium for highly sensitive determination. Analytical Chemistry, 2016, 88: 4192-4195.

[38] Zhang Y J, Cai Y, Yu Y L, Wang J H. A miniature optical emission spectrometric system in a lab-on-valve for sensitive determination of cadmium. Analytica Chimica Acta, 2017, 976: 45-51.

[39] Greda K, Jamroz P, Pohl P. Ultrasonic nebulization atmospheric pressure glow discharge-preliminary study. Spectrochimica Acta Part B: Atomic Spectroscopy, 2016, 121: 22-27.

[40] Doroski T A, Webb M R. Signal enhancement in solution-cathode glow discharge-optical emission spectrometry via low molecular weight organic compounds. Spectrochimica Acta Part B: Atomic Spectroscopy, 2013, 88: 40-45.

[41] Xiao Q, Zhu Z L, Zheng H T, He H Y, Huang C Y, Hu S H. Significant sensitivity improvement of alternating current driven-liquid discharge by using formic acid medium for optical determination of elements. Talanta, 2013, 106: 144-149.

[42] Zhang Z, Wang Z, Li Q, Zou H J, Shi Y. Determination of trace heavy metals in environmental and biological samples by solution cathode glow discharge-atomic emission spectrometry and addition of ionic surfactants for improved

sensitivity. Talanta, 2014, 119: 613-619.

[43] Wang Z, Gai R Y, Zhou L, Zhang Z. Design modification of a solution-cathode glow discharge-atomic emission spectrometer for the determination of trace metals in titanium dioxide. Journal of Analytical Atomic Spectrometry, 2014, 29: 2042-2049.

[44] Doroski T A, King A M, Fritz M P, Webb M R. Solution-cathode glow discharge-optical emission spectrometry of a new design and using a compact spectrograph. Journal of Analytical Atomic Spectrometry, 2013, 28: 1090-1095.

[45] Jiang X, Xu X L, Hou X D, Long Z, Tian Y F, Jiang X M, Xu F J, Zheng C B. A novel capillary microplasma analytical system: Interface-free coupling of glow discharge optical emission spectrometry to capillary electrophoresis. Journal of Analytical Atomic Spectrometry, 2016, 31: 1423-1429.

[46] Leng A Q, Lin Y, Tian Y F, Wu L, Jiang X M, Hou X D, Zheng C B. Pump-and valve-free flow injection capillary liquid electrode discharge optical emission spectrometry coupled to a droplet array platform. Analytical Chemistry, 2017, 89: 703-710.

[47] Jamroz P, Greda K, Dzimitrowicz A, Swiderski K, Pohl P. Sensitive determination of Cd in small-volume samples by miniaturized liquid drop anode atmospheric pressure glow discharge optical emission spectrometry. Analytical Chemistry, 2017, 89: 5729-5733.

[48] He Q, Zhu Z L, Hu S H, Zheng H T, Jin L L. Elemental determination of microsamples by liquid film dielectric barrier discharge atomic emission spectrometry. Analytical Chemistry, 2012, 84: 4179-4184.

[49] Resano M, Vanhaecke F, de Loos-Vollebregt M T C. Electrothermal vaporization for sample introduction in atomic absorption, atomic emission and plasma mass spectrometry: A critical review with focus on solid sampling and slurry analysis. Journal of Analytical Atomic Spectrometry, 2008, 23: 1450-1475.

[50] Hanna S N, Jones B T. A review of tungsten coil electrothermal vaporization as a sample introduction technique in atomic spectrometry. Applied Spectroscopy Reviews, 2011, 46: 624-635.

[51] Zhong G P, Luo H, Zhou Z D, Hou X D. Molybdenum, platinum, and tantalum metal atomizers or vaporizers in analytical atomic spectrometry. Applied Spectroscopy Reviews, 2004, 39: 475-507.

[52] Zou Z R, Deng Y J, Hu J, Jiang X M, Hou X D. Recent trends in atomic fluorescence spectrometry towards miniaturized instrumentation: A review. Analytica Chimica Acta, 2018, 1019: 25-37.

[53] Hou X D, Jones B T. Tungsten devices in analytical atomic spectrometry. Spectrochimica Acta Part B: Atomic Spectroscopy, 2002, 57: 659-688.

[54] Jiang X M, Chen Y, Zheng C B, Hou X D. Electrothermal vaporization for universal liquid sample introduction to dielectric barrier discharge microplasma for portable atomic emission spectrometry. Analytical Chemistry, 2014, 86: 5220-5224.

[55] Deng Y J, Hu J, Li M T, He L, Li K, Hou X D, Jiang X M. Interface-free integration of electrothermal vaporizer and point discharge microplasma for miniaturized optical emission spectrometer. Analytica Chimica Acta, 2021, 1163: 338502.

[56] Deng Y J, Li K, Hou X D, Jiang X M. Flow injection hydride generation and on-line W-coil trapping for electrothermal vaporization dielectric barrier discharge atomic emission spectrometric determination of trace cadmium. Talanta, 2021, 233: 122516.

[57] Zheng H T, Ma J Z, Zhu Z L, Tang Z Y, Hu S H. Dielectric barrier discharge micro-plasma emission source for the determination of lead in water samples by tungsten coil electro-thermal vaporization. Talanta, 2015, 132: 106-111.

[58] Li N, Wu Z C, Wang Y Y, Zhang J, Zhang X N, Zhang H N, Wu W N, Gao J, Jiang J. Portable dielectric barrier bischarge-atomic emission spectrometer. Analytical Chemistry, 2017, 89: 2205-2210.

[59] Jiang J, Li Z J, Wang Y Y, Zhang X N, Yu K, Zhang H, Zhang J, Gao J, Liu X Y, Zhang H N, Wu W H, Li N. Rapid determination of cadmium in rice by portable dielectric barrier discharge-atomic emission spectrometer. Food Chemistry, 2020, 310: 125824.

[60] Zhu Z L, Liu J X, Zhang S C, Na X, Zhang X R. Determination of Se, Pb, and Sb by atomic fluorescence spectrometry using a new flameless, dielectric barrier discharge atomizer. Spectrochimica Acta Part B: Atomic Spectroscopy, 2008, 63: 431-436.

[61] Xing Z, Kuermaiti B, Wang J, Han G J, Zhang S C, Zhang X R. Simultaneous determination of arsenic and antimony by hydride generation atomic fluorescence spectrometry with dielectric barrier discharge atomizer. Spectrochimica Acta Part B: Atomic Spectroscopy, 2010, 65: 1056-1060.

[62] Zhu Z L, He H Y, He D, Zheng H T, Zhang C X, Hu S H. Evaluation of a new dielectric barrier discharge excitation source for the determination of arsenic with atomic emission spectrometry. Talanta, 2014, 122: 234-239.

[63] Na N, Zhao M X, Zhang S C, Yang C D, Zhang X R. Development of a dielectric barrier discharge ion source for ambient mass spectrometry. Journal of the American Society for Mass Spectrometry, 2007, 18: 1859-1862.

[64] Xing Z, Wang J, Zhang S C, Zhang X R. Determination of bismuth in solid samples by hydride generation atomic fluorescence spectrometry with a dielectric barrier discharge atomizer. Talanta, 2009, 80: 139-142.

[65] Zhu Z L, Liu J X, Zhang S C, Na X, Zhang X R. Evaluation of a hydride generation-atomic fluorescence system for the determination of arsenic using a dielectric barrier discharge atomizer. Analytica Chimica Acta, 2008, 607: 136-141.

[66] Zhang Y, Ma J, Na X, Shao Y, Liu J, Mao X, Chen G, Tian D, Qian Y. A portable and field optical emission spectrometry coupled with microplasma trap for high sensitivity analysis of arsenic and antimony simultaneously. Talanta, 2020, 218: 121161.

[67] Burhenn S, Kratzer J, Franzke J. Temporal evolution of tellurium emission lines in a capillary dielectric barrier discharge after hydride generation. Spectrochimica Acta Part B: Atomic Spectroscopy, 2020, 171: 105936.

[68] Burhenn S, Kratzer J, Svoboda M, Klute F D, Michels A, Veza D, Franzke J. Spatially and temporally resolved detection of arsenic in a capillary dielectric barrier discharge by hydride generation high-resolved optical emission spectrometry. Analytical Chemistry, 2018, 90: 3424-3429.

[69] Li M T, Deng Y J, Zheng C B, Jiang X M, Hou X D. Hydride generation-point discharge microplasma-optical emission spectrometry for the determination of trace As, Bi, Sb and Sn. Journal of Analytical Atomic Spectrometry, 2016, 31: 2427-2433.

[70] Li M T, Li K, He L, Zeng X L, Wu X, Hou X D, Jiang X M. Point discharge microplasma optical emission spectrometer: Hollow electrode for efficient volatile hydride/mercury sample introduction and 3D-printing for compact instrumentation. Analytical Chemistry, 2019, 91: 7001-7006.

[71] He L, Li P X, Li K, Lin T, Luo J, Hou X D, Jiang X M. Cross double point discharge as enhanced excitation source for highly sensitive determination of arsenic, mercury and lead by optical emission spectrometry. Journal of Analytical Atomic Spectrometry, 2021, 36: 1193-1200.

[72] Zhu Z L, Chan G C-Y, Ray S J, Zhang X R, Hieftje G M. Microplasma source based on a dielectric barrier discharge for the determination of mercury by atomic emission spectrometry. Analytical Chemistry, 2008, 80: 8622-8627.

[73] Huang C C, Li Q, Mo J M, Wang Z. Ultratrace determination of tin, germanium, and selenium by hydride generation coupled with a novel solution-cathode glow discharge-atomic emission spectrometry method. Analytical Chemistry, 2016, 88: 11559-11567.

[74] Cheng J Q, Li Q, Zhao M Y, Wang Z. Ultratrace Pb determination in seawater by solution-cathode glow discharge-atomic emission spectrometry coupled with hydride generation. Analytica Chimica Acta, 2019, 1077: 107-115.

[75] Guo X H, Peng X X, Li Q, Mo J M, Du Y P, Wang Z. Ultra-sensitive determination of inorganic arsenic valence by solution cathode glow discharge-atomic emission spectrometry coupled with hydride generation. Journal of Analytical Atomic Spectrometry, 2017, 32: 2416-2422.

[76] Peng X X, Wang Z. Ultrasensitive determination of selenium and arsenic by modified helium atmospheric pressure glow discharge optical emission spectrometry coupled with hydride generation. Analytical Chemistry, 2019, 91: 10073-10080.

[77] Greda K, Jamroz P, Jedryczko D, Pohl P. On the coupling of hydride generation with atmospheric pressure glow discharge in contact with the flowing liquid cathode for the determination of arsenic, antimony and selenium with optical emission spectrometry. Talanta, 2015, 137: 11-17.

[78] Zhu Z L, Yang C, Yu P W, Zheng H T, Liu Z F, Xing Z, Hu S H. Determination of antimony in water samples by hydride generation coupled with atmospheric pressure glow discharge atomic emission spectrometry. Journal of Analytical Atomic Spectrometry, 2019, 34: 331-337.

[79] Yang C, He D, Zhu Z L, Peng H, Liu Z F, Wen G J, Bai J H, Zheng H T, Hu S H, Wang Y X. Battery-operated atomic emission analyzer for waterborne arsenic based on atmospheric pressure glow discharge excitation source. Analytical Chemistry, 2017, 89: 3694-3701.

[80] Guo X M, Sturgeon R, Mester Z, Gardner G. UV vapor generation for determination of selenium by heated quartz tube atomic absorption spectrometry. Analytical Chemistry, 2003, 75: 2092-2099.

[81] Guo X M, Sturgeon R, Mester Z, Gardner G. Vapor generation by UV irradiation for sample introduction with atomic spectrometry. Analytical Chemistry, 2004, 76: 2401-2405.

[82] He Y H, Hou X D, Zheng C B, Sturgeon R. Critical evaluation of the application of photochemical vapor generation in analytical atomic spectrometry. Analytical Bioanalytical Chemistry, 2007, 388: 769-774.

[83] Sturgeon R. Photochemical vapor generation: A radical approach to analyte introduction for atomic spectrometry. Journal of Analytical Atomic Spectrometry, 2017, 32: 2319-2340.

[84] Cai Y, Li S H, Dou S, Yu Y L, Wang J H. Metal carbonyl vapor generation coupled with dielectric barrier discharge to avoid plasma quench for optical emission spectrometry. Analytical Chemistry, 2015, 87: 1366-1372.

[85] He H Y, Zhu Z L, Zheng H T, Xiao Q, Jin L L, Hu S H. Dielectric barrier discharge micro-plasma emission source for the determination of thimerosal in vaccines by photochemical vapor generation. Microchemical Journal, 2012, 104: 7-11.

[86] Zhang S, Luo H, Peng M T, Tian Y F, Hou X D, Jiang X M, Zheng C B. Determination of Hg, Fe, Ni, and Co by miniaturized optical emission spectrometry integrated with flow injection photochemical vapor generation and point discharge. Analytical Chemistry, 2015, 87: 10712-10718.

[87] Zhang S, Tian Y F, Yin H, Su Y B, Wu L, Hou X D, Zheng C B. Continuous and inexpensive monitoring of nonpurgeable organic carbon by coupling high-efficiency photo-oxidation vapor generation with miniaturized point-discharge optical emission spectrometry. Environmental Science & Technology, 2017, 51: 9109-9117.

[88] Mo J M, Li Q, Guo X H, Zhang G X, Wang Z. Flow injection photochemical vapor generation coupled with miniaturized solution-cathode glow discharge atomic emission spectrometry for determination and speciation analysis of mercury. Analytical Chemistry, 2017, 89: 10353-10360.

[89] Han B J, Jiang X M, Hou X D, Zheng C B. Miniaturized dielectric barrier discharge carbon atomic emission spectrometry with online microwave-assisted oxidation for determination of total organic carbon. Analytical Chemistry, 2014, 86: 6214-6219.

[90] Li K, Chen H J, Chen Z M, He L, Hou X D, Jiang X M. Miniaturized TOC analyzer using dielectric barrier discharge for catalytic oxidation vapor generation and point discharge optical emission spectrometry. Analytica Chimica Acta, 2021,

1172: 338683.

[91] Yu Y L, Dou S, Chen M L, Wang J H. Iodine excitation in a dielectric barrier discharge micro-plasma and its determination by optical emission spectrometry. Analyst, 2013, 138: 1719-1725.

[92] Yu Y L, Cai Y, Chen M L, Wang J H. Development of a miniature dielectric barrier discharge-optical emission spectrometric system for bromide and bromate screening in environmental water samples. Analytica Chimica Acta, 2014, 809: 30-36.

[93] Zhang D J, Cai Y, Chen M L, Yu Y L, Wang J H. Dielectric barrier discharge-optical emission spectrometry for the simultaneous determination of halogens. Journal of Analytical Atomic Spectrometry, 2016, 31: 398-405.

[94] Williams M, Piepmeier E H. Commercial tungsten filament atomizer for analytical atomic spectrometry. Analytical Chemistry, 1972, 44: 1342-1344.

[95] Hanna S N, Jones B T. A review of tungsten coil electrothermal vaporization as a sample introduction technique in atomic spectrometry. Applied Spectroscopy Reviews, 2011, 46: 624-635.

[96] 吴鹏, 温晓东, 吕弋, 侯贤灯. 钨丝在原子吸收光谱分析中的应用. 分析化学, 2006, 34: 278-282.

[97] Sanford C, Thomas S, Jones B. Portable, battery-powered, tungsten coil atomic absorption spectrometer for lead determinations. Applied Spectroscopy, 1996, 50: 174-181.

[98] Batchelor J, Thomas S, Jones B. Determination of cadmium with a portable, battery-powered tungsten coil atomic absorption spectrometry. Applied Spectroscopy, 1998, 52: 1086-1091.

[99] 章诒学, 宋友才, 李增昌, 谢惠, 杨洪涛, 赵刚. 便携式原子吸收光谱仪的研究与探讨. 光学仪器, 2012, 34: 79-83.

[100] Donati G L, Pharr K E, Calloway C P Jr., Nobrega J A, Jones B T. Determination of Cd in urine by cloud point extraction-tungsten coil atomic absorption spectrometry. Talanta, 2008, 76: 1252-1255.

[101] 范广宇, 蒋小明, 郑成斌, 侯贤灯, 徐开来. 固相萃取-钨丝电热原子吸收光谱分析法测定水样中的银. 光谱学与光谱分析, 2011, 31: 1946-1949.

[102] Cao G M, Xu F J, Wang S L, Xu K L, Hou X D, Wu P. Gold nanoparticle-based colorimetric assay for selenium detection via hydride generation. Analytical Chemistry, 2017, 89: 4695-4700.

[103] Ding Y G, Liu Y, Chen Y L, Huang Y, Gao Y. Photochemical vapor generation for colorimetric speciation of inorganic selenium. Analytical Chemistry, 2019, 91: 3508-3515.

[104] Huang K, Xu K L, Zhu W, Yang L, Hou X D, Zheng C B. Hydride generation for headspace solid-phase extraction with CdTe quantum dots immobilized on paper for sensitive visual detection of selenium. Analytical Chemistry, 2016, 88: 789-795.

[105] Xiong J, Xu K L, Hou X D, Wu P. AuNCs-catalyzed hydrogen selenide oxidation: Mechanism and application for headspace fluorescent detection of Se(IV). Analytical Chemistry, 2019, 91: 6141-6148.

[106] Huang K, Dai R, Deng W Q, Guo S J, Deng H, Wei Y, Zhou F L, Long Y, Li J, Yuan X, Xiong X L. Gold nanoclusters immobilized paper for visual detection of zinc in whole blood and cells by coupling hydride generation with headspace solid phase extraction. Sensors and Actuators B: Chemical, 2018, 255: 1631-1639.

第二部分　进样与联用技术

第8章 化学蒸气发生法

（胡　静[①]　侯贤灯[①]*）

▶ 8.1　化学蒸气发生法概述 / 198

▶ 8.2　氢化物发生 / 198

▶ 8.3　光化学蒸气发生 / 201

▶ 8.4　其他化学蒸气发生法 / 206

▶ 8.5　展望 / 208

[①]四川大学，成都。*通讯作者联系方式：houxd@scu.edu.cn

本章导读

- 首先简单介绍化学蒸气发生法概念及其主要特点。
- 对氢化物发生法，主要讨论过渡金属和贵金属的氢化物发生原理、联用技术和分析应用。
- 重点介绍光化学蒸气发生及其分析应用，包括光化学发生体系、光催化剂和增敏剂的应用。
- 简要介绍其他化学蒸气发生法，包括电化学蒸气发生、烷基化合物发生、卤化物发生、羰基化合物发生、螯合物发生、氧化物发生和等离子体诱导化学蒸气发生。
- 对化学蒸气发生相关研究的发展方向进行了展望。

8.1 化学蒸气发生法概述

化学蒸气发生（chemical vapor generation，CVG）是利用化学反应将凝聚相中的分析元素（或其化合物）转化为挥发性物种的一种重要的原子光谱分析进样技术。化学蒸气发生法的主要特点是：①进样效率高（甚至可接近100%），远远优于传统气动雾化进样法，气相产物的易解离特性也使原子/离子化效率大大提高，从而显著地改善了原子光谱分析法的检出限和灵敏度；②分析元素生成化学蒸气后与可能引起干扰的样品基体分离，因此可以消除或者减少来自基体的光谱/质谱干扰。1969年，Holak首次利用Zn/HCl还原As生成砷化氢（Marsh反应），并利用原子吸收光谱进行了As检测，建立了氢化物发生-原子光谱分析法。随后，分析化学工作者致力于研究新的更为有效的化学蒸气发生法，包括氢化物发生、光化学蒸气发生、电化学蒸气发生、等离子体诱导化学蒸气发生等，并在原子吸收光谱分析法、原子发射光谱分析法、原子荧光光谱分析法和原子质谱分析法中获得广泛应用。本章将重点介绍新型氢化物发生和光化学蒸气发生在分析应用方面的最新进展。

8.2 氢化物发生

8.2.1 传统氢化物发生

如上所述，金属-酸还原体系最早用于氢化物发生（hydride generation，HG）[1]，但该法反应速度慢，可产生的氢化物种类少，难以实现自动化，因此未得到普遍应用。1972年，Braman等[2]提出使用硼氢化盐（tetrahydroborate，THB）代替金属作为还原剂进行AsH_3和SbH_3的发生和直流辉光光谱检测，被认为是氢化物发生技术发展的里程碑。THB反应可在室温下迅速进行，且适用的元素范围更广（As、Sb、Bi、Ge、Sn、Pb、Se、Te、Hg、Zn、Cd等），一经提出即得到了广泛的应用[3-5]。

硼氢化盐还原生成氢化物的过程，过去曾认为分两步进行，即$NaBH_4$先与酸反应生成H·（新生态氢），再与待测元素离子反应生成挥发性氢化物。然而，Laborda等[6]对该"新生态氢"机理提出了质疑，因为从热力学的角度来说，无论是BH_4^-还是Zn都无法将H^+还原成所谓"新生态氢"。分析元素与硼烷络合物（analyte-borane complex，ABC）直接发生反应生成氢化物是目前普遍认可的氢化物发生机理[7]，反应过程如下：

$$L_3B-H + EY_n \underset{A}{\rightleftharpoons} [L_3B\cdots H\cdots EY_{n-1}] \xrightarrow{B} L_3B-Y + Y_{n-1}EH$$

其中，反应 A 为还原剂 L_3B-H（L=H、NH_2、CN 等配体）和分析元素配合物 EY_n（Y=H_2O、OH^-、Cl^-等）反应生成硼络合物中间体；反应 B 为氢原子转移到待测元素的原子上形成氢化物。

8.2.2 过渡金属和贵金属的氢化物发生

除了常见的氢化物和冷原子蒸气（通常简称冷蒸气，如汞冷蒸气）发生元素，人们还发现了更多新的可氢化物发生的元素，大大拓展了硼氢化盐-酸还原体系的适用范围。Sturgeon 等首次通过盐酸或硝酸酸化的铜溶液与 THB 反应产生铜的挥发性物质[8]；随后，一系列贵金属（Au、Ag、Pd、Pt、Ru、Rh、Ir、Os 等）和过渡金属（Ni、Co、Cr、Fe、Cu、Mn、Ti、Zr、Mo、Sc、V 等）的化学蒸气也被原子化后检测出来[9,10]。

为了证明检测到的信号是待测元素通过 CVG 反应生成的挥发性物质所产生的，而不是待测元素被反应产生的气溶胶携带进入检测器的结果，通常将 Be 或 Mg 这种不会产生挥发性物质的元素加入过渡金属或贵金属待测液中作为参比进行检测。Be 或 Mg 产生的信号与待测元素有差别则说明它们不是以相同的传输方式进入检测器的。研究证明，过渡金属或贵金属应该是以挥发性物种的形式存在，而以气溶胶形式进入检测器的部分非常少。

然而，这些过渡金属和贵金属的气态产物的本质是什么，目前仍未确定。这方面的论述也大都建立在假设和实验现象的直接观测的基础上，一般认为，这些金属被硼氢化盐还原生成了不同的中间产物，如氢化物、自由原子、纳米颗粒等。

$$L_nM \rightarrow (L_{n-1}MH \rightarrow L_{n-2}MH_2 \rightarrow \cdots \rightarrow LMH_{n-1}) \rightarrow MH_n \rightarrow M^0 \rightarrow M_m^0 \rightarrow 纳米粒子$$

研究人员曾尝试将 CVG 反应生成的过渡金属和贵金属的气态产物在低温下捕集并传输至质谱仪中进行检测，但是没有成功，猜测气态物质可能是由金属原子组成，在低温捕集过程中团聚在捕集器内，无法再以挥发态传输。值得一提的是，最近 Musil 等将 Ag 的挥发性物质收集后进行透射电镜表征，检测到了（8±2）nm 的银纳米颗粒[11]，确证银在该反应中生成了银纳米粒子，并以气态的形式传输至检测器中。在过渡金属和贵金属 CVG 反应机理方面，Sturgeon 课题组进行了广泛深入的研究[12]，Pohl 等也对该领域的工作做过较为全面的综述[13]。

总的来说，这些元素化学蒸气的发生效率不高且稳定性较差，在传输过程中易分解损失，能否检测到过渡金属和贵金属在化学蒸气发生反应中生成的挥发性物种取决于产物自身的稳定性和气液分离的快慢，这也在很大程度上限制了其在分析领域的应用。在 Sturgeon 等报道铜的氢化物发生后，人们发现加入邻菲罗啉可以大大提高铜的氢化物发生效率，可用于环境和生物样品中痕量铜的测定[14]。当盐酸或硝酸介质中含有 10% 的乙酸时，也可明显提高铜的蒸气发生效率。将少量的二乙基二硫代氨基甲酸钠（DDTC）加入盐酸酸化的 Au 溶液时，能够促进挥发性金化合物的生成。DDTC 还可以选择性地提高六价铬的化学蒸气发生效率，而对三价铬则没有类似的增敏效果，结合 $KMnO_4$ 氧化和差减法，成功地建立了一种无需色谱分离的无机 Cr 形态分析新方法[15]。目前，大量的研究均表明，邻菲罗啉、DDTC、Triton X-100、十六烷基三甲基溴化铵（CTAB）、离子液体等试剂的加入可以提高 Cu、Au、Ag 等过渡金属和贵元素的化学蒸气发生效率，但是增敏机理尚不清楚。一般认为，络合物使过渡金属和贵金属更容易与 THB 反应生成挥发性化合物，或者使得它们的挥发物更加稳定。严秀平课题组[16]曾详细研究了八种离子液体对 Au 的蒸气发生的增感效应，发现增敏效果与离子液体所含阴阳离子的本质有关：含有短链烷烃的阳离子的离子液体增感效果较好，阳离子的烷烃链较长或支链较多的离子液体增感效果不佳；而对于含有相同阳离子的离子液体，阴离子为 Br⁻时增感效果最佳。推测离子液体的加入改变了氢化物发生反应环境的物理性质，反应中产生的细小均匀的气泡将气态产物与其他物质隔离，进而消除了它们带来的干扰。为提高气液分离的效率，郑成斌等设计了薄膜式化学蒸气发生装置[17]，当含铜溶液与硼氢化钠溶液在薄膜反应器表面混合反应时，生成的挥发性铜化合物可以进行快速的气液分离，因此显著地提高了铜的石墨炉原子吸收光谱分析法的检测灵敏度。

8.2.3 氢化物发生的分析应用

氢化物发生法的发展促进了原子光（质）谱及其联用技术在地质、环境、食品和组学等多个学科领域的应用。一方面，由于进样效率高，分析物与基体有效分离，氢化物发生与各种原子光（质）谱以及一些非原子光谱分析技术的联用，可以在提高元素分析灵敏度的同时对源自基体的光（质）谱干扰加以消除；氢化物易于原子化，以氢化物的形式进样，能够提高小型化原子光谱分析仪的激发能力；另一方面，在实际样品分析中，通常还面临目标分析物含量低、检测灵敏度不足和干扰的问题，发展合适的样品前处理、分离富集以及干扰消除的方法，也是该领域的重要研究课题。

目前，多种商品化的原子光谱/质谱分析仪器已配备相应的氢化物发生进样系统，以提高氢化物元素的分析性能。对于氯离子含量较高的样品，喷雾进样时容易形成多原子离子($^{40}Ar^{35}Cl^+$)，干扰 ^{75}As 的 ICP-MS 检测；氢化物发生进样时，氯的干扰大幅降低，使得普通四极杆 ICP-MS 的检测能力可媲美高分辨 ICP-MS（喷雾进样）[18]。近些年，以介质阻挡放电（dielectric barrier discharge, DBD）和尖端放电（point discharge, PD）等微等离子体为原子化器或激发源的小型原子光谱分析仪备受关注，但由于它们抗水汽的能力较低，影响了其分析液体样品中金属离子的潜力。将氢化物发生用作小型原子吸收、原子荧光和原子发射光谱分析仪的进样手段[19-21]，可以显著提高这些微等离子体的原子化和激发能力。此外，氢化物发生与非原子光谱分析技术的联用也展现出了独特的优势，例如，利用金属氢化物与碲化镉量子点或纳米金之间的显色反应[22,23]，建立基于原位氢化物发生和目视比色检测的方法，可以达到降低背景干扰和提高灵敏度的目的。

尽管氢化物发生本身就是一种分离富集技术，但当样品溶液中的干扰元素（如过渡金属离子、硝酸根等）超过一定浓度时，这些干扰离子或干扰离子反应生成的金属纳米小颗粒会抑制氢化物的生成。针对此种情况，可以通过：①加入掩蔽剂，控制过渡金属沉淀的形成；②选择性地富集样品中的待测元素；③采取新的氢化物发生模式或装置；④将样品稀释，使干扰元素的浓度降到一个"安全"的范围，再进行气相富集，也能够起到减少液相干扰和提高分析方法灵敏度的作用。

对于 Cu、Co、Ni、Fe 等过渡金属离子，加入 EDTA、硫脲、羟胺、抗坏血酸等络合剂，与其形成稳定的络合物，是一种很好的消除干扰的方法。例如，加入硫脲可以降低铜对硒氢化物发生的干扰，加入抗坏血酸可以减少过渡金属和贵金属对 As 测定的干扰。在样品中被测元素含量低于检出限或基体较复杂的情况下，则可以考虑分离与富集的方法，如固相萃取（solid phase extraction, SPE）、液液萃取（liquid-liquid extraction, LLE）、共沉淀（co-precipitation）、浊点萃取（cloud point extraction, CPE）等。SPE 具有萃取速度快、富集倍数大、有机溶剂消耗少等优点，结合碳纳米管[24]、二氧化钛[25]、金属有机骨架化合物（metal organic framework, MOF）[26]等纳米材料丰富的表面化学性质对 As、Se、Cd 等元素进行选择性富集，是近年来的一个研究热点。

由于样品前处理大多在酸性条件下完成，所以采用"酸性模式"（即待测元素存在于酸性溶液中）

发生氢化物更为方便和常见，但在这种模式下，过渡金属和贵金属会对氢化物发生造成明显的干扰；将样品溶液调节至碱性，使干扰元素转化为沉淀，然后在"碱性模式"（待测元素溶解于碱性的还原剂中）下与酸性溶液混合反应后生成氢化物[27,28]，可以显著地减小干扰，但是需要避免胶状沉淀对待测元素的夹带和吸附。dos Santos 等发现，汞在碱性溶液中也能直接被还原生成冷蒸气，可用于生物样品中总汞的检测[29]。除此之外，氢化物发生还可以在"非水介质"中进行：将 Sb 萃取到氯仿中，再加入硼氢化钠和冰乙酸溶液时，可以在有机相中进行氢化物发生[30]。将 Hg、As、Sb 萃取至离子液体中，再与固体硼氢化钠、氯化亚锡和氢化铝锂反应，可以直接将其中的元素还原成气态产物后再进行检测[31]，不仅简化了流程，同时也达到了富集待测元素和降低干扰的目的。同样地，通过设计新的氢化物反应装置，能够尽可能地缩短反应时间和加快气液分离，降低过渡金属或贵金属元素的干扰，前面介绍过的薄膜式气液分离器以及可移动还原床氢化物发生器[32]均是基于该原理。

对氢化物进行气相富集，能够有效地降低气相干扰和检出限，实现超痕量的元素分析。常见于报道的气相分离富集方法包括液氮冷却捕集、溶液吸收、热表面原位捕集和放电捕集等[33]。液氮冷却捕集法是将氢化物收集在浸泡于液氮中的"U"形管中，捕集后再加热释放进行检测。该方法是 Holak 为解决 Zn/HCl 体系中氢化物发生较慢的问题时提出来的，随着 THB 体系的出现其一度被淘汰，但因该法富集倍数高，现在超痕量分析中仍占有一席之地[34]。由于仅有氢化物被捕集，大量氢气直接被排空，可以提高等离子体焰的稳定性。采取程序升温的办法，还可以根据氢化物沸点的不同实现同一元素（如砷）氢化物的形态分析以及不同元素氢化物的分离分析[35]。溶液吸收法是利用吸收液将氢化物富集于溶液中，如 Ag-DDC、$AgNO_3$、$KI-I_2$、Ce^{4+}-KI 溶液等。然而，吸收富集法不能在线完成，操作较麻烦，还会将吸收液引入原子化器，造成吸收液基体的干扰。热表面原位捕集法是将氢化物捕集在加热的表面（如 Pd/Pt/Ir 等金属[36,37]、钨丝[38,39]、石墨管[40,41]、金丝[42]等），再升温释放后进行检测的一种技术，多用于电热原子吸收光谱分析（ETAAS）中。2014 年的一项研究表明[43]，在 O_2 条件下 DBD 的内表面也可以捕获 Bi 的氢化物，而在关闭 O_2 通入氢化物发生空白溶液产生的 H_2 时可以释放 Bi。该方法随后被应用于 As、Se、Pb 和 Sb 的氢化物的 DBD 原位捕集和释放，大大地提高了原子光谱分析的灵敏度[44]；其原理是：待测元素在 O_2 条件下以氧化物的形态被捕获在石英表面，在 H_2 条件下原子化后以"原子簇"形式传输至原子光谱分析仪中进行原子化和激发。

8.3 光化学蒸气发生

光化学蒸气发生（photochemical vapor generation，PCVG）是近年来出现的一种新型化学蒸气发生方法，2003 年由 Sturgeon 等提出并用于硒的原子吸收光谱法测定[45]。光化学蒸气发生通常以低分子量有机化合物（low molecular weight organic compound，LMWOC）为光化学反应介质，在紫外光的照射下，使分析物转化为挥发性或半挥发性物质，气液分离后进入原子光谱/质谱分析仪中进行检测。经质谱检测证明，光化学蒸气发生的产物可以是元素蒸气、氢化物、甲基化合物、乙基化合物和羰基化合物等，这与传统的氢化物发生存在很大的区别。除此之外，低分子量有机化合物还具有稳定性好、反应绿色、环境友好、不产生大量的氢气等诸多优点。

经过近 20 年的发展，光化学蒸气发生的适用元素范围不仅覆盖了传统氢化物发生元素，还包括一些过渡金属（铁、钴、镍、钼、钨、铜等）、贵金属（钌、铑、钯、锇、铱、铂）和非金属（碘、溴、氯等）元素，已然成为一种新型高效且普适性很强的化学蒸气发生法。

8.3.1 光化学蒸气发生装置

典型的光化学蒸气发生体系（装置）如图 8-1 所示，待测物与 LMWOC 以一定的流速被泵入光化学反应器中，光照（通常是紫外光）后的气液混合物在载气的带动下进入气液分离器，经气液分离后的化学蒸气进入原子光（质）谱仪中进行检测。

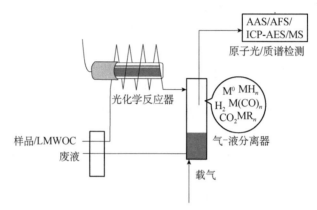

图 8-1　光化学蒸气发生的装置图

光化学反应器在很大程度上决定了蒸气发生的效率。与仪器联用时，应用较多的是流通式的光化学反应器，一般采用汞蒸气放电灯周围缠绕螺旋状石英管或聚四氟乙烯管的形式。将紫外光蒸气发生与雾化室集成一体作为 ICP-MS 的样品引入模块[46]，使元素在 1%～5% 的甲酸存在下雾化并经紫外辐射后测定，不同元素的检测信号可以提高 2～40 倍不等。薄膜式的光化学反应器[47]既具有高效光化学蒸气发生的特点，又实现了高效的即时气液分离。由于发生效率和气液分离速度的提高，抗干扰能力明显增强，多种元素的检出限改善了两个数量级。将石英管烧结在低压汞灯体内部作为紫外光化学反应器[48]，样品溶液流经反应器时可以受到充分且均匀的紫外光照，PCVG 效率进一步大幅提升。此外，我们还将紫外 LED 和连续光源作为辐照光源[49]，并考察了 UV 强度、辐射波长与 PCVG 效率之间的关系[50]。

8.3.2 光催化剂和增敏剂在光化学蒸气发生中的应用

许多元素[如 Se(Ⅳ)、Hg、Ni、Co、Fe、I 等]可以在紫外辐照低分子量有机化合物时高效地生成挥发性的单质或化合物，而有些元素需要光催化剂或增敏剂的促进才能获得较高的 PCVG 效率。

Se(Ⅵ)直接光化学蒸气发生的效率很低，但在含有纳米 TiO_2（一种常见的光化学催化剂）的甲酸溶液中，可以直接反应生成 SeH_2。研究表明，在日光或紫外光照射下，TiO_2 被光激活并生成具有高氧化还原性质的电子和空穴，能够降解或还原附着于其表面的有机物和无机物（如高价态的金属离子），如图 8-2 所示。升高溶液的温度和使用纳米 TiO_2 可以进一步提高 Se(Ⅵ) 的 PCVG 效率[51]。通过贵金属（如 Ag）负载[52]或半导体（如 C_3N_4）复合[53]，还可以调控纳米 TiO_2 的催化性能，促进 Se(Ⅵ)→H_2Se 以及 Hg(Ⅱ)→Hg^0 的转化。将 MOF 用于 PCVG 也可以提高 Se(Ⅵ) 的 PCVG 效率[54-56]。但是，除了 Se、Hg、Bi[57]和 Te[58]等少数几个元素外，纳米 TiO_2 和 MOF 材料对其他元素的催化性能都不太理想。

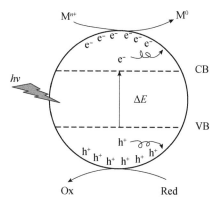

图 8-2 光照下纳米材料表面发生的反应

CB：导带，VB：价带，ΔE：能带间隙

在早期的研究中，人们发现共存离子可能对 PCVG 产生正干扰[59]，但是直到最近才意识到，可以将其作为增敏剂加入分析物的低分子量有机酸溶液中，以提高分析灵敏度。例如，氯化铵对溴化物和溴酸盐的 PCVG 有很高的增敏作用[60]；又如，Pb 在直接光化学蒸气发生时，灵敏度较低，而加入 Ni(Ⅱ)、Co(Ⅱ) 等过渡金属离子后可将其灵敏度提高数千倍[61]。Fe(Ⅲ) 对 As(Ⅲ)[62]、Cu(Ⅱ) 对 Cl(-Ⅰ)[63]均存在类似的增强作用。目前，离子型增敏剂已被广泛地应用于光化学蒸气发生研究中，使得化学蒸气发生的适用元素范围和原子光（质）谱分析性能不断拓展和提高，表 8-1 总结了光催化剂和增敏剂在 PCVG 中的应用。

表 8-1 光催化剂和增敏剂在光化学蒸气发生中的应用

元素	光反应介质	催化剂	产物	参考文献
Se(Ⅵ)	HCOOH	Ag-TiO$_2$ ZrO$_2$ CAU-1 UiO-66(Zr/Ti) PCN-224(Zr/Ti) UiO-66(Ce/Ti)	H$_2$Se	[51,52, 54-56]
Te(Ⅵ)	CH$_3$COOH +HCOOH	Fe(Ⅲ)/TiO$_2$	—	[58]
Bi(Ⅲ)	CH$_3$COOH +HCOOH	Fe-BTC	(CH$_3$)$_3$Bi	[57]
Br(-Ⅰ)	CH$_3$COOH	NH$_4$Cl	CH$_3$Br	[60]
Pb(Ⅱ)	HCOOH	Co(Ⅱ)/Ni(Ⅱ)	—	[61]
As(Ⅲ)	CH$_3$COOH +HCOOH	Fe(Ⅲ)	—	[62]
Cl(-Ⅰ)	CH$_3$COOH	Cu(Ⅱ)	CH$_3$Cl	[63]
F(-Ⅰ)	CH$_3$COOH	Cu(Ⅱ)	CH$_3$F	[64]
Se(Ⅵ)	CH$_3$COOH	Cd(Ⅱ)	H$_2$Se	[65]
Bi(Ⅲ/Ⅴ)	CH$_3$COOH + HCOOH	Fe(Ⅲ)	(CH$_3$)$_3$Bi	[66]
Mo(Ⅵ)	HCOOH	Fe(Ⅲ)	—	[67]
Mo(Ⅵ)	HCOOH	Cu(Ⅱ)+Co(Ⅱ)	—	[68]
W(Ⅵ)	HCOOH	Cd(Ⅱ)	—	[69]
Tl(Ⅰ)	HCOOH	Co(Ⅱ)	—	[70]
As(Ⅲ)/As(Ⅴ)	CH$_3$COOH	Cd(Ⅱ)	(CH$_3$)$_3$As	[71]
Te(Ⅳ)	HCOOH	Co(Ⅱ)	—	[72]
Cd(Ⅱ)	HCOOH	Fe(Ⅱ)	—	[73]

"—" 表示未检测到。

8.3.3 光化学蒸气发生的机理

光化学蒸气发生机理的研究对拓展其应用范围、降低干扰、提高 PCVG 效率、探索新型 PCVG 催化剂及增敏剂等都具有重要意义。Sturgeon 等对光化学蒸气发生机理进行过深入的探讨[74-76]，但大多还是处于猜测、初步实验验证和讨论阶段。

通过冷肼捕集反应生成的挥发性物质并将其引入到 GC-MS 中检测，是研究机理的主要手段。PCVG 产物受不同有机酸影响（表 8-2）：在甲酸中，Se(IV)发生产物有 SeCO 和 H_2Se；在乙酸和丙二酸中，发生产物主要为二甲基硒；在丙酸中，发生产物为二乙基硒[77]。当甲酸为光化学反应介质时，有 AsH_3 生成；当使用乙酸、丙酸、丁酸等单独酸或混合酸时，都生成烷基砷化合物[78]。在甲酸、乙酸和丙酸中，汞和甲基汞的 PCVG 产物都是 Hg^0。基于此原理，可以直接利用白酒[79]、醋[80]、天然水[81]或废水[82]中的有机质，建立基质辅助的光化学蒸气发生法。过渡金属 Fe(III)、Co(II)、Ni(II)的 PCVG 产物多为相应的羰基化合物[83-85]。而卤素的光化学蒸气发生产物则为相应的烷基卤化物[60,63,86]。

表 8-2 元素光化学蒸气发生的产物和介质之间的关系

元素	低分子量有机酸类型			参考文献
	甲酸	乙酸	丙酸	
Hg(II)	Hg^0	Hg^0	Hg^0	[87,88]
CH_3HgCl	Hg^0	Hg^0	Hg^0	[87,88]
Se(IV)	$H_2Se/SeCO$	$Se(CH_3)_2$	$Se(C_2H_5)_2$	[77]
Se(VI)	—	—	—	
As(III/V)	AsH_3	$As(CH_3)_3$	$As(C_2H_5)_3$	[78]
I(-I)	HI	CH_3I	C_2H_5I	[86]
Br(-I)	HBr	CH_3Br	C_2H_5Br	[60]
Cl(-I)	HCl	CH_3Cl	C_2H_5Cl	[63]
Ni(II)	$Ni(CO)_4$	$Ni(CO)_4$	—	[83]
Fe(II/III)	$Fe(CO)_5$	$Fe(CO)_5$	—	[85]
Co(II)	$Co(CO)_4H_2$	—	—	[84]

"—"表示未检测到。

一般认为，有无纳米 TiO_2 等光催化剂的反应机理是不一样的。以 Se(VI)为例，在纳米 TiO_2 或 MOF 存在的情况下，PCVG 是表面吸附和电子转移的过程（图 8-2）：在光照之前，Se(VI)及 $HCOO^-$ 等首先被吸附到光催化剂的表面；当紫外光辐射光催化剂产生光生电子和空穴时，$HCOO^-$ 被空穴或氧化性自由基（如·OH）消耗，阻止其与光生电子复合，同时自身被矿化成二氧化碳和水；光生电子将表面吸附的 Se(VI)还原成单质硒；随着 Se(VI)消耗完毕，溶液中没有能够俘获电子的物质，而光生空穴却一直被 $HCOO^-$ 消耗，大量产生的电子聚集在 TiO_2 表面使其电负性增加，进一步将 Se^0 还原成 H_2Se。

$$TiO_2 \xrightarrow{h\nu} e_{cb}^- + h_{vb}^+$$

$$SeO_4^{2-} + 3H^+ + 2e_{cb}^- \longrightarrow HSeO_3^- + H_2O$$

$$HSeO_3^- + 5H^+ + 4e_{cb}^- \longrightarrow Se^0 + 3H_2O$$

$$Se^0 + 2H^+ + 2e_{cb}^- \longrightarrow H_2Se$$

$$HCOO^- + 2h_{vb}^+ \longrightarrow CO_2 + H^+$$

$$HCOO^- + \cdot OH \longrightarrow \cdot COO^- + H_2O$$

对于没有 TiO$_2$ 等光催化剂存在的光化学蒸气发生体系，通常假设为自由基反应机理，即低分子量有机酸在紫外光照下裂解产生 ·H、·R、·COO$^-$ 等还原性自由基，进而引发一系列的自由基反应，最终生成挥发或半挥发性的氢化物、烷基化合物或羰基化合物。在过渡金属离子存在的情况下，紫外光照时会发生配体向金属电荷转移（ligand to metal charge transfer，LMCT），促进低分子量有机酸的氧化、脱羧和烷基转移[89,90]并使金属还原：

$$M\text{-}O_2CR^{n+} \longrightarrow M^{(n-1)+} + \cdot O_2CR \longrightarrow \cdot R + CO_2$$

$$M^{(n-1)+} \xrightarrow{\cdot H/\cdot CO_2^-/e_{(aq)}^-} M^{(n-2)+} \longrightarrow M^0$$

在过渡金属离子增敏的光化学蒸气发生体系中，Fe(Ⅲ)、Cu(Ⅱ)、Co(Ⅱ)等离子或者它们的混合物可能会改变溶液的光吸收（表现为新的 LMCT 吸收带）以及低分子有机酸的裂解，最终影响 PCVG 反应的产物[75]。近期，我们采用电子顺磁共振波谱（electron paramagnetic resonance，EPR）分析了 PCVG 体系中的自由基种类及含量与反应条件之间的关系[68]：没有过渡金属离子存在时，紫外光照甲酸溶液产生的主要为羟基（·OH）和羧基自由基（·CO$_2^-$），当甲酸溶液中存在 Cu(Ⅱ)和 Co(Ⅱ)离子时，产生的主要是还原性的 ·CO$_2^-$（·OH 可以忽略不计），证明过渡金属离子和自由基的变化对 PCVG 产物产生了直接的影响。

Sturgeon 等[76]从热力学的角度综述了不同元素的 PCVG 机理。他们认为，无论体系中有光催化剂与否，元素氧化还原对的能斯特电位与还原性自由基（e$_{(aq)}^-$、·H、·R、·CO$_2^-$）的氧化还原势或光催化材料的价带导带电位之间的关系是元素能否进行光化学还原和蒸气发生的关键。

8.3.4　光化学蒸气发生的分析应用

作为一种新型化学蒸气进样方法，光化学蒸气发生在复杂基质样品（超）痕量元素及其形态分析方面展示出诱人的优势和应用前景。目前，PCVG 已用于天然水体、地质样品、食品饮料、沉积物、生物组织、汽油、中草药、尿液等不同基质样品中多种元素及其形态的分析。

某些元素的光化学蒸气发生会受到基体或样品消解时引入的阴离子（如 NO$_3^-$）的干扰，需要采取合适的方法加以消除，也可以使用其他定量和校准方法，如基质匹配标准加入法或同位素稀释法[61,91-93]。对于有光催化材料参与的反应，可以利用纳米材料的吸附性质，在 PCVG 反应前对样品中的待分析物进行分离富集，同时发挥其基体分离和催化的双重功能，近期报道的 Fe-BTC MOF 在检测土壤、沉积物和合金中 Bi 的应用即是基于此原理[57]。此外，Lopes 等[94]还系统地研究了 NO$_3^-$ 的干扰机制以及多种消除 NO$_3^-$ 干扰的方法[氨基磺酸、硫酸肼、硫代硫酸钠、V(Ⅲ)、光还原、硝酸试剂]，其中体积分数为 2%的碱性硫酸肼效果最好。

在形态分析方面，利用光源或光催化剂性质的差异，可以进行简单的无需色谱分离的形态分析。例如，紫外光照射下 Hg(Ⅱ)和 MeHg 均能进行光化学蒸气发生，而可见光条件下甲酸仅能使 Hg(Ⅱ)蒸气化[87]；UV-C(254 nm)条件下甲酸可以使 Hg(Ⅱ)和 MeHg 同时进行 PCVG 反应，而 UV-B(311 nm)仅对 Hg(Ⅱ)起作用[95]。通过简单的差减法，即可实现波长调控的无需色谱分离的汞形态分析。此外，王秋泉[52,96]、阴永光[88,97,98]和孙毓璋[99,100]等课题组在光化学蒸气发生与 HPLC-AFS/ICPMS 联用分析方面开展了许多重要的工作。一方面，由于产生氢气少，PCVG 能够减小分析过程中等离子体的波动，提高分析结果的精密度；另一方面，色谱分离可以将基体中的干扰成分与待分析物分离，进而消除其对光化学蒸气发生的影响。

8.4 其他化学蒸气发生法

8.4.1 电化学蒸气发生

电化学蒸气发生（electrochemical vapor generation，ECVG）是通过电化学反应将待测元素转化成易挥发的单质或化合物的一种蒸气发生方式，由于气态产物多为氢化物，也称作电化学氢化物发生（EHG）[101]。1977年，Rigin首次将ECVG作为原子光谱分析的进样方法对砷和锡进行了测定，但由于该法采用间歇式反应器，氢化物发生效率低，未能被广泛采用。黄本立等[102]报道了基于薄层平板电解池的流动注射电化学氢化物发生装置，明显地改善了As、Sb、Se氢化物的发生效率，同时大大地降低了基体干扰。后来的研究也大都沿用了这一电解池模型，现在见于报道的电化学蒸气发生元素包括As、Sb、Bi、Cd、Hg、Ge、Se、Sn、Te、Zn、Tl等11种元素。

电化学氢化物的生成，一般认为经过了四步：待测元素扩散至阴极材料的表面；待测元素被电化学还原至元素态，并沉积在阴极表面；元素态进一步还原，形成氢化物；氢化物扩散至溶液中。其中，水（或水合氢离子）还原生成氢气与氢化物的生成构成竞争性反应。电解池的设计、阴极材料和电解液的选择对氢化物的发生起着关键作用。一般认为，氢过电位较高的电极材料（如Hg、Zn、Pb等）有利于产生氢化物。采用聚苯胺[103]、半胱氨酸[104]、谷胱甘肽[105]修饰电极或铅-锡合金[106]作为ECVG阴极，不仅提高了电化学氢化物发生的效率，同时还可以根据砷、汞元素不同形态化合物在电极上的电化学蒸气发生行为的差异，实现元素的选择性蒸气发生和形态分析。

电化学蒸气发生的主要优点是：不需要THB还原试剂，仅通过电子转移将待测物还原成具有挥发性的氢化物或原子蒸气；过渡金属离子的干扰小于硼氢化物-酸还原体系。但是，电化学蒸气发生也有自身的缺点，比如获得重现性良好的电极表面非常困难，氢化物发生元素之间的相互干扰比硼氢化物更为严重等，这点也是容易理解的。

8.4.2 烷基化合物发生

金属烷基化合物一般具有高的热稳定性及较低的沸点，非常适合与气相色谱联用进行分析。常用的烷基化试剂包括格氏试剂和烷基硼酸盐。格氏试剂（R—MgX，R=CH$_3$、C$_2$H$_5$、C$_3$H$_7$，X=Cl、Br、I）与有机金属化合物（R_n'-M）可生成具有挥发性的金属烷基化合物（R_n'-M-R），曾被广泛用于有机锡、有机汞和烷基铅化合物的测定。由于格氏试剂具有极高的化学活性，遇水易分解，需先将待测物萃取至无水溶剂中，操作十分烦琐，目前已基本被淘汰。四烷基硼酸盐（NaBEt$_4$、NaBPr$_4$、NaBPh$_4$）是一类可以在水相中进行烷基化的试剂，避免了格氏试剂烷基化需要在无水条件下完成所带来的不便。相对于THB来说，NaBEt$_4$具有更高的灵敏度和抗干扰能力，在对Pb和Cd进行乙基化后进样，可获得优异的分析性能。Sturgeon等发现NaBEt$_4$还可以提高Au的化学蒸气发生效率[107]。近年来，该烷基化方法广泛用于烷基铅、有机锡、硒及化合物、汞及其烷基化合物的衍生化和（超）痕量分析[108-110]。结合气相色谱和原子光（质）谱的时间分辨和光（质）谱分辨能力，还可以进行多元素多形态的同时分析[111]，但是需要选择合适的烷基化试剂，使不同形态的烷基化产物可以在气相色谱中分离开来。例如，无机汞和乙基汞的乙基化产物均为Et$_2$Hg，可以改用丙基或苯基硼酸盐进行衍生化。酸度对于烷基硼酸盐的衍生化效率的影响十分显著，需要仔细考察酸度的影响。此外，在进行多形态分析时，还要考虑形态之间的相互转化，必要时可以采取同位素稀释法（isotope dilution，ID）提高分析结果的精密度[112]。

除了常见的金属及有机金属化合物，阴离子（如 Cl^-、Br^-、I^-、F^-、SCN^-、CN^-、S^{2-}、NO_3^-、NO_2^- 等）也可以采用烷基化试剂衍生化后进行测定。D'Ulivo 和 Pagliano 等在这方面做了很多有意义的研究工作[113-115]，他们通过与三烷基氧鎓盐（$Et_3O^+BF_4^-$、$Et_3O^+BF_4^-$）生成乙基化合物，采用顶空 GC-MS 对海水及污水中的多种阴离子进行了高灵敏的测定。

8.4.3 卤化物发生

卤化物（氟化物、氯化物、溴化物）发生的主要目的是获得具有挥发性的金属或非金属以及有机金属卤化物。早在 1923 年，人们将 Ge 与 6 mol/L 盐酸反应并通过蒸馏得到了 $GeCl_4$。卤化物的生成过程与氢化物发生比较相似，可以通过待测元素与高浓度的卤酸（HX，X=F，Cl，Br）或待测元素与卤素的碱金属或碱土金属盐在浓硫酸存在下反应而制得。卤化物发生研究较多的元素是 Ge、Hg、As、Sb、Sn 和 Si。

卤化物发生的主要优点是避免使用相对昂贵的 $NaBH_4/KBH_4$，同时由于高浓度酸的存在，过渡金属和贵金属无法还原成金属纳米小颗粒，因此在卤化物发生中不存在严重的过渡金属或贵金属干扰。例如，将含锗试样和高浓度的 HCl 泵入间歇或流动反应器产生具有挥发性的 $GeCl_4$[116]，然后引入 $Ar-H_2$ 原子化器进行原子化并用非色散原子荧光光谱仪进行检测，几乎不受共存离子的干扰。Mester 等[117]通过 GC-MS 发现在无机砷、甲基砷、二甲基砷的氢化物发生过程中有砷和有机胂的氯化物的存在，向砷酸盐或亚砷酸盐溶液中加入更高浓度（>5 mol/L）的 HCl 时，在溶液上方顶空气体中检测到挥发性三氯化砷，证明在氢化物发生的过程中，当盐酸浓度较大时，产物从氢化物逐渐转变为氯化物。此外，在氯含量较高的介质中，三丁基锡（TBT）和甲基汞（CH_3Hg）也可以转化为具有更高挥发性的 TBT-Cl 和 CH_3HgCl。

由于卤化物蒸气发生的效率不高，需使用高浓度的卤酸，容易对分析仪器造成腐蚀，因此该方法在分析化学中的应用并不普遍。

8.4.4 羰基化合物发生

当一氧化碳通过均匀细小的金属镍粉时会有 $Ni(CO)_4$ 气体生成，这一发现为制造高纯镍粉和 $Ni(CO)_4$ 催化有机合成化学的研究做出了巨大的贡献，也为 Ni 的羰基化合物发生和原子光谱检测提供了一种 CVG 进样方法。样品中的镍经处理后都以离子形式存在，一般需要用强的还原剂（如 THB）还原成单质形式再与 CO 气体反应生成 $Ni(CO)_4$。

传统羰基化合物发生的研究一直不是很活跃，其中最主要的原因可能是需要剧毒气体 CO，这对实验人员的身体健康造成了很大的危险。幸运的是，通过紫外光化学蒸气发生，也可以在甲酸介质中产生 $Ni(CO)_4$ 和 $Fe(CO)_5$[118,119]等羰基化合物。

8.4.5 螯合物发生和氧化物发生

螯合物发生是指用螯合剂与金属离子反应形成具有挥发性的金属化合物蒸气。常用的螯合剂有乙酰丙酮、三氟乙酰丙酮、二硫代氨基乙酸钠、二硫代磷酸盐等。过去常用液液萃取（LLE）使螯合物与基体分离，再引入气相色谱分离后用原子光（质）谱法进行检测。但 LLE 要用到大量的有机试剂，操作烦琐且污染环境，逐渐被固相萃取（SPE）和固相微萃取（solid phase microextraction，SPME）所替代。在 SPE 或 SPME 中，螯合物被捕集在 SPME 或 SPE 柱表面，然后在气相进样口热脱附后进行分离检测[120]。此外，有部分学者还研究了通过流动注射进行在线螯合物发生的可能性。将样品与 Na-DDTC 溶液在线混合生成螯合物，可以采用非色散原子荧光法对 Zn[121]、Au[122]、Ni[123]、Cd[124]等元素进行测定，部分元素的分析性能接近甚至优于氢化物发生法。在实际样品分析中，螯合物发生可与氢化物发生、烷基化合物发生、光化学蒸气发生等技术相辅相成，扩大了化学蒸气发生的适用元素范围。

近年来，氧化物发生主要集中在将 Os 以氧化物形式从基体中分离出来的一种化学蒸气发生法，通常需要强氧化剂（如 H_2O_2、浓 HNO_3）才能将 Os 氧化至最高价态的易挥发的 OsO_4。最近，高英等[125]采用紫外光照 5% HNO_3 也检测到了 Os 的挥发性物质，可能是由于 HNO_3 在光照下产生了氧化性更强的自由基从而将 Os 氧化和蒸气化。此外，用 Ce(Ⅳ)作氧化剂，也能够将 Ru(Ⅲ)转化成具有挥发性的 RuO_4，从而实现蒸气进样，使火焰原子吸收法检测 Ru 的灵敏度比喷雾进样时提高了 60 倍。

8.4.6 等离子体诱导化学蒸气发生

等离子体诱导化学蒸气发生是利用等离子体中的电子或自由基等活性粒子引发氧化还原反应，将待测元素由液态离子直接氧化或还原为挥发性的气态原子或分子，从而实现元素的蒸气发生。其最大优点是避免了化学氧化还原试剂的使用，同时装置简单，易于与便携式原子光谱仪联用。目前报道的主要是基于介质阻挡放电等离子体[126]和液体阴极放电等离子体[127]诱导的汞冷原子蒸气发生，关于等离子体诱导化学蒸气发生的详细介绍可进一步参考文献[128]或本书其他相关章节。

8.5 展　　望

经过数十年的研究和探索，CVG 法在分析对象、蒸气发生体系、检测技术和应用领域等方面都得到了很大的丰富。得益于新的 CVG 体系的建立和发展，可以进行化学蒸气发生的元素越来越多，不断突破人们对于元素化学性质的认知。化学蒸气发生与各种原子光（质）谱分析仪器的结合，很好地发挥了它在提高进样效率和等离子体激发能力以及降低光（质）谱干扰方面的优势。增敏剂、掩蔽剂、干扰机理和蒸气发生机理的研究，以及化学蒸气发生反应器和反应模式的创新，为进一步提高蒸气发生效率、降低基体干扰和复杂样品分析提供了理论和技术支撑。化学蒸气发生法与各种分离、联用技术相结合，也在形态分析特别是组学分析中展现出广阔的应用前景。未来的化学蒸气发生相关研究可以从以下几个方面重点开展工作：

（1）借助先进的分析检测手段及技术，对化学蒸气发生过程的中间产物和终产物进行鉴定，确定其存在的化学物种形式，进而揭示化学蒸气发生反应的机理。从 CVG 机理研究的具体技术来看，理论计算和中间态物种的实验鉴定是必由之路，因此检测自由基和其他瞬态中间体的顺磁共振及超快光谱技术将是理想的选择。高分辨质谱则是检测最终气态产物的有力工具。

（2）虽然能够产生化学蒸气的元素范围已经很广，但其中只有少部分能够用于实际样品分析。针对化学蒸气发生效率低的元素，发展有效的增敏剂和干扰消除方法，是值得探索的研究方向；寻找更多高效的 PCVG 催化剂，特别是基于过渡金属离子的均相催化剂将是重要发展方向。

（3）加快仪器部件和仪器小型化的研制。虽然氢化物和光化学蒸气的发生方法具有较高的灵敏度且适用元素范围广，但仅有前者常见于部分商品化的原子光（质）谱仪器中。CVG和微等离子体的巧妙结合或许可以产生新的微型原子光谱分析仪。

（4）化学蒸气发生法本质上是一种物质组分的分离方法。待测组分与复杂样品基体的有效分离使得检测步骤中相关干扰更少，因此对检测器的要求降低。我们坚信，CVG与原子光谱/质谱以外的其他检测技术的联用的研究值得进一步拓展。

（5）进一步推进化学蒸气发生在食品科学、环境科学和组学等领域的应用，充分发挥化学蒸气发生进样技术的固有优势，可为解决相关学科领域的实际问题提供技术和方法学支撑。

致谢：感谢国家自然科学基金区域创新发展联合基金项目（U21A20283）的资助。

参 考 文 献

[1] Holak W. Gas-sampling technique for arsenic determination by atomic absorption spectrophotometry. Analytical Chemistry, 1969, 41 (12): 1712-1713.

[2] Braman R S, Foreback C C, Justen L L. Direct volatilization spectral emission type detection system for nanogram amounts of arsenic and antimony. Analytical Chemistry, 1972, 44 (13): 2195-2199.

[3] 陶锐, 周宏刚. 氢化物发生-原子吸收分光光度法测定食品中铅. 分析化学, 1985, 13 (4): 283-285.

[4] 姚金玉, 胡庆兰. 氢化物发生-原子吸收分光光度法测定砷和硒. 分析化学, 1977, 5 (5): 407.

[5] Thompson K C, Thomerson D R. Atomic-absorption studies on determination of antimony, arsenic, bismuth, germanium, lead, selenium, tellurium and tin by utilizing generation of covalent hydrides. Analyst, 1974, 99 (1182): 595-601.

[6] Laborda F, Bolea E, Baranguan M T, Castillo J R. Hydride generation in analytical chemistry and nascent hydrogen: When is it going to be over? Spectrochimica Acta Part B: Atomic Spectroscopy, 2002, 57 (4): 797-802.

[7] D'Ulivo A, Dědina J, Mester Z, Sturgeon R E, Wang Q, Welz B. Mechanisms of chemical generation of volatile hydrides for trace element determination (IUPAC Technical Report). Pure and Applied Chemistry, 2011, 83 (6): 1283.

[8] Sturgeon R E, Liu J, Boyko V J, Luong V T. Determination of copper in environmental matrices following vapor generation. Analytical Chemistry, 1996, 68 (11): 1883-1887.

[9] Feng Y L, Lam J W, Sturgeon R E. Expanding the scope of chemical vapor generation for noble and transition metals. Analyst, 2001, 126 (11): 1833-1837.

[10] Cerutti S, Escudero L A, Gasquez J A, Olsina R A, Martinez L D. On-line preconcentration and vapor generation of scandium prior to ICP-OES detection. Journal of Analytical Atomic Spectrometry, 2011, 26 (12): 2428-2433.

[11] Musil S, Kratzer J, Vobecký M, Hovorka J, Benada O, Matoušek T. Chemical vapor generation of silver for atomic absorption spectrometry with the multiatomizer: Radiotracer efficiency study and characterization of silver species. Spectrochimica Acta Part B: Atomic Spectroscopy, 2009, 64 (11): 1240-1247.

[12] Feng Y L, Sturgeon R E, Lam J W, D'Ulivo A. Insights into the mechanism of chemical vapor generation of transition and noble metals. Journal of Analytical Atomic Spectrometry, 2005, 20 (4): 255-265.

[13] Pohl P, Prusisz B. Chemical vapor generation of noble metals for analytical spectrometry. Analytical and Bioanalytical Chemistry, 2007, 388 (4): 753-762.

[14] He L, Zhu X F, Wu L, Hou X D. Determination of trace copper in biological samples by on-line chemical vapor

generation-atomic fluorescence spectrometry. Atomic Spectroscopy, 2008, 29 (3): 93-98.

[15] Zou W, Li C H, Hu J, Hou X D. Selective determination of Cr(Ⅵ) and non-chromatographic speciation analysis of inorganic chromium by chemical vapor generation-inductively coupled plasma mass spectrometry. Talanta, 2020, 218: 121128.

[16] Zhang C, Li Y, Wu P, Jiang Y, Liu Q, Yan X P. Effects of room-temperature ionic liquids on the chemical vapor generation of gold: Mechanism and analytical application. Analytica Chimica Acta, 2009, 650 (1): 59-64.

[17] Zheng C B, Sturgeon R E, Hou X D. Thin film hydride generation: Determination of ultra-trace copper by flow injection in situ hydride trapping graphite furnace AAS. Journal of Analytical Atomic Spectrometry, 2010, 25 (7): 1159-1165.

[18] Klaue B, Blum J D. Trace analyses of arsenic in drinking water by inductively coupled plasma mass spectrometry: High resolution versus hydride generation. Analytical Chemistry, 1999, 71 (7): 1408-1414.

[19] Zhu Z L, Zhang S C, Lv Y, Zhang X R. Atomization of hydride with a low-temperature, atmospheric pressure dielectric barrier discharge and its application to arsenic speciation with atomic absorption spectrometry. Analytical Chemistry, 2006, 78 (3): 865-872.

[20] Yu Y L, Du Z, Chen M L, Wang J H. Atmospheric-pressure dielectric-barrier discharge as a radiation source for optical emission spectrometry. Angewandte Chemie International Edition, 2008, 47 (41): 7909-7912.

[21] Xing Z, Wang J, Zhang S C, Zhang X R. Determination of bismuth in solid samples by hydride generation atomic fluorescence spectrometry with a dielectric barrier discharge atomizer. Talanta, 2009, 80 (1): 139-142.

[22] Huang K, Xu K L, Zhu W, Yang L, Hou X D, Zheng C B. Hydride generation for headspace solid-phase extraction with CdTe quantum dots immobilized on paper for sensitive visual detection of selenium. Analytical Chemistry, 2016, 88 (1): 789-795.

[23] Cao G M, Xu F J, Wang S L, Xu K L, Hou X D, Wu P. Gold nanoparticle-based colorimetric assay for selenium detection via hydride generation. Analytical Chemistry, 2017, 89 (8): 4695-4700.

[24] Wu H, Wang X C, Liu B, Liu Y L, Li S S, Lu J S, Tian J Y, Zhao W F, Yang Z H. Simultaneous speciation of inorganic arsenic and antimony in water samples by hydride generation-double channel atomic fluorescence spectrometry with on-line solid-phase extraction using single-walled carbon nanotubes micro-column. Spectrochimica Acta Part B: Atomic Spectroscopy, 2011, 66 (1): 74-80.

[25] Deng D Y, Zhou J R, Ai X, Yang L, Hou X D, Zheng C B. Ultrasensitive determination of selenium by atomic fluorescence spectrometry using nano-TiO_2 pre-concentration and in situ hydride generation. Journal of Analytical Atomic Spectrometry, 2012, 27 (2): 270-275.

[26] Zou Z R, Wang S L, Jia J, Xu F J, Long Z, Hou X D. Ultrasensitive determination of inorganic arsenic by hydride generation-atomic fluorescence spectrometry using Fe_3O_4@ZIF-8 nanoparticles for preconcentration. Microchemical Journal, 2016, 124 578-583.

[27] Wickstrom T, Lund W, Bye R. Hydride generation atomic absorption spectrometry from alkaline solutions: Determination of selenium in copper and nickel materials. Journal of Analytical Atomic Spectrometry, 1991, 6 (5): 389-391.

[28] 郭小伟, 郭旭明. 从碱性溶液中发生氢化物的原子荧光/原子吸收光谱法测定 Te^{4+} 及 Te^{6+}. 光谱学与光谱分析, 1996, 16 (3): 88-92.

[29] dos Santos E J, Herrmann A B, Frescura V L A, Sturgeon R E, Curtius A J. A novel approach to cold vapor generation for the determination of mercury in biological samples. Journal of the Brazilian Chemical Society, 2008, 19: 929-934.

[30] Aznárez J, Palacios F, Ortega M S, Vidal J C. Extraction-atomic-absorption spectrophotometric determination of antimony by generation of its hydride in non-aqueous media. Analyst, 1984, 109 (2): 123-125.

[31] Wen X D, Gao Y, Wu P, Tan Z Q, Zheng C B, Hou X D. Chemical vapor generation from an ionic liquid using a solid reductant: Determination of Hg, As and Sb by atomic fluorescence spectrometry. Journal of Analytical Atomic Spectrometry, 2016, 31 (2): 415-422.

[32] Tian X D. Movable reduction bed hydride generator coupled with inductively coupled plasma optical emission spectrometry for the determination of some hydride forming elements. Analyst, 1998, 123 (4): 627-632.

[33] 刘腾鹏, 刘美彤, 刘霁欣, 毛雪飞, 钱永忠. 气相富集原子光谱分析技术研究进展. 分析测试学报, 2019, 38(6): 744-754.

[34] Matoušek T, Wang Z, Douillet C, Musil S, Stýblo M. Direct speciation analysis of arsenic in whole blood and blood plasma at low exposure levels by hydride generation-cryotrapping-inductively coupled plasma mass spectrometry. Analytical Chemistry, 2017, 89 (18): 9633-9637.

[35] Chen G Y, Lai B H, Mao X F, Chen T W, Chen M M. Continuous arsine detection using a peltier-effect cryogenic trap to selectively trap methylated arsines. Analytical Chemistry, 2017, 89 (17): 8678-8682.

[36] Doidge P S, Sturman B T, Rettberg T M. Hydride generation atomic absorption spectrometry with *in situ* pre-concentration in a graphite furnace in the presence of palladium. Journal of Analytical Atomic Spectrometry, 1989, 4 (3): 251-255.

[37] Soo H L, Kyung Hoon J, Dong S L. Determination of mercury in environmental samples by cold vapour generation and atomic-absorption spectrometry with a gold-coated graphite furnace. Talanta, 1989, 36 (10): 999-1003.

[38] Cankur O, Ertaş N, Ataman O Y. Determination of bismuth using on-line preconcentration by trapping on resistively heated W coil and hydride generation atomic absorption spectrometry. Journal of Analytical Atomic Spectrometry, 2002, 17 (6): 603-609.

[39] Kula İ, Arslan Y, Bakırdere S, Ataman O Y. A novel analytical system involving hydride generation and gold-coated W-coil trapping atomic absorption spectrometry for selenium determination at $ng \cdot L^{-1}$ level. Spectrochimica Acta Part B: Atomic Spectroscopy, 2008, 63 (8): 856-860.

[40] Infante H G, Sánchez M L F, Sanz-Medel A. Ultratrace determination of cadmium by atomic absorption spectrometry using hydride generation with *in situ* preconcentration in a palladium-coated graphite atomizer. Journal of Analytical Atomic Spectrometry, 1996, 11 (8): 571-575.

[41] Chang C C, Jiang S J. Determination of Hg and Bi by electrothermal vaporization inductively coupled plasma mass spectrometry using vapor generation with *in situ* preconcentration in a platinum-coated graphite furnace. Analytica Chimica Acta, 1997, 353 (2): 173-180.

[42] Guo X M, Guo X W. Determination of ultra-trace amounts of selenium by continuous flow hydride generation AFS and AAS with collection on gold wire. Journal of Analytical Atomic Spectrometry, 2001, 16 (12): 1414-1418.

[43] Kratzer J, Boušek J, Sturgeon R E, Mester Z, Dědina J. Determination of bismuth by dielectric barrier discharge atomic absorption spectrometry coupled with hydride generation: Method optimization and evaluation of analytical performance. Analytical Chemistry, 2014, 86 (19): 9620-9625.

[44] Liu M T, Liu J X, Mao X F, Na X, Ding L, Qian Y Z. High sensitivity analysis of selenium by ultraviolet vapor generation combined with microplasma gas phase enrichment and the mechanism study. Analytical Chemistry, 2020, 92 (10): 7257-7264.

[45] Guo X M, Sturgeon R E, Mester Z, Gardner G J. UV vapor generation for determination of selenium by heated quartz tube atomic absorption spectrometry. Analytical Chemistry, 2003, 75 (9): 2092-2099.

[46] Sturgeon R E, Willie S N, Mester Z. UV/spray chamber for generation of volatile photo-induced products having enhanced sample introduction efficiency. Journal of Analytical Atomic Spectrometry, 2006, 21 (3): 263-265.

[47] Zheng C B, Sturgeon R E, Brophy C, Hou X D. Versatile thin-film reactor for photochemical vapor generation. Analytical Chemistry, 2010, 82 (7): 3086-3093.

[48] Qin D Y, Gao F, Zhang Z H, Zhao L Q, Liu J X, Ye J P, Li J W, Zheng F X. Ultraviolet vapor generation atomic fluorescence spectrometric determination of mercury in natural water with enrichment by on-line solid phase extraction. Spectrochimica Acta Part B: Atomic Spectroscopy, 2013, 88: 10-14.

[49] Hou X L, Ai X, Jiang X M, Deng P C, Zheng C B, Lv Y. UV light-emitting-diode photochemical mercury vapor generation for atomic fluorescence spectrometry. Analyst, 2012, 137 (3): 686-690.

[50] Zou Z R, Tian Y F, Zeng W, Hou X D, Jiang X M. Effect of variable ultraviolet wavelength and intensity on photochemical vapor generation of trace selenium detected by atomic fluorescence spectrometry. Microchemical Journal, 2018, 140: 189-195.

[51] Zheng C B, Wu L, Ma Q, Lv Y, Hou X D. Temperature and nano-TiO_2 controlled photochemical vapor generation for inorganic selenium speciation analysis by AFS or ICP-MS without chromatographic separation. Journal of Analytical Atomic Spectrometry, 2008, 23 (4): 514-520.

[52] Li H M, Luo Y C, Li Z X, Yang L M, Wang Q Q. Nanosemiconductor-based photocatalytic vapor generation systems for subsequent selenium determination and speciation with atomic fluorescence spectrometry and inductively coupled plasma mass spectrometry. Analytical Chemistry, 2012, 84 (6): 2974-2981.

[53] Xu F J, Jiang X M, Hu J, Zhang J Y, Zou Z R, Hou X D, Yan H J. Nano g-C_3N_4/TiO_2 composite: A highly efficient photocatalyst for selenium (Ⅵ) photochemical vapor generation for its ultrasensitive AFS determination. Microchemical Journal, 2017, 135: 158-162.

[54] Jia J, Long Z, Zheng C B, Wu X, Hou X D. Metal organic frameworks CAU-1 as new photocatalyst for photochemical vapour generation for analytical atomic spectrometry. Journal of Analytical Atomic Spectrometry, 2015, 30 (2): 339-342.

[55] Tu J P, Zeng X L, Xu F J, Wu X, Tian Y F, Hou X D, Long Z. Microwave-induced fast incorporation of titanium into UiO-66 metal-organic frameworks for enhanced photocatalytic properties. Chemical Communications, 2017, 53 (23): 3361-3364.

[56] He J H, Zhang Y J, He J, Zeng X L, Hou X D, Long Z. Enhancement of photoredox catalytic properties of porphyrinic metal-organic frameworks based on titanium incorporation via post-synthetic modification. Chemical Communications, 2018, 54 (62): 8610-8613.

[57] Jia Y T, Mou Q, Yu Y, Shi Z M, Huang Y, Ni S J, Wang R L, Gao Y. Reduction of interferences using Fe-containing metal-organic frameworks for matrix separation and enhanced photochemical vapor generation of trace bismuth. Analytical Chemistry, 2019, 91 (8): 5217-5224.

[58] He H Y, Peng X H, Yu Y, Shi Z M, Xu M, Ni S J, Gao Y. Photochemical vapor generation of tellurium: Synergistic effect from ferric ion and nano-TiO_2. Analytical Chemistry, 2018, 90 (9): 5737-5743.

[59] Zheng C B, Sturgeon R E, Hou X D. UV photochemical vapor generation and in situ preconcentration for determination of ultra-trace nickel by flow injection graphite furnace atomic absorption spectrometry. Journal of Analytical Atomic Spectrometry, 2009, 24 (10): 1452-1458.

[60] Sturgeon R E. Detection of bromine by ICP-oa-ToF-MS following photochemical vapor generation. Analytical Chemistry, 2015, 87 (5): 3072-3079.

[61] Gao Y, Xu M, Sturgeon R E, Mester Z, Shi Z M, Galea R, Saull P, Yang L. Metal ion-assisted photochemical vapor generation for the determination of lead in environmental samples by multicollector-ICPMS. Analytical Chemistry, 2015, 87 (8): 4495-4502.

[62] Wang Y L, Lin L L, Liu J X, Mao X F, Wang J H, Qin D Y. Ferric ion induced enhancement of ultraviolet vapour

generation coupled with atomic fluorescence spectrometry for the determination of ultratrace inorganic arsenic in surface water. Analyst, 2016, 141 (4): 1530-1536.

[63] Hu J, Sturgeon R E, Nadeau K, Hou X D, Zheng C B, Yang L. Copper ion assisted photochemical vapor generation of chlorine for its sensitive determination by sector field inductively coupled plasma mass spectrometry. Analytical Chemistry, 2018, 90 (6): 4112-4118.

[64] Sturgeon R E, Pagliano E. Evidence for photochemical synthesis of fluoromethane. Journal of Analytical Atomic Spectrometry, 2020, 35 (9): 1720-1726.

[65] Xu F J, Zou Z R, He J, Li M T, Xu K L, Hou X D. *In situ* formation of nano-CdSe as a photocatalyst: Cadmium ion-enhanced photochemical vapour generation directly from Se(Ⅵ). Chemical Communications, 2018, 54 (38): 4874-4877.

[66] Yu Y, Jia Y T, Shi Z M, Chen Y L, Ni S J, Wang R L, Tang Y R, Gao Y. Enhanced photochemical vapor generation for the determination of bismuth by inductively coupled plasma mass spectrometry. Analytical Chemistry, 2018, 90 (22): 13557-13563.

[67] Šoukal J, Sturgeon R E, Musil S. Efficient photochemical vapor generation of molybdenum for ICPMS detection. Analytical Chemistry, 2018, 90 (19): 11688-11695.

[68] Hu J, Chen H J, Hou X D, Jiang X M. Cobalt and copper ions synergistically enhanced photochemical vapor generation of molybdenum: Mechanism study and analysis of water samples. Analytical Chemistry, 2019, 91 (9): 5938-5944.

[69] Vyhnanovský J, Sturgeon R E, Musil S. Cadmium assisted photochemical vapor generation of tungsten for detection by inductively coupled plasma mass spectrometry. Analytical Chemistry, 2019, 91 (20): 13306-13312.

[70] Xu T, Hu J, Chen H J. Transition metal ion Co(Ⅱ)-assisted photochemical vapor generation of thallium for its sensitive determination by inductively coupled plasma mass spectrometry. Microchemical Journal, 2019, 149: 103972.

[71] Zhou J, Deng D Y, Su Y Y, Lv Y. Determination of total inorganic arsenic in water samples by cadmium ion assisted photochemical vapor generation-atomic fluorescence spectrometry. Microchemical Journal, 2019, 146: 359-365.

[72] Zeng W, Hu J, Chen H J, Zou Z R, Hou X D, Jiang X M. Cobalt ion-enhanced photochemical vapor generation in a mixed acid medium for sensitive detection of tellurium (Ⅳ) by atomic fluorescence spectrometry. Journal of Analytical Atomic Spectrometry, 2020, 35 (7): 1405-1411.

[73] Nováková E, Horová K, Červený V, Hraníček J, Musil S. UV photochemical vapor generation of Cd from a formic acid based medium: Optimization, efficiency and interferences. Journal of Analytical Atomic Spectrometry, 2020, 35 (7): 1380-1388.

[74] Sturgeon R E, Grinberg P. Some speculations on the mechanisms of photochemical vapor generation. Journal of Analytical Atomic Spectrometry, 2012, 27 (2): 222-231.

[75] Sturgeon R E. Photochemical vapor generation: A radical approach to analyte introduction for atomic spectrometry. Journal of Analytical Atomic Spectrometry, 2017, 32 (12): 2319-2340.

[76] Leonori D, Sturgeon R E. A unified approach to mechanistic aspects of photochemical vapor generation. Journal of Analytical Atomic Spectrometry, 2019, 34 (4): 636-654.

[77] Guo X M, Sturgeon R E, Mester Z, Gardner G J. Photochemical alkylation of inorganic selenium in the presence of low molecular weight organic acids. Environmental Science & Technology, 2003, 37 (24): 5645-5650.

[78] Guo X M, Sturgeon R E, Mester Z, Gardner G J. Photochemical alkylation of inorganic arsenic Part 1. Identification of volatile arsenic species. Journal of Analytical Atomic Spectrometry, 2005, 20 (8): 702-708.

[79] Li Y, Zheng C B, Ma Q, Wu L, Hu C W, Hou X D. Sample matrix-assisted photo-induced chemical vapor generation: A reagent free green analytical method for ultrasensitive detection of mercury in wine or liquor samples. Journal of

Analytical Atomic Spectrometry, 2006, 21 (1): 82-85.

[80] Liu Q Y. Direct determination of mercury in white vinegar by matrix assisted photochemical vapor generation atomic fluorescence spectrometry detection. Spectrochimica Acta Part B: Atomic Spectroscopy, 2010, 65 (7): 587-590.

[81] Xia H, Liu X, Huang K, Gao Y, Gan L, He C L, Hou X D. Matrix-assisted UV-photochemical vapor generation for AFS determination of trace mercury in natural water samples: A green analytical method. Spectroscopy Letters, 2010, 43 (7-8): 550-554.

[82] Liu L W, Zheng H L, Yang C, Xiao L, Zhangluo Y L, Ma J Y. Matrix-assisted photochemical vapor generation for the direct determination of mercury in domestic wastewater by atomic fluorescence spectrometry. Spectroscopy Letters, 2014, 47 (8): 604-610.

[83] Guo X M, Sturgeon R E, Mester Z, Gardner G. UV photosynthesis of nickel carbonyl. Applied Organometallic Chemistry, 2004, 18 (5): 205-211.

[84] Grinberg P, Mester Z, Sturgeon R E, Ferretti A. Generation of volatile cobalt species by UV photoreduction and their tentative identification. Journal of Analytical Atomic Spectrometry, 2008, 23 (4): 583-587.

[85] Grinberg P, Sturgeon R E, Gardner G. Identification of volatile iron species generated by UV photolysis. Microchemical Journal, 2012, 105: 44-47.

[86] Grinberg P, Mester Z, D'Ulivo A, Sturgeon R E. Gas chromatography-mass spectrometric identification of iodine species arising from photo-chemical vapor generation. Spectrochimica Acta Part B: Atomic Spectroscopy, 2009, 64 (7): 714-716.

[87] Zheng C B, Li Y, He Y H, Ma Q, Hou X D. Photo-induced chemical vapor generation with formic acid for ultrasensitive atomic fluorescence spectrometric determination of mercury: Potential application to mercury speciation in water. Journal of Analytical Atomic Spectrometry, 2005, 20 (8): 746-750.

[88] Yin Y G, Liu J F, He B, Gao E L, Jiang G B. Photo-induced chemical vapour generation with formic acid: Novel interface for high performance liquid chromatography-atomic fluorescence spectrometry hyphenated system and application in speciation of mercury. Journal of Analytical Atomic Spectrometry, 2007, 22 (7): 822-826.

[89] Bideau M, Claudel B, Faure L, Rachimoellah M. Homogeneous and heterogeneous photoreactions of decomposition and oxidation of carboxylic acids. Journal of Photochemistry, 1987, 39 (1): 107-128.

[90] Carraher J M, Pestovsky O, Bakac A. Transition metal ion-assisted photochemical generation of alkyl halides and hydrocarbons from carboxylic acids. Dalton Transactions, 2012, 41 (19): 5974-5980.

[91] Gao Y, Sturgeon R E, Mester Z, Hou X, Yang L. Multivariate optimization of photochemical vapor generation for direct determination of arsenic in seawater by inductively coupled plasma mass spectrometry. Analytica Chimica Acta, 2015, 901: 34-40.

[92] Gao Y, Sturgeon R E, Mester Z, Hou X D, Zheng C B, Yang L. Direct determination of trace antimony in natural waters by photochemical vapor generation ICPMS: Method optimization and comparison of quantitation strategies. Analytical Chemistry, 2015, 87 (15): 7996-8004.

[93] Gao Y, Li S Z, He H Y, Li T L, Yu T, Liu R, Ni S J, Shi Z M. Sensitive determination of osmium in natural waters by inductively coupled plasma mass spectrometry after photochemical vapor generation. Microchemical Journal, 2017, 130: 281-286.

[94] Lopes G S, Sturgeon R E, Grinberg P, Pagliano E. Evaluation of approaches to the abatement of nitrate interference with photochemical vapor generation. Journal of Analytical Atomic Spectrometry, 2017, 32 (12): 2378-2390.

[95] Chen G Y, Lai B H, Mei N, Liu J X, Mao X F. Mercury speciation by differential photochemical vapor generation at UV-B *vs.* UV-C wavelength. Spectrochimica Acta Part B: Atomic Spectroscopy, 2017, 137: 1-7.

[96] Yin Y M, Liang J, Yang L M, Wang Q Q. Vapour generation at a UV/TiO$_2$ photocatalysis reaction device for determination and speciation of mercury by AFS and HPLC-AFS. Journal of Analytical Atomic Spectrometry, 2007, 22 (3): 330-334.

[97] Yin Y G, Liu J F, He B, Shi J B, Jiang G B. Simple interface of high-performance liquid chromatography-atomic fluorescence spectrometry hyphenated system for speciation of mercury based on photo-induced chemical vapour generation with formic acid in mobile phase as reaction reagent. Journal of Chromatography A, 2008, 1181 (1): 77-82.

[98] Yin Y G, Liu J F, He B, Shi J B, Jiang G B. Mercury speciation by a high performance liquid chromatography-atomic fluorescence spectrometry hyphenated system with photo-induced chemical vapour generation reagent in the mobile phase. Microchimica Acta, 2009, 167 (3): 289.

[99] Chen K J, Hsu I h, Sun Y C. Determination of methylmercury and inorganic mercury by coupling short-column ion chromatographic separation, on-line photocatalyst-assisted vapor generation, and inductively coupled plasma mass spectrometry. Journal of Chromatography A, 2009, 1216 (51): 8933-8938.

[100] Tsai Y N, Lin C H, Hsu I H, Sun Y C. Sequential photocatalyst-assisted digestion and vapor generation device coupled with anion exchange chromatography and inductively coupled plasma mass spectrometry for speciation analysis of selenium species in biological samples. Analytica Chimica Acta, 2014, 806: 165-171.

[101] 李淑萍, 郭旭明, 黄本立, 胡荣宗, 王秋泉, 李彬. 电化学氢化物发生法的进展及其在原子光谱分析中的应用. 分析化学评述与进展, 2001, 29 (8): 967-970.

[102] Lin Y H, Wang X R, Yuan D X, Yang P Y, Huang B L, Zhuang Z X. Flow injection-electrochemical hydride generation technique for atomic absorption spectrometry. Journal of Analytical Atomic Spectrometry, 1992, 7 (2): 287-291.

[103] Jiang X J, Gan W, Wan L Z, Zhang H C, He Y Z. Determination of mercury by electrochemical cold vapor generation atomic fluorescence spectrometry using polyaniline modified graphite electrode as cathode. Spectrochimica Acta Part B: Atomic Spectroscopy, 2010, 65 (2): 171-175.

[104] Zhang W B, Yang X A, Dong Y P, Xue J J. Speciation of inorganic-and methyl-mercury in biological matrixes by electrochemical vapor generation from an L-cysteine modified graphite electrode with atomic fluorescence spectrometry detection. Analytical Chemistry, 2012, 84 (21): 9199-9207.

[105] Yang X A, Lu X P, Liu L, Chi M B, Hu H H, Zhang W B. Selective determination of four arsenic species in rice and water samples by modified graphite electrode-based electrolytic hydride generation coupled with atomic fluorescence spectrometry. Talanta, 2016, 159: 127-136.

[106] Arbab-Zavar M H, Chamsaz M, Yousefi A, Ashraf N. Electrochemical hydride generation of thallium. Talanta, 2009, 79 (2): 302-307.

[107] Xu S K, Sturgeon R E. Flow injection chemical vapor generation of Au using a mixed reductant. Spectrochimica Acta Part B: Atomic Spectroscopy, 2005, 60 (1): 101-107.

[108] Centineo G, Gonzalez E B, Sanz-Medel A. Multielemental speciation analysis of organometallic compounds of mercury, lead and tin in natural water samples by headspace-solid phase microextraction followed by gas chromatography-mass spectrometry. Journal of Chromatography A, 2004, 1034 (1-2): 191-197.

[109] Lin Y, Yang Y, Li Y X, Yang L, Hou X D, Feng X B, Zheng C B. Ultrasensitive speciation analysis of mercury in rice by headspace solid phase microextraction using porous carbons and gas chromatography-dielectric barrier discharge optical emission spectrometry. Environment Science & Technology, 2016, 50 (5): 2468-2476.

[110] Xiao Q, Hu B, He M. Speciation of butyltin compounds in environmental and biological samples using headspace single drop microextraction coupled with gas chromatography-inductively coupled plasma mass spectrometry. Journal of Chromatography A, 2008, 1211 (1): 135-141.

[111] Hu J, Pagliano E, Hou X D, Zheng C B, Yang L, Mester Z. Sub-ppt determination of butyltins, methylmercury and inorganic mercury in natural waters by dynamic headspace in-tube extraction and GC-ICPMS detection. Journal of Analytical Atomic Spectrometry, 2017, 32 (12): 2447-2454.

[112] Yang L, Colombini V, Maxwell P, Mester Z, Sturgeon R E. Application of isotope dilution to the determination of methylmercury in fish tissue by solid-phase microextraction gas chromatography-mass spectrometry. Journal of Chromatography A, 2003, 1011 (1–2): 135-142.

[113] D'Ulivo A, Pagliano E, Onor M, Pitzalis E, Zamboni R. Vapor generation of inorganic anionic species after aqueous phase alkylation with trialkyloxonium tetrafluoroborates. Analytical Chemistry, 2009, 81 (15): 6399-6406.

[114] Pagliano E, Meija J, Sturgeon R E, Mester Z, D'Ulivo A. Negative chemical ionization GC/MS determination of nitrite and nitrate in seawater using exact matching double spike isotope dilution and derivatization with triethyloxonium tetrafluoroborate. Analytical Chemistry, 2012, 84 (5): 2592-2596.

[115] Pagliano E, Meija J, Ding J F, Sturgeon R E, D'Ulivo A, Mester Z. Novel ethyl-derivatization approach for the determination of fluoride by headspace gas chromatography/mass spectrometry. Analytical Chemistry, 2013, 85 (2): 877-881.

[116] Guo X W, Guo X M. Interference-free atomic spectrometric method for the determination of trace amounts of germanium by utilizing the vaporization of germanium tetrachloride. Analytica Chimica Acta, 1996, 330 (2-3): 237-243.

[117] Mester Z, Sturgeon R E. Detection of volatile arsenic chloride species during hydride generation: A new prospectus. Journal of Analytical Atomic Spectrometry, 2001, 16 (5): 470-474.

[118] Zheng C B, Sturgeon R E, Brophy C S, He S P, Hou X D. High-yield UV-photochemical vapor generation of iron for sample introduction with inductively coupled plasma optical emission spectrometry. Analytical Chemistry, 2010, 82 (7): 2996-3001.

[119] Li M T, Deng Y J, Jiang X M, Hou X D. UV photochemical vapor generation-nitrogen microwave induced plasma optical emission spectrometric determination of nickel. Journal of Analytical Atomic Spectrometry, 2018, 33(6): 1086-1091.

[120] Yang L, Mester Z, Abranko L, Sturgeon R E. Determination of total chromium in seawater by isotope dilution sector field ICPMS using GC sample introduction. Analytical Chemistry, 2004, 76 (13): 3510-3516.

[121] Duan X C, Sun R, Fang J L. On-line continuous generation of zinc chelates in the vapor phase by reaction with sodium dithiocarbamates and determination by atomic fluorescence spectrometry. Spectrochimica Acta Part B: Atomic Spectroscopy, 2017, 128: 11-16.

[122] Duan X C, Sun R, Fang J L. Gold determination in geological samples by chelate vapor generation at room temperature coupled with atomic fluorescence spectrometry. Spectrochimica Acta Part B: Atomic Spectroscopy, 2017, 133: 21-25.

[123] Ma G P, Duan X C, Sun J S. Chelate vapor generation using ethanol as efficiency-enhancing reagent for nickel determination in samples by atomic fluorescence spectrometry. Spectrochimica Acta Part B: Atomic Spectroscopy, 2018, 149: 57-61.

[124] Sun R, Ma G P, Duan X C, Sun J S. Determination of cadmium in seawater by chelate vapor generation atomic fluorescence spectrometry. Spectrochimica Acta Part B: Atomic Spectroscopy, 2018, 141: 22-27.

[125] Gao Y, Li S Z, He H Y, Li T L, Yu T, Liu R, Ni S J, Shi Z M. Sensitive determination of osmium in natural waters by inductively coupled plasma mass spectrometry after photochemical vapor generation. Microchemical Journal, 2017, 130, 281-286.

[126] Wu X, Yang W L, Liu M G, Hou X D, Zheng C B. Vapor generation in dielectric barrier discharge for sensitive detection of mercury by inductively coupled plasma optical emission spectrometry. Journal of Analytical Atomic

Spectrometry, 2011, 26 (6): 1204-1209.

[127] Zhu Z L, Chan G C Y, Ray S J, Zhang X R, Hieftje G M. Use of a solution cathode glow discharge for cold vapor generation of mercury with determination by ICP-atomic emission spectrometry. Analytical Chemistry, 2008, 80 (18): 7043-7050.

[128] Liu X, Zhu Z L, Xing P J, Zheng H T, Hu S H. Plasma induced chemical vapor generation for atomic spectrometry: A review. Spectrochimica Acta Part B: Atomic Spectroscopy, 2020, 167: 105822.

第9章　微等离子体原子光谱仪进样技术

(于永亮[①]，王建华[①*])

- 9.1 概述 / 219
- 9.2 原子发射光谱中的微等离子体激发源 / 220
- 9.3 原子吸收/原子荧光光谱中的微等离子体原子化器与捕集器 / 232
- 9.4 质谱中的微等离子体离子源 / 234
- 9.5 微等离子体诱导化学反应 / 236
- 9.6 总结与展望 / 239

① 东北大学，沈阳。*通讯作者联系方式：jianhuajrz@mail.neu.edu.cn

第9章 微等离子体原子光谱仪进样技术

本章导读

- 微等离子体作为原子发射光谱的激发源,可直接检测气体、液体和固体样品。
- 微等离子体作为原子吸收光谱或原子荧光光谱中非热原子化器,可实现蒸气发生进样方式的原子化,以及对气相分析物的富集功能。
- 微等离子体作为常压解吸离子源,结合质谱检测,可实现深度剖面分析、元素分析以及成像分析。
- 微等离子体诱导化学反应,可实现多种元素的蒸气发生进样,以及形态转化与物质分解。

9.1 概 述

原子光谱分析具有优异的灵敏度、选择性和准确度,在涉及金属/类金属元素相关的分析中发挥着不可替代的作用[1],广泛应用于元素组成分析、元素形态分析、金属组学、金属蛋白质组/金属酶组学研究[2-4]。原子光谱分析法,包括原子发射光谱(atomic emission spectrometry,AES)、原子吸收光谱(atomic absorption spectrometry,AAS)、原子荧光光谱(atomic fluorescence spectrometry,AFS)和电感耦合等离子体质谱(inductively coupled plasma-mass spectrometry,ICP-MS),是环境科学、生命科学、食品安全等领域中金属元素检测的最重要技术。然而,原子光谱分析技术趋于成熟,其进一步发展面临瓶颈,在依据新原理新方法的仪器创制方面缺少新突破,在适应环境科学研究中越来越多的现场原位分析方面受到了局限,其分析能力亟待进一步开发与提高。

常规的原子光谱仪器主要采用传统的热原子化与热激发方式(火焰、石墨炉、ICP)[5,6],尽管其可充分满足待测组分原子化与激发的需要,但数千度的高温不仅需要大功率供电设备,且相应仪器难以小型化。因此,发展新型的激发源与原子化模式[7],是破解原子光谱仪器小型化难题的关键。目前,各种微等离子体在原子光谱分析系统中的应用,加快了便携式仪器的发展,在痕量元素分析领域显示出巨大的发展潜力[8-10]。

微等离子体是被限制在一个有限空间(至少有一个维度在毫米或亚毫米范围内)的等离子体[11],其尺寸使它们接近大多数气体Paschen曲线的最小电位,形成的微等离子体非常稳定[12]。目前,基于微等离子体的各种放电结构已被应用于原子光谱分析领域,如辉光放电(glow discharge,GD)、介质阻挡放电(dielectric barrier discharge,DBD)和尖端放电(point discharge,PD)等[10,13]。在适当的电压下,无论有无介电层(如玻璃、石英、陶瓷或聚合物层)存在于两个电极之间,均很容易放电产生微等离子体。微等离子体中所存在的大量高能电子会与周围气体分子发生碰撞,产生各种自由基和离子,将极大地促进待测物质的解离、激发和电离[10,13,14]。这些特性使微等离子体广泛应用于原子光谱分析的各个领域,表现为四种典型的应用:①由于微等离子体具有激发分子或原子的能力,可作为激发源应用于发射光谱(optical emission spectrometry,OES)仪器中,这为研制可直接检测气体、液体和固体样品的便携式仪器提供了可能。②利用微等离子体对分子物质的强大解离能力,可作为AAS和AFS系统中蒸气发生进样的非热原子化器,以及实现对气相分析物的富集功能。③微等离子体具有良好的电离能力,作为常压解吸离子源,结合质谱检测,可实现深度剖面分析、元素分析以及成像分析。④由于微等离子体中存在大量的活性物质如自由基和电子,可促进各种化学反应的发生,实现多种元素的蒸气发生进样,以及形态转化与物质分解。

本章重点介绍基于上述四种典型应用下的微等离子体原子光谱仪进样技术。

9.2 原子发射光谱中的微等离子体激发源

9.2.1 挥发性物质随载气引入微等离子体 OES

微等离子体作为 OES 的激发源可以很容易地进行气体样品分析，该方法已应用于气体介质中低含量污染物或分析物的测定。例如，通过氮在 337 nm 处的特征发射线，DBD-OES 可以检测纯氩气中的微量氮杂质含量[15]；丙酮作为一种挥发性的有机化合物，通过氦气将其引入 DBD-OES 中，可对其在 519 nm 处的特征发射线进行检测[16]；Franzke 等设计了一种基于微空心阴极放电产生微等离子体的方式，用于卤代烃的 OES 检测，通过 912 nm 处的特征发射线，可得到氦气中氯代烃的检出限为 $27×10^{-9}(V/V)$[17]。

在某些情况下，由于环境空气进入微等离子体中会产生复杂的光谱干扰，致使很难选择出待测组分清晰的特征发射线[18]。为此，选择密闭的 DBD 微等离子体可用于隔绝环境空气，克服环境空气所带来的光谱干扰，已成功用于氩气中氨和硫化氢的检测[19,20]。通过测定 326.2 nm 处和 365.06 nm 处的特征发射线，可得到氨和硫化氢的检出限分别为 $0.37×10^{-6}(V/V)$ 和 1.4 mg/m³。段忆翔等在一个小型陶瓷基板芯片（20×10×1 mm³）上构建了一种封闭的氮气 GD 微等离子体发生装置，如图 9-1 所示[21]。该装置不仅克服了环境空气对微等离子体稳定性的影响，而且不需要使用惰性气体。虽然有机样品在氮气微等离子体中的检出限在数百皮克水平，但该装置具有良好的重现性、便携性和低功耗等特点，可用于现场实时检测，具有潜在的应用前景。

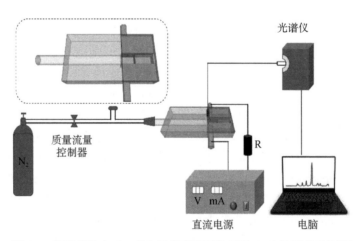

图 9-1 基于芯片内 GD 产生氮微等离子体用于 OES 检测有机物[21]

侯贤灯等提出了一种小型化 PD-OES 系统作为区分各种挥发性有机硫化合物（volatile organic sulfur compound，VOSC）的多通道光学检测器[22]。利用加热的方式，将 VOSC 从样品中挥发出来，通过载气引入 PD-OES 检测。根据 257.6 nm 处的 CS 自由基分子发射线，193.1 nm 和 247.8 nm 处的 C 原子发射线，231.5 nm 和 384.8 nm 处的 C_2 和 CN 自由基分子发射线，作为五个光学通道，对来自 10 个典型 VOSC 的 95 个未知样品进行了分类，准确度可达 98.9%。同样是利用加热的方式，对于从食品酸化液中释放的 SO_2[23]，可利用 DBD 对其进行激发，通过检测 301.9 nm 处的特征发射线，实现对食品中 SO_2 的测定，检出限为 0.01 mg/L。该方法简单、准确、成本低，适用于食品样品中总 SO_2 的测定。

9.2.2 微等离子体 OES 作为气相色谱检测器

微等离子体 OES 除了直接检测气体样品外,还可将其作为气相色谱(gas chromatography,GC)的检测器,发挥色谱分离与光谱分辨的互补优势。基于 DBD-OES 的 GC 检测器[24]可将卤代烃通过微等离子体激发产生分子发射光谱,电荷耦合器件(charge coupled device,CCD)在 258 nm、292 nm 和 342 nm 处分别测定含氯、溴、碘的碳氢化合物,使得在色谱上不能完全分离的组分,通过不同的光谱检测波长实现进一步分离,如图 9-2 所示。CCl_4、$CHCl_3$、CH_2Cl_2、CH_3I、CH_3CH_2I 和 $(CH_3)_2CHI$ 的检出限分别为 0.07 mg/mL、0.06 mg/mL、0.3 mg/mL、0.04 mg/mL、0.05 mg/mL 和 0.02 mg/mL。在此基础上,进一步发展了基于 DBD-OES 的通用、灵敏 GC 检测器,可应用于所有挥发性的含碳化合物[25]。通过将 DBD 装置置于加热套内,使微等离子体工作温度提高至 300℃,以增强 193.0 nm 处的碳原子发射。通过 GC 分离甲醛、乙酸乙酯、甲醇、乙醇、1-丙醇、1-丁醇和 1-戊醇后,用 DBD-OES 对 193.0 nm 处碳特征发射线进行检测,获得的绝对检出限为 0.12~0.28 ng。此外,氢火焰离子化检测器难以检测到的一些含碳化合物,如 HCHO、CO 和 CO_2,也可通过 DBD-OES 实现检测。

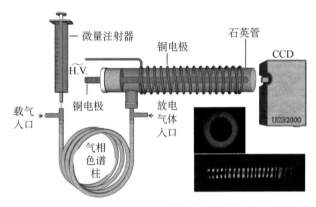

图 9-2 DBD-OES 作为卤代烃的多通道 GC 检测器[24]

通过将顶空固相微萃取(headspace solid phase microextraction,HSPME)与 PD-OES 相结合[26],如图 9-3 所示,Hg 和 Pb 经四乙基硼酸钠(NaBEt$_4$)衍生、固相微萃取后,将其加热释放并引入 PD 微等离子体中激发,得到 Hg 和 Pb 的检出限分别为 0.001 μg/L 和 0.003 μg/L,与光化学蒸气发生相比,Hg 和 Pb 的检出限至少降低了 100 倍。HSPME 将采样、分离与预富集合为一步,不仅简化了实验步骤,而且消除了基质干扰,将其与色谱技术联用,可进一步拓宽其应用,实现物质的形态分析。郑成斌等使用 NaBPh$_4$ 衍生化试剂,将不同形态的 Hg 转化为相应的挥发性物质,利用多孔碳进行 HSPME,随后将浓缩后的 Hg 形态通过 GC-DBD-OES 测定[27],Hg^{2+}、CH_3Hg^+ 和 $CH_3CH_2Hg^+$ 的检出限分别为 0.16 ng、0.24 ng 和 0.34 ng。当使用 PD 代替 DBD 时,可提供更高的激发能,通过微等离子体 OES 测定碳的灵敏度至少能提高 10 倍。因此,后续进一步采用 GC-PD-OES 与 HSPME 联用分析了人发中的 Hg 形态。将头发中提取的 Hg 用 NaBEt$_4$ 转化为相应的挥发性物质,随后进行 HSPME 和 GC-PD-OES 测定,Hg^{2+} 和 MeHg 的检出限分别低至 0.035 ng 和 0.10 ng,相对标准偏差小于 3.5%[28]。在环境分析化学中,GC-PD-OES 除了应用于某些金属和含碳化合物的分析外,还可通过有机硅氧烷中的硅原子发射(251.6 nm 和 288.2 nm),实现各种环状挥发性甲基硅氧烷的分析[29]。在最佳条件下,GC-PD-OES 对六甲基环三硅氧烷、八甲基环四硅氧烷、十甲基环五硅氧烷和十二甲基环六硅氧烷的检出限分别为 0.2 mg/L、0.04 mg/L、0.03 mg/L 和 0.02 mg/L(以

Si 计)。

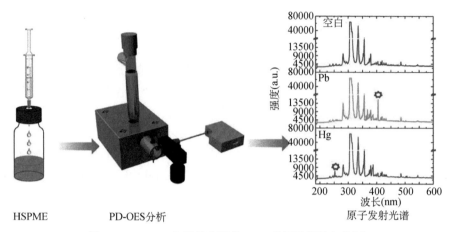

图 9-3 HSPME 与微等离子体 OES 联用检测汞和铅[26]

9.2.3 流动溶液进样-微等离子体 OES

溶液进样是最常用的 OES 进样方式。起先，水溶液中的待测物是通过电解液阴极放电（electrolyte cathode discharge，ELCAD）来实现微等离子体 OES 检测[30]，样品溶液不仅作为分析物，而且参与微等离子体放电。为了改善 ELCAD 装置中微等离子体的稳定性，获得更好的分析结果，各种形式的液体电极放电不断涌现，包括常压辉光放电（atmospheric pressure glow discharge，APGD）[10]、液体阴极辉光放电（solution cathode glow discharge，SCGD）[31]、液体进样大气压辉光放电（liquid sampling-atmospheric pressure glow discharge，LS-APGD）[32]、液体电极介质阻挡放电（liquid electrode-dielectric barrier discharge，LE-DBD）[33]等，从液体样品中释放分析物进入微等离子体的机理通常为阴极溅射、热蒸发、类电喷雾过程和挥发性物质的化学生成[33-36]。

毛细管可用于微量溶液样品的引入，通过注射泵推动溶液从毛细管的一端以一定流速流入管内，在毛细管的另一端形成微等离子体检测 OES，如在针状 Pt 阳极和电解液流动的石英毛细管阴极之间产生 GD 微等离子体，可用于水样中 K、Na、Ca、Mg 和 Zn 的同时 OES 测定[37]。Franzke 等基于毛细管中流动液体表面和钨电极之间形成的微等离子体，通过 OES 评价了碱金属、碱土金属、过渡金属和稀有金属等 23 种元素的分析性能[33]，在 20 μL/min 的低进样流速下，所得 Li 和 Bi 的检出限分别为 0.016 mg/L 和 41 mg/L。汪正等进一步开发了基于 SCGD-OES 的便携式元素分析仪器，通过重新设计，将样品引入、微等离子体激发源和 OES 检测整合到一个小型且高度集成的仪器中，并在笔记本电脑端对其进行控制，利用仪器自身电池供电，实现自动检测和分析。该系统可直接分析液体样品，而无需进行复杂的预处理或元素预浓集，同时分析 Cd、Hg 和 Pb 的检出限在 33～253 mg/L（流动注射模式）和 7～92 mg/L 之间（连续流模式），在小型化微等离子体 OES 仪器创制方面取得了新突破[38]。

自从 2001 年 Marcus 等改进了 ELCAD 系统，发现流动电解质溶液作为阳极也可以产生稳定的微等离子体，至此液体阳极辉光放电（solution anode glow discharge，SAGD）逐渐得到发展[39]。对于 SAGD-AES 装置，使用钛棒作为阴极，石英管中流动的溶液作为阳极，无需冷却系统和稀有气体环境。该装置可高灵敏测定溶液中的 Ag、Bi、Cd、Hg、Pb、Tl 和 Zn[40]，检出限分别为 0.03 μg/L、1.98 μg/L、0.07 μg/L、1.7 μg/L、1.35 μg/L、0.02 μg/L 和 0.12 μg/L。当采用惰性气体射流辅助溶液阳极放电时，可改善微等离子体组成，提供更好的激发条件。为此，Greda 等利用气体射流辅助下的流动液体阳极大气压辉光放电（flowing liquid anode-atmospheric pressure glow discharge，FLA-APGD）作为发射光谱的激发源，对 50 种元素进行了测试，尽管仅识别出 Ag、Cd、Hg、In、Pb、Tl 和 Zn 的发射线，检出限分别为 0.001 μg/L、0.006 μg/L、0.16 μg/L、0.093 μg/L、0.076 μg/L、0.007 μg/L 和 0.018 μg/L，但由于溶液表面被电子轰击，可将元素转化为挥发性物质，在微等离子体激发下产生发射线，与没有气体射流辅助的 FLA-APGD 结构相比，灵敏度提高了 10 倍，而且具有更高的放电稳定性和测量精度（<2%）[41]。在 FLA-APGD 中，金属 OES 增强的原因是由于从溶液中释放了更多的分析物进入微等离子体中，即高的传输效率，并且微等离子体中水蒸气浓度较小，有利原子激发，如图 9-4 所示[42]。将 LS-APGD 微等离子体与单色成像光谱仪联用[43]，可获得特定分析物和背景物质在特定波长下的空间发射轮廓。通过研究分析物在微等离子体中的发射位置以及等离子体气流对这些发射轮廓的影响，发现大部分分析物的发射发生在溶液电极的尖端，即溶质被引入微等离子体的位置处，而各种背景物质的发射则发生在整个微等离子体中，这对了解微等离子体的空间发射特性以及性能优化具有帮助。

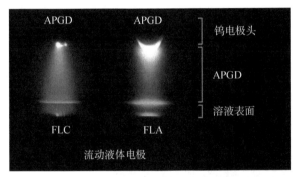

图 9-4　FLC-APGD 与 FLA-APGD 的图像对比[42]

将微等离子体激发源与外界环境隔离，可减少环境气体对分析结果的影响，也是提高 OES 检测灵敏度的一个思路。例如，使用硼硅酸盐玻璃圆筒保护 LS-APGD 微等离子体的放电区域，使其免受环境气氛的干扰，可显著降低氮背景光谱，从而改善了 Ag、Pb、Cd 的检出限。这种采用玻璃护套降低背景干扰的方式，与光谱采集后的背景扣除方式具有可比性[44]。通过对 LS-APGD-OES 的参数进行详细的研究[45,46]，如光学采样位置、酸类型、样品基质、溶液流速、放电电流、样品位移、溶液电极间分离角度和气体流速对铜作为模型基质和分析物的发射响应[45]，发现溶液电极与目标表面的入射角对微等离子体的发射有较大影响，说明鞘层气体的动量对物质的解吸或传输起到一定作用，且溶液流量、电极距离与发射之间存在正相关。Webb 等对 SCGD 进行了成像实验[47]，用激光散射观察了从溶液表面以一定角度范围喷射的液滴轨迹，用干涉滤光片拍摄了各种分析物发射线和背景发射带，并研究了低分子量有机化合物（如 HCOOH 和 CH_3CH_2OH）和非离子表面活性剂（如 Triton X-45 和 Triton X-405）对发射的空间分布影响。其中，低分子量有机化合物会影响某些物质的发射，但不会改变微等离子体结构；较轻的表面活性剂会影响某些背景物质的发射，但不会显著影响分析物的发射或微等离子体结构；较重的表面活性剂（Triton X-405）会影响某些物质的发射以及微等离子体结构。通过研究添加低分子量有机化合物（酸、盐和醇）对 FLA-APGD-OES 测定 In 发射的影响[48]，发现向待测溶液中添加少量 CH_3OH 或 C_2H_5OH(0.5%，V/V)，可使 In 451.1 nm 原子线的强度提高 30 倍。这是由于分析物从液体样品到微等离子体的传输效率提高（从 19% 增至 35%），以及原子化/激发条件的改善。另外，添加醇类也会大大降低基体效应。在 SCGD 系统中[49]，也发现了类似的现象，与不添加化学改性剂相比，含 0.15% 十六烷基三甲基溴化铵和 18% 甲酸的 Cd 溶液发射强度分别提高了 2.4 倍和 2.7 倍。

9.2.4 溶液雾化进样-微等离子体 OES

为了进一步提高微等离子体 OES 的检测灵敏度，溶液样品可采用气动雾化的进样方式，在微型气动雾化器的喷嘴上产生 DBD 微等离子体，将 DBD 能量聚焦在有限空间上，使喷雾中的金属组分发生原子化/激发，如图 9-5 所示[50]。在 3 μL/s 进样流速下，对 80 μL 样品溶液进行雾化，DBD-OES 测定镉的检出限可低至 1.5 μg/L。Greda 等利用小尺寸氢气喷嘴阳极和流动液体阴极之间产生的 APGD，比较了超声雾化进样下与常规流动溶液进样下的微等离子体 OES 分析性能[51]，发现在超声雾化的进样条件下，Ca、Cd、In、K、Li、Mg、Mn、Na 和 Sr 的分析线强度可最大提升 35 倍，检出限在 0.08 μg/L（Li）和 52 μg/L（Mn）之间。

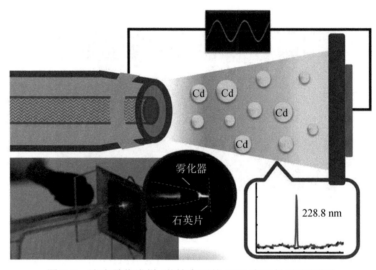

图 9-5　溶液雾化进样-微等离子体 OES 高灵敏测定镉[50]

9.2.5 液滴进样-微等离子体 OES

液滴电极放电可以简化放电结构，实现无泵无阀操作[7,52,53]。比如液膜 DBD 激发源由一个铜电极、一个钨丝电极以及位于它们之间的一块载玻片组成[52]，该载玻片作为电介质阻挡层和样品板。当含有 1 mol/L 硝酸的样品溶液沉积在载玻片表面上时，会形成一层薄液膜。当在电极上施加交流高压（峰值电压 3.7 kV，频率 30 kHz）时，钨丝电极的尖端和液膜表面之间将产生微等离子体，通过 DBD-OES 实现痕量元素分析。该系统无需使用注射器或蠕动泵，所需样品量≤80 μL，且通过样品阵列可实现高通量分析。又如基于基质辅助微等离子体 OES 系统直接进行溶液元素分析[53]，其用一张滤纸承载样品，当微等离子体射流与滤纸基质表面直接作用时，从微等离子体源吸收能量，并将热量传递到滤纸表面的分析物，从而促进原子化和激发，可观察到多种元素的基质辅助微等离子体 OES。只需要 1 μL 样品，即可测得 Ba、Cu、Eu、In、Mn、Ni、Rh 和 Y，检出限低至皮克水平。再如基于简单的液滴阳极 APGD 系统[6]，通过固体针状钨阴极和置于石墨盘阳极上的液滴之间产生微等离子体，使用 50 μL 的液滴上样，镉的检出限为 0.20~0.40 μg/L，如图 9-6 所示。

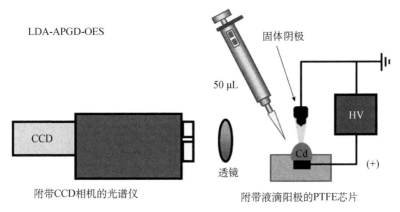

图 9-6　液滴阳极 APGD-OES 灵敏测定微量样品中的镉[6]

后续进一步发展了基于悬滴电极产生的 APGD[54]，当液滴引入微等离子体区域时，可直接观察到放电的自燃，实现了极低的样品溶液消耗（0.4 mL/min）和低放电功率（低于 50 W）。在此基础上，开发了一种毛细管液体电极微等离子体 OES 系统[7]，用于液滴阵列采样分析，如图 9-7 所示。由于毛细管内壁对液体固有的附着力和微等离子体诱导溶液汽化所产生的吸力，样品溶液可在无泵的情况下自动引入到激发源中，样品通量可提高至 90 个样本/h。通过采样时间控制样品量至纳升水平，Cd 和 Hg 的检出限分别为 30 μg/L 和 75 μg/L。利用原位解吸 LS-APGD 对三种不同基板［载玻片、聚四氟乙烯（PTFE）薄板和涂有聚二甲基硅氧烷（PDMS）的载玻片］上的液滴残留进行分析[55]，发现残留物在涂有 PDMS 的玻璃上空间分布均匀，与其他两个基板相比，这种均一性使 OES 检出限改善了约 20 倍（22 μL）和 4 倍（2 μL），该方法可应用于法医、考古等领域。

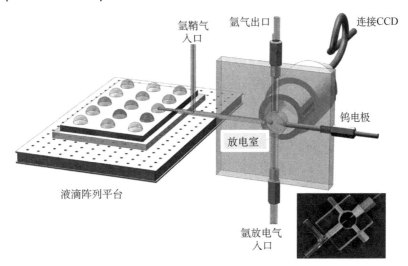

图 9-7　液滴阵列平台与毛细管液体电极放电 OES 联用[7]

9.2.6　适于溶液进样微等离子体 OES 联用的微分离系统

微分离系统的引入将有助于进一步提升微等离子体 OES 的分析性能以及扩展其应用范围。针对复杂生命基体样品的痕量元素分析，于永亮等建立了一种在线微电渗析-液体电极放电-发射光谱（μED-LED-OES）系统[4]，血清样品中的钾离子通过 μED 在线提取，随后被引入到微等离子体区域，实现 OES 定量分析。该系统每次分析所需的血清样品量仅为 20 μL，且可较好地消除血清基质干扰，蛋白质去除效率高达 99%。整个分析过程，包括进样、样品预处理和检测，可在 90 s 内完成。他们还将气动雾化进样 DBD-OES 作为阀上实验室（lab on valve，LOV）样品预处理平台的痕量元素检测方法[56]，通过微珠注射样品预处理、柱后衍生、溶液雾化和 DBD-OES 检测多功能集成于一体，实现了复杂环境样品的基体分离、目标分析物的富集与高灵敏检测，重金属镉的检出限可低至 0.06 μg/L。将强阳离子交换柱用于 LS-APGD-OES 系统在线样品预富集[57]，可提升 LS-APGD 激发源的分析性能，降低硝酸盐/硝酸溶液中银的检出限。考虑到液体电极 GD 不仅可以诱导微等离子体，而且可以使毛细管形成电泳（capillary electrophoresis，CE），因此将直流 GD-OES 与 CE 相结合[58]，构建了新型毛细管微等离子体 OES 分析系统应用于元素形态分析，分析物形态分离与分析信号检测集成于一体，汞、镉、铬、钠和有机或无机组分的检出限在 0.5~500 pg 范围内，如图 9-8 所示。

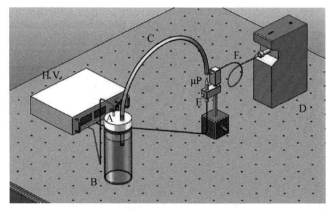

图 9-8　CE 与直流 GD-OES 联用[58]

9.2.7　蒸气发生进样-微等离子体 OES

样品溶液的基体组成对微等离子体 OES 的分析性能有很大影响，大量的水分进入微等离子体中也会降低其激发能力，甚至使其熄灭[59]。因此，较理想的样品引入方式是将液体样品中的待测分析物转化为"纯净"或"干燥"的挥发性物质[60]，如化学蒸气发生（chemical vapor generation）。其中，冷蒸气发生（cold vapor generation，CVG）和氢化物发生（hydride generation，HG）是最为常见的化学蒸气发生方式，应用范围较广，可作为 Hg、Cd、As、Sb、Bi、Se、Te、Ge、Sn、Pb 等元素的样品引入方式。2008 年，王建华[61]和 Hieftje[62]等首次利用 DBD 作为激发源，CVG 作为进样方式，建立了微型化 OES 分析系统，实现了痕量汞的准确测定。传统的 HG 进样方式会伴随有大量氢气的产生，严重干扰微等离子体 OES。为了解决这一问题，发展了一种同轴圆柱型 DBD 微等离子体激发源，该激发源对 H_2 具有良好的耐受性，产生的氢化物引入微等离子体激发源中，可实现对痕量砷的测定，检出限为 4.8 μg/L[63]。侯贤灯等构建了一种小型化的 PD 微等离子体激发源[60]，由于 HG 过程产生的氢气通常消耗微等离子体的激发能量，但在这里可使微等离子体的背景发射最小化，从而使更多的分析物发射谱线具有更少的背景光谱干扰，该 HG-PD-OES 系统测得的 As、Bi、Sb 和 Sn 的检出限分别为 7 μg/L、1 μg/L、5 μg/L 和 2 μg/L。将 HG 作为 SCGD 的进样方式，如图 9-9 所示[64]，使用空心钛管作为阳极和气体进样口，待测元素通过 HG 转化为气态氢化物，并被引入 SCGD 的阳极区域，通过 OES 直接检测，Sn、Ge、Se 的检出限分别为 0.8 μg/L、0.5 μg/L 和 0.2 μg/L，灵敏度得到显著提高。朱振利等开发了一种新型低功率和低氩气消耗量（<8 W，100 mL/min）的微型原子发射光谱仪[65]，采用 HG 进样方式，封闭的直流 APGD 微等离子体作为激发源，用于测定水样中痕量砷，检出限为 0.25 μg/L。将 HG 与 SCGD-AES 相结合[66]建立了一种测定海水中痕量铅的新方法。在 SCGD 中，空

心钛管引入气体作为放电阳极；HG 技术不仅可以将 Pb 从海水基质中分离出来，并减少基质对检测的干扰，还可显著提高 SCGD-AES 的原子化效率；添加 1% $H_2C_2O_4$ 作为掩蔽剂和 3%甲酸以增加 Pb 挥发。该方法测定铅的检出限为 0.17 μg/L，与单独的 SCGD-AES 相比，降低了两个数量级。将化学蒸气发生作为电容耦合氩微波微等离子体(capacitively coupled argon microwave plasma，μCMP)-OES 系统的进样方式[67]，同时测定氢化物形成元素和挥发性汞，获得的 Hg、As、Sb 和 Se 检出限分别为 3.0 ng/mL、1.4 ng/mL、1.5 ng/mL 和 3.8 ng/mL。由于蒸气发生进样方式和相应接口的限制，导致分析物在传输过程中会产生一定的损失。为了克服这一问题，对 PD 的样品引入接口进行改进[68]，采用中空电极作为放电电极，通过 HG 将含有分析物的化学蒸气传输至中空电极，以高的样品引入效率进入微等离子体中激发。由于气态分析物直接从电极内部扩散到微等离子体的中心，而非传统的外部扩散进入微等离子体，从而充分参与等微离子体中的相互作用和激发，实现高激发效率和高稳定性。此外，通过 3D 打印技术制造一些部件，可使整个光谱系统更加紧凑，已成功用于检测 As、Bi、Ge、Hg、Pb、Sb、Se 和 Sn，检出限分别为 2.5 μg/L、0.44 μg/L、1.6 μg/L、0.10 μg/L、2.8 μg/L、1.5 μg/L、31 μg/L 和 0.24 μg/L，相对标准偏差均小于 4%。为了进一步提高检测灵敏度，通过限制两个空心管 APGD 电极之间的空间体积[69]，将微等离子体限制在有限的空间内，成功通过物理方法提高了其激发能力。在所设计的 HG-APGD 中，50 ng/mL Se 和 As 的灵敏度提高了 3 倍以上，所获得的 Se 和 As 检出限分别为 0.13 ng/mL 和 0.087 ng/mL。将钨阴极和空心钛管阳极之间产生的低功率（～10 W）APGD 激发源与 HG 进样方式相结合[70]，使用小型 CCD 光谱仪可实现水中锑的灵敏测定，检出限和定量限分别为 0.14 mg/L 和 0.5 mg/L。后续又提出了一种新颖的间断气流技术[71]，将 APGD 激发源与 HG 进样方式相结合，高灵敏地测定了水样中痕量砷和锑。与 APGD 中常规连续气流相比，间断气流可显著增强 As 和 Sb 发射信号达 1～2 个数量级，所获得的检出限分别为 0.02 μg/L 和 0.003 μg/L。这种增强是由于气流中断后增加了微等离子体中分析物含量而引起的，且增强因子随气体中断的时间间隔增加而增加。

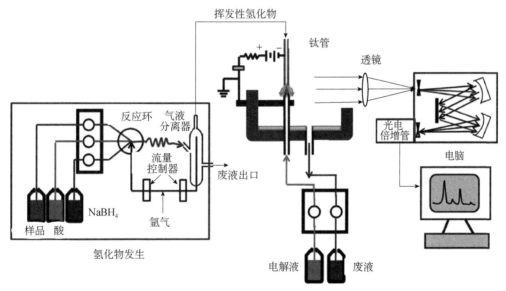

图 9-9　HG-SCGD-OES 测定痕量锡、锗和硒[64]

由于同一元素不同价态的 HG 效率不同，HG 还可作为简单的价态分析方法。将 HG 与 SCGD-OES 相结合[72]，可实现非色谱的 Sb 形态分析。在 HCl 介质中用柠檬酸掩蔽 Sb(Ⅴ)，选择性地将 Sb(Ⅲ)转化为挥发性气态物质，用 0.3% L-半胱氨酸在 0.01 mol/L HCl 中将 Sb(Ⅴ)还原为 Sb(Ⅲ)，进而测定总锑，所获得的Sb(Ⅲ)和总锑的检出限分别为0.39 ng/mL和0.36 ng/mL。通过总锑与Sb(Ⅲ)之差计算出Sb(Ⅴ)的浓度。将连续流化学蒸气发生-GD-AES作为液相色谱（liquid chromatograph，LC）检测器[73]，有机锡（OTs）经LC柱分离后，通过化学蒸气发生转化为挥发性气体，由GD微等离子体激发产生锡在317.66 nm处的原子发射线，并由CCD光谱仪记录，测定三甲基氯化锡和二甲基二氯化锡的检出限分别为0.59 μg/L和0.93 μg/L。该方法可简单、高效的同时测定食品中不同OTs含量。对于一些特定的元素，氧化蒸气发生也可作为微等离子体OES的进样方式，如氯化物、溴化物和碘化物与氧化剂在线反应后，将形成挥发性的单质蒸气，通过氩气引入DBD-OES中检测[74-76]。环境水中的总有机碳可通过微波辅助过硫酸盐氧化，将有机化合物转化为挥发性二氧化碳，引入DBD-OES中，通过193.0 nm的碳原子发射进行检测，检出限为0.01 mg/L，如图9-10所示[77]。

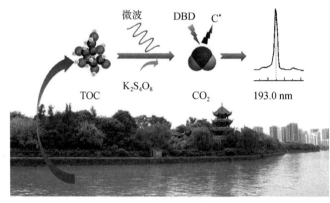

图9-10 在线微波辅助氧化-DBD-OES测定总有机碳[77]

微等离子体激发源不但可用于激发蒸气组分产生原子发射光谱，同样也能用于激发蒸气组分获得分子发射光谱。通过抗坏血酸和盐酸在线还原亚硝酸盐为挥发性的一氧化氮蒸气[78]，利用APGD激发源对NO激发，在237.0 nm处产生特征分子发射线，据此测定亚硝酸盐的检出限为0.26 μg/mL。同理，为进一步实现水质中硝酸盐和铵盐的APGD-OES测定[79]，通过$TiCl_3$将样品中的硝酸盐在线还原为NO，铵盐则通过与NaBrO反应在线氧化为N_2，随后将产生的NO和N_2通过氩气引入到APGD激发源中，对237.0 nm处NO和337.1 nm处N_2的分子发射进行检测，所获得的硝酸盐和铵盐的检出限分别为0.24 mg/L和0.20 mg/L。将DBD-OES作为LC的检测器[80]，发展了一种测定二硫代氨基甲酸酯（DTC）杀菌剂的方法。DTC在微波辅助水解反应器中被高效转化为CS_2，然后通过DBD-OES检测257.49 nm处的特征分子发射，检出限在0.1~1.0 μg/mL。液相色谱的DBD-OES检测器可检测DTCs的每种化学物质，而传统方法则无法区分DTCs的亚类。

光化学蒸气发生（photochemical vapor generation，PVG）作为新型蒸气发生进样方式，在保留原有HG优点的基础上，还具有简单、受过渡金属离子干扰小等优势，同时避免了微等离子体猝灭和大量还原性试剂的使用，成为一种很有发展前景的蒸气发生进样技术。传统氢化物形成元素[81-86]、过渡金属（Ni、Co、Cu、Fe、Mo、W）[87-89]、贵金属（Ag、Au、Rh、Pd、Pt、Os、Ir、Tl）[90,91]、非金属（Cl、I、S）[92]等都可在低分子量羧酸（甲酸、乙酸、丙酸）的存在下，经紫外光（UV）照射生成挥发性物质[93]，并通过微等离子体OES分析测定。例如，镍与HCOOH在石英UV反应器中生成挥发性的羰基镍蒸气[94]，经氩气引入DBD微等离子体中激发，可获得232.0 nm处镍的特征发射线，检出限为1.3 μg/L，如图9-11所示。采用PVG-PD-OES测定汞、铁、镍和钴[95]，检出限分别为

0.10 μg/L、10 μg/L、0.20 μg/L 和 4.5 μg/L。将流动注射 PVG 与 SCGD-OES 相结合[96]，当采用 8 W/254 nm 紫外灯照射 60 s 时，Hg(Ⅱ)和甲基汞(MeHg)都可转化为 Hg(0)，从而实现总汞(T-Hg)的 OES 测定；当采用 4 W/365 nm 紫外灯照射 20 s 时，只有 Hg(Ⅱ)可还原为 Hg(0)，通过调整紫外光的波长、功率以及照射时间，即可实现汞形态的 OES 测定。通过从 T-Hg 中减去 Hg(Ⅱ)计算 MeHg 的浓度，Hg(Ⅱ)和 MeHg 的检出限均为 0.2 μg/L。郑成斌等开发了一种小型化在线不可吹除有机碳（nonpurgeable organic carbon，NPOC）分析系统[97]，利用高效 UV 氧化蒸气发生，可将有机物转化为 CO_2，利用 PD-OES 检测 193.0 nm 处的碳发射线，得到了 0.05 mg/L（以 C 计）的检出限，已成功应用于各种水样中 NPOC 的灵敏准确测定。该系统克服了常规化学需氧量或总有机碳分析仪的许多缺点，如分析时间长，使用昂贵和有毒化学品，产生二次有毒废物，需要大型、耗电且昂贵的仪器，以及难以实现连续在线监测。

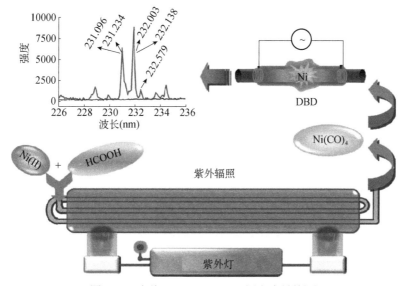

图 9-11　在线 PVG-DBD-OES 测定痕量镍[94]

基于蒸气发生进样方式的微等离子体 OES 除了对特定分析物的定量分析外，还可用于某些特定纳米材料的物理化学性质研究。例如，碳纳米材料的物理化学性质与其表面羧基的含量密切相关，然而常规羧基分析测定的灵敏度低，且没有与基质匹配的标准物质，但是利用微等离子体 OES 成功实现了对碳纳米材料表面羧基的准确定量[98]。邻苯二甲酸氢钾作为定量多壁碳纳米管（MWCNTs）/石墨烯（G）/氧化石墨烯（GO）上羧基的标准样品，使用浓盐酸将 MWCNTs、G 或 GO 上的羧基转化为羧酸，然后与碳酸氢钠反应生成 CO_2，并将其引入 PD-OES 中，检测碳原子发射线。由于 PD-OES 对 CO_2 的检测灵敏度高，基于 10 mg MWCNTs/G/GO 样品量，羧基的检出限为 0.1 μmol/g。GO 的光稳定性极大地影响了其光学性能，但由于仪器和方法的限制，其光稳定性，尤其是痕量水平的光稳定性尚未清楚。将 PVG-PD-OES 作为 GO 降解的跟踪方式，如图 9-12 所示，在 UV 照射和二元光催化剂（H_2O_2、Fe^{3+} 和 TiO_2 NPs）作用下，用 PD-OES 对 GO 光降解产生的挥发性气态含碳化合物进行检测，得到 GO（以 C 计）的检出限为 87.5 μg/L，线性范围为 0.5~10 mg/L[99]。该方法可应用于 GO 产品的实时质量控制，以及预测天然水生环境中 GO 的转化。

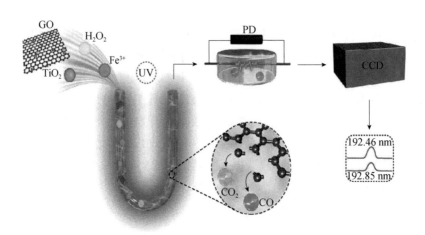

图 9-12 在线 PVG-PD-OES 原位跟踪痕量氧化石墨烯的光降解[99]

9.2.8 电热蒸发进样-微等离子体 OES

电热蒸发（electrothermal evaporation，ETV）样品引入方式可以显著提高微等离子体的原子化/激发能力，尤其适合微量体积的溶液样品，为此设计了一个紧凑的 ETV-DBD-AES 装置[100]，如图 9-13 所示。将 10 μL 样品溶液加到钨丝线圈上，通过执行钨线圈的加热程序，首先除去样品溶剂和基质，然后利用钨线圈提供的额外能量将蒸发/原子化的分析物直接引入 DBD 微等离子体中，并简单加热 DBD，实现了进一步原子化/激发。这将显著提高 DBD 微等离子体的稳定性，并节省了其对分析物重新原子化/激发所需的能量，从而提高了检测灵敏度，获得的汞和锌检出限分别为 0.8 μg/L 和 24 μg/L。在此基础上，进一步研制了一个由电池供电的小型钨线圈 ETV-DBD-OES 设备[5]，主要由微型 ETV、DBD 和光信号采集单元组成，质量仅为 4.5 kg，由 24 V DC 电池供电，最大功耗为 37 W，进样量为 3 μL。该装置可在 3 min 内分析 Hg、Zn、Pb、Ag、Cd、Au、Cu、Mn、Fe、Cr、As，检出限范围为 0.16~11.65 μg/L。将 ETV 作为 APGD-AES 的进样方式，建立了一种快速、灵敏测定稻米中镉的方法[101]。在分析过程中，数毫克（1~15 mg）大米样品先经过电热蒸发脱水和热解，其中镉从残留物中蒸发，并被捕集在冷的钨丝线圈上。然后，加热钨丝线圈将所捕获的 Cd 释放出来，并通过 Ar/H_2 流引入 APGD 微等离子体中激发。整个过程仅需要 3 min，使用 10 mg 大米样品所获得的 Cd（228.8 nm）的检出限为 2.6 pg，即 0.26 μg/kg。

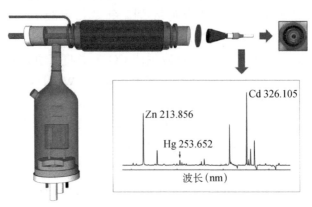

图 9-13 便携式 ETV-DBD-AES 装置[100]

9.2.9 固体剥蚀进样-微等离子体 OES

GD-OES 已成为固体样品元素成像分析与深度剖面分析的商品化仪器[102-105]。利用大气压微等离子体作为激发源，直接分析固体物质是一项很有挑战性的工作。利用 LS-APGD 微等离子体作为解吸/激发源对三种样品（金属薄膜、干溶液残留物和大块材料）进行了验证，将固体表面解吸释放的物质在微等离子体中激发，通过 OES 进行检测，如图 9-14 所示[106]。尽管实验结果基本上是定性的，但可能会在需要元素分析的领域中找到应用，包括金属块、土壤和溶液样品。利用低功率大气压脉冲火花放电(pulsed spark discharge，PSD)-OES[107]，建立了一个新的表面形貌和元素成像分析系统，这种 PSD-OES 系统可以很好地激发 Fe、Cr、Zn、Cu 等原子线。当电极非常靠近样品表面时，发射强度受电极和样品表面之间的距离影响很大，从而使某些样品（如硬币）的表面形貌成像成为可能；对于平坦的样品表面，当距离设置约为 0.5 mm 时，发射强度对微小的距离变化不敏感，因此可以进行元素成像，如识别铜板上的锌条，空间分辨率约为几十微米。该方法在空间分辨率和扫描速度方面仍有很大的提升空间。基于大气压射频（radio frequency，rf）-GD-OES 可实时测量空气颗粒相中的多种元素浓度[108]。该方法将气溶胶颗粒沉积在阴极尖端上，然后使用 rf-GD 剥蚀、原子化和激发颗粒物质，通过光谱仪记录最终的原子发射，以进行元素定性和定量，检出限为 0.055~1.0 ng，重现性为 5%~28%。对 rf-GD 光谱特征的空间分析表明，分析物的激发发生在样品电极附近的区域；对 rf-GD 光谱特征的时间分析表明，收集到的颗粒不断被烧蚀，193 ng 蔗糖颗粒被完全烧蚀的时间约为 2 s。进一步研究 PSD 微等离子体进行气溶胶分析的时空动力学[109]，发现在微等离子体的早期演化过程中，来自碳颗粒的原子发射信号靠近阴极表面区域，C I 和 C II 原子发射分别在 11 μs 和 6 μs 的延迟时间达到峰值强度，峰值发射强度发生在阴极表面上方 0.5~1.3 mm 之间，这将有助于优化该技术进行气溶胶分析。将 APGD 微等离子体作为熔融盐电解过程中的阳极[110]，代替传统的贵金属阳极和碳阳极，可避免电极的腐蚀。通过微等离子体阳极在熔融盐中引发电荷转移反应，以 Ag/Ag^+ 氧化还原作为研究对象，实现了 Ag 的电沉积，电流效率达到 90% 以上，并且微等离子体作为 OES 激发源，还可对熔融盐中银离子浓度进行原位实时测定，为熔盐电化学打开了新的方向。将 APGD 作为激光剥蚀直接固体进样的二次激发源[111]，用于高灵敏测定复杂基质土壤样品中的 Zn、Pb 和 Cd 元素，检出限分别为 0.68 mg/kg、2.71 mg/kg 和 0.31 mg/kg，与激光诱导击穿光谱法相比，锌的检出限降低了一个数量级以上，且设备运行成本极低。

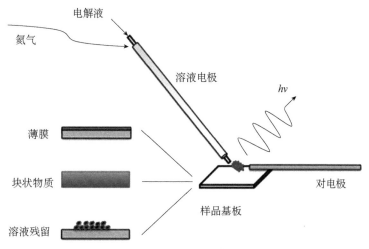

图 9-14　LS-APGD-OES 对三种样品的解吸/激发[106]

9.3 原子吸收/原子荧光光谱中的微等离子体原子化器与捕集器

9.3.1 蒸气发生进样-微等离子体原子化器

原子化器是 AAS 和 AFS 仪器的重要组成部件，其主要功能是使待测分析物分解并原子化。对于 HG 进样方式，商品化的 AAS 或 AFS 仪器通常采用热原子化方式，如外部加热石英管原子化器或火焰石英管原子化器，即当反应产生的氢气和分析物蒸气的混合物通过原子化器时，实现原子化。这种热原子化方式通常温度较高（900℃）[112,113]，使得激发光源和检测器无法足够接近原子化器，并且如此高的原子化温度势必使得仪器的运行功率较大，从而阻碍了仪器的小型化[114]。因此，人们致力于开发非热原子化器来代替传统热原子化器。

微等离子体是由气体放电所产生的非热等离子体，即电子温度远远大于气体温度，其所产生的自由基或离子的化学性质非常活跃，对促进氧化还原以及分子分解起着重要作用。因此，其可以很方便地作为挥发性组分的原子化方式，应用于 AAS 和 AFS 中。张新荣等首先验证了 DBD 微等离子体原子化器用于 HG-AAS 的可行性[114]，如图 9-15 所示，DBD 微等离子体能够实现 AsH_3 在常压下原子化形成 As^0。与传统的热原子化方式（火焰或电热）相比，DBD 具有体积小、电能消耗少（≤5 W）、原子化温度低（~70℃）的显著特点，将 HPLC 和 HG-DBD-AAS 联用可实现砷的形态分析，As(Ⅲ)、As(Ⅴ)、MMA 和 DMA 的检出限分别为 1.0 μg/L、11.8 μg/L、2.0 μg/L 和 18.0 μg/L。随后，他们又进一步扩展了 HG-DBD-AAS 应用范围，实现了对 Se、Sb、Sn 氢化物发生进样 DBD-AAS 检测[115]。为了适应 HG-AFS 原子化的需要，他们还设计了新型 DBD 微等离子体原子化器，其由两个同心石英管组成，内管用于构建 DBD 微等离子体原子化区域，外管用于通入屏蔽气以防止外部空气进入微等离子体原子化区域，可获得与传统的热原子化方式相当的灵敏度，As(Ⅲ)的检出限为 0.04 μg/L[116]。目前，HG-DBD-AFS 的检测范围已涵盖 Te[117]、Bi[118]等常规 HG-AFS 可检测元素。王建华等将 DBD 原子化器集成在一个 LOV 系统中[119]，建立了微型化 AFS 系统，成功地实现了痕量砷的测定，该微型化 AFS 系统可提供与常规 AFS 仪器相当的分析性能。目前，基于等离子体射流[120]、平面型 DBD[112,121,122]、圆柱型 DBD[117]、θ 结构[123]的各种微等离子体原子化器已被报道用于 Cd、As、Sb、Bi、Se 和甲基汞的原子化。其中，等离子体射流解吸原子化为薄层色谱与 AFS 相结合进行汞形态分析提供了一个新的接口[120]。等离子体射流可以快速实现薄层色谱表面上汞形态的解吸和原子化，从而很容易地实现 AFS 对薄层色谱分离出的汞形态直接检测，Hg^{2+}、MeHg 和 PhHg 的检出限分别为 0.51 pg、0.29 pg 和 0.34 pg。

为了对原子化过程有更深入的了解，通过使用各种探针对平面型 DBD 原子化器中的 Bi 进行研究，包括 AAS 监测沿光程分布的自由原子、实时直接分析（direct analysis in real time，DART）-轨道阱质谱仪识别由氢化物发生和原子化器产生的物质结构[124]。得到的数据表明，在 DBD 原子化器的中心部分之外基本没有游离的 Bi 原子，表明其具有高反应活性。通过 ICP-MS 定量分析，在 DBD 或石英管原子化器范围外的气相分析物馏分小于 10%，而在两个原子化器内表面上残留的分析物约为 90%。沉积在原子化器中的 Bi 含量高表明游离 Bi 原子具有高反应活性，这与在 DBD 的物理边界之外几乎不存在任何游离 Bi 原子的事实相符。通过研究其他氢化物形成元素（As、Sb、Se）对分别使用 DBD 和石英管原子化器 AAS 测定 Bi 的干扰程度，发现与石英管原子化器相比，DBD 可更好地耐受 Se 和 Sb 的干扰。

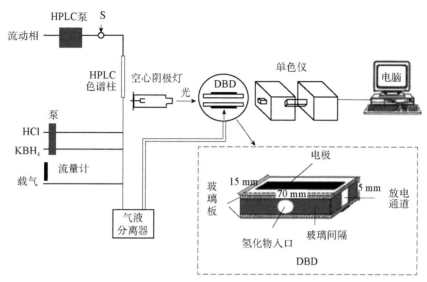

图 9-15 DBD 原子化器在 HG-AAS 测定砷形态中的应用[114]

9.3.2 蒸气发生进样-微等离子体捕集器

Kratzer 等研究发现[112]，在氩气中混合部分氧气的情况下，铋烷产生了一些记忆效应，这使得微等离子体作为捕集器用于分析物预浓集成为可能。之后，他们在平面型石英 DBD 中实现了 As 氢化物的预浓集和原子化[113]，如图 9-16 所示，在 Ar 微等离子体放电中加入 7 mL/min O_2，可将 DBD 原子化器中的 As 定量保留，O_2 关闭后，分析物就完全释放并原子化，预浓集效率达到 100%，并且 300 s 预浓集时间就可将 As 的检出限降低到 0.01 ng/mL。刘霁欣等开发了一种适用于 AFS 分析的 DBD 微等离子体捕集器用于 As 的捕集/释放[125]。当捕集 As 氢化物时，在 600 mL/min Ar 中混合 40 mL/min O_2 作为载气，9.2 kV 放电电压下可使 As 完全捕集在 DBD 原子化器中。当释放 As 时，用 200 mL/min H_2 代替 O_2，调节放电电压为 9.5 kV，可使 As 完全释放并原子化，该方法可得到 As 检出限为 1.0 ng/L 和 8 倍富集因子。使用多种探针研究外部加热的火焰石英管原子化器和 DBD 原子化器中 SeH_2 的原子化[126]，定量分析原子化器内表面上的 Se 沉积物，并用放射性 ^{75}Se 示踪剂通过放射自显影显示其分布，发现在外部加热的火焰石英管原子化器中，约 15% Se 沉积在光路两端的狭窄区域，相反，25%~40% Se 均匀沉积在整个 DBD 光路区域。进一步研究 DBD 微等离子体捕集器用于 Sb 捕集/释放的可行性[127]，发现在 300 s 捕集时间内，预浓集效率和检出限分别为 103%±2%和 0.02 ng/mL。将 PVG 与 DBD 捕集器联用[128]，在 AFS 测定之前对分析物进行气相富集，在含 O_2 的气氛下定量捕集（~100%）在 DBD 管内石英表面上，并在含 H_2 的气氛下释放（~100%），分析灵敏度（峰高）提高了 16 倍，Se 的绝对检出限为 4 pg（进样量为 1.2 mL），线性范围为 0.05~50 μg/L。在此过程中，由 PVG 产生的 H_2Se 和 SeCO 都以 SeO_2 或亚硒酸盐的形式被捕集在 DBD 石英管表面，然后以原子形式释放并传输到 AFS 检测区。

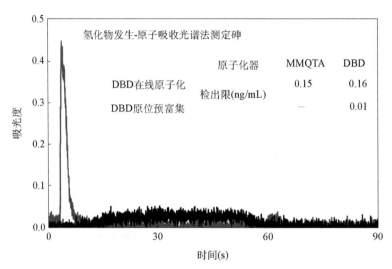

图 9-16 AAS 检测 DBD 中砷烷的预浓集与原子化[113]

9.4 质谱中的微等离子体离子源

自从电喷雾电离（electrospray ionization，ESI）和 DART 作为离子源引入 MS 分析以来，常压解吸电离-质谱（ambient desorption ionization-mass spectrometry，ADI-MS）一直受到人们的关注[12]。基于电晕放电、DBD 和 GD 原理设计了许多微等离子体 ADI[9,12,129-131]，它可实现传统 MS 的 ESI 功能[132]，或作为 ICP-MS 的固体进样方式[133]。微等离子体离子源的一个显著优点是，它可以直接从样品表面剥离化合物，几乎不需要预处理，可作为 ADI。到目前为止，微等离子体离子源已被广泛应用于爆炸物[134]、农药[135-137]、环境污染物[138,139]、代谢物、微生物和血液[129,140]的 MS 分析。

9.4.1 基于微等离子体离子源的固体进样

对于 ICP-MS 来说，微等离子体可作为 ADI，用于深度剖面分析或元素分析[141]。张四纯等报道了低温等离子体（low temperature plasma，LTP）探针与 ICP-MS 相结合用于薄涂层深度剖面分析的新方法[142]，如图 9-17 所示。在环境条件下，将石英毛细管中放电产生直径几十微米的 LTP 探针作用于样品表面，通过烧蚀样品转化为气溶胶，然后通过载气流传输到 ICP-MS 进行检测分析。扫描电子显微镜照片显示 LTP 探针剥蚀后的痕迹为直径小于 10 μm 的孔，横向分辨率约 200 μm，已在硅板上成功完成了 100 nm 单层样品和多层样品（100 nm Al/250 nm SiO$_2$/100 nm Au/50 nm Cr）的深度剖析，可作为现有深度剖面分析方法如 GD-MS/OES、AES 和 SIMS 的补充技术。他们还利用 LTP 探针作为 ADI[143]，实现了对书画艺术品的非破坏性原位 MS 成像分析，且无需喷洒有机试剂。其利用气体温度控制等离子体温度，通过调节毛细管尺寸、放电气体流速和表面扫描速率得到了令人满意的分辨率（250 μm，可通过内径更小的毛细管进一步提高）。杭纬等将脉冲微放电装置与 ICP-MS 相结合，开发了一种用于超薄层深度剖面分析技术。其以钨针为阳极，样品为阴极，在 50 μm 的放电间隙中形成了局部微等离子体，可实现亚毫米横向尺度和 0.6 nm 深度的剥蚀[144]。与直流微放电相比，脉冲微放电在样品的烧蚀和发射方面具有优势[133]。由于瞬时能量增强，脉冲微放电更适合直接固体进样。光谱仪检测发现，在直流模式下只能获取 N$_2$ 的常见发射线，而在脉冲模式下可获得样品的原子和离子线。结合

ICPMS 分析，脉冲微放电对不同基质样品直接固体进样时，在标准曲线线性和检出限方面表现出比直流放电更好的性能。在 MS 各种电离源中，微秒脉冲辉光放电（microsecond pulsed-glow discharge，MP-GD）和缓冲气体辅助激光电离（buffer-gas-assisted laser ionization，BGA-LI）源具有直接定量分析固体的潜力，而无需使用标准参考物质。杭纬等将这两种电离源与正交飞行时间 MS 相结合[145]，评估了这两种电离源的应用潜力，并提出了一种简单的方法来获得定量结果，即如果几乎没有干扰，并且元素峰值电流与它们的浓度成正比，则每种元素的摩尔浓度等于其离子电流占总离子流中的比例。

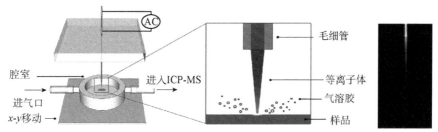

图 9-17　LTP 与 ICP-MS 联用对纳米涂层的深度剖析[142]

9.4.2　基于微等离子体离子源的溶液进样

Shelley 等研究发现在 SCGD 过程中[132]，分析物由流动液体引入，在微等离子体附近被蒸发和电离，通过引入含有金属盐的酸性溶液可在 MS 中产生元素离子以及 H_2O、OH^- 和 NO_3^- 加合物，Cu、Cd、Cs、Pb 和 U 检出限范围从 0.1~4 μg/L，线性范围超过 4 个数量级，相对标准偏差在 5%~16% 之间，如图 9-18 所示。该研究表明 SCGD-MS 能提供元素信息。张新荣等将 LTP 引入反应溶液表面[146]，实现了分析物的解吸和电离，无需任何样品预处理，就可通过 MS 实时监测正在进行的化学反应。这种新的反应监测技术可为有机化学提供用于反应监测和机理阐明的简单方法。为了提供一种能对生物过程介质和细胞培养液进行痕量金属分析的 MS 系统，将 LS-APGD 作为单四极杆质谱仪的离子源[147]，代替标准 ESI 源进行元素分析以及有机化合物的测定，25 μg/mL 的多元素测试溶液产生的信号强度>6×10⁷ AU。尽管 LS-APGD 微等离子体已显示出作为元素/同位素/分子离子源的潜力，但主要采用捕集式质量分析器检测。为此将 LS-APGD 与标准三重四极杆质谱相结合[148]，能够影响许多 MS/MS 技术，并能够扫描捕获型 MS 仪器通常不允许的低质量元素。对于 50 μL 测试元素的进样，其检出限为 0.99~38 ng/mL（即绝对质量为 0.05~2 ng）。进一步将 LS-APGD 离子源与轨道阱质谱仪联用消除谱峰干扰[149]，实现了铀的超高分辨率同位素分析。新的离子源外壳和集成控制系统将增强微等离子体的灵敏度，使元素/同位素分析的分辨率提高了 5 倍。与光束型仪器完全不同，它提供了高质量分辨率的优势，而铀分析的灵敏度、精度或动态范围均不受损失。

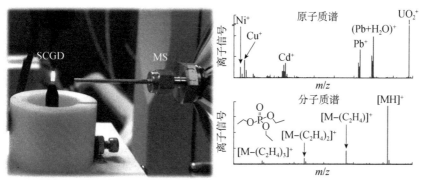

图 9-18　SCGD 作为原子和分子 MS 的多功能离子源[132]

9.5 微等离子体诱导化学反应

9.5.1 微等离子体诱导蒸气发生进样

微等离子体诱导蒸气发生是一种有前途的新型气相生成技术。众所周知，微等离子体中的自由基、电子和高能带电粒子或激发态粒子可以促进各种化学反应发生，从而将分析物转化为其挥发性物质[150]。自 2008 年 Hieftje 等发现 SCGD 形成的自由基可将汞还原成蒸气后[151]，微等离子体已经成功诱导产生了砷、碲[152]、铊、铟[41]、锑、银、硒、铋[153]、铅[154]、锌[155]、碘[156]、锇[157]、汞[158]、硫柳汞[159]和镉挥发物[160]。目前，微等离子体诱导蒸气发生进样方式主要由 GD 和 DBD 实现，可以有效地将溶解的离子转化为挥发性物质。

单液滴溶液电极辉光放电（single drop-solution electrode glow discharge，SD-SEGD）诱导蒸气发生可作为微量（5~20 μL）样品的进样方式[161,162]，如图 9-19 所示，当液滴表面和钨电极之间产生 SD-SEGD 时，会立即生成锌和镉的挥发性物质，并从液相中分离出来，然后通过载气引入原子光谱检测器中测定。同理，Swiderski 等将悬滴阴极大气压辉光放电（hanging drop cathode-atmospheric pressure glow discharge，HDC-APGD）作为 ICP-OES 的样品引入方式[163]，在减少样品消耗（0.56 mL/min）的同时，提高了分析物从溶液到 ICP 的传输效率（通常>80%）。与传统的气动雾化进样方式相比，待测元素的发射线强度平均高出 2 倍，已成功用于测定碱金属、碱土金属、过渡金属和非金属。郑成斌等将毛细管液体电极放电(capillary liquid electrode discharge，CLED)-OES 和微等离子体诱导蒸气发生(microplasma-induced vapor generation，PIVG)-AFS 集成于一体[164]，通过毛细管采样可显著降低样品消耗量，借助固有的毛细管驱动力和微等离子体中溶液汽化产生的驱动力，样品溶液将被自动输送到微等离子体区域中，待测分析物被微等离子体激发的同时，会诱导产生挥发性物质，可使同一样品分别经过 OES 与 AFS 测定，两种原子光谱法相结合可扩大检测的线性范围，汞和镉的检出限分别为 0.03 μg/L 和 0.04 μg/L，线性范围分别达到 0.001~100 mg/L 和 0.001~40 mg/L。为了灵敏地定量鱼体内汞的分布，将多壁碳纳米管辅助基质固相分散（multiwall carbon nanotubes assisted matrix solid-phase dispersion，MWCNTs-MSPD）与 SD-SEGD 诱导冷蒸气发生相结合作为 AFS 进样方式[162]，通过 MWCNTs-MSPD 将有限样品中的汞有效提取到 100 μL 洗脱液里，经 SD-SEGD 冷蒸气发生将不同形态的汞转化为 Hg^0，并进一步转移到 AFS 进行测定。由于避免了样品制备而导致的分析物稀释，可显著提高灵敏度，在 1 mg 样品量时，汞的检出限为 0.01 μg/L（0.2 pg）。采用微等离子体诱导蒸气发生进样时，如果样品溶液中含有甲酸（1%）、2-巯基乙醇或非离子表面活性剂，或简单地将样品溶液转化为气溶胶，则微等离子体诱导蒸气发生效率都将大大提高[152,165,166]。另外，SAGD 微等离子体也可诱导蒸气发生进样[150]，如支持电解质（盐酸、pH=3.2）中的镉和锌离子可在微等离子体-液体界面处转化为挥发性物质，不需要其他还原剂，其蒸气发生效率比常规的电化学氢化物发生体系和 HCl-KBH_4 体系要高得多，AFS 检测镉和锌的检出限可分别低至 0.003 μg/L 和 0.3 μg/L。

邢志等将 DBD 微等离子体用于诱导氢化物发生[152]，成功实现了微量样品中 As、Te、Sb 和 Se 的 HG-AFS 检测。微等离子体由混入少量氢气的氩气流经 DBD 石英管产生，样品溶液通过雾化器转化为气溶胶，引入微等离子体中反应生成氢化物，通过 AFS 检测 As、Te、Sb 和 Se 的绝对检出限分别为 0.6 ng、1.0 ng、1.4 ng 和 1.2 ng。将薄膜 DBD 诱导冷蒸气发生与 DBD-OES 联用[167]，并

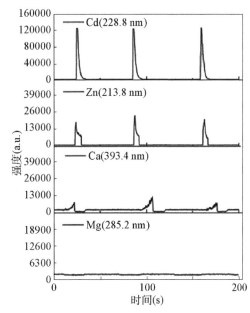

图 9-19　SD-SEGD 诱导蒸气发生检测锌和镉[161]

将其集成在一个 10.5 cm×8.0 cm×1.2 cm 的聚甲基丙烯酸甲酯板中，可获得一种小型汞分析仪。在 DBD 反应器中引入含或不含甲酸的标准溶液或样品溶液，形成薄膜液体，在微等离子体作用下产生汞蒸气，随后从液相中分离出来，并引入 DBD-OES 中检测，得到的检出限分别为 0.20 μg/L 和 2.6 μg/L。基于流动薄膜 DBD 诱导蒸气发生作为 HPLC 与 AFS 在线联用的接口[168]，可实现对无机汞（Hg^{2+}）、甲基汞（MeHg）和乙基汞（EtHg）的形态分析。在不使用其他化学试剂的情况下，使用 DBD 微等离子体可一步完成有机汞形态的分解和 Hg^{2+} 的还原，Hg^{2+}、MeHg 和 EtHg 的检出限分别为 1.6 μg/L、0.42 μg/L 和 0.75 μg/L。基于液体喷雾 DBD 微等离子体诱导蒸气发生进样方式[154] 用于 ICP-MS 测定铅，如图 9-20 所示。将样品溶液转化为气溶胶，同时与气动雾化器喷嘴处产生的 DBD 微等离子体混合，由于细雾滴中的分析物与微等离子体的相互作用增强，极大地促进了 Pb 蒸气的生成，通过 ICP-MS 测定 Pb 的检出限为 0.003 μg/L。利用同样的进样方式，通过 AFS 测定了大米样品中的痕量镉[169]，获得的检出限为 0.01 μg/L。在此基础上，基于液体喷雾 DBD 微等离子体诱导蒸气发生进样方式[153]，通过 ICP-MS 进一步实现了微量液体样品（20 μL）中 Se、Ag、Sb、Pb 和 Bi 的同时灵敏测定，获得的检出限分别为 10 ng/L（200 fg）、2 ng/L（40 fg）、5 ng/L（100 fg）、4 ng/L（80 fg）和 3 ng/L（60 fg）。仅需使用甲酸，溶液样品中 Se、Ag、Sb、Pb 和 Bi 离子就可以高效地同时转化为挥发性物质，无需预还原步骤（对于 Sb 和 Se）或预氧化步骤（对于 Pb），且避免了使用不稳定和昂贵还原剂的麻烦，样品通量高达每小时 180 个样。与传统的气动雾化 ICP-MS 相比，基于液体喷雾 DBD 微等离子体诱导蒸气发生进样方式可以轻松地将灵敏度提高至少 14 倍。将液体喷雾 DBD 微等离子体诱导蒸气发生的元素范围进一步扩展[170]，还可实现 Sc、Y、La、Ce、Pr、Nd、Sm、Eu、Gd、Tb、Dy、Ho、Er、Tm、Yb 和 Lu 16 种稀土元素（rare-earth element，REE）的蒸气发生进样。将其作为 ICP-MS 的进样方式，与传统溶液雾化进样方式相比，REE 检出限改善了 4～10 倍，在 0.002～0.422 ng/L 之间，并且未发现其他金属离子的明显干扰。与超声雾化、微同心雾化、膜去溶剂化和电热蒸发进样方式相比，液体喷雾 DBD 微等离子体诱导蒸气发生进样方式不仅运行成本低、功耗低，且操作简便，可为环境样品中稀土元素的测定提供一种绿色高效的进样方式。

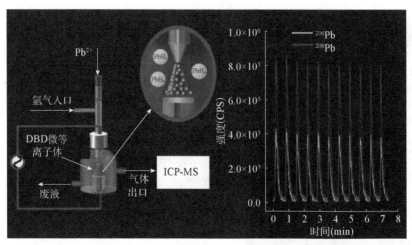

图 9-20　液体喷雾 DBD 微等离子体诱导蒸气发生 ICP-MS 测定铅[154]

9.5.2　微等离子体诱导形态转化与物质分解

微等离子体中富含的电子、自由基和激发态粒子等可使分析物在其中发生一系列的氧化还原反应。利用这一特点，将 PD 微等离子体作为 H_2S 转化为 SO_2 的反应器[171]，效率高达 95%，通过将该反应器与 AFS 联用，使用锌空心阴极灯激发 SO_2 产生荧光，实现对 H_2S 和 SO_2 的形态分析，以及对硫化物和亚硫酸盐的分析测定，如图 9-21 所示。具体而言，当微等离子体关闭时，仅对亚硫酸盐产生的 SO_2 进行检测；当微等离子体开启时，将使硫化物产生的 H_2S 转化为 SO_2，进而检测 SO_2 总含量，只需进行简单的差减即可实现形态分析，硫化物和亚硫酸盐的检出限均为 7.7 μmol/L。

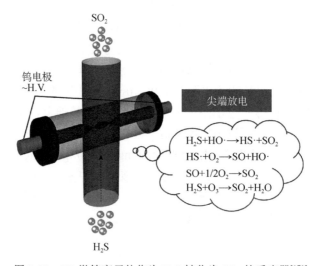

图 9-21　PD 微等离子体作为 H_2S 转化为 SO_2 的反应器[171]

利用同轴 DBD 反应器设计了的固体进样汞分析仪，可以消除来自复杂固体样品的气相干扰[172]。该仪器包括电热蒸发器，用于分解挥发性有机化合物的 DBD 反应器，消除基体干扰的金线圈汞阱和 AFS 检测器，可在环境温度下在线分解多达 12 mg 的干水产品粉末，汞检出限为 0.5 μg/kg。此外，在线 DBD 反应器仅消耗 40 W 的功率，明显低于商用汞分析仪的功率（>300 W）；包括样品预处理，整个分析可在 5 min 内完成。该方法可避免使用化学试剂，更加简便、绿色和安全。使用双层同轴 DBD 反应器消解样品[173]，可实现大米样品中痕量元素的灵敏测定，如图 9-22 所示。针对实际样品研究了三种 DBD 消解体系，包括 H_2O-DBD、H_2O_2-DBD 和 HNO_3-DBD 消解体系，其中 H_2O-DBD 消解体系在无需任何其他化学物质的条件下即可对样品进行消解，而后通过 ICP-MS 检测，Mg、Mn、Zn、Cd、Cr、Co 和 As 的检出限在 0.01～0.35 ng/g 范围内。

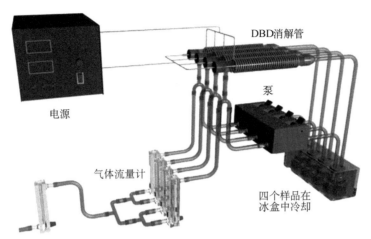

图 9-22　使用 DBD 微等离子体消解样品[173]

9.6　总结与展望

本章概述了微等离子体在原子光谱分析痕量物质中的应用，涵盖了微等离子体激发源、微等离子体原子化器与捕集器、微等离子体离子源与微等离子体诱导化学反应四方面的研究进展。基于微等离子体的原子发射光谱分析，由于其装置体积小、操作简单、功耗低，在便携式痕量分析仪器的研制中发挥了重要作用。目前，基于微等离子体的分析技术研究仍处于起步阶段，已报道的微等离子体原子光谱分析方法，大部分工作主要集中在对元素总量的分析。在面对各种特定的复杂分析任务时，迫切需要开发适于微等离子体原子光谱分析的微分离技术，从而提高分析性能，并扩展形态分析的应用范围。基于微等离子体的原子光谱方法用于深度剖面分析和元素成像正成为新的焦点，通过优化微等离子体放电稳定性、减小等离子体束的尺寸以及增强与样品表面的相互作用，可使其在灵敏度、空间分辨率和扫描速度方面有更大的提升空间。微等离子体诱导蒸气发生作为一种新颖且高效的样品引入方法，这方面还有待进一步研究，以便阐明相关机理，并扩大其应用范围。

参 考 文 献

[1] Wang X H, Zhang S D, Xu Z Y, Lin J Y, Huang B L, Hang W. Atomic spectrometry in China: Past and present. Journal of Analytical Atomic Spectrometry, 2015, 30(4): 852-866.

[2] Butler O T, Cairns W R L, Cook J M, Davidson C M. Atomic spectrometry update: A review of advances in environmental analysis. Journal of Analytical Atomic Spectrometry, 2017, 32(1): 11-57.

[3] Li C H, Long Z, Jiang X M, Wu P, Hou X D. Atomic spectrometric detectors for gas chromatography. TrAC Trends in Analytical Chemistry, 2016, 77: 139-155.

[4] Liu S, Cai Y, Yu Y L, Wang J H. A miniature liquid electrode discharge-optical emission spectrometric system integrating microelectrodialysis for potassium screening in serum. Journal of Analytical Atomic Spectrometry, 2017, 32(9): 1739-1745.

[5] Li N, Wu Z C, Wang Y Y, Zhang J, Zhang X N, Zhang H N, Wu W H, Gao J, Jiang J. Portable dielectric barrier discharge-atomic emission spectrometer. Analytical Chemistry, 2017, 89(4): 2205-2210.

[6] Jamroz P, Greda K, Dzimitrowicz A, Swiderski K, Pohl P. Sensitive determination of Cd in small-volume samples by miniaturized liquid drop anode atmospheric pressure glow discharge optical emission spectrometry. Analytical Chemistry, 2017, 89(11): 5729-5733.

[7] Leng A Q, Lin Y, Tian Y F, Wu L, Jiang X M, Hou X D, Zheng C B. Pump-and valve-free flow injection capillary liquid electrode discharge optical emission spectrometry coupled to a droplet array platform. Analytical Chemistry, 2017, 89(1): 703-710.

[8] Marcus R K, Manard B T, Quarles C D. Liquid sampling-atmospheric pressure glow discharge (LS-APGD) microplasmas for diverse spectrochemical analysis applications. Journal of Analytical Atomic Spectrometry, 2017, 32(4): 704-716.

[9] Brandt S, Klute F D, Schutz A, Franzke J. Dielectric barrier discharges applied for soft ionization and their mechanism. Analytica Chimica Acta, 2017, 951: 16-31.

[10] Pohl P, Jamroz P, Swiderski K, Dzimitrowicz A, Lesniewicz A. Critical evaluation of recent achievements in low power glow discharge generated at atmospheric pressure between a flowing liquid cathode and a metallic anode for element analysis by optical emission spectrometry. TrAC Trends in Analytical Chemistry, 2017, 88: 119-133.

[11] Meng F Y, Yuan X, Li X M, Liu Y, Duan Y X. Microplasma-based detectors for gas chromatography: Current status and future trends. Applied Spectroscopy Reviews, 2014, 49(7): 533-549.

[12] Zeiri O M, Storey A P, Ray S J, Hieftje G M. Microplasma-based flowing atmospheric-pressure afterglow (FAPA) source for ambient desorption-ionization mass spectrometry. Analytica Chimica Acta, 2017, 952: 1-8.

[13] Yu Y L, Zhuang Y T, Wang J H. Advances in dielectric barrier discharge-optical emission spectrometry for the analysis of trace species. Analytical Methods, 2015, 7(5): 1660-1666.

[14] Brandt S, Schütz A, Klute F D, Kratzer J, Franzke J. Dielectric barrier discharges applied for optical spectrometry. Spectrochimica Acta Part B: Atomic Spectroscopy, 2016, 123: 6-32.

[15] Li W, Jiang X M, Xu K L, Hou X D, Zheng C B. Determination of ultratrace nitrogen in pure argon gas by dielectric barrier discharge-molecular emission spectrometry. Microchemical Journal, 2011, 99(1): 114-117.

[16] Yang T, Gao D X, Yu Y L, Chen M L, Wang J H. Dielectric barrier discharge micro-plasma emission spectrometry for the detection of acetone in exhaled breath. Talanta, 2016, 146: 603-608.

[17] Meyer C, Demecz D, Gurevich E L, Marggraf U, Jestel G, Franzke J. Development of a novel dielectric barrier microhollow cathode discharge for gaseous atomic emission spectroscopy. Journal of Analytical Atomic Spectrometry, 2012, 27(4): 677-681.

[18] Greda K, Swiderski K, Jamroz P, Pohl P. Reduction of spectral interferences in atmospheric pressure glow discharge optical emission spectrometry. Microchemical Journal, 2017, 130: 7-13.

[19] Wu Z C, Chen M L, Li P, Zhu Q Q, Wang J H. Dielectric barrier discharge non-thermal micro-plasma for the excitation and emission spectrometric detection of ammonia. Analyst, 2011, 136(12): 2552-2557.

[20] Wu Z C, Jiang J, Li N. Cold excitation and determination of hydrogen sulfide by dielectric barrier discharge molecular emission spectrometry. Talanta, 2015, 144: 734-739.

[21] Meng F Y, Duan Y X. Nitrogen microplasma generated in chip-based ingroove glow discharge device for detection of

organic fragments by optical emission spectrometry. Analytical Chemistry, 2015, 87(3): 1882-1888.

[22] Li M T, Huang S X, Xu K L, Jiang X M, Hou X D. Miniaturized point discharge-radical optical emission spectrometer: A multichannel optical detector for discriminant analysis of volatile organic sulfur compounds. Talanta, 2018, 188: 378-384.

[23] Qian B, Zhao J, He Y, Peng L X, Ge H L, Han B J. Miniaturized dielectric barrier discharge-molecular emission spectrometer for determination of total sulfur dioxide in food. Food Chemistry, 2020, 317: 126437.

[24] Li W, Zheng C B, Fan G Y, Tang L, Xu K L, Lv Y, Hou X D. Dielectric barrier discharge molecular emission spectrometer as multichannel GC detector for halohydrocarbons. Analytical Chemistry, 2011, 83(13): 5050-5055.

[25] Han B J, Jiang X M, Hou X D, Zheng C B. Dielectric barrier discharge carbon atomic emission spectrometer: Universal GC detector for volatile carbon-containing compounds. Analytical Chemistry, 2014, 86(1): 936-942.

[26] Zheng C B, Hu L G, Hou X D, He B, Jiang G B. Headspace solid-phase microextraction coupled to miniaturized microplasma optical emission spectrometry for detection of mercury and lead. Analytical Chemistry, 2018, 90(6): 3683-3691.

[27] Lin Y, Yang Y, Li Y X, Yang L, Hou X D, Feng X B, Zheng C B. Ultrasensitive speciation analysis of mercury in rice by headspace solid phase microextraction using porous carbons and gas chromatography-dielectric barrier discharge optical emission spectrometry. Environmental Science & Technology, 2016, 50(5): 2468-2476.

[28] Yang Y, Tan Q, Lin Y, Tian Y F, Wu L, Hou X D, Zheng C B. Point discharge optical emission spectrometer as a gas chromatography (GC) detector for speciation analysis of mercury in human hair. Analytical Chemistry, 2018, 90(20): 11996-12003.

[29] Yang Y, Wang Y, Hou X L, Lin Y, Yang L, Hou X D, Zheng C B. Can low-temperature point discharge be used as atomic emission source for sensitive determination of cyclic volatile methylsiloxanes? Analytica Chimica Acta, 2020, 1124: 121-128.

[30] Cserfalvi T, Mezei P. Direct solution analysis by glow discharge: Electrolyte-cathode discharge spectrometry. Journal of Analytical Atomic Spectrometry, 1994, 9(3): 345-349.

[31] Yang C, Wang L, Zhu Z L, Jin L L, Zheng H T, Belshaw N S, Hu S H. Evaluation of flow injection-solution cathode glow discharge-atomic emission spectrometry for the determination of major elements in brines. Talanta, 2016, 155: 314-320.

[32] Konegger-Kappel S, Manard B T, Zhang L X, Konegger T, Marcus R K. Liquid sampling-atmospheric pressure glow discharge excitation of atomic and ionic species. Journal of Analytical Atomic Spectrometry, 2015, 30(1): 285-295.

[33] Krahling T, Michels A, Geisler S, Florek S, Franzke J. Investigations into modeling and further estimation of detection limits of the liquid electrode dielectric barrier discharge. Analytical Chemistry, 2014, 86(12): 5822-5828.

[34] Schwartz A J, Ray S J, Elish E, Storey A P, Rubinshtein A A, Chan G C, Pfeuffer K P, Hieftje G M. Visual observations of an atmospheric-pressure solution-cathode glow discharge. Talanta, 2012, 102: 26-33.

[35] Maksimov A I, Titov V A, Khlyustova A V. Electrolyte-as-cathode glow discharge emission and the processes of solution-to-plasma transport of neutral and charged species. High Energy Chemistry, 2004, 38(3): 196-199.

[36] Decker C G, Webb M R. Measurement of sample and plasma properties in solution-cathode glow discharge and effects of organic additives on these properties. Journal of Analytical Atomic Spectrometry, 2016, 31(1): 311-318.

[37] Yu J, Yang S Y, Sun D X, Lu Q F, Zheng J D, Zhang X M, Wang X. Simultaneously determination of multi metal elements in water samples by liquid cathode glow discharge-atomic emission spectrometry. Microchemical Journal, 2016, 128: 325-330.

[38] Peng X X, Guo X H, Ge F, Wang Z. Battery-operated portable high-throughput solution cathode glow discharge optical

emission spectrometry for environmental metal detection. Journal of Analytical Atomic Spectrometry, 2019, 34(2): 394-400.

[39] Marcus R K, Davis W C. An atmospheric pressure glow discharge optical emission source for the direct sampling of liquid media. Analytical Chemistry, 2001, 73(13): 2903-2910.

[40] Yuan M L, Peng X X, Ge F, Li Q, Wang K, Yu D G, Wang Z. Simplified design for solution anode glow discharge atomic emission spectrometry device for highly sensitive detection of Ag, Bi, Cd, Hg, Pb, Tl, and Zn. Microchemical Journal, 2020, 155: 104785.

[41] Greda K, Gorska M, Welna M, Jamroz P, Pohl P. In-situ generation of Ag, Cd, Hg, In, Pb, Tl and Zn volatile species by flowing liquid anode atmospheric pressure glow discharge operated in gaseous jet mode-Evaluation of excitation processes and analytical performance. Talanta, 2019, 199: 107-115.

[42] Greda K, Swiderski K, Jamroz P, Pohl P. Flowing liquid anode atmospheric pressure glow discharge as an excitation source for optical emission spectrometry with the improved detectability of Ag, Cd, Hg, Pb, Tl, and Zn. Analytical Chemistry, 2016, 88(17): 8812-8820.

[43] Hall K A, Paing H W, Webb M R, Marcus R K. Monochromatic spatial imaging of the liquid sampling-atmospheric pressure glow discharge: Effects of gas flow on spatial profiles of analyte and background species. Spectrochimica Acta Part B: Atomic Spectroscopy, 2019, 154: 33-42.

[44] Paing H W, Hall K A, Marcus R K. Sheathing of the liquid sampling-atmospheric pressure glow discharge microplasma from ambient atmosphere and its implications for optical emission spectroscopy. Spectrochimica Acta Part B: Atomic Spectroscopy, 2019, 155: 99-106.

[45] Paing H W, Marcus R K. Parametric evaluation of ambient desorption optical emission spectroscopy utilizing a liquid sampling-atmospheric pressure glow discharge microplasma. Journal of Analytical Atomic Spectrometry, 2017, 32(5): 931-941.

[46] Hall K A, Marcus R K. Parametric optimization and spectral line selection for liquid sampling-atmospheric pressure glow discharge-optical emission spectroscopy. Journal of Analytical Atomic Spectrometry, 2019, 34(12): 2428-2439.

[47] Moon D E, Weeb M R. Imaging studies of emission and laser scattering from a solution-cathode glow discharge. Journal of Analytical Atomic Spectrometry, 2020, 35(9): 1859-1867.

[48] Greda K, Burhenn S, Pohl P, Franzke J. Enhancement of emission from indium in flowing liquid anode atmospheric pressure glow discharge using organic media. Talanta, 2019, 204: 304-309.

[49] Lu Q F, Feng F F, Yu J, Yin L, Kang Y J, Luo H, Sun D X, Yang W. Determination of trace cadmium in zinc concentrate by liquid cathode glow discharge with a modified sampling system and addition of chemical modifiers for improved sensitivity. Microchemical Journal, 2020, 152: 104308.

[50] Cai Y, Zhang Y J, Wu D F, Yu Y L, Wang J H. Nonthermal optical emission spectrometry: Direct atomization and excitation of cadmium for highly sensitive determination. Analytical Chemistry, 2016, 88(8): 4192-4195.

[51] Greda K, Jamroz P, Pohl P. Ultrasonic nebulization atmospheric pressure glow discharge: Preliminary study. Spectrochimica Acta Part B: Atomic Spectroscopy, 2016, 121: 22-27.

[52] He Q, Zhu Z L, Hu S H, Zheng H T, Jin L L. Elemental determination of microsamples by liquid film dielectric barrier discharge atomic emission spectrometry. Analytical Chemistry, 2012, 84(9): 4179-4184.

[53] Yuan X, Zhan X F, Li X M, Zhao Z J, Duan Y X. Matrix-assisted plasma atomization emission spectrometry for surface sampling elemental analysis. Scientific Reports, 2016, 6: 19417.

[54] Świderski K, Pohl P, Jamróz P. A miniaturized atmospheric pressure glow microdischarge system generated in contact with a hanging drop electrode: A new approach to spectrochemical analysis of liquid microsamples. Journal of

Analytical Atomic Spectrometry, 2019, 34(6): 1287-1293.

[55] Paing H W, Marcus R K. Investigation of hydrophobic substrates for solution residue analysis utilizing an ambient desorption liquid sampling-atmospheric pressure glow discharge microplasma. Analyst, 2018, 143(6): 1417-1425.

[56] Zhang Y J, Cai Y, Yu Y L, Wang J H. A miniature optical emission spectrometric system in a lab-on-valve for sensitive determination of cadmium. Analytica Chimica Acta, 2017, 976: 45-51.

[57] Hall K A, Jiang L W, Marcus R K. Demonstration of a novel ion-exchange column for pre-concentration of silver ions in optical emission spectroscopy utilizing a liquid-sampling atmospheric pressure glow discharge microplasma. Journal of Analytical Atomic Spectrometry, 2017, 32(12): 2463-2468.

[58] Jiang X, Xu X L, Hou X D, Long Z, Tian Y F, Jiang X M, Xu F J, Zheng C B. A novel capillary microplasma analytical system: interface-free coupling of glow discharge optical emission spectrometry to capillary electrophoresis. Journal of Analytical Atomic Spectrometry, 2016, 31(7): 1423-1429.

[59] MoB K K, Reinsberg K G, Broekaert J A C. Study of a direct current atmospheric pressure glow discharge in helium with wet aerosol sample introduction systems. Journal of Analytical Atomic Spectrometry, 2014, 29(4): 674-680.

[60] Li M T, Deng Y J, Zheng C B, Jiang X M, Hou X D. Hydride generation-point discharge microplasma-optical emission spectrometry for the determination of trace As, Bi, Sb and Sn. Journal of Analytical Atomic Spectrometry, 2016, 31(12): 2427-2433.

[61] Yu Y L, Du Z, Chen M L, Wang J H. Atmospheric-pressure dielectric-barrier discharge as a radiation source for optical emission spectrometry. Angewandte Chemie-International Edition, 2008, 47(41): 7909-7912.

[62] Zhu Z L, Chan G C, Ray S J, Zhang X R, Hieftje G M. Microplasma source based on a dielectric barrier discharge for the determination of mercury by atomic emission spectrometry. Analytical Chemistry, 2008, 80(22): 8622-8627.

[63] Zhu Z L, He H Y, He D, Zheng H T, Zhang C X, Hu S H. Evaluation of a new dielectric barrier discharge excitation source for the determination of arsenic with atomic emission spectrometry. Talanta, 2014, 122: 234-239.

[64] Huang C C, Li Q, Mo J M, Wang Z. Ultratrace Determination of tin, germanium, and selenium by hydride generation coupled with a novel solution-cathode glow discharge-atomic emission spectrometry method. Analytical Chemistry, 2016, 88(23): 11559-11567.

[65] Yang C, He D, Zhu Z L, Peng H, Liu Z F, Wen G J, Bai J H, Zheng H T, Hu S H, Wang Y X. Battery-operated atomic emission analyzer for waterborne arsenic based on atmospheric pressure glow discharge excitation source. Analytical Chemistry, 2017, 89(6): 3694-3701.

[66] Cheng J Q, Li Q, Zhao M Y, Wang Z. Ultratrace Pb determination in seawater by solution-cathode glow discharge-atomic emission spectrometry coupled with hydride generation. Analytica Chimica Acta, 2019, 1077: 107-115.

[67] Matusiewicz H, Ślachciński M. Trace determination of Hg together with As, Sb, Se by miniaturized optical emission spectrometry integrated with chemical vapor generation and capacitively coupled argon microwave miniplasma discharge. Spectrochimica Acta Part B: Atomic Spectroscopy, 2017, 133: 52-59.

[68] Li M T, Li K, He L, Zeng X L, Wu X, Hou X D, Jiang X M. Point discharge microplasma optical emission spectrometer: Hollow electrode for efficient volatile hydride/mercury sample introduction and 3D-printing for compact instrumentation. Analytical Chemistry, 2019, 91(11): 7001-7006.

[69] Peng X X, Wang Z. Ultrasensitive determination of selenium and arsenic by modified helium atmospheric pressure glow discharge optical emission spectrometry coupled with hydride generation. Analytical Chemistry, 2019, 91(15): 10073-10080.

[70] Zhu Z L, Yang C, Yu P W, Zheng H T, Liu Z F, Xin Z, Hu S H. Determination of antimony in water samples by hydride

generation coupled with atmospheric pressure glow discharge atomic emission spectrometry. Journal of Analytical Atomic Spectrometry, 2019, 34(2): 331-337.

[71] Yang C, Chan G C, He D, Liu Z F, Deng Q S, Zheng H T, Hu S H, Zhu Z L. Highly sensitive determination of arsenic and antimony based on an interrupted gas flow atmospheric pressure glow discharge excitation source. Analytical Chemistry, 2019, 91(3): 1912-1919.

[72] Zhao M Y, Peng X X, Yang B C, Wang Z. Ultra-sensitive determination of antimony valence by solution cathode glow discharge optical emission spectrometry coupled with hydride generation. Journal of Analytical Atomic Spectrometry, 2020, 35(6): 1148-1155.

[73] Qian B, Zhao J, He Y, Peng L X, Ge H L, Han B J. A liquid chromatography detector based on continuous-flow chemical vapor generation coupled glow discharge atomic emission spectrometry: Determination of organotin compounds in food samples. Journal of Chromatography A, 2019, 1608: 460406.

[74] Yu Y L, Dou S, Chen M L, Wang J H. Iodine excitation in a dielectric barrier discharge micro-plasma and its determination by optical emission spectrometry. Analyst, 2013, 138(6): 1719-1725.

[75] Yu Y L, Cai Y, Chen M L, Wang J H. Development of a miniature dielectric barrier discharge-optical emission spectrometric system for bromide and bromate screening in environmental water samples. Analytica Chimica Acta, 2014, 809: 30-36.

[76] Zhang D J, Cai Y, Chen M L, Yu Y L, Wang J H. Dielectric barrier discharge-optical emission spectrometry for the simultaneous determination of halogens. Journal of Analytical Atomic Spectrometry, 2016, 31(2): 398-405.

[77] Han B J, Jiang X M, Hou X D, Zheng C B. Miniaturized dielectric barrier discharge carbon atomic emission spectrometry with online microwave-assisted oxidation for determination of total organic carbon. Analytical Chemistry, 2014, 86(13): 6214-6219.

[78] Zheng H T, Guan X D, Mao X F, Zhu Z L, Yang C, Qiu H O, Hu S H. Determination of nitrite in water samples using atmospheric pressure glow discharge microplasma emission and chemical vapor generation of NO species. Analytica Chimica Acta, 2018, 1001: 100-105.

[79] Zhu Z L, Guan X D, Zheng H T, Yang C, Xing Z, Hu S H. Determination of nitrate and ammonium ions in water samples by atmospheric pressure glow discharge microplasma molecular emission spectrometry coupled with chemical vapour generation. Journal of Analytical Atomic Spectrometry, 2018, 33(12): 2153-2159.

[80] Han B J, Li Y, Qian B, He Y, Peng L X, Yu H M. A novel liquid chromatography detector based on a dielectric barrier discharge molecular emission spectrometer with online microwave-assisted hydrolysis for determination of dithiocarbamates. Analyst, 2018, 143(12): 2790-2798.

[81] Wang Y L, Lin L L, Liu J X, Mao X F, Wang J H, Qin D Y. Ferric ion induced enhancement of ultraviolet vapour generation coupled with atomic fluorescence spectrometry for the determination of ultratrace inorganic arsenic in surface water. Analyst, 2016, 141(4): 1530-1536.

[82] Xu F J, Zou Z R, He J, Li M T, Xu K L, Hou X D. *In situ* formation of nano-CdSe as a photocatalyst: Cadmium ion-enhanced photochemical vapour generation directly from Se(Ⅵ). Chemical Communications, 2018, 54(38): 4874-4877.

[83] Zhen Y F, Yu Y, Zhang A G, Gao Y. Matrix-assisted photochemical vapor generation for determination of trace bismuth in Fe Ni based alloy samples by inductively coupled plasma mass spectrometry. Microchemical Journal, 2019, 151: 104242.

[84] Nováková E, Horová K, Červený V, Hraníček J, Musil S. UV photochemical vapor generation of Cd from a formic acid based medium: Optimization, efficiency and interferences. Journal of Analytical Atomic Spectrometry, 2020, 35(7): 1380-1388.

[85] Gao Y, Xu M, Sturgeon R E, Mester Z, Shi Z M, Galea R, Saull P, Yang L. Metal ion-assisted photochemical vapor generation for the determination of lead in environmental samples by multicollector-ICPMS. Analytical Chemistry, 2015, 87(8): 4495-4502.

[86] Zeng W, Hu J, Chen H J, Zou, Z R, Hou X D, Jiang X M. Cobalt ion-enhanced photochemical vapor generation in a mixed acid medium for sensitive detection of tellurium (Ⅳ) by atomic fluorescence spectrometry. Journal of Analytical Atomic Spectrometry, 2020, 35(7): 1405-1411.

[87] Šoukal J, Sturgeon R E, Musil S. Efficient photochemical vapor generation of molybdenum for ICPMS detection. Analytical Chemistry, 2018, 90(19): 11688-11695.

[88] Hu J, Chen H J, Hou X D, Jiang X M. Cobalt and copper ions synergistically enhanced photochemical vapor generation of molybdenum: Mechanism study and analysis of water samples. Analytical Chemistry, 2019, 91(9): 5938-5944.

[89] Vyhnanovsky J, Sturgeon R E, Musil S. Cadmium assisted photochemical vapor generation of tungsten for detection by inductively coupled plasma mass spectrometry. Analytical Chemistry, 2019, 91(20): 13306-13312.

[90] Guerrero M M L, Alonso E V, Torres A G D, Pavon J M C. Simultaneous determination of traces of Pt, Pd, Os, Ir, Rh, Ag and Au metals by magnetic SPE ICP OES and in situ chemical vapour generation. Journal of Analytical Atomic Spectrometry, 2017, 32(11): 2281-2291.

[91] Xu T, Hu J, Chen H J. Transition metal ion Co(Ⅱ)-assisted photochemical vapor generation of thallium for its sensitive determination by inductively coupled plasma mass spectrometry. Microchemical Journal, 2019, 149: 103972.

[92] Hu J, Sturgeon R E, Nadeau K, Hou X D, Zheng C B, Yang L. Copper ion assisted photochemical vapor generation of chlorine for its sensitive determination by sector field inductively coupled plasma mass spectrometry. Analytical Chemistry, 2018, 90(6): 4112-4118.

[93] Guo X M, Sturgeon R E, Mester Z, Gardner G J. Vapor generation by UV irradiation for sample introduction with atomic spectrometry. Analytical Chemistry, 2004, 76(8): 2401-2405.

[94] Cai Y, Li S H, Dou S, Yu Y L, Wang J H. Metal carbonyl vapor generation coupled with dielectric barrier discharge to avoid plasma quench for optical emission spectrometry. Analytical Chemistry, 2015, 87(2): 1366-1372.

[95] Zhang S, Luo H, Peng M, Tian Y F, Hou X D, Jiang X M, Zheng C B. Determination of Hg, Fe, Ni, and Co by miniaturized optical emission spectrometry integrated with flow injection photochemical vapor generation and point discharge. Analytical Chemistry, 2015, 87(21): 10712-10718.

[96] Mo J M, Li Q, Guo X H, Zhang G X, Wang Z. Flow injection photochemical vapor generation coupled with miniaturized solution-cathode glow discharge atomic emission spectrometry for determination and speciation analysis of mercury. Analytical Chemistry, 2017, 89(19): 10353-10360.

[97] Zhang S, Tian Y F, Yin H L, Su Y B, Wu L, Hou X D, Zheng C B. Continuous and inexpensive monitoring of nonpurgeable organic carbon by coupling high-efficiency photo-oxidation vapor generation with miniaturized point-discharge optical emission spectrometry. Environmental Science & Technology, 2017, 51(16): 9109-9117.

[98] Yang R, Lin Y, Liu B Y, Su Y B, Tian Y F, Hou X D, Zheng C B. Simple universal strategy for quantification of carboxyl groups on carbon nanomaterials: Carbon dioxide vapor generation coupled to microplasma for optical emission spectrometric detection. Analytical Chemistry, 2020, 92(5): 3528-3534.

[99] Tan Z Q, Wang B W, Yin Y G, Liu Q, Li X, Liu J F. In situ tracking photodegradation of trace graphene oxide by the online coupling of photoinduced chemical vapor generation with a point discharge optical emission spectrometer. Analytical Chemistry, 2020, 92(1): 1549-1556.

[100] Jiang X M, Chen Y, Zheng C B, Hou X D. Electrothermal vaporization for universal liquid sample introduction to dielectric barrier discharge microplasma for portable atomic emission spectrometry. Analytical Chemistry, 2014,

86(11): 5220-5224.

[101] Deng Q S, Yang C, Zheng H T, Liu J X, Mao X F, Hu S H, Zhu Z L. Direct determination of cadmium in rice by solid sampling electrothermal vaporization atmospheric pressure glow discharge atomic emission spectrometry using a tungsten coil trap. Journal of Analytical Atomic Spectrometry, 2019, 34(9): 1786-1793.

[102] Kroschk M, Usala J, Addesso T, Gamez G. Glow discharge optical emission spectrometry elemental mapping with restrictive anode array masks. Journal of Analytical Atomic Spectrometry, 2016, 31(1): 163-170.

[103] Vega C G D, Alberts D, Chawla V, Mohanty G, Utke I, Michler J, Pereiro R, Bordel N, Gamez G. Use of radiofrequency power to enable glow discharge optical emission spectroscopy ultrafast elemental mapping of combinatorial libraries with nonconductive components: Nitrogen-based materials. Analytical and Bioanalytical Chemistry, 2014, 406: 7533-7538.

[104] Kodalle T, Greiner D, Brackmann V, Prietze K, Scheu A, Bertram T, Reyes-Figueroa P, Unold T, Abou-Ras D, Schlatmann R, Kaufmann C A, Hoffmann V. Glow discharge optical emission spectrometry for quantitative depth profiling of CIGS thin-films. Journal of Analytical Atomic Spectrometry, 2019, 34(6): 1233-1241.

[105] Lobo L, Fernández B, Aranaz M, Lorenzo A F, Martín-Carbajo J I, Pereiro R. Pulsed radiofrequency glow discharge time-of-flight mass spectrometry: Depth profile analysis of multilayers on conductive and non-conductive substrates. Spectrochimica Acta Part B: Atomic Spectroscopy, 2020, 168: 105865.

[106] Marcus R K, Paing H W, Zhang L X. Conceptual demonstration of ambient desorption-optical emission spectroscopy using a liquid sampling-atmospheric pressure glow discharge microplasma source. Analytical Chemistry, 2016, 88(11): 5579-5584.

[107] Tian Y F, Wu X, Jiang X M, Hou X D. Surface morphological and elemental imaging by low power atmospheric pulsed spark discharge optical emission spectrometry. Microchemical Journal, 2013, 110: 140-145.

[108] Zheng L, Kulkarni P. Rapid elemental analysis of aerosols using atmospheric glow discharge optical emission spectroscopy. Analytical Chemistry, 2017, 89(12): 6551-6558.

[109] Zheng L, Kulkarni P, Diwakar P. Spatial and temporal dynamics of a pulsed spark microplasma used for aerosol analysis. Spectrochimica Acta Part B: Atomic Spectroscopy, 2018, 144: 55-62.

[110] Wei G Y, Liu X G, Lu Y X, Wang Z, Liu S, Ye G, Chen J. Microplasma anode meeting molten salt electrochemistry: Charge transfer and atomic emission spectral analysis. Analytical Chemistry, 2018, 90(22): 13163-13166.

[111] Ge F, Gao L, Peng X X, Li Q, Zhu Y F, Yu J, Wang Z. Atmospheric pressure glow discharge optical emission spectrometry coupled with laser ablation for direct solid quantitative determination of Zn, Pb, and Cd in soils. Talanta, 2020, 218: 121119.

[112] Kratzer J, Bousek J, Sturgeon R E, Mester Z, Dědina J. Determination of bismuth by dielectric barrier discharge atomic absorption spectrometry coupled with hydride generation: Method optimization and evaluation of analytical performance. Analytical Chemistry, 2014, 86(19): 9620-9625.

[113] Novak P, Dedina J, Kratzer J. Preconcentration and atomization of arsane in a dielectric barrier discharge with detection by atomic absorption spectrometry. Analytical Chemistry, 2016, 88(11): 6064-6070.

[114] Zhu Z L, Zhang S C, Lv Y, Zhang X R. Atomization of hydride with a low-temperature, atmospheric pressure dielectric barrier discharge and its application to arsenic speciation with atomic absorption spectrometry. Analytical Chemistry, 2006, 78(3): 865-872.

[115] Zhu Z L, Zhang S C, Xue J H, Zhang X R. Application of atmospheric pressure dielectric barrier discharge plasma for the determination of Se, Sb and Sn with atomic absorption spectrometry. Spectrochimica Acta Part B: Atomic Spectroscopy, 2006, 61(8): 916-921.

[116] Zhu Z L, Liu J X, Zhang S C, Na X, Zhang X R. Evaluation of a hydride generation-atomic fluorescence system for the determination of arsenic using a dielectric barrier discharge atomizer. Analytica Chimica Acta, 2008, 607(2): 136-141.

[117] Xing Z, Kuermaiti B, Wang J, Han G J, Zhang S C, Zhang X R. Simultaneous determination of arsenic and antimony by hydride generation atomic fluorescence spectrometry with dielectric barrier discharge atomizer. Spectrochimica Acta Part B: Atomic Spectroscopy, 2010, 65(12): 1056-1060.

[118] Xing Z, Wang J, Zhang S C, Zhang X R. Determination of bismuth in solid samples by hydride generation atomic fluorescence spectrometry with a dielectric barrier discharge atomizer. Talanta, 2009, 80(1): 139-142.

[119] Yu Y L, Du Z, Chen M L, Wang J H. A miniature lab-on-valve atomic fluorescence spectrometer integrating a dielectric barrier discharge atomizer demonstrated for arsenic analysis. Journal of Analytical Atomic Spectrometry, 2008, 23(4): 493-499.

[120] Liu Z F, Zhu Z L, Zheng H T, Hu S H. Plasma jet desorption atomization-atomic fluorescence spectrometry and its application to mercury speciation by coupling with thin layer chromatography. Analytical Chemistry, 2012, 84(23): 10170-10174.

[121] Duben O, Boušek J, Dědina J, Kratzer J. Dielectric barrier discharge plasma atomizer for hydride generation atomic absorption spectrometry: Performance evaluation for selenium. Spectrochimica Acta Part B: Atomic Spectroscopy, 2015, 111: 57-63.

[122] Yu Y L, Gao F, Chen M L, Wang J H. A miniaturized long-optical path atomic absorption spectrometer with dielectric barrier discharge as atomizer for mercury and methylmercury. Acta Chimica Sinica, 2013, 71(08): 1121-1124.

[123] Navarre E C, Goldberg J M. Design and characterization of a theta-pinch imploding thin film plasma source for atomic emission spectrochemical analysis. Applied Spectroscopy, 2011, 65(1): 26-35.

[124] Kratzer J, Zelina O, Svoboda M, Sturgeon R E, Mester Z, Dedina J. Atomization of bismuthane in a dielectric barrier discharge: A mechanistic study. Analytical Chemistry, 2016, 88(3): 1804-1811.

[125] Mao X F, Qi Y H, Huang J W, Liu J X, Chen G Y, Na X, Wang M, Qian Y Z. Ambient-temperature trap/release of arsenic by dielectric barrier discharge and its application to ultratrace arsenic determination in surface water followed by atomic fluorescence spectrometry. Analytical Chemistry, 2016, 88(7): 4147-4152.

[126] Kratzer J, Musil S, Marschner K, Svoboda M, Matousek T, Mester Z, Sturgeon R E, Dedina J. Behavior of selenium hydride in heated quartz tube and dielectric barrier discharge atomizers. Analytica Chimica Acta, 2018, 1028: 11-21.

[127] Zurynkova P, Dedina J, Kratzer J. Trace determination of antimony by hydride generation atomic absorption spectrometry with analyte preconcentration/atomization in a dielectric barrier discharge atomizer. Analytica Chimica Acta, 2018, 1010: 11-19.

[128] Liu M T, Liu J X, Mao X F, Na X, Ding L, Qian Y Z. High sensitivity analysis of selenium by ultraviolet vapor generation combined with microplasma gas phase enrichment and the mechanism study. Analytical Chemistry, 2020, 92(10): 7257-7264.

[129] Martínez-Jarquín S, Winkler R. Low-temperature plasma (LTP) jets for mass spectrometry (MS): Ion processes, instrumental set-ups, and application examples. TrAC: Trends in Analytical Chemistry, 2017, 89: 133-145.

[130] Chen J, Tang F, Guo C A, Zhang S C, Zhang X R. Plasma-based ambient mass spectrometry: A step forward to practical applications. Analytical Methods, 2017, 9(34): 4908-4923.

[131] Na N, Zhao M X, Zhang S C, Yang C D, Zhang X R. Development of a dielectric barrier discharge ion source for ambient mass spectrometry. Journal of the American Society for Mass Spectrometry, 2007, 18(10): 1859-1862.

[132] Schwartz A J, Williams K L, Hieftje G M, Shelley J T. Atmospheric-pressure solution-cathode glow discharge: A

versatile ion source for atomic and molecular mass spectrometry. Analytica Chimica Acta, 2017, 950: 119-128.

[133] Li W F, Yin Z B, Cheng X L, Huang W, Li J F, Huang B F. Pulsed microdischarge with inductively coupled plasma mass spectrometry for elemental analysis on solid metal samples. Analytical Chemistry, 2015, 87(9): 4871-4878.

[134] Hagenhoff S, Franzke J, Hayen H. Determination of peroxide explosive TATP and related compounds by dielectric barrier discharge ionization-mass spectrometry (DBDI-MS). Analytical Chemistry, 2017, 89(7): 4210-4215.

[135] Mirabelli M F, Wolf J C, Zenobi R. Atmospheric pressure soft ionization for gas chromatography with dielectric barrier discharge ionization-mass spectrometry (GC-DBDI-MS). Analyst, 2017, 142(11): 1909-1915.

[136] Mirabelli M F, Gionfriddo E, Pawliszyn J, Zenobi R. A quantitative approach for pesticide analysis in grape juice by direct interfacing of a matrix compatible SPME phase to dielectric barrier discharge ionization-mass spectrometry. Analyst, 2018, 143(4): 891-899.

[137] Zheng K Y, Dolan M J, Haferl P J, Badiei H, Jorabchi K. Atmospheric-pressure dielectric barrier discharge as an elemental ion source for gas chromatographic analysis of organochlorines. Analytical Chemistry, 2018, 90(3): 2148-2154.

[138] Huba A K, Mirabelli M F, Zenobi R. High-throughput screening of PAHs and polar trace contaminants in water matrices by direct solid-phase microextraction coupled to a dielectric barrier discharge ionization source. Analytica Chimica Acta, 2018, 1030: 125-132.

[139] Dong J L, Qian R, Zhuo S J, Yu P F, Chen Q, Li Z Q. Development and application of a porous cage carrier method for detecting trace elements in soils by direct current glow discharge mass spectrometry. Journal of Analytical Atomic Spectrometry, 2019, 34(11): 2244-2251.

[140] Ma X X, Ouyang Z. Ambient ionization and miniature mass spectrometry system for chemical and biological analysis. TrAC: Trends in Analytical Chemistry, 2016, 85: 10-19.

[141] Harper J D, Charipar N A, Mulligan C C, Zhang X R, Cooks R G, Ouyang Z. Low-temperature plasma probe for ambient desorption ionization. Analytical Chemistry, 2008, 80(23): 9097-9104.

[142] Xing Z, Wang J, Han G J, Kuermaiti B, Zhang S C, Zhang X R. Depth profiling of nanometer coatings by low temperature plasma probe combined with inductively coupled plasma mass spectrometry. Analytical Chemistry, 2010, 82(13): 5872-5877.

[143] Liu Y Y, Ma X X, Lin Z Q, He M J, Han G J, Yang C D, Xing Z, Zhang S C, Zhang X R. Imaging mass spectrometry with a low-temperature plasma probe for the analysis of works of art. Angewandte Chemie-International Edition, 2010, 49(26): 4435-4437.

[144] Cheng X L, Li W F, Hang W, Huang B L. Depth profiling of nanometer thin layers by pulsed micro-discharge with inductively coupled plasma mass spectrometry. Spectrochimica Acta Part B: Atomic Spectroscopy, 2015, 111: 52-56.

[145] Hang L, Xu Z Y, Yin Z B, Hang W. Approaching standardless quantitative elemental analysis of solids: Microsecond pulsed glow discharge and buffer-gas-assisted laser ionization time-of-flight mass spectrometry. Analytical Chemistry, 2018, 90(22): 13222-13228.

[146] Ma X X, Zhang S C, Lin Z Q, Liu Y Y, Xing Z, Yang C D, Zhang X R. Real-time monitoring of chemical reactions by mass spectrometry utilizing a low-temperature plasma probe. Analyst, 2009, 134(9): 1863-1867.

[147] Hoegg E D, Patel B A, Napoli W N, Richardson D D, Marcus R K. Proof-of-concept: Interfacing the liquid sampling-atmospheric pressure glow discharge ion source with a miniature quadrupole mass spectrometer towards trace metal analysis in cell culture media. Journal of Analytical Atomic Spectrometry, 2018, 33: 2015-2020.

[148] Williams T J, Marcus R K. Coupling of the liquid sampling-atmospheric pressure glow discharge (LS-APGD)

ionization source with a commercial triple-quadrupole mass spectrometer. Journal of Analytical Atomic Spectrometry, 2019, 34(7): 1468-1477.

[149] Hoegg E D, Godin S, Szpunar J, Lobinski R, Koppenaal D W, Marcus R K. Coupling of an atmospheric pressure microplasma ionization source with an Orbitrap Fusion Lumos Tribrid 1M mass analyzer for ultra-high resolution isotopic analysis of uranium. Journal of Analytical Atomic Spectrometry, 2019, 34(7): 1387-1395.

[150] Liu X, Liu Z F, Zhu Z L, He D, Yao S Q, Zheng H T, Hu S H. Generation of volatile cadmium and zinc species based on solution anode glow discharge induced plasma electrochemical processes. Analytical Chemistry, 2017, 89(6): 3739-3746.

[151] Zhu Z L, Chan G C, Ray S J, Zhang X R, Hieftje G M. Use of a solution cathode glow discharge for cold vapor generation of mercury with determination by ICP-atomic emission spectrometry. Analytical Chemistry, 2008, 80(18): 7043-7050.

[152] Yang M, Xue J, Li M, Han G J, Xing Z, Zhang S C, Zhang X R. Low temperature hydrogen plasma assisted chemical vapor generation for atomic fluorescence spectrometry. Talanta, 2014, 126: 1-7.

[153] Liu X, Zhu Z L, Bao Z Y, He D, Zheng H T, Liu Z F, Hu S H. Simultaneous sensitive determination of selenium, silver, antimony, lead, and bismuth in microsamples based on liquid spray dielectric barrier discharge plasma-induced vapor generation. Analytical Chemistry, 2019, 91(1): 928-934.

[154] Liu X, Zhu Z L, Li H L, He D, Li Y T, Zheng H T, Gan Y Q, Li Y X, Belshaw, N S, Hu S H. Liquid spray dielectric barrier discharge induced plasma-chemical vapor generation for the determination of lead by ICPMS. Analytical Chemistry, 2017, 89(12): 6827-6833.

[155] Zhu Z L, Liu L, Li Y X, Peng H, Liu Z F, Guo W, Hu S H. Cold vapor generation of Zn based on dielectric barrier discharge induced plasma chemical process for the determination of water samples by atomic fluorescence spectrometry. Analytical and Bioanalytical Chemistry, 2014, 406(29): 7523-7531.

[156] Zhu Z L, He Q, Shuai Q, Zheng H T, Hu S H. Solution cathode glow discharge induced vapor generation of iodine for determination by inductively coupled plasma optical emission spectrometry. Journal of Analytical Atomic Spectrometry, 2010, 25(9): 1390-1394.

[157] Zhu Z L, Huang C Y, He Q, Xiao Q, Liu Z F, Zhang S C, Hu S H. On line vapor generation of osmium based on solution cathode glow discharge for the determination by ICP-OES. Talanta, 2013, 106: 133-136.

[158] Wu X, Yang W L, Liu M G, Hou X D, Zheng C B. Vapor generation in dielectric barrier discharge for sensitive detection of mercury by inductively coupled plasma optical emission spectrometry. Journal of Analytical Atomic Spectrometry, 2011, 26(6): 1204-1209.

[159] Wu Q J, Zhu Z L, Liu Z F, Zheng H T, Hu S H, Li L B. Dielectric barrier discharge-plasma induced vaporization for the determination of thiomersal in vaccines by atomic fluorescence spectrometry. Journal of Analytical Atomic Spectrometry, 2012, 27(3): 496-500.

[160] Zhu Z L, Wu Q J, Liu Z F, Liu L, Zheng H T, Hu S H. Dielectric barrier discharge for high efficiency plasma-chemical vapor generation of cadmium. Analytical Chemistry, 2013, 85(8): 4150-4156.

[161] Li Z A, Tan Q, Hou X D, Xu K L, Zheng C B. Single drop solution electrode glow discharge for plasma assisted-chemical vapor generation: Sensitive detection of zinc and cadmium in limited amounts of samples. Analytical Chemistry, 2014, 86(24): 12093-12099.

[162] Chen Q, Lin Y, Tian Y F, Wu L, Yang L, Hou X D, Zheng C B. Single-drop solution electrode discharge-induced cold vapor generation coupling to matrix solid-phase dispersion: A robust approach for sensitive quantification of total

mercury distribution in fish. Analytical Chemistry, 2017, 89(3): 2093-2100.

[163] Swiderski K, Welna M, Greda K, Pohl P, Jamroz P. Hanging drop cathode-atmospheric pressure glow discharge as a new method of sample introduction for inductively coupled plasma-optical emission spectrometry. Analytical and Bioanalytical Chemistry, 2020, 412(18): 4211-4219.

[164] Xia S A, Leng A Q, Lin Y, Wu L, Tian Y F, Hou X D, Zheng C B. Integration of flow injection capillary liquid electrode discharge optical emission spectrometry and microplasma-induced vapor generation: A system for detection of ultratrace Hg and Cd in a single drop of human whole blood. Analytical Chemistry, 2019, 91(4): 2701-2709.

[165] He Q, Zhu Z L, Hu S H, Jin L L. Solution cathode glow discharge induced vapor generation of mercury and its application to mercury speciation by high performance liquid chromatography-atomic fluorescence spectrometry. Journal of Chromatography A, 2011, 1218(28): 4462-4467.

[166] Li Y X, Zhu Z L, Zheng H T, Jin L L, Hu S H. Significant signal enhancement of dielectric barrier discharge plasma induced vapor generation by using non-ionic surfactants for determination of mercury and cadmium by atomic fluorescence spectrometry. Journal of Analytical Atomic Spectrometry, 2016, 31(2): 383-389.

[167] Leng A Q, Tian Y F, Wang M X, Wu L, Xu K L, Hou X D, Zheng C B. A sensitive and compact mercury analyzer by integrating dielectric barrier discharge induced cold vapor generation and optical emission spectrometry. Chinese Chemical Letters, 2017, 28(2): 189-196.

[168] Liu Z F, Xing Z, Li Z Y, Zhu Z L, Ke Y Q, Jin L L, Hu S H. The online coupling of high performance liquid chromatography with atomic fluorescence spectrometry based on dielectric barrier discharge induced chemical vapor generation for the speciation of mercury. Journal of Analytical Atomic Spectrometry, 2017, 32(3): 678-685.

[169] Liu X, Zhu Z L, Bao Z Y, Zheng H T, Hu S H. Determination of trace cadmium in rice by liquid spray dielectric barrier discharge induced plasma-chemical vapor generation coupled with atomic fluorescence spectrometry. Spectrochimica Acta Part B: Atomic Spectroscopy, 2018, 141: 15-21.

[170] He Q, Wang X X, He H J, Zhang J. A feasibility study of rare-earth element vapor generation by nebulized film dielectric barrier discharge and its application in environmental sample determination. Analytical Chemistry, 2020, 92(3): 2535-2542.

[171] Yu H M, Hu J, Jiang X M, Hou X D, Tian Y F. Point discharge microplasma reactor for high efficiency conversion of H_2S to SO_2 for speciation analysis of sulfide and sulfite using molecular fluorescence spectrometry. Analytica Chimica Acta, 2018, 1042: 79-85.

[172] Liu T P, Liu M T, Liu J X, Mao X F, Zhang S S, Shao Y B, Na X, Chen G Y, Qian Y Z. On-line microplasma decomposition of gaseous phase interference for solid sampling mercury analysis in aquatic food samples. Analytica Chimica Acta, 2020, 1121: 42-49.

[173] Luo Y J, Yang Y, Lin Y, Tian Y F, Wu L, Yang L, Hou X D, Zheng C B. Low-Temperature and atmospheric pressure sample digestion using dielectric barrier discharge. Analytical Chemistry, 2018, 90(3): 1547-1553.

第10章 单纳米颗粒 ICP-MS 分析技术

(胡建煜[①],刘 睿[①],吕 弋[①*],张新荣[②])

▶ 10.1 单纳米颗粒与 ICP-MS 分析 / 252

▶ 10.2 单纳米颗粒 ICP-MS 在环境分析中的应用 / 255

▶ 10.3 单纳米颗粒 ICP-MS 在生物分析中的应用 / 257

▶ 10.4 单纳米颗粒 ICP-MS 分析技术的前景与展望 / 261

▶ 10.5 总结 / 264

①四川大学分析测试中心,成都,②清华大学分析中心,北京。*通迅作者联系方式: lvy@scu.edu.cn

本章导读

- ICP-MS 成为对单个纳米颗粒进行定量检测和理化性质分析的重要方法。
- 在环境分析领域，痕量纳米颗粒在水体和生物体中的不同形态迁移转换可以被单纳米颗粒 ICP-MS 表征和示踪。
- 在生物分析中，基于单纳米粒子标记的均相、异相和多组分分析取得重要进展，为临床检测提供新思路。
- 单纳米颗粒 ICP-MS 分析技术方兴未艾，从单细胞到单分子分析，强大的技术手段如何助力解决前沿分析领域的问题与挑战。

纳米颗粒广泛参与到环境的迁移转化、生物体的代谢过程中，对痕量纳米颗粒的检测和表征无论是对于环境化学还是生物医学都有重要的意义。此外，含金属元素的纳米颗粒含有大量的金属同位素，而电感耦合等离子质谱仪（ICP-MS）是一种对痕量金属元素最有效、最强大的检测技术之一，两者结合构建的单纳米颗粒 ICP-MS 分析技术可以实现对单个纳米颗粒的表征和高灵敏检测[1]。本章简述了单纳米颗粒 ICP-MS 分析技术在纳米颗粒分析方面的发展，以及近年来其在环境分析和生物分析方面的应用，并对单纳米颗粒 ICP-MS 分析技术的现状和发展方向进行了讨论。

10.1　单纳米颗粒与 ICP-MS 分析

在痕量金属元素检测领域，电感耦合等离子质谱仪（ICP-MS）是最为强大的技术手段之一。ICP 所提供的强电离源可以提供平均 6000～8000 K 的高温，在此温度下，含有金属元素的样品可以被迅速地气化，样品中的金属同位素也迅速地失去电子、被离子化，再进入质量分析器中进行检测。大多数金属稳定同位素在 ICP-MS 中都有非常高的灵敏度，检出限低于 pg/mL 水平，同时拥有 9 个数量级的线性范围和极低的基质效应。这些优势促成了一系列基于金属元素标记、利用对样品中的金属同位素进行超痕量定量分析的 ICP-MS 分析方法的建立。其中，基于稀土元素的大环化合物标记[2,3]、多聚稀土元素大环高分子化合物标记[4]和金属纳米粒子标记[5,6]成为研究的热门，纳米粒子中所内含的大量金属同位素为提高分析方法灵敏度提供了有力的支持。

近年来，随着工程纳米材料的大规模应用和纳米粒子标记在生物分析中崭露头角，对单个纳米颗粒的分析和表征面临新的挑战和要求。尽管纳米材料在环境和生物分析中扮演了越来越重要的角色，但由于可靠的检测手段的匮乏使得对于纳米材料在人类健康和环境生态中的监测和分析，复杂基质中单个纳米颗粒的迁移转化，高灵敏检测依然面临巨大的挑战。

与常规的分析物相比，纳米材料的检测不仅需要得到其浓度和物质的量信息，还需要通过可靠的分析手段对其尺寸大小、理化性质进行检测。而这样的纳米材料被广泛应用于工程纳米材料、环境中的纳米结构分析、纳米生物材料标记分析等课题中，其形态和元素构成的多样性、环境和基质的复杂性，以及痕量样品对检出限和线性范围提出的要求都增添了单纳米颗粒分析的难度。

基于这样的背景，单纳米颗粒 ICP-MS 分析技术引起了研究人员的兴趣。单纳米颗粒 ICP-MS 分析的概念最初由 Degueldre 等提出，证明了单颗粒 ICP-MS 可以用于分析胶体和微粒悬浮液[7]。顾名思义，这一分析技术的特点在于实现了每一次数据采集的过程只对单个纳米颗粒进行测量和分析。

在单粒子模式下,一个单个纳米颗粒被引入 ICP,同样的,它也会被快速气化和离子化,形成等离子体中的一团气态离子云——这一个单纳米颗粒所形成的离子云可以被视作一次信号脉冲。为了满足单粒子分析的要求,充分稀释的纳米颗粒悬浮液和适当的数据采集频率都是保证单颗粒纳米粒子分析的前提条件。当纳米粒子通过常规雾化过程进入等离子体时,其流通量需要足够低才可以获得一个单独的纳米颗粒的信号,如图 10-1 所示,不同数据采集频率下,每一个驻留时间中脉冲信号和纳米颗粒数量的对应关系也是不同的。通过快速的数据采集,[>10^4 Hz,驻留时间<100 μs,图 10-1(a)],每一个颗粒所产生的瞬态信号都可以被准确地记录。而当采集频率较低[驻留时间在 ms 级别,图 10-1(b)、(c)]时,几个纳米颗粒合并的信号则会被仪器误认为是一次脉冲信号。

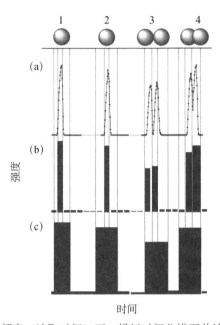

图 10-1　在不同的数据采集频率(读取时间)下,模拟时间分辨下单纳米粒子 ICP-MS 的检测信号
(a) 10 000 Hz (100 μs);(b) 1000 Hz (1 ms);(c) 100 Hz (10 ms)

通过快速的数据采集就可以记录单个纳米粒子产生的瞬态信号,包括平均尺寸、粒度分布、纳米颗粒数量浓度和质量浓度、团聚程度和化学成分都可以被单纳米颗粒 ICP-MS 分析技术准确测量。某种程度上来说,对其他常见纳米粒子表征手段(动态光散射、透射电子显微镜等)形成了挑战。

在单纳米颗粒 ICP-MS 分析中,纳米颗粒的尺寸检出限是量化其分析性能的关键参数之一。西班牙萨拉戈萨大学环境科学研究所(IUCA)的 Laborda 教授团队提出了 ICP-MS 对单纳米颗粒尺寸检出限的计算公式[公式(10-1)][8]。其中,K_{ICPMS} 代表 ICP-MS 的检测效率,由引入 ICP 的原子和实际检测到的原子相比而得,被电离过程、雾化和进样器效率以及通过质谱仪器传输效果影响;K_M($=AN_A/M_w$)表示待测元素同位素的贡献度(A 为待测元素同位素丰度,N_A 为阿伏伽德罗常数,M_w 为所测元素分子量);X_{NP} 表示待测元素在纳米粒子中的质量分数(单元素金属纳米粒子的 X_{NP} 即为 1);ρ 为纳米颗粒的密度。

$$\text{LOD}_{size}=\left(\frac{18\sigma_B}{\pi\rho X_{NP} K_{ICPMS} K_M}\right)^{1/3} \tag{10-1}$$

因此,在单纳米颗粒 ICP-MS 检测中,为了尽可能减小尺寸检出限、提高分析性能,在其他参数都是常数的前提下,以下两个因素的提高就成了关键:

（1）每个纳米粒子中所包含的同位素尽可能多，形成足够大离子团。如图10-1（a）所示，产生足够大的脉冲信号，被检测器识别。

（2）提高ICP-MS传输的效率。与传统模式相比，在传统雾化器中，单颗粒模式的雾化效率较普通模式更低（一般低5%左右）。

在保证样品悬浮液中单颗粒进样的浓度前提下，对纳米粒子的检测与其进样后产生的可以与基线区分开的脉冲信号密切相关。如图10-1（a）所示，可以与背景区分开的最小脉冲信号强度决定了最小可以检测多大尺寸的纳米粒子，对尺寸检出限的计算产生了很大的影响。公式（10-1）也表明，ICP-MS的检测效率对于提高尺寸检出限也至关重要。对ICP进样系统进行改良，提高转换效率，使更多纳米粒子进入等离子体被电离形成脉冲信号也是提高检出限的有效方式——这对于电离电势较低的元素的尺寸检出限提高非常重要。废液收集法相比于粒子频率和尺寸法来说，可以更准确地测得ICP-MS在单纳米颗粒分析中的转换效率，后两者的测量结果往往会偏高[9]。如表10-1所示，一般的，对于单纳米颗粒ICP-MS可以检测到20 nm以上的金属纳米粒子。另外，雾化室对大颗粒的去除效果和在等离子体中的不完全雾化效果也决定了单纳米颗粒ICP-MS能对纳米粒子进行检测的尺寸上限[10]。由于较小颗粒纳米粒子的聚集，一般会要求将最大可检测的尺寸限制在1～5 μm。图10-2总结了一系列常见元素纳米粒子在ICP-MS中的尺寸检出限[11]。

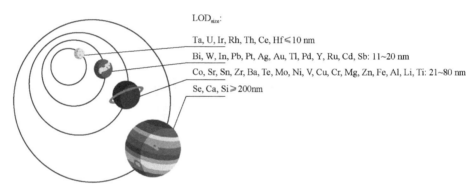

图10-2 常见元素的纳米粒子在单纳米颗粒ICP-MS中的尺寸检出限（LOD_{size}）

对于在等离子体中容易形成多原子分子的元素（如Si、Ti、Fe等），或以溶解的形式存在的元素，无论是作为样品的一部分还是污染物存在都会影响基线的稳定和强度，对最低可分辨的信号及尺寸检出限产生负面影响。因此对于这样的样品分析，提升尺寸检出限就需要进一步提高驻留时间。

单纳米颗粒ICP-MS分析技术在诸多分析领域都取得了可喜的应用进展，表10-1总结了近年来一系列代表性的利用单纳米颗粒ICP-MS对纳米粒子进行检测的报道。在环境分析中，金属纳米颗粒的形成和富集常常伴随着有害的环境问题，单纳米颗粒ICP-MS可以对环境中含量极低的纳米粒子进行表征和分析，研究环境中元素的迁移转换规律；而在生物分析中，生物体本征存在有纳米粒子参与的生命过程，单纳米颗粒ICP-MS可以对这些过程进行高灵敏准确定量追踪，此外，利用纳米颗粒的标记还可以实现基于ICP-MS的均相分析和多组分生物分析，具有很大的发展空间。

表10-1 近年来一些典型的单纳米颗粒ICP-MS检测报道

ICP-MS类型	纳米粒子	LOD	纳米粒子直径	参考文献
ICP-MS 单颗粒模式	单元素和双元素（核壳）纳米颗粒	—	Au 60 nm，Ag 60 nm和AuAg 60 nm	[12]
	稀土氧化物	—	平均（35±3）nm	[13]
	环境水样中的金银纳米粒子	$8×10^4$ particles/L	分别20 nm和19 nm	[14]

续表

ICP-MS 类型	纳米粒子	LOD	纳米粒子直径	参考文献
ICP-MS 单颗粒模式	环境水样中的金纳米粒子	2.2 ng/L	—	[15]
	湖水中的银纳米粒子	—	（33.8±0.4）nm～（46.6±0.4）nm	[16]
	银纳米粒子	0.09 µg/L	18～55 nm	[17]
	环境水样中的硫化银纳米粒子	0.068 µg/L	（4.3±0.9）nm～（17.5±3.0）nm	[18]
	植物摄取的铂纳米粒子	0.002 mg/g	平均 67.1 nm	[19]
	被拟南芥吸收的银纳米粒子	10 nm	吸收前 12.84 nm/吸收后 20.70 nm	[20]
	银纳米粒子	30 nm	30 nm、40 nm、50 nm 和 80 nm	[21]
	TiO_2 纳米粒子	10 ng/L	21 nm 和 50 nm	[22]
	银纳米粒子	0.3 ng/L		[23]
	血浆中的银纳米粒子	13 ng/L	（9.2±3.2）nm～（99.0±5.1）nm	[24]
	饮用水处理中的氧化锌和二氧化铈纳米粒子	0.20 µg/L、0.10 µg/L	35～40 nm、18～20 nm	[25]
	镧系纳米探针	0.5 ng/mL	13.6～16.2 nm	[26]
	NIST 金纳米粒子	20 nm	30 nm 和 60 nm	[27]
	金纳米粒子	0.8～1.0 µg/L	5 nm 和 10 nm/20 nm 和 50 nm	[28]
	金、银、二氧化钛、二氧化硅纳米粒子	Au NPs：1 ng/L SiO_2：0.1 mg/L	20 nm（Au 和 Ag）、50（TiO_2）和 200 nm（SiO_2）	[29]
	镧系元素掺杂的超顺磁氧化铁（SPIO）纳米粒子	—	33.54 nm、33.47 nm、35.57 nm、34.84 nm	[30]
	金银复合的氧化石墨烯纳米片	—	13.3 nm 和 48.8 nm in	[31]
	工程纳米颗粒（ENPs）	0.27～0.55 µg/L	5 nm、20 nm、50 nm	[32]
LA-ICP-MS 激光烧蚀电感耦合等离子体质谱仪	金纳米粒子	10 ng/g	13 nm 和 50 nm	[33]
	水蚤和斑马鱼体内的纳米粒子（Al_2O_3、Ag 和 Au）	500～1000 fg Al、35.7 fg Ag、12.5 fg Au/50 µm	（35±11）nm～（503±367）nm	[34]
	SiO_2 纳米粒子	—	（31±5）nm、（57±20）nm	[35]
ICP-SF-MS 扇形磁场电感耦合等离子体质谱仪	氧化铁纳米粒子	2.4 µg/L	（24.7±1.5）nm	[36]
	分散在环境水样中的银纳米粒子	～10 nm & 2.7 ng/L	15～18 nm	[37]
Mass Cytometer 质谱流式细胞仪	单细胞水平 Au NPs	单细胞上 10 Au NPs	3 nm 核心尺寸	[38]
	单细胞水平 $NaHoF_4$ 纳米粒子	单细胞水平 20 个纳米颗粒	10.1～13.0 nm	[39]

10.2 单纳米颗粒 ICP-MS 在环境分析中的应用

纳米材料在现代工业中被广泛使用，在其生命周期中无可避免地被释放到环境中，引起了人们对环境中的纳米颗粒对生命健康的严重隐忧[40]。研究表明，环境中的纳米颗粒毒理性与其粒径大小密切相关。然而环境中的纳米颗粒含量少、基质复杂，因此建立一种高灵敏的准确定量方式就尤为重要。基于此，单纳米颗粒 ICP-MS 受到了研究人员的青睐，解决了一系列环境中的纳米颗粒（包括金纳米粒子，银纳米粒子，二氧化钛、二氧化硅和稀土元素纳米粒子等）的分析难题[41]。如前文所说，在 ICP 对纳米粒子的离子化过程中，每一个瞬态信号都在极短的停留时间内被测量，因此在这样毫秒级的积分时间内，每一个纳米粒子的噪声很低，由此带来了很高的信噪比优势。

为了研究环境中的纳米粒子在植物体中的代谢过程，法国生物无机与环境分析实验室（aLaboratoire de Chimie Analytique Bio-inorganique et Environnement，LCABIE）的 Jiménez-Lamana 课题组利用单纳米颗粒 ICP-MS 研究了植物对铂纳米粒子（PtNPs）的摄取和代谢过程[19]。通过单纳米颗粒 ICP-MS 提供的纳米颗粒尺寸、颗粒数量等信息，研究人员从植物组织中检出了 Pt NPs 的摄入和聚集，没有发现降解生成的 Pt 离子。

此外，单纳米颗粒 ICP-MS 还被广泛应用于对于实际水样中的低浓度的纳米粒子监测中，为环境水样中纳米颗粒的法规标准制定提供了可靠的数据。奥地利维也纳大学的 Hofmann 教授课题组将包括单纳米颗粒 ICP-MS、普通四极杆 ICP-MS 和 ICP-OES 多种检测手段结合，研究了防晒产品中释放的二氧化钛纳米材料对维也纳多瑙河古湖的释放[42]。他们在 12 个月持续的时间内从湖泊中收集了悬浮颗粒物（SPM），在距离游泳区域 50 m 的位置和 1 m 深的位置进行采样。虽然在透射电镜（TEM）中可以明显观察到纳米颗粒在防晒产品大量使用的夏天明显增多，但为了进一步区分防晒用品中添加的 TiO_2 和天然纳米颗粒，研究人员使用 ICP-MS 对人工合成的 TiO_2 纳米颗粒中其他元素（Al、V、Ga、Y、Nb、Eu、Ho、Er、Tm、Yb 和 Ta）的元素比例进行了检测，结合单纳米颗粒 ICP-MS 分析来确定局部背景值，用于表示特征性的 TiO_2 纳米颗粒成分。检测结果表明，与春季的背景值相比，夏季的 TiO_2 纳米颗粒有轻微增加，并在秋天回落，证明释放的纳米材料在水体中的停留时间很短。

环境中的 Ag NPs 由于其广泛的工业应用和显著的环境效应，近年来受到了研究人员的广泛关注。利用单纳米颗粒 ICP-MS 检测，一系列对 Ag NPs 在环境中的迁移转换的论文被报道。瑞典哥德堡大学的 Hasellöv 教授在 2012 年基于单纳米颗粒 ICP-MS 对 Ag NPs 的检测进行了优化，实现了对最小直径为 20 nm 的纳米颗粒进行检测[43]；美国科罗拉多矿业大学的 Higgins 教授跟踪了 60 nm 和 100 nm 两种工业上制备的银纳米颗粒在实验室和自然水体中的 Ag 离子释出过程，建立了动力学曲线[44]。实际样品中的基质往往更加复杂，而单纳米颗粒 ICP-MS 也可以从污水处理厂的废水中对纳米颗粒进行检测和表征。Mitrano 等从处理前后的污水样品中检测到了 ng/L 级别的溶解银和纳米银。Tuoriniemi 等在废水样品中检测到了含有银、铈和钛的纳米颗粒，其浓度在每毫升 2000～30 000 个，这也与欧洲的预期浓度相符合。

由于银纳米颗粒在工业生产中的大规模添加和使用，它们在生命周期中暴露和释放到环境中的可能也大大增加。Farkas 等对市售的银纳米粒子洗衣机的排出液进行了检测，通过普通模式下的 ICP-MS 发现其排出液中银的含量在 11 μg/L，通过单纳米颗粒 ICP-MS 检测和 TEM 的共同检测，发现废水中大多数银纳米颗粒都在 20 nm 以下；但是也存在少量较大的在 60～100 nm 的颗粒[45]。由于单纳米颗粒 ICP-MS 优秀的检出限，Coleman 等以无脊椎动物带丝蚓作为模型，研究了 Ag NPs 在生物体内的沉积、吸收和富集过程。在 48 小时的净化过程后，单纳米颗粒 ICP-MS 都可以在带丝蚓体内测到 70 nm 的 Ag NPs 沉积[46]。

在此基础上，为了解决复杂基质对单纳米颗粒 ICP-MS 检测的影响，监测自然界中 Ag NPs 和 Ag 离子的转换过程，中国科学院生态环境研究中心的江桂斌院士和刘景富研究员开发了基于中空纤维流场流分离技术和 ICP-MS 检测联用的分析方法[47]。中空纤维流场流分离技术可以实现对不同粒径的 Ag NPs 进行分离，微量浓缩柱实现对 Ag 离子的富集，两者与紫外/ICPMS 双检测器联用，可以实现在线对 Ag NPs 在环境中的整个降解转化过程进行检测和追踪。通过这一系统的准确表征，研究人员发现痕量银离子通过光还原形成小颗粒银纳米离子的过程和高浓度银离子形成大颗粒 Ag NPs 的过程具有显著差异，更好地诠释了 Ag NPs 在自然界的迁移转化过程。

单纳米颗粒 ICP-MS 由于其对痕量纳米粒子的高灵敏检测能力，在环境分析中的纳米颗粒检测中扮演了不可或缺的角色，为多种纳米粒子的检测提供了尺寸、浓度和成分信息。尽管已经有大量的论文报道，但环境样品的复杂基质，小尺寸的纳米颗粒、纳米团簇检测的难度等仍然对环境分析中单纳米颗粒 ICP-MS 的进一步应用提出了挑战。

10.3 单纳米颗粒 ICP-MS 在生物分析中的应用

生物体中的纳米粒子同样存在代谢的过程，其富集和消解都极大程度地影响了纳米粒子对生物体的毒理作用，并且浓度过高的富集会对组织器官产生不可逆的损伤。通过利用单纳米颗粒 ICP-MS 直接去除组织中的纳米颗粒进行检测，为纳米粒子在生物体中的摄入和转化过程研究打开了一扇新的大门。2013 年 Higgins 教授首次报道了一种将单颗粒 ICP-MS 与组织提取结合来对环境相关的生物组织中的工业金属纳米粒子（metallic engineering nanoparticles）摄入过程进行检测的研究。研究人员将生物组织暴露在纳米粒子中，利用四甲基氢氧化铵（TMAH）将生物组织进行提取，并从牛肉提取物中检测到了 100 nm 的金、纳米粒子和 60 nm 的银纳米粒子[48]。

进一步的，基于 ICP-MS 的生物分析在过去几年中也取得了巨大的成果，并成功应用到了对生物小分子[49]、蛋白质[50]、核酸[3]和细胞[51]的高灵敏、多组分绝对定量中。ICP-MS 本身对金属同位素标记的检测拥有极低的检出限（pg/mL），同时拥有 9 个数量级的线性范围和极低的基质效应，此外还有可以裨益多组分检测的高分辨率质谱线。在使用纳米颗粒取代金属同位素做标记之后，由于纳米颗粒内含大量金属原子，检测的灵敏度得到了进一步提升。传统的积分模式下，纳米颗粒标记需要先用酸消解，成为金属离子溶液后才可以进入 ICP-MS 进行检测，检出限通常停留在 ng/mL 级别。消解的步骤和检出限限制了这一类方法在生物分析中的应用，尤其在疾病的早期诊断课题上。

将纳米粒子作为生物分析中的标记，利用单纳米颗粒分析策略进行检测，则有可能建立起更灵敏的生物分析方法[52]。因此，本节重点关注了这一类利用单纳米颗粒 ICP-MS 检测、基于单独的纳米粒子标记计数的生物分析方法。

10.3.1 单纳米颗粒 ICP-MS 异相分析

近年来有一系列基于单纳米颗粒标记的 ICP-MS 的生物分析平台被开发和报道。通常生物分子之间的识别会发生在固液表面或溶液相中，因此根据生物识别反应的环境不同将生物分析分为均相和异相分析两类。异相分析中，游离的待测标记生物分子与固定的捕获物反应，再通过物理方式将标记纳米颗粒与固定相分离，利用 ICP-MS 对纳米标记进行检测。为了实现这样的物理分离，微量滴定板、磁性纳米颗粒、生物膜或组织切片常被用作候选的基质材料。将捕获基团（抗体、核酸等）通过物理吸附或化学键连接的方式固定在基质上，对待测生物分子进行捕获后再与带有纳米颗粒标记的探针进行特异性反应，探针与基质分离后即可以实现高灵敏检测。此类异相分析是最为常用的方案，在提供较低的检出限的同时可以最大限度发挥 ICP-MS 的优秀分析性能。

清华大学张新荣教授课题组于 2009 年首次发表了基于时间分辨（time-resolved analysis，TRA）ICP-MS 质谱检测的单纳米颗粒生物分析研究[53]。他们将纯化后的抗体过夜孵育标记在微量滴定板上，经过 PBS 洗涤和 BSA 封闭后进行竞争免疫反应（图 10-3 右）。标记在二抗上的金纳米颗粒可以通过 10 min 的超声处理从微量滴定板上分离——与酸解离相比，超声的方法避免了纳米颗粒聚集，拥有更好的分析性能。在单纳米颗粒 ICP-MS 中，利用金纳米颗粒的脉冲信号进行定量，α-甲胎蛋白（AFP）作为一种重要的肿瘤标志物，被选作示例待测物，在 TRA 模式下 20 nm、45 nm 和 80 nm 的金纳米颗粒瞬态信号频率和 AFP 浓度直接呈现出了很好的线性相关性，通过计算得出检出限为

0.016 μg/L。除了竞争免疫反应外，为了获得更高的灵敏度，该论文还对传统的夹心免疫法（sandwich-type bioassay）进行了研究。在对比了传统模式和单颗粒模式下 ICP-MS 对金纳米粒子标记的抗体的检测性能之后，该研究证明单颗粒模式下，以金纳米粒子标记的抗体检测 IgG 为例，1 min 的数据采集时间内单颗粒模式取得了 0.02 pg/mL 的检出限，与传统积分模式相比提高了约一个数量级。这证明利用单纳米颗粒 ICP-MS 可以建立更高灵敏度的生物分析方法。但是值得注意的是，单纳米颗粒的 TRA 模式检测的准确度和线性范围都比常规积分模式稍差。Au NPs 的聚集会影响检测的准确性，这也导致了频率计数的不确定性。

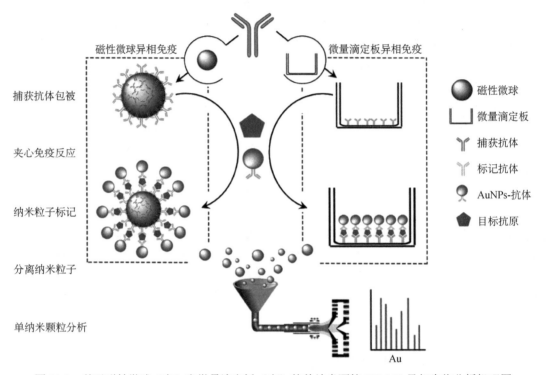

图 10-3　基于磁性微球（左）和微量滴定板（右）的单纳米颗粒 ICP-MS 异相生物分析机理图

除了使用微量滴定板作为基底之外，基于磁性微球的单纳米颗粒 ICP-MS 异相生物分析方法也被开发（图 10-3 左）[54]。该研究也使用了传统的夹心免疫结构，用 ZnSe 量子点（QDs）作为纳米标记，通过 EDC/NHS 的反应连接在二抗上。另外将待测癌坯抗原（CEA）对应的一抗包被在氨基修饰的磁性微球上，用于捕获待测 CEA。再使用 QDs 修饰的二抗靶标 CEA 形成夹心免疫结构。最后记录下单纳米颗粒 ICP-MS 中由标记在抗体上的纳米粒子的电离引起的瞬态信号，瞬时信号的频率与纳米标签的浓度直接相关，并且可以通过频率来对纳米标签抗体的浓度进行定量。在异相免疫反应后，CEA 的检测限为 0.006 ng/mL，该方法也应用于人血清样品检测中，获得了 98%～102%的回收率。

在单纳米颗粒异相分析方法中，由于 ICP-MS 是对纳米粒子直接进行分析，因此将标记的纳米粒子从基底上分离的步骤至关重要。常用的分离方法主要是超声法，将纳米粒子从基底上通过超声消解的方式剥离。但金属纳米粒子性质活泼，超声条件较为苛刻，易发生团聚和变性。因此，为了从原理上解决这一问题，一种利于光诱导剪切位点的分离方法被开发出来[75]。研究人员基于磁性微球基底，在磁性微球和捕获探针 DNA 序列中间直接修饰了光敏位点。在免疫反应结束后，将"磁性微球-探针-纳米粒子"复合物暴露在 325 nm 的紫外光下 30 min，光敏位点即发生断裂，而标记的纳米粒子脱离磁性微球，分散在上清液中。这样的分离方法避免了超声分离可能对纳米粒子的物理性质带来的破坏，实现了更准确的单纳米颗粒分析。该方法被应用于对大肠杆菌 O157 16S rRNA 的绝对定量中，获得了 10 fmol/L 的检出限。

10.3.2 单纳米颗粒 ICP-MS 均相分析

均相生物分析是一类利用在生物反应过程中，待测物与标记分子反应形成生物复合物时，标记试剂本身发生的物理或化学性质变化来对待测物进行检测的分析方法。由于这种性质的变化是生物反应发生时自发产生，因此与异相分析相比不需要额外的分离步骤。均相反应由于其操作简便，环境友好成为生物分析中不可或缺的形式。利用单纳米颗粒 ICP-MS 进行均相生物分析已经成为研究热点，贡献了一大批优秀的研究成果。

在核酸的高灵敏检测课题上，张新荣教授课题组开发出了一种基于单纳米颗粒标记的"一步法"均相 DNA 生物分析法[55]。将 DNA 探针 1 和探针 2 分别固定在柠檬酸盐保护的 Au NPs 上，在缓冲液中将 Au NPs-DNA 1 和 Au NPs-DNA 2 与目标 DNA 杂交。这时，由于杂交反应的发生，Au NPs-DNA 1 和 Au NPs-DNA 2 倾向于形成二聚体、三聚体或更大的聚集体。这一过程引起了 Au NPs 数减少以及纳米粒子大小的增加。通过单纳米颗粒 ICP-MS 对 Au NPs 的数量检测，可以对目标 DNA 进行定量；而单个 Au NPs 的信号强度增加也证明了在 DNA 杂交过程引发的 Au NPs 聚集现象。通过这样简单的"一步法"策略，可以实现对 DNA 的高灵敏检测，获得 1 pmol/L 的检出限，同时借助这一思路，也可以进一步推广设计一系列基于单纳米颗粒 ICP-MS 检测的均相生物分析方法。

吕弋课题组在这一思路上进一步延伸，开发一种基于 Au NPs 标记的双参数、自验证的均相免疫方法（图 10-4）[5]。免疫反应中，修饰有 Au NPs 的特异性癌胚抗原（CEA）抗体与含有 CEA 靶点的样品混合后，引发了 Au NPs 自聚集的发生。该研究利用单纳米颗粒 ICP-MS 实现了两种可以互相验证的双参数检测——通过 Au NPs 聚集后数量的减少引发的计数信号频率的减少，以及聚集后的单个颗粒粒径的增加引发的脉冲信号强度增加，两种参数共同定量，互相验证，在两种参数下都实现了 pmol/L 级别的检出限，实现了单纳米颗粒下的高灵敏准确定量。

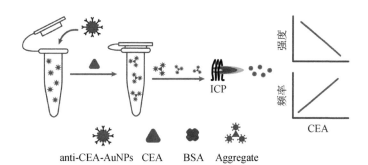

图 10-4 单纳米颗粒 ICP-MS 均相生物分析机理图：以自验证双参数 Au NPs 标记检测 CEA 为例

由于核酸的杂交反应在结构变化上提供了很多可能性，结合均相单纳米颗粒标记，也为核酸分析策略的设计提供了思路源泉，但传统利用 Au NPs 实现的单纳米颗粒 ICP-MS 均相生物分析一直被 Au NPs 的稳定性所掣肘，在盐溶液中极易发生聚集的特性导致这一方法很难推广开来。为了解决这一问题，湖南大学的蒋健晖教授课题组设计了一种基于链式杂交反应（HCR）的单纳米颗粒均相核酸分析方法[56]。研究人员将传统的 Au NPs 与连接 DNA（linker-DNA，L-DNA）通过巯基相连，设计了一种超稳定的球状核酸标记（USNA）。采用这一特殊设计的原因在于传统的 Au NPs 很难克服在 HCR 中所必需的高浓度盐环境，因此对 Au NPs 进行官能化成了必须解决的问题。该研究将巯基化的 L-DNA 连接在 Au NPs 上，提供高盐溶液中的稳定性。另外，将 L-DNA 经过杂交形成双链结构，双链的刚性将 DNA 推离 Au NPs 表面，从而降低了 Au NPs 表面的 DNA 密度，为 HCR 的高效率杂交提供了空间位点。在两个发卡 DNA H1、H2 的辅助下，目标 DNA 可以诱发 HCR 的发生，链式杂交的产物超分子通过每一个 H1 单元预留的互补段与 USNA 结合，最终形成一个粒径明显增大的纳米团，在 ICP-MS 的脉冲强度信号中表现出突然增高的脉冲。由于 HCR 本身具有信号放大的作用，这一方法可以获得 3 fmol/L 的检出限。由于 USNA 在常见的核酸缓冲液中都具有优秀的稳定性，因此极大地拓宽了单纳米粒子均相分析的应用范围。

10.3.3　单纳米颗粒 ICP-MS 多组分分析

在生物分析中，相关的生物标志物往往不止一种，因此多组分生物分子的同时分析就成了重要的研究方向。不同于传统的分析方法（荧光染料分子标记法、电化学标签标记法等）在多组分标记时带来光谱重叠干扰，ICP-MS 本身对超过 100 种金属稳定同位素都可以实现高分辨率高灵敏检测，因此利用不同元素的纳米颗粒实现多组分标记成为可能。加拿大多伦多大学 Tanner 教授课题组分别用 Au NPs 和 Eu 元素的大环化合物标记了 IgG 和 FLAG-BAP 两种蛋白，实现了元素大环化合物和纳米颗粒同时的多组分标记分析[50]；吕弋课题组也开发了利用 Au NPs 和 Ag NPs 两种纳米粒子分别标记 AFP 及其亚型 AFP-3 实现多组分检测的研究。这一系列多组分的纳米粒子标记策略都证明，利用 ICP-MS 优秀的同位素分辨率，结合纳米粒子信号放大的效果，可以实现对多个疾病标志物的同时检测，对分析的准确度大大提高[57]。

结合单纳米颗粒 ICP-MS 检测，多组分纳米粒子标记的生物分析变得更加容易实现。张新荣教授课题组的研究展示了一种使用金、银、铂三种纳米粒子对 DNA 进行标记，对三种与临床疾病相关的核酸标志物（HIV、HAV 和 HBV）进行多组分检测的方法[6]。氨基修饰的三种捕获探针通过羧氨反应连接在羧基修饰的 96 孔板上，捕获探针与待测 DNA 的部分互补；待测 DNA 的另一部分与三种分别修饰了 Au NPs、Ag NPs 和 Pt NPs 的探针 DNA 互补。通过这样一个"捕获 DNA-目标 DNA-探针 DNA-纳米粒子标记"的"夹心杂交"结构，在 2 s 的数据采集时间内，无需额外的放大和扩增过程就可以获得 1 pmol/L 的检出限，实现高灵敏度、高通量的多组分 DNA 检测。广西师范大学的邓必阳教授课题组则开发了一种使用金纳米颗粒（Au NPs）、ZnSe 量子点（QDs）和银纳米颗粒（Ag NPs）同时对细胞角蛋白片段抗原 21-1（CYFRA21-1）、癌胚抗原（CEA）和碳水化合物抗原（CA15-3）三种抗原进行单颗粒多组分异相免疫分析的方法[58]。

以上所提到的两篇报道都是基于异相单纳米颗粒生物分析设计的多组分检测，为了最大限度地发挥单纳米颗粒 ICP-MS 的优势，我们也可以将单组分均相免疫的机理推广开来，如图 10-5 所示，设计一种基于多种金属元素纳米粒子标记的多种疾病标志物同时检测的均相免疫分析方法。对于常见的疾病标志物来说，通常是多种标志物的含量变化共同指示疾病发展程度。因此多组分的检测相较于一种标志物定量分析来说，不仅对于微量样品的利用率更高，并且对疾病的诊断意义更大。利用单纳米颗粒 ICP-MS 同位素分辨率高的优势，我们可以合成多种纳米粒子，分别对几种抗体进行标记，将元素纳米粒子标记的抗体与血浆样品混合，识别多种待测抗原。利用在抗原识别下纳米粒子标记结合粒径产生变化，根据脉冲信号的强度对抗原的浓度进行定量分析。这样的均相体系可以直接被 ICP-MS 的单颗粒模式检测，利用多组分检测的优势，对疾病的诊断、治疗周期的监测具有更实际的意义。

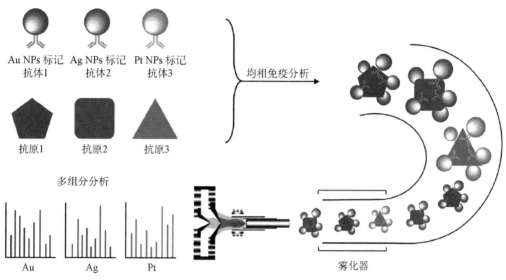

图 10-5　单纳米颗粒 ICP-MS 多组分生物分析机理图

Au NPs、Ag NPs、Pt NPs 标记多种疾病标志物多组分定量分析

与飞行时间质量检测器（time of flight，TOF）相结合，加拿大多伦多大学 Tanner 教授设计的基于 ICP-TOF-MS 的质谱流式细胞仪（mass cytometry，CyTOF），得益于飞行时间质谱仪强大的多元素同时分析能力，可以同时对多种同位素进行分析[59]。普通 ICP-MS 通常使用四极杆质量分析器，在多种同位素同时分析时需要 500~300 μs 的稳定时间——这几乎与单纳米粒子在载气中的保留时间相同，因此普通四极杆 ICP-MS 很难实现对不同同位素的纳米粒子的同时分析。TOF 的优势在于其是利用不同荷质比的同位素在可调电场中的飞行时间差距来进行区分的，因此对不同同位素的同时分析切换更迅速，拥有天然优势。CyTOF 也被认为是开启后荧光-流式细胞仪时代的强大分析仪器。

Mitchell 等就利用 CyTOF 在单细胞层面对细胞对多种纳米颗粒的摄入进行了研究[60]。在人们传统的认知中，CyTOF 的检测都是基于元素的大环化合物标记，而多伦多大学的 Winnik 教授创新性地将镧系纳米粒子（如 $NaHoF_4$ 纳米粒子）标记取代元素大环化合物标签，引入 CyTOF 的检测方法中[39]。该研究用直径 700 nm 的水性凝胶模拟了实际的细胞，测试了不同的纳米颗粒是否可以用作每个细胞上多达 100 个蛋白质的检测。链霉亲和素（Sav）被选做示例蛋白质，通过共价偶联的方式修饰在模拟凝胶上，再通过生物素和牛血清蛋白（BSA）调控封闭，将凝胶表面的 Sav 调控至和细胞表面蛋白质同一水平。研究人员合成了两种生物素化的探针，分别是传统的大环化合物金属聚合物（MCP）——每个探针中含有 50 个 Tb^{3+} 离子；以及端基生物素化的 $NaHoF_4$ 纳米粒子，每个探针含有 15 000 个 Ho 原子。生物素和 Sav 可以特异性结合，BSA 没有这一特性，被选用作为非特异性对比。MCP 探针在 CyTOF 中有效地定量检测了水性凝胶表面的 Sav（每个凝胶表面 20 000 个 Sav）。但对于 Sav 含量更低的凝胶样品，与非特异性 BSA 相比，MCP 探针则很难给出有显著区分度的信号，而此时 $NaHoF_4$ 纳米粒子探针的信号比 MCP 强两个数量级，可以测到每个凝胶表面只有 100~500 个 Sav 的程度。通过这一研究，利用 CyTOF 质谱流式细胞仪进行多组分单纳米颗粒分析成为可能。

10.4　单纳米颗粒 ICP-MS 分析技术的前景与展望

得益于 ICP-MS 对无机元素出色的分析性能，大量元素的纳米颗粒已经通过单纳米颗粒 ICP-MS

成功地进行了分析和检测，这些纳米颗粒也为高灵敏的生物分析提供了候选的标签。尽管在分析方法的设计和仪器的改良上都取得了巨大成功，但现阶段的单纳米颗粒分析仍然存在着一系列的挑战，制约着单纳米颗粒 ICP-MS 在实际应用中的发展。

（1）为了实现多种元素的纳米颗粒同时分析，检测仪器必须对多种同位素的瞬态信号进行同时分析，这需要快速扫描的质谱仪参与[61]——结合电感耦合等离子体飞行时间质谱仪（ICP-TOF-MS）和多检测器质谱仪（MC-ICP-MS）更能实现这一功能，提供更强大的多组分同时分析能力。

（2）目前的方法上，对小粒径的纳米颗粒的检测能力依然不尽如人意，需要进一步优化仪器设计来提高检测的灵敏度。以等效球直径来表示检出限，单个同位素的 LOD 从 10 nm（Ta、U、Ir、Rh、Th、Ce 和 Hf）到 200 nm（Se、Ca 和 Si）不等。而对于多种元素同位素构成的纳米粒子，基于尺寸的 LOD 则更差。

（3）纳米粒子标签长时间稳定性较差，不利于长期储存。在高浓度的盐溶液中也易发生团聚和变性。

单纳米颗粒 ICP-MS 是一种特点鲜明、具有潜力的强大分析工具，在更加优化其分析性能之后，我们相信它将有可能应用到更多前沿的科学领域，解决更多繁复的分析问题。

10.4.1 无标记分析

现有的单纳米颗粒 ICP-MS 分析方法多是基于纳米颗粒标记建立的，尽管单纳米颗粒 ICP-MS 为其快速高通量检测提供了便利的条件和灵敏度，但纳米颗粒预先合成方法的烦琐、短暂的保存时间和不稳定性，仍然增加了这一类纳米颗粒标记法的应用难度。在生物分析中，基于生物分子模板法的自上而下自合成纳米颗粒，因其可以在原子和分子水平精确控制纳米晶体生长，为原位无标记反应提供了条件而在近年来受到了广泛关注。在许多可用的生物分子中，DNA 因其合成便利、性质稳定、高度可调控等性能得到了深入研究，成为一种强大的对金属纳米颗粒进行原位合成的模板。这种通过还原金属离子将金属纳米结构沉积在 DNA 骨架上的方法，已经合成了一系列金属纳米结构，例如 Au、Pt、Pd、Cu、Co、Fe、Ni、Te、Rh 和各种合金等，已经被成功应用到了一系列基于金属纳米粒子的无标记分析中[62]。

例如，自 2004 年 Dickson 等首次报道可以利用 12 个碱基组成的单链 DNA（ssDNA）作为模板合成银纳米团簇（Ag NCs）以来[63]，利用 Ag NCs 特有的荧光性能，基于 DNA 模板合成 Ag NCs 的无标记分析就已经应用到了诸如 DNA、miRNA、生物小分子、生物酶等物质的检测中，建立了一系列便捷的分析方法[64-69]。

将无标记分析和单纳米颗粒 ICP-MS 结合，对原位自合成的纳米颗粒进行直接的均相分析，可以最大限度地释放单纳米颗粒 ICP-MS 在方法设计的便捷性上的优势，提供更高通量、更高灵敏的检测。囿于 Ag NCs 的粒径过小，引入尺度在 5~20 nm 的铜纳米粒子（Cu NPs）作为无标记传感单元更为合适。基于双链 DNA（dsDNA）和 PolyT 单链 DNA（PolyT-ssDNA）分别合成 Cu NPs 的方法在早年间分别由德国科学家 Mokhir 与中国科学家王柯敏等提出[70-72]。通过控制模板中胸腺嘧啶（T）的数量可以调控合成 5~20 nm 的 Cu NPs。利用 Cu NPs 的荧光性质和内含的大量 Cu 元素同位素，可以进行无标记的荧光、ICP-MS 分析。吕弋课题组将两种模式结合，设计研发了一种基于 PolyT-CuNPs 双模态的无标记检测法，用于了三硝基甲苯（TNT）的高灵敏检测中[73]。

通过提升单纳米颗粒 ICP-MS 的检出性能，降低尺寸检出限，则可以实现对此类基于 DNA 模板原位合成的纳米粒子的单颗粒分析，利用原位合成的纳米颗粒的脉冲信号对待测生物分子进行定量，可以实现更为快速、更高通量和灵敏度的生物分析方法。

10.4.2 单细胞分析

近年来，对于单个细胞的分析一直是生物分析的前沿，对于单细胞水平上多种类型的生物标志物的检测，不仅可以对细胞进行功能分型，还对细胞状态、疾病诊断有着巨大价值。利用纳米颗粒设计标记进行单个细胞水平的检测也成为单纳米颗粒标记生物分析的热点之一。武汉大学的胡斌教授课题组就设计了一种掺杂 Cs 元素的多核 MMNPs 和 CdSe/ZnS 量子点共同标记，用于对 HepG2 细胞的识别。CdSe/ZnS 量子点可以被用作荧光成像和质谱定量的双功能探针，提供更加准确的检测结果，并且可以获得 61 个 HepG2 细胞的检出限[51]。该方法可以通过 ICP-MS 进行细胞计数的比率检测，也可以通过荧光成像来显示细胞。在量子点与 EpCAM 抗体偶联，EpCAM 可以特异性捕获溶解的人血中的 HepG2 细胞，量子点的荧光特性可以在细胞成像中对细胞进行可视化检测，另外量子点中的 Cd 元素同位素也可以用于 ICP-MS 检测。另外在 MMNPs 中掺杂的 Cs 元素也可以用作比率型内标，使用 $^{114}Cd/^{133}Cs$ 的信号比代替 ^{114}Cd 进行定量来补偿和抑制由颗粒损失引起的测量误差和信号波动。厦门大学的王秋泉教授则构建了一种具有精确原子数元素标记的病毒状纳米颗粒（VLNP）标记策略，证明了这种纳米材料标记可以实现基于 ICP-MS 的对细胞膜分子介导的细胞计数的可行性[74]。因为生物安全性、实验室易于生产、单分散性良好，尤其是对化学耐受性高等特点，典型的噬菌体 MS2 被作为 VLNP 的实例，以证明具有元素标记的 VLNPs 可以产生信号扩增的功能。MS2 的衣壳由 180 个分子质量为 13 728 Da 的蛋白质单体组成，每种蛋白质单体均具有七个反应性氨基，并具有球形的构型，其外径和内径分别为 27 nm 和 21 nm。此外，衣壳中有 32 个直径为 18 Å 的孔，提供了可以进入内表面的通道。研究人员首先通过去除 MS2 的 RNA 基因组来制备空的 MS2 衣壳，并通过酰化反应和点击反应将元素复合标签共轭到获得的叠氮 MS2 衣壳表面上，以获得元素标记的 MS2。靶标位点则以鼻咽癌 KB 细胞上的叶酸受体（FR）为例，证明了基于 ICP-MS 的元素标记的 MS2 对细胞膜分子介导的细胞计数的有效性。

在单细胞分析领域，利用 CyTOF 质谱流式细胞仪这一强大的分析手段，美国麻省理工学院的 Irvine 教授团队设计了一种两亲性表面配体包被的 Au NPs，用于靶向对淋巴结中的髓样树突状细胞。他们证明基于飞行时间质谱检测，可以实现最小 3 nm 的 Au NPs 核的纳米标记高灵敏分析[38]。相较于普通的单纳米颗粒 ICP-MS 检测，质谱流式细胞仪实现了更小粒径的单颗粒纳米粒子的活体检测，这也为更广阔的生物应用提供了平台。在传统的生物检测中，该研究证明了质谱流式细胞仪对单个细胞中纳米颗粒的分析能力，其灵敏度比流式细胞仪大几个数量级。对于大规模细胞计数，基于质谱流式细胞仪的单纳米颗粒分析解决了组织和细胞中的自发问题，证明了将单纳米颗粒检测与基于抗体的细胞分型结合可以建立快速可靠的体内生物分析方案。结合前文所述的镧系纳米粒子标记的单分子质谱流式细胞仪分析，我们可以设计一种基于质谱流式细胞仪的单细胞分析方法（图 10-6）。利用不同的纳米粒子结合特异性抗体，分别靶标每一个细胞上的不同抗原。再通过配制合适浓度的细胞悬浮液，保证在雾化器中形成单细胞液滴进入 ICP 中，在质量检测器中获得不同元素的频率信号，实现对单个细胞表面抗原的定量。这一方法利用了质谱流式细胞仪单细胞和高分辨率多同位素同时分析的能力，同时也结合了单纳米颗粒信号放大的作用，可以实现对单个细胞多种抗原的高灵敏检测。

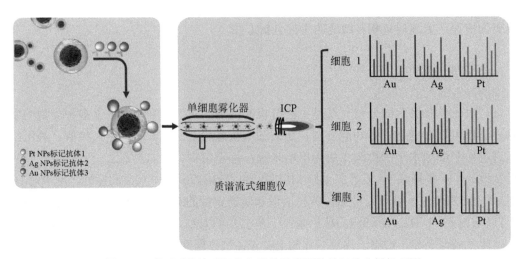

图 10-6 基于质谱流式细胞仪的单纳米颗粒单细胞分析机理图

10.4.3 单分子分析

生物学或临床医学上通常都需要强大的分析手段来对生物分子进行准确而灵敏的定量研究。例如，许多与疾病相关的生物标志物的血清浓度通常为 $10^{-16} \sim 10^{-12}$ mol/L。这样低浓度的分析任务对许多常规的检测方法（如酶联免疫法，通常具有高于 10^{-12} mol/L 的检出限）提出了巨大挑战。因此，提高对极低含量的待测物检测灵敏度的新型生物分析方法引起了科研人员的兴趣，而其中，单分子分析策略在各种超灵敏生物分析方法中具有巨大的潜力。单分子分析是一类对 $1/N_A(1.66×10^{-24})$ mol 或 1.66 ymol 物质的量的待测物进行检测的方法，通过单独检测单个分子，在获得超灵敏检测性能的同时随浓度降低保持信噪比。

但分子分析的主流方法之一是通过纳米颗粒直接计数。第一个用于单分子定量分析的方法是由 Bayley 等开发的，用于计数金属离子和有机分子的纳米孔道[8]。纳米孔的设计非常类似于细胞膜中的孔道，为了测量跨膜的电导，电极被放置在纳米孔的两侧，利用待测物通过纳米孔时产生的位阻导致电导率降低来定量检测，并且通过纳米孔的修饰，来选择性地对待测物进行检测。这样的电阻脉冲信号变化可以反映待测物的浓度。同样的，用荧光标记替代电导检测，利用荧光脉冲进行单分子荧光计数也可以用于单分子定量分析。

与这些方法相比，单纳米颗粒 ICP-MS 具有更出色的分辨率，不同同位素的谱线互不干扰，可实现多组分的检测。与荧光计数法类似，利用单个纳米颗粒带来的脉冲信号变化，也可以实现在单分子层面上的超灵敏检测，同时避免了荧光光谱重叠，在多组分检测上更具优势。

10.5 总　　结

单纳米颗粒 ICP-MS 分析技术作为一种强大的金属纳米材料分析和表征手段，尽管已经得到了数十年的发展，仍然方兴未艾。单纳米颗粒 ICP-MS 分析技术在环境中元素纳米粒子的形态示踪和代谢研究中发挥重要的作用，在生物分析方法中建立了一系列成功的均相、异相和多组分标记方法。单纳米颗粒 ICP-MS 分析技术有非常巨大的发展潜力，质谱流式细胞仪（CyTOF）、激光烧蚀等离子体质谱仪（LA-ICP-MS）等可以实现在单细胞和单分子层面的分析，在超高空间分辨率和超灵敏的纳米粒子标记分析中扮演重要角色。

参 考 文 献

[1] Liu R, Zhang S X, Wei C, Xing Z, Zhang S X, Zhang X R. Metal stable isotope tagging: renaissance of radioimmunoassay for multiplex and absolute quantification of biomolecules. Accounts of Chemical Research, 2016, 49: 775-783.

[2] Yan X W, Yang L M, Wang Q Q. Lanthanide-coded protease-specific peptide-nanoparticle probes for a label-free multiplex protease assay using element mass spectrometry: A proof-of-concept study. Angewandte Chemie-International Edition, 2011, 50: 5130-5133.

[3] Han G J, Zhang S C, Xing Z, Zhang X R. Absolute and relative quantification of multiplex DNA assays based on an elemental labeling strategy. Angewandte Chemie-International Edition, 2013, 52: 1466-1471.

[4] Zhang X D, Zhang G H, Herrera I, Kinach R, Ornatsky O, Baranov V, Nitz M, Winnik M A. Polymer-based elemental tags for sensitive bioassays. Angewandte Chemie-International Edition, 2007, 46: 6111-6114.

[5] Huang Z L, Wang C Q, Liu R, Su Y Y, Lv Y. Self-validated homogeneous immunoassay by single nanoparticle in-depth scrutinization. Analytical Chemistry, 2020, 92: 2876-2881.

[6] Zhang S X, Han G J, Xing Z, Zhang S C, Zhang X R. Multiplex DNA assay based on nanoparticle probes by single particle inductively coupled plasma mass spectrometry. Analytical Chemistry, 2014, 86: 3541-3547.

[7] Degueldre C, Favarger P Y. Colloid analysis by single particle inductively coupled plasma-mass spectroscopy: A feasibility study. Colloids and Surfaces A: Physicochemical and Engineering Aspects, 2003, 217: 137-142.

[8] Braha O, Walker B, Cheley S, Kasianowicz J J, Song L Z, Gouaux J E, Bayley H. Designed protein pores as components for biosensors. Chemistry & Biology, 1997, 4: 497-505.

[9] Pace H E, Rogers N J, Jarolimek C, Coleman V A, Higgins C P, Ranville J F. Determining transport efficiency for the purpose of counting and sizing nanoparticles via single particle inductively coupled plasma mass spectrometry. Analytical Chemistry, 2011, 83: 9361-9369.

[10] Goodall P, Foulkes M E, Ebdon L. Slurry nebulization inductively-coupled plasma spectrometry: The fundamental parameters discussed. Spectrochimica Acta Part B: Atomic Spectroscopy, 1993, 48: 1563-1577.

[11] Lee S Y, Bi X Y, Reed R B, Ranville J F, Herckes P, Westerhoff P. Nanoparticle size detection limits by single particle ICP-MS for 40 elements. Environmental Science & Technology, 2014, 48: 10291-10300.

[12] Merrifield R C, Stephan C, Lead J R. Single-particle inductively coupled plasma mass spectroscopy analysis of size and number concentration in mixtures of monometallic and bimetallic (core-shell) nanoparticles. Talanta, 2017, 162: 130-134.

[13] Fréchette-Viens L, Hadioui M, Wilkinson K J. Practical limitations of single particle ICP-MS in the determination of nanoparticle size distributions and dissolution: Case of rare earth oxides. Talanta, 2017, 163: 121-126.

[14] Yang Y, Long C L, Li H P, Wang Q, Yang Z G. Analysis of silver and gold nanoparticles in environmental water using single particle-inductively coupled plasma-mass spectrometry. Science of the Total Environment, 2016, 563-564: 996-1007.

[15] Liu Y, He M, Chen B B, Hu B. Ultra-trace determination of gold nanoparticles in environmental water by surfactant assisted dispersive liquid liquid microextraction coupled with electrothermal vaporization-inductively coupled plasma-mass spectrometry. Spectrochimica Acta Part B: Atomic Spectroscopy, 2016, 122: 94-102.

[16] Jimenez-Lamana J, Slaveykova V I. Silver nanoparticle behaviour in lake water depends on their surface coating. Science of the Total Environment, 2016, 573: 946-953.

[17] Sötebier C A, Weidner S M, Jakubowski N, Panne U, Bettmer J. Separation and quantification of silver nanoparticles and silver ions using reversed phase high performance liquid chromatography coupled to inductively coupled plasma mass spectrometry in combination with isotope dilution analysis. Journal of Chromatography A, 2016, 1468: 102-108.

[18] Zhou X X, Liu J F, Yuan C G, Chen Y S. Speciation analysis of silver sulfide nanoparticles in environmental waters by magnetic solid-phase extraction coupled with ICP-MS. Journal of Analytical Atomic Spectrometry, 2016, 31: 2285-2292.

[19] Jiménez-Lamana J, Wojcieszek J, Jakubiak M, Asztemborska M, Szpunar J. Single particle ICP-MS characterization of platinum nanoparticles uptake and bioaccumulation by *Lepidium sativum* and *Sinapis alba* plants. Journal of Analytical Atomic Spectrometry, 2016, 31: 2321-2329.

[20] Bao D P, Zhen G O, Zhong C. Characterization of silver nanoparticles internalized by arabidopsis plants using single particle ICP-MS analysis. Frontiers in Plant Science, 2016, 7: 32.

[21] Sötebier C A, Kutscher D J, Rottmann L, Jakubowski N, Panne U, Bettmer J. Combination of single particle ICP-QMS and isotope dilution analysis for the determination of size, particle number and number size distribution of silver nanoparticles. Journal of Analytical Atomic Spectrometry, 2016, 31: 2045-2052.

[22] Soto-Alvaredo J, Dutschke F, Bettmer J, Montes-Bayón M, Pröfrock D, Prange A. Initial results on the coupling of sedimentation field-flow fractionation (SdFFF) to inductively coupled plasma-tandem mass spectrometry (ICP-MS/MS) for the detection and characterization of TiO_2 nanoparticles. Journal of Analytical Atomic Spectrometry, 2016, 31: 1549-1555.

[23] Huang K, Xu K L, Tang J, Yang L, Zhou J R, Hou X D, Zheng C B. Room temperature cation exchange reaction in nanocrystals for ultrasensitive speciation analysis of silver ions and silver nanoparticles. Analytical Chemistry, 2015, 87: 6584-6591.

[24] Roman M, Rigo C, Castillo-Michel H, Munivrana I, Vindigni V, Mičetić I, Benetti F, Manodori L, Cairns W R L. Hydrodynamic chromatography coupled to single-particle ICP-MS for the simultaneous characterization of AgNPs and determination of dissolved Ag in plasma and blood of burn patients. Analytical and Bioanalytical Chemistry, 2016, 408: 5109-5124.

[25] Donovan A R, Adams C D, Ma Y F, Stephan C, Eichholz T, Shi H L. Detection of zinc oxide and cerium dioxide nanoparticles during drinking water treatment by rapid single particle ICP-MS methods. Analytical and Bioanalytical Chemistry, 2016, 408: 5137-5145.

[26] Feng D, Tian F, Qin W J, Qian X H. A dual-functional lanthanide nanoprobe for both living cell imaging and ICP-MS quantification of active protease. Chemical Science, 2016, 7: 2246-2250.

[27] Bustos A M, Petersen E J, Possolo A, Winchester M R. Post hoc interlaboratory comparison of single particle ICP-MS size measurements of NIST gold nanoparticle reference materials. Analytical Chemistry, 2015, 87: 8809-8817.

[28] Matczuk M, Anecka K, Scaletti F, Messori L, Keppler B K, Timerbaev A R, Jarosz M. Speciation of metal-based nanomaterials in human serum characterized by capillary electrophoresis coupled to ICP-MS: A case study of gold nanoparticles. Metallomics, 2015, 7: 1364-1370.

[29] Peters R, Herrera-Rivera Z, Undas A, van der Lee M, Marvin H, Bouwmeester H, Weigel S. Single particle ICP-MS combined with a data evaluation tool as a routine technique for the analysis of nanoparticles in complex matrices. Journal of Analytical Atomic Spectrometry, 2015, 30: 1274-1285.

[30] Elias A, Crayton S H, Warden-Rothman R, Tsourkas A. Quantitative comparison of tumor delivery for multiple targeted nanoparticles simultaneously by multiplex ICP-MS. Scientific Reports, 2014, 4: 5840.

[31] Zhou X Y, Dorn M, Vogt J, Spemann D, Yu W, Mao Z W, Estrela-Lopis I, Donath E, Gao C Y. A quantitative study of the intracellular concentration of graphene/noble metal nanoparticle composites and their cytotoxicity. Nanoscale, 2014, 6: 8535-8542.

[32] Franze B, Engelhard C. Fast separation, characterization, and speciation of gold and silver nanoparticles and their ionic counterparts with micellar electrokinetic chromatography coupled to ICP-MS. Analytical Chemistry, 2014, 86: 5713-5720.

[33] Elci S G, Yan B, Kim S T, Saha K, Jiang Y, Klemmer G A, Moyano D F, Tonga G Y, Rotello V M, Vachet R W. Quantitative imaging of 2 nm monolayer-protected gold nanoparticle distributions in tissues using laser ablation inductively-coupled plasma mass spectrometry (LA-ICP-MS). Analyst, 2016, 141: 2418-2425.

[34] Böhme S, Stärk H J, Kühnel D, Reemtsma T. Exploring LA-ICP-MS as a quantitative imaging technique to study nanoparticle uptake in *Daphnia magna* and zebrafish (*Danio rerio*) embryos. Analytical and Bioanalytical Chemistry, 2015, 407: 5477-5485.

[35] Drescher D, Zeise I, Traub H, Guttmann P, Seifert S, Büchner T, Jakubowski N, Schneider G, Kneipp J. *In situ* characterization of SiO_2 nanoparticle Biointeractions using BrightSilica. Advanced Functional Materials, 2014, 24: 3765-3775.

[36] Meermann B, Wichmann K, Lauer F, Vanhaecke F, Ternes T A. Application of stable isotopes and AF4/ICP-SFMS for simultaneous tracing and quantification of iron oxide nanoparticles in a sediment-slurry matrix. Journal of Analytical Atomic Spectrometry, 2016, 31: 890-901.

[37] Newman K, Metcalfe C, Martin J, Hintelmann H, Shaw P, Donard A. Improved single particle ICP-MS characterization of silver nanoparticles at environmentally relevant concentrations. Journal of Analytical Atomic Spectrometry, 2016, 31: 2069-2077.

[38] Yang Y S, Atukorale P U, Moynihan K D, Bekdemir A, Rakhra K, Tang L, Stellacci F, Irvine D J. High-throughput quantitation of inorganic nanoparticle biodistribution at the single-cell level using mass cytometry. Nature Communications, 2017, 8: 14069.

[39] Lin W J, Hou Y, Lu Y J, Abdelrahman A I, Cao P P, Zhao G Y, Tong L, Qian J S, Baranov V, Nitz M, Winnik M A. A high-sensitivity lanthanide nanoparticle reporter for mass cytometry: Tests on microgels as a proxy for cells. Langmuir, 2014, 30: 3142-3153.

[40] Klaine S J, Koelmans A A, Horne N, Carley S, Handy R D, Kapustka L, Nowack B, von der Kammer F. Paradigms to assess the environmental impact of manufactured nanomaterials. Environmental Toxicology and Chemistry, 2012, 31: 3-14.

[41] Laborda F, Bolea E, Jiménez-Lamana J. Single particle inductively coupled plasma mass spectrometry for the analysis of inorganic engineered nanoparticles in environmental samples. Trends in Environmental Analytical Chemistry, 2016, 9: 15-23.

[42] Gondikas A P, von der Kammer F, Reed R B, Wagner S, Ranville J F, Hofmann T. Release of TiO_2 nanoparticles from sunscreens into surface waters: A one-year survey at the old Danube recreational Lake. Environmental Science & Technology, 2014, 48: 5415-5422.

[43] Tuoriniemi J, Cornelis G, Hassellöv M. Size discrimination and detection capabilities of single-particle ICPMS for environmental analysis of silver nanoparticles. Analytical Chemistry, 2012, 84: 3965-3972.

[44] Mitrano D M, Ranville J F, Bednar A, Kazor K, Hering A S, Higgins C P. Tracking dissolution of silver nanoparticles at environmentally relevant concentrations in laboratory, natural, and processed waters using single particle ICP-MS (spICP-MS). Environmental Science-Nano, 2014, 1: 248-259.

[45] Farkas J, Peter H, Christian P, Urrea J, Hassellöv M, Tuoriniemi J, Gustafsson S, Olsson E, Hylland K, Thomas K V.

Characterization of the effluent from a nanosilver producing washing machine. Environment International, 2011, 37: 1057-1062.

[46] Coleman J G, Kennedy A J, Bednar A J, Ranville J F, Laird J G, Harmon A R, Hayes C A, Gray E P, Higgins C P, Lotufo G, Steevens J A A. Comparing the effects of nanosilver size and coating variations on bioavailability, internalization, and elimination, using Lumbriculus variegatus. Environmental Toxicology and Chemistry, 2013, 32: 2069-2077.

[47] Tan Z Q, Yin Y G, Guo X R, Made M, Moon M H, Liu J F, Jiang G B. Tracking the transformation of nanoparticulate and ionic silver at environmentally relevant concentration levels by hollow fiber flow field-flow fractionation coupled to ICPMS. Environmental Science & Technology, 2017, 51: 12369-12376.

[48] Gray E P, Coleman J G, Bednar A J, Kennedy A J, Ranville J F, Higgins C P. Extraction and analysis of silver and gold nanoparticles from biological tissues using single particle inductively coupled plasma mass spectrometry. Environmental Science & Technology, 2013, 47: 14315-14323.

[49] Wang C Q, Zhao X, Liu R, Zhong Z J, Hu J Y, Lv Y. Isotopic core-satellites enable accurate and sensitive bioassay of adenosine triphosphate. Chemical Communications, 2019, 55: 10665-10668.

[50] Quinn Z A, Baranov V I, Tanner S D, Wrana J L. Simultaneous determination of proteins using an element-tagged immunoassay coupled with ICP-MS detection. Journal of Analytical Atomic Spectrometry, 2002, 17: 892-896.

[51] Yang B, Chen B B, He M, Hu B. Quantum dots labeling strategy for "counting and visualization" of HepG2 cells. Analytical Chemistry, 2017, 89: 1879-1886.

[52] Hu J Y, Deng D Y, Liu R, Lv Y. Single nanoparticle analysis by ICPMS: A potential tool for bioassay. Journal of Analytical Atomic Spectrometry, 2018, 33: 57-67.

[53] Hu S H, Liu R, Zhang S C, Huang Z, Xing Z, Zhang X R. A new strategy for highly sensitive immunoassay based on single-particle mode detection by inductively coupled plasma mass spectrometry. Journal of the American Society for Mass Spectrometry, 2009, 20: 1096-1103.

[54] Cao Y P, Mo G C, Feng J S, He X X, Tang L F, Yu C H, Deng B Y. Based on ZnSe quantum dots labeling and single particle mode ICP-MS coupled with sandwich magnetic immunoassay for the detection of carcinoembryonic antigen in human serum. Analytica Chimica Acta, 2018, 1028: 22-31.

[55] Han G J, Xing Z, Dong Y H, Zhang S C, Zhang X R. One-step homogeneous DNA assay with single-nanoparticle detection. Angewandte Chemie-International Edition, 2011, 50: 3462-3465.

[56] Li B R, Tang H, Yu R, Jiang J H. Single-nanoparticle ICPMS DNA assay based on hybridization-chain-reaction-mediated spherical nucleic acid assembly. Analytical Chemistry, 2020, 92: 2379-2382.

[57] Li Z Y, Li H M, Deng D Y, Liu R, Lv Y. Mass spectrometric assay of alpha-fetoprotein isoforms for accurate serological evaluation. Analytical Chemistry, 2020, 92: 4807-4813.

[58] Cao Y P, Feng J S, Tang L F, Mo G C, Mo W M, Deng B Y. Detection of three tumor biomarkers in human lung cancer serum using single particle inductively coupled plasma mass spectrometry combined with magnetic immunoassay. Spectrochimica Acta Part B: Atomic Spectroscopy, 2020, 166: 105797.

[59] Bandura D R, Baranov V I, Ornatsky O I, Antonov A, Kinach R, Lou X D, Pavlov S, Vorobiev S, Dick J E, Tanner S D. Mass cytometry: Technique for real time single cell multitarget immunoassay based on inductively coupled plasma time-of-flight mass spectrometry. Analytical Chemistry, 2009, 81: 6813-6822.

[60] Mitchell A J, Ivask A, Ju Y. Quantitative measurement of cell-nanoparticle interactions using mass cytometry. Methods in Molecular Biology (Clifton, N.J.), 2019, 1989: 227-241.

[61] Tanner M, Günther D. Short transient signals, a challenge for inductively coupled plasma mass spectrometry: A review. Analytica Chimica Acta, 2009, 633: 19-28.

[62] Chen Z W, Liu C Q, Cao F F, Ren J S, Qu X G. DNA metallization: Principles, methods, structures, and applications. Chemical Society Reviews, 2018, 47: 4017-4072.

[63] Ritchie C M, Johnsen K R, Kiser J R, Antoku Y, Dickson R M, Petty J T. Ag nanocluster formation using a cytosine oligonucleotide template. Journal of Physical Chemistry C, 2007: 175-181.

[64] Zhang J P, Li C, Zhi X, Ramón G A, Liu Y L, Zhang C L, Pan F, Cui D X. Hairpin DNA-templated silver nanoclusters as novel beacons in strand displacement amplification for MicroRNA detection. Analytical Chemistry, 2016, 88: 1294-1302.

[65] CaoQ, Teng Y, Yang X, Wang J, Wang E K. A label-free fluorescent molecular beacon based on DNA-Ag nanoclusters for the construction of versatile biosensors. Biosensors and Bioelectronics, 2015, 74: 318-321.

[66] Zhu Y, Hu X C, Shi S, Gao R R, Huang H L, Zhu Y Y, Lv X Y, Yao T M. Ultrasensitive and universal fluorescent aptasensor for the detection of biomolecules (ATP, adenosine and thrombin) based on DNA/Ag nanoclusters fluorescence light-up system. Biosensors and Bioelectronics, 2016, 79: 205-212.

[67] Zhang M, Guo S M, Li Y R, Zuo P, Ye B C. A label-free fluorescent molecular beacon based on DNA-templated silver nanoclusters for detection of adenosine and adenosine deaminase. Chemical Communications, 2012, 48: 5488-5490.

[68] Zhang L B, Zhu J B, Zhou Z X, Guo S J, Li J, Dong S J, Wang E K. A new approach to light up DNA/Ag nanocluster-based beacons for bioanalysis. Chemical Science, 2013, 4: 4004.

[69] Juul S, Obliosca J M, Liu C, Liu Y L, Chen Y, Imphean D M, Knudsen B R, Ho Y P, Leong K W, Yeh H C. Nanocluster beacons as reporter probes in rolling circle enhanced enzyme activity detection. Nanoscale, 2015, 7: 8332-8337.

[70] Peng J, Ling J, Tan Y Y, Jing C J, Li X J, Chen L Q, Cao Q E. Poly(thymine)-templated selective formation of fluorescent copper nanoparticles. Angewandte Chemie-International Edition, 2013, 52: 9719-9722.

[71] Rotaru A, Dutta S, Jentzsch E, Gothelf K, Mokhir A. Selective dsDNA-templated formation of copper nanoparticles in solution. Angewandte Chemie-International Edition, 2010, 49: 5665-5667.

[72] Liu G Y, Shao Y, Peng J, Dai W, Liu L L, Xu S J, Wu F, Wu X H. Highly thymine-dependent formation of fluorescent copper nanoparticles templated by ss-DNA. Nanotechnology, 2013, 24: 345502.

[73] Hu J Y, Wang C Q, Liu R, Su Y Y, Lv Y. Poly(thymine)-CuNPs: Bimodal methodology for accurate and selective detection of TNT at sub-PPT levels. Analytical Chemistry, 2018, 90: 14469-14474.

[74] Yuan R, Ge F C, Liang T, Zhou Y, Yang L M, Wang Q Q. A virus-like element-tagged nanoparticle ICPMS signal multiplier: membrane biomarker mediated cell counting. Analytical Chemistry, 2019, 91: 4948-4952.

第11章 微流控芯片–等离子体质谱联用技术用于细胞中痕量元素及其形态分析

(胡 斌[①]*，何 蔓[①]，陈贝贝[①])

▶ 11.1 引言 / 271
▶ 11.2 基于微流控芯片的微萃取技术 / 272
▶ 11.3 基于等离子体质谱的单细胞分析 / 282
▶ 11.4 总结与展望 / 294

①武汉大学，武汉。*通讯作者联系方式：binhu@whu.edu.cn

第11章 微流控芯片-等离子体质谱联用技术用于细胞中痕量元素及其形态分析

本章导读

- 首先介绍细胞中痕量元素及形态分析的重要性、对相关分析方法的要求及存在的问题。
- 基于微流控芯片在细胞分析中的突出优势,重点介绍基于微流控芯片的液相微萃取、固相微萃取技术及其在细胞中痕量元素及形态分析中的应用。
- 重点介绍基于等离子体质谱的单细胞分析,包括选择合适的进样系统、质量分析器、检测器,构建液滴微流控芯片、液滴分裂微流控芯片、聚焦微流控芯片等微流体装置,以及微流体技术结合等离子体质谱用于单细胞中痕量元素及形态分析的研究现状。
- 对面向细胞内痕量元素及形态分析的微流控芯片-等离子体质谱联用技术的发展方向进行了展望。

11.1 引 言

细胞是生命体形态结构和生理活动的基本单位,主要成分是水、无机离子、糖类、脂质、核酸和蛋白质等[1];微量金属/类金属在细胞的组成结构和生理功能中有着举足轻重的地位[2],如参与DNA/RNA的转录和翻译,生物大分子的修饰和代谢以及能量的传输和转换等过程。细胞内约三分之一的蛋白质会与锌、铜、铁等辅因子结合;硒和碘等元素则是细胞内蛋白质、多肽的重要组分之一。外界环境也会使细胞内的元素发生变化,影响人体健康。如Cd、Hg、Pb等可通过食物链、饮用水、大气等途径累积在生物体内,造成生物体重金属中毒;含有Pt、Bi、As等元素的药物已被用于癌症、胃溃疡、白血病等疾病的治疗;此外,随着元素标记和生物探针技术的发展,含金属的络合物、聚合物及纳米粒子(nanoparticles,NPs)也被应用于人体内的生化分析。可见,细胞中痕量元素的含量及形态可以作为人体健康和细胞生长状态的"指示剂",获取胞内痕量元素及其形态的准确信息具有十分重要的意义。目前,细胞内痕量元素及形态分析主要面临三个方面的挑战[3]:①细胞的样品量稀少,需要微型化的分析平台;②细胞内目标物的绝对含量低,需要十分灵敏的检测技术;③细胞的基质较为复杂,需要合适的样品前处理技术。

目前,可用于痕量元素及元素形态分析的仪器分析手段包括电感耦合等离子体质谱(inductively coupled plasma-mass spectrometry,ICP-MS)[3]、二次离子质谱(secondary ion mass spectrometry,SIMS)[4]、同步辐射X射线显微荧光技术(synchrotron X-ray fluorescence microscopy,SXRF)[5]以及荧光检测法等。其中,ICP-MS具有灵敏度高、线性范围宽、可多元素/同位素同时分析、易于与各种分离技术联用等优点。但是,当ICP-MS用于细胞样品分析时,通常面临着细胞样品(太少)不能满足常规溶液雾化进样的要求、细胞样品基质复杂和目标元素/形态含量低于ICP-MS检出限等问题。因此,在ICP-MS分析前需要辅以微型化的样品前处理技术进行基质分离和目标分析物富集。另外,常规的细胞分析消耗的细胞数目一般是$10^3 \sim 10^6$个细胞,在检测细胞群的总信号值后,以平均值代表每个细胞的响应水平。但是,由于细胞异质性的广泛存在,常规的细胞分析常常导致部分亚种群细胞的表现决定了整个种群的表现,或是整个种群的表现掩盖了某一亚种群特点的问题[6]。在单个细胞水平开展痕量元素及其形态分析研究,对于从分子水平上揭示痕量元素及其形态在细胞生理活动中的作用机制,具有十分重要的意义。

微流控芯片(microfluidic chip)通过合适的设计和微机电加工来实现分析检测体系的微型化、自动化、集成化与便携化[7],是分析化学、微机电加工、计算机、电子学、材料科学及生物医学交叉形成的一个新的化学分析体系。作为一种化学分析工具，微流控芯片在细胞样品的分析中表现出突出的优势[8,9]：①减少了细胞样品与试剂的使用量（μL 或 nL 级）；②其可集成化的特点能有效提高样品的通量、减少分析时间；③利用其高空间分辨率，能实现对细胞样品的微区操纵和分析，有助于实现单细胞分析；④微流控芯片是集成化系统，对人为操作要求较少，且易于实现自动化，能有效减少样品污染并实现无人化分析；⑤微流控芯片具有便携化特点，可实现"个人化"的化学分析[3]。借助微流控芯片这一微型化的操纵、分析平台可以解决 ICP-MS 分析细胞样品时存在的样品消耗量过大的问题；通过芯片强大的细胞操纵能力可以实现少数细胞甚至单细胞中元素及元素形态的分析。此外，在微流控芯片体系中集成合适的样品前处理技术后，可在不增加细胞样品消耗量的前提下实现细胞样品的基质去除及目标分析物的富集，从而实现少量细胞/单细胞中高灵敏、高选择性、高通量的痕量元素及其形态分析。

11.2 基于微流控芯片的微萃取技术

基于微流控芯片的微萃取技术[10]可以分为液相微萃取和固相微萃取两大类。

11.2.1 基于微流控芯片的液相微萃取技术

液相微萃取主要基于目标分析物在样品溶液相和萃取相之间的分配平衡达到去除基质和富集目标分析物的目的[11]。微流控芯片具有微米级的通道尺寸，微通道内的雷诺系数较小，极易在微通道内形成层流；基于此，可在芯片中实现基于层流的液-液微萃取[12]。微流控芯片是微机电加工的集成系统，在该系统中集成聚合物薄膜极为方便，基于液膜萃取的前处理体系在微流控芯片上也能较容易的实现。此外，通过调控两种不相溶液体的流速、线速度和黏度等参数，能在芯片中得到离散的微液滴，进一步增加两相液体间的物质交换速率，改善萃取性能[13]。目前，用于痕量元素及其形态分析的芯片液相微萃取主要有层流和液滴两种萃取形式[12]。

1. 基于层流的液相微萃取

一般来说，微流体平台中的流体速度 U 不超过 cm/s 量级，通道尺寸 l 不超过亚 mm 量级，即雷诺数 $Re<10(Re=\rho Ul/\mu)$，故流体可以表现出稳定的层流状态。基于多相层流的芯片液相扩散分离体系最早由 Weigl 等[14]提出：在芯片的微米级通道内，溶液表现出稳定的层流状态，多相流动的液体在通道内可以平行同向流动，形成不相混合的多相，从而实现多相液体间物质的交换。虽然有机相能萃取疏水性金属螯合物，但会造成 ICP 功率的不稳定以及采样锥口的碳沉积。采用电热蒸发（electrothermal vaporization，ETV）作为 ICP 的进样技术，可以避免有机相引入 ICP，实现萃取相中目标元素的直接检测。研究工作者[15]构建了基于芯片层流的两相液相微萃取体系（图 11-1）：在水相中加入二乙基二硫代氨基甲酸钠（DDTC）作为痕量元素（Cu、Zn、Cd、Hg、Pb 和 Bi）的螯合试剂，辛醇为有机相，DDTC 与目标元素形成螯合物后从水相被萃取至辛醇中，在芯片出口端接收流出的辛醇，直接引入 ETV-ICP-MS 即可检测水相样品中目标元素的含量。为了尽可能延长萃取时间并增加两相间的接触面积，设计了由 21 个长的直通道（5 cm）及 20 个弧形弯曲连接通道组成的层流萃取通道，层流通道宽度为 200 μm，在芯片通道始端和尾端有两个流入通道及两个流出通道，其中有机相的流入、流出通道宽度为 35 μm，水相的流入、流出通道宽度为 165 μm。在最优的条件下，该分析系统对 Cu、Zn、Cd、Hg、Pb 和 Bi 的检出限为 6.6～89.3 pg/mL；将该方法用于 HepG2 和 Jurkat

T 细胞中痕量 Cu、Zn、Cd、Hg、Pb 和 Bi 的检测，其回收率为 86.6%～119%。该工作有 3 个明显的特点：①萃取有机相直接引入 ETV-ICP-MS，避免了反萃取步骤；②液相微萃取芯片与 ETV-ICP-MS 两者之间的"微微"结合，样品利用效率高（全部萃取有机相被引入 ETV-ICP-MS）；③DDTC 既是待测元素进行液相微萃取的络合剂，又是 ETV-ICP-MS 测定中的化学改进剂。但是，由于两相层流的相接触面积有限且混合较为困难，该萃取体系的萃取效率不甚理想（3.4%～20.2%）。

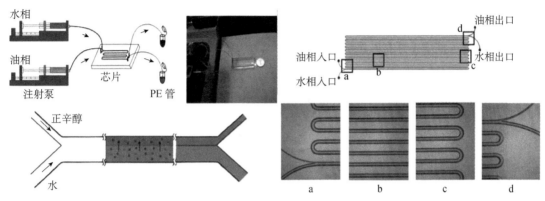

图 11-1　基于芯片层流的液相微萃取体系[15]

2. 基于液滴的液相微萃取

液滴，是一种液体在另外一种不相溶流体中的不连续分布形式，体积通常为 1 pL～10 nL。液滴较大的比表面积和剪切力与两相流速差诱导产生的内部循环流[16,17]，有助于加快反应速率，提高萃取效率；液滴微萃取体系的萃取效率通常优于常规分散相液相微萃取[18]和层流微萃取[19]的萃取效率。此外，增加微通道的弯曲长度，扰乱液滴内部原本上下对称的漩涡，促进混合，也能在一定程度上改善萃取效率。

采用 ICP-MS 作为后续检测手段时，在基于芯片层流的微萃取体系中，通常在 Y 型出口的两端分别收集有机相和水相；而在基于芯片液滴的微萃取体系中必须收集出口的混合相，通过离心操作才能分相，步骤较为烦琐。为了将液滴微萃取平台与 ICP-MS 联用，如何在微通道内实现快速的油水分离，是亟待解决的问题。目前主要有以下三种方法：①通道改性；②膜过滤分离；③基于密度差和毛细力的分相通道。

（1）通道改性：可采用透明胶带改变分离通道的粗糙度（即亲疏水性），该芯片被用于分离唾液水相与萃取了可卡因的四氯乙烯单滴[20]。另有研究工作者[21]将聚乙烯吡咯烷酮和甲基丙烯酸羟乙酯通过层流的方式引进通道内，使通道的上下表面分别具有疏水性和亲水性。当液滴通过改性通道时，能自动分成两相层流，但总流速被限制在 0.5～2 μL/min 范围内，不满足高富集倍数（EF）的要求。

（2）膜过滤分离：利用水油两相对于聚四氟乙烯疏水性多孔膜的不同润湿性及毛细力，达到两相分离的目的[22]。有研究工作者[23]采用含疏水膜的相分离器，利用泵产生负压以及毛细作用，将水油两相分离。这种方式需要额外的膜分离器以及形成相应负压的泵，并且膜的使用会造成目标分析物的一定损失。

（3）基于密度差和毛细力的分相通道：在亚克力树脂材质的分离器上制作渐宽的三角形通道，可迅速地完成基于密度差的相分离；但是其芯片的通道设计或者制作步骤都较为复杂[24]。

有研究工作者[25]基于通道改性的方式设计并构建了新型的液滴微萃取芯片用于油水快速分相［图 11-2（A）］，结合适用于分析微升级样品的 ETV 进样技术和高灵敏的 ICP-MS 检测技术，建立了液滴微萃取芯片与 ETV-ICP-MS 联用分析细胞样品中痕量 Cd、Hg、Pb 和 Bi 的新方法。采用流动聚焦结构产生水包油型液滴，以 1-辛醇作为液滴萃取相，以 DDTC 作为络合剂和 ETV 中的化学改性剂，将目标元素从水相萃取至 1-辛醇液滴中，在芯片的出口端设计了简易的油水分相通道，利用 Comsol

Multiphysics 3.5a 有限元分析软件模拟了微流控芯片分相区油水两相的变化情况。以水平集法作为相界面的处理方法，将表面张力视为水平集变量的梯度与化学势的乘积，并作为体积力代入 Navier-Stokes 方程和连续性方程组，从而实现对两相流动状态的求解。油水分相机理如图 11-2（B）所示，液滴到达孔 A 处时，因通道表面性质的变化，流速会减缓滞留；当后一个液滴追上前面的液滴时，会与之发生碰撞融合；随后，油相会渐渐填满整个不锈钢棒底部；最后，融合后的油相沿着孔 A 侧的通道壁连续流出，形成两相层流，从而达到油水分离的目的。与基于层流的液相微萃取芯片相比，以液滴作为萃取相，大大增加了水相和油相间的接触面积，显著提高了 Cd、Hg、Pb 和 Bi 的萃取效率（95.7%、38.9%、36.6%和 24.9%）。

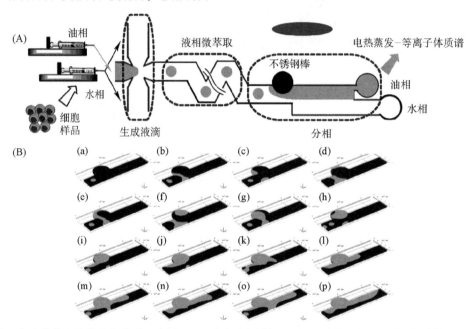

图 11-2 基于液滴微萃取的微流控芯片示意图（A）；油水分相模拟图（B）；油水分相过程的详细模拟图（a）～（p）[25]

11.2.2 基于微流控芯片的固相微萃取技术

与液相微萃取相比，固相微萃取可以利用修饰有功能基团的萃取材料[26]实现更具选择性的富集和更高的萃取效率，且更加绿色环保[27]。固相微萃取中可能存在固定相不均匀的问题，如何将萃取材料固定在芯片中、确保萃取的重现性是构建基于芯片固相微萃取体系的关键。目前，将固相微萃取整合在微流控芯片中主要有三种方法：①与填充柱类似，在通道内装填粒度均一的固定相材料；②在通道内原位合成或嵌入毛细管整体柱；③与开管柱类似，通入溶液在芯片通道壁上均匀地涂覆固定相。

1. 基于芯片的填充柱固相微萃取

在通道内加工围堰式、栅栏式、楔石式等柱塞[28]，或将微颗粒或微膜[29]紧实地固定在芯片内，可得到微流体平台中的填充柱。然而，在微米级的通道中制作围堰和锥形通道对芯片的微机电加工技术的要求较高；体状材料和微米级填充材料比表面积较小，反应动力学和吸附容量较为有限。磁性纳米粒子（MNPs）具有良好的磁导向性、生物相容性和表面易修饰等优点。在外加磁场的作用下，MNPs 可以自组装于芯片微通道中形成填充柱，无需柱塞等微结构，加工简单，且能弥补芯片中固相材料难以更换和再生的不足。因此，将比表面积大和材料表面含有丰富功能基团的新型磁性纳米材料与芯片填充技术联用，极大地改善了芯片微型化样品前处理的分析性能。胡斌等[30]首次将 γ-MPTS 修饰的纳米磁性硅球（Fe_3O_4@SiO_2@γ-MPTS）填充柱用于芯片磁填充柱固相微萃取，建立了基于微流控芯片平台的磁固相微萃取与 ETV-ICP-MS 联用用于细胞中痕量元素分析的新方法。该萃取芯片如图 11-3 所示：$Fe_3O_4@SiO_2@\gamma$-MPTS 磁性纳米粒子从芯片入口 A5 导入，并在外磁体的作用下被固定于长 2.5 cm、宽 500 μm 的芯片通道内形成磁填充微柱；细胞悬浮液及细胞破膜液从芯片入口 A1 和 A2 导入，在混合区内完成细胞破膜，随后进入磁固相萃取区；解吸液（0.2 mol/L HNO_3 和 2%硫脲）自 A4 入口导入进行目标分析物的解吸，收集解吸液后导入 ETV-ICP-MS 进行检测。在该芯片上设计并制作了微气阀控制系统用于控制芯片通道的开合。基于 Cd 和 Se 与巯基之间的强亲和力，$Fe_3O_4@SiO_2@\gamma$-MPTS 填充的芯片磁固相微萃取柱被用于细胞中 Cd 和 Se 的同时萃取；结合 ICP-MS 检测，该方法被用于考察硒化镉量子点孵育的细胞内 Cd 和 Se 的含量变化[31]。此外，基于 $Fe_3O_4@SiO_2@\gamma$-MPTS 的磁固相填充柱还被用于细胞中痕量元素形态的分析：以 $Fe_3O_4@SiO_2@\gamma$-MPTS 作为填充柱的阵列芯片与 Micro HPLC-ICP-MS 在线联用（图 11-4），被用于研究无机汞和甲基汞分别孵育 HepG2 细胞后细胞中汞形态的变化[32]；利用该方法检测了甲基汞和硒代胱氨酸共孵育 HepG2 细胞中甲基汞和无机汞的含量，同时结合 ETV-ICP-MS 测定细胞中 Hg 和 Se 总量，并通过尺寸排阻色谱/反相 HPLC-ICP-MS 等分析手段测定孵育/清除培养基和胞液中 Hg 和 Se 形态的含量，研究了甲基汞和硒代胱氨酸之间的拮抗机制。实验结果表明，硒代胱氨酸促进了细胞对甲基汞的吸收，但却以形成 MeHg-Cys 复合物的形式，减小了甲基汞对细胞的毒性[33]。

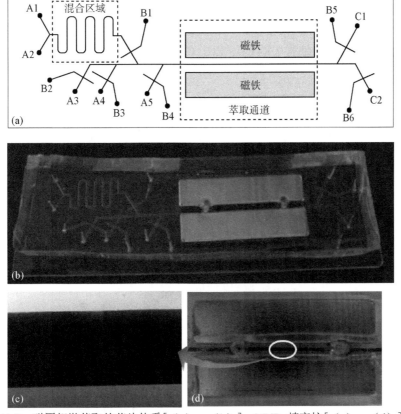

图 11-3　磁固相微萃取的芯片体系[（a）～（b）]；MNPs 填充柱[（c）～（d）][30]

除了上述 γ-MPTS 修饰的纳米磁性硅球作为磁固相填充材料外，基于磺酸基和氨基修饰的纳米磁性硅球作为磁固相萃取材料也被用于痕量元素及形态的分析中。以聚苯乙烯磺酸钠改性的磁性纳米颗粒（Fe_3O_4@PSS）作为填充柱材料，先利用磺酸基吸附呈阳离子形式的硒氨基酸，并以 Na_2CO_3 作为解吸剂，通过高效液相色谱 (HPLC)-ICP-MS 分析了富硒酵母细胞中的五种硒形态[34]。随后，设计了磁固相萃取阵列芯片（图 11-5），在 5 个微萃取通道内填充氨基改性的纳米磁性颗粒

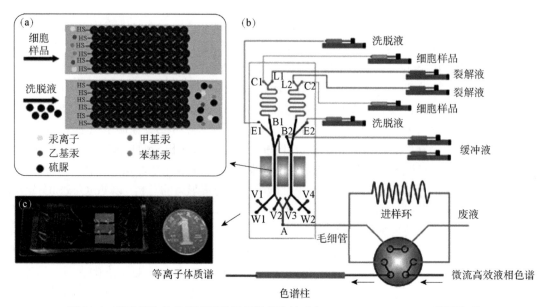

图 11-4 基于芯片的在线阵列磁固相微萃取技术与 MicroHPLC-ICP-MS 联用体系

(a) 磁固相微萃取机理；(b) 在线阵列磁固相微萃取技术与 MicroHPLC-ICP-MS 体系设计图；(c) 微流控芯片制作图[32]

($Fe_3O_4@SiO_2@APTES$)。通过计算机程序控制气阀的开关，依次完成上样、解吸、清洗和再生等步骤，与 ICP-MS 联用在线检测了细胞中的 Cu、Zn、Cd、Hg、Pb 和 Bi[35]。

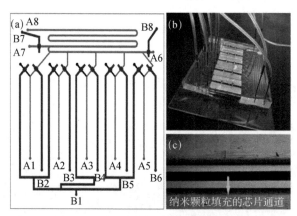

图 11-5 阵列磁固相微萃取技术系统

(a) 芯片设计图；(b) 芯片制作图；(c) $Fe_3O_4@SiO_2@APTES$ 填充柱[35]

上述工作中使用的磁固相萃取材料种类有限（$Fe_3O_4@SiO_2@\gamma$-MPTS、$Fe_3O_4@SiO_2@APTES$ 和 $Fe_3O_4@PSS$），探索吸附容量高、吸附动力学快以及比表面积大的新型磁性材料与芯片磁固相萃取体系，对改善细胞等生物样品中痕量元素及形态分析方法的分析性能具有非常重要的意义。

磁性多孔有机聚合物（MOPs）具有比表面积大、多孔性和良好的稳定性等优点。研究工作者[36]通过利用条件温和的偶氮反应，一步制备巯基功能化的 MOPs 材料，并将巯基功能化的 MOPs 填充进芯片通道内作为萃取柱；制作了含有 8 个微萃取通道的磁固相微萃取（MSPME）阵列芯片，并将其与 ICP-MS 在线联用检测了人发、尿样和细胞样品中 Pt、Au 和 Bi。该方法的样品通量达到 7/h，与 $Fe_3O_4@SiO_2@APTES$ 材料相比，该方法具有吸附容量高及萃取动力学较快等优点。此外，基于磁性金属有机框架材料（MFC）为模板合成的材料在保持原有多孔结构的同时，具有更大的比表面积（比表面积高达 5000 m^2/g）和稳定性[21]。在后续的工作[37]中以 $Fe_3O_4@SiO_2$ 磁性纳米粒子为磁源，制备了纳米磁性 Zr-MFC 复合材料；采用溶剂辅助配体交换，二巯基丁二酸为交换配体，与 MFC 中的对苯二甲酸构架单元进行交换制备了巯基功能化的 MFC-SH 材料。将所合成的两种材料在磁场的作用下分别填充到微流控芯片通道内构建了磁固相萃取双通道（MFC 和 MFC-SH），与 ICP-MS 在线联用应用于细胞样品中四种砷形态的分析（图 11-6）；MFC-SH 填充柱定量吸附 As(Ⅲ)，而 MFC 定量吸附 As(Ⅴ)、DMA 和 MMA，之后通过顺序洗脱策略后 ICP-MS 在线检测；该方法具有较大的吸附容量和较快的吸附解吸动力学。

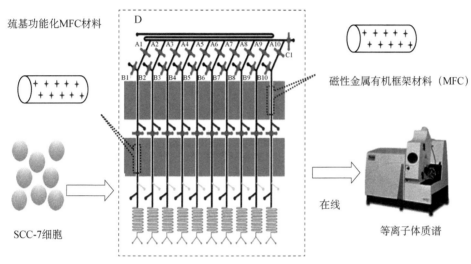

图 11-6　在线双柱阵列磁固相萃取-ICP-MS 体系的示意图[37]

2. 基于芯片的整体柱固相微萃取

基于微流控芯片的整体柱材料可以提供更好的机械稳定性和更大的比表面积，能进一步改善固相萃取体系的性能。根据制作方式的不同，芯片整体柱固相微萃取可以分为原位聚合整体柱和嵌入式整体柱[38]。

原位聚合整体柱芯片是将含有单体、化学功能基和制孔剂等混合溶液引进通道内，再通过光/热引发聚合反应，生成整体柱材料。通过调节溶液的组分和比例，可以控制整体柱的选择性和孔径，丰富了固定相的种类。研究工作者采用光引发原位聚合反应，在 PDMS 芯片中制备了氨基改性的聚(甲基丙烯酸缩水甘油酯-三丙烯酸丙烷三甲醇酯)[poly(GMA-TRIM)]整体柱，之后采用乙二胺对整体柱进行改性使整体材料带上氨基基团用于 HepG2 细胞中 Bi 的萃取[39]。但是，原位合成整体柱面临着在聚合过程中易发生柱床收缩使整体柱材料脱落的问题。

嵌入式芯片整体柱是在芯片外制备整体柱材料后，以嵌入的方式将整体材料集成到芯片上的制备方法；与原位聚合法构建芯片整体柱相比，嵌入式制备方法不必局限于溶胶-凝胶和光引发的聚合方式，可选择的聚合物整体材料种类更多，后续的材料改性更加便利。研究工作者先在通道外采用热引发制备氨基改性的聚(甲基丙烯酸缩水甘油酯-乙二醇二甲基丙烯酸酯)毛细管整体柱，在已完成光刻的硅版上固定毛细管整体柱的方形石英条模板（1000 μm×650 μm×650 μm），再倒入 PDMS 制作芯片。将制备好的毛细管整体柱截成 1 cm 后，涂上适量未固化的 PDMS 并嵌入模具形成的通道中，以保证接口不漏液[40]。此外，还将表面离子印迹技术与毛细管有机无机杂化整体柱相结合制备了聚(γ-甲基丙烯酰氧基三甲氧基硅烷@Gd^{3+}表面离子印迹)毛细管整体柱，并将 8 根整体柱嵌入微流控芯片通道中构建了以芯片为载体的毛细管整体柱阵列微萃取平台，其与 ICP-MS 在线联用，建立了一种高度自动化、可多样品同时分析的新方法用于生物样品中痕量 Gd 的分析（图 11-7）；结果表明，印迹整体柱具有吸附容量高（1584 μg/m）、选择性高（对 La^{3+} 和 Ce^{3+} 相对选择性系数分别为 24.0 和 9.9）的特点，在复杂生物样品痕量元素分析中具有良好的应用潜力[41]。

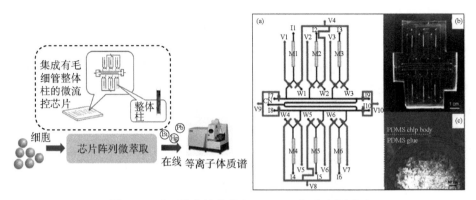

图 11-7　阵列整体柱微萃取-ICP-MS 体系示意图[40]

3. 基于芯片的开管柱固相微萃取

固定相常以涂渍法,通过化学键交联、键合或物理吸附等作用力与通道内壁结合。有研究工作者[42]采用 NaOH 饱和溶液将聚甲基丙烯酸甲酯(PMMA)芯片内壁改性上羧基,实现了微量透析液中 Mn、Co、Ni、Cu 和 Pb 的微萃取和富集。为了省去固定相活化和再生步骤,该课题组接着在羧基改性的通道内通入 50%丙烯酰胺修饰上乙烯基,然后加入单体二氯乙烯和引发剂偶氮二异丁腈,在紫外光的照射下,改性上聚氯乙烯,利用 C—Cl 的强电负性萃取了环境水样中的 Mn、Co、Ni、Cu、Cd 和 Pb[43]。后续建立了含 8 个阵列环形萃取通道的高通量芯片-ICP-MS 痕量元素分析体系[44]。此外,有人在芯片通道内通入 2% $HAuCl_4$ 溶液,利用聚二甲基硅氧烷(PDMS)交联反应后残留的 Si—H 键还原 $HAuCl_4$ 溶液,从而在芯片通道内改性上 Au NPs 得到了 Au NPs 涂覆的芯片并应用于水样中 Hg^{2+} 的萃取[45]。

综上所述,基于微流控芯片的微萃取技术利用了芯片本身的优势,具有微型化、集成化和阵列化的优点,通过对萃取体系的精心设计和微机电精密加工,可针对不同目标分析物实现基质分离和分析物富集等过程。其中液相微萃取技术中的层流微萃取的相接触面积有限且混合较为困难,元素的萃取效率有待提高;液滴微萃取体系具有较大的比表面积和内部循环流,可提高层流微萃取的萃取效率,将液滴微萃取体系与非原位的 ICP-MS 检测技术联用时,必须在微通道内实现快速的油水分相。

在固相萃取中,可通过在萃取材料中修饰特定的功能基团,提高萃取方法的选择性和萃取效率。萃取材料是基于微流体平台的固相微萃取体系的核心。填充柱式微流体平台具有较高的吸附容量和富集倍数。传统填充柱式微流体平台存在柱塞的制作较为复杂等问题,基于磁性纳米材料的芯片填充柱具有功能基团丰富、材料易于更换与再生、芯片加工简单等优势。整体柱微流体平台无需制作复杂的柱塞,吸附容量大,吸附/解吸动力学快。整体柱材料可在芯片通道内原位聚合而成,但需要注意脱壁等问题;也可以将毛细管整体柱嵌入至芯片通道中,毛细管整体柱的材料类型更丰富,但是在嵌入的过程中需要注意接口死体积等问题。开管柱式微流体平台的制备步骤较为简单且柱压小,但所制备的固定相涂层不仅负载容量较低,且随着使用次数的增加,涂层变得不均匀,影响方法的重现性。

11.2.3　基于微流控芯片的微萃取技术在细胞中痕量元素及形态分析中的应用

微流控芯片具有分析速度快、空间分辨率高、样品/试剂消耗少等优点,为细胞分析提供了一个微型化和集成化的平台。在芯片上集成合适的样品前处理技术,有助于去除样品基质和预浓缩细胞内的目标分析物,以及对单个细胞进行封装并实现在线分析。将芯片上的微型化微萃取技术与 ICP-MS 检测相结合,是一种高灵敏度、高选择性、高通量分析痕量元素及其形态的分析方法。

表 11-1 列举了采用微流控芯片-ICP-MS 技术进行细胞内痕量元素及其形态分析的方法。可以看到,采用微流体技术与 ICP-MS 检测相结合的方法,大大降低了细胞的消耗,提高了方法的分析灵敏度;在线分析系统的构建进一步改善了样品通量;液滴芯片的设计有助于单细胞中痕量元素的分析。

第11章 微流控芯片-等离子体质谱联用技术用于细胞中痕量元素及形态分析

表 11-1 基于微流控芯片-ICP-MS 体系在细胞中痕量元素及形态的分析

方法	元素/化合物	细胞类型	消耗的细胞个数	样品体积(mL)	LOD (ng/L)	富集倍数(EF)	样品通量(/h)	参考文献
基于芯片的 MSPME-ETV-ICP-MS	Cd/Hg/Pb	HepG2	5 000	0.5	0.72/0.86/1.12	45/41.6/48.7	—	[30]
基于芯片的 MSPME-ICP-MS	Cd/Se	HepG2	500	0.5	2.2/21	23/21	—	[31]
基于芯片的在线[阵列] MSPME-ICP-MS	Cu/Zn/Cd/Hg/Pb/Bi	HepG2/Jurkat T/MCF7	60 000/300 000/80 000	0.5	49/43/4.2/6.1/13/18	69/69/66/69/61/59	5	[35]
基于芯片的 MCME-ICP-MS	Bi	HepG2	600	0.1	210	3.33	—	[39]
基于芯片的在线[阵列] MCME-ICP-MS	Hg/Pb/Bi	HepG2	30 000	0.25	23/12/13	12.5	16	[40]
基于芯片的 LPME-ETV-ICP-MS	Cu/Zn/Cd/Hg/Pb/Bi	HepG2/Jurkat T	50 000/300 000	0.28	89.3/59.8/7.5/21.6/19.2/6.6	8.54/11.84/8.06/1.35/3.43/6.35	—	[15]
基于液滴芯片的 LPME-ETV-ICP-MS	Cd/Hg/Pb/Bi	HepG2/Hela	350 000	0.35	2.5/3.9/5.5/3.4	33.5/13.6/12.8/8.7	—	[25]
基于芯片的 MSPME-HPLC-ICP-MS	SeCys₂/MeSeCys/SeMet/GluMeSeCys/SeEt	Selenium-enriched yeast	800	0.3	57/117/116/149/95	19/11/8/13/9	—	[34]
基于芯片的在线[阵列] MSPME-HPLC-ICP-MS	Hg^{2+}/$MeHg^+$/$EtHg^+$/$PhHg^+$	HepG2	2 500	0.1	18.8/12.8/17.4/41.8	9.5/9.9/9.4/9.6	3	[32]
基于芯片的在线[阵列] MSPME-ICP-MS	Pt/Au/Bi	Hela	500	0.5	8.6/4.4/3.4	15.2/15.7/15.0	7	[36]
基于芯片的在线[阵列] MSPME-ICP-MS	As(Ⅲ)/As(Ⅴ)/DMA/MMA	SCC-7	8 000	0.25	7.1/4.8/6.3/3.8	24/23/24/25	7	[37]

1. 细胞中痕量元素的分析

将基于微流控芯片的微萃取技术和 ICP-MS 联用，洗脱液可直接导入 ICP-MS 进行测定；由于样品体积有限，在 ICP-MS 中使用微型化进样技术与基于微流控芯片的微萃取技术更匹配。作为一种微量进样技术，ETV 具有样品消耗低、可以直接引入有机相等优点，选择合适的化学改性剂和升温程序还可以减少基体效应[46]。将基于芯片的 MSPME 与 ETV-ICP-MS 相结合，在细胞消耗个数为 5000 个 HepG2 细胞时即可实现 HepG2 细胞中微量 Cd、Hg 和 Pb 的定量分析[30]。结果表明，单个细胞中目标金属离子的含量在 fg/亚 fg 级。为初步研究 Cd、Hg 和 Pb 在 HepG2 细胞的积累情况，比较了控制组 HepG2 细胞与在培育过程中分别添加了 100 μg/L 和 500 μg/L 的 Cd、Hg 和 Pb 的 HepG2 细胞中 Cd、Hg 和 Pb 的测定值。芯片磁固相萃取-ETV-ICP-MS 和直接 ETV-ICP-MS 测定结果均表明，在培养基中添加了 Cd、Hg 和 Pb 的 HepG2 细胞中 Cd、Hg 和 Pb 的测定值均较控制组 HepG2 细胞的测定值高，特别是 Cd 和 Hg，其测定值有数量级的增加，说明 HepG2 细胞对重金属离子 Cd、Hg 和 Pb 具有不同程度的富集能力。另外，在实验中还比较了芯片磁固相萃取-ETV-ICP-MS 和直接 ETV-ICP-MS 检测 HepG2 细胞中 Cd、Hg 和 Pb 的测定值的差异，直接 ETV-ICP-MS 测定 Cd、Hg 和 Pb 的含量均远大于芯片磁固相萃取-ETV-ICP-MS 的测定值，推测是 γ-MPTS 改性纳米磁硅球固相填充柱主要是基于其端巯基与 Cd、Hg 和 Pb 离子之间的相互作用实现对其的萃取富集，而 Cd、Hg 和 Pb 进入 HepG2 细胞中主要与金属硫蛋白等生物分子结合，因此并非所有存在于 HepG2 细胞中的 Cd、Hg 和 Pb 都能被 γ-MPTS 改性纳米磁硅球固相填充柱吸附，导致芯片磁固相萃取-ETV-ICP-MS 的测定值偏低。随后，根据类似的吸附机制，在外磁场的作用下，将巯基改性的纳米磁性硅球填充于芯片的通道内形成磁固相微萃取柱，利用 Cd 和 Se 与巯基之间的强亲和力，仅需消耗 500 个细胞即可实现细胞内 Cd 和 Se 的同时定量；通过超滤法验证了该方法所萃取到的主要是 Cd 的小分子/离子形态[31]。

整体柱微萃取方面，研究者在 PDMS 芯片中原位合成了 poly(GMATRIM)聚合整体柱，并与 ICP-MS 联用用于 HepG2 细胞中痕量 Bi 的分析；该方法单次分析仅需消耗约 600 个细胞。结果表明，细胞内的 Bi 含量在 pg 级，细胞内 Bi 元素多以分子形态存在，该方法所萃取的 Bi 形态为小分子形态络合物[39]。

此外，基于芯片的液相微萃取技术也被用于细胞内痕量元素的分析。研究者[15]将芯片层流两相液相微萃取用于细胞中痕量重金属（Cu、Zn、Cd、Hg、Pb 和 Bi）的分析检测。结果表明：Jurkat T 细胞和 HepG2 细胞的消耗量分别为 300 000 和 5 000 个时可实现目标元素的定量分析，其中 HepG2 细胞的目标元素含量为 2.53～172 fg/cell，而 Jurkat T 细胞的目标元素含量为 0.10～12.2 fg/cell。随后，选择 1-辛醇作为液滴萃取相，DDTC 作为 Cd、Hg、Pb 和 Bi 的络合剂和 ETV 进样中的化学改进剂，建立了液滴微萃取芯片-ETV-ICP-MS 分析生物和细胞样品中痕量重金属的新方法。该方法以液滴作为萃取相，大大增加了相接触面积，显著提高了体系的萃取效率和萃取动力学。10 min 即可完成对 Cd、Hg、Pb 和 Bi 的萃取富集，样品溶液的体积仅需 350 μL，减少了细胞样品的消耗。且细胞分析结果表明，细胞内待测元素含量均在亚 fg/cell 或 fg/cell 级别，加标回收率在 83.5%～112%范围内[25]。

然而，采用上述的离线分析体系通常存在着样品易丢失、易污染、操作较为烦琐以及仪器普及率的问题。为了提高检测通量和避免离线操作可能引进的污染，设计了 MSPME 阵列芯片，在 5 个微萃取通道内填充氨基改性的 MNPs 并与 ICP-MS 联用在线检测了 Jurkat T、HepG2 和 MCF-7 三种细胞中的 Cu、Zn、Cd、Hg、Pb 和 Bi[35]。结果表明，单个细胞内的六种目标分析物的含量在 fg 级，对实际样品加标所得的回收率在 83.8%～117%之间。同时对三种细胞样品的酸消解样品和超声离心处理后的上清液中的六种元素的含量进行了检测，结果显示，超声离心处理后的上清液中六种元素的含量高于采用在线阵列磁固相萃取体系的检测结果，推测是由于六种元素的有机金属形态（例如金属结合蛋白）没有被萃取。而细胞消解检测到的元素含量也高于超声离心处理后的上清液中的六种元素的含量，表明在细胞的膜结构中也含有这六种金属。为了研究在该方法中可被萃取的六种元素的形态，分别对原始细胞样品、经磁固相萃取通道吸附后的细胞样品残液、六种金属离子在含 2%（m/v）硫脲的 0.5 mol/L HNO_3 中的标准溶液和细胞样品萃取后的解吸液中的六种元素形态进行了 SEC/RP-HPLC-ICP-MS 分析。结果表明该方法所萃取的元素形态应为自由离子和/或包含六种元素的小分子形态。随后，设计并制作含有 8 个微萃取通道的 MSPME 阵列芯片与 ICP-MS 在线联用应用于人发、尿样和细胞样品中 Pt、Au 和 Bi 的分析；该方法具有更多的阵列通道和更快的流速，样品通量高达 7/h。未孵育的 HeLa 细胞裂解后未检测到任何元素；未孵育的细胞消解后未检测到 Pt 和 Au 元素，检测到 Bi 的含量为 1.72 fg/cell。采用 0.05 mg/L Pt、Au 和 Bi 孵育 HeLa 细胞 24 h 后；细胞内的目标元素含量相比于未孵育时，检测到细胞内三种目标元素的含量在 fg 级，说明 HeLa 细胞对金属离子 Pt、Au 和 Bi 具有一定程度的富集能力[36]。此外，将毛细管整体柱与微流控芯片结合，并与 ICP-MS 在线联用对细胞等生物样品中的痕量元素 Hg、Pb 和 Bi 进行检测[40]；该方法采用阵列模式，样品通量达到 17/h；且细胞分析结果表明，当细胞消耗量为 30 000 个时，计算得到单个 HepG2 细胞中 Hg、Pb 和 Bi 的含量在亚 fg 级。

2. 细胞中痕量元素形态的分析

将 MSPME 芯片与 HPLC-ICP-MS 联用可分析细胞中的元素形态，研究工作者[34]通过磁性纳米粒子所带的磺酸功能基与目标硒氨基酸之间发生的离子交换作用实现了硒氨基酸的吸附，建立了 MSPE-HPLC-ICP-MS 用于富硒酵母中硒氨基酸形态分析的新方法。以 $SeCys_2$、SeMet、甲基硒代半胱氨酸（MeSeCys）、γ-谷酰基-甲基硒代半胱氨酸（GluMeSeCys）和硒代乙硫氨酸（SeEt）为目标分析物，检出限为 21.5～89.9 ng/L，富集倍数在 10～92 之间。结果表明，富硒酵母细胞的总 Se 量在 1290～1597 μg/g 范围内，而其中被提取出的硒氨基酸含量范围在 19.9～48.4 μg/g 之间。对样品进行了加标回收实验，回收率在 78.8%～106%范围内。随后，构建了以巯基改性 MNPs 作为填充柱的阵列芯片与 HPLC-ICP-MS 的在线联用体系，研究了无机汞和甲基汞分别孵育 HepG2 细胞后汞形态的变化，发现 HepG2 细胞中 $MeHg^+$ 的脱甲基反应。其中，细胞中总汞含量的分析结果表明，$MeHg^+$ 孵育细胞比 Hg^{2+} 孵育细胞更易积累汞；Hg^{2+} 和 $MeHg^+$ 在不同孵育条件下 HepG2 细胞中汞形态分析结果表明，HepG2 细胞对 Hg^{2+} 和 $MeHg^+$ 毒性具有不同的抵御机制[32]。该作者还利用该方法检测了甲基汞和硒代胱氨酸共孵育 HepG2 细胞中甲基汞和无机汞的含量；基于 HPLC-ICP-MS 的综合分析平台，从分子水平上系统地研究了 $SeCys_2$ 对 MeHg 细胞毒性的抑制作用，提出了 MeHg 进入和排出细胞的可能途径，揭示了 $SeCys_2$ 抑制 MeHg 细胞毒性的分子机理：$SeCys_2$ 的加入促进了 HepG2 细胞对 $MeHg^+$ 的摄取；MeHg-GSH 是 $MeHg^+$ 的清除形态；HepG2 细胞中 MeHg 大量转化为络合形态（MeHg-GSH 和 MeHg-Cys）是 $SeCys_2$ 减缓 MeHg 细胞毒性的关键因素[33]。此外，集成有细胞破膜区、微萃取区和微阀的阵列微流控磁固相萃取双通道（MFC 和 MFC-SH）与 ICP-MS 在线联用代替 HPLC 分析了 As(Ⅲ)孵育 SCC-7 细胞中的 As(Ⅴ)、DMA、MMA 和 As(Ⅲ)形态[37]，基于该芯片的阵列设计、样品体积和流速等实验条件，10 个阵列通道的萃取时间为 10 min，4 种砷形态的顺序洗脱时间为 7 min，该体系的样品通量为 7/h。实验结果表明，在未孵育细胞的裂解液中并没有检测到任何砷形态，未孵育细胞的消解液中总砷含量是 0.13 fg/cell；用 10 μmol/L As(Ⅲ)孵育细胞 3 h 后，细胞裂解后检测到胞内四种目标砷形态的含量分别为 0.26 fg/cell、0.023 fg/cell、0.037 fg/cell 和 0.089 fg/cell。

11.3 基于等离子体质谱的单细胞分析

由于细胞个体的基因、蛋白质在生长过程中的随机表达以及外界条件微小变化的累积，细胞会存在巨大的差异，即细胞异质性，因此单细胞分析技术日益受到重视。单细胞分析最大的困难来源于细胞尺寸小、目标分析物绝对含量少、动态的生理过程使细胞基质更为复杂且目标分析物的浓度也呈现出较宽的变化范围等[47]。为了正确地描述细胞群体，还需要快速检测足够多数量的细胞样品。因此，单细胞分析方法需要表征技术的分析体积小、灵敏度高且分析通量高。目前，已经发展出一些针对单细胞的技术手段。例如，能量色散型荧光X射线[48]和质子激发X射线荧光光谱[49]是研究超积累细胞内（比如植物的叶肉细胞）金属/准金属含量和分布的经典方法。但是，这些检测手段的空间分辨率和定量能力仍不能满足检测非超积累细胞中痕量元素的要求。流式细胞仪是目前使用最广泛的单细胞分析方法，它结合了水动力学聚焦和荧光标记策略，可以快速分析细胞胞内或表面的目标物，细胞尺寸也可以由散射光信号得到；但是其应用范围和定量准确性受限于荧光探针的种类和光谱干扰[50]。同步辐射X射线荧光和吸收光谱也是强有力的检测手段，它能提供几十纳米至几微米的空间分辨率和元素的配位信息，具备在亚细胞水平上进行元素分析的能力[51]，已经被应用于单颗藻类、真菌、植物和动物细胞的分析，但照射时间过长或者强度过大的X射线可能会改变元素的化学形态，并且仪器成本昂贵、较为烦琐的制样过程以及样品通量低等都是不容忽视的问题[52]。ICP-MS具有强大的元素分析能力，基于ICP-MS的单细胞分析技术发展迅猛[46]。

Haraguchi等[53]将单个鲑鱼卵细胞酸消解后，通过电感耦合等离子体光谱/质谱检测了细胞内的78种元素含量，首次实现了单细胞水平上的全元素分析。但鲑鱼卵细胞的尺寸是在mm级别，并不适用于绝大多数μm级的哺乳动物类细胞。虽然ICP-MS的检出限非常低，但依然满足不了单细胞分析的检测要求。随着仪器技术的发展，时间分辨（time-resolved analysis，TRA）-ICP-MS得到了研究者们的密切关注。如图11-8所示，TRA-ICP-MS比常规检测模式能显著提高检测细胞悬浮液的信噪比[54]。假设细胞密度为10^6/mL，进样流速为0.3 mL/min，雾化效率约为1%，则引入ICP的进样速率大约为3000个cell/min，即每20 ms引入1个细胞。而离子云在ICP中的移动速率为20~25 m/s，在ICP的停留时间约2 ms[55]。因此，通过控制采样时间（dwell time，t_{dwell}）和细胞密度，理论上可以保证在任何时间内进入ICP的细胞仅为1个。此时采用TRA模式采集数据，即可获得单细胞的信号峰。根据信号峰的个数和强度值，便可同时完成细胞的计数和定量。于是，TRA-ICP-MS被越来越多地应用于单细胞分析[56]。

Li等[57]率先采用TRA-ICP-MS检测暴露铀化合物后的单个枯草杆菌。实验发现，完整的枯草杆菌悬浮液会出现铀跳峰，而10 μg/L铀离子溶液则是稳定的基线。还发现完整细胞中铀、钙和镁元素的响应值均低于相应浓度的离子溶液。利用HPLC先确定细胞内大部分的含铀化合物是铀与DNA或蛋白质的大分子结合态。然后固定炬管位置，调节雾化气流速考察信号强度的变化。实验发现，高雾化气流速能增大铀离子溶液的信号强度，却使细菌悬浮液的信号强度降低。作者认为，在低流速的情况下，细菌在ICP的停留时间更加充足，提高了电离效率，导致信号强度增加，即电离效率的差异，导致完整病菌的信号强度低于溶液。

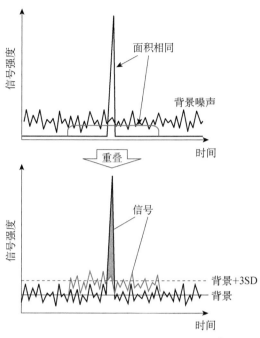

图 11-8 瞬间信号和连续信号的比较[54]

采用 TRA-ICP-MS 检测 Cr^{3+} 孵育后小球藻单细胞中的 Mg、Cu、Mn 和 Cr，出现了相应元素的跳峰。而引入没有孵育 Cr^{3+} 的细胞和 Cr^{3+} 溶液，则没有 Cr 跳峰[58]。如图 11-9 所示，该实验还发现了跳峰个数与细胞密度呈正比且跳峰强度与细胞密度无关。与 Mg 离子溶液相比，作者以平均粒径约 100 nm 的 MgO NPs 作为定量标准，计算得到单细胞中的 Mg 含量与酸消解结果更为吻合。

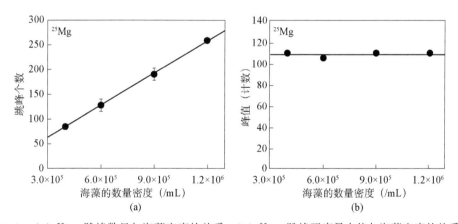

图 11-9 （a）^{25}Mg 跳峰数目与海藻密度的关系；（b）^{25}Mg 跳峰强度最大值与海藻密度的关系[58]

目前，单细胞分析技术被用于不同细胞系的研究。利用抗癌药物姜黄素钴配合物 ([Co(tpa)(-cur)](ClO$_4$)$_2$) 既有 Co 元素又有姜黄素这一特点，通过 ICP-MS 和荧光分光光度计，评价了 MCF-7 和 HepG2 细胞摄取姜黄素钴配合物和姜黄素的行为差异。实验发现，随着孵育浓度的增加，HepG2 对姜黄素钴配合物的摄取越多，并逐渐趋于稳定。然而，在相同浓度范围内，MCF-7 对姜黄素钴配合物则呈指数型增长且摄取量低于 HepG2。作者认为，两者的差异可能是由不同的细胞膜结构/组成造成的[59]。此外，在单细胞水平上研究 HeLa 细胞摄取 Cr(Ⅲ) 与 Cr(Ⅵ) 的行为差异。实验发现，细胞摄取的 Cr 含量与孵育 Cr(Ⅲ) 的浓度呈线性关系，却与 Cr(Ⅵ) 呈指数相关，并且在相同的孵育浓度下，细胞摄取 Cr(Ⅵ) 的含量明显高于 Cr(Ⅲ)[60]。

综上所述，TRA-ICP-MS 单细胞分析通常是：①只有引入细胞悬浮液且检测细胞中含量较多的元素，才会出现单细胞的跳峰信号；②在一定范围内，跳峰个数与细胞密度和跳峰强度与元素含量都呈正相关性，即 TRA-ICP-MS 能够同时实现细胞的计数和定量。这种方式存在着以下不足：①较低的细胞检测率，降低了检测通量；②存在单个跳峰来源于多个细胞的可能性，降低了结果的可靠性；③基于 TRA-ICP-MS 的单细胞研究还较为浅显，没有体现出个体信息的重要性；④TRA-ICP-MS 具有的固有缺陷，比如有损分析、不具备空间分辨和形态分析能力等。于是，在后续的工作中，研究者们从优化 ICP-MS 仪器配置、构建微流体平台和更深入的实际应用等方面进行了改进。

11.3.1 基于 ICP-MS 仪器配置的优化

进样系统：ICP-MS 主要分析液体样品，通过雾化器产生微小的气溶胶，才能保证待测样品在 ICP 中更有效的电离，因此进样系统被认为是 ICP-MS 中的关键环节。溶液雾化进样系统的检测率（detection efficiency，DE）被定义为到达检测器的粒子个数占引进粒子个数的百分比，影响分析方法的速度和可靠性；传输效率（transport efficiency，TE）通过"排废法"计算得到，被定义为样品总体积与排废出的样品体积之差除以样品总体积。因此，DE 与 TE 有关，而 TE 受限于雾化器和雾化室的性能。一般来说，进样流速为 1 mL/min 的常规雾化器（conventional nebulizer，CN）的 TE 为 1%～3%；当进样流速为 0.1 mL/min 时，TE 可以达到 10%～30%[57]。可见，匹配低流速的高效雾化器（high-performance nebulizer，HPN）可以显著提高样品的 DE，比如 ESI 公司的 PFA MicroFlow 型微同心雾化器、Meinhard 公司的 HEN 和 DIHEN 型、日本产业技术综合研究所的 HPCN 型等微同心雾化器以及 Burgener 公司的 Ari Mist 系列、X/NX/SC-175 和 ENYA Mist 型等平行流雾化器。如图 11-10 所示，Groombridge 等[61]以常规玻璃同心雾化器和 50 mL 气旋雾化室作为进样系统，发现在雾化室出口处有细胞沉积，细胞的 DE 只有 1.8%；换成自制的 HPN 后，细胞的 DE 提高至 9%；再将气旋雾化室换成有鞘流气的小体积雾化室，细胞的 DE 能被提高到 75%。Corte-Rodríguez 等[62]自制了小体积雾室并与商品化 HPN 联用，使细胞的 DE 达到 25%。

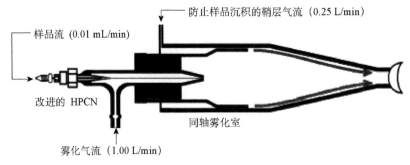

图 11-10　HECIS 示意图[61]

如图 11-11 所示，可拆卸式细胞高效进样系统包括自制 HPN、缠绕镍铬金属丝的单通道雾化室（可控温 80～200℃）[63]。与常规进样系统相比，该进样系统使细胞的 DE 提高了 10 倍。此外，较高的雾室温度促进了样品的去溶剂化，^{115}In 在 140℃时的灵敏度比 25℃提高了 2 倍。

第 11 章 微流控芯片-等离子体质谱联用技术用于细胞中痕量元素及其形态分析

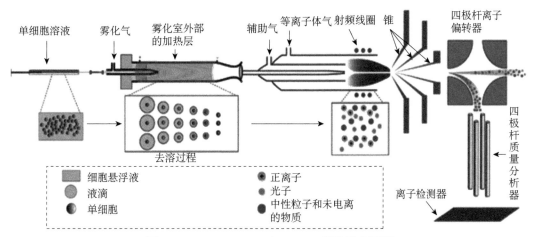

图 11-11 TRA-ICP-MS 与 HECIS 联用示意图[63]

气压促进式 HPN 和雾化室，如图 11-12 所示[64]。与常规进样系统对比，该体系的 LOD、灵敏度、精密度和耐盐量都较好，而氧化物和双电荷干扰以及记忆效应都与常规模式处于相当水平。在流速为 7.5 μL/min 时，该进样系统的 TE 可达到约 100%。

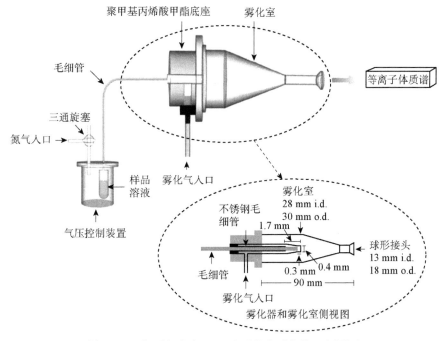

图 11-12 气压促进式 HPN 和雾化室系统的示意图[64]

与溶液雾化不同，激光烧蚀（laser ablation，LA）属于固体进样方式。通过高能激光的照射使微区样品受热解离成气溶胶，再由载气（氦气或氩气）运送至 ICP 中进行解离、原子化和离子化等过程，因此 LA 可以显著提高样品的传输效率[65]。此外，它还具有空间分辨能力，不仅可以提供元素的浓度信息，还可以实现元素的分布成像。目前激光器的光斑也已经能达到亚 μm 级[66]至 μm 级[67]范围，满足单细胞分析的要求。为了获取更好的分辨率，通常使用体积较大的细胞作为研究模型，比如巨噬细胞和成纤维细胞等。此外，还存在一些不足：①LA-ICP-MS 缺乏能模拟细胞基质和尺寸的定量方法；②细胞间的距离小于激光器光斑或细胞重叠造成较低的检测通量。

为了模拟细胞基质的高碳量，研究者们开发了三种定量标准：①在硝酸纤维膜上掺杂 NPs[68] 或标准溶液的液滴[69]；②以商品化喷墨打印机制备与细胞大小和含碳量相似的标准液滴[70]；③在阵列微孔板上加入目标元素、内标和明胶混合溶液的液滴[71]。

为了提高检测通量，研究工作者[72]制作了 PDMS 阵列微孔板。先利用密度大的 CO_2 挤走微孔中的空气，再利用 CO_2 的水溶性，促进细胞悬浮液进入微孔。静置 30 min 后，稍微倾斜微孔板，小心吸取多余的细胞悬浮液。在 500 个微孔中，约有 60%的微孔捕获的是单个细胞，约 4%捕获了多个细胞。与常规方法相比，避免了细胞重叠，使检测通量提高了约 5 倍。另外，有研究者[73]利用压电点样仪和 CellenONE 单细胞分离平台，将单细胞有序引进玻璃片的微格，使捕获率达到 99%，检测通量提高到 550 个 cell/h。

随着技术的日臻发展，LA-ICP-MS 不仅能实现细胞的二维平面成像[74]，还能对细胞进行三维立体结构的表征[75]。除了检测细胞内的痕量元素[75]，还能结合元素标记技术，表征单细胞内的 DNA 和蛋白质[76]。此外，结合常规 ICP-MS 和单颗粒-ICP-MS[77]技术，还可以交叉验证结果的准确性。

质量分析器和检测器：已有研究表明，t_{dwell} 越小，TRA-ICP-MS 中的信噪比越高且 LOD 越低，因此在 TRA-ICP-MS 分析中，t_{dwell} 是很重要的参数。ICP-MS 主要有三类质量分析器——四极杆（quadrupole，Q）、扇形磁场（sector field，SF）和飞行时间（time-of-flight，TOF）。其中，ICP-SF-MS 和 ICP-TOF-MS 的 t_{dwell} 分别是 100-μs 和 10-μs 级别，都已经成功应用到了单细胞分析之中；值得一提的是，由 ICP-TOF-MS 发展出来的质谱流式细胞仪，借助于快速的扫描速度，可以实现单细胞内多达 50 个元素的同时检测。因此利用元素标签代替传统的荧光探针分析单细胞的免疫表型，具有广阔的应用前景[78]。Bandura 等[79]利用 20 种稀土元素聚合物标签同时分析了单细胞表面的 20 种抗原。Bendall 等[78]通过同时分析单细胞的 34 个表面标志物和胞内信号蛋白，将人骨髓细胞样品分成了 30 个亚型，并且研究了在外界刺激下，这些亚群细胞的相应变化，从而更加精确地描述细胞表型。有关质谱流式细胞仪的技术细节和最新进展可以参考文献[79]，此技术有助于更深入地理解细胞的生理过程和疾病机制。

但是这些质量分析器的价格较为昂贵，普及度都没有 ICP-QMS 高。受限于原先的技术水平，ICP-QMS 最初的 t_{dwell} 最小只能达到 10 ms。Miyashita 等[80]以函数信号发生器和电子脉冲计数器来直接读取 ICP-QMS 的数据，由于没有死时间，使 t_{dwell} 可以降低至 0.1 ms。如图 11-13 所示，^{31}P 在 t_{dwell}=0.1 ms 时的信噪比比 t_{dwell}=10 ms 的信噪比提高了约 13 倍，并且还降低了一个跳峰信号来源于多个细胞的概率。随着科学技术的进步，目前商品化 ICP-QMS 的 t_{dwell} 已经能达到 100-μs 级别，比如安捷伦公司推出的 Agilent 7900（t_{dwell}=100 μs）和德国耶拿公司推出的 PlasmaQuant®MS（t_{dwell}=50 μs）。自 2010 年起，珀金埃尔默公司相继推出串联了四极杆离子偏转器的 NexION 300/350/2000 系列，ICP-MS 的数据采集能力高达每秒采集 100 000 个数据点，即 t_{dwell}=10 μs。此外，为了进一步减少基质离子的干扰，安捷伦公司的 Agilent 8800/8900（t_{dwell}=100 μs）、赛默飞公司的 iCAP Q™（t_{dwell}=100 μs）和珀金埃尔默公司的 NexION® 5000（t_{dwell}=10 μs）等型号的三重四极杆 ICP-MS 在近几年也纷纷面世。因此，ICP-QMS 逐渐成为单细胞研究中极具性价比的检测仪器。

11.3.2 微流体平台的优化

单细胞分析依赖于对细胞的操纵，主要包括细胞的定位与捕捉、细胞的分选及融合等。微流体

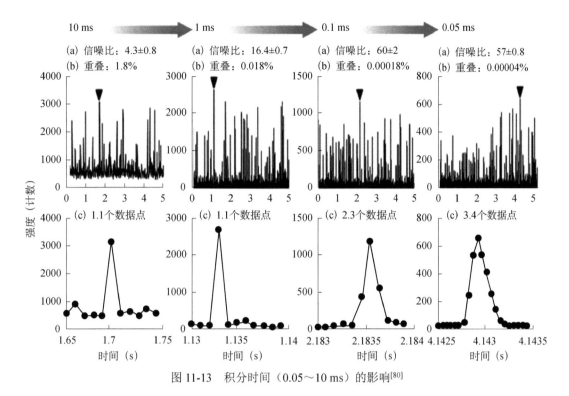

图 11-13 积分时间（0.05～10 ms）的影响[80]

这一微型化的操纵平台，具有与细胞尺寸近似的空间尺寸；不断发展的微机电加工技术使得多种微尺度结构在微流体芯片/装置上得以制作和应用，这均为在微流体平台上实现单细胞获取和分析提供了可能[81-83]。

在微流体平台上对目标细胞进行分离和获取通常采用几种策略[84-86]（图 11-14）：基于液滴的单细胞分离[87]，即将单个细胞通过液滴隔离开来，这一方法具有高通量的优点，但仅适用于已纯化过的单种群细胞；水动力学捕获，即采用流体力学方法进行细胞捕获，这一方法能将目标单细胞从基质（例如多种细胞）中通过尺寸、可形变性等特点分离出来，但对芯片设计和制作的要求较高；磁分离，采用细胞磁标签对目标细胞进行捕获，特异性好，但需要额外的抗体及磁性标签；声捕获，是基于声表面波控制粒子运动的细胞控制方法，能准确地通过粒径实现细胞定位，但可能对细胞活力等生理性质具有负面影响；介电泳捕获，基于细胞（可极化粒子）在不均匀电场中介电常数不同而具有不同运动特性实现细胞分离，通过对电场的调控能简单地对目标细胞进行选择，但在长期使用时难以解决产热的问题；光学捕获[88]，也称光学镊子，采用激光束产生的光学势阱实现对细胞的控制，该方法的适用性更为广泛，但需要昂贵的光学体系。使用最为广泛的微流控单细胞捕获方法是基于液滴的单细胞分离[87]，即将单个细胞通过液滴隔离开来，这一方法具有高通量的优点，十分适用于与 ICP-MS 联用用于分析单细胞中的痕量元素，解决气动雾化-时间分辨 ICP-MS 细胞分析体系中无法完全保证单细胞信号采集的问题。因此，利用微液滴发生器或微流控芯片/微流体装置等微流体平台产生封装单细胞的液滴或形成单行排列的细胞流，可以有效降低此概率。

基于瑞利原理，微液滴发生器（micro droplet generator，μDG）的射流在受到电压产生的机械振动所传递的压力时会变得不稳定，从而克服表面张力和黏滞力并喷出液滴。利用 μDG 产生封装单个富硒酵母细胞的液滴，并以 He 作为去溶剂气，可将液滴引进 ICP-SF-MS（图 11-15），使细胞的 DE 可以达到 100%，并且引入的基质少，能有效降低质谱干扰[89,90]。但是 μDG 也存在一些不足，比如对氩气流速的要求较为苛刻、价格较贵、液滴的尺寸范围与喷头有关，尤其是压电作为一种被动产生液滴的方式，它对溶液的 pH、黏度和含盐量都很敏感。

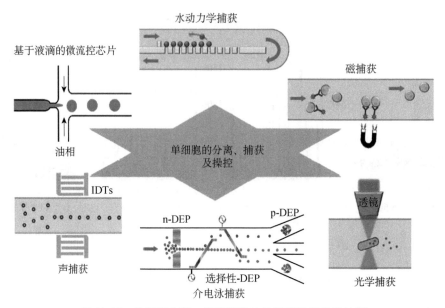

图 11-14　单细胞分离、捕获和操控的微流控芯片设计[84]

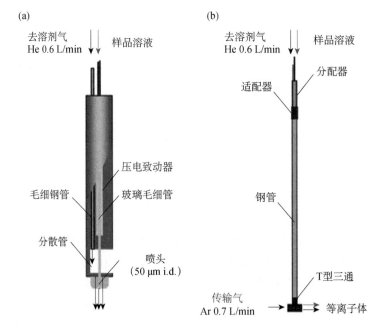

图 11-15　(a) 微液滴发生器；(b) 去溶剂化示意图[90]

液滴微流控芯片一般产生 W/O 型液滴，即液滴的生成通常需要油相作为连续相以实现液滴的分离，Verboket 等[91]在十字聚焦的液滴微流控芯片中采取使用高挥发性有机相全氟己烷作为液滴连续相，并在液滴离开芯片后通过加热和自制的膜去溶装置去除有机溶剂（图 11-16），有效避免了有机相对 ICP-MS 稳定性的干扰，并采用该分析体系实现了对单个红细胞中铁元素的检测。然而，由于该体系中的大量附件是自制的，且该分析体系组成较为复杂，操作复杂。因此，通过寻找更为简便的液滴芯片-ICP-MS 单细胞分析体系十分必要。

胡斌等[92]利用醇类具有黏度大、含碳量低的优点，以正己醇作为有机相，并利用 HPN 和雾室加氧技术降低了有机相的不利影响；以含 1% Span 80 的正己醇作为有机相，细胞悬浮液作为水相，通过简单的流动聚焦结构产生 25 μm 的液滴封装单个 HepG2 细胞（图 11-17）；再通过玻璃毛细管

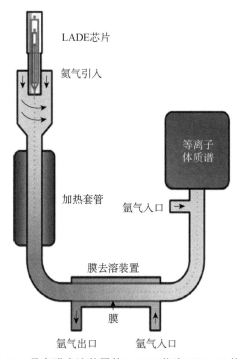

图 11-16 具有膜去溶装置的 LADE 芯片-ICP-MS 体系[91]

（75 μm i.d.×365 μm o.d.）将芯片中生成的包裹单细胞的液滴通过微流雾化器直接导入 ICP-MS，实现单个 HepG2 细胞中 Zn 的分析。该芯片上液滴的生成频率为 $3×10^6$～$6×10^6$ min^{-1}，细胞引入频率为 2500 min^{-1}，方法具有较高的样品通量[92]。常规 PDMS 芯片一般是准 2D 芯片；在高度为几十微米的微通道中，液滴不可避免地会与通道表面接触，所以生成液滴时需要考虑通道的表面性质；且 PDMS 的耐压能力往往比较有限（0.3～0.5 MPa）。针对这一问题，后续的工作[93]基于 PEEK 头、PEEK 管、石英毛细管和四通阀等全商品化部件，构建了具有同轴结构和可视化特性的 3D 液滴微流体装置（图 11-18）。与其他 3D 液滴微流体装置相比，在该液滴微流体装置中，可以观察液滴的生成过程；与准 2D PDMS 芯片相比，该液滴微流体装置无须等待键合后的 PDMS 芯片进行老化，并且不需要黏结额外的毛细管，便能直接与 ICP-MS 联用，大大简化了接口设计和操作步骤，有利于提高细胞的检测率。该装置简单、成本低廉、制作方便，不需要二次加工且耐高压。在此基础上，通过对玻璃毛细管的疏水化改性以及设计可控的气路系统，以 ICP-MS 的工作气体（氩气）作为连续相，构建了对 ICP-MS 仪器更友好的气包水型液滴微流体装置，完全避免了有机相造成的 ICP 功率的不稳定和采样锥口的碳沉积等问题。这种设计不需要膜去溶装置或具有加氧技术的 ICP-MS 和铂锥等昂贵仪器，便可形成封装单细胞的液滴，将一个信号峰来源于多个细胞的概率降低至 $6.1×10^{-3}$；细胞的检测率为 17.6%，该装置具有更普遍的适用性[94]。另外有研究工作者[95]设计了具有十字交叉结构的四通阀，在十字交叉处产生包裹单细胞的液滴，单个液滴封装单细胞的概率小于 0.005%。

上述工作是将封装单细胞的液滴直接引入 ICP-MS，只能实现单细胞中痕量元素总量的分析。为了进一步获取单细胞中不同元素形态的含量信息，研究工作者[96]设计并制作了集成有十字型通道液滴生成区、细胞裂解区和 T 型通道液滴分裂区的集成化芯片，建立了液滴分裂微流控芯片与 ICP-MS 检测体系在线联用分析方法并用于单细胞中 FePt-Cys NPs 的降解行为分析。利用 T 型通道的设计将芯片中生成的封装了单细胞的液滴进行均匀分裂，在梯度磁场的作用下，未降解的磁性 FePt-Cys 在分裂时保留在靠近永磁体的子液滴中且通过废液口排出，远离永磁铁一端的子液滴里则是均相的溶液，里面包含单细胞中因 FePt-Cys NPs 降解形成的游离 Fe 和 Pt；将远离永磁铁的出口端与 ICP-MS 雾化

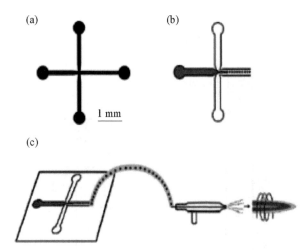

图 11-17　(a) 液滴芯片设计图；(b) 液滴生成；(c) 在线液滴芯片-ICP-MS 单细胞分析示意图[92]

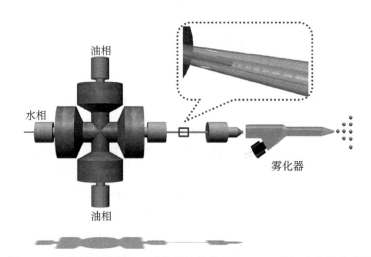

图 11-18　可视化特性的 3D 液滴微流体装置-ICP-MS 单细胞分析体系[93]

器连接，即可实现单细胞内游离 Fe 和 Pt 的在线分析。另外，移除永磁铁并堵住一侧出口，则液滴不发生裂分直接引入 ICP-MS，可以实现单细胞摄取 FePt-Cys 总量的在线分析（图 11-19）。

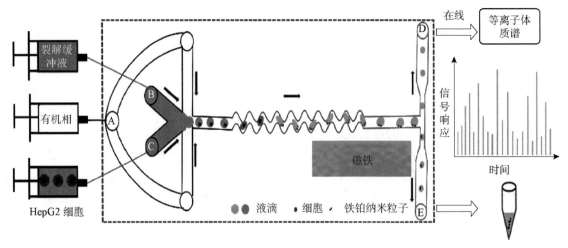

图 11-19　液滴分裂微流控芯片-ICP-MS 体系[96]

为了避免烦琐的前端设计并降低对仪器的要求,还可以通过流场力使细胞提前在微通道内有序排列,提高样品的分析通量。有研究者[97]设计了螺旋形通道与 TRA-ICP-MS 联用,通过含有周期性微柱螺旋形通道的设计可以在 100~800 μL/min 流速范围内实现细胞的单一位置的聚焦。该方法操作方便、能连续且快速地对单细胞进行排列并分析,样品通量高(16 000 cells/min)。在后续的工作中,将玻璃毛细管缠绕于聚氨酯圆柱,利用螺旋结构的曲率效应产生迪恩涡,对管内的速度场和压力场施加作用力,从而使单细胞有序排列(图 11-20);细胞的检测率达 42.1%[98]。该方法既有较好的时空分辨率[(41.55±17.46) μm,(0.97±0.41) ms],又有较高的通量(40 000 cells/min)。

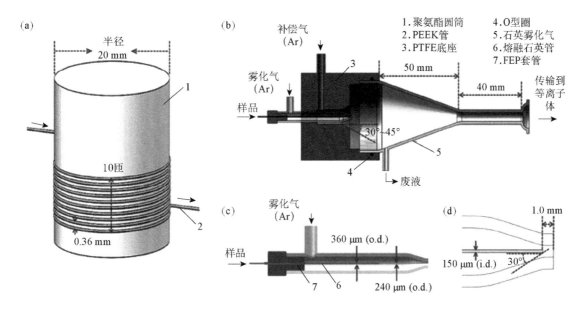

图 11-20　螺旋通道的单细胞聚焦体系[98]

(a)3D 螺旋阵列;(b)雾化器;(c)同心雾化器和同轴雾化室;(d)雾化器喷嘴尖端设计图

另外,有研究者[99]设计了直-弯曲-直通道(图 11-21)进行单细胞的聚焦,并将其与 ICP-MS 联用用于单细胞分析。作者使用具有良好热分解性能的 NH_4HCO_3 缓冲液作为鞘流促使细胞的聚焦,避免了油相的使用,细胞检出率高达 70%,样品通量为 25 000/min。

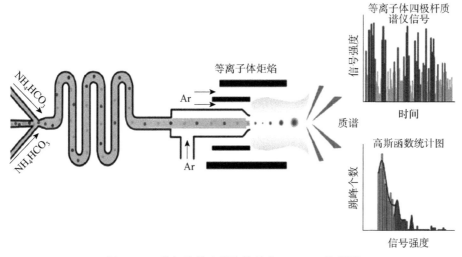

图 11-21　单细胞排序微流控芯片-ICP-MS 体系[99]

由流通池、可观察细胞完整性的显微镜、HPN和雾化室[64]组成的单细胞进样体系，如图11-22所示[100]。在流通池中间引入细胞悬浮液并在两侧引入鞘流，促使形成单行排列的细胞流。实验优化了毛细管内径，使细胞的检测率达到100%。

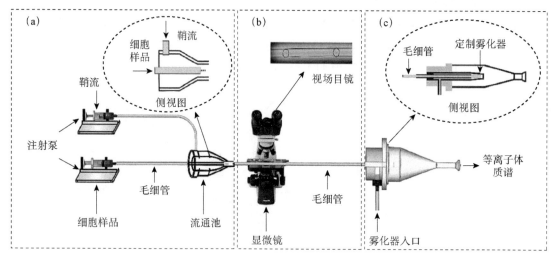

图11-22　单细胞分析体系[100]

(a) 流通池；(b) 显微镜视觉校准系统；(c) 雾化系统

11.3.3 微流体技术结合ICP-MS用于单细胞中痕量元素及形态分析

随着方法学的发展和完善，TRA-ICP-MS开始逐渐被应用于探究更有意义的对象体系，通过分析个体的差异信息，可得到常规方法无法获取的结果。

单细胞中内源性元素的分析：将液滴微流控芯片-TRA-ICP-MS在线联用体系用于孵育ZnO NPs后的HepG2的分析[92]。结果表明了HepG2细胞的异质性，单个细胞中Zn的含量分布符合高斯分布，单个细胞中Zn含量为21.7 fg，与使用常规的酸消解ICP-MS检测方法的测定结果（25.0 fg）相符。此外，通过引入鞘流促使形成单行排列的细胞流并与ICP-MS联用，单细胞分析结果表明，单个红细胞中Cu含量为0.20～0.40 fg，展现了细胞间的异质性，且与微波消解结果相一致（0.266 fg/cell）[100]。

单细胞摄取外源性金属离子的分析：基于周期性微柱螺旋形通道芯片与TRA-ICP-MS联用，研究者[97]从单细胞水平探究了Cd^{2+}和Cu^{2+}的拮抗作用；结果表明，经100 μg/L Cd^{2+}和100 μg/L Cu^{2+}共孵育MCF-7、bEnd3和HepG2三种细胞12 h，竞争率分别达到12.8%、4.81%和10.43%，且单细胞的数据明显反映出细胞间的异质性。

单细胞摄取外源性纳米颗粒的分析：研究者[92]将液滴微流控芯片-TRA-ICP-MS联用体系用于单个HepG2细胞摄取/吸附ZnO NPs的研究：在同一个细胞群中，部分细胞不会摄取/吸附ZnO NPs，部分细胞会摄取/吸附1个ZnO NPs，部分细胞会摄取/吸附2个ZnO NPs，还有部分细胞会摄取/吸附3个ZnO NPs，该结果也同时展现了单个细胞在摄取/吸附ZnO NPs时的异质性。与使用细胞悬浮液进行直接TRA-ICP-MS分析相比，所建立的方法有效地避免了多个细胞存在于一个气溶胶中导致获取错误信息的问题，对细胞异质性研究更为有利。

除了上述对单个细胞摄取ZnO NPs的差异性研究之外，有研究者[95]还对Au NPs孵育MCF-7细胞中的Au进行了单细胞分析；结果表明，随着Au NPs孵育浓度的增加，单个细胞内Au跳峰强度增加；通过对两个样品的单细胞跳峰进行高斯拟合发现，单个细胞内的Au NPs含量分别为2.27 fg和

3.77 fg；而酸消解的结果表明单个细胞内的 Au NPs 含量为 2.38 fg，更进一步说明了细胞的异质性。为了探单细胞对不同 Au NPs 摄取行为的差异，笔者对 HeLa 细胞摄取不同粒径 Au NPs@柠檬酸和 Au NPs@DNA（15 nm、30 nm、60 nm）的行为差异进行了分析。与 Au NPs@柠檬酸相比，细胞摄取 Au NPs@DNA 的含量较为一致且摄取了 Au NPs@DNA 的细胞占比也较高，推测是由于与易团聚的 Au NPs@柠檬酸相比，Au NPs@DNA 在培养基中相对稳定。实验发现，在低浓度和较短时间的孵育条件下，细胞摄取 Au NPs@DNA 比 Au NPs@柠檬酸多；而当浓度和孵育时间增加时，细胞摄取 Au NPs@柠檬酸又比 Au NPs@DNA 多，并且 Au NPs@DNA 达到摄取平台所需的时间较长，这些差异说明两者可能存在不同的内在化通道。该工作先利用低温抑制细胞吸收 Au NPs，发现细胞膜吸附 Au NPs@柠檬酸的含量大于 Au NPs@DNA，推测因为 Au NPs@DNA 的 Zeta 电势更负，与细胞的排斥力更大。随后，加入了网格蛋白抑制剂（氯丙嗪），发现细胞摄取 15 nm、30 nm Au NPs@DNA 的含量明显减少，说明细胞摄取这两种 Au NPs 主要依靠网格蛋白的内吞作用[101]。此外，基于螺旋通道内的迪恩力作用促使单细胞有序排列，研究者[98]对 Au NPs 孵育的 K562 细胞进行了分析；结果表明，随着 Au NPs 孵育浓度的增加，K562 细胞跳峰强度增加，且强度分布更多样化，很好地说明了 K562 细胞对 Au NPs 摄取行为的差异性。为了探究不同种类细胞对 Au NPs 摄取行为的差异性，有研究者[99]将直-弯曲-直通道的设计与 ICP-MS 联用用于 HeLa 和巨噬细胞两种细胞对 Au NPs 摄取差异的研究。结果表明，两种细胞对 Au NPs 的摄取均显示出细胞间的异质性，且 HeLa 单细胞摄取行为的差异更为明显，而巨噬细胞对 Au NPs 的摄取含量约为 HeLa 细胞摄取量的 20 倍。

关于单细胞对外源性纳米颗粒的摄取研究，有研究者[93]基于同轴结构的可视化 3D 液滴微流体装置与 TRA-ICP-MS 联用，对 HepG2 细胞中的 Ag NPs 进行了分析。结果表明，细胞摄取 Ag NPs 所体现出的细胞异质性比 Ag^+ 明显，推测与孵育 Ag NPs 的培养基是非均相溶液有关；用 Ag^+ 孵育 HepG2 细胞 6 h 后，细胞群中几乎所有细胞都摄取了 Ag^+；用 Ag NPs 孵育 HepG2 细胞 12 h 后，才使几乎所有细胞都摄取了 Ag NPs。孵育 Ag^+ 或 Ag NPs 后，细胞摄取 Ag 的含量均随孵育时间延长而不断增加，24 h 时都还未到达吸收平台，且细胞摄取到的 Ag^+ 含量略小于 Ag NPs。

为了探究纳米颗粒在单个细胞中的降解行为，研究工作者[96]构建了液滴分裂微流控芯片，对细胞摄取并降解 FePt-Cys 磁性纳米粒子的行为进行了研究。随着孵育时间的增加，细胞摄取和降解 FePt-Cys 的占比增加；孵育 6 h 时，几乎所有的细胞都摄取了 FePt-Cys NPs；而在孵育 6 h 时，只有在 60%的细胞中 FePt-Cys NPs 发生降解；当孵育时间达到 18 h 时，几乎在所有的细胞中 FePt-Cys NPs 都发生了降解。此外，在细胞内 Pt 的释放率高于 Fe 的释放率，推测释放的 Pt 对细胞产生了更强的毒性并造成了细胞凋亡。该方法无需复杂光刻技术，具有制备简单和高通量分析的优点，并具有从单细胞水平上对不同元素形态进行分析监测的能力。

微流体平台有利于操纵单细胞，可集成细胞分选、样品前处理等多个功能模块，具有实现细胞形态分析以及获取更多信息的应用潜力。随着方法学的日臻成熟，借助 TRA-ICP-MS 技术能获取许多细胞群体实验无法提供的信息。比如，设计单因素的差异性研究，考察不同的细胞对于相同外界刺激呈现出的元素含量、分布函数或摄取/清除动力学的差异，可以为辨别细胞的活性、周期、亚群分型和筛查癌细胞提供新的技术角度；也可以选择形态、尺寸或表面修饰物不同的同元素物质作为外界刺激物，考察细胞对于不同外界刺激物表现出的行为差异，结合刺激物本身的特定差异，探讨刺激物与细胞之间的作用机制。借助微流体平台和其他单细胞检测技术（如 LA-ICP-MS、流式细胞仪、ESI-MS 等），探究元素/元素的空间分布与细胞尺寸/细胞周期/细胞组分的含量和形态变化（如内源性元素、DNA、脂质、蛋白质）等各项生理指标之间的联系，绘制元素指纹识别谱库，将有助于破解细胞生命活动的内在机制[102]。比如，张新荣等[103]将微流控芯片液滴萃取技术和 Pico-ESI-MS 联用用于单个正常星形胶质细胞和胶质母细胞瘤癌细胞中磷脂质的分析，从单细胞水平提供一定的代谢组学信息。随后,他们将微流控液滴芯片微萃取技术和 ESI-MS 联用,用于单个 K562 细胞中 67 种代谢小分子的分析,从单细胞水平探索了 2-脱氧-D-葡萄糖刺激 K562 细胞后,细胞内葡萄糖-磷酸、2-脱氧-D-葡萄糖-磷酸和核糖-磷酸等含量的变化,从单细胞水平探究药物代谢的机理[104]。

11.4 总结与展望

痕量元素在细胞的生理活动中扮演着重要角色,开展细胞中痕量元素及其形态分析有助于从分子水平上揭示痕量元素在细胞生理活动中的作用机制,具有十分重要的意义。尽管 ICP-MS 这一元素特异性检测器能高灵敏地测定痕量元素,但细胞中的痕量元素分析依然面临着细胞样品量极少、细胞基质复杂和目标元素/形态含量极低等问题。本章重点介绍了微流控芯片分析系统与 ICP-MS 结合用于细胞中痕量元素及其形态分析,总结了基于芯片的样品微萃取技术在该领域的发展和应用;此外,本章还总结了基于微流控芯片的单细胞分析方法及其在细胞异质性方面的研究,以及基于微流控芯片-ICP-MS 分析体系在细胞内痕量元素及形态分析方面的应用。

尽管微流控芯片分析系统与 ICP-MS 及其联用技术的结合展现出了对细胞样品中痕量元素及其形态的强大分析能力和应用前景,作为一个新兴的交叉研究领域,该分析技术的应用还需更广泛深入的研究,比如,探索更多选择性好、吸附容量高、反应动力学快的新型固相萃取材料并应用于微流控芯片固相微萃取,进一步提高分析体系的灵敏度、选择性和样品通量,降低细胞的消耗个数;将微流控芯片-ICP-MS 分析体系与生物分析技术、多重联用技术结合,从分子水平上探明目标元素在胞内的迁移转化途径,找寻目标元素的形态转变与其所调控生理功能之间的联系,探究痕量元素的生理作用机制;引入元素标签对细胞中关键生物标志物进行标记,采用液滴芯片-时间分辨-ICP-MS 单细胞分析技术,实现单个细胞水平的生物标志物及痕量元素的同时检测,实现细胞摄取行为与生物指标的综合性分析等。

参 考 文 献

[1] Schmid A, Kortmann H, Dittrich P S, Blank L M. Chemical and biological single cell analysis. Current Opinion in Biotechnology, 2010, 21 (1): 12-20.

[2] Mounicou S, Szpunar J, Lobinski R. Metallomics: The concept and methodology. Chemical Society Reviews, 2009, 38 (4): 1119-1138.

[3] Wang H, He M, Chen B B, Hu B. Advances in ICP-MS-based techniques for trace elements and their species analysis in cells. Journal of Analytical Atomic Spectrometry, 2017, 32 (9): 1650-1659.

[4] Lanni E J, Rubakhin S S, Sweedler J V. Mass spectrometry imaging and profiling of single cells. Journal of Proteomics, 2012, 75 (16): 5036-5051.

[5] Roudeau S, Carmona A, Perrin L, Ortega R. Correlative organelle fluorescence microscopy and synchrotron X-ray chemical element imaging in single cells. Analytical and Bioanalytical Chemistry, 2014, 406 (27): 6979-6991.

[6] Musat N, Halm H, Winterholler B, Hoppe P, Peduzzi S, Hillion F, Horreard F, Amann R, Jorgensen B B, Kuypers M M

M. A single-cell view on the ecophysiology of anaerobic phototrophic bacteria. PNAS, 2008, 105 (46): 17861-17866.

[7] Manz A, Graber N, Widmer H M. Miniaturized total chemical-analysis systems: A novel concept for chemical sencing. Sensors and Actuators B: Chemical, 1990, 1 (1-6): 244-248.

[8] Bhagat A A S, Bow H, Hou H W, Tan S J, Han J, Lim C T. Microfluidics for cell separation. Medical & Biological Engineering & Computing, 2010, 48 (10): 999-1014.

[9] Chao T C, Ros A. Microfluidic single-cell analysis of intracellular compounds. Journal of the Royal Society Interface, 2008, 5: S139-S150.

[10] Wang H, Liu X L, Nan K, Chen B B, He M, Hu B. Sample pre-treatment techniques for use with ICP-MS hyphenated techniques for elemental speciation in biological samples. Journal of Analytical Atomic Spectrometry, 2017, 32 (1): 58-77.

[11] Hu B, He M, Chen B B, Xia L B. Liquid phase microextraction for the analysis of trace elements and their speciation. Spectrochimica Acta Part B: Atomic Spectroscopy, 2013, 86: 14-30.

[12] Atencia J, Beebe D J. Controlled microfluidic interfaces. Nature, 2005, 437 (7059): 648-655.

[13] Kokosa J M. Recent trends in using single-drop microextraction and related techniques in green analytical methods. TrAC Trends in Analytical Chemistry, 2015, 71: 194-204.

[14] Weigl B H, Yager P. Silicon-microfabricated diffusion-based optical chemical sensor. Sensors and Actuators B: Chemical, 1997, 39 (1-3): 452-457.

[15] Wang H, Wu Z Q, Zhang Y, Chen B B, He M, Hu B. Chip-based liquid phase microextraction combined with electrothermal vaporization-inductively coupled plasma mass spectrometry for trace metal determination in cell samples. Journal of Analytical Atomic Spectrometry, 2013, 28 (10): 1660-1665.

[16] Mary P, Studer V, Tabeling P. Microfluidic droplet-based liquid-liquid extraction. Analytical Chemistry, 2008, 80 (8): 2680-2687.

[17] Stone Z B, Stone H A. Imaging and quantifying mixing in a model droplet micromixer. Physics of Fluids, 2005, 17 (6): 063103.

[18] Xu J H, Tan J, Li S W, Luo G S. Enhancement of mass transfer performance of liquid-liquid system by droplet flow in microchannels. Chemical Engineering Journal, 2008, 141 (1-3): 242-249.

[19] Morales M C, Zahn J D. Droplet enhanced microfluidic-based DNA purification from bacterial lysates via phenol extraction. Microfluidics and Nanofluidics, 2010, 9 (6): 1041-1049.

[20] Wagli P, Chang Y C, Homsy A, Hvozdara L, Herzig H P, de Rooij N F. Microfluidic droplet-based liquid-liquid extraction and on-chip IR spectroscopy detection of cocaine in human saliva. Analytical Chemistry, 2013, 85 (15): 7558-7565.

[21] Logtenberg H, Lopez-Martinez M J, Feringa B L, Browne W R, Verpoorte E. Multiple flow profiles for two-phase flow in single microfluidic channels through site-selective channel coating. Lab on a Chip, 2011, 11 (12): 2030-2034.

[22] Kralj J G, Sahoo H R, Jensen K F. Integrated continuous microfluidic liquid-liquid extraction. Lab on a Chip, 2007, 7 (2): 256-263.

[23] Launiere C A, Gelis A V. High precision droplet-based microfluidic determination of americium (Ⅲ) and lanthanide (Ⅲ) solvent extraction separation kinetics. Industrial & Engineering Chemistry Research, 2016, 55 (7): 2272-2276.

[24] Tamagawa O, Muto A. Development of cesium ion extraction process using a slug flow microreactor. Chemical Engineering Journal, 2011, 167 (2-3): 700-704.

[25] Yu X X, Chen B B, He M, Wang H, Tian S, Hu B. Facile design of phase separation for microfluidic droplet-based liquid phase microextraction as a front end to electrothermal vaporization-ICPMS for the analysis of trace metals in cells.

Analytical Chemistry, 2018, 90 (16): 10078-10086.

[26] Hu B, He M, Chen B B. Nanometer-sized materials for solid-phase extraction of trace elements. Analytical and Bioanalytical Chemistry, 2015, 407 (10): 2685-2710.

[27] Giordano B C, Burgi D S, Hart S J, Terray A. On-line sample pre-concentration in microfluidic devices: A review. Analytica Chimica Acta, 2012, 718: 11-24.

[28] AlSuhaimi A O, McCreedy T. Microchip based sample treatment device interfaced with ICP-MS for the analysis of transition metals from environmental samples. Arabian Journal of Chemistry, 2011, 4 (2): 195-203.

[29] Hsu K C, Sun C C, Ling Y C, Jiang S J, Huang Y L. An on-line microfluidic device coupled with inductively coupled plasma mass spectrometry for chromium speciation. Journal of Analytical Atomic Spectrometry, 2013, 28 (8): 1320-1326.

[30] Chen B B, Heng S J, Peng H Y, Hu B, Yu X X, Zhang Z L, Pang D W, Yue X, Zhu Y. Magnetic solid phase microextraction on a microchip combined with electrothermal vaporization-inductively coupled plasma mass spectrometry for determination of Cd, Hg and Pb in cells. Journal of Analytical Atomic Spectrometry, 2010, 25 (12): 1931-1938.

[31] Yu X X, Chen B B, He M, Wang H, Hu B.Chip-based magnetic solid phase microextraction coupled with ICP-MS for the determination of Cd and Se in HepG2 cells incubated with CdSe quantum dots. Talanta, 2018, 179: 279-284.

[32] Wang H, Chen B B, Zhu S, Yu X X, He M, Hu B.Chip-based magnetic solid-phase microextraction online coupled with microHPLC-ICPMS for the determination of mercury species in cells. Analytical Chemistry, 2016, 88 (1): 796-802.

[33] Wang H, Chen B B, He M, Yu X X, Hu B. Selenocystine against methyl mercury cytotoxicity in HepG2 cells. Scientific Reports, 2017, 7.

[34] Chen B B, Hu B, He M, Huang Q, Zhang Y, Zhang X. Speciation of selenium in cells by HPLC-ICP-MS after (on-chip) magnetic solid phase extraction. Journal of Analytical Atomic Spectrometry, 2013, 28 (3): 334-343.

[35] Wang H, Wu Z K, Chen B B, He M, Hu B. Chip-based array magnetic solid phase microextraction on-line coupled with inductively coupled plasma mass spectrometry for the determination of trace heavy metals in cells. Analyst, 2015, 140 (16): 5619-5626.

[36] Chen Z N, Chen B B, He M, Wang H, Hu B. A porous organic polymer with magnetic nanoparticles on a chip array for preconcentration of platinum (Ⅳ), gold (Ⅲ) and bismuth (Ⅲ) prior to their on-line quantitation by ICP-MS. Microchimica Acta, 2019, 186(2). DOI: 10.1007/s00604-018-3139-1.

[37] Chen Z N, Chen B B, He M, Hu B. Magnetic metal-organic framework composites for dual-column solid-phase microextraction combined with ICP-MS for speciation of trace levels of arsenic. Microchimica Acta, 2020, 187 (1). DOI: 10.1007/s00604-019-4055-8.

[38] Vazquez M, Paull B. Review on recent and advanced applications of monoliths and related porous polymer gels in micro-fluidic devices. Analytica Chimica Acta, 2010, 668 (2): 100-113.

[39] Zhang J, Chen B B, Wang H, Huang X, He M, Hu B. Chip-based monolithic microextraction combined with ICP-MS for the determination of bismuth in HepG2 cells. Journal of Analytical Atomic Spectrometry, 2016, 31 (7): 1391-1399.

[40] Zhang J, Chen B B, Wang H, He M, Hu B. Facile chip-based array monolithic microextraction system online coupled with ICPMS for fast analysis of trace heavy metals in biological samples. Analytical Chemistry, 2017, 89 (12): 6878-6885.

[41] Ou X X, He M, Chen B B, Wang H, Hu B. Microfluidic array surface ion-imprinted monolithic capillary microextraction chip on-line hyphenated with ICP-MS for the high throughput analysis of gadolinium in human body fluids. Analyst, 2019, 144 (8): 2736-2745.

[42] Shih T T, Chen W Y, Sun Y C. Open-channel chip-based solid-phase extraction combined with inductively coupled

plasma-mass spectrometry for online determination of trace elements in volume-limited saline samples. Journal of Chromatography A, 2011, 1218 (16): 2342-2348.

[43] Shih T T, Hsu I H, Chen S N, Chen P H, Deng M J, Chen Y, Lin Y W, Sun Y C. A dipole-assisted solid-phase extraction microchip combined with inductively coupled plasma-mass spectrometry for online determination of trace heavy metals in natural water. Analyst, 2015, 140 (2): 600-608.

[44] Shih T T, Hsieh C C, Luo Y T, Su Y A, Chen P H, Chuang Y C, Sun Y C. A high-throughput solid-phase extraction microchip combined with inductively coupled plasma-mass spectrometry for rapid determination of trace heavy metals in natural water. Analytica Chimica Acta, 2016, 916: 24-32.

[45] Hsu K C, Lee C F, Tseng W C, Chao Y Y, Huang Y L. Selective and eco-friendly method for determination of mercury (II) ions in aqueous samples using an on-line AuNPs-PDMS composite microfluidic device/ICP-MS system. Talanta, 2014, 128: 408-413.

[46] He M, Chen B B, Wang H, Hu B. Microfluidic chip-inductively coupled plasma mass spectrometry for trace elements and their species analysis in cells. Applied Spectroscopy Reviews, 2019, 54 (3): 250-263.

[47] Armbrecht L, Dittrich P S. Recent advances in the analysis of single cells. Analytical Chemistry, 2017, 89 (1): 2-21.

[48] Kupper H, Zhao F J, McGrath S P. Cellular compartmentation of zinc in leaves of the hyperaccumulator *Thlaspi caerulescens*. Plant Physiology, 1999, 119 (1): 305-311.

[49] Vogel-Mikus K, Simcic J, Pelicon P, Budnar M, Kump P, Necemer M, Mesjasz-Przybylowicz J, Przybylowicz W J, Regvar M. Comparison of essential and non-essential element distribution in leaves of the Cd/Zn hyperaccumulator Thlaspi praecox as revealed by micro-PIXE. Plant Cell and Environment, 2008, 31 (10): 1484-1496.

[50] Vanhecke D, Rodriguez-Lorenzo L, Clift M J D, Blank F, Petri-Fink A, Rothen-Rutishauser B. Quantification of nanoparticles at the single-cell level: An overview about state-of-the-art techniques and their limitations. Nanomedicine, 2014, 9 (12): 1885-1900.

[51] Rashkow J T, Patel S C, Tappero R, Sitharaman B. Quantification of single-cell nanoparticle concentrations and the distribution of these concentrations in cell population. Journal of the Royal Society Interface, 2014, 11 (94): 20131152.

[52] Wang H J, Wang M, Wang B, Meng X Y, Wang Y, Li M, Feng W Y, Zhao Y L, Chai Z F. Quantitative imaging of element spatial distribution in the brain section of a mouse model of Alzheimer's disease using synchrotron radiation X-ray fluorescence analysis. Journal of Analytical Atomic Spectrometry, 2010, 25 (3): 328-333.

[53] Haraguchi H, Ishii A, Hasegawa T, Matsuura H, Umemura T. Metallomics study on all-elements analysis of salmon egg cells and fractionation analysis of metals in cell cytoplasm. Pure and Applied Chemistry, 2008, 80 (12): 2595-2608.

[54] Iwai T, Shigeta K, Aida M, Ishihara Y, Miyahara H, Okino A. A transient signal acquisition and processing method for micro-droplet injection system inductively coupled plasma mass spectrometry (M-DIS-ICP-MS). Journal of Analytical Atomic Spectrometry, 2015, 30 (7): 1617-1622.

[55] Jorabchi K, Kahen K, Gray C, Montaser A. *In situ* visualization and characterization of aerosol droplets in an inductively coupled plasma. Analytical Chemistry, 2005, 77 (5): 1253-1260.

[56] Mueller L, Traub H, Jakubowski N, Drescher D, Baranov V I, Kneipp J. Trends in single-cell analysis by use of ICP-MS. Analytical and Bioanalytical Chemistry, 2014, 406 (27): 6963-6977.

[57] Li F M, Armstrong D W, Houk R S. Behavior of bacteria in the inductively coupled plasma: Atomization and production of atomic ions for mass spectrometry. Analytical Chemistry, 2005, 77 (5): 1407-1413.

[58] Ho K S, Chan W T. Time-resolved ICP-MS measurement for single-cell analysis and on-line cytometry. Journal of Analytical Atomic Spectrometry, 2010, 25 (7): 1114-1122.

[59] Sun Q X, Wei X, Zhang S Q, Chen M L, Yang T, Wang J H. Single cell analysis for elucidating cellular uptake and

transport of cobalt curcumin complex with detection by time-resolved ICPMS. Analytica Chimica Acta, 2019, 1066: 13-20.

[60] Wei X, Hu L L, Chen M L, Yang T, Wang J H. Analysis of the distribution pattern of chromium species in single cells. Analytical Chemistry, 2016, 88 (24): 12437-12444.

[61] Groombridge A S, Miyashita S, Fujii S, Nagasawa K, Okahashi T, Ohata M, Umemura T, Takatsu A, Inagaki K, Chiba K. High sensitive elemental analysis of single yeast cells (Saccharomyces cerevisiae) by time-resolved inductively-coupled plasma mass spectrometry using a high efficiency cell introduction system. Analytical Sciences, 2013, 29 (6): 597-603.

[62] Corte-Rodríguez M, Alvarez-Fernandez Garcia R, Blanco E, Bettmer J, Montes-Bayon M. Quantitative evaluation of cisplatin uptake in sensitive and resistant individual cells by single-cell ICP-MS (SC-ICP-MS). Analytical Chemistry, 2017, 89 (21): 11491-11497.

[63] Wang H, Wang M, Wang B, Zheng L, Chen H, Chai Z, Feng W. Interrogating the variation of element masses and distribution patterns in single cells using ICP-MS with a high efficiency cell introduction system. Analytical and Bioanalytical Chemistry, 2017, 409 (5): 1415-1423.

[64] Cao Y, Deng B, Yan L, Huang H. An environmentally-friendly, highly efficient, gas pressure-assisted sample introduction system for ICP-MS and its application to detection of cadmium and lead in human plasma. Talanta, 2017, 167: 520-525.

[65] Gunther D, Hattendorf B. Solid sample analysis using laser ablation inductively coupled plasma mass spectrometry. TrAC-Trends in Analytical Chemistry, 2005, 24 (3): 255-265.

[66] Becker J S, Gorbunoff A, Zoriy M, Izmer A, Kayser M. Evidence of near-field laser ablation inductively coupled plasma mass spectrometry (NF-LA-ICP-MS) at nanometre scale for elemental and isotopic analysis on gels and biological samples (vol 21, pg 19, 2006). Journal of Analytical Atomic Spectrometry, 2007, 22 (2): 222.

[67] Wu B, Niehren S, Becker J S. Mass spectrometric imaging of elements in biological tissues by new BrainMet technique-laser microdissection inductively coupled plasma mass spectrometry (LMD-ICP-MS). Journal of Analytical Atomic Spectrometry, 2011, 26 (8): 1653-1659.

[68] Drescher D, Giesen C, Traub H, Panne U, Kneipp J, Jakubowski N. Quantitative imaging of gold and silver nanoparticles in single Eukaryotic cells by laser ablation ICP-MS. Analytical Chemistry, 2012, 84 (22): 9684-9688.

[69] Mueller L, Herrmann A J, Techritz S, Panne U, Jakubowski N. Quantitative characterization of single cells by use of immunocytochemistry combined with multiplex LA-ICP-MS. Analytical and Bioanalytical Chemistry, 2017, 409 (14): 3667-3676.

[70] Wang M, Zheng L N, Wang B, Chen H Q, Zhao Y L, Chai Z F, Reid H J, Sharp B L, Feng W Y. Quantitative analysis of gold nanoparticles in single cells by laser ablation inductively coupled plasma-mass spectrometry. Analytical Chemistry, 2014, 86 (20): 10252-10256.

[71] Van Malderen S J M, Vergucht E, De Rijcke M, Janssen C, Vincze L, Vanhaecke F. Quantitative determination and subcellular imaging of Cu in single cells via laser ablation-ICP-mass spectrometry using high-density microarray gelatin standards. Analytical Chemistry, 2016, 88 (11): 5783-5789.

[72] Zheng L N, Sang Y B, Luo R P, Wang B, Yi F T, Wang M, Feng W Y. Determination of silver nanoparticles in single cells by microwell trapping and laser ablation ICP-MS determination. Journal of Analytical Atomic Spectrometry, 2019, 34 (5): 915-921.

[73] Löhr K, Borovinskaya O, Tourniaire G, Panne U, Jakubowski N. Arraying of single cells for quantitative high throughput laser ablation ICP-TOF-MS. Analytical Chemistry, 2019, 91 (18): 11520-11528.

[74] Managh A J, Edwards S L, Bushell A, Wood K J, Geissler E K, Hutchinson J A, Hutchinson R W, Reid H J, Sharp B L.

Single cell tracking of gadolinium labeled CD4$^{(+)}$ T cells by laser ablation inductively coupled plasma mass spectrometry. Analytical Chemistry, 2013, 85 (22): 10627-10634.

[75] Van Malderen S J M, Van Acker T, Laforce B, De Bruyne M, de Rycke R, Asaoka T, Vincze L, Vanhaecke F. Three-dimensional reconstruction of the distribution of elemental tags in single cells using laser ablation ICP-mass spectrometry via registration approaches. Analytical and Bioanalytical Chemistry, 2019, 411 (19): 4849-4859.

[76] Van Acker T, Buckle T, Van Malderen S J M, van Willigen D M, van Unen V, van Leeuwen F W B, Vanhaecke F. High-resolution imaging and single-cell analysis via laser ablation-inductively coupled plasma-mass spectrometry for the determination of membranous receptor expression levels in breast cancer cell lines using receptor-specific hybrid tracers. Analytica Chimica Acta, 2019, 1074: 43-53.

[77] Hsiao I L, Bierkandt F S, Reichardt P, Luch A, Huang Y J, Jakubowski N, Tentschert J, Haase A. Quantification and visualization of cellular uptake of TiO_2 and Ag nanoparticles: Comparison of different ICP-MS techniques. Journal of Nanobiotechnology, 2016, 14: 50.

[78] Bendall S C, Simonds E F, Qiu P, Amir E-a D, Krutzik P O, Finck R, Bruggner R V, Melamed R, Trejo A, Ornatsky O I, Balderas R S, Plevritis S K, Sachs K, Pe'er D, Tanner S D, Nolan G P. Single-cell mass cytometry of differential immune and drug responses across a human hematopoietic continuum. Science, 2011, 332 (6030): 687-696.

[79] Bandura D R, Baranov V I, Ornatsky O I, Antonov A, Kinach R, Lou X, Pavlov S, Vorobiev S, Dick J E, Tanner S D. Mass cytometry: Technique for real time single cell multitarget immunoassay based on inductively coupled plasma time-of-flight mass spectrometry. Analytical Chemistry, 2009, 81 (16): 6813-6822.

[80] Miyashita S, Groombridge A S, Fujii S, Minoda A, Takatsu A, Hioki A, Chiba K, Inagaki K. Highly efficient single-cell analysis of microbial cells by time-resolved inductively coupled plasma mass spectrometry. Journal of Analytical Atomic Spectrometry, 2014, 29 (9): 1598-1606.

[81] Yin H, Marshall D. Microfluidics for single cell analysis. Current Opinion in Biotechnology, 2012, 23 (1): 110-119.

[82] White A K, Heyries K A, Doolin C, VanInsberghe M, Hansen C L. High-throughput microfluidic single-cell digital polymerase chain reaction. Analytical Chemistry, 2013, 85 (15): 7182-7190.

[83] Zhang X, Wei X, Wei Y, Chen M L, Wang J H. The up-to-date strategies for the isolation and manipulation of single cells. Talanta, 2020, 218: 121147.

[84] Lo S J, Yao D J. Get to understand more from single-cells: Current studies of microfluidic-based techniques for single-cell analysis. International Journal of Molecular Sciences, 2015, 16 (8): 16763-16777.

[85] Valizadeh A, Khosroushahi A Y. Single-cell analysis based on lab on a chip fluidic system. Analytical Methods, 2015, 7 (20): 8524-8533.

[86] Shields C W, Reyes C D, Lopez G P. Microfluidic cell sorting: A review of the advances in the separation of cells from debulking to rare cell isolation. Lab on a Chip, 2015, 15 (5): 1230-1249.

[87] Kang D K, Ali M M, Zhang K, Pone E J, Zhao W. Droplet microfluidics for single-molecule and single-cell analysis in cancer research, diagnosis and therapy. TrAC Trends in Analytical Chemistry, 2014, 58: 145-153.

[88] Huang N T, Zhang H L, Chung M T, Seo J H, Kurabayashi K. Recent advancements in optofluidics-based single-cell analysis: Optical on-chip cellular manipulation, treatment, and property detection. Lab on a Chip, 2014, 14 (7): 1230-1245.

[89] Shigeta K, Koellensperger G, Rampler E, Traub H, Rottmann L, Panne U, Okino A, Jakubowski N. Sample introduction of single selenized yeast cells (*Saccharomyces cerevisiae*) by micro droplet generation into an ICP-sector field mass spectrometer for label-free detection of trace elements. Journal of Analytical Atomic Spectrometry, 2013, 28 (5): 637-645.

[90] Shigeta K, Traub H, Panne U, Okino A, Rottmann L, Jakubowski N. Application of a micro-droplet generator for an

ICP-sector field mass spectrometer-optimization and analytical characterization. Journal of Analytical Atomic Spectrometry, 2013, 28 (5): 646-656.

[91] Verboket P E, Borovinskaya O, Meyer N, Guenther D, Dittrich P S. A new microfluidics-based droplet dispenser for ICPMS. Analytical Chemistry, 2014, 86 (12): 6012-6018.

[92] Wang H, Chen B B, He M, Hu B. A facile droplet-chip-time-resolved inductively coupled plasma mass spectrometry online system for determination of zinc in single cell. Analytical Chemistry, 2017, 89 (9): 4931-4938.

[93] Yu X X, Chen B B, He M, Wang H, Hu B. 3D Droplet-based microfluidic device easily assembled from commercially available modules online coupled with ICPMS for determination of silver in single cell. Analytical Chemistry, 2019, 91 (4): 2869-2875.

[94] Yu X X, Chen B B, He M, Hu B. Argon enclosed droplet based 3D microfluidic device online coupled with time-resolved ICPMS for determination of cadmium and zinc in single cells exposed to cadmium ion. Analytical Chemistry, 2020, 92 (19): 13550-13557.

[95] Wei X, Zheng D H, Cai Y, Jiang R, Chen M L, Yang T, Xu Z R, Yu Y L, Wang J H. High-throughput/high-precision sampling of single cells into ICP-MS for elucidating cellular nanoparticles. Analytical Chemistry, 2018, 90(24): 14543-14550.

[96] Chen Z N, Chen B B, He M, Hu B. A droplet-splitting microchip online coupled with time-resolved-ICPMS for analysis of released Fe and Pt in single cells treated with FePt nanoparticles. Analytical Chemistry, 2020, 92 (18): 12208-12215.

[97] Zhang X, Wei X, Men X, Jiang Z, Ye W Q, Chen M L, Yang T, Xu Z R, Wang J H. Inertial-force-assisted, high-throughput, droplet-free, single-cell sampling coupled with ICP-MS for real-time cell analysis. Analytical Chemistry, 2020, 92 (9): 6604-6612.

[98] Wei X, Zhang X, Guo R, Chen M L, Yang T, Xu Z R, Wang J H. A spiral-helix (3D) tubing array that ensures ultrahigh-throughput single-cell sampling. Analytical Chemistry, 2019, 91 (24): 15826-15832.

[99] Zhou Y, Chen Z, Zeng J, Zhang J, Yu D, Zhang B, Yan X, Yang L, Wang Q. Direct infusion ICP-qMS of lined-up single-cell using an oil-free passive microfluidic system. Analytical Chemistry, 2020, 92 (7): 5286-5293.

[100] Cao Y, Feng J, Tang L, Yu C, Mo G, Deng B. A highly efficient introduction system for single cell-ICP-MS and its application to detection of copper in single human red blood cells. Talanta, 2020, 206: 120174.

[101] Wang H, Chen B B, He M, Li X T, Chen P Y, Hu B. Study on uptake of gold nanoparticles by single cells using droplet microfluidic chip-inductively coupled plasma mass spectrometry. Talanta, 2019, 200: 398-407.

[102] Yao H, Zhao H S, Zhao X, Pan X Y, Feng J X, Xu F J, Zhang S C, Zhang X R. Label-free mass cytometry for unveiling cellular metabolic heterogeneity. Analytical Chemistry, 2019, 91 (15): 9777-9783.

[103] Zhang X C, Zang Q C, Zhao H S, Ma X X, Pan X Y, Feng J X, Zhang S C, Zhang R P, Abliz Z, Zhang X R. Combination of droplet extraction and pico-ESI-MS allows the identification of metabolites from single cancer cells. Analytical Chemistry, 2018, 90 (16): 9897-9903.

[104] Feng J X, Zhang X C, Huang L, Yao H, Yang C D, Ma X X, Zhang S C, Zhang X R. Quantitation of glucose-phosphate in single cells by microwell-based nanoliter droplet microextraction and mass spectrometry. Analytical Chemistry, 2019, 91 (9): 5613-5620.

第 12 章　激光剥蚀等离子体质谱分析

（胡兆初[①]*，张　文[①]，罗　涛[①]，郭　伟[①]，张晨西[①]，廖秀红[①]）

▶ 12.1　引言 / 302

▶ 12.2　激光剥蚀元素分馏机理及其在元素分析中的应用进展 / 303

▶ 12.3　激光在副矿物 U-Th-Pb 定年中的应用 / 304

▶ 12.4　激光在放射性同位素分析中的应用 / 305

▶ 12.5　激光在稳定同位素分析中的应用 / 309

▶ 12.6　激光在单个流体包裹体分析中的应用 / 311

▶ 12.7　激光分析样品前处理技术 / 313

▶ 12.8　激光剥蚀液体样品进样技术 / 316

▶ 12.9　LA-ICP-MS 专业数据处理软件 / 317

▶ 12.10　展望 / 319

---

①中国地质大学（武汉），武汉。*通讯作者联系方式：zchu@vip.sina.com

本章导读

- 介绍了激光剥蚀元素分馏机理及其在元素分析中的应用进展。
- 重点介绍了激光在副矿物 U-Th-Pb 定年中的应用；在放射性同位素和稳定同位素分析中的应用；在流体包裹体分析中的应用。
- 重点介绍了激光整体分析样品前处理技术；激光剥蚀液体样品进样技术；LA-ICP-MS 专业数据处理软件。
- 对激光剥蚀等离子体质谱进行了展望。

12.1 引 言

激光剥蚀等离子体质谱（LA-ICP-MS；其中 ICP-MS 包括四极杆等离子体质谱、多接收杯等离子体质谱等；图 12-1）可对固体样品进行微区原位元素和同位素准确精细分析，研究化学组成在微米尺度上的分布和分配规律，完成在以往整体分析方法下难以完成的工作，从而被认为是地球化学分析方法研究领域最激动人心的重要进展之一。自 1985 年，Gray[1] 发表第一篇关于 LA-ICP-MS 应用的文章以来，该技术已为地球科学、环境科学、材料科学的创新研究提供了重要支撑手段[2-5]。在过去几十年中，激光剥蚀系统发展快速，显著提高了激光剥蚀性能。激光器有两个重要的发展方向，一是向短波长的方向发展，如深紫外波长纳秒激光器，它改善了样品对激光能量的吸收效率而提高了剥蚀性能，极大地促进了 LA-ICP-MS 在各类样品分析中的广泛应用，使其成为固体样品微区分析的主流技术之一。二是向短脉宽的方向发展，如超短脉宽飞秒激光，它改变了样品吸收激光能量的方式，改善了电介质和透明材料的剥蚀性能，大大降低了激光剥蚀过程中的热效应，从源头上缓解了元素分馏和基体效应，为 LA-ICP-MS 技术的发展创造了新的契机。中国是 LA-ICP-MS 技术应用的大国。据 Web of Science 检索的数据表明，2007 年开始，中国采用 LA-ICP-MS 技术每年发表的文章数量已超过美国成为世界第一，并且继续呈现出线性快速增长的趋势。2019 年，中国作者发表的与 LA-ICP-MS 技术相关的论文数量占到了世界的 40%。地球科学是 LA-ICP-MS 技术最重要的应用领域，以 LA-ICP-MS 分析为主要手段获得的地球科学等领域研究成果非常丰硕。例如，对锆石不同结构部位 U-Pb 年龄、Hf 同位素和微量元素的分析可确定其年龄和成因；对珊瑚生长环带的微量元素分析可示踪所对应的古环境变迁等；对微小实验产物的分析，确定元素的分配系数；对单个流体包裹体成矿元素的分析，确定矿床的成因；等等。本章总结了 LA-ICP-MS 在元素、放射性同位素、稳定同位素、副矿物定年、流体包裹体等方面近几年的主要进展。

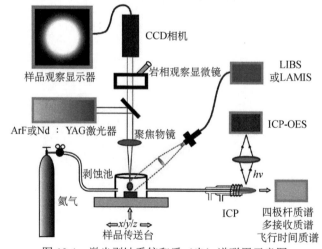

图 12-1 激光剥蚀系统和质（光）谱联用示意图

12.2 激光剥蚀元素分馏机理及其在元素分析中的应用进展

元素分馏是激光剥蚀等离子体质谱准确定量测定存在的一个主要问题。其中，基体效应导致的元素分馏行为的变化是当前元素分馏研究中的一个热点和难点，在实际样品分析工作中也是最需要和最难克服的。所谓的基体效应，本节指的是基体对分析物信号的增强和抑制作用。实际工作中被测量的样品，其基体成分往往是变化的，这种基体成分的变化将直接影响分析结果。而且往往这种基体效应导致的元素分馏在高空间分辨率分析时会被显著放大。任何影响激光剥蚀样品气溶胶产生、传输及离子化的过程均可能产生元素分馏效应，进而影响分析结果的准确度和精密度。研究表明，激光波长越短越有利于样品的能量吸收，剥蚀产生的气溶胶颗粒平均尺寸越小，则分馏效应越小。随着激光剥蚀深度逐渐增加，激光剥蚀产生的"down-hole"分馏效应也越显著，不同样品基体的"down-hole"分馏行为不同，如硅酸盐成分的参考物质（USGS 玻璃和 MPI-DING 玻璃）在高空间分辨率条件下观察到碱金属和过渡族元素相对于 Ca 元素有显著的分馏；而对于 NIST 系列玻璃则无明显的分馏行为[6]。激光脉冲宽度是影响分馏效应的另一重要参数，当激光脉冲宽度减小至飞秒级时可有效减小激光与样品相互作用过程中的热效应[7]，剥蚀产生的气溶胶颗粒更能代表样品的原始组成。飞秒激光被认为是可以减小元素分馏效应的有效手段[8]。需要指出的是，飞秒激光剥蚀过程中产生的元素分馏程度与剥蚀环境紧密相关[9]。在氦气环境下，激光剥蚀直径越大获得的挥发性元素分馏效应越显著，而在小束斑剥蚀条件下则无明显分馏效应[10]。在氦-氩混合气剥蚀环境下，飞秒激光在不同剥蚀束斑条件下均无明显的元素分馏现象[9]。对 193 nm 准分子激光研究表明，样品池内不同采样点处的载气流速显著影响剥蚀所产生的气溶胶颗粒尺寸大小[11,12]，在载气流速较低区域剥蚀产生的大尺寸气溶胶颗粒在等离子体中无法完全离子化是导致元素分馏的重要因素[12]。可以通过降低载气流速[12]或引入氮气、水蒸气等活性基体[13]的方式来增加等离子体温度，提高等离子体对基体的耐受能力，最终降低离子化过程中所产生元素分馏。LA-ICP-MS 分析过程中的元素分馏行为与分析基体的物理化学性质有关，因此采用与待测样品基体匹配的标准物质校正分析更能确保获得准确的 LA-ICP-MS 分析结果。

由于激光剥蚀技术样品进样量小（小于 1 μg），仪器灵敏度是制约 LA-ICP-MS 分析结果的另一重要因素。使用氦气替代氩气作为样品载气已被广泛使用，它可以使 193 nm 激光剥蚀的信号灵敏度提高约 2～5 倍[14]，极大地提高了 LA-ICP-MS 的分析性能。对其增敏机制，最近的研究结果表明该信号增敏不仅仅是因为氦气的使用提高了小尺寸气溶胶颗粒的传输效率，更主要是因为在氦气环境下剥蚀产生的小尺寸气溶胶颗粒在 He-Ar 等离子体环境中具有更高的挥发及离子化效率[13]。通过向等离子体中引入氢气[15]、氮气[16]、水蒸气、甲烷、甲醇、乙醇溶液[14,17-19]等活性基体也可显著增加仪器灵敏度，可观察到约 2～3 倍的信号增强。通过仪器关键硬件的研发也有利于提高分析的信号强度和稳定性，如通过改进磁质谱仪的锥接口[20-23]或改进接口真空泵提高其真空度[24,25]可观察到显著的信号增敏。Hu 等研发的"线性"和"波形"激光剥蚀信号匀化装置可很好地抑制四极杆质谱顺序扫描时产生的光谱螺纹效应，从而显著提高激光剥蚀信号的稳定性[26,27]。采用线性信号平滑装置，成功使激光低剥蚀频率（1 Hz）时信号稳定性提高 11 倍，极大提升了 U-Pb 年龄分析的空间分辨能力[26]。各种增敏技术和关键硬件的应用为提高 LA-ICP-MS 分析的空间分辨能力及开展超低含量样品的元素及同位素比值分析提供了保障。

在进行地质样品元素含量分析时，通常使用 USGS、MPI-DING 以及 NIST 系列玻璃作为标准物质以校正分析过程中的仪器漂移和元素分馏效应[28-30]。但 NIST 系列玻璃与天然地质样品在化学组成有一定差别，其剥蚀特性及元素分馏行为与天然地质样品也不一致[6]，选择基体匹配的外标物质和合适的内标元素是 LA-ICP-MS 准确进行元素含量分析的关键。目前可进行 LA-ICP-MS 元素分析的地质样品矿物种类繁多，如硅酸盐矿物[31,32]、氧化物[8]、碳酸盐矿物[33]、硫化物[34]和熔流体包裹体[35]等。除了对矿物样品进行微区分析以外，学者们还开展了对地质全岩样品进行 LA-ICP-MS 整体分析的方法研究。一般采用粉末压片法[36]和熔融玻璃法[37-39]来制备激光剥蚀尺度上均匀的具有代表性的用于 LA-ICP-MS 整体分析的全岩样品。通过使用高能红外激光直接熔融粉末样品制备熔融玻璃更是一种简便快捷、绿色环保的样品制备方法[40]。此外，LA-ICP-MS 还可直接用于液体样品中元素含量的整体分析[41]，这种方法进样量少，极大降低了与水溶剂相关（氧化物、氢氧化物）的离子干扰[41]。

LA-ICP-MS 分析技术也广泛应用于动植物组织等生物样品中的元素定量分析及元素分布（即生物成像）分析。由于生物样品的特殊性，冷冻剥蚀池的研制是直接激光剥蚀生物组织获得准确定量结果和高质量生物成像的前提[42]。快速冲洗剥蚀池的研制可显著提高生物成像的横向分辨率，如 Van Malderen 等通过研制清洗时间为 6 ms 的剥蚀池可将横向空间分辨率提高至 0.3 μm [43]。为获得准确的分析结果，通常在组织匀浆中添加待分析元素以制备基体匹配的校正标样，如 Reifschneider 等在蛋黄匀浆中添加元素 Tm，加热干燥后黏合在玻璃上以进行老鼠组织 Tm 的生物成像分析[44]。使用生物组织标准物质粉末压片制靶也可制备用于基体匹配的分析标样，如采用骨头粉末压片以校正骨头和牙齿中 Mn 的分布[45]。除采用基体匹配的外标校正外，元素 C、Ir、Rh 等通常被选择为内标进行校正[46,47]，如 Moraleja 等将 Ir 添加在样品表面作为内标元素分析药物在老鼠肾脏组织中的分布情况[47]。此外，也有研究采用同位素稀释法准确分析生物组织中的元素含量，如将富含 ^{57}Fe 的溶液与剥蚀产生的样品气溶胶颗粒混合引入 ICP 等离子体以定量分析羊脑组织中的 Fe 元素分布[48]。

材料样品中的微量元素含量对其产品质量监测有重要意义，LA-ICP-MS 技术还广泛应用于高分子聚合物[49]、钢铁合金[50]、半导体[51]、陶艺[52]等样品中的微量元素检测。随着激光剥蚀系统的发展，如激光束匀化技术的应用以及飞秒激光剥蚀的出现提高了层状材料检测的深度分析能力[52]。同时，深紫外纳秒激光和飞秒激光的出现也减小了不同材料样品间的基体差异，如 Mateo 等采用 LA-ICP-MS 分析技术同时开展了钢铁、高分子聚合物和陶瓷等复杂材料的元素分析[52]。

12.3　激光在副矿物 U-Th-Pb 定年中的应用

同位素年代学研究可提供各种地质体形成的准确时间，为研究各类地质事件及追溯地质演化历史提供了精确的年代信息。副矿物年代学研究是 LA-ICP-MS 在地质学研究中最重要的应用之一。

从 Feng 等[53]和 Fryer 等[54]最初开展锆石 LA-ICP-MS U-Pb 定年工作以来，副矿物 U-Th-Pb 定年测试方法也取得了快速发展和广泛应用。如对锆石样品可成功实现横向（<10 μm）[55]或纵向上（0.6 μm）[56]的超高空间分辨率分析。此外，其他含 U 副矿物的 U-Pb 年代学分析方法也日趋成熟，除了常见的榍石[57,58]、独居石[59]、磷灰石[60,61]、磷钇矿[62]、斜锆石[63,64]、金红石[65,66]、锡石[67]、沥青铀矿[68]、钙钛矿[69]、氟碳铈矿[70]等。近年来，学者们逐渐开发了新的 U-Pb 定年矿物体系，如方解石[71]、石榴石[72,73]、黑钨矿[74]和孔雀石[75]等副矿物的 U-Pb 年代学分析。尽管众多的矿物相成功用于 U-Th-Pb 年代学分析，但在进行 LA-ICP-MS 副矿物 U-Pb 定年分析时产生的元素分馏及基体效应一直影响定年结果的准确度与精密度。由于存在元素分馏及基体效应，LA-ICP-MS 副矿物定年分析时常用的方法是选择与待测样品基体匹配的标样来校正分析过程中产生的元素分馏、质量歧视及仪器漂移。但由于与基体匹配、成分均一的副矿物年龄标样极度缺乏，这也严重制约了副矿物 U-Pb 年代学

的广泛应用和发展。因此，寻找和刻画适用于不同种类副矿物的基体匹配矿物标样一直是学者们的研究热点[71,76-80]，为开展 LA-ICP-MS 副矿物 U-Th-Pb 年龄准确分析提供了基础。除了研发基体匹配的矿物标样，学者对不同副矿物间 U-Pb 年龄分析时的基体效应也进行了研究。如在线扫描剥蚀模式下，分别采用锆石标样做外标分析磷灰石[60]、榍石[81]以及褐帘石[82]的 U-Th-Pb 年龄，获得了部分满足地质应用精度的数据结果。通过优化仪器条件，McFarlane 等[83]利用玻璃标样 NIST 610 做外标，成功分析了褐帘石的 U-Pb 年龄。也有研究表明，通过经验公式校正 "down-hole" 分馏效应的方式也可成功实现以锆石为外标非基体匹配校正褐帘石的 U-Pb 年龄[84]。Luo 等建立了一种简单有效的水蒸气辅助激光剥蚀非基体匹配副矿物 U-Pb 定年方法，使非基体匹配副矿物校正 U-Pb 定年的准确度从 8%～24%提高到小于 1%～2%，成功地实现了以广泛使用的 NIST 610 玻璃或锆石为外标校正分析其他副矿物（锆石、独居石、磷钇矿、榍石和黑钨矿）的 U-Pb 年龄[74,85]（图 12-2）。

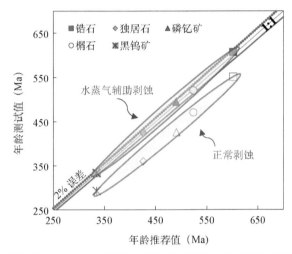

图 12-2 正常剥蚀或水蒸气辅助激光剥蚀模式下非基体匹配校正分析副矿物
（锆石、独居石、磷钇矿、榍石和黑钨矿）U-Pb 年龄结果

另外，有些学者还分别采用激光剥蚀采样技术联合 MC-ICP-MS 开展了独居石 Sm-Nd 定年[86,87]以及激光剥蚀联合 ICP-MS/MS 技术开展了辉钼矿 Re-Os 定年[88]和富钾矿物 Rb-Sr 定年[89]，通过在反应池内引入反应气（O_2 或 CH_4）结合 ICP-MS/MS 实现了 ^{187}Re 和 ^{187}Os、^{87}Rb 和 ^{87}Sr 的分离以及 ^{187}Re/^{187}Os、^{87}Rb/^{87}Sr 比值的准确测试，最终成功构筑 Re-Os 或 Rb-Sr 等时线年龄。LA-ICP-MS 副矿物定年技术蓬勃发展，为探讨成岩成矿、地质演化历史等重要地质问题提供了重要的技术支撑。

12.4 激光在放射性同位素分析中的应用

放射性同位素地球化学利用放射性同位素的衰变进行地质体计时研究，或根据放射性衰变子体同位素组成进行同位素示踪分析，是同位素地球化学中重要的研究领域，主要包括 Rb-Sr、Sm-Nd、Lu-Hf、Re-Os 和 U-Th-Pb 等体系。放射性衰变子体同位素组成变化受放射性母体的衰变常数、含量及衰变时间等参数影响，通常变化较小，需要利用高精度的质谱仪才能将其准确识别。长久以来，高精度放射性同位素分析多采用热电离质谱（TIMS）和多接收电感耦合等离子体质谱（MC-ICP-MS）完成。近年来，激光剥蚀系统与 MC-ICP-MS（LA-MC-ICP-MS）联用开展微区原位放射性同位素分析工作受到了广泛关注，我国分析学家在该领域也取得了较好的研究成果。

12.4.1 激光微区 Sr 同位素分析

地质样品的 Sr 同位素组成被广泛应用于同位素示踪研究，分析对象包括各类高 Sr 低 Rb 的样品，比如碳酸盐矿物（方解石）、磷酸盐矿物（磷灰石）、硅酸盐矿物（斜长石和单斜辉石）。LA-MC-ICP-MS 微区原位 Sr 同位素分析主要受限于较多的质谱干扰影响、Sr 含量、基体匹配参考物质缺乏等问题。Rb 和 Kr 的同质异位素干扰、稀土元素 Er 和 Yb 的双电荷干扰、Ca 和 Ar 相关的多原子离子干扰等共同影响了 Sr 同位素分析测试的准确度和精密度。中国科学院地质与地球物理研究所吴福元院士团队在我国最早开展了 LA-MC-ICP-MS 微区 Sr 同位素分析工作，先后建立了钙钛矿、异性石、磷灰石和氟碳铈矿的 LA-MC-ICP-MS 分析测试技术[80,90-92]。中国地质大学（武汉）地质过程与矿产资源国家重点实验室（GPMR 实验室）团队针对低 Sr 的单斜辉石（<100 μg/g）建立了微区 Sr 同位素分析方法[93]，并将最新的飞秒激光剥蚀系统引入到硅酸盐矿物的 Sr 同位素微区分析中，凭借飞秒激光对透明矿物的剥蚀优势，进一步提升低 Sr 矿物的 Sr 同位素分析精度和准确度[94]。

LA-MC-ICP-MS 微区原位 Sr 同位素在示踪地幔中再循环碳酸盐、示踪岩浆房复杂的开放体系演化过程等重要地球科学问题中起到重要作用。比如 Yu 等利用 LA-MC-ICP-MS 分析了华南桐庐花岗岩及包体中斜长石的 Sr 同位素组成的核幔边变化[95]。斜长石完整记录了岩浆混合和地壳同化混染事件，从核部到幔部，斜长石的 $^{87}Sr/^{86}Sr$ 比值减小（低至 0.7075），而边部又有极高的 Sr 同位素比值（高至 0.7117）（图 12-3）。这些环带记录了岩浆演化的时间顺序，即岩浆混合事件早于同化混染事件。热的镁铁质岩浆注入原先存在的长英质岩浆房不仅增加了岩浆房温度，同时可能引起围岩的强烈破碎进入岩浆，两者结合又加强了同化混染过程。LA-ICP-MS 微区 Rb-Sr 等时线定年技术是目前微区 Sr 同位素分析的一个前沿领域。主要通过三重四极杆的碰撞反应池技术解决 Rb 对 Sr 同位素的干扰问题，使高 Rb 样品的激光微区 Sr 同位素组成准确分析成为可能。但是该技术在我国尚未见到相关报道，值得开展相关研究。

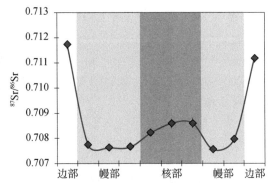

图 12-3　LA-MC-ICP-MS 分析发现斜长石存在复杂的 Sr 同位素分布环带

改自文献[95]

12.4.2 激光微区 Nd 同位素分析

Sm-Nd 同位素体系在同位素地球化学和同位素地质年代学中被广泛应用。激光微区 Sm-Nd 同位素测定的主要对象是轻稀土元素富集矿物，比如磷灰石、独居石、榍石、钙钛矿、异性石、氟碳铈矿和铁锰结核等。激光微区 Nd 同位素分析的主要问题是 ^{144}Sm 对 ^{144}Nd 的同质异位素干扰。研究早

期缺乏正确的 ^{144}Sm 干扰校正策略,直到 2006 年国际上才首次建立了 LA-MC-ICP-MS 磷灰石和榍石 Nd 同位素分析方法[96]。我国科学家紧随其后,中国科学院地质与地球物理研究所杨岳衡等首先建立了钙钛矿、独居石、榍石和磷灰石等矿物的激光微区 Nd 同位素分析方法[97],随后对一系列副矿物开展了 Nd 同位素标准物质研发工作[90-92]。此外,中国地质科学院矿产资源研究所和中国地质大学(武汉)也先后建立了 LA-MC-ICP-MS 天然副矿物 Nd 同位素分析方法[87,98]。

微区原位 Nd 同位素分析可以和 Sr 同位素资料结合,对岩浆源区进行示踪。比如研究金伯利岩中钙钛矿的 Nd 同位素组成,探讨金伯利岩初始岩浆同位素组成[90]。但是利用 Sm-Nd 同位素开展微区年代学研究的工作还较少。主要原因,一方面在于高 Sm 样品的 Nd 同位素分析会受到较为严重的同质异位素干扰影响,另一方面 ^{147}Sm/^{144}Nd 的比值测试常常存在基体效应问题,而微区分析缺乏基体匹配标样是一个长期存在的问题。因此,开发 Nd 同位素分析参考物质,建立准确的 Sm 干扰校正策略将是未来 LA-MC-ICP-MS 开展 Sm-Nd 同位素地质年代学的研究方向。

12.4.3 激光微区 Hf 同位素分析

Lu-Hf 同位素体系被广泛应用于沉积岩物源研究、示踪岩浆源区或为岩浆过程提供证据,并且可以通过 Lu-Hf 同位素构筑高精度的等时线,获得变质岩或地幔岩石的等时线年龄。锆石是 LA-MC-ICP-MS 的主要 Hf 同位素分析对象。锆石具有高 Hf 低 Lu 的特点,其 ^{176}Hf/^{177}Hf 比值基本代表了形成时代体系的 Hf 同位素组成。而且锆石是一种普遍存在且抗风化能力强的矿物,可以保存古老的地质信息。锆石具有很高的 Hf 同位素体系封闭温度,可以在较高的变质过程中保存原始的 Hf 同位素组成。结合锆石的多维地球化学信息,比如 U-Pb 年龄、O 同位素、微量元素及 Ti 温度计,可以为地质学家提供丰富的地球化学信息。

我国锆石 Hf 同位素微区分析始于 2003 年中国科学院广州地球化学研究所李献华研究团队[99],其后中国科学院地质与地球物理研究所[100]、西北大学大陆动力学国家重点实验室[101]、中国地质大学(武汉)GPMR 实验室等[102]科研院校相继建立了锆石 LA-MC-ICP-MS 微区 Hf 同位素分析方法。近年来,全国大量新建实验室或第三方实验室也可开展锆石激光微区 Hf 同位素分析测试,该技术已经成为一项较为普遍的地球化学分析项目。我国研究人员也在锆石 Hf 同位素参考物质研制方面取得一定进展,先后推出了 Penglai、Qinghu、BB、SA01 等 Hf 同位素分析的参考物质[76,103-105]。

目前锆石 Hf 同位素分析受限于高 Yb/Hf 比值(Yb/Hf >20%)锆石的 ^{176}Yb 的同质异位素干扰。尽管经过十余年的调查研究,准确扣除 Yb 干扰依然是一个难题。幸运的是,根据 Fisher 等统计的 12 440 颗锆石中,只有大约 17%锆石的 ^{176}Yb/^{177}Hf 超过 5%,达到 Yb/Hf >20%的锆石在自然界中并不常见[106]。因此在大多数情况下,目前建立的 Yb 干扰校正策略可以满足分析需求。

12.4.4 激光微区 Os 同位素分析

亲铁、亲硫和亲有机性的 Re-Os 同位素体系可以直接用于金属硫化物、超基性岩和富含有机质样品的定年,而且地幔岩石中 Os 同位素比值受后期流体交代作用影响较小,可以很好地示踪地幔演化特征。Re 和 Os 元素在岩石中含量很低,而且存在"块金效应",因此 Re-Os 同位素分析长期以来都采用全岩分析方式测定,激光微区 Re-Os 同位素分析的相关报道比较少见。中国地质大学(武汉)GPMR 实验室研究组开展了微区原位硫化物中 Os 同位素测定方法研究[107]。通过合成人工硫化物 Os 同位素参考物质、优化 Re 干扰校正策略和新型多离子计数器信号采集模式,成功建立了 LA-MC-ICP-MS 微区原位 Os 同位素分析方法。

12.4.5 激光微区 Pb 同位素分析

Pb 同位素组成在自然界中具有较大的变化范围。由于 Pb 同位素独特的地球化学性质，使其在同位素地质年代学和放射性同位素示踪中受到长期的关注。采用 U-Pb 和 Th-Pb 衰变体系可以对含 U-Th 的地质体或者矿物进行年代学研究。此外，不含 U、Th 的矿物（如硫化物和长石）中的普通 Pb 同位素组成可用于示踪初始物质来源。U-Th-Pb 定年方法在其他章节有具体介绍，此处仅讨论 LA-MC-ICP-MS 测试低铀矿物的普通铅同位素研究。LA-MC-ICP-MS 分析 Pb 同位素的主要问题是大多数矿物的 Pb 含量较低，很难准确测定低丰度的 ^{204}Pb。即使采用高灵敏度的离子计数器接收 ^{204}Pb 信号，在常规测试条件下对 Pb 含量低于 10 μg/g 样品的测试精度>0.1%。此外，高汞样品中 ^{204}Hg 或者工作气体中的 Hg 对低 Pb 含量样品的干扰严重影响了 ^{204}Pb 的准确测试。激光剥蚀过程和等离子体离子化过程中同位素分馏效应的准确刻画和基体匹配标准物质缺乏等问题也进一步限制了微区原位高精度 Pb 同位素分析方法的发展。中国地质大学（武汉）GPMR 实验室研究组采用法拉第杯和离子计数器结合的检测器结构建立了 LA-MC-ICPMS 铅同位素分析方法[108]。对 ATHO-G（5.67 μg/g）、KL2-G（2.07 μg/g）和 ML3B-G（1.38 μg/g）等玻璃参考物质，获得的 ^{20x}Pb/^{204}Pb 的准确度达到 0.09%，精密度在 0.14%～0.59%之间（2RSD）。西北大学大陆动力学国家重点实验室研究组 Chen 等使用飞秒激光分析了国际标准玻璃样品中的 Pb 同位素比值[109]。对于 Pb 含量约为 10 μg/g 样品，^{20x}Pb/^{204}Pb 测试精度达到 0.05%，Pb 含量约 2 μg/g 样品，^{20x}Pb/^{204}Pb 测试精度优于 0.3%。但是为了获得足够高的 Pb 信号强度，需要采用大束斑、高剥蚀频率（>20 Hz）的线扫模式分析样品。此外，该课题组还对青铜器、硫化物和硅酸盐岩石全岩 Pb 同位素组成开展了研究[110-113]。

在矿床地质领域，硫化物的 Pb 同位素组成分析的难题在于低温热液环境下形成的硫化物常含有较高含量且变化范围很大的 Hg，能否准确扣除 ^{204}Hg 对 ^{204}Pb 的干扰将直接影响 ^{20x}Pb/^{204}Pb 比值测定的准确度。中国地质大学（武汉）GPMR 实验室研究组设计了镀金激光剥蚀在线除汞装置，用于除去样品中的 Hg[27]。另外，该课题组发现激光剥蚀产生的 Hg 主要是气态的，提出利用气体交换设备将汞蒸气从气溶胶颗粒物中置换，从而实现 Hg 的完全移除，而且该设备不会影响 Pb 同位素分析的信号强度[114]。经过汞移除处理后，富汞硫化物 Pb 同位素比值的测试精密度和准确度得到了显著提高。

LA-MC-ICP-MS 普通铅同位素分析主要被用于以长石为代表的硅酸岩矿物，以方铅矿、黄铁矿为代表的硫化物矿物，或硅酸岩矿物中的熔体包裹体等，开展岩浆源区和地质过程示踪研究，地球化学省的调查研究和岩浆包裹体的成因解释。Zhang 等对华北克拉通房山岩体镁铁质包体长石斑晶进行了微区 Pb 同位素组成分析。分析结果显示房山岩体中镁铁质包体的斜长石斑晶的 Pb 同位素组成具有不均匀性，表明房山岩体形成于一个开放体系，岩浆演化具有复杂、多阶段的特点[108]。中国科学院广州地球化学研究所同位素地球化学国家重点实验室研究组建立了微区熔体包裹体 Pb 同位素测试方法[115]。该研究组利用熔体包裹体的 Pb 同位素组成开展了岩浆演化和构造环境示踪等方面的研究。中国地质大学（武汉）地球科学学院研究组研究了长江上游地区表壳岩系碎屑钾长石 Pb 同位素，用于重建长江上游水系演化过程，为长江水系物源示踪研究提供了很好的基础[116]。

12.4.6 小结

我国 LA-MC-ICP-MS 放射性同位素分析技术始于 2003 年前后，且全国仅少数高校和科研院所拥有相关仪器和少量分析方法。近 10 年来，我国大量购置新型多接收等离子体质谱仪和新型激光剥蚀系统，使该学科发展有了极好的物质基础。目前我国新建立的 Sr-Nd-Hf-Os-Pb 等放射性同位素分析测试技术已经达到国际一流水平，并且提供了大量潜在的副矿物参考物质，为利用 LA-MC-ICP-MS 开展微区原位放射性同位素示踪研究提供了技术保障。但是目前我国在微区原位放射性同位素等时线定年领域与国际同行相比还存在差距，比如国际上利用碰撞反应池技术解决干扰问题，实现副矿物 Rb-Sr 等时线定年的研究工作已经较为成熟，并获得很好的应用成果。但是相关研究在我国开展较少。此外，进一步提高仪器灵敏度，进而提高激光微区分析空间分辨能力和改善分析测试精密度同样是值得继续研究的课题。

12.5 激光在稳定同位素分析中的应用

稳定同位素地球化学除了传统的氢、碳、氮、氧、硫等同位素体系外，还包括了近年来受到广泛关注的金属稳定同位素体系，比如锂、镁、硅、钙、铁、铜、锌等。多接收电感耦合等离子体质谱仪（MC-ICP-MS）的问世是推动金属稳定同位素测试技术及应用快速发展的关键。在过去十余年，金属稳定同位素地球化学成为地球科学中最活跃的分支。稳定同位素测试可分为全岩整体分析和微区原位分析。全岩整体分析主要由气体同位素质谱、MC-ICP-MS 和热电离质谱（TIMS）完成。微区原位分析技术主要有激光剥蚀-质谱联用技术（包括 LA-ICP-MS、LA-MC-ICP-MS）和二次离子质谱（SIMS）。LA-MC-ICP-MS 是目前可开展微区原位稳定同位素分析项目最多的仪器系统，包括锂[117,118]、硼[119,120]、碳[121]、镁[122-124]、硅[124,125]、硫[126-128]、氯[129]、钙[130]、钒[131]、铁[122]、镍[132]、铜[133,134]、锆[135]、锡[136]等多种稳定同位素的高精度准确分析。我国近年来在该领域取得了长足的进展，开发了除氯、钒、镍和锡之外其他的激光微区稳定同位素分析，其中 LA-MC-ICP-MS 分析地质样品的碳稳定同位素比值和锆稳定同位素比值均为国际首创。本节就我国激光微区稳定同位素分析的重要研究成果进行简要概述。

12.5.1 改善同位素分馏和基体效应影响

LA-MC-ICP-MS 影响稳定同位素比值测定准确度最大的因素来自于仪器同位素质量分馏[137]。放射性同位素体系利用自身的稳定同位素比值作为内标校正仪器质量分馏的方式无法被应用于稳定同位素分析，再结合 LA-MC-ICP-MS 中存在两个主要的同位素质量分馏源：激光剥蚀过程以及等离子体电离和离子传输过程，使得激光微区稳定同位素的质量分馏校正成为难题，而且当目标矿物与参考物质之间存在较大基体差异时常出现基体效应导致质量分馏校正失败。研究人员从不同的方向探讨了 LA-MC-ICP-MS 的质量分馏来源，并试图降低基体效应。在激光剥蚀系统方面，飞秒激光剥蚀系统被认为是当前最适合稳定同位素分析的激光剥蚀系统。相对于传统的纳秒级激光器，飞秒激光大大降低了激光剥蚀过程中产生的热分馏效应，尽可能地实现了化学计量剥蚀，消除了在激光剥蚀过程中出现的同位素基体效应。Fu 等利用 fs-LA-MC-ICP-MS 建立了可以同时测定硫化银、黄铜矿、黄铁矿和闪锌矿的激光微区 S 同位素分析方法[126]。研究发现飞秒激光剥蚀硫化物产生了更细粒的气溶胶颗粒、更高的灵敏度以及更好的分析测试精密度。Lin 等对比了飞秒激光和纳秒激光分析电气石和 NIST610 玻璃的 Li 同位素组成，并发现纳秒激光在剥蚀两种不同物质的过程中会导致 Li 同位素组成约 4‰ 的偏差，而飞秒激光可以克服这种基体效应偏差[138]。

ICP 电离过程和离子传输过程被认为是更严重的质量分馏效应和基体效应来源。我国在利用化学辅助技术（水蒸气、氮气等）解决基体效应问题方面有较好的研究进展。比如 Lin 等[138]报道了向激光剥蚀池和等离子体中加入水蒸气，实现了硅酸盐玻璃对电气石样品的 Li 同位素非基体匹配校正[133]。Zhang 等同样采用湿等离子体模式，成功建立了硅酸盐矿物的微区原位 Si 同位素分析测试技术[125]。此外，氮气辅助技术在消除氧化物和氢化物干扰方面显示了极好的效果，比如 Fu 等利用氮气消除 $^{32}S^1H$ 对 ^{33}S 的干扰，成功建立同时测定 $^{34}S/^{32}S$ 和 $^{33}S/^{32}S$ 的激光微区分析方法[126]。Zhang 等的研究发现氮气辅助技术在激光微区 Zr 同位素分析中可以提高 Zr 元素灵敏度 2～3 倍[135]。

12.5.2 参考物质研制工作进展

尽管现有的研究在改善激光剥蚀性能、抑制基体效应等方面有大量创新工作，但是微区原位分析依然需要基体匹配参考物质进行数据质量监控。基体匹配的参考物质需要具有与待测样品一致的激光剥蚀行为，准确地反映待测样品在等离子体、锥界面、离子传输通道中的分馏行为。但是，目前固体参考物质研发速度远远落后于分析技术的进步，无法满足地质学家们对不同基体样品的测试需求。基体匹配的硫化物参考物质缺乏已经成为制约微区原位稳定同位素分析技术发展和应用的最重要影响因素之一。我国近年来在研制基体匹配同位素参考物质方面取得一定进展。

天然矿物作为基体匹配参考物质是最佳选择。理想的天然矿物参考物质应具有稳定的晶体结构，主量元素、目标元素和干扰元素含量接近实际样品，而且有足够数量以满足全球微区分析实验的长期使用。满足上述条件的天然矿物非常罕见。中国科学院广州地球化学研究所推出了部分天然硫化物作为潜在的微区原位硫同位素参考物质，分别是磁黄铁矿 YP136[139]、黄铜矿 HTS4-6[140]。

除了寻找天然矿物，人工合成参考物质也是重要的研究对象。利用纳米级粉末颗粒研制稳定同位素参考物质的工作已经在国内外推广。Feng 等采用水热合成法制备的硫化铁纳米颗粒（600 nm）具有均匀的金属元素含量分布（<3%）和硫-铅同位素组成[141]。Fu 等采用玛瑙研磨仪获得的黄铁矿和闪锌矿超细粉末显示了均匀的硫同位素分布，两个样品的 $\delta^{34}S$ 测试不确定度分别是 0.28‰和 0.15‰，满足实际分析需求[126]。Bao 等将天然黄铁矿、黄铜矿和闪锌矿分别掺杂少量方铅矿，然后将样品研磨成直径数微米的超细粉末，压片样品表现了均匀的硫同位素和铅同位素分布[128]。

粉末压片法在硫化物微量元素和硫-铅同位素参考物质研制中的成功应用证明了其有效性，但是该技术存在一些固有的缺陷。比如粉末压片机械强度弱于天然矿物或者玻璃，导致粉末压片样品的剥蚀速率大于天然样品，会造成参考物质与实际样品测试信号强度不匹配，而且容易引起元素或同位素的深度分馏效应。其次粉末压片样品较难保存，纳米级硫化物粉末容易氧化和受潮膨胀，在干燥箱或真空箱中易开裂破碎。因此，Chen 等尝试将高温（高压）熔融法用于制备硫化物硫同位素参考物质，在 1000℃熔融获得的黄铜矿样品显示了均匀的硫同位素组成[127]。

12.5.3 小结

激光微区稳定同位素分析研究工作已经开展了 10 余年，但是其发展速度相对较慢，国际上建立了成熟微区分析方法的实验室依然较少。其主要原因在于质量分馏校正方案存在不足，基体效应问题尚未彻底解决和长期缺乏基体匹配参考物质。我国在激光微区稳定同位素分析方面的开发工作还是主要集中在高校和科研院所的微区分析实验室，尽管已经有较多研究成果，但是距离充分满足目前地球科学研究的需求还有较大距离，其中一个最需要解决的问题就是提供种类更为丰富和高质量的同位素固体参考物质。

12.6 激光在单个流体包裹体分析中的应用

12.6.1 分析技术简介

流体包裹体是经历成岩成矿过程并被捕获封存于主矿物缺陷之中，并经受复杂的地质过程而保存至今的古流体，特别是分析未被包封改造的单个流体包裹体，已成为地质学研究中"将今论古"的重要依据。LA-ICP-MS 技术已成为单个流体包裹体中元素、同位素化学成分分析的最重要的工具之一[142,143]。

国内外多个实验室对单个流体包裹体 LA-ICP-MS 分析方法进行了持续的研究，并取得了突破性的贡献[142-152]。然而 LA-ICP-MS 单个流体包裹体成分分析的失败率高达 50%以上，从样品选取到定量分析均存在多方面制约。如何成功分析单个流体包裹体？首先，最佳单个流体包裹体选取须严格地控制流体包裹体大小、形状、深度、独立性及宿主物理性质等[6]。其次，常用于激光剥蚀成分分析的石英宿主流体包裹体，由于石英易产生灾难性剥蚀，需要采用一定的剥蚀技巧，如分段剥蚀[142]或直接剥蚀[148]，由小到大快速变束斑剥出包裹体中的流体，通过初始小束斑改性石英表面，达成控制剥蚀或隔离表面污染，大束斑直接带出整个包裹体，有利于缩小积分区间获得有效的定量数据。其次，由于单个流体包裹体均为微米级，体积小、元素含量低、信号持续时间有限等特点，质谱分析难度较高。小体积/双体积剥蚀池/原位提取剥蚀池[12-14,153-155]，可缩短气溶胶洗出时间，这对流体包裹体信号获得高信噪比、低检出限非常有利；而低温剥蚀池可提前冻住流体包裹体内的流动成分，减少包裹体打开瞬间的崩裂溅射损失，对富液态 CO_2 包裹体可达到最佳控制剥蚀[156,157]。对低浓度微体积的样品，质谱仪器的检测灵敏度制约了其研究发展，Harlaux 等对比了三类质谱的流体包裹体分析能力，包括四极杆质谱（QMS）、双聚焦磁质谱（SFMS）、时间飞行质谱（TOFMS）[158]，特别是可实现准同时分析的 TOFMS，将包裹体信号持续时间有限的短板改善到极致，分析精度得到极大提高；拥有高灵敏度 SFMS 对低含量元素特别有效，但较慢的扫描时间限制了质谱分析元素数量[35]；最常规 QMS 成本低、性能稳定、易操作，通过折中考虑元素数量和驻留时间，达到了测试大部分元素的目的。最后，由于没有完全匹配的外部标准，以及包裹体内标元素定量误差较大，流体包裹体成分分析不确定性往往高达 20%以上。硅酸盐熔融玻璃成为包裹体广泛使用定量外标，内标元素多用 Na、Cl 元素，但通过显微测温获得包裹体内盐度并结合质谱信号比例计算内标 Na 的浓度时的经验公式[146,159]或经验模型[149,160]，无法涵盖各类流体包裹体成分特征以计算准确内标浓度，其产生的误差甚至高达 30%以上。近期郭伟等[151]提出了采用 LA-ICP-OES-MS 联用方式，流体包裹体一次进样同时分析的思路，由基体效应小的 OES 测试获取内标元素，可避免显微测温的额外步骤和经验公式计算误差，但其实用性仍需进一步验证。

12.6.2 应用研究

利用 LA-ICP-MS 获得的单个流体包裹体中的元素含量或比值，可有效地追踪热液流体在成岩成矿过程中的演化和特征、成矿元素迁移和沉淀、成矿流体来源等，为矿床学综合研究提供了十分重要的途径。

12.6.2.1 成矿元素迁移和沉淀

多期次流体包裹体不同的元素浓度，展示了古流体元素的分离分配关系和演变趋势，因此研究元素行为变化是探讨成矿过程的基础。针对元素在不同相中的迁移和分配行为，Heinrich 等[161]给出了一个明确的思路，即气相和液相中流体包裹体元素富集差异，推测富集在液相中的元素与 Cl 形成络合物，而富集在气相中的元素与 S 形成络合物。Chang 等[162]通过研究藏东玉龙斑岩多期岩浆热液 Cu-Mo 矿床多期矿脉中流体包裹体的 LA-ICP-MS 成分测量结果，展现了 Cu、Mo 元素在卤水相和气相流体包裹体中的行为，其中 Mo 在相分离前的早期析出沉淀，高盐度卤水携带 Cu 元素于后期堆积成矿。作者结合流体包裹体相变化和 Mo/Cu、As/K、Cs/K 比值变化及其他元素浓度对比，建立了该 Cu-Mo 矿床岩浆热液流体形成与演化模型。因此，元素在不同流体相中的迁移行为改变了流体成分，也影响了部分元素沉淀。对于流体中元素的沉淀，最具代表性的文章是 Audétat 等[163]通过 LA-ICP-MS 分析的澳大利亚 Yanlee Load 锡矿中的石英单晶流体包裹体，同一颗石英中获得完美的各世代流体包裹体组合，直接展示了不同流体演化阶段的元素成分和温度压力的变化关系，后期 B、Sn 元素浓度直线下降几个数量级匹配了电气石和锡矿的沉淀并分析出大气降水混合导致的元素析出沉淀的作用机制。近期 Liu 等[164]测定了中国西南北衙金矿多期脉石流体包裹体的金属元素浓度，通过早期、晚期元素比值对比（如 Rb、Cs 与 Na+K 浓度比值），排除了外部流体稀释情况[165]，给出了流体降温降压冷却的沉淀机制。

12.6.2.2 成矿流体来源

成矿流体来源是矿床地质学研究的基础问题之一。简单来说，成矿流体中元素富集程度，指示了源于某一特征的富某元素热液流体或岩浆。如王莉娟等[166]采用 LA-ICP-MS 分析了大井矿床锡铜矿体和黄岗梁矿床锡矿体中萤石及石英单个流体包裹体中 Rb/Sr、Sn/Cu、Na/K 比值，认为大井矿床富铜流体可能来源于深源的基性岩浆，而富锡流体来源于浅源的花岗岩浆。通常只有极少数元素，特别是卤素 Cl、Br、I 等，即使在广泛的流-岩相互作用中也能保留源的原始成分特征。因为卤素在水溶液中的浓度较高，而在大多数硅酸盐矿物中的丰度相对较低，它们行为在很大程度上是不变的，所以矿物包裹的古流体中这些元素具有一定的来源指示意义。Stoffell 等[167]利用 LA-ICP-MS 研究了北美 Tri-State 和 Northern Arkansas 两个地区 MVT 型铅锌矿的成矿流体，其中石英及闪锌矿流体包裹体中 Cl 和 Br 含量落在蒸发浓缩的海水趋势线上，即该成矿流体主要来源于蒸发浓缩的海水，而闪锌矿流体中 Br 含量更高，则暗示析出闪锌矿的流体源于更高程度浓缩海水。Leisen 等[168]详细地讨论了如何准确测试流体包裹体中 Cl、Br 卤素元素，并对不同地质体环境低盐度（阿尔卑斯石英脉）到高盐度（加拿大阿萨巴斯卡盆地铀矿、法国比利牛斯山脉滑石矿床）的流体包裹体进行了 Cl、Br、Na 浓度分析。元素含量比值结果清晰显示出不同流体包裹体中的古流体可能性来源：蒸发海水或溶解盐岩（图 12-4）。然而 Br 相对于主要卤素 Cl 的富集和耗竭可能是海水蒸发和卤石溶解以外的过程造成的，仅根据 Br、Cl、Na 及其比例不能明确地解决这一问题。众所周知，I 是一种亲生物元素，强烈富集在海洋有机物中，Fusswinkel 等[169]结合流体包裹体中的 Br/Cl、I/Cl 比值对成矿流体来源提供必要的约束，很好地解释了成矿流体与富含有机质的沉积岩不同程度的流-岩相互作用，为有机质流体来源提供了扎实的依据。

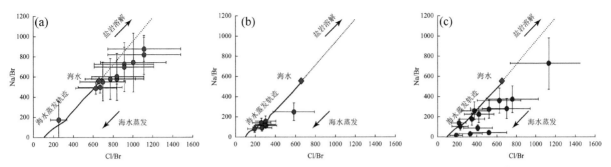

图 12-4 由低盐度到高盐度三类流体包裹体 LA-ICP-MS 测试的 Na/Br、Cl/Br 比值关系图[168]

(a) 阿尔卑斯石英脉流体包裹体；(b) 加拿大阿萨巴斯卡盆地铀矿流体包裹体；
(c) 法国比利牛斯山脉滑石矿床流体包裹体。

12.6.2.3 成矿流体特征及演化

成矿流体中元素含量丰富，不可避免地会产生独特的成形现象、演化痕迹或矿化类型，依靠流体包裹体元素成分测试，可获取特征流体信息，对探讨成矿机制及找矿提供了强有力的证据。Gomes 等[170]研究了巴西 Espinhaço 山脉东部铁矿床形成的氧化物化学和流体包裹体，LA-ICP-MS 流体包裹体成分测试显示不同元素的富集和亏损情况，暗示变质流体参与交代事件的演化，各演化过程决定了铁矿中铁的升级换代和高品位矿体的形成。Li 等[171]分析了中国长江中下游磁铁矿-磷灰石矿床中具有极高盐度的流体包裹体成分，发现流体包裹体含有特征的 Cl/Br、Na/K 和 Na/Br 比值，以及高 S/B 和 Ca/Na 比值，即它们与蒸发岩密切相关，综合研究表明岩浆渗出的流体上升过程中吸收了大量三叠纪蒸发岩，蒸发岩同化产生的高盐度盐水提高了溶解和运输 Fe 的能力。据此认为含蒸发岩的沉积盆地比无蒸发岩的盆地更有成矿前景。Chen 等[172]测定了华北克拉通北部大苏计斑岩钼矿多期岩脉中的流体包裹体成分，发现流体中 Sr 和 Na、Ca，K 和 Pb、Ba 含量呈正相关关系，这与斜长石、钾长石分解现象相吻合，说明绢云母化过程对成矿流体组成有较大影响。最近，Zhang 等[173]研究了西藏荣那 Cu-(Au)矿床斑岩到浅层热液过渡行为，LA-ICP-MS 流体包裹体元素的比值如 K/Na 和 Pb/Na，在除含明矾石脉外的所有包裹体类型中几乎保持一致，尽管它们的盐度变化很大；而高硫化超热静脉流体包裹体中的 Cu 浓度通常高于与斑岩有关的静脉，这类浅层低温热液脉流体包裹体中铜的高浓度和常见的硫化物置换特征表明，低温热液成矿流体可能与斑岩期硫化物的溶解和再沉淀有关。作者认为岩浆深部有单相流体出溶，经过沸腾和冷却，形成斑岩蚀变和成矿作用。因此，通过独特的成矿流体元素变化，匹配相应的矿化现象，阐释成矿过程和寻找具有经济价值的矿床，成为流体包裹体研究的重要目的。

12.7 激光分析样品前处理技术

LA-ICP-MS 是一种环境友好的"绿色"分析技术，它直接将固体样品导入 ICP 质谱检测设备。目前该技术主要用于固体微区原位微量元素和同位素准确精细分析，是地质分析领域的重要发展方向。但是该技术目前无法较好地直接应用于岩石、沉积物或土壤等全岩地质样品中的主微量元素含量和同位素的测定。其主要原因是地质样品基体复杂且非常不均一，LA-ICP-MS 是一种微区分析技术，每次分析进样量非常少（一般每次小于 1 μg），剥蚀的样品量太小无法代表实际样品的化学组成。因而提高样品的均一性（代表性）是从根本上解决这些问题的唯一途径。激光整体分析对样品制备技术主要提出了两点要求：一是足够均一，激光微区采样可以代表样品的整体组成；二是具有较好的机械稳定性，表面平整，使激光与固体样品间的相互作用具有较好的重现性。基于这两点要求，主要形成了粉末压片法、熔融玻璃制备技术，以及溶胶-凝胶法等样品固化技术。

12.7.1 粉末压片法

粉末压片法是 X 射线荧光光谱（XRF）测定主量元素常用的样品制备技术，该方法简便、易操作，可以方便地加入内标元素便于后续的数据校正，因此广泛地应用于 LA-ICP-MS 和 LIBS 中[174-177]。但是，粉末压片法通常存在样品不够均一、样品颗粒间黏合力较差，以及激光对不同样品的剥蚀量相差很大等问题，早期的研究无法获取准确的分析结果（偏差高达 20%~40%）[178]。针对这些问题，很多研究聚焦于黏合剂的使用和超细粉末的制备两个方面。

12.7.1.1 黏合剂

粉末压片法最早使用的黏合剂有聚氯乙烯粉末（PVC）、聚乙烯醇（PVA）、硼酸（H_3BO_3）等[1,178,179]，用来增强压片样品的机械稳定性。另外黏合剂还可以作为标准样品和待测样品的基体统一[180]，消除不同基体剥蚀行为带来的误差。特别是对于极易受基体影响的 LIBS 方法，相比于聚乙烯醇、银、铝和淀粉等材料，KBr 基质可以获得较好的信号强度和剥蚀坑形貌[181]。Yao 等[177]发现黏合剂 KBr 添加的百分比不同也会对分析结果产生影响，当 KBr 占比为 60%（质量分数）时，激光诱导等离子体激发温度的相对标准偏差最小（4.26%），基体效应最小。

12.7.1.2 超细粉末

近年来关于制备超细粉末压片的研究越来越多[176,182-184]，减小粉末粒径不仅可以增强样品颗粒间的凝聚力，使样品更加均一，还使剥蚀产物更容易被传输到等离子体中充分离子化，从而降低检出限[185]。

Wu 等[176]通过湿磨法制备了一系列岩石标准样品的超细粉末，经过 45 min 的研磨，花岗岩样品 GZG G1RF 的粒径为 d_{50}=1.2 μm、d_{90}=5.5 μm，研磨过程中出现了 6.6%的样品损耗，除了 SiO_2 受到玛瑙容器的污染，并未发现其他元素的明显污染和丢失。不同于物理研磨的方式，合适的化学手段可以破坏矿物颗粒的物理结构，将地质样品彻底粉碎和均匀化。Zhang 等[36]发现不同的硅酸盐岩石样品经过 NH_4HF_2 处理后可以得到晶粒形貌和尺寸一致、元素分布均一的超细粉末（d_{80}<8.5 μm），对超细粉末进行压片（参见图 12-5）可以方便地实现内标元素的添加和不同样品、标样的批量处理，减小基体效应。

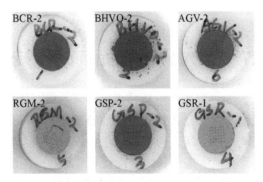

图 12-5 一系列硅酸盐岩石标准物质经 NH_4HF_2 处理后所得超细粉末压片样品

12.7.2 熔融玻璃法

在极高的温度下将粉末样品熔融成玻璃无疑是获取均匀样品最有效的方法,同时玻璃的稳定和致密性使得激光采样的行为更具有重现性,从而提高分析结果的准确性。熔融玻璃法包括直接熔融法[38,39,186-188]和助熔剂熔融法[189-191]。

12.7.2.1 直接熔融

早期的玻璃制备方法来源于电子探针(EPMA)的制样技术[23,192],利用高熔点金属带(W、Mo和Ir)加热样品熔融成玻璃状态,通常需要1300~1800℃的高温,使得样品中冷凝温度较低的元素(Pb、Cs、Ge、Sn等)容易在熔融过程中挥发丢失[24,193]。对熔融玻璃制备的优化集中在如何抑制挥发性元素丢失的问题上,将装有样品的密闭容器(BN坩埚、内置Mo腔的石墨管、铂金坩埚)置于马弗炉等高温环境下进行熔融,可以有效地抑制挥发性元素的丢失[37-39]。但是,对于富含难熔副矿物(如花岗岩、花岗闪长岩等)的样品即使是在高温长时间(1600℃,35 min)的熔融条件下,锆石也并未被完全熔解,需要进一步将玻璃粉碎再重熔[3]。另外,科马提岩、橄榄岩在淬火的过程中容易结晶出橄榄石和其他矿物,导致样品不均一;高SiO_2含量的样品在熔融状态下黏度过高不容易混匀;这些阻碍了直接熔融法的广泛应用[194]。最新的样品熔融方法采用高能量脉宽红外激光器直接对样品进行照射熔融[40],激光器可以提供3000 K以上的温度,可以迅速将包含难熔副矿物的地质样品熔融成均一的玻璃(图12-6),每个样品的熔融过程仅需要几十到几百毫秒,且全过程不需要使用任何化学试剂和样品容器装置,绿色环保。但是由于过高的温度,不可避免地造成了易挥发元素的丢失。

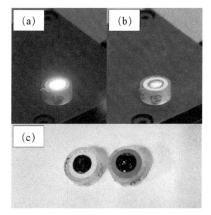

图 12-6 高能量激光器熔融玻璃样品制备

(a)激光照射熔融过程;(b)迅速冷却过程;(c)熔融所得玻璃样品

12.7.2.2 助熔剂熔融

偏硼酸锂和硼酸锂助熔剂的加入可以降低体系的熔点,从而有效地减少了易挥发元素的丢失。但是助熔剂的加入不仅使待测样品被稀释,还引入了杂质,因此助熔剂空白和元素检出限是衡量此种方法优劣性的重要标准[191,195,196]。将助熔剂和样品的比例减小(减小至2~3∶1)是改善这一问题的有效途径[191]。

助熔剂(偏硼酸锂、硼酸锂、金属氧化物等)的加入可以改变熔体的组成,防止科马提岩和橄榄岩在退火过程中形成橄榄石晶体,降低流纹岩和花岗岩熔体的黏度,使玻璃更加均匀[190,197,198]。Zhang等[190]以人工合成的钠长石作为助熔剂与橄榄石样品混合,在1500~1550℃温度下熔融10~

15 min，可以准确地测定玻璃样品中的多种痕量元素，第一系列过渡金属元素与溶液方法的测试值偏差在 10%以内。同时，助熔剂的加入使制得样品的基体组成更为相近，特别是对于 LIBS 方法，减弱了测试中的粒径效应和基体干扰[199,200]。

12.7.2.3 其他样品固化方法

除了使用高压将粉末压制成片和利用高温将样品熔融的方法，还有报道通过加入金属醇盐、有机玻璃等物质使样品在液体介质中固化成稳定的样品靶的技术。

溶胶-凝胶法是由金属醇盐、水、共溶剂或催化剂等形成胶体悬浮液，金属醇盐基团水解、连续聚合形成颗粒，这些颗粒继续聚集形成网状结构包裹住液体，干燥后可以获得均一的干凝胶样品[201,202]。该方法易于加入粉末状样品和内标，可以快速有效地形成便于激光分析的固体样品。采用同样的制样方法对参考物质进行处理作为校正的标准样品，无需基体匹配。前人的研究中尝试过 Zr、Al、Si 等多种干凝胶，Zr 干凝胶制备更快速、容易，且更为均一[201]。而 Al 干溶胶因为其更低的光谱背景更适合于 LIBS 的样品分析中[203]。Li 等[204]选用和待测植物样品基体一致的琼脂作为固化剂，加入内标元素制备了一系列用于激光分析的基体匹配校正标准，减小了样品与标准物质间的基体效应和元素分馏效应。

将溶于溶剂的有机玻璃和待测样品混合，加热或自然风干后也可以得到稳定的样品靶用于激光分析[205,206]。Zhu 等[206]先将内标元素 Cs 混入有机玻璃［聚甲基丙烯酸甲酯（PMMA）］，优化温度、样品和有机玻璃的比例等参数后，得到 11 种主微量元素均一的样品。

随着样品前处理技术方法的逐渐成熟，激光剥蚀进样技术已经成功地应用在了地质、生物、环境等样品的整体元素和同位素分析中。但相对于常规基于溶液的分析而言，还有非常大的差距，研究快速制备均一和稳定的适合激光分析的前处理技术仍然是接下来的研究中需要重点关注的内容。

12.8 激光剥蚀液体样品进样技术

人们对激光剥蚀进样技术的关注点主要聚焦在对固体物质样品的分析应用上。即使是在采用 LA-ICP-MS 进行液体样品分析的报道中，人们往往也是倾向于先将样品通过一些前处理技术（如：干燥/沉降化[207,208]、胶体化[204]、PVA 成膜[209]等）转化成固体物质，随后再利用 LA 将其引入 ICP-MS 进行分析。激光与液体直接相互作用的研究报道在 ICP-MS 分析中极为少见。不过，在激光诱导击穿光谱（LIBS）中关于 LA 与液体相互作用的研究却是相对比较常见，这可能是因为该技术在发展的最初就被认为是一项可以对任何物理状态的样品进行直接测试的技术[210]。实际上，LA 作为液体样品引入技术的相关报道可以追溯至 1993 年。相比于溶液雾化进样，激光剥蚀进样的样品引入量（通过改变激光束斑、频率等）控制起来很方便，即信号强度极为可控，这被认为有利于进一步拓宽 ICP-MS 浓度检测的线性范围；同时基于样品消耗量少且不涉及废液排放的优势，该项进样技术被认为对含放射性污染物质溶液的检测有重要意义[211]。随后人们提出，在 LA-ICP-MS 对固体样品的分析测试中，可通过 LA 剥蚀标准溶液作为外标来实现定量校正[142,212-216]。众所周知的是，由于 LA-ICP-MS 测试中元素分馏和基体效应的存在，定量校正时采用基体匹配的固体标样来进行分析校正一直是主流做法；但这对于那些成分复杂多变的样品来说（例如流体包裹体），获得基体匹配的固体标样显然是非常困难的。液体标样的有效替代无疑可以极大程度地提高测试的便利性，拓宽 LA-ICP-MS 分析样品的类型范围，随后该方法也被应用于沥青[217]和金刚石[218]中微量元素测试的校正。但自此之后，学者们对 LA 作为一种液体样品引入技术的发掘似乎就此停滞。直至 2019 年，新的研究表明基于 LA 进样的微量特性，其在降低溶剂相关的多原子离子干扰（质谱干扰）和抑制基体效应（非质谱干扰）方面皆表现优越[41]。相比于传统的溶液雾化进样 ICP-MS，LA-ICP-MS 分析溶液样品时氧化物及氢氧化物的产率可降低 1~2 个数量级，稀释倍数 80~2000 倍的变化和溶液介质酸浓度 2%~30%变化都被证明不会对测试结果的准确度造成明显的影响[41]。研究显示 LA-ICP-MS 在面向液体样品时元素分馏效应和基体效应明显减弱，非基体匹配校正也更具可行性。利用激光剥蚀进样技术耐基体等的特点，LA-ICP-MS 还被成功直接用于红酒这类复杂有机样品中多元素含量分析[219]。LA-ICP-MS 分析液体样品时所需要的装置与常规的固体样品分析装置几乎没有差别，唯一需要额外配备的是容纳液体样品的微型特氟龙孔槽。以红酒样品为例[219]（图 12-7）：取出的红酒样品在滴加内标溶液并摇匀之后，采用移液枪上样并用实验室专用的封口膜封口制备成液体靶，之后便转移进入剥蚀池。在正式采集数据之前，将激光束斑聚焦到封口膜上将其烧穿，使气溶胶得以从被击穿的孔上被剥蚀池中的载气流吹扫进入 ICP-MS 分析。这是一种直接、简便、绿色、高效的方法。

总体来说，LA 在作为 ICP-MS 的一项面向于液体样品的进样方法时，具有样品消耗量少且易控制、传输效率高、溶剂相关的质谱干扰弱、基体耐受度高、记忆效应弱等优势[41,211,215,219]。目前关于 LA-ICP-MS 直接面向液体样品的报道还非常少，诸多性质和特有的优点都还有待进一步探索与发掘。面对广大的液体样品测试的市场需求，LA-ICP-MS 向该方向的进军无疑可以更大程度地拓宽其应用范围。

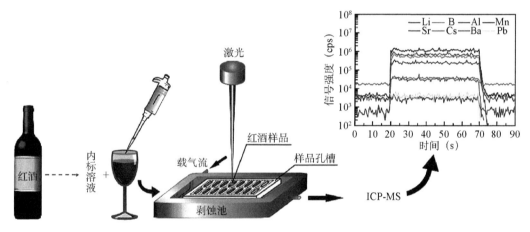

图 12-7　红酒滴入内标溶液后被直接引入 LA-ICP-MS 中进行多种元素的含量分析

12.9　LA-ICP-MS 专业数据处理软件

LA-ICP-MS 的数据处理过程比溶液分析技术更为复杂，涉及信号提取、背景扣除、校正方案选择、干扰校正、信号或同位素比值漂移校正等过程。而且 LA-ICP-MS 分析效率高，数据量大，使数据处理工作不仅是专业性较强的脑力劳动，也是需要消耗大量时间的体力劳动。针对 LA-ICP-MS 的数据特点，开发特定的、专业的数据处理软件对推广 LA-ICP-MS 分析技术具有重要的意义。根据 LA-ICP-MS 的具体用途，数据处理软件的功能被划分为元素定量分析、副矿物 U-Pb 定年、同位素分析和元素面扫描（mapping）等，其中元素定量分析还考虑是否可用于包裹体的元素定量分析。

目前全球微区实验室使用的 LA-ICP-MS 数据处理软件并不相同，一般通过以下途径获得：①购买专业综合性的数据处理软件；②索要免费的、开源的数据处理软件；③实验室自行编写符合自己研究需要的处理软件，包括一些小型的基于 Excel 的数据处理列表。表 12-1 总结了 14 款目前常见的 LA-ICP-MS 元素定量、U-Pb 定年和同位素分析的相关软件[220-232]。部分软件具有综合性的数据处理功能，比如 Glitter、ICPMSDataCal 和 Iolite。大部分软件则是针对某一项功能而编写的实验室内部软件，比如 SILLS 被用于流体包裹体和熔体包裹体的微量元素定量分析，Iso-Compass 是专业开展微区同位素组成分析的软件。这些软件通常都采用可视化的数据表达形式，通过预先设计数据处理模板，实现数据的自动批量化处理，并导出最终的数据报告。商业数据处理软件通常具有更好的商业仪器数据兼容性，更优秀的软件使用流畅度，而实验室自行开发的软件通常为开源代码，提供了潜在的可扩展性和实用性，当然这需要用户具有较好的编程能力。

表 12-1 部分 LA-ICP-MS 数据处理软件汇编

软件	平台	功能	费用	参考文献或链接
Glitter	Windows	元素定量、U-Pb 定年	4400 USD	http://www.glitter-gemoc.com/
ICPMSDataCal	VBA	元素定量（包裹体）、U-Pb 定年、同位素比值	免费	[31]
Iolite	C++和 Python	元素定量、U-Pb 定年、同位素比值、元素面扫描	3000/3500 USD	[221]
LAMTRACE	LOTUS 1-2-3	元素定量、U-Pb 定年	免费	[228]
LAMDATE-LAMTOOL	VB	U-Pb 定年、同位素比值	免费	[229]
SILLS	Matlab	元素定量（包裹体）	免费	[230]
AMS	Java	元素定量（包裹体）	免费	[232]
Lars-C	Delphi 3	元素定量	免费	[231]
TERMITE	R 语言	元素定量	免费	[224]
Latools	Python	元素定量	免费	[226]
Vizualage	DRS for Iolite (Igor Pro)	U-Pb 定年	免费	[222]
ET_Redux	Java 1.8	U-Pb 定年	免费	[223]
IsoFishR	R 语言	Sr 同位素比值	免费	[227]
Iso-Compass	C#	同位素比值	免费	[225]

专业 LA-ICP-MS 数据处理软件的开发工作一直未曾停止，更快的数据处理速度、更可靠的数据校正策略和误差传递方案对 LA-ICP-MS 的技术发展起到了很大的促进作用。而过去 10 年，LA-ICP-MS 最为重要的应用发展之一是元素 mapping 技术（imaging 技术）。LA-ICP-MS 元素 mapping 的优势在于较低的检出限、快速多元素获取能力、更大范围的元素 mapping（厘米级）。而 LA-ICP-MS 元素 mapping 产生的大量数据涉及更为复杂的处理过程，包括信号的分割提取、坐标尺度转化、背景扣除、元素定量校正、干扰校正、数据成图、图像处理等工作，对数据处理软件提出了更高的要求。过去 10 年，无论是微区实验室自行开发 mapping 软件或者商业仪器公司的与设备捆绑的 mapping 软件超过了 40 余个。由于本书篇幅限制，在此不列出软件的名称，可以参考 Weiskirchen 等[233]关于 LA-ICP-MS 元素 mapping 软件的综述。

尽管国际上有较多的 LA-ICP-MS 数据处理软件，但是我国自行开发的相关软件很少。目前被广泛使用的软件仅有中国地质大学（武汉）刘勇胜教授开发的 ICPMSDataCal 综合数据处理软件，可以完成微区单矿物和包裹体的微量元素定量、副矿物 U-Pb 定年和同位素比值分析等功能[220]。该课题组在 2020 年推出了专门针对 LA-MC-ICP-MS 的同位素数据处理软件 Iso-Compass[225]。元素 mapping 软件仅见合肥工业大学汪方跃教授的 LIMS 报道[234]。

从全球激光微区分析领域来看，百花齐放的数据处理软件其实并不利于最终测试数据的相互比对工作。不同实验室的数据处理软件不可避免地存在数据校正策略、数据提取和筛选以及误差传递方式的差异，容易导致不同软件校正后的数据存在一定的系统偏差。最重要的是越来越多复杂且不同思路的误差传递方案使数据之间的比对变得困难。因此，未来推行一种或者几种经过相互验证的软件作为标准的 LA-ICP-MS 数据处理软件，实现一致的全球激光微区分析数据处理流程，将使激光微区分析测试数据的可靠性、可重复性得到进一步的提升。

12.10 展　　望

激光剥蚀等离子体质谱已成为最重要的微区原位元素和同位素分析技术之一。随着地球科学等学科向精细研究发展，高时间和空间分辨率（如定年、原位成像分析）、高准确度和精度（如同位素分析）和高灵敏度和选择性（如低含量样品测定）的分析检测要求对激光剥蚀等离子体质谱专家提出了新的挑战。如何进一步提高仪器的检测性能是微区分析地球化学当前发展的一个主要挑战。这也是一个永恒不变的主题，因为人们对高分析性能的追求是永无止境的。虽然目前我国大部分的 LA-ICP-MS 仪器设备都还依靠国外引进，但科研仪器和工具研制的重要性现已得到了广泛共识。

急需开展微区元素和同位素分析标准物质的研制。标准物质是国家测量体系的重要组成部分，是科学研究和分析测试必备的物质条件，也是开发新技术和新方法不可或缺的材料。我国已拥有大量国家一级、二级地质矿物标准物质。但用于单矿物微区（微米级）元素-同位素分析的国家标准物质极少。采用基体匹配的外标进行校正是克服激光剥蚀等离子体质谱分析中存在的主要问题（基体效应和元素分馏）最简捷有效的方法。目前微区元素和同位素分析标准物质供需矛盾突出。这与我国微区元素和同位素分析大国和强国的地位极不相称。研发高质量的国际通行的微区元素和同位素分析标准物质势在必行。

综上所述，与以往分析测试手段进步带来了地球科学等研究的突破一样，可以预见，LA-ICP-MS 新技术、新方法的开发必将极大地拓宽我们的工作领域，为更多问题的深入研究提供可能，也必将为推动地球科学等学科成果的获得提供技术源泉。

致谢：感谢国家杰出青年科学基金（41725013）、国家自然科学基金专项项目（42142005）和湖北省创新研究群体（2020CFA045）项目的资助。

参 考 文 献

[1] Gray A L. Solid sample introduction by laser ablation for inductively coupled plasma source mass spectrometry. Analyst, 1985, 110: 551-556.

[2] Limbeck A, Galler P, Bonta M, Bauer G, Nischkauer W, Vanhaecke F. Recent advances in quantitative LA-ICP-MS analysis: Challenges and solutions in the life sciences and environmental chemistry. Analytical and Bioanalytical Chemistry, 2015, 407(22): 6593-6617.

[3] Woodhead J D, Horstwood M S A, Cottle J M. Advances in isotope ratio determination by LA-ICP-MS. Elements, 2016, 12(5): 317-322.

[4] Jenner F E, Arevalo R D, Jr. Major and trace element analysis of natural and experimental igneous systems using LA-ICP-MS. Elements, 2016, 12(5): 311-316.

[5] Pozebon D, Scheffler G L, Dressler V L. Recent applications of laser ablation inductively coupled plasma mass spectrometry (LA-ICP-MS) for biological sample analysis: A follow-up review. Journal of Analytical Atomic Spectrometry, 2017, 32(5): 890-919.

[6] Hu Z C, Liu Y S, Chen L, Zhou L, Li M, Zong K Q, Zhu L Y, Gao S. Contrasting matrix induced elemental fractionation in NIST SRM and rock glasses during laser ablation ICP-MS analysis at high spatial resolution. Journal of Analytical Atomic Spectrometry, 2011, 26(2): 425-430.

[7] Russo R E, Mao X, Gonzalez J J, Mao S S. Femtosecond laser ablation ICP-MS. Journal of Analytical Atomic Spectrometry, 2002, 17(9): 1072-1075.

[8] Li Z, Hu Z C, Günther D, Zong K Q, Liu Y S, Luo T, Zhang W, Gao S, Hu S H. Ablation characteristic of ilmenite using uv nanosecond and femtosecond lasers: Implications for non-matrix-matched quantification. Geostandards and Geoanalytical Research, 2016, 40(4): 477-491.

[9] Luo T, Ni Q, Hu Z C, Zhang W, Shi Q H, Günther D, Liu Y S, Zong K Q, Hu S H. Comparison of signal intensities and elemental fractionation in 257 nm femtosecond LA-ICP-MS using He and Ar as carrier gases. Journal of Analytical Atomic Spectrometry, 2017, 32(11): 2217-2225.

[10] Li Z, Hu Z C, Liu Y S, Gao S, Li M, Zong K Q, Chen H H, Hu S H. Accurate determination of elements in silicate glass by nanosecond and femtosecond laser ablation ICP-MS at high spatial resolution. Chemical Geology, 2015, 400: 11-23.

[11] Hu Z C, Liu Y S, Gao S, Hu S H, Dietiker R, Günther D. A local aerosol extraction strategy for the determination of the aerosol composition in laser ablation inductively coupled plasma mass spectrometry. Journal of Analytical Atomic Spectrometry, 2008, 23(9): 1192-1203.

[12] Luo T, Wang Y, Hu Z C, Günther D, Liu Y S, Gao S, Li M, Hu S H. Further investigation into ICP-induced elemental fractionation in LA-ICP-MS using a local aerosol extraction strategy. Journal of Analytical Atomic Spectrometry, 2015, 30(4): 941-949.

[13] Günther D, Heinrich C A. Enhanced sensitivity in laser ablation-ICP mass spectrometry using helium-argon mixtures as aerosol carrier. Journal of Analytical Atomic Spectrometry, 1999, 14(9): 1363-1368.

[14] Luo T, Hu Z C, Zhang W, Günther D, Liu Y S, Zong K Q, Hu S H. Reassessment of the influence of carrier gases He and Ar on signal intensities in 193 nm excimer LA-ICP-MS analysis. Journal of Analytical Atomic Spectrometry, 2018, 33(10): 1655-1663.

[15] Guillong M, Heinrich C A. Sensitivity enhancement in laser ablation ICP-MS using small amounts of hydrogen in the carrier gas. Journal of Analytical Atomic Spectrometry, 2007, 22(12): 1488-1494.

[16] Hu Z C, Gao S, Liu Y S, Hu S H, Chen H H, Yuan H L. Signal enhancement in laser ablation ICP-MS by addition of nitrogen in the central channel gas. Journal of Analytical Atomic Spectrometry, 2008, 23(8): 1093-1101.

[17] Liu S H, Hu Z C, Günther D, Ye Y H, Liu Y S, Gao S, Hu S H. Signal enhancement in laser ablation inductively coupled plasma-mass spectrometry using water and/or ethanol vapor in combination with a shielded torch. Journal of Analytical Atomic Spectrometry, 2014, 29(3): 536-544.

[18] Floor G H, Millot R, Iglesias M, Négrel P. Influence of methane addition on selenium isotope sensitivity and their spectral interferences. Journal of Mass Spectrometry, 2011, 46(2): 182-188.

[19] Fliegel D, Frei C, Fontaine G, Hu Z C, Gao S, Günther D. Sensitivity improvement in laser ablation inductively coupled

plasma mass spectrometry achieved using a methane/argon and methanol/water/argon mixed gas plasma. Analyst, 2011, 136(23): 4925-4934.

[20] Hu Z C, Liu Y S, Gao S, Liu W G, Zhang W, Tong X R, Lin L, Zong K Q, Li M, Chen H H. Improved *in situ* Hf isotope ratio analysis of zircon using newly designed X skimmer cone and jet sample cone in combination with the addition of nitrogen by laser ablation multiple collector ICP-MS. Journal of Analytical Atomic Spectrometry, 2012, 27(9): 1391-1399.

[21] Xu L, Hu Z C, Zhang W, Yang L, Liu Y S, Gao S, Luo T, Hu S H. *In situ* Nd isotope analyses in geological materials with signal enhancement and non-linear mass dependent fractionation reduction using laser ablation MC-ICP-MS. Journal of Analytical Atomic Spectrometry, 2015, 30(1): 232-244.

[22] He T, Ni Q, Miao Q, Li M. Effects of cone combinations on the signal enhancement by nitrogen in LA-ICP-MS. Journal of Analytical Atomic Spectrometry, 2018, 33(6): 1021-1030.

[23] Gou L F, Jin Z D, Deng L, He M Y, Liu C Y. Effects of different cone combinations on accurate and precise determination of Li isotopic composition by MC-ICP-MS. Spectrochimica Acta Part B: Atomic Spectroscopy, 2018, 146: 1-8.

[24] Yuan H L, Bao Z A, Chen K Y, Zong C L, Chen L, Zhang T. Improving the sensitivity of a multi-collector inductively coupled plasma mass spectrometer via expansion-chamber pressure reduction. Journal of Analytical Atomic Spectrometry, 2019, 34(5): 1011-1017.

[25] Wu C C, Burger M, Günther D, Shen C C, Hattendorf B. Highly-sensitive open-cell LA-ICPMS approaches for the quantification of rare earth elements in natural carbonates at parts-per-billion levels. Analytica Chimica Acta, 2018, 1018: 54-61.

[26] Hu Z C, Liu Y S, Gao S, Xiao S Q, Zhao L S, Günther D, Li M, Zhang W, Zong K Q. A "wire" signal smoothing device for laser ablation inductively coupled plasma mass spectrometry analysis. Spectrochimica Acta Part B: Atomic Spectroscopy, 2012, 78: 50-57.

[27] Hu Z C, Zhang W, Liu Y S, Gao S, Li M, Zong K Q, Chen H H, Hu S H. "Wave" signal-smoothing and mercury-removing device for laser ablation quadrupole and multiple collector ICPMS analysis: Application to lead isotope analysis. Analytical Chemistry, 2015, 87(2): 1152-1157.

[28] Jochum K P, Willbold M, Raczek I, Stoll B, Herwig K. Chemical characterisation of the usgs reference glasses GSA-1G, GSC-1G, GSD-1G, GSE-1G, BCR-2G, BHVO-2G and BIR-1G using EPMA, ID-TIMS, ID-ICP-MS and LA-ICP-MS. Geostandards and Geoanalytical Research, 2005, 29(3): 285-302.

[29] Jochum K P, Stoll B, Herwig K, Willbold M, Hofmann A W, Amini M, Aarburg S, Abouchami W, Hellebrand E, Mocek B, Raczek I, Stracke A, Alard O, Bouman C, Becker S, Dücking M, Brätz H, Klemd R, Bruin D d, Canil D, Cornell D, Hoog C-J d, Dalpë C, Danyushevsky L, Eisenhauer A, Gao Y, Snow J E, Groschopf N, Günther D, Latkoczy C, Guillong M, Hauri E H, Höfer H, Lahaye Y, Horz K, Jacob D E, Kasemann S A, Kent A J R, Ludwig T, Zack T, Mason P R D, Meixner A, Rosner M, Misawa K, Nash B P, Pfänder J, Premo W R, Sun W D, Tiepolo M, Vannucci R, Vennemann T, Wayne D, Woodhead J D. MPI-DING reference glasses for in situ microanalysis: New reference values for element concentrations and isotope ratios. Geochemistry, Geophysics, Geosystems, 2006, 7(2), Q02008. DOI: 10.1029/2005GC001060.

[30] Jochum K P, Weis U, Stoll B, Kuzmin D, Yang Q, Raczek I, Jacob D E, Stracke A, Birbaum K, Frick D A. Determination of reference values for NIST SRM 610-617 glasses following ISO guidelines. Geostandards and Geoanalytical Research, 2011, 35(4): 397-429.

[31] Liu Y S, Hu Z C, Gao S, Günther D, Xu J, Gao C Q, Chen H H. *In situ* analysis of major and trace elements of anhydrous minerals by LA-ICP-MS without applying an internal standard. Chemical Geology, 2008, 257(1): 34-43.

[32] Batanova V G, Thompson J M, Danyushevsky L V, Portnyagin M V, Garbe-Schönberg D, Hauri E, Kimura J I, Chang Q, Senda R, Goemann K. New olivine reference material for in situ microanalysis. Geostandards and Geoanalytical Research, 2019, 43(3): 453-473.

[33] Chen L, Liu Y S, Hu Z C, Shan G, Zong K Q, Chen H H. Accurate determinations of fifty-four major and trace elements in carbonate by LA-ICP-MS using normalization strategy of bulk components as 100%. Chemical Geology, 2011, 284(3-4): 283-295.

[34] Cook N J, Ciobanu C L, Pring A, Skinner W, Shimizu M, Danyushevsky L, Saini-Eidukat B, Melcher F. Trace and minor elements in sphalerite: A LA-ICPMS study. Geochimica et Cosmochimica Acta, 2009, 73(16): 4761-4791.

[35] Wälle M, Heinrich C A. Fluid inclusion measurements by laser ablation sector-field ICP-MS. Journal of Analytical Atomic Spectrometry, 2014, 29(6): 1052-1057.

[36] Zhang W, Hu Z C, Liu Y S, Yang W W, Chen H H, Hu S H, Xiao H Y. Quantitative analysis of major and trace elements in NH_4HF_2-modified silicate rock powders by laser ablation-inductively coupled plasma mass spectrometry. Analytica Chimica Acta, 2017, 983: 149-159.

[37] Zhu L Y, Liu Y S, Hu Z C, Hu Q H, Tong X R, Zong K Q, Chen H H, Gao S. Simultaneous determination of major and trace elements in fused volcanic rock powders using a hermetic vessel heater and LA-ICP-MS. Geostandards and Geoanalytical Research, 2013, 37(2): 207-229.

[38] Bao Z A, Yuan H L, Zong C L, Liu Y S, Chen K Y, Zhang Y. Simultaneous determination of trace elements and lead isotopes in fused silicate rock powders using a boron nitride vessel and fsLA-(MC)-ICP-MS. Journal of Analytical Atomic Spectrometry, 2016, 31(4): 1012-1022.

[39] He Z W, Huang F, Yu H M, Xiao Y, Wang F, Li Q, Xia Y, Zhang X. A flux-free fusion technique for rapid determination of major and trace elements in silicate rocks by LA-ICP-MS. Geostandards and Geoanalytical Research, 2016, 40(1): 5-21.

[40] Zhang C X, Hu Z C, Zhang W, Liu Y S, Zong K Q, Li M, Chen H H, Hu S H. Green and fast laser fusion technique for bulk silicate rock analysis by laser ablation-inductively coupled plasma mass spectrometry. Analytical Chemistry, 2016, 88(20): 10088-10094.

[41] Liao X H, Hu Z C, Luo T, Zhang W, Liu Y S, Zong K Q, Zhou L, Zhang J F. Determination of major and trace elements in geological samples by laser ablation solution sampling-inductively coupled plasma mass spectrometry. Journal of Analytical Atomic Spectrometry, 2019, 34(6): 1126-1134.

[42] Hamilton J, Gorishek E, Mach P, Sturtevant D, Ladage M, Suzuki N, Padilla P, Mittler R, Chapman K, Verbeck G. Evaluation of a custom single Peltier-cooled ablation cell for elemental imaging of biological samples in laser ablation-inductively coupled plasma-mass spectrometry (LA-ICP-MS). Journal of Analytical Atomic Spectrometry, 2016, 31(4): 1030-1033.

[43] Van Malderen S J, Van Elteren J T, Vanhaecke F. Development of a fast laser ablation-inductively coupled plasma-mass spectrometry cell for sub-μm scanning of layered materials. Journal of Analytical Atomic Spectrometry, 2015, 30(1): 119-125.

[44] Reifschneider O, Wentker K S, Strobel K, Schmidt R, Masthoff M, Sperling M, Faber C, Karst U. Elemental bioimaging of thulium in mouse tissues by laser ablation-ICPMS as a complementary method to heteronuclear proton magnetic resonance imaging for cell tracking experiments. Analytical Chemistry, 2015, 87(8): 4225-4230.

[45] Praamsma M L, Parsons P J. Characterization of calcified reference materials for assessing the reliability of manganese determinations in teeth and bone. Journal of Analytical Atomic Spectrometry, 2014, 29(7): 1243-1251.

[46] Deiting D, Börno F, Hanning S, Kreyenschmidt M, Seidl T, Otto M. Investigation on the suitability of ablated carbon as

an internal standard in laser ablation ICP-MS of polymers. Journal of Analytical Atomic Spectrometry, 2016, 31(8): 1605-1611.

[47] Moraleja I, Esteban-Fernández D, Lázaro A, Humanes B, Neumann B, Tejedor A, Mena M L, Jakubowski N, Gómez-Gómez M M. Printing metal-spiked inks for LA-ICP-MS bioimaging internal standardization: Comparison of the different nephrotoxic behavior of cisplatin, carboplatin, and oxaliplatin. Analytical Bioanalytical Chemistry, 2016, 408(9): 2309-2318.

[48] Douglas D N, O'reilly J, O'connor C, Sharp B L, Goenaga-Infante H. Quantitation of the Fe spatial distribution in biological tissue by online double isotope dilution analysis with LA-ICP-MS: A strategy for estimating measurement uncertainty. Journal of Analytical Atomic Spectrometry, 2016, 31(1): 270-279.

[49] İzgi B, Kayar M. Determination of bromine and tin compounds in plastics using laser ablation inductively coupled plasma mass spectrometry (LA-ICP-MS). Talanta, 2015, 139: 117-122.

[50] Luo T, Wang Y, Li M, Zhang W, Chen H H, Hu Z C. Determination of major and trace elements in alloy steels by nanosecond and femtosecond laser ablation ICP-MS with non-matrix-matched calibration. Atomic Spectroscopy, 2020, 41(1): 11-19.

[51] Cakara A, Bonta M, Riedl H, Mayrhofer P H, Limbeck A. Development of a multi-variate calibration approach for quantitative analysis of oxidation resistant Mo-Si-B coatings using laser ablation inductively coupled plasma mass spectrometry. Spectrochimica Acta Part B: Atomic Spectroscopy, 2016, 120: 57-62.

[52] Mateo M P, Garcia C C, Hergenröder R. Depth analysis of polymer-coated steel samples using near-infrared femtosecond laser ablation inductively coupled plasma mass spectrometry. Analytical Chemistry, 2007, 79(13): 4908-4914.

[53] Feng R, Machado N, Ludden J. Lead geochronology of zircon by LaserProbe-Inductively coupled plasma mass spectrometry (LP-ICPMS). Geochimica et Cosmochimica Acta, 1993, 57(14): 3479-3486.

[54] Fryer B J, Jackson S E, Longerich H P. The application of laser ablation microprobe-inductively coupled plasma-mass spectrometry (LAM-ICP-MS) to in situ (U)-Pb geochronology. Chemical Geology, 1993, 109(1): 1-8.

[55] Mukherjee P K, Souders A K, Sylvester P J. Accuracy and precision of U-Pb zircon geochronology at high spatial resolution (7~20 μm spots) by laser ablation-ICP-single-collector-sector-field-mass spectrometry. Journal of Analytical Atomic Spectrometry, 2019, 34(1): 180-192.

[56] Corbett E P, Simonetti A, Shaw P, Corcoran L, Crowley Q G, Hoare B C. Shallow sampling by multi-shot laser ablation and its application within U-Pb zircon geochronology. Chemical Geology, 2020, 554, 119568.

[57] El Korh A. Ablation behavior and constraints on the U-Pb and Th-Pb geochronometers in titanite analyzed by quadrupole inductively coupled plasma mass spectrometry coupled to a 193 nm excimer laser. Spectrochimica Acta Part B: Atomic Spectroscopy, 2013, 86: 75-87.

[58] Sun J F, Yang J H, Wu F Y, Xie L W, Yang Y H, Liu Z C, Li X H. *In situ* U-Pb dating of titanite by LA-ICPMS. Chinese Science Bulletin, 2012, 57(20): 2506-2516.

[59] Paquette J L, Tiepolo M. High resolution (5 μm) U-Th-Pb isotope dating of monazite with excimer laser ablation (ELA)-ICPMS. Chemical Geology, 2007, 240(3-4): 222-237.

[60] Chew D M, Sylvester P J, Tubrett M N. U-Pb and Th-Pb dating of apatite by LA-ICPMS. Chemical Geology, 2011, 280(1): 200-216.

[61] Thomson S N, Gehrels G E, Ruiz J, Buchwaldt R. Routine low-damage apatite U-Pb dating using laser ablation-multicollector-ICPMS. Geochemistry, Geophysics, Geosystems, 2012, 13(2), Q0AA21. DOI: 10.1029/2011GC003928.

[62] Liu Z C, Wu F Y, Guo C L, Zhao Z F, Yang J H, Sun J F. *In situ* U-Pb dating of xenotime by laser ablation (LA)-ICP-MS. Chinese Science Bulletin, 2011, 56(27): 2948-2956.

[63] Wohlgemuth-Ueberwasser C C, Söderlund U, Pease V, Nilsson M K. Quadrupole LA-ICP-MS U/Pb geochronology of baddeleyite single crystals. Journal of Analytical Atomic Spectrometry, 2015, 30(5): 1191-1196.

[64] Ibanez-Mejia M, Gehrels G E, Ruiz J, Vervoort J D, Eddy M P, Li C. Small-volume baddeleyite (ZrO$_2$) U-Pb geochronology and Lu-Hf isotope geochemistry by LA-ICP-MS: Techniques and applications. Chemical Geology, 2014, 384: 149-167.

[65] Xia X P, Ren Z Y, Wei G J, Zhang L, Sun M, Wang Y. *In situ* rutile U-Pb dating by laser ablation-MC-ICPMS. Geochemical Journal, 2013, 47(4): 459-468.

[66] Zack T, Stockli D F, Luvizotto G L, Barth M G, Belousova E, Wolfe M R, Hinton R W. *In situ* U-Pb rutile dating by LA-ICP-MS: ^{208}Pb correction and prospects for geological applications. Contributions to Mineralogy Petrology, 2011, 162(3): 515-530.

[67] Yuan S D, Peng J, Hao S, Li H, Geng J, Zhang D. *In situ* LA-MC-ICP-MS and ID-TIMS U-Pb geochronology of cassiterite in the giant Furong tin deposit, Hunan Province, South China: New constraints on the timing of tin-polymetallic mineralization. Ore Geology Reviews, 2011, 43(1): 235-242.

[68] Zong K Q, Chen J, Hu Z C, Liu Y S, Li M, Fan H, Meng Y. *In-situ* U-Pb dating of uraninite by fs-LA-ICP-MS. Science China Earth Sciences, 2015, 58(10): 1731-1740.

[69] Wu F Y, Mitchell R H, Li Q L, Sun J, Liu C Z, Yang Y H. *In situ* UPb age determination and SrNd isotopic analysis of perovskite from the Premier (Cullinan) kimberlite, South Africa. Chemical Geology, 2013, 353: 83-95.

[70] Yang Y H, Wu F Y, Li Y, Yang J H, Xie L W, Liu Y, Zhang Y B, Huang C. *In situ* U-Pb dating of bastnaesite by LA-ICP-MS. Journal of Analytical Atomic Spectrometry, 2014, 29(6): 1017-1023.

[71] Roberts N M W, Rasbury E T, Parrish R R, Smith C J, Horstwood M S A, Condon D J. A calcite reference material for LA-ICP-MS U-Pb geochronology. Geochemistry, Geophysics, Geosystems, 2017, 18(7): 2807-2814.

[72] Yang Y H, Wu F Y, Yang J H, Mitchell R H, Zhao Z F, Xie L W, Huang C, Ma Q, Yang M, Zhao H. U-Pb age determination of schorlomite garnet by laser ablation inductively coupled plasma mass spectrometry. Journal of Analytical Atomic Spectrometry, 2018, 33(2): 231-239.

[73] Deng X D, Li J W, Luo T, Wang H Q. Dating magmatic and hydrothermal processes using andradite-rich garnet U-Pb geochronometry. Contributions to Mineralogy and Petrology, 2017, 172(9): 71.

[74] Luo T, Deng X D, Li J W, Hu Z C, Zhang W, Liu Y S, Zhang J. U-Pb geochronology of wolframite by laser ablation inductively coupled plasma mass spectrometry. Journal of Analytical Atomic Spectrometry, 2019, 34(7): 1439-1446.

[75] Kahou Z S, Brichau S, Poujol M, Duchêne S, Campos E, Leisen M, D'abzac F-X, Riquelme R, Carretier S. First U-Pb LA-ICP-MS *in situ* dating of supergene copper mineralization: Case study in the Chuquicamata mining district, Atacama Desert, Chile. Mineralium Deposita, 2020: 1-14.

[76] Huang C, Wang H, Yang J H, Ramezani J, Yang C, Zhang S B, Yang Y H, Xia X P, Feng L J, Lin J. SA01-A proposed zircon reference material for microbeam U-Pb age and Hf-O isotopic determination. Geostandards and Geoanalytical Research, 2020, 44(1): 103-123.

[77] Eddy M P, Ibañez-Mejia M, Burgess S D, Coble M A, Cordani U G, Desormeau J, Gehrels G E, Li X, Maclennan S, Pecha M. GHR 1 zircon-A new Eocene natural reference material for microbeam U-Pb geochronology and Hf isotopic analysis of zircon. Geostandards and Geoanalytical Research, 2019, 43(1): 113-132.

[78] Wiedenbeck M, Allé P, Corfu F, Griffin W L, Meier M, Oberli F, Quadt A V, Roddick J C, Spiegel W. Three natural zircon standards for U-Th-Pb, Lu-Hf, trace element and REE analysis. Geostandards Newsletter, 1995, 19(1): 1-23.

[79] Zhang L, Wu J L, Tu J R, Wu D, Li N, Xia X P, Ren Z Y. RMJG rutile: A new natural reference material for microbeam U-Pb dating and Hf isotopic analysis. Geostandards and Geoanalytical Research, 2019, 44(1): 133-145.

[80] Yang Y H, Wu F Y, Li Q L, Rojas-Agramonte Y, Yang J H, Li Y, Ma Q, Xie L W, Huang C, Fan HR. *In situ* U-Th-Pb dating and Sr-Nd isotope analysis of bastnäsite by LA-(MC)-ICP-MS. Geostandards and Geoanalytical Research, 2019, 43(4): 543-565.

[81] Storey C D, Jeffries T E, Smith M. Common lead-corrected laser ablation ICP-MS U-Pb systematics and geochronology of titanite. Chemical Geology, 2006, 227(1): 37-52.

[82] Darling J R, Storey C D, Engi M. Allanite U-Th-Pb geochronology by laser ablation ICPMS. Chemical Geology, 2012, 292-293: 103-115.

[83] McFarlane C R M. Allanite U Pb geochronology by 193 nm LA ICP-MS using NIST610 glass for external calibration. Chemical Geology, 2016, 438: 91-102.

[84] Burn M, Lanari P, Pettke T, Engi M. Non-matrix-matched standardisation in LA-ICP-MS analysis: General approach, and application to allanite Th-U-Pb dating. Journal of Analytical Atomic Spectrometry, 2017, 32(7): 1359-1377.

[85] Luo T, Hu Z C, Zhang W, Liu Y S, Zong K Q, Zhou L, Zhang J F, Hu S H. Water vapor-assisted "universal" nonmatrix-matched analytical method for the *in situ* U-Pb dating of zircon, monazite, titanite, and xenotime by laser ablation-inductively coupled plasma mass spectrometry. Analytical Chemistry, 2018, 90(15): 9016-9024.

[86] Phillips S E, Hanchar J M, Miller C F, Fisher C M, Lancaster P J, Darling J R. High-spatial-resolution isotope geochemistry of monazite (U-Pb & Sm-Nd) and zircon (U-Pb & Lu-Hf) in the Old Woman and North Piute Mountains, Mojave Desert, California. EGU General Assembly Conference Abstracts, 2014: 8647.

[87] Xu L, Yang J H, Ni Q, Yang Y H, Hu Z C, Liu Y S, Wu Y B, Luo T, Hu S H. Determination of Sm-Nd isotopic compositions in fifteen geological materials using laser ablation MC-ICP-MS and application to monazite geochronology of metasedimentary rock in the North China Craton. Geostandards and Geoanalytical Research, 2018, 42(3): 379-394.

[88] Hogmalm K J, Dahlgren I, Fridolfsson I, Zack T. First *in situ* Re-Os dating of molybdenite by LA-ICP-MS/MS. Mineralium Deposita, 2019, 54(6): 821-828.

[89] Zack T, Hogmalm K J. Laser ablation Rb/Sr dating by online chemical separation of Rb and Sr in an oxygen-filled reaction cell. Chemical Geology, 2016, 437: 120-133.

[90] Yang Y H, Wu F Y, Wilde S A, Liu X M, Zhang Y B, Xie L W, Yang J H. *In situ* perovskite Sr-Nd isotopic constraints on the petrogenesis of the Ordovician Mengyin kimberlites in the North China Craton. Chemical Geology, 2009, 264(1-4): 24-42.

[91] Wu F Y, Yang Y H, Marks M a W, Liu Z C, Zhou Q, Ge W C, Yang J S, Zhao Z F, Mitchell R H, Markl G. *In situ* U-Pb, Sr, Nd and Hf isotopic analysis of eudialyte by LA-(MC)-ICP-MS. Chemical Geology, 2010, 273(1-2): 8-34.

[92] Yang Y H, Wu F Y, Yang J H, Chew D M, Xie L W, Chu Z Y, Zhang Y B, Huang C. Sr and Nd isotopic compositions of apatite reference materials used in U-Th-Pb geochronology. Chemical Geology, 2014, 385: 35-55.

[93] Tong X R, Liu Y S, Hu Z C, Chen H H, Zhou L, Hu Q H, Xu R, Deng L X, Chen C F, Yang L, Gao S. Accurate determination of Sr isotopic compositions in clinopyroxene and silicate glasses by LA-MC-ICP-MS. Geostandards and Geoanalytical Research, 2016, 40(1): 85-99.

[94] Zhang W, Hu Z C, Liu Y S, Wu T, Deng X D, Guo J L, Zhao H. Improved *in situ* Sr isotopic analysis by a 257 nm femtosecond laser in combination with the addition of nitrogen for geological minerals. Chemical Geology, 2018, 479: 10-21.

[95] Yu K Z, Liu Y S, Hu Q H, Ducea M N, Hu Z C, Zong K Q, Chen H H. Magma recharge and reactive bulk assimilation in enclave-bearing granitoids, Tonglu, south China. Journal of Petrology, 2018, 59(5): 795-824.

[96] Foster G L, Vance D. *In situ* Nd isotopic analysis of geological materials by laser ablation MC-ICP-MS. Journal of Analytical Atomic Spectrometry, 2006, 21(3): 288-296.

[97] Yang Y H, Sun J F, Xie L W, Fan H R, Wu F Y. *In situ* Nd isotopic measurement of natural geological materials by LA-MC-ICPMS. Chinese Science Bulletin, 2008, 53(7): 1062-1070.

[98] Xu L, Hu Z C, Zhang W, Yang L, Liu Y S, Gao S, Luo T, Hu S H. *In situ* Nd isotope analyses in geological materials with signal enhancement and non-linear mass dependent fractionation reduction using laser ablation MC-ICP-MS. Journal of Analytical Atomic Spectrometry, 2014, 30(1): 232-244.

[99] 李献华, 梁细荣, 韦刚健, 刘颖. 锆石 Hf 同位素组成的 LAM-MC-ICPMS 精确测定. 地球化学, 2003, 32(1): 86-90.

[100] 徐平, 吴福元, 谢烈文, 杨岳衡. U-Pb 同位素定年标准锆石的 Hf 同位素. 科学通报, 2004, 49(14): 1403-1410.

[101] Yuan H L, Gao S, Dai M N, Zong C L, Günther D, Fontaine G H, Liu X M, Diwu C R. Simultaneous determinations of U-Pb age, Hf isotopes and trace element compositions of zircon by excimer laser-ablation quadrupole and multiple-collector ICP-MS. Chemical Geology, 2008, 247(1-2): 100-118.

[102] Hu Z C, Liu Y S, Gao S, Liu W G, Zhang W, Tong X R, Lin L, Zong K Q, Li M, Chen H H, Zhou L, Yang L. Improved *in situ* Hf isotope ratio analysis of zircon using newly designed X skimmer cone and jet sample cone in combination with the addition of nitrogen by laser ablation multiple collector ICP-MS. Journal of Analytical Atomic Spectrometry, 2012, 27(9): 1391-1399.

[103] 李献华, 唐国强, 龚冰, 杨岳衡, 侯可军, 胡兆初, 李秋立, 刘宇, 李武显. Qinghu(清湖)锆石: 一个新的 U-Pb 年龄和 O, Hf 同位素微区分析工作标样. 科学通报, 2013, 58(20): 1954-1961.

[104] Li X H, Long W G, Li Q L, Liu Y, Zheng Y F, Yang Y H, Chamberlain K R, Wan D F, Guo C H, Wang X C, Tao H. Penglai zircon megacrysts: A potential new working reference material for microbeam determination of Hf-O isotopes and U-Pb age. Geostandards and Geoanalytical Research, 2010, 34(2): 117-134.

[105] Huang C, Wang H, Yang J H, Xie L W, Yang Y H, Wu S T. Further characterization of the BB zircon via SIMS and MC-ICP-MS for Li, O, and Hf isotopic compositions. Minerals, 2019, 2019(12): 774-790.

[106] Fisher C M, Vervoort J D, Hanchar J M. Guidelines for reporting zircon Hf isotopic data by LA-MC-ICPMS and potential pitfalls in the interpretation of these data. Chemical Geology, 2014, 363(1): 125-133.

[107] Zhu L Y, Liu Y S, Ma T T, Lin J, Hu Z C, Wang C. In situ measurement of Os isotopic ratios in sulfides calibrated against ultra-fine particle standards using LA-MC-ICP-MS. Journal of Analytical Atomic Spectrometry, 2016, 31(7): 1414-1422.

[108] Zhang W, Hu Z C, Yang L, Liu Y S, Zong K Q, Xu H J, Chen H H, Gao S, Xu L. Improved Inter-calibration of faraday cup and ion counting for *in situ* Pb isotope measurements using LA-MC-ICP-MS: Application to the study of the origin of the Fangshan pluton, north china. Geostandards and Geoanalytical Research, 2015, 39(4): 467-487.

[109] Chen K Y, Yuan H L, Bao Z A, Zong C L, Dai M N. Precise and accurate *in situ* determination of lead isotope ratios in NIST, USGS, MPI-DING and CGSG glass reference materials using femtosecond laser ablation MC-ICP-MS. Geostandards and Geoanalytical Research, 2014, 38(1): 5-21.

[110] Yuan H L, Chen K Y, Bao Z A, Zong C L, Dai M N, Chao F, Cong Y. Determination of lead isotope compositions of geological samples using femtosecond laser ablation MC-ICPMS. Chinese Science Bulletin, 2013, 58(32): 3914-3921.

[111] Yuan H L, Yin C, Liu X, Chen K Y, Bao Z A, Zong C L, Dai M N, Lai S C, Wang R, Jiang S Y. High precision *in-situ* Pb isotopic analysis of sulfide minerals by femtosecond laser ablation multi-collector inductively coupled plasma mass spectrometry. Science China Earth Sciences, 2015, 58(10): 1713-1721.

[112] Bao Z, Zhang H F, Yuan H L, Liu Y, Chen K Y, Zong C L. Flux-free fusion technique using a boron nitride vessel and rapid acid digestion for determination of trace elements by ICP-MS. Journal of Analytical Atomic Spectrometry, 2016, 31(11): 2261-2271.

[113] 陈开运, 范超, 袁洪林, 包志安, 宗春蕾, 戴梦宁, 凌雪, 杨颖. 飞秒激光剥蚀-多接收电感耦合等离子质谱原位微区分析青铜中铅同位素组成——以古铜钱币为例. 光谱学与光谱分析, 2013, 32(5): 1342-1349.

[114] Zhang W, Hu Z C, Günther D, Liu Y S, Ling W L, Zong K Q, Chen H H, Gao S. Direct lead isotope analysis in Hg-rich sulfides by LA-MC-ICP-MS with a gas exchange device and matrix-matched calibration. Analytica Chimica Acta, 2016, 948: 9-18.

[115] Zhang L, Ren Z Y, Nichols A R L, Zhang Y H, Zhang Y, Qian S P, Liu J Q. Lead isotope analysis of melt inclusions by LA-MC-ICP-MS. Journal of Analytical Atomic Spectrometry, 2014, 29: 1393-1405.

[116] Zhang Z, Tyrrell S, Li C, Daly J S, Sun X, Li Q. Pb isotope compositions of detrital K-feldspar grains in the upper-middle Yangtze River system: Implications for sediment provenance and drainage evolution. Geostandards and Geoanalytical Research, 2014, 15(7): 2765-2779.

[117] Steinmann L K, Oeser M, Horn I, Seitz H-M, Weyer S. *In situ* high-precision lithium isotope analyses at low concentration levels with femtosecond-LA-MC-ICP-MS. Journal of Analytical Atomic Spectrometry, 2019, 34: 1447-1458.

[118] Lin J, Liu Y S, Tong X R, Zhu L Y, Zhang W, Hu Z C. Improved *in situ* Li isotopic ratio analysis of silicates by optimizing signal intensity, isotopic ratio stability and intensity matching using ns-LA-MC-ICP-MS. Journal of Analytical Atomic Spectrometry, 2017, 32(4): 834-842.

[119] Kimura J, Chang Q, Ishikawa T, Tsujimori T. Influence of laser parameters on isotope fractionation and optimisation of lithium and boron isotope ratio measurements using laser ablation-multiple Faraday collector-inductively coupled plasma mass spectrometry. Journal of Analytical Atomic Spectrometry, 2016, 31(11): 2305-2320.

[120] Lin L, Hu Z C, Yang L, Zhang W, Liu Y S, Gao S, Hu S H. Determination of boron isotope compositions of geological materials by laser ablation MC-ICP-MS using newly designed high sensitivity skimmer and sample cones. Chemical Geology, 2014, 386(0): 22-30.

[121] Chen W, Lu J, Jiang S Y, Zhao K D, Duan D F. *In situ* carbon isotope analysis by laser ablation MC-ICP-MS. Analytical Chemistry, 2017, 89(24): 13415-13421.

[122] Oeser M, Weyer S, Horn I, Schuth S. High-precision Fe and Mg isotope ratios of silicate reference glasses determined *in situ* by femtosecond LA-MC-ICP-MS and by solution nebulisation MC-ICP-MS. Geostandards and Geoanalytical Research, 2014, 38(3): 311-328.

[123] Xie L W, Yin Q Z, Yang J H, Wu F Y, Yang Y H. High precision analysis of Mg isotopic composition in olivine by laser ablation MC-ICP-MS. Journal of Analytical Atomic Spectrometry, 2011, 26(9): 1773-1780.

[124] Janney P E, Richter F M, Mendybaev R A, Wadhwa M, Georg R B, Watson E B, Hines R R. Matrix effects in the analysis of Mg and Si isotope ratios in natural and synthetic glasses by laser ablation-multicollector ICPMS: A comparison of single-and double-focusing mass spectrometers. Chemical Geology, 2011, 281(1-2): 26-40.

[125] Zhang C X, Zhao H, Zhang W, Luo T, Li M, Zong K Q, Liu Y S, Hu Z C. A high performance method for the accurate and precise determination of silicon isotopic compositions in bulk silicate rock samples using laser ablation MC-ICP-MS. Journal of Analytical Atomic Spectrometry, 2020. DOI: 10.1039/d0ja00036a.

[126] Fu J L, Hu Z C, Zhang W, Yang L, Liu Y S, Li M, Zong K Q, Gao S, Hu S H. *In situ* sulfur isotopes (δ^{34}S and δ^{33}S) analyses in sulfides and elemental sulfur using high sensitivity cones combined with the addition of nitrogen by laser ablation MC-ICP-MS. Analytica Chimica Acta, 2016, 911: 14-26.

[127] Chen L, Chen K Y, Bao Z A, Liang P, Sun T T, Yuan H L. Preparation of standards for *in situ* sulfur isotope measurement in sulfides using femtosecond laser ablation MC-ICP-MS. Journal of Analytical Atomic Spectrometry, 2017, 32(1): 107-116.

[128] Bao Z A, Chen L, Zong C L, Yuan H L, Chen K Y, Dai M N. Development of pressed sulfide powder tablets for *in situ*

sulfur and lead isotope measurement using LA-MC-ICP-MS. International Journal of Mass Spectrometry, 2017, 421: 255-262.

[129] Toyama C, Kimura J, Chang Q, Vaglarov B S, Kuroda J. A new high-precision method for determining stable chlorine isotopes in halite and igneous rock samples using UV-femtosecond laser ablation multiple Faraday collector inductively coupled plasma mass spectrometry. Journal of Analytical Atomic Spectrometry, 2015, 30(10): 2194-2207.

[130] Zhang W, Hu Z C, Liu Y S, Feng L P, Jiang H S. *In situ* calcium isotopic ratio determination in calcium carbonate materials and calcium phosphate materials using laser ablation-multiple collectorinductively coupled plasma mass spectrometry. Chemical Geology, 2019, 522: 16-25.

[131] Schuth S, Horn I, Brüske A, Wolff P E, Weyer S. First vanadium isotope analyses of V-rich minerals by femtosecond laser ablation and solution-nebulization MC-ICP-MS. Ore Geology Reviews, 2017, 81: 1271-1286.

[132] Weyrauch M, Oeser M, Bruske A, Weyer S. *In situ* high-precision Ni isotope analysis of metals by femtosecond-LA-MC-ICP-MS. Journal of Analytical Atomic Spectrometry, 2017, 32(7): 1312-1319.

[133] Lazarov M, Horn I. Matrix and energy effects during *in-situ* determination of Cu isotope ratios by ultraviolet-femtosecond laser ablation multicollector inductively coupled plasma mass spectrometry. Spectrochimica Acta Part B: Atomic Spectroscopy, 2015, 111: 64-73.

[134] Resano M, Aramendia M, Rello L, Calvo M L, Berail S, Pecheyran C. Direct determination of Cu isotope ratios in dried urine spots by means of fs-LA-MC-ICPMS. Potential to diagnose Wilson's disease. Journal of Analytical Atomic Spectrometry, 2013, 28(1): 98-106.

[135] Zhang W, Wang Z C, Frédéric M, Inglis E, Tian S Y, Li M, Liu Y S, Hu Z C. Determination of Zr isotopic ratios in zircons using laser-ablation multiple-collector inductively coupled-plasma mass-spectrometry. Journal of Analytical Atomic Spectrometry, 2019, 34: 1800-1809.

[136] Schulze M, Ziegerick M, Horn I, Weyer S, Vogt C. Determination of tin isotope ratios in cassiterite by femtosecond laser ablation multicollector inductively coupled plasma mass spectrometry. Spectrochimica Acta Part B: Atomic Spectroscopy, 2017, 130: 26-34.

[137] Zhang W, Hu Z C. A critical review of isotopic fractionation and interference correction methods for isotope ratio measurements by laser ablation multi-collector inductively coupled plasma mass spectrometry. Spectrochimica Acta Part B: Atomic Spectroscopy, 2020, 171: 105929.

[138] Lin J, Liu Y S, Hu Z C, Chen W, Zhang C X, Zhao K D, Jin X Y. Accurate analysis of Li isotopes in tourmalines by LA-MC-ICP-MS under "wet" conditions with nonmatrix-matched calibration. Journal of Analytical Atomic Spectrometry, 2019, 34: 1145-1153.

[139] Li R, Xia X P, Yang S H, Chen H Y, Yang Q. Off-mount calibration and one new potential pyrrhotite reference material for sulfur isotope measurement by secondary ion mass spectrometry. Geostandards and Geoanalytical Research, 2019, 43(1): 177-187.

[140] Li R, Xia X P, Chen H Y, Wu N P, Zhao T P, Lai C, Yang Q, Zhang Y Q. A potential new chalcopyrite reference material for secondary ion mass spectrometry sulfur isotope ratio analysis. Geostandards and Geoanalytical Research, 2020, 44(3): 485-500.

[141] Feng Y T, Zhang W, Hu Z C, Liu Y S, Chen K, Fu J L, Xie J Y, Shi Q H. Development of sulfide reference materials for *in situ* platinum group elements and S-Pb isotope analyses by LA-(MC)-ICP-MS. Journal of Analytical Atomic Spectrometry, 2018, 33(12): 2172-2183.

[142] Günther D, Audetat A, Frischknecht R, A. Heinrich C. Quantitative analysis of major, minor and trace elements in fluid inclusions using laser ablation-inductively coupled plasmamass spectrometry. Journal of Analytical Atomic

Spectrometry, 1998, 13(4): 263-270.

[143] Heinrich C A, Pettke T, Halter W E, Aigner-Torres M, Horn I. Quantitative multi-element analysis of minerals, fluid and melt inclusions by laser-ablation inductively-coupled-plasma mass-spectrometry. Geochimica et Cosmochimica Acta, 2003, 67(18): 3473-3497.

[144] Lin X, Guo W, Jin L L, Hu S H. Review: Elemental analysis of individual fluid inclusions by laser ablation-ICP-MS. Atomic Spectroscopy, 2020, 41(1): 1-10.

[145] 胡圣虹, 胡兆初, 刘勇胜, 罗彦, 林守麟, 高山. 单个流体包裹体元素化学组成分析新技术——激光剥蚀电感耦合等离子体质谱(LA-ICP-MS). 地学前缘, 2001, (4): 434-440.

[146] Allan M M, Yardley B W D, Forbes L J, Shmulovich K I, Banks D A, Shepherd T J. Validation of LA-ICP-MS fluid inclusion analysis with synthetic fluid inclusions. American Mineralogist, 2005, 90(11-12): 1767-1775.

[147] Pettke T. Analytical protocols for element concentration and isotope ratio measurements in fluid inclusions by LA-(MC-)ICP-MS. Mineralogical Association on Canada, 2008, 40:189-217.

[148] Pettke T, Oberli F, Audétat A, Guillong M, Simon A C, Hanley J J, Klemm L M. Recent developments in element concentration and isotope ratio analysis of individual fluid inclusions by laser ablation single and multiple collector ICP-MS. Ore Geology Reviews, 2012, 44: 10-38.

[149] Leisen M, Dubessy J, Boiron M-C, Lach P. Improvement of the determination of element concentrations in quartz-hosted fluid inclusions by LA-ICP-MS and Pitzer thermodynamic modeling of ice melting temperature. Geochimica et Cosmochimica Acta, 2012, 90: 110-125.

[150] 蓝廷广, 胡瑞忠, 范宏瑞, 毕献武, 唐燕文, 周丽, 毛伟, 陈应华. 流体包裹体及石英LA-ICP-MS分析方法的建立及其在矿床学中的应用. 岩石学报, 2017, 33(10): 3239-3262.

[151] 郭伟, 林贤, 胡圣虹. 单个流体包裹体LA-ICP-MS分析及应用进展. 地球科学, 2020, 45(4): 1362-1374.

[152] Pan J-Y, Ni P, Wang R-C. Comparison of fluid processes in coexisting wolframite and quartz from a giant vein-type tungsten deposit, South China: Insights from detailed petrography and LA-ICP-MS analysis of fluid inclusions. American Mineralogist, 2019, 104(8): 1092-1116.

[153] Liu Y S, Hu Z C, Yuan H L, Hu S H, Cheng H H. Volume-optional and low-memory (VOLM) chamber for laser ablation-ICP-MS: Application to fiber analyses. Journal of Analytical Atomic Spectrometry, 2007, 22(5): 582-585.

[154] Lindner H, Autrique D, Pisonero J, Günther D, Bogaerts A. Numerical simulation analysis of flow patterns and particle transport in the HEAD laser ablation cell with respect to inductively coupled plasma spectrometry. Journal of Analytical Atomic Spectrometry, 2010, 25(3): 295-304.

[155] Hu Z C, Liu Y S, Gao S, Hu S H, Dietiker R, Günther D. A local aerosol extraction strategy for the determination of the aerosol composition in laser ablation inductively coupled plasma mass spectrometry. Journal of Analytical Atomic Spectrometry, 2008, 23(9): 1192-1203.

[156] Albrecht M, Derrey I T, Horn I, Schuth S, Weyer S. Quantification of trace element contents in frozen fluid inclusions by UV-fs-LA-ICP-MS analysis. Journal of Analytical Atomic Spectrometry, 2014, 29(6): 1034-1041.

[157] Jian W, Albrecht M, Lehmann B, Mao J W, Horn I, Li Y H, Ye H S, Li Z Y, Fang G G, Xue Y S. UV-fs-LA-ICP-MS Analysis of CO_2-rich fluid inclusions in a frozen state: Example from the Dahu Au-Mo deposit, Xiaoqinling region, Central China. Geofluids, 2018, 3692180.

[158] Harlaux M, Borovinskaya O, Frick D A, Tabersky D, Gschwind S, Richard A, Günther D, Mercadier J. Capabilities of sequential and quasi-simultaneous LA-ICPMS for the multi-element analysis of small quantity of liquids (pL to nL): Insights from fluid inclusion analysis. Journal of Analytical Atomic Spectrometry, 2015, 30(9): 1945-1969.

[159] Heinrich C A, Ryan C G, Mernagh T P, Eadington P J. Segregation of ore metals between magmatic brine and vapor; A

fluid inclusion study using PIXE microanalysis. Economic Geology, 1992, 87(6): 1566-1583.

[160] Steele-Macinnis M, Ridley J, Lecumberri-Sanchez P, Schlegel T U, Heinrich C A. Application of low-temperature microthermometric data for interpreting multicomponent fluid inclusion compositions. Earth-Science Reviews, 2016, 159: 14-35.

[161] Heinrich C A, Günther D, Audétat A, Ulrich T, Frischknecht R. Metal fractionation between magmatic brine and vapor, determined by microanalysis of fluid inclusions. Geology, 1999, 27(8): 755-758.

[162] Chang J, Li J W, Audétat A. Formation and evolution of multistage magmatic-hydrothermal fluids at the Yulong porphyry Cu-Mo deposit, eastern Tibet: Insights from LA-ICP-MS analysis of fluid inclusions. Geochimica et Cosmochimica Acta, 2018, 232: 181-205.

[163] Audétat A, Günther D, Heinrich C A. Formation of a magmatic-hydrothermal ore deposit: Insights with LA-ICP-MS analysis of fluid inclusions. Science, 1998, 279(5359): 2091-2094.

[164] Liu H Q, Bi X W, Lu H Z, Hu R Z, Lan T G, Wang X S, Huang M L. Nature and evolution of fluid inclusions in the Cenozoic Beiya gold deposit, SW China. Journal of Asian Earth Sciences, 2018, 161(AUG.1): 35-56.

[165] Wilkinson J J, Stoffell B, Wilkinson C C, Jeffries T E, Appold M S. Anomalously metal-rich fluids form hydrothermal ore deposits. Science, 2009, 323(5915): 764-767.

[166] 王莉娟, 王玉往, 王京彬, 朱和平, Günther D. 内蒙古大井锡多金属矿床流体成矿作用研究:单个流体包裹体组分 LA-ICP-MS 分析证据. 科学通报, 2006, (10): 1203-1210.

[167] Stoffell B, Appold M S, Wilkinson J J, Mcclean N A, Jeffries T E. Geochemistry and evolution of Mississippi Valley-type mineralizing brines from the Tri-State and northern Arkansas Districts determined by LA-ICP-MS microanalysis of fluid inclusions. Economic Geology, 2008, 103(7): 1411-1435.

[168] Leisen M, Boiron M C, Richard A, Dubessy J. Determination of Cl and Br concentrations in individual fluid inclusions by combining microthermometry and LA-ICPMS analysis: Implications for the origin of salinity in crustal fluids. Chemical Geology, 2012, 330-331: 197-206.

[169] Fusswinkel T, Giehl C, Beermann O, Fredriksson J R, Garbe-Schönberg D, Scholten L, Wagner T. Combined LA-ICP-MS microanalysis of iodine, bromine and chlorine in fluid inclusions. Journal of Analytical Atomic Spectrometry, 2018, 33(5): 768-783.

[170] Gomes S D, Berger S, Figueiredo E Silva R C, Hagemann S G, Rosière C A, Banks D A, Lobato L M, Hensler A S. Oxide chemistry and fluid inclusion constraints on the formation of itabirite-hosted iron ore deposits at the eastern border of the southern Espinhaço Range, Brazil. Ore Geology Reviews, 2018, 95: 821-848.

[171] Li W T, Audétat A, Zhang J. The role of evaporites in the formation of magnetite-apatite deposits along the Middle and Lower Yangtze River, China: Evidence from LA-ICP-MS analysis of fluid inclusions. Ore Geology Reviews, 2015, 67: 264-278.

[172] Chen P W, Zeng Q D, Zhou T C, Wang Y B, Yu B, Chen J Q. Evolution of fluids in the Dasuji porphyry Mo deposit on the northern margin of the North China Craton: Constraints from Microthermometric and LA-ICP-MS analyses of fluid inclusions. Ore Geology Reviews, 2019, 104: 26-45.

[173] Zhang X N, Li G M, Qin K Z, Lehmann B, Li J X, Zhao J X. Porphyry to epithermal transition at the Rongna Cu-(Au) deposit, Tibet: Insights from H-O isotopes and fluid inclusion analysis. Ore Geology Reviews, 2020, 123: 103585.

[174] 吴石头, 许春雪, 陈宗定, 王亚平, 白金峰. LA-ICP-MS 结合超高压粉末压片技术快速分析碳酸盐岩中 Mg, Ca, Sr, Ba 和轻稀土元素. 分析试验室, 2019, 38(9): 1089-1094.

[175] 姜劲锋, 徐鸿志, 郭伟, 刘先国, 胡圣虹. 粉末压饼 LA-ICP-MS 测定土壤样品中微量元素. 分析试验室, 2007, (11): 20-24.

[176] Wu S T, Karius V, Schmidt B C, Simon K, Wörner G. Comparison of ultrafine powder pellet and flux-free fusion glass for bulk analysis of granitoids by laser ablation-inductively coupled plasma-mass spectrometry. Geostandards and Geoanalytical Research, 2018, 42(4): 575-591.

[177] Yao S C, Zhao J B, Xu J L, Lu Z M, Lu J D. Optimizing the binder percentage to reduce matrix effects for the LIBS analysis of carbon in coal. Journal of Analytical Atomic Spectrometry, 2017, 32(4): 766-772.

[178] Perkins W T, Pearce N, Jeffries T E. Laser ablation inductively coupled plasma mass spectrometry: A new technique for the determination of trace and ultra-trace elements in silicates. Geochimica et Cosmochimica Acta, 1993, 57(2): 475-482.

[179] Morrison C. Laser ablation-inductively coupled plasma-mass spectrometry: An investigation of elemental responses and matrix effects in the analysis of geostandard materials. Chemical Geology, 1995, 119(1-4): 13-29.

[180] Holá M, Mikuška P, Hanzlíková R, Kaiser J, Kanický V. Tungsten carbide precursors as an example for influence of a binder on the particle formation in the nanosecond laser ablation of powdered materials. Talanta, 2010, 80(5): 1862-1867.

[181] Gondal M, Hussain T, Yamani Z, Baig M. The role of various binding materials for trace elemental analysis of powder samples using laser-induced breakdown spectroscopy. Talanta, 2007, 72(2): 642-649.

[182] Jantzi S C, Almirall J R. Characterization and forensic analysis of soil samples using laser-induced breakdown spectroscopy (LIBS). Analytical and Bioanalytical Chemistry, 2011, 400(10): 3341-3351.

[183] Jantzi S C, Almirall J R. Elemental analysis of soils using laser ablation inductively coupled plasma mass spectrometry (LA-ICP-MS) and laser-induced breakdown spectroscopy (LIBS) with multivariate discrimination: Tape mounting as an alternative to pellets for small forensic transfer specimens. Applied Spectroscopy, 2014, 68(9): 963-974.

[184] Peters D, Pettke T. Evaluation of Major to ultra trace element bulk rock chemical analysis of nanoparticulate pressed powder pellets by LA-ICP-MS. Geostandards and Geoanalytical Research, 2017, 41(1): 5-28.

[185] Mukherjee P K, Khanna P P, Saini N K. Rapid determination of trace and ultra trace level elements in diverse silicate rocks in pressed powder pellet targets by LA-ICP-MS using a matrix-independent protocol. Geostandards and Geoanalytical Research, 2014, 38(3): 363-379.

[186] 朱律运, 刘勇胜, 胡兆初, 高山, 王晓红, 田滔. 玄武岩全岩元素含量快速、准确分析新技术: 双铱带高温炉与LA-ICP-MS 联用法. 地球化学, 2011, (5): 407-417.

[187] 包志安, 贺国芬, 陈开运, 宋佳瑶, 袁洪林, 柳小明. 高温熔融研制岩石标准玻璃方法初步研究: 以玄武岩为例. 地球化学, 2011, (3): 223-236.

[188] 包志安, 袁洪林, 陈开运, 宋佳瑶, 戴梦宁, 宗春蕾. 高温熔融研制安山岩玻璃标准物质初探. 岩矿测试, 2011, (5): 521-527.

[189] 胡圣虹, 罗彦, 刘勇胜, 林守麟, 高山. $Li_2B_4O_7$ 熔融玻璃 LA-ICP-MS 测定岩石中痕量稀土元素. 武汉: 第九届全国稀土分析化学学术报告会, 2001.

[190] Zhang S Y, Zhang H L, Hou Z, Ionov D A, Huang F. Rapid determination of trace element compositions in peridotites by LA-ICP-MS using an albite fusion method. Geostandards and Geoanalytical Research, 2019, 43(1): 93-111.

[191] Yu Z S, Norman M D, Robinson P. Major and trace element analysis of silicate rocks by XRF and laser ablation ICP-MS using lithium borate fused glasses: Matrix effects, instrument response and results for international reference materials. Geostandards and Geoanalytical Research, 2003, 27(1): 67-89.

[192] Nicholls I A. A direct fusion method of preparing silicate rock glasses for energy-dispersive electron microprobe analysis. Chemical Geology, 1974, 14(3): 151-157.

[193] Stoll B, Jochum K P, Herwig K, Amini M, Flanz M, Kreuzburg B, Kuzmin D, Willbold M, Enzweiler J. An automated

iridium-strip heater for LA-ICP-MS bulk analysis of geological samples. Geostandards and Geoanalytical Research, 2008, 32(1): 5-26.

[194] Nehring F, Jacob D E, Barth M G, Foley S F. Laser-ablation ICP-MS analysis of siliceous rock glasses fused on an iridium strip heater using MgO dilution. Microchimica Acta, 2008, 160(1-2): 153-163.

[195] Günther D, V. Quadt A, Wirz R, Cousin H, Dietrich V J. Elemental analyses using laser ablation-inductively coupled plasma-mass spectrometry (LA-ICP-MS) of geological samples fused with $Li_2B_4O_7$ and calibrated without matrix-matched standards. Mikrochimica Acta, 2001, 136(3): 101-107.

[196] Leite T D F, Escalfoni R, Da Fonseca T C O, Miekeley N. Determination of major, minor and trace elements in rock samples by laser ablation inductively coupled plasma mass spectrometry: Progress in the utilization of borate glasses as targets. Spectrochimica Acta Part B: Atomic Spectroscopy, 2011, 66(5): 314-320.

[197] Monsels D A, Van Bergen M J, Mason P R D. Determination of trace elements in bauxite using laser ablation-inductively coupled plasma-mass spectrometry on lithium borate glass beads. Geostandards and Geoanalytical Research, 2018, 42(2): 239-251.

[198] Malherbe J, Claverie F, Alvarez A, Fernandez B, Pereiro R, Molloy J L. Elemental analyses of soil and sediment fused with lithium borate using isotope dilution laser ablation-inductively coupled plasma-mass spectrometry. Analytica Chimica Acta, 2013, 793: 72-78.

[199] Aguilera J A, Aragón C. Analysis of rocks by CSigma laser-induced breakdown spectroscopy with fused glass sample preparation. Journal of Analytical Atomic Spectrometry, 2017, 32(1): 144-152.

[200] Carvalho A a C, Alves V C, Silvestre D M, Leme F O, Oliveira P V, Nomura C S. Comparison of fused glass beads and pressed powder pellets for the quantitative measurement of Al, Fe, Si and Ti in bauxite by laser-induced breakdown spectroscopy. Geostandards and Geoanalytical Research, 2017, 41(4): 585-592.

[201] Horner N S, Beauchemin D. The use of sol-gels as solid calibration standards for the analysis of soil samples by laser ablation coupled to inductively coupled plasma mass spectrometry. Journal of Analytical Atomic Spectrometry, 2014, 29(4): 715-720.

[202] Klemm W, Bombach G. A simple method of target preparation for the bulk analysis of powder samples by laser ablation inductively coupled plasma mass spectrometry (LA-ICP-MS). Fresenius' Journal of Analytical Chemistry, 2001, 370(5): 641-646.

[203] Brouard D, Gravel J-F Y, Viger M L, Boudreau D. Use of sol-gels as solid matrixes for laser-induced breakdown spectroscopy. Spectrochimica Acta Part B: Atomic Spectroscopy, 2007, 62(12SI): 1361-1369.

[204] Li Y T, Guo W, Hu Z C, Jin L L, Hu S H, Guo Q H. Method development for direct multielement quantification by LA-ICP-MS in food samples. Journal of Agricultural and Food Chemistry, 2018, 67(3): 935-942.

[205] Pakieła M, Wojciechowski M, Wagner B, Bulska E. A novel procedure of powdered samples immobilization and multi-point calibration of LA ICP MS. Journal of Analytical Atomic Spectrometry, 2011, 26(7): 1539-1543.

[206] Zhu Y B, Hioki A, Chiba K. Quantitative analysis of the elements in powder samples by LA-ICP-MS with PMMA powder as the binder and Cs as the internal standard. Journal of Analytical Atomic Spectrometry, 2013, 28(2): 301-306.

[207] Yang L, Sturgeon R E, Mester Z. Quantitation of trace metals in liquid samples by dried-droplet laser ablation inductively coupled plasma mass spectrometry. Analytical Chemistry, 2005, 77(9): 2971-2977.

[208] Foltynová P, Bednařík A, Kanický V, Preisler J. Diode laser thermal vaporization ICP MS with a simple tubular cell for determination of lead and cadmium in whole blood. Journal of Analytical Atomic Spectrometry, 2014, 29(9): 1585-1590.

[209] Dos Santos Augusto A, Sperança M A, Andrade D F, Pereira-Filho E R. Nutrient and contaminant quantification in

solid and liquid food samples using laser-ablation inductively coupled plasma-mass spectrometry (LA-ICP-MS): Discussion of calibration strategies. Food Analytical Methods, 2017, 10(5): 1515-1522.

[210] Winefordner J D, Gornushkin I B, Correll T, Gibb E, Smith B W, Omenetto N. Comparing several atomic spectrometric methods to the super stars: Special emphasis on laser induced breakdown spectrometry, LIBS, a future super star. Journal of Analytical Atomic Spectrometry, 2004, 19(9): 1061-1083.

[211] Prabhu R K, Vijayalakshmi S, Mahalingam T R, Viswanathan K S, Mathews C K. Laser vaporization inductively coupled plasma mass spectrometry: A technique for the analysis of small volumes of solutions. Journal of Analytical Atomic Spectrometry, 1993, 8(4): 565-569.

[212] Ghazi A M, Mccandless T E, Vanko D A, Ruiz J. New quantitative approach in trace elemental analysis of single fluid inclusions: Applications of laser ablation inductively coupled plasma mass spectrometry (LA-ICP-MS). Journal of Analytical Atomic Spectrometry, 1996, 11(9): 667-674.

[213] Moissette A, Shepherd T J, Chenery S R. Calibration strategies for the elemental analysis of individual aqueous fluid inclusions by laser ablation inductively coupled plasma mass spectrometry. Journal of Analytical Atomic Spectrometry, 1996, 11(3): 177-185.

[214] Boue-Bigne F, J. Masters B, S. Crighton J, L. Sharp B. A calibration strategy for LA-ICP-MS analysis employing aqueous standards having modified absorption coefficients. Journal of Analytical Atomic Spectrometry, 1999, 14(11): 1665-1672.

[215] Günther D, Frischknecht R, Müschenborn H-J, Heinrich C A. Direct liquid ablation: A new calibration strategy for laser ablation-ICP-MS microanalysis of solids and liquids. Fresenius' Journal of Analytical Chemistry, 1997, 359(4): 390-393.

[216] Masters B J, Sharp B L. Universal calibration strategy for laser ablation inductively coupled plasma mass spectrometry based on the use of aqueous standards with modified absorption coefficients. Analytical Communications, 1997, 34(9): 237-239.

[217] Mossman D J, Jackson S E, Gauthier-Lafaye F. Trace element and isotopic analysis by laser ablation ICP-MS of ore deposit bitumens: A test case with uranium ores from Oklo, Gabon. Energy Sources, 2001, 23(9): 809-822.

[218] Rege S, Jackson S, Griffin W L, Davies R M, Pearson N J, O'reilly S Y. Quantitative trace-element analysis of diamond by laser ablation inductively coupled plasma mass spectrometry. Journal of Analytical Atomic Spectrometry, 2005, 20(7): 601.

[219] Liao X H, Luo T, Zhang S H, Zhang W, Zong K Q, Liu Y S, Hu Z C. Direct and rapid multi-element analysis of wine samples in their natural liquid state by laser ablation ICPMS. Journal of Analytical Atomic Spectrometry, 2020, 35(6): 1071-1079.

[220] Liu Y S, Hu Z C, Gao S, Günther D, Xu J, Gao C G, Chen H H. In situ analysis of major and trace elements of anhydrous minerals by LA-ICP-MS without applying an internal standard. Chemical Geology, 2008, 257(1-2): 34-43.

[221] Paton C, Hellstrom J, Paul B, Woodhead J, Hergt J. Iolite: Freeware for the visualisation and processing of mass spectrometric data. Journal of Analytical Atomic Spectrometry, 2011, 26: 2508-2518.

[222] Petrus J A, Kamber B S. VizualAge: A novel approach to laser ablation ICP-MS U-Pb geochronology data reduction. Geostandards and Geoanalytical Research, 2012, 36(3): 247-270.

[223] McLean N M, Bowring J F, Gehrels G. Algorithms and software for U-Pb geochronology by LA-ICPMS. Geochemistry, Geophysics, Geosystems, 2016, 17(7): 2480-2496.

[224] Mischel S A, Mertz-Kraus R, Jochum K P, Scholz D. TERMITE: An R script for fast reduction of laser ablation inductively coupled plasma mass spectrometry data and its application to trace element measurements. Rapid

Communications in Mass Spectrometry, 2017, 31(13): 1079-1087.

[225] Zhang W, Hu Z C, Liu Y S. Iso-Compass: New freeware software for isotopic data reduction of LA-MC-ICP-MS. Journal of Analytical Atomic Spectrometry, 2020, 35: 1087-1096.

[226] Branson O, Fehrenbacher J S, Vetter L, Sadekov A Y, Eggins S M, Spero H J. *LAtools*: A data analysis package for the reproducible reduction of LA-ICPMS data. Chemical Geology, 2019, 504: 83-95.

[227] Willmes M, Ransom K M, Lewis L S, Denney C T, Glessner J J G, Hobbs J A. IsoFishR: An application for reproducible data reduction and analysis of strontium isotope ratios ($^{87}Sr/^{86}Sr$) obtained via laser-ablation MC-ICP-MS. PLOS ONE, 2018, 13(9): e0204519.

[228] Jackson S. LAMTRACE data reduction software for LA-ICP-MS. Laser Ablation ICP-MS in the Earth Sciences: Current Practices and Outstanding Issues, 2008, 40: 305-307.

[229] Košler J, Forst L, Sláma J. LamDate and LamTool: Spreadsheet-based data reduction for laser ablation ICP-MS. Laser Ablation ICP-MS in the Earth Sciences: Current Practices and Outstanding Issues, 2008, 40: 315-317.

[230] Guillong M, Meier D L, Allan M M, Heinrich C A, Yardley B W D. SILLS: A matlab-based program for the reduction of laser ablation ICP-MS data of homogeneous materials and inclusions. Laser Ablation ICP-MS in the Earth Sciences: Current Practices and Outstanding Issues, 2008, 40: 328-333.

[231] Gebel A. Laser ablation data reduction software for concentration measurements-Lars-C. Laser Ablation ICP-MS in the Earth Sciences: Current Practices and Outstanding Issues, 2008, 40: 341-342.

[232] Mutchler S R, Fedele L, Bodnar R J. Analysis management system (AMS) for reduction of laser ablation ICP-MS data. Laser Ablation ICP-MS in the Earth Sciences: Current Practices and Outstanding Issues, 2008, 40: 318-327.

[233] Weiskirchen R, Weiskirchen S, Kim P, Winkler R. Software solutions for evaluation and visualization of laser ablation inductively coupled plasma mass spectrometry imaging (LA-ICP-MSI) data: A short overview. Journal of Cheminformatics, 2019, 11(1): 16.

[234] 汪方跃, 葛粲, 宁思远, 聂利青, 钟国雄, White N C. 一个新的矿物面扫描分析方法开发和地质学应用. 岩石学报, 2017, 33(11): 3422-3436.

第13章 激光剥蚀等离子体质谱法深度剖析技术

(张鼎文[①],张丽萍[②],靳兰兰[①],郭 伟[①],胡圣虹[①*])

▶ 13.1 引言 / 336

▶ 13.2 深度剖析技术的优势与局限 / 336

▶ 13.3 LA-ICP-MS 深度剖析技术原理 / 337

▶ 13.4 深度剖析性能的影响因素 / 340

▶ 13.5 LA-ICP-MS 深度剖析的应用 / 351

▶ 13.6 展望 / 354

---

①中国地质大学(武汉),武汉;②中国地质科学院郑州矿产综合利用研究所,郑州。*通讯作者联系方式:shhu@cug.edu.cn

本章导读

- 主要介绍了激光剥蚀电感耦合等离子体质谱深度剖析技术。
- 简单介绍了 LA-ICP-MS 深度剖析技术的基本原理及相较于其他深度剖析技术的优势、深度分辨率的计算方法。
- 系统介绍了激光与物质相互作用,如激光剥蚀参数、剥蚀池的载气、激光在样品上的聚焦情况、样品厚度和剥蚀池的几何形状等因素对多层涂层材料深度剖析和深度分辨率的影响。
- 最后介绍了 LA-ICP-MS 深度剖析技术在陶瓷涂层、金属镀层材料、玻璃和聚合物涂层分析中的应用实效。

13.1 引 言

由于涂层材料能够满足使用均质材料而不能满足的关键需求,因而受到了越来越多的关注[1]。不同组成和厚度的涂层材料具有特殊的性能,如通过使用单一或多组分薄层可显著改善材料的热、电、光和磁学性质,以及材料的机械硬度、表面附着力、催化性能和耐腐蚀性等[2],由此研发生产出各种高级涂层(如硬涂层、光学涂层和耐腐涂层等)以及高新材料(如纳米材料、复合物材料、光电材料等)[3]。这些材料优越的物理、化学性能很大程度上取决于涂层组成(化学计量)、材料纯度、厚度、不同层间的界面和涂层元素的空间分布[1],因此需要发展一种快速准确的、检测限低(μg/g~ng/g水平)和空间分辨率高(可分辨 nm~μm 范围)的表面和深度剖析技术来表征这些高性能涂层和界面中元素的空间分布[1,4,5],实现对元素组成的二维空间的定性定量描述,以帮助解决许多基础性研究问题,并提供工业流程的生产控制和质量保证[6]。微小范围的深度剖析可对深度方向(微量)元素组成进行描述,能够识别样品中的浓度梯度或定位异构固体内部的边界。研究固体样品中微量元素随着深度的空间分辨变化对很多微区分析应用也十分重要[7]。因此,建立 nm 到 μm 尺度涂层材料等层状样品的深度剖析方法是十分重要而具有挑战性的研究工作[8]。

13.2 深度剖析技术的优势与局限

样品的深度剖析是指分析固体样品组成或杂质元素由表层到深部随深度变化的空间分布情况。固体样品的表层可以通过光子、电子或离子与材料最外层原子层的相互作用进行检测[6]。迄今为止,已有多种技术成功用于多层材料的深度剖析,如扫描电镜与能量色散 X 射线光谱仪结合技术(energy dispersive X-ray spectroscopy,SEM-EDX)、俄歇电子光谱(Auger electron spectroscopy,AES)、X 射线光电子能谱(X-ray photoelectron spectroscopy,XPS)、透射电子显微镜(transmission electron microscope,TEM)、原子探针断层扫描(atom probe tomography,APT)、二次离子质谱(secondary ion mass spectroscopy,SIMS)和辉光放电光谱/质谱(glow discharge optical emission spectrometry/mass spectrometry,GD-OES/GD-MS)、激光剥蚀电感耦合等离子体质谱法(laser ablation inductively coupled plasma mass spectrometry,LA-ICP-MS)等,这些技术之间由于检出限的差异、定量校准、深度分辨率和检测时间等限制了其检测能力和应用的局限性[9]。为了评价这些方法,必须综合考虑分析时间、精确度和深度分辨率，以及多元素同时分析的能力和方法灵活性等[10]。

基于其与物质作用的原理，深度剖析技术可分为以下两大类。

（1）光子作用于物质技术：如 XPS 被用作材料第一原子层的深度剖析方法[4]，通过溅射可以在几纳米到几微米的范围内取样，广泛应用于对薄膜的分析，但不适合用于超过几微米距离的微量分析或深度剖析，且基体效应严重、分析耗时、可检测元素种类受限[11]。TEM/SEM-EDX 同样只能分析样品表层化学成分信息[12]，但是对于标准样品的要求十分严格，需要基体匹配标样定量，并且空间分辨率较差。APT 可实现纳米尺寸范围内样品三维方向原子尺寸重建，但仪器昂贵、分析耗时，通常仅限于导电样品分析[13]。基于激光剥蚀的分析技术（如 LIBS、LA-ICP-MS）有许多优点：可以进行微区分析和深度剖析，空间分辨率较高（深度 0.1~10 μm，横向 10~100 μm），且无需复杂的样品前处理，对样品的形状和尺寸限制少，可分析导电和不导电样品，即使单个激光脉冲采样也能得到全部化学组成分析信息[14]。激光剥蚀技术如 LA-ICP-MS、LA-TOF-MS 和 LIBS 等已经成功应用于不同类型涂层和层状样品的深度剖析中[15]。

（2）电子作用于物质技术：包括 AES、SIMS、GD-OES 和 GD-MS，这些技术已经广泛用于不同厚度涂层材料的研究。AES 与 XPS 类似，空间分辨率高但存在严重的基体效应。SIMS 可实现原子层尺度的深度剖析，可分析导电或不导电材料，静态或动态模式下选择不同的溅射源可得到薄层中元素或是分子的信息分布[16]，适合于低浓度层和掺杂分析。但 SIMS 基体效应严重、需要相关的干扰校正，还需要超高真空、分析耗时且仪器和维护费用昂贵[4]。GD-OES 和 GD-MS 作为常用的深度剖析方法之一，其具有纳米级深度分辨率、易于操作、高灵敏度，可实现无基体匹配下的定量分析，但其横向分辨率只有毫米级，对样品的形状和尺寸也有特定要求[17]。

综上所述，LA-ICP-MS 已经成功应用于不同涂层材料的深度剖析，可实现纳米至微米尺度的厚度分析，采用不连续取样可实现微米级区域样品分析，也可通过线扫和点阵剥蚀分析厘米区域样品，这也是其他技术实现不了的[18]。虽然它并不能完全取代其他的深度剖析技术[17,19]，与 TOFSIMS、GD-OES/MS 或 SIMS 相比，其深度分辨率存在局限性，但目前文献报道的 LA-ICP-MS 对于 CdTe 光伏器件的深度分辨率已达到 50 nm[5]，有望成为极具潜力的深度剖析技术。

13.3　LA-ICP-MS 深度剖析技术原理

LA-ICP-MS 仪器由激光剥蚀系统和质谱分析仪器组成，其中激光剥蚀系统由高能量激光器、激光光束传输系统、高分辨率的电荷耦合器件（charge-coupled device，CCD）摄像头观测系统和剥蚀池组成。图 13-1 为 LA-ICP-MS 的工作原理图，将固体样品固定在剥蚀池中，激光束聚焦在样品表面进行剥蚀（可以线扫或单点分析），激光剥蚀产生的气溶胶由载气从剥蚀池传输到 ICP 等离子体中电离，质谱系统分析检测。关于深度剖析的研究相对较少，随着激光系统的改进，更高的激光能量、更均匀的激光能量分布、更精确的剥蚀参数，LA-ICP-MS 深度剖析技术得以迅速发展[10]。目前 LA-ICP-MS 深度剖析中广泛使用的是纳秒脉冲的 Nd：YAG 或 ArF 准分子激光器，以及飞秒脉冲的 Ti 蓝宝石激光[1]。

LA-ICP-MS 进行深度剖析时，理想的状态是激光在样品表面产生一个"完美"的剥蚀坑，即垂直的坑壁和平坦的坑底[21]。激光作用在样品表面，剥蚀坑向深层延伸，代表不同深度的样品化学组成信息被 ICP-MS 检测，用对应剥蚀深度的激光脉冲数来计算平均剥蚀速率，用剥蚀速率、剥蚀时间和脉冲数进行计算得出深度，从而得到样品组成成分的深度分布信息[22]。在大多数激光系统中随着深度增加，很难得到一个理想的剥蚀坑[6]，由于剥蚀过程中元素分馏的影响，即样品中不同元素剥蚀

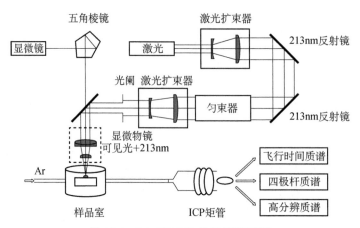

图 13-1　LA-ICP-MS 仪器原理图[20]

蒸发和传输过程的差异而导致非计量剥蚀[23]，因此，深度分析时剥蚀坑的形状、非代表取样、不均匀的剥蚀速率、气溶胶传输过程中的混合和信号拖尾、质谱检测方面都会影响深度分析的性能。理想的激光剥蚀系统得到的时间分辨图（信号强度-时间图）理论上开始剥蚀时上层信号迅速上升并进入平台区，在两层交界区上层信号开始下降。下层信号开始上升，在下层信号达到最大值进入平台区时，上层信号降到背景值。而在实际的两层玻璃深度剖析时（如图 13-2 所示），当下层玻璃 K、Sr 信号开始上升时，上层玻璃 Pb、Zr 信号还未开始下降；当下层玻璃信号开始到达最大值时，上层玻璃信号依然存在（下降中或下降到某一平台），这导致交界处的深度轮廓不能准确描述真实的界面信息[6]，这种现象被称为边界扩散或混合效应。

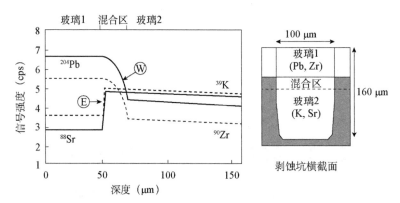

图 13-2　两层玻璃样品中 K、Sr、Zr 和 Pb 元素随深度的 ICP-MS 信号响应

图中 Ⓔ 和 Ⓦ 分别表示混合区增强和衰减信号，实验是由 266 nm Nd：YAG 激光在斑径 100 μm、2GW/cm² 的参数下获得[6]

Coedo 等[24]使用 266 nm Nd：YAG 激光对钢基质的 Cu 涂层进行深度剥蚀，发现从 Cu 涂层向钢基质剥蚀时，Fe 信号上升快 Cu 信号下降缓慢；从材料另一侧，即钢基质向 Cu 涂层剥蚀时，发现各层信号的变化趋势与前者相同（图 13-3）。涂层-基质信号的混合现象主要与激光剥蚀过程，气溶胶传输过程的元素分馏有关，与材料本身的化学分散和元素的挥发性无关。气溶胶在剥蚀池中洗出慢、气溶胶颗粒的再沉积、激光冲击对样品表面造成的破坏都会导致气溶胶传输过程的元素分馏，因此使用气溶胶快速洗出的小体积剥蚀池可以减少内部样品记忆效应[6]。

界面信号混合效应受多种因素的影响，包括仪器条件、激光与物质相互作用和样品物理化学特性[1,8]实际的深度剖析图中，界面元素信号上升和下降越迅速，深度分辨结果就越好[10,20]。评估深度剖析性能的关键参数是深度分辨率（ΔZ），它可以用来衡量一个具有明显分界面的准确分辨程度[1]。

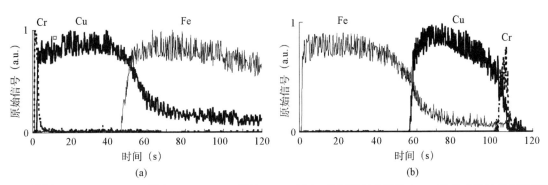

图 13-3　采用 266 nm Nd∶YAG 激光 LA-ICP-MS 获得 80 μm 厚 Cu 涂层样品（表面<1 μm 的 Cr 薄层）的深度轮廓[24]

(a) 从 Cr 层向基质钢剥蚀；(b) 从基质钢向 Cr 层剥蚀

Hofmann[25]提出了深度分辨率的定义和评价方法。在此基础上，Mateo 等[4,8]和 Plotnikov 等[10]分别报道了深度分辨率的计算方法（图 13-4）。Mateo 等认为深度分辨率是指全部测量信号的 16%~84% 之间的深度范围[图 13-4（a）]，通过等式（13-1）~式（13-3）计算得到：

$$\text{AAR} = \frac{D}{P_{50}} \tag{13-1}$$

$$\Delta P = P_{84} - P_{16} \tag{13-2}$$

$$\Delta Z = \Delta P \times \text{AAR} \tag{13-3}$$

式中，ΔP 是归一化信号最大值 84%（P_{84}）至 16%（P_{16}）的脉冲数；AAR 是平均剥蚀速率（μm/pulse）；D 是涂层厚度（μm）；P_{50} 是到达涂层归一化信号的 50% 对应的脉冲数。

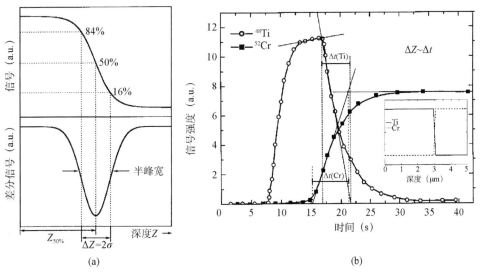

图 13-4　深度分辨率的两种计算方式示意图

(a) 通过归一化信号的 16% 与 84% 区间对应的剥蚀深度计算[8]；(b) 通过层界面交界处切线的最大斜率的方式计算[10]

Plotnikov 等[10]的 ΔZ 计算方法是通过原始信号的深度轮廓图[图 13-4（b）]，涂层/基质界面区域内瞬时信号正切的最大斜率计算的，通过等式计算：

$$\Delta Z = \Delta t \times f \times R \tag{13-4}$$

式中，Δt 是达到最大信号所需时间（s）；f 是激光重复率（Hz）；R 是每脉冲的剥蚀速率（μm/pulse）。

从以上计算公式可以看出采用低剥蚀速率可获得更好的深度分辨率，有部分文献直接使用深度剥蚀的平均速率来表征深度分辨率，但 Margetic 等[2]认为这样表示深度分辨率是不合适的，只有探测合适厚度的多层材料时，才能获得可靠的激光剥蚀深度分辨率。

13.4 深度剖析性能的影响因素

激光剥蚀的深度分辨率是影响深度剖析性能很重要的因素，激光剥蚀速率过大、不均匀的激光光能量分布，以及不规则的剥蚀坑都会严重影响激光剥蚀的深度分辨率[21]。为此众多学者通过不断地改进激光系统以得到更稳定的激光能量、更好的光束轮廓（平头）、更合适的剥蚀参数、更好的成像和软件/计算机完全控制的系统[10]。采用紫外和深紫外波长的激光器可改善样品的剥蚀行为，通过光束均质器将高斯轮廓激光转变成平头激光，以改善剥蚀坑的形状，显著提高深度分辨率；与纳秒激光相比，脉冲宽度更短的飞秒激光热效应更小，更低的剥蚀速率和最小的元素分馏可得到更好的深度分辨结果[18]。下面将从激光与物质相互作用、激光剥蚀参数（如激光波长、脉冲宽度、激光能量、激光斑径和激光重复频率）、剥蚀池载气组成、激光在样品上的聚焦情况、样品厚度和剥蚀池的几何形状等方面进行阐述。

13.4.1 激光与物质的相互作用

激光与物质相互作用十分复杂，受到表面加热、蒸发、分解、激发等作用的影响[6]。应用于深度剖析的红宝石激光器、Nd：YAG 激光器和准分子激光器与物质作用，即激光剥蚀过程是光化学过程和光热过程共同作用的，这些过程包括碰撞、热、电子、流体和分层等行为。激光辐射到材料表面会发生吸收、反射和透射，由于不同材料理化性质的差异，材料对会有以下五种主要机制吸收激光能量：杂质的吸收、逆韧致吸收、多光子的电离、光致电离和空穴吸收[26]。材料粒子获得的激光能量转化为电子的激发能、动能以及晶格能，这些能量在各自自由度之间的分布并不均匀，导致材料粒子通过不断碰撞让能量保持平衡，碰撞行为最后在宏观表现为材料温度的升高。许多学者从材料热效应温度表达式推导出剥蚀深度的计算公式[27]，假设材料表面材料吸收激光能量发生汽化后，表面粒子温度能维持在汽化温度不变。根据能量守恒定理，可以得到单脉冲剥蚀厚度 Z_t 计算公式为

$$Z_t = \frac{(1-R)(E-E_{\text{vth}})}{C_1\rho_1\Delta T + L_v} \tag{13-5}$$

式中，ρ_1 是金属的液态密度，C_1 是液态比热容，ΔT 是变化的最大温度差，L_v 是金属的汽化潜热，E_{vth} 是脉冲激光的汽化能量阈值，E 是激光能量，R 是金属的表面反射率。当激光能量小于激光汽化能量阈值时（$E<E_{\text{vth}}$），材料吸收的能量不足以使粒子从表面蒸发出去。

根据倪晓昌等[28]关于 E_{vth} 与样品达到蒸发态的时间 τ_v 关系的研究，证实材料汽化阈值 E_{vth} 不仅受激光脉冲参数的影响，还与材料自身的光学、热学性质相关，依此推导出激光剥蚀材料厚度与时间的关系式为：

$$Z_t = \frac{(1-R)^2 I_0(t-\tau_v)}{C_1\rho_1\Delta T + L_v} \tag{13-6}$$

式中，I_0 为激光入射到材料表面（$Z=0$）的平均功率密度。

目前用于深度剖析的激光剥蚀系统主要有纳秒激光和飞秒激光，这两种激光的剥蚀过程存在根

本的差异，超短脉冲飞秒激光（$10^{14}\sim10^{15}$ W/cm²）的激光能量比短脉冲纳秒激光（$10^9\sim10^{10}$ W/cm²）大得多，如图 13-5 所示，纳秒激光脉冲会导致激光光斑周围形成"热影响区"，热影响区从剥蚀坑壁向材料内部可延伸 40 μm[29]，这是由于纳秒激光脉冲宽度较大，热辐射扩散到固体样品中造成的。热影响区的材料经过固态，液态，等离子体态的转变，这一过程会产生裂纹和沉积层，并对邻近的结构造成破坏，因此，激光-物质相互作用过程中挥发性元素优先蒸发，气溶胶在材料表面再沉积都会引起分馏效应。与纳秒激光系统相比，飞秒激光的脉冲宽度可达 10 ps，低于被加热电子吸收脉冲能量转移到晶格所需的时间（即热弛豫时间）。飞秒激光有效地降低了热效应，并能够提供高深度分辨率和高灵敏度的化学成分信息[30]。

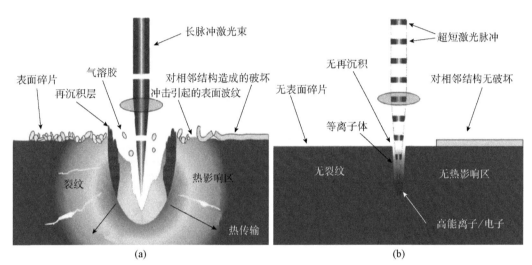

图 13-5　激光与物质的相互作用[30]
（a）纳秒脉冲；（b）飞秒脉冲

13.4.2　激光剥蚀系统

影响深度分辨率的激光器参数主要有激光波长、脉冲宽度和激光能量分布，这些因素直接影响剥蚀坑形状和底部粗糙度。不平整的剥蚀坑形状和较大的表面粗糙度都会导致不同水平层的样品材料被同时剥蚀[21]，从而导致样品层间的信号混合，致使深度剖析信号失真[18]。

不同脉冲宽度和波长的激光器与物质作用的机理不同，长波长宽脉冲激光剥蚀过程以热效应为主，样品以蒸气和熔融态两种形式被除去，样品的熔化导致交界层的改变和混合，激光剥蚀深度分辨率降低[图 13-6（a）]。短波长窄脉冲的激光具有更高的激光能量，剥蚀过程从热效应转变为光效应为主，激光与物质的作用效率更高，与图 13-6（b）进行对比，390 nm 紫外激光剥蚀产生的剥蚀坑周边的热效应显著降低。ArF 准分子激光器的波长为 193 nm 的深紫外区[31]，并使用光束均质器将高斯光束转换为均匀能量分布的平头光束，匀质化的激光能量分布在样品表面，避免坑壁与坑底被同时剥蚀，得到较为理想的剥蚀坑，Bleiner 等[3]第一个使用 193 nm ArF 激光对 Ti 涂层进行深度剥蚀分析，得到最佳剥蚀速率 0.2 μm/pluse；Gutierrez-Gonzalez 等[5]使用 193 nm ArF 准分子激光测定了 CdTe 光伏器件的层厚，获得了与 GD-TOFMS 和 SIMS 一致的结果，并成功用于 Ti 涂层[32]、Zn 涂层[33]、层状高新材料[34]等分析。

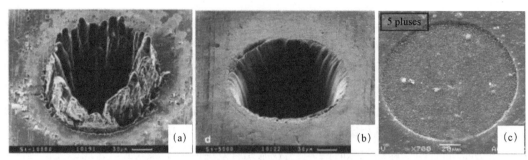

图 13-6 钢箔导电材料在（a）780 nm-3.3 ns-LA、1 mJ、4.2 J/cm² 激光脉冲和（b）390 nm-250 fs-LA、0.5 mJ、2.5 J/cm² 产生的剥蚀坑[35]；（c）热成像材料在 193 nm ArF 准分子激光（匀束）100 mJ、3 Hz、120 μm 产生的剥蚀坑[34]

虽然飞秒系统能减小激光剥蚀中的热效应，但高斯能量分布的激光束会导致不均匀的剥蚀坑形状[图 13-7（a）、（b）]，降低了深度分辨率。为改善飞秒激光的能量分布，通常采用孔径成像或分级反射镜等技术。Samek 等[15]采用液晶阵列（LCD）技术将红外飞秒高斯激光变为平头激光，有效地减小了激光剥蚀过程的热效应，但能量的传输效率过低不足以有效地剥蚀样品。Kaser 等[36]报道了基于孔径辅助衍射和光透镜重组光束的均匀化技术，采用两阶段的傅里叶光学处理器，第一阶段让激光初始光束在圆形孔径衍射形成艾里斑（Airy disk），第二阶段，衍射的光束在透镜的焦平面上转化为均匀的平头光束。利用该装置，可以得到圆柱形的剥蚀坑[图 13-7（c）、（d）]，进而提高了 fs-LAICP-MS 在深度剖析方面的分析能力。

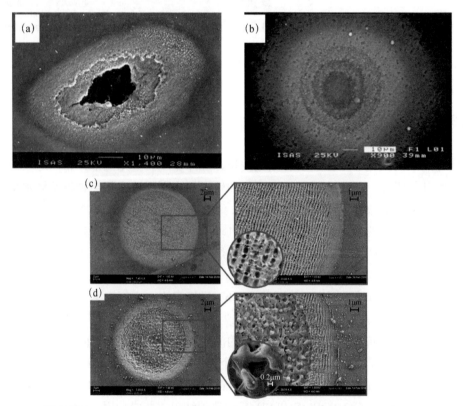

图 13-7 775 nm 飞秒激光（170～200 fs）对（a）Cu-Ag 涂层（0.9 J/cm²、80 shots）和（b）TiN-TiAlN 多层样品（0.35 J/cm²、100 shots）进行深度剖析的剥蚀坑形貌图[2]；在 1 J/cm² 的激光能量下，（c）5 个脉冲和（d）25 个脉冲后，NIST 2135c 上剥蚀坑的 SEM 显微照片[36]

激光系统的改善，如波长更短和脉冲宽度更窄的激光、匀质化的激光能量分布，使得 LA-ICP-MS 成功应用于各类固体样品的深度剖析。深度剖析的性能还取决于激光剥蚀参数，如激光能量、激光斑径和激光重复频率，其会影响气溶胶颗粒的大小和剥蚀坑的几何形状。不平整的剥蚀坑形状、不同粒径的气溶胶颗粒和元素分馏效应都会影响分析精度[37]，选择正确的激光系统和合适的激光剥蚀参数可有效地降低分析误差[38]。

13.4.2.1 激光能量

激光辐射强度（irradiance，W/cm^2）或能量（energy，J/cm^2）的大小影响着分析区每个脉冲剥蚀的物质的量[1,5]。深度剖析往往需要采用较低的激光能量以获得最小的剥蚀速率，但激光能量要大于材料的剥蚀阈值，并确保分析元素信号高于 ICP-MS 检出限。而实际分析中，在满足这些条件时又会发生部分元素的优先蒸发，导致分馏效应增强[4,5]。

不少学者对激光能量与深度剖析影响因素（剥蚀坑形状、钻孔时间和深度分辨率）的关系做了大量研究。如图 13-8 所示，Mason 等[6]使用 193 nm 激光对比了 0.2 GW/cm^2 和 1.4 GW/cm^2 的激光能量对剥蚀坑形状的影响，当采用相对低的激光能量剥蚀时，随着剥蚀深度的增加，剥蚀坑的斑径逐渐减小，剥蚀坑很难保持均匀的几何形状，这是由于激光聚焦位置远离初始焦点导致激光能量损失，不能有效地剥蚀材料。Plotnikov 等[10]研究了激光能量对 TiN 涂层钻孔时间的影响，发现激光能量在 0.3～2.5 mJ 范围内，钻孔时间随能量的升高而降低，在 1.5 mJ（约 2.5 GW/cm^2）时得到了最低的标准偏差。研究表明剥蚀速率随着激光能量的增加而加快，过低的能量则会导致信号的不稳定性和分析时间延长。Coedo 等[24]使用 266 nm Nd∶YAG 激光对标准样品 NIST 1361b 和 NIST 1362b 进行深度剖析，探讨了激光能量与平均剥蚀速率和深度分辨率的关系，获得的结论与上述一致（图 13-9）。

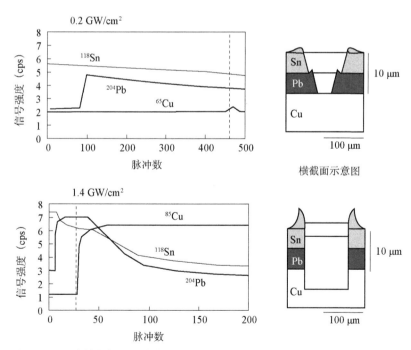

图 13-8 两种激光能量剥蚀多层金属样品时 ICP-MS 的响应和剥蚀坑横截面[6]

13.4.2.2 激光斑径

LA-ICP-MS 深度剖析的准确性不仅取决于激光能量，还取决于激光斑径的选择。Mason 等[6]使用平头 193 nm（25 ns）ArF 准分子激光 LA-ICP-MS 对两层玻璃进行了深度剖析（图 13-10），使用

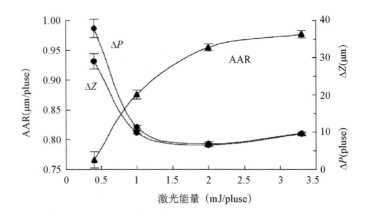

图 13-9 激光脉冲能量对 AAR、ΔP 和 ΔZ 的影响，激光重复频率为 1 Hz[24]

120 μm 与 60 μm 斑径得到的深度轮廓图的区别很小，但还是可观察到两者的区别，区别可能是由于较小的剥蚀坑仅有中心光束剥蚀到样品上，并形成一个平坦的图像平面，也可能受采样区域材料厚度变化的影响。较大的斑径能更好地对平面层样品进行剥蚀，但往往是以牺牲了交界区深度分辨率为代价，实验最终选取 120 μm 斑径。Gutierre-Gonzalez 等[5]使用平头 193 nm（4 ns）ArF 准分子激光 LA-ICP-MS 对 CdTe 光电材料进行深度剖析[图 13-11（a）]，实验证实 65 μm 和 160 μm 斑径条件下平均剥蚀速率与激光脉冲数和剥蚀材料本身相关，击穿 CdTe、CdS、SnO_2 层分别需要 10、22、33 个脉冲，而与激光斑径无关。尽管匀束 193 nm 激光得到的剥蚀坑轮廓并不是完全平坦，剥蚀坑边缘的剥蚀量更多，剥蚀坑底的粗糙度较大，随着深度增加，剥蚀坑的凸度变强，而剥蚀到 SnO_2 层时凸度变缓，反映出材料组成对剥蚀过程的影响。从元素分布图[图 13-11（b）、（c）]来看，160 μm 斑径下取样量更大，混合更明显，因此实验选取 65 μm 斑径。Plotnikov 等[10]固定激光能量，通过改变斑径来研究 TiN 涂层深度剥蚀时光束的边缘效应，发现使用较小的斑径（60 μm），剥蚀产生气溶胶的量小、信号低、相对标准偏差（RSD）大，采用较大的斑径（250 μm），多余的激光能量会被材料吸收，产生的熔融和热效应导致 RSD 增大，因此选择适中的（120 μm）斑径为最佳条件。

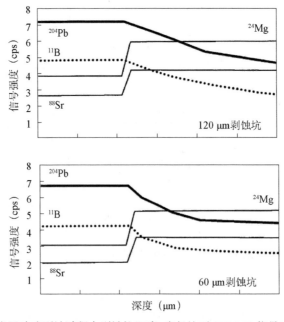

图 13-10 多层玻璃剥蚀过程中剥蚀坑深度-直径比对 ICP-MS 信号响应的影响[6]

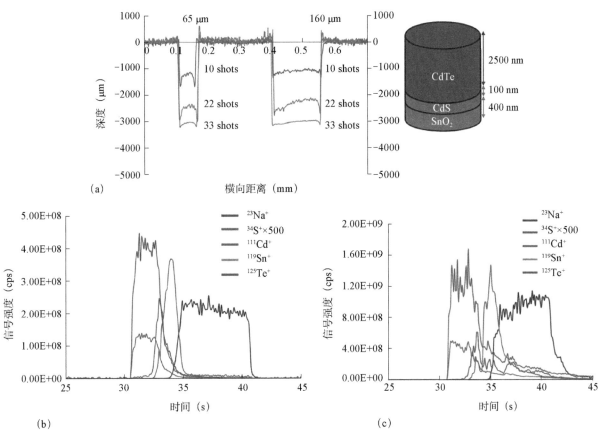

图 13-11 （a）斑径在 65 μm 和 160 μm 下得到的剥蚀坑轮廓图；斑径（b）65 μm、（c）160 μm 得到的多层 CdTe 光伏电池的深度轮廓，激光重复频率为 10 Hz[5]

13.4.2.3 激光脉冲重复频率

激光脉冲重复频率是控制钻孔时间的重要参数之一，钻孔时间与频率成反比，虽然脉冲重复频率与每脉冲剥蚀速率和深度分辨率之间没有发现明显相关性[10,39]，但选取较小的重复率可有效地减少每脉冲之间的信号混合，提高深度分辨率。Pisonero 等[17]采用紫外 fs-LA-ICP-MS 对 Cr/Ni 金属涂层进行深度剖析，激光重复频率分别为 1 Hz 和 4 Hz 时得到 Cr 元素的信号轮廓（图 13-12），发现在 4 Hz 重复率下信号混合严重，而在 1 Hz 条件下可清晰分辨每个脉冲产生的信号。实验在 0.4 J/cm²、1 Hz 条件下获得的深度分辨率可达 290 nm。

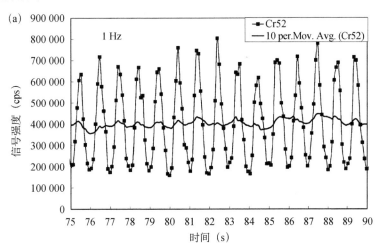

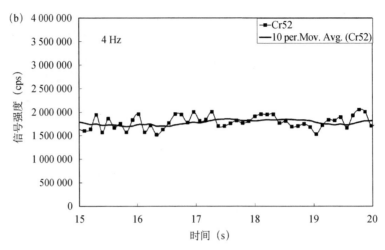

图 13-12　不同的激光重复频率 [1 Hz（a）和 4 Hz（b）] 获得的 ^{52}Cr 信号轮廓（设置激光能量为 0.4 J/cm^2）[17]

13.4.3　剥蚀池的载气

载气的性质与流速严重影响气溶胶颗粒的传输效率[40]。LA-ICP-MS 通常使用 Ar 和 He 作为载气[41]。Mank 等[7]使用 266 nm Nd：YAG 激光 LA-ICP-MS 对 NIST 610 进行深度剖析，研究在 He 和 Ar 条件下对深度剖析的影响 [图 13-13（a）]，发现相同剥蚀条件下，分析元素的分馏系数变化规律相似，而 He 作为载气时元素 Y 的信号强度是 Ar 的 2～3 倍，剥蚀穿透的深度比 Ar 提高 20%～30%，单个脉冲剥蚀深度（0.27 μm）大于 Ar（0.22 μm）。两种载气下剥蚀坑周边颗粒沉积的形貌完全不同，在 Ar 中剥蚀时观察到严重的再沉积现象。Mason 等[6]使用平头 193 nm ArF 激光 LA-ICP-MS 对多层金属涂层（5 μm Sn/5 μm Pb/Cu）进行深度剖析 [图 13-13（b）]，结果表明对于 5 μm 涂层样品两种载气中剥蚀信号的差异并不明显，但对于 25 μm 的涂层样品可明显地观察到在 He 中各层元素信号降低更迅速，激光剥蚀的深度分辨率更好。

不同激光系统的情况不同，266 nm 激光剥蚀产生的气溶胶受剥蚀载气的影响较小，而 193 nm 激光使用 He 作为载气可显著提高信号灵敏度[42]。Luo 等[43]使用 257 nm fs-LA-ICP-MS 分别研究了 He、Ar 和 He-Ar 混合气下的信号强度和元素分馏。实验表明，采用 He 代替 Ar 气氛剥蚀，样品中难熔元素（如稀土元素）的灵敏度增加了 1.05～1.20 倍，挥发性元素的信号强度增加了 1.5～3.0 倍；在 He 中剥蚀坑周围沉积的气溶胶颗粒更少，气溶胶传输效率更高；在 He-Ar 混合气中元素信号强度与在 He 中相近，同时各元素分馏系数比在 He 或 Ar 中获得的分馏系数更接近 1。

13.4.4　激光在样品上的聚焦情况

激光在样品上聚焦区域的不同会导致分析区激光能量和斑径的变化，进而影响深度剖析性能。Coedo 等[24]采用 266 nm Nd：YAG 激光研究在相同脉冲能量下，不同聚焦位置对深度剖析的影响，如图 13-14（a）所示，散焦越大，剥蚀坑直径越大，辐射量越低。焦点在样品上方与下方的激光能量和剥蚀坑形状的变化趋势并不相同，相同工作距离聚焦在样品上方产生的激光能量更低，剥蚀坑直径更大，这是由于当聚焦点在样品上方时，激光光子的能量被焦点上的 Ar（剥蚀载气）吸收，形成二次等离子体，到达样品表面的激光能量降低。此外，该实验还研究了在最佳化的能量范围内 ΔZ 随 AAR 和 ΔP 的变化 [图 13-14（b）]。

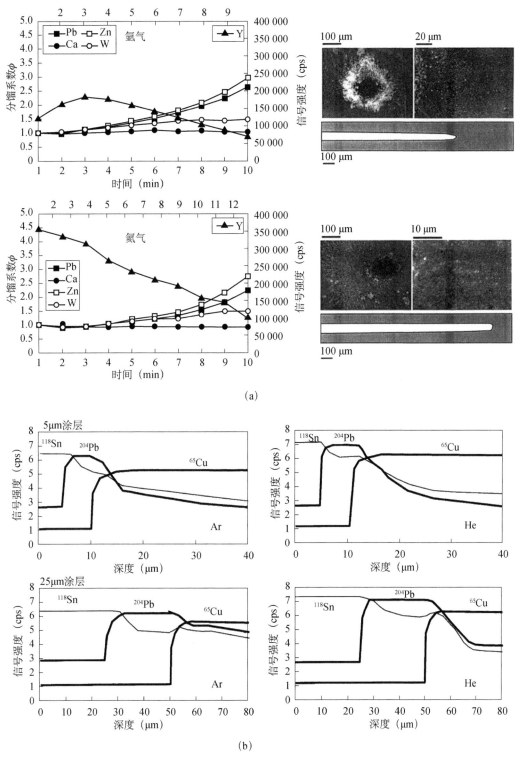

图 13-13 不同载气条件对深度剖析的影响

(a) Ar 和 He 中的剥蚀坑形貌图和 ICP-MS 信号响应与时间的关系图[7];(b) 剥蚀池载气对不同厚度的多层材料剥蚀过程中 ICP-MS 信号响应的影响[6],实验结果是在激光能量 1.4 GW/cm², 斑径 120 μm 的条件下得到的

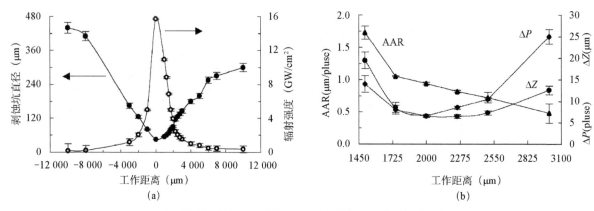

图 13-14　激光在样品表面聚焦点对深度剖析的影响[24]

（a）在 2 mJ/pulse 脉冲能量下，不同聚焦位置对剥蚀坑直径和激光能量的影响；（b）工作距离与平均剥蚀速率、深度分辨率的关系，工作距离是指激光镜头到样品表面的距离和激光焦距的差值，工作距离<0 是聚焦在样品上方，工作距离>0 是聚焦在样品下方

随着剥蚀坑向深层延伸，激光聚焦点的偏离会引起剥蚀坑直径变窄。在理想状态下，通过调整仪器样品平台在 Z 轴方向上的移动速率来保证样品移动速率与激光剥蚀速率完全相等，然而随着剥蚀坑深度增加，气溶胶难以被载气吹出，导致信号衰减，灵敏度大幅降低。周韵等[14]优化激光频率和样品台移动速率，选择大于剥蚀速率的移动速率进行深度剖析，发现剥蚀坑底的半径随深度加深而减小，最终形成一个倒圆锥形。Mank 等[7]采用 266 nm Nd：YAG 激光 LA-ICP-MS 对 NIST 610 深度剖析，比较移动平台和固定平台两种模式下，元素分馏和剥蚀坑深度轮廓的差异（图 13-15），由于移动平台的不稳定性导致不规则剥蚀坑形状和气溶胶颗粒再沉积。因此，改变激光聚焦点的方法并不适合深度剖面分析。

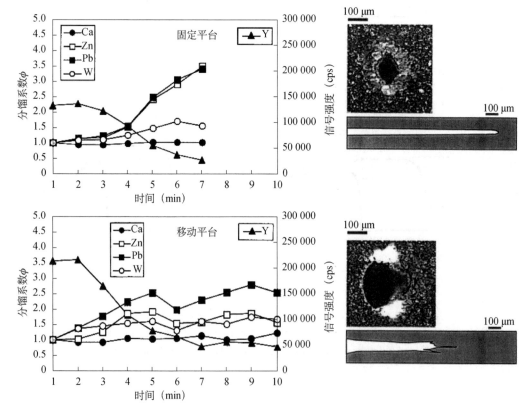

图 13-15　两种模式下信号强度和分馏系数随时间的变化和激光剥蚀坑的横截面图[7]

13.4.5 样品厚度

样品厚度也严重影响着深度分辨率，且厚度越薄的涂层其深度分辨率越高。Mason 等[6]使用 193 nm（25 ns）匀束准分子激光研究了不同厚度涂层样品在剥蚀到达边界后涂层信号拖尾（信号随相对脉冲数变化）的影响，发现涂层越薄信号下降越陡，混合效应越小。Plotnikov 等[10]使用 266 nm Nd：YAG 激光对不同厚度的 Ti 涂层样品进行了深度剖析，研究了钻孔时间与涂层厚度的关系，结果表明涂层越薄钻孔时间越短，相同厚度的 TiN 钻孔时间最长，对比 TiN 和 TiC 材料，其物理性质如密度、熔点、反射率和显微硬度等参数都会影响激光剥蚀行为[图 13-16（a）]。Coedo 等[24]使用 266 nm Nd：YAG 激光系统对 8 个不同厚度（从 5.9 μm 到 199 μm）的铜涂层标准 NIST 1361b 和 NIST 1362b 进行深度剖析[图 13-16（b）]，结果表明涂层越薄平均剥蚀速率越大，对于较厚涂层，随着剥蚀坑深度增加，激光能量的衰减导致平均剥蚀速率降低。

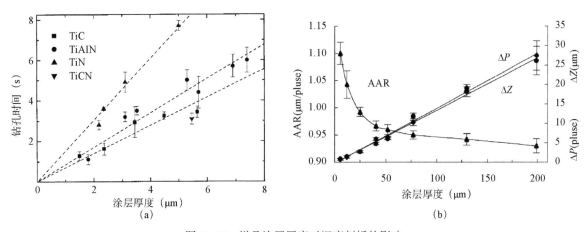

图 13-16　样品涂层厚度对深度剖析的影响

（a）不同材料钻孔时间与涂层厚度的关系[10]；（b）AAR、ΔP 和 ΔZ 与涂层厚度的关系[24]

13.4.6 剥蚀池几何形状和尺寸

气溶胶传输过程中脉冲信号混合和信号拖尾是影响 LA-ICP-MS 深度剖析分辨率的因素[19]，因为不同层的样品材料会因混合效应同时进入等离子体中电离[21]。剥蚀池体积会直接影响 ICP-MS 瞬时信号的时间衰减和形状变化[17,19,23]。为了避免连续激光剥蚀中层间信号的混合，需要采用较低的激光重复率以及气溶胶快速洗出的剥蚀池，以使每个脉冲产生的信号能够快速衰减[19]。使用小体积剥蚀池、优化空气动力学条件和气体混合物（Ar-He）会增加信号轮廓的陡度，可得到更好的深度分辨率[10]。

剥蚀池的改进通常包括对其几何形状和尺寸、气体动力学、出入口喷嘴，以及剥蚀池到 ICP 间管路等的改进。Arrowsmith 等[44]最早通过改变气体流动方向设计了小体积剥蚀池（体积都在 1～3 cm³），洗出时间为 1～3 s。Devos 等[45]设计了一种开放剥蚀池分析大体积的考古银器。Pisonero 等[46]设计了基于"文丘里效应"的气溶胶提取剥蚀池"HEAD cell"，洗出时间小于 0.5 s，将气溶胶颗粒尺寸变得更小，有效减少了分馏效应，提高了电离效率，离子信号的稳定性和重复率显著提高[图 13-17（a）]。Garcia 等[47]发现入射喷嘴会导致瞬时信号变宽，混合加剧，而出口喷嘴对信号形状无影响，其设计的 0.6 cm³ "烟囱状"剥蚀池信号洗出时间为 0.8 s，但存在双峰现象。Liu 等[48]设计的体积开放、低记忆效应剥蚀池，信号洗出时间为 0.4 s，有效地减少了元素的记忆效应，还成功实现了对光纤中 Pb 等元素分布的分析。Muller 等[49]设计的锥形双室剥蚀池"HelEx cell"，气溶胶洗出时间小于 1.5 s，并成功应用于金属涂层的深度剖析。Wagner 等[50]也设计了一种适合于文物分析的开放剥蚀池（洗出时间约 3 s）。Leach 等[51]探讨了剥蚀池体积对 ICP-MS 瞬时信号的影响，结果表明最小化的剥蚀池体积可有效地减小瞬时信号的峰宽；Gurevich 等[52]基于对气溶胶混合过程的研究，提出了一种毫秒级洗出时间的剥蚀池，有效抑制了气体的湍流和剥蚀层流的影响，可准确区分 20 Hz 重复频率单个激光脉冲产生的信号。Wang 等[53]设计的低分散"管状"剥蚀池实现了 30 ms 的洗出时间，即使激光频率 20～30 Hz 也能区分单个脉冲的信号和抑制气溶胶拖尾的影响 [图 13-17（b）]。Gundlach-Graham[54]在上述剥蚀池的基础上进行了改进，并将样品与剥蚀池底部小孔的距离从 350 μm 缩小至 275 μm，信号洗出时间降低至 10 ms。Van Malderen[55]设计的半开放管状剥蚀池，剥蚀池内直径 1.2 mm，长 372 mm，洗出时间约为 6 ms，可区分 200～300 Hz 下的脉冲信号。Douglas 等[56]设计的双体积低分散剥蚀池（Sniffer cell），洗出时间降低至 4.9 ms，有效地提高信号的灵敏度。

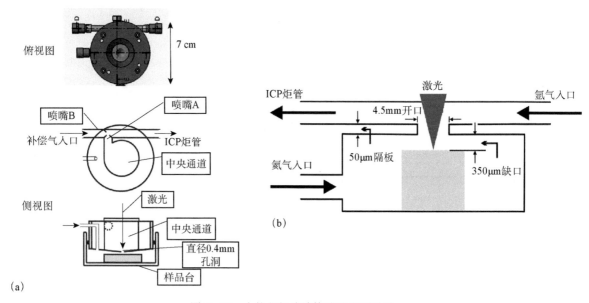

图 13-17　小体积气溶胶快速洗出剥蚀池

（a）高效气溶胶分散剥蚀池"HEAD cell"[46]；（b）低分散"管状"剥蚀池[53]

深度剖析中采用快速洗出的剥蚀池可减少剥蚀过程中气溶胶的混合。Pisonero 等[17]使用"HEAD cell"结合 265 nm Ti 蓝宝石飞秒激光分析了 500 nm Cr 薄层，得到的深度分辨率小于 300 nm；Gutierrez-Gonzalez 等[5]采用 HelEx 双室剥蚀池对多层 CdTe 光电材料进行深度剖析，通过对比不同激光剥蚀频率（1 Hz 和 5 Hz）的深度剖析轮廓，证实采用低的激光剥蚀频率能有效避免气溶胶传输过程的混合，提高了深度分辨率（图 13-18）。Hattendorf 等[57]采用 HelEx 双室剥蚀池和 193 nm 激光，采取线扫描对不同厚度 Nd 层（0.5 nm、1 nm、3 nm、6 nm）样品进行深度剖析，得到不同厚度 Nd 层的信号响应与层厚的关系，成功测定了亚纳米级 Nd 金属层的厚度；采用空白样品信号的 10 倍标准偏差（以峰高计算），获得厚度的理论检出限可达 0.0002 nm，即使在背景噪声严重的情况下也可达 0.025 nm，证实了 LA-ICP-MS 具有准确测定"原子"层厚度的潜力。

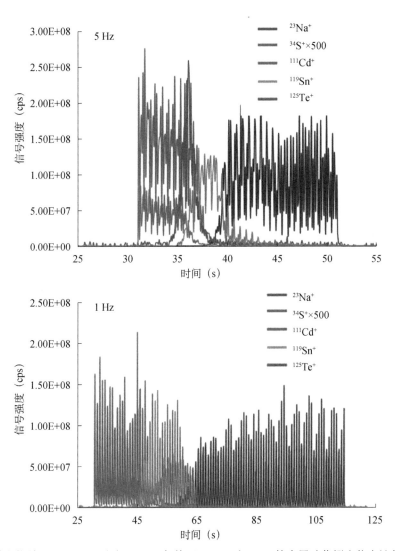

图 13-18 在激光能量 1.03 J/cm²、斑径 65 μm 条件下，5 Hz 和 1 Hz 的多层碲化镉光伏电池的深度剖析图[5]

13.5 LA-ICP-MS 深度剖析的应用

13.5.1 陶瓷涂层

陶瓷涂层具有优异的耐磨、耐腐蚀、耐高温和高热阻等优点，已成功应用于金属切削工具、涡轮发动机、轴承等耐磨耐蚀部件中。Zr、Ti 的氮化物和氧化物都属于这类涂层，通常使用化学气相沉积（CVD）或物理气相沉积（PVD）等技术制备，其厚度为微米级。Shaheen 等[58]使用 193 nm ArF 准分子激光器对氧化铝陶瓷进行深度剖析，研究激光参数对平均剥蚀速率的影响，结果表明 10 Hz 和 100 Hz 的脉冲重复频率下，剥蚀速率不变，脉冲重复频率增加至 300 Hz 时，剥蚀速率略有下降，在 5.1 J/cm²、10.1 J/cm²、15.2 J/cm² 的激光能量，50 μm 的激光斑径下得到氧化铝材料的平均剥蚀速率分别为 51 nm/pluse、78 nm/pluse、94 nm/pluse。Torrisi 等[59]采用 Nd:YAG 激光 ICP-MS 对由青铜和银合金构成的古钱币和古彩色陶瓷进行深度剖析,成功测定了西西里彩色陶器由 Fe/C、Cu/C、Fe/O、Cu/O 构成的彩色层的元素组成和厚度。

13.5.2 金属镀层材料

金属镀层材料在各种领域有着重要的用途,如机械零件和工具的硬保护层、镜头和反射镜的光学涂层、电极涂层、金属-有机骨架薄膜等。表 13-1 简要介绍了部分前人的工作情况。

表 13-1 金属镀层材料深度剖析的研究

作者	年份	激光器类型	工作进展
Mason 等[6]	2001	266 nm 和 193 nm 激光	对 5~50 μm 厚的多层金属涂层样品进行深度剖析时,发现剥蚀过程中交界层间信号的混合是影响深度分辨率的因素,确定了降低混合效应、提高分析准确度的最佳条件
Mateo 等[8]	2001	308 nm XeCl 准分子激光	报道了深度分辨率的计算方法,研究激光能量对镍铜镀层黄铜样品的平均剥蚀速率和深度分辨率的影响,采用适中的激光能量可获得最好的深度分辨率
Coedo 等[24]	2005	266 nm 纳秒紫外 Nd:YAG 激光	对已知厚度的铜涂层的钢材料深度剖析,研究激光能量和涂层厚度对深度分辨率的影响,采用激光脉冲能量为 2 mJ/pluse 时可获得 6.5 μm 深度分辨率
Pisonero 等[17]	2007	Ti 基蓝宝石 265 nm 紫外飞秒激光	对 Cr 薄层的 Ni 基材料进行了深度剖析,使用激光匀质器和小体积的气溶胶快速洗出剥蚀池,测得剥蚀速率约 10 nm/pulse,深度分辨率小于 300 nm
Kaser 等[36]	2018	Ti 基蓝宝石 400 nm 紫外飞秒激光	对 NIST 2135c 标准参考材料(由 9 个交替的 Cr/Ni 金属层构成,Si 基,Cr 层和 Ni 层厚度分别为 57 nm 和 56 nm)进行分析,得到最佳的深度分辨率为 49 nm

13.5.3 玻璃

多层玻璃材料普遍用于大型平板显示器、电视屏幕等电子产品中,使用无损伤或最小化破坏快速检测的技术来表征这些层的化学组成,对产品的加工生产、性能优化及质量监控是十分重要的。

van Elteren 等[11]使用 193 nm ArF 激光 LA-ICP-MS 在 1.79 mm×1.79 mm 的平面上,使用 80 μm 斑径、7 J/cm² 能量、1 Hz 频率、单点阵列剥蚀,每个点 50 个脉冲,间距 90 μm,获得各元素的 3D 分布图,进而探讨了中世纪玻璃制品的风化降解机制。他们还采用 LA-ICP-MS 技术结合去卷积方法研究了腐蚀玻璃中 Mn 的 3D 分布情况[60]。Gutierrez-Gonzalez 等[5]使用 193 nm ArF 激光 LA-ICP-MS 和 HelEx 剥蚀池对多层 CdTe 光电材料进行深度剖析(图 13-19),研究激光参数对深度剖析的影响,发现 1.03 J/cm² 能量下剥蚀速率最小,为 50 nm/pulse,在 1 Hz、65 μm 条件下获得了 100 nm CdS 层的深度剖析信号,与 GD-TOFMS 和 SIMS 的结果相比取得了良好的一致性。

13.5.4 聚合物涂层材料

聚合物涂层材料在工业与日常消费品中普遍存在,能够显著增加材料的耐腐耐磨等物理化学性能。Mateo 等[4]采用 795 nm 飞秒激光对聚合物涂层材料(热镀锌钢板基底上镀有含 Ti 聚合物层和含 Sr-Cr 聚合物层)进行深度剖析,实验使用液晶阵列将激光束变为平头激光,并在较低的激光重复率(0.2 Hz)和 1.6 J/cm² 能量下得到深度分辨率分别为 240nm(热镀锌钢板涂层,见图 13-20)和 2.3 μm(含 Ti、Sr-Cr 聚合物涂层,见图 13-21),优于 GD-OES 得到的深度分辨率。

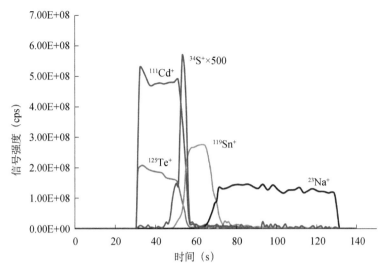

图 13-19 CdTe 光电镀层的 LA-ICP-MS 深度剖析图[5]

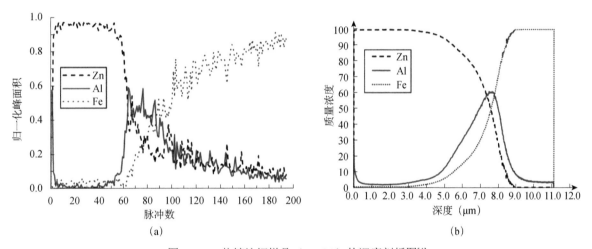

图 13-20 热镀锌钢样品（HDGS）的深度剖析图[4]
（a）fs-LA-ICP-MS；（b）GD-OES

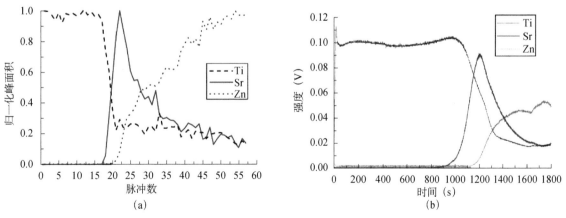

图 13-21 含 Ti、Sr-Cr 聚合物涂层的深度轮廓图[4]
（a）fs-LA-ICP-MS；（b）GD-OES

13.6 展望

LA-ICP-MS 已成为固体材料分析的有力手段，被广泛应用于材料科学、地球科学、生命科学、环境科学、考古和法医科学等各类样品的整体分析、微区分析和深度剖析或空间元素分布。LA-ICP-MS 深度剖析的性能主要取决于激光特征（波长、脉冲宽度和光束形状）和剥蚀参数（如激光能量、剥蚀斑径、剥蚀载气等），优化激光剥蚀条件，LA-ICP-MS 可提供小样品或多层样品特定截面的浓度渐变信息，用于确定新型合成材料的纯度和化学计量组成，从而进行质量控制，其显示出对高级多层材料进行快速定量深度剖析的巨大潜力。未来发展可使用小剥蚀池、优化空气动力学条件和气体混合物增强气溶胶的传输效率，快速冲洗和低分散的剥蚀池有利于减少不同脉冲气溶胶混合引起的信号叠加，改善多层样品交界模糊的现象以得到更好的深度分辨率；使用飞秒激光器可提高激光与物质作用效率，有效改善剥蚀坑轮廓和非计量剥蚀，提供较低的剥蚀速率，降低元素分馏效应；飞行时间质谱检测器的使用可进行快速多元素信息的同时捕获，缩短元素分析时间，实现纳米级甚至单个原子层水平的深度分辨率和多元素快速成像分析，为多层复合涂层功能材料的表征与鉴别，提供了一种新的分析技术。

参 考 文 献

[1] Farinas J C, Coedo A G, Dorado T. Influence of relative abundance of isotopes on depth resolution for depth profiling of metal coatings by laser ablation inductively coupled plasma mass spectrometry. Talanta, 2010, 81(1-2): 301-308.

[2] Margetic V, Bolshov M, Stockhaus A, Niemax K, Hergenröder R. Depth profiling of multi-layer samples using femtosecond laser ablation. Journal of Analytical Atomic Spectrometry, 2001, 16(6): 616-621.

[3] Bleiner D, Plotnikov A, Vogt C, Wetzig K, Gunther D. Depth profile analysis of various titanium based coatings on steel and tungsten carbide using laser ablation inductively coupled plasma-"time of flight" mass spectrometry. Fresenius' Journal of Analytical Chemistry, 2000, 368(2-3): 221-226.

[4] Mateo M P, Garcia C C, Hergenröder R. Depth analysis of polymer-coated steel samples using near-infrared femtosecond laser ablation inductively coupled plasma mass spectrometry. Analytical Chemistry, 2007, 79(13): 4908-4914.

[5] Gutierrez-Gonzalez A, Gonzalez-Gago C, Pisonero J, Tibbetts N, Menendez A, Velez M. Capabilities and limitations of LA-ICP-MS for depth resolved analysis of CdTe photovoltaic devices. Journal of Analytical Atomic Spectrometry, 2015 30(1): 191-197.

[6] Mason P R D, Mank A J G. Depth-resolved analysis in multi-layered glass and metal materials using laser ablation inductively coupled plasma mass spectrometry (LA-ICP-MS). Journal of Analytical Atomic Spectrometry, 2001,16(12): 1381-1386.

[7] Mank A J G, Mason P R D. A critical assessment of laser ablation ICP-MS as an analytical tool for depth analysis in silica-based glass samples. Journal of Analytical Atomic Spectrometry, 1999, 14(8): 1143-1153.

[8] Mateo M P, Vadillo J M, Laserna J J. Irradiance-dependent depth profiling of layered materials using laser-induced plasma spectrometry. Journal of Analytical Atomic Spectrometry, 2001, 16(11): 1317-1321.

[9] Grimaudo V, Moreno-Garcia P, Riedo A, Meyer S, Tulej M, Neuland M B, Mohos M, Gutz C, Waldvogek S R, Wurz P, Broekmann P. Toward three-dimensional chemical imaging of ternary Cu-Sn-Pb alloys using femtosecond laser ablation/ionization mass spectrometry. Analytical Chemistry, 2017, 89(3): 1632-1641.

[10] Plotnikov A, Vogt C, Hoffmann V, Taschner C, Wetzig K. Application of laser ablation inductively coupled plasma quadrupole mass spectrometry (LA-ICP-QMS) for depth profile analysis. Journal of Analytical Atomic Spectrometry, 2001, 16(11): 1290-1295.

[11] van Elteren J T, Izmer A, Sala M, Orsega E F, Selih V S, Panighello S, Vanhaecke F. 3D laser ablation-ICP-mass spectrometry mapping for the study of surface layer phenomena: A case study for weathered glass. Journal of Analytical Atomic Spectrometry, 2013, 28(7): 994-1004.

[12] Li Y B, Qian G J, Li J, Gerson A R. Kinetics and roles of solution and surface species of chalcopyrite dissolution at 650 mV. Geochimica Et Cosmochimica Acta, 2015, 161: 188-202.

[13] Galindo R E, Gago R, Duday D, Palacio C. Towards nanometric resolution in multilayer depth profiling: A comparative study of RBS, SIMS, XPS and GDOES. Analytical And Bioanalytical Chemistry, 2010, 396(8): 2725-2740.

[14] 周韵, 楚民生, 龚思维, 吉静, 郅惠博, 吴益文. 激光剥蚀-电感耦合等离子体原子发射光谱法分析金属材料镀层. 理化检验 (化学分册), 2013, 49(11): 1305-1308.

[15] Samek O, Hommes V, Hergenröder R, Kukhlevsky S V. Femtosecond pulse shaping using a liquid-crystal display: Applications to depth profiling analysis. Review of Scientific Instruments, 2005, 76(8): 1-4.

[16] Chang C J, Chang H Y, You Y W, Liao H Y, Kuo Y T, Kao W L, Yen G J, Tsai M H, Shyue J J. Parallel detection, quantification, and depth profiling of peptides with dynamic-secondary ion mass spectrometry (D-SIMS) ionized by C_{60}^+-Ar^+ co-sputtering. Analytica Chimica Acta, 2012, 718(5): 64-69.

[17] Pisonero J, Koch J, Walle M, Hartung W, Spencer N D, Gunther D. Capabilities of femtosecond laser ablation inductively coupled plasma mass spectrometry for depth profiling of thin metal coatings. Analytical Chemistry, 2007, 79(6): 2325-2333.

[18] Limbeck A, Bonta M, Nischkauer W. Improvements in the direct analysis of advanced materials using ICP-based measurement techniques. Journal of Analytical Atomic Spectrometry, 2016, 32(2): 212-232.

[19] Pisonero J, Gunther D. Femtosecond laser ablation inductively coupled plasma mass spectrometry: Fundamentals and capabilities for depth profiling analysis. Mass Spectrometry Reviews, 2008, 27(6): 609-623.

[20] 张勇, 贾云海, 陈吉文, 沈学静, 刘英, 赵雷, 李冬玲, 韩鹏程, 赵振, 樊万伦, 王海舟. 激光烧蚀-电感耦合等离子体质谱技术在材料表面微区分析领域的应用进展. 光谱学与光谱分析, 2014, 34(8): 2238-2243.

[21] Bleiner D, Belloni F, Doria D, Lorusso A, Nassisi V. Overcoming pulse mixing and signal tailing in laser ablation inductively coupled plasma mass spectrometry depth profiling. Journal of Analytical Atomic Spectrometry, 2005, 20(12): 1337-1343.

[22] Vadillo J M, Laserna J J. Laser-induced plasma spectrometry: Truly a surface analytical tool. Spectrochimica Acta Part B:Atomic Spectroscopy, 2004, 59(2): 147-161.

[23] Russo R E, Mao X L, Borisov O V, Liu H C. Influence of wavelength on fractionation in laser ablation ICP-MS. Journal of Analytical Atomic Spectrometry, 2000, 15(9): 1115-1120.

[24] Coedo A G, Dorado T, Padilla I, Farinas J C. Depth profile analysis of copper coating on steel using laser ablation inductively coupled plasma mass spectrometry. Journal of Analytical Atomic Spectrometry, 2005, 20(7): 612-620.

[25] Hofmann S. From depth resolution to depth resolution function: Refinement of the concept for delta layers, single layers and multilayers. Surface And Interface Analysis, 1999, 27(9): 825-834.

[26] 周益春, 段祝平, 解伯民. 强激光破坏机制研究进展. 力学与实践, 1995, (1): 10-18.

[27] Chmel A E. Fatigue laser-induced damage in transparent materials. Materials Science and Engineering B: Solid State Materials for Advanced Technology, 1997, 49(3): 175-190.

[28] 倪晓昌, 王清月. 飞秒、皮秒激光烧蚀金属表面的有限差分热分析. 中国激光, 2004, (3): 277-280.

[29] Harzic R L, Huot N, Audouard E, Jonin C, Laporte P, Valette S, Fraczkiewicz A, Fortunier R. Comparison of heat-affected zones due to nanosecond and femtosecond laser pulses using transmission electronic microscopy. Applied Physics Letters, 2002, 80(21): 3886-3888.

[30] Fernandez B, Claverie F, Pecheyran C, Donard O F X. Direct analysis of solid samples by fs-LA-ICP-MS. TrAC Trends in Analytical Chemistry, 2007, 26(10): 951-966.

[31] 王岚, 杨理勤, 王亚平, 冯亮, 陈雪, 陈占生. 激光剥蚀电感耦合等离子体质谱微区分析进展评述. 地质通报, 2012, 31(4): 637-645.

[32] Cui Y, Moore J F, Milasinovic S, Liu Y M, Gordon R J, Hanley L. Depth profiling and imaging capabilities of an ultrashort pulse laser ablation time of flight mass spectrometer. Review of Scientific Instruments, 2012, 83(9).1-8.

[33] Ales H, Vítezslav O, Karel N, Detlef G, Viktor K, Gunther D, Kanicky V. Feasibility of depth profiling of Zn-based coatings by laser ablation inductively coupled plasma optical emission and mass spectrometry using infrared Nd：YAG and ArF* lasers. Spectrochimica Acta Part B: Atomic Spectroscopy, 2005, 60(3): 307-318.

[34] Balcaen L I L, Lenaerts J, Moens L, Vanhaecke F. Application of laser ablation inductively coupled plasma (dynamicreaction cell) mass spectrometry for depth profiling analysis of high-tech industrial materials. Journal of Analytical Atomic Spectrometry, 2005, 20(5)417-423.

[35] Momma C, Chichkov B N, Nolte S, Alvensleben F V, Tunnermann A, Welling H, Wellegehausen B. Short-pulse laser ablation of solid targets. Optics Communications, 1996, 129(1): 134-142.

[36] Kaser D, Hendriks L, Koch J, Koch J, Gunther, D. Depth profile analyses with sub 100-nm depth resolution of a metal thin film by femtosecond-laser ablation-inductively coupled plasma-time-of-flight mass spectrometry. Spectrochimica Acta Part B: Atomic Spectroscopy, 2018, 149: 176-183.

[37] Hergenroder R. A model of non-congruent laser ablation as a source of fractionation effects in LA-ICP-MS. Journal of Analytical Atomic Spectrometry, 2006, 21(5): 505-516.

[38] Bian Q, Garcia C C, Koch J, Niemax K. Non-matrix matched calibration of major and minor concentrations of Zn and Cu in brass, aluminium and silicate glass using NIR femtosecond laser ablation inductively coupled plasma mass spectrometry. Journal of Analytical Atomic Spectrometry, 2006, 21(2): 187-191.

[39] Kanicky V, Kuhn H R, Guenther D. Depth profile studies of ZrTiN coatings by laser ablation inductively coupled plasma mass spectrometry. Analytical and Bioanalytical Chemistry, 2004, 380(2): 218-226.

[40] Jan K, Simon E J, Yang Z, Wirth R. Effect of oxygen in sample carrier gas on laser-induced elemental fractionation in U-Th-Pb zircon dating by laser ablation ICP-MS. Journal of Analytical Atomic Spectrometry, 2014, 29(5): 832-840.

[41] Gunther D, Heinrich C A. Comparison of the ablation behaviour of 266 nm Nd：YAG and 193 nm ArF excimer lasers for LA-ICP-MS analysis. Journal of Analytical Atomic Spectrometry, 1999, 14(9): 1369-1374.

[42] Horn I, Gunther D. The influence of ablation carrier gasses Ar, He and Ne on the particle size distribution and transport efficiencies of laser ablation-induced aerosols: Implications for LA-ICP-MS. Applied Surface Science, 2003, 207(1-4): 144-157.

[43] Luo T, Ni Q, Hu Z C, Zhang W, Shi Q H, Gunther D, Liu Y S, Zong K Q, Hu S H. Comparison of signal intensities and elemental fractionation in 257 nm femtosecond LA-ICP-MS using He and Ar as carrier gases. Journal of Analytical Atomic Spectrometry, 2017, 32(11): 2217-2225.

[44] Arrowsmith P, Hughes S K. Entrainment and transport of laser ablated plumes for subsequent elemental analysis. Applied

Spectroscopy, 1988, 42(7): 1231-1239.

[45] Devos W, Moor C, Lienemann P. Determination of Impurities in antique silver objects for authentication by laser ablation inductively coupled plasma mass spectrometry (LAICPMS). Journal of Analytical Atomic Spectrometry, 1999, 14(4): 621-626.

[46] Pisonero J, Fliegel D, Gunther D. High efficiency aerosol dispersion cell for laser ablation-ICP-MS. Journal of Analytical Atomic Spectrometry, 2006, 21(9): 922-931.

[47] Garcia C C, Lindner H, Niemax K. Transport efficiency in femtosecond laser ablation inductively coupled plasma mass spectrometry applying ablation cells with short and long washout times. Spectrochimica Acta Part B: Atomic Spectroscopy, 2007, 62(1): 13-19.

[48] Liu Y S, Hu Z C, Yuan H L, Hu S H, Cheng H H. Volume-optional and low-memory (VOLM) chamber for laser ablation-ICPMS: Application to fiber analyses. Journal of Analytical Atomic Spectrometry, 2007, 22(5): 582-585.

[49] Muller W, Shelley M, Miller P, Broude S. Initial performance metrics of a new custom-designed ArF excimer LA-ICP-MS system coupled to a two-volume laser-ablation cell. Journal of Analytical Atomic Spectrometry, 2009, 24(2):209-214.

[50] Wagner B, Jędral W. Open ablation cell for LA-ICP-MS investigations of historic objects. Journal of Analytical Atomic Spectrometry, 2011, 26(10):2058-2063.

[51] Leach A M, Hieftje G M. Factors affecting the production of fast transient signals in single shot laser ablation inductively coupled plasma mass spectrometry. Applied Spectroscopy, 2002, 56(1): 62-69.

[52] Gurevich E L, Hergenroder R. A simple laser ICP-MS ablation cell with wash-out time less than 100 ms. Journal of Analytical Atomic Spectrometry, 2007, 22(9): 1043-1050.

[53] Wang H A O, Grolimund D, Giesen C, Borca C N, Shaw-Stewart J R H, Bodenmiller B, Gunther D. Fast chemical imaging at high spatial resolution by laser ablation inductively coupled plasma mass spectrometry. Analytical Chemistry, 2013, 85(21): 10107-10116.

[54] Gundlach-Graham A, Burger M, Allner S, Schwarz G, Wang H A O, Gyr L, Grolimund D, Hattendorf B, Gunther D. High-speed, high-resolution, multielemental laser ablation-inductively coupled plasma-time-of-flight mass spectrometry imaging: Part I. Instrumentation and two-dimensional imaging of geological samples. Analytical Chemistry, 2015, 87(16): 8250-8258.

[55] Van Malderen S J M, van Elteren J T, Vanhaecke F. Development of a fast laser ablation-inductively coupled plasma-mass spectrometry cell for sub-μm scanning of layered materials. Journal of Analytical Atomic Spectrometry, 2015, 30(1): 119-125.

[56] Douglas D N, Managh A J, Reid H J, Sharp B L. High-speed. integrated ablation cell and dual concentric injector plasma torch for laser ablation-inductively coupled plasma mass spectrometry. Analytical Chemistry, 2015, 87(22): 11285-11294.

[57] Hattendorf B, Pisonero J, Gunther D, Bordel N. Thickness determination of subnanometer layers using laser ablation inductively coupled plasma mass spectrometry. Analytical Chemistry, 2012, 84(20): 8771-8776.

[58] Shaheen M E, Gagnon J E, Fryer B J. Experimental studies on ablation characteristics of alumina after irradiation with a 193-nm ArF excimer laser. European Physical Journal Plus, 2021,136(1):1-17.

[59] Torrisi A, Cutroneo M, Castrizio E D, Torrisi L. Laser ablation coupled to mass quadrupole spectrometry for analysis in the cultural heritage. Plasma Physics by Laser and Applications 2013 Conference. Bristol: IOP Publishing Ltd, 2014, 508:1-6.

[60] Van Malderen S J M, van Elteren J T, Vanhaecke F. Submicrometer imaging by laser ablation-inductively coupled plasma mass spectro metryvia signal and image deconvolution approaches. Analytical Chemistry, 2015, 87(12):6125-6132.

第 14 章 色谱-原子光谱/质谱联用技术

(何 滨[①], 阴永光[①], 刘丽红[①], 史建波[①], 胡立刚[①]*)

▶ 14.1 引言 / 359

▶ 14.2 气相色谱-原子光谱/质谱联用技术 / 359

▶ 14.3 液相色谱-原子光谱/质谱联用技术 / 362

▶ 14.4 毛细管电泳-电感耦合等离子体质谱联用技术 / 372

▶ 14.5 展望 / 379

[①]中国科学院生态环境研究中心, 北京。*通讯作者联系方式: lghu@rcees.ac.cn

第14章 色谱-原子光谱/质谱联用技术

本章导读

- 介绍了气相色谱-原子光谱/质谱联用技术的发展及相关样品前处理技术。
- 介绍了液相色谱-原子光谱/质谱联用技术接口的发展及多维色谱联用技术。
- 介绍了毛细管电泳-电感耦合等离子体质谱联用技术接口、雾室、雾化器的发展。

14.1 引　言

传统的金属元素检测只针对元素的总量，但在环境和生命体中许多金属元素以不同的价态和化学形态存在，这些价态和化学形态不同的金属化合物在环境中的迁移转化规律、生物可利用性以及生理活性和毒性也具有显著差异。因此，仅以元素总量为依据的研究已不能满足现代环境科学、食品安全、医药卫生和生命科学发展的需要，元素的形态分析往往比总量分析更为重要。因此，发展高灵敏度和高选择性的元素形态分析联用技术和方法，对研究重金属的环境化学行为和生物生理作用有着非常重要的意义。

随着色谱、原子光谱和质谱以及样品前处理技术的发展，具有高分辨率和高灵敏度的色谱-原子光谱/质谱联用技术成为金属化合物形态分析的主要手段。各种色谱技术如气相色谱（GC）、液相色谱（LC）、离子色谱（IC）和毛细管电泳（CE）等被应用于不同形态金属化合物的分离，而原子吸收光谱法（AAS）、原子荧光光谱法（AFS）以及电感耦合等离子体质谱法（ICP-MS）等元素选择性检测器可以对特定元素进行响应，成为金属化合物形态分析中的主要检测方法。

14.2 气相色谱-原子光谱/质谱联用技术

14.2.1 气相色谱-原子吸收联用技术（GC-AAS）

1966年，Kolb等[1]提出采用原子吸收光谱仪作为GC检测器，通过将气相色谱分离柱直接连接到原子吸收光谱仪的燃烧器上实现了汽油中四甲基铅和四乙基铅的在线联用分析，分析过程中液体待测样品在GC进样口注入，在气化室气化后由载气带入色谱柱，被分析物经分离后由空气-乙炔气带入AAS燃烧器原子化。与传统的气相色谱检测器相比，气相色谱与原子吸收光谱法相结合有利于高选择性、高灵敏地检测金属，适用于易挥发性金属化合物的痕量分析。火焰原子吸收法（FAAS）虽然操作简单、运行成本低廉，但原子化效率低，待测原子在测量光路停留时间短，其测定灵敏度低于电热原子吸收法（ETAAS）。使用加热石英管作为GC-AAS的原子化器，对Me_4Pb、Me_3EtPb、Me_2Et_2Pb、$MeEt_3Pb$、Et_4Pb五种有机铅化合物进行分离测定，灵敏度比火焰原子化器可提高1000倍[2]，该技术还可应用于有机锡和有机铅的分析[3-5]。在GC与原子吸收光谱联用系统中，联用仪器主要由GC、联用接口和原子吸收光谱仪三大部分组成。理想的设计是将气相色谱流出物直接导入原子化器进行原子化并测定，以减少样品损失，避免峰展宽，降低死体积，提高测定灵敏度。这样简单的联用接口适合于测定分子量较小、结构对称、挥发性强的有机金属化合物，而对于分子量较大、挥发性较弱的有机金属化合物的分离测定，联用系统的接口必须保持一定温度使待测样品在接口中维持气化状态或原子化。在保温接口的研制方面，中国科学院生态环境研究中心江桂斌研究组较早开展了这方面的工作，研制了一种可控温加热联用接口，该接口由内径为 1 mm 的不锈钢管，外套内径 4 mm、长 8 cm 的石英管组成，石英管上绕有镍铬电阻丝，加热温度可由变压器调控，部分易裂解的有机金属化合物如有机汞在接口部位就可发生热裂解而被原子化[6]。他们还发明了 T 型电热石英原子化器，可直接将色谱分离组分以最佳传送方式导入吸收光路，克服了组分稀释和峰变宽问题，从而实现了五种有机汞在线的高灵敏度测定，并用于多种环境和生物样品中有机汞的形态分析[6,7]。为了更好地分析有机锡化合物，他们还对 T 型电热石英炉原子化器进行了改进，缩短了接口部分的长度，避免了色谱流出物在接口部分被吸附、分解，极大地改善了色谱峰形；同时延长了原子化器部分石英管的长度并在管外环绕镍铬电阻丝控制原子化器温度在 600～900℃，使有机锡化合物能够原子化并停留较长时间，从而提高了测定灵敏度，成功分离测定了 7 种不同形态的有机锡化合物[8]。Matousek 等[9]则描述了一种带有多原子化器的氢化物发生-冷阱捕集-气相色谱-原子吸收光谱（HG-CT-GC-AAS）自动化系统，他们采用在 Tris 缓冲溶液中选择性地将 iAs(Ⅲ)、MMA(Ⅲ)、DMA(Ⅲ) 和 TMA(Ⅴ)衍生成相应的氢化物，而在加入了 L-半胱氨酸的缓冲溶液中将三价和五价的 iAs、MMA、DMA 等效转化成氢化物，再用 GC-AAS 分离，从而测定了不同价态的 iAs、MMA、DMA 及 TMA(Ⅴ)。这种方法可以对复杂生物基质(如细胞培养系统)中无机砷的代谢产物进行高通量的形态分析，而无需样品预处理，从而保持了样品中三价和五价形态的分布。Feng 等[10]则应用 HG-GC-AAS 方法测定了海洋沉积物样品中的有机锡化合物，在 1 mol/L HCl 浸提液中加入 8-羟基喹啉，大大改善了甲苯对三丁基锡和三苯基锡的萃取效率。基于原子吸收光谱的气相色谱联用系统可以检测到复杂样品中 ppt 水平的多种有机金属化合物，广泛适用于样品中极性差别不大、在一定的加热温度下有挥发性但热稳定、原子化温度较低的金属化合物，如烷基汞、硒、锡、锗和铅等的形态分析。使用高强度的短弧氙灯作为原子吸收的连续光源以及高分辨率的双光栅单色器和灵敏的电荷耦合检测器可实现原子吸收的多元素同时测定[11]，将 GC 与连续光源原子吸收光谱（CSAAS）联用，在背景校正条件下分离测定有机锡的检测限为 6～22 pg，与线源原子吸收检测器（LSAAS）相比检测灵敏度提高 5～10 倍，即使在无背景校正时，CSAAS 的灵敏度也比 LSAAS 高 2 倍[12]。

14.2.2 气相色谱-原子荧光联用技术（GC-AFS）

原子荧光光谱法（AFS）是一种灵敏度高、选择性强的元素测定方法，与 AAS 相比，虽然其可测定的元素有限，但在灵敏度、线性范围和抗光谱干扰方面的优势已经在理论上[13,14]和实验中[15]得到证实。Van Loon 等最先提出 GC-AFS 联用技术，他们用非色散原子荧光光谱法，采用氮屏蔽的空气-乙炔火焰原子化器，使流出物通过燃烧器底部的一个端口进入火焰中进行测定[16]。该方法操作简便，且与原子吸收相比，大多数元素的检测限可以降低 1～3 个数量级。Dulivo 等[17]用毛细管 GC 与多通道非色散原子荧光仪联用技术分离并同时检测了烷基硒、烷基铅和烷基锡，其中 Sn 和 Pb 检出限与 GC-AAS 相当，Se 的检出限降低了 15 倍，表明该联用系统在对有机金属化合物的检测灵敏度上有一定优势。Armstrong 等[18]对气相色谱和 AFS、ICP-MS 联用在有机汞形态分析中的应用进行了比较，发现两种检测器均具有很高的灵敏度和选择性，AFS 和 ICP-MS 的检出限分别为 0.9 pg 和 0.25 pg（以 Hg 计），尽管 ICP-MS 可进行多元素和同位素分析，但 AFS 运行费用低且操作简单。适用于 GC-AAS 的接口，例如加热套管、热裂解装置等，也适用于 GC 和 AFS 的联用。Cai 等采用以电热裂解器为接口的 GC-AFS 对甲基汞和乙基汞进行了分析，GC 分离后的有机汞可在接口处 800℃ 下分解为 Hg^0 从而被 AFS 测定，无机汞经格氏试剂衍生后也可用该联用系统分析，1 L 水样经巯基棉富集后有机汞的检测限可低至 0.02 ng/L[19,20]。在色谱分离前采用乙基化衍生和吹扫捕集技术，水样中甲基汞的检测限可降至 0.016 ng/L[21]。Cai 等[22]还比较了三种联用技术，即 GC-AFS、GC-MIP-AES（气相色谱-微波诱导等离子体-原子发射光谱）和 GC-MS 对经四乙基硼酸钠和四苯基硼酸钠衍生后的甲基汞和乙基汞分析性能，结果表明，GC-AFS 和 GC-AES 均比 CG-MS 更适用于大多数环境样品中汞的测定，其检出限为 pg 水平。采用氢化物发生-顶空固相微萃取对待测组分进行分离富集可提高 GC-AFS 测定水样中丁基锡方法的灵敏度，一丁基锡（MBT）、二丁基锡（DBT）和三丁基锡（TBT）的检测限可分别降至 10 ng/mL、0.2 ng/mL、0.1 ng/mL[23]。为改善 GC 分离中某些组分的拖尾，提高分离效率，还可对 GC-AFS 联用技术中尾吹气及尾吹气流量进行优化，采用 400 mL/min 氩气替代氮气进行尾气吹扫，对甲基汞（MMC）和乙基汞（EMC）进行了很好的分离检测[24]。Gorecki 等[25]配备了带阀门和氩气加热器的 GC-AFS 系统，以消除固相微萃取（SPME）纤维吸附水分所造成的信号下降。Gomez-Ariza 使用改进的 AFS 检测器，即在检测器中加入石英流池，增加了检测器中汞原子的浓度，将 Hg^{2+} 和 MeHg 的测定灵敏度提高了约 2 倍[26]。

14.2.3 气相色谱-原子发射光谱联用（GC-AES）

原子发射光谱法（AES）由于检测动态范围宽、灵敏度和选择性好，可实现多元素同时测定，也可作为特效检测器与 GC 联用。McCormack 等首次建立了 GC-AES 联用系统[27]，并被用于农药中有机磷[28]和鱼体中有机汞的分析[29]，色谱流出物可被引入氩/氦等离子体中并被原子化和激发而被检测，其联用接口相对简单。由于原子发射谱线较窄、重叠概率和光谱干扰较少，AES 几乎可以测量所有的元素，但需采用背景校正技术以消除因原子化不完全而残留的分子所产生的光谱发射带对原子发射光谱的影响。随着 AES 技术的发展，各种类型的等离子体激发光源取代了早期的火焰、弧光放电以及火花放电等技术，侯贤灯等对各种等离子体激发技术和放电激发技术进行了总结[30]。新的激发光源技术大大提高了光源的稳定性，降低了基体干扰，提高了测定灵敏度和重现性。

14.2.4 气相色谱-电感耦合等离子体质谱联用技术（GC-ICP-MS）

电感耦合等离子体质谱仪（ICP-MS）具有选择性高、检出限低、可多元素同时测定等优点，其独有的同位素稀释分析还能有效地提高测定的准确性和精密度。1986 年 Van Loon[31]首次报道将 ICP-MS 作为 GC 检测器以来，GC-ICP-MS 联用技术获得了广泛的应用。GC 的高分辨率和 ICP-MS 的高灵敏度使 GC-ICP-MS 成为环境样品中金属和非金属元素形态分析的有力手段。对 ICP-MS 来说，GC 分离后的气相进样是高效且无干扰的进样方法。由于高温的氩气等离子体火焰能够很容易地将待测化合物原子化并电离，GC-ICP-MS 接口加热套管的温度只要达到维持待测物气化即可，不需要很高的热裂解温度。应用稳定同位素稀释技术还可以提高 ICP-MS 检测的特异性和准确度。Baxter 等用 ^{198}Hg 标记的甲基汞反相同位素稀释法分析了血浆和血清中甲基汞。他们采用二氯甲烷萃取、反萃取入水溶液，然后经乙基化衍生、吹扫捕集、热解吸等前处理技术以及气相色谱分离和 ICP-MS 特异性检测，分离测定了正常人血浆和血清中的甲基汞浓度，检测限可低至 0.03 ng/mL[32]。利用同位素稀释技术可提高水、底泥和生物样品中甲基汞的分析准确性和精密度，马旭等采用自制的 T 型三通将经 GC 分离并热裂解的甲基汞直接导入 ICP-MS 的炬管，这种接口设计缩短了 GC 与 ICP-MS 的传输距离，同时经 T 型管导入 Ar 气为进样辅助气提高了 ICP-MS 的进样速度以及分析的灵敏度，^{198}Hg 和 ^{202}Hg 的绝对检测限分别为 4.3 pg 和 8.1 pg，测定的相对标准偏差小于 3.6%[33]。使用 GC-ICP-MS 联用分析技术，结合高效的萃取和富集等样品前处理技术，可以得到低于 fg 级的绝对检出限[34]。

14.2.5 气相色谱联用技术中的样品前处理技术

在进行形态分析前必须保证在待测形态不发生变化的前提下将待测物全部从样品基质中分离出来，对样品中的痕量和超痕量待测物还需进行富集以满足分析仪器的检测限。在样品的前处理上，传统的液-液萃取仍是常用的分离富集方法[35-38]，该方法需使用大量有机溶剂，操作烦冗且容易造成样品损失，重现性差，实际操作中通常会根据待测形态及样品基质的性质进行改进。

在采用气相色谱进行形态分离时通常要求被分离组分具有易挥发性和热稳定性，而许多金属离子和有机金属化合物形态是极性的，挥发性差，在进气相色谱分离前还需进行衍生，通过氢化反应、乙基化[39-44]、格氏试剂衍生[20,35,45,46]等方法将待测形态转化成分子量相对较小的氢化物或结构相对对称的乙基化、丙基化或苯基化衍生物，以增加待测形态的挥发性。自从 Craig 等[47]证明了甲基汞氢化物的存在后，氢化物发生法被广泛用于甲基汞[44,48,49]和其他有机汞[50,51]的分离测定。Craig 研究小组首先用 NaBEt$_4$ 对 MeHg$^+$ 进行了衍生[41]，此后这一方法被广泛用于有机汞的测定[52]。但由于衍生产物均为二乙基汞，NaBEt$_4$ 衍生法不适用于无机汞和乙基汞的同时测定。由于格氏试剂的多样性，格氏化衍生反应种类繁多，适用性广，缺点是格氏反应必须在有机溶液中进行。

在富集技术方面，除了传统的液-液萃取浓缩，还发展了固相萃取（SPE）[53-55]、固相微萃取（SPME）[56-58]、吹扫捕集（purge and trap）[59-61]和冷阱捕获（cryogenic trapping，CT）[62-64]等富集技术。固相微萃取（SPME）是 20 世纪 80 年代发展起来的一项新型的无溶剂化样品前处理技术，它集萃取、浓缩、解吸于一体，与气相色谱联用可快速、有效地分析环境样品中痕量挥发和半挥发性有机化合物[65,66]。对于极性的和挥发性差的有机金属化合物，则可通过衍生化反应提高它们的挥发性[67,68]。何滨等采用氢化物发生顶空固相微萃取（HG-HSSPME）技术分离富集无机汞和有机汞化合物，并用 GC-AAS 对土壤、底泥和生物体中的有机汞进行了分析[56,57]。Davis 等[69]使用涂覆聚二甲基硅氧烷（PDMS）的二氧化硅纤维进行 SPME，然后使用 GC-ICP-MS，以 0.5 g SRM 1566b 牡蛎组织样本为基础，获得了 4.2 pg/g(as Hg) 的甲基汞 LOD。吹扫捕集技术适用于从液体或固体样品中萃取沸点低于 200℃、溶解度小于 2% 的挥发性或半挥发性有机物[70]。

14.3 液相色谱-原子光谱/质谱联用技术

相较于气相色谱-原子光谱/质谱联用，液相色谱与原子光谱/质谱的联用更为直接。与气相色谱不同，液相色谱分离通常在室温下或略高的温度下进行，因此无需调节液相色谱和原子光谱/质谱之间流路的温度[71]。通常，液相色谱的流速与 ICP-MS 进样速度相匹配，其联用通常直接采用气动雾化将液相色谱流出物引入 ICP 系统。液相色谱与原子荧光、原子吸收的联用通常采用氢化物/冷蒸气发生的接口。

液相色谱-原子光谱/质谱联用技术的分析对象十分广泛，涵盖了从无机离子（如三价砷、五价砷）、有机金属（如甲基汞）、分子（如砷糖、砷脂、多溴联苯醚）到生物大分子（如蛋白质）及其络合物（如汞-蛋白络合物）、纳米颗粒（如 CdSe 量子点、纳米银）的多种分析物。随着 ICP-MS/MS 技术的发展，其对 S、P 等元素的检测灵敏度显著提高，采用液相色谱-ICP-MS/MS 联用，可直接通过 S、P 定量对多肽[72]、蛋白质[73,74]与 DNA[73]等进行分析，也可通过元素标记实现生物大分子的检测[75]。

需要特别指出的是，对于一些挥发性形态，可采用化学吸收法将其转化为非挥发的水溶性形态，进一步采用液相色谱-原子光谱/质谱联用技术进行定量分析,从而可以定量出原始的挥发性形态。例如,对于挥发性砷化合物,用1%硝酸银浸渍硅胶管吸附之后采用5%沸稀硝酸洗脱与双氧水氧化,可将挥发性的砷化氢、一甲基砷化氢、二甲基砷化氢、三甲基砷转化为水溶性的五价砷、一甲基砷酸、二甲基砷酸、三甲基砷氧化物,之后可用阴离子交换色谱-ICP-MS进行定量[76]。而对于挥发性的硒化物[77,78]与硫化物[78],则可以采用70% HNO_3 水溶液吸收后进行分析。

针对不同分析物的物理化学特性,可选择不同的液相色谱技术对其进行分离。常用于与原子光谱/质谱联用的液相色谱技术主要有反相色谱、正相色谱、离子交换色谱、离子对色谱、尺寸排阻色谱、亲水作用色谱等,其中又以尺寸排阻色谱、反相色谱与离子交换色谱使用最为广泛。近年来,低流速液相色谱如毛细管与纳流液相色谱的应用也日益增多。

以上色谱分离技术各有其优劣。尺寸排阻色谱是分离不同分子量分析物的利器,但其色谱效率较低,柱回收率较低。离子交换色谱与离子对色谱在分离荷电离子方面具有独特的优势,但通常使用高盐流动相,可能带来盐沉积的问题。这一问题可通过采用挥发性盐作为流动相来解决。正相色谱与反相色谱的流动相常含有一定比例有机溶剂,可能导致ICP的不稳定与碳沉积。流动相中高有机相带来的问题可通过雾化室冷却、加氧等措施来缓解。微型液相色谱流动相流速低,因此可显著降低盐或有机溶剂带来的ICP干扰问题,但低流速流动相也导致须对雾化器与雾化室进行改进,以提高雾化与样品引入效率[79]。

下面分别就液相色谱-原子光谱/质谱的联用系统接口、多维液相色谱-原子光谱/质谱联用、原子光谱/质谱与其他互补技术的同时检测、液相色谱-原子光谱/质谱联用中的样品预富集技术等方面进行介绍。

14.3.1 液相色谱-原子光谱/质谱联用的接口

液相色谱-原子光谱/质谱在线联用的关键是将液相色谱分离的样品高效转移至光谱/质谱仪中。在线联用接口装置通常包括雾化接口与冷蒸气/氢化物发生接口两大类。其中,冷蒸气/氢化物发生接口广泛应用于液相色谱与原子荧光、原子吸收等的在线联用。而雾化与冷蒸气/氢化物发生两类接口在液相色谱与ICP-OES、ICP-MS的在线联用中均得到广泛应用。

14.3.1.1 液相色谱-原子光谱/质谱联用的雾化接口

在液相色谱-原子光谱/质谱联用系统的雾化接口中,色谱流出物在雾化器中雾化形成气溶胶,在雾化室中较大粒径气溶胶颗粒被分离去除,小粒径气溶胶颗粒进入ICP中。因此,雾化器和雾化室是实现死体积小、进样效率高的关键。理想的雾化器和雾化室应满足:产生的气溶胶平均粒径小,分布范围窄;传输效率高;雾化器与雾化室死体积小,峰展宽小;色谱流动相与雾化器适用流速相匹配。在传统的液相色谱中,其流动相流速通常为0.5~1.5 mL/min。这一流速与常规的ICP-MS样品引入流速相匹配,因此可较为方便地将色谱流出物直接连接ICP-MS的常规雾化器。常用的雾化器有同心雾化器、交叉流雾化器、沟槽雾化器(Babington)、平行流路雾化器(Burgner)[80]、超声雾化器[81-83]等,其中以同心雾化器使用最为广泛。常用雾化室有旋流雾化室、Scott雾化室、圆锥形雾化室等。在这些常规的雾化器/雾化室接口中,样品传递效率通常较低(2%~10%)。对于液相色谱流动相中高有机相的情况,可通过雾化室低温制冷(如<-10℃)、降低有机溶剂挥发来解决。在某些情况下,加热雾化室也是选择之一,这不仅有助于去除有机溶剂,降低ICP的负载,还可显

著提高检测灵敏度[84,85]。但在另一些情况下，加热则降低分析物的灵敏度，且这种变化依赖于分析物形态[86]。雾化室温度控制对灵敏度的影响可参见综述[87]。此外也可通过柱后加氧降低采样锥的碳沉积[87]。

近年来，小口径液相色谱的使用日益广泛，其具有流动相消耗低、废液产生少、分辨率高、ICP基质负载低的优点。但当其流动相流速降至几十 µL/min 甚至更低时，传统雾化器和雾化室的雾化与传输效率不能满足要求，死体积过大，柱后峰展宽严重。因此需针对小口径液相色谱低流速的特点对其雾化接口进行改进。目前，低流速雾化器主要包括：微同心雾化器（micro-concentric nebulizer, MCN）[88]、高效雾化器（high-efficiency nebulizer, HEN）、振荡毛细管雾化器（oscillating capillary nebulizer, OCN）、电喷雾雾化器（electrospray nebulizer）等（表 14-1）。微同心雾化器设计流速为 50~500 µL/min，其可匹配小体积雾室，降低死体积[89,90]。在流速<0.2 mL/min 时，微同心雾化器的检出限优于传统同心雾化器[88,91]。通过进一步降低微同心雾化器的毛细管与出气口内径，可进一步使其适用更低流速（如 0.5~7.5 µL/min[92]），提高雾化效率。高效雾化器的使用流速较微同心雾化器更低，通常为 10~150 µL/min[93]。以上雾化器的样品传递效率仍然较低，直接注射雾化器（direct-injection nebulizer, DIN）采用了全新的设计以解决这一问题[94]。实际上，直接注射雾化器是一种将喷雾口直接位于 ICP 焰基下的一种微同心雾化器，其设计流速约 100 µL/min 或更低，无需雾化室，色谱流出物可 100%进入 ICP[94-97]。研究显示，与传统雾化器以及微同心雾化器相比，采用直接注射雾化器时，微孔液相色谱-ICP-MS 柱效与灵敏度均有增加[98]。但直接注射雾化器操作较为复杂，易于折断。因此，在直接注射雾化器与高效雾化器的基础上进一步发展了直接注射高效雾化器（direct-injection high-efficiency nebulizer, DIHEN）。相较于其他微型雾化器，直接注射高效雾化器在峰宽、分辨率、峰对称性、检出限等方面性能优异[99]。近年来，在以上雾化器的基础上，又设计出许多新型雾化器，其具有优异的雾化性能与样品传输效率（表 14-1）。这些雾化器大多仍基于同心雾化器的结构，进一步优化了毛细管参数与位置，其流速可降至 nL/min。

表 14-1　液相色谱-原子光谱/质谱联用中微雾化器的设计与应用

雾化器	设计图	备注	参考文献
微同心雾化器		流速<10~100 µL/min	[100,101]
高效雾化器		流速 10~150 µL/min，便于安装与操作	[93]

第14章 色谱-原子光谱/质谱联用技术

续表

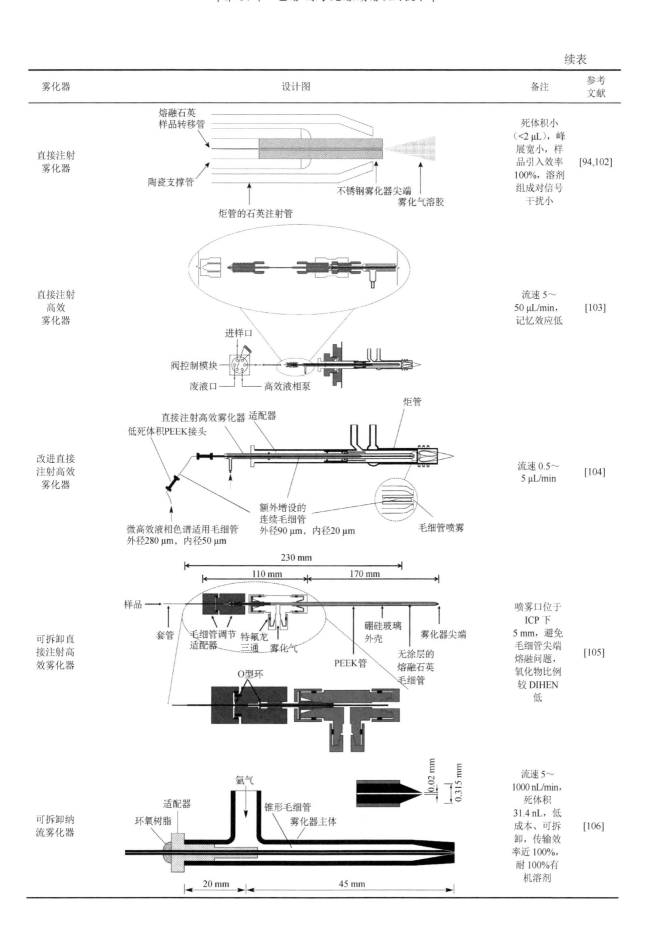

雾化器	设计图	备注	参考文献
直接注射雾化器		死体积小（<2 μL），峰展宽小，样品引入效率100%，溶剂组成对信号干扰小	[94,102]
直接注射高效雾化器		流速5～50 μL/min，记忆效应低	[103]
改进直接注射高效雾化器		流速0.5～5 μL/min	[104]
可拆卸直接注射高效雾化器		喷雾口位于ICP下5 mm，避免毛细管尖端熔融问题，氧化物比例较DIHEN低	[105]
可拆卸纳流雾化器		流速5～1000 nL/min，死体积31.4 nL，低成本、可拆卸，传输效率近100%，耐100%有机溶剂	[106]

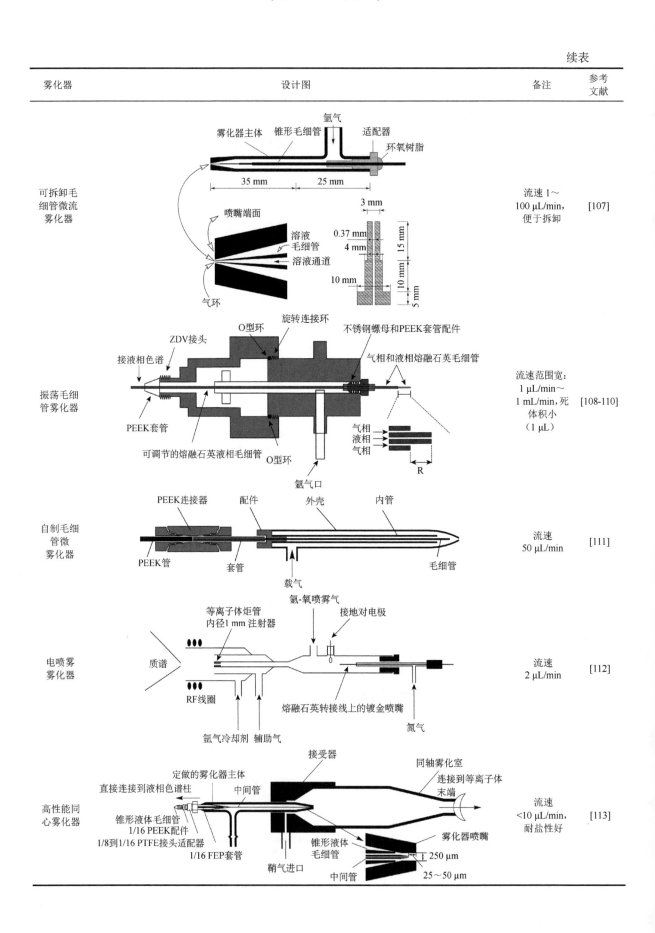

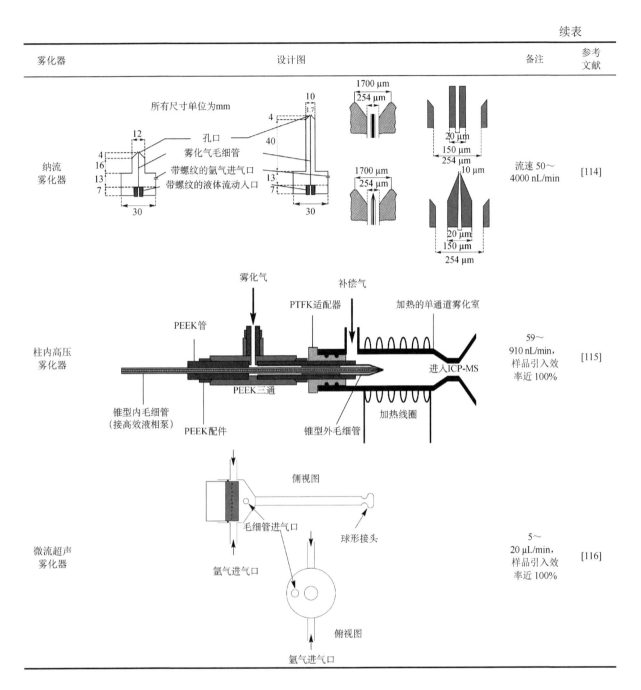

14.3.1.2 液相色谱-原子荧光联用的冷蒸气/氢化物发生接口

液相色谱与原子荧光、原子吸收等光谱技术的联用,以冷蒸气/氢化物发生接口为主。液相色谱流出物在线转化为元素冷蒸气/氢化物,经气液分离后进一步原子化并进行原子荧光/吸收检测。由于液相色谱与原子荧光/吸收联用的冷蒸气/氢化物发生接口类似,且目前的研究以原子荧光检测为主流,这里仅就液相色谱-原子荧光联用的冷蒸气/氢化物发生接口进行总结介绍。

针对不同的分析物,可采用不同的冷蒸气/氢化物发生策略。对于汞而言,无机汞及不同形态烷基汞均可与 $KBH_4/NaBH_4$ 反应,生成零价汞与氢化烷基汞,经气液分离后,氢化烷基汞可在氩氢火焰中[117]或经紫外降解[118]转化为零价汞,从而进一步进行原子荧光检测。但更为常用的方式是将有机汞经氧化降解转化为二价汞,经 $KBH_4/NaBH_4$、$SnCl_2$ 还原为零价汞后检测。有机汞的氧化过程可通过在线混合氧化剂如 $KBr-KBrO_3$、Fe^{3+} 等实现[119-121]，也可采用紫外、紫外微波等氧化方式[122,123]，或者氧化剂与能量方式相结合（如 $K_2S_2O_8$+紫外、$K_2S_2O_8$+微波）。特别地，当流动相中存在半胱氨酸时，其与 $KBH_4/NaBH_4$ 的混合可直接将有机汞转化为零价汞，因此该接口避免了紫外灯、微波探针等的使用，也无需在线混合氧化剂[124]。近年来，光化学蒸气发生作为一种新型的蒸气引入方式使用日益广泛。光化学蒸气发生也可作为液相色谱-原子荧光联用的接口[125]。在紫外光照下，柱后混合的甲酸可直接将有机汞转化为零价汞[126]，也可采用 TiO_2 等光催化材料进一步提高转化效果[120,127,128]。考虑到不同汞形态的液相色谱流动相中会加入缓冲盐，因此可将流动相中的甲酸-甲酸铵同时作为缓冲盐与柱后光化学蒸气发生试剂，这一接口无需添加任何柱后试剂与蠕动泵等，得到了极大简化[129]。类似地，流动相中的甲酸与巯基乙醇也可以作为溶液阴极辉光放电蒸气发生的增强剂[130]。介质阻挡放电化学蒸气发生接口的使用也可极大减少柱后化学试剂的使用[131]。

对于易于氢化物发生的一些砷、硒、锑等元素形态（如无机砷、甲基砷）而言，可直接柱后 $KBH_4/NaBH_4$ 混合后使其转化为挥发性的氢化物，进而在氩氢火焰中原子化后检测。但某些砷、硒、锑元素形态（如砷糖）的氢化物发生效率较差，通常将其氧化后再进行氢化物发生。类似于汞，可采用的氧化剂包括 $K_2S_2O_8$、$KBr-KBrO_3$ 等，能量方式包括紫外、微波、加热，或其与氧化剂的复合。某些形态如六价硒、五价锑的直接 $KBH_4/NaBH_4$ 还原效率不高，可在 $KBH_4/NaBH_4$ 还原之前增加预还原步骤（还原剂 KI[132]、TiO_2[133]、半胱氨酸[134]），使其转化为更易氢化物发生的四价硒与三价锑。基于甲酸的光化学蒸气发生也可用于硒形态分析的液相色谱-原子荧光联用接口，固定化 $Ag-TiO_2/ZrO_2$ 光催化剂的使用可进一步提升光化学蒸气发生效率与检测灵敏度[135]。双阳极电化学氢化物发生作为柱后接口也可避免 $KBH_4/NaBH_4$ 等不稳定还原试剂的使用[136-138]。需要特别指出的是，在砷、硒、锑检测中，如未采用 $KBH_4/NaBH_4$ 或其浓度较低时，产生的氢气不足以支持稳定的氩氢火焰，可采用柱后补充氢气的方法解决这一问题。

14.3.1.3 液相色谱-ICP-MS 联用的冷蒸气/氢化物发生接口

如上所述，液相色谱可较为方便地通过雾化器接口与 ICP-MS 实现联用。但在某些情况下，这些联用也存在一些缺陷：①流动相中较高比例的有机溶剂引入等离子体会导致高反射功率、采样锥碳沉积以及显著的记忆效应，严重时会直接导致等离子体熄火[139]；②样品基质对检测的干扰，如高浓度的非挥发性盐导致采样锥盐沉积、氯离子形成的 $ArCl^+$ 对砷测定的干扰[140]、BrH^+ 对硒测定的干扰[141]；③样品引入效率不足，导致灵敏度较低[142]。采用柱后冷蒸气/氢化物发生作为液相色谱与 ICP-MS 联用的接口可降低有机溶剂引入量，降低基质干扰，提高分析物的引入效率，提高检测灵敏度[142]。

一般地，液相色谱流出物与酸混合后，进一步与硼氢化钠混合，可在酸性条件下实现元素形态的柱后冷蒸气/氢化物发生。在此过程中，可直接利用 ICP-MS 雾化器促进元素形态与硼氢化钠的高效混合，雾化室作为冷蒸气/氢化物的气液分离器[139,143-145]。但这种利用雾化室作为气液分离器的接口可能引入氢化物发生过程生成的盐（如 NaCl），从而造成 ICP-MS 进样系统的盐沉积。通常，不同形态元素的冷蒸气/氢化物发生效率存在较大差异[146]。例如五价砷与六价硒的氢化物发生效率低于三价砷与四价硒[147]。可在柱后氢化物发生之前，采用预还原的方法将其转化为氢化物发生效率较高的低价态，常用的预还原剂有半胱氨酸[147-149]、$Na_2S_2O_4$[150,151]。此外，砷糖、硒糖等有机形态的直接氢化物发生效率极低[152-154]。这一现象虽然有助于进行复杂样品中高氢化物生成能力的元素形态分析（如三甲基硒[153,155]、无机砷[146,156]），但也给低氢化物生成能力的元素形态分析带来不便。由于不同元素形态的氢化物发生效率不同导致其检测灵敏度差异较大，在采用氢化物光化学蒸气发生接口时，不能采用归一化法对未知元素形态进行定量分析。为提升氢化物光化学蒸气发生的效率，可在冷蒸气/氢化物发生之前引入在线消解装置，将难以冷蒸气/氢化物发生的元素形态氧化降解为易于冷蒸气/氢化物发生的元素形态。在线氧化中，常用的氧化剂有过硫酸钾[157-160]、KBrO$_3$-HBr[161-163]等，通常同时采用微波[161,163]、紫外光照[157-160]等手段促进氧化反应的发生，其中又以紫外光照应用最为广泛。高效光氧化反应器（high-efficiency photooxidation reactor）的独特设计可引入185 nm真空紫外的有效辐照，可无需使用氧化剂快速实现氧化过程（如砷甜菜碱转化为砷酸盐仅需3.5 s）[140]，因此该方法对于未知砷化合物的高灵敏检测具有极大优势。此外，在紫外光降解中，可采用二氧化钛等纳米材料进一步提升降解性能[150,164]。

采用硼氢化钠作为冷蒸气/氢化物发生的还原剂，会导致氢气等副产物生成，氢气可能进一步导致ICP稳定性问题并加剧背景噪声[147]。此外过硫酸钾、KBrO$_3$-HBr等氧化剂的使用也导致管路易于老化，需定期更换。为克服这些问题，光化学蒸气发生作为一种新型的化学蒸气发生手段被用于液相色谱-ICP-MS联用的接口。光化学蒸气生成通常采用甲酸作为柱后反应试剂，具有集氧化/还原为一体、产生氢气少、试剂用量少及装置简便等显著的优点。类似地，可采用二氧化钛悬浮液[165-167]以及固定化二氧化钛[168]、纳米金修饰二氧化钛[151]、纳米银修饰二氧化钛[135]、纳米氧化锆[135]等光催化材料进一步提升分析性能。目前，光化学蒸气发生已成功用于液相色谱-ICP-MS联用系统汞[166,169]、砷[151]、硒[135,165,167,168]的分离检测。相对于传统氢化物发生（使用硼氢化钠），二氧化钛促进的光化学蒸气发生可显著提高Se(Ⅳ)的样品引入效率与灵敏度[167,168]。

14.3.2 多维液相色谱-原子光谱/质谱联用

色谱在复杂混合物分离分析中发挥了极大的作用。但对于样品中成百上千性质各异的分析物，一维色谱往往难以达到分离要求，需要使用二维甚至三维以上的色谱分离技术。通常，多维液相色谱技术将不同选择性（正交性）的液相分离柱串联组合，以增加系统的分离能力。对于多维液相色谱-原子光谱/质谱联用系统，第一维色谱通常为尺寸排阻色谱，第二维色谱形式多样，可以为亲水相互作用色谱[170,171]、阴离子交换色谱[172-176]、阳离子交换色谱[177,178]、亲和色谱[179,180]、反相色谱[181-184]、反相离子对色谱[185]、正相色谱[171]、多孔石墨碳色谱[186]等，也可以是其他非液相色谱分离技术，如毛细管电泳[187]、凝胶电泳[188]。目前报道的多维液相色谱-原子光谱/质谱联用系统，其色谱分离可多至三维[189,190]。多维液相色谱的关键是两种色谱分离系统之间的切换。目前，在多维液相色谱-原子光谱/质谱联用系统中，多维液相色谱之间的联用仍以离线为主。但随着技术的发展，在线联用的报道日益增多[174,179,184,189,191]。多维液相色谱在线联用时，通常采用中心切割（heart cut）改变流动相流路，将第一维未分离的组分导入第二维色谱柱进行第二次分离[174,184]。最近，报道了全二维液相色谱（comprehensive two-dimensional liquid chromatography）与ICP-MS/MS在石油产品中的硫、钒和镍化合物分析中的应用，尺寸排阻色谱与反相色谱的串联使得系统可同时得到分析物的分子量与疏水性信息[183]。

14.3.3 原子光谱/质谱与其他互补技术的同时检测

原子光谱/质谱作为元素特异性的检测可提供各种形态的元素识别与定量信息，其可与其他互补技术联用，提供更多分析物的信息。例如，液相色谱-紫外可见光谱-原子光谱/质谱串联检测，可同时提供被分析物的紫外可见光谱信息，在溶解有机质[192,193]、蛋白质[194,195]、DNA[196]等分析中应用十分广泛。此外，原子光谱/质谱与电喷雾离子化-质谱的并行检测，可同时提供分析物的元素结构信息，在未知元素形态的识别（如代谢产物）上发挥着重要的作用[197-201]。例如，通过柱后Ba的加入可将F⁻在氩等离子体中转化为可被ICP-MS/MS检测的[BaF]⁺，从而实现液相色谱分离的全氟化合物的F

特异性检测；结合 ESI-MS/MS 结构分析，可进行样品中未知有机氟化合物的非靶标筛查[202]。

14.3.4 液相色谱-原子光谱/质谱联用中的样品预富集技术

在某些环境与生物基质中元素形态浓度较低，为实现其液相色谱-原子光谱/质谱联用分析，进样之前采用合适的预富集步骤可提高元素形态的检测能力。预富集包括离线与在线富集两类。

液相色谱-原子光谱/质谱联用的离线富集技术见表 14-2 所示，主要包括固相萃取、液相萃取两大类方法。固相萃取中，除了较为传统的固相萃取柱富集外，分散固相萃取、搅拌棒吸附萃取、固相微萃取技术也得到应用。液相萃取中，浊点萃取、分散液液微萃取以及中空纤维液相微萃取使用较为广泛。在以上萃取过程中，可向萃取介质中加入修饰剂改性，以提高待分析元素形态的萃取性能。例如在汞的固相萃取与液相微萃取等预富集过程中，通常加入二乙基二硫代氨基甲酸盐[203-205]、双硫腙[206-208]等改性试剂，提高汞形态在固相与有机相中的保留或分配，以提高富集倍数与回收率。此外，在固相萃取中，也可通过共价键合巯基[209]或直接使用含硫固定相[210]以提高其对元素形态的富集能力。以上离线富集方法已广泛应用于汞、硒、砷、锡等元素的形态分析。特别地，采用基于 Triton X-114 浊点萃取方法，可实现 Ag_2S、Ag、ZnS、ZnO 等纳米颗粒的同时萃取，而加入双(对磺酰苯基)苯基膦可选择性溶解 Ag 与 ZnO 纳米颗粒，而保持 Ag_2S 和 ZnS 纳米颗粒大小和形貌不变，经 LC-ICP-MS 测定，可实现 ng/L 量级 Ag_2S 和 ZnS 纳米颗粒的选择性分析[211]。

表 14-2 液相色谱-原子光谱/质谱联用的离线富集技术

元素形态	离线富集模式	参考文献
无机汞与有机汞	修饰剂改性 C_{18} 固相萃取	[203,204,206]
	磁分散固相萃取	[208,212-214]
	巯基石墨烯分散固相萃取	[209]
	MoS_2 纳米片分散固相萃取	[210]
	分散液液微萃取	[215-220]
	浊点萃取	[205,221,222]
	中空纤维液相微萃取	[207,223,224]
无机硒与硒代氨基酸	磁分散固相萃取	[225]
	芯片磁固相萃取	[226]
	阴离子交换固相萃取	[227]
	磁分散固相萃取+中空纤维液相微萃取	[228]
	中空纤维液相微萃取	[223]
无机砷与甲基砷	阴离子交换固相萃取	[229]
	MnO_2 固相萃取	[230]
	管内中空纤维固相微萃取	[231]
苯基砷化合物	搅拌棒吸附萃取	[232]
	分散液液微萃取	[233]
	中空纤维液液微萃取	[234]
有机锡	固相萃取	[235,236]
	搅拌棒吸附萃取	[237]
	磁分子印迹聚合物分散固相萃取	[238]
	固相微萃取	[239]
Ag_2S 与 ZnS 纳米颗粒	浊点萃取	[211]

液相色谱-原子光谱/质谱联用的在线富集技术见表 14-3 所示。在线富集技术以在线固相萃取为主，其可较为方便地实现自动化。可将液相色谱进样系统的定量环更换为固相萃取小柱，实现大体积的上样富集。上样体积从 1 mL 以下至 200 mL 不等，洗脱液可全部或部分地引入分离检测系统，富集倍数可达几十倍至上千倍。根据分析物的特性，进行商品化或自制固相萃取小柱的选择。对于阴离子，如无机砷、无机硒、六价铬，多采用基于阴离子交换机制的在线固相萃取。对于疏水性物种，如含硒多肽、维生素 B_{12} 等，可直接采用基于疏水作用的 C_{18} 在线固相萃取模式。对于无机汞或有机汞，其富集机制形式多样，固相萃取柱可采用阳离子交换固定相，也可将固定相（C_{18}、阴离子交换树脂、硅球、多壁碳纳米管、氧化石墨烯等）以化学吸附或键合的方式修饰上含硫基团，从而具备吸附无机汞或有机汞的能力。除此之外，也有一些报道将液相色谱的分离柱同时作为固相富集柱，实现柱上富集（on-column preconcentration）。在柱上富集过程中，载流驱动样品溶液经过分离柱，样品中的分析物（待测形态）在柱头位置保留并富集，更换流动相后进行分析物的洗脱、分离与检测[240-243]。例如，以 IC-Pak A HC 阴离子交换柱作为富集柱与分离柱，上样 5 mL 样品后，以 10 mmol/L NH_4NO_3 为流动相，可实现溴离子与溴酸根的富集、分离与检测[243]；以多孔石墨碳为富集柱与分离柱，以七氟丁酸为样品载流上样（1 mL），240 mmol/L 甲酸（含 1%甲醇，pH 2.6）为流动相，可实现无机硒与硒代氨基酸的柱上富集、分离与检测[240]。

表 14-3 液相色谱-原子光谱/质谱联用的在线富集技术

元素形态	在线富集模式	进样量（mL）	富集倍数	参考文献
	二乙基二硫代氨基甲酸盐修饰 C_{18} 在线固相萃取	—	—	[244]
	在线芯片磁固相微萃取	0.1	~10	[245]
	吡咯烷二硫代氨基甲酸钠修饰 C_{18} 在线固相萃取	5 或 10	—	[246]
		58.5	751~953	[247]
	吡咯烷二硫代氨基甲酸铵修饰 C_{18} 在线固相萃取	7	50	[248]
		100	160~400	[249]
无机汞与有机汞	3-巯基-1-丙烷磺酸钠修饰阴离子交换在线固相萃取	6	1025~1108	[250]
	聚 L-蛋氨酸修饰多壁碳纳米管在线固相萃取	20	190	[251]
	巯基功能化硅球在线固相萃取	5	830~916	[252]
		5	50	[253]
	巯基与硫脲功能化硅球在线固相萃取	10~200	—	[254-256]
	阳离子交换在线固相萃取	30	1250	[257]
	4-苯基-3-氨基-硫脲修饰氧化石墨烯在线固相萃取	10	1794~1963	[258]
无机砷与甲基砷	基于膦改性聚合物微球的在线固相萃取	25×3（连续进样 3 次）	28~30	[259]
Se(Ⅵ)、Se(Ⅳ)	阴离子交换在线固相萃取	10-100	—	[260]
含硒多肽	C_{18} 在线固相萃取	—	100	[261]
Cr(Ⅵ)、Cr(Ⅲ)	N,N-双(2-氨基乙基)乙烷-1,2-二胺功能化聚合物在线固相萃取	30×3	105、128	[262]
Cr(Ⅵ)	四丁基溴化铵修饰 C_{18} 在线固相萃取	2	20	[263]
维生素 B_{12}	C_{18} 在线固相微萃取	—	—	[264]

14.3.5 液相色谱-原子光谱/质谱联用中的多元素同时检测

采用原子光谱/质谱作为液相色谱的检测器进行多元素同时检测，不仅可加快分析速度、减少试剂耗费，还可通过元素比例信息推测相应的元素形态信息。目前，关于原子荧光与原子吸收光谱作为液相色谱的检测器进行多元素同时检测的报道较少。可通过柱后串联双原子荧光检测器[265]或采用双通道原子荧光检测器[266]实现多种元素形态的同时分析。ICP-AES[267,268]与ICP-MS[269]在多元素同时检测方面具有与生俱来的优势。目前，采用原子光谱/质谱作为液相色谱的检测器进行多元素同时检测仍以ICP-MS为主。液相色谱-ICP-MS可用于多种元素无机态或小分子形态的同时分析，如As/Se[270-272]、As/Cr[273]、As/Cr/Sb[274-276]、As/Cr/Se[277]、As/Se/Sb/Te[278]、Hg/As[279]、Hg/Pb[280]、Hg/Se[281]、Hg/As/Pb[282]、Br/I[272]等。同时，对于金属结合蛋白，多元素同时分析提供了更加丰富的蛋白与金属结合的信息[283-289]。特别地，对于蛋白及金属纳米颗粒，多元素同时检测可提供分析物的元素组成信息。对植物蛋白的分析表明，植物的汞、硒共暴露后，可检出含有S、Se、Hg信号的共流出，提示汞与蛋白上的巯基或硒氢基的结合可能与硒-汞拮抗相关[290]。对于量子点（同时检测Cd/Te/S[291]、Cd/Zn[292,293]、Cd/Se/S[294]）和其他纳米颗粒（同时检测Ag/S[295]、Au/Ag[295]），多元素检测可提供纳米颗粒组成信息，并指示纳米颗粒的溶解等转化过程。同时应该注意到，多元素同时检测虽然具有多项优势，但无机态或小分子形态的多元素同时分析也对多种元素形态的同时提取、净化以及分离提出了挑战。

14.4　毛细管电泳-电感耦合等离子体质谱联用技术

14.4.1　CE-ICP-MS联用技术的接口

CE-ICP-MS联用技术结合了两者的优点，具有分离效率高、分离快速、灵敏度高、样品和缓冲盐需求量小等优势，已被广泛应用于金属元素形态分析、金属蛋白分析、金属纳米颗粒检测、金属和化合物相互作用研究等方面。与GC、LC等分离技术不同，CE不能与ICP-MS直接联用，接口的设计是CE-ICP-MS联用技术成功的关键因素。

14.4.1.1　CE-ICP-MS联用接口的需求

在CE-ICP-MS联用系统中，接口需要满足以下几个条件。首先需要解决CE进样速率与ICP-MS雾化器流速的匹配问题。CE的进样速率很低（通常为0.1~1 μL/min），而ICP-MS常用的标准同心雾化器的样品进样速率为100~1000 μL/min，CE与ICP-MS的进样速率不匹配，在毛细管中容易产生负压和吸力，进而产生层流，降低CE的分离效率，产生峰展宽。因此联用系统的接口需要对两者的流速进行匹配，减少负压，保证CE的分离效率。其次，CE的进样量很少（<100 nL），ICP-MS常用的雾化器的样品雾化效率较低，且样品容易被外加的鞘流液高度稀释，导致分析的灵敏度降低，接口连接部分要保证较高的样品进样效率和雾化效率。再次，由于毛细管末端接口处的死体积，样品容易发生扩散，降低了CE的分离效率，因此接口设计要尽量减少系统的死体积。最后，要保持CE毛细管内的直流电。由于毛细管的末端直接与ICP-MS进样系统连接，无法与电极连接，因此需要在接口处设计导电连接元件，保证毛细管末端有持续稳定的电连接[296]。

14.4.1.2 CE-ICP-MS 接口的主要类型

目前，CE-ICP-MS 联用技术主要有 3 种接口：鞘流液接口、无鞘液接口和氢化物发生接口[297]。

（1）无鞘液接口。无鞘液接口技术将毛细管直接与雾化器连接，Deng 等利用铂金属丝缠绕在 CE 毛细管的末端实现电路的导通，将毛细管直接连接到雾化器中，毛细管的末端固定到距离雾化器出口 1 mm 的位置（如图 14-1 所示）[298]。无鞘液接口技术的优势是结构简单，不需要引入鞘流液，但是无法实现 CE 与 ICP 流速的匹配，很难消除雾化器吸力作用下造成的毛细管层流和峰展宽，因此应用非常有限。

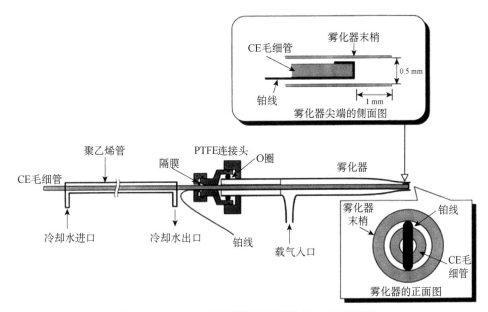

图 14-1　CE-ICP 联用系统的无鞘液接口示意图[298]

（2）氢化物发生接口。氢化物发生接口技术是将 CE 毛细管与氢化物发生器连接，CE 电泳液被氢化和衍生化后传输到 ICP 中。Magnuson 等最早设计了 CE-ICP-MS 联用系统的氢化物发生接口，CE 电泳液分别与补充液、HCl 和 NaBH$_4$ 混合，最后进入到气液分离器中并被载气传输到 ICP-MS 中进行检测[299]。Tian 等发展了一种 CE-ICP-AES 联用系统的可移动的还原床氢化物发生器接口，该接口操作简单，不需要使用气液分离器就可以实现微量体积样品的氢化物发生和传输[图 14-2（a）][300]。随后，氢化物发生接口还被应用于 CE 与 ICP-MS[图 14-2（b）][301]、ICP-OES[图 14-2（c）][302] 和 ICP-AES[图 14-2（d）][303]等联用系统的接口，但是这种接口只适用于几种元素的检测，应用范围较小。

（3）鞘流液接口。鞘流液接口技术是最常用的 CE-ICP-MS 联用系统的接口。鞘液接口技术是通过外加一路鞘流液，对 CE 毛细管的流速进行补充，使得整体流速与 ICP-MS 雾化器的样品引入流速相匹配，消除雾化器自喷对毛细管产生的吸力和层流。典型的鞘流液接口一般是采用一个 T 形三通管或十字形的四通管将毛细管、铂电极、鞘流液和雾化器连接起来，如图 14-3 所示。

利用鞘流液接口连接的 CE-ICP-MS 系统如图 14-4 所示，毛细管从 T 形或十字形的接口穿过，连接在 ICP-MS 雾化器的样品入口处；在垂直方向上通过自吸[305]、注射泵[306]或者蠕动泵[307]注入一路鞘流液，消除雾化器自喷雾所产生的负压和在毛细管产生的吸力，避免层流的产生；将铂电极接地，鞘流液在接口处与毛细管出口端混合时，可以保证毛细管末端与接口的电连接。

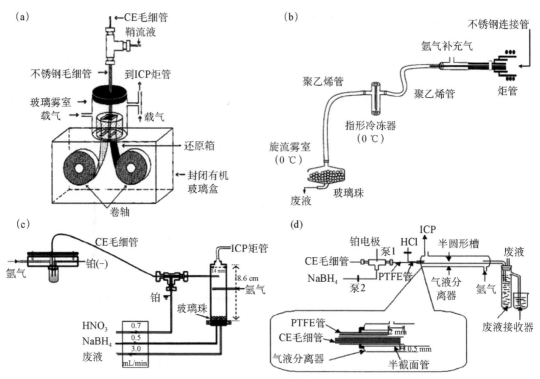

图 14-2 CE 与 ICP 联用系统的氢化物发生接口示意图

(a) CE-ICP-MS 系统可移动的还原床氢化物发生器接口[300];(b) CE-ICP-MS 联用系统氢化物发生接口[301];
(c) CE-ICP-OES 氢化物发生接口[302];(d) CE-ICP-AES 氢化物发生接口[303]

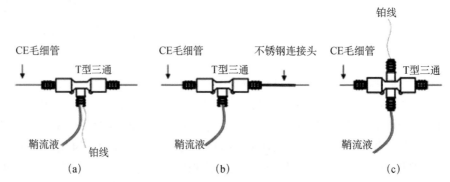

图 14-3 CE-ICP-MS 联用技术常见接口设计[304]

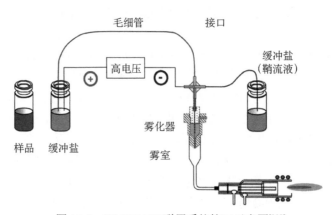

图 14-4 CE-ICP-MS 联用系统接口示意图[308]

14.4.2 雾化器和雾室

在鞘液接口中，CE 电泳液与鞘流液混合并传输到雾化器中进行雾化，同时平衡雾化器的负压产生的层流；另一方面，鞘流液的引入会引起样品的稀释，降低检测的灵敏度。因此，雾化器的雾化效率和自吸速率在很大程度上决定了联用系统的分析性能。发展高效和低自吸速率的雾化器是 CE-ICP 接口的关键。目前，常用的商业化雾化器一般有微同心雾化器（microconcentric nebulizer，MCN）[309-311]、微量同心雾化器（MicroMist nebulizer，MMN）[312-315]、高效雾化器（high efficiency nebulizer，HEN）[316-318]、超声雾化器[116]、Burgener Mira Mist 雾化器[319-321]、交叉雾化器[322]、流动聚焦雾化器[323]等（图 14-5）。

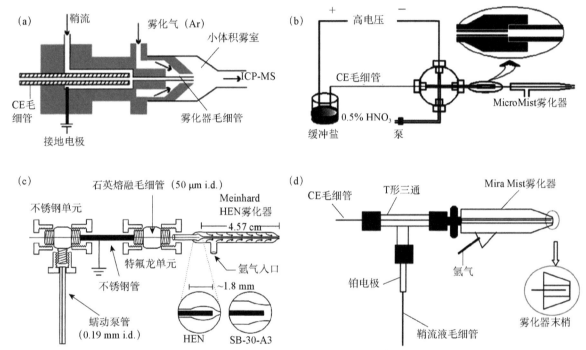

图 14-5　典型的 CE-ICP-MS 联用系统的鞘流液接口示意图

（a）CETAC CEI-100[324]；（b）MicroMist 雾化器接口[326]；（c）高效雾化器（HEN）[316]；（d）Burgener Mira Mist 雾化器[320]

CETAC 公司的同心雾化器 CEI-100 型作为 CE-ICP-MS 的商用接口[结构图如图 14-5（a）所示][324,325]，不仅具有雾化器的功能，同时具备了一个非常小体积（5 mL）的雾室的功能，死体积较小；并且在自吸模式下鞘流液的流速可以达到 2～12 μL/min，能有效消除毛细管内的层流且不会引起样品的过度稀释，广泛应用于 CE-ICP-MS 系统的连接。MicroMist 雾化器是最常用的雾化器，通过与 T 形三通管或十字形的四通管的结合，可以实现 CE-ICP-MS 的连接。除此之外，还可以通过在 CE 毛细管的入口端增加一个外加的负压来抵消雾化器产生的吸力，同时可以使用较低流速的鞘流液或者不使用鞘流液，以减少对样品的稀释，提高灵敏度。

Yang 等[311]利用微量雾化器设计了一种基于鞘液的 CE-ICP-MS 接口，由一根不锈钢毛细管（600 μm 内径，800 μm 外径）、聚乙烯管、两个蠕动泵、一个脉冲控制和一个 PEEK 三通组件组成[图 14-6（a）]。CE 毛细管的末端直接插入不锈钢毛细管中，不锈钢毛细管与蠕动泵 1 和 CE 电源的阴极连接，实现 CE 毛细管的电流回路。CE 毛细管的流出液首先与蠕动泵 1 的补充液混合，然后在 PEEK 三通处被蠕动泵 2 的超纯水传输到雾化器和 ICP-MS 中。在整个分离过程中，蠕动泵 1 将 CE 毛细管与雾化器隔离，避免了自吸效应引起的层流。随后，作者对接口进行了进一步的优化，将不锈钢的毛细管放置在一个聚乙烯管中，CE 毛细管从聚乙烯管中穿过进入不锈钢管中，聚乙烯管的另一端放在超纯水中

[图14-6（b）]。在检测过程中，不锈钢毛细管中一直充满补充液，保持稳定的电连接。该接口有效抑制了CE毛细管的层流，可以对分离时间在20 s以上的不同形态化合物进行检测分析。

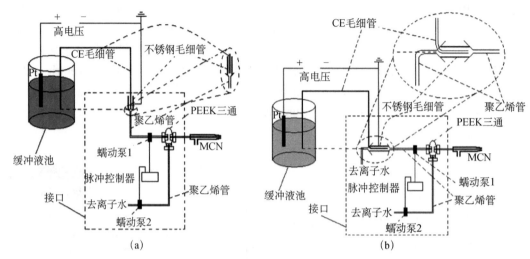

图14-6　基于补充液方式的CE-ICP-MS接口示意图（a）[311]和改进后的接口（b）[307]

除了使用商品化的雾化器之外，CE-ICP-MS鞘流液接口还有使用喷雾装置改装设计的接口[327]。Liu等将商业化的喷雾针组件进行改装后，作为连接CE与ICP-MS的接口（如图14-7所示）。喷雾针接口采用不锈钢双层同轴设计，内层为不锈钢毛细管，外层为不锈钢管。该喷雾针作为ICP-MS的雾化器被直接安装在雾室基座上。毛细管柱从喷雾针的内层不锈钢管穿过，毛细管的末端在喷雾针头处向外伸出0.1 mm，内层不锈钢管与CE毛细管的间隙中加入鞘流液，外层不锈钢管中加入载气。不锈钢喷雾针的外壳接地，确保毛细管两端的闭合电流回路。由于毛细管的流速极低，难以将被分析物有效雾化，在喷雾针的内层不锈钢管中引入低速、稳定的鞘流液，确保样品的有效雾化，同时润湿毛细管的出口端以实现毛细管两端的电流回路。通过使用该接口，喷雾针替代了商品化的MicroMist雾化器，化合物经过毛细管分离后，不需要经过传输直接被载气雾化，消除了毛细管末端与雾化器之间的死体积，因此保持了CE的高分离效率。由于毛细管的末端直接进入雾室，化合物被雾化后被全部传输到ICP-MS中进行检测，提高了进样效率。

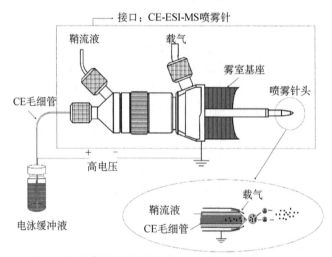

图14-7　由喷雾针组件改装后的CE-ICP-MS接口示意图[327]

此外，实验室自制的微雾化器也被应用于CE与ICP-MS的接口，包括直接进样高效雾化器（direct injection high-efficiency nebulizer，DIHEN）[328-330]、高效交叉流动微量雾化器（high-efficiency cross-flow micro nebulizer，HECFMN）[331]、可拆卸毛细管微流雾化器（demountable capillary microflow nebulizer，d-CMN）[107,332]以及基于芯片的交叉流动雾化器[333]等（参见图14-8）。Bendahl等[328]采用了一种直接进样高效雾化器作为CE-ICP-MS的接口[图14-8（a）]。该接口由PEEK接口、高硼硅玻璃管、高硼硅玻璃样品毛细管（外径1.4 mm，内径0.95 mm）、铂电极等组成，CE毛细管分别从PEEK接口和样品毛细管中穿过，鞘流液的注射针头和铂电极在同一个管路，同时起到鞘流液和接地导电的作用。通过调节，样品毛细管在距离喷嘴0.5 mm的位置，可以使雾化器保持在一个恒定的4～5 bar的气体背景压力下，实现雾化器鞘流液的自吸效应；CE毛细管在超过喷嘴0.1 mm的位置，可以使雾化后的气溶胶完全进入ICP-MS炬管中心。该雾化器省去了雾室，将样品直接传输到ICP炬管的位置，与传统雾化器和雾室相比，降低了样品传输过程中的扩散和碰撞损失，提高了样品的雾化和传输效率。

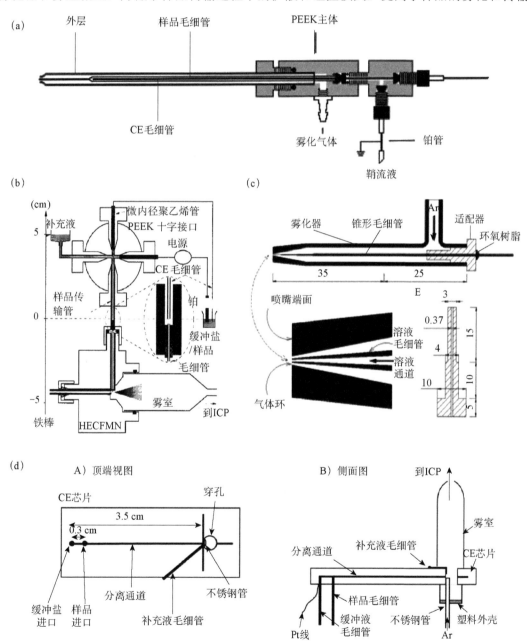

图14-8 CE-ICP-MS系统的接口示意图

（a）DIHEN[328]；（b）HECFMN[331]；（c）d-CMN[107]；（d）基于芯片的交叉流动雾化器[333]

在 CE-ICP-MS 联用系统中，雾室是影响 CE 分离效率的另一重要因素，雾室的体积、温度等因素均会影响电泳峰的展宽和拖尾程度[324]。研究发现，圆锥形的雾室（体积 35~60 mL）的样品传输效率比传统双通道 Scott-type 雾室（体积 90~150 mL）的雾化效率要高[334]。Tangen 等通过比较 Scott-type 雾室和直接进样（省略雾室）的进样方式发现，雾室会引起显著的峰高降低和峰展宽，同时由于传输距离的增加，使用双通道雾室时的最大峰高值出峰时间延迟 3.7 s[335]。Liu 等设计了直接进样的全耗型雾室，该接口中，喷雾针、雾室与炬管的中心线在同一条直线上（图 14-9），最大限度地降低了系统的死体积，显著减少了样品传输距离。化合物被载气雾化后，在该接口的中心线上被直线传输到炬管中进行离子化，减少了样品碰撞损失，降低了样品扩散，提高了样品传输效率，从而提高了检测的灵敏度[336]。通过比较发现，使用全耗型雾室作为接口时，砷化合物的峰高比使用标准双通道雾室时显著增高，不同化合物的检出限降低 2~4 倍，分离度提高 20%。另外，小体积的雾室可以降低冲洗时间，最终会改善峰展宽和拖尾的现象。除了雾室的体积外，雾化室的温度也会影响样品的传输效率，通常情况下，高温能够促进气溶胶的去溶剂化效率，进而提高样品的传输效率[332]。常用的冷却式雾室温度是 2℃，能够使样品气溶胶更均匀，减少水蒸气向等离子体的传输，传输效率一般是 1%~2%；而通过采用加热的单通道雾室（90℃），样品传输效率可以高达 100%[337]。

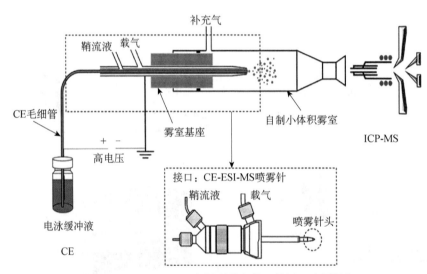

图 14-9 CE-ICP-MS 全耗型雾室接口装置示意图[336]

毛细管电泳是一种非常高效的分离方式，但是由于其检测灵敏度较低，限制了 CE 的应用。通过设计高效的接口，搭建 CE-ICP-MS 联用系统，能显著提高方法检测的灵敏度，该联用系统已成为元素形态分析非常重要的工具。但是与 HPLC 相比，CE 的进样量极低、检出限高、接口复杂、样品传输过程损失大等特点使其应用范围受到一定的限制。因此，未来 CE-ICP-MS 联用技术的发展趋势将主要集中在提高接口效率、提高 CE 富集能力、拓宽联用系统的应用研究等方面。首先，CE-ICP-MS 的接口影响了联用系统检测的灵敏度和稳定性。目前商业化的 CE-ICP-MS 接口比较少，仍有改进和优化的空间，开发进样效率高、死体积小、样品传输效率高、操作简单的一体化接口将能提高 CE-ICP-MS 联用系统的稳定性和可靠性。其次，CE 的进样体积很低（纳升级），开发样品的在线富集方法将有利于 CE 灵敏度的提高。通过简单高效的接口，结合 CE 高效分离和 ICP-MS 灵敏检测的技术优势，将为元素多维形态分析和相关研究提供强有力的技术支持。

14.5 展　　望

色谱-原子光谱/质谱联用技术为金属与类金属形态分析技术的发展提供了坚实的基础，但由于人们对元素在环境和生物体中的存在形式以及对其性质了解甚少，可供利用的标准物很少，使得色谱-原子光谱/质谱联用技术在元素的全形态分析中的应用受到了限制。高灵敏的有机质谱分析方法，在不需要标准物的条件下就能够确定复杂基质中未知形态元素的组成和化合物结构，可帮助人们获取那些尚不了解但在环境和生命活动中起重要作用的元素形态。ICP-MS 是一种高选择的元素特异性检测器，可以预见，未来色谱-ICP-MS 和色谱-有机质谱联用技术将在化学与生物转化过程的未知元素形态筛选上发挥重要作用。此外，联用仪器的小型化是今后有害元素分析的一个迫切需求，也是进行现场监测与流行病学调查所必需的。

参 考 文 献

[1] Kolb B, Kemmner G, Schleser F H, Wiedekin E. Elementspezifische anzeige gaschromatographisch getrennter metallverbindungen mittels atom-absorptions-spektroskopie (AAS). Zeitschrift Fur Analytische Chemie Fresenius, 1966, 221 (SEP): 166-175.

[2] Chau Y K, Wong P T S, Saitoh H. Determination of tetraalkyl lead compounds in the atmosphere. Journal of Chromatographic Science, 1976, 14(3): 162-164.

[3] Chau Y K, Wong P T S, Bengert G A. Determination of methyltin (Ⅳ) and tin (Ⅳ) species in water by gas chromatography/atomic absorption spectrophotometry. Analytical Chemistry, 1982, 54(2): 246-249.

[4] Ahmad I, Chau Y K, Wong P T S, Carty A J, Taylor L. Chemical alkylation of lead (Ⅱ) salts to tetraalkyllead (Ⅳ) in aqueous solution. Nature, 1980, 287(5784): 716-717.

[5] Chau Y K, Wong P T S, Kramar O. The determination of dialkyllead, trialkyllead, tetraalkyllead and lead (Ⅱ) ions in water by chelation/extraction and gas chromatography/atomic absorption spectrometry. Analytica Chimica Acta, 1983, 146(FEB): 211-217.

[6] Jiang G B, Ni Z M, Wang S R, Han H B. Determination of organomercurials in air by gas chromatography-atomic absorption spectrometry. Journal of Analytical Atomic Spectrometry, 1989, 4(4): 315-318.

[7] Jiang G B, Ni Z M, Wang S R, Han H B. Organic mercury speciation in fish by capillary gas chromatography interfaced with atomic absorption spectrometry. Fresenius Zeitschrift Fur Analytische Chemie, 1989, 334(1): 27-30.

[8] 徐福正, 江桂斌, 韩恒斌. 气相色谱与原子吸收联用及其在有机锡化合物形态分析中的应用. 分析化学, 1995, 23(11): 1308-1312.

[9] Matousek T, Hernandez-Zavala A, Svoboda M, Langrova L, Adair B M, Drobna Z, Thomas D J, Styblo M, Dedina J. Oxidation state specific generation of arsines from methylated arsenicals based on L-cysteine treatment in buffered media for speciation analysis by hydride generation-automated cryotrapping-gas chromatography-atomic absorption spectrometry with the multiatomizer. Spectrochimica Acta Part B: Atomic Spectroscopy, 2008, 63(3): 396-406.

[10] Feng Y L, Narasaki H. Speciation of organotin compounds in marine sediments by capillary column gas chromatography-

atomic absorption spectrometry coupled with hydride generation. Analytical and Bioanalytical Chemistry, 2002, 372(2): 382-386.

[11] Welz B, Becker-Ross H, Florek S, Heitmann U, Vale M G R. High-resolution continuum-source atomic absorption spectrometry: What can we expect? Journal of the Brazilian Chemical Society, 2003, 14(2): 220-229.

[12] Van D N, Radziuk B, Frech W. A comparison between continuum- and line source AAS for speciation analysis of butyl- and phenyltin compounds. Journal of Analytical Atomic Spectrometry, 2006, 21(7): 708-711.

[13] Winefordner J D, Elser R C. Atomic fluorescence spectrometry. Analytical Chemistry, 1971, 43(4): 24A-42A.

[14] West C D. Relative effect of molecular absorption on atomic absorption and atomic fluorescence. Analytical Chemistry, 1974, 46(6): 797-799.

[15] Thompson K C, Reynolds G D. The atomic-fluorescence determination of mercury by the cold vapour technique. Analyst, 1971, 96(1148): 771-775.

[16] Van Loon J C, Lichwa J, Radziuk B. Non-dispersive atomic fluorescence spectroscopy, a new detector for chromatography. Journal of Chromatography, 1977, 136(2): 301-305.

[17] Dulivo A, Papoff P. Simultaneous detection of alkylselenide, alkyllead and alkyltin compounds by gas chromatography using a multi-channel non-dispersive atomic fluorescence spectrometric detector and a miniature flame as the atomiser. Journal of Analytical Atomic Spectrometry, 1986, 1(6): 479-484.

[18] Armstrong H L, Corns W T, Stockwell P B, O'Connor G, Ebdon L, Evans E H. Comparison of AFS and ICP-MS detection coupled with gas chromatography for the determination of methylmercury in marine samples. Analytica Chimica Acta, 1999, 390(1-3): 245-253.

[19] Cai Y, Jaffe R, Alli A, Jones R D. Determination of organomercury compounds in aqueous samples by capillary gas chromatography-atomic fluorescence spectrometry following solid-phase extraction. Analytica Chimica Acta, 1996, 334(3): 251-259.

[20] Cai Y, Jaffe R, Jones R. Ethylmercury in the soils and sediments of the Florida Everglades. Environmental Science & Technology, 1997, 31(1): 302-305.

[21] Hurley J P, Benoit J M, Babiarz C L, Shafer M M, Andren A W, Sullivan J R, Hammond R, Webb D A. Influences of watershed characteristics on mercury levels in Wisconsin rivers. Environmental Science & Technology, 1995, 29(7): 1867-1875.

[22] Cai Y, Monsalud S, Jaffe R, Jones R D. Gas chromatographic determination of organomercury following aqueous derivatization with sodium tetraethylborate and sodium tetraphenylborate: Comparative study of gas chromatography coupled with atomic fluorescence spectrometry, atomic emission spectrometry and mass spectrometry. Journal of Chromatography A, 2000, 876(1-2): 147-155.

[23] Shi J-B, Jiang G-B. Application of gas chromatography-atomic fluorescence spectrometry hyphenated system for speciation of butyltin compounds in water samples. Spectroscopy Letters, 2011, 44(6): 393-398.

[24] 史建波, 廖春阳, 王亚伟, 江桂斌. 气相色谱和原子荧光联用测定生物和沉积物样品中甲基汞. 光谱学与光谱分析, 2006, 26(2): 336-339.

[25] Gorecki J, Diez S, Macherzynski M, Kalisinska E, Golas J. Improvements and application of a modified gas chromatography atomic fluorescence spectroscopy method for routine determination of methylmercury in biota samples. Talanta, 2013, 115: 675-680.

[26] Gomez-Ariza J L, Lorenzo F, Garcia-Barrera T. Guidelines for routine mercury speciation analysis in seafood by gas chromatography coupled to a home-modified AFS detector. Application to the Andalusian coast (south Spain). Chemosphere, 2005, 61(10): 1401-1409.

[27] McCormack A J, Tong S C, Cooke W D. Sensitive selective gas chromatography detector based on emission spectrometry of organic compounds. Analytical Chemistry, 1965, 37(12): 1470-1476.

[28] Bache C A, Lisk D J. Determination of organophosphorus insecticide residues using the emission spectrometric detector. Analytical Chemistry, 1965, 37(12): 1477-1480.

[29] Bache C A, Lisk D J. Gas chromatographic determination of organic mercury compounds by emission spectrometry in a helium plasma. Application to the analysis of methylmercuric salts in fish. Analytical Chemistry, 1971, 43(7): 950-952.

[30] Li C H, Long Z, Jiang X M, Wu P, Hou X D. Atomic spectrometric detectors for gas chromatography. TrAC-Trends in Analytical Chemistry, 2016, 77: 139-155.

[31] Van Loon J C, Alcock L R, Pinchin W H, French J B. Inductively coupled plasma source mass spectrometry: A new element/isotope specific mass spectrometry detector for chromatography. Spectroscopy Letters, 1986, 19(10): 1125-1135.

[32] Baxter D C, Faarinen M, Osterlund H, Rodushkin I, Christensen M. Serum/plasma methylmercury determination by isotope dilution gas chromatography-inductively coupled plasma mass spectrometry. Analytica Chimica Acta, 2011, 701(2): 134-138.

[33] Ma X, Yin Y G, Shi J B, Liu J F, Jiang G B. Species-specific isotope dilution-GC-ICP-MS for accurate and precise measurement of methylmercury in water, sediments and biological tissues. Analytical Methods, 2014, 6(1): 164-169.

[34] Lobinski R, Adams F C. Speciation analysis by gas chromatography with plasma source spectrometric detection. Spectrochimica Acta Part B: Atomic Spectroscopy, 1997, 52(13): 1865-1903.

[35] Bulska E, Baxter D C, Frech W. Capillary column gas chromatography for mercury speciation. Analytica Chimica Acta, 1991, 249(2): 545-554.

[36] Tomiyasu T, Nagano A, Sakamoto H, Yonehara N. Differential determination of organic mercury and inorganic mercury in sediment, soil and aquatic organisms by cold-vapor atomic absorption spectrometry. Analytical Sciences, 1996, 12(3): 477-481.

[37] Medina I, Rubi E, Mejuto M C, Cela R. Speciation of organomercurials in marine samples using capillary electrophoresis. Talanta, 1993, 40(11): 1631-1636.

[38] 庞秀言, 梁淑轩. 气相色谱-原子吸收联用技术测定人体体液中烷基汞. 色谱, 1997, 15(2): 130-132.

[39] Tseng C M, de Diego A, Pinaly H, Amouraoux D, Donard O F X. Cryofocusing coupled to atomic absorption spectrometry for rapid and simple mercury speciation in environmental matrices. Journal of Analytical Atomic Spectrometry, 1998, 13(8): 755-764.

[40] Fischer R, Rapsomanikis S, Andreae M O. Determination of methylmercury in fish samples using GC/AA and sodium tetraethylborate derivatization. Analytical Chemistry, 1993, 65(6): 763-766.

[41] Ashby J, Clark S, Craig P J. Methods for the production of volatile organometallic derivatives for application to the analysis of environmental samples. Journal of Analytical Atomic Spectrometry, 1988, 3(5): 735-736.

[42] Tseng C M, deDiego A, Martin F M, Donard O F X. Rapid and quantitative microwave-assisted recovery of methylmercury from standard reference sediments. Journal of Analytical Atomic Spectrometry, 1997, 12(6): 629-635.

[43] Rapsomanikis S, Craig P J. Speciation of mercury and methylmercury compounds in aqueous samples by chromatography atomic-absorption spectrometry after ethylation with sodium tetraethylborate. Analytica Chimica Acta, 1991, 248(2): 563-567.

[44] Gerbersmann C, Heisterkamp M, Adams F C, Broekaert J A C. Two methods for the speciation analysis of mercury in fish involving microwave-assisted digestion and gas chromatography atomic emission spectrometry. Analytica Chimica Acta, 1997, 350(3): 273-285.

[45] Bulska E, Emteborg H, Baxter D C, Frech W, Ellingsen D, Thomassen Y. Speciation of mercury in human whole blood

by capillary gas chromatography with a microwave-induced plasma emission detector system following complexometric extraction and butylation. Analyst, 1992, 117(3): 657-663.

[46] Emteborg H, Baxter D C, Frech W. Speciation of mercury in natural waters by capillary gas chromatography with a microwave-induced plasma emission detector following preconcentration using a dithiocarbamate resin microcolumn installed in a closed flow injection system. Analyst, 1993, 118(8): 1007-1013.

[47] Craig P J, Mennie D, Ostah N, Donard O F X, Martin F. Novel methods for derivatization of mercury (II) and methylmercury (II) compounds for analysis. Analyst, 1992, 117(4): 823-824.

[48] Tseng C M, DeDiego A, Martin F M, Amouroux D, Donard O F X. Rapid determination of inorganic mercury and methylmercury in biological reference materials by hydride generation, cryofocusing, atomic absorption spectrometry after open focused microwave-assisted alkaline digestion. Journal of Analytical Atomic Spectrometry, 1997, 12(7): 743-750.

[49] Weber J H. Speciation of methylarsenic, methyl- and butyltin, and methylmercury compounds and their inorganic analogues by hydride derivatization. TrAC Trends in Analytical Chemistry, 1997, 16(2): 73-78.

[50] Puk R, Weber J H. Determination of mercury (II), monomethylmercury cation, dimethylmercury and diethylmercury by hydride generation, cryogenic trapping and atomic absorption spectrometric detection. Analytica Chimica Acta, 1994, 292(1-2): 175-183.

[51] Craig P J, Garraud H, Laurie S H, Mennie D, Stojak G H. Nuclear magnetic resonance and mass spectra of organomercury hydrides and deuterides, part II. Journal of Organometallic Chemistry, 1994, 468(1-2): 7-11.

[52] Krystek P, Favaro P, Bode P, Ritsema R. Methyl mercury in nail clippings in relation to fish consumption analysis with gas chromatography coupled to inductively coupled plasma mass spectrometry: A first orientation. Talanta, 2012, 97: 83-86.

[53] Lee Y H, Mowrer J. Determination of methylmercury in natural waters at the sub-nanograms per litre level by capillary gas chromatography after adsorbent preconcentration. Analytica Chimica Acta, 1989, 221(2): 259-268.

[54] Lansens P, Meuleman C, Leermakers M, Baeyens W. Determination of methylmercury in natural waters by headspace gas chromatography with microwave-induced plasma detection after preconcentration on a resin containing dithiocarbamate groups. Analytica Chimica Acta, 1990, 234(2): 417-424.

[55] Emteborg H, Baxter D C, Sharp M, Frech W. Evaluation, mechanism and application of solid-phase extraction using a dithiocarbamate resin for the sampling and determination of mercury species in humic-rich natural waters. Analyst, 1995, 120(1): 69-77.

[56] He B, Jiang G B. Analysis of organomercuric species in soils from orchards and wheat fields by capillary gas chromatography on-line coupled with atomic absorption spectrometry after in situ hydride generation and headspace solid phase microextraction. Fresenius' Journal of Analytical Chemistry, 1999, 365(7): 615-618.

[57] He B, Jiang G B, Ni Z M. Determination of methylmercury in biological samples and sediments by capillary gas chromatography coupled with atomic absorption spectrometry after hydride derivatization and solid phase microextraction. Journal of Analytical Atomic Spectrometry, 1998, 13(10): 1141-1144.

[58] Cai Y, Monsalud S, Furton K G, Jaffe R, Jones R D. Determination of methylmercury in fish and aqueous samples using solid-phase microextraction followed by gas chromatography atomic fluorescence spectrometry. Applied Organometallic Chemistry, 1998, 12(8-9): 565-569.

[59] Horvat M, Liang L, Bloom N S. Comparison of distillation with other current isolation methods for the determination of methyl mercury compounds in low level environmental samples: Part 2 water. Analytica Chimica Acta, 1993, 282(1): 153-168.

[60] Liang L, Horvat M, Bloom N S. An improved speciation method for mercury by GC/CVAFS after aqueous phase ethylation and room temperature precollection. Talanta, 1994, 41(3): 371-379.

[61] Pereiro I R, Wasik A, Lobinski R. Purge-and-trap isothermal multicapillary gas chromatographic sample introduction accessory for speciation of mercury by microwave-induced plasma atomic emission spectrometry. Analytical Chemistry, 1998, 70(19): 4063-4069.

[62] Bloom N. Determination of picogram levels of methylmercury by aqueous phase ethylation, followed by cryogenic gas chromatography with cold vapour atomic fluorescence detection. Canadian Journal of Fisheries and Aquatic Sciences, 1989, 46(7): 1131-1140.

[63] Stoichev T, Martin-Doimeadios R C R, Tessier E, Amouroux D, Donard O F X. Improvement of analytical performances for mercury speciation by on-line derivatization, cryofocussing and atomic fluorescence spectrometry. Talanta, 2004, 62(2): 433-438.

[64] Ritsema R, Donard O F X. On-line speciation of mercury and methylmercury in aqueous samples by chromatography-atomic fluorescence spectrometry after hydride generation. Applied Organometallic Chemistry, 1994, 8(7-8): 571-575.

[65] Tong W, Link A, Eng J K, Yates J R. Identification of proteins in complexes by solid phase microextraction multistep elution capillary electrophoresis tandem mass spectrometry. Analytical Chemistry, 1999, 71(13): 2270-2278.

[66] Buchholz K D, Pawliszyn J. Optimization of solid-phase microextraction conditions for determination of phenols. Analytical Chemistry, 1994, 66(1): 160-167.

[67] Cai Y, Bayona J M. Determination of methylmercury in fish and river water samples using in situ sodium tetraethylborate derivatization following by solid-phase microextraction and gas chromatography-mass spectrometry. Journal of Chromatography A, 1995, 696(1): 113-122.

[68] Moens L, DeSmaele T, Dams R, VandenBroeck P, Sandra P. Sensitive, simultaneous determination of organomercury, -lead, and -tin compounds with headspace solid phase microextraction capillary gas chromatography combined with inductively coupled plasma mass spectrometry. Analytical Chemistry, 1997, 69(8): 1604-1611.

[69] Davis W C, Vander Pol S S, Schantz M M, Long S E, Day R D, Christopher S J. An accurate and sensitive method for the determination of methylmercury in biological specimens using GC-ICP-MS with solid phase microextraction. Journal of Analytical Atomic Spectrometry, 2004, 19(12): 1546-1551.

[70] Bellar T A, Lichtenberg J J. Determining volatile organics at microgram-per litre levels by gas chromatography. Journal American Water Works Association, 1974, 66(12): 739-744.

[71] Popp M, Hann S, Koellensperger G. Environmental application of elemental speciation analysis based on liquid or gas chromatography hyphenated to inductively coupled plasma mass spectrometry: A review. Analytica Chimica Acta, 2010, 668(2): 114-129.

[72] Schaumloeffel D, Giusti P, Preud'Homme H, Szpunar J, Lobinski R. Precolumn isotope dilution analysis in nanoHPLC-ICPMS for absolute quantification of sulfur-containing peptides. Analytical Chemistry, 2007, 79(7): 2859-2868.

[73] Gong J, Solivio M J, Merino E J, Caruso J A, Landero-Figueroa J A. Developing ICP-MS/MS for the detection and determination of synthetic DNA-protein crosslink models via phosphorus and sulfur detection. Analytical and Bioanalytical Chemistry, 2015, 407(9): 2433-2437.

[74] Gronbaek-Thorsen F, Sturup S, Gammelgaard B, Moller L H. Development of a UPLC-IDA-ICP-MS/MS method for peptide quantitation in plasma by Se-labelling, and comparison to S-detection of the native peptide. Journal of Analytical Atomic Spectrometry, 2019, 34(2): 375-383.

[75] Kretschy D, Koellensperger G, Hann S. Elemental labelling combined with liquid chromatography inductively coupled plasma mass spectrometry for quantification of biomolecules: A review. Analytica Chimica Acta, 2012, 750: 98-110.

[76] Mestrot A, Uroic M K, Plantevin T, Islam M R, Krupp E M, Feldmann J, Meharg A A. Quantitative and qualitative trapping of arsines deployed to assess loss of volatile arsenic from paddy soil. Environmental Science & Technology. 2009, 43(21): 8270-8275.

[77] Winkel L, Feldmann J, Meharg A A. Quantitative and qualitative trapping of volatile methylated selenium species entrained through nitric acid. Environmental Science & Technology, 2010, 44(1): 382-387.

[78] Vriens B, Ammann A A, Hagendorfer H, Lenz M, Berg M, Winkel L H E. Quantification of methylated selenium, sulfur, and arsenic in the environment. PloS One, 2014, 9(7): e102906.

[79] Grotti M, Terol A, Todoli J L. Speciation analysis by small-bore HPLC coupled to ICP-MS. TrAC-Trends in Analytical Chemistry, 2014, 61: 92-106.

[80] Yanes E G, Miller-Ihli N J. Use of a parallel path nebulizer for capillary-based microseparation techniques coupled with an inductively coupled plasma mass spectrometer for speciation measurements. Spectrochimica Acta Part B: Atomic Spectroscopy, 2004, 59(6): 883-890.

[81] Gammelgaard B, Jons O. Comparison of an ultrasonic nebulizer with a cross-flow nebulizer for selenium speciation by ion-chromatography and inductively coupled plasma mass spectrometry. Journal of Analytical Atomic Spectrometry, 2000, 15(5): 499-505.

[82] Falter R, Wilken R D. Determination of carboplatinum and cisplatinum by interfacing HPLC with ICP-MS using ultrasonic nebulisation. Science of the Total Environment, 1999, 225(1-2): 167-176.

[83] Yang K L, Jiang S J. Determination of selenium compounds in urine samples by liquid chromatographyinductively coupled plasma mass spectrometrywith an ultrasonic nebulizer. Analytica Chimica Acta, 1995, 307(1): 109-115.

[84] Brennan R G, Rabb S A, Jorabchi K, Rutkowski W F, Turk G C. Heat-assisted argon electrospray interface for low-flow rate liquid sample introduction in plasma spectrometry. Analytical Chemistry, 2009, 81(19): 8126-8133.

[85] Paredes E, Grotti M, Mermet J M, Luis Todoli J. Heated-spray chamber-based low sample consumption system for inductively coupled plasma spectrometry. Journal of Analytical Atomic Spectrometry, 2009, 24(7): 903-910.

[86] Grotti M, Ardini F, Terol A, Magi E, Luis Todoli J. Influence of chemical species on the determination of arsenic using inductively coupled plasma mass spectrometry at a low liquid flow rate. Journal of Analytical Atomic Spectrometry, 2013, 28(11): 1718-1724.

[87] Leclercq A, Nonell A, Todoli Torro J L, Bresson C, Vio L, Vercouter T, Chartier F. Introduction of organic/hydro-organic matrices in inductively coupled plasma optical emission spectrometry and mass spectrometry: A tutorial review. Part II. Practical considerations. Analytica Chimica Acta, 2015, 885: 57-91.

[88] Marchante-Gayon J M, Thomas C, Feldmann I, Jakubowski N. Comparison of different nebulisers and chromatographic techniques for the speciation of selenium in nutritional commercial supplements by hexapole collision and reaction cell ICP-MS. Journal of Analytical Atomic Spectrometry, 2000, 15(9): 1093-1102.

[89] Polec K, Perez-Calvo M, Garcia-Arribas O, Szpunar J, Ribas-Ozonas B, Lobinski R. Investigation of metal complexes with metallothionein in rat tissues by hyphenated techniques. Journal of Inorganic Biochemistry, 2002, 88(2): 197-206.

[90] Malavolta M, Piacenza F, Basso A, Giacconi R, Costarelli L, Pierpaoli S, Mocchegiani E. Speciation of trace elements in human serum by micro anion exchange chromatography coupled with inductively coupled plasma mass spectrometry. Analytical Biochemistry, 2012, 421(1): 16-25.

[91] Ackley K L, Sutton K L, Caruso J A. A comparison of nebulizers for microbore LC-ICP-MS with mobile phases containing methanol. Journal of Analytical Atomic Spectrometry, 2000, 15(9): 1069-1073.

[92] Schaumloffel D, Encinar J R, Lobinski R. Development of a sheathless interface between reversed-phase capillary HPLC and ICPMS via a microflow total consumption nebulizer for selenopeptide mapping. Analytical Chemistry, 2003,

75(24): 6837-6842.

[93] Pergantis S A, Heithmar E M, Hinners T A. Microscale flow-injection and microbore high performance liquid chromatography coupled with inductively coupled plasma mass spectrometry via a high-efficiency nebulizer. Analytical Chemistry, 1995, 67(24): 4530-4535.

[94] Wiederin D R, Smith F G, Houk R S. Direct injection nebulization for inductively coupled plasma mass spectrometry. Analytical Chemistry, 1991, 63(3): 219-225.

[95] Shum S C K, Pang H M, Houk R S. Speciation of mercury and lead compounds by microbore column liquid chromatography-inductively coupled plasma mass spectrometry with direct injection nebulization. Analytical Chemistry, 1992, 64(20): 2444-2450.

[96] Powell M J, Boomer D W, Wiederin D R. Determination of chromium species in environmental samples using high pressure liquid chromatography direct injection nebulization and inductively coupled plasma mass spectrometry. Analytical Chemistry, 1995, 67(14): 2474-2478.

[97] Chao W S, Jiang S J. Determination of organotin compounds by liquid chromatography-inductively coupled plasma mass spectrometry with a direct injection nebulizer. Journal of Analytical Atomic Spectrometry, 1998, 13(12): 1337-1341.

[98] Sun Y C, Lee Y S, Shiah T L, Lee P L, Tseng W C, Yang M H. Comparative study on conventional and low-flow nebulizers for arsenic speciation by means of microbore liquid chromatography with inductively coupled plasma mass spectrometry. Journal of Chromatography A, 2003, 1005(1-2): 207-213.

[99] Stefanka Z, Koellensperger G, Stingeder G, Hann S. Down-scaling narrowbore LC-ICP-MS to capillary LC-ICP-MS: A comparative study of different introduction systems. Journal of Analytical Atomic Spectrometry, 2006, 21(1): 86-89.

[100] Tangen A, Trones R, Greibrokk T, Lund W. Microconcentric nebulizer for the coupling of micro liquid chromatography and capillary zone electrophoresis with inductively coupled plasma mass spectrometry. Journal of Analytical Atomic Spectrometry, 1997, 12(6): 667-670.

[101] Woller A, Garraud H, Boisson J, Dorthe A M, Fodor P, Donard O F X. Simultaneous speciation of redox species of arsenic and selenium using an anion-exchange microbore column coupled with a micro-concentric nebulizer and an inductively coupled plasma mass spectrometer as detector. Journal of Analytical Atomic Spectrometry, 1998, 13(2): 141-149.

[102] Shum S C K, Houk R S. Elmental speciation by aion-exchange and size exclusion chromatography with detection by inductively coupled plasma mass spectrometry with direct injection nebulization. Analytical Chemistry, 1993, 65(21): 2972-2976.

[103] Acon B W, McLean J A, Montaser A. A direct injection high efficiency nebulizer interface for microbore high-performance liquid chromatography-inductively coupled plasma mass spectrometry. Journal of Analytical Atomic Spectrometry, 2001, 16(8): 852-857.

[104] Wind M, Eisenmenger A, Lehmann W D. Modified direct injection high efficiency nebulizer with minimized dead volume for the analysis of biological samples by micro- and nano-LC-ICP-MS. Journal of Analytical Atomic Spectrometry, 2002, 17(1): 21-26.

[105] Westphal C S, Kahen K, Rutkowski W E, Acon B W, Montaser A. Demountable direct injection high efficiency nebulizer for inductively coupled plasma mass spectrometry. Spectrochimica Acta Part B: Atomic Spectroscopy, 2004, 59(3): 353-368.

[106] Shen L, Sun J, Cheng H, Liu J, Xu Z, Mu J. A demountable nanoflow nebulizer for sheathless interfacing nano-high performance liquid chromatography with inductively coupled plasma mass spectrometry. Journal of Analytical Atomic

Spectrometry, 2015, 30(9): 1927-1934.

[107] Cheng H, Yin X, Xu Z, Wang X, Shen H. A simple and demountable capillary microflow nebulizer with a tapered tip for inductively coupled plasma mass spectrometry. Talanta, 2011, 85(1): 794-799.

[108] B'Hymer C, Sutton K L, Caruso J A. Comparison of four nebulizer-spray chamber interfaces for the high-performance liquid chromatographic separation of arsenic compounds using inductively coupled plasma mass spectrometric detection. Journal of Analytical Atomic Spectrometry, 1998, 13(9): 855-858.

[109] Hoang T T, May S W, Browner R F. Developments with the oscillating capillary nebulizer-effects of spray chamber design, droplet size and turbulence on analytical signals and analyte transport efficiency of selected biochemically important organoselenium compounds. Journal of Analytical Atomic Spectrometry, 2002, 17(12): 1575-1581.

[110] Wang L Q, May S W, Browner R F, Pollock S H. Low-flow interface for liquid chromatography inductively coupled plasma mass spectrometry speciation using an oscillating capillary nebulizer. Journal of Analytical Atomic Spectrometry, 1996, 11(12): 1137-1146.

[111] Takasaki Y, Sakagawa S, Inagaki K, Fujii S-I, Sabarudin A, Umemura T, Haraguchi H. Development of salt-tolerance interface for an high performance liquid chromatography/inductively coupled plasma mass spectrometry system and its application to accurate quantification of DNA samples. Analytica Chimica Acta, 2012, 713: 23-29.

[112] Raynor M W, Dawson G D, Balcerzak M, Pretorius W G, Ebdon L. Electrospray nebulisation interface for micro-high performance liquid chromatography inductively coupled plasma mass spectrometry. Journal of Analytical Atomic Spectrometry, 1997, 12(9): 1057-1064.

[113] Inagaki K, Fujii S-I, Takatsu A, Chiba K. High performance concentric nebulizer for low-flow rate liquid sample introduction to ICP-MS. Journal of Analytical Atomic Spectrometry, 2011, 26(3): 623-630.

[114] Rappel C, Schaumloeffel D. Improved nanonebulizer design for the coupling of nanoHPLC with ICP-MS. Journal of Analytical Atomic Spectrometry, 2010, 25(12): 1963-1968.

[115] Cheng H, Zhang W, Wang Y, Liu J. Interfacing nanoliter liquid chromatography and inductively coupled plasma mass spectrometry with an in-column high-pressure nebulizer for mercury speciation. Journal of Chromatography A, 2018, 1575: 59-65.

[116] Tarr M A, Zhu G X, Browner R F. Microflow ultrasonic nebulizer for inductively coupled plasma-atomic emission spectrometry. Analytical Chemistry, 1993, 65(13): 1689-1695.

[117] Yin Y G, Wang Z H, Peng J F, Liu J F, He B, Jiang G B. Direct chemical vapour generation-flame atomization as interface of high performance liquid chromatography-atomic fluorescence spectrometry for speciation of mercury without using post-column digestion. Journal of Analytical Atomic Spectrometry, 2009, 24(11): 1575-1578.

[118] Huang K, Xu K, Hou X, Jia Y, Zheng C, Yang L. UV-induced atomization of gaseous mercury hydrides for atomic fluorescence spectrometric detection of inorganic and organic mercury after high performance liquid chromatographic separation. Journal of Analytical Atomic Spectrometry, 2013, 28(4): 510-515.

[119] Bramanti E, Lomonte C, Galli A, Onor M, Zamboni R, Raspi G, D'Ulivo A. Characterization of denatured metallothioneins by reversed phase coupled with on-line chemical vapour generation and atomic fluorescence spectrometric detection. Journal of Chromatography A, 2004, 1054(1-2): 285-291.

[120] Bramanti E, Lomonte C, Onor M, Zamboni R, D'Ulivo A, Raspi G. Mercury speciation by liquid chromatography coupled with on-line chemical vapour generation and atomic fluorescence spectrometric detection (LC-CVGAFS). Talanta, 2005, 66(3): 762-768.

[121] Zhang X, Ji D, Zhang Y, Lu Y, Fu J, Wang Z. Fe^{3+}-catalyzed degradation of organic mercury as a simple post-column interface for the speciation of mercury by high-performance liquid chromatography-catalytic cold vapor-atomic

fluorescence spectrometry. Journal of Analytical Atomic Spectrometry, 2020, 35(4): 693-700.

[122] Acosta G, Spisso A, Fernandez L P, Martinez L D, Pacheco P H, Gil R A. Determination of thimerosal in pharmaceutical industry effluents and river waters by HPLC coupled to atomic fluorescence spectrometry through post-column UV-assisted vapor generation. Journal of Pharmaceutical and Biomedical Analysis, 2015, 106: 79-84.

[123] Angeli V, Ferrari C, Longo I, Onor M, D'Ulivo A, Bramanti E. Microwave-assisted photochemical reactor for the online oxidative decomposition and determination of p-hydroxymercurybenzoate and its thiolic complexes by cold vapor generation atomic fluorescence detection. Analytical Chemistry, 2011, 83(1): 338-343.

[124] Wang Z H, Yin Y G, He B, Shi J B, Liu J F, Jiang G B. L-Cysteine-induced degradation of organic mercury as a novel interface in the HPLC-CV-AFS hyphenated system for speciation of mercury. Journal of Analytical Atomic Spectrometry, 2010, 25(6): 810-814.

[125] Yin Y, Liu J, Jiang G. Photo-induced chemical-vapor generation for sample introduction in atomic spectrometry. TrAC-Trends in Analytical Chemistry, 2011, 30(10): 1672-1684.

[126] Tang L, Chen F, Yang L, Wang Q. The determination of low-molecular-mass thiols with 4-(hydroxymercuric)benzoic acid as a tag using HPLC coupled online with UV/HCOOH-induced cold vapor generation AFS. Journal of Chromatography B: Analytical Technologies in the Biomedical and Life Sciences, 2009, 877(28): 3428-3433.

[127] Yin Y, Liang J, Yang L, Wang Q. Vapour generation at a UV/TiO_2 photocatalysis reaction device for determination and speciation of mercury by AFS and HPLC-AFS. Journal of Analytical Atomic Spectrometry, 2007, 22(3): 330-334.

[128] de Quadros D P C, Campanella B, Onor M, Bramanti E, Borges D L G, D'Ulivo A. Mercury speciation by high-performance liquid chromatography atomic fluorescence spectrometry using an integrated microwave/UV interface. Optimization of a single step procedure for the simultaneous photo-oxidation of mercury species and photo-generation of Hg^0. Spectrochimica Acta Part B: Atomic Spectroscopy, 2014, 101: 312-319.

[129] Yin Y, Liu J, He B, Shi J, Jiang G. Simple interface of high-performance liquid chromatography-atomic fluorescence spectrometry hyphenated system for speciation of mercury based on photo-induced chemical vapour generation with formic acid in mobile phase as reaction reagent. Journal of Chromatography A, 2008, 1181(1-2): 77-82.

[130] He Q, Zhu Z, Hu S, Jin L. Solution cathode glow discharge induced vapor generation of mercury and its application to mercury speciation by high performance liquid chromatography-atomic fluorescence spectrometry. Journal of Chromatography A, 2011, 1218(28): 4462-4467.

[131] Liu Z, Xing Z, Li Z, Zhu Z, Ke Y, Jin L, Hu S. The online coupling of high performance liquid chromatography with atomic fluorescence spectrometry based on dielectric barrier discharge induced chemical vapor generation for the speciation of mercury. Journal of Analytical Atomic Spectrometry, 2017, 32(3): 678-685.

[132] Simon S, Barats A, Pannier F, Potin-Gautier M. Development of an on-line UV decomposition system for direct coupling of liquid chromatography to atomic-fluorescence spectrometry for selenium speciation analysis. Analytical and Bioanalytical Chemistry, 2005, 383(4): 562-569.

[133] Liang J, Wang Q Q, Huang B L. Electrochemical vapor generation of selenium species after online photolysis and reduction by UV-irradiation under nano TiO_2 photocatalysis and its application to selenium speciation by HPLC coupled with atomic fluorescence spectrometry. Analytical and Bioanalytical Chemistry, 2005, 381(2): 366-372.

[134] Quiroz W, Olivares D, Bravo M, Feldmann J, Raab A. Antimony speciation in soils: Improving the detection limits using post-column pre-reduction hydride generation atomic fluorescence spectroscopy (HPLC/pre-reduction/HG-AFS). Talanta, 2011, 84(2): 593-598.

[135] Li H, Luo Y, Li Z, Yang L, Wang Q. Nanosemiconductor-based photocatalytic vapor generation systems for subsequent selenium determination and speciation with atomic fluorescence spectrometry and inductively coupled plasma mass

spectrometry. Analytical Chemistry, 2012, 84(6): 2974-2981.

[136] Shen-Tu C, Fan Y, Hou Y, Wang K, Zhu Y. Arsenic species analysis by ion chromatography-bianode electrochemical hydride generator-atomic fluorescence spectrometry. Journal of Chromatography A, 2008, 1213(1): 56-61.

[137] 申屠超, 侯逸众, 范云场, 朱岩. 离子色谱-双阳极电化学氢化物发生-原子荧光光谱法测定 I 型牙髓失活材料中的砷形态. 分析化学, 2009, 37(2): 263-266.

[138] 侯逸众, 范云场, 朱岩, 陈梅兰, 申屠超. 离子色谱-双阳极电化学氢化物发生-原子荧光光谱法测定当归中 Sb(Ⅲ)和 Sb(Ⅴ). 分析试验室, 2009, 28(10): 38-40.

[139] Tu Q, Johnson W, Buckley B. Mercury speciation analysis in soil samples by ion chromatography, post-column cold vapor generation and inductively coupled plasma mass spectrometry. Journal of Analytical Atomic Spectrometry, 2003, 18(7): 696-701.

[140] Nakazato T, Tao H. A high-efficiency photooxidation reactor for speciation of organic arsenicals by liquid chromatography-hydride generation-ICPMS. Analytical Chemistry, 2006, 78(5): 1665-1672.

[141] Darrouzes J, Bueno M, Simon S, Pannier F, Potin-Gautier M. Advantages of hydride generation interface for selenium speciation in waters by high performance liquid chromatography-inductively coupled plasma mass spectrometry coupling. Talanta, 2008, 75(2): 362-368.

[142] Arslan Y, Yildirim E, Gholami M, Bakirdere S. Lower limits of detection in speciation analysis by coupling high-performance liquid chromatography and chemical-vapor generation. TrAC-Trends in Analytical Chemistry, 2011, 30(4): 569-585.

[143] Wan C C, Chen C S, Jiang S J. Determination of mercury compounds in water samples by liquid chromatography inductively coupled plasma mass spectrometry with an *in situ* nebulizer/vapor generator. Journal of Analytical Atomic Spectrometry, 1997, 12(7): 683-687.

[144] Chiou C S, Jiang S J, Danadurai K S K. Determination of mercury compounds in fish by microwave-assisted extraction and liquid chromatography-vapor generation-inductively coupled plasma mass spectrometry. Spectrochimica Acta Part B: Atomic Spectroscopy, 2001, 56(7): 1133-1142.

[145] de Souza S S, Campiglia A D, Barbosa F, Jr. A simple method for methylmercury, inorganic mercury and ethylmercury determination in plasma samples by high performance liquid chromatography-cold-vapor-inductively coupled plasma mass spectrometry. Analytica Chimica Acta, 2013, 761: 11-17.

[146] Nakazato T, Taniguchi T, Tao H, Tominaga M, Miyazaki A. Ion-exclusion chromatography combined with ICP-MS and hydride generation-ICP-MS for the determination of arsenic species in biological matrices. Journal of Analytical Atomic Spectrometry, 2000, 15(12): 1546-1552.

[147] Chen L W L, Lu X, Le X C. Complementary chromatography separation combined with hydride generation-inductively coupled plasma mass spectrometry for arsenic speciation in human urine. Analytica Chimica Acta, 2010, 675(1): 71-75.

[148] Hwang C J, Jiang S J. Determination of arsenic compounds in water samples by liquid chromatography inductively coupled plasma mass spectrometry with an in-situ nebulizer-hydride generator. Analytica Chimica Acta, 1994, 289(2): 205-213.

[149] 干宁, 李榕生, 李天华, 王峰, 徐伟民. 高效液相色谱-氢化物发生-动态反应池-电感耦合等离子体质谱法检测废水中 4 种砷形态. 冶金分析, 2009, 29(4): 14-19.

[150] Sun Y C, Chen Y J, Tsai Y N. Determination of urinary arsenic species using an on-line nano-TiO_2 photooxidation device coupled with microbore LC and hydride generation-ICP-MS system. Microchemical Journal, 2007, 86(1): 140-145.

[151] Lin C-H, Chen Y, Su Y-A, Luo Y-T, Shih T-T, Sun Y-C. Nanocomposite-coated microfluidic-based photocatalyst-

assisted reduction device to couple high-performance liquid chromatography and inductively coupled plasma-mass spectrometry for online determination of inorganic arsenic species in natural water. Analytical Chemistry, 2017, 89(11): 5892-5900.

[152] Gallagher P A, Wei X Y, Shoemaker J A, Brockhoff C A, Creed J T. Detection of arsenosugars from kelp extracts via IC-electrospray ionization-MS-MS and IC membrane hydride generation ICP-MS. Journal of Analytical Atomic Spectrometry, 1999, 14(12): 1829-1834.

[153] Kuehnelt D, Kienzl N, Juresa D, Francesconi K A. HPLC/vapor generation/ICPMS of selenium metabolites relevant to human urine-selective determination of trimethylselenonium ion. Journal of Analytical Atomic Spectrometry, 2006, 21(11): 1264-1270.

[154] Marschner K, Musil S, Miksik I, Dedina J. Investigation of hydride generation from arsenosugars: Is it feasible for speciation analysis? Analytica Chimica Acta, 2018, 1008: 8-17.

[155] Kuehnelt D, Juresa D, Kienzl N, Francesconi K A. Marked individual variability in the levels of trimethylselenonium ion in human urine determined by HPLC/ICPMS and HPLC/vapor generation/ICPMS. Analytical and Bioanalytical Chemistry, 2006, 386(7-8): 2207-2212.

[156] Petursdottir A H, Gunnlaugsdottir H, Joerundsdottir H, Mestrot A, Krupp E M, Feldmann J. HPLC-HG-ICP-MS: A sensitive and selective method for inorganic arsenic in seafood. Analytical and Bioanalytical Chemistry, 2012, 404(8): 2185-2191.

[157] Wei H Y, Brockhoff-Schwegel C A, Creed J T. A comparison of urinary arsenic speciation via direct nebulization and on-line photo-oxidation-hydride generation with IC separation and ICP-MS detection. Journal of Analytical Atomic Spectrometry, 2001, 16(1): 12-19.

[158] Gomez-Ariza J L, Sanchez-Rodas D, Giraldez I, Morales E. Comparison of biota sample pretreatments for arsenic speciation with coupled HPLC-HG-ICP-MS. Analyst, 2000, 125(3): 401-407.

[159] Rubio R, Padro A, Alberti J, Rauret G. Determination of arsenic speciation by liquid chromatography hydride generation inductively coupled plasma atomicemission spectrometrywith online UV photoxoidation. Analytica Chimica Acta, 1993, 283(1): 160-166.

[160] Dagnac T, Padro A, Rubio R, Rauret G. Speciation of arsenic in mussels by the coupled system liquid chromatography UV irradiation hydride generation inductively coupled plasma mass spectrometry. Talanta, 1999, 48(4): 763-772.

[161] LaFuente J M G, Dlaska M, Sanchez M L F, Sanz-Medel A. Organic and inorganic selenium speciation in urine by on-line vesicle mediated high-performance liquid chromotography-focused microwave digestion-hydride generation-inductively coupled plasma mass spectrometry. Journal of Analytical Atomic Spectrometry, 1998, 13(5): 423-429.

[162] Clough R, Belt S T, Fairman B, Catterick T, Evans E H. Uncertainty contributions to single and double isotope dilution mass spectrometry with HPLC-CV-MC-ICP-MS for the determination of methylmercury in fish tissue. Journal of Analytical Atomic Spectrometry, 2005, 20(10): 1072-1075.

[163] LaFuente J M G, Marchante-Gayon J M, Sanchez M L F, Sanz-Medel A. Urinary selenium speciation by high-performance liquid chromatography-inductively coupled plasma mass spectrometry: Advantages of detection with a double-focusing mass analyser with a hydride generation interface. Talanta, 1999, 50(1): 207-217.

[164] Tsai M W, Sun Y C. On-line coupling of an ultraviolet titanium dioxide film reactor with a liquid chromatography/hydride generation/inductively coupled plasma mass spectrometry system for continuous determination of dynamic variation of hydride-and nonhydride-forming arsenic species in very small microdialysate samples. Rapid Communications in Mass Spectrometry, 2008, 22(2): 211-216.

[165] Sun Y C, Chang Y C, Su C K. On-line HPLC-UV/nano-TiO$_2$-ICPMS system for the determination of inorganic

selenium species. Analytical Chemistry, 2006, 78(8): 2640-2645.

[166] Chen K J, Hsu I H, Sun Y C. Determination of methylmercury and inorganic mercury by coupling short-column ion chromatographic separation, on-line photocatalyst-assisted vapor generation, and inductively coupled plasma mass spectrometry. Journal of Chromatography A, 2009, 1216(51): 8933-8938.

[167] Shih T T, Hsu I H, Wu J F, Lin C H, Sun Y C. Development of chip-based photocatalyst-assisted reduction device to couple high performance liquid chromatography and inductively coupled plasma-mass spectrometry for determination of inorganic selenium species. Journal of Chromatography A, 2013, 1304: 101-108.

[168] Shih T-T, Lin C-H, Hsu I H, Chen J-Y, Sun Y-C. Development of a titanium dioxide-coated microfluidic-based photocatalyst-assisted reduction device to couple high-performance liquid chromatography with inductively coupled plasma mass spectrometry for determination of inorganic selenium species. Analytical Chemistry, 2013, 85(21): 10091-10098.

[169] 徐进勇, 王彤, 陈杜军, 叶隆慧, 倪师军. 光诱导蒸气发生-高效液相色谱-电感耦合等离子体质谱联用测定汞形态. 分析化学, 2012, 40(1): 169-172.

[170] Kinska K, Bierla K, Godin S, Preud'homme H, Kowalska J, Krasnodebska-Ostrega B, Lobinski R, Szpunar J. A chemical speciation insight into the palladium(Ⅱ) uptake and metabolism by *Sinapis alba*. Exposure to Pd induces the synthesis of a Pd-histidine complex. Metallomics, 2019, 11(9): 1498-1505.

[171] Far J, Preud'homme H, Lobinski R. Detection and identification of hydrophilic selenium compounds in selenium-rich yeast by size exclusion-microbore normal-phase HPLC with the on-line ICP-MS and electrospray Q-TOF-MS detection. Analytica Chimica Acta, 2010, 657(2): 175-190.

[172] Chan Q, Caruso J A. A metallomics approach discovers selenium-containing proteins in selenium-enriched soybean. Analytical and Bioanalytical Chemistry, 2012, 403(5): 1311-1321.

[173] Bouyssiere B, Knispel T, Ruhnau C, Denkhaus E, Prange A. Analysis of nickel species in cytosols of normal and malignant human colonic tissues using two dimensional liquid chromatography with ICP-sector field MS detection. Journal of Analytical Atomic Spectrometry, 2004, 19(1): 196-200.

[174] Yun Z, Li L, Liu L, He B, Zhao X, Jiang G. Characterization of mercury-containing protein in human plasma. Metallomics, 2013, 5(7): 821-827.

[175] Pedrero Z, Encinar J R, Madrid Y, Camara C. Identification of selenium species in selenium-enriched *Lens esculenta* plants by using two-dimensional liquid chromatography-inductively coupled plasma mass spectrometry and Se-77 selenomethionine selenium oxide spikes. Journal of Chromatography A, 2007, 1139(2): 247-253.

[176] Garcia-Sartal C, Taebunpakul S, Stokes E, del Carmen Barciela-Alonso M, Bermejo-Barrera P, Goenaga-Infante H. Two-dimensional HPLC coupled to ICP-MS and electrospray ionisation (ESI)-MS/MS for investigating the bioavailability *in vitro* of arsenic species from edible seaweed. Analytical and Bioanalytical Chemistry, 2012, 402(10): 3359-3369.

[177] McSheehy S, Pohl P, Lobinski R, Szpunar J. Complementarity of multidimensional HPLC-ICP-MS and electrospray MS-MS for speciation analysis of arsenic in algae. Analytica Chimica Acta, 2001, 440(1): 3-16.

[178] Casal S G, Far J, Bierla K, Ouerdane L, Szpunar J. Study of the Se-containing metabolomes in Se-rich yeast by size-exclusion-cation-exchange HPLC with the parallel ICP-MS and electrospray orbital ion trap detection. Metallomics, 2010, 2(8): 535-548.

[179] Arias-Borrego A, Callejon-Leblic B, Rodriguez-Moro G, Velasco I, Gomez-Ariza J L, Garcia-Barrera T. A novel HPLC column switching method coupled to ICP-MS/QTOF for the first determination of selenoprotein P(SELENOP) in human breast milk. Food Chemistry, 2020, 321: 126692.

[180] Garcia-Sevillano M A, Garcia-Barrera T, Gomez-Ariza J L. Simultaneous speciation of selenoproteins and selenometabolites in plasma and serum by dual size exclusion-affinity chromatography with online isotope dilution inductively coupled plasma mass spectrometry. Analytical and Bioanalytical Chemistry, 2014, 406(11): 2719-2725.

[181] McSheehy S, Pohl P, Szpunar J, Potin-Gautier M, Lobinski R. Analysis for selenium speciation in selenized yeast extracts by two-dimensional liquid chromatography with ICP-MS and electrospray MS-MS detection. Journal of Analytical Atomic Spectrometry, 2001, 16(1): 68-73.

[182] Kaewkhomdee N, Mounicou S, Szpunar J, Lobinski R, Shiowatana J. Characterization of binding and bioaccessibility of Cr in Cr-enriched yeast by sequential extraction followed by two-dimensional liquid chromatography with mass spectrometric detection. Analytical and Bioanalytical Chemistry, 2010, 396(3): 1355-1364.

[183] Bernardin M, Le Masle A, Bessueille-Barbier F, Lienemann C-P, Heinisch S. Comprehensive two-dimensional liquid chromatography with inductively coupled plasma mass spectrometry detection for the characterization of sulfur, vanadium and nickel compounds in petroleum products. Journal of Chromatography A, 2020, 1611: 460605.

[184] Galvez L, Rusz M, Jakupec M A, Koellensperger G. Heart-cut 2DSEC-RP-LC-ICP-MS as a screening tool in metal-based anticancer research. Journal of Analytical Atomic Spectrometry, 2019, 34(6): 1279-1286.

[185] Gergely V, Kubachka K M, Mounicou S, Fodor P, Caruso J A. Selenium speciation in *Agaricus bisporus* and *Lentinula edodes* mushroom proteins using multi-dimensional chromatography coupled to inductively coupled plasma mass spectrometry. Journal of Chromatography A, 2006, 1101(1-2): 94-102.

[186] Lindemann T, Hintelmann H. Identification of selenium-containing glutathione S-conjugates in a yeast extract by two-dimensional liquid chromatography with inductively coupled plasma MS and nanoelectrospray MS/MS detection. Analytical Chemistry, 2002, 74(18): 4602-4610.

[187] Mounicou S, McSheehy S, Szpunar J, Potin-Gautier M, Lobinski R. Analysis of selenized yeast for selenium speciation by size-exclusion chromatography and capillary zone electrophoresis with inductively coupled plasma mass spectrometric detection (SEC-CZE-ICP-MS). Journal of Analytical Atomic Spectrometry, 2002, 17(1): 15-20.

[188] Verola Mataveli L R, Zezzi Arruda M A. Expanding resolution of metalloprotein separations from soybean seeds using 2D-HPLC-ICP-MS and SDS-PAGE as a third dimension. Journal of Proteomics, 2014, 104: 94-103.

[189] Garcia-Sevillano M A, Garcia-Barrera T, Gomez-Ariza J L. Development of a new column switching method for simultaneous speciation of selenometabolites and selenoproteins in human serum. Journal of Chromatography A, 2013, 1318: 171-179.

[190] McSheehy S, Pannier F, Szpunar J, Potin-Gautier M, Lobinski R. Speciation of seleno compounds in yeast aqueous extracts by three-dimensional liquid chromatography with inductively coupled plasma mass spectrometric and electrospray mass spectrometric detection. Analyst, 2002, 127(2): 223-229.

[191] Miyayama T, Ogra Y, Suzuki K T. Separation of metallothionein isoforms extracted from isoform-specific knockdown cells on two-dimensional micro high-performance liquid chromatography hyphenated with inductively coupled plasma-mass spectrometry. Journal of Analytical Atomic Spectrometry, 2007, 22(2): 179-182.

[192] Wrobel K, Sadi B B M, Wrobel K, Castillo J R, Caruso J A. Effect of metal ions on the molecular weight distribution of humic substances derived from municipal compost: Ultrafiltration and size exclusion chromatography with spectrophotometric and inductively coupled plasma-MS detection. Analytical Chemistry, 2003, 75(4): 761-767.

[193] Heumann K G, Rottmann L, Vogl J. Elemental speciation with liquid chromatography inductively coupled plasma mass spectrometry. Journal of Analytical Atomic Spectrometry, 1994, 9(12): 1351-1355.

[194] Dean J R, Munro S, Ebdon L, Crews H M, Massey R C. Studies of metalloprotein species by directly coupledhigh performance liquid chromatography inductively coupled plasma mass spectrometry. Journal of Analytical Atomic

Spectrometry, 1987, 2(6): 607-610.

[195] Kannamkumarath S S, Wrobel K, Wuilloud R G. Studying the distribution pattern of selenium in nut proteins with information obtained from SEC-UV-ICP-MS and CE-ICP-MS. Talanta, 2005, 66(1): 153-159.

[196] Siethoff C, Feldmann I, Jakubowski N, Linscheid M. Quantitative determination of DNA adducts using liquid chromatography electrospray ionization mass spectrometry and liquid chromatography high-resolution inductively coupled plasma mass spectrometry. Journal of Mass Spectrometry, 1999, 34(4): 421-426.

[197] Takahashi K, Ogra Y. Identification of the biliary selenium metabolite and the biological significance of selenium enterohepatic circulation. Metallomics, 2020, 12(2): 241-248.

[198] Nearing M M, Koch I, Reimer K J. Complementary arsenic speciation methods: A review. Spectrochimica Acta Part B: Atomic Spectroscopy, 2014, 99: 150-162.

[199] Meermann B, Sperling M. Hyphenated techniques as tools for speciation analysis of metal-based pharmaceuticals: Developments and applications. Analytical and Bioanalytical Chemistry, 2012, 403(6): 1501-1522.

[200] Ellis J L, Conklin S D, Gallawa C M, Kubachka K M, Young A R, Creed P A, Caruso J A, Creed J T. Complementary molecular and elemental detection of speciated thioarsenicals using ESI-MS in combination with a xenon-based collision-cell ICP-MS with application to fortified NIST freeze-dried urine. Analytical and Bioanalytical Chemistry, 2008, 390(7): 1731-1737.

[201] Raab A, Feldmann J. Biological sulphur-containing compounds: Analytical challenges. Analytica Chimica Acta, 2019, 1079: 20-29.

[202] Jamari N L A, Dohmann J F, Raab A, Krupp E M, Feldmann J. Novel non-targeted analysis of perfluorinated compounds using fluorine-specific detection regardless of their ionisability (HPLC-ICPMS/MS-ESI-MS). Analytica Chimica Acta, 2019, 1053: 22-31.

[203] 戴礼洪, 刘潇威, 王迪, 蒋梦. 固相萃取-高效液相色谱-原子荧光光谱法联用测定水中无机汞(Ⅱ)及有机汞. 理化检验(化学分册), 2015, 51(8): 1178-1182.

[204] Xiong X, Qi X, Liu J, Wang J, Wu C. Comparison of modifers for mercury speciation in water by solid phase extraction and high performance liquid chromatography-atomic fluorescence spectrometry. Analytical Letters, 2014, 47(14): 2417-2430.

[205] Chen H, Chen J, Jin X, Wei D. Determination of trace mercury species by high performance liquid chromatography-inductively coupled plasma mass spectrometry after cloud point extraction. Journal of Hazardous Materials, 2009, 172(2-3): 1282-1287.

[206] Yin Y G, Chen M, Peng J F, Liu J F, Jiang G B. Dithizone-functionalized solid phase extraction-displacement elution-high performance liquid chromatography-inductively coupled plasma mass spectrometry for mercury speciation in water samples. Talanta, 2010, 81(4-5): 1788-1792.

[207] Wang Z, Xu Q, Li S, Luan L, Li J, Zhang S, Dong H. Hollow fiber supported ionic liquid membrane microextraction for speciation of mercury by high-performance liquid chromatography-inductively coupled plasma mass spectrometry. Analytical Methods, 2015, 7(3): 1140-1146.

[208] Li L, Bi R, Wang Z, Xu C, Li B, Luan L, Chen X, Xue F, Zhang S, Zhao N. Speciation of mercury using high-performance liquid chromatography-inductively coupled plasma mass spectrometry following enrichment by dithizone functionalized magnetite-reduced graphene oxide. Spectrochimica Acta Part B: Atomic Spectroscopy, 2019, 159: 105653.

[209] Li L, Wang Z, Zhang S, Wang M. Directly-thiolated graphene based organic solvent-free cloud point extraction-like method for enrichment and speciation of mercury by HPLC-ICP-MS. Microchemical Journal, 2017, 132: 299-307.

[210] Gao X, Dai J, Zhao H, Zhu J, Luo L, Zhang R, Zhang Z, Li L. Synthesis of MoS$_2$ nanosheets for mercury speciation analysis by HPLC-UV-HG-AFS. RSC Advances, 2018, 8(33): 18364-18371.

[211] Zhou X-X, Jiang L-W, Wang D-J, He S, Li C-J, Yan B. Speciation analysis of Ag$_2$S and ZnS nanoparticles at the ng/L level in environmental waters by cloud point extraction coupled with LC-ICPMS. Analytical Chemistry, 2020, 92(7): 4765-4770.

[212] Zhang S, Luo H, Zhang Y, Li X, Liu J, Xu Q, Wang Z. *In situ* rapid magnetic solid-phase extraction coupled with HPLC-ICP-MS for mercury speciation in environmental water. Microchemical Journal, 2016, 126: 25-31.

[213] He Y, He M, Nan K, Cao R, Chen B, Hu B. Magnetic solid-phase extraction using sulfur-containing functional magnetic polymer for high-performance liquid chromatography-inductively coupled plasma-mass spectrometric speciation of mercury in environmental samples. Journal of Chromatography A, 2019, 1595: 19-27.

[214] Zhu S, Chen B, He M, Huang T, Hu B. Speciation of mercury in water and fish samples by HPLC-ICP-MS after magnetic solid phase extraction. Talanta, 2017, 171: 213-219.

[215] Liu Y-M, Zhang F-P, Jiao B-Y, Rao J-Y, Leng G. Automated dispersive liquid-liquid microextraction coupled to high performance liquid chromatography-cold vapour atomic fluorescence spectroscopy for the determination of mercury species in natural water samples. Journal of Chromatography A, 2017, 1493: 1-9.

[216] Song X, Ye M, Tang X, Wang C. Ionic liquids dispersive liquid-liquid microextraction and HPLC-atomic fluorescence spectrometric determination of mercury species in environmental waters. Journal of Separation Science, 2013, 36(2): 414-420.

[217] Leng G, Chen W, Wang Y. Speciation analysis of mercury in sediments using ionic-liquid-based vortex-assisted liquid-liquid microextraction combined with high-performance liquid chromatography and cold vapor atomic fluorescence spectrometry. Journal of Separation Science, 2015, 38(15): 2684-2691.

[218] Leng G, Yin H, Li S, Chen Y, Dan D. Speciation analysis of mercury in sediments using vortex-assisted liquid-liquid microextraction coupled to high-performance liquid chromatography-cold vapor atomic fluorescence spectrometry. Talanta, 2012, 99: 631-636.

[219] Jia X, Han Y, Wei C, Duan T, Chen H. Speciation of mercury in liquid cosmetic samples by ionic liquid based dispersive liquid-liquid microextraction combined with high-performance liquid chromatography-inductively coupled plasma mass spectrometry. Journal of Analytical Atomic Spectrometry, 2011, 26(7): 1380-1386.

[220] Jia X, Han Y, Liu X, Duan T, Chen H. Speciation of mercury in water samples by dispersive liquid-liquid microextraction combined with high performance liquid chromatography-inductively coupled plasma mass spectrometry. Spectrochimica Acta Part B: Atomic Spectroscopy, 2011, 66(1): 88-92.

[221] Yu L P. Cloud point extraction preconcentration prior to high-performance liquid chromatography coupled with cold vapor generation atomic fluorescence spectrometry for speciation analysis of mercury in fish samples. Journal of Agricultural and Food Chemistry, 2005, 53(25): 9656-9662.

[222] Chen J, Chen H, Jin X, Chen H. Determination of ultra-trace amount methyl-, phenyl- and inorganic mercury in environmental and biological samples by liquid chromatography with inductively coupled plasma mass spectrometry after cloud point extraction preconcentration. Talanta, 2009, 77(4): 1381-1387.

[223] Moreno F, Garcia-Barrera T, Gomez-Ariza J L. Simultaneous speciation and preconcentration of ultra trace concentrations of mercury and selenium species in environmental and biological samples by hollow fiber liquid phase microextraction prior to high performance liquid chromatography coupled to inductively coupled plasma mass spectrometry. Journal of Chromatography A, 2013, 1300: 43-50.

[224] Chen B, Wu Y, Guo X, He M, Hu B. Speciation of mercury in various samples from the micro-ecosystem of East Lake

by hollow fiber-liquid-liquid-liquid microextraction-HPLC-ICP-MS. Journal of Analytical Atomic Spectrometry, 2015, 30(4): 875-881.

[225] Cao Y, Yan L, Huang H, Deng B. Selenium speciation in radix puerariae using ultrasonic assisted extraction combined with reversed phase high performance liquid chromatography-inductively coupled plasma-mass spectrometry after magnetic solid-phase extraction with 5-sulfosalicylic acid functionalized magnetic nanoparticles. Spectrochimica Acta Part B: Atomic Spectroscopy, 2016, 122: 172-177.

[226] Chen B, Hu B, He M, Huang Q, Zhang Y, Zhang X. Speciation of selenium in cells by HPLC-ICP-MS after (on-chip) magnetic solid phase extraction. Journal of Analytical Atomic Spectrometry, 2013, 28(3): 334-343.

[227] Bueno M, Potin-Gautier M. Solid-phase extraction for the simultaneous preconcentration of organic (selenocystine) and inorganic Se(Ⅳ), Se(Ⅵ) selenium in natural waters. Journal of Chromatography A, 2002, 963(1-2): 185-193.

[228] Guo X, He M, Nan K, Yan H, Chen B, Hu B. A dual extraction technique combined with HPLC-ICP-MS for speciation of seleno-amino acids in rice and yeast samples. Journal of Analytical Atomic Spectrometry, 2016, 31(2): 406-414.

[229] Gomez M, Camara C, Palacios M A, LopezGonzalvez A. Anionic cartridge preconcentrators for inorganic arsenic, monomethylarsonate and dimethylarsinate determination by on-line HPLC-HG-AAS. Fresenius' Journal of Analytical Chemistry, 1997, 357(7): 844-849.

[230] Tian Y, Chen M-L, Chen X-W, Wang J-H, Hirano Y, Sakamoto H, Shirasaki T. Arsenic preconcentration via solid phase extraction and speciation by HPLC-gradient hydride generation atomic absorption spectrometry. Journal of Analytical Atomic Spectrometry, 2011, 26(1): 133-140.

[231] Chen B, Hu B, He M, Mao X, Zu W. Synthesis of mixed coating with multi-functional groups for in-tube hollow fiber solid phase microextraction-high performance liquid chromatography-inductively coupled plasma mass spectrometry speciation of arsenic in human urine. Journal of Chromatography A, 2012, 1227: 19-28.

[232] Mao X, Chen B, Huang C, He M, Hu B. Titania immobilized polypropylene hollow fiber as a disposable coating for stir bar sorptive extraction-high performance liquid chromatography-inductively coupled plasma mass spectrometry speciation of arsenic in chicken tissues. Journal of Chromatography A, 2011, 1218(1): 1-9.

[233] Yang Y, Liu Z, Chen H, Li S. Low-density solvent-based dispersive liquid-liquid microextraction followed by HPLC-ICP-MS for speciation analysis of phenylarsenics in lake water. International Journal of Environmental Analytical Chemistry, 2019, 99(1): 87-100.

[234] Guo X, Chen B, He M, Hu B, Zhou X. Ionic liquid based carrier mediated hollow fiber liquid liquid liquid microextraction combined with HPLC-ICP-MS for the speciation of phenylarsenic compounds in chicken and feed samples. Journal of Analytical Atomic Spectrometry, 2013, 28(10): 1638-1647.

[235] Fairman B, Catterick T, Wheals B, Polinina E. Reversed-phase ion-pair chromatography with inductively coupled plasma-mass spectrometry detection for the determination of organo-tin compounds in waters. Journal of Chromatography A, 1997, 758(1): 85-92.

[236] Schulze G, Lehmann C. Separation of monobutyltin, dibutyltinand tributyltincompounds by isocratic ion-exchange liquid chromatography coupled with hydride generation atomic absorption spectrometric determination. Analytica Chimica Acta, 1994, 288(3): 215-220.

[237] Mao X, Fan W, He M, Chen B, Hu B. C-18-coated stir bar sorptive extraction combined with HPLC-ICP-MS for the speciation of butyltins in environmental samples. Journal of Analytical Atomic Spectrometry, 2015, 30(1): 162-171.

[238] Yang H, Zhang H, Zhu X Y, Chen S D, Liu L, Pan D. Determination of tributyltin in seafood based on magnetic molecularly imprinted polymers coupled with high-performance liquid chromatography-inductively coupled plasma mass spectrometry. Journal of Food Quality, 2017, 7405475.

[239] Ugarte A, Unceta N, Sampedro M C, Goicolea M A, Gomez-Caballero A, Barrio R J. Solid phase microextraction coupled to liquid chromatography-inductively coupled plasma mass spectrometry for the speciation of organotin compounds in water samples. Journal of Analytical Atomic Spectrometry, 2009, 24(3): 347-351.

[240] Dauthieu M, Bueno M, Darrouzes K, Gilon N, Potin-Gautier M. Evaluation of porous graphitic carbon stationary phase for simultaneous preconcentration and separation of organic and inorganic selenium species in "clean" water systems. Journal of Chromatography A, 2006, 1114(1): 34-39.

[241] Abbas-Ghaleb K, Gilon N, Cretier G, Mermet J M. Preconcentration of selenium compounds on a porous graphitic carbon column in view of HPLC-ICP-AES speciation analysis. Analytical and Bioanalytical Chemistry, 2003, 377(6): 1026-1031.

[242] Mester Z, Fodor P. Selenium speciation with on-column preconcentration high-performance liquid chromatography atomic fluorescence spectrometry using ultrasonic nebulization technique. Analytica Chimica Acta, 1999, 386(1-2): 89-97.

[243] Pantsar-Kallio M, Manninen P K G. Optimizing ion chromatography-inductively coupled plasma mass spectrometry for speciation analysis of arsenic, chromium and bromine in water samples. International Journal of Environmental Analytical Chemistry, 1999, 75(1-2): 43-55.

[244] 殷学锋, 刘梅. 在线固相萃取富集-液相色谱分离冷原子吸收联机测定不同形态汞. 分析化学, 1996, 24(11): 1248-1252.

[245] Wang H, Chen B, Zhu S, Yu X, He M, Hu B. Chip-based magnetic solid-phase microextraction online coupled with microHPLC-ICPMS for the determination of mercury species in cells. Analytical Chemistry, 2016, 88(1): 796-802.

[246] Falter R, Ilgen G. Coupling of the RP C_{18} preconcentration HPLC-UV-PCO system with atomic fluorescence detection for the determination of methylmercury in sediment and biological tissue. Fresenius' Journal of Analytical Chemistry, 1997, 358(3): 407-410.

[247] Yin X F, Frech W, Hoffmann E, Ludke C, Skole J. Mercury speciation by coupling cold vapour atomic absorption spectrometry with flow injection on-line preconcentration and liquid chromatographic separation. Fresenius' Journal of Analytical Chemistry, 1998, 361(8): 761-766.

[248] Carneado S, Pero-Gascon R, Ibanez-Palomino C, Lopez-Sanchez J F, Sahuquillo A. Mercury(II) and methylmercury determination in water by liquid chromatography hyphenated to cold vapour atomic fluorescence spectrometry after online short-column preconcentration. Analytical Methods, 2015, 7(6): 2699-2706.

[249] Sarzanini C, Sacchero G, Aceto M, Abollino O, Mentasti E. Simultaneous determination of methyl-mercury, ethyl-mercury, phenyl-mercury,andinorganic mercury bycold vapor atomic absorption spectrometry withonlinechromatographic separation. Journal of Chromatography, 1992, 626(1): 151-157.

[250] Cheng H, Wu C, Shen L, Liu J, Xu Z. Online anion exchange column preconcentration and high performance liquid chromatographic separation with inductively coupled plasma mass spectrometry detection for mercury speciation analysis. Analytica Chimica Acta, 2014, 828: 9-16.

[251] Londonio A, Emir Hasuoka P, Pacheco P, Andres Gil R, Smichowski P. Online solid phase extraction-HPLC-ICP-MS system for mercury and methylmercury preconcentration using functionalised carbon nanotubes for their determination in dietary supplements. Journal of Analytical Atomic Spectrometry, 2018, 33(10): 1737-1744.

[252] Cheng H, Wu C, Liu J, Xu Z. Thiol-functionalized silica microspheres for online preconcentration and determination of mercury species in seawater by high performance liquid chromatography and inductively coupled plasma mass spectrometry. RSC Advances, 2015, 5(25): 19082-19090.

[253] Shade C W, Hudson R J M. Determination of MeHg in environmental sample matrices using Hg-thiourea complex ion

chromatography with on-line cold vapor generation and atomic fluorescence spectrometric detection. Environmental Science & Technology, 2005, 39(13): 4974-4982.

[254] Brombach C-C, Gajdosechova Z, Chen B, Brownlow A, Corns W T, Feldmann J, Krupp E M. Direct online HPLC-CV-AFS method for traces of methylmercury without derivatisation: A matrix-independent method for urine, sediment and biological tissue samples. Analytical and Bioanalytical Chemistry, 2015, 407(3): 973-981.

[255] Brombach C-C, Chen B, Corns W T, Feldmann J, Krupp E M. Methylmercury in water samples at the pg/L level by online preconcentration liquid chromatography cold vapor-atomic fluorescence spectrometry. Spectrochimica Acta Part B: Atomic Spectroscopy, 2015, 105: 103-108.

[256] Brombach C C, Ezzeldin M F, Chen B, Corns W T, Feldmann J, Krupp E M. Quick and robust method for trace determination of MeHg in rice and rice products without derivatisation. Analytical Methods, 2015, 7(20): 8584-8589.

[257] Jia X Y, Gong D R, Han Y, Wei C, Duan T C, Chen H T. Fast speciation of mercury in seawater by short-column high-performance liquid chromatography hyphenated to inductively coupled plasma spectrometry after on-line cation exchange column preconcentration. Talanta, 2012, 88: 724-729.

[258] Yang S, Zhang D, Cheng H, Wang Y, Liu J. Graphene oxide as an efficient adsorbent of solid-phase extraction for online preconcentration of inorganic and organic mercurials in freshwater followed by HPLC-ICP-MS determination. Analytica Chimica Acta, 2019, 1074: 54-61.

[259] Jia X, Gong D, Wang J, Huang F, Duan T, Zhang X. Arsenic speciation in environmental waters by a new specific phosphine modified polymer microsphere preconcentration and HPLC-ICP-MS determination. Talanta, 2016, 160: 437-443.

[260] Cai Y, Cabanas M, Fernandezturiel J L, Abalos M, Bayona J M. On line preconcentration of selenium(IV) and selenium(VI) in aqueous matrices followed by liquid chromatography inductively coupled plasma-mass spectrometry determination. Analytica Chimica Acta, 1995, 314(3): 183-192.

[261] Giusti P, Schaumloffel D, Preud'homme H, Szpunar J, Lobinski R. Selenopeptide mapping in a selenium-yeast protein digest by parallel nanoHPLC-ICP-MS and nanoHPLC-electrospray-MS/MS after on-line preconcentration. Journal of Analytical Atomic Spectrometry, 2006, 21(1): 26-32.

[262] Jia X, Gong D, Xu B, Chi Q, Zhang X. Development of a novel, fast, sensitive method for chromium speciation in wastewater based on an organic polymer as solid phase extraction material combined with HPLC-ICP-MS. Talanta, 2016, 147: 155-161.

[263] Posta J, Alimonti A, Petrucci F, Caroli S. On-line separation and preconcentration of chromium species in seawater. Analytica Chimica Acta, 1996, 325(3): 185-193.

[264] Bishop D P, Blanes L, Wilson A B, Wilbanks T, Killeen K, Grimm R, Wenzel R, Major D, Macka M, Clarke D, Schmid R, Cole N, Doble P A. Microfluidic high performance liquid chromatography-chip hyphenation to inductively coupled plasma-mass spectrometry. Journal of Chromatography A, 2017, 1497: 64-69.

[265] Gomez-Ariza J L, Lorenzo F, Garcia-Barrera T. Simultaneous determination of mercury and arsenic species in natural freshwater by liquid chromatography with on-line UV irradiation, generation of hydrides and cold vapor and tandem atomic fluorescence detection. Journal of Chromatography A, 2004, 1056(1-2): 139-144.

[266] 王振华, 何滨, 史建波, 阴永光, 江桂斌. 液相色谱-双通道原子荧光检测联用法同时测定砷和硒的形态. 色谱, 2009, 5: 203-208.

[267] Manley S A, Byrns S, Lyon A W, Brown P, Gailer J. Simultaneous Cu-, Fe-, and Zn-specific detection of metalloproteins contained in rabbit plasma by size-exclusion chromatography-inductively coupled plasma atomic emission spectroscopy. Journal of Biological Inorganic Chemistry, 2009, 14(1): 61-74.

[268] Gailer J, Buttigieg G A, Denton M B. Simultaneous arsenic-and selenium-specific detection of the dimethyldiselenoarsinate anion by high-performance liquid chromatography-inductively coupled plasma atomic emission spectrometry. Applied Organometallic Chemistry, 2003, 17(8): 570-574.

[269] Marcinkowska M, Baralkiewicz D. Multielemental speciation analysis by advanced hyphenated technique-HPLC/ICP-MS: A review. Talanta, 2016, 161: 177-204.

[270] Iserte L O, Roig-Navarro A F, Hernandez F. Simultaneous determination of arsenic and selenium species in phosphoric acid extracts of sediment samples by HPLC-ICP-MS. Analytica Chimica Acta, 2004, 527(1): 97-104.

[271] Castillo A, Roig-Navarro A F, Pozo O J. Capabilities of microbore columns coupled to inductively coupled plasma mass spectrometry in speciation of arsenic and selenium. Journal of Chromatography A, 2008, 1202(2): 132-137.

[272] Cheng H Y, Zhang W W, Wang Y C, Liu J H. Graphene oxide as a stationary phase for speciation of inorganic and organic species of mercury, arsenic and selenium using HPLC with ICP-MS detection. Microchimica Acta, 2018, 185(9): 425.

[273] Marcinkowska M, Komorowicz I, Baralkiewicz D. Study on multielemental speciation analysis of Cr(VI), As(III) and As(V) in water by advanced hyphenated technique HPLC/ICP-DRC-MS. Fast and reliable procedures. Talanta, 2015, 144: 233-240.

[274] Marcinkowska M, Komorowicz I, Baralkiewicz D. New procedure for multielemental speciation analysis of five toxic species: As(III), As(V), Cr(VI), Sb(III) and Sb(V) in drinking water samples by advanced hyphenated technique HPLC/ICP-DRC-MS. Analytica Chimica Acta, 2016, 920: 102-111.

[275] Marcinkowska M, Lorenc W, Baralkiewicz D. Study of the impact of bottles material and color on the presence of As^{III}, As^{V}, Sb^{III}, Sb^{V} and Cr^{VI} in matrix-rich mineral water-Multielemental speciation analysis by HPLC/ICP- DRC-MS. Microchemical Journal, 2017, 132: 1-7.

[276] Lorenc W, Markiewicz B, Kruszka D, Kachlicki P, Baralkiewicz D. Study on speciation of As, Cr, and Sb in bottled flavored drinking water samples using advanced analytical techniques IEC/SEC-HPLC/ICP-DRC-MS and ESI-MS/MS. Molecules, 2019, 24(4): 668.

[277] Martinez-Bravo Y, Roig-Navarro A F, Lopez F J, Hernandez F. Multielemental determination of arsenic, selenium and chromium(VI) species in water by high-performance liquid chromatography-inductively coupled plasma mass spectrometry. Journal of Chromatography A, 2001, 926(2): 265-274.

[278] Guerin T, Astruc M, Batel A, Borsier M. Multielemental speciation of As, Se, Sb and Te by HPLC-ICP-MS. Talanta, 1997, 44(12): 2201-2208.

[279] Gomez-Ariza J, Lorenzo F, Garcia-Barrera T. Comparative study of atomic fluorescence spectroscopy and inductively coupled plasma mass spectrometry for mercury and arsenic multispeciation. Analytical and Bioanalytical Chemistry, 2005, 382(2): 485-492.

[280] Chang L-F, Jiang S-J, Sahayam A C. Speciation analysis of mercury and lead in fish samples using liquid chromatography-inductively coupled plasma mass spectrometry. Journal of Chromatography A, 2007, 1176(1-2): 143-148.

[281] Li Y-F, Chen C, Li B, Wang Q, Wang J, Gao Y, Zhao Y, Chai Z. Simultaneous speciation of selenium and mercury in human urine samples from long-term mercury-exposed populations with supplementation of selenium-enriched yeast by HPLC-ICP-MS. Journal of Analytical Atomic Spectrometry, 2007, 22(8): 925-930.

[282] Zhang D, Yang S, Ma Q, Sun J, Cheng H, Wang Y, Liu J. Simultaneous multi-elemental speciation of As, Hg and Pb by inductively coupled plasma mass spectrometry interfaced with high-performance liquid chromatography. Food Chemistry, 2020, 313: 126119.

[283] Bayon M M, Cabezuelo A B S, Gonzalez E B, Alonso J I G, Sanz-Medel A. Capabilities of fast protein liquid

chromatography coupled to a double focusing inductively coupled plasma mass spectrometer for trace metal speciation in human serum. Journal of Analytical Atomic Spectrometry, 1999, 14(6): 947-951.

[284] Ferrarello C N, Bayon M M, de la Campa R F, Sanz-Medel A. Multi-elemental speciation studies of trace elements associated with metallothionein-like proteins in mussels by liquid chromatography with inductively coupled plasma time-of-flight mass spectrometric detection. Journal of Analytical Atomic Spectrometry, 2000, 15(12): 1558-1563.

[285] St Remy R R D, Sanchez M L F, Sastre J B L, Sanz-Medel A. Multielemental distribution patterns in premature human milk whey and pre-term formula milk whey by size exclusion chromatography coupled to inductively coupled plasma mass spectrometry with octopole reaction cell. Journal of Analytical Atomic Spectrometry, 2004, 19(9): 1104-1110.

[286] Wuilloud R G, Kannamkumarath S S, Caruso J A. Multielemental speciation analysis of fungi porcini (*Boletus edulis*) mushroom by size exclusion liquid chromatography with sequential on-line UV-ICP-MS detection. Journal of Agricultural and Food Chemistry, 2004, 52(5): 1315-1322.

[287] Wuilloud R G, Kannamkumarath S S, Caruso J A. Speciation of nickel, copper, zinc, and manganese in different edible nuts: A comparative study of molecular size distribution by SEC-UV-ICP-MS. Analytical and Bioanalytical Chemistry, 2004, 379(3): 495-503.

[288] Rodriguez-Cea A, Arias A R L, Fernandez de la Campa M R, Moreira J C, Sanz-Medel A. Metal speciation of metallothionein in white sea catfish, *Netuma barba*, and pearl cichlid, *Geophagus brasiliensis*, by orthogonal liquid chromatography coupled to ICP-MS detection. Talanta, 2006, 69(4): 963-969.

[289] Alvarado G, Murillo M. Multielemental fractionation in human peripheral blood mononuclear cells by size exclusion liquid chromatography coupled to UV and ICP-MS detection. Journal of Chromatographic Science, 2010, 48(9): 697-703.

[290] McNear D H, Jr., Afton S E, Caruso J A. Exploring the structural basis for selenium/mercury antagonism in *Allium fistulosum*. Metallomics, 2012, 4(3): 267-276.

[291] 李惠玲, 胡月, 孟佩俊, 张雪莹, 谢韵漪, 黄沛力. 碲化镉量子点稳定性测定的体积排阻高效液相色谱-电感耦合等离子体质谱法. 中华劳动卫生职业病杂志, 2017, 35(3): 217-220.

[292] Paydary P, Larese-Casanova P. Water chemistry influences on long-term dissolution kinetics of CdSe/ZnS quantum dots. Journal of Environmental Sciences, 2020, 90: 216-233.

[293] Paydary P, Larese-Casanova P. Separation and quantification of quantum dots and dissolved metal cations by size exclusion chromatography-ICP-MS. International Journal of Environmental Analytical Chemistry, 2015, 95(15): 1450-1470.

[294] Peng L, He M, Chen B, Qiao Y, Hu B. Metallomics study of CdSe/ZnS quantum dots in HepG2 cells. ACS Nano, 2015, 9(10): 10324-10334.

[295] Zhou X-X, Liu J-F, Jiang G-B. Elemental mass size distribution for characterization, quantification and identification of trace nanoparticles in serum and environmental waters. Environmental Science & Technology, 2017, 51(7): 3892-3901.

[296] Kannamkumarath S S, Wrobel K, Wrobel K, B'Hymer C, Caruso J A. Capillary electrophoresis-inductively coupled plasma-mass spectrometry: An attractive complementary technique for elemental speciation analysis. Journal of Chromatography A, 2002, 975(2): 245-266.

[297] Cheng H Y, Li P, Liu J H, Ye M Y. Coupling electrophoretic separation with inductively coupled plasma spectroscopic detection: Interfaces and applications from elemental speciation, metal-ligand interaction to indirect determination. Journal of Analytical Atomic Spectrometry, 2016, 31(9): 1780-1810.

[298] Deng B Y, Chan W T. Simple interface for capillary electrophoresis-inductively coupled plasma atomic emission

spectrometry. Journal of Chromatography A, 2000, 891(1): 139-148.

[299] Magnuson M L, Creed J T, Brockhoff C A. Speciation of arsenic compounds in drinking water by capillary electrophoresis with hydrodynamically modified electroosmotic flow detected through hydride generation inductively coupled plasma mass spectrometry with a membrane gas-liquid separator. Journal of Analytical Atomic Spectrometry, 1997, 12(7): 689-695.

[300] Tian X-D, Zhuang Z-X, Chen B. Movable reduction bed hydride generation system as an interface for capillary zone electrophoresis and inductively coupled plasma atomic emission spectrometry for arsenic speciation analysis. Analyst, 1998, 123(5): 899-903.

[301] Richardson D D, Kannamkumarath S S, Wuilloud R G, Caruso J A. Hydride generation interface for speciation analysis coupling capillary electrophoresis to inductively coupled plasma mass spectrometry. Analytical Chemistry, 2004, 76(23): 7137-7142.

[302] Alfredo Suárez C, Fernanda Giné M. A reactor/phase separator coupling capillary electrophoresis to hydride generation and inductively coupled plasma optical emission spectrometry (CE-HG-ICP OES) for arsenic speciation. Journal of Analytical Atomic Spectrometry, 2005, 20(12): 1395-1397.

[303] Deng B, Feng J, Meng J. Speciation of inorganic selenium using capillary electrophoresis-inductively coupled plasma-atomic emission spectrometry with on-line hydride generation. Analytica Chimica Acta, 2007, 583(1): 92-97.

[304] Álvarez-Llamas G, Fernández de laCampa M D R, Sanz-Medel A. ICP-MS for specific detection in capillary electrophoresis. TrAC Trends in Analytical Chemistry, 2005, 24(1): 28-36.

[305] Møller C, Stürup S, Hansen H R, Gammelgaard B. Comparison of two CE-ICP-MS interfaces and quantitative measurements of carboplatin in plasma samples using an internal standard. Journal of Analytical Atomic Spectrometry, 2009, 24(9): 1208.

[306] Suárez C A, Araújo G C L, Giné M F, Kakazu M H, Sarkis J E S. Sequential injection analysis (SIA) for arsenic speciation by capillary electrophoresis hyphenated to inductively coupled plasma sector field mass spectrometry (CE-ICP-SFMS). Spectroscopy Letters, 2009, 42(6-7): 376-382.

[307] Yang G, Xu J, Zheng J, Xu X, Wang W, Xu L, Chen G, Fu F. Speciation analysis of arsenic in *Mya arenaria* Linnaeus and *Shrimp* with capillary electrophoresis-inductively coupled plasma mass spectrometry. Talanta, 2009, 78(2): 471-476.

[308] Schaumloffel D. Capillary liquid separation techniques with ICP MS detection. Analytical and Bioanalytical Chemistry, 2004, 379(3): 351-354.

[309] Lee T H, Jiang S J. Determination of mercury compounds by capillary electrophoresis inductively coupled plasma mass spectrometry with microconcentric nebulization. Analytica Chimica Acta, 2000, 413(1-2): 197-205.

[310] Day J A, Sutton K L, Soman R S, Caruso J A. A comparison of capillary electrophoresis using indirect UV absorbance and ICP-MS detection with a self-aspirating nebulizer interface. Analyst, 2000, 125(5): 819-823.

[311] Yang G, Xu X, Wang W, Xu L, Chen G, Fu F. A new interface used to couple capillary electrophoresis with an inductively coupled plasma mass spectrometry for speciation analysis. Electrophoresis, 2008, 29(13): 2862-2868.

[312] Alvarez-Llamas G, de la Campa M R F, Sanchez M L F, Sanz-Medel A. Comparison of two CE-ICP-MS interfaces based on microflow nebulizers: Application to cadmium speciation in metallothioneins using quadrupole and double focusing mass analyzers. Journal of Analytical Atomic Spectrometry, 2002, 17(7): 655-661.

[313] Li B-H, Yan X-P. Short-column CE coupled with inductively coupled plasma MS for high-throughput speciation analysis of chromium. Electrophoresis, 2007, 28(9): 1393-1398.

[314] Sun J, He B, Yin Y, Li L, Jiang G. Speciation of organotin compounds in environmental samples with semi-permanent

coated capillaries by capillary electrophoresis coupled with inductively coupled plasma mass spectrometry. Analytical Methods, 2010, 2(12): 2025-2031.

[315] Möser C, Kautenburger R, Philipp Beck H. Complexation of europium and uranium by humic acids analyzed by capillary electrophoresis-inductively coupled plasma mass spectrometry. Electrophoresis, 2012, 33(9-10): 1482-1487.

[316] Kinzer J A, Olesik J W, Olesik S V. Effect of laminar flow in capillary electrophoresis: Model and experimental results on controlling analysis time and resolution with inductively coupled plasma mass spectrometry detection. Analytical Chemistry, 1996, 68(18): 3250-3257.

[317] Sutton K L, B'hymer C, Caruso J A. Ultraviolet absorbance and inductively coupled plasma mass spectrometric detection for capillary electrophoresis: A comparison of detection modes and interface designs. Journal of Analytical Atomic Spectrometry, 1998, 13(9): 885-891.

[318] Lu Q, Bird S M, Barnes R M. Interface for capillary electrophoresis and inductively coupled plasma mass spectrometry. Analytical Chemistry, 1995, 67(17): 2949-2956.

[319] Meermann B, Bartel M, Scheffer A, Trumpler S, Karst U. Capillary electrophoresis with inductively coupled plasma-mass spectrometric and electrospray time of flight mass spectrometric detection for the determination of arsenic species in fish samples. Electrophoresis, 2008, 29(12): 2731-2737.

[320] Kang J, Kutscher D, Montes-Bayón M, Blanco-González E, Sanz-Medel A. Enantioselective determination of thyroxine enantiomers by ligand-exchange CE with UV absorbance and ICP-MS detection. Electrophoresis, 2009, 30(10): 1774-1782.

[321] Graser C-H, Banik N I, Bender K A, Lagos M, Marquardt C M, Marsac R, Montoya V, Geckeis H. Sensitive redox speciation of iron, neptunium, and plutonium by capillary electrophoresis hyphenated to inductively coupled plasma sector field mass spectrometry. Analytical Chemistry, 2015, 87(19): 9786-9794.

[322] Li J X, Umemura T, Odake T, Tsunoda K. A high-efficiency cross-flow micronebulizer interface for capillary electrophoresis and inductively coupled plasma mass spectrometry. Analytical Chemistry, 2001, 73(24): 5992-5999.

[323] Kovachev N, Aguirre M Á, Hidalgo M, Simitchiev K, Stefanova V, Kmetov V, Canals A. Elemental speciation by capillary electrophoresis with inductively coupled plasma spectrometry: A new approach by Flow Focusing® nebulization. Microchemical Journal, 2014, 117: 27-33.

[324] Sonke J E, Salters V J M. Capillary electrophoresis-high resolution sector field inductively coupled plasma mass spectrometry. Journal of Chromatography A, 2007, 1159(1-2): 63-74.

[325] Schaumloffel D, Prange A. A new interface for combining capillary electrophoresis with inductively coupled plasma-mass spectrometry. Fresenius' Journal of Analytical Chemistry, 1999, 364(5): 452-456.

[326] Li B H. Rapid speciation analysis of mercury by short column capillary electrophoresis on-line coupled with inductively coupled plasma mass spectrometry. Analytical Methods, 2011, 3(1): 116-121.

[327] Liu L, He B, Yun Z, Sun J, Jiang G. Speciation analysis of arsenic compounds by capillary electrophoresis on-line coupled with inductively coupled plasma mass spectrometry using a novel interface. Journal of Chromatography A, 2013, 1304: 227-233.

[328] Bendahl L, Gammelgaard B, Jons O, Farver O, Hansen S H. Interfacing capillary electrophoresis with inductively coupled plasma mass spectrometry by direct injection nebulization for selenium speciation. Journal of Analytical Atomic Spectrometry, 2001, 16(1): 38-42.

[329] Tangen A, Lund W, Josefsson B, Borg H. Interface for the coupling of capillary electrophoresis and inductively coupled plasma mass spectrometry. Journal of Chromatography A, 1998, 826(1): 87-94.

[330] Liu Y, Lopezavila V, Zhu J J, Wiederin D R, Beckert W F. capillary electrophoresis coupled online with

inductively-coupled plasma-mass spectrometry for elemental epeciation. Analytical Chemistry, 1995, 67(13): 2020-2025.

[331] Li J, Umemura T, Odake T, Tsunoda K I. A high-efficiency cross-flow micronebulizer interface for capillary electrophoresis and inductively coupled plasma mass spectrometry. Analytical Chemistry, 2001, 73(24): 5992-5999.

[332] Cheng H, Liu J, Yin X, Shen H, Xu Z. Elimination of suction effect in interfacing microchip electrophoresis with inductively coupled plasma mass spectrometry using porous monolithic plugs. Analyst, 2012, 137(13): 3111-3118.

[333] Hui A Y N, Wang G, Lin B, Chan W-T. Interface of chip-based capillary electrophoresis-inductively coupled plasma-atomic emission spectrometry (CE-ICP-AES). Journal of Analytical Atomic Spectrometry, 2006, 21(2): 134-140.

[334] Olesik J W, Kinzer J A, Olesik S V. Capillary electrophoresis inductively coupled plasma spectrometry for rapid elemental speciation. Analytical Chemistry, 1995, 67(1): 1-12.

[335] Tangen A, Lund W. Capillary electrophoresis-inductively coupled plasma mass spectrometry interface with minimised dead volume for high separation efficiency. Journal of Chromatography A, 2000, 891(1): 129-138.

[336] Liu L, Yun Z, He B, Jiang G. Efficient interface for online coupling of capillary electrophoresis with inductively coupled plasma-mass spectrometry and its application in simultaneous speciation analysis of arsenic and selenium. Analytical Chemistry, 2014, 86(16): 8167-8175.

[337] Cheng H, Xu Z, Liu J, Wang X, Yin X. A microfluidic system for introduction of nanolitre sample in inductively coupled plasma mass spectrometry using electrokinetic flow combined with hydrodynamic flow. Journal of Analytical Atomic Spectrometry, 2012, 27(2): 346-353.

第15章　原子光谱分析中样品制备技术

（钟燕辉[①]，肖小华[①]，胡玉斐[①]，李攻科[①]*）

▶ 15.1　固体分析法 / 404

▶ 15.2　试液制备法 / 406

▶ 15.3　相分离法 / 409

▶ 15.4　场辅助样品制备法 / 419

▶ 15.5　样品制备/原子光谱分析在线联用技术 / 426

▶ 15.6　总结与展望 / 430

---

①中山大学，广州。*通讯作者联系方式：cesgkl@mail.sysu.edu.cn

第15章 原子光谱分析中样品制备技术

本章导读

- 综述了近十年来原子光谱分析中样品制备技术的研究进展。
- 对比了各种样品制备技术在原子光谱分析应用中的优点和局限。
- 讨论了原子光谱分析元素总量及形态时样品制备技术的差异。
- 展望了原子光谱分析中样品制备技术的发展趋势。

原子光谱主要分为：原子发射光谱（AES）、原子吸收光谱（AAS）、原子荧光光谱（AFS）、X射线荧光光谱（XRF）和等离子体质谱法（ICP-MS）。原子光谱主要适用于无机元素微量及痕量分析，所测样品包括固体和液体，其样品制备技术对测定误差影响很大。原子光谱分析仪器种类繁多，根据仪器所需样品状态（液体或固体），采取不同样品制备方法[1]。结合各种样品制备技术，原子光谱技术在地质、冶金、机械、化工、农业、食品、轻工、生物医药、环境保护、材料科学等领域得到广泛应用。

样品制备过程是原子光谱分析中耗时最长的一个环节，也是后续检测的关键，往往决定测定结果的准确度。样品制备目的主要有以下5点：样品种类繁多、复杂程度高，需消除基体或共存物质干扰[2,3]；浓缩、富集和稀释调节样品浓度，使其达到最佳检测范围[4]；不适合后续分离或检测的物质，不能完全消解法，需进行介质置换，如元素形态分析[5]；保护仪器，避免仪器系统污染，延长仪器使用寿命[6,7]；缩短样品制备时间、减少试剂消耗[8]。

样品制备技术是将分析物从样品基体中转移到定性和定量评价的过程。这个过程包括破坏样品和不破坏样品[9]。对于不破坏样品的方法，属于直接分析法，仅需简单的处理过程，几乎不消耗系统能量，也无需引入外部能量，是简单、快速、不费时费力的分析法[10]。大多数样品制备法都需要破坏样品，将待测元素从样品中溶出。溶解稀释和消解是样品混乱度增大（熵增）过程，熵增是热力学第二定律的自发过程，以开放系统的耗散结构牺牲能量为代价而进行[11]。相分离过程使样品中分析物从混乱到有序，为熵减的过程。通过引入外部能量，如微波、超声、热场、电场和光能等提高样品制备效率、缩短消解或萃取时间，使热力学第二定律的非自发过程最大限度地快速进行[12]。样品制备装置与分析仪器联用技术能自动或半自动处理样品，提高分析速度和灵敏度，是现代仪器分析发展的重要方向之一[13]。以下综述了原子光谱分析中样品制备技术（图15-1）的现状及进展，旨在为原子光谱分析样品制备技术的发展提供参考。

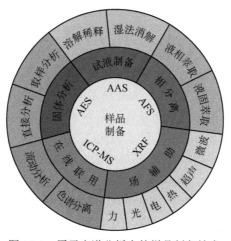

图15-1 原子光谱分析中的样品制备技术

15.1 固体分析法

固体分析法能直接使用仪器分析极少量的固体样品，从中获得有关元素的组成信息[14]。在原子光谱测定化学元素的特定领域，固体分析法相比湿化学法有几个优点：简单和快速的样品预处理；降低污染和分析物损失风险；由于样品未被稀释，分析物的可检测性更强；减少了危险试剂的使用和废物产生；消耗样品量小；能获得高通量分析结果[10]。

固体分析在原子光谱分析领域得到了广泛应用，固体样品直接分析法使用的原子光谱仪有较强激发光源，直接用光源激发固体样品中待测物，如 X 射线荧光光谱（XRF）、激光诱导击穿光谱（LIBS）技术和辉光放电质谱（GD-MS）等。固体取样法一般利用电热蒸发（ETV）、石墨炉（GF）原子化和激光烧蚀（LA）等，直接将样品中分析物引入检测器。

15.1.1 直接分析法

XRF 所测样品包括固体、粉末和液体，但样品中不能含水、油和挥发性成分，更不能有腐蚀溶剂，其样品制备技术对测定误差影响很大。XRF 是一种快速、无损的多元素分析技术，用于测定元素百分比含量。虽然 XRF 检出限比其他原子光谱法（如 ICP-MS、AES 和 AAS）高，但其便携性及易操作性使其成为现场分析中的有力工具，用于食品、药品、化妆品和香料[15]的检测，硬化水泥膏固化后新鲜断裂表面的检测[16]，手机印刷电路板[17]的检测。

便携式 XRF 技术在考古学和地质学中有着广泛的应用空间[18]。例如辅助确定中国青花瓷起源时间和地点[19]、研究壁画颜料[20]、土地利用类型和土壤退化[21]分析研究以及碳酸盐露头锤击新鲜岩屑的分析[22]。便携式 XRF 技术具有现场测量优势，同时能提供可靠定量结果。

对于不能直接分析的样品，通常需要将样品直接沉积在石英玻璃样品载体上，干燥后用 XRF 测定，如头发[23]、燃料油和原油[24,25]、尿液和血浆[26]、浮游生物标本[27]、电子烟内部液体样品[28]、动物组织[29]、防晒产品[30]、指甲油[31]和氧化硅催化剂[32]。XRF 具有高通用性、快速同时多元素检测，样品制备简单，分析速度快的优点，因此在急诊医学、法医鉴定和质量监控中应用广泛。

激光诱导击穿光谱（LIBS）技术使用高能激光光源，在分析材料表面形成高强度激光光斑，使样品激发发光，通过检测系统分析。该技术无需对样品进行烦琐化学处理，对样品破坏小，具有快速实时、可远程监测等特点[33]。LIBS 无损分析技术可用于文物化学成分的研究，如故宫出土的陶瓷[34]、古迹自然沉积物[35]和混凝土[36]等。

由于基体干扰、异质性和样品粒度影响，复杂样品定量分析校准是 LIBS 中最困难的问题之一。Carvalho 等[37]提出了硼酸盐熔合校准高硅含量样品的方法。熔融法制备样品以最大限度地减少基体组分强吸收效应，提高测定结果准确度，实现了 LIBS 复杂样品的定量分析。

对于不能用 LIBS 直接分析的样品，通常制备成微丸或粉末。Casado-Gavalda 等[38]将动物组织制备成直径 1.3 cm 的球团，LIBS 能成功地绘制出球团内 Cu 含量。LIBS 能提供样品空间信息，分析异质性样品。球团颗粒会降低检测的灵敏度，黏结剂的排放也可能引起干扰。Suyanto 等[39]使用亚目支撑微网架制备粉末样品，LIBS 可显示样品中所有主要元素的发射线。由于丸状样品基体效应明显，使用粉末样品获得的光谱质量更优越，具有更好的强度。

LIBS 和 XRF 的联合使用能提供多层样品元素组成的必要信息。Pospíšilová 等[40]采用 LIBS 联合

XRF深度解析分析多层历史画架绘画模型样品。LIBS深度剖面法能区分不同材料组成层和估计其厚度，但容易产生拖尾效应。X射线信号取决于材料层厚度和特定层对X射线辐射吸收，X射线能更好地穿透有机材料，引起信号衰减以及信号放大，在测量中会观察到重元素存在。两种方法都有局限性，但是通过比较可得到相应信息。

辉光放电既可作为光源也可作为离子源被应用到固态样品含量和深度分析。Di Sabatino等[41]采用辉光放电质谱（GD-MS）分析了太阳电池的硅杂质。

15.1.2 取样分析法

激光烧蚀（LA）是一种常用的固体取样法。Measures等[42]最早将LA技术结合发射光谱法测定元素含量。使用激光器进行一次激光轰击，在固体样品上产生一个直径很小深坑，使等离子体气流过样品表面，消耗的样品经一导管导入等离子炬，样品在焰炬中电离进行光谱分析。LA使用大功率、高重复激光器，大面积取样[43]，采样均匀性好，且不受样品种类限制，可实现样品微小区域的分析。

LA通常作为ICP-MS固体样品引入装置，这项技术经历快速发展并得到广泛应用[44]。如测定人眼组织结构中元素分布[45]，木乃伊头发样品中As[46]，植物叶片中Ca、K和Mg[47]，玉米根横截面中Hg[48]和药物产品中元素杂质[49]，人造玻璃中Pb同位素[50]，废弃高分子材料中金属元素[51]等。LA-ICP-MS可在痕量和超痕量浓度水平上进行相对快速、精准和空间分辨的元素和同位素比值测量，且样品制备简单。

LA作为样品引入系统的优点是样品制备简单并具高空间分辨率，对于不能直接引入系统的样品，可采用简单方法将样品制备成干燥固体。干滴法只需在过滤器表面上放置液体样品，将溶液转化为一个小的、薄的干燥点，作为LA的分析对象[52]。Hsieh等[53]用LA-ICP-MS在滤膜上直接分析干燥血液样品中多种元素。Papaslioti等[54]将植物和食品材料压制成微球，使用LA-ICP-MS进行元素分析。与液体样品引入系统对比，LA样品处理量更大、样品制备简单、符合绿色化学要求，避免了危险试剂使用。

电热蒸发（ETV）是一种有效的直接固体取样技术，将样品和残渣一起放入自动取样器中，在测量前，样品基体在ETV炉中以热解步骤直接除去。它与ICP-OES、ICP-MS和AAS结合，无需大量样品制备，就能快速分析多种元素。ETV-ICP-OES成功用于多类型样品的分析，如生物样品[55]、食品[56]、化石燃料[57]、塑料[58,59]和药用粉末[60]等。Mello等[61]采用ETV-ICP-MS直接分析石墨中微量稀土元素，避免烦琐制样步骤，减少干扰。ETV-ICP-MS分析高纯铜材料时，使用CHF_3作为卤化试剂，避免熔融样品从ETV到等离子体系统的高基质输入，充分实现了基体分离[62]，该法省时经济，在很大程度上实现了自动化。

将ETV与原子光谱技术结合，如电热原子吸收光谱（ETAAS），可分析整个样品中微量元素和主要元素，制备样品的工作量小，分析时间短，从而实现连续测量大量样品。GF是非火焰原子化器，是ETV中广为应用的一种。在石墨炉原子吸收光谱（GFAAS）中样品通常以液体溶液的形式引入，但GFAAS也可直接分析固体样品。Santos等[63]采用GFAAS直接测定了99.5%纯度聚酰亚胺中Cr、Cu、Mn、Na和Ni，其结果与微波辅助消解和中子活化分析的结果吻合。Resano等[64]使用GFAAS直接测定了大型水蚤样品中Zn，单个样本测量时间为2 min，可实现无脊椎动物的无污染分析，其唯一样品的准备步骤是干燥。GFAAS直接测定固体样品，不需要复杂样品制备过程，样品用量少。

GFAAS直接分析所用样品量通常很少，对于小质量样品，固体材料均匀性较低，得到的精度较差。通过引入高分辨率连续光源石墨炉原子吸收光谱（HR-CS-GFAAS），扩展了GFAAS直接分析固体的适用范围和优势。与线源原子吸收法相比，高分辨率连续光源原子吸收法具有较高灵敏度、有效的背景校正、快速选择波长的优点。HR-CS-GFAAS 能直接同时测定面粉中 Cu 和 Fe[65]，空气中可吸入颗粒物中的 Cl 和 Br[66]，电气和电子设备聚合物中 Cr 和 Sb[67]，茶叶中 Cd 和 Cr[68]，冻干蔬菜样品中 Cd、Cr 和 Cu[69]。Castilho 等[70]比较了直接固体样品分析、微波辅助酸浸出和超声辅助酸萃取玻璃纤维过滤器上的空气颗粒物，采用 HR-CS-GFAAS 测定样品中 Cu、Mo 和 Sb，发现 3 种方法测定结果无显著性差异，但固体分析法缩短了分析时间。Gonzalez-Alvarez 等[71]用萃取剂萃取海水样品中微量 Ag，采用 GH-CS-SS-GFAAS 直接分析含萃取剂和分析物的搅拌棒。HR-CS-GFAAS 直接分析固体样品，简便快速、灵敏准确，且样品消耗量少，样品制备步骤简单。

15.2　试液制备法

大部分原子光谱仪器适合液体样品引入，因此在原子光谱分析中样品多以液体形式进行分析。将样品转化为易于处理的液体，方便进一步精确分配和准备校准系列[72]。研究人员使用各种形式的溶解稀释和消解将各种状态的样品转化为易于分析的液体。

15.2.1　溶解稀释法

基于液体样品的原子光谱分析，稀释是减少基体干扰的常用方法。根据样品和分析仪器的实际情况，使用不同试剂溶解或稀释样品，以减小基体或共存物质干扰，使其浓度达到最佳检测范围。溶解稀释法是一种简单快速的待测元素溶出技术，对比消解法，有独特优势。采用消解法及稀释法处理红酒[73]、果汁[74,75]、人类胎盘[76]、鲜鱼[77]和鱼油[78]的测定结果无显著差异。消解法可能导致样品污染，影响检测结果的准确度，而简单稀释能避免这个问题。如稀释法可提高 ICP-MS 测定全血中 Ti 的特异性[79]。血浆和全血样品制备时用 Triton X-100、抗泡剂 B 和 L-半胱氨酸稀释，可抑制氢化物发生原子吸收光谱（HG-AAS）检测过程中氢化物发生器的过度泡沫[80]。溶解稀释法与消解法相比，试剂用量少、污染风险小、快速可靠、分析时间短。

溶解稀释法基于相似相容的原理，针对不同的样品须使用相应的溶剂。如液体饮料样品用水[81,82]或酸[83-85]稀释；血液样品用水[79]或稀酸[86]稀释；乙醇燃料用稀酸稀释[87]；蜂蜜用非离子表面活性剂（Triton X-405）和稀酸稀释[88]；药材样品用 N,N-二甲基甲酰胺稀释[89]；生物组织用中性磷酸盐缓冲生理盐水[90]、甲酸[91]和苄基三甲基氢氧化物溶液[92]稀释；化妆品用无水乙醇和琼脂溶液稀释[93]。GFAAS 直接同时测定稀释样品，需要用化学基体改进剂配合溶剂稀释样品，如 Pd(NO$_3$)$_2$ 和钨钌基体改进剂配合稀酸稀释牛奶[94]；Pd(NO$_3$)$_2$ 配合水稀释细胞[95]；Pd(NO$_3$)$_2$ 和 Mg(NO$_3$)$_2$ 配合 Triton X-100 和稀酸稀释血液[96]。

难溶于酸的样品，若能在合适分散剂中保持稳定，可将样品制备成悬浮液直接引入检测系统。如部分金刚石纳米颗粒可自发形成胶体溶液[97]；聚乙烯亚胺为分散剂（pH=4.0）可制备流动性在 30%（m/v）以下的陶瓷碳化硅料浆[98]；聚丙烯酰胺是制备氮化硅悬浮体的良好分散剂[99]。全反射 X 射线荧光光谱法（TXRF）常用于直接测定悬浮液样品，如小麦粉[100]、氧化铜锌矿[101]和煤灰[102]。悬浮液直接引入法为难溶样品检测提供了一种多元素、快速简单和相对成本低的分析方法。

用溶剂稀释原油样品直接引入原子光谱是一种省时省力的元素分析法[103]，研究报道正丙醇[78]、二甲苯[104-107]、乙醇[108-110]、甲醇[109]、异辛烷[111]、甲基氢氧化铵[112]、N-甲基吡咯烷酮和 HNO$_3$（2%，v/v）[113]均可作为油脂样品稀释剂。Doyle 等[114]用甲苯稀释原油样品后，放置在一张滤纸的中心，该滤纸夹在两个聚丙烯薄膜箔之间，并连接到 XRF 直接测定原油中 S、Ca、Fe、Ni、V，避免了复杂的样品制备。

与溶解稀释法相似的微乳液制备法可用于制备油类样品。将油类样品制备成纳米微乳直接分析，所得到的微乳液是透明的，它是油和水的混合物，使用表面活性剂和低分子量的溶剂为共溶剂[115]。其优点是制备方便、稳定性良好、黏度低、样品处理量大、避免使用有毒溶剂等。油脂样品的微乳液一般由油脂样品、Triton X-100、HNO_3 和有机试剂组成，有机试剂包括丙醇[116-120]、乙醇[121]、丁醇[121]、二甲苯[122,123]，或直接将样品、水和丙醇制备成微乳液[124]。根据所检测的元素和测定仪器选择不同的物料配比制备不同的微乳液。特定顺序制备微乳液能保证乳液稳定性，Aranda 等[125]在生物柴油混合物中加入 0.25 mL 浓硝酸，保持 15 min 后加入 Triton X-100 和异丙醇，放入超声浴中 15 min，再加入超纯水，将得到的溶液直接注入 GFAAS 系统中分析，与微波封闭酸消解法的结果吻合。

基于乳状液和微乳体系的粗植物油样品制备简单快速，使用低毒溶剂和最少的制备操作步骤。Bohrer 等[126,127]采用硬脂酸辛酯为油相，Tween-80 和 Triton X-100 为表面活性剂制备微乳，通过增加表面活性剂-水外相体积，相应减小内相体积改变水油比，便于萃取巧克力中不同金属元素。类似方法还可用于鸡蛋中金属元素测定[128]。与其他样品制备法相比，该法更简单，所需试剂少，且重复性好，经济实用。表 15-1 列出了溶解稀释法的应用进展，针对不同样品选择合适的溶剂。

表 15-1 溶解稀释法应用进展

样品	溶解或稀释试剂	分析仪器	参考文献
食品	HCl 和 Cs 溶液	HR-CS-GFAAS	[73]
	HNO_3（2%，v/v）	ICP-AES/OES	[74, 75]
	水	GFAAS	[82]
	HNO_3（0.5 mol/L）	FAAS/FAES	[83]
	王水	ICP-OES	[84]
	HNO_3（1%，v/v）	GFAAS	[85]
	HCl 和 Triton X-405	GD-OES	[88]
	N,N-二甲基甲酰胺	ICP-AES	[89]
	HNO_3（1%，v/v）	GFAAS	[94]
	Triton X-100（1%，v/v）	TXRF	[100]
动物组织或体液	Triton X-100（1%，v/v）	TXRF	[76]
	水	ICP-MS	[79]
	HNO_3（0.1%，v/v）	ICP-MS	[86]
	Triton X-100、硝酸和水	HR-CS-GFAAS	[96]
	$Pd(NO_3)_2$ 和水	GFAAS	[95]
	中性磷酸盐缓冲生理盐水	ICP-AES	[90]
	甲酸	ICP-MS	[91]
	苯基三甲基氢氧化物溶液（40%，v/v）	GFAAS	[92]
化妆品	无水乙醇和琼脂溶液	HG-AFS	[93]
固体无机材料或废料	水	ICP-AES	[97]
	聚乙烯亚胺（2%，v/v）	ICP-OES	[98]
	聚丙烯酸胺	ICP-OES	[99]
	乙二醇	TXRF	[101]
	HNO_3（1%，v/v）	TXRF	[102]

续表

样品	溶解或稀释试剂	分析仪器	参考文献
油类样品	正丙醇	HR-CS-GFAAS	[78]
	HNO_3（0.1 mol/L）	HR-CS-GFAAS	[87]
	甲醇或乙醇	WC-AES	[109]
	甲基氢氧化铵	ICP-MS	[112]
	N-甲基吡咯烷酮和 HNO_3（2%, v/v）	FAAS、ICP-MS 和 ETAAS	[113]
	二甲苯	ICP-OES/MS 和 GFAAS	[104-107]
	异辛烷	ICP-OES	[111]
	甲苯	XRF	[114]

在稀释法中，确定最佳稀释因子需烦琐耗时的离线样品制备，因为稀释对发射谱线和基质干扰的影响不同。Cheung 等[129]针对这一问题，使用 HPLC 泵在线混合样品和稀释剂，并引入 ICP-AES 雾化器。对校准标准品和含基质样品进行线性梯度稀释。将两条发射线（来自同一或不同元素）的信号比值作为稀释因子的函数，分析者不仅能识别是否存在基体干扰，还能确定克服干扰所需的最佳稀释因子。Donati 等[130]提出标准稀释分析（SDA）法，可应用于大多数接受液体样品的仪器技术，并能同时实现两个波长检测。它结合了传统的标准加入法和内标法，校正了由样品大小、方向或仪器参数变化而引起的基体效应和波动。单样品的 SDA 结合 ICP-MS 分析仅需 200 s。既不需要准备一系列标准溶液，也不需要构造通用校准图。该法通过将两种溶液组合在一个容器中进行：第一种溶液含 50%的样品和 50%的标准混合物；第二种含 50%的样品和 50%的溶剂。当第一种溶液被第二种溶液稀释时，实时收集数据。得到的数据用于绘制二维图，y 轴为分析物/内部标准信号，x 轴为内部标准浓度的倒数。样品中分析物的浓度由该图的斜率和截距的比值决定。SDA 的准确度和精密度都优于外标法、标准加入法和内标法[131]。SDA 能与多种原子光谱技术结合，测定实际样品[132]。SDA 为复杂基体样品分析提供了一种更简单、更快和更准确的方法。

金属结合蛋白的分析需谨慎小心样品制备操作，以确保金属蛋白复合物保持其自然状态，并在样品分析期间保持金属元素形态。研究表明，改变溶剂极性、pH 值、离子强度以及蛋白质离子疏水性对金属在蛋白质的保留有直接影响[133]。

15.2.2 湿法消解

原子光谱分析法中，消解是将样品中有机质破坏，使样品消解成试液的过程，广泛应用于元素总量分析。场辅助消解法应用广泛，后续将单独介绍，这里主要讨论试剂消解法。李攻科等[134]用常温消解和密闭微波消解处理口服液样品，研究表明，常温消解方式采用 HNO_3、$HClO_4$ 和 H_2O_2 混合物消解剂，可不加基体改进剂直接用 GFAAS 测定试液。常温试剂消解法比场辅助消解法仍有优势。采用开放式冷却体系湿法消解，可避免蒸汽、分析物及试剂的损失，也可通过加热减少消解时间[135]。半封闭系统加快了消解过程，避免了污染及分析物挥发的损失，提高了分析的准确度。

强酸溶液是最常用的消解试剂[136]，消解液直接定容时，酸浓度过高不利于原子光谱测定，需除掉多余的酸（赶酸）。HNO_3 对许多样品有良好的消解效果，如石笋[137]、肉类[135]、咖啡[138]和茶叶[139]。而在高压灰化设备辅助消解过程中，过氧化物对样品的消解没有任何显著的改善，因此 HNO_3 为最佳消解试剂[140]。浓 HNO_3 是环境不友好试剂，不符合绿色化学的理念。消解液中加入其他辅助溶剂可减少 HNO_3 消耗，如 H_2O_2 和 HNO_3 消解细胞[95]、HNO_3 和 Triton X-100 消解水稻[141]。H_2O_2 与 HNO_3 联合用于样品消解，在不降低消解效率的前提下，可使全脂奶粉和牛肝脏消解中通常使用的 HNO_3 分别减少 14 倍和 9.3 倍[142]。Thomaidis 等[143]用 HNO_3 消解虾样品过程中使用柠檬酸提高了方法灵敏度，降低了检出限。

HNO_3 和 HF 混合物常用于消解玻璃样品[144]。消解溶液 HNO_3、H_2O_2 和 HF 可用于消解葡萄干样品[145]。与单用 HNO_3 或 HF 消解植物样品中硅质物质相比，H_2O_2 提高了消解液的消解能力。HF 是十分危险的试剂，在赶酸过程中使用不当会对操作人员造成伤害。Mutsuga 等[146]使用不含 HF 的 KOH 和 H_3BO_3 溶液消解食品添加剂中 TiO_2 和某些硅酸盐中 SiO_2。因此，碱有望代替 HF 作为硅质物质消解试剂。

HCl 也常应用于样品消解中。例如，浓 HCl 消解金属容器[147]，HNO_3 与少量 HCl 混合消解口红[148]，用稀 HCl 消解食品萃取所有形态的 Fe[149]，用 $CuBr_2$ 和 HCl 溶液使有机汞从土壤中释放出来[150]。

根据样品中测定的分析物，使用不同消解试剂。针对地质样品中金属元素，Niedzielski 等[151]使用氢氟酸消解样品测定元素总浓度，王水萃取法测定类总元素浓度，盐酸萃取法测定酸浸馏分中元素含量。

油类样品的消解试剂一般是有机试剂、表面活性剂和疏水溶剂混合物。Azcarate 等[152]用二甲苯、Triton X-114 和 H_2O 消解旧润滑油。该法样品制备简单，能降低污染风险和分析物损失，对磨损金属的定量分析有显著改善，是一种直接快速的样品制备法。

对于蛋白质样品，酶是高效又环保的消解试剂。模拟胃液和模拟肠液能有效释放蛋白质样品中 Ca、Cu、Fe、Mg 和 Zn 等元素[153]。酶水解过的肉类样品，能直接用水萃取其中 As[154,155]。酶解法使用少量溶剂或不使用溶剂，是绿色简单的样品制备法。

消解溶液不但能破碎样品，还能萃取样品中目标分析物。以 Na_2CO_3（0.1 mol/L）处理红茶、绿茶和草本茶可萃取总 Cr[156]，酸性过硫酸盐萃取土壤悬浮液中总 P[157]，四甲基氢溴酸溶液能直接萃取生物样品中无机汞[158]。Aranha 等[159]用四甲基氢化铵溶液萃取肉类样品中 Cd，仅需在室温下稳定、均匀搅拌 10 min。该法比传统煅烧法所需样品量减少 20 倍，样品制备时间也减少 500 倍，这非常有利于常规实验室的分析。

15.3 相 分 离 法

根据目标物的物理、化学和生物特性，相分离法可实现复杂样品中目标物分离和富集，是使分析物从原始样品随机状态到高度有序的预检测状态，属于熵减过程。相分离法一般包括液-液萃取法和液-固萃取法。液-液萃取的原理是相分配，液-固萃取的原理是吸附或离子交换[160]。本节总结了近十年相分离技术在原子光谱分析样品制备中的应用进展。

15.3.1 液-液萃取法

液-液萃取的原理是相分配，用溶剂分离和萃取液体混合物中组分的过程。液-液萃取法又称溶剂萃取或抽提。在液体混合物中加入与其不相混溶（微溶）的溶剂，利用其组分在溶剂中不同溶解度而达到分离或萃取目的。一般根据样品和所测定的元素选择不同的萃取溶剂。

Tanase 等[161]用水溶液萃取法萃取加工食品中 K 和 Na。Barros 等[162]用 HNO_3（1%，w/v）或纯水溶液萃取生物柴油中 Na、K、Ca 和 Mg。选择合适的温度条件和样品状态可提高萃取率[163,164]。

在液氮中保存的新鲜脑和肝组织，用 Tris-HCl 缓冲液低温萃取，与同一样品的酸消解进行比较，脑内 Mn 的萃取率从 17%提高到 26%，肝内 Mn 的萃取率从 28%提高到 44%[165]。冷冻法不能提高从植物中萃取分析物的萃取率。Jedynak 等[166]使用不同预处理法制备芥末植物样品，分别是有无使用液氮处理新鲜、冷冻和干燥植物，最适合的样品预处理是不使用液氮萃取干燥植物材料。因此，处理不同样品的气氛和温度条件是不同的。Alava 等[167]强调样品的颗粒大小影响 As 的萃取率，当使用水为萃取溶剂时，全粒大米的萃取效率是粉状大米的 75%。

在萃取前对样品进行合适的处理，能使元素保持其原始形态。Bluemlein 等[168]推荐以新鲜植物作为植物螯合蛋白复合物的物种形成分析研究，因为常见的冻干过程会促进 As-植物螯合肽复合物的解体。当无法进行新鲜植物分析时，应在分析前进行冷冻（80℃）保存物种。Zhang 等[169]萃取冻干或磨碎植物时，没有检测到植物螯合肽复合物的存在。而使用新鲜植物和采用温和萃取条件，则可观察到其色谱峰。

样品制备是物种形成的关键步骤之一，在此过程中可能会导致样品损失、污染、物种间相互转换，甚至是对分析物的低萃取[170]。浓 HNO_3 能使 Cu 从二硫代氨基甲酸络合物中释放出来[171]。对于样品中不同物种形态萃取，有几种溶剂以及它们的混合物可供选择，包括甲醇、水、硝酸、甲酸、酶解液、磷酸缓冲液和三乙胺-碳酸缓冲液[172-174]。相关辅助萃取法也很多，如搅拌和加热[172]、超声[164]、压力[175,176]或微波[176]，甚至这些方法的组合来完成[177,178]。这里重点介绍溶剂萃取法，场辅助萃取法在后面详细介绍。

稀酸溶液为萃取液能保留样品中元素不同形态。酸类溶剂被用于萃取植物组织中与植物螯合素相连接的 As(Ⅲ)，这是由于复合物的稳定性会随 pH 值降低而增加。如甲酸（1%，w/v，pH=2.2）可作为低温萃取测定 As-植物螯合肽复合物的溶剂[168]。H_2SO_4（1.0 mol/L）为萃取液能同时保留 As(Ⅲ)和 As(Ⅴ)[179]。王水萃取无机 As[As(Ⅲ, Ⅴ)]和有机 As[二甲基砷酸盐（DMA）、单甲基砷酸盐（MMA）]时，As(Ⅲ)被氧化为 As(Ⅴ)，两种有机 As 形态同时被保存[180]。

蛋白质萃取液是两种 As(Ⅴ)和 As(Ⅲ)的较好萃取剂，且蛋白质萃取液的形态和数量可随样品类型而变化。Rahman 等[181]使用水、甲醇、$NH_4H_2PO_4$ 和蛋白质萃取液萃取蔬菜样品中 As(Ⅲ)和 As(Ⅴ)。虽然 $NH_4H_2PO_4$ 萃取的 As(Ⅲ)和 As(Ⅴ)的量大致相同，但 As(Ⅲ)用蛋白质萃取液萃取的量是前者的两倍，萃取剂对 As(Ⅲ)的增溶作用存在差异，蛋白质可能与植物螯合素络合。

低共熔溶剂萃取法是一种新型绿色、快速的液相萃取法。低共熔溶剂必须包含氢键受体和氢键供体，如氯化胆碱-草酸[182,183]、氯化锌-乙酰胺[184]、氯化胆碱-柠檬酸[185]和苹果酸-脯氨酸低共熔溶剂[186]。低共熔溶剂萃取法通常需要外部能量辅助，提高萃取能力，如微波[182]、超声[183]和涡流[185]等，或引入非离子表面活性剂（Triton X-114）[184]，提高相转移比。Zounr 等[187]采用空气辅助低共熔溶剂萃取 Pb，以 4-(2-噻唑拉唑)间苯二酚为络合剂，在氯化胆碱苯酚低共熔溶剂中加入含分析物的水溶液，吸取混合物，用注射器注射 9 次，得到混浊溶液。低共熔溶剂萃取结合原子光谱技术能快速地进行元素形态分析。Akramipour 等[188]采用低共熔萃取溶剂从血液样品中萃取 Se(Ⅳ)-二乙基二硫代磷酸络合物。将 Se(Ⅵ)转化为 Se(Ⅳ)，再用 GFAAS 分析总无机 Se。

离子液体和超分子溶剂也是常用的萃取剂。Amjadi 等[189]用离子液体微萃取样品中 Sn。该法将 Sn(Ⅳ)-APDC 络合物萃取到小体积的 1-己基-3-甲基六氟膦酸铵离子液体中，经相分离后，用 FAAS 测定最终溶液中富集的分析物。Fang 等[190]使用硫醇功能化离子液体选择性吸附 Cd(Ⅱ)。Moradi 等[191]将超分子溶剂微萃取结合 GFAAS 测定水、尿、化妆品和食品中总 Se。两种形式的超分子溶剂由①1-辛醇的倒六角聚集（棒状胶束）和②辛酸在四氢呋喃的反胶束组成，将 Se(Ⅳ)与 APDC 络合，得到疏水络合物，并萃取到超分子溶剂相。

传统液-液萃取法需要消耗大量样品和溶剂，也耗费大量时间[192]，因此发展了各种液相萃取技术，如破乳萃取法、浊点萃取法和分散液-液微萃取法等。

15.3.1.1 破乳萃取法

破乳萃取法（EIEB）是指乳状液完全破坏，成为不相溶的两相。破乳实质上是消除乳状液稳定化条件，使分散的液滴聚集、分层的过程。

破乳过程是 EIEB 萃取中最关键的过程，通常采用加热[193,194]或离心[195]破乳，使溶液形成分离良好的两相或三相。Bakircioglu 等[196]用 EIEB 萃取食用油中 Cd、Cr、Cu、Fe、Mn、Ni、Pb 和 Zn 等元素。EIEB 的第一步是形成稳定的油包水乳剂的食用油，即食用油、Triton X-114 和 HNO_3 混合溶液。在乳剂形成后，用热水浴使乳剂破裂，形成三相：①上相，只含食用油的有机相；②中间相，含萃取金属的酸性水相；③下相，富含表面活性剂的相。与超声萃取和湿法消解对比，从 EIEB 中得到的金属元素含量平均值较高[197,198]。

乳化试剂是 EIEB 中关键的试剂，Cassella 等[199]研究了 Triton X-100、Triton X-114、Tween-20 和十二烷基硫酸钠对 EIEB 的影响。研究表明，采用 Triton X-100 和 Triton X-114 制备乳剂对金属萃取率更高。HNO_3 和正丙醇适用于生物柴油的常规分析[200]。选用合适的乳化试剂能提高分析的稳定性和灵敏度。

样品/萃取剂体积比也影响萃取效果。de Sousa 等[193]研究了 EIEB 对原油中 Ni 和 V 的萃取效果，研究表明，样品体积小于 3 mL 时，两种分析物的萃取率都较高。Fernandes 等[201]采用破乳萃取结合 FAAS 测定润滑油中 Ca、Mg、Zn，样品用甲苯（矿物油和半合成油）和乙苯（合成油）稀释，溶剂比例始终为 40%（v/v），当萃取液中 HNO_3 和 Triton X-114 浓度分别为 4.2 mol/L 和 3%（m/v）时萃取率最高。

15.3.1.2 浊点萃取法

浊点萃取法（CPE）是液-液萃取技术，它不使用挥发性有机溶剂，环境友好。以中性表面活性剂胶束水溶液的溶解性和浊点现象为基础，改变实验参数引发相分离，将疏水性物质与亲水性物质分离。该法已成功地应用于金属离子、金属螯合物、生物大分子分离与纯化。

CPE 在金属分离纯化过程中，无机元素与络合剂形成螯合物，再萃取到表面活性剂相。例如，二乙基二硫代磷酸盐络合 As 和 Hg[202,203]、8-羟基喹啉络合 V[204]、钙羧酸络合 Co[205]、二乙基二硫代氨基甲酸酯络合 Se(Ⅳ)[206]，V(Ⅴ)和 V(Ⅳ)在少量氧化石墨烯（GO）存在下被定量转移到富表面活性剂相[207]。

外场辅助能提高 CPE 效率，到达浊点后，产生的临界胶束在外场辅助作用下均匀地分散在水相中。Altunay 等[208]采用超声辅助 CPE 预富集食品和饮料中 Sb(Ⅲ)和 Sn(Ⅳ)。在十六烷基三甲基铵（既可提高灵敏度，又可辅助配体和离子的结合）存在下，Sb(Ⅲ)和 Sn(Ⅳ)与 2-(2,4-二羟基苯基)-3,5,7-羟色胺-4-桑色素形成稳定三元复合物，被萃取到聚乙烯乙二醇-单对壬基苯基醚胶束萃取剂中。Gürkan 等[209]采用超声辅助 CPE 对干果和蔬菜样品中微量 Ag^+ 和 Cd^{2+} 进行预浓缩，在 pH 值为 6.0 时，碘化钾过量，Ag^+ 和 Cd^{2+} 与红色染料 T 形成三元复合物，三元复合物被定量萃取到聚(乙基乙二醇-单对壬基苯基醚)胶束相，再采用 FAAS 测定。他们也采用该技术萃取牛奶、蔬菜和食品中微量 V 和 Mo，V(Ⅴ)和 Mo(Ⅵ)稳定阴离子草酸配合物与[9-(二乙胺)苯并[a]苯噁嗪-5-亚基]铵离子缔合，溶液 pH 值为 4.5，将形成的离子缔合物萃取到聚氧乙烯壬基苯基醚胶束相，再采用 FAAS 测定[210]。Khan 等[211]使用超声辅助 CPE 用于同时预浓缩不同类型胆结石患者血清中 Pb 和 Cd。超声辅助 CPE 是一种新型绿色的方法，与传统 CPE 相比，简化并加速了萃取过程。

15.3.1.3 分散液-液微萃取

样品制备是耗时且容易引入分析误差的过程,传统液-液萃取法操作烦琐、消耗大量的有毒有机溶剂。分散液-液微萃取(DLLME)是在均相液-液微萃取和CPE基础上形成的一种简单、快速的微萃取技术。具有有机溶剂用量少、操作简单、成本低、回收率高、富集倍数高、对环境友好特点。在DLLME萃取过程中,螯合剂与目标分析物形成络合物,萃取剂以细小的液滴均匀地分散到水相中,增大其与目标化合物的接触面积,快速达到平衡状态,短时间内实现分析物分离富集。根据样品和分析物,选择合适螯合剂、萃取溶剂和分散液。Marguí等[212]对比了中空纤维液相微萃取(HFLPME)和DLLME结合TXRF测定水中无机锑含量,采用DLLME处理的样品可实现低浓度范围Sb(Ⅲ)和Sb(Ⅴ)的测定。表15-2列出了DLLME法应用研究进展。

表15-2 DLLME法应用研究进展

样品	分析物	螯合剂	萃取溶剂	分散剂	参考文献
水样	Zn、Cu	5-Br-PADAP	四氯化碳	甲醇	[213, 214]
	Rh	5-Br-PADAP	$NaNO_3$	甲醇	[215, 216]
	Co	1-(2-吡啶偶氮)-2-萘酚	癸酸反胶束超分子团	四氢呋喃	[217]
	Co^{2+}、Cu^{2+}、Ni^{2+}、Zn^{2+}	二乙基二硫代氨基甲酸钠(DDTC)	1,2-DBE	DMSO	[218]
饮料	Cu、Fe	DDTC	二氯苯	甲醇	[216]
	Cr	双(2-甲氧基苯甲醛)乙烯二亚胺	1-硝基-3-甲基咪唑鎓盐	乙醇	[219]
	V	N-苯甲酰-N-苯基羟胺	1-己基-3-甲基咪唑四氟硼酸	—	[220]
	Cd	8-羟基喹啉	1-十二醇	甲醇	[221]
	Mo	4-氨基-3-羟基-1-萘磺酸	1-十一醇	—	[222]
植物样品	Mn、Cd	氯化1-丁基-3-甲基咪唑	间苯二酚	—	[223]
	Mn(Ⅱ)、Mn(Ⅶ)	1,3-二丁基咪唑六氟磷酸盐	氨氯化铵	甲醇	[224]
	Mo	8-羟基喹啉	1-十一醇	—	[225]
海产品	Hg	1,5-二苯基-3-甲硫醇	四氯化碳	丙酮	[214]
油类	Ca、Mg、Na、K	HNO_3	HNO_3	异丙醇	[226, 227]
	Ni			正丙醇	[228]
道路粉尘	Pd	—	四氯化碳	乙醇	[229]

螯合剂影响元素的选择性测定,需根据待测分析物和样品状态选择螯合剂。DLLME萃取油脂样品中元素,通常使用反向DLLME,HNO_3在油脂样品萃取中起着重要作用,将金属离子从生物柴油[226,227]和氢化植物油[228]转移到萃取溶液中,以有机金属分子或有机复合物形式结合在油性基质中。与油脂中元素测定的溶剂稀释或消解常规法相比,该法相对简单、成本低、试剂消耗量小。采用离子液体为螯合剂可进一步减小有机溶剂消耗量[213,215,223],如用1,3-二丁基咪唑六氟磷酸盐室温离子液体在溶液pH为10.0和7.0时,分别螯合Mn(Ⅱ)和Mn(Ⅶ)[224]。在无配体DLLME法中,只需萃取溶剂和分散剂,无需任何螯合剂[229]。

超分子DLLME法结合了DLLME与超分子预浓缩和反向胶束的优点。Jafarvand等[217]利用超分子DLLME法萃取环境水样中痕量Co^{2+},以1-(2-吡啶偶氮)-2-萘酚为络合剂吸附Co^{2+},并分散在四氢呋喃水混合物中,利用癸酸制成的反胶束超分子团萃取剂对Co^{2+}络合物进行微萃取。

由于等离子体对常用有机溶剂耐受性有限,DLLME联用ICP-AES会出现问题(溶剂黏度大和挥发性高)。1-十一醇、1-丁基-3-甲基-咪唑六氟磷酸和氯甲烷是DLLME中常见的萃取溶剂。在1-

十一醇和1-丁基-3-甲基-咪唑六氟磷酸萃取物中，进行简单甲醇稀释即可解决这个问题[230]。

以离子液体为萃取溶剂可提高DLLME萃取率，如采用$NaPF_6$改性的1-硝基-3-甲基咪唑嗡盐萃取水和食品中微量Cr[219]、$NaPF_6$改性的1-己基-3-甲基咪唑四氟硼酸超声辅助萃取水和牛奶中V[220]。磁性离子液体从水溶液中萃取$Se(Ⅳ)$和2,3-二氨基萘络合物后，施加外部磁场，目标物被磁性离子液体富集[231]。

分散剂能分散萃取剂，提高萃取效率。引入外场可加快萃取剂分散，如采用超声[220]和涡流[223,225]辅助分散萃取剂，无需外加分散剂，减少了有机溶剂消耗。Sorouraddin等[218]采用加热辅助DLLME，快速高效地从大容量水样中同时萃取微量Co^{2+}、Cu^{2+}、Ni^{2+}、Zn^{2+}。将含分析物和螯合剂水相温度调高，快速注入分散剂和萃取剂，混合物置于冰水浴中，离心分离，采用GFAAS测定沉淀相中富集物。

Liang等[232]发展了置换DLLME结合GFAAS测定Hg的方法。置换DLLME法基于两步DLLME。首先，Cu^{2+}与吡咯烷二硫代氨基甲酸酯（PDC）反应形成Cu-PDC，DLLME法萃取，用分散剂将沉积物相分散到含Hg^{2+}的样品溶液中，再进行DLLME。由于Hg-PDC的稳定性高于Cu-PDC，因此Hg^{2+}从预萃取的Cu-PDC络合物中取代Cu^{2+}，进入沉积相。预浓缩Hg^{2+}过程中，同时存在的PDC络合物稳定性较低的金属离子无法从Cu-PDC络合物中取代Cu^{2+}，消除了干扰。

15.3.1.4 其他液-液萃取法

超临界流体萃取具有萃取率高、产品纯度好、流程简单、能耗低等优点，绿色环保，几乎不使用有机溶剂，可用于分离金属离子[233]。在氟化-二酮螯合剂2,2-二甲基-6,6,7,7,8,8,8-七氟-3,5-辛二酮存在下，用CO_2成功萃取尿液中Ga[234]和蚀刻废水中In[235]。

溶剂棒微萃取（SBME）可将待测分析物预富集在微体积受体溶液中，选择性高、灵敏简便。三辛基甲基氯化铵离子液体与$CdCl_n^{(n-2)-}$发生离子交换，采用三相SBME系统分离富集海水样品中Cd。萃取步骤如下：采用毛细管纤维制成溶剂棒，并用热尖端和平头镊子密封其中一个纤维端。在光纤腔内填充受体溶液，热封光纤另一端，形成SBME系统。再将其浸入含载体的有机溶液中浸渍纤维孔隙，用超纯水冲洗纤维壁中过量有机溶液。最后，从样品中取出纤维，回收受体溶液，FAAS测定Cd[236]。López-López等[237]以离子液体溶解在含十二烷基-1-醇的煤油中作为受体溶液，富集海水中Ag。

固体-液体萃取是一种简单、方便和绿色的样品制备法。de Oliveira等[238]用固体-液体萃取结合FAAS测定香肠中Na和K。样品制备主要包括用超纯水和涡流搅拌从样品中萃取分析物。

单液滴萃取法环境友好，消耗样品和试剂较少。Neri等[239]研究了单液滴微萃取法萃取水样中$Cu(Ⅱ)$，采用1-(2-噻唑)-2-萘酚络合铜，有利于溶剂传质，单液滴机械化注入的辅助装置，改善了萃取系统的性能。

中空纤维液相微萃取（HFLPME）在环境分析中应用引起了人们的广泛关注。Es'haghi等[240]使用商业50 μL Hamilton微注射器，采用HFLPME萃取水样中Cu^{2+}。先将中空纤维浸渍在溶剂中进行壁浸渍，再用乙醇洗涤中空纤维外部将溶剂去除。将20倍的有机溶剂抽到微注射器中，注入纤维腔内，使萃取剂充满纤维通道。将其浸入含分析物供体水溶液中，搅拌溶液。萃取结束后，将溶剂缩回注射器，以FAAS分析目标物。此外，Frentiu等[241]提出了双液-液萃取法萃取海产品中甲基汞（MeHg），样品先在有机溶剂中萃取，再将有机相转移到L-半胱氨酸溶液中进行定量测定。

15.3.2 液-固萃取法

液-固萃取的原理是吸附或离子交换，萃取过程发生在固相与液相之间，目标物或杂质经吸附或离子交换保留在固体材料上，以此富集或分离目标物。主要包括固相萃取法、分散固相萃取法及固相微萃取法等。

15.3.2.1 固相萃取法

固相萃取（SPE）由液固萃取和液相色谱技术相结合发展而来，主要用于样品分离、纯化和浓缩。SPE 使用选择性高的固体吸附剂，可实现分析物快速分离，方法简单、成本较低、有机溶剂使用少，有较高的重复使用率和预浓缩因子，适合用于发展自动化检测技术。

大部分 SPE 技术的原理是基于固体吸附，在 SPE 过程中，选择合适的吸附剂是获得良好回收率和高富集率的关键因素。研究人员合成了各种类型的纳米材料、高分子聚合物和复合材料等固体吸附材料，选择性吸附目标物。表 15-3 列出了针对不同目标分析物的固体吸附材料进展。

表 15-3 针对不同目标分析物的固体吸附材料

	固体吸附材料	目标分析物	参考文献
聚合物复合材料	多壁碳纳米管/聚(2-氨基噻吩)纳米复合材料	Cd^{2+} 和 Pb^{2+}	[242]
	聚丙烯胺枝状大分子接枝多壁碳纳米管杂化材料	Au^{3+} 和 Pd^{2+}	[243]
	聚邻苯乙醇胺/多壁碳纳米管复合材料	Pb^{2+}	[244]
	聚-2-(5-甲基异噁唑)甲基丙烯酰胺-羧基-2-丙烯酰胺-2-甲基-1-丙磺酸-二乙烯基苯	$Cr(III)$	[245]
	聚乙烯亚胺包覆碳点	$Cr(VI)$	[246]
	含亚氨基二乙酸的螯合纤维	$Nd(III)$	[247]
	硫基吸附材料	MeHg	[248]
	功能化离子液体涂层聚四氟乙烯管	Hg	[249]
	含 S 和 N 的杂链聚合物	Pt、Pd 和 Rh	[250]
	GO-聚(2,6-二氨基吡啶)复合材料	Cd^{2+}	[251]
	螯合二胺双模板印迹吸附剂	Fe^{3+}	[252]
生物材料	没食子酸改性纤维素	Al^{3+}	[253]
	西藏雪莲花粉末	稀土元素	[254]
	辣木种子粉末	Cd	[255]
改性无机材料	硝酸镍浸渍活性炭热改性吸附剂	As	[256]
		Se	[257]
	杂化胺功能化的二氧化钛/二氧化硅纳米颗粒	Pb^{2+}、Cu^{2+} 和 Zn^{2+}	[258]
	富硫醇多面体寡聚倍半氧硅氧烷	Hg^{2+} 和 CH_3Hg^+	[259]
	生物形态多孔纳米钛酸钙	Ni^{2+}	[260]
	三元硅胶负载硫化铜纳米复合材料	$Cr(VI)$	[261]
	镍铝层状双氢氧化物纳米吸附剂	$Cr(VI)$ 和 $Mn(VII)$	[262]
	巯基十二酸修饰的 TiO_2 核-Au 壳纳米粒子	Cd	[263]
	1-(2-噻唑拉唑)-2-萘酚固定化纳米 TiO_2	Ni^{2+}	[264]
	1-甲基-3-丁基咪唑溴铵离子液体修饰的 SiO_2	Pb^{2+}	[265]
	1-丁基-3-甲基咪唑六氟磷酸盐离子液体修饰的 SiO_2	Cd	[266]
	1,3-二(正丁基)六氟磷酸咪唑离子液体修饰的 SiO_2	Pb	[267]
	$SiO_2/Nb_2O_5/ZnO$	Co^{2+}	[268]

续表

固体吸附材料		目标分析物	参考文献
改性无机材料	2,2-吡啶-SBA-15 介孔杂化材料	Cu^{2+}	[269]
	硅胶支撑的热解木质素	稀土元素	[270]

聚合物高分子链能吸附金属离子，原理是金属离子共享导电聚合物的=N—和—S—基团电子对。纳米材料和高分子聚合物复合材料对金属具有很高的亲和力，这是由于复合材料中存在良好的吸附位点（S 和 N）[242]。吸附材料对目标离子的耐受性一般超过其他离子[244]。聚乙烯亚胺包覆碳点不仅能选择性吸附痕量 Cr(Ⅵ)，且在 FAAS 中也是一种信号增强剂[246]。

采用固相萃取结合原子光谱法可实现元素的形态分析。如采用聚-2-(5-甲基异噁唑)甲基丙烯酰胺-羧基-2-丙烯酰胺-2-甲基-1-丙磺酸-二乙烯基苯同时吸附 Cr(Ⅲ)和 Cr(Ⅵ)，Cr(Ⅲ)在 pH=1.5～4.5 时定量洗脱，而 Cr(Ⅵ)无法被洗脱[245]。王建华等[259]使用富硫醇多面体寡聚倍半氧硅氧烷吸附 Hg^{2+} 和 CH_3Hg^+，用硫脲（2%，v/v）为洗脱剂回收 Hg。在冷雾化模式下利用 AFS 定量测定 Hg^{2+}，在热雾化模式下定量总 Hg。

镍铝层状双氢氧化物（LDH）纳米吸附剂结合 FAAS 对 Cr、Mn 进行形态分析。以 Ni-Al(NO_3^-) LDH 吸附 Cr(Ⅵ)和 Mn(Ⅶ)氧阴离子，并在 pH=6.0 时与 LDH 层间的 NO_3^- 交换，而 Cr(Ⅲ)和 Mn(Ⅱ)阳离子通过填充柱而不保留。分别采用 H_2O_2 和 KIO_4 的酸性溶液将 Cr(Ⅲ)和 Mn(Ⅱ)预氧化成 Cr(Ⅵ)和 Mn(Ⅶ)。该法成功于 Cr 和 Mn 总含量、Cr(Ⅵ)和 Mn(Ⅶ)以及 Cr(Ⅲ)和 Mn(Ⅱ)的测定[262]。

目标物的洗脱也是固相萃取的重要过程。洗脱溶液一般是稀 HNO_3[260,271]、HCl[265,272]、硫脲[273]。洗脱步骤使材料可重复利用，但也存在解吸收率不可恢复、非预期随机污染和空白值不一致等风险，可能影响分析的准确度、精密度和富集因子。Baysal 等[263]将吸附 Cd 的巯基十二酸修饰的 TiO_2 核-Au 壳纳米粒子浆液直接进入 GFAAS 分析。采用常规间歇式收集吸附剂上的 Cr(Ⅲ)，液固分离后，吸附剂料浆直接导入 GFAAS 分析，消除了洗脱过程中存在的问题。

基于离子识别的分离技术因其对靶离子高选择性而受到广泛关注。离子印迹聚合物制备的一般过程为：以待测目标离子和络合剂的螯合物为模板分子，制备新型离子印迹聚合物。如 Ni^{2+} 印迹聚合物[272,274]、Zn^{2+} 印迹聚合物[275]、Ru^{3+} 印迹聚合物[273]、K^+ 选择性冠醚[271]、Cu^{2+} 印迹聚合物[276]、Cd^{2+} 和 Pb^{2+} 印迹聚合物[277]、Pb^{2+} 印迹聚合物[278,279]、Pd^{2+} 印迹聚合物[280]和 Tl^+ 印迹聚合物[281]等。离子印迹聚合物对目标离子的选择性高于具有相同电荷和相似离子半径的竞争性金属离子。

用离子液体改性纳米材料可简单制备高选择性、高吸附量的固相吸附剂，广泛应用于各种样品中元素的萃取分析。如 Amjadi 等[264]采用 1-十六烷基-3-甲基咪唑溴化铵将 1-(2-噻唑拉唑)-2-萘酚固定在纳米 TiO_2 上制备高容量固相吸附剂，用于预浓缩痕量 Ni^{2+}。Cheng 等[267]采用双硫腙螯合 1,3-二(正丁基)六氟磷酸咪唑修饰 TiO_2 制备固相材料，富集食品中微量 Pb。离子液体修饰的 SiO_2 材料在特定 pH 条件下能保留目标离子，在 HCl（1.0 mol/L）溶液中完全洗脱[265,266]。因此，该类材料能重复多次利用，有良好的发展潜力。

离子交换树脂也是常用的固相萃取材料。如 Zn^{2+}（1%，w/w）改性 Dowex 1X8 树脂在 pH=7.0～8.0 的条件下，使用磷酸二氢钾/氢氧化钠缓冲溶液，富集 Zn^{2+}，用 HNO_3（0.1 mol/L）溶液洗脱[282]。离子交换大孔树脂 Amberlite XAD-16 与凝胶强离子交换树脂固相萃取柱结合 FAAS 测定蜂蜜中 Ca 和 Mg 的浓度[283]。反相和阳离子交换的两柱固相萃取法结合 FAAS 测定甜菜根汁中 Cu、Fe、Mn 和 $Zn^{[284]}$。离子交换柱的优点是富集能力强，重复利用次数多。

离子交换柱还可实现不同元素形态的分离。Issa 等[285]利用强碱阴离子交换树脂和杂种树脂结合

ICP-MS 测定水中 As 的浓度。水的酸度对 As 形态的控制起重要作用,将 pH 值调整到小于 8.0,强碱阴离子交换树脂保留 As(Ⅴ)并通过 As(Ⅲ)。在 pH=5~11,与杂种树脂结合的氧化铁颗粒能吸附所有无机 As。Smolíkova 等[286]采用羧酸阳离子交换树脂制备梯度扩散树脂凝胶,结合 ETAAS 测定水生生物中 As(Ⅲ)、As(Ⅴ)、DMA 和 MMA。在 NaCl（10 g/L）和 NaOH（10 g/L）的存在下,采用微波辅助树脂凝胶萃取 As 形态。新树脂凝胶具有均匀凝胶结构、良好重现性和较高吸附能力,可长期使用。改变固相萃取的 pH 值,或改性树脂,制备新型离子交换柱,可实现元素形态分离。

基质固相分散萃取法（MSPD）的原理是将固相萃取材料与样品一起研磨,得到半干状态混合物并将其作为填料装柱,用溶剂淋洗柱子,将各种待测物洗脱下来。Wang 等[287]将聚乙烯亚胺改性凹凸棒石材料作为 MSPD 固相载体,用于萃取海产品中 Cd。Chen 等[288]将样品与多壁碳纳米管混合,再转移到注射器中,洗脱液注入注射器中与样品反应,最后将洗脱液推出注射器,收集萃取液,可将少量样品中 Hg 有效萃取到洗脱液中。

15.3.2.2 分散固相萃取法

分散固相萃取（DSPE）是一种新的样品制备技术,具有简单安全、低溶剂用量、自动化等优点。DSPE 的吸附剂直接添加到含目标物的样品溶液中,而不形成柱。DSPE 促进目标物和吸附剂之间的相互作用,缩短样品制备时间。金属在固体材料上富集的机理可能包括吸附和还原。小尺寸固体材料提供大的表面积以实现高效吸附和还原反应的目的[289]。DSPE 法环境友好,仅使用少量纳米材料,不需要有机溶剂进行萃取,也不需要制备填料预浓缩柱。

在 DSPE 过程中,选择合适的吸附剂是获得良好回收率和高富集率的关键因素。磁性材料在外加磁场作用下,吸附剂很容易地从复杂基体中分离。由于相分离简单,磁性材料被广泛应用于 DSPE 中[290]。磁性离子印迹聚合物结合了磁性材料易分离及离子印迹聚合物选择性高的优点,在 DSPE 中应用广泛。Dahaghi 等[291]以 $Fe_3O_4@SiO_2$ 为核心、靛红为配体、4-乙烯基吡啶为功能单体,制备磁性 Cu^{2+} 离子印迹聚合物。He 等[292]以磁性碳纳米管/Fe_3O_4 复合材料为核心、3-丙基铵三乙氧基硅烷为功能单体、四乙基硅酸盐为交联剂,Cu^{2+} 为模板合成磁性离子印迹聚合物。Qi 等[293]采用溶胶-凝胶技术制备磁性 Cr(Ⅵ)印迹纳米颗粒,偏氟咪唑和 3-氨基丙基三乙氧基硅烷分别为有机功能单体和共单体,甲基丙氧基丙基三甲氧基硅烷为偶联剂,在有机相和无机相中形成稳定共价键。

石墨烯（GO）材料也是一类应用广泛的分散固相萃取材料,未经修饰的石墨烯材料吸附量大、选择性较低。Zawisza 等[294]用 GO 预浓缩水样中痕量 Co^{2+}、Ni^{2+}、Cu^{2+}、Zn^{2+} 和 Pb^{2+},还用 GO 预浓缩葡萄酒样品的 Co^{2+}、Ni^{2+} 和 Cu^{2+}[295]。金属离子的配合物被吸附在分散的石墨烯上,吸附结束后过滤,用 EDXRF 直接测定固体,避免洗脱过程。Al_2O_3 负载在 GO 上的复合材料,选择性预浓缩 As(Ⅴ)和 Cr(Ⅲ)。该复合材料在 pH=5 时选择性吸附砷酸盐,在 pH=6 时选择性吸附铬(Ⅲ)[296]。Sitko 等[297]用接枝 3-巯基丙基三甲氧基硅烷在 GO 表面合成吸附剂,该吸附剂在砷酸盐存在下对亚砷酸盐具有选择性,可用于 As 形态检测。由于石墨烯纳米片具有褶皱结构和在水中分散性好等特点,是 DSPE 进行快速、简单的预浓缩和测定重金属离子的理想材料,对石墨烯材料进行一定的修饰还能提高其选择性。表 15-4 总结了 DSPE 固体材料研究进展。

DSPE 富集目标物一般需要加入络合物,在特定 pH 值下,与目标离子形成螯合物,调整 pH 值将目标物从固体材料中洗脱[312]。在碱性介质中,Pb^{2+} 与 1-(2-吡啶偶氮)-2-萘酚试剂形成络合物,再定量萃取到十六烷基三甲基溴化铵磁性纳米粒子表面。吸附剂经磁分离后,用 HCl 的甲醇溶液（0.5%,w/v）洗脱[301]。洗脱剂一般为酸性溶液、水[304]和 EDTA[310]。直接测定固体材料中目标物,可避免洗脱步骤,但限制了材料的重复使用。Skorek 等[313]使用多壁碳纳米管吸附 Ni^{2+}、Co^{2+}、Cu^{2+} 和 Pb^{2+} 与吡啶二硫代氨基甲酸铵的螯合物后,用 EDXRF 直接测定。

表 15-4　DSPE 的固体吸附材料

分散固相萃取的固体材料	分析物	参考文献
Fe_3O_4@SiO_2@Cu^{2+}印迹聚合物	Cu^{2+}	[291]
2-氨基苯并噻唑功能化磁性纳米粒子	Cd^{2+}、Cu^{2+}、Ni^{2+}	[298]
$MnFe_2O_4$ 和磁性水铝镍石-铝硅酸盐吸附剂	Pb^{2+}	[299]
十六烷基三甲基溴化铵包覆的磁性纳米粒子	Pb^{2+}	[300]
十六烷基三甲基溴化铵磁性纳米粒子	Pb^{2+}	[301]
双硫腙功能化磁性金属有机骨架	Pb^{2+}	[302]
磁性纳米 Fe/Fe_2O_3	Cd、Pb、Ni、Cr 和 As	[289]
2-吡啶羧基甲醛硫代氨基脲基修饰的磁性 GO	Hg^{2+}	[303]
Fe_3O_4/羟基磷灰石/石墨烯量子点	Cu	[304]
SiO_2 包覆的磁铁矿纳米颗粒	Zn^{2+}	[305]
席夫碱改性氧化硅铁磁性颗粒	Pb^{2+}、Cd^{2+} 和 Cu^{2+}	[306]
二苯卡巴腙/十二烷基硫酸钠固定化 Fe_3O_4	Cd^{2+}	[307]
琼脂糖包覆磁性纳米颗粒	Pd^{2+}	[308]
MoS_2 和 Fe_3O_4 复合材料	Pb^{2+} 和 Cu^{2+}	[309]
磁性金属有机骨架 MOF-199 纳米颗粒	Cd^{2+}、Pb^{2+} 和 Ni^{2+}	[310]
磁性丙烯胺修饰的 GO-聚(醋酸乙烯酯-共二乙烯基苯)	Pb^{2+}、Cd^{2+}、Cu^{2+}、Ni^{2+} 和 Co^{2+}	[311]
Ag 纳米粒子与壳聚糖形成聚电解质复合物	Cr^{3+}	[312]
Al_2O_3 负载在 GO 材料上的复合材料	As(V) 和 Cr(III)	[296]
3-巯基丙基三甲氧基硅烷修饰的 GO	亚砷酸盐	[297]
多壁碳纳米管	Ni^{2+}、Co^{2+}、Cu^{2+} 和 Pb^{2+}	[313]
Au^{3+}印迹硫氰基功能化二氧化硅材料	Au^{3+}	[314]
三聚氰胺浸渍的多壁碳纳米管	Pb^{2+}、Cd^{2+} 和 Ni^{2+}	[315]
TiO_2	Hg^{2+} 和 CH_3Hg^+	[310, 316]

外场的引入可加快固体材料分散到样品溶液中的过程,并加速固体材料富集目标物。如超声浴[304,305,315-317]、机械搅拌[305,314]和磁搅拌[308]既能将材料分散到整个溶液中,又能辅助固体材料预富集目标离子。

常用分散溶剂如乙醇[309]和甲醇[318],将固体材料分散到溶液。Cunha 等[319]提出微乳化分散磁固相萃取。该法结合了磁性纳米粒子和微乳液的特性,成功应用于从汽油中萃取 Cu(II)。功能化的磁性纳米粒子被分散在汽油、缓冲液和 1-丙醇的微乳液中。微乳液的高比表面积保证 Cu(II) 与络合剂的有效接触,从而快速(约 40 s)预浓缩 Cu(II)。Farahani 等[320]使用铁磁流体分散固相萃取法富集环境和食品样品的 Cu。铁磁流体包含磁性吸附剂和离子液体载体,用注射器将铁磁流体快速注射到水样,倒转试管,萃取剂完全分散在样品溶液,形成半不透明状态,分析物的疏水复合物被磁性颗粒吸附,实现快速萃取。与其他磁性固相萃取法相比,该法操作简便、萃取过程短[321]。此外,该法对于难以观察的萃取相和不透明或深色样品的分离非常有效。

对于复杂样品,需要结合多种相分离技术去除基体或萃取目标物以消除干扰。Chen 等[322]应用 DLLME 和 DSPE 预浓缩/分离目标物,结合 ETV-ICP-MS 分析 Sb 的形态。在 DSPE 中,二氧化钛纳米纤维吸附 Sb。不进行预氧化或预还原的情况下在含 Sb(III)的上水相(pH=5.0)和含 Sb(V)的洗脱溶液(pH=0.5)中分别加入螯合剂、乙醇(分散剂)和 $CHCl_3$(萃取剂)进行 DLLME 预浓缩。Sadeghi 等[323]将 DSPE 与超声辅助 DLLME、GFAAS 联用测定食品及环境样品中 Pb 和 Cd。

15.3.2.3 固相微萃取法

固相微萃取（SPME）可用于富集样品中微量元素。Rahmi 等[324]直接在商用注射器中制备高分子聚合物材料，pH=5.0 时，以 0.9 mL 的硝酸（2 mol/L）为洗脱液，单步萃取 27 种元素，结合 ICP-MS 测定，回收率超过 80%。Shahdousti 等[325]采用聚吡咯包覆纤维的顶空 SPME，富集 1,2-二氨基苯衍生化后的 Se(Ⅳ)，转化为吡烯醇形式，通过离子迁移谱分析。Stanisz 等[249]采用离子液体涂层聚四氟乙烯管制备 SPME 材料，萃取土壤样品中 Hg。在离子液体涂层之前，对聚四氟乙烯管的表面进行机械加工，使其更加粗糙，便于离子液体在表面沉积。采用镀锡聚四氟乙烯管直接浸没萃取汞。胡斌等[326]提出用刚果红改性单壁碳纳米管制备熔融石英毛细管 SPME 涂层，萃取人头发中微量 La、Eu、Dy 和 Y，再用氟化辅助 ETV-ICP-OES 测定，头发样品分析的回收率为 93%～105%。

微固相萃取（μ-SPE）技术是 SPE 的微型化装置。Maya 等[327]报道的自动微固相萃取法，实现了磁性金属有机骨架（MOF）材料自动微固相萃取。以含 Fe_3O_4 纳米粒子的亚微米 MOF 晶体为基础制备一种杂化材料，将其保留在微型磁棒表面，磁棒放置在自动双向注射泵的注射器内，通过自动激活/失激活磁搅拌，使 MOF 混合材料得以分散和随后的磁回收。该技术适用于小尺寸吸附材料，且能与大多数检测器和分离技术连接。

15.3.2.4 其他固相萃取法

共沉淀法可用于分离和预浓缩样品中痕量金属离子。Gouda[328]使用 4-(2-羟基亚苄基氨基)-1,2-二氢-2,3-二甲基-1-苯基吡唑-5-酮为有机共沉淀剂使水和食物样品中 Cr^{3+}、Cu^{2+}、Fe^{3+}、Pb^{2+} 和 Zn^{2+} 发生共沉淀。用缓冲液调溶液 pH=7.0，静置 10 min 后，离心分离。沉淀物溶解于硝酸，用 FAAS 测定目标物。

搅拌棒吸附萃取（SBSE）结合高效液相色谱(HPLC)-ICP-MS 可用于分析生物样品中硒代胱氨酸、甲硒基-半胱氨酸、硒硫氨酸、硒寡肽（谷氨酰-硒-半胱氨酸和硒二谷胱甘肽）[329]。有别于传统有机聚合物搅拌棒萃取涂料（如聚二甲硅氧烷），部分磺化聚苯乙烯-TiO_2 涂层的萃取原理是阳离子交换作用，直接萃取高极性硒氨酸和硒寡肽。该 SBSE 也适用于萃取环境及生物样品中阳离子化合物。

Chen 等[330]采用聚合物单体为萃取相，研制了便携式多通道针尖微萃取的现场样品制备装置。根据所研究分析物的化学性质，原位合成有富集功能和高渗透性单体。将三个装有吸附剂的针尖安装在三个注射器上，并连接到一个螺杆电机驱动注射器，精确调节吸附和解吸阶段的流量。由于吸附剂单体中含多功能性基团，对农药（氨基甲酸盐、三唑）和重金属离子（Cd^{2+}、Pb^{2+}、Cu^{2+}）的协同萃取性能良好。该装置结构简单方便，结合 FAAS 能检测金属离子，如图 15-2 所示。

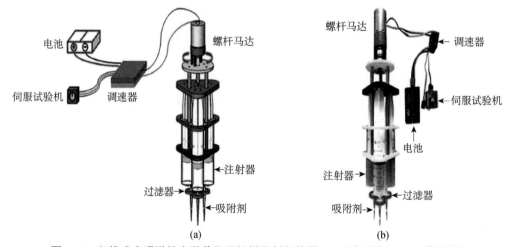

图 15-2 便携式多通道针尖微萃取现场样品制备装置（a）原理图和（b）模型[330]

引自 2020 American Chemical Society

Huang 等[331]提出搅拌饼吸附萃取法，搅拌饼的制作过程非常简单。首先，通过单体的原位聚合成整体，将整体材料插入注射器筒中，在萃取过程中磁搅拌整体材料。由于整体饼在搅拌过程中不与容器壁接触，没有萃取介质摩擦损失，搅拌饼能使用 1000 h 以上。随后他们以 3-(1-乙基咪唑-3-基)丙基甲基丙烯酰胺溴和二甲基丙烯酸乙烯原位聚合制备新型聚合物离子液体吸附剂，作为搅拌饼吸附萃取介质萃取环境水样中痕量 Sb，结合 HG-AFS 测定[332]。

样品种类繁多、复杂程度高，可能含多目标物。此外，这些目标物性质可能非常相似且含量水平低。为消除基体干扰，必须在仪器分析前完成目标物的分离富集。利用相分离法能选择性富集某些元素的形态，因此，研制新型相分离溶剂或材料是元素形态分析研究的发展方向之一。

15.4 场辅助样品制备法

引入外部能量是加快样品制备过程中能量交换和传质的有效方法。引入外场，在消解过程中可减少试剂用量，如采用超声和微波辅助萃取沉积颗粒物时，在稀释的王水中可达到王水萃取的效果，是一种绿色环保方法[333]。常用的场辅助样品制备技术包括微波场、超声场、电场、热场、光场和力场等。

15.4.1 微波场

微波通过对样品体加热，加速消解及传质过程[334]，因此，微波辅助样品制备技术具有能量传递快、选择性加热、环境友好、溶剂消耗少、快速高效、高通量等优点。

15.4.1.1 微波湿法制备

常用的微波辅助样品制备技术是微波辅助萃取和微波辅助湿法消解，缩短制样时间、提高制备效率。Leme 等[335]分别采用块状蒸煮酸矿化法、微波辅助酸消解法萃取蜂蜜样品中无机元素，得出微波辅助酸消解法是适合的制备法。Muller 等[336]在稀 HNO_3 溶液中微波辅助消解药用植物，用 ICP-MS 测定 As、Cd、Hg 和 Pb。相比之下，干灰化法会导致 Pb 明显损失、Hg 完全损失。用微波辅助消解法可同时用 HNO_3（4 mol/L）消解 8 个样品，消解物碳含量低于 320 mg/L。微波辅助消解可减少试剂消耗和废物产生，绿色环保。在封闭微波辅助消解系统中完成样品处理，样品不与外界接触，降低了污染风险。Cindric 等[337]将干燥的苹果在开放式和封闭式微波辅助消解系统中处理，用 ICP-AES 测定，发现封闭微波辅助消解系统测定结果较准确[145]。

针对样品和分析物，需采用不同酸化溶液与微波辅助技术共同处理样品。Martins 等[338]用微波辅助消解茶叶、面粉和小麦，萃取其中金属元素，每种样品所需酸化溶剂不同。李攻科等[339]用微波辅助消解石蒜样品，用 GFAAS 测定 Cd、Cr 和 Pb，在 $NH_4H_2PO_4$ 基体改进剂存在下，可提高待测元素的稳定性。Fernandez 等[340]采用微波辅助萃取鱼类样品中 Cd、Hg、Pb，结果表明用浓 HNO_3 和 H_2O_2 混合物萃取 Hg，用 $HClO_4$ 萃取 Pb 和 Cd 的效果最好。表 15-5 列出了各类样品微波湿法制备技术的研究进展。

微波不但能促进液相萃取和湿法消解，还能促进固体材料吸附目标物。如微波辅助促进没食子酸改性纤维素吸附 Al^{3+}，吸附效果显著高于传统回流法[253]。微波辅助湿法样品制备技术可用于元素形态的分析。如在微酸性（pH=4.50）和低浓度四氢硼酸盐（0.1%，w/v）条件下形成不同形态无机 As，以水为溶剂微波辅助萃取，为 As 萃取提供了合适的环境，同时也防止形态间相互转化[353]。Pasias

表 15-5 微波湿法样品制备进展

样品	主要分析物	消解/萃取溶液	参考文献
食品	Cr	HNO_3（2 mol/L）和 H_2O_2（30%, v/v）	[156]
	12 种矿物元素	HNO_3（2 mol/L）	[335]
	Cu、Fe、Zn	HNO_3 和 H_2O_2	[341]
	12 种矿物元素	HNO_3 和 V_2O_5 混合物	[342]
	人体必需矿物元素	HNO_3（69%, v/v）、H_2O_2（30%, v/v）和纯水	[343]
	12 种矿物元素	HNO_3 和 H_2O_2	[344]
	Ca、Mg、Na 和 K	HNO_3（67%, v/v）和纯水	[345]
	As 和 Hg	HNO_3 和 H_2O_2	[346]
药品	As、Cd、Hg 和 Pb	HNO_3（4 mol/L）	[336]
	9 种矿物元素	HNO_3（7 mol/L）和 H_2O_2（30%, v/v）	[163]
	As、Cd、Hg 和 Pb	逆王水	[347]
动物组织	Pb	HNO_3（7.00 mol/L）	[348]
	Cd、Hg、Pb	HNO_3 和 H_2O_2，$HClO_4$	[340]
	Hg	HCl 和 $SnCl_2$	[349]
	Se	HNO_3 和 H_2O_2	[350]
	As	HNO 和 H_2O_2	[351]
灰烬	Si	HF	[352]

等[354]介绍了 ETAAS 测定大米和米粉食品中总 As、总无机 As、As(III)和 As(V)的三种样品制备法。样品在 HNO_3（65%, v/v）溶液中微波辅助消解测定总 As，样品在 HNO_3（1 mol/L）和 EDTA 中溶解取上清液测定总无机 As，样品在 HNO_3（1 mol/L）溶液中微波辅助消解，再用 HNO_3（1 mol/L）和 EDTA 溶液溶解，并将溶液 pH 值调到 4.8~4.9，离心过滤测定 As(III)。这三种制备法巧妙地利用消解法测定 As 形态，避免萃取和消解过程中分析物的损失和转化。

Nobrega 等[355]基于单反应室设计的超短波微波，发展了单反应室微波辅助消解法，利用简单的小瓶和架子在高压和高温下消解有机样品。这种超短波振荡器有一个微波电源（1500 W），将微波能量传输到腔室（1 L），根据容器体积大小，这个腔室能容纳 5 个、15 个或 22 个容器。可直接控制反应室中每个样品的压力和温度。

Krzyzaniak 等[356]采用单反应室微波辅助消解法结合 ICP-OES 和 ICP-MS 测定了碳纳米管吸附的微量元素。单反应室法可使用稀 HNO_3 溶液进行高效消解，也可运行混合批次样品来增大样品量。该法简单易于操作，所有的样品制备步骤都是采用单容器进行的，不需要任何容器组装和拆卸，也不需要溶液转移，克服了传统封闭式微波消解法的局限。Anschau 等[357]仅用 H_2O_2 在单反应室系统中消解食品。Yang 等[358]研究了高压燃烧强酸消解、高压高温强酸炸弹消解和微波单反应室辅助酸消解处理原油样品，发现微波单反应室辅助酸消解是最好的方法，19 个分析物的平均回收率从 93%到 113%。由于单反应室微波辅助消解技术满足快速制样、吞吐量大和更大样品容量（每次消解 1.2 g），因此能检测低丰度元素。

15.4.1.2 微波诱导燃烧法

微波诱导燃烧（MIC）法是干灰化法的替代法，在传统密闭容器湿法消解系统的石英容器内放置一个小石英架，微波辐射启动燃烧，样品用氧在几分钟内进行快速完全消解，只需稀酸吸收分析物[359]，短时间内消解样品，避免了元素挥发损失，又可获得与原子光谱仪器兼容的低残酸溶液。

MIC 消解普通植物[360]和动物[361]样品，与微波辅助湿法消解对比，两种样品制备法均适用于样品消解，而 MIC 法的优势不仅在于能用稀 HNO_3 为吸收溶液，且具有较高消解效率和较低检出限。Barin 等[362]使用微波诱导燃烧法、微波湿法消解和干灰化法结合 ICP-MS 测定三环活性药物中 As、Cd、Hg 和 Pb。微波辅助密闭容器湿消解法对卡马西平、盐酸阿米替林和盐酸咪丙嗪的消解效果不佳，所有物质均有残渣。高压干灰化消解系统需高温和长时间消解，并降低样品量，且 Hg 在干灰化消解中完全挥发。采用 MIC 法消解三环类原药，所有物质均能高效消解。此外，采用 MIC 法能快速消解浓酸难以消解的样品，如石墨[363]、头发[364]、高分子聚合物等[365,366]。

尽管 MIC 对一些有机样品具有较高消解率，但它不适用于易燃液体，这是因为不可预测的快速反应可能会产生危险。Dalla Nora 等[367]提出用石英羊毛为阻燃剂控制挥发性燃料在封闭系统中燃烧速率，进一步通过 ICP-MS 测定微量元素，采用该法可获得较高消解率（99%）。

微波消解原油样品的条件苛刻[368]。Pereira 等[369]使用 MIC 法测定轻、重质原油中稀土元素。将原油样品插入聚碳酸酯胶囊，用氧气和硝酸铵（6 mol/L）为点火器进行燃烧。与传统原油消解法（微波-紫外消解和中子活化分析）相比，MIC 法能用稀酸为吸收液，获得了更低检出限，减少了 ICP-MS 测定稀土元素的干扰。

MIC 法消解的样品量大、消解效率高、试剂消耗少，是很有发展前景的方法。Henn 等[370]研制了单模微波探针便携式微波辅助固体样品分析装置。样品直接在含微晶纤维素颗粒的石英载体上称重，与 NH_4NO_3 溶液浸润，并将其引入微波系统燃烧。氧气连续流过石英管，将燃烧产物带到 FAAS 仪器的火焰炉喷嘴。微波照射一直持续到样品点火（5 s），再关闭微波并收集信号。微波辅助固体取样系统便携，适于与传统原子光谱仪联用。

15.4.2 超声辅助样品制备法

超声辅助分散、萃取、消解是一种简易、快速、绿色的场辅助法，易于在实验室进行常规分析。通过对比研究超声辅助消解与传统酸消解处理草药[371]、茶叶[372]和土壤[373]，发现超声辅助样品制备是快速环保的方法。表 15-6 列出了超声辅助湿法样品制备技术的研究进展。

表 15-6 超声辅助湿法样品制备进展

	样品	超声时间（min）	萃取溶液	参考文献
无机环境样品	玻璃微珠	60	HNO_3 和 HF	[144]
	土壤	15	水和 Triton 溶液（1%，v/v）	[373]
	矿山尾矿	5/6	HNO_3、HCl 和 H_2O_2	[383]
	道路粉尘	10	水和吡咯烷二硫代氨基甲酸铵	[384]
	空气颗粒物	30	HCl（4.0 mol/L）	[385]
	土壤和沉积物	1/12	甘油（20%，v/v）水溶液	[386]
食品	粉状植物	10	HCl（2.0 mol/L）、HNO_3（2.0 mol/L）和表面活性剂	[372]
	茶叶	15	王水	[387]
	有机蔬菜	10/3	HCl（0.10 mol/L）	[388]
	谷物	10	Triton X-100 和 8-羟基喹啉-5-磺酸	[389]
	中药	10	8-羟基喹啉(1%，v/v)	[390]
	猪饲料	5	HCl（0.10 mol/L）	[391]
	速溶咖啡	15	王水	[377, 392]
	玉米粉	30	HNO_3（0.014 mol/L）和 Triton X-100（1%，v/v）	[393]
	大米	30	HNO_3（2.0 mol/L）	[394]
	椰奶	10	HNO_3（1.0 mol/L）	[395]

续表

	样品	超声时间（min）	萃取溶液	参考文献
食品	乳制品	20	HCl (2.0 mol/L)	[396]
	淀粉	17	HCl 和 HNO₃	[397]
	酸奶	1	HNO₃ (0.6 mol L)或 HCl (0.2 mol/L)	[398]
	骨头	2	甘油 (10%，v/v)和 HNO₃ (5.0%，v/v)	[399]
		3	HNO₃ (3%，v/v)和 HCl (2%，v/v)	[400]
	生物组织	3	HCl (2 mol/L)	[401]
		5	水	[402]
工业制品或原料	化妆品	30	己烷水溶液(2.5%，v/v)	[403]
	磷肥	20	氯化镧、HCl 和硫脲溶液	[404]
	原油	30	乙烷、Triton X-100 和纯水	[405]
	油料	10	HNO₃ (1.4 mol/L)	[406]

　　超声辅助样品制备法可在常温常压下进行，安全简便。dos Santos 等[144]选择超声辅助消解玻璃微珠，确保了准确测定并减少了 HF 的挥发。Saleh 等[374]提出超声辅助萃取处理猪饲料样品，其结果与矿化样品吻合。超声前离心显著提高了部分微量元素的萃取率（12%~44%），提高了精密度和准确度[375]。大米样品在 60℃的封闭系统中煮 20 min，再超声处理 30 min，结果与微波辅助消解吻合[376]。超声辅助样品制备可减少人工取样和操作成本，且避免了产生有毒残留物。

　　超声还能提高溶剂对目标分析物的萃取率。Depoi 等[203]证明超声对 Hg^{2+} 的萃取非常有效。对比微波酸化消解，超声辅助消解萃取速溶咖啡中无机 As and Se 的效果最佳[377]。异戊醇为消泡剂，超声技术能提高醋中 Hg 的萃取率[378]。在 DLLME[220]或 DSPE[303]中，超声辅助制备技术通过增强萃取剂或固体材料在溶液中分散，促进了目标分析物的快速萃取。

　　超声波辅助样品处理技术装置主要有超声浴和超声探头。Machado 等[379]对超声浴和超声探头两种超声装置进行了对比研究，超声浴可同时制备多个样品，且污染风险较小。超声探针直接插入样品瓶，可对混合物进行超声搅拌，适用于微量分析，样品和试剂消耗少[380]。Zhang 等[381]证明在样品制备过程中提高超声波频率不仅能加快萃取，也能有效改善 Hg 萃取率，特别是有机汞。他们构建了快速有效的超声浴和超声探头组合设备以制备样品。20 kHz 超声探头和 40 kHz 超声浴条件下对有机汞的萃取率最好。Zhang 等[382]使用超声浴和超声探头组合制备样品，与单独使用超声探头或超声浴相比，组合法对有机汞的萃取率至少提高 40%。

15.4.3　电场辅助样品制备法

　　电场辅助样品制备技术主要包括电化学氧化还原、电化学富集和电泳分离。电化学氧化还原法简单方便、选择性高。该法将溶解的金属元素在电极表面进行电化学氧化或还原，再用适当方法剥离。Zu 等[187]采用自制电化学流动池电解还原生成 Hg 蒸气，再用 AFS 进行检测，阴极电极是玻璃碳棒，阳极是铂。Zawisza 等[407]提出电场辅助氧化多壁碳纳米管吸附痕量 Cr、Mn、Co、Ni、Cu 和 Zn。

　　电化学富集可降低原子光谱仪检出限。Hutton 等[408]用微波等离子体化学气相沉积技术合成大面积无支撑的硼掺杂金刚石并以其为电极材料，通过阴极电沉积在电极上富集目标元素，用 XRF 分析。电极材料拥有广泛的溶剂窗口、对 XRF 光束的透明度好和能生产机械强度大的无支撑薄膜，可较好地应用于电沉积和 XRF 检测。电沉积和检测原理如图 15-3 所示。

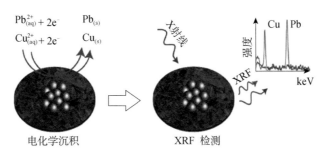

图 15-3　电沉积和 XRF 检测金属元素的原理示意图[408]

引自 2014 American Chemical Society

电渗析富集离子溶质在电场作用下将分析物从样品溶液定量转移到受体溶液,该法可在数秒内富集离子溶质。由于离子定量转移,富集系数(样品的浓度与得到的受体溶液的浓度之比)仅取决于样品溶液与受体溶液的流速比[409]。该法可同时实现基体分离和富集。Allegretta 等[410]用蒸馏水清洗蚯蚓,用纸干燥后放入皮氏培养皿中,施加 5 V 电压,持续 3 s,萃取体腔液。

凝胶电泳结合原子光谱技术可用于分析金属结合蛋白。Zahler[411]使用凝胶电泳结合同步 XRF 成像技术同时检测多种蛋白质和金属离子。如图 15-4(a)所示,样品制备采用标准凝胶电泳和电印迹技术,用非变性聚丙烯酰胺凝胶电泳分离蛋白质混合物,再将样品转移到聚偏二氟乙烯膜上,有效固定蛋白质和金属。图 15-4(b)所示为同步 X 射线荧光技术检测蛋白质和金属,每个元素以特征能量发射荧光 X 射线,测定其强度。Feng 等[412]利用 LA-ICP-MS 测定凝胶电泳分离后的金属结合蛋白。样品制备采用标准凝胶电泳和电印迹技术。该技术广泛适用,无需专门设备,有一定发展前景。

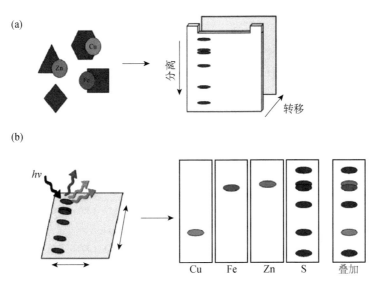

图 15-4　蛋白质凝胶的 XRF 成像[411]

(a)标准凝胶电泳和电印迹技术制备样品;(b)同步 X 射线荧光技术对蛋白质和金属进行定量

引自 2010 American Chemical Society

电膜萃取基于电离化合物的电动力学迁移,其驱动力是施加在支撑液膜上的电势。其中一个电极放置在样品水溶液,另一个电极放置在光纤腔内的水溶液受体中。Boutorabi 等[413]采用管内电膜萃取结合 FAAS 选择性测定环境样品中 Cr(Ⅵ),该法以薄聚丙烯放在一个管作为支持膜溶剂,将小体积水受体溶液与水供体溶液分离。管内电膜萃取的有机溶剂消耗量低,萃取过程简单可重复,是一种环保、自动化的萃取装置。研究表明,利用电驱动作为绿色辅助能量在样品制备领域是一种有效的方法。

15.4.4 热场辅助样品制备法

加热可加快样品制备速度。研究表明，复杂样品在酸化溶液中加热，更容易分解[414]。Silva 等[415]采用低温湿法消解涂料样品，在 40℃下使用 HNO₃ 去除有机溶剂（20 min），在 120℃下使用 HCl 和 HF 分散块状样品（3 h）。该法使用不同步骤消除不同基质对测定影响。

水浴加热法和回流加热法是最常用的加热法。Pereira 等[424]将有机质土壤燃烧后进行热水解反应，用于卤素测定。该体系可完全消解有机质含量最高可达 400 mg 的土壤和有机质含量最低 100 mg 的土壤。样品制备过程不超过 12 min（样品燃烧约 1.8 min，热水解约 10 min），快速获得卤素的定量结果。Soares 等[425]用羟化四甲基铵处理口红样品，将混合物放入 60℃水浴中浸泡 60 min，再引入 GFAAS 测定总 Pb 含量。该法与微波辅助分解法的结果一致。Pereira 等[426]将回流加热法应用于果汁样品的制备。de Oliveira 等[427]用室温开放体系、回流加热和超声辅助消解稻米，结果表明回流加热法有效简单、安全可靠。此外，与微波辅助消解法相比，密闭加热消解系统能保留挥发性元素如 Cd 和 Se[428]。

在均相液-液微萃取技术中，利用温度辅助预浓缩水中重金属阳离子。环己胺为络合剂和萃取剂，样品和环己胺混合物摇晃形成均匀溶液，再向溶液中加入 NaCl；再次振荡后，将试管放入恒温（70℃）水浴中。高温下环己胺的溶解度较低，形成浑浊溶液。用离心法在水相上方收集阳离子-环己胺配合物微滴[429]。

高温熔融是另一种固体样品消解法。Whitty-Leveille 等[430]使用 LiBO₂ 为助溶剂结合高温（1050℃）消解的矿物样品能直接进入 ICP-MS/MS 测定。Amosova 等[431]用 110 mg 的岩石样品，偏硼酸锂和 LiBr 溶液为助溶剂，在铂坩埚电炉高温（1100℃）融合。该融合技术可从同一样品中同时测定 35 种元素。

将样品制备成易于分析的固体，需要合适的制备程序，如高温制球[432]、烘箱干灰化[433]和热板干燥[434]。Schmidt 等[435]研究了干燥条件对鱼组织中 Hg（Hg^{2+} 和 CH_3Hg^+）形态的影响，结果表明，在 100℃以上干燥温度，某些 CH_3Hg^+ 会发生损失和向 Hg^{2+} 转化。因此，在元素形态分析中，必须选择合适的样品制备温度，制备成易于分析的样品状态（固体或液体），避免目标分析物的损失。

表 15-7 列出了针对不同样品特点，采用加热辅助酸消解法的研究进展。

表 15-7　热场辅助酸解样品制备技术研究进展

样品	加热温度和时间	酸化溶液	参考文献
食品	150℃，46 min	HNO₃（65%，v/v）和 H₂O₂（30%，v/v）	[416]
	200℃，1 h	HNO₃ 和 H₂O₂	[417]
	120℃，2 h	HNO₃（65%，v/v）和 H₂O₂（30%，v/v）	[418]
	80℃，40 min	HNO₃：H₂O₂（2∶1）	[419]
动物组织	60℃，5 min	羟化四甲基铵、Triton X-100 和有机硅消泡剂	[420]
	90℃，1 h	浓 HNO₃ 和 H₂O₂（2 mol/L）	[421]
日用品	100℃，180 min	稀 HNO₃、H₂O₂ 和 Triton X-100（5%，v/v）	[422]
	90℃，2 h	浓 HNO₃	[423]

15.4.5 光场辅助样品制备法

光辅助化学蒸气生成法（PVG）是分析化学的一个新兴研究领域，是样品制备的重要方法。该法仪器简单、绿色环保、过渡金属干扰小、成本低，在原子光谱分析中得到广泛应用。

紫外光化学蒸气发生（UV-PVG）与等离子体微炬联用的显著特点是使用单一试剂（甲酸）从固体样品中萃取Hg[436]，可用于测定海产品中Hg形态[241]，是一种简单、绿色、灵敏的痕量Hg测定法。Liu等[437]往含Ni的样品溶液中加入甲酸，紫外线照射，产生羰基镍，随后被引入AFS系统。Potes等[438]将鱼样品溶解在氢氧化四甲基铵中，加入正丙醇和水，混合物在UV辐射中产生Hg挥发性化合物被带到GFAAS进行分析。他们还研究了PVG耦合GFAAS测定硒的方法。PVG在流动注射模式下进行，在UV照射下产生挥发性Se通过气液分离器从冷凝相中分离，并引入石墨炉中测定[439]。

UV辅助氧化样品制备能减少酸消耗，是绿色环保的方法。dos Santos等[440]使用UV-H_2O_2光氧化制备果汁样品，结合FAAS测定Cu，与酸消解得到的结果吻合。UV光解是一种可靠、有效的样品消解法。Dash等[441]用UV光解法辅助异硫蓝矿化，H_2O_2在紫外光影响下提供羟基自由基，加入HNO_3产生NO·使样品溶解并保持氧化条件。虽然矿化时间较长，但在光分解系统中一次处理12个样品，显著缩短了样品平均制备时间，有利于大量样品的分析。

红外加热辅助样品制备技术利用电磁辐射传热原理，以直接传热方式，使红外辐射材料对红外光能量有效转换并被加热物质分子振动所吸收，而达到加热、干燥等目的，具有快速节能、热效率高、无污染等特点。红外辅助消解样品制备为复杂样品分析提供了快速而经济的替代法，能容纳相对大量的样品，提高了样品均匀性，并提高了检测能力。Lopes等[442]将大豆样品与HNO_3（65%，w/w）一起放入红外辅助加热装置中，可完成大豆样品消解。他们还将红外酸消解法结合FAAS分析亚麻籽中Cu、Fe、Mn和Zn，测定结果与传统微波辅助消解的结果一致[443]。红外灯蒸发使样品预先碳化（不灰化）可将元素检出限降低一个数量级。Zaksas等[444]将血液和石墨粉放在红外灯中蒸发（不灰化）进行全血样品制备，此外，他们也将该法用于动物器官的直接多元素分析[445]，为生物医学研究提供了大量生物样本的制备方法。

Zhang等[446]提出聚焦红外光波灰化（FILA）的新型干灰化法，用于消解植物样品。他们在干法灰化装置中采用高性能红外石英管为加热元件，在加热管表面镀一层金，反射和聚焦红外光，增强了加热效果。自行设计可流通的石英灰化管保证氧气穿透整个样品层，提高了样品灰化率，样品在1 min内迅速从室温加热到900℃，大多数样品能在0.5 h内完成灰化。Zhang等[447]将FILA系统应用于明胶胶囊灰化，整个灰化过程只需15 min，可同时处理12个样品。FILA与微波辅助消解法、常规灰化法、湿酸消解法、马弗炉灰化法的结果吻合。整个红外灰化装置体积仅为普通马弗炉体积的五分之一，方法快速高效，环境友好。

15.4.6 力场辅助样品制备法

样品制备过程中，力场辅助产生压力能更好地消解样品。Pereira等[448]提出在氧压下用稀酸进行微波辅助消解运动补充剂，与无氧常规微波辅助消解法相比，新方法降低了分析物的检出限，且使用稀酸减少了操作风险，试剂消耗量少，符合绿色化学理念。Kiyataka等[140]用高压灰化设备（HPA）在封闭系统中消解高密度聚乙烯酸奶包装材料，将样品切成小块，用去离子水冲洗，放入高压、封闭的石英分解容器。此外，HPA也可用于消解聚对苯二甲酸乙二醇酯[449]和聚丙烯[450]包装材料。

此外，采用力场将样品制备成各种易于分析的形状，如高速气流粉末分散[451]、交叉撕裂和研磨[452]。Andrey等[453]使用液压机（10 t）将可可粉制成球团。Oliveira等[454]将可可粉或巧克力粉用液压机（25 t）制成球团。每个样品只制备一个颗粒，颗粒化样品两面的分析总时间为150 s，这是一种快速简便方法。表15-8列出了热场、光场和力场在制备固体分析样品中的研究进展。

表 15-8 场辅助固体分析样品制备进展

场	制备程序	样品	分析仪器	参考文献
热场	高温熔融	矿物	ICP-MS/MS	[430]
	高温熔融	岩石	ICP-MS/MS	[431]
	高温制球	粉状碳化硅	LA-ICP-MS	[432]
	烘箱干灰化	植物叶片	HR-CS-FAAS	[433]
	热板干燥	小麦粉	TXRF	[434]
光场	红外灯蒸发	全血	ICP-AES	[444]
	红外灯蒸发	动物器官	ICP-AES	[445]
力场	高速气流分散	干粉	ICP-AES	[451]
	交叉撕裂和研磨	电子废料	ICP-OES、XRF 和 CV-AFS	[452]
	高压制球	可可粉	XRF	[453, 454]

15.4.7 多场协同辅助样品制备法

为进一步提高样品制备效率，缩短分析时间，多场辅助样品制备技术受到广泛关注。Hartwig 等[455]使用微波辅助紫外消解（MAWD-UV）系统萃取人造黄油中 Ni、Pd 和 Pt。仅用 10 mL HNO_3（4 mol/L）消解 500 mg 人造黄油，且消解率高于 98%，减少了 ICP-MS 测定的干扰。Mello 等[456]采用 MAWD-UV 处理聚合物，紫外光提高消解率，允许使用稀酸消解。Souza 等[457]用 MAWD-UV 萃取原油中无机物质。Oliveira 等[458]使用 MAWD-UV 处理石油焦，仅用稀 HNO_3 和 H_2O_2 溶液，60 min 加热时间，就能消解 500 mg 石油焦。MAWD-UV 可消解常规微波辅助酸消解无法完全消解的样品。

Lopes 等[459]使用红外辐射与微波辐射（IR-MW）增加有机样品消解量。结果与常规微波辅助辐射分解结果一致。IR-MW 系统易于实施，使用商用红外光灯，并在微波消解容器中使用红外辐射，采用 IR-MW 系统处理饲料样品可达到更好精度，使用少量硝酸可完全消解大量样品或富含有机化合物的样品。

加速溶剂萃取法是在提高温度和压力的条件下，用有机溶剂萃取的自动化法。Cui 等[460]以甲醇和水混合物为萃取剂，采用加速溶剂萃取法萃取动物肝脏样品中氨基砷、硝苯砷酸和罗沙肿。

外场辅助样品制备技术如微波和超声辅助消解或萃取法，能减少溶剂用量，加快样品的消解或目标物的萃取，适合于样品总元素分析及形态分析，如李攻科等[461]以 HNO_3 和 H_2O_2 微波辅助消解法获取烟叶中重金属总量，以超声水萃取法进行重金属初级形态分析。场的引入可选择性富集目标物，如电沉积法与凝胶电泳分离法等，在样品元素形态分析或生物蛋白结合金属分析中有广阔应用前景。

15.5 样品制备/原子光谱分析在线联用技术

样品制备与原子光谱分析技术联用，通过自动或半自动处理样品，提高了分析的精密度、准确度和速度。其中色谱分离法及流动分析法与原子光谱分析联用技术发展最快。色谱法的分离功能提高了原子光谱技术元素形态分析的能力，流动分析法解决了静态样品制备稳定性差的问题。

15.5.1 色谱分离/原子光谱分析在线联用技术

色谱法与原子光谱联用技术实现了元素的形态分析。在实际应用过程中一般先用预富集法，如

SPE[248,462]和SBSE[463]富集目标物，再通过色谱分离技术分离元素形态进入检测系统。表15-9列出了高效液相色谱（HPLC）、液相色谱（LC）及气相色谱（GC）与原子光谱联用技术研究进展。

表15-9 色谱法与原子光谱联用进展

分离方法	样品	分析物	参考文献
LC	全血	As(Ⅲ)、As(Ⅴ)、MMA、DMA 和 AsB	[464]
	血液	氯	[465]
	食用菌	AsB、As(Ⅲ)、DMA、MMA、As(Ⅴ) 和 AsC	[466]
	海鲜	MeHg	[467]
	鱼虾	AsB、As(Ⅲ)、DMA、MMA 和 As(Ⅴ)	[468]
	铜	Th 和 U	[469]
	肥料	As(Ⅲ)和As(Ⅴ)	[470]
GC	血液	MeHg、Et-Hg 和 Ino-Hg	[471]
	血浆和血清	MeHg	[472]
	红酒	MeHg 和 Ino-Hg	[473]
HPLC	环境水样	MeHg	[248]
	尿样	靶甲状腺素	[463]
	RNA	磷	[474]
	铁补充剂	As(Ⅲ)、As(Ⅴ)、Cr(Ⅲ)和Cr(Ⅳ)	[475]
	海产品	MeHg、Et-Hg 和 Ino-Hg	[476]
	海洋动物	AsB、As(Ⅲ)、DMA、MMA 和 As(Ⅴ)	[477]
	全血和关节积液	Cr(Ⅲ)和 Cr(Ⅵ)	[478]
	血清	Al-Cit 和 Al-Tf	[479]
	鱼	MeHg、Et-Hg 和 Ino-Hg	[480]
	血浆和血清	MeHg、Et-Hg 和 Ino-Hg	[481]
	头发	MeHg、Et-Hg 和 Ino-Hg	[482]
	熊猫和猪骨	As(Ⅴ)、As(Ⅲ)和有机 As	[483]
	水样和鱼	Hg^{2+}、$MeHg^+$和$PhHg^+$	[462]

LC是以液体为流动相的色层分离分析法，与原子光谱分析技术联用，涉及的分离原理有吸附色谱法[464]和离子色谱交换法[465,466]等，与人工分离相比，LC自动分离系统的使用减少了研究人员的工作时间（约80%）[469]。

GC是以气体为流动相的色层分离分析法。Rodrigues 等[471]采用 GC-ICP-MS 同时测定血液样品中 MeHg、乙基汞（Et-Hg）和 Ino-Hg。与 LC-ICP-MS 测定血液样品 Hg 形态比较，结果吻合。GC 结合 ICP-MS 可在样品中直接测量 MeHg，无需预富集步骤[472]。使用 GC-ICP-MS 无需烦琐的清理步骤，缩短了分析时间。

HPLC以液体为流动相，采用高压输液系统，将不同极性的单一溶剂或不同比例的混合溶剂、缓冲液等流动相泵入装有固定相的色谱柱，在柱内各成分被分离后引入检测器分析。HPLC-ICP-MS 在线元素检测技术已应用于生物样品元素形态分析[484]。Chen 等[479]用 3-[(3-胆酰胺基丙基)-二甲基铵]-1-丙磺酸盐动态包覆 C_{18} 柱构建了 HPLC-UV-ICP-MS 系统，在人血清直接注射系统中进行 Al 的形态分析，4 min 即可分离小分子铝络合物（柠檬酸铝）和大分子铝蛋白复合物（转铁蛋白）。高压泵系统使 HPLC 分离法的样品和溶剂消耗量少，样品制备时间短，提高了测定的灵敏度且扩宽了检测线性范围。

色谱法与原子光谱分析联用技术是有效的元素形态分析法,但仍需解决色谱与检测器接口的匹配问题,使之成为集分离与检测一体化的在线技术。在元素形态分析过程中,需使用合适流动相防止元素形态间相互转换,此外,应关注色谱-原子光谱技术在生物科学中的应用,如金属蛋白形态的形成、含杂原子的分子或其他标记外源金属的化合物,分析微量元素的生物学功能。

15.5.2　流动分析/原子光谱在线联用技术

流动分析技术是流动的样品制备技术,在一个连续(在线)封闭系统中处理样品。该法通常能提高样品吞吐量,试剂和样品消耗以及废物产生量通常较低。流动分析联用技术主要有流动注射(FI)技术和连续流动技术。

15.5.2.1　流动注射/原子光谱分析在线联用

当需要进行大量分析时,预浓缩和样品引入是重要的步骤。FI 是自动化引入样品的方法,既可提供快速的萃取和富集能力,又能提高分析方法的精密度,因此被广泛应用于样品的制备。

Chantada-Vázquez 等[485]采用 FI-ICP-MS 直接分析人血清样品中的多种无机元素。血清样品(200 μL)用 HNO_3(1%, w/v)直接稀释至 2.0 mL,再放入自动进样器,加载到定量环,用蠕动泵将负载的样品注入检测器(图 15-5)。该方法对样品容量要求低,所需样品制备最少,有较高采样率(每次重复分析需 2.50 min)。

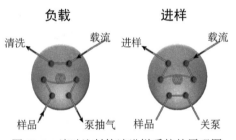

图 15-5　流动注射快速进样系统的原理图

FI 通常结合固相萃取柱,连续萃取或预富集分析物,一般使用聚四氟乙烯管连接蠕动泵、固相萃取柱、样品导入系统和检测器,蠕动泵运送试剂、洗脱液、还原剂和样品,固相萃取柱分离或富集目标物,样品导入系统将样品以气体或液体形式导入检测器。Parodi 等[486]用氧化碳纳米管微柱富集水样中 Hg^{2+},洗脱后的 Hg^{2+} 在系统中被还原为 Hg 蒸气,用气液分离器将 Hg 蒸气引入检测器。Puanngam 等[487]用 2-(3-(2-氨基乙基硫)-丙基硫)乙醇修饰的硅胶预富集 Hg^{2+},洗脱后的 Hg^{2+} 被还原为 Hg 蒸气,Hg 蒸气被重新捕获在镀金钨丝上,在干燥的 Ar 气流中直接电加热钨丝,Hg 释放进入检测器。Chen 等[288]用 MSPD 法(图 15-6)将少量样品中 Hg 有效萃取到 100 μL 洗脱液,再通过单滴溶液电极辉光放电诱导冷蒸气发生转化为 Hg 蒸气,并进一步运输到 AFS 进行测定。测定步骤为:在蠕动泵的作用下,萃取液最初通过六通阀被导向 20 μL 样品环;启动阀门,通过载气推动溶液形成液滴,液滴挂在钢管一端。在 60 V 电压下产生并维持微等离子体 10 s,再将 Hg 转化为蒸气,再扫入 AFS 进行检测。该方法具有样品稀释量小、空白低、样品导入效率高、灵敏度高、毒性化学物质和样品消耗小等优点。

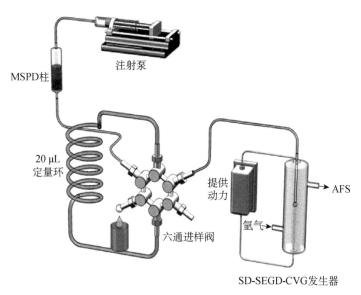

图 15-6　SD-SEGD-CVG 实验装置示意图[288]

引自 American Chemical Society

选择合适固相萃取材料能提高原子光谱分析法的选择性。如采用功能化磁性纳米粒子微柱选择性吸附 Au、Pd 和 Pt[488]，印迹聚合物微柱选择性吸附 Cd^{2+}[489]，Dowex 1X8 树脂微型玻璃柱富集 Zn^{2+}[490]，非极性固相萃取柱吸附海藻中无机 As[491]，聚合物树脂强阳离子交换柱固相萃取天然水样中痕量 Cd(Ⅱ)、Pd(Ⅱ)和 Cu(Ⅱ)[492]。FI-SPE/原子光谱法实现了在线测定样品中无机元素。

FI 系统可同时测定样品中多种元素形态。Karunasagar 等[493]用聚苯胺微柱预富集鱼组织中超痕量 Ino-Hg 和 MeHg。在 pH<3 时，只有 Ino-Hg 能被吸附，pH=7 时，MeHg 和 Ino-Hg 均能被吸附。分别用 HCl（2%，v/v）或 HCl（2%，v/v）和硫脲（0.02%，v/v）混合溶剂选择性洗脱两种 Hg。Tarley 等[494]采用 $SiO_2/Al_2O_3/TiO_2$ 和[3-(2-氨基乙基酰胺)丙基]三甲氧基硅烷功能化的硅胶（SiO_2/AAPTMS）双柱流动注射系统分别富集水样中 Cr(Ⅲ)和 Cr(Ⅵ)，采用相同洗脱液，按顺序将 Cr(Ⅲ)和 Cr(Ⅵ)分别洗脱，并用 FAAS 测定。与色谱法分离技术相比，FI 与原子光谱联用进行元素形态分析的仪器装置成本低、操作简单灵活。

Zierhut 等[495]提出一种无试剂全自动 FI 分析系统，与 AFS 联用测定 Hg，使用活性纳米金捕集器直接预富集天然水中 Hg，吸附的 Hg 通过加热（700℃）从收集器中释放出来，Ar 气流将 Hg 蒸气输送到内置金捕集器中，重新收集 Hg。Hg 蒸气在通往内置金捕集器的过程中，气流经过气液分离器去除水。不需要任何试剂进行物种转换、预浓缩或解吸，降低了污染风险，减少了试剂和时间消耗。

Cheng 等[496]采用直接分析微流控样品导入系统结合 ICP-MS 测定黄酒中 Cd 和 Pb。该系统由集成微流控芯片、流动注射分析的八路多功能阀、注射泵和 ICP-MS 仪器的蠕动泵组成。AlSuhaimi 等[497]构建了一个适用于 ICP-MS 监控样品制备的微芯片 SPE 装置。利用标准光刻法和湿化学蚀刻法制造三通道玻璃微流控装置，每个微通道都可填充 SPE 材料。微流控装置与 ICP-MS 仪器通过一个小流量同心喷雾管连接，并与输送样品和试剂的分流阀连接。该装置可用于现场样品的远程小型化处理。

FI 与原子光谱仪器联用技术的装置成本低，操作方法灵活简便，提高了样品吞吐量，减少了试剂的消耗。FI 与 SPE 结合可提高选择性，与微流控芯片结合可应用于现场分析，有良好的发展前景。

15.5.2.2　连续流动技术/原子光谱分析在线联用

常规离线样品制备，多个步骤分立进行，费时费力易引起误差。连续流动技术克服了这些缺点，它的批处理系统操作简单，价格低廉，与原子光谱仪器在线联用测定分析物，方便应用于常规分析实验室。

溶液-阴极辉光放电光谱法（SCGD-OES）利用三个蠕动泵的简单系统来实现在线溶液混合。该法能被计算机控制和自动化，使标准增加的校准能够迅速地在线执行。三蠕动泵系统可获得与高效液相色谱泵系统相同的效果，提高了 SCGD-OES 分析的样品吞吐量，减少离线了样品制备时间。该法还可与其他流动溶液在线取样仪器，如 AAS、ICP-OES 和 ICP-MS 等联用[498]。

流动间歇萃取系统批处理操作简单，价格低廉，可通过 ICP-OES 在线测定植物叶片[499]和动物组织[500]中宏观和微量元素，适合于常规分析实验室。Cunha 等[501]采用单相液-液萃取与 GFAAS 联用测定醋样品中 Cr(Ⅵ)。该法采用间流歇分析仪，自动化促进单相溶液的形成/破裂、反应和萃取。

在线连续浸出法具有样品制备快速简便、污染风险低、可实时获取浸出数据等优点[502]。Horner 等[502]采用连续在线浸出法释放水稻中各种形态 As。Cunha 等[503]用活塞推进间歇流分析仪对汽油和石脑油进行低稀释自动制备微乳。在流动金属分析中，使用微乳液能提高分析物稳定性和乳化的均匀性，从而提高样品制备的便利性。

梯度扩散薄膜（DGT）技术以菲克扩散第一定律为理论基础，在特定时间内穿过特定厚度的扩散膜的某一离子进行定量化测量计算而获得该离子浓度。DGT 是无损原位被动采样技术，检测重金属活性肽不引入其他离子，不改变介质环境，结果稳定可靠。Desaulty 等[504]用 DGT 被动采样器与 ICP-MS 联用测量了水溶液中 Pb 和 Zn 同位素比值。DGT 装置可实现水的同位素组成随时间的推移而整合，并原位预浓缩金属。这一工作促进了样品野外采集和同位素分析的发展。

连续流动系统除了应用于样品制备，在样品引入装置中同样发挥着重要作用。Yang 等[505]将在线同位素稀释技术与 LA-ICP-MS 相结合测定 p 型硅晶片中硼。利用常规雾化系统，将激光烧蚀样品气溶胶与连续提供富硼气溶胶在线混合。两气溶胶流混合后，硼同位素比值迅速变化，由 ICP-MS 系统记录，并根据同位素稀释原理进行定量。在线固体分析法不需要内部或外部固体标准物质，能准确地定量元素浓度。

Romero 等[506]使用 TXEF 分析无机 As 和无机 Sb 形态，该法选择性生成 AsH_3 和 SbH_3，再转移到含固定化钯纳米颗粒的石英反光镜上。随着 AsH_3 和 SbH_3 生成，在不同反应条件下选择性捕集 As(Ⅲ)和 Sb(Ⅲ)。柠檬酸/柠檬酸缓冲介质（pH=4.5）中选择性测定 As(Ⅲ)，高酸性介质（pH<2）选择性测定 Sb(Ⅲ)。此外，KI 抗坏血酸混合物将 As(Ⅴ)和 Sb(Ⅴ)预还原为 As(Ⅲ)和 Sb(Ⅲ)，同时测定总 As 和总 Sb。

Zhang 等[507]采用钨线圈捕集器、多孔碳蒸发器和镍铬线圈在线灰化炉组成的固体取样装置结合 ICP-MS 测定样品中 Cd。该法的双气路系统由载气和辅助气路组成，并由单独气体流量控制器控制。将样品称重到采样器中，在线灰化装置干燥和捣碎样品。取样器将捣碎的样品残渣带入蒸发器，密封载气管路，将蒸发的 Cd 由载气清洗出来，并在室温下捕获在钨线圈捕集器上。加热（2000℃），将捕获的 Cd 从钨线圈捕集器中释放出来，随后由载气将 Cd 带入 ICP-MS 测定。

连续流动技术减少了样品制备步骤，使样品制备过程在线省时。连续流动技术能批处理样品，与原子光谱仪器联用可在线测定样品中分析物。连续流动技术还将样品引入装置与检测器连接，提高了检测方法的选择性、精密度和准确度。

15.6 总结与展望

样品制备是原子光谱分析中耗时费力的环节，也是后续检测的关键，决定元素检测结果的准确度、精密度及速度，分析检测的绝大多数时间花费在将分析物从原始基体转移到可分析状态上，加速并简化样品制备过程能显著缩短分析时间。本章介绍了近年来原子光谱中样品制备的研究进展，综述了5种样品制备法，包括固体分析法、溶液制备法、相分离法、场辅助样品制备法及样品制备/原子光谱分析联用技术。这些样品制备法密不可分，互相结合。固体分析法可能需用场辅助法将样品加工成方便分析的状态，绝大部分原子光谱仪是以液体形式导入样品，溶液制备法是大部分样品制备的基础。相分离法可去除样品中杂质，萃取目标分析物。场辅助样品制备法，场的引入加快了样品制备速度。样品制备与原子光谱分析联用通常需要使用基本预处理法，使样品制备更加高效。

原子光谱分析元素总量的样品制备方法发展较完善，而有关元素形态分析的样品制备方法研究虽已取得进步，但在方法学及应用上仍存在许多问题，需进一步研究。现阶段元素形态分析研究在很大程度上依赖于能否得到不同形态元素的标准物，但人们对环境、生物体中元素存在形式及它们的性质研究不够，可供利用的标准物很少，因此限制了方法学发展[508]。未来应发展新方法，如高灵敏有机质谱分析法，在不需要标准物的条件下确定复杂基体中未知元素形态组成、化合物结构，获得那些人们尚不了解、但在环境和生命活动中起着重要作用的元素形态。在元素形态研究中，人们的注意力大多集中在环境和生命体系中相对稳定化合物的积累和迁移上，对其转换、代谢过程产生的中间产物研究不多。

原子光谱分析中样品制备技术发展方向：为满足更复杂、更小数量分析样品要求，必须在样品制备技术原理上有所创新；发展新技术提供样品中元素的全部形态，并保持样品中原有元素形态不变；评价样品制备法的效率、灵敏度和准确性应受到重视；仪器的小型化、在线联用技术是今后原子光谱分析样品制备的迫切需求；标准化操作程序和标准参考物质也是今后努力的重要方向；此外，还应考虑样品制备过程对环境造成的危害，发展绿色样品制备技术。

参 考 文 献

[1] 邓勃, 李玉珍, 刘明钟. 实用原子光谱分析. 北京：化学工业出版社, 2013: 171-206.

[2] 黄怡淳, 丁炜炜, 张卓旻, 李攻科. 食品安全分析样品前处理——快速检测联用方法研究进展. 色谱, 2013, 31(7): 613-619.

[3] Gao Y, Shi Z, Long Z, Wu P, Zheng C, Hou X. Determination and speciation of mercury in environmental and biological samples by analytical atomic spectrometry. Microchemical Journal, 2012, 103: 1-14.

[4] Stanisz E, Werner J, Zgola-Grzeskowiak A. Liquid-phase microextraction techniques based on ionic liquids for preconcentration and determination of metals. TrAC Trends in Analytical Chemistry, 2014, 61: 54-66.

[5] 周朗君, 古君平, 施文庄, 韩冰, 李攻科, 胡玉玲. 食品中重金属元素形态分析前处理与检测研究进展. 食品安全质量检测学报, 2014, 5(5): 1261-1269.

[6] Pohl P, Stecka H. Elemental composition of white refined sugar by instrumental methods of analysis. Critical Reviews in Analytical Chemistry, 2011, 41 (2): 100-113.

[7] 吴瑶庆, 孟昭荣. 无机元素原子光谱分析样品预处理技术. 北京：中国纺织出版社, 2019: 2.

[8] 郑国经. 分析化学手册. 3A. 原子光谱分析. 第三版. 北京：化学工业出版社, 2016: 9.

[9] 赵娟. 原子光谱分析及其在冶金工业中的应用. 北京：化学工业出版社, 2020: 183-187.

[10] Machado R C, Andrade D F, Babos D V, Castro J P, Costa V C, Speranca M A, Garcia J A, Gamela R R, Pereira-Filho E R. Solid sampling: Advantages and challenges for chemical element determination: A critical review. Journal of

Analytical Atomic Spectrometry, 2020, 35 (1): 54-77.

[11] Prigogine I, Lefever R. Stability and self-organization in open systems. Chichester: Wiley-Interscience, 1975.

[12] Xia L, Yang J, Su R, Zhou W, Zhang Y, Zhong Y, Huang S, Chen Y, Li G. Recent progress in fast sample preparation techniques. Analytical Chemistry, 2020, 92 (1): 34-48.

[13] Yu Y L, Jiang Y, Chen M L, Wang J H. Lab-on-valve in the miniaturization of analytical systems and sample processing for metal analysis. TrAC Trends in Analytical Chemistry, 2011, 30 (10): 1649-1658.

[14] Kurfurst U. Solid Sample Analysis: Direct and slurry sampling using GF-AAS and ETV-ICP. Berlin: Springer, 1998.

[15] Guimaraes D, Praamsma M L, Parsons P J. Evaluation of a new optic-enabled portable X-ray fluorescence spectrometry instrument for measuring toxic metals/metalloids in consumer goods and cultural products. Spectrochimica Acta Part B: Atomic Spectroscopy, 2016, 122: 192-202.

[16] Bran-Anleu P, Caruso F, Wangler T, Pomjakushina E, Flatt R J. Standard and sample preparation for the micro XRF quantification of chlorides in hardened cement pastes. Microchemical Journal, 2018, 141: 382-387.

[17] Carvalho R R V, Coelho J A O, Santos J M, Aquino F W B, Carneiro R L, Pereira-Filho E R. Laser-induced breakdown spectroscopy (LIBS) combined with hyperspectral imaging for the evaluation of printed circuit board composition. Talanta, 2015, 134: 278-283.

[18] Steiner A E, Conrey R M, Wolff J A. PXRF calibrations for volcanic rocks and the application of in-field analysis to the geosciences. Chemical Geology, 2017, 453: 35-54.

[19] De Pauw E, Tack P, Verhaeven E, Bauters S, Acke L, Vekemans B, Vincze L. Microbeam X-ray fluorescence and X-ray absorption spectroscopic analysis of Chinese blue-and-white kraak porcelain dating from the Ming dynasty. Spectrochimica Acta Part B: Atomic Spectroscopy, 2018, 149: 190-196.

[20] Donais M K, George D, Duncan B, Wojtas S M, Daigle A M. Evaluation of data processing and analysis approaches for fresco pigment studies by portable X-ray fluorescence spectrometry and portable Raman spectroscopy. Analytical Methods, 2011, 3 (5): 1061-1071.

[21] Melquiades F L, Andreoni L F S, Thomaz E L. Discrimination of land-use types in a catchment by energy dispersive X-ray fluorescence and principal component analysis. Applied Radiation and Isotopes, 2013, 77: 27-31.

[22] Quye-Sawyer J, Vandeginste V, Johnston K J. Application of handheld energy-dispersive X-ray fluorescence spectrometry to carbonate studies: Opportunities and challenges. Journal of Analytical Atomic Spectrometry, 2015, 30 (7): 1490-1499.

[23] Borgese L, Zacco A, Bontempi E, Pellegatta M, Vigna L, Patrini L, Riboldi L, Rubino F M, Depero L E. Use of total reflection X-ray fluorescence (TXRF) for the evaluation of heavy metal poisoning due to the improper use of a traditional ayurvedic drug. Journal of Pharmaceutical and Biomedical Analysis, 2010, 52 (5): 787-790.

[24] Cinosi A, Andriollo N, Pepponi G, Monticelli D. A novel total reflection X-ray fluorescence procedure for the direct determination of trace elements in petrochemical products. Analytical and Bioanalytical Chemistry, 2011, 399(2): 927-933.

[25] Pedrozo-Penafiel M J, Doyle A, Mendes L A N, Tristao M L B, Saavedra A, Aucelio R Q. Methods for the determination of silicon and aluminum in fuel oils and in crude oils by X-ray fluorescence spectrometry. Fuel, 2019, 243: 493-500.

[26] Telgmann L, Holtkamp M, Kuennemeyer J, Gelhard C, Hartmann M, Klose A, Sperling M, Karst U. Simple and rapid quantification of gadolinium in urine and blood plasma samples by means of total reflection X-ray fluorescence (TXRF). Metallomics, 2011, 3 (10): 1035-1040.

[27] Woelfl S, Ovari M, Nimptsch J, Neu T R, Mages M. Determination of trace elements in freshwater rotifers and ciliates by total reflection X-ray fluorescence spectrometry. Spectrochimica Acta Part B: Atomic Spectroscopy, 2016, 116: 28-33.

[28] Kamilari E, Farsalinos K, Poulas K, Kontoyannis C G, Orkoula M G. Detection and quantitative determination of heavy metals in electronic cigarette refill liquids using total reflection X-ray fluorescence spectrometry. Food and Chemical Toxicology, 2018, 116: 233-237.

[29] Gruber A, Mueller R, Wagner A, Colucci S, Spasic M V, Leopold K. Total reflection X-ray fluorescence spectrometry for trace determination of iron and some additional elements in biological samples. Analytical and Bioanalytical Chemistry, 2020. DOI: 10.1007/s00216-020-02614-8.

[30] Bairi V G, Lim J H, Quevedo I R, Mudalige T K, Linder S W. Portable X-ray fluorescence spectroscopy as a rapid screening technique for analysis of TiO_2 and ZnO in sunscreens. Spectrochimica Acta Part B: Atomic Spectroscopy, 2016, 116: 21-27.

[31] Melquiades F L, da Silva A M A. Identification of sulphur in nail polish by pattern recognition methods combined with portable energy dispersive X-ray fluorescence spectral data. Analytical Methods, 2016, 8 (19): 3920-3926.

[32] Pouzar M, Kratochvil T, Capek L, Smolakova L, Cernohorsky T, Krejcova A, Hromadko L. Quantitative LIBS analysis of vanadium in samples of hexagonal mesoporous silica catalysts. Talanta, 2011, 83 (5): 1659-1664.

[33] Owolabi T O, Gondal M A. Development of hybrid extreme learning machine based chemo-metrics for precise quantitative analysis of LIBS spectra using internal reference pre-processing method. Analytica Chimica Acta, 2018, 1030: 33-41.

[34] Li Y, Zhu J, Ji L, Shan Y, Jiang S, Chen G, Sciau P, Wang W, Wang C. Study of arsenic in famille rose porcelain from the Imperial Palace of Qing Dynasty, Beijing, China. Ceramics International, 2018, 44 (2): 1627-1632.

[35] Anderson E, Almond M J, Matthews W. Analysis of wall plasters and natural sediments from the Neolithic town of catalhoyuk (Turkey) by a range of analytical techniques. Spectrochimica Acta Part A: Molecular and Biomolecular Spectroscopy, 2014, 133: 326-334.

[36] Bonta M, Eitzenberger A, Burtscher S, Limbeck A. Quantification of chloride in concrete samples using LA-ICP-MS. Cement and Concrete Research, 2016, 86: 78-84.

[37] Carvalho A A C, Cozer L A, Luz M S, Nunes L C, Roche F R P, Nomura C S. Multi-energy calibration and sample fusion as alternatives for quantitative analysis of high silicon content samples by laser-induced breakdown spectrometry. Journal of Analytical Atomic Spectrometry, 2019, 34 (8): 1701-1707.

[38] Casado-Gavalda M P, Dixit Y, Geulen D, Cama-Moncunill R, Cama-Moncunill X, Markiewicz-Keszycka M, Cullen P J, Sullivan C. Quantification of copper content with laser induced breakdown spectroscopy as a potential indicator of offal adulteration in beef. Talanta, 2017, 169: 123-129.

[39] Suyanto H, Lie T J, Kurniawan K H, Kagawa K, Tjia M O. Practical soil analysis by laser induced breakdown spectroscopy employing subtarget supported micro mesh as a powder sample holder. Spectrochimica Acta Part B: Atomic Spectroscopy, 2017, 137: 59-63.

[40] Pospíšilová E, Novotny K, Porizka P, Hradil D, Hradilova J, Kaiser J, Kanicky V. Depth-resolved analysis of historical painting model samples by means of laser-induced breakdown spectroscopy and handheld X-ray fluorescence. Spectrochimica Acta Part B: Atomic Spectroscopy, 2018, 147: 100-108.

[41] Di Sabatino M. Detection limits for glow discharge mass spectrometry (GDMS) analyses of impurities in solar cell silicon. Measurement, 2014, 50: 135-140.

[42] Measures R M, Drewell N, Kwong H S. Atomic lifetime measurements obtained by the use of laser ablation and selective excitation spectroscopy. Physical Review A Genenal Physics (USA), 1977, 16 (3): 1093-1097.

[43] Harilal S S, Brumfield B E, LaHaye N L, Hartig K C, Phillips M C. Optical spectroscopy of laser-produced plasmas for standoff isotopic analysis. Applied Physics Reviews, 2018, 5 (2): 32.

[44] Konz I, Fernandez B, Luisa Fernandez M, Pereiro R, Sanz-Medel A. Laser ablation ICP-MS for quantitative biomedical applications. Analytical and Bioanalytical Chemistry, 2012, 403 (8): 2113-2125.

[45] Konz I, Fernandez B, Luisa Fernandez M, Pereiro R, Gonzalez H, Alvarez L, Coca-Prados M, Sanz-Medel A. Gold internal standard correction for elemental imaging of soft tissue sections by LA-ICP-MS: Element distribution in eye microstructures. Analytical and Bioanalytical Chemistry, 2013, 405 (10): 3091-3096.

[46] Byrne S, Amarasiriwardena D, Bandak B, Bartkus L, Kane J, Jones J, Yanez J, Arriaza B, Cornejo L. Were Chinchorros exposed to arsenic? Arsenic determination in Chinchorro mummies' hair by laser ablation inductively coupled plasma-mass spectrometry (LA-ICP-MS). Microchemical Journal, 2010, 94 (1): 28-35.

[47] Ranulfi A C, Senesi G S, Caetano J B, Meyer M C, Magalhaes A B, Villas-Boas P R, Milori D M B P. Nutritional characterization of healthy and *Aphelenchoides besseyi* infected soybean leaves by laser-induced breakdown spectroscopy (LIBS). Microchemical Journal, 2018, 141: 118-126.

[48] Debeljak M, van Elteren JT, Vogel-Mikus K. Development of a 2D laser ablation inductively coupled plasma mass spectrometry mapping procedure for mercury in maize (*Zea mays* L.) root cross-sections. Analytica Chimica Acta, 2013, 787: 155-162.

[49] Pluhacek T, Rucka M, Maier V. A direct LA-ICP-MS screening of elemental impurities in pharmaceutical products in compliance with USP and ICH-Q3D. Analytica Chimica Acta, 2019, 1078: 1-7.

[50] Sjastad K E, Simonsen S L, Andersen T. Studies of SRM NIST glasses by laser ablation multicollector inductively coupled plasma source mass spectrometry (LA-ICP-MS). Journal of Analytical Atomic Spectrometry, 2012, 27 (6): 989-999.

[51] Stehrer T, Heitz J, Pedarnig J D, Huber N, Aeschlimann B, Guenther D, Scherndl H, Linsmeyer T, Wolfmeir H, Arenholz E. LA-ICP-MS analysis of waste polymer materials. Analytical and Bioanalytical Chemistry, 2010, 398 (1): 415-424.

[52] Do T M, Hsieh H F, Chang W C, Chang E E, Wang C F. Analysis of liquid samples using dried-droplet laser ablation inductively coupled plasma mass spectrometry. Spectrochimica Acta Part B: Atomic Spectroscopy, 2011, 66 (8): 610-618.

[53] Hsieh H F, Chang W S, Hsieh Y K, Wang C F. Using dried-droplet laser ablation inductively coupled plasma mass spectrometry to quantify multiple elements in whole blood. Analytica Chimica Acta, 2011, 699 (1): 6-10.

[54] Papaslioti E M, Parviainen A, Roman Alpiste M J, Marchesi C, Garrido C J. Quantification of potentially toxic elements in food material by laser ablation-inductively coupled plasma-mass spectrometry (LA-ICP-MS) via pressed pellets. Food Chemistry, 2019, 274: 726-732.

[55] Detcheva A, Barth P, Hassler J. Calibration possibilities and modifier use in ETV-ICP-OES determination of trace and minor elements in plant materials. Analytical and Bioanalytical Chemistry, 2009, 394 (5): 1485-1495.

[56] Li Y T, Jiang S J, Sahayam A C. Electrothermal vaporization inductively coupled plasma mass spectrometry for the determination of Cr, Cd, Hg, and Pb in honeys. Food Analytical Methods, 2017, 10 (2): 434-441.

[57] Vogt T, Bauer D, Neuroth M, Ottto M. Quantitative multi-element analysis of argonne premium coal samples by ETV-ICP-OES: A highly efficient direct analytical technique for inorganics in coal. Fuel, 2015, 152: 96-102.

[58] Boerno F, Richter S, Deiting D, Jakubowski N, Panne U. Direct multi-element analysis of plastic materials via solid sampling electrothermal vaporization inductively coupled plasma optical emission spectroscopy. Journal of Analytical Atomic Spectrometry, 2015, 30 (5): 1064-1071.

[59] Kojima N, Mizoguchi Y, Tanabe K, Iida Y, Hashimoto B, Uchihara H, Ohshita Y, Okamoto Y. Simple and convenient analytical method for the direct determination of chlorine species by ETV-ICP-AES using tungsten boat furnace vaporiser and exchangeable small sample cuvettes. Polymer Testing, 2017, 59: 262-267.

[60] Santos C M M, Nunes M A G, Costa A B, Pozebon D, Duarte F A, Dressler V L. Multielement determination in

medicinal plants using electrothermal vaporization coupled to ICP-OES. Analytical Methods, 2017, 9 (23): 3497-3504.

[61] Mello P A, Pedrotti M F, Cruz S M, Muller E I, Dressler V L, Flores E M M. Determination of rare earth elements in graphite by solid sampling electrothermal vaporization-inductively coupled plasma mass spectrometry. Journal of Analytical Atomic Spectrometry, 2015, 30 (10): 2048-2055.

[62] Hassler J, Barth P, Richter S, Matschat R. Determination of trace elements in high-purity copper by ETV-ICP-OES using halocarbons as chemical modifiers. Journal of Analytical Atomic Spectrometry, 2011, 26 (12): 2404-2418.

[63] Santos R F, Carvalho G S, Duarte F A, Bolzan R C, Flores E M M. High purity polyimide analysis by solid sampling graphite furnace atomic absorption spectrometry. Spectrochimica Acta Part B: Atomic Spectroscopy, 2017, 129: 42-48.

[64] Briceno J, Belarra M A, De Schamphelaere K A C, Vanblaere S, Janssen C R, Vanhaecke F, Resano M. Direct determination of Zn in individual *Daphnia magna* specimens by means of solid sampling high-resolution continuum source graphite furnace atomic absorption spectrometry. Journal of Analytical Atomic Spectrometry, 2010, 25 (4): 503-510.

[65] dos Santos L O, Brandao G C, dos Santos A M P, Ferreira S L C, Lemos V A. Direct and simultaneous determination of copper and iron in flours by solid sample analysis and high-resolution continuum source graphite furnace atomic absorption apectrometry. Food Analytical Methods, 2017, 10 (2): 469-476.

[66] de Gois J S, Almeida T S, Alves J C, Araujo R G O, Borges D L G. Assessment of the halogen content of brazilian inhalable particulate matter (PM_{10}) using high resolution molecular absorption spectrometry and electrothermal vaporization inductively coupled plasma mass spectrometry, with direct solid sample analysis. Environmental Science & Technology, 2016, 50 (6): 3031-3038.

[67] Duarte A T, Dessuy M B, Vale M G R, Welz B. Determination of chromium and antimony in polymers from electrical and electronic equipment using high-resolution continuum source graphite furnace atomic absorption spectrometry. Analytical Methods, 2013, 5 (24): 6941-6946.

[68] Borges A R, Bazanella D N, Duarte A T, Zmozinski A V, Vale M G R, Welz B. Development of a method for the sequential determination of cadmium and chromium from the same sample aliquot of yerba mate using high-resolution continuum source graphite furnace atomic absorption spectrometry. Microchemical Journal, 2017, 130: 116-121.

[69] Pozzatti M, Borges A R, Dessuy M B, Vale M G R, Welz B. Determination of cadmium, chromium and copper in vegetables of the Solanaceae family using high-resolution continuum source graphite furnace atomic absorption spectrometry and direct solid sample analysis. Analytical Methods, 2017, 9 (2): 329-337.

[70] Castilho I N B, Welz B, Vale M G R, de Andrade J B, Srnichowski P, Shaltout A A, Colares L, Carasek E. Comparison of three different sample preparation procedures for the determination of traffic-related elements in airborne particulate matter collected on glass fiber filters. Talanta, 2012, 88: 689-695.

[71] Gonzalez-Alvarez RJ, Bellido-Milla D, Pinto J J, Moreno C. Determination of silver in seawater by the direct analysis of solvent bars by high resolution continuum source solid sampling graphite furnace atomic absorption spectrometry. Journal of Analytical Atomic Spectrometry, 2018, 33 (11): 1925-1931.

[72] 辛仁轩. 等离子体发射光谱分析. 北京：化学工业出版社, 2018: 17.

[73] Boschetti W, Rampazzo R T, Dessuy M B, Vale M G R, de Oliveira Rios A, Hertz P, Manfroi V, Celso P G, Ferrao M F. Detection of the origin of Brazilian wines based on the determination of only four elements using high-resolution continuum source flame AAS. Talanta, 2013, 111: 147-155.

[74] Cindric I J, Zeiner M, Kroeppl M, Stingeder G. Comparison of sample preparation methods for the ICP-AES determination of minor and major elements in clarified apple juices. Microchemical Journal, 2011, 99 (2): 364-369.

[75] Szymczycha-Madeja A, Welna M. Evaluation of a simple and fast method for the multi-elemental analysis in commercial fruit juice samples using atomic emission spectrometry. Food Chemistry, 2013, 141 (4): 3466-3472.

[76] Margui E, Ricketts P, Fletcher H, Karydas A G, Migliori A, Leani J J, Hidalgo M, Queralt I, Voutchkov M. Total reflection X-ray fluorescence as a fast multielemental technique for human placenta sample analysis. Spectrochimica Acta Part B: Atomic Spectroscopy, 2017, 130: 53-59.

[77] Zmozinski A V, Passos L D, Damin I C F, Espirito Santo M A B, Vale M G R, Silva M M. Determination of cadmium and lead in fresh fish samples by direct sampling electrothermal atomic absorption spectrometry. Analytical Methods, 2013, 5 (22): 6416-6424.

[78] Schneider M, Pereira E R, de Quadros D P C, Welz B, Carasek E, de Andrade J B, Menoyo J d C, Feldmann J. Investigation of chemical modifiers for the determination of cadmium and chromium in fish oil and lipoid matrices using HR-CS GF AAS and a simple 'dilute-and-shoot' approach. Microchemical Journal, 2017, 133: 175-181.

[79] Koller D, Bramhall P, Devoy J, Goenaga-Infante H, Harrington C F, Leese E, Morton J, Nunez S, Rogers J, Sampson B, Powell J J. Analysis of soluble or titanium dioxide derived titanium levels in human whole blood: Consensus from an inter-laboratory comparison. Analyst, 2018, 143 (22): 5520-5529.

[80] Matousek T, Wang Z, Douillet C, Musil S, Styblo M. Direct speciation analysis of arsenic in whole blood and blood plasma at low exposure levels by hydride generation-cryotrapping-inductively coupled plasma mass spectrometry. Analytical Chemistry, 2017, 89 (18): 9633-9637.

[81] Zacharia A, Zhuravlev A, Chebotarev A, Arabadji M. Graphite "filter furnace" atomizer with Pd-Mg chemical modifier for direct analysis of foods using electrothermal atomic absorption spectrometry. Food Analytical Methods, 2015, 8 (3): 668-677.

[82] Ivanova-Petropulos V, Balabanova B, Bogeva E, Frentiu T, Ponta M, Senila M, Gulaboski R, Irimie F D. Rapid determination of trace elements in macedonian grape brandies for their characterization and safety evaluation. Food Analytical Methods, 2017, 10 (2): 459-468.

[83] Pohl P, Kalinka M, Pieprz M. Development of a very simple and fast analytical methodology for FAAS/FAES measurements of Ca, K, Mg and Na in red beetroot juices along with chemical fractionation of Ca and Mg by solid phase extraction. Microchemical Journal, 2019, 147: 538-544.

[84] Szymczycha-Madeja A, Welna M, Pohl P. Simple and fast sample preparation procedure prior to multi-element analysis of slim teas by ICP-OES. Food Analytical Methods, 2014, 7 (10): 2051-2063.

[85] Ivanova-Petropulos V, Jakabova S, Nedelkovski D, Pavlik V, Balazova Z, Hegedus O. Determination of Pb and Cd in macedonian wines by electrothermal atomic absorption spectrometry (ETAAS). Food Analytical Methods, 2015, 8 (8): 1947-1952.

[86] Zhang D, Wang X, Liu M, Zhang L, Deng M, Liu H. Quantification of strontium in human serum by ICP-MS using alternate analyte-free matrix and its application to a pilot bioequivalence study of two strontium ranelate oral formulations in healthy Chinese subjects. Journal of Trace Elements in Medicine and Biology, 2015, 29: 69-74.

[87] Almeida J S, Souza O C C O, Teixeira L S G. Determination of Pb, Cu and Fe in ethanol fuel samples by high-resolution continuum source electrothermal atomic absorption spectrometry by exploring a combination of sequential and simultaneous strategies. Microchemical Journal, 2018, 137: 22-26.

[88] Greda K, Jamroz P, Dzimitrowicz A, Pohl P. Direct elemental analysis of honeys by atmospheric pressure glow discharge generated in contact with a flowing liquid cathode. Journal of Analytical Atomic Spectrometry, 2015, 30 (1): 154-161.

[89] Tu Q, Wang T, Antonucci V. High-efficiency sample preparation with dimethylformamide for multi-element determination in pharmaceutical materials by ICP-AES. Journal of Pharmaceutical and Biomedical Analysis, 2010, 52 (2): 311-315.

[90] Kollander B, Andersson M, Pettersson J. Fast multi-element screening of non-digested biological materials by slurry

introduction to ICP-AES. Talanta, 2010, 80 (5): 2068-2075.

[91] Liu R, Xu M, Shi Z, Zhang J, Gao Y, Yang L. Determination of total mercury in biological tissue by isotope dilution ICP-MS after UV photochemical vapor generation. Talanta, 2013, 117: 371-375.

[92] Ferreira K S, Ferreira W A, Mendonca Gomes J M, Correa-Junior J D, Donnici C L, Borba da Silva J B. Use of fast alkaline solubilisation to determine copper in bovine liver, fish tissues (salmon), and rolled oats by graphite furnace atomic absorption spectrometry using aqueous calibration. Microchemical Journal, 2016, 124: 350-355.

[93] Liu S, Zhang P, Liu H, Wang W, Lian K. Determination of trace mercury in cosmetics by suspension-sampling hydride generation atomic fluorescence spectrometry. Asian Journal of Chemistry, 2013, 25 (13): 7315-7318.

[94] Freschi G P G, Fortunato F M, Freschi C D, Gomes Neto J A. Simultaneous and direct determination of As, Bi, Pb, Sb, and Se and Co, Cr, Cu, Fe, and Mn in milk by electrothermal atomic absorption spectrometry. Food Analytical Methods, 2012, 5 (4): 861-866.

[95] Polgari Z, Ajtony Z, Kregsamer P, Streli C, Mihucz V G, Reti A, Budai B, Kralovanszky J, Szoboszlai N, Zaray G. Microanalytical method development for Fe, Cu and Zn determination in colorectal cancer cells. Talanta, 2011, 85 (4): 1959-1965.

[96] Wojciak-Kosior M, Szwerc W, Strzemski M, Wichlacz Z, Sawicki J, Kocjan R, Latalski M, Sowa I. Optimization of high-resolution continuum source graphite furnace atomic absorption spectrometry for direct analysis of selected trace elements in whole blood samples. Talanta, 2017, 165: 351-356.

[97] Volkov D S, Proskurnin M A, Korobov M V. Elemental analysis of nanodiamonds by inductively-coupled plasma atomic emission spectroscopy. Carbon, 2014, 74: 1-13.

[98] Wang Z, Zhang J, Qiu D, Zou H, Qu H, Chen Y, Yang P. Preparation of a high-concentration nm-size ceramic silicon carbide slurry for the ICP-OES determination of ultra-trace impurities in a sample. Journal of Analytical Atomic Spectrometry, 2010, 25 (9): 1482-1484.

[99] Wang Z, Zhang J, Zhang G, Qiu D, Yang P. Direct determination of trace impurities in high-purity silicon nitride by axial viewed inductively coupled plasma optical emission spectrometry using a slurry nebulization technique. Journal of Analytical Atomic Spectrometry, 2015, 30 (4): 909-915.

[100] Stosnach H. Analytical determination of selenium in medical samples, staple food and dietary supplements by means of total reflection X-ray fluorescence spectroscopy. Spectrochimica Acta Part B: Atomic Spectroscopy, 2010, 65 (9-10): 859-863.

[101] Sharanov P Y, Volkov D S, Alov N V. Quantification of elements in copper-zinc ores at micro-and macro-levels by total reflection X-ray fluorescence and inductively coupled plasma atomic emission spectrometry. Analytical Methods, 2019, 11 (29): 3750-3756.

[102] Haberl J, Fromm S, Schuster M. Digestions *vs.* suspensions: The influence of sample preparation on precision and accuracy in total-reflection X-ray fluorescence analysis by the example of waste incineration fly ash. Spectrochimica Acta Part B: Atomic Spectroscopy, 2019, 154: 82-90.

[103] Poirier L, Nelson J, Gilleland G, Wall S, Berhane L, Lopez-Linares F. Comparison of preparation methods for the determination of metals in petroleum fractions (1000°F+) by microwave plasma atomic emission spectroscopy. Energy Fuels, 2017, 31 (8): 7809-7815.

[104] Poirier L, Nelson J, Leong D, Berhane L, Hajdu P, Lopez-Linares F. Application of ICP-MS and ICP-OES on the determination of nickel, vanadium, iron, and calcium in petroleum crude oils via direct dilution. Energy Fuels, 2016, 30 (5): 3783-3790.

[105] de Souza J R, Duyck C B, Fonseca T C O, Saint'Pierre T D. Multielemental determination in oil matrices diluted in

xylene by ICP-MS with a dynamic reaction cell employing methane as reaction gas for solving specific interferences. Journal of Analytical Atomic Spectrometry, 2012, 27 (8): 1280-1286.

[106] de Albuquerque FI, Duyck CB, Fonseca T C O, Saint'Pierre T D. Determination of As and Se in crude oil diluted in xylene by inductively coupled plasma mass spectrometry using a dynamic reaction cell for interference correction on Se-80. Spectrochimica Acta Part B-Atomic Spectroscopy, 2012, 71-72: 112-116.

[107] Kowalewska Z. Direct determination of nickel in xylene solutions of raw material for catalytic cracking with application of graphite furnace atomic absorption spectrometry. Analytical Methods, 2013, 5 (1): 192-201.

[108] Barros A I, de Oliveira A P, de Magalhaes M R L, Villa R D. Determination of sodium and potassium in biodiesel by flame atomic emission spectrometry, with dissolution in ethanol as a single sample preparation step. Fuel, 2012, 93 (1): 381-384.

[109] Dancsak S E, Silva S G, Nobrega J A, Jones B T, Donati G L. Direct determination of sodium, potassium, chromium and vanadium in biodiesel fuel by tungsten coil atomic emission spectrometry. Analytica Chimica Acta, 2014, 806: 85-90.

[110] Quadros D P C, Rau M, Idrees M, Chaves E S, Curtius A J, Borges D L G. A simple and fast procedure for the determination of Al, Cu, Fe and Mn in biodiesel using high-resolution continuum source electrothermal atomic absorption spectrometry. Spectrochimica Acta Part B: Atomic Spectroscopy, 2011, 66 (5): 373-377.

[111] Gazulla M F, Rodrigo M, Orduna M, Ventura M J, Andreu C. High precision measurement of silicon in naphthas by ICP-OES using isooctane as diluent. Talanta, 2017, 164: 563-569.

[112] Navarre E C, Bright L K, Osterhage A, Lada B. Development of an *N*-methyl pyrrolidone based method of analysis for lead in paint. Analytical Methods, 2012, 4 (12): 4295-4302.

[113] Savio M, Ortiz M S, Almeida C A, Olsina R A, Martinez L D, Gil R A. Multielemental analysis in vegetable edible oils by inductively coupled plasma mass spectrometry after solubilisation with tetramethylammonium hydroxide. Food Chemistry, 2014, 159: 433-438.

[114] Doyle A, Saavedra A, Tristao M L B, Aucelio R Q. Determination of S, Ca, Fe, Ni and V in crude oil by energy dispersive X-ray fluorescence spectrometry using direct sampling on paper substrate. Fuel, 2015, 162: 39-46.

[115] Burguera J L, Burguera M. Pretreatment of oily samples for analysis by flow injection-spectrometric methods. Talanta, 2011, 83 (3): 691-699.

[116] Amais R S, Garcia E E, Monteiro M R, Nobrega J A. Determination of Ca, Mg, and Zn in biodiesel microemulsions by FAAS using discrete nebulization. Fuel, 2012, 93 (1): 167-171.

[117] Nunes LS, Barbosa J T P, Fernandes A P, Lemos V A, dos Santos W N L, Korn M G A, Teixeira L S G. Multi-element determination of Cu, Fe, Ni and Zn content in vegetable oils samples by high-resolution continuum source atomic absorption spectrometry and microemulsion sample preparation. Food Chemistry, 2011, 127 (2): 780-783.

[118] Lobo F A, Goveia D, Oliveira A P, Romao L P C, Fraceto L F, Dias Filho N L, Rosa A H. Development of a method to determine Ni and Cd in biodiesel by graphite furnace atomic absorption spectrometry. Fuel, 2011, 90 (1): 142-146.

[119] Leite C C, Zmozinski A V, Vale M G R, Silva M M. Determination of Fe, Cr and Cu in used lubricating oils by ET AAS using a microemulsion process for sample preparation. Analytical Methods, 2015, 7 (8): 3363-3371.

[120] Pessoa H M, Hauser-Davis R A, de Campos R C, Ribeiro de Castro E V, Weitzel Dias Carneiro M T, Brandao G P. Determination of Ca, Mg, Sr and Ba in crude oil samples by atomic absorption spectrometry. Journal of Analytical Atomic Spectrometry, 2012, 27 (9): 1568-1573.

[121] Roveda L M, Raposo Jr J L. Internal standardization in dispersion dystems: An efficient application to determine Mg in crude vegetable oils by FS-FAAS. Food Analytical Methods, 2019, 12 (5): 1111-1120.

[122] Carballo S, Teran J, Soto R M, Carlosena A, Andrade J M, Prada D. Green approaches to determine metals in lubricating oils by electrothermal atomic absorption spectrometry (ETAAS). Microchemical Journal, 2013, 108: 74-80.

[123] Quadros D P C, Chaves E S, Lepri F G, Borges D L G, Welz B, Becker-Ross H, Curtius A J. Evaluation of Brazilian and Venezuelan crude oil samples by means of the simultaneous determination of Ni and V as their total and non-volatile fractions using high-resolution continuum source graphite furnace atomic absorption spectrometry. Energy Fuels, 2010, 24(11): 5907-5911.

[124] de Jesus A, Zmozinski A V, Vieira M A, Ribeiro A S, da Silva M M. Determination of mercury in naphtha and petroleum condensate by photochemical vapor generation atomic absorption spectrometry. Microchemical Journal, 2013, 110: 227-232.

[125] Aranda P R, Gasquez J A, Olsina R A, Martinez L D, Gil R A. Method development for Cd and Hg determination in biodiesel by electrothermal atomic absorption spectrometry with emulsion sample introduction. Talanta, 2012, 101: 353-356.

[126] Ieggli C V S, Bohrer D, do Nascimento P C, de Carvalho L M. Determination of sodium, potassium, calcium, magnesium, zinc and iron in emulsified chocolate samples by flame atomic absorption spectrometry. Food Chemistry, 2011, 124(3): 1189-1193.

[127] Ieggli C V S, Bohrer D, do Nascimento P C, de Carvalho L M, Gobo L A. Determination of aluminum, copper and manganese content in chocolate samples by graphite furnace atomic absorption spectrometry using a microemulsion technique. Journal of Food Composition and Analysis, 2011, 24(3): 465-468.

[128] Ieggli C V S, Bohrer D, do Nascimento P C, de Carvalhoa L M, Garcia S C. Determination of sodium, potassium, calcium, magnesium, zinc, and iron in emulsified egg samples by flame atomic absorption spectrometry. Talanta, 2010, 80(3): 1282-1286.

[129] Cheung Y, Schwartz A J, Hieftje G M. Use of gradient dilution to flag and overcome matrix interferences in axial-viewing inductively coupled plasma-atomic emission spectrometry. Spectrochimica Acta Part B: Atomic Spectroscopy, 2014, 100: 38-43.

[130] Jones W B, Donati G L, Calloway Jr C P, Jones B T. Standard dilution analysis. Analytical Chemistry, 2015, 87(4): 2321-2327.

[131] Goncalves D A, Jones B T, Donati G L. The reversed-axis method to estimate precision in standard additions analysis. Microchemical Journal, 2016, 124: 155-158.

[132] Goncalves D A, McSweeney T, Santos M C, Jones B T, Donati G L. Standard dilution analysis of beverages by microwave-induced plasma optical emission spectrometry. Analytica Chimica Acta, 2016, 909: 24-29.

[133] Quarles Jr C D, Randunu K M, Brumaghim J L, Marcus R K. Metal retention in human transferrin: Consequences of solvent composition in analytical sample preparation methods. Metallomics, 2011, 3 (10): 1027-1034.

[134] 范华均, 李攻科, 黎蔚波. 石墨炉原子吸收光谱法测定中药口服液中的铬铅镉. 分析试验室, 2005, 24(3): 36-39.

[135] Oreste E Q, de Oliveira R M, Nunes A M, Vieira M A, Ribeiro A S. Sample preparation methods for determination of Cd, Pb and Sn in meat samples by GFAAS: Use of acid digestion associated with a cold finger apparatus versus solubilization methods. Analytical Methods, 2013, 5 (6): 1590-1595.

[136] Pozebon D, Dressler V L, Alexandre Marcelo M C, de Oliveira T C, Ferrao M F. Toxic and nutrient elements in yerba mate (*Ilex paraguariensis*). Food Additives & Contaminants Part B: Surveillance, 2015, 8(3): 215-220.

[137] Fortes F J, Vadillo I, Stoll H, Jimenez-Sanchez M, Moreno A, Laserna J J. Spatial distribution of paleoclimatic proxies in stalagmite slabs using laser-induced breakdown spectroscopy. Journal of Analytical Atomic Spectrometry, 2012, 27(5): 868-873.

[138] Szymczycha-Madeja A, Pohl P, Welna M, Stelmach E, Jedryczko D. The evaluation of the suitability of different alternative sample preparation procedures prior to the multi-elemental analysis of brews of ground roasted and instant coffees by FAAS and ICP-OES. Food Research International, 2016, 89: 958-966.

[139] Szymczycha-Madeja A, Welna M, Pohl P. Comparison and validation of different alternative sample preparation procedures of tea infusions prior to their multi-element analysis by FAAS and ICP-OES. Food Analytical Methods, 2016, 9 (5): 1398-1411.

[140] Bizzi C A, Flores E L M, Nobrega J A, Oliveira J S S, Schmidt L, Mortari S R. Evaluation of a digestion procedure based on the use of diluted nitric acid solutions and H_2O_2 for the multielement determination of whole milk powder and bovine liver by ICP-based techniques. Journal of Analytical Atomic Spectrometry, 2014, 29 (2): 332-338.

[141] Kiyataka P H M, Dantas S T, Lima Pallone J A. Method for assessing lead, cadmium, mercury and arsenic in high-density polyethylene packaging and study of the migration into yoghurt and simulant. Food Additives and Contaminants Part A: Chemistry Analysis Control Exposure & Risk Assessment, 2014, 31 (1): 156-163.

[142] Scheffler G L, Dressler V L, Pozebon D. Rice slurry analysis using mixed-gas plasma and axially viewed ICP-OES. Food Analytical Methods, 2014, 7 (7): 1415-1423.

[143] Pasias I N, Pappa C, Katsarou V, Thomaidis N S, Piperaki E A. Alternative approaches to correct interferences in the determination of boron in shrimps by electrothermal atomic absorption spectrometry. Spectrochimica Acta Part B: Atomic Spectroscopy, 2014, 92: 23-28.

[144] dos Santos E J, Herrmann A B, Prado S K, Fantin E B, dos Santos V W, Miranda de Oliveira A V, Curtius A J. Determination of toxic elements in glass beads used for pavement marking by ICP-OES. Microchemical Journal, 2013, 108: 233-238.

[145] Fang Y L, Zhang A, Wang H, Li H, Zhang Z W, Chen S X, Luan L Y. Health risk assessment of trace elements in Chinese raisins produced in Xinjiang province. Food Control, 2010, 21 (5): 732-739.

[146] Mutsuga M, Sato K, Hirahara Y, Kawamura Y. Analytical methods for SiO_2 and other inorganic oxides in titanium dioxide or certain silicates for food additive specifications. Food Additives and Contaminants Part A: Chemistry Analysis Control Exposure & Risk Assessment, 2011, 28 (4): 423-427.

[147] Dessuy M B, de Jesus R M, Brandao G C, Ferreira S L C, Vale M G R, Welz B. Fast sequential determination of antimony and lead in pewter alloys using high-resolution continuum source flame atomic absorption spectrometry. Food Additives and Contaminants Part A: Chemistry Analysis Control Exposure & Risk Assessment, 2013, 30 (1): 202-207.

[148] Metarwiwinit S, Mukdasai S, Poonsawat C, Srijaranai S. A simple dispersive-micro-solid phase extraction based on a colloidal silica sorbent for the spectrophotometric determination of Fe(II) in the presence of tetrabutylammonium bromide. New Journal of Chemistry, 2018, 42 (5): 3401-3408.

[149] Niedzielski P, Zielinska-Dawidziak M, Kozak L, Kowalewski P, Szlachetka B, Zalicka S, Wachowiak W. Determination of iron species in samples of iron-fortified food. Food Analytical Methods, 2014, 7 (10): 2023-2032.

[150] Fernandez-Martinez R, Rucandio I. A simplified method for determination of organic mercury in soils. Analytical Methods, 2013, 5 (16): 4131-4137.

[151] Niedzielski P, Kozak L, Wachelka M, Jakubowski K, Wybieralska J. The microwave induced plasma with optical emission spectrometry (MIP-OES) in 23 elements determination in geological samples. Talanta, 2015, 132: 591-599.

[152] Mariela Azcarate S, Paradiso Langhoff L, Manuel Camina J, Savio M. A green single-tube sample preparation method for wear metal determination in lubricating oil by microwave induced plasma with optical emission spectrometry. Talanta, 2019, 195: 573-579.

[153] Menezes E A, Oliveira A F, Franca C J, Souza G B, Nogueira A R A. Bioaccessibility of Ca, Cu, Fe, Mg, Zn, and crude protein in beef, pork and chicken after thermal processing. Food Chemistry, 2018, 240: 75-83.

[154] Liu Q, Peng H, Lu X, Le XC. Enzyme-assisted extraction and liquid chromatography mass spectrometry for the determination of arsenic species in chicken meat. Analytica Chimica Acta, 2015, 888: 1-9.

[155] Yilmaz E. Use of hydrolytic enzymes as green and effective extraction agents for ultrasound assisted-enzyme based hydrolytic water phase microextraction of arsenic in food samples. Talanta, 2018, 189: 302-307.

[156] Mandiwana K L, Panichev N, Panicheva S. Determination of chromium (Ⅵ) in black, green and herbal teas. Food Chemistry, 2011, 129 (4): 1839-1843.

[157] de Sousa W V, da Silva F L F, Gouveia S T, Matos W O, Ribeiro L P D, Lopes G S. Infrared radiation applied as a heating source in milk sample preparation for the determination of trace elements by inductively coupled plasma-optical emission spectroscopy. Revista Virtual de Quimica, 2017, 9 (6): 2226-2236.

[158] Wu Y, Lee Y I, Wu L, Hou X. Simple mercury speciation analysis by CVG-ICP-MS following TMAH pre-treatment and microwave-assisted digestion. Microchemical Journal, 2012, 103: 105-109.

[159] Aranha T S C P, Oliveira A, Queiroz H M, Cadore S. A fast alkaline treatment for cadmium determination in meat samples. Food Control, 2016, 59: 447-453.

[160] 刘震. 现代分离科学. 北京: 化学工业出版社, 2017: 279-289.

[161] Tanase C M, Griffin P, Koski K G, Cooper M J, Cockell K A. Sodium and potassium in composite food samples from the Canadian total diet study. Journal of Food Composition and Analysis, 2011, 24 (2): 237-243.

[162] Barros A I, de Oliveira A P, Gomes Neto J A, Dalla Villa R. Liquid-liquid extraction as a sample preparation procedure for the determination of Na, K, Ca, and Mg in biodiesel. Analytical Methods, 2017, 9 (36): 5395-5399.

[163] Pereira Junior J B, Dantas K G F. Evaluation of inorganic elements in cat's claw teas using ICP-OES and GF AAS. Food Chemistry, 2016, 196: 331-337.

[164] Mir K A, Rutter A, Koch I, Smith P, Reimer K J, Poland J S. Extraction and speciation of arsenic in plants grown on arsenic contaminated soils. Talanta, 2007, 72 (4): 1507-1518.

[165] dos Santos A M P, Oliveira A C, Souza A S, de Jesus R M, Ferreira S L C. Determination and evaluation of the mineral composition of Chinese cabbage (*Beta vulgaris*). Food Analytical Methods, 2011, 4 (4): 567-573.

[166] Jedynak L, Kowalska J, Kossykowska M, Golimowski J. Studies on the uptake of different arsenic forms and the influence of sample pretreatment on arsenic speciation in white mustard (*Sinapis alba*). Microchemical Journal, 2010, 94 (2): 125-129.

[167] Alava P, Van de Wiele T, Tack F, Du Laing G. Extensive grinding and pressurized extraction with water are key points for effective and species preserving extraction of arsenic from rice. Analytical Methods, 2012, 4 (5): 1237-1243.

[168] Bluemlein K, Raab A, Feldmann J. Stability of arsenic peptides in plant extracts: Off-line versus on-line parallel elemental and molecular mass spectrometric detection for liquid chromatographic separation. Analytical and Bioanalytical Chemistry, 2009, 393 (1): 357-366.

[169] Zhang W H, Cai Y, Downum K R, Ma L Q. Arsenic complexes in the arsenic hyperaccumulator pteris vittata (Chinese brake fern). Journal of Chromatography A, 2004, 1043 (2): 249-254.

[170] Dong Y, Gao M, Liu X, Qiu W, Song Z. The mechanism of polystyrene microplastics to affect arsenic volatilization in arsenic-contaminated paddy soils. Journal of Hazardous Materials, 2020, 398: 122896.

[171] Thompson C M, Ellwood M J, Wille M. A solvent extraction technique for the isotopic measurement of dissolved copper in seawater. Analytica Chimica Acta, 2013, 775: 106-113.

[172] Jedynak L, Kowalska J, Harasimowicz J, Golimowski J. Speciation analysis of arsenic in terrestrial plants from arsenic

contaminated area. Science of the Total Environment, 2009, 407 (2): 945-952.

[173] Pizarro I, Gomez M, Camara C, Palacios M A. Arsenic speciation in environmental and biological samples-extraction and stability studies. Analytica Chimica Acta, 2003, 495 (1-2): 85-98.

[174] Schmidt A C, Haufe N, Otto M. A systematic study on extraction of total arsenic from down-scaled sample sizes of plant tissues and implications for arsenic species analysis. Talanta, 2008, 76 (5): 1233-1240.

[175] Yuan C G, Jiang G B, He B. Evaluation of the extraction methods for arsenic speciation in rice straw, *Oryza sativa* L., and analysis by HPLC-HG-AFS. Journal of Analytical Atomic Spectrometry 2005, 20 (2): 103-110.

[176] Larios R, Fernandez-Martinez R, LeHecho I, Rucandio I. A methodological approach to evaluate arsenic speciation and bioaccumulation in different plant species from two highly polluted mining areas. Science of the Total Environment, 2012, 414: 600-607.

[177] Narukawa T, Inagaki K, Kuroiwa T, Chiba K. The extraction and speciation of arsenic in rice flour by HPLC-ICP-MS. Talanta, 2008, 77 (1): 427-432.

[178] Taebunpakul S, Liu C, Wright C, McAdam K, Heroult J, Braybrook J, Goenaga-Infante H. Determination of total arsenic and arsenic speciation in tobacco products: From tobacco leaf and cigarette smoke. Journal of Analytical Atomic Spectrometry, 2011, 26 (8): 1633-1640.

[179] Ferreira S L C, dos Santos W N L, dos Santos I F, Junior M M S , Silva L O B, Barbosa U A, de Santana F A, Queiroz A F dS. Strategies of sample preparation for speciation analysis of inorganic antimony using hydride generation atomic spectrometry. Microchemical Journal, 2014, 114: 22-31.

[180] Welna M, Pohl P, Szymczycha-Madeja A. Non-chromatographic speciation of inorganic arsenic in rice by hydride generation inductively coupled plasma optical emission spectrometry. Food Analytical Methods, 2019, 12 (2): 581-594.

[181] Rahman F, Chen Z L, Naidu R. A comparative study of the extractability of arsenic species from silverbeet and amaranth vegetables. Environmental Geochemistry and Health, 2009, 31: 103-113.

[182] Ghanemi K, Navidi M A, Fallah-Mehrjardi M, Dadolahi-Sohrab A. Ultra-fast microwave-assisted digestion in choline chloride-oxalic acid deep eutectic solvent for determining Cu, Fe, Ni and Zn in marine biological samples. Analytical Methods, 2014, 6 (6): 1774-1781.

[183] Matong J M, Nyaba L, Nomngongo P N. Determination of As, Cr, Mo, Sb, Se and V in agricultural soil samples by inductively coupled plasma optical emission spectrometry after simple and rapid solvent extraction using choline chloride-oxalic acid deep eutectic solvent. Ecotoxicology and Environmental Safety, 2017, 135: 152-157.

[184] Ali J, Tuzen M, Kazi T G. Green and innovative technique develop for the determination of vanadium in different types of water and food samples by eutectic solvent extraction method. Food Chemistry, 2020, 306: 125638.

[185] Ataee M, Ahmadi-Jouibari T, Noori N, Fattahi N. The speciation of inorganic arsenic in soil and vegetables irrigated with treated municipal wastewater. RSC Advances, 2020, 10 (3): 1514-1521.

[186] Elik A, Demirbas A, Altunay N. Developing a new and simple natural deep eutectic solvent based ultrasonic-assisted microextraction procedure for determination and preconcentration of As and Se from rice samples. Analytical Methods, 2019, 11 (27): 3429-3438.

[187] (a) Zounr R A, Tuzen M, Khuhawar M Y. A simple and green deep eutectic solvent based air assisted liquid phase microextraction for separation, preconcentration and determination of lead in water and food samples by graphite furnace atomic absorption spectrometry. Journal of Molecular Liquids, 2018, 259: 220-226.（b）Zu W, Wang Z. Ultra-trace determination of methylmercuy in seafood by atomic fluorescence spectrometry coupled with electrochemical cold vapor generation. Journal of Hazardous Materials, 2016, 304: 467-473.

[188] Akramipour R, Golpayegani M R, Ghasemi M, Noori N, Fattahi N. Development of an efficient sample preparation

method for the speciation of Se(Ⅳ)/Se(Ⅵ) and total inorganic selenium in blood of children with acute leukemia. New Journal of Chemistry, 2019, 43 (18): 6951-6958.

[189] Amjadi M, Manzoori J L, Hamedpour V. Optimized ultrasound-assisted temperature-controlled ionic liquid microextraction coupled with FAAS for determination of tin in canned foods. Food Analytical Methods, 2013, 6 (6): 1657-1664.

[190] Fang G, Zhang J, Lu J, Ma L, Wang S. Preparation, characterization, and application of a new thiol-functionalized ionic liquid for highly selective extraction of Cd(Ⅱ). Microchimica Acta, 2010, 171 (3-4): 305-311.

[191] Moradi M, Kashanaki R, Borhani S, Bigdeli H, Abbasi N, Kazemzadeh A. Optimization of supramolecular solvent microextraction prior to graphite furnace atomic absorption spectrometry for total selenium determination in food and environmental samples. Journal of Molecular Liquids, 2017, 232: 243-250.

[192] Hashemi B, Zohrabi P, Kim K H, Shamsipur M, Deep A, Hong J. Recent advances in liquid-phase microextraction techniques for the analysis of environmental pollutants. TrAC Trends in Analytical Chemistry, 2017, 97: 83-95.

[193] de Sousa J M, Cassella R J, Lepri F G. Evaluation of extraction induced by emulsion breaking for Ni and V extraction from off-shore Brazilian crude oils. Energy Fuels, 2019, 33 (11): 10435-10441.

[194] Trevelin A M, Marotto R E S, de Castro E V R, Brandao G P, Cassella R J, Carneiro M T W D. Extraction induced by emulsion breaking for determination of Ba, Ca, Mg and Na in crude oil by inductively coupled plasma optical emission spectrometry. Microchemical Journal, 2016, 124: 338-343.

[195] Caldas L F S, Brum D M, de Paula C E R, Cassella R J. Application of the extraction induced by emulsion breaking for the determination of Cu, Fe and Mn in used lubricating oils by flame atomic absorption spectrometry. Talanta, 2013, 110: 21-27.

[196] Bakircioglu D, Kurtulus Y B, Yurtsever S. Comparison of extraction induced by emulsion breaking, ultrasonic extraction and wet digestion procedures for determination of metals in edible oil samples in turkey using ICP-OES. Food Chemistry, 2013, 138 (2-3): 770-775.

[197] Bakircioglu D, Topraksever N, Kurtulus Y B. Separation/preconcentration system based on emulsion-induced breaking procedure for determination of cadmium in edible oil samples by flow injection-flame atomic absorption spectrometry. Food Analytical Methods, 2015, 8 (9): 2178-2184.

[198] Bakircioglu D, Topraksever N, Kurtulus Y B. Determination of zinc in edible oils by flow injection FAAS after extraction induced by emulsion breaking procedure. Food Chemistry, 2014, 151: 219-224.

[199] Cassella R J, Brum D M, Robaina N F, Lima C F. Extraction induced by emulsion breaking: A model study on metal extraction from mineral oil. Fuel, 2018, 215: 592-600.

[200] Antunes G A, dos Santos H S, da Silva Y P, Silva M M, Piatnicki C M S, Samios D. Determination of iron, copper, zinc, aluminum, and chromium in biodiesel by flame atomic absorption spectrometry using a microemulsion preparation method. Energy Fuels, 2017, 31(3): 2944-2950.

[201] Fernandes A, Vinhal J O, Bourdot Dutra A J, Cassella R J. Study of the extraction of Ca, Mg and Zn from different types of lubricating oils (mineral, semi-synthetic and synthetic) employing the emulsion breaking strategy. Microchemical Journal, 2019, 145: 1112-1118.

[202] Depoi F dS, de Oliveira TC, de Moraes D P, Pozebon D. Preconcentration and determination of As, Cd, Pb and Bi using different sample introduction systems, cloud point extraction and inductively coupled plasma optical emission spectrometry. Analytical Methods, 2012, 4 (1): 89-95.

[203] Depoi F dS, Bentlin F R S, Pozebon D. Methodology for Hg determination in honey using cloud point extraction and cold vapour-inductively coupled plasma optical emission spectrometry. Analytical Methods, 2010, 2 (2): 180-185.

[204] Khan S, Kazi T G, Baig J A, Kolachi N F, Afridi H I, Wadhwa S K, Shah A Q, Kandhro G A, Shah F. Cloud point extraction of vanadium in pharmaceutical formulations, dialysate and parenteral solutions using 8-hydroxyquinoline and nonionic surfactant. Journal of Hazardous Materials, 2010, 182 (1-3): 371-376.

[205] Ulusoy H I, Gurkan R, Demir O, Ulusoy S. Micelle-mediated extraction and flame atomic absorption spectrometric method for determination of trace cobalt ions in beverage samples. Food Analytical Methods, 2012, 5 (3): 454-463.

[206] Wen S, Zhu X, Wei Y, Wu S. Cloud point extraction-inductively coupled plasma mass spectrometry for separation/analysis of aqueous-exchangeable and unaqueous-exchangeable selenium in tea samples. Food Analytical Methods, 2013, 6 (2): 506-511.

[207] Lopez-Garcia I, Jose Marin-Hernandez J, Hernandez-Cordoba M. Graphite furnace atomic absorption spectrometric determination of vanadium after cloud point extraction in the presence of graphene oxide. Spectrochimica Acta Part B: Atomic Spectroscopy, 2018, 143: 42-47.

[208] Altunay N, Gurkan R, Yildirim E. A new ultrasound assisted-cloud point extraction method for the determination of trace levels of tin and antimony in food and beverages by flame atomic absorption apectrometry. Food Analytical Methods, 2016, 9 (10): 2960-2971.

[209] Gürkan R, Altunay N, Yildirim E. Combination of ultrasonic-assisted cloud point extraction with flame AAS for preconcentration and determination of trace amounts of silver and cadmium in dried nut and vegetable samples. Food Analytical Methods, 2016, 9 (11): 3218-3229.

[210] Gurkan R, Korkmaz S, Altunay N. Preconcentration and determination of vanadium and molybdenum in milk, vegetables and foodstuffs by ultrasonic-thermostatic-assisted cloud point extraction coupled to flame atomic absorption spectrometry. Talanta, 2016, 155: 38-46.

[211] Khan M, Kazi T G, Afridi H I, Bilal M, Akhtar A, Ullah N, Khan S, Talpur S. Application of ultrasonically modified cloud point extraction method for simultaneous enrichment of cadmium and lead in sera of different types of gallstone patients. Ultrasonics Sonochemistry, 2017, 39: 313-320.

[212] Marguí E, Sague M, Queralt I, Hidalgo M. Liquid phase microextraction strategies combined with total reflection X-ray spectrometry for the determination of low amounts of inorganic antimony species in waters. Analytica Chimica Acta, 2013, 786: 8-15.

[213] Pytlakowska K, Sitko R. Energy-dispersive X-ray fluorescence spectrometry combined with dispersive liquid-liquid microextraction for simultaneous determination of zinc and copper in water samples. Analytical Methods, 2013, 5 (21): 6192-6199.

[214] Bidari A, Ganjali M R, Assadi Y, Kiani A, Norouzi P. Assay of total mercury in commercial food supplements of marine origin by means of DLLME/ICP-AES. Food Analytical Methods, 2012, 5 (4): 695-701.

[215] Molaakbari E, Mostafavi A, Afzali D. Ionic liquid ultrasound assisted dispersive liquid-liquid microextraction method for preconcentration of trace amounts of rhodium prior to flame atomic absorption spectrometry determination. Journal of Hazardous Materials, 2011, 185 (2-3): 647-652.

[216] Seeger T S, Rosa F C, Bizzi C A, Dressler V L, Flores E M M, Duarte F A. Feasibility of dispersive liquid-liquid microextraction for extraction and preconcentration of Cu and Fe in red and white wine and determination by flame atomic absorption spectrometry. Spectrochimica Acta Part B: Atomic Spectroscopy, 2015, 105: 136-140.

[217] Jafarvand S, Shemirani F. Supramolecular-based dispersive liquid-liquid microextraction: A novel sample preparation technique for determination of inorganic species. Microchimica Acta, 2011, 173 (3-4): 353-359.

[218] Sorouraddin S M, Nouri S. Simultaneous temperature-assisted dispersive liquid-liquid microextraction of cobalt, copper, nickel and zinc ions from high-volume water samples and determination by graphite furnace atomic absorption

spectrometry. Analytical Methods, 2016, 8 (6): 1396-1404.

[219] Zeeb M, Ganjali M R, Norouzi P. Preconcentration and trace determination of chromium using modified ionic liquid cold-induced aggregation dispersive liquid-liquid microextraction: Application to different water and food samples. Food Analytical Methods, 2013, 6 (5): 1398-1406.

[220] Zeeb M, Mirza B, Zare-Dorabei R, Farahani H. Ionic liquid-based ultrasound-assisted *in situ* solvent formation microextraction combined with electrothermal atomic absorption spectrometry as a practical method for preconcentration and trace determination of vanadium in water and food samples. Food Analytical Methods, 2014, 7 (9): 1783-1790.

[221] Wu Q, Wu C, Wang C, Lu X, Li X, Wang Z. Sensitive determination of cadmium in water, beverage and cereal samples by a novel liquid-phase microextraction coupled with flame atomic absorption spectrometry. Analytical Methods, 2011, 3 (1): 210-216.

[222] Tuzen M, Shemsi A M, Bukhari A A. Vortex-assisted solidified floating organic drop microextraction of molybdenum in beverages and food samples coupled with graphite furnace atomic absorption spectrometry. Food Analytical Methods, 2017, 10 (1): 219-226.

[223] Eskina V V, Dalnova O A, Filatova D G, Baranovskaya V B, Karpov Y A. Separation and preconcentration of platinum-group metals from spent autocatalysts solutions using a hetero-polymeric S, N-containing sorbent and determination by high-resolution continuum source graphite furnace atomic absorption spectrometry. Talanta, 2016, 159: 103-110.

[224] Wen S, Zhu X. Speciation analysis of Mn(Ⅱ)/Mn(Ⅶ) in tea samples using flame atomic absorption spectrometry after room temperature ionic liquid-based dispersive liquid-liquid microextraction. Food Analytical Methods, 2014, 7 (2): 291-297.

[225] Barros J A V A, Angel Aguirre M, Kovachev N, Canals A, Nobrega J A. Vortex-assisted dispersive liquid-liquid microextraction for the determination of molybdenum in plants by inductively coupled plasma optical emission spectrometry. Analytical Methods, 2016, 8 (4): 810-815.

[226] Nogueira da Silva K R, Greco A dS, Corazza M Z, Raposo J L Jr. Feasibility of dispersive liquid-liquid microextraction to determine Ca, Mg, K, and Na in biodiesel by atomic spectrometry. Analytical Methods, 2018, 10 (26): 3284-3291.

[227] Lourenco E C, Eyng E, Bittencourt P R S, Duarte F A, Picoloto R S, Flores E L M. A simple, rapid and low cost reversed-phase dispersive liquid-liquid microextraction for the determination of Na, K, Ca and Mg in biodiesel. Talanta, 2019, 199: 1-7.

[228] Kalschne D L, Canan C, Beato M O, Leite O D, Moraes Flores E L. A new and feasible analytical method using reversed-phase dispersive liquid-liquid microextraction (RP-DLLME) for further determination of Nickel in hydrogenated vegetable fat. Talanta, 2020, 120409.

[229] Mohammadi S Z, Afzali D, Taher M A, Baghelani Y M. Determination of trace amounts of palladium by flame atomic absorption spectrometry after ligandless-dispersive liquid-liquid microextraction. Microchimica Acta, 2010, 168 (1-2): 123-128.

[230] Martinez D, Torregrosa D, Grindlay G, Gras L, Mora J. Coupling dispersive liquid-liquid microextraction to inductively coupled plasma atomic emission spectrometry: An oxymoron? Talanta, 2018, 176: 374-381.

[231] Wang X, Chen P, Cao L, Xu G, Yang S, Fang Y, Wang G, Hong X. Selenium speciation in rice samples by magnetic ionic liquid-based up-and-down-shaker-assisted dispersive liquid-liquid microextraction coupled to graphite furnace atomic absorption spectrometry. Food Analytical Methods, 2017, 10 (6): 1653-1660.

[232] Liang P, Yu J, Yang E, Mo Y. Determination of mercury in food and water samples by displacement-dispersive liquid-liquid microextraction coupled with graphite furnace atomic absorption spectrometry. Food Analytical Methods, 2015, 8 (1): 236-242.

[233] Rezaee M, Yamini Y, Faraji M. Evolution of dispersive liquid-liquid microextraction method. Journal of Chromatography A, 2010, 1217(16): 2342-2357.

[234] Wu C C, Liu H M, Determination of gallium in human urine by supercritical carbon dioxide extraction and graphite furnace atomic absorption spectrometry. Journal of Hazardous Materials, 2009, 163(2-3): 1239-1245.

[235] Liu H M, Wu C C, Lin Y H, Chiang C K. Recovery of indium from etching wastewater using supercritical carbon dioxide extraction. Journal of Hazardous Materials, 2009, 172(2-3): 744-748.

[236] Herce-Sesa B, Lopez-Lopez J A, Moreno C. Ionic liquid solvent bar micro-extraction of $CdCl_n^{(n-2)-}$ species for ultra-trace Cd determination in seawater. Chemosphere, 2018, 193: 306-312.

[237] López-López J A, Herce-Sesa B, Moreno C. Solvent bar micro-extraction with graphite atomic absorption spectrometry for the determination of silver in ocean water. Talanta, 2016, 159: 117-121.

[238] de Oliveira T L, de Oliveira A P, Villa R D. Solid-liquid extraction as a clean sample preparation procedure for determination of Na and K in meat products. Journal of Food Composition and Analysis, 2017, 62: 164-167.

[239] Neri T S, Rocha D P, Munoz R A A, Coelho N M M, Batista A D. Highly sensitive procedure for determination of Cu(II) by GF AAS using single-drop microextraction. Microchemical Journal, 2019, 147: 894-898.

[240] Es'haghi Z, Azmoodeh R. Hollow fiber supported liquid membrane microextraction of Cu^{2+} followed by flame atomic absorption spectroscopy determination. Arabian Journal of Chemistry, 2010, 3 (1): 21-26.

[241] Covaci E, Senila M, Ponta M, Darvasi E, Petreus D, Frentiu M, Frentiu T. Methylmercury determination in seafood by photochemical vapor generation capacitively coupled plasma microtorch optical emission spectrometry. Talanta, 2017, 170: 464-472.

[242] Nabid M R, Sedghi R, Bagheri A, Behbahani M, Taghizadeh M, Oskooie H A, Heravi M M. Preparation and application of poly(2-amino thiophenol)/MWCNTs nanocomposite for adsorption and separation of cadmium and lead ions via solid phase extraction. Journal of Hazardous Materials, 2012, 203: 93-100.

[243] Behbahani M, Gorji T, Mahyari M, Salarian M, Bagheri A, Shaabani A. Application of polypropylene amine dendrimers (POPAM)-grafted MWCNTs hybrid materials as a new sorbent for solid-phase extraction and trace determination of gold(III) and palladium(II) in food and environmental samples. Food Analytical Methods, 2014, 7 (5): 957-966.

[244] Ghorbani-Kalhor E, Behbahani M, Abolhasani J, Khanmiri R H. Synthesis and characterization of modified multiwall carbon nanotubes with poly(N-phenylethanolamine) and their application for removal and trace detection of lead ions in food and environmental samples. Food Analytical Methods, 2015, 8 (5): 1326-1334.

[245] Sacmaci S, Kartal S, Yilmaz Y, Sacmaci M, Soykan C. A new chelating resin: Synthesis, characterization and application for speciation of chromium(III)/(VI) species. Chemical Engineering Journal, 2012, 181: 746-753.

[246] Liu Y, Hu J, Li Y, Wei H P, Li X S, Zhang X H, Chen S M, Chen X Q. Synthesis of dolyethyleneimine capped carbon dots for preconcentration and slurry sampling analysis of trace chromium in environmental water samples. Talanta, 2015, 134: 16-23.

[247] Sadeghi H B, Panahi H A, Abdouss M, Esmaiilpour B, Nezhati M N, Moniri E, Azizi Z. Modification and characterization of polyacrylonitrile fiber by chelating ligand for preconcentration and determination of neodymium ion in biological and environmental samples. Journal of Applied Polymer Science, 2013, 128 (2): 1125-1130.

[248] Brombach C C, Chen B, Corns W T, Feldmann J, Krupp E M. Methylmercury in water samples at the pg/L level by online preconcentration liquid chromatography cold vapor-atomic fluorescence spectrometry. Spectrochimica Acta Part B: Atomic Spectroscopy, 2015, 105: 103-108.

[249] Stanisz E, Werner J, Matusiewicz H. Task specific ionic liquid-coated PTFE tube for solid-phase microextraction prior to chemical and photo-induced mercury cold vapour generation. Microchemical Journal, 2014, 114: 229-237.

[250] Evans E H, Pisonero J, Smith C M M, Taylor R N. Atomic spectrometry update: Review of advances in atomic spectrometry and related techniques. Journal of Analytical Atomic Spectrometry, 2016, 31 (5): 1057-1077.

[251] Kojidi M H, Aliakbar A. A graphene oxide based poly(2,6-diaminopyridine) composite for solid-phase extraction of Cd(Ⅱ) prior to its determination by FAAS. Microchimica Acta, 2017, 184 (8): 2855-2860.

[252] Xie F, Liu G, Wu F, Guo G, Li G. Selective adsorption and separation of trace dissolved Fe(Ⅲ) from natural water samples by double template imprinted sorbent with chelating diamines. Chemical Engineering Journal, 2012, 183: 372-380.

[253] Mortada W I, Kenawy I M M, Abou El-Reash Y G, Mousa A A. Microwave assisted modification of cellulose by gallic acid and its application for removal of aluminium from real samples. International Journal of Biological Macromolecules, 2017, 101: 490-501.

[254] Zhang Q, He M, Chen B, Hu B. Preparation, characterization and application of *Saussurea tridactyla* Sch-Bip as green adsorbents for preconcentration of rare earth elements in environmental water samples. Spectrochimica Acta Part B: Atomic Spectroscopy, 2016, 121: 1-10.

[255] Alves V N, Mosquetta R, Melo Coelho N M, Bianchin J N, Di Pietro Roux K C, Martendal E, Carasek E. Determination of cadmium in alcohol fuel using *Moringa oleifera* seeds as a biosorbent in an on-line system coupled to FAAS. Talanta, 2010, 80 (3): 1133-1138.

[256] Dobrowolski R, Otto M. Preparation and evaluation of Ni-loaded activated carbon for enrichment of arsenic for analytical and environmental purposes. Microporous and Mesoporous Materials, 2013, 179: 1-9.

[257] Dobrowolski R, Otto M. Preparation and evaluation of Fe-loaded activated carbon for enrichment of selenium for analytical and environmental purposes. Chemosphere, 2013, 90 (2): 683-690.

[258] Rajabi M, Barfi B, Asghari A, Najafi F, Aran R. Hybrid amine-functionalized titania/silica nanoparticles for solid-phase extraction of lead, copper, and zinc from food and water samples: Kinetics and equilibrium studies. Food Analytical Methods, 2015, 8 (4): 815-824.

[259] Wang W, Chen M, Chen X, Wang J. Thiol-rich polyhedral oligomeric silsesquioxane as a novel adsorbent for mercury adsorption and speciation. Chemical Engineering Journal, 2014, 242: 62-68.

[260] Zhang D, Wang M, Ren G, Song E. Preparation of biomorphic porous calcium titanate and its application for preconcentration of nickel in water and food samples. Materials Science & Engineering C: Materials for Biological Applications, 2013, 33 (8): 4677-4683.

[261] Zhan M, Yu H, Li L, Nguyen D T, Chen W. Detection of hexavalent chromium by copper sulfide nanocomposites. Analytical Chemistry, 2019, 91 (3): 2058-2065.

[262] Abdolmohammad-Zadeh H, Sadeghi G H. A nano-structured material for reliable speciation of chromium and manganese in drinking waters, surface waters and industrial wastewater effluents. Talanta, 2012, 94: 201-208.

[263] Baysal A, Akman S, Demir S, Kahraman M. Slurry sampling electrothermal atomic absorption spectrometric determination of chromium after separation/enrichment by mercaptoundecanoic acid modified gold coated TiO_2 nanoparticles. Microchemical Journal, 2011, 99 (2): 421-424.

[264] Amjadi M, Samadi A. Modified ionic liquid-coated nanometer TiO_2 as a new solid phase extraction sorbent for preconcentration of trace nickel. Colloids and Surfaces A: Physicochemical and Engineering Aspects, 2013, 434: 171-177.

[265] Ayata S, Bozkurt S S, Ocakoglu K. Separation and preconcentration of Pb(Ⅱ) using ionic liquid-modified silica and its determination by flame atomic absorption spectrometry. Talanta, 2011, 84 (1): 212-215.

[266] Liang P, Peng L. Ionic liquid-modified silica as sorbent for preconcentration of cadmium prior to its determination by

flame atomic absorption spectrometry in water samples. Talanta, 2010, 81 (1-2): 673-677.

[267] Cheng J, Ma X, Wu Y. Silica gel chemically modified with ionic liquid as novel sorbent for solid-phase extraction and preconcentration of lead from beer and tea drink samples followed by flame atomic absorption spectrometric determination. Food Analytical Methods, 2014, 7 (5): 1083-1089.

[268] Diniz K M, Gorla F A, Ribeiro E S, Olimpio do Nascimento M B, Correa R J, Teixeira Tarley C R, Segatelli M G. Preparation of $SiO_2/Nb_2O_5/ZnO$ mixed oxide by sol-gel method and its application for adsorption studies and on-line preconcentration of cobalt ions from aqueous medium. Chemical Engineering Journal, 2014, 239: 233-241.

[269] Imran K, Harinath Y, Naik B R, Kumar N S, Seshaiah K. A new hybrid sorbent 2, 2′-pyridil functionalized SBA-15 (pyl-SBA-15) synthesis and its applications in solid phase extraction of Cu(II) from water samples. Journal of Environmental Chemical Engineering, 2019, 7 (3): 103170.

[270] Maraschi F, Speltini A, Tavani T, Gulotta M G, Dondi D, Milanese C, Prato M. Silica-supported pyrolyzed lignin for solid-phase extraction of rare earth elements from fresh and sea waters followed by ICP-MS detection. Analytical and Bioanalytical Chemistry, 2018, 410 (29): 7635-7643.

[271] Rajabi H R, Shamsipur M, Pourmortazavi S M. Preparation of a novel potassium ion imprinted polymeric nanoparticles based on dicyclohexyl 18C6 for selective determination of K^+ ion in different water samples. Materials Science & Engineering C: Materials for Biological Applications, 2013, 33 (6): 3374-3381.

[272] Singh D K, Mishra S. Synthesis, characterization and analytical applications of Ni(II)-ion imprinted polymer. Applied Surface Science, 2010, 256 (24): 7632-7637.

[273] Zambrzycka E, Kiedysz U, Wilczewska A Z, Lesniewska B, Godlewska-Zykiewicz B. A novel ion imprinted polymer as a highly selective sorbent for separation of ruthenium ions from environmental samples. Analytical Methods, 2013, 5 (12): 3096-3105.

[274] Abbasi S, Roushani M, Khani H, Sahraei R, Mansouri G. Synthesis and application of ion-imprinted polymer nanoparticles for the determination of nickel ions. Spectrochimica Acta Part A: Molecular and Biomolecular Spectroscopy, 2015, 140: 534-543.

[275] Shamsipur M, Rajabi H R, Pourmortazavi S M, Roushani M. Ion imprinted polymeric nanoparticles for selective separation and sensitive determination of zinc ions in different matrices. Spectrochimica Acta Part A: Molecular and Biomolecular Spectroscopy, 2014, 117: 24-33.

[276] Shamsipur M, Besharati-Seidani A, Fasihi J, Sharghi H. Synthesis and characterization of novel ion-imprinted polymeric nanoparticles for very fast and highly selective recognition of copper(II) ions. Talanta, 2010, 83 (2): 674-681.

[277] Barciela-Alonso M C, Plata-Garcia V, Rouco-Lopez A, Moreda-Pineiro A, Bermejo-Barrera P. Ionic imprinted polymer based solid phase extraction for cadmium and lead pre-concentration/determination in seafood. Microchemical Journal, 2014, 114: 106-110.

[278] Behbahani M, Bagheri A, Taghizadeh M, Salarian M, Sadeghi O, Adlnasab L, Jalali K. Synthesis and characterisation of nano structure lead(II) ion-imprinted polymer as a new sorbent for selective extraction and preconcentration of ultra trace amounts of lead ions from vegetables, rice, and fish samples. Food Chemistry, 2013, 138 (2-3): 2050-2056.

[279] Fayazi M, Taher M A, Afzali D, Mostafavi A, Ghanei-Motlagh M. Synthesis and application of novel ion-imprinted polymer coated magnetic multi-walled carbon nanotubes for selective solid phase extraction of lead(II) ions. Materials Science & Engineering C: Materials for Biological Applications, 2016, 60: 365-373.

[280] Jiang Y, Kim D. Synthesis and selective adsorption behavior of Pd(II)-imprinted porous polymer particles. Chemical Engineering Journal, 2013, 232: 503-509.

[281] Fayazi M, Ghanei-Motlagh M, Taher M A, Ghanei-Motlagh R, Salavati MR. Synthesis and application of a novel

nanostructured ion-imprinted polymer for the preconcentration and determination of thallium(I) ions in water samples. Journal of Hazardous Materials, 2016, 309: 27-36.

[282] De Martino M G, Macarovscha G T, Cadore S. The use of zincon for preconcentration and determination of zinc by flame atomic absorption spectrometry. Analytical Methods, 2010, 2 (9): 1258-1262.

[283] Sergiel I, Pohl P. Determination of the total content of calcium and magnesium and their bioavailability in ripened bee honeys. Journal of Agricultural and Food Chemistry, 2010, 58 (13): 7497-7501.

[284] Pohl P, Pieprz M, Dzimitrowicz A, Jamroz P, Szymczycha-Madeja A, Welna M. New green determination of Cu, Fe, Mn, and Zn in beetroot juices along with their chemical fractionation by solid-phase extraction. Molecules, 2019, 24 (20): 3645.

[285] Ben Issa N, Rajakovic-Ognjanovic V N, Jovanovic B M, Rajakovic L V. Determination of inorganic arsenic species in natural waters: Benefits of separation and preconcentration on ion exchange and hybrid resins. Analytica Chimica Acta, 2010, 673 (2): 185-193.

[286] Smolíková V, Pelcova P, Ridoskova A, Hedbavny J, Grmela J. Development and evaluation of the iron oxide-hydroxide based resin gel for the diffusive gradient in thin films technique. Analytica Chimica Acta, 2020, 1102: 36-45.

[287] Wang T, Chen Y, Ma J, Jin Z, Chai M, Xiao X, Zhang L, Zhang Y. A polyethyleneimine-modified attapulgite as a novel solid support in matrix solid-phase dispersion for the extraction of cadmium traces in seafood products. Talanta, 2018, 180: 254-259.

[288] Chen Q, Lin Y, Tian Y, Wu L, Yang L, Hou X, Zheng C. Single-drop solution electrode discharge-induced cold vapor generation coupling to matrix solid-phase dispersion: A robust approach for sensitive quantification of total mercury distribution in fish. Analytical Chemistry, 2017, 89 (3): 2093-2100.

[289] Karatapanis A E, Petrakis D E, Stalikas C D. A layered magnetic iron/iron oxide nanoscavenger for the analytical enrichment of ng·L^{-1} concentration levels of heavy metals from water. Analytica Chimica Acta, 2012, 726: 22-27.

[290] Giakisikli G, Anthemidis A N. Magnetic materials as sorbents for metal/metalloid preconcentration and/or separation: A review. Analytica Chimica Acta, 2013, 789: 1-16.

[291] Dahaghin Z, Mousavi H Z, Boutorabi L. Application of magnetic ion-imprinted polymer as a new environmentally-friendly nonocomposite for a selective adsorption of the trace level of Cu(II) from aqueous solution and different samples. Journal of Molecular Liquids, 2017, 243: 380-386.

[292] He H, Xiao D, He J, Li H, He H, Dai H, Peng J. Preparation of a core-shell magnetic ion-imprinted polymer via a sol-gel process for selective extraction of Cu(II) from herbal medicines. Analyst, 2014, 139 (10): 2459-2466.

[293] Qi X, Gao S, Ding G, Tang A-N. Synthesis of surface Cr(VI)-imprinted magnetic nanoparticles for selective dispersive solid-phase extraction and determination of Cr(VI) in water samples. Talanta, 2017, 162: 345-353.

[294] Zawisza B, Sitko R, Malicka E, Talik E. Graphene oxide as a solid sorbent for the preconcentration of cobalt, nickel, copper, zinc and lead prior to determination by energy-dispersive X-ray fluorescence spectrometry. Analytical Methods, 2013, 5 (22): 6425-6430.

[295] Pytlakowska K. Graphene-based preconcentration system prior to energy dispersive X-ray fluorescence spectrometric determination of Co, Ni, and Cu ions in wine samples. Food Analytical Methods, 2016, 9 (8): 2270-2279.

[296] Baranik A, Gagor A, Queralt I, Margui E, Sitko R, Zawisza B. Determination and speciation of ultratrace arsenic and chromium species using aluminium oxide supported on graphene oxide. Talanta, 2018, 185: 264-274.

[297] Sitko R, Janik P, Zawisza B, Talik E, Margui E, Queralt I. Green approach for ultratrace determination of divalent metal ions and arsenic species using total-reflection X-ray fluorescence spectrometry and mercapto-modified graphene oxide nanosheets as a novel adsorbent. Analytical Chemistry, 2015, 87 (6): 3535-3542.

[298] Bagheri H, Asgharinezhad A A, Ebrahimzadeh H. Determination of trace amounts of Cd(II), Cu(II), and Ni(II) in food samples using a novel functionalized magnetic nanosorbent. Food Analytical Methods, 2016, 9 (4): 876-888.

[299] Kardar Z S, Beyki M H, Shemirani F. Takovite-aluminosilicate@MnFe$_2$O$_4$ nanocomposite, a novel magnetic adsorbent for efficient preconcentration of lead ions in food samples. Food Chemistry, 2016, 209: 241-247.

[300] Farajzadeh M A, Yadeghari A. Extraction and preconcentration of nickel, cadmium, cobalt, and lead cations using dispersive solid phase extraction performed in a narrow-bore tube. Journal of Industrial and Engineering Chemistry, 2018, 59: 377-387.

[301] Faraji M, Shariati S, Yamini Y, Adeli M. Preconcentration of trace amounts of lead in water samples with cetyltrimethylammonium bromide coated magnetite nanoparticles and its determination by flame atomic absorption spectrometry. Arabian Journal of Chemistry, 2016, 9: S1540-S1546.

[302] Wang Y, Xie J, Wu Y, Ge H, Hu X. Preparation of a functionalized magnetic metal-organic framework sorbent for the extraction of lead prior to electrothermal atomic absorption spectrometer analysis. Journal of Materials Chemistry A, 2013, 1 (31): 8782-8789.

[303] Keramat A, Zare-Dorabei R. Ultrasound-assisted dispersive magnetic solid phase extraction for preconcentration and determination of trace amount of Hg (II) ions from food samples and aqueous solution by magnetic graphene oxide (Fe$_3$O$_4$@ GO/2-PTSC): Central composite design optimization. Ultrasonics Sonochemistry, 2017, 38: 421-429.

[304] Sricharoen P, Limchoowong N, Areerob Y, Nuengmatcha P, Techawongstien S, Chanthai S. Fe$_3$O$_4$/hydroxyapatite/graphene quantum dots as a novel nano-sorbent for preconcentration of copper residue in Thai food ingredients: Optimization of ultrasound-assisted magnetic solid phase extraction. Ultrasonics Sonochemistry, 2017, 37: 83-93.

[305] Abdolmohammad-Zadeh H, Hassanlouei S, Zamani-Kalajahi M. Preparation of ionic liquid-modified SiO$_2$@Fe$_3$O$_4$ nanocomposite as a magnetic sorbent for use in solid-phase extraction of zinc(II) ions from milk and water samples. RSC Advances, 2017, 7 (38): 23293-23300.

[306] Bagheri H, Afkhami A, Saber-Tehrani M, Khoshsafar H. Preparation and characterization of magnetic nanocomposite of Schiff base/silica/magnetite as a preconcentration phase for the trace determination of heavy metal ions in water, food and biological samples using atomic absorption spectrometry. Talanta, 2012, 97: 87-95.

[307] Mirabi A, Dalirandeh Z, Rad A S. Preparation of modified magnetic nanoparticles as a sorbent for the preconcentration and determination of cadmium ions in food and environmental water samples prior to flame atomic absorption spectrometry. Journal of Magnetism and Magnetic Materials, 2015, 381: 138-144.

[308] Safdarian M, Hashemi P, Adeli M. One-step synthesis of agarose coated magnetic nanoparticles and their application in the solid phase extraction of Pd(II) using a new magnetic field agitation device. Analytica Chimica Acta, 2013, 774: 44-50.

[309] Baghban N, Yilmaz E, Soylak M. A magnetic MoS$_2$-Fe$_3$O$_4$ nanocomposite as an effective adsorbent for dispersive solid-phase microextraction of lead(II) and copper(II) prior to their determination by FAAS. Microchimica Acta, 2017, 184 (10): 3969-3976.

[310] Ghorbani-Kalhor E. A metal-organic framework nanocomposite made from functionalized magnetite nanoparticles and HKUST-1 (MOF-199) for preconcentration of Cd(II), Pb(II), and Ni(II). Microchimica Acta, 2016, 183(9): 2639-2647.

[311] Khan M, Yilmaz E, Sevinc B, Sahmetlioglu E, Shah J, Jan M R, Soylak M. Preparation and characterization of magnetic allylamine modified graphene oxide-poly(vinyl acetate-*co*-divinylbenzene) nanocomposite for vortex assisted magnetic solid phase extraction of some metal ions. Talanta, 2016, 146: 130-137.

[312] Djerahov L, Vasileva P, Karadjova I. Self-standing chitosan film loaded with silver nanoparticles as a tool for selective

determination of Cr(Ⅵ) by ICP-MS. Microchemical Journal, 2016, 129: 23-28.

[313] Skorek R, Zawisza B, Margui E, Queralt I, Sitko R. Dispersive micro solid-phase extraction using multiwalled carbon nanotubes for simultaneous determination of trace metal ions by energy-dispersive X-ray fluorescence spectrometry. Applied Spectroscopy, 2013, 67 (2): 204-209.

[314] Dobrzynska J, Dabrowska M, Olchowski R, Dobrowolski R. An ion-imprinted thiocyanato-functionalized mesoporous silica for preconcentration of gold(Ⅲ) prior to its quantitation by slurry sampling graphite furnace AAS. Microchimica Acta, 2018, 185 (12): 564.

[315] Fahimirad B, Asghari A, Rajabi M. A novel nanoadsorbent consisting of covalently functionalized melamine onto MWCNT/Fe_3O_4 nanoparticles for efficient microextraction of highly adverse metal ions from organic and inorganic vegetables: Optimization by multivariate analysis. Journal of Molecular Liquids, 2018, 252: 383-391.

[316] Krawczyk M, Stanisz E. Ultrasound-assisted dispersive micro solid-phase extraction with nano-TiO_2 as adsorbent for the determination of mercury species. Talanta, 2016, 161: 384-391.

[317] Asghari A, Parvari S M, Hemmati M, Rajabi M. Statistical evaluation of three kinds of sonochemically-prepared magnetic conductive polymer nanocomposites for ultrasound-assisted ligandless uptake of some deleterious metal ions in vegetable samples. Journal of Molecular Liquids, 2018, 268: 867-874.

[318] Behbahani M, Hassanlou P G, Amini M M, Omidi F, Esrafili A, Farzadkia M, Bagheri A. Application of solvent-assisted dispersive solid phase extraction as a new, fast, simple and reliable preconcentration and trace detection of lead and cadmium ions in fruit and water samples. Food Chemistry, 2015, 187: 82-88.

[319] Cunha F A S, Ferreira D T S, Andrade W C R, Fernandes J P A, Lyra W S, Pessoa A G G, Ugulino de Araujo M C. Macroemulsion-based dispersive magnetic solid phase extraction for preconcentration and determination of copper(Ⅱ) in gasoline. Microchimica Acta, 2018, 185 (2): 99.

[320] Farahani M D, Shemirani F, Ramandi N F, Gharehbaghi M. Ionic liquid as a ferrofluid carrier for dispersive solid phase extraction of copper from food samples. Food Analytical Methods, 2015, 8 (8): 1979-1989.

[321] Gharehbaghi M, Farahani M D, Shemirani F. Dispersive magnetic solid phase extraction based on an ionic liquid ferrofluid. Analytical Methods, 2014, 6 (23): 9258-9266.

[322] Chen S, Zhu S, Lu D. Dispersive micro-solid phase extraction combined with dispersive liquid-liquid microextraction for speciation analysis of antimony by electrothermal vaporization inductively coupled plasma mass spectrometry. Spectrochimica Acta Part B: Atomic Spectroscopy, 2018, 139: 70-74.

[323] Sadeghi M, Rostami E, Kordestani D, Veisi H, Shamsipur M. Simultaneous determination of ultra-low traces of lead and cadmium in food and environmental samples using dispersive solid-phase extraction (DSPE) combined with ultrasound-assisted emulsification microextraction based on the solidification of floating organic drop (UAEME-SFO) followed by GFAAS. RSC Advances, 2017, 7 (44): 27656-27667.

[324] Rahmi D, Takasaki Y, Zhu Y, Kobayashi H, Konagaya S, Haraguchi H, Umemura T. Preparation of monolithic chelating adsorbent inside a syringe filter tip for solid phase microextraction of trace elements in natural water prior to their determination by ICP-MS. Talanta, 2010, 81 (4-5): 1438-1445.

[325] Shahdousti P, Alizadeh N. Headspace-solid phase microextraction of selenium(Ⅳ) from human blood and water samples using polypyrrole film and analysis with ion mobility spectrometry. Analytica Chimica Acta, 2011, 684 (1-2): 67-71.

[326] Wu S, Hu C, He M, Chen B, Hu B. Capillary microextraction combined with fluorinating assisted electrothermal vaporization inductively coupled plasma optical emission spectrometry for the determination of trace lanthanum, europium, dysprosium and yttrium in human hair. Talanta, 2013, 115: 342-348.

[327] Maya F, Palomino Cabello C, Manuel Estela J, Cerda V, Turnes Palomino G. Automatic in-syringe dispersive microsolid phase extraction using magnetic-metal organic frameworks. Analytical Chemistry, 2015, 87 (15): 7545-7549.

[328] Gouda A A. A new coprecipitation method without carrier element for separation and preconcentration of some metal ions at trace levels in water and food samples. Talanta, 2016, 146: 435-441.

[329] Mao X, Hu B, He M, Chen B. High polar organic-inorganic hybrid coating stir bar sorptive extraction combined with high performance liquid chromatography-inductively coupled plasma mass spectrometry for the speciation of seleno-amino acids and seleno-oligopeptides in biological samples. Journal of Chromatography A, 2012, 1256: 32-39.

[330] Chen L, Wang Z, Pei J, Huang X. Highly permeable monolith-based multichannel in-tip microextraction apparatus for simultaneous field sample preparation of pesticides and heavy metal ions in environmental waters. Analytical Chemistry, 2020, 92 (2): 2251-2257.

[331] Huang X J, Chen L L, Lin F H, Yuan D X. Novel extraction approach for liquid samples: Stir cake sorptive extraction using monolith. Journal of Separation Science, 2011, 34 (16-17): 2145-2151.

[332] Zhang Y, Mei M, Ouyang T, Huang X. Preparation of a new polymeric ionic liquid-based sorbent for stir cake sorptive extraction of trace antimony in environmental water samples. Talanta, 2016, 161: 377-383.

[333] Mimura A M S, Ferreira C C M, Silva J C J. Fast and feasible sample preparation methods for extraction of trace elements from deposited particulate matter samples. Analytical Methods, 2017, 9 (3): 490-499.

[334] Rutkowska M, Owczarek K, de la Guardia M, Plotka-Wasylka J, Namiesnik J. Application of additional factors supporting the microextraction process. TrAC Trends in Analytical Chemistry, 2017, 97: 104-119.

[335] Leme A B P, Bianchi S R, Carneiro R L, Nogueira A R A. Optimization of sample preparation in the determination of minerals and trace elements in honey by ICP-MS. Food Analytical Methods, 2014, 7 (5): 1009-1015.

[336] Muller A L H, Muller E I, Barin J S, Flores E M M. Microwave-assisted digestion using diluted acids for toxic element determination in medicinal plants by ICP-MS in compliance with United States pharmacopeia requirements. Analytical Methods, 2015, 7 (12): 5218-5225.

[337] Juranovic Cindric I, Krizman I, Zeiner M, Kampic S, Medunic G, Stingeder G. ICP-AES determination of minor and major elements in apples after microwave assisted digestion. Food Chemistry, 2012, 135 (4): 2675-2680.

[338] Martins C A, Cerveira C, Scheffler G L, Pozebon D. Metal Determination in tea, wheat, and wheat flour using diluted nitric acid, high-efficiency nebulizer, and axially viewed ICP-OES. Food Analytical Methods, 2015, 8 (7): 1652-1660.

[339] 范华均, 李攻科, 黎蔚波. 石墨炉原子吸收光谱法测定中药口服液中的铬铅镉. 分析试验室, 2005, 3: 36-39.

[340] Herrero Fernandez Z, Valcarcel Rojas L A, Montero Alvarez A, Estevez Alvarez J R, dos Santos Junior J A, Pupo Gonzalez I, Rodriguez Gonzalez M, Alberro Macias N, Lopez Sanchez D, Hernandez Torres D. Application of cold vapor-atomic absorption (CVAAS) spectrophotometry and inductively coupled plasma-atomic emission spectrometry methods for cadmium, mercury and lead analyses of fish samples. Validation of the method of CVAAS. Food Control, 2015, 48: 37-42.

[341] Acar O, Tunceli A, Turker A R. Comparison of wet and microwave digestion methods for the determination of copper, iron and zinc in some food samples by FAAS. Food Analytical Methods, 2016, 9 (11): 3201-3208.

[342] Gillermina Ceballos-Magana S, Marcos Jurado J, Muniz-Valencia R, Alcazar A, de Pablos F, Jesus Martin M. Geographical authentication of tequila according to its mineral content by means of support vector machines. Food Analytical Methods, 2012, 5 (2): 260-265.

[343] Frentiu T, Ponta M, Darvasi E, Frentiu M, Cordos E. Analytical capability of a medium power capacitively coupled plasma for the multielemental determination in multimineral/multivitamin preparations by atomic emission spectrometry. Food Chemistry, 2012, 134 (4): 2447-2452.

[344] Larrea-Marin M T, Pomares-Alfonso M S, Gomez-Juaristi M, Sanchez-Muniz F J, Rodenas de la Rocha S. Validation of an ICP-OES method for macro and trace element determination in *Laminaria* and *Porphyra* seaweeds from four different countries. Journal of Food Composition and Analysis, 2010, 23 (8): 814-820.

[345] Chekri R, Noel L, Millour S, Vastel C, Kadar A, Sirot V, Leblanc J C, Guerin T. Calcium, magnesium, sodium and potassium levels in foodstuffs from the second French Total Diet Study. Journal of Food Composition and Analysis, 2012, 25 (2): 97-107.

[346] 李蓉, 杨伟, 杨春花, 周美丽, 范晓旭. 微波消解-原子荧光光谱法同时测定香辛料中砷、汞含量. 食品安全质量检测学报, 2018, 9(22): 5906-5910.

[347] da Silva C S, Pinheiro F C, Britto do Amaral C D, Nobrega J A. Determination of As, Cd, Hg and Pb in continuous use drugs and excipients by plasma-based techniques in compliance with the United States Pharmacopeia requirements. Spectrochimica Acta Part B: Atomic Spectroscopy, 2017, 138: 14-17.

[348] de Sousa R A, Sabarense C M, Prado G L P, Metze K, Cadore S. Lead biomonitoring in different organs of lead intoxicated rats employing GF AAS and different sample preparations. Talanta, 2013, 104: 90-96.

[349] Frentiu T, Butaciu S, Ponta M, Senila M, Darvasi E, Frentiu M, Petreus D. Determination of total mercury in fish tissue using a low-cost cold vapor capacitively coupled plasma microtorch optical emission microspectrometer: Comparison with direct mercury determination by thermal decomposition atomic absorption spectrometry. Food Analytical Methods, 2015, 8 (3): 643-648.

[350] Martins C T, Almeida C M M, Alvito P C. Selenium content of raw and cooked marine species consumed in portugal. Food Analytical Methods, 2011, 4 (1): 77-83.

[351] Ngah C W Z C W, Yahya M A. Optimisation of digestion method for determination of arsenic in shrimp paste sample using atomic absorption spectrometry. Food Chemistry, 2012, 134 (4): 2406-2410.

[352] Le Blond J S, Strekopytov S, Unsworth C, Williamson B J. Testing a new method for quantifying Si in silica-rich biomass using HF in a closed vessel microwave digestion system. Analytical Methods, 2011, 3 (8): 1752-1758.

[353] Lehmann E L, Fostier A H, Arruda M A Z. Hydride generation using a metallic atomizer after microwave-assisted extraction for inorganic arsenic speciation in biological samples. Talanta, 2013, 104: 187-192.

[354] Pasias I N, Thomaidis N S, Piperaki E A. Determination of total arsenic, total inorganic arsenic and inorganic arsenic species in rice and rice flour by electrothermal atomic absorption spectrometry. Microchemical Journal, 2013, 108: 1-6.

[355] Nobrega J A, Pirola C, Fialho L L, Rota G, de Campos Jordao C E K M A, Pollo F. Microwave-assisted digestion of organic samples: How simple can it become? Talanta, 2012, 98: 272-276.

[356] Krzyzaniak S R, Iop G D, Holkem A P, Flores E M M, Mello P A. Determination of inorganic contaminants in carbon nanotubes by plasma-based techniques: Overcoming the limitations of sample preparation. Talanta, 2019, 192: 255-262.

[357] Anschau K F, Enders M S P, Senger C M, Duarte F A, Dressler V L, Muller E I. A novel strategy for medical foods digestion and subsequent elemental determination using inductively coupled plasma optical emission spectrometry. Microchemical Journal, 2019, 147: 1055-1060.

[358] Yang W, Casey J F, Gao Y. A new sample preparation method for crude or fuel oils by mineralization utilizing single reaction chamber microwave for broader multi-element analysis by ICP techniques. Fuel, 2017, 206: 64-79.

[359] Flores E M D, Barin J S, Paniz J N G, Medeiros J A, Knapp G. Microwave-assisted sample combustion: A technique for sample preparation in trace element determination. Analytical Chemistry, 2004, 76 (13): 3525-3529.

[360] Barin J S, Pereira J S F, Mello P A, Knorr C L, Moraes D P, Mesko M F, Duarte F A. Focused microwave-induced combustion for digestion of botanical samples and metals determination by ICP-OES and ICP-MS. Talanta, 2012, 94: 308-314.

[361] Maciel J V, Knorr C L, Flores E M M, Mueller E I, Mesko M F, Primel E G, Duarte F A. Feasibility of microwave-induced combustion for trace element determination in *Engraulis anchoita* by ICP-MS. Food Chemistry, 2014, 145: 927-931.

[362] Barin J S, Tischer B, Picoloto R S, Antes F G, da Silva F E B, Paula F R, Flores E M M. Determination of toxic elements in tricyclic active pharmaceutical ingredients by ICP-MS: A critical study of digestion methods. Journal of Analytical Atomic Spectrometry, 2014, 29 (2): 352-358.

[363] Cruz S M, Schmidt L, Dalla Nora F M, Pedrotti M F, Bizzi C A, Barin J S, Flores E M M. Microwave-induced combustion method for the determination of trace and ultratrace element impurities in graphite samples by ICP-OES and ICP-MS. Microchemical Journal, 2015, 123: 28-32.

[364] Novo D L R, Pereira R M, Henn A S, Costa V C, Moraes Flores E M, Mesko M F. Are there feasible strategies for determining bromine and iodine in human hair using interference-free plasma based-techniques? Analytica Chimica Acta, 2019, 1060: 45-52.

[365] Krzyzaniak S R, Santos R F, Nora F M D, Cruz S M, Flores E M M, Mello P A. Determination of halogens and sulfur in high-purity polyimide by IC after digestion by MIC. Talanta, 2016, 158: 193-197.

[366] Mello P A, Diehl L O, Oliveira J S S, Muller E I, Mesko M F, Flores E M M. Plasma-based determination of inorganic contaminants in waste of electric and electronic equipment after microwave-induced combustion. Spectrochimica Acta Part B: Atomic Spectroscopy, 2015, 105: 95-102.

[367] Dalla Nora F M, Cruz S M, Giesbrecht C K, Knapp G, Wiltsche H, Bizzi CA, Barin J, Flores E M M. A new approach for the digestion of diesel oil by microwave-induced combustion and determination of inorganic impurities by ICP-MS. Journal of Analytical Atomic Spectrometry, 2017, 32 (2): 408-414.

[368] dos Anjos S L, Alves J C, Rocha Soares S A, Araujo R G O, de Oliveira O M C, Queiroz A F S, et al. Multivariate optimization of a procedure employing microwave-assisted digestion for the determination of nickel and vanadium in crude oil by ICP-OES. Talanta, 2018, 178: 842-846.

[369] Pereira J S F, Pereira L S F, Mello P A, Guimaraes R C L, Guarnieri R A, Fonseca T C O, Flores E M M. Microwave-induced combustion of crude oil for further rare earth elements determination by USN-ICP-MS. Analytica Chimica Acta, 2014, 844: 8-14.

[370] Henn A S, Frohlich A C, Pedrotti M F, Duarte F A, Paniz J N G, Flores E M M, Bizzi C A. Microwave-assisted solid sampling system for Hg determination in polymeric samples using FF-AAS. Microchemical Journal, 2019, 147: 463-468.

[371] Siriangkhawut W, Sittichan P, Ponhong K, Chantiratikul P. Quality assessment of trace Cd and Pb contaminants in Thai herbal medicines using ultrasound-assisted digestion prior to flame atomic absorption spectrometry. Journal of Food and Drug Analysis, 2017, 25 (4): 960-967.

[372] Bezerra M A, Castro J T, Macedo R C, da Silva D G. Use of constrained mixture design for optimization of method for determination of zinc and manganese in tea leaves employing slurry sampling. Analytica Chimica Acta, 2010, 670 (1-2): 33-38.

[373] Bilo F, Borgese L, Cazzago D, Zacco A, Bontempi E, Guarneri R, Bernardello M, Attuati S, Lazo P, Depero L E. TXRF analysis of soils and sediments to assess environmental contamination. Environmental Science and Pollution Research, 2014, 21 (23): 13208-13214.

[374] Saleh M A D, Padilha P M, Hauptli L, Berto DA. The ultra-sonication of minerals in swine feed. Journal of Animal Science and Biotechnology, 2015, 6:32.

[375] Bastos Silva L O, da Silva D G, Leao D J, Matos G D, Costa Ferreira S L. Slurry sampling for the determination of

mercury in rice using cold vapor atomic absorption spectrometry. Food Analytical Methods, 2012, 5 (6): 1289-1295.

[376] Saleh M A D, Berto D A, Padilha P M. Ultrasound-assisted extraction of Na and K from swine feed and its application in a digestibility assay: A green analytical procedure. Ultrasonics Sonochemistry, 2013, 20 (6): 1353-1358.

[377] Welna M, Szymczycha-Madeja A, Pohl P. Improvement of determination of trace amounts of arsenic and selenium in slim coffee products by HG-ICP-OES. Food Analytical Methods, 2014, 7 (5): 1016-1023.

[378] Silva Junior M M, Bastos Silva L O, Leao D J, Lopes dos Santos W N, Welz B, Costa Ferreira S L. Determination of mercury in alcohol vinegar samples from Salvador, Bahia, Brazil. Food Control, 2015, 47: 623-627.

[379] Machado I, Bergmann G, Piston M. A simple and fast ultrasound-assisted extraction procedure for Fe and Zn determination in milk-based infant formulas using flame atomic absorption spectrometry (FAAS). Food Chemistry, 2016, 194: 373-376.

[380] Alvarez-Vazquez M A, Bendicho C, Prego R. Ultrasonic slurry sampling combined with total reflection X-ray spectrometry for multi-elemental analysis of coastal sediments in a ria system. Microchemical Journal, 2014, 112: 172-180.

[381] Zhang X, Li Y, Li G, Hu C. Preparation of Fe/activated carbon directly from rice husk pyrolytic carbon and its application in catalytic hydroxylation of phenol. RSC Advances, 2015, 5 (7): 4984-4992.

[382] Zhang W, Xue J, Yang X, Wang S. Determination of inorganic and total mercury in seafood samples by a new ultrasound-assisted extraction system and cold vapor atomic fluorescence spectrometry. Journal of Analytical Atomic Spectrometry, 2011, 26 (10): 2023-2029.

[383] Khan A H, Shang J Q, Alam R. Optimization of sample preparation method of total sulphur measurement in mine tailings. International Journal of Environmental Science and Technology, 2014, 11 (7): 1989-1998.

[384] Hsu W H, Jiang S J, Sahayam A C. Determination of Pd, Rh, Pt, Au in road dust by electrothermal vaporization inductively coupled plasma mass spectrometry with slurry sampling. Analytica Chimica Acta, 2013, 794: 15-19.

[385] Correia F O, Almeida T S, Garcia R L, Queiroz A F S, Smichowski P, da Rocha G O, Araujo R G O. Sequential determination and chemical speciation analysis of inorganic As and Sb in airborne particulate matter collected in outdoor and indoor environments using slurry sampling and detection by HG AAS. Environmental Science and Pollution Research, 2019, 26 (21): 21416-21424.

[386] Husakova L, Urbanova I, Safrankova M, Sidova T. Slurry sampling high-resolution continuum source electrothermal atomic absorption spectrometry for direct beryllium determination in soil and sediment samples after elimination of SiO interference by least-squares background correction. Talanta, 2017, 175: 93-100.

[387] Szymczycha-Madeja A, Welna M, Pohl P. Determination of essential and non-essential elements in green and black teas by FAAS and ICP-OES simplified-multivariate classification of different tea products. Microchemical Journal, 2015, 121: 122-129.

[388] Federici Padilha C dC, de Moraes P M, Garcia L dA, Costa Pozzi C M, Pereira Lima G P, Serra Valente J P, Alves Jorge S M, Padilha P dM. Evaluation of Cu, Mn, and Se in vegetables using ultrasonic extraction and GFAAS quantification. Food Analytical Methods, 2011, 4 (3): 319-325.

[389] Huang S Y, Jiang S J. 8-Hydroxyquinoline-5-sulfonic acid as the modifier for the determination of trace elements in cereals by slurry sampling electrothermal vaporization ICP-MS. Analytical Methods, 2010, 2 (9): 1310-1315.

[390] Lin M L, Jiang S J. Determination of As, Cd, Hg and Pb in herbs using slurry sampling electrothermal vaporisation inductively coupled plasma mass spectrometry. Food Chemistry, 2013, 141 (3): 2158-2162.

[391] Saleh M A D, Padilha P M, Hauptli L, Berto D A. The ultra-sonication of minerals in swine feed. Journal of Animal Science and BioTechnology, 2015, 6:32.

[392] Szymczycha-Madeja A, Welna M, Pohl P. Simplified multi-element analysis of ground and instant coffees by ICP-OES and FAAS. Food Additives and Contaminants Part A: Chemistry Analysis Control Exposure & Risk Assessment, 2015, 32 (9): 1488-1500.

[393] daSilva D G, Matos G D, dos Santos A M P, Fiuza RdP, de Jesus R M. A procedure using slurry sampling for the determination of manganese in corn flour by ET AAS. Analytical Methods, 2011, 3 (11): 2625-2629.

[394] dos Santos W N L, Cavalcante D D, Macedo S M, Nogueira J S, da Silva E G P. Slurry sampling and HG-AFS for the determination of total arsenic in rice samples. Food Analytical Methods, 2013, 6 (4): 1128-1132.

[395] Santos D C M B, Carvalho L S B, Lima D C, Leao D J, Teixeira L S G, Korn M G A. Determination of micronutrient minerals in coconut milk by ICP-OES after ultrasound-assisted extraction procedure. Journal of Food Composition and Analysis, 2014, 34 (1): 75-80.

[396] Brandao G C, Matos G D, Ferreira S L C. Slurry sampling and high-resolution continuum source flame atomic absorption spectrometry using secondary lines for the determination of Ca and Mg in dairy products. Microchemical Journal, 2011, 98 (2): 231-233.

[397] Amorim F A C, Costa V C, Guedes W N, de Sa I P, dos Santos M C, da Silva E G P, Lima D D C. Multivariate optimization of method of slurry sampling for determination of iron and zinc in starch samples by flame atomic absorption spectrometry. Food Analytical Methods, 2016, 9 (6): 1719-1725.

[398] de Andrade C K, Klack de Brito P M, dos Anjos V E, Quinaia S P. Determination of Cu, Cd, Pb and Cr in yogurt by slurry sampling electrothermal atomic absorption spectrometry: A case study for Brazilian yogurt. Food Chemistry, 2018, 240: 268-274.

[399] Husakova L, Sidova T, Ibrahimova L, Svizelova M, Mikysek T. Direct determination of lead in bones using slurry sampling and high-resolution continuum source electrothermal atomic absorption spectrometry. Analytical Methods, 2019, 11 (9): 1254-1263.

[400] Costas M, Lavilla I, Gil S, Pena F, de la Calle I, Cabaleiro N, Bendicho C. Evaluation of ultrasound-assisted extraction as sample pre-treatment for quantitative determination of rare earth elements in marine biological tissues by inductively coupled plasma-mass spectrometry. Analytica Chimica Acta, 2010, 679 (1-2): 49-55.

[401] Costas-Rodriguez M, Lavilla I, Bendicho C. Assessment of ultrasound-assisted extraction as sample pre-treatment for the measurement of lead isotope ratios in marine biological tissues by multicollector inductively coupled plasma-mass spectrometry. Spectrochimica Acta Part B: Atomic Spectroscopy, 2011, 66 (6): 483-488.

[402] Yi Y Z, Jiang S J, Sahayam A C. Palladium nanoparticles as the modifier for the determination of Zn, As, Cd, Sb, Hg and Pb in biological samples by ultrasonic slurry sampling electrothermal vaporization inductively coupled plasma mass spectrometry. Journal of Analytical Atomic Spectrometry, 2012, 27 (3): 426-431.

[403] de la Calle I, Menta M, Klein M, Seby F. Screening of TiO_2 and Au nanoparticles in cosmetics and determination of elemental impurities by multiple techniques (DLS, SP-ICP-MS, ICP-MS and ICP-OES). Talanta, 2017, 171: 291-306.

[404] de Jesus R M, Silva L O B, Castro J T, de Azevedo Neto A D, de Jesus R M, Ferreira S L C. Determination of mercury in phosphate fertilizers by cold vapor atomic absorption spectrometry. Talanta, 2013, 106: 293-297.

[405] Luz M S, Oliveira P V. Niobium carbide as permanent modifier for silicon determination in petrochemical products by emulsion-based sampling GF AAS. Fuel, 2014, 116: 255-260.

[406] Dias Peronico V C, Raposo J L Jr. Ultrasound-assisted extraction for the determination of Cu, Mn, Ca, and Mg in alternative oilseed crops using flame atomic absorption spectrometry. Food Chemistry, 2016, 196: 1287-1292.

[407] Zawisza B, Sitko R. Electrochemically assisted sorption on oxidized multiwalled carbon nanotubes for preconcentration of Cr, Mn, Co, Ni, Cu and Zn from water samples. Analyst, 2013, 138 (8): 2470-2476.

[408] Hutton L A, O'Neil G D, Read T L, Ayres Z J, Newton M E, Macpherson J V. Electrochemical X-ray fluorescence spectroscopy for trace heavy metal analysis: Enhancing X-ray fluorescence detection capabilities by four orders of magnitude. Analytical Chemistry, 2014, 86 (9): 4566-4572.

[409] Ohira S I, Yamasaki T, Koda T, Kodama Y, Toda K. Electrodialytic in-line preconcentration for ionic solute analysis. Talanta, 2018, 180: 176-181.

[410] Allegretta I, Porfido C, Panzarino O, Fontanella M C, Beone G M, Spagnuolo M, Terzano R. Determination of As concentration in earthworm coelomic fluid extracts by total-reflection X-ray fluorescence spectrometry. Spectrochimica Acta Part B: Atomic Spectroscopy, 2017, 130: 21-25.

[411] Zahler N H. Gel Electrophoresis and X-ray Fluorescence: A powerful combination for the analysis of protein metal binding. ACS Chemical Biology, 2010, 5 (6): 541-543.

[412] Feng L, Zhang D, Wang J, Shen D, Li H. A novel quantification strategy of transferrin and albumin in human serum by species-unspecific isotope dilution laser ablation inductively coupled plasma mass spectrometry (ICP-MS). Analytica Chimica Acta, 2015, 884: 19-25.

[413] Boutorabi L, Rajabi M, Bazregar M, Asghari A. Selective determination of chromium (Ⅵ) ions using in-tube electro-membrane extraction followed by flame atomic absorption spectrometry. Microchemical Journal, 2017, 132: 378-384.

[414] Lima D C, dos Santos A M P, Araujo R G O, Scarminio I S, Bruns R E, Ferreira S L C. Principal component analysis and hierarchical cluster analysis for homogeneity evaluation during the preparation of a wheat flour laboratory reference material for inorganic analysis. Microchemical Journal, 2010, 95 (2): 222-226.

[415] Silva F L F, Duarte T A O, Melo L S, Ribeiro L P D, Gouveia S T, Lopes G S, Matos W O. Development of a wet digestion method for paints for the determination of metals and metalloids using inductively coupled plasma optical emission spectrometry. Talanta, 2016, 146: 188-194.

[416] Oliveira S S, Alves C N, Boa Morte E S, Santos Junior A dF, Oliveira Araujo R G, Muniz Batista Santos D C. Determination of essential and potentially toxic elements and their estimation of bioaccessibility in honeys. Microchemical Journal, 2019, 151: 104221.

[417] Pohl P, Dzimitrowicz A, Jamroz P, Greda K. Development and optimization of simplified method of fast sequential HR-CS-FAAS analysis of apple juices on the content of Ca, Fe, K, Mg, Mn and Na with the aid of response surface methodology. Talanta, 2018, 189: 182-189.

[418] da Silva D G, Junior M M S, Silva L O B, Portugal L A, Matos G D, Ferreira S L C. Determination of cadmium in rice by electrothermal atomic absorption spectrometry using aluminum as permanent modifier. Analytical Methods, 2011, 3 (11): 2495-2500.

[419] Astolfi M L, Marconi E, Protano C, Vitali M, Schiavi E, Mastromarino P, Canepari S. Optimization and validation of a fast digestion method for the determination of major and trace elements in breast milk by ICP-MS. Analytica Chimica Acta, 2018, 1040: 49-62.

[420] Campillo N, Munoz-Delgado E, Lopez-Garcia I, Baeza-Albarracin Y, Hernandez-Cordoba M. Suspensions of biological tissues in alkaline medium for the determination of copper, manganese and cobalt by electrothermal atomic absorption spectrometry. Microchimica Acta, 2010, 171 (1-2): 71-79.

[421] Ferreira Damin I C, Zmozinski A V, Borges A R, Rodrigues Vale M G, da Silva M M. Determination of cadmium and lead in fresh meat by slurry sampling graphite furnace atomic absorption spectrometry. Analytical Methods, 2011, 3 (6): 1379-1385.

[422] Batista E F, Augusto AdS, Pereira-Filho E R. Determination of Cd, Co, Cr, Cu, Ni and Pb in cosmetic samples using a simple method for sample preparation. Analytical Methods, 2015, 7 (1): 329-335.

[423] Lewen N, Nugent D. The use of inductively coupled plasma-atomic emission spectroscopy (ICP-AES) in the determination of lithium in cleaning validation swabs. Journal of Pharmaceutical and Biomedical Analysis, 2010, 52 (5): 652-655.

[424] Pereira L S F, Pedrotti M F, Vecchia P D, Pereira J S F, Flores E M M. A simple and automated sample preparation system for subsequent halogens determination: Combustion followed by pyrohydrolysis. Analytica Chimica Acta, 2018, 1010: 29-36.

[425] Soares A R, Nascentes C C. Development of a simple method for the determination of lead in lipstick using alkaline solubilization and graphite furnace atomic absorption spectrometry. Talanta, 2013, 105: 272-277.

[426] Pereira C C, de Souza A O, Oreste E Q, Vieira M A, Ribeiro A S. Evaluation of the use of a reflux system for sample preparation of processed fruit juices and subsequent determination of Cr, Cu, K, Mg, Na, Pb and Zn by atomic spectrometry techniques. Food Chemistry, 2018, 240: 959-964.

[427] de Oliveira R M, Nascimento Antunes A C, Vieira M A, Medina A L, Ribeiro A S. Evaluation of sample preparation methods for the determination of As, Cd, Pb, and Se in rice samples by GF AAS. Microchemical Journal, 2016, 124: 402-409.

[428] Miranda K, Vieira A L, Bechlin M A, Fortunato F M, Virgilio A, Ferreira E C, Gomes Neto J A. Determination of Ca, Cd, Cu, Fe, K, Mg, Mn, Mo, Na, Se, and Zn in foodstuffs by atomic spectrometry after sample preparation using a low-cost closed-vessel conductively heated digestion system. Food Analytical Methods, 2016, 9 (7): 1887-1894.

[429] Sorouraddin S M, Farajzadeh M A, Okhravi T. Cyclohexylamine as extraction solvent and chelating agent in extraction and preconcentration of some heavy metals in aqueous samples based on heat-induced homogeneous liquid-liquid extraction. Talanta, 2017, 175: 359-365.

[430] Whitty-Leveille L, Turgeon K, Bazin C, Lariviere D. A comparative study of sample dissolution techniques and plasma based instruments for the precise and accurate quantification of REEs in mineral matrices. Analytica Chimica Acta, 2017, 961: 33-41.

[431] Amosova A A, Panteeva S V, Chubarov V M, Finkelshtein A L. Determination of major elements by wavelength-dispersive X-ray fluorescence spectrometry and trace elements by inductively coupled plasma mass spectrometry in igneous rocks from the same fused sample (110 mg). Spectrochimica Acta Part B: Atomic Spectroscopy, 2016, 122: 62-68.

[432] Zhou H, Wang Z, Zhu Y, Li Q, Zou H J, Qu H Y, Chen Y R, Du Y P. Quantitative determination of trace metals in high-purity silicon carbide powder by laser ablation inductively coupled plasma mass spectrometry without binders. Spectrochimica Acta Part B: Atomic Spectroscopy, 2013, 90: 55-60.

[433] Oliveira S R, Gomes Neto J A, Nobrega J A, Jones B T. Determination of macro-and micronutrients in plant leaves by high-resolution continuum source flame atomic absorption spectrometry combining instrumental and sample preparation strategies. Spectrochimica Acta Part B: Atomic Spectroscopy, 2010, 65 (4): 316-320.

[434] Machado I, Mondutey S, Pastorino N, Arce V, Piston M. A green analytical method for the determination of Cu, Fe, Mn, and Zn in wheat flour using total reflection X-ray fluorescence. Journal of Analytical Atomic Spectrometry, 2018, 33 (7): 1264-1268.

[435] Schmidt L, Bizzi C A, Duarte F A, Dressler V L, Flores E M M. Evaluation of drying conditions of fish tissues for inorganic mercury and methylmercury speciation analysis. Microchemical Journal, 2013, 108: 53-59.

[436] Covaci E, Angyus S B, Senila M, Ponta M, Darvasi E, Frentiu M, Frentiu T. Eco-scale non-chromatographic method for mercury speciation in fish using formic acid extraction and UV-Vis photochemical vapor generation capacitively coupled plasma microtorch optical emission spectrometry. Microchemical Journal, 2018, 141: 155-162.

[437] Liu L, Deng H, Wu L, Zheng C, Hou X. UV-induced carbonyl generation with formic acid for sensitive determination of nickel by atomic fluorescence spectrometry. Talanta, 2010, 80 (3): 1239-1244.

[438] Potes M dL, Kolling L, de Jesus A, Dessuy M B, Rodrigues Vale M G, da Silva M M. Determination of mercury in fish by photochemical vapor generation graphite furnace atomic absorption spectrometry. Analytical Methods, 2016, 8 (46): 8165-8172.

[439] Potes M dL, Nakadi F V, Grasel Frois C F, Rodrigues Vale M G, da Silv M M. Investigation of the conditions for selenium determination by photochemical vapor generation coupled to graphite furnace atomic absorption spectrometry. Microchemical Journal, 2019, 147: 324-332.

[440] Brandao G C, Aureliano M D O, da Silva Sauthier M C, dos Santos W N L. Photo-oxidation using UV radiation as a sample preparation procedure for the determination of copper in fruit juices by flame atomic absorption spectrometry. Analytical Methods, 2012, 4 (3): 855-858.

[441] Dash K, Venkateswarlu G, Thangavel S, Rao S V, Chaurasia S C. Ultraviolet photolysis assisted mineralization and determination of trace levels of Cr, Cd, Cu, Sn, and Pb in isosulfan blue by ICP-MS. Microchemical Journal, 2011, 98 (2): 312-316.

[442] Campos V M, Silva F L F, Oliveira J P S, Ribeiro L P D, Matos W O, Lopes G S. Investigation of a rapid infrared heating assisted mineralization of soybean matrices for trace element analysis. Food Chemistry, 2019, 280: 96-102.

[443] Oliveira J P S, Silva F L F, Monte R J G, Matos W O, Lopes G S. A new approach to mineralization of flaxseed (*Linum usitatissimum* L.) for trace element analysis by flame atomic absorption spectrometry. Food Chemistry, 2017, 224: 335-341.

[444] Zaksas N P, Gerasimov V A, Nevinsky G A. Simultaneous determination of Fe, P, Ca, Mg, Zn, and Cu in whole blood by two-jet plasma atomic emission spectrometry. Talanta, 2010, 80 (5): 2187-2190.

[445] Zaksas N P, Nevinsky G A. Solid sampling in analysis of animal organs by two-jet plasma atomic emission spectrometry. Spectrochimica Acta Part B: Atomic Spectroscopy, 2011, 66 (11-12): 861-865.

[446] Zhang N, Li Z, Zheng J, Yang X, Shen K, Zhou T, Zhang Y. Multielemental analysis of botanical samples by ICP-OES and ICP-MS with focused infrared lightwave ashing for sample preparation. Microchemical Journal, 2017, 134: 68-77.

[447] Zhang N, Zheng J. Rapid sample preparation with multi-channel focused infrared micro-ashing prior to determination of chromium in gelatin capsules by electrothermal atomic absorption spectrometry. Analytical Methods, 2018, 10 (8): 920-925.

[448] Pereira J S F, Pereira L S F, Mello P A, Guimaraes R C L, Guarnieri R A, Fonseca T C O, Flores E M M. Microwave-induced combustion of crude oil for further rare earth elements determination by USN-ICP-MS. Analytica Chimica Acta, 2014, 844: 8-14.

[449] Kiyataka P H M, Dantas S T, Albino A C, Lima Pallone J A. Antimony assessment in PET bottles for soft drink. Food Analytical Methods, 2018, 11 (1): 1-9.

[450] Kiyataka P H M, Dantas S T, Pallone J A L. Method for analysis and study of migration of lead, cadmium, mercury and arsenic from polypropylene packaging into ice cream and simulant. Food Analytical Methods, 2015, 8 (9): 2331-2338.

[451] Shao Y. A novel device for powder sample introduction into the inductively coupled plasma and proposed strategy of powder sample preparation. Spectrochimica Acta Part B: Atomic Spectroscopy, 2010, 65 (11): 967-972.

[452] Wienold J, Recknagel S, Scharf H, Hoppe M, Michaelis M. Elemental analysis of printed circuit boards considering the ROHS regulations. Waste Management, 2011, 31 (3): 530-535.

[453] Andrey D, Dufrier J P, Perring L. Analytical capabilities of energy dispersive X-Ray fluorescence for the direct quantification of iron in cocoa powder and powdered cocoa drink. Spectrochimica Acta Part B: Atomic Spectroscopy,

2018, 148: 137-142.

[454] Oliveira L B, dos Santos W P C, Teixeira L S G, Korn M G A. Direct analysis of cocoa powder, chocolate powder, and powdered chocolate drink for multi-element determination by energy dispersive X-ray fluorescence spectrometry. Food Analytical Methods, 2020, 13 (1): 195-202.

[455] Hartwig C A, Pereira R M, Novo D L R, Teixeira Oliveira D T, Mesko M F. Green and efficient sample preparation method for the determination of catalyst residues in margarine by ICP-MS. Talanta, 2017, 174: 394-400.

[456] Iop G D, Krzyzaniak S R, Silva J S, Flores E M M, Costa A B, Mello P A. Feasibility of microwave-assisted ultraviolet digestion of polymeric waste electrical and electronic equipment for the determination of bromine and metals (Cd, Cr, Hg, Pb and Sb) by ICP-MS. Journal of Analytical Atomic Spectrometry, 2017, 32 (9): 1789-1797.

[457] Souza J P, Barela P S, Kellermann K, Santos M F P, Moraes D P, Pereira J S F. Microwave-assisted ultraviolet digestion: An efficient method for the digestion of produced water from crude oil extraction and further metal determination. Journal of Analytical Atomic Spectrometry, 2017, 32 (12): 2439-2446.

[458] Oliveira J S S, Picoloto R S, Bizzi C A, Mello P A, Barin J S, Flores E M M. Microwave-assisted ultraviolet digestion of petroleum coke for the simultaneous determination of nickel, vanadium and sulfur by ICP-OES. Talanta, 2015, 144: 1052-1058.

[459] Dantas A N S, Matos W O, Gouveia ST, Lopes G S. The combination of infrared and microwave radiation to quantify trace elements in organic samples by ICP-OES. Talanta, 2013, 107: 292-296.

[460] Cui J, Xiao Y-b, Dai L, Zhao X, Wang Y. Speciation of organoarsenic species in food of animal origin using accelerated solvent extraction (ASE) with determination by HPLC-hydride generation-atomic fluorescence spectrometry (HG-AFS). Food Analytical Methods, 2013, 6 (2): 370-379.

[461] 古君平，胡静，周朗君，陈静夷，胡玉玲，李攻科，曾尊祥. 原子吸收光谱法测定烟叶中的重金属总量及形态分析. 分析测试学报，2015，34(1): 111-114.

[462] Zhu S, Chen B, He M, Huang T, Hu B. Speciation of mercury in water and fish samples by HPLC-ICP-MS after magnetic solid phase extraction. Talanta, 2017, 171: 213-219.

[463] Fan W, Mao X, He M, Chen B, Hu B. Stir bar sorptive extraction combined with high performance liquid chromatography-ultraviolet/inductively coupled plasma mass spectrometry for analysis of thyroxine in urine samples. Journal of Chromatography A, 2013, 1318: 49-57.

[464] Ito K, Palmer C D, Steuerwald A J, Parsons P J. Determination of five arsenic species in whole blood by liquid chromatography coupled with inductively coupled plasma mass spectrometry. Journal of Analytical Atomic Spectrometry, 2010, 25 (8): 1334-1342.

[465] Schwan A M, Martin R, Goessler W. Chlorine speciation analysis in blood by ion chromatography-inductively coupled plasma mass spectrometry. Analytical Methods, 2015, 7 (21): 9198-9205.

[466] Chen S, Guo Q, Liu L. Determination of arsenic species in edible mushrooms by high-performance liquid chromatography coupled to inductively coupled plasma mass spectrometry. Food Analytical Methods, 2017, 10 (3): 740-748.

[467] Zmozinski A V, Carneado S, Ibanez-Palomino C, Sahuquillo A, Fermin Lopez-Sanchez J, da Silva M M. Method development for the simultaneous determination of methylmercury and inorganic mercury in seafood. Food Control, 2014, 46: 351-359.

[468] Schmidt L, Landero J A, Santos R F, Mesko M F, Mello P A, Flores E M M, Caruso J A. Arsenic speciation in seafood by LC-ICP-MS/MS: Method development and influence of culinary treatment. Journal of Analytical Atomic Spectrometry, 2017, 32 (8): 1490-1499.

[469] Arnquist I J, di Vacri M L, Hoppe E W. An automated ultraclean ion exchange separation method for the determinations

of Th-232 and U-238 in copper using inductively coupled plasma mass spectrometry. Nuclear Instruments & Methods in Physics Research Section A: Accelerators Spectrometers Detectors and Associated Equipment, 2020, 965: 163761.

[470] 黄均明, 柴刚, 韩岩松, 保万魁, 刘红芳, 刘蜜, 王旭. 液相色谱-原子荧光光谱法测定水溶肥料中无机砷含量. 中国土壤与肥料, 2020, (4): 252-257.

[471] Rodrigues J L, Alvarez C R, Farinas N R, Berzas Nevado J J, Barbosa F Jr, Rodriguez Martin-Doimeadios R C. Mercury speciation in whole blood by gas chromatography coupled to ICP-MS with a fast microwave-assisted sample preparation procedure. Journal of Analytical Atomic Spectrometry, 2011, 26 (2): 436-442.

[472] Baxter D C, Faarinen M, Osterlund H, Rodushkin I, Christensen M. Serum/plasma methylmercury determination by isotope dilution gas chromatography-inductively coupled plasma mass spectrometry. Analytica Chimica Acta, 2011, 701 (2): 134-138.

[473] Dressler V L, Moreira Santos C M, Antes F G, Stum Bentlin F R, Pozebon D, Moraes Flores E M. Total mercury, inorganic mercury and methyl mercury determination in red wine. Food Analytical Methods, 2012, 5 (3): 505-511.

[474] Tu Q, Guidry E N, Meng F, Wang T, Gong X. A high-throughput flow injection inductively coupled plasma mass spectrometry method for quantification of oligonucleotides. Microchemical Journal, 2016, 124: 668-674.

[475] Araujo-Barbosa U, Pena-Vazquez E, Carmen Barciela-Alonso M, Costa Ferreira S L, Pinto dos Santos A M, Bermejo-Barrera P. Simultaneous determination and speciation analysis of arsenic and chromium in iron supplements used for iron-deficiency anemia treatment by HPLC-ICP-MS. Talanta, 2017, 170: 523-529.

[476] Batista B L, Rodrigues J L, de Souza S S, Oliveira Souza V C, Barbosa F Jr. Mercury speciation in seafood samples by LC-ICP-MS with a rapid ultrasound-assisted extraction procedure: Application to the determination of mercury in Brazilian seafood samples. Food Chemistry, 2011, 126 (4): 2000-2004.

[477] Schmidt L, Landero J A, Novo D L R, Duarte F A, Mesko M F, Caruso J A, Moraes Flores E M. A feasible method for As speciation in several types of seafood by LC-ICP-MS/MS. Food Chemistry, 2018, 255: 340-347.

[478] Pechancova R, Pluhacek T, Gallo J, Milde D. Study of chromium species release from metal implants in blood and joint effusion: Utilization of HPLC-ICP-MS. Talanta, 2018, 185: 370-377.

[479] Chen H, Du P, Chen J, Hu S, Li S, Liu H. Separation and preconcentration system based on ultrasonic probe-assisted ionic liquid dispersive liquid liquid microextraction for determination trace amount of chromium(Ⅵ) by electrothermal atomic absorption spectrometry. Talanta, 2010, 81 (1-2): 176-179.

[480] Doker S, Bosgelmez I I. Rapid extraction and reverse phase-liquid chromatographic separation of mercury(Ⅱ) and methylmercury in fish samples with inductively coupled plasma mass spectrometric detection applying oxygen addition into plasma. Food Chemistry, 2015, 184: 147-153.

[481] de Souza S S, Campiglia A D, Barbosa F Jr. A simple method for methylmercury, inorganic mercury and ethylmercury determination in plasma samples by high performance liquid chromatography-cold-vapor-inductively coupled plasma mass spectrometry. Analytica Chimica Acta, 2013, 761: 11-17.

[482] de Souza S S, Rodrigues J L, de Oliveira Souza V C, Barbosa F Jr. A fast sample preparation procedure for mercury speciation in hair samples by high-performance liquid chromatography coupled to ICP-MS. Journal of Analytical Atomic Spectrometry, 2010, 25 (1): 79-83.

[483] Yu H, Du H, Wu L, Li R, Sun Q, Hou X. Trace arsenic speciation analysis of bones by high performance liquid chromatography-inductively coupled plasma mass spectrometry. Microchemical Journal, 2018, 141: 176-180.

[484] Bishop D P, Hare D J, Clases D, Doble P A. Applications of liquid chromatography-inductively coupled plasma-mass spectrometry in the biosciences: A tutorial review and recent developments. Trac-Trends in Analytical Chemistry, 2018, 104: 11-21.

[485] Chantada-Vazquez P M, Herbello-Hermelo P, Bermejo-Barrera P, Moreda-Pineiro A. Discrete sampling based-flow injection as an introduction system in ICP-MS for the direct analysis of low volume human serum samples. Talanta, 2019, 199: 220-227.

[486] Parodi B, Londonio A, Polla G, Savio M, Smichowski P. On-line flow injection solid phase extraction using oxidised carbon nanotubes as the substrate for cold vapour-atomic absorption determination of Hg(Ⅱ) in different kinds of water. Journal of Analytical Atomic Spectrometry, 2014, 29 (5): 880-885.

[487] Puanngam M, Dasgupta P K, Unob F. Automated on-line preconcentration of trace aqueous mercury with gold trap focusing for cold vapor atomic absorption spectrometry. Talanta, 2012, 99: 1040-1045.

[488] Ye J, Liu S, Tian M, Li W, Hu B, Zhou W, Jia Q. Preparation and characterization of magnetic nanoparticles for the on-line determination of gold, palladium, and platinum in mine samples based on flow injection micro-column preconcentration coupled with graphite furnace atomic absorption spectrometry. Talanta, 2014, 118: 231-237.

[489] Gawin M, Konefal J, Trzewik B, Walas S, Tobiasz A, Mrowiec H, Witek E. Preparation of a new Cd(Ⅱ)-imprinted polymer and its application to determination of cadmium(Ⅱ) via flow-injection-flame atomic absorption spectrometry. Talanta, 2010, 80 (3): 1305-1310.

[490] Peixoto R R A, Macarovscha G T, Cadore S. On-line preconcentration and determination of zinc using zincon and flame atomic absorption spectrometry. Food Analytical Methods, 2012, 5 (4): 814-820.

[491] Zhang W, Qi Y, Qin D, Liu J, Mao X, Chen G, Wei C, Qian Y. Determination of inorganic arsenic in algae using bromine halogenation and on-line nonpolar solid phase extraction followed by hydride generation atomic fluorescence spectrometry. Talanta, 2017, 170: 152-157.

[492] Anthemidis A N, Xidia S, Giakisikli G. Study of bond Elut® Plexa™ PCX cation exchange resin in flow injection column preconcentration system for metal determination by flame atomic absorption spectrometry. Talanta, 2012, 97: 181-186.

[493] Krishna M V B, Chandrasekaran K, Karunasagar D. On-line speciation of inorganic and methyl mercury in waters and fish tissues using polyaniline micro-column and flow injection-chemical vapour generation-inductively coupled plasma mass spectrometry (FI-CVG-ICP-MS). Talanta, 2010, 81 (1-2): 462-472.

[494] Tarley C R T, Lima G F, Nascimento D R, Assis A R S, Ribeiro E S, Diniz K M, Bezerra M A, Segatelli M G. Novel on-line sequential preconcentration system of Cr(Ⅲ) and Cr(Ⅵ) hyphenated with flame atomic absorption spectrometry exploiting sorbents based on chemically modified silica. Talanta, 2012, 100: 71-79.

[495] Zierhut A, Leopold K, Harwardt L, Schuster M. Analysis of total dissolved mercury in waters after on-line preconcentration on an active gold column. Talanta, 2010, 81 (4-5): 1529-1535.

[496] Cheng H, Liu J, Xu Z, Yin X. A micro-fluidic sub-microliter sample introduction system for direct analysis of Chinese rice wine by inductively coupled plasma mass spectrometry using external aqueous calibration. Spectrochimica Acta Part B: Atomic Spectroscopy, 2012, 73: 55-61.

[497] AlSuhaimi A O, McCreedy T. Microchip based sample treatment device interfaced with ICP-MS for the analysis of transition metals from environmental samples. Arabian Journal of Chemistry, 2011, 4 (2): 195-203.

[498] Schwartz A J, Ray S J, Hieftje G M. Automatable on-line generation of calibration curves and standard additions in solution-cathode glow discharge optical emission spectrometry. Spectrochimica Acta Part B: Atomic Spectroscopy, 2015, 105: 77-83.

[499] Marques T L, Nobrega J A. Fast and simple flow-batch extraction procedure for screening of macro and micronutrients in dried plant leaves by ICP-OES. Microchemical Journal, 2017, 134: 27-34.

[500] Marques T L, Nobrega J A. Application of a flow-batch extraction system for on-line determination of minerals in

animal foods by inductively coupled plasma optical emission spectrometry. Food Analytical Methods, 2018, 11 (4): 1243-1249.

[501] Cunha F A S, Pereira A S G, Fernandes J P A, Lyra W S, Araujo M C U, Almeida L F. Automated single-phase liquid-liquid extraction for determination of Cr(Ⅵ) using graphite furnace atomic absorption spectrophotometry without wet digestion of samples. Food Analytical Methods, 2017, 10 (4): 921-930.

[502] Horner N S, Beauchemin D. A simple method using on-line continuous leaching and ion exchange chromatography coupled to inductively coupled plasma mass spectrometry for the speciation analysis of bio-accessible arsenic in rice. Analytica Chimica Acta, 2012, 717: 1-6.

[503] Cunha F A S, Sousa R A, Harding D P, Cadore S, Almeida L F, Araujo M C U. Automatic microemulsion preparation for metals determination in fuel samples using a flow-batch analyzer and graphite furnace atomic absorption spectrometry. Analytica Chimica Acta, 2012, 727: 34-40.

[504] Desaulty A M, Meheut M, Guerrot C, Berho C, Millot R. Coupling DGT passive samplers and multi-collector ICP-MS: A new tool to measure Pb and Zn isotopes composition in dilute aqueous solutions. Chemical Geology, 2017, 450: 122-134.

[505] Yang C K, Chi P H, Lin Y C, Sun Y C, Yang M H. Development of an on-line isotope dilution laser ablation inductively coupled plasma mass spectrometry (LA-ICP-MS) method for determination of boron in silicon wafers. Talanta, 2010, 80 (3): 1222-1227.

[506] Romero V, Vilas L, Lavilla I, Bendicho C. Speciation of inorganic As and Sb in natural waters by total reflection X-ray fluorescence following selective hydride generation and trapping onto quartz reflectors coated with nanostructured Pd. Journal of Analytical Atomic Spectrometry, 2017, 32: 1705-1712.

[507] Zhang Y, Mao X, Liu J, Wang M, Qian Y, Gao C, Qi Y. Direct determination of cadmium in foods by solid sampling electrothermal vaporization inductively coupled plasma mass spectrometry using a tungsten coil trap. Spectrochimica Acta Part B-Atomic Spectroscopy, 2016, 118: 119-126.

[508] 孙绮旋, 魏星, 刘珣, 杨婷, 陈明丽, 王建华. 原子光谱/元素质谱在生命分析中的应用进展. 光谱学与光谱分析, 2019, 39(5): 1340-1345.

第三部分 分析应用

第 16 章　金属组学发展现状

(严雪婷[①]，周　莹[①]，李洪艳[①]，孙红哲[①]*)

▶ 16.1　金属组学研究内容及技术　/ 468

▶ 16.2　金属蛋白质组学研究　/ 475

▶ 16.3　金属组学在金属药物研究中的应用　/ 479

▶ 16.4　展望　/ 482

[①]香港大学，香港。*通讯作者联系人方式：hsun@hku.hk

本章导读

- 金属组学是一门新兴的前沿交叉学科,其目标是阐明生物体中尤其是细胞与组织内全部金属离子及其金属配合物的形态结构、功能及其生物分子机制。
- 金属蛋白质组学是金属组学中一个重要的领域,主要研究一个蛋白质组内所有金属蛋白的结构、金属的配位环境、蛋白质功能和构效关系。
- 金属组学分析技术大大推进了金属药物机理研究,不仅揭示了潜在药物靶点及金属药物作用通路,还将促进新型金属药物的设计合成和发展。
- 在介绍金属(蛋白质)组学概念、研究方法的基础上,分析展望金属组学的发展前景。

16.1 金属组学研究内容及技术

16.1.1 金属组学概述

继基因组学、蛋白质组学和代谢组学等热门主流组学之后,牛津大学 Robert J. P. Williams 教授和日本名古屋大学 Hiroki Haraguchi 教授相继提出了新的组学研究方向——"金属组(metallome)"/"金属组学(metallomics)",它是系统研究生命体内自由或络合的全部金属元素的分布、含量、化学形态及其功能的一门综合学科[1]。金属组学最重要的研究目标是阐明生物体系中与金属离子结合的生物分子的生理作用和功能。在金属组学中,细胞、器官或生物组织中的金属及类金属蛋白、酶和其他含有金属的生物分子以及游离金属离子的集合称为金属组[2]。金属组学的发展进程见图 16-1。

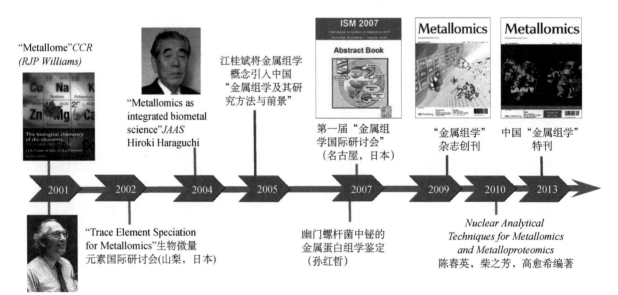

图 16-1 金属组学发展进程

一些微量金属元素在生物体液和器官中可以与不同的蛋白分子结合形成金属蛋白,当金属蛋白在生物细胞和器官中发挥催化或调控生化反应、影响生理功能时被称为"金属酶"。在生命体系中约有三分之一的蛋白质为金属蛋白或金属酶,不同种类的金属离子和金属酶参与了生物体生命周期中基因(DNA、RNA)和蛋白质的合成与代谢过程[3]。金属组学研究生物体系中的金属组,不仅包括金属与蛋白质、酶和基因的相互作用和功能关系，还包括金属与小分子配体或其他代谢产物之间的作用关系。

早期的金属组学研究主要是围绕元素在环境及生物体的分布及所造成的影响，研究的是元素的总量，而现代金属组学研究关注金属离子和类金属在细胞内的分布及形态，与基因组、蛋白质组和代谢组之间的关系，并从分子生物学的角度出发对其与生物大分子的复杂体，特别是金属蛋白、金属酶进行结构、功能和活性的分析。

金属组学研究不仅将解答生物体内金属与生物分子间的相互作用，而且将为金属蛋白质等生命物质所承担的生物功能和作用机制开辟新的研究途径。可以说，金属组学是形态分析发展到较高水平的必然产物，只有真正掌握了金属或类金属在生物体内存在的全部信息，才能清楚地了解金属在生命过程中的作用。

16.1.2 金属组学分析技术及应用

近二十年来，金属组学发展十分迅速，金属组学研究技术也在不断进步和完善，目前常用的金属组学研究技术主要包括金属组定量分析技术，如中子活化（NAA）技术、等离子体（plasma）技术、质谱流式细胞技术等；金属组学分布研究技术，如激光剥蚀-电感耦合等离子体质谱法（LA-ICP-MS）、同步辐射X射线荧光成像（SR-XRF）、二次离子质谱（SIMS）等技术；金属组学形态分析技术，如高效液相色谱-电感耦合等离子体质谱联用技术（HPLC-ICP-MS）和凝胶电泳-同步辐射X射线荧光成像技术（GE-SRXRF）；金属组学结构分析技术，如分子质谱学技术（MALDI-TOF-MS或ESI-MS）、X射线吸收光谱、穆斯堡尔谱、核磁共振等，以及其他分析技术（图16-2）[4]。

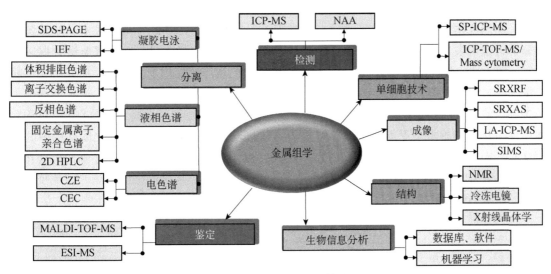

图16-2 金属组学研究技术

16.1.2.1 金属组定量分析技术

中子活化分析（neutron activation analysis，NAA）是活化分析中最重要的一种方法，用反应堆、加速器或同位素中子源产生的中子作为轰击粒子的活化分析方法，是确定物质元素成分的定性和定量的分析方法[5]。NAA具有很高的灵敏度和准确性，可同时测定多达30种元素的含量，对元素周期表中大多数元素的分析灵敏度可达$10^{-6} \sim 10^{-13}$ g/g[6]，因此在环境、生物、地学、材料、考古、法学等微量金属元素分析中都有广泛应用。Koeman等利用NAA技术对预处理后的海洋哺乳动物组织中甲基汞含量进行分析，结果表明90%以上的汞残留附着在组织中[7]。

基于等离子体技术的分析仪器主要有早期的电感耦合等离子体原子发射光谱（inductively coupled plasma-atomic emission spectrometry，ICP-AES）和后期的电感耦合等离子体质谱（inductively coupled plasma mass spectrometry，ICP-MS）。ICP-AES 和 ICP-MS 具有多元素同时检测与高灵敏度的优点，即使在复杂的分子环境中，也能对低浓度下的多种元素进行同时定量检测[8,9]。相对于 ICP-AES，ICP-MS 具有更高的灵敏度，对金属的检测限可达 pg/g 或 ng/L 水平[10]，比 ICP-AES 低 2～3 个数量级，而且 ICP-MS 的进样方式更加多元化，除了液体和气体样品可以直接进样，固体样品也可以通过激光烧蚀后直接进行检测[11]。由于 ICP-MS 具有操作方便、数据采集快、灵敏度高等优点，通过单独使用或与其他仪器联用已广泛应用于复杂生物样品和环境样品中金属元素的分析检测。Sanchez 等利用 ICP-MS 分析了人体尿液中 As、Cd、Co、Cu、Mn、Mo、Ni、Pb、Se、Zn 等 15 种元素的含量和相互关系，结果表明地下水可能是人体尿液中砷、钼和钨的来源，槟榔可能是镉暴露的来源[12]。

质谱流式细胞技术（mass cytometry，CyTOF）是基于质谱原理实现单细胞水平金属分析的一个新的应用方向[图 16-3（a）][13]。CyTOF 主要采用金属同位素标记的特异性抗体来标记细胞生物分子（蛋白质或核酸），标记好的细胞以单个细胞的形式依次进入 ICP-MS，进而被等离子体炬离子化为一个个独立的离子云，离子云中的金属标签随后被质谱定量检测出来，从而得到单个细胞对应的生物分子信息。CyTOF 既继承了传统单细胞流式细胞仪高速分析的特点，又具有质谱检测的高分辨能力。金属同位素（主要是镧系元素）标签的引入突破了传统荧光流式通道数量局限，能同时实现理论上超过 100 种细胞生物分子的同步检测，有助于我们对单细胞进行更全面的表型、信号通路及功能研究。同时，和传统基于荧光信号的流式细胞仪相比，镧系元素低的生物丰度赋予 CyTOF 更低的背景信号，同时避免了荧光光谱重叠所带来的信号干扰。DNA 条形码（DNA barcoding）技术的引入[14]，使得 CyTOF 可以实现多个样本同时检测，大大减小了由不同检测时间、操作及仪器差异所带来的干扰，有利于不同样本间更准确的定量比较，同时缩短了检测时间，降低了检测成本。随着各类高维度数据分析方法的发展（viSNE[15]、SPADE[16]、Citrus[17]等），能帮助我们对流式质谱数据进行更深入的挖掘。鉴于 CyTOF 在单细胞分析中的独特优势，其已被广泛应用于生物医学研究中（疾病早期诊断、生物标志物的发现、药物筛选、免疫学、衰老等相关研究[18-22]）。近年来，流式质谱成像系统（LA-CyTOF）的引入[图 16-3（b）][23,24]，大大扩宽了肿瘤和组织中原位分析通道数目，有助于我们在组织微环境下从亚细胞水平更加全面地了解复杂的细胞表型及其在组织微环境空间结构中的相互关系。LA-CyTOF 的出现，为探索免疫和免疫介导疾病在微环境中的相互关系提供了新的可视化研究方法。目前流式质谱成像系统已被应用于肿瘤细胞精准分型[24,25]、循环肿瘤细胞检测[26]、免疫系统与肿瘤关系[27]、糖尿病相关研究及病毒[28-30]等。质谱流式细胞技术的广泛推广将有利于生物医学的发展。

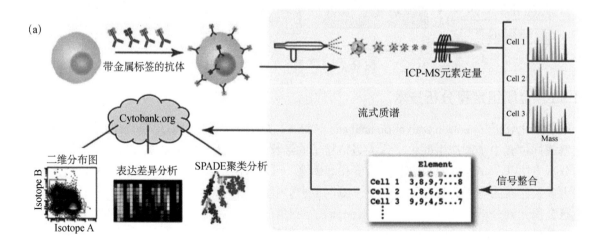

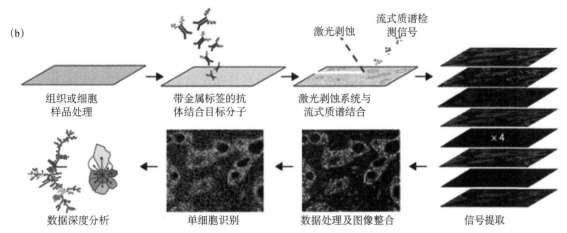

图 16-3 质谱流式细胞技术（a）[13]；流式质谱成像系统（b）[23,24]

16.1.2.2 金属组分布研究技术

生物体系中的金属组通常参与到众多不同生物功能中，研究这些生物活性金属元素在生物组织中的空间分布不仅可以提供金属元素的吸收、转运、蓄积和代谢过程等信息，还能揭示外源性金属作用于生物的目标器官或靶标位点，为金属元素的生物功能研究提供依据和线索，同时还可以为金属间多元素相互作用研究提供有用信息。目前研究金属组分布的主要技术包括：激光剥蚀-电感耦合等离子体质谱（laser ablation inductively coupled plasma mass spectrometry，LA-ICP-MS）和同步辐射微区 X 射线荧光（synchrotron radiation micro-X-ray fluorescence spectroscopy，SR-μXRF）。

LA-ICP-MS 技术主要通过脉冲激光束与样品相互作用，将产生的气溶胶引入到等离子体炬中激发、电离，经质谱分析器依据其荷质比分离和收集，进而确定未知试样组分和含量。LA-ICP-MS 仅需极小样品量（微克级）即可分析，激光直接气化固态样品避免了烦琐的前处理和分析过程中的损失，同时具有较高分辨率（2～5 μm）和较低检出限（10^{-9}），使其在固体材料研究中越来越受到人们的关注[31]。Barst 等利用 LA-ICP-MS 在鱼类组织结构中对汞沉积物和巨噬细胞中心进行共定位，发现了汞生物积累和免疫应答之间的潜在关联[32]。近年来，LA-ICP-MS 技术也被成功应用于单细胞中元素成像研究，Van Malderen 等通过加入醋酸铀和铱类化合物对 HeLa 细胞进行染色，经切片成像后获得了细胞中铀和铱的二维元素分布图像[33]。

SR-μXRF 采用由加速器产生的同步辐射作光源进行 X 射线荧光分析，具有强度高、准确性好和能量范围宽等优点，通过聚焦系统可将入射 X 射线光斑聚焦到微米甚至是纳米级别，大大提高了 XRF 的分辨率和灵敏度[34]。SR-μXRF 是目前研究金属组分布最热门的技术之一，具有样品取样量少、分析速度快、可作微区三维扫描分析给出被测元素的分布特征等优点，已被广泛用于生物样品如组织或细胞中微量元素的分布测定。高愈希等利用 SR-μXRF 技术对模式动物线虫暴露纳米铜后体内元素分布进行了研究，发现暴露后铜在体内含量增加且分布也发生了改变[35]。柴之芳团队利用 SR-XRF 技术揭示了大鼠在神经退行性疾病发展过程中脑内 Fe、Cu、Zn 等金属元素的微区分布，同时结合金纳米颗粒免疫标记技术，实现了金属元素与淀粉样蛋白的共成像[36,37]。Korbas 等利用 SR-XRF 技术研究了有机汞暴露斑马鱼之后的金属元素分布情况，直观地提供了微米空间分辨率层面上的分部信息，揭示了汞的视觉损伤可能与汞在斑马鱼眼部富集相关，而不仅仅是由于汞的神经毒性诱导产生[38]。

二次离子质谱（secondary ion mass spectrometry，SIMS）也可用于金属组的分布研究，通过高能量的一次离子束轰击样品表面，使样品表面的原子或原子团吸收能量而从表面发生溅射产生二次离子，通过质量分析器收集、分析这些二次离子，就可以得到关于样品表面信息的图谱。SIMS 可同时检测所有元素，检测限为 ng/kg 级别，分辨率可达到 10 μm，可对材料或者生物组织进行微区成分分析[39]。但 SIMS 对样品具有局部破坏性，且经常受到原子、离子以及基质干扰，进而可能会影响 SIMS 图像的定量化和信息深度。在生命科学领域，SIMS 可用于单细胞可视化分析[40]，获得药物在细胞内的吸收、分布、代谢等信息，还可用于研究药物在组织或者细胞中的定位[41]，对于提高药物的靶向性以及合理设计药物具有重要意义。近年来，随着 NanoSIMS 技术的兴起，实现了亚细胞水平上多元素的同时成像[42,43]。利用 NanoSIMS 的高灵敏度和分辨率，各种金属药物（如铂[44,45]、钌[46]、金[47]、钆[48]等）在细胞内的分布及作用机制被广泛研究。半胱氨酸硫醇是很多癌症相关蛋白中的药物靶点，通过 NanoSIMS 结合电子显微镜成像技术首次证明了金(Ⅲ)卟啉Ⅸ二甲酯（AuMesoIX）独特的硫醇靶向特性及其抗癌活性，AuMesoIX 可以作用于半胱氨酸残基并抑制抗肿瘤靶标蛋白的活性[图 16-4（a）]，而且 AuMesoIX 在小鼠体内也表现出抗肿瘤活性[49]。同时，利用 NanoSIMS 研究一些生命必需元素（如钙、铜、铁、锌等[50-52]）在细胞中的动态变化与生物活性的关系也被广泛报道[图 16-4（b）]。

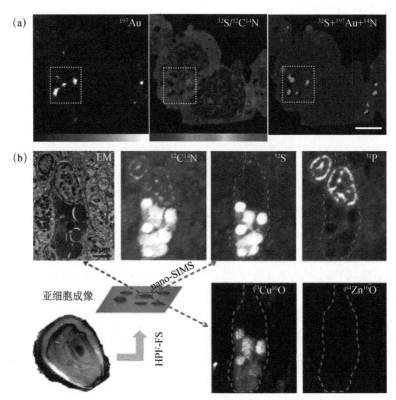

图 16-4　NanoSIMS 在金属成像中的应用[49,52]

16.1.2.3　金属组形态分析技术

金属元素的形态分析在环境和生物分析中都是极其重要的，因为不同金属元素在生物体内的作用及其代谢过程在很大程度上取决于金属元素存在的化学形态，而不仅仅是金属元素的总量。金属元素在生物体内的含量可直接决定其生物功能或生物毒性，但同一种金属元素的不同化学形态，也可能显示出不同生物毒性或生物利用度。例如水俣病的元凶甲基汞，在经微生物作用之前作为无机汞是没有明显毒性作用的，但是通过微生物的转化和生物放大作用进一步富集于鱼体，最终被人摄入之后引起了汞中毒。另一种常见金属元素——铬，三价铬是生物体所必需的微量元素，可维持正常葡萄糖代谢过程，而六价铬则被认为是可引起肺癌和皮肤癌的致癌物质。上述例子可以看出，仅仅根据被测金属元素的总量结果来评价其对环境和生物体系的影响是很不充分的，也不能满足现代科学研究的要求。这就对金属组学的发展提出了新的挑战，金属组的形态分析应运而生，为环境分析、生物化学分析、金属药物设计等提供了更充分的研究手段。从金属元素的形态水平上进行研究，可以更加深入地了解金属离子与生物分子在生命过程中的相互作用，阐明金属元素在生物体内的代谢过程、生理生化功能、毒性效应及作用机制。

金属组的形态分析需要多种技术手段联合使用，通常需要高效能的分离技术达到各形态的有效分离，再利用选择性强、灵敏度高的元素特征检测器实现超痕量形态的测定，色谱-光谱联用技术已成为金属组学研究的主要方法[53]。目前，常用的金属组形态分析方法主要有高效液相色谱-电感耦合等离子体质谱联用技术（HPLC-ICP-MS）、凝胶电泳-同步辐射 X 射线荧光成像技术（GE-SR-XRF）和柱状凝胶电泳-电感耦合等离子体质谱（GE-ICP-MS）等。

生物样品中金属组形态分析之前通常需要先进行样品前处理，即在保证样品中金属组形态不发生变化的前提下从固体或者液体中提取到易于后续进样分析的液相或气相中。金属组的样品前处理通常采用湿消化法，将酸、碱或者酶等加入样品中，使待测成分释放出来进入后续检测装置，但此法使用过程中会涉及腐蚀或危险溶剂，且可能会引入对后续仪器产生损害的试剂，因此使用溶剂的选择上需特别注意。针对不同类型的待测样品可能还会用到干灰化法或微波消解等方法，实际操作中要根据样品的组成及后续的分离检测仪器选择合适的前处理方法。

样品中金属组经提取后还需进一步分离才能进行元素形态分析，常用分离手段主要有色谱技术和电泳技术两大类。色谱技术包括高效液相色谱（high performance liquid chromatography，HPLC）、固相萃取（solid-phase extraction，SPE）、固相微萃取（solid phase microextraction，SPME）、压力液相萃取（pressurized liquid chromatography，PLE）以及气相色谱（gas chromatography，GC）[54]。电泳技术包括 1D/2D 凝胶电泳（1D/2D-GE）、毛细管电泳（capillary electrophoresis，CE）、毛细管区带电泳（capillary zone electrophoresis，CZE）、毛细管等速电泳（capillary isotachophoresis，CITP）、毛细管等电聚焦（capillary isoelectric focusing，CIEF）、胶束电动毛细管色谱（micellar electrophoresis capillary chromatography，MECC）、毛细管凝胶电泳（capillary gel electrophoresis，CGE）以及亲和毛细管电泳（affinity capillary electrophoresis，ACE）[55]。

基于色谱技术的分离方法效率通常较低，更适用于小分子化合物和基质较简单的样品分离，对于较复杂的样品可利用多维色谱进行分离。喻宏伟等建立了在线 HPLC-ICP-MS 分离检测系统，并应用于生物样品中硒的化学形态分析[56]。李玉峰等利用 HPLC-ICP-MS 研究了汞暴露人群补充硒后尿液中的汞与硒的形态变化[57]。HPLC-ICP-MS 还可与同位素稀释技术联合使用同时检测牛肉和奶制品中的 Cr(Ⅲ)和 Cr(Ⅵ)[58]。除了检测单一金属元素的不同形态外，HPLC-ICP-MS 还可同时检测多种金属元素形态，Araujo-Barbosa 等利用 HPLC-ICP-MS 对微波萃取后的补铁剂中的砷和铬元素形态进行了检测，样品中的 As(Ⅲ)、As(Ⅴ)、Cr(Ⅲ)和 Cr(Ⅵ)等金属形态可以在 5 min 内完成分离和检测[59]。

电泳技术由于其独特的分辨率优势更适用于大分子化合物和复杂样品的分离，且在蛋白质组学研究中已获得广泛应用。电泳技术结合 SR-XRF 构成一种准在线联用技术，可用于研究生物样品中的金属分布。利用 SR-XRF 的元素选择性分析模式对电泳分离后蛋白条带进行扫描测定，可在相应位置测得多种元素的含量。陈春英等利用该方法原位测定了人体肝脏组织中的微量元素，在人肝胞液中发现了分子质量为 29.5 kDa 的含锌蛋白[60]。二维凝胶电泳是目前分辨率最高的蛋白质分离技术，2-DE-SR-XRF 技术可实现更为复杂样品的分离和检测，赵甲亭等利用 2-DE-SR-XRF 技术对富硒酵母中的含硒蛋白进行了全定量分析，其检出限可达 0.2 μg/g，所有含硒蛋白的总硒量为 125.56 μg/g[61]。Lima 等采用 2-DE（IEF/SDS-PAGE）结合 SR-XRF 对罗非鱼肝脏组织及血浆中的金属蛋白进行研究，检测发现肝脏的提取蛋白中含 Ca、Fe 和 Zn 元素的蛋白数分别为 12、6 和 8；血浆蛋白含 Mn 和 Zn

元素的蛋白数分别为 4 和 6[62,63]。

此外，基于电泳技术的柱状凝胶电泳和毛细管电泳也可以和 ICP-MS 在线联用检测生物样本中的金属形态。柱状凝胶电泳与电感耦合等离子体质谱在线联用系统是对生物分子进行分离、检测、定量的有效工具，目前该系统已经成功应用于检测纳米金颗粒[64]、金属蛋白中的铁元素[65]、气溶胶中的碘[66]、核苷酸与顺铂的反应[67]以及多种生物样本中的金属蛋白。

16.1.2.4　金属组结构分析技术

表征金属组在环境样本或生物体内的结构信息可以直观地反映金属的富集归趋、结合位点、作用方式等，为了科学直观地展示金属在样本中的局域结构细节，必须要用到高分辨率、高准确性和高兼容性的结构分析技术。目前，常用的金属组结构分析方法主要有分子质谱学技术、基于辐射光技术及核磁共振技术，分子质谱学技术会在金属蛋白质组学部分详细介绍，这里主要介绍后两种技术。基于辐射光的方法可获得原子水平的结构信息，其光源包括 X 射线、γ 射线及中子流等。

X 射线吸收光谱（X-ray absorption spectroscopy，XAS）是利用 X 射线入射到样品前后信号变化来分析材料元素组成、电子态及微观结构等信息的光谱学手段。当 X 射线穿过样品时，由于样品对 X 射线的吸收，光的强度会发生衰减，这种衰减与样品的组成及结构密切相关。XAS 方法通常具有元素分辨性，几乎对所有原子都具有响应性，对固体（晶体或非晶）、液体、气体等各类样品都可以进行相关测试。基于同步辐射光源，当 X 射线经过样品时所激发的光电子被周围配位原子所散射，致使 X 射线吸收强度随能量发生振荡，研究这些振荡信号可以得到该体系的电子和几何局域结构，这种振荡结构称为 X 射线吸收精细结构（X-ray absorption fine structure，XAFS），主要包括 X 射线吸收近边结构（X-ray absorption near edge structure，XANES）和扩展 X 射线吸收精细结构（extend X-ray absorption fine structure，EXAFS）[68]。XANES 是指吸收边附近约 50 eV 范围内的精细结构，该方法的特点是振荡剧烈，谱采集时间短，对价态、未占据电子态和电荷转移等化学信息敏感，对温度依赖性弱，可用于高温原位化学实验，可快速鉴别元素的化学种类[69]。EXAFS 则指吸收边高能侧 50~1000 eV 范围出现的振荡，该方法可以得到中心原子与配位原子的键长、配位数、无序度等信息，但对立体结构并不敏感[70]。Harris 等采用 EXAFS 技术发现了海鱼中汞的主要形态为硫化甲基汞，其硫配基可能来源于半胱氨酸或蛋白质[71]。李玉峰等利用 XAS 对汞长期暴露人群的头发及血液样本进行了原位研究，结果表明汞主要与硫元素结合，且在头发和血液中也以不同配位形式存在[72]。

X 射线具有很强的穿透能力，能在晶体中产生衍射花样，对衍射花样进行分析可以确定对应的晶体结构，使得 X 射线衍射成为研究物质结构的主要手段之一。单晶 X 射线衍射（single crystal X-ray diffraction，SXRD）是利用单晶体对 X 射线形成衍射效应进而对物质进行内部原子在空间分布状况的结构分析方法。将具有一定波长的 X 射线照射到结晶性物质上时，X 射线因在结晶内遇到规则排列的原子或离子而发生散射，散射的 X 射线在某些方向上相位得到加强，从而显示与结晶结构相对应的特有的衍射现象[73]。SXRD 是研究晶体三维结构的强有力工具，分辨率可达 0.15~2 nm，该技术的特点在于可以获得元素存在的化合物状态、原子间相互结合的方式，从而可进行价态分析，但由于需要获得单晶，因此实际应用中受到一定限制。

穆斯堡尔谱（Mössbauer spectroscopy）是基于固体中的某些放射性原子核发射或吸收 γ 射线进行分析的技术[74]。该方法的主要特点是分辨率高，灵敏度高，抗干扰能力强，对试样无破坏，实验技术较为简单，试样的制备技术也不复杂，样本可以是粉末、超细小颗粒，甚至是冷冻的溶液，最终可提供样品中金属活性位点的多种物理和化学信息。但该技术的主要的缺点是：只有有限数量（42 种元素）的核有穆斯堡尔效应，且许多还必须在低温下或在具有制备源条件的实验室内进行，室温下

只有 ^{57}Fe 和 ^{119}Sn 等少数的穆斯堡尔核得到了充分的应用[75]。目前生物化学领域应用最多的是研究与铁元素相关的生物大分子或化合物的结构及变化。例如，Bonková 等使用穆斯堡尔谱分析了人和马脾脏组织冻干后的粉状样品中铁元素的形态[76]。Lindahl 课题组利用穆斯堡尔谱结合 EPR 和其他分析技术对老鼠器官中的铁元素形态进行了一系列研究[77-80]。虽然穆斯堡尔谱被证实是生物样品中铁元素形态分析的一种有效手段，但是该技术的应用仍然受低温、外加磁场以及生物分子中较低含铁量等因素的限制。

中子衍射技术是随着反应堆的出现而发展起来的一种研究物质结构的方法，采用中子作为光源的衍射技术与 X 射线衍射原理十分相似，主要不同之处在于 X 射线与电子相互作用，而中子与原子核相互作用。单晶中子衍射（single-crystal neutron diffraction，SCND）可用于测定晶体中的同位素、氢元素和原子序数接近的元素的位置，通常以 X 射线衍射的初步结果为基础，进一步提供晶体中难以用 X 射线等手段测定原子的位置信息[81]。但是，由于中子数的通量远小于 X 射线的通量，因此所需实验周期较长且对晶体样品要求较高。

核磁共振（nuclear magnetic resonance，NMR）是基于化学位移理论发展起来的，可用于测定生物大分子诸如蛋白质、DNA、RNA 等与金属离子、金属药物及其他化合物相互作用后的化学成分和分子结构，可以提供样品的三维结构和相互作用动力学信息，其分辨率与单晶 X 射线衍射相当，可以达到 0.15～2 nm[82]。NMR 技术最大的优点是可以直接在溶液中测定，无需晶体形式样品，实验条件更接近于生物大分子的生理环境，是一种无损测量技术，适用于分析不能结晶或者结晶后构象可能会改变的蛋白质。因此，NMR 技术成为除 X 射线以外最主要的金属组结构测定方法之一。Riccardo 等系统综述了 NMR 技术用于解析过渡金属元素 Cu(Ⅱ)/Cu(I)、Fe(Ⅲ)/Fe(Ⅱ)、Zn(Ⅱ) 及 Ni(Ⅱ) 与蛋白质相互作用所形成复合物的分子结构，阐述了 NMR 可提供金属配位、结构、动力学以及结合位点等重要生物信息[83]。

16.1.3　金属组学发展现状

金属组学研究目前正处于蓬勃发展的阶段，金属组学以及金属组学研究技术的不断发展正将该领域相关研究推向一个新的高度。近年来，金属组学专辑期刊的发行以及《金属组学》（*Metallomics*）期刊的成立为金属组学的发展提供了一个良好的平台。目前，多技术手段联合应用平台的建立和完善为金属组学与其他学科的交叉发展创造了更多的可能，金属组学技术越来越多地应用于生物化学、环境科学、分子生物学、毒理学、纳米材料学及医学等学科当中，不仅揭示了金属、类金属元素在生物体系中的吸收、转运、功能及代谢机理，同时也促进了各学科向更深入的方向发展。

16.2　金属蛋白质组学研究

16.2.1　金属蛋白质组学介绍

近年来，随着蛋白质组学与金属组学研究的交叉发展，一门全新的交叉学科金属蛋白质组学（metalloproteomics）逐渐形成。生物体系中大多数的微量金属元素通常与蛋白质结合，人体内有三分之一的蛋白与金属元素结合形成金属蛋白（包括金属酶）。金属蛋白质组（metalloproteome）

包括一个蛋白质组中所有的金属蛋白或具有金属结合位点的所有蛋白质，而与之对应的金属蛋白质组学主要研究一个蛋白质组内所有金属蛋白质的结构、金属的配位环境、蛋白质功能和构效关系[84]。

金属蛋白质组学研究可以使我们了解生物体内金属蛋白的分布、组成、功能、结构以及合成过程，还包括这些蛋白质翻译后的过程，如修饰、组装、转运、金属吸收等以及微配位环境对这些过程的影响，了解微量金属元素在生物体内的存在方式、作用机理、生物效应及其与相关疾病的关联，可以为我们在疾病诊断、金属药物的研发中提供理论依据[85]。

16.2.2　金属蛋白质组学研究技术

金属蛋白质组研究与蛋白质组的研究思路相似，首先利用全息样品制备技术从各种培养的细胞、细菌、生理或病理条件下的器官、组织中提取蛋白质，针对不同样品特征采用适当的分离技术将蛋白质分离，表征金属蛋白序列、结构以及功能，分析各个蛋白质中的微量元素及其相应功能，其中，生物样品中蛋白质结构的表征和金属蛋白的种类、形态分析是非常重要的研究步骤[86]。

16.2.2.1　金属蛋白质组学提取技术

与传统蛋白质组学方法比较，金属蛋白质组学更关注于与金属相结合的蛋白质，因此金属蛋白的提取分离方法要有别于常用的蛋白质组学方法，为了防止在提取过程中金属从蛋白质中脱落，需要在样品的整个前处理过程保持温和的条件。金属蛋白的提取主要包括生物组织/细胞破碎和蛋白质提取。根据生物样品来源和形态的不同，破碎方法分为温和、中等以及高强度三种不同模式，根据后续蛋白的分离检测条件以及样品属性决定所用裂解液或蛋白提取液。

在金属蛋白的提取过程中，须针对生物样品的不同属性和特点选择适合的破碎方法，柴之芳等详细综述了各种组织或细胞常用的破碎方法[87]。易裂解的细胞和微生物一般选择温和的裂解方式，而生物固体组织或者有较厚细胞壁的微生物一般选择高强度的破碎方式。蛋白酶或者普通裂解液通常用于普通细胞的裂解，一般需要在冰上操作，以保证蛋白活性。超声破碎是一种强有力的机械破碎方法，常用于组织或细菌样品的破碎，通过超声将裂解液中的组织或细胞裂解从而获得溶出蛋白，但是由于超声功率比较大，往往会产生较多热量和泡沫，不仅会使蛋白质变性，还可能造成金属的脱落，因此，超声破碎过程一般在冰浴中进行，仪器间歇工作，以降低溶液温度。对于较大的固体组织也可以利用振荡研磨进行破碎，通过在裂解液中加入钢珠或玻璃珠振荡研磨，以达到蛋白溶出的目的，但该方法也不能完全保证蛋白复合体的完整性。金属蛋白的提取过程中还应注意避免使用金属器具，防止在此过程中引入外来金属。

16.2.2.2　金属蛋白质组学分离技术

金属蛋白的分离是金属蛋白质组学研究中的一个重要环节，金属蛋白常用的分离技术与蛋白质组学分离技术基本相同，但是考虑金属蛋白的特殊性，在分离的过程中需特别关注与蛋白相结合的金属。目前，金属蛋白常用的分离技术主要有高效液相色谱、毛细管电泳以及凝胶电泳等[88]。

高效液相色谱是一种常用的蛋白质分离方法，通过后续与原子吸收光谱或质谱在线联用可以用于分离检测金属蛋白。高效液相色谱按照不同的分离机理主要可分为离子交换色谱、体积排阻色谱、反相色谱和亲和色谱等。Szpunar等报道了高效液相色谱在分离金属蛋白中的应用[89]。高效液相色谱在分离金属蛋白时对金属的影响较小，但是采用单一技术分离复杂蛋白样品时，分离分辨率较低，因此通常采用两种及以上液相色谱串联以提高分辨率。由于不同的液相色谱所采用的流动相有所差异，且流动相所含组分可能会对后续与之相连的质谱产生影响，因此在色谱柱的选择上需要考虑较多因素[90]。

固定化金属离子亲和色谱（immobilized metal affinity chromatography，IMAC）也称金属螯合亲和色谱（metal chelate affinity chromatography，MCAC），1975 年由 Porath 等提出，将过渡金属离子[Cu(II)、Zn(II)、Ni(II)]通过与组氨酸或半胱氨酸特异性地结合在固相基质上形成稳定的复合物，对不同靶分子进行亲和富集，随后以竞争性洗脱方式实现蛋白的分离、富集、纯化[91]。

毛细管电泳是基于化合物的分子荷质比差异来实现分离的，具有快速、高效、所需样品量少的特点，毛细管电泳的高分离能力使之可以用来分离极性和带电荷的化合物，因此也是分离金属和类金属蛋白的有效技术手段。毛细管电泳通常与传统 UV 检测器相连用于检测蛋白，但是 UV 检测器灵敏度较差，缺乏选择性，并不适用于复杂生物样品的分析。近年来发展的 CE-ICP-MS 联用技术，具有高灵敏度和选择性，已逐渐成为金属生物分子形态分析的有效工具[92]。毛细管电泳还可以与 ICP-MS 和 ESI-MS 并联体系联合使用，该联用体系结合了毛细管电泳的高分离能力、ICP-MS 的高灵敏度、ESI-MS 的高表征能力，可对化合物定量的同时也获得具体的结构信息，该联用系统已成为金属组学研究的主要方法。

电泳是生物大分子以及各种复合物分离的经典技术，常用于蛋白质和核酸的分离。用于蛋白质分离的电泳技术主要有两大类：等电聚焦电泳（isoelectric focusing electrophoresis，IEF）和聚丙烯酰胺凝胶电泳（polyacrylamide gel electrophoresis，PAGE）。等电聚焦电泳和聚丙烯酰胺凝胶电泳具有样品用样量少、分辨率高等优点，除了被应用于蛋白质组的分离外，也常被应用于金属蛋白组的分离，通过将两种电泳结合，可同时分离上千种蛋白质。

聚丙烯酰胺凝胶电泳可分为变性凝胶电泳（SDS-PAGE）与非变性凝胶电泳（Native-PAGE），都以聚丙烯酰胺凝胶为电泳支持物，根据凝胶的电荷效应与分子筛效应，将具有不同电荷和不同分子量的蛋白质分开。SDS-PAGE 性能稳定、分辨率高、分离重复性好，但是 SDS 会改变蛋白的构象，从而影响金属与蛋白的结构，在电泳的过程中可能会使金属蛋白中的金属脱落，因此 SDS-PAGE 不能很好地适应金属蛋白的分离。非变性凝胶电泳则可以保持蛋白质的原始性状，对金属蛋白的影响较小，但是分辨率会略差于 SDS-PAGE。为了更好地分离检测金属蛋白，Bertrand 等先采用 SDS-PAGE 对金属蛋白进行分离，得到明显的蛋白条带后，逐渐减少凝胶中 SDS 含量，从而使凝胶中检测到的金属含量逐渐增多，根据检测到金属的位置结合 SDS-PAGE 上条带位置即可确定其中的金属蛋白[93]。

16.2.2.3 金属蛋白质组学检测技术

为了确定金属蛋白所含有的金属元素，一般在蛋白分离之后会进行金属元素的检测。通常样品中所含有的元素浓度都较低，需要灵敏度足够高的检测器，常用的金属检测器主要有核分析技术、火焰原子吸收光谱（flame atomic absorption spectrometry，FAAS）和电热原子吸收光谱（electrothermal atomic absorption spectrometry，ETAAS）以及电感耦合等离子体质谱。

1. 核分析技术

现代核技术具有灵敏度高、准确度好、本底效应低等优点，先进的核分析技术与各种化学或生化分离技术结合可用于生物样品中微量元素的形态分析，分子活化分析（molecular activation analysis，MAA）、质子激发 X 射线荧光分析（proton induced X-ray emission，PIXE）、同步辐射 X 射线荧光分析等技术已广泛用于生物体系中金属蛋白的形态分析[94]。

分子活化分析是将传统的中子活化分析与特效的物理或化学方法相结合来实现分析的目的,主要用于生物、环境以及地学中微量元素化学特征的研究。柴之芳等建立了用于生物样品中微量元素形态分析的 MAA 方法,并且研究了植物、动物以及人类对稀土元素的吸收,探究了稀土元素与生物大分子的结合[95]。Behne 等用中子活化分析方法研究了哺乳动物体内的含硒蛋白及氨基酸,共检测出了四种特异性的硒蛋白,为其生物效应提供了依据[96]。

质子激发 X 射线荧光分析利用原子受质子激发后产生的特征 X 射线的能量和强度来进行物质定性和定量分析,是一种非破坏性的鉴定手段。该方法是一种多元素微量分析技术,灵敏度较高,可与色谱分离技术联用[97]。PIXE 更容易检测出蛋白质条带中的微量元素,因此常用于凝胶电泳分离之后金属蛋白的分析。

同步辐射 X 射线荧光技术使用同步辐射光源替代了传统的 X 射线光源,因此具有多元素分析、灵敏度高、非破坏性等优点,常用于电泳分离蛋白之后对凝胶上的金属元素进行检测,但是由于该技术的成本较高,其应用受到一定限制。

2. 火焰原子吸收光谱和电热原子吸收光谱

原子吸收光谱(atomic absorption spectroscopy,AAS)是基于待测元素的基态原子对特征谱线的吸收而建立的元素分析方法,广泛应用于生物样品中金属元素的定量分析。电热原子吸收光谱具有较低的检出限(μg/L 数量级),可以用于痕量元素的定量,但成本较高、相对于 ICP-MS 耗时较长。近年来,激光技术逐渐应用到原子吸收分析,用激光代替空心阴极灯光源,使原子吸收在痕量、超痕量领域有了更大的应用空间。

3. 电感耦合等离子体质谱

电感耦合等离子体质谱是以等离子体作为离子源的一种质谱型分析技术。与核分析技术相比,ICP-MS 主要用于多元素同时测定,而且灵敏度更高,检出限在 ng/L 量级,成本也较低。ICP-MS 进样系统可以液体自动进样,使其能够与高效液相色谱和毛细管电泳等分离技术在线联用,目前 ICP-MS 检测技术已被应用于分析金属和生物大分子蛋白复合物[98,99]、硒蛋白[100]、金属硫蛋白以及金属/半金属结合的多糖[101]等。激光剥蚀固体进样系统与 ICP-MS 联用的新分析技术产生于 20 世纪 80 年代后期[102],LA-ICP-MS 可以直接对凝胶或者固体组织上的元素信息进行原位检测而无需复杂的前处理步骤,具有原位、实时、高灵敏度、多元素同时测定等优点[103,104]。

16.2.2.4 金属蛋白质组学鉴定技术

生物大分子的结构鉴定对于其作用和功能的确定十分重要,质谱技术常被用于分析小分子和大分子化合物的结构,例如蛋白质、多肽和低聚核苷酸等[105-107]。质谱检测通过测量离子荷质比(电荷-质量比),从而得到其分子量和结构的信息。常用的生物分子电离技术有基质辅助激光解吸电离(matrix assisted laser desorption ionization,MALDI)和电喷雾电离(electrospray ionization,ESI)。在一些情况下,配体的鉴定可以通过电离技术(MALDI 和 ESI)和质谱技术的协同作用实现,从而产生了基质辅助激光解吸电离-飞行时间质谱和电喷雾质谱联用技术。

基质辅助激光解吸电离-飞行时间质谱(MALDI-TOF-MS)是近年来发展起来的一种新型的软电离生物质谱,对于凝胶分离的蛋白质鉴定起着十分重要的作用,特别是 MALDI 技术的发展使质谱可以被用于常规蛋白质鉴定。通过解析多肽片段的断裂模式可以得到氨基酸序列的信息,将通过质谱得到的多肽或者多肽片段的序列根据所对应物种进行搜库,根据谱库的比对结果进而得到目标蛋白质信息。MALDI-TOF-MS 的一个主要优点就是可以分析复杂基质中的生物分子，生物分子的多肽片段不受样品中存在的盐和缓冲试剂干扰。MALDI-TOF-MS 的另外一个优点是具有较好的检测灵敏度，可以鉴定低浓度的蛋白质，而且可测定的生物大分子的分子质量高达 600 kDa。Rigueira 等将金属检测技术（GFAAS）与蛋白质鉴定技术（MALDI-TOF-MS）结合成功检测出大豆中存在的金属结合蛋白[108]。

电喷雾电离质谱（ESI-MS）通过将溶液中的离子转变为气相离子而进行质谱分析。电喷雾电离质谱不仅可以用于无机物的检测分析，还可以用于有机金属离子复合物和生物大分子的检测。在电喷雾电离质谱中高分子量的分子会带有多个电荷，这种多电荷离子的产生大大扩展了普通质谱仪分析的质量范围。电喷雾电离质谱还具有很高的灵敏度，能与各种色谱技术联用，用于复杂样品体系的分析。

16.2.3 金属蛋白质组学发展现状

金属蛋白质组学研究可揭示金属蛋白的合成、修饰、转运等一系列生物过程，了解金属元素的生物效应、作用机理及其相关疾病的发病机制，为金属化合物的生态与健康评估以及金属药物的使用和设计提供了更多的理论依据，越来越多金属蛋白质组学的综合研究，极大地推动了金属生物学信息数据库的构建。近年来，各种金属蛋白质组学技术与研究方法正在不断改进和完善，多种核分析技术、质谱技术联合使用是很有潜力的综合研究方法，特别是中子科学的发展和多组学平台的建立将为金属蛋白质组学研究提供强有力的技术支持，极大地促进了金属元素的生物过程和生物学效应研究。

16.3 金属组学在金属药物研究中的应用

金属离子作为生命体的基本组成部分，在生命过程中起着不可替代的生理作用。金属离子也常应用于医疗领域作为疾病诊断和治疗试剂，典型的癌症治疗药物顺铂及其他铂类药物在无机药物发展和癌症研究中起到了至关重要的作用[109]。从远古时代开始使用的抗菌金属银到近现代的抗菌药物铋以及除铂以外的其他抗癌金属药物，如钌、锗、锡、镓、砷、金等都逐渐发展起来，并被应用于临床疾病的治疗当中[110]。

尽管金属药物的研发和使用受到越来越广泛的关注，但其生物学作用的分子靶点在很大程度上仍是未知的[111]。传统的金属药物研究侧重于金属与生物体的某些靶标蛋白或其他生物分子之间的相互作用，不能全面系统地提供细胞对金属药物的反应。但实际上金属药物通常同时作用于多个生物靶点，因此需要搭建综合平台来系统全面地揭示金属药物的生物活性及分子作用机制[112,113]。目前，新兴的组学方法已经被用来探索组成有机体细胞的各种类型分子的作用、关系和行为，综合多组学方法可以揭示金属药物的作用机制[114]。

典型的生物学技术平台通常是由众多组学方法，如转录组学、蛋白质组学、代谢组学等组成，它能够获得药物作用后生物样本中基因、蛋白质及小分子代谢产物的定性和定量结果，但这些组学方法无法获得药物结合靶标蛋白的信息[115,116]。考虑到金属药物通常与生物大分子（蛋白质、DNA、RNA）结合并影响其生物功能，因此，系统地检测金属药物结合蛋白对理解其在生物学中的作用尤为重要。在这种情况下，金属组学和金属蛋白质组学的出现，结合其他组学技术共同提供了大量有效的实验工具，有助于为金属药物的作用机理和研发设计提供系统完善的依据。金属组学和金属蛋白质组学可以提供金属的特异性信息，可实现对金属药物靶标配体/蛋白的识别、鉴定以及药物作用通路的探究[117,118]。目前，在金属蛋白质组学和其他组学交叉领域，已发展和建立了一系列高通量、高灵敏度的分析技术平台，用于解析金属药物处理后细胞或组织中金属与蛋白质形成的复杂混合物[119]。

  目前细菌耐药性的问题正变得越来越严重，缺乏新的抗生素成为医学界面临的最严峻挑战之一，具有广谱抗菌性的金属化合物一直是药物研发的关注重点。金属铋具有很好的抗菌性，铋类化合物作为抑菌药物已有上百年历史，本章作者所在实验室利用 IMAC/2D-GE、ICP-MS 与 MALDI-TOF-MS 等联用技术系统分析了铋在幽门螺杆菌内的结合蛋白，揭示了金属铋在生物体内的分子靶点及其作用机制[图 16-5（a）][120,121]。综合利用金属蛋白质组学、代谢组学与转录组学等技术，课题组发现 Ga(Ⅲ)会与铜绿假单胞菌中的 RNA 聚合酶结合进而抑制 RNA 合成，从而降低细菌代谢率和能量的利用，而外源性补充乙酸盐可进一步增强 Ga(Ⅲ)的抗菌活性，这一协同效应在动物模型中进一步得到了验证[122]。银的抗菌性及其作用机制虽然已有很多报道，但有关其生物大分子的作用靶点仍不是很明确。在前期金属蛋白质组学技术的基础之上，实验室又进一步建立了 LC-GE-ICP-MS 联用技术，成功分离检测了 34 种大肠杆菌中的银结合蛋白［图 16-5（b）］，生物信息学分析发现这些银结合蛋白主要参与到三羧酸（TCA）循环、乙醛酸循环、糖酵解、翻译、氧化应激反应和胞内 pH 调节等生物过程，通过多组学技术的综合分析发现银主要作用于糖酵解和 TCA 循环中的多种酶，进而影响乙醛酸循环通路，抑制细胞氧化应激反应，最终造成细菌损伤和死亡[123]。对银离子结合蛋白甘油醛-3-磷酸脱氢酶（GAPDH）进行结构研究，发现银离子通过与 GAPDH 中 Cys149 活性位点结合进而影响其生物活性，上述结果从分子生物学层面阐释了金属药物的抗菌机制［图 16-5（b）][124]。

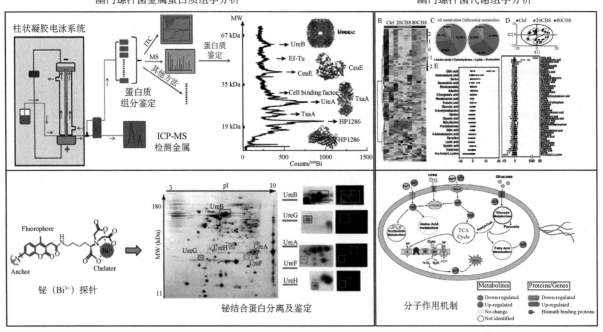

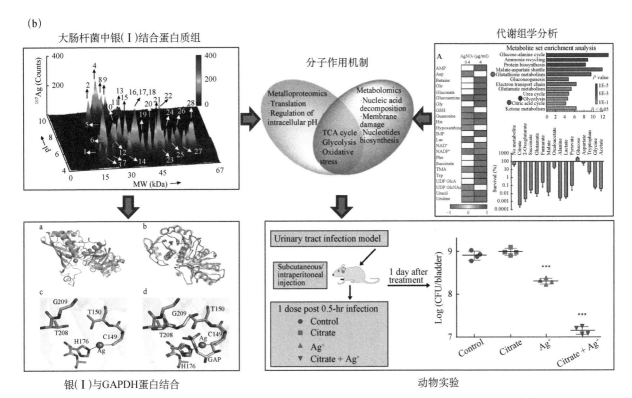

图 16-5　多组学技术在铋、银药物结合蛋白和作用机理研究中的应用

传统铂化疗药物往往伴产生严重副作用和药物抗性，因此不同金属抗癌药物的分子机制和新型金属药物设计研发成为金属药物研究的重点方向。乐晓春等利用砷亲和色谱柱和质谱技术在人肺癌细胞 A549 中进行靶向蛋白质组学分析，分离鉴定出 50 种核蛋白以及 24 种膜/细胞器组分蛋白[125]，该团队还设计了一种叠氮化标记砷化合物（p-azidophenylarsenoxide，PAzPAO）在细胞中可检测出 48 种砷结合蛋白[126]，这些结合蛋白的发现揭示了砷类抗癌药物的更多潜在作用靶点。支志明团队利用光亲和标记、点击化学、蛋白质组学、NMR 以及荧光猝灭等技术确定了卟啉金(Ⅲ)类抗癌化合物的直接分子靶点是分子伴侣 Hsp60[127]；该团队还进一步确定了抗肿瘤双(N-杂环卡宾)铂(Ⅱ)复合物在肿瘤细胞中的作用靶点并非传统认为的 DNA，而是胞内蛋白天冬酰胺合成酶[128]。Meier 等通过综合蛋白质组学技术发现钌类抗癌药物新的潜在作用靶点的分子靶点——网蛋白，该蛋白靶点在体外敲除模型中也得到了验证[129]。

钌化合物是铂类抗癌药物的最佳替代物，对正常细胞具有较轻的毒副作用，Klose 等利用 LA-ICP-MS 研究了两种抗癌药物（有机钌化合物和锇化合物）在肿瘤小鼠体内的组织分布情况（图 16-6），肝脏、肾脏、肌肉以及肿瘤组织中均可检测到锇和钌元素，锇化合物主要分布在组织边缘，而钌则重点富集在组织和肿瘤的内部[130]。

金属药物在生物体系内的作用机理是十分复杂的，通过整合多组学技术手段有助于对金属药物的生物活性研究提供系统性的认知，可在分子水平上揭示金属药物吸收、转运、分布及作用机制。金属组学和金属蛋白质组学在分析技术上的发展极大地推进了金属药物机理的研究，此外单细胞分析和人工智能的迅速崛起也将为金属药物的设计合成提供更多帮助。

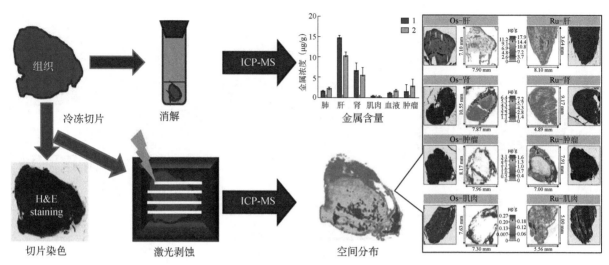

图 16-6　LA-ICP-MS 用于锇和钌化合物生物分布研究[130]

16.4　展　　望

金属组学从形成到发展虽然只有短短十几年，但现代分析技术、生物技术和生物信息学工具的迅速发展，结合环境科学、分析化学、无机化学、蛋白质组学和毒理学等方法，为金属组学研究提供了大量的数据资料和可用的关键技术，金属组学与其他学科的交叉合作也极大地促进了分析技术手段的进步。目前在金属组学研究中，无论是方法学还是机理研究仍有很多方面需要深入完善，在未来的研究中，应继续向该领域引入创新性的技术，如基于深度学习的人工智能方法[131]，通过多技术融合最终建立综合一体化的金属生物学分析平台。近年来，金属组学研究也衍生出一些新兴分支组学，例如环境金属组学、纳米金属组学及金属病理组学，这些创新交叉学科的发展不仅为金属元素的生物活性研究提供了更多的数据信息，同时也在金属相关的生态环境系统和人类健康风险研究中发挥了重要作用。我国金属组学研究一直紧跟国际金属组学研究步伐，某些研究方面已起到引领作用，随着越来越多学者在交叉领域合作的加深和大科学装置的建设、使用，我国金属组学研究将会在国际上占有更加重要的地位。

致谢：感谢香港研究资助局、香港创新科技署、国家自然科学基金委员会、香港大学和叶志成·范港喜基金会的支持。感谢本课题组葛瑞光、孙雪松、杨楠、夏炜、胡立刚、黎佑芷、阳娅、洪祎璠、汪旻稷、都秀波、巢艾伦、张苑茵、王海波、王宇传、江南、程天凡、徐小晗以及我们的合作者郝权、陈春英、李玉锋、柴之芳、江桂斌和 Joanna Szpunar、Ryszard Lobinski 对金属（蛋白质）组学的发展所做出的贡献。

参 考 文 献

[1] Haraguchi H. Metallomics as integrated biometal science. Journal of Analytical Atomic Spectrometry, 2004, 19: 5-14.

[2] Shi W, Chance M R. Metallomics and metalloproteomics. Cellular and Molecular Life Sciences, 2008, 65: 3040-3048.
[3] Degtyarenko K. Bioinorganic motifs: Towards functional classification of metalloproteins. Bioinformatics, 2000, 16: 851-864.
[4] Chen B W, Hu L G, He B, Luan T G, Jiang G B. Environmetallomics: Systematically investigating metals in environmentally relevant media. TrAC-Trends in Analytical Chemistry, 2020, 126: 115875.
[5] De Soete D, Gijbels R, Hoste J. Neutron Activation Analysis. New York: Wiley, 1972.
[6] 中国科学院高能物理研究所. 中子活化分析在环境学、生物学和地学中的应用. 北京: 原子能出版社, 1992.
[7] Koeman J H, Peeters W H M, Koudstaal-Hol C H M. Mercury-selenium correlations in marine mammals. Nature, 1973, 245(1): 385-386.
[8] Szpunar J. Advances in analytical methodology for bioinorganic speciation analysis. metallomics, metalloproteomics and heteroatom-tagged proteomics and metabolomics. Analyst, 2005, 130: 442-465.
[9] Rodriguez-Moro G, Ramirez-Acosta S, Arias-Borrego A, Garcia-Barrera T, Gomez-Ariza J L. Environmental metallomics. Metallomics, 2018, 1055: 39-66.
[10] Profrock D, Prange A. Inductively coupled plasma-mass spectrometry (ICPMS) for quantitative analysis in environmental and life sciences: A review of challenges, solutions, and trends. Applied Spectroscopy, 2012, 66: 843-868.
[11] Durrant S F, Ward N I. Laser ablation-inductively coupled plasma-mass spectrometry (LA-ICP-MS) for the multielemental analysis of biological materials: A feasibility study. Food Chemistry, 1994, 49(3): 317-323.
[12] Sanchez T R, Slavkovich V, LoIacono N, van Geen A, Ellis T, Chillrud S N, Balac O, Islam T, Parvez F, Ahsan H, Graziano J H, Navas-Acien A. Urinary metals and metal mixtures in Bangladesh: Exploring environmental sources in the health effects of arsenic longitudinal study (HEALS). Environment International, 2018, 121: 852-860.
[13] Bendall S C, Simonds E F, Qiu P, El-ad D A, Krutzik P O, Finck R, Bruggner R V, Melamed R, Trejo A, Ornatsky O I. Single-cell mass cytometry of differential immune and drug responses across a human hematopoietic continuum. Science, 2011, 332 (6030): 687-696.
[14] Bodenmiller B, Zunder E R, Finck R, Chen T J, Savig E S, Bruggner R V, Simonds E F, Bendall S C, Sachs K, Krutzik P O. Multiplexed mass cytometry profiling of cellular states perturbed by small-molecule regulators. Nature Biotechnology, 2012, 30(9): 858-867.
[15] Amir E-a D, Davis K L, Tadmor M D, Simonds E F, Levine J H, Bendall S C, Shenfeld D K, Krishnaswamy S, Nolan G P, Pe'er D. viSNE enables visualization of high dimensional single-cell data and reveals phenotypic heterogeneity of leukemia. Nature Biotechnology, 2013, 31 (6): 545-552.
[16] Qiu P, Simonds E F, Bendall S C, Gibbs K D, Bruggner R V, Linderman M D, Sachs K, Nolan G P, Plevritis S K. Extracting a cellular hierarchy from high-dimensional cytometry data with SPADE. Nature Biotechnology, 2011, 29 (10): 886-891.
[17] Bruggner R V, Bodenmiller B, Dill D L, Tibshirani R J, Nolan G P. Automated identification of stratifying signatures in cellular subpopulations. Proceedings of the National Academy of Sciences of the United States of America, 2014, 111(26): 2770-2777.
[18] Simoni Y, Chng M H Y, Li S, Fehlings M, Newell E W. Mass cytometry: A powerful tool for dissecting the immune landscape. Current Opinion in Immunology, 2018, 51: 187-196.
[19] Atkuri K R, Stevens J C, Neubert H. Mass cytometry: A highly multiplexed single-cell technology for advancing drug development. Drug Metabolism and Disposition, 2015, 43 (2): 227-233.
[20] Baca Q, Cosma A, Nolan G, Gaudilliere B. The road ahead: Implementing mass cytometry in clinical studies, one cell at a time. Cytometry Part B: Clinical Cytometry, 2017, 92 (1): 10-11.

[21] Bengsch B, Ohtani T, Herati R S, Bovenschen N, Chang K-M, Wherry E J. Deep immune profiling by mass cytometry links human T and NK cell differentiation and cytotoxic molecule expression patterns. Journal of Immunological Methods, 2018, 453: 3-10.

[22] Mrdjen D, Pavlovic A, Hartmann F J, Schreiner B, Utz S G, Leung B P, Lelios I, Heppner F L, Kipnis J, Merkler D. High-dimensional single-cell mapping of central nervous system immune cells reveals distinct myeloid subsets in health, aging, and disease. Immunity, 2018, 48 (2): 380-395.

[23] Pantanowitz L, Prefer F, Wilbur D C. Advanced imaging technology applications in cytology. Diagnostic Cytopathology, 2019, 47 (1): 5-14.

[24] Giesen C, Wang H A, Schapiro D, Zivanovic N, Jacobs A, Hattendorf B, Schüffler P J, Grolimund D, Buhmann J M, Brandt S. Highly multiplexed imaging of tumor tissues with subcellular resolution by mass cytometry. Nature Methods, 2014, 11 (4): 417-422.

[25] Wagner J, Rapsomaniki M A, Chevrier S, Anzeneder T, Langwieder C, Dykgers A, Rees M, Ramaswamy A, Muenst S, Soysal S D, Jacobs A, Windhager J, Silina K, van den Broek M, Dedes K J. Rodríguez Martínez M, Weber W P, Bodenmiller B. A single-cell atlas of the tumor and immune ecosystem of human breast cancer. Cell, 2019, 177(5): 1330-1345.

[26] Gerdtsson E, Pore M, Thiele J-A, Gerdtsson A S, Malihi P D, Nevarez R, Kolatkar A, Velasco C R, Wix S, Singh M. Multiplex protein detection on circulating tumor cells from liquid biopsies using imaging mass cytometry. Convergent Science Physical Oncology, 2018, 4 (1): 015002.

[27] Chevrier S, Levine J H, Zanotelli V R T, Silina K, Schulz D, Bacac M. An immune atlas of clear cell renal cell carcinoma. Cell, 2017, 169: 736-749.

[28] Wang W, Su B, Pang L, Qiao L, Feng Y, Ouyang Y, Guo X, Shi H, Wei F, Su X. High-dimensional immune profiling by mass cytometry revealed immunosuppression and dysfunction of immunity in COVID-19 patients. Cellular & Molecular Immunology, 2020: 1-3.

[29] Leng Z, Zhu R, Hou W, Feng Y, Yang Y, Han Q, Shan G, Meng F, Du D, Wang S. Transplantation of ACE2-mesenchymal stem cells improves the outcome of patients with COVID-19 pneumonia. Aging and Disease, 2020, 11 (2): 216-228.

[30] Ouyang Y, Yin J, Wang W, Shi H, Shi Y, Xu B, Qiao L, Feng Y, Pang L, Wei F. Down-regulated gene expression spectrum and immune responses changed during the disease progression in COVID-19 patients. Clinical Infectious Diseases, 2020, 71(16): 2052-2060.

[31] Becker J S. Imaging of metals in biological tissue by laser ablation inductively coupled plasma mass spectrometry (LA-ICP-MS): State of the art and future developments. Journal of Mass Spectrometry, 2013, 48: 255-268.

[32] Barst B D, Gevertz A K, Chumchal M M, Smith J D, Rainwater T R, Drevnick P E, Hudelson K E, Hart H, Verbeck G F, Roberts A P. Laser ablation ICP-MS co-localization of mercury and immune response in fish. Environmental Science & Technology, 2011, 45(20): 8982-8988.

[33] Van Malderen S J M, Van Acker T, Laforce B. Three-dimensional reconstruction of the distribution of elemental tags in single cells using laser ablation ICP-mass spectrometry via registration approaches. Analytical and Bioanalytical Chemistry, 2019, 411: 4849-4859.

[34] Chen C, Chai Z, Gao Y. Nuclear Analytical Techniques for Metallomics and Metalloproteomics. Cambridge: RSC Publishing, 2010.

[35] Gao Y, Liu N, Chen C. Mapping technique for biodistribution of elements in a model organism, *Caenorhabditis elegans*, after exposure to copper nanoparticles with microbeam synchrotron radiation X-ray fluorescence. Journal of Analytical Atomic Spectrometry, 2008, 23: 1121-1124.

[36] Wang H J, Wang M, Wang B, Meng X Y, Wang Y, Li M, Feng W Y, Zhao Y L and Chai Z F. Quantitative imaging of element spatial distribution in the brain section of a mouse model of Alzheimer's disease using synchrotron radiation X-ray fluorescence analysis. Journal of Analytical Atomic Spectrometry, 2010, 25: 328-333.

[37] Wang H, Wang M, Wang B, Li M, Chen H, Yu X, Zhao Y, Feng W, Chai Z. The distribution profile and oxidation states of biometals in APP transgenic mouse brain: dyshomeostasis with age and as a function of the development of Alzheimer's disease. Metallomics, 2012, (4): 289-296.

[38] Korbas M, Blechinger S R, Krone P H, Pickering I J, George G N. Localizing organomercury uptake and accumulation in zebrafish larvae at the tissue and cellular level. Proceedings of the National Academy of Sciences of the United States of America, 2008, 105: 12108-12112.

[39] David Briggs, Alan Brown, Vickerman J C, Adams F. Handbook of static secondary ion mass spectrometry (SIMS). Analytica Chimica Acta, 1990, 236: 509-510.

[40] Shao C, Zhao Y, Wu K, Jia F, Luo Q, Liu Z, Wang F. Correlated secondary ion mass spectrometry-laser scanning confocal microscopy imaging for single cell-principles and applications. Chinese Journal of Analytical Chemistry, 2018, 46(7): 1005-1016.

[41] Wu K, Jia F, Zheng W, Luo Q, Zhao Y, Wang F. Visualization of metallodrugs in single cells by secondary ion mass spectrometry imaging. Journal of Biological Inorganic Chemistry, 2017, 22: 653-661.

[42] Lee R F S, Theiner S, Meibom A, Koellensperger G, Keppler B K, Dyson P J. Application of imaging mass spectrometry approaches to facilitate metal-based anticancer drug research. Metallomics, 2017, 9: 365-381.

[43] Massonnet P, Heeren R M A. A concise tutorial review of TOF-SIMS based molecular and cellular imaging. Journal of Analytical Atomic Spectrometry, 2019, 34: 2217-2228.

[44] Legin A A, Schintlmeister A, Jakupec M A, Galanski M, Lichtscheidl I, Wagner M, Keppler B K. NanoSIMS combined with fluorescence microscopy as a tool for subcellular imaging of isotopically labeled platinum-based anticancer drugs. Chemical Science, 2014, 5: 3135-3143.

[45] Lee R F S, Riedel T, Escrig S, Maclachlan C, Knott G W, Davey C A, Johnsson K, Meibom A, Dyson P J. Differences in cisplatin distribution in sensitive and resistant ovarian cancer cells: A TEM/NanoSIMS study. Metallomics, 2017, 9: 1413-1420.

[46] Lee R F S, Escrig S, Croisier M, Clerc-Rosset S, Knott G W, Meibom A, Davey C A, Johnsson K, Dyson P J. NanoSIMS analysis of an isotopically labelled organometallic ruthenium(Ⅱ) drug to probe its distribution and state in vitro. Chemical Communications, 2015, 51: 16486-16489.

[47] Wedlock L E, Kilburn M R, Cliff J B, Filgueira L, Saunders M, Berners-Price S J. Visualising gold inside tumour cells following treatment with an antitumour gold(Ⅰ) complex. Metallomics, 2011, 3: 917-925.

[48] Duane R S, Daniel R L, Subhash C. Subcellular SIMS imaging of gadolinium isotopes in human glioblastoma cells treated with a gadolinium containing MRI agent. Applied Surface Science, 2004, 231-232: 457-461.

[49] Tong K C, Lok C N, Wan P K, Hu D, Fung Y M E, Chang X Y, Huang S, Jiang H, Che C M. An anticancer gold(Ⅲ)-activated porphyrin scaffold that covalently modifies protein cysteine thiols. Proceedings of the National Academy of Sciences of the United States of America, 2020, 117: 1321-1329.

[50] Hong-Hermesdorf A, Miethke M, Gallaher S, Kropat J, Dodani S C, Chan J, Barupala D, Domaille D W, Shirasaki D I, Loo J A, Weber P K, Pett-Ridge J, Stemmler T L, Chang C J, Merchant S S. Subcellular metal imaging identifies dynamic sites of Cu accumulation in Chlamydomonas. Nature Chemical Biology, 2014, 10: 1034-1042.

[51] Dauphas S, Delhaye T, Lavastre O, Corlu A, Guguen-Guillouzo C, Ababou-Girard S, Geneste S. Localization and quantitative analysis of antigen-antibody binding on 2D substrate using imaging NanoSIMS. Analytical Chemistry,

2008, 80: 5958-5962.

[52] Weng N Y, Jiang H, Wang W X. *In situ* subcellular imaging of copper and zinc in contaminated oysters revealed by nanoscale secondary ion mass spectrometry. Environmental Science & Technology, 2017, 51: 14426-14435.

[53] Sanz-Medel A. Trace element analytical speciation in biological systems: Importance, challenges and trends. Spectrochimica Acta Part B: Atomic Spectroscopy, 1988, 53(2): 197-211.

[54] 江桂斌. 环境样品前处理技术. 北京：化学工业出版社, 2004：171-190.

[55] Righettu P G. Capillary Electrophoresis in Analytical Biotechnology. Florida: CRC Press, 1966: 20-23.

[56] 喻宏伟, 陈春英, 高愈希, 李柏, 柴之芳. 高效液相色谱-电感耦合等离子体质谱法分析生物样品中硒的化学形态. 分析化学, 2006, 34(6): 749-753.

[57] Li Y F, Chen C Y, Li B, Wang Q, Wang J X, Gao Y X, Zhao Y L, Chai Z F. Simultaneous speciation of selenium and mercury in human urine samples from long-term mercury-exposed populations with supplementation of selenium-enriched yeast by HPLC-ICP-MS. Journal of Analytical Atomic Spectrometry, 2007, 22(8): 925-930.

[58] Saraiva M, Chekri R, Leufroy A, Guérin T, Sloth J J, Jitaru P. Development and validation of a single run method based on species specific isotope dilution and HPLC-ICP-MS for simultaneous species interconversion correction and speciation analysis of Cr(III)/Cr(VI) in meat and dairy products. Talanta, 2021, 222: 121538.

[59] Araujo-Barbosa U, Pena-Vazquez E, Barciela-Alonso M C, Costa Ferreira S L, Pinto dos Santos A M, Bermejo-Barrera P. Simultaneous determination and speciation analysis of arsenic and chromium in iron supplements used for iron-deficiency anemia treatment by HPLC-ICP-MS. Talanta, 2017, 170: 523-529.

[60] 陈春英, 章佩群, 柴之芳, 李光城, 黄宇营. 同步辐射X荧光分析法原位测定人肝金属蛋白中的微量元素. 同步辐射装置用户科技论文集, 2000, 1: 361-365.

[61] Zhao J, Pu Y, Gao Y, Peng X, Li Y, Xu X, Li B, Zhu N, Dong J, Wu G, Li Y-F. Identification and quantification of seleno-proteins by 2-DE-SR-XRF in selenium-enriched yeasts. Journal of Analytical Atomic Spectrometry, 2015, 30: 1408-1413.

[62] Santos F A, Lima P M, Neves R C F, Moraes P M, Pérez A A, Silva M A O, Arruda M A Z, Castro G R, Padilha P M. Metallomic study of plasma samples from *Nile tilapia* using SR-XRF and GFAAS after separation by 2D PAGE: Initial results. Microchimica Acta, 2011, 173: 43-49.

[63] Lima P M, Neves R C F, Santos F A, Pérez C A, Silva M A O da, Arruda M A Z, Castro G R, Padilha P M. Analytical approach to the metallomic of *Nile tilapia* (*Oreochromis niloticus*) liver tissue by SRXRF and FAAS after 2D-PAGE separation: Preliminary results. Talanta, 2010, 82: 1052-1056.

[64] Helfrich A, Brüchert W, Bettmer J. Size characterisation of Au nanoparticles by ICP-MS coupling techniques. Journal of Analytical Atomic Spectrometry, 2006, 21: 431-434.

[65] Anorbe M G, Messerschmidt J, Feldmann I, Jakubowski N. On-line coupling of gel electrophoresis (GE) and inductively coupled plasma-mass spectrometry (ICP-MS) for the detection of Fe in metalloproteins. Journal of Analytical Atomic Spectrometry, 2007, 22: 917-924.

[66] Brüchert W, Helfrich A, Zinn N, Klimach T. Gel electrophoresis coupled to inductively coupled plasma-mass spectrometry using species-specific isotope dilution for iodide and iodate determination in aerosols. Analytical Chemistry, 2007, 79: 1714-1719.

[67] Brüchert W, Krüger R, Tholey A, Montes-Bayón M, Bettmer J. A novel approach for analysis of oligonucleotide-cisplatin interactions by continuous elution gel electrophoresis coupled to isotope dilution inductively coupled plasma mass spectrometry and matrix-assisted laser desorption/ionization mass spectrometry. Electrophoresis, 2005, 29: 1451-14591.

[68] 麦振洪. 同步辐射光源及其应用. 北京: 科学出版社, 2013.

[69] Rehr J J, Ankudinov A L. Progress in the theory and interpretation of XANES. Coordination Chemistry Reviews, 2005, 249(1-2): 131-140.

[70] 王其武, 刘文汉. X 射线吸收精细结构及其应用. 北京: 科学出版社, 1994.

[71] Harris H H, Pickering I J, George G N. The chemical form of mercury in fish. Science, 2003, 301: 1203-1203.

[72] Li Y F, Chen C, Li B, Li W, Qu L, Dong Z, Nomura M, Gao Y, Zhao J, Hu W, Zhao Y, Chai Z. Mercury in human hair and blood samples from people living in Wanshan mercury mine area, Guizhou, China: An XAS study. Journal of Inorganic Biochemistry, 2008, 102: 500-506.

[73] 周公度. 晶体和准晶体的衍射. 北京: 北京大学出版社, 1999.

[74] Sharma V K, Klingelhöfer G, Nishida T. Mössbauer Spectroscopy: Applications in Chemistry, Biology, and Nanotechnology. New York: Wiley, 2013.

[75] Gütlich P, Bill E, Trautwein A X. Mössbauer Spectroscopy and Transition Metal Chemistry: Fundamentals and Applications. Berlin: Springer-Verlag, 2011.

[76] Bonková I, Miglierini M, Bujdoš M. Analysis of iron-binding compounds in biological tissues by Mössbauer spectrometry. AIP Conference Proceedings, 2016, 1781(1): 020011.

[77] Holmes-Hampton G P, Chakrabarti M, Cockrell A L, McCormick S P, Abbott L C, Lindahl L S, Lindahl P A. Changing iron content of the mouse brain during development. Metallomics, 2012, 4 (8): 761-770.

[78] Chakrabarti M, Cockrell A L, Park J, McCormick S P, Lindahl L S, Lindahl P A. Speciation of iron in mouse liver during development, iron deficiency, IRP2 deletion and inflammatory hepatitis. Metallomics, 2015, 7 (1): 93-101.

[79] Chakrabarti M, Barlas M N, McCormick S P, Lindahl L S, Lindahl P A. Kinetics of iron import into developing mouse organs determined by a pup-swapping method. Journal of Biological Chemistry, 2015, 290 (1): 520-528.

[80] Wofford J D, Chakrabarti M, Lindahl P A. Mössbauer spectra of mouse hearts reveal age-dependent changes in mitochondrial and ferritin iron levels. Journal of Biological Chemistry, 2017, 292 (13): 5546-5554.

[81] Hazemann I, Dauvergne M T, Blakeley M P, Meilleur F, Haertlein M, Van Dorsselaer A, Mitschler A, Myles D A, Podjarny A. High-resolution neutron protein crystallography with radically small crystal volumes: Application of perdeuteration to human aldose reductase. Acta Crystallographica Section D: Biological Crystallography, 2005, 61(10): 1413-1417.

[82] Keeler J. Understanding NMR Spectroscopy. Second ed. Chichester: John Wiley, 2010.

[83] Riccardo D R, Slawomir P, Henryk K, Daniela V. NMR investigations of metal interactions with unstructured soluble protein domains. Coordination Chemistry Reviews, 2014, 269: 1-12.

[84] Shi W, Chance M R. Metallomics and metalloproteomics. Cellular and Molecular Life Sciences, 2008, 65: 3040-3048.

[85] Bettmer J. Metalloproteomics: A challenge for analytical chemists. Analytical and Bioanalytical Chemistry, 2005, 383: 370-371.

[86] Gao Y, Chen C Y, Chai Z F. Advanced nuclear analytical techniques for metalloproteomics. Journal of Analytical Atomic Spectrometry, 2007, 22: 856-866.

[87] Peng X, Zhang J, Gao Y, Chai Z F. Techniques for extraction and separation of metalloproteins. Progress in Chemistry, 2012, 24: 834-843.

[88] Jiang G B, He B. Methodology and foreground of metallomics. Bulletin of National Natural Science Foundation of China, 2005, 3: 151-155.

[89] Szpunar J. Bio-inorganic speciation analysis by hyphenated techniques. Analyst, 2005, 130: 442-465.

[90] Li X, Chen X G, Kong L, Zhou H F. Multidimensional liquid chromatography and its applications in life science.

Chinese Bulletin of Life Science, 2003, 15: 95-100.

[91] Porath J, Carlsson J, Olsson I, Belfrage G. Metal chelate affinity chromatography, a new approach to protein fractionation. Nature, 1975, 258: 598-599.

[92] Timerbaev A R, Pawlak K, Aleksenko S S, Foteeva L S, Matczuk M, Jarosz M. Advances of CE-ICP-MS in speciation analysis related to metalloproteomics of anticancer drugs. Talanta, 2012, 102: 164-170.

[93] Bertrand M, Weber G, Schoefs B. Metal determination and quantification in biological material using particle-induced X-ray emission. TrAC-Trends in Analytical Chemistry, 2003, 22: 254-262.

[94] 高愈希, 陈春英, 柴之芳. 先进核分析技术在金属蛋白质组学研究中的应用. 核化学与放射化学, 2008, 30: 1-16.

[95] Chen C, Zhang P, Chai Z. Distribution of some rare earth elements and their binding species with proteins in human liver studied by instrumental neutron activation analysis combined with biochemical techniques. Analytica Chimica Acta, 2001, 439: 19-27.

[96] Behne D, Hammel C, Pfeifer H, Rothlein D, Gessner H, Kyriakopoulos A. Speciation of selenium in the mammalian organism. Analyst, 1998, 123: 871-873.

[97] Garman E F, Grime G W. Elemental analysis of proteins by microPIXE. Progress In Biophysics & Molecular Biology, 2005, 89: 173-205.

[98] Inagaki K, Mikuriya N, Morita S, Haraguchi H, Nakahara Y, Hattori M, Kinosita T, Saito H. Speciation of protein-binding zinc and copper in human blood serum by chelating resin pre-treatment and inductively coupled plasma mass spectrometry. Analyst, 2000, 125: 197-203.

[99] Manley S A, Byrns S, Lyon A, Brown P, Gailer J. Simultaneous Cu-, Fe-, and Zn-specific detection of metalloproteins contained in rabbit plasma by size-exclusion chromatography-inductively coupled plasma atomic emission spectroscopy. Journal of Biological Inorganic Chemistry, 2009, 14: 61-74.

[100] Encinar J R, Ouerdane L, Buchmann W, Tortajada J, Lobinski R, Szpunar J. Identification of water-soluble selenium-containing proteins in selenized yeast by size-exclusion-reversed-phase HPLC/ICPMS followed by MALDI-TOF and electrospray Q-TOF mass spectrometry. Analytical Chemistry, 2003, 75: 3765-3774.

[101] Sanz-Medel A, Montes-Bayon M, Saanchez M L F. Trace element speciation by ICP-MS in large biomolecules and its potential for proteomics. Analytical and Bioanalytical Chemistry, 2003, 377: 236-247.

[102] Pozebon D, Scheffler G L, Dressler V L, Nunes M A G. Review of the applications of laser ablation inductively coupled plasma mass spectrometry (LA-ICP-MS) to the analysis of biological samples. Journal of Analytical Atomic Spectrometry, 2014, 29: 2204-2228.

[103] Becker J S, Lobinski R, Becker J S. Metal imaging in non-denaturating 2D electrophoresis gels by laser ablation inductively coupled plasma mass spectrometry (LA-ICP-MS) for the detection of metalloproteins. Metallomics, 2009, 1: 312-316.

[104] Becker J S, Zoriy M, Wu B, Matusch A, Becker J S. Imaging of essential and toxic elements in biological tissues by LA-ICP-MS. Journal of Analytical Atomic Spectrometry, 2008, 23: 1275-1280.

[105] Politis A, Schmidt C. Structural characterisation of medically relevant protein assemblies by integrating mass spectrometry with computational modeling. Journal of Proteomics, 2018, 175: 34-41.

[106] Gomez-Ariza J L, Garcia-Barrera T, Lorenzo F, Bernal V, Villegas M J, Oliveira V. Use of mass spectrometry techniques for the characterization of metal bound to proteins (metallomics) in biological systems. Analytica Chimica Acta, 2004, 524: 15-22.

[107] McDonald W H, Ohi R, Miyamoto D T, Mitchison T J, Yates J R. Comparison of three directly coupled HPLC MS/MS strategies for identification of proteins from complex mixtures: single-dimension LC-MS/MS, 2-phase MudPIT, and

3-phase MudPIT. International Journal of Mass Spectrometry, 2002, 219: 245-251.

[108] Rigueira L M B, Lana D A P D, dos Santos D M, Pimenta A M, Augusti R, Costa L M. Identification of metal-binding to proteins in seed samples using RF-HPLC-UV, GFAAS and MALDI-TOF-MS. Food Chemistry, 2016, 211: 910-915.

[109] Mjos K D, Orvig C. Metallodrugs in medicinal inorganic chemistry. Chemical Reviews, 2014, 114: 4540-4563.

[110] Franz K J, Metzler-Nolte N. Introduction: Metals in medicine. Chemical Reviews, 2019, 119: 727-729.

[111] Barry N P, Sadler P J. Exploration of the medical periodic table: towards new targets. Chemical Communications, 2013, 49: 5106-5131.

[112] Sun X, Tsang C N, Sun H. Identification and characterization of metallodrug binding proteins by (metallo)proteomics. Metallomics, 2009, 1: 25-31.

[113] De Almeida A, Oliveira B L, Correia J D G, Soveral G, Casini A. Emerging protein targets for metal-based pharmaceutical agents: An update. Coordination Chemistry Reviews, 2013, 257: 2689-2704.

[114] Li H, Wang R, Sun H. Systems approaches for unveiling the mechanism of action of bismuth drugs: New medicinal applications beyond *Helicobacter pylori* infection. Accounts of Chemical Research, 2019, 52: 216-227.

[115] Rabilloud T, Lescuyer P. Proteomics in mechanistic toxicology: History, concepts, achievements, caveats, and. potential. Proteomics, 2015, 15: 1051-1074.

[116] Graves P R, Haystead T A J. Molecular biologist's guide to proteomics. Microbiology and Molecular Biology Reviews, 2002, 66: 39-63.

[117] Wang Y, Li H, Sun H. Metalloproteomics for unveiling the mechanism of action of metallodrugs. Inorganic Chemistry, 2019, 58: 13673-13685.

[118] Wang Y, Wang H, Li H, Sun H. Metallomic and metalloproteomic strategies in elucidating the molecular mechanisms of metallodrugs. Dalton Transactions, 2015, 44: 437-447.

[119] Wang H, Zhou Y, Xu X, Li H, Sun H. Metalloproteomics in conjunction with other omics for uncovering the mechanism of action of metallodrugs: Mechanism-driven new therapy development. Current Opinion in Chemical Biology, 2020, 55: 171-179.

[120] Han B, Zhang Z, Xie Y, Hu X, Wang H, Xia W, Wang Y, Li H, Wang Y, Sun H. Multi-omics and temporal dynamics profiling reveal disruption of central metabolism in *Helicobacter pylori* on bismuth treatment. Chemical Science, 2018, 9: 7488-7497.

[121] Hu L, Cheng T, He B, Li L, Wang Y, Lai Y T, Jiang G, Sun H. Identification of metal-associated proteins in cells by using continuous-flow gel electrophoresis and inductively coupled plasma mass spectrometry. Angewandte Chemie-International Edition, 2013, 52: 4916-4920.

[122] Yan H, Wang N, Weinfeld M, Cullen W R, Le X C. Identification of arsenic-binding proteins in human cells by affinity chromatography and mass spectrometry. Analytical Chemistry, 2009, 81: 4144-4152.

[123] Wang H, Yan A, Liu Z, Yang X, Xu Z, Wang Y, Wang R, Koohi-Moghadam M, Hu L, Xia W, Tang H, Wang Y, Li H, Sun H. Deciphering molecular mechanism of silver by integrated omic approaches enables enhancing its antimicrobial efficacy in *E. coli*. PLoS Biology, 2019, 17: e3000292.

[124] Wang H, Wang M, Yang X, Xu X, Hao Q, Yan A, Hu M, Lobinski R, Li H, Sun H. Antimicrobial silver targets glyceraldehyde-3-phosphate dehydrogenase in glycolysis of *E. coli*. Chemical Science, 2019, 10: 7193-7199.

[125] Yan H, Wang N, Weinfeld M, Cullen W R, Le X C. Identification of arsenic-binding proteins in human cells by affinity chromatography and mass spectrometry. Analytical Chemistry, 2009, 81: 4144-4152.

[126] Yan X, Li J, Liu Q, Peng H, Popowich A, Wang Z, Li X F, Le X C. *p*-Azidophenylarsenoxide: An arsenical "bait" for the in situ capture and identification of cellular arsenic-binding proteins. Angewandte Chemie-International Edition,

2016, 55: 14051-14056.

[127] Hu D, Liu Y, Lai Y T, Tong K C, Fung Y M, Lok C N, Che C M. Anticancer gold(Ⅲ) porphyrins target mitochondrial chaperone Hsp60. Angewandte Chemie-International Edition, 2016, 55: 1387-1391.

[128] Hu D, Yang C, Lok C N, Xing F, Lee P Y, Fung Y M E, Jiang H, Che C M. An antitumor bis(N-heterocyclic carbene) platinum(Ⅱ) complex that engages asparagine synthetase as an anticancer target. Angewandte Chemie-International Edition, 2019, 58: 10914-10918.

[129] Meier S M, Kreutz D, Winter L, Klose M H M, Cseh K, Weiss T, Bileck A, Alte B, Mader J C, Jana S, Chatterjee A, Bhattacharyya A, Hejl M, Jakupec M A, Heffeter P, Berger W, Hartinger C G, Keppler B K, Wiche G, Gerner C. An organoruthenium anticancer agent shows unexpected target selectivity for plectin. Angewandte Chemie-International Edition, 2017, 56: 8267-8271.

[130] Klose M H M, Theiner S, Kornauth C, Meier-Menches S M, Heffeter P, Berger W, Koellensperger G, Keppler B K. Bioimaging of isosteric osmium and ruthenium anticancer agents by LA-ICP-MS. Metallomics, 2018, 10: 388-396.

[131] Koohi-Moghadam M, Wang H, Wang Y, Yang X, Li H, Wang J, Sun H. Predicting disease-associated mutation of metal binding sites in proteins using a deep learning approach. Nature Machine Intelligence, 2019, 1: 561-567.

第 17 章 金属组学：高通量分析技术

（李玉锋[①]，陈春英[②]*）

▶ 17.1　金属组学简介 / 492

▶ 17.2　金属组学高通量分析方法 / 494

▶ 17.3　总结与展望 / 499

[①]中国科学院高能物理研究所，北京；[②]国家纳米科学中心，北京。*通讯作者联系方式：chenchy@nanoctr.cn

本章导读

- 金属组学是综合研究体系内全部元素的分布、含量、化学种态及功能的学科。
- 金属组学已形成了纳米金属组学、环境金属组学、医学金属组学等分支方向。
- XRF、PIXE、EDX、NAA、ICP-OES/MS 及 LIBS 等可实现高通量定量分析。
- XRF、PIXE、EDX、SIMS、LA-ICP-MS 及 LIBS 等可实现高通量分布研究。
- 形态及结构分析通常为低通量技术，而 HTXAS 等高通量方法正在逐渐成熟。

我们所处的物质世界，是由不同元素组成的。碳、氢、氧、氮及磷五种元素是构成糖类、脂类、蛋白质和核酸等生命物质的基础元素，占生物体质量的 95% 以上。其他元素（特别是金属及类金属元素）的含量虽然低，却也是生命存在的重要物质基础，它们作为活性中心或结构中心参与构成了体内 30% 左右的蛋白质和 50%～70% 的酶。比如，DNA 及 RNA 的合成离不开 DNA 和 RNA 聚合酶（含锌酶）的催化；在体内扮演不同角色的过氧化氢酶、固氮酶、铁传递蛋白及细胞色素等蛋白质（酶）的活性中心均含有数量不同的铁原子等[1]。而一些非生命必需金属元素如汞、镉及铅等可引起严重的环境健康问题，同样值得大力关注。

17.1　金属组学简介

17.1.1　金属组学的定义

鉴于金属及类金属元素在生命体系中的重要性，2002 年，日本学者 Haraguchi 提出了金属组学（metallomics）这一学科方向，即综合研究生命体内，特别是一种细胞内，自由或络合的全部金属原子的分布、含量、化学种态及其功能。与之相对应，细胞、器官或组织中金属及类金属蛋白质、酶或其他含有金属的生物分子及游离金属及类金属离子的集合则称之为"金属组（metallome）"[2,3]，即生物体系中的所有含金属物种。

2010 年，国际理论与应用化学联合会（IUPAC）发布了对金属组和金属组学的定义[4]：金属组是指生物体系中所有的金属或类金属的种态和/或总量，即①金属组可以通过测定能够代表生物体系（或其组成部分）或特定部位的生物样品来确定；②金属组可以通过如下具有不同程度近似的方法来表征，比如体系内所有元素的浓度、所有与特定配体（蛋白质或代谢物）结合的金属络合物的总量或特定元素的所有物种（比如铜金属组）等；③金属组只能通过确定热力学平衡常数描述。

金属组学研究生物体系中金属组，包括研究金属离子或其他金属种态与基因、蛋白质、代谢物质及其他生物分子的相互作用和功能关系。①金属组学重点关注生物体系中的金属元素（例如，铜、锌、铁、锰、钼、镍、钙、镉、铅、汞、铀）或类金属元素（例如，砷、硒、锑）等；②金属组学旨在阐明一系列元素含量或形态与基因组的联系。这种联系可以是统计性的（某一特定基因存在可导致某一元素含量的富集）、结构性的（某一金属蛋白质的序列与某一基因相关）或功能性的（某一生物配体的存在是由于某一基因编码造成的）；③金属组学是一种综合性研究方法。

17.1.2 金属组学的研究内容

金属组学的研究内容包括但不局限于如下内容[2,5,6]：①生物体液、细胞、器官中金属组的含量与分布分析；②生物体系中元素的形态分析；③金属组的结构分析；④应用模型络合物阐释金属组反应机理；⑤金属蛋白质和金属酶的鉴定；⑥生物分子和金属的代谢研究；⑦基于多元素分析的健康和疾病的医学诊断；⑧化学治疗中的无机药物的设计；⑨地球上生命体系的化学演变；⑩医学、环境科学、食品科学、农业、毒理学、生物地球化学中的金属辅助的生物学等。

17.1.3 金属组学的发展

金属组学的提出得到了学术界的广泛关注。*Journal of Analytical Atomic Spectrometry* 于 2004 年和 2007 年分别发行了两期"金属组学"专刊[7]。2009 年 *Metallomics* 杂志正式发行，这是金属组学研究领域的第一种金属组学专业期刊，为金属组学的发展提供了一个良好的平台。2007 年，第一届国际金属组学会议于日本举办，来自世界各地的约 400 名专家学者与会。之后每隔一到两年，国际金属组学会议分别于美国、德国、西班牙、中国、德国及波兰举办，其中中国科学院高能物理研究所和清华大学联合举办了第五届国际金属组学会议（http://metallomics.antpedia.com）。第八届金属组学国际会议于 2022 年在日本举办。国际上第一个金属组学研究中心——美洲金属组学研究中心于美国西西那提大学成立。2018 年，伦敦金属组学平台（London Metallomics Facility，https://www.clustermarket.com/london-metallomics-facility）于伦敦国王学院成立。

我国学者充分关注了金属组学这一学科方向。江桂斌[8]、金钦汉[9]等于 2005 年对金属组学的概念进行了介绍并对金属组学的研究方法特别是金属组的分离与检测方法进行了总结。李玉锋、陈春英等[10]于 2010 年在金属组学研究基础上提出了纳米金属组学（nanometallomics），旨在系统研究金属相关纳米材料在体内的吸收、转化、代谢、排泄及毒性以及它们与基因、蛋白质等生物分子的相互作用[11]。2013 年，由香港大学孙红哲组织的"中国金属组学研究进展"专辑发表于 *Metallomics* 杂志[12]。该专辑共收录了 21 篇我国科学家关于金属组学研究的文章，内容涵盖了金属组学在生物学、生物化学、环境科学、分子生物学、毒理学及医学中应用等。胡立刚、江桂斌等[13,14]于 2016 年提出了环境金属组学（environmetallomics），旨在系统研究金属污染物在生物体内和所处环境中的含量、分布、形态、传输和转化等[15]。此外，金属组学的其他分支方向，如医学金属组学（medimetallomics）[16-19]、疾病金属组学（pathometallomics）[20-22]、地学金属组学（geometallomics）[23]、农业金属组学（agrometallomics）[24-26]、放射金属组学（radiometallomics）[27]、考古金属组学（archaeometallomics）[28]、计量金属组学（metrometallomics）[29]、材料金属组学[30]及稀土金属组学（REEtallomics）[31-33]等已经或正在陆续提出。

在学术研讨会方面，2008 年，金属组学与金属蛋白质组学研讨会于北京举行，来自法国 Pau 大学、丹麦国家食品研究所及中国香港大学、北京大学医学部、复旦大学、厦门大学、中国科学院生态环境研究中心、国家纳米科学中心以及中国科学院高能物理研究所等多家单位的科研人员报告了各自在金属组学领域的研究进展。2019 年 1 月，2019 金属组学研讨会于北京召开。2020 年，北京金属组学平台（Beijing Metallomics Facility，https://indico.ihep.ac.cn/event/11261/）正式成立，该平台联合了北京地区 13 家高等院校和科研单位，旨在充分发挥北京地区金属组学研究相关机构优势，加强合作与交流，推动我国金属组学研究的发展。2020 年 12 月，2020 金属组学研讨会暨北京金属组学平台年会及原子光谱沙龙年会于北京召开。

17.2 金属组学高通量分析方法

"工欲善其事，必先利其器"。金属组学研究离不开高通量分析方法的支持，包括高通量元素定量技术、高通量元素分布技术及高通量形态及结构分析技术等[34,35]。本章系统总结了金属组学研究中的各种分析技术，如基于 X 射线、质子流、电子流、中子流、等离子体、γ 射线及激光的原子光谱技术等，并对这些方法的特点和局限性作简单说明。

17.2.1 高通量元素定量技术

高通量元素定量技术包括基于 X 射线、质子流、电子流、中子流、等离子体以及激光的多元素分析技术等[36]。

X 射线荧光分析（X-ray fluorescence analysis，XRF）技术是一种典型的多元素分析技术。当 X 射线与原子发生碰撞时，逐出一个内层电子使整个原子体系处于激发态，激发态原子寿命约为 10^{-12}～10^{-14} s，如果较外层的电子跃入内层空穴所释放的能量以辐射形式放出，便产生特征 X 射线荧光，而荧光强度与相应元素的含量有一定的关系，据此可以进行元素定量分析[37]。XRF 中的 X 射线若采用同步辐射光，则可实现绝对检测限 10^{-12}～10^{-15} g，相对检测限为 μg/g，甚至可达 ng/g 水平[38,39]。若采用高能质子流作为激发源而进行荧光分析，则称之为质子激发 X 射线荧光（proton induced X-ray emission，PIXE）分析，PIXE 具有比 XRF 更高的灵敏度[40]。利用高速电子流作为激发源而进行荧光分析，通常称为能量色散型 X 射线光谱（energy dispersive X-ray spectrometry，EDX）分析，该技术通常与电子显微镜联用，可实现局部微小区域的多元素定性和定量分析，但其检测限较差，在 g/kg 水平[41]。

中子活化分析（neutron activation analysis，NAA）是另一种重要的多元素定量分析技术，它可以同时测定样品中多达 30 种元素的含量。它的绝对检测限可低至 10^{-6}～10^{-13} g[42]。中子活化分析的基本原理是，样品中的稳定核素（^{A}Z）在入射中子流的作用下发生中子捕获反应，生成放射性核素（^{A+1}Z），^{A+1}Z 具有一定的半衰期，它发生衰变时会发射 β 粒子和 γ 射线。利用高分辨的 γ 射线谱仪检测样品受中子照射后产生的 γ 射线的强度即可实现对样品中元素定性或定量分析。中子活化分析技术的一个最大优点是基质干扰非常少，这是因为无论是探针（中子流）还是分析信号（γ 射线）都与样品中的其他元素不发生作用。另外，由于不涉及样品消化或溶解，可最大限度地避免试剂或实验室的污染[42]。

基于电感耦合等离子体技术的仪器如电感耦合等离子体发射光谱（ICP-OES）及电感耦合等离子体质谱（ICP-MS）可进行高效、高灵敏度的多元素定量分析。相对而言，ICP-OES 测定灵敏度较低、选择性也较差，而 ICP-MS 的检测限可达 pg/g 级，比 ICP-OES 高 2～3 个数量级[43]。对 ICP-MS 而言，液体和气体样品可以直接进样，对固体样品也可通过激光烧蚀（laser ablation，LA）技术进行直接测试[44]。由于 LA-ICP-MS 具有数据采集快、样品前处理少及灵敏度高等优点，目前已广泛应用于复杂的生物样品、环境样品及地质样品中元素测定。

激光诱导击穿光谱（laser-induced breakdown spectroscopy，LIBS）是近几十年随着激光技术以及光谱仪器的发展而兴起的痕量元素探测手段。用聚焦后的一束高能激光照射到样品材料上，靶表面焦点处的温度迅速上升（一般为几十到几百个纳秒），使得作用区内的样品材料开始熔化，并高速向外喷射，喷射过程中熔融粒子通过吸收激光能量和碰撞过程进一步被分解、激发和电离，从而形成等离子体。对等离子体进行分析,可以实现对样品的元素成分定性及其含量进行半定量和定量分析[45]。由于超短脉冲激光聚焦后能量密度较高,可以将任何物态(固态、液态、气态)的样品激发形成等离子体,因此理论上 LIBS 技术可以分析任何物态的样品,仅受激光的功率以及摄谱仪、检测器的灵敏度和波长范围的限制。此外,几乎所有的元素被激发形成等离子体后都会发出特征谱线,因此,LIBS 可以分析大多数的元素,检测限可低至 μg/g 量级[46]。

XRF、PIXE、EDX、NAA、ICP-OES、ICP-MS 及 LIBS 都是优异的多元素分析技术,XRF、PIXE、EDX、NAA 与 LIBS 可实现固体样品的原位、非/微破坏性分析,但需要注意的是,中子活化处理后的样品仍可能存在放射性,因此需要专业人员进行操作[47]。

在获得样品中不同元素的含量后,可以对其进行大数据处理,从而有助于开发疾病诊断或治疗的新方法[20]。例如,乳腺癌是女性人群中发病率仅次于肺癌的癌症,研究乳腺癌组织、癌旁组织及正常组织的多元素水平发现,与正常组织相比,肿瘤组织内 Ca、Cu 及 Zn 水平明显降低,而通过多变量判别分析可实现对癌症组织的正确识别率达 87%[48]。对男性前列腺癌的研究发现,癌症组织 S、K、Ca、Fe、Zn 及 Rb 的含量明显降低[49]。多元素分析同样可以用于环境污染物的溯源。例如,对机场附近的大气颗粒物进行元素分析并溯源,发现 Zn、Pb 主要来自燃料燃烧,Cu 主要来自飞机尾气排放而 Cl 主要来自公路融雪剂中的盐分[50]。此外,人们发现鱼类、海洋哺乳动物、海鸟及人类体内硒与汞含量具有正相关,而这种相关性的发现有助于人们进一步研究硒与汞的协同或拮抗作用[51]。

17.2.2 高通量元素分布技术

除了元素含量外,元素的空间分布研究同样非常重要,其有助于了解元素在生物体内转移与分布等。多元素分布的技术包括:①基于质谱技术的 LA-ICP-MS 与二次离子质谱(SIMS)等;②基于荧光分析的 SR-XRF、PIXE 及 EDX 等;③基于激光技术的 LIBS 等。虽然放射性同位素和稳定同位素标记技术也可以研究元素在体内的转移与分布,但该技术通常局限于研究特定元素,因此未归为高通量元素分布技术[52,53]。

质谱技术最大的优势在于检测限低,LA-ICP-MS 的检测限为亚 mg/kg 级,分辨率可达到 μm 量级[41]。目前 LA-ICP-MS 已广泛应用于地学[54,55]、叶子、蛇尾、年轮、鼠脑、人脑切片及牙齿等生物样品中的元素分布研究[56,57]。SIMS 是用质谱法分析由几千电子伏能量的一次离子打到样品靶上溅射产生的正、负二次离子的方法。其主要优点是可以分析包括氢、氦在内的几乎全部元素(不易离子化的稀有气体元素除外)及其同位素,检测限为 ng/kg 水平,空间分辨率可达 nm 量级[58,59]。SIMS 是一种局部破坏性技术,不适合一些活的生物样品的检测[60],但也有人利用 SIMS 研究了大豆根系、葡萄籽、动物组织及细胞等样品中金属组的分布[61,62]。也正是由于这种局部破坏性特点,可以通过逐层扫描的方式实现微区面成分分析和深度剖析,两者结合即可完成三维微区分析。

SR-XRF 微区分析技术是通过聚焦系统,如 K-B 镜(Kirkpatrick-Baez mirror system)、Fresnel 波带片(Fresnel zone plate)的调节,将 X 射线光斑大小调节到微米甚至是纳米尺寸[63],这样就可以研究亚细胞甚至细胞器水平上元素分布特征,同时还能开展二维和三维元素分布研究[36,64-66]。例如,利用 200×200 nm^2 分辨率的 SRXRF,发现细胞内的铜主要分布于线粒体和高尔基体等细胞器中[67]。Carmona 等利用 K-B 镜将同步光源聚焦至 90 nm 后研究了多巴胺能细胞株中金属组的分布[68]。他们发现,多巴胺囊泡中的铁主要为颗粒状,分布于神经突及神经末梢。另一方面,他们发现,不同细胞的神经突均蓄积铜、锌甚至是铅元素。而利用纳米分辨率的 X 射线束流,甚至可以观察到 Pt 在染色体上的分布情况[69]。Gao 等[70]对暴露于纳米铜的线虫体内元素分布进行了研究,发现纳米铜暴露后,线虫体内铜、钾元素含量明显增加,而且铜在体内的分布也发生改变。He 等[71]研究了稀土长期暴露6个月后对大鼠脑中微量元素分布的影响，发现大鼠暴露稀土氯化镧和氯化镱后，稀土La和Yb有明显的蓄积，提示稀土可通过血脑屏障进入大脑，同时脑中Mg、Ca、Fe、Cu、Zn、Mn、Co等微量元素的分布也发生了变化。此外，稀土在血清、肝脏和骨头中均有蓄积，而且Mg、Ca、Fe、Cu、Zn、Mn、Co等元素的分布也发生了变化。对样品进行线扫描（一维）或面扫描（二维）可得到细胞、组织或器官切片中金属组分布的信息[72,73]，但这些信息均基于对样品表面扫描的结果。要获得样品的立体信息，就需要对其进行三维扫描。利用共聚焦系统可实现样品的三维扫描，其深度分辨率可达10～40 μm，而检测限可达亚mg/kg级[74]。此外，同步辐射微束荧光CT（SR-XRF computerized tomography）同样可实现样品的三维扫描，其空间分辨率为μm甚至更低[75-78]。三维SR-XRF技术已用于水草[79]、人体骨关节[80]及古陶瓷等文物[81,82]的研究。

利用微区扫描PIXE技术也可研究动植物组织及人血细胞、肿瘤细胞中多元素的分布[83-85]。同时，利用共聚焦技术，三维微区扫描PIXE也已研制成功，其空间分辨率可达4 μm[86,87]。

透射电镜（TEM）或扫描电镜（SEM）与X射线能谱联用技术（TEM/SEM-EDX）已用于测定拟南芥的根、茎，小鼠肝组织及锥体神经元中金属的空间分布，空间分辨率可达亚nm量级，特别是利用基于球差校正技术的电子显微镜[88-90]。

利用微区扫描LIBS技术同样可以研究元素的分布情况。陈启蒙等[91]以100 μm分辨率对天然铁矿石样品表面14 mm×11 mm范围内进行面扫描分析，得到其表面Ca、Al、Ti和Mn四种元素的分布情况。杨春等[92]利用LIBS技术对焊缝熔合区的元素成分分布进行了分析，在熔合区域清楚观察到了元素浓度梯度分布曲线。张建华等[93]已经实现了横向空间分辨率为0.9 μm的LIBS分析方法。

总之，LA-ICP-MS、SIMS、SR-XRF、PIXE、SEM/TEM-EDX及LIBS均为优异的多元素分布研究技术（表17-1），它们的检测限通常在mg/kg水平且空间分辨率可达nm水平。与商业化的LA-ICP-MS、SIMS、SEM/TEM-EDX及LIBS等仪器相比，依托于大型装置的SR-XRF及PIXE的机时有限，故其应用也受到了一定限制。

表17-1　几种高通量元素分布研究技术

方法	激发源	可探测元素	维数	空间分辨率	参考文献
LA-ICP-MS	激光	几乎所有元素	1D，2D	μm	[41]
SIMS	离子束	几乎所有元素	1～3D	10 nm	[59]
SR-XRF	X射线	氧以后的元素	1～3D	nm	[69]
PIXE	质子流	钠以后的元素	1～3D	4 μm	[86]
EM-EDX	电子流	C、N、O、F、Mn、Fe、Ni、Cu等	1D，2D	亚nm	[90]
LIBS	激光	几乎所有元素	1～3D	μm	[93]

17.2.3　高通量元素形态及结构分析技术

元素形态及结构分析通常涉及样品中元素形态的提取（酸、碱提取，酶解等）、分离（GC、LC、EC及GE等）、检测（ICP-MS及NAA等）、鉴定（MALDI-TOF-MS及ESI-MS等）及结构分析（PX、XAS、NS等）。上述各步骤需要高通量分析手段才能满足金属组学研究的高通量分析要求。前面已经提到ICP-MS、XRF和NAA等技术具有多元素同时检测能力，是高通量元素分析的有力手段，但这些技术均不能直接给出形态信息，因而形态分析中需要事先提取并分离各种形态。

元素形态分离通常基于：①色谱技术，如液-固萃取中的HPLC、固相萃取（solid-phase extraction，SPE）和固相微萃取（solid-phase microextraction，SPME），液-液萃取中的溶剂萃取，压力液相萃取

（pressurized liquid extraction，PLE）及液-气萃取中的气相色谱（GC）和吹扫技术等[94]；②电泳技术，如 1-D、2-D 凝胶电泳（GE），毛细管电泳（CE），毛细管区带电泳（CZE），毛细管凝胶电泳（CGE），胶束电动毛细管（MEKC）及毛细管阵列电泳（CAE）等[95]。

基于色谱技术的分离方法通常效率较低。比如，无论是 HPLC 还是 GC，都需要单次进样，而多个 SPE 柱的同时应用使得批量分离成为可能，此外柱切换技术也是一种提高分离效率的方法[96]。

基于电泳技术的分离方法已在金属蛋白质研究中获得应用。凝胶电泳技术（GE），尤其是 2D 凝胶电泳技术是分离蛋白质的有力方法，结合 ICP-MS 及 NAA 等技术可实现对金属蛋白分析。将 GE 和 XRF 结合也可用于金属蛋白分析[97]。高愈希等[98]探索了金属蛋白中的金属定量测定问题，该法测定对 Fe、Cu、Zn 检出限分别为 2.43 μg/g、1.12 μg/g 和 0.96 μg/g；测定蛋白条带内 Fe 和 Zn 的回收率分别为 90.4%和 115.7%。该联用技术可用于生物样品中微量元素的化学形态分析，同时给出蛋白质的微量元素组成和等电点等信息[99]。Zhao 等[100,101]利用该法对富硒酵母中含硒蛋白及暴露于汞的水稻根部的蛋白进行了定量分析，总计发现了 157 种含硒蛋白，所有含硒蛋白的总硒含量为 125.56 μg/g，而汞暴露的水稻根部含汞蛋白分子质量为 15～25 kDa。

Nielsen 等[102]利用 GE 与 LA-ICP-MS 联用技术，研究了人血清与钴（Co）结合的情况，发现 Co 主要与血清白蛋白、α_2-巨球蛋白、α_1-脂蛋白、β_1-脂蛋白、α_2-抗胰蛋白酶及结合珠蛋白等蛋白结合。Szökefalvi-Nagy 等[103-105]用 SDS-PAGE 处理了一种来源于紫色硫光合作用菌（*Thiocapsa roseopersicina*）的氢化酶（HiPIP，内含 4 个 Fe—S 原子簇），凝胶干燥后，用 PIXE 技术分析了条带内 Fe、Ni 的相对含量，所用蛋白含量少于 1 μg。但 GE 难以与 ICP-MS 或 NAA 等金属组检测器在线联用。利用电泳技术结合 SR-XRF、LA-ICPMS 或 PIXE 进行金属蛋白研究的最大难点是保证与蛋白结合的金属离子不会在样品处理过程中脱落或被其他杂质离子取代，因此需要采取非蛋白变性分离技术，避免凝胶中有其他杂质离子，还要保证染色液中没有杂质离子[106]。与之相比较，毛细管电泳（CE）可以很容易与其他检测器连接，从而实现自动化操作。目前 CE 与 ICP-MS 连接可用于分析检测金属卟啉、环境样品中锑的形态、维生素 B_1 等[15,107]。

经上述步骤分离后的元素形态，若是已知稳定的化合物，如甲基汞、有机硒、砷或锡化合物等，可通过与标准样品的保留时间比对而得以确认。例如，喻宏伟等[108]建立了在线分离分析生物样品中含硒化合物的联用方法，用反相离子对高效液相色谱分离含硒化合物，用电感耦合等离子体质谱（ICP-MS）在线分析硒含量，实现了对硒酸钠、亚硒酸钠、硒代胱氨酸、硒代胱胺、硒代蛋氨酸、硒代尿素和三甲基硒离子等 7 种生物体常见的含硒小分子的分析。考虑到生物样品的复杂性，在色谱图中经常可发现无法指认的色谱峰，这就需要应用分子质谱学技术，如 MALDI-TOF-MS 或 ESI-MS 来确认或确定其分子信息。MALDI-TOF-MS 通常难以与上述分离技术在线联用，而 ESI-MS 可直接与 HPLC 等分离技术在线连接。目前，HPLC-ICP-MS 与 ESI-TOF-MS 联用已成功鉴定了富硒酵母中的多种含硒化合物[109]。

除形态分析外，通常还需要对所得感兴趣的物质进行结构分析。单晶 X 射线衍射谱（protein X-ray diffraction，PX）是研究晶体三维结构的强有力工具，其分辨率可达 0.15～2 nm，而中子激发技术，如单晶中子衍射（SCND）可为单晶 X 射线衍射谱提供有力的补充，它可以研究氢等轻元素原子[110]。由于需要获得生物大分子的单晶，X 射线或者中子衍射的实际应用受到一定的限制。小角 X 射线散射（SAXS）及小角中子散射（SANS）均可以得到固态或液态生物样品的结构信息，不需要获得样品的单晶，其分辨率可低至 1 nm[111,112]，而中子全散射方法能覆盖从亚原子量级到几个纳米量级的尺度[113]。近年来，冷冻电镜技术的发展为研究物质结构提供了另一个强有力的工具，可实现生物分子"近原子级"的分辨率[114]。

核磁共振谱（NMR）同样可以给出样品三维结构信息，其分辨率与单晶 X 射线衍射相媲美，达到 0.15～2 nm[115]。NMR 最大的优势是它可在溶液中直接测定，无需晶体形式的样品。NMR 测定晶体结构时可获得一系列（10～50 个）能够符合晶胞结构的信息，而单晶 X 射线衍射获得的是一个或只有几个符合条件的结构。因此，对 NMR 获得的晶体结构常常需要进行再确认。

穆斯堡尔谱是基于晶格束缚中的原子核发射和吸收 γ 光子时，原子核无反冲发射和共振吸收 γ 射线的现象[116]，可用来研究正常和病理条件下含铁生物大分子、模型化合物以及药物中铁的电子结构，在病理过程和环境因素作用条件下含铁生物大分子的定性、定量变化，为从分子水平进一步理解疾病的发生发展过程提供信息[117-120]。利用 ^{57}Fe 穆斯堡尔谱与快速冷冻猝灭法联合使用研究生物化学反应过程中 Fe 局域结构的变化，可以捕捉反应的活性中间体并对其中铁的局域结构细节进行表征，从而研究非血红素二铁酶和 α-酮戊二酸双加氧酶的氧激活作用，以了解含铁酶的反应机理[121]。穆斯堡尔谱方法的优点是分辨率高，灵敏度高，抗干扰能力强，对试样无破坏，设备和测量简单，可同时提供金属活性位点的多种物理和化学信息，不要求对蛋白质样品进行特别的纯化和结晶处理。其局限性首先表现在只有有限数量的核素（如 ^{57}Fe、^{57}Co、^{119}Sn、^{153}Sm 及 ^{197}Au 等）具有穆斯堡尔效应，且许多还必须在低温以及外加磁场下进行，这使它在应用上受到限制[122]。

X 射线吸收谱（XAS），特别是 X 射线吸收精细结构光谱（EXAFS）无需晶体样品即可获得样品的局域结构信息，可以原位探测吸收原子的 2～3 个邻近配位壳层，获得目标元素的电子结构信息和化学结构信息，其空间分辨率可达 0.1～1 pm[123]，所以 XAS 已成为研究微观结构重要的手段之一。Harris 等[124]采用 EXAFS 技术研究了海鱼肌肉中汞的分布特征，发现汞主要以硫化甲基汞形式存在，含硫配基可能来自半胱氨酸或蛋白质。Li 等[125,126]利用 XAS 原位研究了长期汞暴露人群头发及血液中汞的形态，发现汞主要与硫结合，头发中汞为三配位，而血液中汞为四配位。

XAS 还可对由单晶 X 射线衍射谱所得到的结构尤其是局域结构进行精修，从而获得更精确的结构信息[127]。Corbet 等[128]用 XAS 方法测定了离体的 MoFe 蛋白和 ADP·AlF4（MoFe 蛋白和 Fe 蛋白的稳定复合体）中固氮酶 FeMo 辅因子的金属局部结构。二者中 Mo 的局域结构没有明显不同，高质量的 Mo K-和 L-边 XANES 数据为研究固氮酶循环中的其他中间状态提供了基础，用偏振单晶 XAS 方法还可研究 FeMo 辅因子的其他状态。研究表明该技术完全可以用来为晶体衍射技术尚无法探明的 FeMo 辅因子中心补充结构细节。还有一些其他研究组也在 PX 与 XAS 技术相互补充方面做了很好的工作[129,130]。为了实现实验设备的相互补充，已经出现了一些建立在同一束线上的 PX 和 XAS 组合设备，如 Daresbury 的 MPW10 束线[128]和 SSRL 的 BL9-3 束线等[131]。

以上所提到的结构分析方法通常为低通量方法。为实现高通量结构分析，Scott 等提出了一种高通量 XAS（HTXAS）方法[132]。他们用高通量技术克隆标记表达和纯化了源自 *Pyrococcus furiosus* 的大约 2200 个开读框架中的 25 个重要基因，收集到专门设计的 25 孔样品架中（样品孔直径 1.5 mm，以 5×5 方式排列于 1 英寸聚碳酸酯支架上），每孔装 0.2～1 mmol/L 蛋白质溶液 3 μL，用 1 mm×1 mm 的同步辐射光分析样品中金属组分布，然后根据元素分布情况用 30 元固体探测器对目标蛋白进行进一步种态分析（XANES）和结构分析（EXAFS）。利用这种方法，他们在这 25 种基因产物中检出了两个含 Ni 蛋白和一个含 Zn 蛋白。利用这一方法对 3879 个纯化蛋白进行分析，发现其中 9%的蛋白含有过渡金属原子，如 Zn、Cu、Ni 及 Mn 等。结合生物信息学方法，还可实现对一系列金属蛋白活性中心的分析[133]。HTXAS 技术仍存在几个瓶颈。例如，如何实现 XAS 微量样品自动进样，如何实现低含量、微量样品的测定与数据的自动化采集及如何对大量数据进行快速处理与分析等。Benfatto 等[134]提出了从 XANES 图谱提取结构信息的新方法，成功地用 MXAN 程序对不同的 XANES 谱进行了拟合，获得了满意的结果。Rehr 等[135]发展了用 Bayes-Turchin 法代替传统的最小二乘法拟合的新方法，有望成为 XAS 的自动分析工具。

17.3 总结与展望

在金属组学研究中，XRF、PIXE、EDX、NAA、ICP-OES、ICP-MS 及 LIBS 技术可实现多元素定量分析，而 SR-XRF、PIXE、SEM/TEM-EDX、SIMS、LA-ICP-MS 及 LIBS 可实现高通量元素分布研究。金属组的形态及结构分析通常为低通量方法，但目前也已有一些方法可实现高通量形态或结构分析方法，如 HTXAS 及相关数据处理技术。

生物信息学是一门数学、统计、计算机与生物医学交叉结合的新兴学科，它已广泛地渗透到医学的各个研究领域中，成为生物医学发展中不可缺少的重要工具[136]。随着各种高通量金属组学技术的发展，将获得大量金属组数据，而处理这些海量数据不可避免地应用到生物信息学技术。因此，生物信息学技术将极大促进金属组学的研究。此外，考虑到机器学习、人工智能技术在大数据处理中的重要作用，它们也必将成为金属组学研究的重要工具[137,138]。

近年来，人造纳米材料[139,140]、大气细/超细颗粒物[141,142]、微米/纳米塑料[143,144]等新兴污染物的环境健康效应也引起了人们极大的关注，考虑到环境污染物暴露势必对机体内微量元素动态平衡造成扰动，因此金属组学方法有望在这些新兴污染物的环境健康效应研究中发挥独特作用。与此同时，随着分析技术的进步，有望实现对研究对象中全元素进行高通量分析，从而金属组学可以进一步升级为元素组学（elementomics）[145]。

参 考 文 献

[1] Mertz W. The essential trace elements. Science, 1981, 213(4514): 1332-1338.

[2] Haraguchi H. Metallomics as integrated biometal science. Journal of Analytical Atomic Spectrometry, 2004, 19(1): 5-14.

[3] Haraguchi H, Matsuura H. Chemical speciation for metallomics, Proceedings of International Symposium on Bio-Trace Elements 2002 (BITRE 2002) Wako, 2003; Enomoto S, Seko Y. The Institute of Physical and Chemical Research (RIKEN): Wako, 2003: 3-8.

[4] Lobinski R, Becker J S, Haraguchi H, Sarkar B. Metallomics: Guidelines for terminology and critical evaluation of analytical chemistry approaches (IUPAC Technical Report). Pure and Applied Chemistry, 2010, 82(2): 490-504.

[5] Mounicou S, Szpunar J, Lobinski R. Metallomics: The concept and methodology. Chemical Society Reviews, 2009, 38(4): 1119-1138.

[6] 葛瑞光, 陈卓, 孙红哲. 金属组学: 研究生命体系中金属离子的前沿交叉学科. 中国科学: B 辑, 2009, 39(7): 590-606.

[7] Koppenaal D W, Hieftje G M. Metallomics: An interdisciplinary and evolving field. Journal of Analytical Atomic Spectrometry, 2007, 22(8): 855-855.

[8] 江桂斌, 何滨. 金属组学及其研究方法与前景. 中国科学基金, 2005, 19(3): 151-155.

[9] 金伟, 牟颖, 金钦汉. 金属组学、代谢组学及其它. 理化检验-化学分册, 2005, 41(4): 296-299.

[10] Li Y-F, Wang L, Zhang L, Chen C, Nuclear-based metallomics in metallic nanomaterials: Nanometallomics//Chen C, Chai Z, Gao Y. Nuclear Analytical Techniques for Metallomics and Metalloproteomics. Cambridge: RSC Publishing, 2010.

[11] Li Y-F, Gao Y, Chai Z, Chen C. Nanometallomics: An emerging field studying the biological effects of metal-related

nanomaterials. Metallomics, 2014, 6(2): 220-232.

[12] Sun H. Metallomics in China. Metallomics, 2013, 5(7): 782-783.

[13] 胡立刚, 何滨, 江桂斌, 环境金属组学//李玉锋, 孙红哲, 陈春英, 柴之芳.金属组学. 北京: 科学出版社, 2016.

[14] Hu L, He B, Wang Y, Jiang G, Sun H. Metallomics in environmental and health related research: Current status and perspectives. Chinese Science Bulletin, 2013, 58(2): 169-176.

[15] Chen B, Hu L, He B, Luan T, Jiang G. Environmetallomics: Systematically investigating metals in environmentally relevant media. Trends in Analytical Chemistry, 2020, 126: 115875.

[16] 闫赖赖, 吕超, 邱锦云, 刘雅琼, 王京宇. 探讨精神分裂症与血清中 10 种元素的关系. 中国生化药物杂志, 2015, 35(6): 55-58.

[17] 文华, 刘多见, 王栋芳, 刘雅琼, 闫赖赖, 曾静, 卢庆彬, 欧阳荔, 王京宇. 慢性轻度不可预见性应激大鼠血液组分中 17 种氧化应激相关元素的测定分析. 卫生研究, 2015, 44(5): 806-812.

[18] 王栋芳, 闫赖赖, 卢庆彬, 文华, 徐瑞平, 张立新, 张钰, 刘多见, 王京宇, 刘雅琼. 食管鳞癌患者血清 15 种非必需微量元素定量分析. 中华肿瘤防治杂志, 2015, 22(9): 649-654.

[19] Guo Z, Sadler P J. Metals in medicine. Angewandte Chemie International Edition, 1999, 38(11): 1512-1531.

[20] Callejón-Leblic B, Gómez-Ariza J L, Pereira-Vega A, García-Barrera T. Metal dyshomeostasis based biomarkers of lung cancer using human biofluids. Metallomics, 2018, 10(10): 1444-1451.

[21] Al-sandaqchi A T, Brignell C, Collingwood J F, Geraki K, Mirkes E M, Kong K, Castellanos M, May S T, Stevenson C W, Elsheikha H M. Metallome of cerebrovascular endothelial cells infected with *Toxoplasma gondii* using μ-XRF imaging and inductively coupled plasma mass spectrometry. Metallomics, 2018, 10(10): 1401-1414.

[22] Ma J, Yan L, Guo T, Yang S, Liu Y, Xie Q, Ni D, Wang J. Association between serum essential metal elements and the risk of schizophrenia in China. Scientific Reports, 2020, 10(1): 10875.

[23] Joannes-Boyau R, Adams J W, Austin C, Arora M, Moffat I, Herries A I R, Tonge M P, Benazzi S, Evans A R, Kullmer O, Wroe S, Dosseto A, Fiorenza L. Elemental signatures of *Australopithecus africanus* teeth reveal seasonal dietary stress. Nature, 2019, 572(7767): 112-115.

[24] Duan G, Liu W, Chen X, Hu Y, Zhu Y. Association of arsenic with nutrient elements in rice plants. Metallomics, 2013, 5(7): 784-792.

[25] Bai X, Li Y, Liang X, Li H, Zhao J, Li Y-F, Gao Y. Botanic metallomics of mercury and selenium: Current understanding of mercury-selenium antagonism in plant with the traditional and advanced technology. Bulletin of Environmental Contamination and Toxicology, 2019, 102(5): 628-634.

[26] Li X, Liu T, Chang C, Lei Y, Mao X. Analytical methodologies for agrometallomics: A critical review. Journal of Agricultural and Food Chemistry, 2021, 69(22): 6100-6118.

[27] Liang Y, Liu Y, Li H, Bai X, Yan X, Li Y-F, Zhao J, Gao Y. Advances of synchrotron radiation-based radiometallomics for the study of uranium. Atomic Spectroscopy, 2021, 42: DOI: 10.46770/AS.2021.105.

[28] Li L, Yan L, Sun H, Zhoua Y, Li Y-F, Feng X. Archaeometallomics as a tool for studying ancient ceramics. Atomic Spectroscopy, 2021, 42: DOI: 10.46770/AS.2021.107.

[29] Pan M, Zang Y, Zhou X, Lu Y, Xiong J, Li H, Feng L. Inductively coupled plasma mass spectrometry for metrometallomics: The study of quantitative metalloproteins. Atomic Spectroscopy, 2021, 42: DOI: 10.46770/AS.2021.104.

[30] Li Q, Cai Z, Fang Y, Wang Z. Matermetallomics: Concept and analytical methodology. Atomic Spectroscopy, 2021, 42: DOI: 10.46770/AS.2021.101.

[31] Ma Y, He X, Zhang P, Zhang Z, Guo Z, Tai R, Xu Z, Zhang L, Ding Y, Zhao Y, Chai Z. Phytotoxicity and biotransformation of La_2O_3 nanoparticles in a terrestrial plant cucumber (*Cucumis sativus*). Nanotoxicology, 2011, 5(4): 743-753.

[32] Zhang Z, Zhao Y, Chai Z. Applications of radiotracer techniques for the pharmacology and toxicology studies of nanomaterials. Chinese Science Bulletin, 2009, 54(2): 173-182.

[33] Ma Y, Xie C, He X, Zhang B, Yang J, Sun M, Luo W, Feng S, Zhang J, Wang G, Zhang Z. Effects of ceria nanoparticles and $CeCl_3$ on plant growth, biological and physiological parameters, and nutritional value of soil grown common bean (*Phaseolus vulgaris*). Small, 2020, 16(21): 1907435.

[34] 李玉锋, 高愈希, 陈春英, 李柏, 赵宇亮, 柴之芳. 金属组学: 高通量分析技术进展与展望. 中国科学 B 辑: 化学, 2009, 39(7): 580-589.

[35] 赵甲亭, 李玉锋, 高愈希, 分析方法学//李玉锋, 孙红哲, 陈春英, 柴之芳. 金属组学. 北京: 科学出版社, 2016.

[36] Wang L, Yan L, Liu J, Chen C, Zhao Y. Quantification of nanomaterial/nanomedicine trafficking *in vivo*. Analytical Chemistry, 2018, 90(1): 589-614.

[37] Potts P J, Ellis A T, Kregsamer P, Streli C, West M, Wobrauschek P. X-ray fluorescence spectrometry. Journal of Analytical Atomic Spectrometry, 1999, 14(11): 1773-1799.

[38] Van Langevelde F, Vis R D. Trace element determinations using a 15-keV synchrotron X-ray microprobe. Analytical Chemistry, 1991, 63(20): 2253-2259.

[39] 吉昂. X 射线荧光光谱三十年. 岩矿测试, 2012, 31(3): 383-398.

[40] Johansson E. PIXE: A novel technique for elemental analysis. Endeavour, 1989, 13(2): 48-53.

[41] Motelica-Heino M, Le Coustumer P, Thomassin J H, Gauthier A, Donard O F X. Macro and microchemistry of trace metals in vitrified domestic wastes by laser ablation ICP-MS and scanning electron microprobe X-ray energy dispersive spectroscopy. Talanta, 1998, 46(3): 407-422.

[42] 中国科学院高能物理研究所. 中子活化分析在环境学、生物学和地学中的应用. 北京: 原子能出版社, 1992.

[43] Thompson M, Walsh J N. Handbook of Inductively Coupled Plasma Spectrometry. Glasgow: Blackie, 1983.

[44] Durrant S F, Ward N I. Laser ablation-inductively coupled plasma-mass spectrometry (LA-ICP-MS) for the multielemental analysis of biological materials: A feasibility study. Food Chemistry, 1994, 49(3): 317-323.

[45] 段忆翔, 林庆宇. 激光诱导击穿光谱分析技术及其应用. 北京: 科学出版社, 2016.

[46] 任佳, 高勋. 双脉冲激光诱导击穿光谱技术及其应用. 应用物理, 2019, 9(2): 79-86.

[47] Chen C, Chai Z, Gao Y. Nuclear Analytical Techniques for Metallomics and Metalloproteomics. Cambridge: RSC Publishing, 2010.

[48] Piacenti da Silva M, Zucchi O L A D, Ribeiro-Silva A, Poletti M E. Discriminant analysis of trace elements in normal, benign and malignant breast tissues measured by total reflection X-ray fluorescence. Spectrochimica Acta B, 2009, 64(6): 587-592.

[49] Leitão R G, Palumbo A, Souza P A V R, Pereira G R, Canellas C G L, Anjos M J, Nasciutti L E, Lopes R T. Elemental concentration analysis in prostate tissues using total reflection X-ray fluorescence. Radiation Physics and Chemistry, 2014, 95: 62-64.

[50] Groma V, Osan J, Alsecz A, Torok S, Meirer F, Streli C, Wobrauschek P, Falkenberg G. Trace element analysis of airport related aerosols using SR-TXRF. Quarterly Journal of the Hungarian Meteorological Service, 2008, 112(2): 83-97.

[51] Koeman J H, Peeters W H M, Koudstaal-Hol C H M, Tjloe P S, deGoeij J J M. Mercury-selenium correlations in marine mammals. Nature, 1973, 245(5425): 385-386.

[52] Zhu M T, Feng W Y, Wang Y, Wang B, Wang M, Ouyang H, Zhao Y L, Chai Z F. Particokinetics and extrapulmonary translocation of intratracheally instilled ferric oxide nanoparticles in rats and the potential health risk assessment. Toxicological Sciences, 2009, 107(2): 342-351.

[53] Lee K J, Nallathamby P D, Browning L M, Osgood C J, Xu X-H N. *In vivo* imaging of transport and biocompatibility of

single silver nanoparticles in early development of zebrafish embryos. ACS Nano, 2007, 1(2): 133-143.

[54] Meurer W P, Claeson D T. Evolution of crystallizing interstitial liquid in an arc-related cumulate determined by LA-ICP-MS-mapping of a large amphibole oikocryst. Journal of Petrology, 2002, 43(4): 607-629.

[55] Devos W, Senn-Luder M, Moor C, Salter C. Laser ablation inductively coupled plasma mass spectrometry (LA-ICP-MS) for spatially resolved trace analysis of early-medieval archaeological iron finds. Fresenius' Journal of Analytical Chemistry, 2000, 366(8): 873-880.

[56] Becker J S, Zoriy M, Przybylski M, Becker J S. High resolution mass spectrometric brain proteomics by MALDI-FTICR-MS combined with determination of P, S, Cu, Zn and Fe by LA-ICP-MS. International Journal of Mass Spectrometry, 2007, 261(1): 68-73.

[57] Kang D, Amarasiriwardena D, Goodman A. Application of laser ablation-inductively coupled plasma-mass spectrometry (LA-ICP-MS) to investigate trace metal spatial distributions in human tooth enamel and dentine growth layers and pulp. Analytical and Bioanalytical Chemistry, 2004, 378(6): 1608-1615.

[58] Briggs B Y D, Brown A, Vickerman J C. Handbook of static secondary ion mass spectrometry (SIMS). Analytical Chemistry, 1988, 60: 1791-1799.

[59] 王涛, 葛祥坤, 范光, 郭冬发. FIB-TOF-SIMS 联用技术在矿物学研究中的应用. 铀矿地质, 2019, (4): 247-252.

[60] 孙立民. 飞行时间二次离子质谱在生物材料和生命科学中的应用(下). 质谱学报, 2014, 35(5): 385-396.

[61] Lazof D B, Goldsmith J G, Rufty T W, Linton R W. Rapid uptake of aluminum into cells of intact soybean root tips (a microanalytical study using secondary ion mass spectrometry). Plant Physiology, 1994, 106(3): 1107-1114.

[62] Chandra S, Morrison G H. Sample preparation of animal tissues and cell cultures for secondary ion mass spectrometry (SIMS) microscopy. Biology of the Cell, 1992, 74(1): 31-42.

[63] Yuan Q-X, Deng B, Guan Y, Zhang K, Liu Y-J. Novel developments and applications of nanoscale synchrotron radiation microscopy. Physics, 2019, 48(4): 205-218.

[64] Schroer C, Kurapova O, Patommel J, Boye P, Feldkamp J, Lengeler B, Burghammer M, Riekel C, Vincze L, Van der Hart A. Hard X-ray nanoprobe based on refractive X-ray lenses. Applied Physical Letters, 2005, 87: 124103.

[65] Yan H, Bouet N, Zhou J, Huang X, Nazaretski E, Xu W, Cocco A P, Chiu W K S, Brinkman K S, Chu Y S. Multimodal hard X-ray imaging with resolution approaching 10 nm for studies in material science. Nano Futures, 2018, 2(1): 011001.

[66] Li Y, Hu W, Zhao J, Chen Q, Wang W, Li B, Li Y-F. Selenium decreases methylmercury and increases nutritional elements in rice growing in mercury-contaminated farmland. Ecotoxicology and Environmental Safety, 2019, 182: 109447.

[67] Yang L, McRae R, Henary M M, Patel R, Lai B, Vogt S, Fahrni C J. Imaging of the intracellular topography of copper with a fluorescent sensor and by synchrotron X-ray fluorescence microscopy. Proceedings of the National Academy of Sciences of the United States of America, 2005, 102(32): 11179-11184.

[68] Carmona A, Cloetens P, Devès G, Bohic S, Ortega R. Nano-imaging of trace metals by synchrotron X-ray fluorescence into dopaminergic single cells and neurite-like processes. Journal of Analytical Atomic Spectrometry, 2008, 23: 1083-1088.

[69] Yan H, Nazaretski E, Lauer K, Huang X, Wagner U, Rau C, Yusuf M, Robinson I, Kalbfleisch S, Li L, Bouet N, Zhou J, Conley R, Chu Y S. Multimodality hard-X-ray imaging of a chromosome with nanoscale spatial resolution. Scientific Reports, 2016, 6: 20112.

[70] Gao Y, Liu N, Chen C, Luo Y, Li Y-F, Zhang Z, Zhao Y, Zhao Y, Iida A, Chai Z. Mapping technique for biodistribution of elements in a model organism, *Caenorhabditis elegans*, after exposure to copper nanoparticles with microbeam synchrotron radiation X-ray fluorescence. Journal of Analytical Atomic Spectrometry, 2008, 23: 1121-1124.

[71] He X, Feng L, Xiao H, Li Z, Liu N, Zhao Y, Zhang Z, Chai Z, Huang Y. Unambiguous effects of lanthanum? Toxicological Letters, 2007, 170(1): 94-96.

[72] Cui L, Zhao J, Chen J, Zhang W, Gao Y, Li B, Li Y-F. Translocation and transformation of selenium in hyperaccumulator plant *Cardamine enshiensis* from Enshi, Hubei, China. Plant and Soil, 2018, 425(1-2): 577-588.

[73] Bai X, Li Y, Liang X, Li H, Zhao J, Li Y-F, Gao Y. Botanic metallomics of mercury and selenium: Current understanding of mercury-selenium antagonism in plant with the traditional and advanced technology. Bulletin of Environmental Contamination and Toxicology, 2019, 102(5): 628-634.

[74] Vincze L, Vekemans B, Brenker F E, Falkenberg G, Rickers K, Somogyi A, Kersten M, Adams F. Three-dimensional trace element analysis by confocal X-ray microfluorescence imaging. Analytical Chemistry, 2004, 76(22): 6786-6791.

[75] Hansel C M, La Force M J, Fendorf S, Sutton S. Spatial and temporal association of As and Fe species on aquatic plant roots. Environmental Science & Technology, 2002, 36(9): 1988-1994.

[76] Yang G-L, Yang M-X, Lv S-M, Tan A-J. The effect of chelating agents on iron plaques and arsenic accumulation in duckweed (*Lemna minor*). Journal of Hazardous Materials, 2021, 419: 126410.

[77] Bleuet P, Gergaud P, Lemelle L, Bleuet P, Tucoulou R, Cloetens P, Susini J, Delette G, Simionovici A. 3D chemical imaging based on a third-generation synchrotron source. Trends in Analytical Chemistry, 2010, 29(6): 518-527.

[78] Laforce B, Masschaele B, Boone M N, Schaubroeck D, Dierick M, Vekemans B, Walgraeve C, Janssen C, Cnudde V, Van Hoorebeke L, Vincze L. Integrated three-dimensional microanalysis combining X-ray microtomography and X-ray fluorescence methodologies. Analytical Chemistry, 2017, 89(19): 10617-10624.

[79] Kanngießer B, Malzer W, Pagels M, Lühl L, Weseloh G. Three-dimensional micro-XRF under cryogenic conditions: A pilot experiment for spatially resolved trace analysis in biological specimens. Analytical and Bioanalytical Chemistry, 2007, 389(4): 1171-1176.

[80] Zoeger N, Streli C, Wobrauschek P, Jokubonis C, Pepponi G, Roschger P, Hofstaetter J, Berzlanovich A, Wegrzynek D, Chinea-Cano E, Markowicz A, Simon R, Falkenberg G. Determination of the elemental distribution in human joint bones by SR micro XRF. X-Ray Spectrometry, 2008, 37(1): 3-11.

[81] Yi L, Qin M, Wang K, Lin X, Peng S, Sun T, Liu Z. The three-dimensional elemental distribution based on the surface topography by confocal 3D-XRF analysis. Applied Physics A, 2016, 122(9): 856.

[82] Mantouvalou I, Wolff T, Hahn O, Rabin I, Lühl L, Pagels M, Malzer W, Kanngiesser B. 3D Micro-XRF for cultural heritage objects: New analysis strategies for the investigation of the Dead Sea scrolls. Analytical Chemistry, 2011, 83(16): 6308-6315.

[83] Tylko G, Mesjasz-Przybylowicz J, Przybylowicz W J. In-vacuum micro-PIXE analysis of biological specimens in frozen-hydrated state. Nuclear Instruments & Methods in Physics Research Section B, 2007, 260(1): 141-148.

[84] Kramer U, Grime G W, Smith J A C, Hawes C R, Baker A J M. Micro-PIXE as a technique for studying nickel localization in leaves of the hyperaccumulator plant *Alyssum lesbiacum*. Nuclear Instruments & Methods in Physics Research Section B, 1997, 130(1): 346-350.

[85] Sie S H, Thresher R E. Micro-PIXE analysis of fish otoliths: Methodology and evaluation of first results for stock discrimination. International Journal of PIXE, 1992, 2(3): 357-379.

[86] Ishii K, Matsuyama S, Watanabe Y, Kawamura Y, Yamaguchi T, Oyama R, Momose G, Ishizaki A, Yamazaki H, Kikuchi Y. 3D-imaging using micro-PIXE. Nuclear Instruments & Methods in Physics Research Section A, 2007, 571(1-2): 64-68.

[87] Ohkura S, Ishii K, Matsuyama S, Terakawa A, Kikuchi Y, Kawamura Y, Catella G, Hashimoto Y, Fujikawa M, Hamada N, Fujiki K, Hatori E, Yamazaki H. *In vivo* 3D imaging of *Drosophila melanogaster* using PIXE-micron-CT. X-Ray

Spectrometry, 2011, 40(3): 191-193.

[88] Kametani K, Nagata T. Quantitative elemental analysis on aluminum accumulation by HVTEM-EDX in liver tissues of mice orally administered with aluminum chloride. Medical Molecular Morphology, 2006, 39(2): 97-105.

[89] Jiang J, Sato S. Detection of calcium and aluminum in pyramidal neurons in the gerbil hippocampal CA1 region following repeated brief cerebral ischemia: X-ray microanalysis. Medical Electron Microscopy, 1999, 32(3): 161-166.

[90] 陈陆, 刘军, 王勇, 张泽. α-Cu_2Se 精细结构的球差校正扫描透射电镜表征. 物理化学学报, 2019, 35(2): 16-21.

[91] 陈启蒙, 杜敏, 郝中骐, 邹孝恒, 李嘉铭, 易荣兴, 李祥友, 陆永枫, 曾晓雁. 激光诱导击穿光谱技术成分面扫描分析与应用. 光谱学与光谱分析, 2016, 36(5): 1473-1477.

[92] 杨春, 张勇, 贾云海, 王海舟. 激光诱导击穿光谱对焊缝熔合区的元素成分分布分析. 光谱学与光谱分析, 2014, (4): 1089-1094.

[93] 张建华. 具有纳米尺度空间分辨的 LA-LIBS 元素显微分析. 广州: 华南理工大学, 2017.

[94] 江桂斌. 环境样品前处理技术. 北京: 化学工业出版社, 2004.

[95] Righetti P G. Capillary Electrophoresis in Analytical Biotechnology. Florida: CRC Press, 1996.

[96] 顾云. 柱切换 HPLC 技术在生物样品测定中的应用. 天津药学, 2003, 3(15): 80-82.

[97] Stone S F, Bernasconi G, Haselberger N, Makarewicz M, Ogris R, Wobrauschek P, Zeisler R, Detection and Determination of Selenoproteins by Nuclear Techniques//Kučera J, Obrusník I, Sabbioni E. Nuclear Analytical Methods in the Life Sciences 1994. Humana Press, 1994, 299-307.

[98] 高愈希, 丰伟悦, 李柏, 章佩群, 何伟, 黄宇营, 柴之芳. 同步辐射 X 荧光法测定电泳分离后蛋白条带内的锌. 核技术, 2004, 27(3): 165-168.

[99] 董元兴, 高愈希, 陈春英, 李柏, 邢丽, 喻宏伟, 何伟, 黄宇营, 柴之芳. 用同步辐射 X-荧光定量测定电泳分离后蛋白条带内的微量元素. 分析化学, 2006, 34(4): 443-446.

[100] Zhao J, Pu Y, Gao Y, Peng X, Li Y, Xu X, Li B, Zhu N, Dong J, Wu G, Li Y-F. Identification and quantification of seleno-proteins by 2-DE-SR-XRF in selenium-enriched yeasts. Journal of Analytical Atomic Spectrometry, 2015, 30: 1408-1413.

[101] Li Y, Zhao J, Li Y-F, Xu X, Zhang B, Liu Y, Cui L, Li B, Gao Y, Chai Z. Comparative metalloproteomic approaches for the investigation proteins involved in the toxicity of inorganic and organic forms of mercury in rice (*Oryza sativa* L.) roots. Metallomics, 2016, 8(7): 663-671.

[102] Neilsen J L, Abildtrup A, Christensen J, Watson P, Cox A, McLeod C W. Laser ablation inductively coupled plasma-mass spectrometry in combination with gel electrophoresis: A new strategy for speciation of metal binding serum proteins. Spectrochimica Acta B, 1998, 53(2): 339-345.

[103] Szökefalvi-Nagy Z, Demeter I, Bagyinka C, Kovács K L. PIXE analysis of proteins separated by polyacrylamide gel electrophoresis. Nuclear Instruments & Methods in Physics Research Section B, 1987, 22(1-3): 156-158.

[104] Bagyinka C, Szökefalvi-Nagy Z, Demeter I, Kovacs K L. Metal composition analysis of hydrogenase from *Thiocapsa roseopersicina* by proton induced X-ray emission spectroscopy. Biochemical and Biophysical Research Communications, 1989, 162(1): 422-426.

[105] Szökefalvi-Nagy Z, Bagyinka C, Demeter I, Kovács K, Le Quynh H. Location and quantification of metal ions in enzymes combining polyacrylamide gel electrophoresis and particle-induced X-ray emission. Biological Trace Element Research, 1990, 26-27(1): 93-101.

[106] Wittig I, Braun H-P, Schagger H. Blue native PAGE. Nature Protocols, 2006, 1(1): 418-428.

[107] 叶美英, 殷学锋. 毛细管电泳和电感耦合等离子体质谱接口技术进展. 光谱学与光谱分析, 2003, 23(1): 89-93.

[108] 喻宏伟, 陈春英, 高愈希, 李柏, 柴之芳. 高效液相色谱-电感耦合等离子体质谱法分析生物样品中硒的化学形

态. 分析化学, 2006, 34(6): 749-753.

[109] Infante H G, O'Connor G, Rayman M, Hearn R, Cook K. Simultaneous identification of selenium-containing glutathione species in selenised yeast by on-line HPLC with ICP-MS and electrospray ionisation quadrupole time of flight (QTOF)-MS/MS. Journal of Analytical Atomic Spectrometry, 2006, 21(11): 1256-1263.

[110] Hazemann I, Dauvergne M T, Blakeley M P, Meilleur F, Haertlein M, Van Dorsselaer A, Mitschler A, Myles D A, Podjarny A. High-resolution neutron protein crystallography with radically small crystal volumes: Application of perdeuteration to human aldose reductase. Acta Crystallographica Section D, 2005, 61(Pt 10): 1413-1417.

[111] Feigin L A, Svergun D I. Structure analysis by small-angle X-ray and neutron scattering. New York: Plenum Press 1987.

[112] 孙小沛, 祝万钱, 徐中民, 秦宏亮. 上海光源时间分辨超小角散射线多层膜单色器的设计. 核技术, 2019, (11): 5-10.

[113] 左太森. 微小角中子散射谱仪设计和中子全散射方法学研究. 东莞: 中国科学院大学, 2017.

[114] 邢崇阳. 厌氧氨氧化细菌内纳米舱的结构与功能研究. 重庆: 中国科学院大学, 2020.

[115] 宋启泽. 核磁共振原理及应用. 北京: 兵器工业出版社, 1992.

[116] Clémancey M, Blondin G, Latour J-M, Garcia-Serres R, Mössbauer Spectroscopy//Fontecilla-Camps J C, Nicolet Y. Metalloproteins: Methods and Protocols. Totowa, NJ: Humana Press, 2014: 153-170.

[117] Heinnickel M, Agalarov R, Svensen N, Krebs C, Golbeck J H. Identification of FX in the heliobacterial reaction center as a [4Fe-4S] cluster with an S=3/2 ground spin state. Biochemistry, 2006, 45(21): 6756-6764.

[118] Oshtrakh M I. Study of the relationship of small variations of the molecular structure and the iron state in iron containing proteins by Mössbauer spectroscopy: Biomedical approach. Spectrochimica Acta Part A, 2004, 60(1): 217-234.

[119] Kamnev A A, Tugarova A V. Sample treatment in Mössbauer spectroscopy for protein-related analyses: Nondestructive possibilities to look inside metal-containing biosystems. Talanta, 2017, 174: 819-837.

[120] Oshtrakh M I, Semionkin V A. Mössbauer spectroscopy with a high velocity resolution: Advances in biomedical, pharmaceutical, cosmochemical and nanotechnological research. Spectrochimica Acta Part A, 2013, 100: 78-87.

[121] Krebs C, Price J C, Baldwin J, Saleh L, Green M T, Bollinger J M. Rapid freeze-quench ^{57}Fe Mössbauer spectroscopy: Monitoring changes of an iron-containing active site during a biochemical reaction. Inorganic Chemistry, 2005, 44(4): 742-757.

[122] Oshtrakh M I. Applications of Mössbauer spectroscopy in biomedical research. Cell Biochemistry and Biophysics, 2019, 77(1): 15-32.

[123] 王其武, 刘文汉. X 射线吸收精细结构及其应用. 北京: 科学出版社, 1994.

[124] Harris H H, Pickering I J, George G N. The chemical form of mercury in fish. Science, 2003, 301(5637): 1203-1203.

[125] Li Y-F, Chen C, Li B, Li W, Qu L, Dong Z, Nomura M, Gao Y, Zhao J, Hu W, Zhao Y, Chai Z. Mercury in human hair and blood samples from people living in Wanshan mercury mine area, Guizhou, China: An XAS study. Journal of Inorganic Biochemistry, 2008, 102(3): 500-506.

[126] 李玉锋, 陈春英, 邢丽, 刘涛, 谢亚宁, 高愈希, 李柏, 瞿丽雅, 柴之芳. 贵州万山汞矿地区人发中汞的含量及其赋存状态的 XAFS 原位研究. 核技术, 2004, 27(12): 899-903.

[127] Cheung K C, Strange R W, Hasnain S S. 3D EXAFS refinement of the Cu site of structural change at the metal centre in an oxidation-reduction process: An integrated approach combining EXAFS and crystallography. Acta Crystallography, 2000, D56: 697-704.

[128] Corbett M C, Tezcan F A, Einsle O, Walton M Y, Rees D C, Latimer M J, Hedman B, Hodgson K O. Mo K- and L-edge

X-ray absorption spectroscopic study of the ADP. AlF4-stabilized nitrogenase complex: Comparison with MoFe protein in solution and single crystal. Journal of Synchrotron Radiation, 2005, 12(1): 28-34.

[129] Hasnain S S, Hodgson K O. Structure of metal centres in proteins at subatomic resolution. Journal of Synchrotron Radiation, 1999, 6(4): 852-864.

[130] Pohl E, Haller J C, Mijovilovich A, Meyer-Klaucke W, Garman E, Vasil M L. Architecture of a protein central to iron homeostasis: Crystal structure and spectroscopic analysis of the ferric uptake regulator. Molecular Microbiology, 2003, 47(4): 903-915.

[131] Latimer M J, Ito K, McPhillips S E, Hedman B. Integrated instrumentation for combined polarized single-crystal XAS and diffraction data acquisition for biological applications. Journal of Synchrotron Radiation, 2005, 12(1): 23-27.

[132] Scott R A, Shokes J E, Cosper N J, Jenney F E, Adams M W W. Bottlenecks and roadblocks in high-throughput XAS for structural genomics. Journal of Synchrotron Radiation, 2005, 12(1): 19-22.

[133] Shi W, Punta M, Bohon J, Sauder J M, D'Mello R, Sullivan M, Toomey J, Abel D, Lippi M, Passerini A, Frasconi P, Burley S K, Rost B, Chance M R. Characterization of metalloproteins by high-throughput X-ray absorption spectroscopy. Genome Research, 2011, 21(6): 898-907.

[134] Benfatto M, Della Longa S, Natoli C R. The MXAN procedure: a new method for analysing the XANES spectra of metalloproteins to obtain structural quantitative information. Journal of Synchrotron Radiation, 2003, 10(1): 51-57.

[135] Rehr J J, Kozdon J, Kas J, Krappe H J, Rossner H H. Bayes-Turchin approach to XAS analysis. Journal of Synchrotron Radiation, 2005, 12(1): 70-74.

[136] Lesk A M. Introduction to Bioinformatics. New York: Oxford University Press, 2002.

[137] Wang C, Steiner U, Sepe A. Synchrotron big data science. Small, 2018, 14(46): 1802291.

[138] 师芸, 马东晖, 吕杰, 李杰, 史经俭. 基于流形光谱降维和深度学习的高光谱影像分类. 农业工程学报, 2020, 36(6): 151-160.

[139] Chen C, Li Y-F, Qu Y, Chai Z, Zhao Y. Advanced nuclear analytical and related techniques for the growing challenges in nanotoxicology. Chemical Society Reviews, 2013, 42(21): 8266-8303.

[140] Wang J, Chen C, Liu Y, Jiao F, Li W, Lao F, Li Y, Li B, Ge C, Zhou G, Gao Y, Zhao Y, Chai Z. Potential neurological lesion after nasal instillation of TiO_2 nanoparticles in the anatase and rutile crystal phases. Toxicological Letters, 2008, 183(1-3): 72-80.

[141] Ding J, Guo J, Wang L, Chen Y, Hu B, Li Y, Huang R, Cao J, Zhao Y, Geiser M, Miao Q, Liu Y, Chen C. Cellular responses to exposure to outdoor air from the Chinese Spring Festival at the air-liquid interface. Environmental Science & Technology, 2019, 53(15): 9128-9138.

[142] Mei M, Song H, Chen L, Hu B, Bai R, Xu D, Liu Y, Zhao Y, Chen C. Early-life exposure to three size-fractionated ultrafine and fine atmospheric particulates in Beijing exacerbates asthma development in mature mice. Partical and Fibre Toxicology, 2018, 15(1): 13.

[143] Li L, Luo Y, Li R, Zhou Q, Peijnenburg W J G M, Yin N, Yang J, Tu C, Zhang Y. Effective uptake of submicrometre plastics by crop plants via a crack-entry mode. Nature Sustainability, 2020, 3: 929-937.

[144] Mitrano D M, Beltzung A, Frehland S, Schmiedgruber M, Cingolani A, Schmidt F. Synthesis of metal-doped nanoplastics and their utility to investigate fate and behaviour in complex environmental systems. Nature Nanotechnology, 2019, 14(4): 362-368.

[145] Li Y-F, Chen C, Qu Y, Gao Y, Li B, Zhao Y, Chai Z. Metallomics, elementomics, and analytical techniques. Pure and Applied Chemistry, 2008, 80(12): 2577-2594.

第18章　砷的环境行为、代谢机理、健康效应

原子光谱及相关形态分析方法应用于砷的研究

（彭汉勇[①②]*，严晓文[③]*，景传勇[②]*，刘晴晴[④]，黄　科[⑤]，孟晓光[⑥]，
王秋泉[③]，赵方杰[⑤]，乐晓春[①]）

▶ 18.1　砷形态分析方法简介 / 509

▶ 18.2　砷形态代谢研究：以洛克沙胂为例 / 511

▶ 18.3　砷的主要环境暴露 / 517

▶ 18.4　砷的健康效应研究：以砷与蛋白质的结合为例 / 522

▶ 18.5　总结与展望 / 529

[①]University of Alberta，加拿大；[②]中国科学院生态环境研究中心，中国；[③]厦门大学，中国；[④]西南大学，中国；[⑤]南京农业大学，中国；[⑥]Stevens Institute of Technology，美国。*通讯作者联系方式：hypeng@rcees.ac.cn（彭汉勇），xwyan@xmu.edu.cn（严晓文），cyjing@rcees.ac.cn（景传勇）

本章导读

- 砷污染及其人群暴露是世界范围内亟待解决的环境健康问题。
- 砷的毒性和环境行为与砷的化学形态密切相关。
- 原子光谱、质谱、色谱及联用技术是形态分析的重要手段。
- 利用形态分析方法鉴定从前未知的砷代谢产物、研究砷与蛋白质的相互作用、探讨砷形态的健康效应。
- 通过砷形态分析研究砷的分布、控制和减少人体暴露于有毒的含砷化合物，促进和保护公共健康。

砷（arsenic, As）是一种在自然界中广泛存在的微量元素，在地壳中的平均浓度约为 5 μg/g（ppm）。尽管平均丰度不高，但由于自然矿化作用，砷在世界某些地区浓度很高。当受到一系列自然过程和人为活动的影响时，砷会逐渐迁移释放到环境中。长期暴露于地下水中的砷被认为是世界上最大的环境健康灾难之一，全球超过 1 亿人处于受砷暴露引发的相关疾病风险中[1,2]。对阿根廷、智利、西孟加拉、印度、巴基斯坦以及我国内蒙古、台湾等地区的流行病学研究发现，长期饮用含高浓度砷的饮用水与肾癌、肺癌、皮肤癌等癌症的发病有显著相关性[3-5]。除了癌症，长期砷暴露还会引起心血管疾病、糖尿病等各种慢性疾病[6-8]。目前，砷已经被世界卫生组织的国际癌症研究机构（International Agency for Research on Cancer）以及美国环境保护署（U.S. Environmental Protection Agency）列为人类致癌物。2011 年发布的重要环境污染物清单把砷排在铅、汞和多氯联苯（PCB）等环境污染物之前，位列第一。

砷的毒性与其化学结构密切相关。无机砷（As^{III}、As^V）和有机胂（MMA^V、DMA^V）之间的毒性差异非常明显，并且三价有机胂（MMA^{III} 和 DMA^{III}）的毒性要高于其五价形态，甚至大于无机砷的毒性[9-13]。另外一些形态，如砷甜菜碱（AsB）、砷胆碱（AsC）和砷糖（arsenosugars）则被认为是无毒的[14-16]。Le 等[17]设计了急性早幼粒细胞白血病（APL）细胞和人白血病（HL-60）细胞的凋亡实验，考察了六种砷形态的毒性（见图 18-1）。这几种砷形态的毒性顺序为氧化苯胂（PAO^{III}）>MMA^{III}≥As^{III}>As^V>MMA^V>DMA^V，其中无机砷比有机胂（单甲基和双甲基胂酸）的毒性强，而三价的有机胂又比无机砷的毒性强，PAO^{III} 的毒性比 DMA^V 高出了四个数量级。由此可见，总砷的含量并不能直接反映砷对人体健康的影响，而砷的形态起着更重要的作用。因此，砷的形态分析能够更直接提供砷的毒性信息。

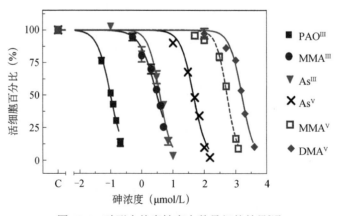

图 18-1　砷形态的毒性存在数量级的差异[17]

目前，生命体中砷形态代谢产物分析主要采用色谱与光谱/质谱联用技术，将高效的分离手段与高灵敏的检测技术相结合。色谱分离技术包括气相色谱（gas chromatography，GC）、毛细管电泳（capillary electrophoresis，CE）和高效液相色谱（high performance liquid chromatography，HPLC）等。其中，HPLC 对样品性质要求低，分离模式多样，与各种仪器联用的接口简单，具有分离效率高、耐受基体能力强、重现性好等优点，是目前与质谱联用最为广泛的分离技术。质谱检测方法（元素特异性检测 ICP-MS 和分子结构特异性检测 ESI-MS 等）在砷形态分析中具有显著的优势，ICP-MS 可以实现痕量/超痕量砷形态的定量分析，ESI-MS 可以实现已知/未知砷形态的定性分析。联用技术能够有效地实现各种手段间的优势互补，是砷形态分析的最有力手段，但是由于砷形态种类的多样性和生物样品的复杂性，使得生命体内砷形态代谢研究依然面临巨大的挑战。

18.1 砷形态分析方法简介

18.1.1 砷形态分析中的前处理技术

实际样品的砷形态往往处于复杂基质中，如水、植物、海产品、大米、血液、唾液、指甲、头发和藻类细胞等环境和生物样品，其赋存形态与检测仪器并不匹配。同时，砷形态的含量低、种类多且性质差异大，直接分析不同样品中砷形态存在很大的困难。因此，针对不同样品，发展砷形态的提取、分离和富集等前处理方法，是砷形态分析的基础。

甲醇/水或乙腈/水是一种较温和的提取砷形态的方法。Pizzaro 等[18]采用甲醇-水作为提取剂，提取鸡肉中砷形态效率达到了 70%~75%，其中包含约 15%的 AsB、50%的 DMA 和 34%的未知砷形态。Sanchez-Rodas 等[19]采用甲醇-水来提取冻干的鸡肉样品中砷形态，并辅以加热和超声等方式提高了提取效率，达到了 80%~100%。

蛋白酶提取的方式能够释放与器官组织中蛋白质结合的砷，对动植物组织中的砷形态提取更有效，例如胃蛋白酶、胰蛋白酶、链霉蛋白酶、脂肪酶等[20]。Sanz 等[21]用含 As^{III} 的饲料喂养鸡，70 天后采集鸡组织样品，采用甲醇/水和蛋白酶提取的方式对冻干的鸡肉样品中砷形态进行了提取，同时对比了适用于不同样品（大米、鱼肉、鸡肉和土壤）的提取方法，只有鸡肉样品需要采用酶提取的方式来提高提取效率。Le 等[22]发展了胃蛋白酶萃取方法（100 mg 胃蛋白酶溶于 5 mL 0.5% HCl 溶液），并用超声辅助进一步提高提取效率，结果表明胃蛋白酶酶解方法提取砷形态的效率为 90%±2%。

在众多砷形态的分离方法中，固相萃取技术（SPE）因具有简单、快速、选择性高、易于在线等诸多优点，得到了广泛应用[23,24]。Mondal 等[25]采用强碱性的阴离子交换材料 AG1 X8 树脂填充柱来选择性地区分无机砷形态。在样品溶液 pH=3 的条件下，As^{III} 以弱酸分子形式存在，而 As^{V} 依然保持着离子状态，因而能够被阴离子交换树脂固定相保留。97%的 As^{III} 能够通过柱子，而 As^{V} 能被完全保留。Ben Issa 等[26]将强碱性阴离子交换树脂与水合氧化铁固体颗粒混合作为固相萃取材料。在 pH≤8 时，阴离子交换树脂只保留 As^{V}，而对 As^{III} 没有保留。当 pH=5~11 时，水合氧化铁能同时保留 As^{V} 和 As^{III}。所制得的材料吸附容量大，强碱性阴离子交换树脂对 As^{V} 的吸附可达到 370 μg/g，而水合氧化铁固体颗粒混合树脂对 As^{III}、As^{V} 的吸附容量分别超过 4150 μg/g、3500 μg/g。Boyaci 等[27]制备了一种同时含有氨基和巯基的生物功能材料，充分利用了砷形态的电负性和砷硫结合能力。当 pH=1~9 时，As^{III} 与巯基结合，能达到定量吸附，而在 pH=3 时，通过与氨基作用实现了对 As^{V} 的定量吸附。

随着材料科学的发展，新的功能性纳米材料，包括纳米纤维、磁性纳米材料、金属氢氧化物沉淀以及纳米氧化钛等，在砷的形态分析中得到了更多的应用。Chen 等[28]开发了碳纳米纤维（CNFs）作为固相萃取材料，采用吡咯烷二硫代氨基甲酸铵（APDC）作为络合试剂。在样品溶液的 pH=1～3 时，As^{III} 能够被选择性地吸附在 CNF 微柱上，而 As^{V} 没有保留。聚丙烯酸酯（PHEMA）微球[29]和聚乙烯亚胺（BPEI）修饰的多壁碳纳米管（MWCNTs）[30]对 As^{V} 的吸附选择性更高，而在 pH 范围为 3～8 时，氨基改性的纳米磁性材料仅保留 As^{V}[31]。

一些新的萃取模式也被应用于砷形态分析，如毛细管微萃取（capillary microextraction，CME）、液相微萃取（liquid phase microextraction，LPME）和搅拌棒吸附萃取（stir bar sorptive extraction，SBSE）等。例如，Hu 等[32]在毛细管微萃取中，通过化学键合或者涂覆的方式对毛细管内壁进行改性，或者制成多孔的整体柱材料。有序介孔 Al_2O_3 整体柱和 N-(2-氨乙基)-3-氨丙基三甲氧基硅烷（AAPTS）改性的氨基整体柱都能够选择性吸附 As^{V}，而 3-巯丙基三甲氧基硅烷（MPTS）改性的巯基整体柱则可保留 As^{III}[33]。Pu 等[34]采用将中空纤维膜套在溶剂棒上，负载有机溶剂后通过搅拌棒萃取溶液中的 APDC-As^{III} 的络合物。Mao 等[35]使用 TiO_2 负载的中空纤维膜搅拌棒，有效萃取了九种砷形态，包括 As^{III}、As^{V}、MMA、DMA、4-氨基苯胂酸(4-APAA)、4-羟基苯胂酸(4-HPAA)、苯胂酸(PA)、4-硝基苯胂酸(4-NPAA)和 3-硝基-4-羟基苯胂酸(3-NHPAA，ROX，洛克沙胂)。

18.1.2 砷形态分析中的色谱分离技术

根据样品种类和砷形态性质的不同，众多色谱分离技术被用于分离样品中的砷形态，包括阴/阳离子交换色谱、反相色谱、离子对色谱、亲水相互作用色谱、尺寸排阻色谱等。为实现多种砷形态的高效分离，需要考虑不同砷形态性质、色谱柱材料和流动相成分、pH、盐浓度等因素的影响，往往还要结合多种分离模式。

由于多数砷形态的 pK_a 差异明显，阴/阳离子交换色谱分离技术能够高效分离多种砷形态，是应用非常广泛的色谱分离方法。Le 等[22]采用强阴离子交换柱（PRP X-110S），实现了对 AsB、As^{III}、DMA^{V}、MMA^{V}、4-AHPAA、As^{V}、3-氨基-4-羟基苯胂酸(3-AHPAA)、N-乙酰基-4-羟基苯胂酸(N-AHAA) 和 ROX 等九种砷形态的基线分离。为了与后续的 ICP-MS 和 ESI-MS 检测相匹配，采用了碳酸氢铵（60 mmol/L）水溶液和 5%甲醇作为流动相，并通过调节流动相 pH 至 8.75 和梯度洗脱方式，解决了阴离子交换柱上对两性离子 AsB 和无机砷 As^{III}（pK_{a1}=9.2）难以分离的难题。

尽管同一类型色谱柱的分离原理相似，但由于不同的商品化色谱柱采用不同功能基团修饰的固定相填料，造成其用于砷形态分析时的分离条件和分析性能会有所差异。例如，Jackson 等[36]评价了三根离子色谱柱（Dionex AS14、AS16 和 AS7）对鸡粪中砷形态的分析性能。六种砷形态 As^{III}、As^{V}、DMA、MMA、4-APAA 和 ROX 在三根色谱柱的检出限均低于 0.5 μg/L，但在 AS16 和 AS7 上的检出限更低（低于 0.05 μg/L）。Grant 等[37]使用 HPLC-ICP-MS 分析了鸡的组织样品中砷形态，评价了两种色谱柱（PRP-X100 和 Dionex AS7）对砷形态的分离效果。ROX 不能从 PRP-X100 洗脱下来，而 Dionex AS7 则可以分离 As^{III}、As^{V}、DMA、MMA 和 ROX。

选取合适的色谱分离技术，不仅要根据砷形态的性质，还要考虑到分析柱上流动相成分与后续分析仪器的匹配，包括流动相的有机相浓度和缓冲盐成分等。Falnoga 等[38]用含有 30 μg/g 无机砷 As^{III} 的饲料喂养鸡，然后采用 Hamilton PRP-X100 阴离子交换柱对砷形态分离后进行 HG-AFS 分析，流动相为 pH=6 的 15 mmol/L 磷酸钾溶液。尽管磷酸盐缓冲体系使用广泛，与生理环境更匹配，但是此类难挥发性缓冲盐并不适用于原子质谱检测。Polatajko 和 Szpunar 等[39]采用甲醇-水的方式对鸡肉样品进行萃取，提取液采用 Dionex AS7 柱分离后进入 ICP-MS 检测，所用的流动相为含有 0.5%甲醇的 0.5 mmol/L 和 50 mmol/L 硝酸溶液。

18.1.3 砷形态分析中的检测技术

砷的测定方法以光谱和质谱为主。原子吸收光谱法（AAS）、原子发射光谱法（ICP-OES）、原子荧光光谱法（AFS）、X射线荧光光谱法（XRF）、电感耦合等离子体质谱法（ICP-MS）等都已成为广泛使用的检测砷技术[40]。

随着砷形态研究的逐渐深入，样品和砷形态种类日趋复杂，多种仪器的联用成为当前砷形态分析的主要手段，将不同仪器的优势整合在一个分析平台，从多个角度获取样品中砷形态的信息，已成为砷形态分析的主角[41,42]（图18-2）。例如，液相色谱-氢化物发生-原子荧光光谱联用系统，将高效液相色谱的分离能力与原子荧光光谱价廉和灵敏结合了起来，使该技术兼顾成本与性能，广泛应用于砷形态的测定与分析中。原子质谱（比如ICP-MS）能高灵敏地提供砷元素形态的定量信息，但由于其离子化温度过高，往往需要辅以分子质谱来提供其结构信息。通过将原子质谱与分子质谱联用能够有效地实现两种手段间的优势互补，经HPLC将砷的不同形态分离纯化后，不管后续的检测器是原子质谱还是分子质谱，砷形态的保留时间均是一致的，结果相互印证。因此，原子/分子质谱联用技术平台所提供的定量与定性信息使得它在元素形态分析上有着广泛的应用，尤其是未知形态的鉴定。

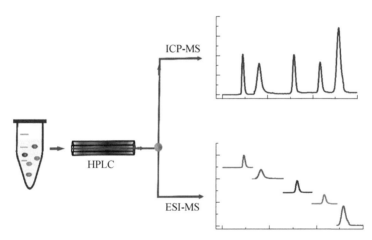

图18-2 HPLC分离后由ICP-MS和ESI-MS同时检测多种砷形态

除此之外，砷形态的现场分析是建立高效、简便、经济的砷污染现场监测预警体系所亟待解决的现实问题。近年来，表面增强拉曼散射（surface enhanced raman scattering，SERS）在砷形态分析检测领域得到越来越多的应用。例如，采用基于$Fe_3O_4@Ag$磁性卫星SERS基底，可实现地下水中三价砷与五价砷的选择性富集与检测[43]。将水样通过附载SERS基底的过滤装置，利用便携拉曼光谱仪采集信号，整个流程耗时1 min，检测性能可满足生活饮用水的水质标准要求。

18.2 砷形态代谢研究：以洛克沙胂为例

各种砷形态的毒性、生物利用性和迁移过程等都与其代谢过程和产物的化学结构密切相关。随着基于色谱分离和光谱/质谱检测技术的砷形态分析方法不断发展，生命体中砷代谢形态的真实面貌才逐渐被揭开。砷元素在生物体内的代谢是进一步了解砷形态毒性的基础，成为科研工作者密切关注的研究方向。

无机砷的逐步甲基化是其在生命体中的一条主要代谢途径[44]。五价砷酸被谷胱甘肽（GSH）还原成三价亚砷酸，在甲基化酶（As3MT）和甲基供体（SAM）的存在下，再进一步被甲基化。通过逐步还原和甲基化过程，无机砷代谢产生的甲基化产物包括单甲基胂酸（MMA^V）、单甲基亚胂酸（MMA^{III}）、二甲基胂酸（DMA^V）、二甲基亚胂酸（DMA^{III}）和三甲基胂氧化物（TMAO^V）等。在大部分哺乳动物中，DMA^{V}是主要的代谢产物[45]，但在老鼠和一些细菌中，DMA^{V}还能进一步被代谢为TMAO^{V}和三甲基亚胂（TMA^{III}）[45]。

不同有机胂形态之间的代谢行为差异较大，代谢产物复杂，对其代谢途径的认识尚有一定的局限性。例如，在海鱼和贝壳类中，砷含量较高，主要包括砷甜菜碱（AsB）和砷糖等形态[46,47]，其中AsB 化学稳定性好，能快速地从人体内排出[14,48,49]，不易在体内代谢[14-16]，但 Le 等[50,51]在食用贝壳和海藻后的人尿液中，发现了以 DMA^{V}为主的多种砷形态，表明砷糖能在人体内被代谢为其他的砷形态[52-54]。有机胂在生物体内的代谢途径尚不明确，仍然亟需更系统、深入的工作来促进我们对有机胂代谢途径的理解。

本节将系统性地研究一种有机胂在动物、细胞和微生物中的代谢途径。从 20 世纪 40 年代开始，洛克沙胂（ROX）作为一种有机胂饲料添加剂被广泛应用于家禽养殖业中[55,56]，通过禽类肉制品进入人体，同时含洛克沙胂的粪便会被用作农作物肥料。因此，一直以来人们对其在环境和公共健康中的影响存在担忧[57,58]，然而以往仅提供砷总量信息不足以评价其潜在危害[59,60]。近年来随着砷形态分析方法的发展，科研工作者从用洛克沙胂喂养鸡的鸡肝中检出了毒性更高的无机砷和一些其他砷形态[61]，包括 As^{III}、As^V、MMA、DMA、3-AHPAA 和一些未知砷形态[35,36,61-69]。Polatajko 和 Szpunar[39]报道了鸡肉中的砷形态含量分别为 106 μg/kg DMA、37 μg/kg 砷甜菜碱（AsB，来自于饲料中的海产品）以及约 15 μg/kg 未知砷形态。在 Pizzaro 等[18]的研究工作中发现了占砷形态总量为 34%的未知砷形态。由于这些未知砷形态尚未被鉴定，ROX 在生物体内的代谢转化始终是一个谜[65,68,70-72]，因此对于其在食品动物中使用始终存在安全性的担忧[73]。

18.2.1 洛克沙胂在鸡体内代谢研究

有机胂化合物，如洛克沙胂（3-硝基-4-羟基苯胂酸）、阿散酸（4-氨基苯胂酸）和硝苯坤酸（4-硝基苯胂酸），作为禽畜饲料添加剂被广泛应用于家禽养殖工业，主要用途是加快鸡的生长速度、抵抗肠道疾病、增加肉鸡重量以及改善鸡肉颜色等[74,75]。美国食品药品监督管理局（FDA）于 1964 年批准使用洛克沙胂作为猪和鸡的饲料添加剂[74]。我国农业部于 1996 年批准了洛克沙胂的使用，依据《兽药管理条例》，洛克沙胂的用量分别为每 1000 kg 饲料添加 50 g，休药期为 5 天，但在蛋鸡产蛋期和肉猪出栏期是禁用的。已有报道表明[61,65,70]，洛克沙胂能够在鸡体内被代谢成多种砷形态，包括毒性较高的无机砷形态。2011 年起，全球最大的洛克沙胂生产商 Pfizer subsidiary Alpharma（US）自动停止生产和销售洛克沙胂[76]。美国和中国相继于 2013 年、2019 年禁止在食品动物中使用洛克沙胂作为饲料添加剂，但是其在很多其他国家和地区仍然被广泛使用。

目前，已有文献对洛克沙胂进行了研究[36,61-64]，但其在鸡体内的代谢过程和代谢产物却依然不太清晰[35,66,67,77-79]。因此，Le 等[80]设计了一个包含 1600 只鸡的喂养实验，系统性地研究了洛克沙胂在鸡体内的代谢。在阿尔伯塔大学家禽养殖中心依照美国食品药品监督管理局规定喂养，整个喂养周期为 35 天。这 1600 只鸡中包含两种不同的商业品种，即 Ross 308（800 只）和 Cobb 500（800 只），每 100 只鸡装在一个笼子里喂养，分为控制组和实验组两组。在实验组中，用含洛克沙胂的饲料进行喂养，给药期为四周，然后停药一周；而在控制组，全程使用不含洛克沙胂的饲料进行喂养。在每一个采样日,分别从 16 个笼子里随机取样,并将鸡的各个器官组织样品分别采集后立刻保存于 −70℃,包括鸡胸肉、肝、肾、肠、皮、血、毛以及粪便等。

将高效液相色谱(HPLC)分离与电感耦合等离子体质谱(ICP-MS)和电喷雾质谱(ESI-MS)检测联用,在砷形态鉴定和定量分析上已经发挥了重要的作用。尽管 ICP-MS 具有灵敏度高、线性范围宽、可同位素分析等诸多优点[16,36,81-84],但是却不能提供分子结构信息。电喷雾质谱(ESI-MS)则能够提供分子量信息,包括三重四极杆质谱(ESI-MS/MS)和飞行时间质谱(ESI-TOF-MS),是鉴定砷形态的有力工具[85-88]。为了研究复杂基体中的砷形态,Le 等[22]建立了 HPLC 与 ICP-MS 和 ESI-MS 联用技术平台,同时定量和定性分析鸡组织样品中的砷形态,为后续工作中研究砷在鸡体内代谢过程及其代谢产物奠定了基础。

以鸡肝为例[80],经过酶萃取提取砷形态后,将建立的质谱联用技术平台用于分析实验组和控制组样品中的砷形态(图 18-3)。HPLC-ICP-MS 的结果显示,在控制组中仅有来自背景暴露的常见的砷形态,而在实验组中检测到 11 种不同的砷形态。通过砷标准品的加标实验,样品中的砷形态保留时间与 8 种所加入的标准品吻合良好,包括 AsB、As^{III}、DMA^V、MMA^V、As^V、3-AHPAA、N-AHPAA 和 ROX。同时,在 HPLC-ESI-MS/MS 的多反应监测(multiple reaction monitoring,MRM)模式中,通过对每种砷形态的两组特征子离子对进行监测,对所检测到的砷形态进一步验证。值得注意的是,仍然有三种未知砷形态不能与现有的砷标准品吻合。

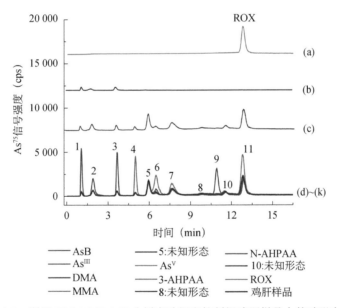

图 18-3 利用 HPLC-ICP-MS 分析实验组和控制组鸡肝样品中的砷形态[80]

(a)ROX 标准品;(b)控制组样品;(c)实验组样品;(d)~(k)分别加入砷标准品后的实验组样品

通过已建立的质谱联用技术平台,分析已有的砷标准品的特征信息,如 As^{III}、As^V、MMA、DMA、3-AHPAA、N-AHPAA 和 ROX。在 ESI-MS/MS 中,这些砷形态能够产生一些相同的含砷碎片离子,如 m/z 91(AsO)、m/z 107(AsO$_2$)、m/z 123(AsO$_3$)或者 m/z 121(AsO$_2$CH$_3$)。利用这些特征碎片离子,可在 ESI-MS/MS 中寻找能够产生这些特征离子的前驱母离子(Prec 模式),并对可能的前驱母离子逐一采集完整的碎片离子信息[子离子增强模式(EPI)]。通过建立信息依赖采集(information dependent acquirement,IDA)分析方法(Prec-IDA-EPI)能够快速扫描并采集所有可能的前驱母离子的分子碎片信息,利用高分辨率质谱 ESI-TOF-MS 采集每个碎片离子精确分子量,得到其所对应的化学组成,并据此逐一分析其可能的分子形态。

此外，含砷化合物的一些其他特征信息也可用于快速扫描和筛选可能的母离子。通过 ICP-MS 对砷元素进行特异性检测，能够提供未知砷形态的色谱保留时间，从而缩小分析范围。由于砷的原子量为 74.9216，高分辨率质谱分析含砷离子碎片会有质量亏损（mass defect），大多数含砷碎片离子的 m/z 都会低于整数值。几种苯胂酸化合物的子离子谱图中均产生了一个与母离子相差 m/z 18 的子离子，即从母离子上丢掉了一个 H_2O，因此，中性丢失（NL 模式）也可用于扫描可能的苯胂酸形态。对于已报道的分子，可将所得到的未知物化学组成在数据库中搜索，快速匹配可能的化合物，常用的数据库包括 ChemIndex、Chemfinder、ChemSpider、ISI 或 Merk Index 等。

经过 ESI-MS/MS（Prec-IDA-EPI）方法分析，通过对含 m/z 107 或 m/z 121 的前驱母离子进行扫描，在与 HPLC-ICP-MS 中一种未知砷形态相匹配的保留时间处，发现了能够同时多种特征含砷碎片离子的前驱母离子 m/z 260，包括含砷特征子离子 m/z 91、m/z 107 和 m/z 121，以及中性丢失离子 m/z 242。高分辨质谱 ESI-TOF-MS 分析（图 18-4），该未知砷形态的精确分子量为 m/z 259.9548，分子组成为 $C_7H_7O_5NAs^-$，与理论值的质量偏差仅有 0.1 ppm。通过对其碎片离子进行分析，确认 m/z 260 的未知砷形态为甲基化的洛克沙胂。

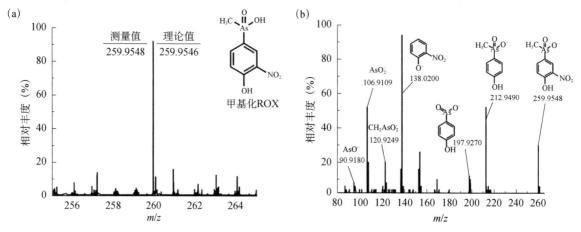

图 18-4　ESI-TOF-MS 鉴定鸡肝中甲基化的洛克沙胂

（a）新砷形态（图中的第 10 个色谱峰）的分子离子峰 m/z 259.9548（$C_7H_7NAsO_5^-$）；（b）m/z 259.9548 的子离子谱图及碎片离子

甲基化的洛克沙胂从未有过报道，也没有相关的标准品进行匹配验证。为了进一步确认其分子结构，通过化学合成的方法，制备了这种新发现的甲基化苯胂酸。通过质谱对合成的甲基化苯胂酸产物进行表征，所得到的离子谱图与样品中的新形态完全吻合。

将合成的甲基化苯胂酸加标到实际样品中，对比加标前后 HPLC-ICP-MS 分析砷形态的谱图，在该未知形态相同的保留时间处，加标后的砷形态信号强度明显增加。同时，对比加标前后 HPLC-ESI-MS/MS（MRM 模式）分析样品的检测结果，甲基化苯胂酸的两组子离子对 m/z 260 → m/z 107 和 m/z 260 → m/z 138 在实际样品中能被检测到，在加标后两组子离子对的信号强度均有明显增加，且峰强度的比值保持不变。以上结果确认了样品中的该未知砷形态就是甲基化的洛克沙胂。

采用同样的方法，鉴定出其他两种未知砷形态并进行了验证。三种未知砷形态均为甲基化的苯胂酸，其化学形态分别为甲基化 3-AHPAA、甲基化 N-AHPAA 和甲基化 ROX，与砷相连的氧被甲基取代，苯环上的取代基分别为氨基、乙酰基和硝基。

为了验证苯胂酸的甲基化是否需要砷甲基化酶的参与，设计了胞内甲基化模拟实验，在与胞内环境相似的溶液中，包含了砷甲基化酶（As3MT/wt）还原性试剂（GSH）和甲基供体（AdoMet）等。加入三价苯胂酸作为底物后，在不同的时间点分别取样，利用 HPLC-ICP-MS 和 ESI-MS 对产物进行检测。与起始溶液相比，经过 30 min 的反应，就已经产生了甲基化苯胂酸。随着时间的增加，甲基化苯胂酸的含量逐渐增加，6 个小时后即达到 36%。苯胂酸能够被甲基化，而这个甲基化过程需要砷甲基化酶的参与。

在鉴定出鸡体内 11 种洛克沙胂的代谢产物后，得到了洛克沙胂的完整代谢途径（图 18-5）。ROX 被还原成 3-AHPAA，然后进一步乙酰化生成 N-AHPAA，而这三种苯胂酸都能被还原成其三价形态，并在甲基化酶的催化下被甲基化，生成三种甲基化的苯胂酸。洛克沙胂在鸡肝中的代谢途径分为两个过程：第一个过程是氧化还原，由于鸡肝中存在大量含硫基的还原性化合物，如 GSH 等，五价的砷容易被还原成三价，而 3-AHPAA 也是洛克沙胂上的硝基被还原的产物；第二个过程是甲基化和乙酰化，从定量的结果上看，乙酰化的效率较低，而主要的代谢反应是甲基化。

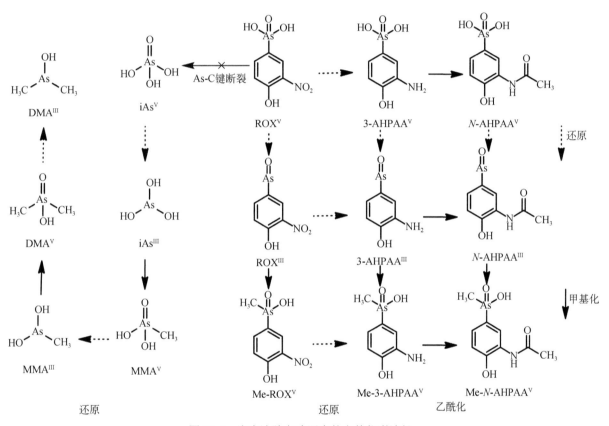

图 18-5 洛克沙胂在鸡肝中的完整代谢途径

ROX 被还原成 3-AHPAA[89]，然后进一步乙酰化生成 N-AHPAA[61]，而这三种苯胂酸都能被还原成其三价形态，并在甲基化酶的催化下被甲基化，生成三种甲基化的苯胂酸。砷-碳键的断裂产生的无机砷也会逐步被甲基化

在鸡体内苯胂酸的甲基化过程中，会产生三价苯胂酸中间体，毒性要明显高于其五价形态。三价洛克沙胂 ROX^{III} 对 T24 细胞的 IC_{50} 为 0.19 μmol/L，而其五价形态 ROX^{V} 的 IC_{50} 为 5.7 mmol/L，差异高达近 30 000 倍。而无机三价砷 As^{III} 对 T24 细胞的 IC_{50} 为 6.9 μmol/L，三价形态 ROX^{III} 的毒性比无机砷的毒性还要高出近 30 倍。尽管洛克沙胂本身的毒性并不高，但其代谢中间产物会产生如此巨大的毒性反差，革新了人们对于洛克沙胂作为家禽饲料添加剂安全性的认识。

18.2.2 洛克沙胂在细胞内代谢

洛克沙胂能否被人代谢以及如何代谢，成为研究洛克沙胂健康效应所亟需回答的问题。以人源细胞为模型来探究洛克沙胂的代谢行为，为研究洛克沙胂对人体健康的影响提供了直接的关联信息。研究者[90]考察了洛克沙胂在两种肝细胞中的代谢，包括人肝癌细胞（HepG2）和人原代肝细胞。将这两种肝细胞分别暴露于 20 μmol/L 洛克沙胂一天后，对胞内的砷形态进行提取并利用 HPLC-ICP-MS 进行分析。在原代肝细胞中则可检测到 ROX、MMA^V、As^{V}和一些未知的砷形态，不同个体提供的原代肝细胞代谢洛克沙胂后的产物浓度差异较大。在 HepG2 细胞中含有 ROX、DMA^V、MMA^V、As^V、3-AHPAA、N-AHPAA 和几种未知的砷形态，经过一天的孵育后，洛克沙胂代谢物占总砷含量的 10%，其中代谢的主要产物是 3-AHPAA。两种肝细胞内砷形态分析结果表明，HepG2 细胞和人原代肝细胞都能将 ROX 进一步代谢。

人肝细胞中存在未知的砷形态，在质谱联用技术平台上对其进行鉴定分析后，发现了一种新的苯胂酸化合物，为巯基化的洛克沙胂，洛克沙胂的砷原子上的羟基被巯基取代。对这种新发现的砷形态的细胞毒性试验表明，巯基化洛克沙胂对肺腺癌细胞（A549）的半致死量浓度（IC$_{50}$）为（380±80）μmol/L，对人膀胱移行细胞癌细胞（T24）的 IC$_{50}$ 为（42±10）μmol/L。洛克沙胂对 A549 和 T24 细胞的 IC$_{50}$ 分别为（9300±1600）μmol/L 和（6800±740）μmol/L。巯基化洛克沙胂比其代谢底物洛克沙胂本身具有更高的毒性。

为了研究洛克沙胂进入细胞的能力，Liu 等[91]采用单层细胞（人结直肠腺癌细胞，Caco-2）模型对 ROX 的积累和运输进行评估，并将它与其他一些砷形态（As^{III}、As^{V}和 AsB）进行比较。当 Caco-2 细胞分别暴露于 3 μmol/L Rox、AsB 和 As^{III} 24 h 后，细胞内砷的蓄积在 12 h 达到最大值。在细胞内蓄积的 As^{III} 占给药量的 13%，而 AsB 只占 3%，ROX 仅占 1%，说明 Caco-2 对 ROX 的吸收非常少。在用 Caco-2 形成的单层细胞膜模拟小肠吸收环境时，ROX 从 Caco-2 单层膜的顶侧到底侧的透过率仅为 As^{III} 的 1/4 和 AsB 的 1/2，充分说明了 ROX 的生物吸收率较其他砷形态低。而与人肝源细胞相比，Caco-2 细胞几乎不代谢 ROX。

洛克沙胂可通过巯基化途径被人类肝细胞代谢，代谢所产生的新形态及其毒性的研究，为研究苯胂酸在人体砷暴露及健康风险上的影响提供依据。此外，不同的砷形态进入细胞的能力以及与胞内生物分子结合能力也有差异。对苯胂酸孵育后的细胞进行裂解提取后，所提取的游离砷形态占总砷的 72%±10%，仍有部分砷代谢产物与生物大分子结合而残留在细胞碎片中。对含砷的生物大分子砷形态进行表征，例如砷结合膜蛋白，将进一步推动苯胂酸对人体健康影响的研究。

18.2.3 微生物转化洛克沙胂

随着洛克沙胂的广泛使用，含洛克沙胂的畜禽粪便作为肥料被大量应用于农业生产中，该过程加剧了农田土壤及水体的砷污染[92,93]。微生物是砷在自然生态系统中代谢的主要媒介[94]，在长期与无机砷共存的过程中，进化出了多种砷代谢途径。常见的微生物对无机砷的代谢途径包括：①砷的氧化还原循环，涵盖三价和五价砷之间的相互转化[95]；②砷的甲基化/去甲基化循环，砷甲基化微生物在 As^{III}-S-腺苷甲硫氨酸甲基转移酶（ArsM）的作用下，将甲基供体上的甲基基团逐步添加至三价砷的砷原子上，生成 MMA、DMA 和 TMA 等一系列甲基砷化合物[96-98]，而环境中的另一些微生物可在砷裂解酶（ArsI）的作用下，断裂甲基砷的 C·As 键，脱去砷原子上的甲基生成无机砷化合物[99]。此外，苯胂类有机胂化合物的微生物代谢是其在环境中降解的主要途径，也是砷生物地球化学循环的重要组成部分[100-102]。

2003 年，Garbarino 等发现堆肥过程中存在厌氧菌转化 ROX 这一现象[103]。截至目前，已发现多株厌氧微生物包括菌株 *Alkaliphilus oremlandii* OhILAs[104,105]、*Shewanella oneidensis* MR-1[106]和 *Shewanella putrefaciens* CN32[107]等具有转化 ROX 的能力。厌氧环境中，*Shewanella oneidensis* MR-1 利

用 ROX 作为电子受体，将其还原为终产物 3-AHPAA[106]；而 *Shewanella putrefaciens* CN32 转化 ROX 的产物中除 3-AHPAA 外，还有少量的还原态 3-AHPAA 和 As^{III}[107]。以往的研究很少关注好氧条件下微生物介导的 ROX 转化过程。Huang 等[108]近期从一份砷污染的稻田土壤中分离到一株好氧 ROX 转化菌 *Enterobacter* sp. CZ-1。经 HPLC-ICP-MS、HPLC-ICP-MS/ESI-MS 和 HPLC-ESI-qTOF-MS 鉴定，发现该菌株转化 ROX 生成 5 种含砷代谢产物，分别是：3-AHPAA、*N*-AHPAA、As^{III}、As^{V} 和一种全新的含硫洛克沙胂衍生物 $AsC_9H_{13}N_2O_6S$，其中 3-AHPAA 和 *N*-AHPAA 是两种最主要的 ROX 代谢产物。*N*-AHPAA 是菌株 CZ-1 乙酰化 3-AHPAA 的产物，性质非常稳定，不能被该菌进一步转化。值得一提的是，*N*-AHPAA 还是临床妇科药物滴维净的主效成分，用于治疗难治性阴道滴虫病[109,110]。研究中，Huang 等[108]提出了三条推测的 ROX 好氧转化途径：①ROX 上的硝基先被还原为氨基生成中间代谢产物 3-AHPAA，后者再经乙酰化反应进一步转化为 *N*-AHPAA（主途径；见图 18-6）；②少量的 3-AHPAA 可非酶转化为 As^{V}，后者再经还原反应转化为 As^{III}；③ROX 还可通过一条未知的转化途径生成新型代谢产物 $AsC_9H_{13}N_2O_6S$，该产物的毒理作用有待进一步研究。另外，为了进一步揭示菌株 *Enterobacter* sp. CZ-1 将 ROX 转化为临床药物 *N*-AHPAA 的分子机制，Huang 等[111]采用基因敲除与回补、异源表达和体外酶学试验等方法，从菌株 CZ-1 体内鉴定出两个负责将 3-AHPAA 转化为 *N*-AHPAA 的酶 NhoA1 和 NhoA2，其中 NhoA1 是菌株 CZ-1 体内主效的乙酰基转移酶，其催化活性为 NhoA2 的 115 倍。该研究成果不仅拓宽了对于乙酰基转移酶功能的认识，同时也为利用生物技术生产临床药物 *N*-AHPAA 提供了可用的基因与酶学材料。

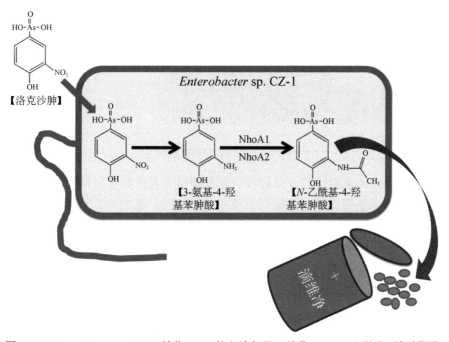

图 18-6 *Enterobacter* sp. CZ-1 转化 ROX 的主途径及乙酰化 3-AHPAA 的分子机制[108]

18.3 砷的主要环境暴露

砷进入到环境中一部分是通过大自然循环，矿石或土壤中的砷溶入水体，或通过微生物代谢，生成气态的甲基化胂化合物，释放到大气中，再经过雨水进入水体和土壤[112]。另外还有一部分来自于人类的开发利用，包括以下方式：矿物燃料燃烧、金属提炼、采矿的废水、肥料、颜料、家畜饲料添加剂、木头防腐剂、杀虫剂以及药物等[113]。Hindmarsh 和 McCurdy 的研究表明人类活动在环境中所产生的砷总量比大自然循环带入的更多[114]。

通过环境暴露，在人体中积累的砷会引发一系列的健康问题，其中也包括了一些癌症相关疾病[115]（图 18-7）。流行病学研究发现一些由砷引起的癌症，包括了膀胱癌、皮肤癌、肝癌、肺癌和肾癌等[116,117]。皮肤癌在孟加拉国和印度得到了确证[118,119]，此外在我国台湾还发现了黑足病[120]。砷在这些地方人体内的积累还会增加其他一些疾病的发病率，包括心脏病[121]、慢性呼吸道疾病[122]、糖尿病[123]等，并会对胎儿造成伤害[124]，影响儿童智力发育[125]。

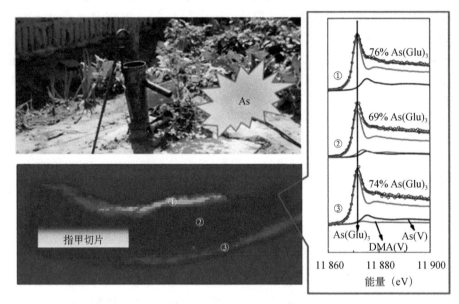

图 18-7　人指甲中砷的累积

砷产生的环境暴露风险来源广泛，最直接且主要的进入人体的途径是摄入含砷饮用水和食物。在发展中国家，数亿人由于长期饮用高砷水而导致健康损伤，成为世界上具有大规模流行病学数据的覆盖范围最广、最严重的地方病之一。世界卫生组织（WHO）对饮用水中砷的含量限量是 10 μg/L。由于地质原因，在很多地方地下水中含有较高浓度的砷，如美国的新墨西哥州、犹他州、亚利桑那州和内华达州等[126]。此外，加拿大、智利、阿根廷、墨西哥、秘鲁、泰国和澳大利亚等国家的一些地区砷含量都超过限定值[120,127,128]，而在孟加拉国的大部分地区饮用水中砷含量都高于 50 μg/L[129]。这种情况在印度尤其严重，更多的人口受砷含量超标的影响[118]。在中国山西、内蒙古、台湾等地区也存在相似的情况[119,120]。由于含砷地下水、土壤、肥料等应用于农业生产，还会造成食品的砷污染，一个人平均每天吸收的砷总含量在 20~300 μg/d。此外，接触和使用含砷制品也增加了砷暴露风险，如木头防腐剂（copper chromated arsenate，CCA）处理过的木制家具[130]和新兴的电子烟[131,132]。

摄入体内的砷形态能够被进一步代谢，例如，无机砷暴露的人尿液中会有其甲基化代谢产物 DMA^V 和 MMA^V，而甲基化中间代谢产物 MMA^{III} 和 DMA^{III} 的毒性比无机砷要高出许多。因此，对砷形态的分析需求是综合且复杂的，要满足三个层面：第一，测定砷总量和各形态含量信息，全面了解砷污染水平；第二，分析砷形态的分布规律，以解释由砷所引发的一系列生物效应及作用机制；

第三，研究特定时间段砷形态变化，并综合多方面的信息，"动态描述"砷形态转化的化学过程。在此基础上，为研究和制定降低砷暴露风险策略提供有力的科学依据。

18.3.1 地下水体中的砷

环境水体中砷的主要存在形式是无机砷（As^{III}、As^V），在细菌、酵母等微生物的作用下，无机砷可以被代谢为有机砷，如 DMA 和 MMA 等，而在海洋中还会有更为复杂的有机砷，如砷胆碱、砷甜菜碱、砷糖和砷磷脂等。水体中的砷形态转化与 pH、氧化还原氛围、磷酸根、碳酸氢根等离子与砷的竞争吸附、含砷硫化物的氧化、含砷铁氧化物的还原释放等众多因素相关，其中电子活度电位（pE）也被证明是高砷地下水成因及演变的关键因素[133,134]。当 pE>-3 时，砷主要以 As^V 存在并吸附在固相表面，是砷的吸附稳定区；在 $-7<pE<-3$ 区间，As^V 还原为 As^{III}，随含砷矿物的溶解一同脱附释放到地下水中，是砷的还原脱附区；当 pE<-7 时，砷与硫化物生成硫化砷矿物，是砷的还原沉淀区。

Jing 等[135,136]研究发现我国地下水砷污染问题面临的首要困境是地下水砷浓度及价态分布并不清晰，仅亟待测定水质的水井就有上百万口，同时高砷地下水的形成机制及演变规律尚不明确。通过对我国山西、内蒙古典型砷污染地区调查发现，34%的被调查农村居民（$n=557$）患有疑似及以上典型砷中毒症状，然而有 54%的村民甚至不清楚自家的饮用井水中是否含砷。对当地 97 个地下水进行水样采集后，经过 AFS 和 ICP-OES 等仪器进行分析。结果表明，大同盆地地下水中砷的平均浓度为 280 μg/L，河套平原的砷平均浓度为 314 μg/L，其中最高浓度分别为 1160 μg/L 和 804 μg/L。两个地区地下水中的溶解态砷约有 73%以 As^{III} 的形式存在，未检测到有机砷。地下水都呈弱碱性，pH 值分别为 8.0（山西）和 8.2（内蒙古）。大同盆地地下水的平均 E_h 值是-113 mV，同时低的 NO_3^- 浓度（0.5 mg/L）及 SO_4^{2-} 浓度（43.0 mg/L）表明，地下水的还原氛围可能是砷释放的主要形成因素。而河套平原地下水的 E_h 为-162 mV，相对高的 NO_3^- 浓度（2.4 mg/L）及 SO_4^{2-} 浓度（174.0 mg/L）也说明，共存离子的竞争吸附是地下水中砷浓度高的原因之一。

在山西、内蒙古地区开展的高砷地下水成因解析也证实了不同氧化还原电位区间对于高砷地下水形成机制及演变规律的关键作用（表 18-1）。大同盆地地下水中的砷浓度与铁、锰浓度及氧化还原电位存在着显著相关性，而河套平原地下水中的砷浓度与铁、锰浓度无显著相关关系。在两个地区的地下水中，砷浓度都与磷酸根浓度有显著的相关性。这可能是由于磷酸根与砷竞争吸附的结果。由于这两个地区都是农业区，地下水中含有磷酸根可能是长期施用磷肥引起的。此外，在两个地区中，砷、铁和锰的浓度都与氧化还原电位有关，三者的高浓度都出现在 pE-4 到-2 范围内。

表 18-1 大同盆地（山西）和河套盆地（内蒙古）地下水的水化学参数

参数	大同					河套				
	最高值	最低值	平均值	中位数	标准偏差	最高值	最低值	平均值	中位数	标准偏差
TotalDissl.As(μg/L)	1160	<1	280	260	242	804	<1	314	304	214
%As(III)	92	N.D.	73	76	14	97	N.D.	73	78	20
Alkalinity (CaCO$_3$ mg/L)	737	252	384	350	124	905	140	542	534	160
Depth (m)	90	20	36	30	16	32	13	23	23	5
EC (mS/cm)	5.2	0.5	1.1	0.7	1	6.3	0.01	2.1	1.7	1.3
ORP (mV)	79	−242	−113	−145	82	65	−205	−162	−174	45
pH	8.5	7.7	8.0	8.0	0.2	8.8	7.4	8.2	8.3	0.3
T (℃)	15.1	10.1	11.7	11.8	0.9	17.5	9.9	12.0	11.8	1.6
TDS (mg/L)	3290	298	694	421	636	3960	8	1339	1080	799

续表

参数	大同					河套				
	最高值	最低值	平均值	中位数	标准偏差	最高值	最低值	平均值	中位数	标准偏差
TOC (mg/L)	17.5	1.4	4.9	4.0	3.4	16.0	1.0	6.3	5.8	3.3
Al (mg/L)	0.22	0.14	0.15	0.14	0.02	0.15	0.08	0.12	0.12	0.02
Ca (mg/L)	40.4	6.4	19.4	17.8	8.7	219.8	3.6	34.3	17.2	39.1
Cl (mg/L)	631.8	7.9	92.7	26.4	162	1644.5	39.0	327.8	238.8	308.0
F (mg/L)	5.6	0.4	1.2	0.8	1.1	3.8	0.3	1.1	0.9	0.7
Fe (mg/L)	0.35	<0.01	0.07	0.10	0.10	0.38	<0.01	0.06	0.02	0.09
K (mg/L)	10.9	0.4	1.0	0.7	1.7	9.3	1.3	3.8	3.2	1.7
Mg (mg/L)	112.5	21.7	39.5	30.4	22.7	264.7	17.2	71.3	45.0	63.9
Mn (mg/L)	0.23	<0.01	0.03	<0.01	0.06	1.16	<0.01	0.04	<0.01	0.15
Na (mg/L)	941.4	40.0	165.9	91.9	177.1	946.4	52.3	327.1	300.4	170.5
NO$_3$ (mg/L)	5.8	<0.1	0.5	<0.1	1.3	20	<0.1	2.4	1.4	3.4
Pb (mg/L)	N.D.[a]	N.D.	N.D.	N.D.	—	N.D.	N.D.	N.D.	N.D.	—
PO$_4^{3-}$ (mg/L)	3.6	<0.1	0.8	0.5	0.9	1.2	<0.1	0.5	0.4	0.2
Si (mg/L)	12.2	0.03	10.1	10.8	2.2	12.3	5.8	8.4	7.8	1.6
SO$_4^{2-}$ (mg/L)	490.2	0.4	43.0	4.6	92.1	973.3	0.4	174.0	92.9	214.2

注：样品数量分别为大同37个和河套62个。a N.D.：未检测到。Pb检出限=40 μg/L。

研究表明两个地区高砷地下水的主要成因可能是在还原条件下，微生物对铁锰氧化物的还原作用，使矿物溶解释放出表面吸附的砷，从而使水体中砷浓度升高。此外，磷酸根的竞争吸附也是导致地下水砷浓度高的另一个原因。而河套平原地下水中砷与铁、锰没有相关性，说明当地地下水砷的释放是由多种过程共同作用引起的。通过对地下水成分与砷浓度的分析，探讨了中国典型地下水砷污染地区——山西及内蒙古地下水中砷浓度高的主要原因，揭示了在当地地下水的还原环境下，铁锰氧化物的还原溶解是释放砷的主要原因。此外，农业磷肥的使用，也是导致当地地下水砷浓度高的一个重要原因。

对山西省山阴县高砷暴露的饮用水、蔬菜及粮食进行了调查研究，并且评估了当地高砷暴露人群的健康状况。当地居民的饮用地下水中砷平均浓度为 168 μg/L（n=113），75%超过国家标准（10 μg/L）。其中 As^{Ⅲ} 平均浓度为 111 μg/L，As^{Ⅴ} 为 57.8 μg/L。蔬菜中砷平均浓度为 1.21 μg/g（n=121，干重），约93%的样品高于我国食品卫生标准所规定的限量值（0.05 μg/g）。其中砷含量较高的有黄瓜、茄子和西红柿。该地蔬菜中砷浓度与浇灌的地下水中砷浓度成正相关，说明长期浇灌含砷地下水很有可能导致蔬菜中砷浓度增高，而采集的粮食（n=22）中，共有 7 个超过了我国食品卫生标准规定的限量值。根据美国 EPA 健康风险评估模型，从膳食角度对饮用水、蔬菜和粮食中砷进行计算，得到该地区人群砷日均摄入量（average daily dose，ADD）为 1.92×10^{-3} mg/(kg·d)，有96%的人摄入量超过了最高摄入量[3×10^{-4} mg/(kg·d)]，ADD 与水中砷浓度相关系数 r 为 0.997（p<0.001），说明人体平均摄入量与水中砷浓度显著相关。除饮用水外，蔬菜和粮食所占的 ADD 分别达到 2.51×10^{-4} mg/(kg·d) 和 3.65×10^{-4} mg/(kg·d)，是人体摄入砷的不可忽略因素。

人组织样品中的砷形态分析是评估砷暴露风险的重要指示物，包括人尿液、指甲、头发等。通过对采集的山阴县人群尿液中砷形态分析表明，总砷平均浓度为 58.6 μg/L，主要形态为 DMA 和 MMA，70%以上的人群尿液砷浓度高于背景值（10 μg/L）。尿砷浓度随饮用水中砷浓度增大而升高，且呈显著相关。当饮用水中砷浓度低于 10 μg/L 时，尿砷与水砷相关性仍然较高（n=31，r_s=0.996，p<0.001）。指甲和头发中平均砷含量分别为 7.1 μg/g 和 3.5 μg/g，75%的指甲样品和59%的头发样品

高于正常水平（指甲为 1.5 μg/g，头发为 1.0 μg/g），指甲和头发中砷与水中砷浓度成正相关（两者 p 均为 0.001）。经美国 EPA 健康风险评估模型对该地区的砷暴露风险进行分析，得到平均危害系数为 6.40，其中有 91%超过正常值（1.0），致癌风险平均值为 2.87×10^{-3}，高于最低致癌风险值（1×10^{-4}）。该地区的高砷地下水已经给当地居民带来了砷暴露风险，通过饮用水和食物进入人体的砷含量均超出限定值，成为影响当地人健康状况的重要因素，也突出了该地区砷污染防控的紧迫性[137-139]。

18.3.2 食品中的砷

环境砷暴露的另一个重要来源是食物。地下水或土壤中的砷通过农作物进入人类的食物链，而人类对砷的开发利用也会带来食品中的砷污染。例如，洛克沙胂（ROX）作为饲料添加剂被广泛用于家禽养殖业中[74,140]，它在家禽中会被代谢成毒性更高的无机砷（As^{III}、As^{V}）[39,141]。残留在家禽体内的洛克沙胂及其代谢产物通过食物进入人体，或者是通过排泄物、皮毛等废弃物排放到环境中，通过生物链累积并最终与人类接触，从而影响人类健康[142]。

为了详细研究洛克沙胂及其代谢产物在家禽中的含量变化，Le 等[80]设计了洛克沙胂喂养实验，通过对 1600 只鸡肉进行 35 天的喂养，监测了鸡肉中洛克沙胂及其代谢产物的含量变化。在鸡肉中，研究人员发现在所有对照组样品中均未检出 ROX，但在所有实验组样品中均检出 ROX。在实验组的鸡肉样品中检测到的无机三价砷（As^{III}，98%）和甲基化砷化物（DMA^{V}，93%；MMA^{V}，100%）比在对照组中检测到的频率（As^{III}，26%；DMA^{V}，92%；MMA^{V}，92%）高。分别对这两组数据两两之间进行双尾 T 检验，As^{III}、DMA^{V}、MMA^{V} 和 ROX 这几种砷形态在实验组中的浓度也显著性地高于对照组（As^{III}，$p\leqslant0.001$；DMA^{V}，$p\leqslant0.001$；MMA^{V}，$p=0.01$；ROX，$p\leqslant0.001$）。在洛克沙胂喂养四周后的实验组鸡肉样品中含有砷甜菜碱 AsB、As^{III}[(38±19) μg/kg]、DMA^{V}[(20±16) μg/kg]、MMA^{V}[(13±5) μg/kg]、ROX[(8±3) μg/kg]和甲基化 ROX[(8±3) μg/kg]，而对照组的鸡肉样品中仅有背景砷信号。

对整个喂养周期的鸡肉样品中砷形态进行定量分析后，可以获取每个砷形态随着洛克沙胂喂养而产生的浓度变化。在给药期间（0~28 天），As^{III}、DMA^{V} 和 MMA^{V} 的浓度与 ROX 的浓度变化趋势相似，到第 28 天达到了最大值。在休药期间（28~35 天），各个砷形态的浓度均快速下降。对这个过程进行药物动力学模型分析，所有砷形态的半衰期均小于 1 天。As^{III} 在鸡胸肉中保留时间最长（1 天），而 DMA^{V} 保留时间最短（0.4 天），ROX 和 MMA^{V} 具有相似的半衰期（0.7 天）。随后砷形态浓度的下降速率都变缓，As^{III}、DMA^{V} 和 MMA^{V} 在最后四天的样品中含量没有明显差异。在休药期最后一天，与对照组中砷含量相比，实验组中 As^{III}（$p=0.01$）、DMA^{V}（$p=0.02$）和 ROX（$p<0.001$）仍然明显高出。尽管有一周的休药期，实验组中残留的砷含量也无法达到与对照组相当的水平。

使用家禽排泄物肥料、含砷杀虫剂、含砷的灌溉用水，都会使得砷在植物中积累。植物能够从土壤中吸收砷，并运输到叶子或者果实。研究发现，大米能够富集环境中的砷，从而增加了砷在大米中的含量，长期食用会产生砷显露风险[124]。在中国，大米是非常重要的主食，消耗量极大，人们从大米中所吸收的砷甚至会高于饮用水。中国对大米中无机砷的限量为 0.15 μg/g[124]，美国大米中砷的平均含量为 0.26 μg/g，比欧洲、印度和孟加拉国的大米砷含量要高出 5 倍[143]。

大米中砷主要以无机砷为存在形态[144]，但也有部分植物能够将无机砷甲基化。Lomax 等[145]研究了三种植物（大米、西红柿和红首蓿）中砷的形态。当这些植物生长的环境中含有有机胂（DMA 和 MMA）时，它们会在植物组织中保留。在不含有机胂的土壤中，这些植物会保留无机砷，其中主要是三价的亚砷酸。在五价砷酸培养环境中，甲基化有机胂并没有在这三种植物中检出，也说明了这三种植物不能进一步代谢无机砷。另一方面，Raab 等[146]和 Xu 等[147]在用水培法栽种的植物中发现了痕量有机胂（MMA 或 DMA），而所用培养溶液中只含有无机砷，这个结果表明了部分植物也具有了甲基化无机砷的能力。砷能在植物中积累，但其甲基化能力却与植物的种类相关。

18.3.3 电子香烟中的砷

电子烟在世界范围内非常流行。因其成分简单，过去电子烟常被人认为比传统香烟"安全"，但最近的研究表明电子烟含有或者产生多种有害物质，砷就是其中之一[131,132]。每 10 口抽吸产生的电子烟气溶胶中，总砷含量约为 $0.001\sim0.01$ μg[148,149]。不同的报告显示电子烟油中总砷浓度为 1.5 μg/L、(0.08 ± 0.04) μg/L、26.7 μg/kg 或 $0.83\sim3.04$ μg/kg[150-152]。

为了获得有关电子烟油和产生的雾化气中砷形态的信息，Song 等根据电子烟的销售信息，从加拿大和中国市场购买了 17 种较有代表性的电子烟油样品[153]。研究人员通过 HPLC-ICPMS 分析，发现电子烟油中存在 As^{III}、As^{V} 和 MMA，检出率分别为 59%、94% 和 47%，浓度范围在 $<0.01\sim8.30$ μg/kg。此外，在雾化气冷凝液中，As^{III}、As^{V} 和 MMA 的检出率分别为 100%、88% 和 13%，浓度范围为 $<0.01\sim4.60$ μg/kg。进一步比较电子烟油和雾化气冷凝液后发现，电子烟油雾化后 As^{III} 的浓度显著性增加。

在此基础上，研究人员估算了由吸入电子烟雾化气中的无机砷所引起的肺癌风险。雾化气中无机砷的中位浓度为 3.4 μg/m^3，根据美国环境保护署（EPA）的无机砷吸入暴露的单位风险因素（4.3×10^{-3} 每 1 μg/m^3），假设电子烟吸入时间占终生的 1%，那么增加的肺癌风险为 1.5×10^{-4}。这是非常高的肺癌风险，比 EPA 的百万分之一的目标高出 150 倍以上。这充分说明电子烟因"危害少"成为戒烟工具，作为传统香烟的替代品，但我们仍不能忽视电子烟的危害。

18.4 砷的健康效应研究：以砷与蛋白质的结合为例

砷的环境暴露对人类健康有着重要的影响。砷易与生物分子相互作用，导致其正常生理功能的丧失，进而干扰生命体的正常活动，例如，降低超氧化物歧化酶（SOD）和谷胱甘肽过氧化物酶（GSH-Px）的活性，对生物膜造成损害；影响血红素的生成，从而影响血红蛋白的形成；改变细胞壁的通透性，损害毛细血管；破坏蛋白质结构，抑制酶的活性；抑制 DNA 损伤的修复。另一方面，砷又很早就被中西医用作药剂治疗疾病，其药用历史可以追溯到 2400 多年前；Hippocrates（公元前 $460\sim$前 375 年）使用雄黄和雌黄（As$_2$S$_3$）来治疗肿瘤和癌性溃疡；我国东汉时期（$25\sim220$ 年）中医使用含有雄黄（As$_4$S$_4$）的药物外敷治疗疮疡和痈肿。砒霜也是一种古老的中药，其化学成分为三氧化二砷，最早载于《开宝本草》："砒霜：疗诸疟，风痰在胸膈，可作吐药。不可久服，伤人。"砷的药用功能至今仍在不断开发和研究中。

砷与生物分子相互作用的分子机制研究是理解其毒性和药效机制的关键。三价砷与巯基具有较高的亲和力，容易与蛋白质分子上的巯基结合，抑制蛋白质活性，导致蛋白质构象变化，影响蛋白质的正常功能[154-158]（图 18-8）。不同三价砷形态与巯基结合的化学计量比有所不同，从而具有不同的蛋白质结合能力，如 As^{III}、MMA^{III} 和 DMA^{III} 可以分别与三个、二个和一个巯基结合[159]。砷与蛋白质分子的结合虽然是生物体内砷的代谢和解毒途径，但是也会导致一些富含巯基的金属硫蛋白、血红蛋白和角蛋白等生物分子的生理功能丧失产生毒性[160,161]。砷与巯基的结合是一种可逆相互作用，一般情况下砷与蛋白质的解离常数在 nmol/L～μmol/L 水平[162]，其稳定性受浓度、溶液组成和 pH 值、巯基个数、溶液氧化还原氛围的影响[163,164]。

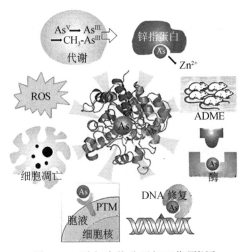

图 18-8　砷与生物分子相互作用[154]

ROS：活性氧；ADME：药物代谢；PTM：蛋白质翻译后修饰

砷与蛋白质巯基结合并抑制其活性是砷产生细胞毒性的主要原因，相互作用的亲和性导致不同砷形态的细胞毒性差异。例如，在细胞线粒体能量代谢中起关键作用的丙酮酸脱氢酶（pyruvate dehydrogenase，PDH）复合物由多种酶和辅助因子组成，硫辛酸是 PDH 复合物中起氧化还原作用的酶辅助因子。研究发现 As^{III}、MMA^{III} 和 DMA^{III} 通过与硫辛酸的邻二硫醇基团结合抑制了 PDH 的活性。MMA^{III} 的抑制活性大于 As^{III}，这与 MMA^{III}-硫辛酸之间的更高的亲和性相关[165-168]。另一类砷结合蛋白是氧化还原蛋白，包括硫氧还蛋白家族和谷胱甘肽还原酶家族蛋白，如硫氧还蛋白、硫氧还蛋白还原酶、硫氧还蛋白过氧化物酶等，它们在维持细胞氧化还原平衡、转录因子调控、砷甲基化等方面发挥了重要作用。这些氧化还原蛋白的共同特点是活性催化位点含有由两个 Cys 组成的邻位巯基。砷活性位点双巯基结合会抑制氧化还原蛋白的活性，打乱细胞氧化还原平衡并干扰细胞的生化过程，引起细胞毒性[169-171]。研究还发现砷抑制多种 DNA 修复蛋白的活性，包括多腺苷二磷酸核糖聚合酶-1（PARP-1）、甲酰胺嘧啶 DNA 糖基化酶（Fpg）；着色性干皮病 A 组蛋白（XPA）等蛋白[172,173]。As^{III}、MMA^{III} 和 DMA^{III} 可以与这些蛋白质的锌指结构的巯基结合，取代锌离子。锌指结构对 DNA 损伤位点的结合起关键作用，缺少了锌离子的 DNA 修复蛋白无法识别并修复 DNA 损伤位点。三价砷可以取代锌指结合的锌离子，其取代能力不仅与浓度相关，也与砷的形态有关，MMA^{III} 和 DMA^{III} 具有比 As^{III} 更强的锌取代能力[174,175]。

砷作为一种最成功的治疗急性早幼粒细胞性白血病（acute promyelocytic leukemia，APL）药物[176-179]，其分子机制也与砷和蛋白巯基结合的性质相关。近年来的研究发现砷与 APL 癌蛋白（PML-RARα）巯基结合并诱导其快速降解是砷治疗 APL 的主要分子机制[180,181]。无机三价砷（As^{III}）可以与 PML-RARα 的 PML 部分半胱氨酸巯基结合，引起 PML-RARα 构象变化、聚集、类泛素化修饰（SUMOylation）和泛素化修饰（ubiquitylation），最终诱导癌症蛋白的降解和 APL 细胞凋亡[182,183]。As^{III} 对少数 APL 患者没有显著效果，这是由于患者体内的 PML-RARα 癌蛋白的 PML 部分产生了突变（A216V，L218P，C212A，C213A，S214L，A216T，L217F）[184]。其中一些位点，例如 Cys212 和 Cys213 是砷的巯基结合位点，其突变将导致 PML 无法与砷结合。相反，若没有这些突变，APL 患者则不产生抗药性[185]。最近一项研究发现 4 μmol/L As^{III} 足以改变正常 PML 蛋白的溶解度并诱导其降解；相反，即使 As^{III} 的浓度达到 50 μmol/L 也无法使 PML（A216V）突变体降解[186]。以上研究结果表明突变导致 PML 构象变化或巯基结合位点消失，从而降低 As^{III} 与 PML 的亲和性，使得砷无法与癌蛋白特异性结合并诱导其降解，是 PML 蛋白突变患者产生 As^{III} 抗药性的可能原因。

研究砷与蛋白质相互作用，捕获和鉴定砷结合蛋白质不仅可以从分子水平上加深对砷的生物作用机制的理解，还有助于寻找生物体内砷作用靶点，推动砷的健康效应研究和砷药物研发。

18.4.1 砷结合蛋白质的直接分析

当小分子的砷形态与蛋白质大分子结合后，结合的砷形态能够在尺寸排除色谱法（SEC）中与其小分子形态分离，同时由于可以选择接近生理条件的溶液作为流动相来保持蛋白质分子的构象和活性，因此 SEC 被广泛用于砷结合蛋白质的分离和表征。早在 1988 年，Brown 等就合成了以 ^{14}C 作为放射示踪剂的砷化合物，并将 SEC 与放射线检测和时间分辨荧光各向异性光谱法结合来研究砷化合物与硫氧还蛋白的相互作用规律[187]。他们发现砷的结合导致蛋白质的水化半径变大，结合速率与砷化合物的极性、刚性和尺寸相关，小尺寸非极性的砷化合物更容易与蛋白质巯基结合。对于复杂样品，单独检测砷化合物的信号无法确定蛋白质的种类，需要更多的互补技术来表征砷结合蛋白。由于血红素在 420 nm 处有强烈吸收，因此在 SEC 分离之后同时检测 ^{74}As 的放射信号和 420 nm 处吸收，Cornelis 等发现老鼠血液中与砷结合的蛋白质的分子质量大约为 60 kDa，在 420 nm 有强吸收的血红蛋白是砷的主要结合靶点[188]。Kimpe 等利用同样的技术进一步确认腹膜注射 ^{74}As 标记砷酸盐的兔子血红细胞中的血红蛋白是砷结合蛋白[189]。以 ^{73}As 作为放射示踪剂，Styblo 使用 SEC 分析了包含大鼠肝细胞质、GSH、S-苷甲硫胺酸 和 1 μmol/L [^{73}As]arsenite 成分的体外砷甲基化体系中砷与蛋白质的相互作用，发现分子质量大约为 1000 kDa、135 kDa 和 38 kDa 的蛋白质可以与砷结合，表明砷与蛋白质的结合在砷甲基化反应过程中发挥关键作用[190]，这一工作为后续的砷甲基化机理研究提供了重要线索。

随着原子光谱/质谱和仪器接口技术的发展，SEC-AAS/-ICPMS 等联用技术发展起来并逐渐取代放射线检测器用于金属结合蛋白的研究。1998 年，Zhang 等将氢化物发生原子吸收光谱法与 SEC 联用（SEC-HGAAS）研究了持续性非卧床腹膜透析患者血液中的砷结合蛋白，发现转铁蛋白是无机砷的主要结合靶点，不同患者血液蛋白结合的砷浓度介于 0.19～0.59 μg/L[191]。作为血液砷结合蛋白工作的延续，Lu 等使用 SEC-ICPMS 分析了不同砷形态与老鼠和人血红蛋白的相互作用亲和力，发现老鼠血红蛋白对砷形态的亲和力是人的 3～6 倍。对喂食含砷饲料老鼠的血红细胞的 SEC-ICPMS 定量结果表明，大部分的砷形态与血红蛋白结合。他们还进一步使用 nano-ESIMS 测定了血红蛋白与砷结合的化学计量比，并发现砷主要与 α 链结合。这一工作证明砷与红细胞内的血红蛋白结合是老鼠血液累积砷形态的分子机制[192]。Naranmandura 等进一步对喂食含砷饲料老鼠血液的砷代谢产物进行形态分析和 SEC-ICPMS 分析，发现 DMA 是老鼠血红蛋白的主要结合砷形态[193]。为了研究砷的代谢机理，Naranmandura 等分析了静脉注射亚砷酸盐（0.5 mg As/kg 体重）的大鼠肝肾砷代谢产物，他们发现需要使用 H$_2$O$_2$ 将结合在蛋白质上的砷释放出来，表明在甲基化过程中与蛋白质结合的砷是+3 价态，而+5 价的砷是代谢终产物而不是代谢中间产物，基于这个实验结果，他们推测在生物甲基化过程中砷保持与蛋白质巯基结合的还原甲基化机理[194,195]。此外，SEC-ICPMS 还被用于研究细胞[196]、心血管组织[197]、肝脏[198]、尿液[199]、小鼠[200,201]等生物样品的砷结合蛋白质研究。

基于分子大小和电荷性质的平板凝胶电泳可更好地分离复杂生物样品中砷结合蛋白质，在生物学研究中得到广泛应用。与激光剥蚀进样(LA)-ICPMS[202]、同步辐射 X 射线分析（SXFS）[203]、放射自显影[204]等检测技术相结合，平板凝胶电泳可用于金属结合蛋白质的分离和检测。但是由于 LA-ICPMS 的灵敏度较低，SXFS 设备稀缺，以及放射自显影技术对人体造成辐射伤害需要专业化实验室等原因限制了平板凝胶电泳在金属结合蛋白分离分析的使用。为了解决这些问题，Hu 等发展了连续流动凝胶电泳-ICPMS 联用技术（GE-ICPMS）用于金属结合蛋白质的在线分离与实时检测[205-207]。他们将凝胶填充于圆筒柱内，在柱出口的径向方向加上流路将分离后的蛋白质连续带入到 ICPMS 中进行实时在线金属检测，使用该技术他们研究了金属铋与幽门螺杆菌蛋白的相互作用并证明 HpSlyD 是一种铋结合蛋白。GE-ICPMS 还被用于研究 *E. coli* 中的金属伴侣蛋白 HpHypA 和 HpHspA，以及 GAPDH 等蛋白质与 Zn^{2+}、Ni^{2+}、Cu^{2+}、Co^{2+} 以及 Ag^+ 等金属离子的相互作用，显示出这一技术在金属结合蛋白研究中的优势[208-210]。Xu 等使用 GE-ICPMS 技术研究了有机胂药物 *S*-二甲基砷-谷胱甘肽 (ZIO-101、darinaparsin) 的作用机理，发现 Histone H3.3 是 ZIO-101 的结合靶点，进一步使用 RT-PCR、蛋白质印迹（Western blotting）、流式细胞仪（flow cytometry）等技术证明砷与 Histone H3.3 的半胱氨酸巯基结合引起了 Trail 诱导白血病细胞凋亡是 ZIO-101 的作用机理[211]。

18.4.2 砷结合蛋白质的捕获与分析

选择性地分离砷结合蛋白质是解析其生物作用靶点的重要前提，主要包括砷亲和色谱和砷的生物素化亲和标签（As-biotin）等方式。砷亲和色谱采用三价砷修饰的固定相作为填料，用于捕获和纯化砷结合蛋白质，结合蛋白质组、蛋白质印迹等蛋白质技术鉴定，能够更全面地分析砷形态与蛋白质作用，并发现生物体内新的砷结合靶点。砷的生物素化亲和标签是另一种捕获砷结合蛋白质的方法，相比于砷亲和柱，As-biotin 的空间位阻较小，容易与处于生理条件下的砷结合蛋白结合，接着可以利用亲和素修饰磁珠捕获纯化 As-biotin 结合蛋白用于后续的蛋白质组学鉴定。

早在 1982 年，Hannestad 等就首先设计了以三价砷为亲和配体的色谱。他们将对氨基苯砷酸直接修饰到 CNBr 活化的 Sepharose 6B 填料上制备了砷亲和色谱，接着研究了洗脱溶液的 pH 值和硫醇洗脱物对结合在亲和柱上的单硫醇和二硫醇的洗脱效果[212]。后来的几个研究小组对砷亲和色谱进行了改进，以用于研究砷与蛋白质之间的相互作用，或用于纯化蛋白质。为了研究三价砷化合物 PAO 对 3T3-L1 脂肪细胞的胰岛素依赖型己糖摄取的影响，Hoffman 和 Lane 将结合了 β-巯基乙醇（β-ME）的 4-氨基苯砷酸修饰到羟基琥珀酰亚胺（NHS）活化的琼脂糖上制备了砷亲和树脂，接着使用含有 β-ME 的 1% Triton X-100 或 SDS 溶液洗脱结合在亲和树脂上的砷结合蛋白质，在 SDS-PAGE 上进一步分离，并通过 4-[^{125}I]碘代苯砷酸对砷结合蛋白质进行放射自显影成像，发现包括胰岛素反应性葡萄糖转运蛋白 GLUT4 和微管蛋白在内的几种来自于 3T3-L1 脂肪细胞的蛋白质能够与砷结合[213]。

利用三价砷与含有邻位巯基的蛋白质（VTPs）具有较高亲和力的性质，砷亲和柱被用于 VTPs 的捕获和纯化。Zhou 等[214]将对氨基苯胂酸（PAPAO）修饰到 NHS 活化的羧甲基-Bio-Gel，接着使用该色谱柱纯化人血浆卵磷脂-胆固醇酰基转移酶（LCAT）。LCAT 是一种二硫基酶，其催化位点含有两个还原型半胱氨酸残基（Cys31 和 Cys184），可以与砷亲和柱结合。使用含有 2,3-二巯基丙烷磺酸的缓冲液可将 LCAT 洗脱，总产率为 11%。Kalef 等在 1993~1997 年间使用砷亲和色谱对白血病细胞的 VTPs 进行了捕获、纯化和鉴定研究[215,216]。他们制备了基于 Sepharose 4B 的亲和树脂，并将其与氨己酰-4-氨基苯砷偶联，从 L1210 鼠白血病淋巴母细胞中纯化 VTPs。用这种亲和色谱法一步纯化了在酵母中合成的 1-三碘甲状腺原氨酸重组大鼠 c-erbAβ1T3 受体，产率为 11%。同样的方法也被 Berleth 和 Boerman 等用于泛素蛋白连接酶(E3)[217]、视黄醇脱氢酶（RoDH）[218]等 VTPs 的纯化。

Winski 和 Carter 使用砷亲和色谱表征大鼠红细胞中的砷结合蛋白。他们使用缓冲液和乙二醇除去非特异性结合蛋白，接着用半胱氨酸、谷胱甘肽和二巯基丙烷磺酸盐（DMPS）三种硫醇逐级洗脱保留在砷亲和柱上的蛋白质，发现用 100 mmol/L 谷胱甘肽洗脱出来的血红蛋白是大鼠红细胞中主要砷结合蛋白[219]。使用相同的砷亲和色谱策略，Menzel 等根据分子量，鉴定出两种砷结合蛋白质分别是微管蛋白和肌动蛋白，他们还发现可以被砷诱导高度表达的血红素加氧酶 1（HO1）蛋白不能与砷亲和柱结合，因为它不含还原型半胱氨酸[220]。Chang 等为了从中国仓鼠卵巢（CHO）和 SA7（抗砷 CHO）细胞中分离出砷调节蛋白，使用了商品化的 PAO-琼脂糖树脂。使用含有 2 mol/L NaCl 的 Tris 缓冲液洗涤除去非特异性结合蛋白后，再依次用 20 mmol/L β-ME 和 20 mmol/L DTT 洗脱特异性结合蛋白。通过氨基酸序列分析鉴定出半乳凝集素 1（Gal-1；在 CHO 细胞的 β-ME 洗脱级分中）、谷胱甘肽 S-转移酶 P 型（GST-P）和硫氧还蛋白过氧化物酶 II（TPX-II）为砷结合蛋白。TPX-II 在 SA7 细胞中优先表达，但不在 CHO 或 SA7N 中表达。相反，Gal-1 在 CHO 和 SA7N 细胞中特异性表达，而在 SA7 细胞中没有发现。Gal-1 和 TPX-II 与 As^{III} 的特异性结合通过免疫共沉淀和蛋白质重组技术进行了验证[221]。

Mizumura 等制备了 As^{V}、As^{III}-GSH 和 As^{III} 三种不同的砷琼脂糖亲和柱来研究肝胞质蛋白与五价砷、谷胱甘肽修饰三价砷和裸露三价砷的相互作用。他们将 NHS 活化的琼脂糖凝胶（Sepharose 4 Fast Flow）与 PAPAO 混合，合成了固定化的 As^{V}-琼脂糖亲和柱。用 GSH 还原后，该凝胶进一步转化为 As^{III}-GSH-琼脂糖亲和柱，As^{III}-GSH 可以在过量 GSH 存在的条件下保持稳定。当溶液中不含有 GSH 时，GSH 从 As^{III}-GSH-琼脂糖亲和色谱上水解解离，得到不含 GSH 的 As^{III}-琼脂糖亲和柱。用含有 GSH、DTT 或 LDS（十二烷基硫酸锂）的缓冲液分别洗脱与不同砷亲和色谱结合的肝细胞质蛋白，并用 SDS-PAGE 和 EZBlue 染色，未发现与五价砷结合的蛋白质。利用 MALDI-MS/MS 鉴定出蛋白二硫键异构酶相关蛋白 5（PDSIRP5）和过氧化物酶 1/增强蛋白（PRX1/EP）与 As^{III}-琼脂糖亲和柱结合。发现过氧化物酶 2、胞质焦磷酸酶、磷酸甘油酸激酶 1 和 KM-102 衍生还原酶样因子（硫氧还蛋白还原酶）这四种蛋白与 As^{III}-GSH-琼脂糖亲和柱特异性结合[222]。He 等将 PAPAO 与 Affigel 偶联以制备 PAO 亲和性，用于探测砷与 Kelch 样环氧氯丙烷相关蛋白 1（Keap1）、核因子 E2 相关因子 2（Nrf2）、金属调节转录因子（MTF1）等蛋白的相互作用[223]。砷亲和材料还被用于从细胞裂解物中分离出腺嘌呤核苷酸转位酶（ANT）和蛋白质酪氨酸磷酸酶（PTP）等蛋白质[224,225]。所有这些鉴定出的蛋白质中都存在半胱氨酸簇作为砷的结合位点。Yan 等开发了一种基于砷亲和色谱靶向蛋白质组学方法，用于鉴定人细胞中的砷结合蛋白质。砷亲和柱是通过 Eupergit C 珠粒与 4-氨基苯基氨酸氧化物（NAPO^{III}）反应合成的。通过逐渐增加 1%SDS 的缓冲液中 DTT 的含量来增加砷结合蛋白质的洗脱强度，接着使用 LC-ESI-MS/MS 对洗脱下来的蛋白质进行鉴定分析。这种方法表明，在大量非特异性蛋白存在的情况下，可以从 A549 人肺癌细胞的细胞核和细胞膜/细胞器中分别鉴定出 50 种和 24 种砷结合蛋白质，证明砷亲和柱对砷结合蛋白质的良好选择性[226]。

蛋白质邻二硫醇的可逆氧化还原在细胞氧化还原信号转导、代谢和基因表达调节过程中发挥着重要作用。Gitler 等介绍了一种通用方法用于鉴定和富集细胞中处于氧化和还原状态的 VTPs。他们首先使用 N-乙基马来酰亚胺（NEM）封闭鼠白血病 L1210 细胞中的还原型邻二硫醇，然后用 DTT 还原氧化的邻二硫醇，这些被还原的邻二硫醇蛋白质接着使用 PAO 亲和色谱进行富集[216]。由于 DTT 可以与砷高亲和力结合，使用 DTT 作为二硫键还原剂会导致砷亲和柱捕获的蛋白质减少。因此 Foley 等使用非硫醇还原剂三(2-羧乙基)膦（TCEP）代替 DTT 来还原蛋白质巯基，接着使用 PAO 亲和柱提高了对大鼠脑的 Triton X-100 可溶性提取物中砷结合蛋白质的捕获效率。他们发现两种丰度最高的蛋白质是白蛋白和磷酸三糖异构酶（TPI），另一种鉴定的蛋白质磷酸酶 2A（PP2Ac）的催化亚基含有可逆氧化的硫醇[227]。使用类似的技术，研究人员还从多形核白细胞（PMN）的细胞胞质中发现蛋白磷酸酶 2B（钙调磷酸酶，CAN）和硫氧还蛋白[228]，以及其他三种蛋白质（谷胱甘肽小号转移酶 P1-1、淋巴细胞胞浆蛋白和丝切蛋白），是砷结合蛋白质[229]。

砷的生物素化亲和标签（As-biotin）采用两端分别修饰了三价砷和生物素（Biotin）的亲和标签，在生理条件下与砷靶标蛋白结合后，通过亲和素修饰磁珠捕获纯化 As-biotin 结合蛋白。1999 年，Moaddel 等使用三种不同长度（2～15 Å）的连接臂将 PAPAO 和 biotin 偶联，合成了三种新型 As-biotin 标签用于 VTPs 的捕获。他们以含有邻位巯基的尼古丁受体为模型研究了 As-biotin 与蛋白质的相互作用，发现当邻位巯基与 As-biotin 结合时，巯基氧化试剂二硫双(2-硝基苯甲酸)（DTNB）无法与巯基发生二硫键交换反应；当加入 2,3-二巯基丙磺酸与 As-biotin 竞争结合后，尼古丁受体的巯基被释放出来从而与 DTNB 反应，说明 As-biotin 与邻位巯基有较强的结合能力。且尼古丁受体与三种不同长度的 As-biotin 结合不影响 biotin 与亲和素的结合，说明合成的 As-biotin 对 VTPs 具有良好的捕获效果[230]。

Zhang 等使用五氟苯酚生物素酯与 PAPAO 偶联制备了 As-biotin 标签，结合 PAGE 分离和 MALDI-TOFMS 鉴定技术从 MCF-7 人类乳腺癌细胞中捕获和鉴定出 50 种砷结合蛋白质，主要分类为代谢酶、结构蛋白和压力响应蛋白。他们进一步研究了砷与其中的 β-微管蛋白和丙酮酸激酶 M2 的结合对其生物活性的影响，确认这两种蛋白是砷结合蛋白质[231]。相同的方法被用于证明砷可以与急性早幼粒细胞白血病（APL）的 PML 和 PML-RARα 蛋白结合，但不与 RARα 结合，说明砷与癌蛋白的结合对于 APL 的治疗效果十分重要[232]。

为了提高 As-biotin 的合成效率，Heredia-Moya 等发明了将 biotin 与 PAPAO 直接偶联的一锅合成法。他们把氯甲酸乙酯与 biotin 和 PAPAO 混合，氯甲酸乙酯将 biotin 的羧基转化为酰氯，随即便可与 PAPAO 的氨基直接偶联合成 As-biotin，产率达到 49%[233]。

细胞表面 VTPs 邻位巯基的氧化、还原或二硫键交换反应可能参与细胞的氧化还原感应和信号传导，Donoghue 等合成了一种不透膜的三价砷-GSH 衍生物以鉴定哺乳动物细胞表面的邻位巯基。由于 GSH 在细胞内部合成并分泌到细胞外，无法重新被细胞摄取，利用这个性质，Donoghue 等将 PAPAO 与 GSH 的巯基偶联，生成 4-(N-(S-谷胱甘肽乙酰基)氨基)-对苯氧化物（p-GSAO）[234]。结果显示 p-GSAO 与二巯基紧密结合，但是与单个巯基结合并不紧密，表明 p-GSAO 对邻位巯基良好的选择性。为了鉴定细胞表面 VTPs，他们进一步将生物素连接到 GSAO 的 γ-谷氨酰胺的伯氨基上，得到 GSAO-B。生物素部分还可防止细胞膜上的 γ-谷氨酰转肽酶切割 γ-谷氨酰胺残基[235]。他们在含有和不含有 2,3-二巯基丙醇（DMP）的条件下，用 GSAO-B 标记细胞，裂解，并与亲和素-琼脂糖孵育，以收集生物素捕获的砷结合蛋白质，接着用含有 50 mmol/L DTT 的缓冲液洗脱结合的蛋白，在 SDS-PAGE 上分离并采用抗体印迹来检测目标蛋白质。最终在牛主动脉内皮（BAE）细胞和人纤维肉瘤（HT1080）细胞的表面分别鉴定出 10 种和 12 种不同蛋白质，两种细胞中都发现了蛋白质二硫键异构酶（PDI）。用 GSAO-B 从细胞裂解液中捕获的蛋白质与从细胞表面中捕获的蛋白质的种类差异较大，证明 GSAO-B 确实不容易穿透细胞膜，对细胞表面的 VTPs 具有很好的选择性。

Dilda 等比较了对氨基苯砷和邻氨基苯砷与 GSH 的偶联产物（p-GSAO 和 o-GSAO）对内皮细胞和癌细胞增殖的抑制效果，发现 o-GSAO 的细胞增殖抑制效果比 p-GSAO 高 50 倍，这是由于 o-GSAO 的细胞摄取和积累效率比 p-GSAO 高 300 倍。将 p-GSAO 和 o-GSAO 与分离出来的线粒体进行孵育，发现和 p-GSAO 一样，o-GSAO 也可以与腺嘌呤核苷酸转移酶（ANT）的邻位巯基结合，但是诱导线粒体通透性效率却有 8 倍的提高[236]。细胞凋亡和坏死的共同特征是细胞膜完整性的丧失，一些不可渗透细胞膜的物质因此也可以加入死亡的细胞。利用这个性质，Park 等在 p-GSAO 上修饰 biotin 和荧光团来研究砷与死亡细胞蛋白的相互作用，发现 p-GSAO 可以快速累积在死亡细胞质中并与 90 kDa 热休克蛋白（Hsp90）、真核翻译延伸因子和纤维蛋白 A 等蛋白质的二硫醇结合。他们还将 ^{111}In 修饰到 p-GSAO 上，用于对小鼠肿瘤细胞的死亡情况进行 SPECT/CT 无创性成像[237]。

近几年，基于 As-biotin 的砷结合蛋白质捕获技术还被用于研究砷的致癌机理。核心组蛋白 H2B 的 120 号位赖氨酸（K120）的泛素化对于打开 30 nm 染色质纤维并促进 DNA 双链断裂（DSB）修复很重要[238]。在环指家族 E3 泛素连接酶中，主要由 RNF20-RNF40 异二聚体负责 H2B 的 K120 的泛素化。Zhang 等通过 As-biotin 和 As^{III} 的体外、体内实验证明砷可以与 RNF20-RNF40 的 RING finger 结构域上的半胱氨酸巯基结合，改变组蛋白表观遗传标记，进而抑制 DNA 的 DSB 修复[239]。基于类似的思路和方法，Jiang 等发现砷也可以与泛素连接酶 Rbx1 E3 的半胱氨酸富集结构结合，进而抑制 Rbx1 E3 的靶点 Nrf2 的泛素化，并激活 Nrf2 诱导的抗氧化信号通路[240]。Lee 等使用 As-biotin 亲和捕获和相关实验发现酵母甘油-3-磷酸脱氢酶（Gpd1）活性位点的 Cys306 是 MMA^{III} 的主要靶点，MMA^{III} 的结合抑制了 Gpd1 的活性，进而导致细胞在非高渗压条件下过量积累甘油[241]。这些工作证明砷与蛋白质巯基结合是砷发挥其生物效应的化学基础。

18.4.3 胞内砷结合蛋白质的原位捕获与鉴定

砷亲和柱方法需要先将细胞裂解以获得细胞内部的蛋白质用于后续的分析，同时面临较大的空间位阻和非特异性吸附等问题，一些砷结合蛋白质在分析过程中由于发生氧化或构象变化而无法被高效率、特异性捕获。尽管 As-biotin 更易与蛋白质结合，但由于其体积相对较大，而且 biotin 基团有可能干扰细胞内部的 biotin 代谢通路[242,243]，因此原位捕获活细胞内部的砷结合蛋白质仍面临着极大的挑战[244]。

为了克服这些问题，Yan 等[245]发展了含有叠氮和三价砷基团的对叠氮苯砷（p-azidophenylarsenoxide，PAzPAO）探针用于活细胞内部砷结合蛋白的原位捕获与鉴定（图 18-9）。由于体积小且疏水性较强，PAzPAO 可以快速进入细胞，其三价砷基团与蛋白质靶点在细胞内部原位结合。将 PAzPAO 孵育的细胞裂解，然后通过生物正交反应将环辛炔修饰的生物素与叠氮基团偶联，接着使用亲和素修饰磁珠特异性捕获砷结合蛋白质，最后使用 DTT 将砷结合蛋白质从磁珠上释放下来进行蛋白质组学鉴定，成功在活细胞中原位捕获和鉴定出 50 多种砷结合蛋白。发现鉴定出的多个高丰度蛋白是活性位点含有邻位双巯基的氧化还原蛋白，表明与氧化还原蛋白结合并干扰细胞氧化还原平衡是砷的重要作用机理。进一步发现并证明砷与糖酵解路径上的甘油醛 3-磷酸脱氢酶（GAPDH）催化位点的半胱氨酸巯基（Cys152）特异性结合，并抑制其体外和体内活性，表明砷可能通过抑制 GAPDH 进而抑制癌细胞的有氧糖酵解，导致癌细胞死亡，GAPDH 是一个潜在的新型砷药物靶点。

随着更为专一、高效的砷结合蛋白质分析技术的发展，鉴定出更多的砷结合蛋白质，系统深入地理解含砷化合物的生物效应，将会在预防、治疗砷相关疾病和开发新型含砷药物上发挥更大的作用。

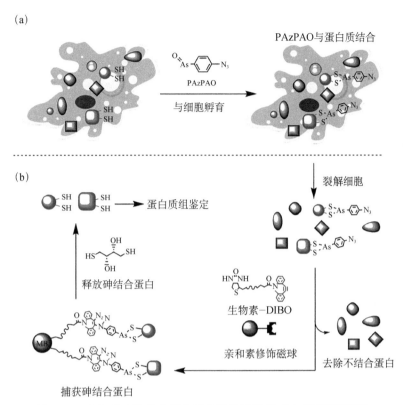

图 18-9 （a）胞内砷结合蛋白质的原位捕获和（b）鉴定流程[245]

18.5 总结与展望

砷形态的代谢、环境暴露与健康效应研究涉及环境影响、食品安全和人体健康影响等众多领域，是世界范围内亟待解决的环境健康问题。基于原子光（质）谱的多种分析仪器联用技术成为砷形态分析的重要手段，为研究砷的毒性和环境行为提供了技术支撑。随着砷形态分析技术不断发展，在环境和生物样品中发现越来越多从前未知的砷形态及其代谢途径，揭示了更多新的砷暴露来源，为进一步开展砷暴露引起健康效应的分子机制研究、评价砷暴露风险和制定含砷化合物使用规范提供了理论基础。通过研究砷的分布，控制和减少人体暴露于含砷化合物，促进和保护公共健康。砷形态的健康效应研究是原子光（质）谱领域当前重要的发展方向之一，未来将主要从以下几个方面来重点开展工作：

（1）基于样品前处理方法、原子光（质）谱与分子质谱联用技术，发展生命体系中砷形态的高灵敏度、高选择性分析技术和综合分析平台，是研究砷环境污染及其对食品安全和人体健康影响等关键问题的重要手段。

（2）对砷形态的环境暴露、生物转化和安全性评估等进行深入研究，包括对其已知形态的探究、未知形态的认知、代谢途径及生物安全性的评价等。

（3）通过对不同砷化合物与生物分子相互作用的研究，探索砷形态产生毒性的分子机理，同时开展含砷药物的作用靶点和功能研究，为系统地研究砷在生物体中的代谢、毒性以及可利用性提供了丰富的信息。

参 考 文 献

[1] Podgorski J, Berg M. Global threat of arsenic in groundwater. Science, 2020, 368: 845-850.

[2] Uppal J S, Zheng Q, Le X C. Arsenic in drinking water-recent examples of updates from southeast asia. Current Opinion in Environmental Science & Health, 2019, 7: 126-135.

[3] Cantor K P, Lubin J H. Arsenic, internal cancers, and issues in inference from studies of low-level exposures in human populations. Toxicology and Applied Pharmacology, 2007, 222: 252-257.

[4] Celik I, Gallicchio L, Boyd K, Lam T K, Matanoski G, Tao X, Shiels M, Hammond E, Chen L, Robinson K A, Caulfield L E, Herman J G, Guallar E, Alberg A J. Arsenic in drinking water and lung cancer: A systematic review. Environmental Research, 2008, 108: 48-55.

[5] Heck J E, Andrew A S, Onega T, Rigas J R, Jackson B P, Karagas M R, Duell E J. Lung cancer in a U.S. Population with low to moderate arsenic exposure. Environmental Health Perspectives, 2009, 117: 1718-1723.

[6] Chen C J, Wang S L, Chiou J M, Tseng C H, Chiou H Y, Hsueh Y M, Chen S Y, Wu M M, Lai M S. Arsenic and diabetes and hypertension in human populations: A review. Toxicology and Applied Pharmacology, 2007, 222: 298-304.

[7] Chen Y, Graziano J H, Parvez F, Liu M, Slavkovich V, Kalra T, Argos M, Islam T, Ahmed A, Rakibuz-Zaman M, Hasan R, Sarwar G, Levy D, van Geen A, Ahsan H. Arsenic exposure from drinking water and mortality from cardiovascular disease in Bangladesh: Prospective cohort study. The British Medical Journal, 2011, 342: d2431.

[8] Naujokas M F, Anderson B, Ahsan H, Aposhian H V, Graziano J H, Thompson C, Suk W A. The broad scope of health effects from chronic arsenic exposure: Update on a worldwide public health problem. Environmental Health Perspectives, 2013, 121: 295-302.

[9] Andrewes P, Kitchin K T, Wallace K. Dimethylarsine and trimethylarsine are potent genotoxins *in vitro*. Chemical Research in Toxicology, 2003, 16: 994-1003.

[10] Lin S, Del Razo L M, Styblo M, Wang C Q, Cullen W R, Thomas D J. Arsenicals inhibit thioredoxin reductase in cultured rat hepatocytes. Chemical Research in Toxicology, 2001, 14: 305-311.

[11] Mass M J, Tennant A, Roop B C, Cullen W R, Styblo M, Thomas D J, Kligerman A D. Methylated trivalent arsenic species are genotoxic. Chemical Research in Toxicology, 2001, 14: 355-361.

[12] Petrick J S, Ayala-Fierro F, Cullen W R, Carter D E, Aposhian H V. Monomethylarsonous acid (MMA[III]) is more toxic than arsenite in chang human hepatocytes. Toxicology and Applied Pharmacology, 2000, 163: 203-207.

[13] Moe B, Peng H Y, Lu X F, Chen B W, Chen L W L, Gabos S, Li X F, Le X C. Comparative cytotoxicity of fourteen trivalent and pentavalent arsenic species determined using real-time cell sensing. Journal of Environmental Sciences, 2016, 49: 113-124.

[14] Kaise T, Watanabe S, Itoh K. The acute toxicity of arsenobetaine. Chemosphere, 1985, 14: 1327-1332.

[15] Leufroy A, Noel L, Dufailly V, Beauchemin D, Guerin T. Determination of seven arsenic species in seafood by ion exchange chromatography coupled to inductively coupled plasma-mass spectrometry following microwave assisted extraction: Method validation and occurrence data. Talanta, 2011, 83: 770-779.

[16] Ackley K L, B'Hymer C, Sutton K L, Caruso J A, Speciation of arsenic in fish tissue using microwave-assisted extraction followed by HPLC-ICP-MS. Journal of Analytical Atomic Spectrometry, 1999, 14: 845-850.

[17] Charoensuk V, Gati W P, Weinfeld M, Le X C. Differential cytotoxic effects of arsenic compounds in human acute

promyelocytic leukemia cells. Toxicology and Applied Pharmacology, 2009, 239: 64-70.

[18] Pizarro I, Gomez M, Camara C, Palacios M A. Arsenic speciation in environmental and biological samples-extraction and stability studies. Analytica Chimica Acta, 2003, 495: 85-98.

[19] Sanchez-Rodas D, Gomez-Ariza J L, Oliveira V. Development of a rapid extraction procedure for speciation of arsenic in chicken meat. Analytical and Bioanalytical Chemistry, 2006, 385: 1172-1177.

[20] Liu Q Q, Peng H Y, Lu X F, Le X C. Enzyme-assisted extraction and liquid chromatography mass spectrometry for the determination of arsenic species in chicken meat. Analytica Chimica Acta, 2015, 888: 1-9.

[21] Sanz E, Munoz-Olivas R, Camara C. Evaluation of a focused sonication probe for arsenic speciation in environmental and biological samples. Journal of Chromatography A, 2005, 1097: 1-8.

[22] Peng H Y, Hu B, Liu Q Q, Yang Z L, Lu X F, Huang R F, Li X F, Zuidhof M J, Le X C. Liquid chromatography combined with atomic and molecular mass spectrometry for speciation of arsenic in chicken liver. Journal of Chromatography A, 2014, 1370: 40-49.

[23] Le X C, Yalcin S, Ma M S. Speciation of submicrogram per liter levels of arsenic in water: On-site species separation integrated with sample collection. Environmental Science & Technology, 2000, 34: 2342-2347.

[24] Yalcin S, Le X C. Speciation of arsenic using solid phase extraction cartridges. Journal of Environmental Monitoring, 2001, 3: 81-85.

[25] Mondal P, Balomajumder C, Mohanty B. Quantitative separation of As(Ⅲ) and As(Ⅴ) from a synthetic water solution using ion exchange columns in the presence of Fe and Mn ions. Clean-Soil Air Water, 2007, 35: 255-260.

[26] Ben Issa N, Rajakovic-Ognjanovic V N, Jovanovic B M, Rajakovic L V. Determination of inorganic arsenic species in natural waters-benefits of separation and preconcentration on ion exchange and hybrid resins. Analytica Chimica Acta, 2010, 673: 185-193.

[27] Boyaci E, Cagir A, Shahwan T, Eroglu A E. Synthesis, characterization and application of a novel mercapto-and amine-bifunctionalized silica for speciation/sorption of inorganic arsenic prior to inductively coupled plasma mass spectrometric determination. Talanta, 2011, 85: 1517-1525.

[28] Chen S Z, Zhan X L, Lu D B, Liu C, Zhu L. Speciation analysis of inorganic arsenic in natural water by carbon nanofibers separation and inductively coupled plasma mass spectrometry determination. Analytica Chimica Acta, 2009, 634: 192-196.

[29] Doker S, Uzun L, Denizli A. Arsenic speciation in water and snow samples by adsorption onto phema in a micro-pipette-tip and gfaas detection applying large-volume injection. Talanta, 2013, 103: 123-129.

[30] Chen M L, Lin Y M, Gu C B, Wang J H. Arsenic sorption and speciation with branch-polyethyleneimine modified carbon nanotubes with detection by atomic fluorescence spectrometry. Talanta, 2013, 104: 53-57.

[31] Huang C Z, Xie W, Li X, Zhang J P. Speciation of inorganic arsenic in environmental waters using magnetic solid phase extraction and preconcentration followed by ICP-MS. Microchimica Acta, 2011, 173: 165-172.

[32] Hu W L, Zheng F, Hu B. Simultaneous separation and speciation of inorganic As(Ⅲ)/As(Ⅴ) and Cr(Ⅲ)/Cr(Ⅵ) in natural waters utilizing capillary microextraction on ordered mesoporous Al_2O_3 prior to their on-line determination by ICP-MS. Journal of Hazardous Materials, 2008, 151: 58-64.

[33] Zheng F, Hu B. Dual silica monolithic capillary microextraction (CME) on-line coupled with ICP-MS for sequential determination of inorganic arsenic and selenium species in natural waters. Journal of Analytical Atomic Spectrometry, 2009, 24: 1051-1061.

[34] Pu X L, Chen B B, Hu B. Solvent bar microextraction combined with electrothermal vaporization inductively coupled plasma mass spectrometry for the speciation of inorganic arsenic in water samples. Spectrochimica Acta Part B: Atomic

Spectroscopy, 2009, 64: 679-684.

[35] Mao X J, Chen B B, Huang C Z, He M, Hu B. Titania immobilized polypropylene hollow fiber as a disposable coating for stir bar sorptive extraction-high performance liquid chromatography-inductively coupled plasma mass spectrometry speciation of arsenic in chicken tissues. Journal of Chromatography A, 2011, 1218: 1-9.

[36] Jackson B P, Bertsch P M. Determination of arsenic speciation in poultry wastes by IC-ICP-MS. Environmental Science & Technology, 2001, 35: 4868-4873.

[37] Grant T D. Assessing the environmental and biological implications of various elements through elemental speciation using inductively coupled plasma mass spectrometry. Electronic Thesis or Dissertation. University of Cincinnati, 2004. https://etd.ohiolink.edu/.

[38] Falnoga I, Stilbilj V, Tusek-Znidaric M, Slejkovec Z, Mazej D, Jacimovic R, Scancar J. Effect of arsenic trioxide on metallothionein and its conversion to different arsenic metabolites in hen liver. Biological Trace Element Research, 2000, 78: 241-254.

[39] Polatajko A, Szpunar J. Speciation of arsenic in chicken meat by anion-exchange liquid chromatography with inductively coupled plasma-mass spectrometry. Journal of AOAC International, 2004, 87: 233-237.

[40] Taylor A, Branch S, Day M P, Patriarca M, White M. Atomic spectrometry update. Clinical and biological materials, foods and beverages. Journal of Analytical Atomic Spectrometry, 2010, 25: 453-492.

[41] Bacon J R, Butler O T, Cairns W R L, Cook J M, Davidson C M, Cavoura O, Mertz-Kraus R. Atomic spectrometry update : A review of advances in environmental analysis. Journal of Analytical Atomic Spectrometry, 2020, 35: 9-53.

[42] Ardini F, Dan G, Grotti M. Arsenic speciation analysis of environmental samples. Journal of Analytical Atomic Spectrometry, 2020, 35: 215-237.

[43] Du J J, Cui J L, Jing C Y. Rapid *in situ* identification of arsenic species using a portable Fe_3O_4@Ag SERS sensor. Chemical Communications, 2014, 50: 347-349.

[44] Challenger F. Biological methylation. Chemical Reviews, 1945, 36: 315-361.

[45] Lu X F, Arnold L L, Cohen S M, Cullen W R, Le X C. Speciation of dimethylarsinous acid and trimethylarsine oxide in urine from rats fed with dimethylarsinic acid and dimercaptopropane sulfonate. Analytical Chemistry, 2003, 75: 6463-6468.

[46] Le X C, Cullen W R, Reimer K J. Determination of urinary arsenic and impact of dietary arsenic intake. Talanta, 1993, 40: 185-193.

[47] Cullen W R, Reimer K J. Arsenic speciation in the environment. Chemical Reviews, 1989, 89: 713-764.

[48] Freeman H C, Uthe J F, Fleming R B, Odense P H, Ackman R G, Landry G, Musial C. Clearance of arsenic ingested by man from arsenic contaminated fish. Bulletin of Environmental Contamination and Toxicology, 1979, 22: 224-229.

[49] Le X C, Cullen W R, Reimer K J. Human urinary arsenic excretion after one-time ingestion of seaweed, crab, and shrimp. Clinical Chemistry, 1994, 40: 617-624.

[50] Le X C, Ma M S. Short-column liquid chromatography with hydride generation atomic fluorescence detection for the speciation of arsenic. Analytical Chemistry, 1998, 70: 1926-1933.

[51] Popowich A, Zhang Q, Le X C. Arsenobetaine: The ongoing mystery. National Science Review, 2016, 3: 451-458.

[52] Francesconi K A, Tanggaard R, McKenzie C J, Goessler W. Arsenic metabolites in human urine after ingestion of an arsenosugar. Clinical Chemistry, 2002, 48: 92-101.

[53] Hansen H R, Raab A, Francesconi K A, Feldmann I. Metabolism of arsenic by sheep chronically exposed to arsenosugars as a normal part of their diet. 1. Quantitative intake, uptake, and excretion. Environmental Science & Technology, 2003, 37: 845-851.

[54] Ma M S, Le X C. Effect of arsenosugar ingestion on urinary arsenic speciation. Clinical Chemistry, 1998, 44: 539-550.

[55] Chapman H D, Johnson Z B. Use of antibiotics and roxarsone in broiler chickens in the USA: Analysis for the years 1995 to 2000. Poultry Science, 2002, 81: 356-364.

[56] Zhao Y P, Cui J L, Fang L P, An Y L, Gan S C, Guo P R, Chen J H. Roxarsone transformation and its impacts on soil enzyme activity in paddy soils: A new insight into water flooding effects. Environmental Research, 2021: 111636.

[57] Hughes M F, Beck B D, Chen Y, Lewis A S, Thomas D J. Arsenic exposure and toxicology: A historical perspective. Toxicological Sciences, 2011, 123: 305-332.

[58] Silbergeld E K, Nachman K. The environmental and public health risks associated with arsenical use in animal feeds. Annals of the New York Academy of Sciences, 2008, 1140: 346-357.

[59] Smith A H, Hopenhaynrich C, Bates M N, Goeden H M, Hertzpicciotto I, Duggan H M, Wood R, Kosnett M J, Smith M T. Cancer risks from arsenic in drinking-water. Environmental Health Perspectives, 1992, 97: 259-267.

[60] Jacobson-Kram D P, Mushak M, Piscator D, Sivulka Chu M. Health Assessment Document for Inorganic Arsenic (Final Report 1984). U.S. Environmental Protection Agency, Washington, D.C., EPA/600/8-83/021F (NTIS PB84190891).

[61] 黄连喜, 何兆桓, 曾芳, 姚丽贤, 周昌敏, 国彬. 液相色谱-氢化物发生-原子荧光联用同时测定洛克沙胂及其代谢物. 分析化学, 2010, 38: 1321-1324.

[62] Kerr K B, Narveson J R, Lux F A. Toxicity of an organic arsenical, 3-nitro-4-hydroxyphenylarsonic acid-residues in chicken tissues. Journal of Agricultural and Food Chemistry, 1969, 17: 1400-1402.

[63] Chiou P W S, Chen K L, Yu B. Effects of roxarsone on performance, toxicity, tissue accumulation and residue of eggs and excreta in laying hens. Journal of the Science of Food and Agriculture, 1997, 74: 229-236.

[64] Sierra-Alvarez R, Cortinas I, Field J A. Methanogenic inhibition by roxarsone (4-hydroxy-3-nitrophenylarsonic acid) and related aromatic arsenic compounds. Journal of Hazardous Materials, 2010, 175: 352-358.

[65] Rosal C G, Momplaisir G M, Heithmar E M. Roxarsone and transformation products in chicken manure: Determination by capillary electrophoresis-inductively coupled plasma-mass spectrometry. Electrophoresis, 2005, 26: 1606-1614.

[66] Liu X P, Zhang W F, Hu Y A, Cheng H F. Extraction and detection of organoarsenic feed additives and common arsenic species in environmental matrices by HPLC-ICP-MS. Microchemical Journal, 2013, 108: 38-45.

[67] Xie D, Mattusch J, Wennrich R. Separation of organoarsenicals by means of zwitterionic hydrophilic interaction chromatography (ZIC®-HILIC) and parallel ICP-MS/ESI-MS detection. Engineering in Life Sciences, 2008, 8: 582-588.

[68] Yao L X, Li G L, Dang Z, Yang B M, He Z H, Zhou C M. Uptake and transport of roxarsone and its metabolites in water spinach as affected by phosphate supply. Environmental Toxicology and Chemistry, 2010, 29: 947-951.

[69] Arai Y, Lanzirotti A, Sutton S, Davis J A, Sparks D L. Arsenic speciation and reactivity in poultry litter. Environmental Science & Technology, 2003, 37: 4083-4090.

[70] 黄连喜, 魏岚, 姚丽贤, 何兆桓, 周昌敏. 洛克沙胂代谢物在土壤中的累积及其植物有效性研究. 农业环境科学学报, 2019, 38: 1079-1088.

[71] Pizarro I, Gomez M M, Fodor P, Palacios M A, Camara C. Distribution and biotransformation of arsenic species in chicken cardiac and muscle tissues. Biological Trace Element Research, 2004, 99: 129-143.

[72] Cortinas I, Field J A, Kopplin M, Garbarino J R, Gandolfi A J, Sierra-Alvarez R. Anaerobic biotransformation of roxarsone and related N-substituted phenylarsonic acids. Environmental Science & Technology, 2006, 40: 2951-2957.

[73] Hileman B. Arsenic in chicken production. Chemical & Engineering News, 2007, 85: 34-35.

[74] 钟国清. 有机胂饲料添加剂的合成和应用. 饲料工业, 1995, 7: 27-28.

[75] Liu Q Q, Lu X F, Peng H Y, Popowich A, Tao J, Uppal J S, Yan X W, Boe D, Le X C. Speciation of arsenic: A review of phenylarsenicals and related arsenic metabolites. TrAC: Trends in Analytical Chemistry, 2017, 104: 171-182.

[76] Madison N J. Pfizer to suspend sale of 3-nitro (roxarsone) in the United States. Pfizer Media Statement, 2011.

[77] Nachman K E, Baron P A, Raber G, Francesconi K A, Navas-Acien A, Love D C. Roxarsone, inorganic arsenic, and other arsenic species in chicken: A U.S.-based market basket sample. Environmental Health Perspectives, 2013, 121: 818-824.

[78] Liu L H, He B, Yun Z J, Sun J, Jiang G B. Speciation analysis of arsenic compounds by capillary electrophoresis on-line coupled with inductively coupled plasma mass spectrometry using a novel interface. Journal of Chromatography A, 2013, 1304: 227-233.

[79] Chen D M, Zhang H H, Tao Y F, Wang Y L, Huang L L, Liu Z L, Pan Y H, Peng D P, Wang X, Dai M H, Yuan Z H. Development of a high-performance liquid chromatography method for the simultaneous quantification of four organoarsenic compounds in the feeds of swine and chicken. Journal of Chromatography B: Analytical Technologies in the Biomedical and Life Sciences, 2011, 879: 716-720.

[80] Peng H Y, Hu B, Liu Q Q, Li J H, Li X F, Zhang H Q, Le X C. Methylated phenylarsenical metabolites discovered in chicken liver. Angewandte Chemie International Edition, 2017, 56: 6773-6777.

[81] Le X C, Cullen W R, Reimer K J. Speciation of arsenic compounds by HPLC with hydride generation atomic absorption spectrometry and inductively coupled plasma mass spectrometry detection. Talanta, 1994, 41: 495-502.

[82] McSheehy S, Szpunar J, Lobinski R, Haldys V, Tortajada J, Edmonds J S. Characterization of arsenic species in kidney of the clam tridacna derasa by multidimensional liquid chromatography-ICPMS and electrospray time-of-flight tandem mass spectrometry. Analytical Chemistry, 2002, 74: 2370-2378.

[83] Gong Z L, Lu X F, Ma M S, Watt C, Le X C. Arsenic speciation analysis. Talanta, 2002, 58: 77-96.

[84] Yin Y G, Liu J F, Jiang G B. Recent advances in speciation analysis of mercury, arsenic and selenium. Chinese Science Bulletin, 2013, 58: 150-161.

[85] Pergantis S A, Wangkarn S, Francesconi K A, Thomas-Oates J E. Identification of arsenosugars at the picogram level using nanoelectrospray quadrupole time-of-flight mass spectrometry. Analytical Chemistry, 2000, 72: 357-366.

[86] McKnight-Whitford A, Chen B W, Naranmandura H, Zhu C, Le X C. New method and detection of high concentrations of monomethylarsonous acid detected in contaminated groundwater. Environmental Science & Technology, 2010, 44: 5875-5880.

[87] Schaumloffel D, Tholey A. Recent directions of electrospray mass spectrometry for elemental speciation analysis. Analytical and Bioanalytical Chemistry, 2011, 400: 1645-1652.

[88] Amayo K O, Raab A, Krupp E M, Gunnlaugsdottir H, Feldmann J. Novel identification of arsenolipids using chemical derivatizations in conjunction with RP-HPLC-ICPMS/ESMS. Analytical Chemistry, 2013, 85: 9321-9327.

[89] Moody J P, Williams R T. The metabolism of 4-hydroxy-3-nitrophenylarsonic acid in hens. Food and Cosmetics Toxicology, 1964, 2: 707-715.

[90] Liu Q Q, Leslie E M, Moe B, Zhang H Q, Douglas D N, Kneteman N M, Le X C. Metabolism of a phenylarsenical in human hepatic cells and identification of a new arsenic metabolite. Environmental Science & Technology, 2018, 52: 1386-1392.

[91] Liu Q Q, Leslie E M, Le X C. Accumulation and transport of roxarsone, arsenobetaine, and inorganic arsenic using the human immortalized caco-2 cell line. Journal of Agricultural and Food Chemistry, 2016, 64: 8902-8908.

[92] D'Angelo E, Zeigler G, Beck E G, Grove J, Sikora F, Arsenic species in broiler (*Gallus gallus domesticus*) litter, soils, maize (*Zea mays* L.), and groundwater from litter-amended fields. Science of the Total Environment, 2012, 438: 286-292.

[93] Jackson B P, Bertsch P M, Cabrera M L, Camberato J J, Seaman J C, Wood C W. Trace element speciation in poultry litter. Journal of Environmental Quality, 2003, 32: 535-540.

[94] Rhine E D, Garcia-Dominguez E, Phelps C D, Young L Y. Environmental microbes can speciate and cycle arsenic. Environmental Science & Technology, 2005, 39: 9569-9573.

[95] Stolz J F, Basu P, Santini J M, Oremland R S. Arsenic and selenium in microbial metabolism. Annual Review of Microbiology, 2006, 60: 107-130.

[96] Huang K, Chen C, Zhang J, Tang Z, Shen Q R, Rosen B P, Zhao F J. Efficient arsenic methylation and volatilization mediated by a novel bacterium from an arsenic-contaminated paddy soil. Environmental Science & Technology, 2016, 50: 6389-6396.

[97] Qin J, Rosen B P, Zhang Y, Wang G J, Franke S, Rensing C. Arsenic detoxification and evolution of trimethylarsine gas by a microbial arsenite s-adenosylmethionine methyltransferase. Proceedings of the National Academy of Sciences of the United States of America, 2006, 103: 2075-2080.

[98] Zhu Y G, Yoshinaga M, Zhao F J, Rosen B P. Earth abides arsenic biotransformations. Annual Review of Earth and Planetary Sciences, 2014, 42: 443-467.

[99] Yoshinaga M, Rosen B P. A C·As lyase for degradation of environmental organoarsenical herbicides and animal husbandry growth promoters. Proceedings of the National Academy of Sciences of the United States of America, 2014, 111: 7701-7706.

[100] Shmakov S, Abudayyeh O O, Makarova K S, Wolf Y I, Gootenberg J S, Semenova E, Minakhin L, Joung J, Konermann S, Severinov K, Zhang F, Koonin E V. Discovery and functional characterization of diverse class 2 CRISPR-Cas systems. Molecular Cell, 2015, 60: 385-397.

[101] Sharma V K, Sohn M. Aquatic arsenic: Toxicity, speciation, transformations, and remediation. Environment International, 2009, 35: 743-759.

[102] Yao L, Huang L, He Z, Zhou C, Lu W, Bai C. Delivery of roxarsone via chicken diet→chicken→chicken manure→soil→rice plant. Science of the Total Environment, 2016: 1152-1158.

[103] Garbarino J R, Bednar A J, Rutherford D W, Beyer R S, Wershaw R L. Environmental fate of roxarsone in poultry litter. I. Degradation of roxarsone during composting. Environmental Science & Technology, 2003, 37: 1509-1514.

[104] Fisher E, Dawson A M, Polshyna G, Lisak J, Crable B, Perera E, Ranganathan M, Thangavelu M, Basu P, Stolz J F. Transformation of inorganic and organic arsenic by *Alkaliphilus oremlandii* sp. Nov. strain OhILAs. Annals of the New York Academy of Sciences, 2008, 1125: 230-241.

[105] Stolz J F, Perera E, Kilonzo B, Kail B, Crable B, Fisher E, Ranganathan M, Wormer L, Basu P. Biotransformation of 3-nitro-4-hydroxybenzene arsonic acid (roxarsone) and release of inorganic arsenic by clostridium species. Environmental Science & Technology, 2007, 41: 818-823.

[106] Chen G W, Ke Z C, Liang T F, Liu L, Wang G. Shewanella oneidensis mr-1-induced Fe(III) reduction facilitates roxarsone transformation. PLOS ONE, 2016, 11: e0154017.

[107] Han J C, Zhang F, Cheng L, Mu Y, Liu D F, Li W W, Yu H Q. Rapid release of arsenite from roxarsone bioreduction by exoelectrogenic bacteria. Environmental Science & Technology Letters, 2017, 4: 350-355.

[108] Huang K, Peng H, Gao F, Liu Q, Lu X, Shen Q, Le X C, Zhao F J. Biotransformation of arsenic-containing roxarsone by an aerobic soil bacterium *Enterobacter* sp. Cz-1. Environmental Pollution, 2019, 247: 482-487.

[109] Chen M Y, Smith N A, Fox E F, Bingham J S, Barlow D. Acetarsol pessaries in the treatment of metronidazole resistant trichomonas vaginalis. International Journal of STD & AIDS, 1999, 10: 277-280.

[110] Kiely C J, Clark A, Bhattacharyya J, Moran G W, Lee J C, Parkes M. Acetarsol suppositories: Effective treatment for refractory proctitis in a cohort of patients with inflammatory bowel disease. Digestive Diseases and Sciences, 2018, 63: 1011-1015.

[111] Huang K, Gao F, Le X C, Zhao F J. *N*-hydroxyarylamine *O*-acetyltransferases catalyze acetylation of 3-amino-4-hydroxyphenylarsonic acid in the 4-hydroxy-3-nitrobenzenearsonic acid transformation pathway of *Enterobacter* sp. strain cz-1. Applied and Environmental Microbiology, 2020, 86: e02050-19.

[112] International Programme on Chemical Safety. Arsenic-Environmental Health Criteria 18. Geneva: World Health Organization, 1981.

[113] Grund S C, Hanusch K, Wolf H U. Arsenic and arsenic compounds in Ullmann's encyclopedia of industrial chemistry. Wiley-VCH Verlag GmbH & Co. KGaA, 2008.

[114] Hindmarsh J T, McCurdy R F, Clinical and environmental aspects of arsenic toxicity. Critical Reviews in Clinical Laboratory Sciences, 1986, 23: 315-347.

[115] Wu M M, Chiou H Y, Wang T W, Hsueh Y M, Wang I H, Chen C J, Lee T C. Association of blood arsenic levels with increased reactive oxidants and decreased antioxidant capacity in a human population of northeastern Taiwan. Environmental Health Perspectives, 2001, 109: 1011-1017.

[116] Abernathy C O, Liu Y P, Longfellow D, Aposhian H V, Beck B, Fowler B, Goyer R, Menzer R, Rossman T, Thompson C, Waalkes M. Arsenic: Health effects, mechanisms of actions, and research issues. Environmental Health Perspectives, 1999, 107: 593-597.

[117] Council N R. Arsenic in the Drinking Water (update). Washington D.C.: National Academy Press, 2001.

[118] Chakraborti D, Mukherjee S C, Pati S, Sengupta M K, Rahman M M, Chowdhury U K, Lodh D, Chanda C R, Chakraborty A K, Basul G K. Arsenic groundwater contamination in middle Ganga plain, Bihar, India: A future danger? Environmental Health Perspectives, 2003, 111: 1194-1201.

[119] Sun G F. Arsenic contamination and arsenicosis in China. Toxicology and Applied Pharmacology, 2004, 198: 268-271.

[120] Council N R. Arsenic in the Drinking Water. Washington D.C.: National Academy Press, 1999.

[121] Tseng C H, Chong C K, Tseng C P, Hsueh Y M, Chiou H Y, Tseng C C, Chen C J. Long-term arsenic exposure and ischemic heart disease in arseniasis-hyperendemic villages in Taiwan. Toxicology Letters, 2003, 137: 15-21.

[122] Hendryx M. Mortality from heart, respiratory, and kidney disease in coal mining areas of appalachia. International Archives of Occupational and Environmental Health, 2009, 82: 243-249.

[123] Kile M L. Environmental arsenic exposure and diabetes. JAMA: The Journal of the American Medical Association, 2008, 300: 845-846.

[124] Gilbert-Diamond D, Cottingham K L, Gruber J F, Punshon T, Sayarath V, Gandolfi A J, Baker E R, Jackson B P, Folt C L, Karagas M R. Rice consumption contributes to arsenic exposure in us women. Proceedings of the National Academy of Sciences, 2011, 108: 20656-20660.

[125] Wasserman G A, Liu X H, Parvez F, Ahsan H, Factor-Litvak P, van Geen A, Slavkovich V, Lolacono N J, Cheng Z Q, Hussain I, Momotaj H, Graziano J H. Water arsenic exposure and children's intellectual function in Araihazar, Bangladesh. Environmental Health Perspectives, 2004, 112: 1329-1333.

[126] Frost F J, Muller T, Petersen H V, Thomson B, Tollestrup K. Identifying us populations for the study of health effects related to drinking water arsenic. Journal of Exposure Analysis and Environmental Epidemiology, 2003, 13: 231-239.

[127] McGuigan C F, Hamula C L A, Huang S, Gabos S, Le X C. A review on arsenic concentrations in canadian drinking water. Environmental Reviews, 2010, 18: 291-307.

[128] Graziano J H and van Geen A. Reducing arsenic exposure from drinking water: Different settings call for different approaches. Environmental Health Perspectives, 2005, 113: A360-A361.

[129] Chakraborti D, Rahman M M, Das B, Murrill M, Dey S, Mukherjee S C, Dhar R K, Biswas B K, Chowdhury U K, Roy S, Sorif S, Selim M, Rahman M, Quamruzzaman Q. Status of groundwater arsenic contamination in bangladesh: A

14-year study report. Water Research, 2010, 44: 5789-5802.

[130] Lew K, Acker J P, Gabos S, Le X C. Biomonitoring of arsenic in urine and saliva of children playing on playgrounds constructed from chromated copper arsenate-treated wood. Environmental Science & Technology, 2010, 44: 3986-3991.

[131] Deng W C, Schofield J R M, Le X C, Li X F. Electronic cigarettes and toxic substances, including arsenic species. Journal of Environmental Sciences, 2020, 92: 278-283.

[132] Liu Q Q, Huang C Z, Le X C. Arsenic species in electronic cigarettes: Determination and potential health risk. Journal of Environmental Sciences, 2020, 91: 168-176.

[133] Jing C Y, Liu S Q, Meng X G. Arsenic remobilization in water treatment adsorbents under reducing conditions: Part I. Incubation study. Science of the Total Environment, 2008, 389: 188-194.

[134] 罗婷, 景传勇. 地下水砷污染形成机制研究进展. 环境化学, 2011, 30: 77-83.

[135] Cui J L, Shi J B, Jiang G B, Jing C Y. Arsenic levels and speciation from ingestion exposures to biomarkers in ShanXi, China: Implications for human health. Environmental Science & Technology, 2013, 47: 5419-5424.

[136] Luo T, Hu S, Cui J L, Tian H X, Jing C Y. Comparison of arsenic geochemical evolution in the Datong basin (ShanXi) and Hetao basin (Inner Mongolia), China. Applied Geochemistry, 2012, 27: 2315-2323.

[137] Ye, L, Liu, W J, Shi Q T, Jing C Y. Arsenic mobilization in spent nZVI waste residue: Effect of *Pantoea* sp. IMH. Environmental Pollution, 2017, 230: 1081-1089.

[138] Luo T, Ye L, Chan T S, Jing C Y. Mobilization of arsenic on nano-TiO_2 in soil columns with sulfate reducing bacteria. Environmental Pollution, 2018, 234: 762-768.

[139] Luo T, Tian H X, Guo Z, Zhuang G Q, Jing C Y. Fate of arsenate adsorbed on nano-TiO_2 in the presence of sulfate reducing bacteria. Environmental Science & Technology, 2013, 47: 10939-10946.

[140] Nigra A E, Nachman K E, Love D C, Grau-Perez M, Navas-Acien A. Poultry consumption and arsenic exposure in the U.S. population. Environmental Health Perspectives, 2017, 125: 370-377.

[141] Conklin S D, Shockey N, Kubachka K, Howard K D, and Carson M C. Development of an ion chromatography-inductively coupled plasma-mass spectrometry method to determine inorganic arsenic in liver from chickens treated with roxarsone. Journal of Agricultural and Food Chemistry, 2012, 60: 9394-9404.

[142] Yang Z L, Peng H Y, Lu X F, Liu Q Q, Huang R F, Hu B, Kachanoski G, Zuidhof M J, and Le X C. Arsenic metabolites, including *N*-acetyl, 4-hydroxy-*m*-arsanilic acid, in chicken litter from a roxarsone-feeding study involving 1600 chickens. Environmental Science & Technology, 2016, 50: 6737-6743.

[143] Williams P N, Price A H, Raab A, Hossain S A, Feldmann J, Meharg A A. Variation in arsenic speciation and concentration in paddy rice related to dietary exposure. Environmental Science & Technology, 2005, 39: 5531-5540.

[144] Liang F, Li Y L, Zhang G L, Tan M G, Lin J, Liu W. Total and speciated arsenic levels in rice from china. Food Additives and Contaminants: Part A—Chemistry, Analysis, Control, Exposure and Risk Assessment, 2010, 27: 810-816.

[145] Lomax C, Liu W J, Wu L Y, Xue K, Xiong J B, Zhou J Z, McGrath S P, Meharg A A, Miller A J, Zhao F J. Methylated arsenic species in plants originate from soil microorganisms. New Phytologist, 2012, 193: 665-672.

[146] Raab A, Ferreira K, Meharg A A, Feldmann J. Can arsenic-phytochelatin complex formation be used as an indicator for toxicity in helianthus annuus? Journal of Experimental Botany, 2007, 58: 1333-1338.

[147] Xu X Y, McGrath S P, Zhao F J. Rapid reduction of arsenate in the medium mediated by plant roots. New Phytologist, 2007, 176: 590-599.

[148] Palazzolo D L, Crow A P, Nelson J M, Johnson R A. Trace metals derived from electronic cigarette (ECIG) generated

aerosol: Potential problem of ECIG devices that contain nickel. Frontiers in Physiology, 2017, 7: 663.

[149] Williams M, Bozhilov K, Ghai S, Talbot P. Elements including metals in the atomizer and aerosol of disposable electronic cigarettes and electronic hookahs. PlOS ONE, 2017, 12: e0175430.

[150] Beauval N, Antherieu S, Soyez M, Gengler N, Grova N, Howsam M, Hardy E M, Fischer M, Appenzeller B M R, Goossens J F, Allorge D, Garcon G, Lo-Guidice J M, Garat A. Chemical evaluation of electronic cigarettes: Multicomponent analysis of liquid refills and their corresponding aerosols. Journal of Analytical Toxicology, 2017, 41: 670-678.

[151] Beauval N, Howsam M, Antherieu S, Allorge D, Soyez M, Garcon G, Goossens J F, Lo-Guidice J M, Garat A. Trace elements in e-liquids-development and validation of an ICP-MS method for the analysis of electronic cigarette refills. Regulatory Toxicology and Pharmacology, 2016, 79: 144-148.

[152] Olmedo P, Goessler W, Tanda S, Grau-Perez M, Jarmul S, Aherrera A, Chen R, Hilpert M, Cohen J E, Navas-Acien A, Rule A M. Metal concentrations in E-cigarette liquid and aerosol samples: The contribution of metallic coils. Environmental Health Perspectives, 2018, 126: 027010.

[153] Song J J, Go Y Y, Mun J Y, Lee S, Im G J, Kim Y Y, Lee J H, Chang J. Effect of electronic cigarettes on human middle ear. International Journal of Pediatric Otorhinolaryngology, 2018, 109: 67-71.

[154] Shen S, Li X F, Cullen W R, Weinfeld M, Le X C. Arsenic binding to proteins. Chemical Reviews, 2013, 113: 7769-7792.

[155] Cullen W R, McBride B C, Reglinski J. The reaction of methylarsenicals with thiols: Some biological implications. Journal of Inorganic Biochemistry, 1984, 21: 179-193.

[156] Lopez S, Miyashita Y, Simons S S, Jr. Structurally based, selective interaction of arsenite with steroid receptors. Journal of Biological Chemistry, 1990, 265: 16039-16042.

[157] Kitchin K T, Wallace K. Arsenite binding to synthetic peptides based on the zn finger region and the estrogen binding region of the human estrogen receptor-alpha. Toxicology and Applied Pharmacology, 2005, 206: 66-72.

[158] Lu M, Wang H, Li X F, Arnold L L, Cohen S M, Le X C. Binding of dimethylarsinous acid to Cys-13α of rat hemoglobin is responsible for the retention of arsenic in rat blood. Chemical Reviews, 2007, 20: 27-37.

[159] Jiang G, Gong Z, Li X F, Cullen W R, Le X C. Interaction of trivalent arsenicals with metallothionein. Chemical Research in Toxicology, 2003, 16: 873-880.

[160] Tang Z, Zhao F J. The roles of membrane transporters in arsenic uptake, translocation and detoxification in plants. Critical Reviews in Environmental Science and Technology, 2021, 51: 2449-2484.

[161] 刘劼, 乐晓春. 砷与蛋白相互作用研究进展. 分析科学学报, 2009, 25: 465-472.

[162] Sapra A, Ramadan D, Thorpe C. Multivalency in the inhibition of oxidative protein folding by arsenic(Ⅲ) species. Biochemistry, 2015, 54: 612-621.

[163] Zhao L, Wang Z, Xi Z, Xu D, Chen S, Liu Y. The reaction of arsenite with proteins relies on solution conditions. Inorganic Chemistry, 2014, 53: 3054-3061.

[164] Li G, Yuan S, Zheng S, Chen Y, Zheng Z, Liu Y, Huang G. The effect of salts in promoting specific and competitive interactions between zinc finger proteins and metals. Journal of the American Society for Mass Spectrometry, 2017, 28: 2658-2664.

[165] Petrick J S, Jagadish B, Mash E A, Aposhian H V. Monomethylarsonous acid (MMA^{Ⅲ}) and arsenite: LD$_{50}$ in hamsters and in vitro inhibition of pyruvate dehydrogenase. Chemical Research in Toxicology, 2001, 14: 651-656.

[166] Samuel S, Kathirvel R, Jayavelu T, Chinnakkannu P. Protein oxidative damage in arsenic induced rat brain: Influence of DL-alpha-lipoic acid. Toxicology Letters, 2005, 155: 27-34.

[167] Bergquist E R, Fischer R J, Sugden K D, Martin B D. Inhibition by methylated organo-arsenicals of the respiratory 2-oxo-acid dehydrogenases. Journal of Organometallic Chemistry, 2009, 694: 973-980.

[168] Spuches A M, Kruszyna H G, Rich A M, Wilcox D E. Thermodynamics of the As(III)-thiol interaction: Arsenite and monomethylarsenite complexes with glutathione, dihydrolipoic acid, and other thiol ligands. Inorganic Chemistry, 2005, 44: 2964-2972.

[169] 陈保卫, Le X C. 中国关于砷的研究进展. 环境化学, 2011, 30: 1936-1943.

[170] Lu J, Chew E H, Holmgren A. Targeting thioredoxin reductase is a basis for cancer therapy by arsenic trioxide. Proceedings of the National Academy of Sciences of the United States of America, 2007, 104: 12288-12293.

[171] Wang Z, Zhang H, Li X F, Le X C. Study of interactions between arsenicals and thioredoxins (human and *E. coli*) using mass spectrometry. Rapid Communications in Mass Spectrometry, 2007, 21: 3658-3666.

[172] Walter I, Schwerdtle T, Thuy C, Parsons J L, Dianov G L, Hartwig A. Impact of arsenite and its methylated metabolites on PARP-1 activity, PARP-1 gene expression and poly(ADP-ribosyl)ation in cultured human cells. DNA Repair (Amst), 2007, 6: 61-70.

[173] Hartwig A, Blessing H, Schwerdtle T, Walter I. Modulation of DNA repair processes by arsenic and selenium compounds. Toxicology, 2003, 193: 161-169.

[174] Zhou X, Sun X, Cooper K L, Wang F, Liu K J, Hudson L G. Arsenite interacts selectively with zinc finger proteins containing C3H1 or C4 motifs. Journal of Biological Chemistry, 2011, 286: 22855-22863.

[175] Schwerdtle T, Walter I, Hartwig A. Arsenite and its biomethylated metabolites interfere with the formation and repair of stable BPDE-induced DNA adducts in human cells and impair XPAZF and FPG. DNA Repair (Amst), 2003, 2: 1449-1463.

[176] 张亭栋. 含砷中药治疗白血病研究——谈谈癌灵1号注射液对白血病的治疗. 中国中西医结合杂志, 1998, 18: 581.

[177] 陈国强, 陈赛娟, 王振义, 陈竺. 氧化砷注射液治疗早幼粒细胞性白血病的机制研究及展望. 中国中西医结合杂志, 1998, 18: 581-582.

[178] Shen Z X, Chen G Q, Ni J H, Li X S, Xiong S M, Qiu Q Y, Zhu J, Tang W, Sun G L, Yang K Q, Chen Y, Zhou L, Fang Z W, Wang Y T, Ma J, Zhang P, Zhang T D, Chen S J, Chen Z, Wang Z Y. Use of arsenic trioxide (As_2O_3) in the treatment of acute promyelocytic leukemia (APS): Ii. Clinical efficacy and pharmacokinetics in relapsed patients. Blood, 1997, 89: 3354-3360.

[179] Niu C, Yan H, Yu T, Sun H P, Liu J X, Li X S, Wu W, Zhang F Q, Chen Y, Zhou L, Li J M, Zeng X Y, Yang R R, Yuan M M, Ren M Y, Gu F Y, Cao Q, Gu B W, Su X Y, Chen G Q, Xiong S M, Zhang T D, Waxman S, Wang Z Y, Chen Z, Hu J, Shen Z X, Chen S J. Studies on treatment of acute promyelocytic leukemia with arsenic trioxide: Remission induction, follow-up, and molecular monitoring in 11 newly diagnosed and 47 relapsed acute promyelocytic leukemia patients. Blood, 1999, 94: 3315-3324.

[180] Lallemand-Breitenbach V, Zhu J, Puvion F, Koken M, Honore N, Doubeikovsky A, Duprez E, Pandolfi P P, Puvion E, Freemont P, de The H. Role of promyelocytic leukemia (PML) sumolation in nuclear body formation, 11s proteasome recruitment, and As_2O_3-induced PML or PML/retinoic acid receptor alpha degradation. Journal of Experimental Medicine, 2001, 193: 1361-1371.

[181] Lallemand-Breitenbach V, Jeanne M, Benhenda S, Nasr R, Lei M, Peres L, Zhou J, Zhu J, Raught B, de The H. Arsenic degrades PML or PML-RARα through a sumo-triggered RNF4/ubiquitin-mediated pathway. Nature Cell Biology, 2008, 10: 547-555.

[182] Zhang X W, Yan X J, Zhou Z R, Yang F F, Wu Z Y, Sun H B, Liang W X, Song A X, Lallemand-Breitenbach V, Jeanne M, Zhang Q Y, Yang H Y, Huang Q H, Zhou G B, Tong J H, Zhang Y, Wu J H, Hu H Y, de The H, Chen S J, Chen Z.

Arsenic trioxide controls the fate of the PML-RARα oncoprotein by directly binding PML. Science, 2010, 328: 240-243.

[183] Kaiming C, Sheng Y, Zheng S, Yuan S, Huang G, Liu Y. Arsenic trioxide preferentially binds to the ring finger protein PML: Understanding target selection of the drug. Metallomics, 2018, 10: 1564-1569.

[184] Lou Y, Ma Y, Sun J, Ye X, Pan H, Wang Y, Qian W, Meng H, Mai W, He J, Tong H, Jin J. Evaluating frequency of PML-RARA mutations and conferring resistance to arsenic trioxide-based therapy in relapsed acute promyelocytic leukemia patients. Annals of Hematology, 2015, 94: 1829-1837.

[185] Goto E, Tomita A, Hayakawa F, Atsumi A, Kiyoi H, Naoe T. Missense mutations in PML-RARA are critical for the lack of responsiveness to arsenic trioxide treatment. Blood, 2011, 118: 1600-1609.

[186] Jiang Y H, Chen Y J, Wang C, Lan Y F, Yang C, Wang Q Q, Hussain L, Maimaitiying Y, Islam K, Naranmandura H. Phenylarsine oxide can induce the arsenite-resistance mutant PML protein solubility changes. International Journal of Molecular Sciences, 2017, 18: 247.

[187] Brown S B, Turner R J, Roche R S, Stevenson K J. Conformational analysis of thioredoxin using organoarsenical reagents as probes. A time-resolved fluorescence anisotropy and size exclusion chromatography study. Biochemistry and Cell Biology, 1989, 67: 25-33.

[188] Cornelis R, de Kimpe J. Elemental speciation in biological fluids. Invited lecture. Journal of Analytical Atomic Spectrometry, 1994, 9: 945-950.

[189] Kimpe J D, Cornelis R, Mees L, Vanholder R. Basal metabolism of intraperitoneally injected carrier-free ^{74}As-labeled arsenate in rabbits. Fundamental and Applied Toxicology, 1996, 34: 240-248.

[190] Styblo M, Thomas D J. Binding of arsenicals to proteins in an *in vitro* methylation system. Toxicology and Applied Pharmacology, 1997, 147: 1-8.

[191] Zhang X, Cornelis R, Kimpe J D, Mees L, Lameire N. Study of arsenic-protein binding in serum of patients on continuous ambulatory peritoneal dialysis. Clinical Chemistry, 1998, 44: 141-147.

[192] Lu M, Wang H, Li X F, Lu X, Cullen W R, Arnold L L, Cohen S M, Le X C. Evidence of hemoglobin binding to arsenic as a basis for the accumulation of arsenic in rat blood. Chemical Research in Toxicology, 2004, 17: 1733-1742.

[193] Naranmandura H, Suzuki K T. Identification of the major arsenic-binding protein in rat plasma as the ternary dimethylarsinous-hemoglobin-haptoglobin complex. Chemical Research in Toxicology, 2008, 21: 678-685.

[194] Chen B, Lu X, Shen S, Arnold L L, Cohen S M, Le X C. Arsenic speciation in the blood of arsenite-treated F344 rats. Chemical Research in Toxicology, 2013, 26: 952-962.

[195] Naranmandura H, Suzuki N, Suzuki K T. Trivalent arsenicals are bound to proteins during reductive methylation. Chemical Research in Toxicology, 2006, 19: 1010-1018.

[196] Alp O, Merino E J, Caruso J A. Arsenic-induced protein phosphorylation changes in HeLa cells. Analytical and Bioanalytical Chemistry, 2010, 398: 2099-2107.

[197] Pizarro I, Gómez M, Cámara C, Palacios M A, Roman-Silva D A. Evaluation of arsenic species-protein binding in cardiovascular tissues by bidimensional chromatography with ICP-MS detection. Journal of Analytical Atomic Spectrometry, 2004, 19: 292-296.

[198] Garcia-Sevillano M A, Garcia-Barrera T, Navarro F, Gomez-Ariza J L. Analysis of the biological response of mouse liver (*Mus musculus*) exposed to As_2O_3 based on integrated-omics approaches. Metallomics, 2013, 5: 1644-1655.

[199] Mandal B K, Suzuki K T, Anzai K, Yamaguchi K, Sei Y. A SEC-HPLC-ICP MS hyphenated technique for identification of sulfur-containing arsenic metabolites in biological samples. Journal of Chromatography B: Analytical Technologies in the Biomedical and Life Sciences, 2008, 874: 64-76.

[200] Gonzalez-Fernandez M, Garcia-Barrera T, Arias-Borrego A, Bonilla-Valverde D, Lopez-Barea J, Pueyo C, Gomez-Ariza J L. Metal-binding molecules in the organs of *Mus musculus* by size-exclusion chromatography coupled with UV spectroscopy and ICP-MS. Analytical and Bioanalytical Chemistry, 2008, 390: 17-28.

[201] Garcia-Sevillano M A, Garcia-Barrera T, Navarro-Roldan F, Montero-Lobato Z, Gomez-Ariza J L. A combination of metallomics and metabolomics studies to evaluate the effects of metal interactions in mammals. Application to *Mus musculus* mice under arsenic/cadmium exposure. Journal of Proteomics, 2014, 104: 66-79.

[202] Tsang C N, Bianga J, Sun H, Szpunar J, Lobinski R. Probing of bismuth antiulcer drug targets in *H. pylori* by laser ablation-inductively coupled plasma mass spectrometry. Metallomics, 2012, 4: 277-283.

[203] Lobinski R, Becker J S, Haraguchi H, Sarkar B. Metallomics: Guidelines for terminology and critical evaluation of analytical chemistry approaches (IUPAC technical report). Pure and Applied Chemistry, 2010, 82: 493-504.

[204] Sevcenco A M, Pinkse M W, Wolterbeek H T, Verhaert P D, Hagen W R, Hagedoorn P L. Exploring the microbial metalloproteome using mirage. Metallomics, 2011, 3: 1324-1330.

[205] Hu L, Cheng T, He B, Li L, Wang Y, Lai Y T, Jiang G, Sun H. Identification of metal-associated proteins in cells by using continuous-flow gel electrophoresis and inductively coupled plasma mass spectrometry. Angewandte Chemie International Edition, 2013, 125: 5016-5020.

[206] Wang Y, Wang H, Li H, Sun H. Metallomic and metalloproteomic strategies in elucidating the molecular mechanisms of metallodrugs. Dalton Transactions, 2015, 44: 437-447.

[207] Xu X, Wang H, Li H, Sun H. Metalloproteomic approaches for matching metals to proteins: The power of inductively coupled plasma mass spectrometry (ICP-MS). Chemistry Letters, 2020, 49: 697-704.

[208] Wang Y, Hu L, Yang X, Chang Y Y, Hu X, Li H, Sun H. On-line coupling of continuous-flow gel electrophoresis with inductively coupled plasma-mass spectrometry to quantitatively evaluate intracellular metal binding properties of metallochaperones *Hp*HypA and *Hp*HspA in *E. coli* cells. Metallomics, 2015, 7: 1399-1406.

[209] Wang H, Wang M, Yang X, Xu X, Hao Q, Yan A, Hu M, Lobinski R, Li H, Sun H. Antimicrobial silver targets glyceraldehyde-3-phosphate dehydrogenase in glycolysis of *E. coli*. Chemical Science, 2019, 10: 7193-7199.

[210] Wang H, Yan A, Liu Z, Yang X, Xu Z, Wang Y, Wang R, Koohi-Moghadam M, Hu L, Xia W, Tang H, Wang Y, Li H, Sun H. Deciphering molecular mechanism of silver by integrated omic approaches enables enhancing its antimicrobial efficacy in *E. coli*. PLOS Biology, 2019, 17: e3000292.

[211] Xu X, Wang H, Li H, Hu X, Zhang Y, Guan X, Toy P H, Sun H. *S*-dimethylarsino-glutathione (Darinaparsin®) targets histone H3.3, leading to trail-induced apoptosis in leukemia cells. Chemical Communications, 2019, 55: 13120-13123.

[212] Hannestad U, Lundqvist P, Sörbo B. An agarose derivative containing an arsenical for affinity chromatography of thiol compounds. Analytical Biochemistry, 1982, 126: 200-204.

[213] Hoffman R D, Lane M D. Iodophenylarsine oxide and arsenical affinity chromatography: New probes for dithiol proteins. Journal of Biochemistry, 1992, 267: 14005-14011.

[214] Zhou G, Jauhiainen M, Stevenson K, Dolphin P J. Human plasma lecithin: Cholesterol acyltransferase. Journal of Chromatography B: Biomedical Sciences and Applications, 1991, 568: 69-83.

[215] Kalef E, Walfish P G, Gitler C. Arsenical-based affinity chromatography of vicinal dithiol-containing proteins: Purification of L1210 leukemia cytoplasmic proteins and the recombinant rat c-erb a beta 1 T3 receptor. Analytical Biochemistry, 1993, 212: 325-334.

[216] Gitler C, Zarmi B, Kalef E. General method to identify and enrich vicinal thiol proteins present in intact cells in the oxidized, disulfide state. Analytical Biochemistry, 1997, 252: 48-55.

[217] Berleth E S, Kasperek E M, Grill S P, Braunscheidel J A, Graziani L A, Pickart C M. Inhibition of ubiquitin-protein

ligase (E3) by mono-and bifunctional phenylarsenoxides. Evidence for essential vicinal thiols and a proximal nucleophile. Journal of Biological Chemistry, 1992, 267: 16403-16411.

[218] Boerman M H, Napoli J L. Effects of sulfhydryl reagents, retinoids, and solubilization on the activity of microsomal retinol dehydrogenase. Archives of Biochemistry and Biophysics, 1995, 321: 434-441.

[219] Winski S L, Carter D E. Interactions of rat red blood cell sulfhydryls with arsenate and arsenite. Journal of Toxicology and Environmental Health, 1995, 46: 379-397.

[220] Menzel D B, Hamadeh H K, Lee E, Meacher D M, Said V, Rasmussen R E, Greene H, Roth R N. Arsenic binding proteins from human lymphoblastoid cells. Toxicology Letters, 1999, 105: 89-101.

[221] Chang K N, Lee T C, Tam M F, Chen Y C, Lee L W, Lee S Y, Lin P J, Huang R N. Identification of galectin I and thioredoxin peroxidase II as two arsenic-binding proteins in Chinese hamster ovary cells. Biochemical Journal, 2003, 371: 495-503.

[222] Mizumura A, Watanabe T, Kobayashi Y, Hirano S. Identification of arsenite-and arsenic diglutathione-binding proteins in human hepatocarcinoma cells. Toxicology and Applied Pharmacology, 2010, 242: 119-125.

[223] He X, Ma Q. Induction of metallothionein I by arsenic via metal-activated transcription factor 1: Critical role of C-terminal cysteine residues in arsenic sensing. Journal of Biological Chemistry, 2009, 284: 12609-12621.

[224] McStay G P, Clarke S J, Halestrap A P. Role of critical thiol groups on the matrix surface of the adenine nucleotide translocase in the mechanism of the mitochondrial permeability transition pore. Biochemical Journal, 2002, 367: 541-548.

[225] Jin Y J, Yu C L, Burakoff S J. Human 70-kDa SHP-1l differs from 68-kDa SHP-1 in its C-terminal structure and catalytic activity. Journal of Biological Chemistry, 1999, 274: 28301-28307.

[226] Yan H, Wang N, Weinfeld M, Cullen W R, Le X C. Identification of arsenic-binding proteins in human cells by affinity chromatography and mass spectrometry. Analytical Chemistry, 2009, 81: 4144-4152.

[227] Foley T D, Stredny C M, Coppa T M, Gubbiotti M A. An improved phenylarsine oxide-affinity method identifies triose phosphate isomerase as a candidate redox receptor protein. Neurochemical Research, 2010, 35: 306-314.

[228] Foley T D, Melideo S L, Healey A E, Lucas E J, Koval J A. Phenylarsine oxide binding reveals redox-active and potential regulatory vicinal thiols on the catalytic subunit of protein phosphatase 2a. Neurochemical Research, 2011, 36: 232-240.

[229] Bogumil R, Ullrich V. Phenylarsine oxide affinity chromatography to identify proteins involved in redox regulation: Dithiol-disulfide equilibrium in serine/threonine phosphatase calcineurin. Methods in Enzymology, 2002, 348: 271-280.

[230] Moaddel R, Sharma A, Huseni T, Jones G S, Hanson R N, Loring R H. Novel biotinylated phenylarsonous acids as bifunctional reagents for spatially close thiols: Studies on reduced antibodies and the agonist binding site of reduced torpedo nicotinic receptors. Bioconjugate Chemistry, 1999, 10: 629-637.

[231] Zhang X, Yang F, Shim J Y, Kirk K L, Anderson D E, Chen X. Identification of arsenic-binding proteins in human breast cancer cells. Cancer Letter, 2007, 255: 95-106.

[232] Jeanne M, Lallemand-Breitenbach V, Ferhi O, Koken M, Le Bras M, Duffort S, Peres L, Berthier C, Soilihi H, Raught B, de Thé H. PML/RARA oxidation and arsenic binding initiate the antileukemia response of As_2O_3. Cancer Cell, 2010, 18: 88-98.

[233] Heredia-Moya J, Kirk K L. An improved synthesis of arsenic-biotin conjugates. Bioorganic & Medicinal Chemistry, 2008, 16: 5743-5746.

[234] Donoghue N, Yam P T, Jiang X M, Hogg P J. Presence of closely spaced protein thiols on the surface of mammalian

cells. Protein Science, 2000, 9: 2436-2445.

[235] Donoghue N, Hogg P J. Characterization of redox-active proteins on cell surface. Methods in Enzymology, 2002, 348: 76-86.

[236] Dilda P J, Decollogne S, Rossiter-Thornton M, Hogg P J. Para to ortho repositioning of the arsenical moiety of the angiogenesis inhibitor 4-(N-(S-glutathionylacetyl)amino)phenylarsenoxide results in a markedly increased cellular accumulation and antiproliferative activity. Cancer Research, 2005, 65: 11729-11734.

[237] Park D, Don A S, Massamiri T, Karwa A, Warner B, MacDonald J, Hemenway C, Naik A, Kuan K T, Dilda P J, Wong J W, Camphausen K, Chinen L, Dyszlewski M, Hogg P J. Noninvasive imaging of cell death using an Hsp90 ligand. Journal of the American Chemical Society, 2011, 133: 2832-2835.

[238] Mailand N, Bekker-Jensen S, Faustrup H, Melander F, Bartek J, Lukas C, Lukas J. RNF8 ubiquitylates histones at DNA double-strand breaks and promotes assembly of repair proteins. Cell, 2007, 131: 887-900.

[239] Zhang F, Paramasivam M, Cai Q, Dai X, Wang P, Lin K, Song J, Seidman M M, Wang Y. Arsenite binds to the ring finger domains of RNF20-RNF40 histone E3 ubiquitin ligase and inhibits DNA double-strand break repair. Journal of the American Chemical Society, 2014, 136: 12884-12887.

[240] Jiang J, Tam L M, Wang P, Wang Y. Arsenite targets the ring finger domain of RBX1 E3 ubiquitin ligase to inhibit proteasome-mediated degradation of Nrf2. Chemical Research in Toxicology, 2018, 31: 380-387.

[241] Lee J, Levin D E. Methylated metabolite of arsenite blocks glycerol production in yeast by inhibition of glycerol-3-phosphate dehydrogenase. Molecular Biology of the Cell, 2019, 30: 2134-2140.

[242] Manthey K C, Griffin J B, Zempleni J. Biotin supply affects expression of biotin transporters, biotinylation of carboxylases and metabolism of interleukin-2 in jurkat cells. Journal of Nutrition, 2002, 132: 887-892.

[243] Zempleni J, Uptake, localization, and noncarboxylase roles of biotin. Annual Review of Nutrition, 2005, 25: 175-196.

[244] Zhang H N, Yang L, Ling J Y, Czajkowsky D M, Wang J F, Zhang X W, Zhou Y M, Ge F, Yang M K, Xiong Q, Guo S J, Le H Y, Wu S F, Yan W, Liu B, Zhu H, Chen Z, Tao S C. Systematic identification of arsenic-binding proteins reveals that hexokinase-2 is inhibited by arsenic. Proceedings of the National Academy of Sciences of the United States of America, 2015, 112: 15084-15089.

[245] Yan X, Li J, Liu Q, Peng H, Popowich A, Wang Z, Li X F, Le X C. p-Azidophenylarsenoxide: An arsenical " bait " for the in situ capture and identification of cellular arsenic-binding proteins. Angewandte Chemie International Edition, 2016, 55: 14051-14056.

第 19 章　原子光谱在汞研究中的应用

（冯新斌[①]*，史建波[②]，李　平[①]，阴永光[②]，付学吾[①]，俞　奔[②]，闫海鱼[①]，孟　博[①]）

▶ 19.1　原子光谱法是汞分析的重要手段 / 545

▶ 19.2　基于原子光谱的汞分析方法 / 545

▶ 19.3　原子光谱在汞研究中的应用进展 / 551

---

①中国科学院地球化学研究所，贵阳；②中国科学院生态环境研究中心，北京。*通讯作者联系方式：fengxinbin@vip.skleg.cn

本章导读

- 简要介绍了汞分析中常用的原子光谱技术，包括原子吸收光谱、原子荧光光谱、原子发射光谱等；
- 介绍了原子光谱及联用技术在汞的新形态分析中的应用，涉及颗粒结合态零价汞、一价汞、硫化汞与硒化汞纳米颗粒、二甲基汞、乙基汞；
- 介绍了天然汞同位素分析方法，涉及总汞和甲基汞同位素前处理和测定方法；
- 重点介绍了原子光谱在汞研究中的应用，涉及原子光谱在大气、水生生态系统和稻田汞研究的应用以及汞同位素技术应用进展。

19.1 原子光谱法是汞分析的重要手段

汞及其化合物不仅是一类毒性很高的持久性有毒污染物，还是一种全球性污染物。汞污染已经成为最重要的全球环境问题之一。我国是目前全球汞使用量和汞排放量最大的国家，环境汞污染形势十分严峻。汞的含量与形态分析是汞生物地球化学循环研究的基础，而原子光谱法是汞分析的最重要手段。近年来，汞同位素分析技术快速发展，为示踪汞污染来源和揭示汞的关键地球化学过程提供了新的有力工具。本章将重点介绍基于原子光谱的汞分析方法以及原子光谱在汞研究中的应用进展。

19.2 基于原子光谱的汞分析方法

19.2.1 汞分析中常用原子光谱技术

环境汞分析一般包括总汞和有机汞（甲基汞）。总汞的测定方法主要包括原子吸收光谱法（atomic absorption spectrometry，AAS）、原子荧光法（atomic fluorescence spectrometry，AFS）、电感耦合等离子体原子发射光谱法（inductive coupled plasma atomic emission spectrometry，ICP-AES）、电感耦合等离子体质谱法（inductive coupled plasma mass spectrometry，ICP-MS）及直接测汞仪检测法。而有机汞一般需经气相色谱（gas chromatography，GC）、高效液相色谱（high performance liquid chromatography，HPLC）等方法进行成分分离后再行检测。

1. 原子吸收光谱法

原子吸收光谱法在国内外已被广泛用于汞的分析，尤其是冷原子吸收法极大地提高了测定的灵敏度。AAS 是基于汞蒸气能够吸收波长 253.7 nm 的共振线，其吸光度值在一定范围内与汞含量成正比，再结合外标法计算出样品汞含量。该方法具有抗干扰能力强、分析速度快、样品用量少、选择性及稳定性好、检出限低、精密度高、操作简便等优点。缺点是标准曲线的动态范围相对较窄，通常小于 2 个数量级。

2. 原子荧光光谱法

原子荧光光谱法，在汞阴极灯照射下，基态汞原子被激发至高能态，再由高能态回到基态发射出特征波长的荧光，荧光强度与汞含量成正比，以荧光强度进行定量分析汞元素含量。该方法灵敏度比原子吸收法更高，具有干扰少、线性测量范围宽、原子化器和测量系统记忆效应小、仪器结构简单、调整方便等优点。

3. 电感耦合等离子体原子发射光谱法

电感耦合等离子体原子发射光谱法利用高频电感耦合等离子体作为光源激发汞元素，使汞原子处于激发态，由激发态返回基态时放出辐射，产生光谱，利用汞元素的特征共振发射线波长进行定性分析，利用光谱强度与样品浓度成正比进行定量分析。该方法的灵敏度和准确度不及ICP-MS，但成本价格较ICP-MS低。

4. 电感耦合等离子体质谱法

电感耦合等离子体质谱法利用高频电感耦合等离子体作为高温离子源，用电场和磁场将运动的离子按它们的质荷比分离，测出离子准确质量数即可进行定量分析。该方法具有高的灵敏度和准确性，样品预处理简单，可以获得同位素比值信息。但其成本高，价格昂贵。

5. 直接测汞仪法

直接测汞仪法（direct mercury analyzer，DMA）基于冷原子吸收与热解处理相结合的原理，无需样品的预处理即可对固、液态样品中的总汞进行检测，含汞废气经吸收液无害化处理后排放。该法准确度、精密度和灵敏度高（检出极限为0.02 ng）、速度快（平均每5 min测定一个样品）、简便可靠。

19.2.2 汞的新形态分析

汞在环境介质中的形态十分复杂，熟知的汞形态包括总汞、零价汞、二价汞、甲基汞、二甲基汞等。实际上，环境中还可能存在多种汞的"新"形态，如颗粒结合态零价汞、一价汞、硫化汞与硒化汞纳米颗粒、乙基汞等。与此同时，对于一些"传统"汞形态如二甲基汞的认识也不断拓展。原子光谱及其联用技术在这些"传统"及"新"的汞形态识别中发挥着重要的作用。

1. 颗粒结合态零价汞

水体溶解性气态汞（以零价汞为主）在汞的水气交换中发挥了关键作用。在环境水体中，除了溶解性气态汞，零价汞还可以与悬浮颗粒物结合，以颗粒结合态零价汞的形态存在[1]。与溶解性气态汞不同的是，与悬浮颗粒物结合的零价汞是不可吹扫的（nonpurgeable）[1]。因此，颗粒结合态零价汞的赋存及其与溶解性气态汞之间的平衡，将影响零价汞的水气交换。研究表明，向湖水中加入零价汞，4小时后其可吹扫零价汞回收率下降至19.6%，过滤实验表明回收率下降与湖水中的悬浮颗粒物有关[1]。同位素示踪（$^{201}Hg^0$、$^{199}Hg^{2+}$）结合热脱附-ICP-MS分析表明，零价汞部分与悬浮颗粒物结合，生成了颗粒结合态零价汞（150 ℃可释放）以及其他结合更强的汞形态或氧化态汞[1]。以磁铁矿颗粒（Fe_3O_4）为悬浮颗粒物模型，热脱附分析显示吸附于颗粒上的汞以零价汞为主[1]。采用同位素稀释法结合热脱附-ICP-MS分析，可以定量环境样品中颗粒结合态零价汞，Everglades上覆水中颗粒结合态零价汞可占总零价汞的45%～80%[1]。这一发现提示须对环境水体中零价汞的赋存审慎评估，进一步揭示颗粒结合态零价汞在水体零价汞循环中的关键作用。

2. 一价汞

理论上，汞的氧化还原过程多为单电子传递，因此一价汞应是汞氧化还原过程重要的中间体。但由于一价汞的不稳定性，目前尚缺乏一价汞的有效表征手段，对环境中一价汞的赋存状态仍缺乏认识。在以往的汞研究中，仅有少数工作涉及样品一价汞的分析。目前报道的一价汞的分析方法有直接质谱法[2]、X射线光电子能谱[3]、X射线吸收精细结构[4]以及拉曼光谱[5]等谱学技术。这些技术的不足之处在于：①提取、衍生、分析过程中存在一价汞转化，难以保证一价汞的稳定性；②灵敏度较低，难以实现环境浓度一价汞的定量分析。热脱附-原子光谱广泛用于固相样品中汞的形态分析。该方法虽有无需样品提取的优势，但在热解吸程序升温过程中，一价汞通常呈现两个峰（温度为80℃与130℃），分别对应于两步热反应（$Hg_2Cl_2 \xrightarrow{\Delta} Hg^0 + HgCl_2$；$HgCl_2 \xrightarrow{\Delta} Hg^0 + Cl_2$）[6,7]。因此，一价汞的热解吸峰往往与其他汞形态（如Hg^0、$HgCl_2$）无法完全分离[8]，该方法对一价汞的定性、定量存在不足。近期研究发现，以巯基乙醇为萃取剂，在温和的振荡萃取条件下，一价汞可保持稳定[9]。采用巯基乙醇提取，结合HPLC-ICP-MS分析，可实现环境固相基质中一价汞的分析。研究发现，提取过程中固相基质组分仍会催化部分一价汞还原为零价汞，所以该方法低估了样品中一价汞的含量[9]。采用这一方法，对多种环境固相基质中的一价汞进行了检测。研究表明，一价汞在汞矿区土壤、植物、大气$PM_{2.5}$、粉煤灰、二价汞暴露的微藻和废弃的荧光灯管中普遍存在[9]。其中，大气$PM_{2.5}$中一价汞占总汞的比例高达30%，这表明一价汞可能对大气中汞形态转化有重要影响。二价汞暴露的微藻中一价汞的赋存也表明环境中可能存在一价汞的生物生成。此外，在煤炭、飞灰、脱硫石膏、荧光灯管、汞矿区植物中还存在未知汞形态[9]。这一研究进一步揭示了环境汞形态的复杂性，为深入理解一价汞在汞氧化还原循环中的作用提供了技术手段。

3. 硫化汞与硒化汞纳米颗粒

汞具有极强的亲硫性，易于形成硫化汞或硒化汞。因此，硫化汞与硒化汞是环境与生物体中汞重要的汇[10-12]。纳米颗粒态是硫化汞与硒化汞重要的存在形式[10,13]。目前，多采用X射线吸收精细结构[14,15]与透射电镜技术[16]对环境与生物样品中的硫化汞与硒化汞进行识别。X射线吸收精细结构可给出硫化汞与硒化汞的定性与定量信息，但由于汞的X射线吸收精细结构特征性不强，因此其定性定量仍存在较大不确定性。此外，X射线吸收精细结构不能给出硫化汞与硒化汞的粒径信息。透射电镜可对硫化汞与硒化汞定性并给出粒径统计信息，但难以给出定量信息。如前所述，热脱附-原子光谱/质谱在环境样品的硫化汞、硒化汞识别中起着重要的作用[17,18]。近期，单颗粒-ICP-MS与场流分离-ICP-MS也进一步用于硫化汞、硒化汞纳米颗粒的识别与粒径表征。以四氢呋喃为载液，非对称流场分离-ICP-MS联用对石油烃凝析气的分析显示，汞主要以纳米颗粒的形式存在，其粒径约为10~50 nm[19]。采用四氢呋喃为溶剂，单颗粒-ICP-MS进一步证实含汞纳米颗粒的存在[19]。扫描透射电镜-元素分析确认含汞纳米颗粒为硫化汞[19]。领航鲸肝脏与大脑中汞、硒浓度均随年龄显著增加，而甲基汞占总汞的比例则随年龄增加而下降[20]。激光剥蚀-ICP-MS对大脑切片的分析显示，成年领航鲸大脑中存在汞、硒共存的"热点"区域，提示存在硒化汞颗粒[20]。单颗粒-ICP-MS分析显示，汞、硒纳米颗粒的颗粒数与粒径存在一致性，提示所检测到的颗粒为硒化汞纳米颗粒[20]。需要特别指出的是，由于含汞纳米颗粒多含有两种及以上元素（如汞和硒），ICP-MS联用系统的多元素分析就显得尤为重要。在单颗粒-ICP-MS分析时，同一颗粒上的多元素检测可提供颗粒组成的重要信息，因此可实现这一功能的单颗粒-ICP-飞行时间（TOF）-MS与单颗粒-MC-ICP-MS将在未来含汞纳米粒分析中发挥重要作用。目前，硫化汞与硒化汞纳米颗粒的环境赋存、生物摄入、转化行为尚不清楚，亟需建立多介质的综合表征技术对其展开研究。

4. 二甲基汞

以往人们认为二甲基汞主要在海洋环境中存在，近期采用吹扫捕集-热脱附-气相色谱-原子荧光在淡水系统如稻田环境中也观测到二甲基汞的广泛存在[21,22]。在汞污染稻田上覆水中溶解性二甲基汞浓度可达0.39~91 pg/L[平均(12±22)pg/L]，与深海和海岸上升流中观测浓度相当[22]。鉴于二甲基汞的高挥发性，这些发现有助于进一步理解汞的生物地球化学循环，并为大气中甲基汞的来源提供了一个可能的解释。未来应进一步加强二甲基汞的生成途径（化学[23,24]与生物[25]生成）与环境行为研究。

5. 乙基汞

人们对环境中的甲基汞研究较多，而对于其他有机汞如乙基汞的环境赋存知之甚少。这可能是由于甲基汞分析所采用的传统乙基化衍生前处理使得后续的气相色谱-原子光谱/质谱分析无法区分乙基汞与二价汞。乙基汞的分析可通过乙基化衍生之外的其他衍生策略来解决。美国佛罗里达 Everglades 湿地底泥与土壤中检测到乙基汞的普遍存在（n.d.~4.91 μg/kg），综合采用不同衍生方法（溴化、丁基化、苯基化）、不同色谱柱（DB-1、DB-17、Rtx-1、Rtx-50、DB-5MS、C_{18}）的 GC-AFS/ICP-MS、HPLC-AFS、GC-质谱联用技术对乙基汞的赋存进行识别与确认[26-28]。类似地，加拿大 Kejimkujik 国家公园湿地土壤[29]、加拿大湿地底泥（0.3~3.7 μg/kg）[30]、苏格兰 Union 运河底泥（n.d.~6.12 μg/kg）[31]、斯洛文尼亚 Idrija 汞矿区土壤（n.d.~17.4 μg/kg）[32]、日本妙高山森林土壤（n.d.~7.7 mg/kg）[33]以及原煤[34,35]中均检测到乙基汞的存在。在美国 Everglades 与加拿大几个湿地环境中，乙基汞仅存在于底泥与土壤中，间隙水与上覆水中并未观测到乙基汞的存在[27,30]。Everglades 底泥/土壤、苏格兰 Union 运河底泥中乙基汞与总汞呈正相关[26,31]，但加拿大几个湿地底泥中乙基汞与总汞无相关性[30]。Everglades 底泥与土壤中乙基汞与甲基汞的平均浓度相当[26,28]。Kejimkujik 国家公园湿地土壤中的乙基汞含量与甲基汞并无相关性，乙基汞含量与岩性相关：与杂砂岩和二长花岗岩相比，黑色硫化物板岩和灰色板岩中的乙基汞含量最高[29]。斯洛文尼亚 Idrija 汞矿区土壤中的乙基汞/甲基汞的垂直分布与 TOC 分布有相似的趋势，提示有机汞生成与有机质降解可能具有一定的联系[32]。日本妙高山森林土壤中乙基汞含量与甲基汞相当，亦与总汞无关，且在上层土壤中含量较高[33]。据推测，妙高山森林土壤中的乙基汞来自于火山喷气孔释放汞的进一步转化[33]。目前，这些环境中乙基汞的来源尚不清楚。人为污染（如西力生的合成与排放[36]）、环境中的化学乙基化（如乙基铅的转烷基化[37]、光化学乙基化[38]）以及生物乙基化（如豌豆幼苗对汞的乙基化[39]）均可能导致乙基汞的环境赋存。对于以上研究区域，大部分地点（苏格兰 Union 运河除外）均远离工业区，因此乙基汞人为直接排放的可能性较小，其很可能来自于环境中汞的乙基化过程。模拟实验显示，汞污染河漫滩土壤中存在汞的乙基化与甲基化，乙基汞与甲基汞生成受氧化还原电位控制[40]。E_h 为 0 mV 左右有利于乙基化，而-50~100 mV 之间则有利于甲基化[40]。对于汞污染河漫滩土壤，生物炭与石膏修复对乙基汞生成影响较小[41]。以上研究可能提示，与甲基汞类似，乙基汞在环境介质中也是普遍存在的，但乙基汞的来源与形成途径仍有待进一步研究。

19.2.3 汞同位素分析

汞在自然界中存在七种稳定同位素，包括 ^{196}Hg、^{198}Hg、^{199}Hg、^{200}Hg、^{201}Hg、^{202}Hg 和 ^{204}Hg。由于汞具有较高的原子质量，相对于传统同位素体系如 C、N、S 等，天然汞同位素分馏值非常小，同时天然样品中的汞含量通常较低。较小的分馏值和较低的含量使得天然样品的汞同位素测定在早期十分困难[42]。多接收（multi-collector，MC）-ICP-MS 具备多个法拉第杯，可同时测定多种同位素，相对于单接收（single-collector）-ICP-MS 或 ICP-TOF-MS 具有更高的检测精度，成为分析天然汞同位素体系的有效工具。

1. 总汞同位素的测定方法

1）样品前处理方法

汞同位素分析的前处理过程要求具有较高的回收率并尽量避免发生额外的同位素分馏。由于不同样品中汞的赋存形态和浓度均有差异，因此针对各种环境和生物样品需要采用不同的前处理方法。

针对汞含量较高的固体样品，湿法消解仍然是目前广泛使用的前处理方法。湿法消解作为测定环境样品总汞含量的前处理方式，广泛应用于沉积物、土壤和生物等固体样品的分析，方法操作简便、回收率高，消解液中汞的形态单一。在汞同位素分析中，通常使用混合酸液如 HNO_3/H_2SO_4[43]或王水[44]消解固体样品，并加入 BrCl 溶液对消解液进行保存，然后使用 MC-ICP-MS 进行测定。对于汞含量较低的固体样品，湿法消解具有一定的局限性。为了满足测试的汞浓度要求，需要消解较多固体样品，这不可避免使用较大量的酸液。由于此时溶液体积大，汞浓度低，且消解液基质复杂，通常还需要使用 $SnCl_2$ 还原配合吹扫捕集的方法对消解液中的汞进行富集。吹扫捕集方法的引入增加了前处理过程的复杂程度，而且可能影响到前处理过程的整体回收率。因此，有研究者提出了一种双步热解处理低浓度固态样品的方法[45,46]，该方法使用氧气作为载气，将固体样品加热至近 1000℃，挥发性物质在高温下被分解氧化，而热解释放的零价汞被氧化性溶液如反王水[46]或酸性高锰酸钾溶液[45]捕集并转化为二价汞。

大气汞具有三种形态，包括气态单质汞、气态氧化汞和颗粒态汞。颗粒态汞使用大气颗粒物采集器采集到滤膜上成为常规的固体样品进行双步热解法前处理。气态单质汞和气态氧化汞作为气态汞，在大气中含量极低，需要使用不同的捕集装置对其进行采集。对于气态单质汞，在早期通常使用金捕集管收集[47]。由于同位素检测需要采集较大量的汞，因此必须延长金捕集管收集大气汞的时间。在此过程中，大气中其他气体如水汽等会对金具有钝化作用从而降低采样效率，并可能引入额外的同位素分馏[48]。目前部分研究采用改性活性炭捕集管[48]收集气态单质汞。相对于金捕集管，改性活性炭减少了钝化风险，同时其较强的吸附作用可以使用比金捕集管更高的采样流速。对于气态氧化汞，可以使用 KCl 改性的滤膜进行采集[49]，随后使用双步热解法进行前处理。总的来说，这些方法都是通过不同的捕集方式将足量的不同形态汞转化为固体样品，然后按照固体样品的方法进行同位素分析。

水体样品中汞同位素分析具有更大的挑战。水体中的汞含量通常很低，且具有复杂的形态如溶解性气态汞、溶解性二价离子态汞、溶解性结合态汞以及悬浮颗粒物结合态汞等。吹扫捕集法可以将溶解性气态汞转化为气态汞后进行捕集；滤膜过滤可以收集悬浮颗粒物结合态汞。针对其他水体汞形态，可使用湿法消解，利用添加酸和 BrCl 溶液将溶解性结合态汞转化为溶解性二价离子态汞，然后采用不同方法进行富集，如受人为源排放污染的高汞含量海水可使用吹扫捕集法进行预富集[50]，或使用离子交换树脂柱富集大气降水[51]及天然海水[52]。

综上所述，随着待测样品汞浓度的降低，这些前处理方法变得更加复杂。低浓度样品的前处理过程中往往伴随着汞的形态转变，存在可能的同位素分馏。这些前处理方法的使用都是为了满足连续流进样测定的需求。然而，近年的部分研究表明，连续流进样系统具有一定的局限性，特别是样品溶液基质的干扰[53]，如挥发性烃类[54]、钨氧化物及氢氧化物[55]。因此，尽管增加了样品前处理的复杂程度，Blum 和 Johnson 仍建议将消解的样品溶液进行一次额外的吹扫捕集处理，并使用统一基质的溶液（如酸性高锰酸钾溶液）收集吹扫气态零价汞[53]。

2）进样系统

早期的进样系统使用金捕集管收集经过前处理后的气态零价汞，随后将其加热使捕集的汞随着载气进入 ICP[56]。然而，此进样方法与气相色谱或高效液相色谱类似，导入 ICP 的含汞气体检测后产生的是瞬时信号。瞬时信号无论在仪器调谐过程中，还是信号处理过程中都具有很大的局限性。受到当时信号处理方式和计算方法的限制，无法精确计算同位素比值，因此也很难获得准确的分馏值。

随后，一种使用冷蒸气发生的连续流进样系统[42]被广泛应用于汞同位素测定中。该系统使用两路载气，首先使用蠕动泵将含二价汞的溶液与 $SnCl_2$ 溶液混合后导入冷蒸气发生器，二价汞被还原为零价汞，经过气液分离后被吹入其中一路载气中，废液被蠕动泵抽离。同时，Tl 标准溶液经过雾化后形成气溶胶，随着另一路载气与含汞载气在进样系统混合，最后导入 ICP 进行测定。

相对于金捕集管热解进样系统，连续流进样系统可以获得连续且稳定的信号，降低了仪器调谐和信号处理的难度。在仪器调谐过程中，更容易观察信号强度的响应变化，可获得高精度的稳定信号。如 Geng 等[57]优化了连续流进样系统，将 1 ng/mL 的二价汞溶液响应信号提升至 1.78 V。响应信号强度越高，对于分馏值的测定就可获得更高的精确度。

部分研究也重新审视了早期的金捕集管与 MC-ICP-MS 联用技术。Berail 等[58]使用双金捕集管系统收集经过吹扫后的气态零价汞，随后将其加热并导入 MC-ICP-MS。由于金捕集管的特性，溶液基质中的可挥发性物质被排除，无法进入 ICP，这样就彻底去除了溶液基质对同位素分析的影响。虽然这种方法测得的数据相对于连续流进样系统而言具有较大的标准偏差，但值得深入研究和优化，并最终应用于低汞含量样品的同位素测定。

3）信号采集、数据处理及标准化

对于连续流进样系统而言，MC-ICP-MS 在样品测定时输出稳定的同位素信号。无论是 Neptune 系列还是 Nu Plasma 系列的 MC-ICP-MS，在进行汞同位素测试时都会将杯结构对应至原子质量 198 amu、199 amu、200 amu、201 amu、202 amu、203 amu 和 205 amu，分别对应样品溶液中 ^{198}Hg、^{199}Hg、^{200}Hg、^{201}Hg、^{202}Hg、^{203}Tl 和 ^{205}Tl。依据仪器法拉第杯结构设定的不同，一些研究还同时测定了高汞浓度溶液样品中的 ^{196}Hg 和 ^{204}Hg。不同质量数的汞同位素信号被转换为与 ^{198}Hg 信号强度的比值，并通过 ^{205}Tl/^{203}Tl 比值进行矫正以消除质谱仪测试过程中产生的质量歧视效应。为了消除质谱仪和进样系统的记忆效应，汞同位素分析中通常采用样品-标准交叉进样的方法（sample-standard bracketing method）。

早期的研究往往使用分馏系数 α、ε 等分析汞同位素分馏过程前后的同位素组成差异。由于天然汞同位素的分馏程度相对较小，同时为了更直观和方便地进行不同体系之间的分馏差异比较，分馏系数 δ 渐渐成为了最为广泛使用的数据表示方式。Blum 和 Bergquist 于 2007 年建议使用 NIST SRM 3133 作为计算 δ 时使用的标准样品，同时使用 UM-Almadèn（现已商业化为 NIST SRM 8610）作为第二标准样品用于评估质谱仪测量的准确度[59]。当分馏系数较小（<10‰）时，使用下面的简化公式对样品中的汞同位素组成进行描述：

$$\delta^{xxx}\mathrm{Hg}(‰) = \left\{ \left[\left(^{xxx}\mathrm{Hg}/^{198}\mathrm{Hg}\right)_{\mathrm{sample}} / \left(^{xxx}\mathrm{Hg}/^{198}\mathrm{Hg}\right)_{\mathrm{NIST\ SRM\ 3133}} \right] - 1 \right\} \times 1000 \quad (19\text{-}1)$$

$$\Delta^{xxx}\mathrm{Hg}(‰) = \delta^{xxx}\mathrm{Hg} - \beta \times \delta^{202}\mathrm{Hg} \quad (19\text{-}2)$$

$$\beta = \begin{cases} 0.252 & (xxx = 199) \\ 0.502 & (xxx = 200) \\ 0.752 & (xxx = 201) \\ 1.493 & (xxx = 204) \end{cases} \quad (19\text{-}3)$$

其中，xxx 代表质量数分别为 199、200、201、202 和 204。δ 和 Δ 分别表示该质量数同位素的质量分馏值和非质量分馏值。

与连续流进样系统不同的是，金捕集管热解进样系统在样品测试过程中无法输出稳定的连续信号，因此也无法直接利用仪器输出的比值数据进行计算。Berail 等[58]使用仪器输出的同位素信号强度，将其比值进行线性拟合。然后将拟合后的比值进行质量歧视矫正和分馏值计算中，缩小了传统使用信号值积分而产生的误差。

2. 甲基汞同位素测定方法

随着仪器本身测量精度的进步和调谐方法的改进，研究者开始关注更加复杂的形态汞同位素分析。在诸多汞形态中，甲基汞是最受关注、环境意义最为重大的汞形态。随着汞同位素分析技术快速发展，利用同位素分馏技术进行汞甲基化及去甲基化机理研究也成为关注的焦点，但首先需要解决的问题就是如何准确测定甲基汞的同位素组成。

为了满足进样系统的需求，需要使用合适的方法对样品中的甲基汞进行分离提纯。通常利用乙基化试剂将含汞溶液中的无机汞和甲基汞进行乙基化衍生，并使用GC[60]或HPLC[61]分离衍生后的甲基乙基汞且将其和少量未完全分离的挥发性含汞化合物富集在Tenax管中。随后加热Tenax管，在不同的温度释放出不同形态的汞，再使用金捕集管接收由甲基乙基汞热解产生的零价汞。然后加热金捕集管，用氧化性溶液如酸性高锰酸钾或反王水等富集释放出的汞，最终实现将样品中的甲基汞分离并转化为含汞溶液。这种前处理方式过程较为复杂，特别是乙基化衍生的步骤对样品中甲基汞的回收率非常关键。对于生物样品（例如鱼肉和大米样品），可以采用$NaBr/CuSO_4$萃取、硫代硫酸钠反萃取的连续提取方法有效分离甲基汞，实现甲基汞同位素的测定[62,63]。

也有研究者将GC与MC-ICP-MS联用进行甲基汞同位素的在线测定[64-66]。同样利用乙基化试剂将样品溶液中的无机汞和甲基汞进行乙基化衍生，然后使用GC分离各形态汞并直接导入MC-ICP-MS进行测定。但是该联用系统产生的是瞬时信号，在测定过程中会产生同位素比值漂移[67]。色谱峰的宽度、积分时间、积分点数量、数据处理方式以及同位素比值漂移的校正都会对测定结果的准确性和精度造成影响。该方法前处理过程相对简单，并可以有效地避免甲基汞离线提纯和富集过程中的回收率问题，但仍需要继续研究和优化以获得满意的甲基汞同位素分析结果。

19.3 原子光谱在汞研究中的应用进展

19.3.1 原子光谱在大气汞研究中的应用

汞在大气中主要以气态单质汞（gaseous elemental mercury，GEM）、气态氧化汞（gaseous oxidized mercury，GOM）、颗粒汞（particulate bound mercury，PBM）以及一些有机形态汞（如单甲基汞和二甲基汞）的形式存在[68,69]。GEM是大气汞最主要的形态，占大气汞总量的90%以上[70-72]。由于浓度非常低，大气汞检测通常采用预富集结合冷原子荧光光谱法（cold vapor atomic fluorescence spectrometry，CVAFS）测定，其中GEM的预富集主要采用金捕汞管和活性炭，GOM的预富集主要采用镀KCl的扩散管、回流喷雾箱、石英棉、尼龙膜和离子交换膜技术，PBM的预富集主要采用膜技术（如石英膜、聚四氟乙烯膜、醋酸纤维膜等），而甲基汞和二甲基汞主要采用Tenax管和回流喷雾箱进行预富集[68,73-79]。预富集的大气汞经化学或热解析法还原为气态单质汞，之后采用CVAFS法进行测定。近年来，研究人员基于预富集和CVAFS法开发了一系列大气汞在线监测设备，包括加拿大Tekran公司生产的Tekran 2537/1130/1135大气汞形态（GEM、GOM和PBM）分析仪以及立陶宛生产的Gardis-1A型、日本Nihon公司生产的Mercury/AM-3型和德国ENVEA GmbH公司生产的UT-3000型等气态汞（GEM+GOM）分析仪[76,80-82]。俄罗斯Lumex公司基于塞曼原子吸收光谱法和高频调制偏振光联合技术，开发不需要预富集而直接测定大气GEM的Lumex RA-915分析仪[83]。

前人在大气汞监测领域开展了大量的比对工作。不同型号的大气汞在线分析仪和基于手动方法监测的大气GEM浓度具有较好的一致性[76,84-86]。而有关大气GOM和PBM浓度的比对通常显示出较大的偏差[84,85,87,88]。当前对大气 GOM 具体包括的形态的认识比较匮乏，GOM 预富集的方法如镀 KCl 扩散管、回流喷雾箱、石英棉、尼龙滤膜和阳离子交换膜对不同形态的 GOM 预富集效率存在一定差异[89]，这是导致 GOM 分析误差的一个重要原因。此外，GOM 的预富集效率还受到臭氧和大气相对湿度等大气环境的影响[90,91]，这也会影响 GOM 的分析精度。在 PBM 测试方面，预富集所采用的滤膜材质和孔径、采样过程 GOM 或者 GEM 的吸附作用以及滤膜采集到 PBM 的解析方法都会对分析精度产生干扰[84,85,88,92]。总体来说，目前国际上有关大气 GEM 浓度的监测可靠性比较高，但在痕量大气汞形态 GOM、PBM 和有机形态汞含量的监测方面还有很大不确定性，因此需要研发更高效的预富集方法，并配合可靠的校正方法来降低其分析误差。

近 20 年来，大气汞特别是 GEM、PBM 和 GOM 的监测得到快速发展。基于全球大气汞监测网络（global mercury observation system，GMOS）的研究显示，大气 GEM 浓度具有显著的南北半球差异，2013 年北半球中高纬度地区 GEM 平均浓度为 1.55 ng/m^3，是热带地区平均值（1.23 ng/m^3）的 1.26 倍和南半球中高纬度地区平均值（0.93 ng/m^3）的 1.67 倍[93]，而热带和南半球中高纬度地区较低的人为源汞排放强度和较高的大气 GEM 氧化清除过程是导致这些地区大气 GEM 浓度偏低的重要原因[70,71,94]。在北半球，由于绝大部分（约 85%）的人为源汞排放主要集中在亚洲地区[95]，导致亚洲地区大气 GEM 浓度普遍高于（约 1.6 倍）北半球其他地区[96]。我国大气 GEM 浓度普遍高于世界其他地区，偏远地区和城市地区 GEM 平均浓度分别为 2.86 ng/m^3 和 9.20 ng/m^3，约是北半球背景值（约 1.5 ng/m^3）的 1.9 倍和 6.1 倍[97]，这与我国较高的人为源和自然源大气汞排放相对应[98,99]。近 20 年来，随着欧美地区对大气汞排放不断控制，欧洲和北美大气 GEM 浓度出现了显著的下降趋势，其中 2010~2015 年期间欧洲和北美地区 GEM 浓度约比 1990~1995 年期间下降了 32%[100]。然而，这种下降趋势在自由对流层和北半球高纬度地区却不太显著，其 2010~2015 年间平均值约比 1990~1995 年间下降了 7%[100]。这表明尽管欧美地区大气汞排放量有所降低，但北半球其他地区的大气汞排放可能仍在持续增加，从而导致北半球总体的大气汞排放量并未出现明显下降趋势，这是和北半球人为源汞排放清单的预测结果是一致的[95]。

北半球大气 GOM 和 PBM 浓度同样具有显著的区域分布特征，但人为源汞排放并不是唯一的影响因素。比如北半球城市地区大气 GOM 和 PBM 的平均浓度分别为 9.9 pg/m^3 和 10.0 pg/m^3，是低海拔偏远地区的 1.4~3.5 倍[101]。和大气 GEM 类似，亚洲地区 GOM 和 PBM 浓度也存在明显偏高现象，其平均浓度是欧洲和北美地区的 1.4~8.1 倍[101]。和大气 GEM 不同的是，在一些高海拔的偏远地区，其大气 GOM 和 PBM 平均浓度分别达到了 12.1 pg/m^3 和 11.0 pg/m^3，略高于城市地区，显著高于低海拔偏远地区（1.6~4.3 倍）。这种现象主要与高海拔地区较强的光化学反应有关。在高海拔地区，受较高辐射强度和大气氧化物浓度以及较低大气相对湿度的影响，GEM 容易转化为 GOM 和 PBM，这是导致这些地区大气 GOM 和 PBM 偏高的重要因素[102-107]。

和大气 GEM、GOM 和 PBM 的监测相比，目前国际上有关大气甲基汞和二甲基汞的研究相对较少。北极地区大气甲基汞和二甲基汞的平均浓度分别为 2.9 pg/m^3 和 3.9 pg/m^3 [108]，略低于瑞典哥德堡城市地区的监测结果（平均值 7.4 pg/m^3）[74]。目前对大气总有机形态汞的来源还不是很清楚，研究推测一些自然过程比如气溶胶的甲基化过程和海洋水体排放可能是重要来源[108]。此外，研究者在垃圾填埋场排气口监测的甲基汞和二甲基汞的平均浓度可达 2.1~77.0 ng/m^3 [75,109]，是城市和偏远地区的 10^3~10^4 倍，表明垃圾填埋是陆地生态系统重要的大气有机形态汞污染来源。

19.3.2 原子光谱在水生生态系统汞研究中的应用

水生生态系统主要包括水体、沉积物和水生生物三个组成要素。水体中的汞一般以单质汞（Hg^0）、

Hg^+和Hg^{2+}三种价态的无机或有机化合物形式存在。天然水体如湖泊、水库和海洋中汞含量低至纳克级，通常总汞（THg）含量约为5.0～10.0 ng/L，甲基汞含量约为0.01～1.0 ng/L，溶解在水体内的单质汞-溶解气态汞（DGM）一般低至10 pg/L。由于汞活跃的理化性质，环境中的汞极易在常温下发生形态转化和迁移，这大大增加了汞形态含量测定的难度。沉积物中汞主要以Hg^{2+}、Hg^0、HgO、HgS、$CH_3Hg(SR)$、$(CH_3Hg)_2S$等形式存在，且在固相和液相间保持平衡。未受污染的沉积物中，THg含量通常为0.01～0.3 mg/kg（干重），其中甲基汞含量约占0.1%～20.0%。水生生物体内汞以甲基汞为主要存在形态。水生生物体内的汞含量差异巨大，一般为0.01～5.00 mg/kg（湿重）[110]。通常在中性偏碱水体的生物体内汞含量较低，而在海洋、富含有机质的酸化湖泊中，鱼体汞含量常常超过世界卫生组织规定的食用水产品标准限值，而且鱼体内的甲基汞含量可达THg含量的98%。

因此，水生生态系统中汞的测定不仅要关注汞含量，而且需要关注汞形态。对于不同的环境介质，采取相应的前处理方式，结合原子光谱法进行测定。

1. 水样中不同形态汞的测定方法

国内外对淡水样品中汞的分析测定方法已比较成熟。无机汞的分析普遍采用美国EPA Method 1631E方法，甲基汞测定采用水相乙基化[111,112]或苯基化衍生结合GC-CVAFS法[113]。水样中汞的形态测定主要是按照化学性质、赋存状态及操作规程综合进行分类下的汞形态进行测定的。通常水样测定的汞形态包括：甲基汞、溶解态甲基汞、颗粒态甲基汞、活性汞、DGM、颗粒态汞、溶解态汞和THg。

水样可以分为原水和经过微孔滤膜过滤的水样，原水用于分析THg、甲基汞、活性汞和DGM，过滤水用于分析溶解态汞和溶解态甲基汞，差量法求得颗粒态汞和颗粒态甲基汞。当原水中颗粒物较多时，也可将滤膜上的颗粒物进行微波消解，测定消解液的颗粒态汞和颗粒态甲基汞。由于天然水体汞含量极低，所有样品分析结果的准确性取决于整个采样和样品分析的所有环节的严格操作，包括采用低空白的试剂、滤膜及采样瓶的除汞。

1）无机汞的测定方法

水样无机汞的测定是在美国EPA Method 1631E水体汞分析方法的基础上加以改进的[114]。DGM需要在水样采集后，立即以300 mL/min的无汞高纯氮气（99.999%）吹扫、金管预富集后带回实验室，在12小时内测定。活性汞测定前需加$SnCl_2$还原成单质汞，用镀金石英管预富集、二次金汞齐-CVAFS法测定；THg测定前24小时加0.5% BrCl氧化，测定前30 min加入0.2%盐酸羟氨溶液去除游离卤素，采用金管预富集、CVAFS测定。

目前常用的检测器包括Tekran 2500、Tekran 2600、Brooksrand Model Ⅲ、Lumex RA 915+和Nippon NIC RA-4300FG+等分析仪，适合痕量天然水体汞的测定。而液相色谱与ICP-MS、F732测汞仪及其他测汞仪更适合测定汞含量较高的样品。

质量控制（QA/QC）：水样测定一般设置5%～10%的野外试剂空白以及平行样。标准参考物质一般采用河水NRCC ORMS-4（总汞含量推荐值为22 ng/L）作为标样，要求回收率为90%～105%。仪器的工作标准曲线线性相关系数$R \geq 0.999$，THg方法的最低检出限为0.2～0.5 ng/L。

2）甲基汞的测定方法

水中甲基汞的检测是在美国EPA Method 1630方法的基础上加以改进的，使用蒸馏、水相乙基化、Tenax管预富集和GC-CVAFS测定[111,115-117]。对水样进行蒸馏的目的是消除基体干扰，蒸馏装置采用Teflon蒸馏瓶。通过对蒸馏瓶加热（控制温度为145℃）使气相中的甲基汞随氮气进入接收瓶（冰浴冷却）。当水样被蒸出80%～85%时立即将接收瓶取出，将蒸馏液于黑暗、室温下放置，在48 h内加入80 μL 1%的四乙基硼化钠使其乙基化20 min，Tenax管预富集，在65～100℃下15% OV-3填充的

GC 色谱柱恒温分离、700～900℃热解为单质汞、CVAFS 测定[118]。甲基汞通常采用 Tekran 2500、Brooksrand Model ⅠⅡ或全自动甲基汞测汞仪 MERX-M 和 Tekran 2700 等高精度、低检出限仪器进行测定。

质量控制（QA/QC）：水样测定需要设置 5%～10%的野外试剂空白和平行样。仪器的工作标准曲线线性相关系数 $R \geqslant 0.999$，45 mL 水样甲基汞的最低检出限为 0.009 ng/L。

2. 沉积物中汞的测定方法

沉积物由液相（即孔隙水）和固相两部分组成。沉积物中汞的形态研究分类也是根据操作步骤划分的，一种分为总汞和甲基汞，另一种按照 Tessier 等[119]的五步连续化学萃取分为不同赋存状态的汞，即可交换态、碳酸盐结合态、铁锰氧化物结合态、有机结合态和残渣态。

沉积物孔隙水中的汞测定方法与水样中汞的分析方法相同。沉积物固相和土壤中无机汞的测定需要先进行消解，消解方法有 HNO_3-H_2SO_4-V_2O_5 法、HNO_3-H_2SO_4-$KMnO_4$ 法、H_2SO_4-$KMnO_4$ 法、HNO_3-$HClO_4$-HF 法、王水法等。所有的方法中王水消解法被认为是最有效和简洁的方法。其次高温热解法是最为简单的 THg 测定方法。本节重点介绍采用王水消解、CVAFS 法测定沉积物中无机汞的方法。

准确称取 0.5 g 湿样或者 0.1 g 干样到比色管中，再加入王水，100～140℃加热 2.5～3 h 进行消解、BrCl 氧化、盐酸羟胺去除游离卤素、$SnCl_2$ 还原，CVAFS 法测定。沉积物土壤消解液中总汞测定所用仪器同水样[120]。采用热解法直接测定沉积物或土壤干样则更为快速、简单，标准参考物质测定的回收率高达 99.6%～102%。高温热解法测定总汞的仪器有 Milestone DMA-80[121]、Lumex RA 915+等[122]。

沉积物或土壤样品中甲基汞的形态分析主要包括三个步骤：萃取、分离和测定。萃取技术有酸萃取、碱萃取及有机溶剂萃取，结合微波和加热。汞分离普遍采用 HPLC 和 GC 技术。HPLC 对于汞的检测限太高，很难满足低汞含量环境样品（比如天然沉积物）中汞的形态分析。而 GC 结合高效预富集方法如镀金石英砂和 Tenax 管预富集样品中的汞，可以大大降低检出限，因此，GC 成为超低含量样品中汞形态的首选分离技术。另一种方法是毛细管电泳，该技术已被广泛研究并成为 HPLC 和 GC 汞形态分析，特别是在研究生物分子与烷基汞之间相互作用方面的方法补充。固体样品中甲基汞测定相对较难，最大问题是基质干扰，因此除了仪器的检测限，样品的前处理尤为重要。国内研究者结合固相微萃取（solid-phase micro-extraction，SPME），采用巯基棉富集-GC-电子捕获（electron capture detector，ECD）法用于环境样品的有机汞分析，也取得良好效果[123]。沉积物或土壤中甲基汞需要采用 HNO_3/$CuSO_4$ 溶液浸提，CH_2Cl_2 萃取并结合水相乙基化，GC-CVAFS 法测定，方法检出限可达 0.006 ng/g[124]。

3. 生物样品汞的形态测定方法

生物样品中汞的形态测定通常分为 THg 和甲基汞。鱼体 THg 测定与沉积物相同，样品需要采用 H_2SO_4/HCl（2∶8，V/V）消解；也可采用 Milestone DMA-80、Lumex RA 915+直接测汞仪高温热解的方法测定。鱼体甲基汞采用碱消解，少量样品（一般<50 μL）进样测定。鱼体 THg 和甲基汞的上机测定与沉积物消解液相同。

19.3.3 原子光谱在稻田汞研究中的应用

无机汞可在特殊的环境条件下被转化成毒性更强的甲基汞，进而在食物链中生物富集和生物放大，对人体健康构成潜在威胁。国际学术界普遍认同，食用鱼类等水产品是人体甲基汞暴露的主要途径[125]。在欧洲和北美地区,人群普遍存在食用鱼类等水产品甲基汞暴露的健康风险问题。然而,Feng 等[126]研究发现我国南方内陆居民甲基汞暴露的主要途径是食用大米。这一发现打破了国际上认为食用鱼等水产品是人体甲基汞暴露主要来源的传统认识。水稻是世界上最重要的粮食作物之一,全球一半以上的人口以稻米为主食。中国水稻种植面积约占全球 50%,近三分之二的人口以稻米为主食。汞污染区稻米富集甲基汞是一个普遍的现象,居民食用汞污染的大米所导致的汞暴露风险不容忽视。国内外学者围绕稻田生态系统汞的生物地球化学循环开展了大量的研究工作,取得了一系列重要研究成果。

1. 稻田土壤中汞的甲基化过程

稻田土壤中无机汞的甲基化过程和甲基汞的去甲基化过程是汞形态转化的重要环节,直接影响土壤甲基汞的含量水平,最终影响甲基汞在稻米的富集程度[127,128]。利用单一富集稳定汞同位素示踪技术结合 GC-ICP-MS,Zhao 等[128]系统测定了贵州万山汞矿区稻田土壤无机汞的甲基化速率和甲基汞的去甲基化速率。研究发现,汞矿区稻田土壤中甲基汞的浓度是甲基化和去甲基化共同作用的结果,净甲基化的潜力在一定程度上可反映稻田土壤甲基汞的浓度水平[127,128]。研究证实,大气沉降的"新汞"比土壤中的"老汞"更容易转化成甲基汞,进而被水稻吸收富集[128,129]。因此,减少大气汞排放是降低水稻甲基汞污染的有效途径。稻田作为一种特殊的间歇性湿地生态系统,土壤存在活跃的甲基化作用,这是稻米富集甲基汞的主要原因之一[127-130]。稻田土壤中无机汞的甲基化和甲基汞的去甲基化非常复杂,受多种物理、化学、生物因素的影响,主要包括甲基化微生物种群丰度及活性[131-133]、有机质含量[134-136]、pH[128,137]、氧化还原电位[133,137]、大气汞沉降[127-130]。

2. 水稻体内汞的来源与分布特征

基于 CVAFS、GC-CVAFS 和 ICP-MS 分析技术,国内外学者对水稻体内汞的来源与分布特征开展了系统的研究工作。发现水稻地上部分无机汞主要来源于大气,根部无机汞主要来源于土壤,而水稻茎部的无机汞同时来源于大气和土壤[130,138,139]。由于水稻不同部位无机汞的来源不同,导致不同污染类型的稻田系统(如土法炼汞区和废弃汞矿区)中水稻各部位无机汞含量分布存在差异,土法炼汞区表现为叶>根>茎>壳>米(糙米),废弃汞矿区为根>叶>茎>壳>米(糙米)[130]。水稻叶片可以直接吸收大气中的气态单质汞,但是大气中的气态单质汞不能通过水稻胚珠维管束途径输送至稻米[129,130,140]。对于整株成熟水稻,大部分无机汞富集在水稻地上部分,叶部富集的无机汞明显高于其他部位。对于不同污染类型的水稻田,无机汞在水稻不同部位的分布比例存在差异[130]。而水稻体内的甲基汞主要来源于土壤,水稻根际土壤甲基汞含量是控制水稻各部位甲基汞富集程度的关键因素[130,138]。因此,对于不同污染类型的稻田系统(土法炼汞区和废弃汞矿区),水稻各部位甲基汞浓度分布特征具有一致性,都表现为米(糙米)>根>壳>茎>叶。对于整株成熟水稻,绝大部分的甲基汞储存在稻米中,而稻米的甲基汞主要赋存于精米[130]。

基于同步辐射 X 射线荧光微区谱学成像技术(synchrotron radiation micro-X-ray fluorescence,SR-μXRF)的研究表明,相对于胚乳,汞(主要为无机汞)和铁、钙、磷、钾等营养元素强烈富集在糙米表层——对应为果皮和糊粉层。基于 X 射线近边吸收谱学分析技术(X-ray absorption near edge structure,XANES),Meng 等[141]研究发现,稻米中的无机汞主要是与半胱氨酸(cysteine)结合,并以植物螯合肽(phytochelatins,PCs)的形式存在。因此,大气中的无机汞被稻米吸收后很难发生运移,致使其主要富集在糙米表层。同样,稻米中的甲基汞也主要与半胱氨酸结合,但与无机汞不同的是,与半胱氨酸结合的甲基汞主要赋存于蛋白质[141]。利用 HPLC-ICP-MS,同样也证实稻米中的甲基汞主要与半胱氨酸结合,以 CH_3Hg-L-cysteine(CH_3HgCys)的形式存在[142]。与半胱氨酸结合的甲基汞(CH_3HgCys)可以直接穿过血脑和胎盘屏障,对人体靶器官(脑)和胎儿造成伤害,这更加凸显了稻米甲基汞的危害性。

3. 水稻无机汞和甲基汞的吸收富集过程

由于稻米中的无机汞主要与半胱氨酸结合（Hg-cysteine），无机汞的这种结合形态在水稻体内具有重要的解毒作用[141,143]。基于反相高效液相色谱联用电感耦合等离子质谱仪/电喷雾离子化质谱仪（RP-HPLC-ICPMS/ESI-MS），Krupp 等[143]发现水稻体内对汞具有解毒作用的 PCs 可以有效螯合 Hg^{2+} 形成 Hg-PCs 化合物，促进 Hg^{2+} 从细胞质向液泡的转运，从而将 Hg^{2+} 滞留在根部，减少向地上部分的运移，因此水稻根表铁膜组织对土壤中的无机汞表现出潜在的"屏障"作用[130,138]。水稻生长期间，根、茎、叶部持续从环境中吸收无机汞进入体内，其中茎和叶可以从大气吸收气态单质汞并在水稻体内转化成二价汞，但绝大部分被固定在水稻植株体内，并未发现明显的运移现象[138]。

水稻对甲基汞的吸收富集过程与无机汞截然不同，土壤中的甲基汞可以自由穿过水稻根表铁膜组织而被根部吸收[129]。水稻在整个生长期间，根部持续从土壤吸收甲基汞并向地上部分运移。在水稻成熟之前，甲基汞主要储存在茎部和叶部；而在水稻成熟期间，茎部和叶部的绝大部分甲基汞被迅速转移至稻米[129]。水稻对甲基汞的富集是吸收-运移-富集的动态变化过程[129]。Meng 等[141]推测，土壤中的甲基汞被水稻根部吸收后，很可能与蛋白质结合并被运移至地上部分（主要为茎和叶），在水稻果实成熟期间随蛋白质一起被转运、富集至稻米。

19.3.4 汞同位素技术应用进展

汞同位素可以用来示踪汞污染来源和揭示汞的关键地球化学过程。过去十多年来，随着 MC-ICP-MS 技术的快速发展，汞同位素地球化学研究取得了引人注目的进展。

1. 源排放特征

我国是全球最大的汞生产、使用和排放国，燃煤、有色金属冶炼（主要是锌冶炼）和水泥生产是主要人为源排放行业[144]。我国煤燃烧排放汞的同位素特征为 $\delta^{202}Hg$=−0.70‰、$\Delta^{199}Hg$=−0.05‰[145]。我国锌矿石的 $\delta^{202}Hg$ 值变化范围较大（−1.87‰～+0.70‰），$\Delta^{199}Hg$ 值为−0.24‰～+0.18‰，不同类型矿石的 $\Delta^{199}Hg$ 值存在显著差异[146]。

2. 大气汞循环

汞是一种全球性污染物，大气汞的长距离迁移转化是全球汞生物地球化学循环的重要环节。我国背景区与城市大气总汞具有不同的汞同位素特征。背景区由于森林植被排放影响具有偏正的 $\delta^{202}Hg$(0.07‰～0.99‰)和偏负 $\Delta^{199}Hg$(−0.24‰～0.09‰)，而城市区域由于工业排放等人为活动影响具有偏负的 $\delta^{202}Hg$(−0.73‰±0.54‰)和近零的 $\Delta^{199}Hg$(0.02‰±0.07‰)[147-149]。我国背景区 PBM 具有显著偏正的 $\Delta^{199}Hg$ 值（=0.27‰±0.22‰～0.66‰±0.32‰），明显高于一次人为源（0‰）和城市 PBM(−0.02‰～0.05‰)，主要受二次颗粒汞来源的影响[150]。我国街道灰尘 $\delta^{202}Hg$ 值为−0.61‰±0.92‰，$\Delta^{199}Hg$ 值为−0.03‰±0.08‰，其主要来源为煤燃烧和工业活动[151]。

3. 森林生态系统汞循环

森林系统占全球陆地总面积的 31%，是全球汞生物地球化学循环最活跃的区域[152]。温带混交林植被生长季夜晚 GEM 发生亏损且 $\delta^{202}Hg$ 值偏重，指示森林植被对大气 GEM 有吸收作用[153]。森林叶片 $\Delta^{199}Hg$ 值随其生长呈下降趋势（−0.16‰至−0.47‰～−0.34‰），叶片释放的 GEM 具有明显偏正的 $\Delta^{199}Hg$，说明植被/大气界面存在汞再释放作用[154]。长白山温带森林 GEM 的 $\delta^{202}Hg$ 和 $\Delta^{199}Hg$ 值与植物生长指数具有显著正相关关系，提示植被影响大气汞同位素特征；而哀牢山亚热带森林冠层的 GEM 的 δ^{202}Hg 与人为源排放量具有显著正相关,表明植被活动和人为源释放是北半球大气汞同位素空间变化的主要驱动因子[148]。青藏高原/亚热带山地森林的土壤与凋落物具有相似的 Δ^{199}Hg 和 Δ^{200}Hg 值,指示凋落物是森林土壤的主要输入汞源[155,156];而森林土壤与降水的 Δ^{199}Hg 值具有显著差异,降水主要通过影响凋落物生物量和降解过程间接影响森林土壤汞库[156]。汞同位素三元混合模型结果显示,冰川退化区土壤汞主要来源于植被演替吸收的大气汞,可达原始冰川汞库的 10 倍,1850 年小冰期以来约 300~400 Mg 汞沉降累积在冰川退缩区,未来进一步的全球变暖-植被格局改变将显著影响全球汞的生物地球化学循环[157]。

4. 水生生态系统汞循环

在水生生态系统中,无机汞可以转化为甲基汞并随着食物链累积放大,对水生生态及人群健康造成危害,因此水生生态系统汞的地球化学循环一直是环境汞研究的重要领域。贵州红枫湖和百花湖沉积物 δ^{202}Hg 分别为−1.67‰~−2.02‰和−0.60‰~−1.10‰,百花湖沉积物的汞主要来自于有机化工厂排放,而汞同位素 MIF 显示红枫湖沉积物汞主要来自流域侵蚀[158]。我国珠江三角洲河口和南海沉积物的汞同位素特征具有显著差异,南海沉积物高 Δ^{199}Hg 值表明部分汞进入沉积物之前经历了显著的 Hg^{2+} 光还原过程,而珠江三角洲河口沉积物 Δ^{199}Hg 值偏低表明河流输入是汞的主要来源[159]。珠江三角洲海洋野生鱼肉 Δ^{199}Hg 值为 0.05‰±0.10‰~0.59‰±0.30‰,Δ^{199}Hg/Δ^{201}Hg 约为 1.26,说明水中的 MeHg 在进入食物链之前发生光降解;草食性、底栖性和肉食性鱼类的 Δ^{199}Hg 值具有显著差异,可以利用汞同位素揭示食物链甲基汞暴露来源[160]。

5. 土壤-农作物系统汞循环

土壤是全球最大的汞库,而农作物是人类基本的食物来源之一。森林、泥炭、草地和苔原生态系统表层土壤 Δ^{199}Hg 和 Δ^{200}Hg 与植被的汞同位素类似,其表层土壤汞 54%±21%、12%±9% 和 34%±18% 分别来自大气 GEM、大气 GOM 以及土壤母质,而农田表层土壤汞主要来自于人为源[161]。贵州省汞矿区、燃煤区域和锌冶炼区表层土壤 δ^{202}Hg 和 Δ^{199}Hg 值具有显著差异,这是由不同汞污染源的混合作用以及自然界同位素分馏过程引起的[162]。万山汞矿区土壤水溶态汞 δ^{202}Hg(0.70‰±0.13‰)、$(NH_4)_2S_2O_3$ 提取态汞 δ^{202}Hg(1.28‰±0.25‰)相对土壤总汞 δ^{202}Hg(−0.02‰±0.16‰)显著偏正,而 Δ^{199}Hg 无显著差异[163]。水稻植株 Δ^{199}Hg 值为−0.29‰~+0.01‰,可以利用 Δ^{199}Hg 二元混合模型计算水稻不同部位大气和土壤汞的相对贡献[164]。玉米地土壤、大气和玉米植株 δ^{202}Hg 和 Δ^{199}Hg 值存在显著差异,玉米植株从大气和土壤吸收汞的过程发生明显的 MDF,而植株地上部分汞主要来源于大气[165]。

参 考 文 献

[1] Wang Y, Li Y, Liu G, Wang D, Jiang G, Cai Y. Elemental mercury in natural waters: Occurrence and determination of particulate Hg(0). Environmental Science & Technology, 2015, 49(16): 9742-9749.

[2] Raofie F, Ariya P A. Product study of the gas-phase BrO-initiated oxidation of Hg^0: Evidence for stable Hg^{1+} compounds. Environmental Science & Technology, 2004, 38(16): 4319-4326.

[3] Zhou Q, Lei Y, Liu Y, Tao X, Lu P, Duan Y, Wang Y. Gaseous elemental mercury removal by magnetic Fe-Mn-Ce sorbent in simulated flue gas. Energy & Fuels, 2018, 32(12): 12780-12786.

[4] Allen P G, Gash A E, Dorhout P K, Strauss S H. XAFS studies of soft-heavy-metal-ion-intercalated M_xMOS_2 (M = Hg^{2+},

Ag$^+$) solids. Chemistry of Materials, 2001, 13(7): 2257-2265.

[5] Richter P W, Wong P T T, Whalley E. Effect of pressure on Raman-spectra of mercurous chloride and bromide. Journal of Chemical Physics, 1977, 67(5): 2348-2354.

[6] Lopez-Anton M A, Yuan Y, Perry R, Maroto-Valer M M. Analysis of mercury species present during coal combustion by thermal desorption. Fuel, 2010, 89(3): 629-634.

[7] Rallo M, Lopez-Anton M A, Meij R, Perry R, Maroto-Valer M M. Study of mercury in by-products from a Dutch co-combustion power station. Journal of Hazardous Materials, 2010, 174(1-3): 28-33.

[8] Windmoller C C, Wilken R D, Jardim W D. Mercury speciation in contaminated soils by thermal release analysis. Water Air and Soil Pollution, 1996, 89(3-4): 399-416.

[9] Wang Y, Liu G, Li Y, Liu Y, Guo Y, Shi J, Hu L, Cai Y, Yin Y, Jiang G. Occurrence of mercurous [Hg(Ⅰ)] species in environmental solid matrices as probed by mild 2-mercaptoethanol extraction and HPLC-ICP-MS analysis. Environmental Science & Technology Letters, 2020, 7(7): 482-488.

[10] Chen Y, Yin Y, Shi J, Liu G, Hu L, Liu J, Cai Y, Jiang G. Analytical methods, formation, and dissolution of cinnabar and its impact on environmental cycle of mercury. Critical Reviews in Environmental Science and Technology, 2017, 47(24): 2415-2447.

[11] Gajdosechova Z, Mester Z, Feldmann J, Krupp E M. The role of selenium in mercury toxicity-current analytical techniques and future trends in analysis of selenium and mercury interactions in biological matrices. TrAC-Trends in Analytical Chemistry, 2018, 104: 95-109.

[12] 张华, 冯新斌, 王祖光, Larssen T. 硒汞相互作用及机理研究进展. 地球与环境, 2013, 41(6): 696-708.

[13] Dang F, Li Z, Zhong H. Methylmercury and selenium interactions: Mechanisms and implications for soil remediation. Critical Reviews in Environmental Science and Technology, 2019, 49(19): 1737-1768.

[14] Manceau A, Wang J, Rovezzi M, Glatzel P, Feng X. Biogenesis of mercury-sulfur nanoparticles in plant leaves from atmospheric gaseous mercury. Environmental Science & Technology, 2018, 52(7): 3935-3948.

[15] Korbas M, O'Donoghue J L, Watson G E, Pickering I J, Singh S P, Myers G J, Clarkson T W, George G N. The chemical nature of mercury in human brain following poisoning or environmental exposure. ACS Chemical Neuroscience, 2010, 1(12): 810-818.

[16] Barnett M O, Harris L A, Turner R R, Stevenson R J, Henson T J, Melton R C, Hoffman D P. Formation of mercuric sulfide in soil. Environmental Science & Technology, 1997, 31(11): 3037-3043.

[17] Feng X B, Lu J Y, Gregoire D C, Hao Y J, Banic C M, Schroeder W H. Analysis of inorganic mercury species associated with airborne particulate matter/aerosols: Method development. Analytical and Bioanalytical Chemistry, 2004, 380(4): 683-689.

[18] Liu X, Wang S, Zhang L, Wu Y, Duan L, Hao J. Speciation of mercury in FGD gypsum and mercury emission during the wallboard production in China. Fuel, 2013, 111: 621-627.

[19] Ruhland D, Nwoko K, Perez M, Feldmann J, Krupp E M. AF4-UV-MALS-ICP-MS/MS, spICP-MS, and STEM-EDX for the characterization of metal-containing nanoparticles in FGD condensates from petroleum hydrocarbon samples. Analytical Chemistry, 2019, 91(1): 1164-1170.

[20] Gajdosechova Z, Lawan M M, Urgast D S, Raab A, Scheckel K G, Lombi E, Kopittke P M, Loeschner K, Larsen E H, Woods G, Brownlow A, Read F L, Feldmann J, Krupp E M. In vivo formation of natural HgSe nanoparticles in the liver and brain of pilot whales. Scientific Reports, 2016, 6: 34361.

[21] Wang Z, Sun T, Driscoll C T, Yin Y, Zhang X. Mechanism of accumulation of methylmercury in rice (Oryza sativa L.) in a mercury mining area. Environmental Science & Technology, 2018, 52(17): 9749-9757.

[22] Wang Z, Sun T, Driscoll C T, Zhang H, Zhang X. Dimethylmercury in floodwaters of mercury contaminated rice paddies. Environmental Science & Technology, 2019, 53(16): 9453-9461.

[23] Jonsson S, Mazrui N M, Mason R P. Dimethylmercury formation mediated by inorganic and organic reduced sulfur surfaces. Scientific Reports, 2016, 6: 27958.

[24] Kanzler C R, Lian P, Trainer E L, Yang X, Govind N, Parks J M, Graham A M. Emerging investigator series: Methylmercury speciation and dimethylmercury production in sulfidic solutions. Environmental Science-Processes & Impacts, 2018, 20(4): 584-594.

[25] Baldi F, Parati F, Filippelli M. Dimethylmercury and dimethylmercury-sulfide of microbial origin in the biogeochemical cycle of Hg. Water Air and Soil Pollution, 1995, 80(1-4): 805-815.

[26] Cai Y, Jaffe R, Jones R. Ethylmercury in the soils and sediments of the Florida Everglades. Environmental Science & Technology, 1997, 31(1): 302-305.

[27] Mao Y, Liu G, Meichel G, Cai Y, Jiang G. Simultaneous speciation of monomethylmercury and monoethylmercury by aqueous phenylation and purge-and-trap preconcentration followed by atomic spectrometry detection. Analytical Chemistry, 2008, 80(18): 7163-7168.

[28] Mao Y, Yin Y, Li Y, Liu G, Feng X, Jiang G, Cai Y. Occurrence of monoethylmercury in the Florida Everglades: Identification and verification. Environmental Pollution, 2010, 158(11): 3378-3384.

[29] Siciliano S D, Sangster A, Daughney C J, Loseto L, Germida J J, Rencz A N, O'Driscoll N J, Lean D R S. Are methylmercury concentrations in the wetlands of Kejimkujik National Park, Nova Scotia, Canada, dependent on geology? Journal of Environmental Quality, 2003, 32(6): 2085-2094.

[30] Holmes J, Lean D. Factors that influence methylmercury flux rates from wetland sediments. Science of the Total Environment, 2006, 368(1): 306-319.

[31] Cavoura O, Brombach C C, Cortis R, Davidson C M, Gajdosechova Z, Keenan H E, Krupp E M. Mercury alkylation in freshwater sediments from Scottish canals. Chemosphere, 2017, 183: 27-35.

[32] Tomiyasu T, Kodamatani H, Imura R, Matsuyama A, Miyamoto J, Akagi H, Kocman D, Kotnik J, Fajon V, Horvat M. The dynamics of mercury near Idrija mercury mine, Slovenia: Horizontal and vertical distributions of total, methyl, and ethyl mercury concentrations in soils. Chemosphere, 2017, 184: 244-252.

[33] Kodamatani H, Katsuma S, Shigetomi A, Hokazono T, Imura R, Kanzaki R, Tomiyasu T. Behavior of mercury from the fumarolic activity of Mt. Myoko, Japan: Production of methylmercury and ethylmercury in forest soil. Environmental Earth Sciences, 2018, 77(13): 478.

[34] Gao E, Jiang G, He B, Yin Y, Shi J. Speciation of mercury in coal using HPLC-CV-AFS system: Comparison of different extraction methods. Journal of Analytical Atomic Spectrometry, 2008, 23(10): 1397-1400.

[35] Lusilao-Makiese J, Tessier E, Amouroux D, Tutu H, Chimuka L, Cukrowska E M. Speciation of mercury in South African coals. Toxicological and Environmental Chemistry, 2012, 94(9): 1688-1706.

[36] Acosta G, Spisso A, Fernandez L P, Martinez L D, Pacheco P H, Gil R A. Determination of thimerosal in pharmaceutical industry effluents and river waters by HPLC coupled to atomic fluorescence spectrometry through post-column UV-assisted vapor generation. Journal of Pharmaceutical and Biomedical Analysis, 2015, 106: 79-84.

[37] Hempel M, Kuballa J, Jantzen E. Discovery of a transalkylation mechanism-identification of ethylmercury[+] at a tetraethyllead-contaminated site using sodiumtetrapropylborate, GC-AED and HPLC-AFS. Fresenius' Journal of Analytical Chemistry, 2000, 366(5): 470-475.

[38] Yin Y, Chen B, Mao Y, Thanh W, Liu J, Cai Y, Jiang G. Possible alkylation of inorganic Hg(II) by photochemical processes in the environment. Chemosphere, 2012, 88(1): 8-16.

[39] Fortmann L C, Gay D D, Wirtz K O. Ethylmercury: Formation in plant tissues and relation to methylmercury formation. Trace Substances in Environmental Health-XI Proceedings of the 11th Annual Conference, 1977: 117-122.

[40] Beckers F, Awad Y M, Beiyuan J, Abrigata J, Mothes S, Tsang D C W, Ok Y S, Rinklebe J. Impact of biochar on mobilization, methylation, and ethylation of mercury under dynamic redox conditions in a contaminated floodplain soil. Environment International, 2019, 127: 276-290.

[41] Beckers F, Mothes S, Abrigata J, Zhao J, Gao Y, Rinklebe J. Mobilization of mercury species under dynamic laboratory redox conditions in a contaminated floodplain soil as affected by biochar and sugar beet factory lime. Science of the Total Environment, 2019, 672: 604-617.

[42] Foucher D, Hintelmann H. High-precision measurement of mercury isotope ratios in sediments using cold-vapor generation multi-collector inductively coupled plasma mass spectrometry. Analytical and Bioanalytical Chemistry, 2006, 384(7-8): 1470-1478.

[43] Estrade N, Carignan J, Donard O F X. Tracing and quantifying anthropogenic mercury sources in soils of northern France using isotopic signatures. Environmental Science & Technology, 2011, 45(4): 1235-1242.

[44] Das R, Salters V J M, Odom A L. A case for *in vivo* mass-independent fractionation of mercury isotopes in fish. Geochemistry Geophysics Geosystems, 2009, 10: Q11012.

[45] Sherman L S, Blum J D, Nordstrom D K, McCleskey R B, Barkay T, Vetriani C. Mercury isotopic composition of hydrothermal systems in the Yellowstone plateau volcanic field and Guaymas basin sea-floor rift. Earth and Planetary Science Letters, 2009, 279(1-2): 86-96.

[46] Sun R, Enrico M, Heimbuerger L E, Scott C, Sonke J E. A double-stage tube furnace-acid-trapping protocol for the pre-concentration of mercury from solid samples for isotopic analysis. Analytical and Bioanalytical Chemistry, 2013, 405(21): 6771-6781.

[47] Sherman L S, Blum J D, Johnson K P, Keeler G J, Barres J A, Douglas T A. Mass-independent fractionation of mercury isotopes in Arctic snow driven by sunlight. Nature Geoscience, 2010, 3(3): 173-177.

[48] Fu X, Heimbuerger L E, Sonke J E. Collection of atmospheric gaseous mercury for stable isotope analysis using iodine- and chlorine-impregnated activated carbon traps. Journal of Analytical Atomic Spectrometry, 2014, 29(5): 841-852.

[49] Rolison J M, Landing W M, Luke W, Cohen M, Salters V J M. Isotopic composition of species-specific atmospheric Hg in a coastal environment. Chemical Geology, 2013, 336: 37-49.

[50] Lin H, Yuan D, Lu B, Huang S, Sun L, Zhang F, Gao Y. Isotopic composition analysis of dissolved mercury in seawater with purge and trap preconcentration and a modified Hg introduction device for MC-ICP-MS. Journal of Analytical Atomic Spectrometry, 2015, 30(2): 353-359.

[51] Chen J, Hintelmann H, Feng X, Dimock B. Unusual fractionation of both odd and even mercury isotopes in precipitation from Peterborough, ON, Canada. Geochimica et Cosmochimica Acta, 2012, 90: 33-46.

[52] Strok M, Baya P A, Hintelmann H. The mercury isotope composition of Arctic coastal seawater. Comptes Rendus Geoscience, 2015, 347(7-8): 368-376.

[53] Blum J D, Johnson M W. Recent developments in mercury stable isotope analysis. Reviews in Mineralogy and Geochemistry, 2017, 82 (1): 733-757.

[54] Georg R B, Newman K. The effect of hydride formation on instrumental mass discrimination in MC-ICP-MS: A case study of mercury (Hg) and thallium (Tl) isotopes. Journal of Analytical Atomic Spectrometry, 2015, 30(9): 1935-1944.

[55] Guo W, Hu S, Wang X, Zhang J, Jin L, Zhu Z, Zhang H. Application of ion molecule reaction to eliminate WO interference on mercury determination in soil and sediment samples by ICP-MS. Journal of Analytical Atomic Spectrometry, 2011, 26(6): 1198-1203.

[56] Xie Q L, Lu S Y, Evans D, Dillon P, Hintelmann H. High precision Hg isotope analysis of environmental samples using gold Trap-MC-ICP-MS. Journal of Analytical Atomic Spectrometry, 2005, 20(6): 515-522.

[57] Geng H, Yin R, Li X. An optimized protocol for high precision measurement of Hg isotopic compositions in samples with low concentrations of Hg using MC-ICP-MS. Journal of Analytical Atomic Spectrometry, 2018, 33(11): 1932-1940.

[58] Berail S, Cavalheiro J, Tessier E, Barre J P G, Pedrero Z, Donard O F X, Amouroux D. Determination of total Hg isotopic composition at ultra-trace levels by on line cold vapor generation and dual gold-amalgamation coupled to MC-ICP-MS. Journal of Analytical Atomic Spectrometry, 2017, 32(2): 373-384.

[59] Blum J D, Bergquist B A. Reporting of variations in the natural isotopic composition of mercury. Analytical and Bioanalytical Chemistry, 2007, 388(2): 353-359.

[60] Qin C, Chen M, Yan H, Shang L, Yao H, Li P, Feng X. Compound specific stable isotope determination of methylmercury in contaminated soil. Science of the Total Environment, 2018, 644: 406-412.

[61] Entwisle J, Malinovsky D, Dunn P J H, Goenaga-Infante H. Hg isotope ratio measurements of methylmercury in fish tissues using hplc with off line cold vapour generation MC-ICPMS. Journal of Analytical Atomic Spectrometry, 2018, 33(10): 1645-1654.

[62] Li P, Du B Y, Maurice L, Laffont L, Lagane C, Point D, Sonke J E, Yin R S, Lin C J, Feng X B. Mercury isotope signatures of methylmercury in rice samples from the Wanshan mercury mining area, China: Environmental implications. Environmental Science & Technology, 2017, 51(21): 12321-12328.

[63] Masbou J, Point D, Sonke J E. Application of a selective extraction method for methylmercury compound specific stable isotope analysis (MeHg-CSIA) in biological materials. Journal of Analytical Atomic Spectrometry, 2013, 28(10): 1621-1628.

[64] Epov V N, Berail S, Jimenez-Moreno M, Perrot V, Pecheyran C, Amouroux D, Donard O F X. Approach to measure isotopic ratios in species using multicollector-ICPMS coupled with chromatography. Analytical Chemistry, 2010, 82(13): 5652-5662.

[65] Epov V N, Rodriguez-Gonzalez P, Sonke J E, Tessier E, Amouroux D, Maurice Bourgoin L, Donard O F X. Simultaneous determination of species-specific isotopic composition of Hg by gas chromatography coupled to multicollector ICPMS. Analytical Chemistry, 2008, 80(10): 3530-3538.

[66] Queipo-Abad S, Rodriguez-Gonzalez P, Garcia Alonso J I. Measurement of compound-specific Hg isotopic composition in narrow transient signals by gas chromatography coupled to multicollector ICP-MS. Journal of Analytical Atomic Spectrometry, 2019, 34(4): 753-763.

[67] Gourgiotis A, Berail S, Louvat P, Isnard H, Moureau J, Nonell A, Manhes G, Birck J L, Gaillardet J, Pecheyran C, Chartier F, Donard O F X. Method for isotope ratio drift correction by internal amplifier signal synchronization in MC-ICPMS transient signals. Journal of Analytical Atomic Spectrometry, 2014, 29(9): 1607-1617.

[68] Bloom N, Fitzgerald W F. Determination of volatile mercury species at the picogram level by low-temperature gas-chromatography with cold-vapor atomic fluorescence detection. Analytica Chimica Acta, 1988, 208(1-2): 151-161.

[69] Lindqvist O, Johansson K, Aastrup M, Andersson A, Bringmark L, Hovsenius G, Hakanson L, Iverfeldt A, Meili M, Timm B. Mercury in the Swedish environment-recent research on causes, consequences and corrective methods. Water Air and Soil Pollution, 1991, 55(1-2): 6-261.

[70] Horowitz H M, Jacob D J, Zhang Y X, Dibble T S, Slemr F, Amos H M, Schmidt J A, Corbitt E S, Marais E A, Sunderland E M. A new mechanism for atmospheric mercury redox chemistry: Implications for the global mercury budget. Atmospheric Chemistry and Physics, 2017, 17(10): 6353-6371.

[71] Holmes C D, Jacob D J, Corbitt E S, Mao J, Yang X, Talbot R, Slemr F. Global atmospheric model for mercury including oxidation by bromine atoms. Atmospheric Chemistry and Physics, 2010, 10(24): 12037-12057.

[72] Selin N E, Jacob D J, Park R J, Yantosca R M, Strode S, Jaegle L, Jaffe D. Chemical cycling and deposition of atmospheric mercury: Global constraints from observations. Journal of Geophysical Research-Atmospheres, 2007, 112(D2): 1-14.

[73] Brosset C, Lord E. Methylmercury in ambient air-method of determination and some measurement results. Water Air and Soil Pollution, 1995, 82(3-4): 739-750.

[74] Lee Y H, Wangberg I, Munthe J. Sampling and analysis of gas-phase methylmercury in ambient air. Science of the Total Environment, 2003, 304(1-3): 107-113.

[75] Feng X B, Tang S L, Li Z G, Wang S F, Liang L. Landfill is an important atmospheric mercury emission source. Chinese Science Bulletin, 2004, 49(19): 2068-2072.

[76] Schroeder W H, Keeler G, Kock H, Roussel P, Schneeberger D, Schaedlich F. International field intercomparison of atmospheric mercury measurement methods. Water Air and Soil Pollution, 1995, 80(1-4): 611-620.

[77] Landis M S, Stevens R K, Schaedlich F, Prestbo E M. Development and characterization of an annular denuder methodology for the measurement of divalent inorganic reactive gaseous mercury in ambient air. Environmental Science & Technology, 2002, 36(13): 3000-3009.

[78] Feng X B, Sommar J, Gardfeldt K, Lindqvist O. Improved determination of gaseous divalent mercury in ambient air using KCl coated denuders. Fresenius' Journal of Analytical Chemistry, 2000, 366(5): 423-428.

[79] Lindberg S E, Stratton W J. Atmospheric mercury speciation: Concentrations and behavior of reactive gaseous mercury in ambient air. Environmental Science & Technology, 1998, 32(1): 49-57.

[80] Urba A, Kvietkus K, Sakalys J, Xiao Z, Lindqvist O. A new sensitive and portable mercury-vapor analyzer GARDIS-1A. Water Air and Soil Pollution, 1995, 80(1-4): 1305-1309.

[81] Osawa T, Ueno T, Fu F F. Sequential variation of atmospheric mercury in Tokai-mura, seaside area of eastern central Japan. Journal of Geophysical Research-Atmospheres, 2007, 112: D19107.

[82] Nakazawa K, Nagafuchi O, Kawakami T, Inoue T, Yokota K, Serikawa Y, Cyio B, Elvince R. Human health risk assessment of mercury vapor around artisanal small-scale gold mining area, Palu city, central Sulawesi, Indonesia. Ecotoxicology and Environmental Safety, 2016, 124: 155-162.

[83] Kim K H, Mishra V K, Hong S. The rapid and continuous monitoring of gaseous elemental mercury (GEM) behavior in ambient air. Atmospheric Environment, 2006, 40(18): 3281-3293.

[84] Ebinghaus R, Jennings S G, Schroeder W H, Berg T, Donaghy T, Guentzel J, Kenny C, Kock H H, Kvietkus K, Landing W, Muhleck T, Munthe J, Prestbo E M, Schneeberger D, Slemr F, Sommar J, Urba A, Wallschlager D, Xiao Z. International field intercomparison measurements of atmospheric mercury species at Mace Head, Ireland. Atmospheric Environment, 1999, 33(18): 3063-3073.

[85] Munthe J, Wangberg I, Pirrone N, Iverfeldt A, Ferrara R, Ebinghaus R, Feng X, Gardfeldt K, Keeler G, Lanzillotta E, Lindberg S E, Lu J, Mamane Y, Prestbo E, Schmolke S, Schroeder W H, Sommar J, Sprovieri F, Stevens R K, Stratton W, Tuncel G, Urba A. Intercomparison of methods for sampling and analysis of atmospheric mercury species. Atmospheric Environment, 2001, 35(17): 3007-3017.

[86] 付学吾, 冯新斌, 张辉. 贵阳市大气气态总汞: Lumex RA-915AM 与 Tekran 2537A 的对比观测. 生态学杂志, 2011, (05): 939-943.

[87] Huang J Y, Miller M B, Weiss-Penzias P, Gustin M S. Comparison of gaseous oxidized Hg measured by KCl-coated denuders, and nylon and cation exchange membranes. Environmental Science & Technology, 2013, 47(13): 7307-7316.

[88] Lynam M M, Keeler G J. Comparison of methods for particulate phase mercury analysis: Sampling and analysis. Analytical and Bioanalytical Chemistry, 2002, 374(6): 1009-1014.

[89] Gustin M S, Amos H M, Huang J, Miller M B, Heidecorn K. Measuring and modeling mercury in the atmosphere: A critical review. Atmospheric Chemistry and Physics, 2015, 15(10): 5697-5713.

[90] Lyman S N, Jaffe D A, Gustin M S. Release of mercury halides from KCl denuders in the presence of ozone. Atmospheric Chemistry and Physics, 2010, 10(17): 8197-8204.

[91] McClure C D, Jaffe D A, Edgerton E S. Evaluation of the KCl denuder method for gaseous oxidized mercury using $HgBr_2$ at an In-Service AMNet Site. Environmental Science & Technology, 2014, 48(19): 11437-11444.

[92] Lynam M M, Keeler G J. Artifacts associated with the measurement of particulate mercury in an urban environment: The influence of elevated ozone concentrations. Atmospheric Environment, 2005, 39(17): 3081-3088.

[93] Sprovieri F, Pirrone N, Bencardino M, D'Amore F, Carbone F, Cinnirella S, Mannarino V, Landis M, Ebinghaus R, Weigelt A, Brunke E G, Labuschagne C, Martin L, Munthe J, Wangberg I, Artaxo P, Morais F, Barbosa H D J, Brito J, Cairns W, Barbante C, Dieguez M D, Garcia P E, Dommergue A, Angot H, Magand O, Skov H, Horvat M, Kotnik J, Read K A, Neves L M, Gawlik B M, Sena F, Mashyanov N, Obolkin V, Wip D, Bin Feng X, Zhang H, Fu X W, Ramachandran R, Cossa D, Knoery J, Marusczak N, Nerentorp M, Norstrom C. Atmospheric mercury concentrations observed at ground-based monitoring sites globally distributed in the framework of the GMOS network. Atmospheric Chemistry and Physics, 2016, 16(18): 11915-11935.

[94] Lamborg C H, Fitzgerald W F, O'Donnell J, Torgersen T. A non-steady-state compartmental model of global-scale mercury biogeochemistry with interhemispheric atmospheric gradients. Geochimica et Cosmochimica Acta, 2002, 66(7): 1105-1118.

[95] Streets D G, Horowitz H M, Lu Z, Levin L, Thackray C P, Sunderland E M. Global and regional trends in mercury emissions and concentrations, 2010—2015. Atmospheric Environment, 2019, 201: 417-427.

[96] Mao H T, Cheng I, Zhang L M. Current understanding of the driving mechanisms for spatiotemporal variations of atmospheric speciated mercury: A review. Atmospheric Chemistry and Physics, 2016, 16(20): 12897-12924.

[97] Fu X W, Zhang H, Yu B, Wang X, Lin C J, Feng X B. Observations of atmospheric mercury in China: A critical review. Atmospheric Chemistry and Physics, 2015, 15(16): 9455-9476.

[98] Wu Q R, Wang S X, Li G L, Liang S, Lin C J, Wang Y F, Cai S Y, Liu K Y, Hao J M. Temporal trend and spatial distribution of speciated atmospheric mercury emissions in China during 1978—2014. Environmental Science & Technology, 2016, 50(24): 13428-13435.

[99] Wang X, Lin C J, Yuan W, Sommar J, Zhu W, Feng X B. Emission-dominated gas exchange of elemental mercury vapor over natural surfaces in China. Atmospheric Chemistry and Physics, 2016, 16(17): 11125-11143.

[100] Zhang Y X, Jacob D J, Horowitz H M, Chen L, Amos H M, Krabbenhoft D P, Slemr F, St Louis V L, Sunderland E M. Observed decrease in atmospheric mercury explained by global decline in anthropogenic emissions. Proceedings of the National Academy of Sciences of the United States of America, 2016, 113(3): 526-531.

[101] Mao H T, Hall D, Ye Z Y, Zhou Y, Felton D, Zhang L M. Impacts of large-scale circulation on urban ambient concentrations of gaseous elemental mercury in New York, USA. Atmospheric Chemistry and Physics, 2017, 17(18): 11655-11671.

[102] Timonen H, Ambrose J L, Jaffe D A. Oxidation of elemental Hg in anthropogenic and marine airmasses. Atmospheric Chemistry and Physics, 2013, 13(5): 2827-2836.

[103] Swartzendruber P C, Jaffe D A, Prestbo E M, Weiss-Penzias P, Selin N E, Park R, Jacob D J, Strode S, Jaegle L. Observations of reactive gaseous mercury in the free troposphere at the Mount Bachelor observatory. Journal of

Geophysical Research-Atmospheres, 2006, 111, D24301.

[104] Fain X, Obrist D, Hallar A G, Mccubbin I, Rahn T. High levels of reactive gaseous mercury observed at a high elevation research laboratory in the Rocky Mountains. Atmospheric Chemistry and Physics, 2009, 9(20): 8049-8060.

[105] Fu X W, Marusczak N, Heimburger L E, Sauvage B, Gheusi F, Prestbo E M, Sonke J E. Atmospheric mercury speciation dynamics at the high-altitude Pic Du Midi observatory, southern France. Atmospheric Chemistry and Physics, 2016, 16(9): 5623-5639.

[106] Shah V, Jaegle L, Gratz L E, Ambrose J L, Jaffe D A, Selin N E, Song S, Campos T L, Flocke F M, Reeves M, Stechman D, Stell M, Festa J, Stutz J, Weinheimer A J, Knapp D J, Montzka D D, Tyndall G S, Apel E C, Hornbrook R S, Hills A J, Riemer D D, Blake N J, Cantrell C A, Mauldin R L. Origin of oxidized mercury in the summertime free troposphere over the southeastern US. Atmospheric Chemistry and Physics, 2016, 16(3): 1511-1530.

[107] Lyman S N, Jaffe D A. Formation and fate of oxidized mercury in the upper troposphere and lower stratosphere. Nature Geoscience, 2012, 5(2): 114-117.

[108] Baya P A, Gosselin M, Lehnherr I, St Louis V L, Hintelmann H. Determination of monomethylmercury and dimethylmercury in the Arctic marine boundary layer. Environmental Science & Technology, 2015, 49(1): 223-232.

[109] Lindberg S E, Southworth G, Prestbo E M, Wallschlager D, Bogle M A, Price J. Gaseous methyl-and inorganic mercury in landfill gas from landfills in Florida, Minnesota, Delaware, and California. Atmospheric Environment, 2005, 39(2): 249-258.

[110] Yan H, Li Q, Yuan Z, Jin S, Jing M. Research progress of mercury bioaccumulation in the aquatic food chain, China: A review. Bulletin of Environmental Contamination and Toxicology, 2019, 102(5): 612-620.

[111] Liang L, Horvat M, Bloom N S. An improved speciation method for mercury by GC/CVAFS after aqueous phase ethylation and room temperature precollection. Talanta, 1994, 41(3): 371-379.

[112] Liang L, Horvat M, Cernichiari E, Gelein B, Balogh S. Simple solvent extraction technique for elimination of matrix interferences in the determination of methylmercury in environmental and biological samples by ethylation-gas chromatography-cold vapor atomic fluorescence spectrometry. Talanta, 1996, 43(11): 1883-1888.

[113] Mao Y, Liu G, Meichel G, Cai Y, Jiang G. Simultaneous speciation of monomethylmercury and monoethylmercury by aqueous phenylation and purge-and-trap preconcentration followed by atomic spectrometry detection. Analytical Chemistry, 2008, 80: 7163-7168.

[114] 闫海鱼. 环境样品中不同形态汞的分析方法建立与贵州百花湖汞的生物地球化学循环特征的初步研究. 贵阳: 中国科学院地球化学研究所, 2005.

[115] Horvat M, Liang L, Bloom N S. Comparison of distillation with other current isolation methods for the determination of methyl mercury compounds in low level environmental samples: Part Ⅱ. Water. Analytica Chimica Acta, 1993, 282(1): 153-168.

[116] Horvat M, Bloom N S, Liang L. Comparison of distillation with other current isolation methods for the determination of methyl mercury compounds in low level environmental samples: Part 1. Sediments. Analytica Chimica Acta, 1993, 281(1): 135-152.

[117] 蒋红梅. 水库对乌江河流汞生物地球化学循环的影响. 贵阳: 中国科学院地球化学研究所, 2005.

[118] 蒋红梅, 冯新斌, 梁琏, 商立海, 闫海鱼, 仇广乐. 蒸馏-乙基化 GC-CVAFS 法测定天然水体中的甲基汞. 中国环境科学, 2004, 24(5): 568-571.

[119] Tessier A, Campbell P G C, Bisson M. Sequential extraction procedure for the speciation of particulate trace metals. Analytical Chemistry, 1979, 51(7): 844-851.

[120] 李仲根, 冯新斌, 何天容, 闫海鱼. 王水水浴消解-冷原子荧光法测定土壤和沉积物中的总汞. 矿物岩石地球化

学通报, 2005, 24(2): 140-143.

[121] Cizdziel J V, Tolbert C, Brown G. Direct analysis of environmental and biological samples for total mercury with comparison of sequential atomic absorption and fluorescence measurements from a single combustion event. Spectrochimica Acta Part B: Atomic Spectroscopy, 2010, 65(2): 176-180.

[122] 王翠萍, 闫海鱼, 刘鸿雁, 冯新斌, 王建旭. 使用 Lumex 测汞仪快速测定固体样品中总汞的方法. 地球与环境, 2010, (3): 126-130.

[123] 何滨, 江桂斌. 固相微萃取毛细管气相色谱-原子吸收联用测定农田土壤中的甲基汞和乙基汞. 岩矿测试, 1999, 18(4): 259-262.

[124] 何天容, 冯新斌, 戴前进, 仇广乐, 商立海, 蒋红梅, Liang L. 萃取-乙基化结合 GC-CVAFS 法测定沉积物及土壤中的甲基汞. 地球与环境, 2004, 32(2): 83-86.

[125] Clarkson T W. Mercury-major issues in environmental-health. Environmental Health Perspectives, 1993, 100: 31-38.

[126] Feng X, Li P, Qiu G, Wang S, Li G, Shang L, Meng B, Jiang H, Bai W, Li Z, Fu X. Human exposure to methylmercury through rice intake in mercury mining areas, Guizhou Province, China. Environmental Science & Technology, 2008, 42(1): 326-332.

[127] Zhao L, Anderson C W N, Qiu G, Meng B, Wang D, Feng X. Mercury methylation in paddy soil: Source and distribution of mercury species at a Hg mining area, Guizhou Province, China. Biogeosciences, 2016, 13(8): 2429-2440.

[128] Zhao L, Qiu G, Anderson C W N, Meng B, Wang D, Shang L, Yan H, Feng X. Mercury methylation in rice paddies and its possible controlling factors in the Hg mining area, Guizhou Province, southwest China. Environmental Pollution, 2016, 215: 1-9.

[129] Meng B, Feng X, Qiu G, Liang P, Li P, Chen C, Shang L. The process of methylmercury accumulation in rice (*Oryza sativa* L.). Environmental Science & Technology, 2011, 45(7): 2711-2717.

[130] Meng B, Feng X, Qiu G, Cai Y, Wang D, Li P, Shang L, Sommar J. Distribution patterns of inorganic mercury and methylmercury in tissues of rice (*Oryza sativa* L.) plants and possible bioaccumulation pathways. Journal of Agricultural and Food Chemistry, 2010, 58(8): 4951-4958.

[131] Liu Y R, Yu R Q, Zheng Y M, He J Z. Analysis of the microbial community structure by monitoring an Hg methylation gene *(hgcA)* in paddy soils along an Hg gradient. Applied and Environmental Microbiology, 2014, 80(9): 2874-2879.

[132] Liu Y R, Zheng Y M, Zhang L M, He J Z. Linkage between community diversity of sulfate-reducing microorganisms and methylmercury concentration in paddy soil. Environmental Science and Pollution Research, 2014, 21(2): 1339-1348.

[133] Wang X, Ye Z, Li B, Huang L, Meng M, Shi J, Jiang G. Growing rice aerobically markedly decreases mercury accumulation by reducing both Hg bioavailability and the production of MeHg. Environmental Science & Technology, 2014, 48(3): 1878-1885.

[134] Zhu H, Zhong H, Fu F, Zeng Z. Incorporation of decomposed crop straw affects potential phytoavailability of mercury in a mining-contaminated farming soil. Bulletin of Environmental Contamination and Toxicology, 2015, 95(2): 254-259.

[135] Zhu H, Zhong H, Evans D, Hintelmann H. Effects of rice residue incorporation on the speciation, potential bioavailability and risk of mercury in a contaminated paddy soil. Journal of Hazardous Materials, 2015, 293: 64-71.

[136] Zhu H, Zhong H, Wu J. Incorporating rice residues into paddy soils affects methylmercury accumulation in rice. Chemosphere, 2016, 152: 259-264.

[137] Rothenberg S E, Feng X. Mercury cycling in a flooded rice paddy. Journal of Geophysical Research-Biogeosciences, 2012, 117: 3003.

[138] Meng B, Feng X, Qiu G, Wang D, Liang P, Li P, Shang L. Inorganic mercury accumulation in rice (*Oryza sativa* L.). Environmental Toxicology and Chemistry, 2012, 31(9): 2093-2098.

[139] Strickman R J, Mitchell C P J. Accumulation and translocation of methylmercury and inorganic mercury in *Oryza sativa*: An enriched isotope tracer study. Science of the Total Environment, 2017, 574: 1415-1423.

[140] Rothenberg S E, Feng X, Dong B, Shang L, Yin R, Yuan X. Characterization of mercury species in brown and white rice (*Oryza sativa* L.) grown in water-saving paddies. Environmental Pollution, 2011, 159(5): 1283-1289.

[141] Meng B, Feng X, Qiu G, Anderson C W N, Wang J, Zhao L. Localization and speciation of mercury in brown rice with implications for Pan-Asian public health. Environmental Science & Technology, 2014, 48(14): 7974-7981.

[142] Li L, Wang F, Meng B, Lemes M, Feng X, Jiang G. Speciation of methylmercury in rice grown from a mercury mining area. Environmental Pollution, 2010, 158(10): 3103-3107.

[143] Krupp E M, Mestrot A, Wielgus J, Meharg A A, Feldmann J. The molecular form of mercury in biota: Identification of novel mercury peptide complexes in plants. Chemical Communications, 2009, (28): 4257-4259.

[144] Zhang L, Wang S X, Wang L, Wu Y, Duan L, Wu Q R, Wang F Y, Yang M, Yang H, Hao J M, Liu X. Updated emission inventories for speciated atmospheric mercury from anthropogenic sources in China. Environmental Science & Technology, 2015, 49: 3185-3194.

[145] Yin R, Feng X, Chen J. Mercury stable isotopic compositions in coals from major coal producing fields in China and their geochemical and environmental implications. Environmental Science & Technology, 2014, 48(10): 5565-5574.

[146] Yin R, Feng X, Hurley J P, Krabbenhoft D P, Lepak R F, Hu R, Zhang Q, Li Z, Bi X. Mercury isotopes as proxies to identify sources and environmental impacts of mercury in sphalerites. Scientific Reports, 2016, 6: 18686.

[147] Fu X, Yang X, Tan Q, Ming L, Lin T, Lin C J, Li X, Feng X. Isotopic composition of gaseous elemental mercury in the boundary layer of east China sea. Journal of Geophysical Research-Atmospheres, 2018, 123(14): 7656-7669.

[148] Fu X, Zhang H, Liu C, Zhang H, Lin C J, Feng X. Significant seasonal variations in isotopic composition of atmospheric total gaseous mercury at forest sites in China caused by vegetation and mercury sources. Environmental Science & Technology, 2019, 53(23): 13748-13756.

[149] Yu B, Fu X, Yin R, Zhang H, Wang X, Lin C J, Wu C, Zhang Y, He N, Fu P, Wang Z, Shang L, Sommar J, Sonke J E, Maurice L, Guinot B, Feng X. Isotopic composition of atmospheric mercury in China: New evidence for sources and transformation processes in air and in vegetation. Environmental Science & Technology, 2016, 50(17): 9262-9269.

[150] Fu X, Zhang H, Feng X, Tan Q, Ming L, Liu C, Zhang L. Domestic and transboundary sources of atmospheric particulate bound mercury in remote areas of China: Evidence from mercury isotopes. Environmental Science & Technology, 2019, 53(4): 1947-1957.

[151] Sun G, Feng X, Yang C, Zhang L, Yin R, Li Z, Bi X, Wu Y. Levels, sources, isotope signatures, and health risks of mercury in street dust across China. Journal of Hazardous Materials, 2020: 122276.

[152] 王训, 袁巍, 冯新斌. 森林生态系统汞的生物地球化学过程. 化学进展, 2017, 29(9): 970-980.

[153] Fu X, Zhu W, Zhang H, Sommar J, Yu B, Yang X, Wang X, Lin C J, Feng X. Depletion of atmospheric gaseous elemental mercury by plant uptake at Mt. Changbai, northeast China. Atmospheric Chemistry and Physics, 2016, 16(20): 12861-12873.

[154] Yuan W, Sommar J, Lin C J, Wang X, Li K, Liu Y, Zhang H, Lu Z, Wu C, Feng X. Stable isotope evidence shows re-emission of elemental mercury vapor occurring after reductive loss from foliage. Environmental Science & Technology, 2019, 53(2): 651-660.

[155] Wang X, Luo J, Yin R, Yuan W, Lin C J, Sommar J, Feng X, Wang H, Lin C. Using mercury isotopes to understand mercury accumulation in the montane forest floor of the eastern Tibetan Plateau. Environmental Science & Technology,

2017, 51(2): 801-809.

[156] Wang X, Yuan W, Lu Z, Lin C J, Yin R, Li F, Feng X. Effects of precipitation on mercury accumulation on subtropical montane forest floor: Implications on climate forcing. Journal of Geophysical Research-Biogeosciences, 2019, 124(4): 959-972.

[157] Wang X, Luo J, Yuan W, Lin C J, Wang F Y, Liu C, Wang G X, Feng X B. Global warming accelerates uptake of atmospheric mercury in regions experiencing glacier retreat. Proceedings of the National Academy of Sciences of the United States of America, 2020, 117(4): 2049-2055.

[158] Feng X, Foucher D, Hintelmann H, Yan H, He T, Qiu G. Tracing mercury contamination sources in sediments using mercury isotope compositions. Environmental Science & Technology, 2010, 44(9): 3363-3368.

[159] Yin R, Feng X, Chen B, Zhang J, Wang W, Li X. Identifying the sources and processes of mercury in subtropical estuarine and ocean sediments using Hg isotopic composition. Environmental Science & Technology, 2015, 49(3): 1347-1355.

[160] Yin R, Feng X, Zhang J, Pan K, Wang W, Li X. Using mercury isotopes to understand the bioaccumulation of Hg in the subtropical pearl river estuary, south China. Chemosphere, 2016, 147: 173-179.

[161] Wang X, Yuan W, Lin C J, Zhang L, Zhang H, Feng X. Climate and vegetation as primary drivers for global mercury storage in surface soil. Environmental Science & Technology, 2019, 53(18): 10665-10675.

[162] Feng X, Yin R, Yu B, Du B. Mercury isotope variations in surface soils in different contaminated areas in Guizhou Province, China. Chinese Science Bulletin, 2013, 58(2): 249-255.

[163] Yin R, Feng X, Wang J, Bao Z, Yu B, Chen J. Mercury isotope variations between bioavailable mercury fractions and total mercury in mercury contaminated soil in Wanshan mercury mine, SW China. Chemical Geology, 2013, 336: 80-86.

[164] Yin R, Feng X, Meng B. Stable mercury isotope variation in rice plants (*Oryza sativa* L.) from the Wanshan mercury mining district, SW China. Environmental Science & Technology, 2013, 47(5): 2238-2245.

[165] Sun G, Feng X, Yin R, Zhao H, Zhang L, Sommar J, Li Z, Zhang H. Corn (*Zea mays* L.): A low methylmercury staple cereal source and an important biospheric sink of atmospheric mercury, and health risk assessment. Environment International, 2019, 131: 104971.

第 20 章　MC-ICP-MS 在生物和环境同位素分析中的应用

（杨学志[①]，王伟超[①]，张璐瑶[①]，令伟博[①]，杨　航[①]，陈子谷[①]，刘　倩[①]*，江桂斌[①]）

▶ 20.1　引言 / 569

▶ 20.2　MC-ICP-MS 的应用原理 / 569

▶ 20.3　MC-ICP-MS 分析方法最新进展 / 571

▶ 20.4　MC-ICP-MS 在环境同位素分析中的应用 / 574

▶ 20.5　MC-ICP-MS 在生物同位素分析中的应用 / 578

▶ 20.6　展望 / 579

[①]中国科学院生态环境研究中心，北京。*通讯作者联系方式：qianliu@rcees.ac.cn

第 20 章 MC-ICP-MS 在生物和环境同位素分析中的应用

本章导读

- 简要介绍 MC-ICP-MS 的应用原理，具体内容包括 MC-ICP-MS 的结构、功能及应用理论基础；
- 介绍 MC-ICP-MS 分析方法的最新研究进展，涉及样品前处理、仪器性能优化、联用技术及标准样品制备四个方面；
- 重点介绍 MC-ICP-MS 在环境同位素分析中的应用，涉及重金属示踪、大气细颗粒物示踪和纳米颗粒物示踪；
- 简要介绍 MC-ICP-MS 在生物同位素分析中的应用，涉及微生物、植物、动物和人体四个方面；
- 展望 MC-ICP-MS 在环境科学和生物学等领域的未来发展方向。

20.1 引 言

质谱技术的进步是高精度稳定同位素分析的重要推动力。20 世纪 90 年代以前，稳定同位素分析主要依靠热电离质谱（thermal ionization mass spectrometry，TIMS），尽管 TIMS 采用了单聚焦质量分析器和多接收器等核心部件，保证了同位素分析的精度和稳定性，但由于采用热电离源的缘故，其应用性受到一定限制，比如容易引入同位素分馏和低电离能元素分析受限等。电感耦合等离子质谱仪（inductively coupled plasma mass spectrometry，ICP-MS）的发明突破了元素浓度分析的瓶颈，实现了多元素浓度的高精度同时分析。相比于热电离源，ICP 具有较高的离子化能力，可以降低离子化过程中的同位素分馏效应和拓宽质谱仪的元素分析范围。1992 年，Walder 等将 ICP 与双聚焦质量分析器和多接收器等核心部件进行耦合，实现了 U 和 Pb 同位素的高精度分析，这一发明标志着多接收器电感耦合等离子体质谱仪（multi-collector inductively coupled plasma mass spectrometry，MC-ICP-MS）的诞生，从此开启了非传统稳定同位素分析的新时代。

非传统稳定同位素通常指的是传统稳定同位素（C、N、O、H 和 S）之外的其他元素，例如金属元素同位素。从 20 世纪 90 年代初开始，非传统稳定同位素技术获得了快速发展，逐渐形成了多元化的非传统稳定同位素应用体系，涉及领域也从最初的地学领域，逐渐扩展到了环境科学和生物学等领域，展现出巨大的发展潜力。为了反映 MC-ICP-MS 的最新发展趋势，本章将重点介绍 MC-ICP-MS 在分析方法和应用中（环境科学和生物学领域）的最新进展，并对未来发展方向做进一步的展望。

20.2 MC-ICP-MS 的应用原理

经过二十多年的发展，MC-ICP-MS 仪器性能得到快速提升，同时，基于 MC-ICP-MS 发展起来的非传统稳定同位素理论体系逐渐完善。本节主要介绍 MC-ICP-MS 的相关应用原理，具体内容包括 MC-ICP-MS 的结构、功能及应用理论基础。

20.2.1 MC-ICP-MS 的结构和功能

如图 20-1 所示，MC-ICP-MS 主要是由进样系统、离子源（ICP）、双聚焦质量分析器（扇形电场和磁场）、离子检测器等系统组成。其中，ICP、双聚焦质量分析器和离子检测器是保证 MC-ICP-MS 实现高精度同位素比值分析的核心部件。

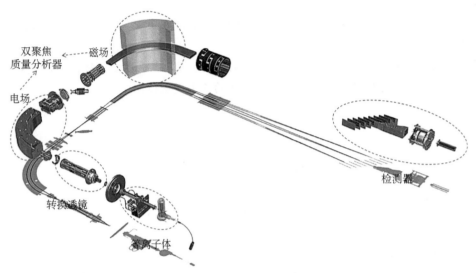

图 20-1　Neptune MC-ICP-MS（Neptune MC-ICP-MS 的基本操作培训手册，2007）

ICP 一般指的是由高频电流激发的氩气等离子体，核心温度可达 8000 K，其提供的能量刚好可使大部分元素电离出一个核外电子。当样品进入 ICP，会被迅速蒸发、解离、原子化及离子化（去掉一个电子），进而形成离子束，之后离子束被提取到后面的真空传输通道中进行分离。相比于其他离子源（如热电离源），ICP 具有较高的离子化效率，可以拓宽质谱仪的元素分析范围和提高同位素分析的灵敏度，因此很快发展为同位素质谱仪的核心部件。但 ICP 也存在一个缺点，即容易产生多原子和双电荷离子干扰（具有与目标离子相近的荷质比），进而严重影响同位素比值的准确测定。为了克服这一点，研究人员提出了很多改进方法来减少干扰离子的形成，比如优化等离子体条件、减少进入等离子体的水气量和降低基质浓度等。此外，进一步提高仪器分辨率，将目标离子与荷质比相近的干扰离子进行彻底分离，也是消除这种干扰效应的有效方法，因此具有高分辨率特性的双聚焦质量分析器进入了人们的视野。

受离子初始能量差异的影响，单聚焦质量分析器（仅有一个磁场）具有较低的质量分辨率，而双聚焦质量分析器可以校正离子的初始能量差异，因此其分辨率显著提高。双聚焦质量分析器是由扇形电场和磁场两部分组成，其中扇形电场是将质量相同而速度不同的离子分离聚焦，起到校正离子初始能量差异的作用，而磁场可将不同荷质比的离子进行分离，使其分别进入不同法拉第杯离子检测器进行同时检测。这种同时实现速度和方向双聚焦的质量分析器，大大提高了 MC-ICP-MS 的质量分辨率。

相比于其他离子检测器（如电子倍增器），法拉第杯可以直接收集带电荷的离子，且具有较大的线性动态范围，因此适合作为同位素分析的离子检测器（元素的不同同位素丰度差异很大）。同位素分析初期，经常使用单一法拉第杯检测器来依次接收不同同位素的离子信号，在这种情况下同位素比值测定极易受到离子源稳定性的影响。后经改进，人们将多个法拉第杯组合成检测器阵列，同时收集检测不同荷质比离子，大大增加了 MC-ICP-MS 的分析稳定性。

20.2.2 MC-ICP-MS 的应用理论基础

目前，MC-ICP-MS 已成为非传统稳定同位素分析的首选仪器，它的使用大大促进了非传统稳定同位素体系和相关理论的建立和发展，反过来非传统稳定同位素理论也已成为 MC-ICP-MS 的应用理论基础。

与放射性同位素相对，稳定同位素是指不发生或极不易发生放射性衰变（半衰期大于 10^{15} 年）的同位素，可以在自然界中稳定存在。通常，稳定同位素又可细分为传统稳定同位素（一般指 C、N、O、H、S）和非传统稳定同位素（除传统稳定同位素以外的其他同位素）。自然界中很多过程，比如氧化还原、沉降、络合、吸附、溶解、蒸发和扩散等，都会导致物质的同位素组成发生变化，这一现象称之为稳定同位素分馏。经过漫长的地球化学过程，自然界中不同储库逐渐具有了特定的同位素组成特征，通过研究这些不同储库的同位素组成特征及稳定同位素分馏机制，可以很好地反推物质的来源和相关的地球化学过程。因此，稳定同位素分馏已被作为来源和过程示踪的有力工具，广泛应用到元素的地球化学循环过程研究中。

稳定同位素分馏发生的机制是稳定同位素之间存在一定的质量差异，进而导致同位素之间存在轻微的物理化学性质差异。目前，自然界中发现的大部分同位素分馏现象都符合质量依赖效应，即发生了质量依赖分馏（mass-dependent fractionation，MDF）。质量依赖分馏又可分为热力学分馏和动力学分馏。其中，热力学分馏通常发生于浓度已达平衡的反应体系中，重同位素更倾向富集在"更强的成键环境"中，比如更低的配位数和更短的键长；而动力学分馏通常发生于浓度未达平衡的反应过程中，轻同位素通常具有更快的反应速率，因此更容易富集于反应产物中。此外，少部分元素在实验室模拟（Ti、Cr、Zn、Sr 和 Mo 等）和自然界（Hg、O、S 和 Fe 中）发现存在非质量依赖效应，即发生了非质量依赖分馏（mass-independent fractionation，MIF）。通常，非质量依赖分馏主要是由核体积效应和磁效应导致的。同位素分馏的程度可以通过比较反应前后物质间同位素组成变化的大小来反映，而物质的同位素组成通常是以参比于同位素标准物质的相对千分差（δ）来表示，即

$$\delta^x \mathrm{E} = \left(\frac{(^x\mathrm{E}/^y\mathrm{E})_{样品}}{(^x\mathrm{E}/^y\mathrm{E})_{标准物质}} - 1 \right) \times 1000$$

其中，E 代表某种化学元素，x 和 y 分别代表该元素两种同位素的质量数。

20.3 MC-ICP-MS 分析方法最新进展

MC-ICP-MS 分析方法的发展促进了非传统稳定同位素技术在多个学科的应用。目前，非传统稳定同位素技术应用领域已从传统的地学领域逐渐扩展到环境科学和生物学等交叉学科领域，相应地，MC-ICP-MS 需要分析的样品也从简单的岩石样品逐渐向低丰度、高有机质含量的生物样品转变，因此，如何提高 MC-ICP-MS 的分析性能和灵敏度，进一步拓展 MC-ICP-MS 的应用范围，一直是该领域的重要研究课题。之前已有一些综述总结了非传统稳定同位素技术在分析方法方面的研究进展，作为补充，本节将重点介绍 MC-ICP-MS 在样品前处理、仪器性能优化、仪器联用技术和标准物质制备四个方面的最新进展。

20.3.1 样品前处理

MC-ICP-MS 分析的灵敏度和准确度受到样品基质效应的严重制约，因此分离纯化样品中的目标元素是高精度同位素比值测定的前提。目前，柱色谱法是进行样品纯化的主流方法，已广泛应用于 50 多种元素的分离纯化，但在应用过程中仍然存在流程复杂、耗时长和耗酸量高等问题。针对这些不足，研究人员针对不同元素的理化性质，不断优化过柱流程，力求在保证回收率的基础上，不断提高过柱效率和降低相关费用。

目前，元素的分离纯化仍以多柱分离法为主，即使用两种及以上的离子交换树脂（比如 BPAH、TRU、AG1-X4、AG1-X8、AGMP-1M、AG50W-X12 等）的组合来去除众多基质元素，从而达到分离富集目标元素的目的。对于基质相对简单的样品来说，将多柱分离法简化为单柱分离法，可以同时实现简化流程、节省时间和降低耗酸量等优点。近期，中国地质大学（武汉）胡兆初课题组[1]通过改进多种浓酸的配比使用方法，针对含 Mo 地质样品成功开发了一种基于 TRU 树脂的单柱 Mo 元素分离方案，达到了简化流程和降低成本的目的。而对于复杂基质的样品而言，多柱分离体系仍然不可或缺。在此情况下，可以根据干扰元素和待测元素之间的化学性质差异，通过优化洗脱酸种类和浓度来实现复杂基质中目标元素的高效分离。近期一些研究针对 Mo[2]、Cd[3,4]、Ti[5]和 Sb[6]等元素在复杂基质中的纯化方案做出了改进，在保证分离效率的基础上使样品回收率接近 100%，保证了 MC-ICP-MS 分析的准确度。此外，值得一提的是，Mahan 等[7]设计了可离心色谱柱，用离心力来代替重力为液体流过色谱柱提供推力，大大节省了过柱时间。

提高 MC-ICP-MS 的痕量分析的能力，有利于拓宽非传统稳定同位素技术的应用领域（尤其是生物学领域），为了解决这一点，预浓缩技术应运而生。目前，常用的预浓缩技术有蒸发、色谱富集、捕集和共沉淀。以 Hg 元素为例，陈玖斌等开发了色谱技术用以富集水体中痕量 Hg 元素，为痕量 Hg 同位素分析提供了方法学支持[8]。对于硅元素而言，尽管自然水体中硅酸根离子以溶解态的形势存在，但受限于低丰度，很难直接进行同位素分析，而基于共沉淀技术（硅酸根离子可与 NaOH 或 NH_4OH 形成沉淀），可以轻松实现溶解态硅酸根的分离富集。此外，不同的捕集技术也被逐渐开发出来，比如金捕集、KMO_4 捕集和酸溶液捕集[9-11]等。未来，针对复杂基质的预浓缩技术仍需进一步开发，以促进 MC-ICP-MS 在环境科学和生物学等新兴领域的应用。

20.3.2 仪器性能优化

MC-ICP-MS 主要是由进样系统、离子源（ICP）、双聚焦质量分析器和离子检测器等核心部件组成，优化提升各个部件的性能是提升仪器整体性能的重要着手点。具体地，进样系统的优化可以提升灵敏度和减少干扰离子的产生，比如将湿法进样改为干法进样，仪器灵敏度一般可以提升数倍以上，同时可以降低氢化物和氧化物等多原子干扰的产生。对于离子源而言，适当降低 ICP 的能量（冷等离子体法），可以抑制 Ar 的电离，从而减少 ArO 等多原子干扰离子的产生。进一步提升双聚焦质量分析器的分辨率（如大型同位素质谱仪 Nu Plasma 1700），可以大大提高仪器整体分辨率，从而可以轻松实现 Si、Fe 和 Ca 等难测元素与干扰离子的分离；此外，加入碰撞反应池单元（如新款同位素质谱仪 Nu sapphire），也是减少干扰离子产生、提升仪器分辨率的有效方式。对于离子检测器部分，通过增加法拉第杯放大器电阻，可以显著提高信号灵敏度。近期的研究在离子检测器方面取得了一些新的进展。

尽管 ICP 电离源保证了很高的离子化效率，但同时也容易引入 $^{40}Ar^+$、$^{40}Ar^{16}O^+$ 等多原子干扰。其中，加入碰撞池和采用冷等离子体等方法已被证明可以降低多原子干扰的影响。近期，特殊设计的盲杯技术（dummy bucket）也展示了降低多原子干扰的能力[12,13]。盲杯技术又称"空"法拉第杯，即使用一只接地状态的金属杯，将其安装在法拉第杯检测器的接地挡板的低质量侧，可以有效减少散逸离子逸出，并将 $^{40}Ar^+$ 的强信号分配给该"空"杯。该技术可以较好地消除 K、Ca、Fe 等元素同位素测试过程中的拖尾现象，提高分析的精度。需要指出的是，目前该技术由于受到造价和 Ar 相关的粒子散射干扰，尚未广泛应用。

MC-ICP-MS 产生的离子束所带的电荷量很小，因此需要借助法拉第杯放大器的高电阻来实现信号放大，从而实现对极小离子束的低噪定量分析。发明初期，MC-ICP-MS 一般使用的是 $10^{10}\sim10^{11}\ \Omega$ 电阻器，即可以实现 $10^{10}\sim10^{11}$ 倍数的信号放大。为了增加信号灵敏度，Nielsen 等用 $10^{12}\ \Omega$ 电阻器代替初始 $10^{11}\ \Omega$ 电阻器，提高了 V 同位素分析的灵敏度[14]；同期，Peters 等基于 $10^{12}\ \Omega$ 电阻器建立了 Pt 同位素的高精度分析方法[15]。为了进一步提升信号灵敏度，近期 Creech 等[16]和 Dellinger 等[17]基于 $10^{13}\ \Omega$ 电阻器分别建立了 Pt 和 Re 同位素的分析方法，灵敏度提升了一个数量级，同时又很好地兼顾了信噪比的问题。

20.3.3 仪器联用技术

发展 MC-ICP-MS 与各种进样系统的联用技术，可以显著提高 MC-ICP-MS 的分析效率和能力，因此联用技术的发展与优化一直是稳定同位素分析的前沿方向。目前，常用的 MC-ICP-MS 联用进样系统有激光剥蚀（LA）、气相色谱（GC）、液相色谱（LC）、冷蒸气发生装置（CVG）等。

LA 是一种直接固体取样技术，将 LA 与 MC-ICP-MS 联用，可以实现固体样品的微区原位同位素分析，从而扩展了 MC-ICP-MS 的应用领域。目前 LA-MC-ICP-MS 已成为地学分析的常用工具，常用的 LA 进样系统有纳秒（ns-LA）和飞秒（fs-LA）两种，其中 fs-LA 的激光瞬时功率更高，可以提供更高的空间和时间分辨率。近期，该联用系统也取得了一些进展，如利用 ns-LA-MC-ICP-MS 实现了微米级陨石原位 Fe 同位素的在线分析[18]，以及用 fs-LA-MC-ICP-MS 精准测定了硅酸岩样品中 Si 同位素的组成[19]。此外，借用梯度扩散薄膜原位被动采样技术（DGT）的使用，可以将 LA-MC-ICP-MS 的分析对象扩展到液体介质。具体地，DGT 采样器将水体中的目标元素富集于树脂层上（平面结构），进而可直接使用 LA-MC-ICP-MS 剥蚀树脂层来进行原位同位素分析[20]。

除了利用 LA 进行微区原位同位素分析外，还常用色质联用技术来实现形态同位素分析。其中，气质联用系统（GC-MC-ICP-MS）常用于分析气态 Hg 同位素[21]、有机化合物中的 S 同位素[22]，以及卤代化合物中的 Cl[23]、Br[24]等卤族同位素等，通过积分瞬时信号从而获得同位素比值。高效液相色谱（HPLC）也适用于在线同位素分析，如用 HPLC-MC-ICP-MS 在线分离 Cr(III)和 Cr(VI)，进而实现了对氧化过程中 Cr 同位素分馏的实时追踪[25]。但受限于 MC-ICP-MS 易受干扰的信号响应状态，以及样品纯化程度的影响，将 GC 或 LC 与 MC-ICP-MS 联用的应用范围比较受限，其分析精度低于常规离线分析，需要进一步优化。

20.3.4 标准物质制备

样品同位素比值的测定需要以稳定可靠的标准物质作为基准进行比较和仪器分馏校正，因此同位素标准物质的制备是稳定同位素分析的基础工作。但是，目前稳定同位素标准品的研发工作仍然存在很多不足，比如很多元素至今未有国际通用标准品、标准物质逐渐被耗尽、标准品的共享性受限（实验室内部研发的标准品）等，进而限制了分析方法的开发与应用。因此，标准物质的制备与共享仍将长期是稳定同位素分析领域的关注重点。近期，黄铜矿标样 HTS4-6[26]、Mg 同位素标准溶液 GSB-Mg[27]、两种 Cu 同位素标准溶液 NWU-Cu-A 和 NWU-Cu-B[28]、天然锆石参考物质 SA01[29]、Nd 同位素标准参考物质 GSB 04-3258-2015[30]以及铁铜锌同位素标准参考溶液 CAGS-Fe、CAGS-Cu 和 CAGS-Zn[31]等被制备出来，缓解了相关元素标准品短缺的问题。

20.4 MC-ICP-MS 在环境同位素分析中的应用

生物地球化学循环过程通常包含复杂的生物、物理和化学反应，容易导致同位素分馏效应，从而使物质的同位素组成发生改变。基于物质的同位素组成和分馏机理可以对物质的来源和迁移转化过程进行示踪。本节主要介绍了 MC-ICP-MS 在重金属示踪、大气细颗粒示踪和纳米颗粒物示踪研究中的最新进展。

20.4.1 重金属示踪最新进展

重金属指的是原子量比较大的金属元素，包括 Ni、Co、Cu、Zn、Cd、Sn、Sb、Hg、Pb 等。重金属从矿石中冶炼精炼出来，投入到人类的生产生活中，如合金制造、油漆生产、电池制造、电镀、皮革加工、燃煤、垃圾焚烧、交通运输等。这些人为过程会释放大量的重金属到环境中，对生态系统和人体健康造成危害[32,33]。污染区域内重金属的来源示踪对于重金属排放控制非常重要。目前常用的溯源手段包括重金属的时间及空间浓度变化观测、不同元素比分析、相对富集因子法等[34-36]。颗粒物中重金属溯源还会用到主成分分析法[37]。近年来随着 MC-ICP-MS 分析方法的飞速发展，非传统稳定同位素比值也逐步被用作环境中重金属来源示踪的工具。

利用非传统稳定同位素比值进行溯源，不仅需要测定环境样品中的同位素比值，还需要获得潜在污染源的同位素组成信息。因此污染源谱同位素库的建立和丰富对利用非传统稳定同位素比值进行溯源应用非常重要。目前重金属铜（Cu）、锌（Zn）、铅（Pb）、汞（Hg）等同位素溯源研究较多，其在环境样品及人为活动源中的同位素测定比较广泛。

以铅元素为例，相比于其他重金属，自然界中铅同位素比值的范围相对更广，不同来源铅同位素变化相对更大，因此更容易区分。环境介质中的铅可能来自人为源（如煤炭燃烧、矿石开采等）或者自然源（如岩石风化等）。仅煤炭这一种环境样品，不同产地的铅同位素比值就有所差异。中国地质大学闭向阳等[38]采集了来自中国 18 个省的主要煤矿中的煤炭，发现从北到南，煤炭中的 ^{206}Pb 含量逐渐升高，中国西南部 ^{206}Pb/^{207}Pb>1.20，东北和西北在 1.19～1.20 之间，北方<1.17，中部和东部分别为 1.17 和 1.18。在另一项更早的调查中[39]，中国煤炭中铅同位素比值 ^{206}Pb/^{207}Pb 分布在 1.14～1.18，这个范围更为集中，这可能和采样量更小以及测量技术有关。在这份调查中，世界上其他地区煤炭中铅同位素比值 ^{206}Pb/^{207}Pb 均值分别为：欧洲 1.19±0.04；南美洲和北美洲 1.21±0.2；亚洲 1.17±0.03；大洋洲 1.20±0.03；南非 1.21±0.02。除了煤炭之外，其他的燃料中铅同位素比值也曾被报道过，例如很早之前使用的含铅汽油 ^{206}Pb/^{207}Pb 和 ^{208}Pb/^{206}Pb 分别为 1.12～1.18 和 2.10～2.15[40]，后来使用的无铅汽油中 ^{206}Pb/^{207}Pb 和 ^{208}Pb/^{206}Pb 大概为 1.08 和 2.11[40]，天然气中 ^{206}Pb/^{207}Pb 和 ^{208}Pb/^{206}Pb 大概为 1.07 和 2.15[41]。含铅矿石的铅同位素测定对于溯源同样意义重大。Sangster 等[42]总结了世界各地主要铅矿的同位素比值，其中喷流沉积矿（SEDEX）的同位素范围最广。中国铅矿的 ^{206}Pb/^{207}Pb 和 ^{208}Pb/^{206}Pb 分布区间分别为 1.02～1.18 和 2.10～2.26。在另一项调查中[38]，中国铅矿（闪锌矿/方铅矿）的同位素比值 ^{206}Pb/^{207}Pb 和 ^{208}Pb/^{206}Pb 主要分布区间为 1.02～1.18 和 2.09～2.31，和前文报道

接近。中国北方的铅矿 $^{206}Pb/^{207}Pb<1.10$，南方>1.14，趋势和煤炭一样[38]。欧洲地质调查局曾测量过欧洲农业土壤的铅同位素分布地图，$^{206}Pb/^{207}Pb$ 范围在 1.12～1.73 之间，平均值为 1.20。北欧的 $^{206}Pb/^{207}Pb$ 高于南欧，且有一条明显的分界线，这条分界线和跨欧洲缝合带相似，结合浓度分布作者认为南欧和北欧之间明显的铅同位素差异主要来自于地质原因和风化条件等[43]。

重金属从源到汇的过程可能会发生同位素分馏效应，进而影响溯源分析的准确性，因此需要对迁移转化过程的分馏效应进行研究和校正。环境过程包括蒸发、沉降、吸附、沉淀、氧化还原、生物作用等都可能导致同位素分馏。高温过程如矿石冶炼等会使得更多的轻同位素先蒸发出来造成比较明显的分馏，例如 Cu、Zn、Fe、Cd 等[44-47]。但是近期有一个关于煤燃烧过程中镉同位素分馏的研究[48]表明，飞灰中的镉同位素反而比煤渣中的镉同位素更重，这主要是由飞灰颗粒在燃烧过程中容易释放，而镉蒸汽冷凝过程中重的镉同位素优先吸附到飞灰颗粒物中导致的，并不是因为蒸发过程导致的分馏。吸附过程会由于溶液条件以及吸附剂表面性质不同造成不同的分馏现象，Guinoiseau 等[49]发现 pH 和离子强度对 Zn^{2+} 吸附到高岭土上的分馏影响较大，这和高岭土上的吸附位点有关。南京大学李伟强等[50]也发现 Zn^{2+} 浓度和 pH 会对 Zn^{2+} 吸附到铝氧化物上的分馏产生影响，通过 EXAFS 表征发现这些因素是通过影响 Zn^{2+} 吸附到铝氧化物上成键进而影响分馏的。一些过渡金属在氧化还原过程中也容易发生同位素分馏，例如微生物异化铁还原过程中产生的 Fe(Ⅱ)的 $\delta^{56}Fe$ 比初始加入的 Fe(Ⅱ)的 $\delta^{56}Fe$ 低。蓝藻细菌光合作用生成的氧气能氧化周围溶液中的 Fe(Ⅱ)，生成的 Fe(Ⅲ)氢氧化物沉淀的 $\delta^{56}Fe$ 变高[51]。溶液中金属与其他有机配体的络合作用会使得不同的金属配合物之间同位素组成有差异。Ryan 等[52]发现 Cu 与不同的有机配体结合形成的配合物相比于初始溶液 $\Delta^{65}Cu$ 为+0.14±0.84‰，Morgan 等[53]发现 Fe 与不同有机配体形成的配合物之间也有显著区别，有机配体与金属之间的结合力是影响同位素分馏的一个重要原因。此外，生物的吸收也会造成环境中同位素分馏，例如细菌和硅藻对 Zn^{2+} 的摄取也会产生分馏，影响海水中 Zn 同位素比值[54]。

近年来，重金属同位素的示踪应用呈逐步增多趋势。Walraven 等[55]对荷兰高速公路旁的土壤和地下水进行铅浓度和同位素比值测定，发现表层土壤 $^{206}Pb/^{207}Pb$ 低于深层土壤，推测表层土壤中的铅主要来自天然源和汽油燃烧排放。且地下水的 $^{206}Pb/^{207}Pb$ 也变低，地下水中的铅污染很可能是酸性土壤中的铅迁移释放造成的，据推测贡献率达到 30%～100%。Takano 等[56]测定了海水颗粒中 Ni 和 Cu 同位素，推测海水颗粒中的 Ni 和 Cu 来自海底沉积物和生物有机粒子。Tu 等[57]根据我国台湾二伦河源头至河口的锌同位素组成和锌空间分布推测该河流中的 Zn 至少有三个来源，包括自然源、电镀废水和金属表面加工工业废水。Kanagaraj 等[58]通过测定印度南部某地区地下水中 Cr 同位素，推测地下水中的 Cr 部分来自制革废水。Kersten 等[59]利用 Tl 同位素证明水泥厂周边土壤中 Tl 污染来自水泥窑粉尘释放。除了环境介质，生物体内污染同位素示踪也有一些报道，例如城市不同地方夹竹桃叶 Zn 同位素比值差异表明不同的污染来源[60]；根据儿童乳齿中 Pb 含量和 Pb 同位素比值可以推测儿童成长过程中周围环境中 Pb 的状态[61]。

除了上述环境过程提到的同位素质量依赖分馏（MDF）现象，自然界中还存在非质量依赖分馏（MIF）现象。汞一共有 7 种稳定同位素（^{196}Hg、^{198}Hg、^{199}Hg、^{200}Hg、^{201}Hg、^{202}Hg、^{204}Hg）。奇数质量数同位素（^{199}Hg、^{201}Hg）MIF 通常发生在水体中无机汞和甲基汞的光还原过程[62]（无机汞 $\Delta^{199}Hg/\Delta^{201}Hg=1.00\pm0.02$，甲基汞 $\Delta^{199}Hg/\Delta^{201}Hg=1.36\pm0.03$），产生的 MIF 可能会沿食物链传播。甲基汞目前在大气、水体、鱼类、植物、北极雪、人类头发中都检测出了奇数质量数 MIF[63-68]。偶数质量数同位素（^{200}Hg、^{204}Hg）MIF 通常发生在大气汞的干湿沉降过程中[64,69-71]，部分可能是由于 Hg(0) 的光氧化引起的，不过降雨过程可能也会引起奇数质量数 MIF[71]。

目前利用 Hg 稳定同位素溯源的工作大多是同时使用 $\delta^{202}Hg$ 和非质量依赖分馏大小（$\Delta^{199}Hg$、$\Delta^{200}Hg$、$\Delta^{201}Hg$、$\Delta^{204}Hg$）来识别不同的污染源。陈玖斌等[71]通过测定加拿大安大略省 8 个淡水湖湖水和沉积物中 Hg 同位素，发现所有淡水湖湖水中汞同位素都呈现明显偏正的 $\varDelta^{199}$Hg 和 $\varDelta^{200}$Hg，而沉积物中呈现明显偏负的 $\varDelta^{199}$Hg 和不显著的 $\varDelta^{200}$Hg。湖水中偏正的 $\varDelta^{200}$Hg 可能来自于当地降水；流域湿地的甲基汞和来自流域风化的 Hg(II)可能是造成湖水 $\varDelta^{199}$Hg 偏正的原因。沉积物中的 Hg 则很有可能来自当地土壤的冲刷输入。中国科学院生态环境研究中心史建波等[72]对中国沿海沉积物中 Hg 含量和同位素组成进行测定，发现海口汞同位素 δ^{202}Hg 比海洋沉积物高，$\varDelta^{199}$Hg 更低，且从北到南 δ^{202}Hg 逐渐降低而 $\varDelta^{199}$Hg 逐渐升高。他们基于汞同位素混合模型和海洋-城市距离提出了一个混合模型，发现海口沉积物和海洋沉积物在汞的来源上存在明显差异，主要的汞源从陆地输入（河流和工业废水排放）逐渐向大气沉降转变。史建波等[73]还发现渤海海洋食物链中，营养级和生物体内 Hg 同位素 MDF 和奇数质量数 MIF 存在明显相关性，生物体内的甲基汞可能来自底泥，且在生物链中传递。生物体内的 Hg 偶数质量数 MIF 暗示大气中 Hg(0)也可能是生物体内的一个来源。他们[74]还曾检测过青藏高原谢格拉山上苔藓、针叶和表层土壤中汞的同位素组成，根据 $\varDelta^{199}$Hg 推测这两种植物中的汞主要来自大气，表层土壤中的 Hg 平均有 87%±9%来自大气。关于 Hg 在各种天然源和人为源中的同位素组成特征、溯源应用、分馏机制等，Li 等[75]和 Kwon 等[76]有更详细的介绍。

新的重金属稳定同位素体系的开发和发展可拓宽重金属同位素示踪应用，例如 Ga、Ge 等[77,78]，目前这些元素的测定方法都得到了发展，但环境示踪应用尚待进一步开发。当单元素稳定同位素比值无法有效示踪时，可以考虑多元素同位素比值[41,79]、元素比[34]或其他参数联合示踪[80]。

20.4.2　大气细颗粒物示踪最新进展

近年来，随着我国社会经济的快速发展，环境污染问题愈发严重，大气污染就是最严峻的污染之一。空气污染物是气体和颗粒物（particulate matter，PM）的复杂组合，特别是细颗粒物（空气动力直径小于 2.5 μm）严重影响了人体健康和生态环境[81]。以最为常见的雾霾为例，$PM_{2.5}$ 就是雾霾天气的主要元凶；同时，大气细颗粒物往往会携带多种金属/非金属组分，如汞、铅、锌、铜、有机质和黑炭等混合组分[82]。要想实现大气污染综合治理，首先必须了解这些细颗粒物的产生来源。

作为地壳中第二丰富的非金属元素，硅元素广泛存在于大气细颗粒物中。由于硅同位素在大气运输过程中不容易产生同位素分馏效应，因此硅同位素可以用来进行大气污染溯源分析。中国科学院生态环境研究中心刘倩等[83]利用硅同位素对大气 $PM_{2.5}$ 的来源进行解析，发现春冬季燃煤排放和工业进程是导致季节性颗粒物浓度暴发和雾霾天频发的主要原因。结合分析实际大气 $PM_{2.5}$ 样品与一次污染源的硅浓度和同位素组成，发现燃煤排放源和工业排放源与样品中硅同位素组成更为接近，从而为追溯 $PM_{2.5}$ 来源建立了科学有效的方法。该团队[84]进一步利用硅元素的化学惰性性质，基于"硅稀释效应"定量评估了大气化学过程中二次源的贡献比重。随着 $PM_{2.5}$ 浓度的增加，硅丰度显著降低。这种稀释效应利用了二次颗粒物生成过程中总的硅浓度保持不变的特性，从而有效定量了二次气溶胶所占比重，为研究气溶胶化学过程提供了一种有效的分析工具。将 Si 同位素和 Si 丰度结合起来，可以形成二维 Si 指纹技术，并对 2013 年以来北京地区 $PM_{2.5}$ 的一次源和二次源的年际变化趋势进行了持续跟踪[85]。结果发现，从 2013~2017 年间（即《大气污染防治行动计划》实施期间），$PM_{2.5}$ 的一次源和二次源贡献都发生了急剧变化。富集轻 Si 同位素的一次源（即燃煤和工业源）的贡献显著下降，证明了污染控制政策对于燃煤和工业源的有效管控。同时，$PM_{2.5}$ 的年平均 Si 丰度从 2013 年的 1.2%上升到 2017 年的 4.6%，说明二次源污染比重显著降低（从 83%降到 42%）。这一结果为未来制定更为有效的污染控制政策、进一步降低北京地区 $PM_{2.5}$ 污染水平提供了数据。

由于较大的环境毒性效应，追踪大气重金属组分汞和铅一直是研究热点。Das 等[86]发现印度加尔各答省工业排放源、垃圾焚烧源以及汽车尾气源排放的 PM_{10} 中汞同位素组成存在明显差异，说明不同人为排放源具有不同的汞同位素指纹特征，为大气颗粒物中汞污染溯源的实现提供了科学基础。陈玖斌等[69]利用汞同位素组成变化有效表征了北京地区 $PM_{2.5}$ 的贡献源组成和气候影响。汞同位素可以有效示踪大气颗粒物相关组分来源，但由于大气化学过程的复杂性，这方面研究还需深入挖掘。铅同位素的研究工作主要集中在人为活动方面，如工业排放、煤的燃烧以及生物质燃烧等。Kayee 等[87]发现泰国地区高浓度细颗粒物气溶胶主要来自于生物质燃烧，且进一步得出了 $^{206}Pb/^{207}Pb$ 与 $PM_{2.5}$ 之间的相关性。而 Lee 等[88]通过分析韩国大田市上空亚洲尘/非亚洲尘中人为/天然源铅同位素组成，发现煤燃烧是主要的人为污染源，分别占亚洲尘/非亚洲尘铅源的 81%和 80%；此外，作为天然来源，气候影响下的阿拉善高原也有部分输入。这些研究工作同样表明铅同位素在解析人类活动对大气影响方面的重要作用。

最近，针对铜、锌、铁、锶、钕等金属稳定同位素的相关溯源工作也多见报道。Gonzalez 等[89]收集了欧洲两个主要城市（巴塞罗那和伦敦）大气颗粒物中锌和铜同位素的时空变化数据后发现，巴塞罗那上空细颗粒物中铜主要来自非尾气排放，而偏轻锌同位素比值特征表明工厂冶炼和燃烧排放是大气中锌的主要贡献源；与之相比，伦敦地区偏重的铜同位素（高达+0.97‰±0.21‰）和较高的浓度表明，化石燃料是主要排放源，而大气锌则主要来自非废气排放。与铜、锌相比，利用铁同位素示踪大气细颗粒物的研究还处于起步阶段，Flament、Majestic 等[90,91]的研究证实了人为活动的铁同位素组成与大气细颗粒物中铁同位素组成的相似性，说明基于铁同位素追溯大气颗粒物中铁的天然来源与人为来源是可行的。锶同位素近年来也逐渐被开发应用于大气细颗粒物污染溯源研究中。由于锶同位素组成（$^{87}Sr/^{86}Sr$）主要受来源影响，因而具有天然的溯源优势。Widory 等[92]通过研究北京地区 $PM_{2.5}$ 和 TSP 中锶同位素组成发现，大气颗粒物中锶主要来自于燃煤燃烧，表明锶同位素同样可以作为人为源的示踪手段。与锶相比，自然界中钕元素具有 7 种稳定同位素，且在实际测量中精度可以达到 0.1‰，是一种潜在的溯源工具。Geagea 等[93]探究了斯特拉斯堡凯尔地区大气颗粒物中钕同位素组成，发现不同来源的 $\delta^{144}Nd$ 具有差异性，说明可以基于钕同位素组成对大气颗粒物中的钕进行溯源。这些非传统稳定同位素的溯源研究，将有助于人们深入探究大气细颗粒物来源，为源头治理提供了科学参考。

20.4.3 纳米颗粒物示踪最新进展

纳米材料广泛应用于医疗、食品、化妆品和涂料等行业，已成为社会发展的重要支撑。但是，纳米材料的广泛使用会对生态环境造成潜在的危害，并进一步威胁到人类的健康安全[94-96]。环境中纳米颗粒的准确识别和示踪对于纳米颗粒物的后续毒性评估具有重要的作用。由于实际环境体系的复杂性和多变性，现有技术手段并不能有效示踪特定纳米颗粒物来源及其迁移转化过程[97]。纳米颗粒物在合成、环境迁移与转化等涉及物理/化学过程中极有可能发生同位素分馏效应，基于 MC-ICP-MS 的同位素分析技术为纳米颗粒溯源及其环境过程示踪提供了可能性。

近年来，基于非传统稳定同位素技术的纳米颗粒溯源方兴未艾。Larner 等[98]和 Laycock 等[99]分别测量了工程纳米颗粒 ZnO 和 CeO_2 中 Zn 和 Ce 的同位素组成，发现与天然物质同位素组成相比，人工纳米颗粒与其并无显著差异，因而无法识别环境中 ZnO 和 CeO_2 纳米颗粒的来源。Yang 等[100]测量了几种商业产品银稳定同位素组成后发现，硫代癸烷修饰的纳米银、膳食补充剂和丝袜中的银同位素组成存在明显差异。刘倩等[101]发现人为源和天然源纳米银在转化过程中具有不同的银同位素分馏效应，这为纳米银的来源识别提供了可能性。这些研究结果展示了 MC-ICP-MS 在纳米银溯源研究中潜在的应用价值。该课题组基于多同位素示踪体系（Si-O）的 SiO_2 纳米颗粒溯源，发现人为源和天然源 SiO_2 纳米颗粒具有可分辨的硅-氧双同位素指纹特征，进一步结合机器学习模型（如 Linear discriminant analysis 模型）给出了人为源和天然源 SiO_2 纳米颗粒的定量判别结果，准确率达 93.3%。此外，硅-氧二维同位素指纹还可以在一定程度上识别工程 SiO_2 纳米颗粒的合成方法和厂家来源，进一步证明了 MC-ICP-MS 在环境纳米颗粒物示踪方面的应用潜力。

此外，利用 MC-ICP-MS 还可以对环境中纳米颗粒的转化过程进行示踪。一方面，天然纳米颗粒物的生成过程可以发生显著的同位素分馏效应。在自然水体中，溶解性 Fe^{2+} 可以生成硫化铁类纳米颗粒（FeS_m），其生长过程伴随显著的铁同位素分馏效应[102]。人为源纳米颗粒在自然水体中的转化也会导致同位素分馏效应，基于此可以示踪人为源纳米颗粒在环境中的转化与归趋。刘倩等[101]报道了纳米银在自然水体中的转化过程可发生显著的银同位素分馏效应，同位素分馏因子达 0.86‰，具体地：该团队通过分析纳米银生成和溶解过程中伴随的银同位素分馏特征，揭示了纳米银在自然环境中的生成和溶解机制，为纳米银的迁移转化研究提供了强有力的示踪工具。基于这一示踪工具，该课题组[103]结合纳米银在不同浓度腐殖酸水体中溶解过程伴随的银同位素分馏特征，进一步揭示了纳米银在实际水体中的稳定机制。

目前，环境中大量纳米颗粒的毒理学机制尚不明确，这一环节被掣肘则与纳米颗粒的环境行为分析相关。MC-ICP-MS 为我们提供了一个有力的手段，基于该仪器的优势，人们可以更好地了解纳米颗粒来源及其环境行为，这对于评估环境中纳米颗粒的分布及毒理机制具有良好的应用前景。

20.5　MC-ICP-MS 在生物同位素分析中的应用

生命体对元素的生物地球化学循环起着重要的作用。元素地球化学循环涉及的一些生物过程，如：生物吸收转化、生物富集、生物氧化还原等，通常会伴随着同位素分馏现象。利用 MC-ICP-MS 对这一分馏现象进行观察可以追溯生物体中相关物质的来源及转化过程。

MC-ICP-MS 在生物同位素方面的早期应用中主要涉及一些同位素组成的观测，如微生物体内不同蛋白质的同位素比值差异[104]；动植物不同组织器官中的同位素分布差异[105,106]；人体不同组织[107]、性别间[108]的同位素差异以及疾病相关的同位素分馏现象[109]等。近年来，随着 MC-ICP-MS 分析方法的不断改进，非传统稳定同位素技术在生物学领域获得了较快的发展，已逐渐从同位素组成测量层次向同位素分馏机理探究方面深入发展。本节将主要根据生物类别（微生物、植物、动物和人体）介绍生物同位素分析的最新进展。

微生物是自然界中存在最广泛的生命体，在物质的生物地球化学循环中扮演着重要的角色，依靠 MC-ICP-MS 获得的同位素信息有助于科学家们更深入了解微生物的相关生命活动。在近期的微生物同位素研究中，科学家们主要关注两个方面：①微生物摄取物质过程产生的同位素分馏效应；②微生物对外界环境中同位素分馏效应的影响。微生物摄取元素的过程中容易产生同位素分馏效应，而不同元素的分馏效果也可能是不一样的。Chen 等[110]发现浮游微生物在吸收铀元素时会产生同位素分馏效应，且该过程受到热力学平衡分馏效应的主导。除了吸收过程，微生物还可以通过间接方式改变外界环境中物质的同位素组成。Miller 等发现抗生素处理可改变老鼠结肠中的铜同位素组成，并且同时改变了肠道中铜转运蛋白的表达。这一现象暗示肠道菌群的变化可以影响到结肠中铜同位素的组成[111]。

植物是生态系统的重要组成部分，对元素的生物地球化学循环也具有重要影响。植物吸收转运相关元素的过程中通常伴随着同位素分馏效应，研究这一分馏效应有助于理解相关元素在植物体中的吸收转运过程和相关机理。近期的研究发现土壤的理化性质不仅会影响植物的吸收效率，而且会影响植物吸收过程中的同位素分馏效应。Arnold 等[112]发现在好氧土壤中，土壤与地上植株的锌同位素分馏量很小；而在厌氧土壤中，水稻植株更倾向于从土壤中吸收富集较轻的锌同位素。此外，土壤的 pH 以及元素在土壤中赋存状态也会影响植株吸收过程中伴随的同位素分馏效应[113]。植物中元素的转运过程同样容易产生同位素分馏效应。Ryan 等近期发现番茄在转运 Cu 的过程中会发生同位素分馏效应，且该分馏效应受到不同组织中的氧化还原过程的影响，进而导致番茄茎部优先富集较重的 ^{65}Cu，而根部和叶子则富集较轻的 ^{63}Cu[114]。

相对于植物而言，动物一般具有更加复杂的形态结构和生理功能（比如摄食、消化、吸收、排泄和运动等），因此相关的元素吸收转运过程通常更为复杂。Jaouen 等利用 MC-ICP-MS 分析了植物、食草动物和食肉动物体内的铁、铜、锌同位素组成，发现食草动物的骨骼相对于植物富集重同位素 ^{66}Zn，而 Fe 和 Cu 没有显示出明显的富集现象；同样，食肉动物和食草动物骨骼内 Fe、Cu 和 Zn 具有不同的同位素分馏效应。这些不同的同位素分馏现象为示踪相关元素在食物链中的传递提供了潜在的工具[115]。此外，有研究指出肠道吸收过程也有可能导致同位素分馏效应[106]。近期，Mahan 等开展了哺乳动物相关的同位素应用研究，他们通过分析哥廷根小型猪体内锌同位素的组成特征发现，相对于血液而言，肝脏、心脏和大脑中富集较轻的锌同位素，揭示了锌同位素在小猪体内不同部位具有不同的富集行为[116]。

随着消解和纯化技术的发展，针对人体样品的同位素前处理方法逐渐完善，进一步加速了稳定同位素技术在人体健康研究中的应用进展。大量文献数据表明多种肿瘤的发生与铜元素密切相关[117-119]，因此人们更早地开始关注肿瘤（如肝癌[120]、结肠癌[108]、乳腺癌[108]等）与铜同位素之间的关系。通常，相对于正常人，癌症患者血液中倾向于富集轻铜同位素。Balter 等[120]在研究肝癌与 Cu 同位素关系时，也发现了肝癌患者血液中富集轻同位素的现象，他们推测这一现象主要是由患者体内环境改变导致的，而非食源影响。近期，陆达伟等[121]将非传统稳定同位素技术应用于真实人体内的细颗粒物溯源研究。他们采集了胸腔积液（患者）和血液样本（正常人和患者），发现这些真实人体样本中存在大量纳米级颗粒物（以金属颗粒为主）。进一步通过体内颗粒物的同位素组成与体内样本和细颗粒物体外污染源（燃烧源）进行对比，推测体内含铁颗粒物以外源性为主，且主要来自于燃烧过程，比如汽车尾气和燃煤燃烧等。

20.6 展望

经过二十多年的实践，MC-ICP-MS 在分析方法开发与实际应用中获得了快速发展，已成为非传统稳定同位素分析的首选仪器。在应用领域方面，MC-ICP-MS 也从最初的地学领域，逐渐扩展到了环境科学和生物学领域。可以预见，随着非传统稳定同位素技术的发展，MC-ICP-MS 将会获得更广阔的舞台，同时，其本身在理论基础、仪器性能、方法开发及应用领域等方面的发展上，仍需获得更多投入，进而可以进一步促进非传统稳定同位素技术的发展。未来，还需要在以下几个方面继续加强研究：

（1）进一步提高仪器分析性能和优化前处理方法，努力克服环境样品同位素分析的难点，比如复杂基质和痕量分析。

（2）进一步扩展及完善 MC-ICP-MS 的应用体系，比如开发新的同位素应用体系和相关应用领域、完善非传统稳定同位素源谱数据库、探究实际环境过程中的同位素分馏影响机制等。

（3）推进 MC-ICP-MS 在环境科学和生物学等相关领域的发展。MC-ICP-MS 在地学领域的应用促进了非传统稳定同位素技术理论体系的发展与完善，未来，推进 MC-ICP-MS 在环境科学和生物学等交叉学科领域的应用，将会为解决相关交叉学科面临的困难（例如示踪来源和过程）提供方法学支持。

参 考 文 献

[1] Feng L P, Zhou L, Hu W F, Zhang W, Li B C, Liu Y S, Hu Z C, Yang L. A simple single-stage extraction method for Mo separation from geological samples for isotopic analysis by MC-ICP-MS. Journal of Analytical Atomic Spectrometry, 2020, 35(1): 145-154.

[2] Fan J J, Li J, Wang Q, Zhang L, Zhang J, Zeng X L, Ma L, Wang Z L. High-precision molybdenum isotope analysis of low-Mo igneous rock samples by MC-ICP-MS. Chemical Geology, 2020, 545: 119648.

[3] Tan D, Zhu J-M, Wang X, Han G, Lu Z, Xu W. High-sensitivity determination of Cd isotopes in low-Cd geological samples by double spike MC-ICP-MS. Journal of Analytical Atomic Spectrometry, 2020, 35(4): 713-727.

[4] Friebel M, Toth E R, Fehr M A, Schönbächler M. Efficient separation and high-precision analyses of tin and cadmium isotopes in geological materials. Journal of Analytical Atomic Spectrometry, 2020, 35(2): 273-292.

[5] He X, Ma J, Wei G, Zhang L, Wang Z, Wang Q. A new procedure for titanium separation in geological samples for ^{49}Ti/^{47}Ti ratio measurement by MC-ICP-MS. Journal of Analytical Atomic Spectrometry, 2020, 35(1): 100-106.

[6] Liu J, Chen J, Zhang T, Wang Y, Yuan W, Lang Y, Tu C, Liu L, Birck J L. Chromatographic purification of antimony for accurate isotope analysis by MC-ICP-MS. Journal of Analytical Atomic Spectrometry, 2020, 35(7): 1360-1367.

[7] Mahan B M, Wu F, Dosseto A, Chung R, Schaefer B, Turner S. SpinChem™: Rapid element purification from biological and geological matrices via centrifugation for MC-ICP-MS isotope analyses: A case study with Zn. Journal of Analytical Atomic Spectrometry, 2020, 35(5): 863-872.

[8] Chen J, Hintelmann H, Dimock B. Chromatographic pre-concentration of Hg from dilute aqueous solutions for isotopic measurement by MC-ICP-MS. Journal of Analytical Atomic Spectrometry, 2010, 25(9): 1402-1409.

[9] Sonke J E, Zambardi T, Toutain J P. Indirect gold trap-MC-ICP-MS coupling for Hg stable isotope analysis using a syringe injection interface. Journal of Analytical Atomic Spectrometry, 2008, 23(4): 569-573.

[10] Gratz L E, Keeler G J, Blum J D, Sherman L S. Isotopic composition and fractionation of mercury in great lakes precipitation and ambient air. Environmental Science & Technology, 2010, 44(20): 7764-7770.

[11] Lin H, Yuan D, Lu B, Huang S, Sun L, Zhang F, Gao Y. Isotopic composition analysis of dissolved mercury in seawater with purge and trap preconcentration and a modified Hg introduction device for MC-ICP-MS. Journal of Analytical Atomic Spectrometry, 2015, 30(2): 353-359.

[12] Li M, Lei Y, Feng L, Wang Z, Belshaw N S, Hu Z, Liu Y, Zhou L, Chen H, Chai X. High-precision Ca isotopic measurement using a large geometry high resolution MC-ICP-MS with a dummy bucket. Journal of Analytical Atomic Spectrometry, 2018, 33(10): 1707-1719.

[13] Li X, Han G, Zhang Q, Miao Z. An optimal separation method for high-precision K isotope analysis by using MC-ICP-MS with a dummy bucket. Journal of Analytical Atomic Spectrometry, 2020, 35(7): 1330-1339.

[14] Nielsen S G, Owens J D, Horner T J. Analysis of high-precision vanadium isotope ratios by medium resolution MC-ICP-MS. Journal of Analytical Atomic Spectrometry, 2016, 31(2): 531-536.

[15] Peters S T M, Münker C, Wombacher F, Elfers B-M. Precise determination of low abundance isotopes (^{174}Hf, ^{180}W and

^{190}Pt) in terrestrial materials and meteorites using multiple collector ICP-MS equipped with 1012 Ω Faraday amplifiers. Chemical Geology, 2015, 413: 132-145.

[16] Creech J B, Schaefer B F, Turner S P. Application of 1013 Ω amplifiers in low-signal plasma-source isotope ratio measurements by MC-ICP-MS: A case study with Pt isotopes. Geostandards and Geoanalytical Research, 2020, 44(2): 223-229.

[17] Dellinger M, Hilton R G, Nowell G M. Measurements of rhenium isotopic composition in low-abundance samples. Journal of Analytical Atomic Spectrometry, 2020, 35(2): 377-387.

[18] González de Vega C, Costas-Rodríguez M, Van Acker T, Goderis S, Vanhaecke F. Nanosecond laser ablation-multicollector Inductively coupled plasma-mass spectrometry for *in situ* Fe isotopic analysis of micrometeorites: Application to micrometer-sized glassy cosmic spherules. Analytical Chemistry, 2020, 92(5): 3572-3580.

[19] Zhang C, Zhao H, Zhang W, Luo T, Li M, Zong K, Liu Y, Hu Z. A high performance method for the accurate and precise determination of silicon isotopic compositions in bulk silicate rock samples using laser ablation MC-ICP-MS. Journal of Analytical Atomic Spectrometry, 2020, 35(9): 1887-1896.

[20] Desaulty A M, Lach P, Perret S. Rapid determination of lead isotopes in water by coupling DGT passive samplers and MC-ICP-MS laser ablation. Journal of Analytical Atomic Spectrometry, 2020, 35(8): 1537-1546.

[21] Dzurko M, Foucher D, Hintelmann H. Determination of compound-specific Hg isotope ratios from transient signals using gas chromatography coupled to multicollector inductively coupled plasma mass spectrometry (MC-ICP/MS). Analytical and Bioanalytical Chemistry, 2008, 393(1): 345.

[22] Amrani A, Sessions A L, Adkins J F. Compound-specific δ^{34}S analysis of volatile organics by coupled GC/multicollector-ICP-MS. Analytical Chemistry, 2009, 81(21): 9027-9034.

[23] Van Acker M R M D, Shahar A, Young E D, Coleman M L. GC/multiple collector-ICPMS method for chlorine stable isotope analysis of chlorinated aliphatic hydrocarbons. Analytical Chemistry, 2006, 78(13): 4663-4667.

[24] Gelman F, Halicz L. High precision determination of bromine isotope ratio by GC-MC-ICPMS. International Journal of Mass Spectrometry, 2010, 289(2): 167-169.

[25] Karasiński J, Nguyen-Marcińczyk C T, Wojciechowski M, Bulska E, Halicz L. Determination of isotope fractionation of Cr(Ⅲ) during oxidation by LC/low-resolution MC-ICPMS. Journal of Analytical Atomic Spectrometry, 2020, 35(3): 560-566.

[26] Huang Q, Reinfelder J R, Fu P, Huang W. Variation in the mercury concentration and stable isotope composition of atmospheric total suspended particles in Beijing, China. Journal of Hazardous Materials, 2019, 383: 121131.

[27] Bao Z, Huang K-J, Xu J, Deng L, Yang S, Zhang P, Yuan H. Preparation and characterization of a new reference standard GSB-Mg for Mg isotopic analysis. Journal of Analytical Atomic Spectrometry, 2020, 35(6): 1080-1086.

[28] Yuan H, Yuan W, Bao Z, Chen K, Huang F, Liu S. Development of two new copper isotope standard solutions and their copper isotopic compositions. Geostandards and Geoanalytical Research, 2017, 41(1): 77-84.

[29] Huang C, Wang H, Yang J H, Ramezani J, Yang C, Zhang S B, Yang Y H, Xia X P, Feng L J, Lin J, Wang T T, Ma Q, He H Y, Xie L W, Wu S T. SA01: A proposed zircon reference material for microbeam U-Pb age and Hf-O isotopic determination. Geostandards and Geoanalytical Research, 2020, 44(1): 103-123.

[30] Li J, Tang S h, Zhu X K, Pan C X. Production and certification of the reference material GSB 04-3258-2015 as a ^{143}Nd/^{144}Nd isotope ratio reference. Geostandards and Geoanalytical Research, 2017, 41(2): 255-262.

[31] Tang S, Zhu X, Li J, Yan B, Li S, Li Z, Wang Y, Sun J. New standard solutions for measurement of iron, copper and zinc isotopic compositions by multi-collector Inductively coupled plasma-mass spectrometry. Rock and Mineral Analysis, 2016, 35(2): 127-133.

[32] Valente A J M. Assessment of heavy metal pollution from anthropogenic activities and remediation strategies: A review. Journal of Environmental Management, 2019, 246: 101-108.

[33] Giripunje M D, Fulke A B, Meshram P U. Remediation techniques for heavy-metals contamination in lakes: A mini-review. Clean-Soil Air Water, 2015, 43(9): 1350-1354.

[34] Sun G X, Wang X J, Hu Q H. Using stable lead isotopes to trace heavy metal contamination sources in sediments of Xiangjiang and Lishui Rivers in China. Environmental Pollution, 2011, 159(12): 3406-3410.

[35] Hu X, Sun Y, Ding Z, Zhang Y, Wu J, Lian H, Wang T. Lead contamination and transfer in urban environmental compartments analyzed by lead levels and Isotopic compositions. Environmental Pollution, 2014, 187: 42-48.

[36] Lee H C, Kim M K, Jo W K. Pb isotopic ratios in airborne PM_{10} of an iron/metal industrial complex area and nearby residential areas: Implications for ambient sources of Pb pollution. Atmospheric Research, 2011, 99(3-4): 462-470.

[37] Cong Z, Kang S, Luo C, Li Q, Huang J, Gao S, Li X. Trace elements and lead isotopic composition of PM_{10} in Lhasa, Tibet. Atmospheric Environment, 2011, 45(34): 6210-6215.

[38] Bi X Y, Li Z G, Wang S X, Zhang L, Xu R, Liu J L, Yang H M, Guo M Z. Lead isotopic compositions of selected coals, Pb/Zn ores and fuels in China and the application for source tracing. Environmental Science & Technology, 2017, 51(22): 13502-13508.

[39] Diaz-Somoano M, Kylander M E, Lopez-anton M A, Suarez-Ruiz I, Martinez-Tarazona M R, Ferrat M, Kober B, Weiss D J. Stable lead isotope compositions in selected coals from around the world and implications for present day aerosol source tracing. Environmental Science & Technology, 2009, 43(4): 1078-1085.

[40] Yang Z P, Wen L U, Xin X, Jun L I. Lead isotope signatures and source identification in urban soil of Changchun City. Journal of Jilin University, 2008, 38(4): 663-669.

[41] Kong H, Teng Y, Song L, Wang J, Zhang L. Lead and strontium isotopes as tracers to investigate the potential sources of lead in soil and groundwater: A case study of the Hun River alluvial fan. Applied Geochemistry, 2018, 97: 291-300.

[42] Sangster D F, Outridge P M, Davis W J. Stable lead isotope characteristics of lead ore deposits of environmental significance. Dossiers Environnement, 2000, 8(2): 115-147.

[43] Reimann C, Flem B, Fabian K, Birke M, Ladenberger A, Ngrel P, Demetriades A, Hoogewerff J. Lead and lead isotopes in agricultural soils of Europe: The continental perspective. Applied Geochemistry, 2012, 27(3): 532-542.

[44] Yin N H, Sivry Y, Benedetti M F, Lens P N L, Van-Hullebusch E D. Application of Zn isotopes in environmental impact assessment of Zn-Pb metallurgical industries: A mini review. Applied Geochemistry, 2016, 64: 128-135.

[45] Moynier F, Vance D, Fujii T, Savage P. The isotope geochemistry of zinc and copper//Teng F Z, Watkins J, Dauphas N. Non-Traditional Stable Isotopes. 2017, vol. 82: 543-600.

[46] Kurisu M, Adachi K, Sakata K, Takahashi Y. Stable isotope ratios of combustion iron produced by evaporation in a steel plant. ACS Earth & Space Chemistry, 2019, 3(4): 588-598.

[47] Zhong Q, Zhou Y, Tsang D C W, Liu J, Zhang Z. Cadmium isotopes as tracers in environmental studies: A review. Science of the Total Environment, 2020, 736: 139585.

[48] Fotios F, Lin M, Engle M A, Leslie R, Geboy N J, Costa M A. Cadmium isotope fractionation during coal combustion: Insights from two U.S. coal-fired power plants. Applied Geochemistry, 2018, 96: 100-112.

[49] Guinoiseau D, Gelabert A, Moureau J, Louvat P, Benedetti M F. Zn isotope fractionation during sorption onto kaolinite. Environmental Science & Technology, 2016, 50(4): 1844-1852.

[50] Gou W, Li W, Ji J, Li W. Zinc isotope fractionation during sorption onto Al oxides: Atomic level understanding from EXAFS. Environmental Science & Technology, 2018, 52(16): 9087-9096.

[51] Swanner E D, Bayer T, Wu W, Hao L, Obst M, Sundman A, Byrne J M, Michel F M, Kleinhanns I C, Kappler A,

Schoenberg R. Iron isotope fractionation during Fe(II) oxidation mediated by the oxygen-producing marine cyanobacterium *Synechococcus* PCC 7002. Environmental Science & Technology, 2017, 51(9): 4897-4906.

[52] Ryan B M, Kirby J K, Degryse F, Scheiderich K, Mclaughlin M J. Copper isotope fractionation during equilibration with natural and synthetic ligands. Environmental Science & Technology, 2014, 48(15): 8620-8626.

[53] Morgan J L L, Wasylenki L E, Nuester J, Anbar A D. Fe isotope fractionation during equilibration of Fe-organic complexes. Environmental Science & Technology, 2010, 44(16): p.6095-6101.

[54] Michael, Koebberich, Derek, Vance. Kinetic control on Zn isotope signatures recorded in marine diatoms. Geochimica Et Cosmochimica Acta, 2017, 210: 97-113.

[55] Walraven N, van Os B J H, Klaver G T, Middelburg J J, Davies G R. The lead (Pb) isotope signature, behaviour and fate of traffic-related lead pollution in roadside soils in The Netherlands. Science of the Total Environment, 2014, 472: 888-900.

[56] Takano S, Liao W H, Tian H A, Huang K F, Sohrin Y. Sources of particulate Ni and Cu in the water column of the northern South China Sea: Evidence from elemental and isotope ratios in aerosols and sinking particles. Marine Chemistry, 2020, 219: 103751.

[57] Tu Y, You C, Kuo T. Source identification of Zn in Erren River, Taiwan: An application of Zn isotopes. Chemosphere, 2020, 248: 126044.

[58] Kanagaraj G, Elango L. Chromium and fluoride contamination in groundwater around leather tanning industries in southern India: Implications from stable isotopes $\delta^{53}Cr/\delta^{52}Cr$, geochemical and geostatistical modelling. Chemosphere, 2018, 220: 943-953.

[59] Kersten M, Xiao T, Kreissig K, Brett A, Coles B J, Rehkaemper M. Tracing anthropogenic thallium in soil using stable isotope compositions. Environmental Science & Technology, 2014, 48(16): 9030-9036.

[60] Martin A, Caldelas C, Weiss D, Aranjuelo I, Navarro E. Assessment of metal immission in urban environments using elemental concentrations and zinc isotope signatures in leaves of *Nerium oleander*. Environmental Science & Technology, 2018, 52(4): 2071-2080.

[61] Shepherd T J, Dirks W, Roberts N M W, Patel J G, Hodgson S, Pless-Mulloli T, Walton P, Parrish R R. Tracing fetal and childhood exposure to lead using isotope analysis of deciduous teeth. 2016, 146: 145-153.

[62] Bergquist B A, Blum J D. Mass-dependent and-independent fractionation of Hg isotopes by photoreduction in aquatic systems. Science, 2007, 318(5849): 417-420.

[63] Blum J D, Sherman L S, Johnson M W. Mercury isotopes in earth and environmental sciences. Annual Review of Earth & Planetary Sciences, 42(1): 249-269.

[64] Enrico M, Le Roux G, Marusczak N, Heimbuerger L E, Claustres A, Fu X, Sun R, Sonke J E. Atmospheric mercury transfer to peat bogs dominated by gaseous elemental mercury dry deposition. Environmental Science & Technology, 2016, 50(5): 2405-2412.

[65] Sherman L S, Blum J D, Johnson K P, Keeler G J, Barres J A, Douglas T A. Mass-independent fractionation of mercury isotopes in Arctic snow driven by sunlight. Nature Geoscience, 2010, 3(3): 173-177.

[66] Li M, Juang C A, Ewald J D, Yin R, Mikkelsen B, Krabbenhoft D P, Balcom P H, Dassuncao C, Sunderland E M. Selenium and stable mercury isotopes provide new insights into mercury toxicokinetics in pilot whales. Science of the Total Environment, 2020, 710: 136325.

[67] Li M, Sherman L S, Blum J D, Grandjean P, Mikkelsen B, Weihe P, Sunderland E M, Shine J P. Assessing sources of human methylmercury exposure using stable mercury isotopes. Environmental Science & Technology, 2014, 48(15): 8800-8806.

[68] Tsui T K, Blum J D, Finlay J C, Balogh S J, Nollet Y H, Palen W J, Power M E. Variation in terrestrial and aquatic sources of methylmercury in stream predators as revealed by stable mercury isotopes. Environmental Science & Technology, 2014, 48(17): 10128-10135.

[69] Huang Q, Chen J, Huang W, Fu P, Guinot B, Feng X, Shang L, Wang Z, Wang Z, Yuan S, Cai H, Wei L, Yu B. Isotopic composition for source identification of mercury in atmospheric fine particles. Atmospheric Chemistry and Physics, 2016, 16(18): 11773-11786.

[70] Blum J D, Johnson M W. Recent developments in mercury stable isotope analysis. Reviews in Mineralogy & Geochemistry, 2017, 82(1): 733-757.

[71] Chen J B, Hintelmann H, Feng X B, Dimock B. Unusual fractionation of both odd and even mercury isotopes in precipitation from Peterborough, ON, Canada. Geochimica Et Cosmochimica Acta, 2012, 90: 33-46.

[72] Meng M, Sun R Y, Liu H W, Yu B, Yin YG, Hu L G, Shi J B, Jiang G B. An integrated model for input and migration of mercury in Chinese coastal sediments. Environmental Science & Technology, 2019, 53(5): 2460-2471.

[73] Meng M, Sun R Y, Liu H W, Yu B, Yin Y G, Hu L G, Chen J B, Shi J B, Jiang G B. Mercury isotope variations within the marine food web of Chinese Bohai Sea: Implications for mercury sources and biogeochemical cycling. Journal of Hazardous Materials, 2020, 384: 121379.

[74] Liu H W, Shao J J, Yu B, Liang Y, Duo B, Fu J J, Yang R Q, Shi J B, Jiang G B. Mercury isotopic compositions of mosses, conifer needles, and surface soils: Implications for mercury distribution and sources in Shergyla Mountain, Tibetan Plateau. Ecotoxicology and Environmental Safety, 2019, 172: 225-231.

[75] Li W, Gou W X, Li W Q, Zhang T Y, Yu B, Liu Q, Shi J B. Environmental applications of metal stable isotopes: Silver, mercury and zinc. Environmental Pollution, 2019, 252(Pt B): 1344-1356.

[76] Kwon S Y, Blum J D, Yin R, Tsui M T K, Yang Y H, Choi J W. Mercury stable isotopes for monitoring the effectiveness of the minamata convention on mercury. Earth-Science Reviews, 2020, 203: 103111.

[77] Yuan W, Saldi G D, Chen J, Zuccolini M V, Birck J L, Liu Y, Schott J. Gallium isotope fractionation during Ga adsorption on calcite and goethite. Geochimica Et Cosmochimica Acta, 2018, 223: 350-363.

[78] Pokrovsky O S, Galy A, Schott J, Pokrovski G S, Mantoura S. Germanium isotope fractionation during Ge adsorption on goethite and its coprecipitation with Fe oxy(hydr)oxides. Geochimica Et Cosmochimica Acta, 2014, 131: 138-149.

[79] Souto-Oliveira C E, Babinski M, Araujo D F, Andrade M F. Multi-isotopic fingerprints (Pb, Zn, Cu) applied for urban aerosol source apportionment and discrimination. Science of the Total Environment, 2018, 626: 1350-1366.

[80] Zhu Z, Li Z, Wang S, Bi X. Magnetic mineral constraint on lead isotope variations of coal fly ash and its implications for source discrimination. Science of the Total Environment, 2020, 713: 136320.

[81] Zhang R, Wang G, Guo S, Zamora M L, Ying Q, Lin Y, Wang W, Hu M, Wang Y. Formation of urban fine particulate matter. Chemical Review, 2015, 115(10): 3803-3855.

[82] Weagle C L, Snider G, Li C, van Donkelaar A, Philip S, Bissonnette P, Burke J, Jackson J, Latimer R, Stone E, Abboud I, Akoshile C, Anh N X, Brook J R, Cohen A, Dong J, Gibson M D, Griffith D, He K B, Holben B N, Kahn R, Keller C A, Kim J S, Lagrosas N, Lestari P, Khian Y L, Liu Y, Marais E A, Martins J V, Misra A, Muliane U, Pratiwi R, Quel E J, Salam A, Segev L, Tripathi S N, Wang C, Zhang Q, Brauer M, Rudich Y, Martin R V. Global sources of fine particulate matter: Interpretation of $PM_{2.5}$ chemical composition observed by SPARTAN using a global chemical transport model. Environmental Science & Technology, 2018, 52(20): 11670-11681.

[83] Lu D, Liu Q, Yu M, Yang X, Fu Q, Zhang X, Mu Y, Jiang G. Natural silicon isotopic signatures reveal the sources of airborne fine particulate matter. Environmental Science & Technology, 2018, 52(3): 1088-1095.

[84] Lu D, Tan J, Yang X, Sun X, Liu Q, Jiang G. Unraveling the role of silicon in atmospheric aerosol secondary formation:

A new conservative tracer for aerosol chemistry. Atmospheric Chemistry and Physics, 2019, 19(5): 2861-2870.

[85] Yang X, Lu D, Tan J, Sun X, Zhang Q, Zhang L, Li Y, Wang W, Liu Q, Jiang G. Two-dimensional silicon fingerprints reveal dramatic variations in the sources of particulate matter in Beijing during 2013~2017. Environmental Science & Technology, 2020, 54(12): 7126-7135.

[86] Das R, Wang X, Khezri B, Webster R D, Sikdar P K, Datta S. Mercury isotopes of atmospheric particle bound mercury for source apportionment study in urban Kolkata, India. Elementa: Science of the Anthropocene, 2016, 4: 1-12.

[87] Kayee J, Sompongchaiyakul P, Sanwlani N, Bureekul S, Wang X, Das R. Metal concentrations and source apportionment of $PM_{2.5}$ in Chiang Rai and Bangkok, Thailand during a biomass burning season. ACS Earth and Space Chemistry, 2020, 4(7): 1213-1226.

[88] Lee P K, Yu S. Lead isotopes combined with a sequential extraction procedure for source apportionment in the dry deposition of Asian dust and non-Asian dust. Environmental Pollution, 2016, 210: 65-75.

[89] Gonzalez R O, Strekopytov S, Amato F, Querol X, Reche C, Weiss D. New insights from zinc and copper isotopic compositions into the sources of atmospheric particulate matter from two major european cities. Environmental Science & Technology, 2016, 50(18): 9816-9824.

[90] Flament P, Mattielli N, Aimoz L, Choel M, Deboudt K, de Jong J, Rimetz-Planchon J, Weis D. Iron isotopic fractionation in industrial emissions and urban aerosols. Chemosphere, 2008, 73(11): 1793-1798.

[91] Majestic B J, Anbar A D, Herckes P. Stable isotopes as a tool to apportion atmospheric iron. Environmental Science & Technology, 2009, 43(12) 4327-4333.

[92] Widory D, Liu X, Dong S. Isotopes as tracers of sources of lead and strontium in aerosols (TSP & $PM_{2.5}$) in Beijing. Atmospheric Environment, 2010, 44(30): 3679-3687.

[93] Geagea M L, Srille P, Gauthier-Lafaye F, Millet M. Tracing of industrial aerosol sources in an urban environment Using Pb, Sr, and Nd isotopes. Environmental Science & Technology, 2008, 42(3): 692-698.

[94] Nowack B, Bucheli T D. Occurrence, behavior and effects of nanoparticles in the environment. Environmental Pollution, 2007, 150(1): 5-22.

[95] Handy R D, von der Kammer F, Lead J R, Hassellov M, Owen R, Crane M. The ecotoxicology and chemistry of manufactured nanoparticles. Ecotoxicology, 2008, 17(4): 287-314.

[96] Klaine, S J, Alvarez P J J, Batley G E, Fernandes T F, Handy R D, Lyon D Y, Mahendra S, McLaughlin M J, Lead J R. Nanomaterials in the environment: Behavior, fate, bioavailability, and effects. Environmental Toxicology and Chemistry, 2008, 27(9): 1825-1851.

[97] Yin Y, Tan Z, Hu L, Yu S, Liu J, Jiang G. Isotope tracers to study the environmental fate and bioaccumulation of metal-containing engineered nanoparticles: Techniques and applications. Chemical Review, 2017, 117(5): 4462-4487.

[98] Larner F, Rehkamper M. Evaluation of stable isotope tracing for ZnO nanomaterials: New constraints from high precision isotope analyses and modeling. Environmental Science & Technology, 2012, 46(7): 4149-4158.

[99] Laycock A, Coles B, Kreissig K, Rehkämper M. High precision $^{142}Ce/^{140}Ce$ stable isotope measurements of purified materials with a focus on CeO_2 nanoparticles. Journal of Analytical Atomic Spectrometry, 2016, 31(1): 297-302.

[100] Yang L, Dabek-Zlotorzynska E, Celo V. High precision determination of silver isotope ratios in commercial products by MC-ICP-MS. Journal of Analytical Atomic Spectrometry, 2009, 24(11): 1564-1569.

[101] Lu D, Liu Q, Zhang T, Cai Y, Yin Y, Jiang G. Stable silver isotope fractionation in the natural transformation process of silver nanoparticles. Nature Nanotechnology, 2016, 11(8): 682-686.

[102] Guilbaud R, Butler I B, Ellam R M, Rickard D. Fe isotope exchange between Fe(Ⅱ)$_{aq}$ and nanoparticulate

mackinawite (FeS$_m$) during nanoparticle growth. Earth and Planetary Science Letters, 2010, 300(1-2): 174-183.

[103] Zhang T, Lu D, Zeng L, Yin Y, He Y, Liu Q, Jiang G. Role of secondary particle formation in the persistence of silver nanoparticles in humic acid containing water under light irradiation. Environmental Science & Technology, 2017, 51(24): 14164-14172.

[104] Zhu X, Guo Y, Williams R, O'nions R, Matthews A, Belshaw N, Canters G, De Waal E, Weser U, Burgess B. Mass fractionation processes of transition metal isotopes. Earth and Planetary Science Letters, 2002, 200(1-2): 47-62.

[105] Guelke M, Von Blanckenburg F. Fractionation of stable iron isotopes in higher plants. Environmental Science & Technology, 2007, 41(6): 1896-1901.

[106] Balter V, Lamboux A, Zazzo A, Telouk P, Leverrier Y, Marvel J, Moloney A P, Monahan F J, Schmidt O, Albarede F. Contrasting Cu, Fe, and Zn isotopic patterns in organs and body fluids of mice and sheep, with emphasis on cellular fractionation. Metallomics, 2013, 5(11): 1470-1482.

[107] Walczyk T, von Blanckenburg F. Natural iron isotope variations in human blood. Science, 2002, 295(5562): 2065-2066.

[108] Telouk P, Puisieux A, Fujii T, Balter V, Bondanese V P, Morel A P, Clapisson G, Lamboux A, Albarede F. Copper isotope effect in serum of cancer patients: A pilot study. Metallomics, 2015, 7(2): 299-308.

[109] Krayenbuehl P-A, Walczyk T, Schoenberg R, von Blanckenburg F, Schulthess G. Hereditary hemochromatosis is reflected in the iron isotope composition of blood. Blood, 2005, 105(10): 3812-3816.

[110] Chen X, Zheng W, Anbar A D. Uranium isotope fractionation (^{238}U/^{235}U) during U(Ⅵ) uptake by freshwater plankton. Environmental Science & Technology, 2020, 54(5): 2744-2752.

[111] Miller K A, Vicentini F A, Hirota S A, Sharkey K A, Wieser M E. Antibiotic treatment affects the expression levels of copper transporters and the isotopic composition of copper in the colon of mice. Proceedings of the National Academy of Sciences of the United States of America, 2019, 116(13): 5955-5960.

[112] Arnold T, Markovic T, Kirk G J D, Schönbächler M, Rehkämper M, Zhao F J, Weiss D J. Iron and zinc isotope fractionation during uptake and translocation in rice (*Oryza sativa*) grown in oxic and anoxic soils. Comptes Rendus Geoscience, 2015, 347(7-8): 397-404.

[113] Li S Z, Zhu X K, Wu L H, Luo Y M. Cu isotopic compositions in Elsholtzia splendens: Influence of soil condition and growth period on Cu isotopic fractionation in plant tissue. Chemical Geology, 2016, 444: 49-58.

[114] Ryan B M, Kirby J K, Degryse F, Harris H, McLaughlin M J, Scheiderich K. Copper speciation and isotopic fractionation in plants: Uptake and translocation mechanisms. New Phytologist, 2013, 199(2): 367-378.

[115] Jaouen K, Pons M L, Balter V. Iron, copper and zinc isotopic fractionation up mammal trophic chains. Earth and Planetary Science Letters, 2013, 374: 164-172.

[116] Mahan B, Moynier F, Jorgensen A L, Habekost M, Siebert J. Examining the homeostatic distribution of metals and Zn isotopes in Gottingen minipigs. Metallomics, 2018, 10(9): 1264-1281.

[117] Majumder S, Chatterjee S, Pal S, Biswas J, Efferth T, Choudhuri S K. The role of copper in drug-resistant murine and human tumors. Biometals, 2009, 22(2): 377-384.

[118] Gupte A, Mumper R J. Elevated copper and oxidative stress in cancer cells as a target for cancer treatment. Cancer Treatment Reviews, 2009, 35(1): 32-46.

[119] Shanbhag V, Jasmer-McDonald K, Zhu S, Martin A L, Gudekar N, Khan A, Ladomersky E, Singh K, Weisman G A, Petris M J. ATP7A delivers copper to the lysyl oxidase family of enzymes and promotes tumorigenesis and metastasis. Proceedings of the National Academy of Sciences of America, 2019, 116(14): 6836-6841.

[120] Balter V, da Costa A N, Bondanese V P, Jaouen K, Lamboux A, Sangrajrang S, Vincent N, Fourel F, Télouk P, Gigou

M. Natural variations of copper and sulfur stable isotopes in blood of hepatocellular carcinoma patients. Proceedings of the National Academy of Sciences of America, 2015, 112(4): 982-985.

[121] Lu D, Luo Q, Chen R, Zhuansun Y, Jiang J, Wang W, Yang X, Zhang L, Liu X, Li F, Liu Q, Jiang G. Chemical multi-fingerprinting of exogenous ultrafine particles in human serum and pleural effusion. Nature Communications, 2020, 11(1): 2567.

第 21 章　LA-ICP-MS 的生物元素成像应用

（徐立宁[①]，王煦栋[②]，徐　明[①,②*]）

▶ 21.1　概述 / 589

▶ 21.2　重金属原位成像 / 589

▶ 21.3　金属类药物原位成像 / 592

▶ 21.4　疾病诊断 / 596

▶ 21.5　蛋白识别 / 600

▶ 21.6　细胞检测 / 605

▶ 21.7　纳米材料原位成像 / 607

▶ 21.8　展望 / 610

①中国科学院生态环境研究中心，北京；②国科大杭州高等研究院，杭州。*通讯作者联系方式：mingxu@rcees.ac.cn

第21章 LA-ICP-MS 的生物元素成像应用

本章导读

- LA-ICP-MS 成像技术可以实现生物组织或细胞内元素的原位成像分析。
- LA-ICP-MS 成像技术与其他分离分析技术相结合，可以实现元素代谢分布特征与生物分子作用机理的解析。
- 基于 LA-ICP-MS 的生物元素成像分析技术与方法仍有待于进一步发展。

21.1 概 述

1985 年，英国萨里大学的 Gray 首次将激光剥蚀（laser ablation，LA）系统与电感耦合等离子体质谱（inductively coupled plasma mass spectrometry，ICP-MS）相结合，应用于原位化学分析[1]。此后的 30 多年，激光剥蚀电感耦合等离子体质谱（LA-ICP-MS）已被广泛应用于固态样品的元素或同位素分析。LA-ICP-MS 的基本原理是利用高能激光束使固态样品熔融和气化，并通过载气将气化后的样品载带至高温等离子体内进行电离，最后通过质谱进行定性与定量分析。

早期研究中，LA-ICP-MS 常被用于地质与环境固体样品中痕量或超痕量元素分析[2,3]。近年，LA-ICP-MS 技术也逐渐被应用于生物样品的元素成像分析，其空间分辨率可低至 1 μm[4]。与其他元素成像技术相比，如高能同步辐射 X 射线荧光光谱（SR-XRF）或二次离子质谱（SIMS），LA-ICP-MS 具有更高的灵敏度、更低的使用成本、更丰富的元素采集信息、更简单的样品制备等诸多优势，可以实现原位、实时、快速、高灵敏的多元素或同位素成像分析[5]。目前，针对生物样品，已经发展出各式各样的 LA-ICP-MS 原位成像定量分析策略，可以利用溶液校准法或固体校准法实现元素或同位素的准确定量分析[6]。因此，作为科学前沿领域，生物样品的 LA-ICP-MS 元素成像分析技术已被广泛应用于环境科学、生命科学、临床医学、材料科学等诸多领域[7-12]。本章将重点介绍近年 LA-ICP-MS 技术在生物元素成像分析方面的最新研究进展。

21.2 重金属原位成像

作为一类重要污染物，重金属可以通过摄入、呼吸、皮肤接触等多种暴露途径进入生物体，发生生物富集。进而，重金属可以通过食物链传递，进入人体，并造成组织器官损伤与致病风险。因此，揭示生物体内重金属的分布特征与生物利用性就极为关键，有助于阐明重金属的暴露水平、健康危害、毒性特征与致病机理。

21.2.1 汞

作为一种典型重金属，汞（Hg）的环境污染问题一直备受关注。为了解斯洛文尼亚汞污染地区食用菌的汞暴露风险，Kavčič 等利用 LA-ICP-MS 分析了蘑菇（*Boletus* spp.和 *Scutiger pes-caprae*）中汞的分布特征与形态[13]。LA-ICP-MS 元素成像结果显示，Hg 主要分布于蘑菇的菌盖和菌褶表层，并且会与硒（Se）结合形成硒化汞（HgSe），为 *S. pes-caprae* 超富集 Hg 与 Se 的现象提供了科学解释。Becker 等则对来源于法国布列塔尼污染水域的网目织纹螺（*Nassarius reticulatus*）体内的 Hg 分布进行了研究[14]。利用 LA-ICP-MS 对 N. reticulatus 的纵向切片进行元素成像，可以实现对 Hg 的 μg/g 水平的检测，并发现 Hg 主要分布于 N. reticulatus 的消化腺内。Karst 等比较了黑腹果蝇（Drosophila melanogaster）对 3 种汞化合物（$HgCl_2$、CH_3HgCl、$C_9H_9HgNaO_2S$）的摄入与体内分布特征[15]。LA-ICP-MS 成像结果显示，经饮食暴露后，Hg^{2+}难以穿越血脑屏障进入脑部，而 2 种有机汞均体现出穿越血脑屏障的能力，会累积于脑部。并且，CH_3HgCl 的脑部累积比 $C_9H_9HgNaO_2S$ 高 3 倍。由于人脑视觉[外侧膝状核（LGN）]和听力[内侧膝状核（MGN）]的中继中心易发生 Hg 累积，可能会干扰神经信号传递，导致神经系统疾病风险。因此，Pamphlett 等对 50 名 20～104 岁神经系统疾病与精神疾病患者脑部的 LGN 和 MGN 切片样品进行了研究[16]。LA-ICP-MS 成像结果显示，一名患者脑部的 LGN 区发生了 Hg 累积（图 21-1），可能是其患有帕金森病（parkinson's disease，PD）的原因之一。此外，在另一项研究中，他们还分析了 50 例 34～69 岁乳腺癌女性患者的乳腺组织样品，利用 LA-ICP-MS 成像证实了患者乳腺组织中 Hg 的存在[17]。

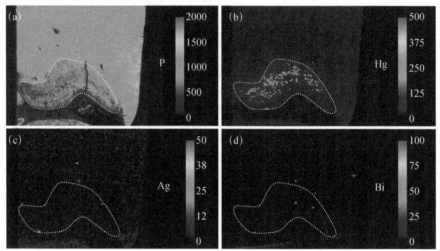

图 21-1　精神系统疾病患者脑部 LNG 区的 LA-ICP-MS 成像[16]
(a)、(b)、(c) 和 (d) 分别为 P、Hg、Ag 和 Bi 的分布

21.2.2　镉

镉（Cd）易于富集在植物和动物体内，具有很强的生物毒性。例如，黄连（Coptis chinensis Franch）是一种传统中医药用植物，经常会受到 Cd 污染，带来潜在的健康风险。Liang 等利用 LA-ICP-MS 成像技术分析了黄连的根茎横截面，发现 Cd 主要分布于黄连的周皮、皮层、木髓和根迹维管束内，与细胞壁结合，说明 Ca^{2+} 通道可能是 Cd 进入植物体的重要途径[18]。Arruda 等发现 Cd 暴露会导致向日葵种子内 Cd 的富集[19]。LA-ICP-MS 成像结果显示，Cd 主要分布于向日葵种子的子叶部分，并且会导致 Cu、Fe、Mn 等必需微量元素的稳态失衡。Basnet 等研究了中国湖南省"锡矿山"地区生产水稻（Oryza sativa L.）子实内 Cd 的分布特征[20]。LA-ICP-MS 成像结果显示，Cd 主要累积于胚乳，并可能与 Zn 的结合位点发生竞争，降低 Zn 含量。Berhard 等成功地利用 LA-ICP-MS 成像技术评价了人体直肠、主动脉、卵巢、胰腺、睾丸、肝脏、肾髓质-皮质缘，以及肾皮质内部 Cd 的空间分布特征[21]。结果显示，在直肠、肾脏和睾丸等参与人体 Cd 的代谢器官和组织中，其呈现不均匀的分布。

21.2.3　其他重金属

为研究微量元素（Cr、Mn、Ni、Cu、Zn、As、Cd、Hg、Pb）在小麦谷粒中的相互作用机制，

Vanhaecke 等基于 LA-ICP-MS 三维重构元素成像技术，分析了成熟小麦（*Triticum aestivum* L.）和黑麦（*Secale cereale* L.）子实内特征元素的空间分布[22-24]。如图 21-2 所示，通过将二维横截面的元素成像结果进行空间重建，就可以获得小麦谷粒中多种微量元素的三维空间分布。结果证实，由于转运与储存机理的不同，Mn 和 Zn 主要富集在谷粒的糊粉层、种皮、折痕的维管组织和胚芽组织中，而 Cr、As、Cd 和 Pb 主要富集于谷粒胚乳内。此外，LA-ICP-MS 技术可实现同位素及同位素比值的原位成像分析。例如，Becker 等利用 LA-ICP-MS 研究了香薷（*Elsholtzia splendens*）叶片内 Cu 的富集过程与分布情况，以及 Cu 暴露对叶片必需元素（K、Mg、Mn、P、S 和 B）的影响[25]。为达到示踪目的，他们采用 ^{65}Cu 进行暴露，并分析检测了 ^{65}Cu / ^{63}Cu 的比值。结果显示，香薷叶主要通过叶柄和主脉进行 Cu 的摄入与累积。在新生叶片中，Cu 暴露会导致 K、Mg、Mn、P 和 S 的累积，但不影响 B 含量。在老叶片中，Cu 暴露会导致 K 和 P 含量降低，而其他必需元素不受影响。

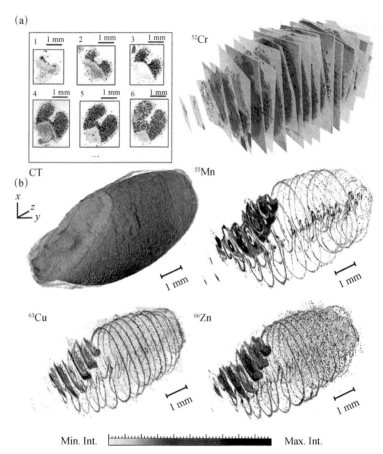

图 21-2 （a）将黑麦的二维 Cr 图像进行三维空间重建的过程；
（b）成熟小麦的显微 CT 图像，以及 ^{55}Mn、^{63}Cu、^{66}Zn 的三维空间分布[22]

虽然 LA-ICP-MS 技术可以提供强大的原位元素分析能力，却无法识别分子的化学结构信息。为弥补该短板，Karst 等将 LA-ICP-MS 与基质辅助激光解析质谱（MALDI-MS）技术相结合，对黑腹果蝇（*Drosophila melanogaster*）体内的砷脂 AsHC 进行了研究（图 21-3）[26]。基于上述策略，既可实现对果蝇体内 As 分布的定量解析，还可实现对 AsHC 分子的定位识别。基于上述方法，他们发现，果蝇幼虫经 AsHC332 暴露后，其体内 As 的分布相对均匀，而成虫体内 As 主要累积于腹部脂肪体附近。AsHC332 还可以穿过血脑屏障，进入果蝇的脑部。此外，通过比较亚砷酸盐或 AsHC332 暴露后的果蝇成虫，发现一旦暴露停止后，亚砷酸盐可以被果蝇从体内快速清除，而 AsHC332 的体内滞留时间较长。除小分子化合物，还可以将 LA-ICP-MS 与免疫组织化学技术相结合，研究重金属与蛋白分子的作用机理。例如，为揭示 Pb 与钙离子通道相关 STIM1 蛋白间的潜在联系，Ishizuka 等利用 LA-ICP-MS 成像技术研究了 Pb 在小鼠肝、肾和脑中的分布[27]。研究结果显示，Pb 在肝脏中均匀分布，但在肾脏和脑部的分布并不均匀。在肾脏中，Pb 更倾向于富集在肾髓质，而非肾皮质。在脑部，Pb 主要富集在海马体、丘脑和下丘脑区域。然而，LA-ICP-MS 成像与免疫组织化学染色的结果显示，Pb 并没有诱导 STIM1 蛋白的特异性表达。

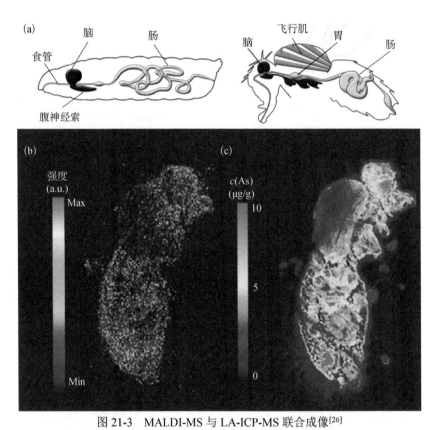

图 21-3　MALDI-MS 与 LA-ICP-MS 联合成像[26]

（a）黑腹果蝇成虫及幼虫体内的器官分布示意图；AsHC332 暴露 3 天后，（b）MALDI-MS 对黑腹果蝇体内 AsHC 的成像，（c）对相同组织切片 As 分布的 LA-ICP-MS 成像

此外，还可将 LA-ICP-MS 与同步辐射 X 射线荧光光谱（SR-XRFS）技术相结合，同时原位定量分析元素，并解析元素的化学价态和形态。例如，Liang 等利用 LA-ICP-MS 进行原位成像，发现 Cr 主要富集于黄连根部与叶柄的维管柱区域，以及根状茎的皮层与外皮层；同时，利用 X 射线近边吸收谱法（XANES）进行形态分析，发现黄连吸收的 Cr(Ⅵ) 被还原为 Cr(Ⅲ)，并且组氨酸与磷酸参与了 Cr 的体内累积过程[28]。尽管 SR-XRFS 也可以实现对生物样本中金属元素的无损原位成像，并获取元素的形态信息，但灵敏度要低于 LA-ICP-MS。例如，Bohle 等比较了 LA-ICP-MS 与 SR-μXRF 对骨骼中 Zn 和 W 的成像结果，发现前者的灵敏度与分辨率显著优于后者，可更好地区分 Zn 和 W 在骨骼中的特异性分布[29]。

21.3　金属类药物原位成像

金属类药物具有独特的药物活性，在肿瘤治疗、组织成像、抗菌抗病毒等方面均有广泛应用，如铂（Pt）类抗肿瘤药物、钆（Gd）类造影剂、银（Ag）抗菌剂等。尽管如此，金属类药物的治病机理尚未被完全阐述，阻碍了新型药物的研制。此外，金属类药物毒副作用的机理也亟待深入揭示，有助于保证金属类药物的使用安全性。因此，已有不少研究利用 LA-ICP-MS 元素成像技术进行金属类药物的机理研究。

21.3.1 铂类抗肿瘤药物

铂类药物主要包括顺铂（cisplatin）、卡铂（carboplatin）、奈达铂（nedaplatin）、奥沙利铂（oxaliplatin）等（图 21-4），具有抑制肿瘤细胞增殖的特性，被广泛用作抗肿瘤药物。例如，顺铂可与 DNA 结合，诱导肿瘤细胞死亡，用于治疗实体肿瘤。然而，由于铂（Pt）的高亲和性，顺铂还可与蛋白分子作用，破坏细胞正常的生理功能，产生毒副作用。在治疗过程中，顺铂主要通过肾脏排泄，因此大概三分之一用药患者会产生肾中毒现象。为探究顺铂的肾脏毒性机理，Gómez-Gómez 等利用 LA-ICP-MS 成像技术研究了顺铂在大鼠肾脏中的累积及肾脏损害，LA-ICP-MS 对肾脏切片中 Pt 的检测限可达 50 fg，空间分辨率达 8 μm[30]。研究结果显示，低剂量顺铂（5 mg/kg）暴露后，Pt 主要累积于肾脏的肾皮质和皮质髓质交界处（图 21-5），对应于近端小管 S3 段。经高剂量（16 mg/kg）暴露后，Pt 在肾脏内的分布会延伸至肾柱，且累积水平升高约 1 个数量级。此外，Pt 的累积还显著地改变了该部位 Cu 和 Zn 的空间分布，导致两者含量降低。Vanhaecke 等将 LA-ICP-MS 的横向扫描分辨率进一步提高至 1 mm，在亚细胞水平进行了元素成像分析[31]。结果显示，Pt 主要累积于肾脏近端小管的上皮细胞，造成上皮细胞的脱落坏死。Moraleja 等比较了顺铂、卡铂和奥沙利铂的肾脏分布与毒性差异[32,33]。LA-ICP-MS 成像结果显示，顺铂和卡铂主要分布于肾脏的皮质区域，而奥沙利铂则均匀分布于整个肾脏。上述发现证实，不同铂类药物的肾脏分布与其代谢和毒性紧密相关。此外，Ciarimboli 等发现，顺铂还可穿过血睾屏障，进入睾丸，导致小鼠睾丸质量下降[34]。LA-ICP-MS 成像结果显示，Pt 主要分布于睾丸白膜与间质区域，少量分布于曲细精管内。而且，睾丸内 Pt 含量较高的区域出现精子减少与形态异常的现象。除肾脏毒性，长期使用顺铂还会对耳蜗造成损伤，并导致 40%～80% 患者的永久性听力丧失。为揭示耳蜗中顺铂的分布特征，Cunningham 等采用 LA-ICP-MS 成像技术研究了人耳蜗的顺铂分布，并观察到 Pt 主要分布于耳蜗的血管纹、蜗神经纤维、耳蜗神经与耳蜗骨间边界及内骨膜（图 21-6），与耳蜗功能损伤直接相关[35]。

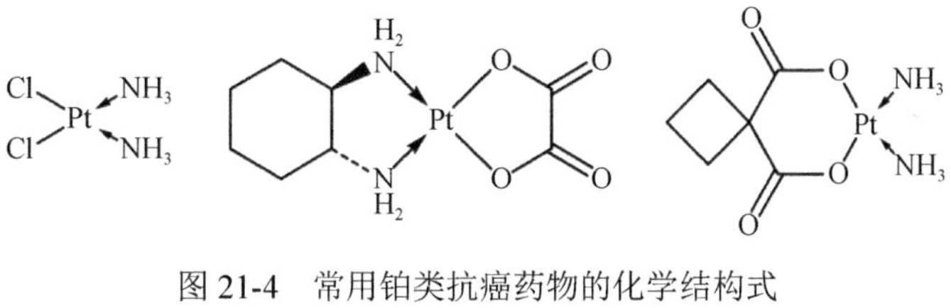

图 21-4　常用铂类抗癌药物的化学结构式

从左到右依次为顺铂、奥沙利铂和卡铂

在实体肿瘤治疗过程中，肿瘤的异质化会对抗肿瘤药物的瘤内分布与药物敏感性产生很大的影响，如降低药物的渗透性与药效。Egger 等利用结肠癌小鼠肿瘤模型评价了奥沙利铂和赛特铂的肿瘤组织分布特征[36]。LA-ICP-MS 成像结果显示，Pt 在肿瘤与癌旁组织的分布确实存在异质性。大多数情况下，恶性肿瘤区域的 Pt 含量要低于周围软组织，如结缔组织与微血管。Karst 等则利用人胆管癌多细胞肿瘤球体评价了顺铂与乙酰丙酮铂的肿瘤渗透性[37]。他们发现，乙酰丙酮铂主要分布于肿瘤球体的增殖区，而顺铂的亲水性较强，能够比较有效地渗透整个肿瘤球体。采取类似的研究策略，

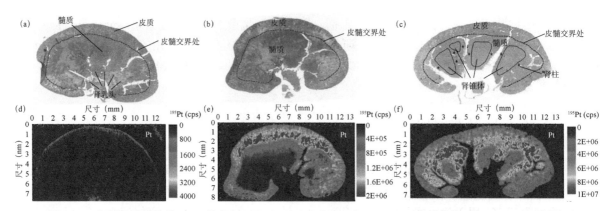

图 21-5 大鼠肾脏切片的 H&E 染色[（a）、（b）、（c）]和 LA-ICP-MS 元素成像[（d）、（e）、（f）][30]

（a，d）对照组、（b，e）5 mg/kg 和（c，f）16 mg/kg 处理组

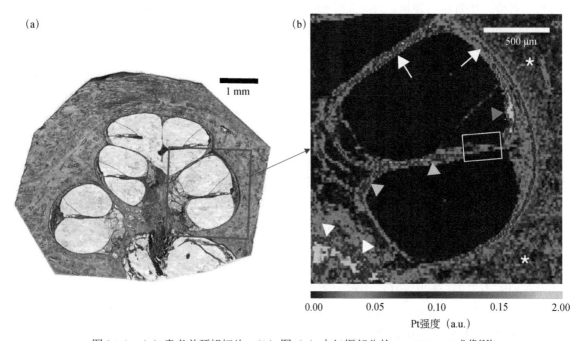

图 21-6 （a）患者前耳蜗切片；（b）图（a）中红框部分的 LA-ICP-MS 成像[35]

绿色、黄色和白色三角分别指示血管纹、蜗神经纤维和耳蜗神经与耳蜗骨间边界；
白色箭头指示耳蜗内骨膜、星号指示耳蜗骨、白色框指示 Corti 器

Koellensperger 等发现，奥沙利铂处理后的 HCT116 人结肠癌多细胞肿瘤球体内，约 20%～25% Pt 分布于肿瘤球体的增殖区，约 30% Pt 分布于静区，而其余 Pt 分布于坏死区[38]。此外，将 LA-ICP-MS 与 MALDI-MS 成像技术相结合，还可以实现对铂类药物与其他小分子化合物的空间分布研究，揭示抗肿瘤药物的潜在作用机制[39]。

21.3.2 钆类造影剂

在磁共振成像（MRI）检查中，钆（Gd）类造影剂（GBCA）主要用于增强器官、血管和组织的影像质量。依据化学结构特征，GBCA 可分为线性分子和大环分子两类。其中，线性 GBCA 更易释放钆离子，而大环 GBCA 相对稳定。目前，国内批准上市的 GBCA 医用药品有钆喷酸葡胺、钆双胺、钆贝葡胺、钆塞酸二钠、钆特酸葡胺、钆特醇、钆布醇等（图 21-7）。例如，钆布醇（Gadovist）已

被应用于临床诊断，为评价钆布醇的体内分布特征，Beck 等同时利用 LA-ICP-MS 元素成像技术与 MALDI-MS 分子成像技术对小鼠心脏、肝脏、肾脏和脑的钆布醇进行研究[40]。研究结果显示，钆布醇在心脏的含量较高，且主要分布于心室。此外，在肾脏与肝脏均可检测到钆布醇，但其脑部含量相对较低。当肾脏功能不全患者使用 GBCA 后，会发生肾性系统性纤维化（NSF），体现为皮肤纤维化，并伴有肺脏、食管、心脏和骨骼肌的纤维化。为揭示 GBCA 导致 NSF 的机理，Karst 等利用 LA-ICP-MS 成像技术研究了一名疑似 NSF 患者的皮肤样品[41]。他们发现，在该名患者的皮肤组织中，Gd 分布不均匀，且 P 的分布特征体现出一致性（图 21-8），表明皮肤组织内存在 $GdPO_4$ 沉积物。此外，进一步研究发现，即便经过 GBCA 治疗 8 年后，患者皮肤组织中依然可以检测到完整的 Gd-HP-DO3A 分子，表明大环 GBCA 在体内具有很强的稳定性。然而，经 2 种商用 GBCA（Omniscan 和 Magnevist）给药 1 个月后，Pascolo 等只在小鼠的门齿中检测到 Gd，并未在肾脏、肝脏、肱骨、背部皮肤检测到 Gd[42]。LA-ICP-MS 元素成像进一步确认了 Gd 在小鼠牙周膜中的沉积，并且 Gd 的沉积与 Fe 的组织分布紧密相关。

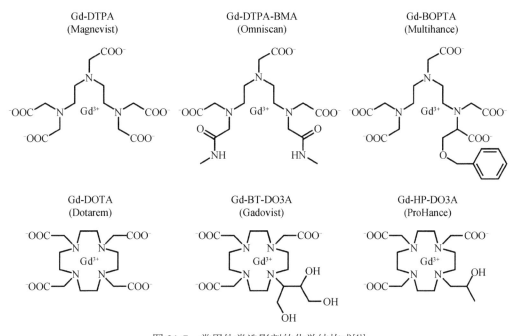

图 21-7　常用钆类造影剂的化学结构式[41]

依次为钆喷酸葡胺、钆双胺、钆贝葡胺、钆特酸葡胺、钆布醇和钆特醇

由于 GBCA 可能会通过血脑屏障进入脑部，因此需要研究并评估 GBCA 的脑部分布特征与机理。Karst 等研究了一名患者脑部不同区域的 GBCA 分布[43]。LA-ICP-MS 成像结果显示，在患者小脑齿状核区域及血管处，可以观察到较强的 Gd 信号，表明 GBCA 可以通过血液传输进入脑部。相较于小脑齿状核区域，Gd 在脑部基底核与额叶区域内的分布较为均一，但脑干区域的 Gd 含量较低。此外，他们还发现，Gd 在脑部的沉积与 P 紧密相关[44]。Richter 等则利用 LA-ICP-MS 研究了犬类脑部钆双胺的分布特征[45]。研究结果显示，在静脉注射 35 个月后，Gd 在小脑深部核团（DCN）内的累积要明显地高于 DCN 周边组织，并且 DCN 与血管处的 Gd 与 Zn 分布存在一致性。

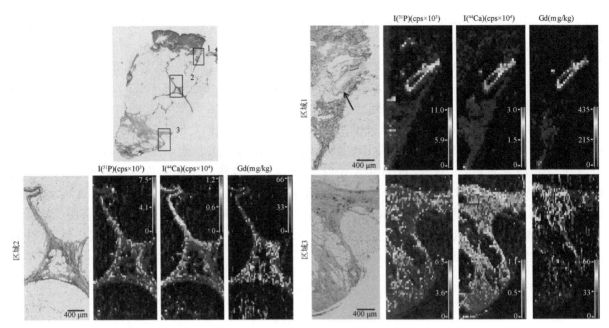

图 21-8　NSF 患者皮肤的显微和 LA-ICP-MS 成像（P、Ca 和 Gd）[41]

区域 1、2、3 为左上图中的黑色方框内区域

21.3.3　其他金属药物

除上述 2 种金属类药物，Ru 类化合物也被视为重要的潜在抗肿瘤药物。Hartinger 等基于 LA-ICP-MS 成像技术研究了小鼠器官与组织内 Ru 配合物（SKP1339）的分布规律，发现肝脏、脾脏和肌肉内 Ru 分布较为均匀，而肾脏皮质内 Ru 含量较髓质更高[46]。Keppler 等则利用 LA-ICP-MS 成像技术比较了具有类似分子结构的 Ru 与 Os 类抗肿瘤药物在肝脏、肾脏、肌肉和肿瘤的分布差异[47]。研究结果显示，2 种药物均主要累积于肾脏皮质内，但前者比后者体现出更好的组织渗透性。此外，LA-ICP-MS 成像技术还被应用于含 Co、In、B 的小分子或生物大分子药物的体内代谢、分布与机理研究[48-50]。

21.4　疾病诊断

由于必需微量元素（essential trace element）几乎参与人体所有的生理过程，因此一旦某些必需微量元素发生稳态失衡，就可能会诱发疾病。作为一种强大的元素分析手段，LA-ICP-MS 元素成像技术有助于实现快速、精准的疾病临床诊断、治疗与机理研究。

21.4.1　神经系统疾病

脑具有非常复杂的组织结构特征。LA-ICP-MS 元素成像可以实现脑组织内痕量元素的原位解析，提供重要的空间分布信息，用于研究脑部损伤与疾病[51,52]，如创伤性脑损伤[53]、创伤性应激障碍[54]、癫痫[55]等。目前，利用 LA-ICP-MS 研究较多的神经系统疾病包括阿尔兹海默症（Alzheimer disease，

AD）与帕金森病（Parkinson disease，PD）。AD 是一种神经系统退行性疾病，患者的脑部会产生 β 淀粉样肽（Amyloid-β，Aβ）的聚集，并发生记忆力衰退现象。2005 年，Mc Leod 等就利用 Eu 和 Ni 标记的单克隆抗体，并应用 LA-ICP-MS 元素成像技术成功地测定了 AD 转基因小鼠脑部组织切片中 Aβ 的沉积特征[56]。他们发现，在 AD 患者脑部，Fe 会加速累积。为评价 Fe 的脑部分布特征，Doble 等利用 LA-ICP-MS 元素成像技术定量分析了 AD 患者脑部额叶皮层白质和灰质的 Fe 分布[57]。研究结果表明，相较于正常人脑，患者脑部额叶皮层的白质和灰质区域均发生明显的 Fe 累积，说明患者脑部的 Fe 代谢发生了异常。为实现精准定量地分析 AD 患者脑组织切片的 Fe、Cu 和 Zn，Feng 等将同位素稀释应用到 LA-ICP-MS 元素成像分析[58]。与 XRF 成像技术相比，该方法不仅能提供元素空间分布信息，还可实现元素绝对定量，更准确地揭示脑部微量元素含量的变化差异（图 21-9）。基于 LA-ICP-MS 元素成像以及免疫组化分析的结果，他们发现，在 AD 大鼠脑组织切片中，Fe、Cu 和 Zn 的分布与 Aβ 沉积存在相关性。近期，Donnelly 等开发了一种新型 Cu 络合物，可以穿过血脑屏障，实现对 AD 患者脑组织切片中 Aβ 样斑块的标记，进而利用 LA-ICP-MS 进行成像定位[59]。

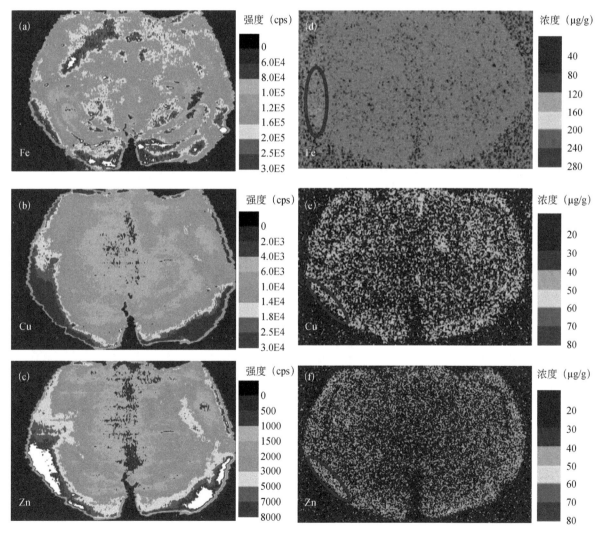

图 21-9 利用 LA-ICP-MS[（a）、（b）、（c）]和 XRF[（d）、（e）、（f）]分别对 AD 小鼠脑组织切片进行元素成像[58]

圆圈为 Fe、Cu 和 Zn 的累积区域

PD 是一种多发于中老年人的神经系统变性疾病，主要的病理特征是脑黑质（SN）多巴胺（dopamine，DA）神经元的变性死亡，进而导致纹状体 DA 含量显著降低。为揭示 PD 的致病机理，Doble 等研究了 6-羟多巴胺（6-OHDA）诱导 PD 小鼠脑部 SN 区域 Fe、Cu、Zn 和 Mn 的变化[60]。LA-ICP-MS 成像结果显示，与对照组相比，PD 小鼠脑部 SN 区域 Fe、Cu、Zn 和 Mn 均不同程度升高，尤其是 Fe 发生了累积。此外，他们还进一步考察了 SN 区域 Fe 与 DA 的分布相关性，发现在单侧 6-OHDA 损伤模型小鼠的 SN 区域，受损侧酪氨酸羟化酶（TH）阳性区域的 Fe 水平显著升高（图 21-10），但 TH 活力降低[61]。上述发现证明，在 PD 发生发展过程中，Fe 和 DA 具有相关性，增加了 SN 区域神经元的氧化损伤。类似地，Becker 等同时采用 LA-ICP-MS 与 MALDI-MS 成像技术研究了 Fe、Cu、Zn、Mn 与脂质分子的分布特征[62]。他们观察到，SN 区域的 Fe、Cu、Mn 含量上升，但单不饱和脂肪酸含量降低。在另一项研究中，该研究组考察了 1-甲基-4-苯基-1,2,3,6-四氢吡啶（MPTP）诱导 PD 模型小鼠脑部 Fe、Cu、Zn 和 Mn 的分布特征。LA-ICP-MS 成像结果显示，在 MPTP 处理 2 小时与 7 天后，脑室周围与齿筋膜区域的 Cu 显著降低。28 天后，该区域的 Cu 恢复至正常水平，或发生过度补偿。在脚间核区域，Fe 水平显著升高，但 SN 区域的 Fe 没有升高[63]。上述发现证明，Fe 和 Cu 在 PD 发生发展过程中起关键作用。近期，Curtis 等研究了 PD 患者脑部嗅球区域的金属元素分布特征[64]。他们发现，与正常人脑部相比，PD 患者脑部嗅球区域 Fe 和 Na 的升高可能与嗅觉功能丧失有关。

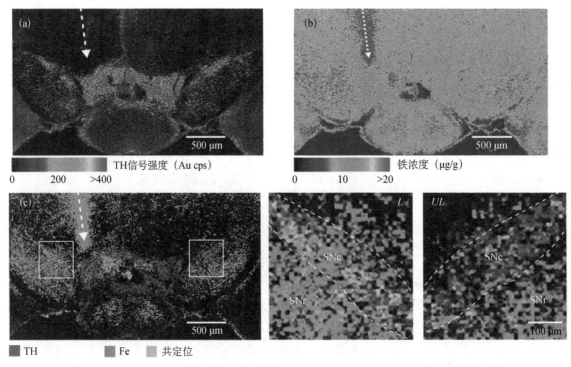

图 21-10　6-OHDA 小鼠脑部损伤模型[61]（a）Fe、（b）TH 和（c）共定位的原位成像

21.4.2　癌症

在癌症发生发展过程中，肿瘤组织的微量元素稳态与代谢会发生异常。Fernández-Sánchez 等采用 LA-ICP-MS 成像技术考察了乳腺癌组织 Ca、Fe、Cu 和 Zn 的变化[65]。LA-ICP-MS 对 Fe、Cu 和 Zn 的检测限分别达到 10～150 ng/g，对 Cd 的检测限达到 2 μg/g。他们发现，相较于非肿瘤组织，乳腺癌肿瘤组织的 Ca、Fe、Cu 和 Zn 含量明显更高。近期，他们又采用 LA-ICP-MS 与 MALDI-MS 成像

技术研究了乳腺癌组织中基质金属蛋白酶（MMP-11）的分布特征[66]。结果表明，相较于正常组织，乳腺癌组织 MMP-11 的表达水平更高。由于 MMP-11 是一种含 Zn 蛋白，因此乳腺癌组织中 Zn 与 MMP-11 的分布呈现出一致性，可用于指示 MMP-11 的分布特征（图 21-11）。据报道，相较于正常组织，前列腺肿瘤中 Zn 及其代谢相关的柠檬酸、天冬氨酸的水平均会降低。为揭示上述现象的潜在机理，Andersen 等采用 LA-ICP-MS 与 MALDI-MS 成像技术检测了前列腺肿瘤组织中的 Zn、$ZnCl_3^-$、柠檬酸、天冬氨酸以及 N-乙酰天冬氨酸的空间分布特征[67]。研究结果表明，与正常上皮组织相比，肿瘤上皮组织的 Zn、柠檬酸和天冬氨酸均显著降低，且存在一定相关性。在正常与肿瘤基质区域，则没有体现出明显的差异。此外，将 LA-ICP-MS 与傅里叶变换红外光谱（FTIR）成像技术结合，可实现同时研究乳腺癌组织中痕量元素和有机分子分布[68]。例如，Miklos 等利用 LA-ICP-MS 成像技术研究了放疗过程中 Mn 对不同类型肿瘤放射敏感性的影响[69]。他们发现，不同肿瘤组织的 Mn 含量介于 0.02～1.15 μg/g 间，含量差别可达 60 倍。其中，放射敏感性最高的肿瘤（睾丸）的 Mn 含量最低，而放射耐受性最强的肿瘤（胶质母细胞瘤、黑色素瘤）的 Mn 含量最高。并且，在 7 种肿瘤类型中，肿瘤总 Mn 含量与放射敏感性存在直接相关性。

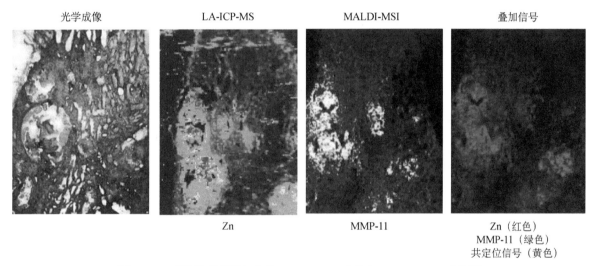

图 21-11　乳腺肿瘤组织切片的 LA-ICP-MS 成像和 MALDI-MSI 成像[66]

21.4.3　肝脏疾病

LA-ICP-MS 元素成像技术在肝脏疾病研究中也被广泛应用[5]。例如，威尔逊病（Wilson disease，WD）是一种以青少年为主的常染色体隐性遗传性疾病。铜转运 ATP 酶基因（Atp7b）的突变引起肝脏细胞的 Cu 代谢紊乱，无法将 Cu 正常地分泌到胆汁中，从而导致各组织器官的 Cu 异常积累，如肝脏、肾脏、脑、角膜等。因此，WD 患者常患有肝脏疾病和比较严重的神经系统缺陷。相较于传统的组织 Cu 染色方法，如罗丹宁染色，LA-ICP-MS 可以在不影响组织样本 Cu 分布的前提下，实现更高分辨率与更低检测限的 Cu 原位成像[70]。Weiskirchen 等采用 LA-ICP-MS 元素成像技术对 WD 小鼠和患者的肝脏组织切片进行了研究[71]。在 Atp7b 基因缺失 WD 小鼠的衰老过程中，肝脏 Cu 的累积具有年龄依赖性，并伴有 Fe 和 Zn 的升高。D-青霉胺（DPA）是一种治疗 WD 的 Cu 螯合剂药物。为揭示 DPA 治疗 WD 副作用的可能机理，Karst 等采用 10 μm 激光 LA-ICP-MS 考察了 DPA 用药对 Atp7b 基因缺失 WD 大鼠与 WD 患者肝脏 Cu 的影响[72]。他们发现，与正常组相比，DPA 可有效降低 WD 个体肝脏内的 Cu 累积。然而，经 DPA 用药后，Cu 在肝脏的空间分布并不均匀，血管周边区域 Cu 分布较低，但其他区域 Cu 含量仍较高，这可能是导致 WD 患者停用 DPA 治疗后产生严重副作用的重要原因之一。近期，他们进一步采用更高分辨率的 4 μm 激光 LA-ICP-MS 对细菌肽铜载体（MB）治疗 WD 大鼠的效果进行了评价。结果显示，经 MB 治疗后，WD 大鼠肝脏组织残余的 Cu 聚集点更小，Cu 浓度更低（图 21-12）。在停止 MB 治疗 7 周后，虽然仍可观察到 Cu 的累积，但没有超过对照组的 Cu 含量，表明 MB 对 WD 的治疗效果具可持续性[73]。此外，LA-ICP-MS 元素成像技术还被应用于 Menkes 综合征（Menkes disease）[74]、血色素沉着病（hemochromatosis）[75]、胆汁淤积性肝病（cholestatic liver disease）[76] 的致病机理研究。

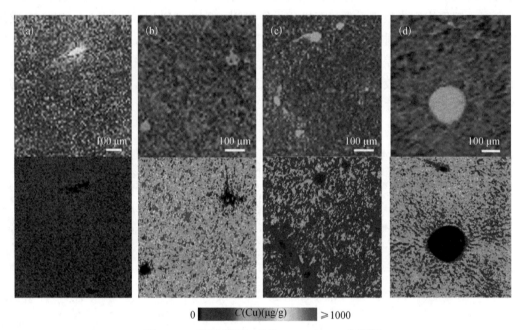

图 21-12　肝脏组织切片的 LA-ICP-MS 成像[73]
（a）正常组、（b）WD 组、（c）62.5 mg/kg 和（d）150 mg/kg MB 处理组

21.5　蛋 白 识 别

蛋白是最为重要的一类生物功能大分子，参与生命活动的所有过程。LA-ICP-MS 元素成像技术既可以应用于金属蛋白分析，也可用于非金属蛋白分析。通过与蛋白分离分析技术（如凝胶电泳、免疫标记、分子质谱）结合，可以进一步扩展 LA-ICP-MS 的应用范围，针对蛋白分子进行研究[77-79]。

21.5.1　凝胶电泳成像法

作为最常用的蛋白分离方法，聚丙烯酰胺凝胶电泳（polyacrylamide gel electrophoresis，PAGE）可以实现生物样本中复杂蛋白组分的定性与定量分析。通常，金属蛋白分子中的金属离子结合域并不稳定，容易被表面活性剂破坏而丢失金属离子，如十二烷基磺酸钠（SDS）。因此，常规 SDS-PAGE 无法保证金属蛋白稳定性，需要使用非变性聚丙烯酰胺凝胶电泳（non-denaturing PAGE）分离金属蛋白，再利用 LA-ICP-MS 进行研究[80]。例如，基于 LA-ICP-MS 元素成像分析，Polatajko 等比较了

SDS-PAGE 和 native-PAGE 对菠菜（*Spinacia oleracea* L.）叶片中 Cd 结合蛋白的分离效果[81]。结果证明，SDS-PAGE 会造成蛋白结合 Cd 的损失，而 native-PAGE 可实现 Cd 结合蛋白的有效分离，更适合研究 Cd 结合蛋白。我们将非变性 PAGE、LA-ICP-MS 与高效液相色谱-电喷雾串联质谱（HPLC-ESI-MS/MS）结合，筛查了海洋 Cu 污染暴露后，牡蛎（*Crassostrea gigas*）鳃与消化腺的 Cu 结合蛋白[82]。结果表明，牡蛎消化腺内的超氧化物歧化酶（Cu/Zn-SOD）和 L-抗坏血酸氧化酶均具有很强的 Cu 结合能力，揭示了牡蛎耐受 Cu 污染的分子机理。为实现更精准的金属蛋白研究，Becker 等采用非变性二维凝胶电泳（non-denaturing 2D electrophoresis，ND-2DE）分离大鼠肾脏中的金属结合蛋白，并利用 LA-ICP-MS 进行元素成像分析和 MALDI-MS 鉴定蛋白种类，最终成功识别了 4 种可能的金属结合蛋白[83]。采用类似的研究策略，我们利用 ND-2DE、LA-ICP-MS 与 HPLC-ESI-MS/MS 筛查了 U 暴露后，克氏原螯虾（*Procambarus clarkii*）肝胰腺内的 U 结合蛋白，并成功鉴定了 6 种重要的 U 结合蛋白（图 21-13）[84]。其中，铁蛋白（ferritin）的 Fe 与 U 信号体现出高度一致性，表明其是 U 的重要靶蛋白。

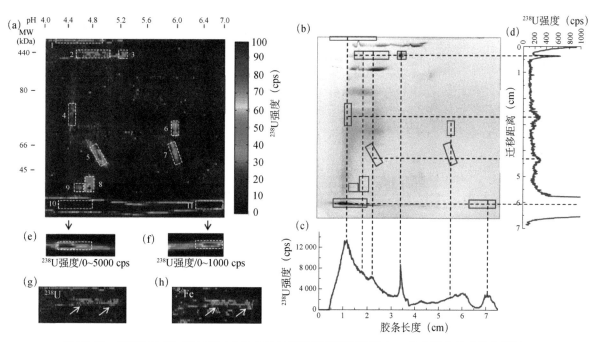

图 21-13　利用 ND-2D-PAGE LA-ICP-MS 对克氏原螯虾肝胰腺内 U 结合蛋白的成像分析[84]

除金属结合蛋白，一些蛋白质分子中还含有非金属微量元素（如 Se、I 等），也可以采用 LA-ICP-MS 进行检测[85-87]。例如，在硒蛋白或含硒蛋白分子中，Se 主要以共价键结合于硒代半胱氨酸（SeCys）或硒代蛋氨酸（SeMet）残基，稳定性好，可应用常规 SDS-PAGE 和 SDS-2DE 分离，再采用 LA-ICP-MS 成像分析。基于上述思路，Bierla 等利用 SDS-2DE 和 LA-ICP-MS 元素成像技术成功地识别了富硒酵母中的 19 种含硒蛋白，并对这些蛋白的蛋氨酸（Met）和半胱氨酸（Cys）的取代程度进行了研究[88]。他们发现，在含硒蛋白中，SeMet/Met 的平均取代率约为 42.9%，而 SeCys/Cys 的平均取代率约为 14.6%。考虑到富硒酵母蛋白的 Cys/Met 丰度比值（2∶1），表明富硒酵母蛋白所含的 10%～15% Se 以 SeCys 的形式存在。Szpunar 等则通过 SDS-2DE 和 LA-ICP-MS 元素成像技术筛查了海藻（*Nori*）中的含碘蛋白，并成功检测到 5 种碘结合蛋白[89]。

除利用内源性元素进行 LA-ICP-MS 成像分析，还可使用外源性元素标签对蛋白分子进行标记识别。例如，Linscheid 等基于点击化学（click chemistry）反应，利用镧系元素（Lu 和 Tm）亲和的 MeCAT 标记热应激处理前后大肠杆菌（*Escherichia coli*）的提取蛋白[90]。通过 SDS-2DE 和 LA-ICP-MS 元素成像比较提取蛋白的 Lu 和 Tm 信号，发现了 29 种异常表达蛋白，检测限优于传统银染法。

21.5.2 免疫印迹成像法

在凝胶电泳的基础上，免疫印迹可利用抗原-抗体的特异性结合，实现了生物样品中目标蛋白的检测。为实现 LA-ICP-MS 元素成像分析，常利用特定的元素标签对抗体（IgG）进行标记，再通过免疫反应标记目标蛋白，最终对蛋白分子进行检测[91]。例如，使用具有 Tb 络合能力的 MeCAT、MAXPAR 或 SCN-DOTA 标记细胞色素 P450（CYPs）抗体后，再通过免疫印迹法（Western blot，WB）分离、标记大鼠肝脏微粒体中的 CYPs，最终 LA-ICP-MS 的检测限可达 fmol 级[92]。Jakubowski 等利用 Eu-DOTA 和 I 标记 CYPs 的两个同工酶 CYP1A1 和 CYP2E1 的抗体，再通过 WB 与 LA-ICP-MS 成功检测了生物样品中 CYP1A1 和 CYP2E1 的变化[93]。进而，他们使用含 5 种镧系元素（Ho、Tm、Lu、Eu 和 Tb）的 DOTA 分别标记 CYP1A1、CYP2B1、CYP2C11、CYP2E1 和 CYP3A1 的抗体，实现同时定量分析大鼠肝脏微粒体中 CYPs 同工酶的变化（图 21-14）[94]。基于免疫印迹原理，他们还采用微阵列法与 LA-ICP-MS，实现了 8 种镧系元素（Er、Nd、Pr、Ho、Tm、Lu、Eu 和 Tb）标记 CYPs（CYP2E1、CYP4A1、CYP3A、CYP2C6、CYP1A1、CYP1A2、Cyt b5 和 CYP2B1/2B2）的同时高通量分析[95]。此外，为达到更好的特异性标记及准确的定量分析，新型抗体元素标签和标记策略还在持续发展[96]。

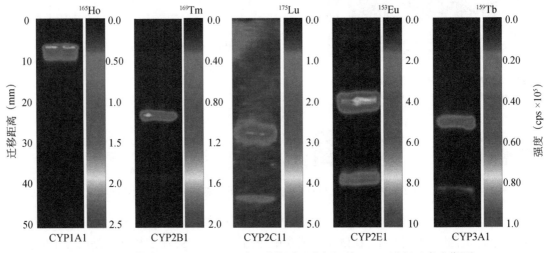

图 21-14　利用 WB LA-ICP-MS 对 8 种镧系元素标记的 CYPs 进行元素成像[94]

21.5.3 组织原位成像法

金属硫蛋白（metallothionein，MT）是一种富含半胱氨酸（30%）的金属蛋白分子，具有重要的生物功能。在人眼球中，视网膜含有最高浓度的 Zn，并主要位于视网膜色素上皮区域。由于 MT 是细胞质中最主要的含 Zn 蛋白，因此可通过 LA-ICP-MS 对 Zn 进行成像分析，研究视网膜的 MT 分布特征[97]。此外，基于免疫标记原理，可使用元素标记的抗体对生物样本中的蛋白进行特异性标记，再通过 LA-ICP-MS 进行原位成像，最终实现对蛋白空间分布特征的研究。例如，Giesen 等分别利用镧系元素（Ho、Tm、Tb）络合的 SCN-DOTA 标记 3 种肿瘤标志物 Her2、CK7 和 MUC1 的抗体，并研究了乳腺癌组织中目标蛋白的分布特征[98]。LA-ICP-MS 成像结果显示，患者乳腺癌组织 MUC1 的表达水平要显著高于 Her2 和 CK7。此外，还可采用纳米颗粒对抗体进行标记，实现目标蛋白的原位分析。Fernandez 等使用 2.7 nm 的水溶性金纳米团簇（AuNCs）标记了 MT1/2 和 MT3 的抗体。之后，利用抗体-抗原特异性作用标记 MT1/2 和 MT3，最后使用 LA-ICP-MS 成像技术进行原位分析[99,100]。他们发现，在视网膜组织中，MT1/2 和 MT3 主要位于内核层、外核层和神经节细胞层。在该研究中，由于采用的 AuNCs 平均含 579 个 Au 原子，显著提升了 LA-ICP-MS 的检测能力，实现了信号放大目的，对 MT1/2 的检测限可达 0.5 ng/g。同时，激光束的光斑可以降低至 4 μm，极大地提高了元素成像的空间分辨率。近期，他们又利用 2.2 nm 的 AuNCs 标记膜铁转运蛋白（FPN）抗体，并利用 LA-ICP-MS 成像技术研究了 AD 患者脑部海马 CA1 区锥体层内 FPN 和 Fe 的分布特征（图 21-15）。研究结果显示，人脑海马区内 Fe 非均匀分布，主要集中于锥体层，并且 Fe 与 FPN 的分布规律一致。此外，AD 患者该区域的 FPN 和 Fe 水平均高于正常人[101]。

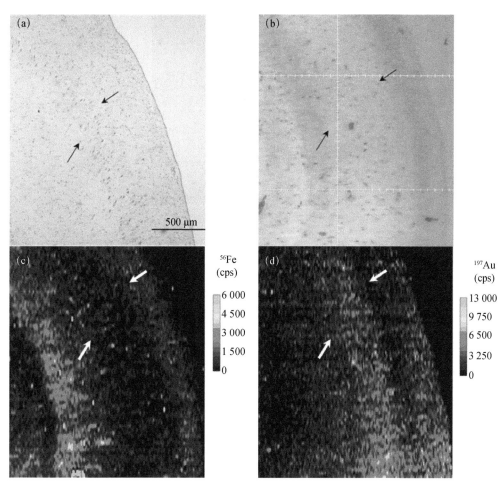

图 21-15　AD 患者脑组织切片的 FPN 免疫组织化学分析[（a）、（b）]和 LA-ICP-MS 元素成像分析[（c）、（d）][101]

基质金属蛋白酶（MMPs）能够降解细胞外基质蛋白，并在肿瘤的侵袭、转移过程中起关键性作用。Fernández-Sánchez 等利用金纳米颗粒（AuNPs）标记 MMP-11 抗体，通过两步免疫法标记研究了患者乳腺癌组织中 MMP-11 的分布[102]。LA-ICP-MS 成像结果显示，在转移癌和非转移癌组织中均可检测到 MMP-11，但非转移癌组织中 MMP-11 的表达水平较低。并且，与非转移癌相比，转移癌组织中 MMP-11 的空间分布更不均匀。除抗体–抗原的免疫亲和反应，还可利用其他生物亲和作用进行蛋白标记。例如，由于 CL 多肽对膜型基质金属蛋白酶-1（MT1-MMP）具有特异亲和性，可用于肿瘤细胞表面 MT1-MMP 的标记。因此，Gao 等设计并制备了一种 CL 多肽表面修饰的 AuNCs，用于原位标记、识别肿瘤组织中的 MT1-MMP（图 21-16）。通过 LA-ICP-MS 对 PC-14 和 A549 肺部肿瘤组织的 MT1-MMP 成像，他们发现 PC-14 肿瘤组织的信号要高于 A549 肿瘤组织，表明 PC-14 肿瘤细胞比 A549 肿瘤细胞表达了更多的 MT1-MMP。此外，通过研究肾癌患者的肿瘤组织，他们发现癌变区域的 MT1-MMP 大量表达，但正常组织区域几乎不表达 MT1-MMP[103]。

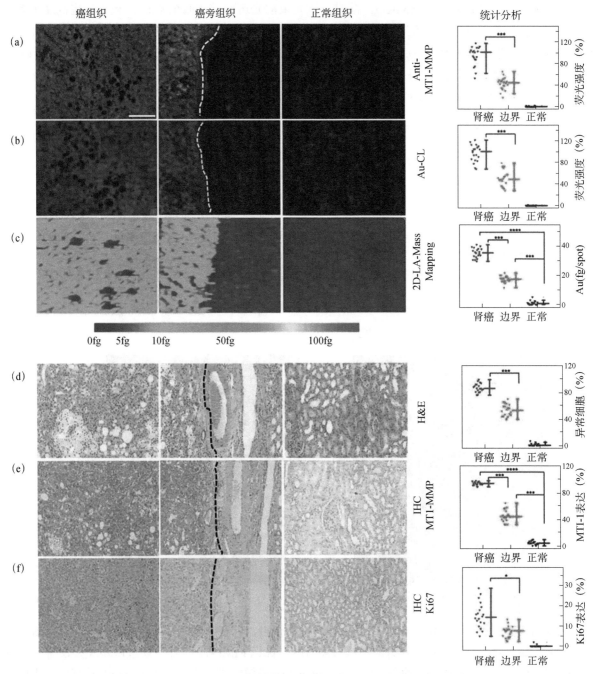

图 21-16 原发性肾癌患者肿瘤组织的 MTI-MMP 免疫荧光成像（a）、金纳米簇探针荧光成像（b）、LA-ICP-MS 元素成像（c）、H&E 染色（d）和免疫组化分析［(e)、(f)］[103]

21.6 细胞检测

近年,作为科学前沿热门领域,单细胞分析受到广泛关注。单细胞分析可以获取细胞在微环境中的准确信息,对于研究细胞行为、分子信号传导、生理与病理过程,以及疾病早期诊断均具有十分重要的科学意义。因此,以 LA-ICP-MS 技术为基础的单细胞研究也得到了越来越多的重视与应用[104-107]。

21.6.1 单细胞成像

受激光光斑、细胞大小,以及灵敏度难于检测单个细胞的元素含量等原因限制,LA-ICP-MS 技术用于单细胞成像还存在诸多难题与挑战。例如,Wang 等使用 LA-ICP-MS 元素成像分析 AuNPs 处理后的 RAW264.7 巨噬细胞,除 Au 与 Mg 的信号,检测不到其他痕量元素信号[108]。Vanhaecke 等利用 LA-ICP-MS 元素成像研究了 Cu(0.5~100 μg/L)暴露后,锥状斯氏藻(*Stropsiella trochoidea*)细胞内的 Cu 累积,发现不同细胞个体对 Cu 的吸收和累积存在很大差异[109]。Costas-Rodríguez 等研究了人神经母细胞瘤 SH-SY5Y 细胞分化前后胞内 Cu 的变化。成像结果表明,细胞分化前 Cu 的胞内分布相对均匀,而细胞分化后的胞内 Cu 沿神经突起方向发生了重新分布[110]。利用 LA-ICP-MS 的同位素检测能力,Fernández 等比较研究了人视网膜色素上皮 HRPEsv 细胞对 2 种富集同位素 Zn 化合物($^{168}ZnSO_4$、^{170}Zn-gluconate)的摄入与分布[111]。结果显示,2 种外源 Zn 化合物的胞内分布并无显著性差异。Koellensperger 等则利用 LA-ICP-MS 的多元素分析检测能力,对使用顺铂患者的血液涂片进行了研究[112]。利用 Fe 和 P 元素含量的差异,他们成功地区分了红细胞与白细胞(图 21-17),另外,还可观察到血细胞外存在 Pt 信号,而红细胞结合的 Pt 只占很小比例。

为提高细胞成像的分辨率与检测灵敏度,使用外源元素标签对细胞进行标记是一种可行的方案,可有效增强细胞检测信号,并降低细胞内源痕量元素的干扰。例如,Herrmann 等分别采用 Ir 化合物(Intercalator-Ir)与 mDOTA(Tm)标记细胞核与细胞整体轮廓[113]。元素成像结果表明,对于小鼠成纤维 NIH-3T3 细胞和人肺上皮 A549 细胞,即便很短的标记时间(<1 h)和很低的试剂浓度(<0.5 μmol/L)也足以实现细胞标记与成像。并且,LA-ICP-MS 对镧系元素(如 Ir、Ho)的检测限可低至 fg 级[114]。Jakubowski 等同时采用 6 种镧系元素标记的抗体(anti-beta-actin-Er、anti-Cdk2-Yb、anti-cyclin B1-Eu、anti-pRb-Nd、anti-pH 3-Lu、anti-Ki67-Dy),以及 mDOTA(Ho)和 Intercalator-Ir 对小鼠成纤维 3T3 细胞的细胞周期进行了研究[115]。他们发现,在细胞 G1、G2 和 M 期,可检测到 *pRb* 和 *Ki67* 表达,而细胞 M 期的 *pH 3* 表达则体现出较大差异。近期,Vanhaecke 等还利用 Intercalator-Ir 标记人类宫颈癌 HeLa 细胞,并利用 LA-ICP-MS 的二维 Ir 元素成像结果进行了细胞三维图像的重建[116]。Buckle 等则制备了一种可配位结合 Ho 或 In 元素的标签分子(hybrid-Cy5-Ac-TZ4011),靶向标记细胞表面的趋化因子受体 4(CXCR4),用于 CXCR4 过表达细胞的 LA-ICP-MS 元素成像[117]。此外,还可利用 LA-ICP-MS 成像技术同时检测 [Tm]DTPA-Lys(Cy5)-Cys(Ac-TZ14011) 和 [Ho]DTPA-Lys(Fluorescein)- Cys(Cetuximab) 分别标记的 CXCR4 和表皮生长因子受体(EGFR),对乳腺癌细胞亚型(MDA-MB-231 X4 和 MDA-MB-468)进行区分和识别[118]。

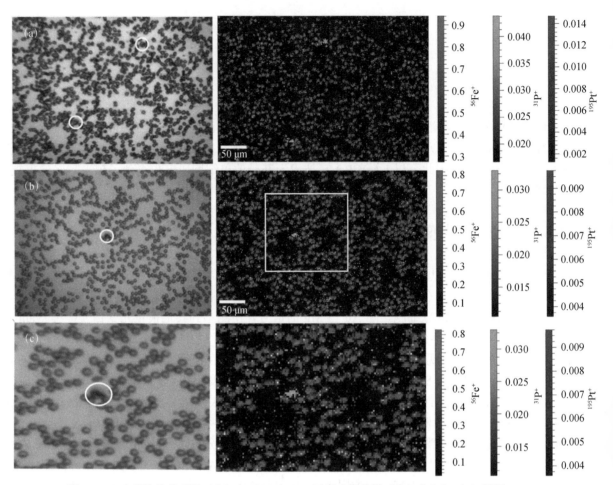

图 21-17　血液涂片的明场（左）与 LA-ICP-MS 元素成像分析（右）[（a）、（b）][112]；
（c）为（b）内白框区域的局部放大，白色圆圈指示白细胞

21.6.2　体内细胞示踪

除体外细胞成像，还可以利用 LA-ICP-MS 元素成像技术对体内的细胞进行示踪和定位，揭示细胞的体内分布特征与迁移行为[119]。例如，巨噬细胞和干细胞常被用于临床治疗，因此揭示这些细胞的体内行为对改善临床治疗效果至关重要。Karst 等利用 DOTMA（Tm）标记骨髓来源的巨噬细胞（BMDMs），并研究了 BMDMs 在局部炎症模型小鼠体内的分布特征[120]。经静脉注射 8 天后，LA-ICP-MS 成像结果显示，Tm 标记的 BMDMs 可以迁移至小鼠皮下炎症部位，并在肝脏中大量分布。Nakada 等使用 Cr 标记小鼠间充质干细胞（mMSCs），对体内注射的 mMSCs 进行示踪[121]。LA-ICP-MS 成像结果表明，经肌肉注射或静脉注射后，Cr 标记的 mMSCs 主要分布于大腿肌肉或肺部，表明肺部毛细血管对 mMSCs 具有截留作用。我们采用 100 nm Au NPs 对人骨髓来源间充质干细胞（hMSCs）进行标记，并利用 LA-ICP-MS 对小鼠肺癌原位模型进行研究。元素成像结果表明，经尾静脉注射 1 天后，Au NPs 标记的 hMSCs 集中分布于血管周围。经过 7 天后，更多 hMSCs 会迁移至肿瘤结节区域（图 21-18）。

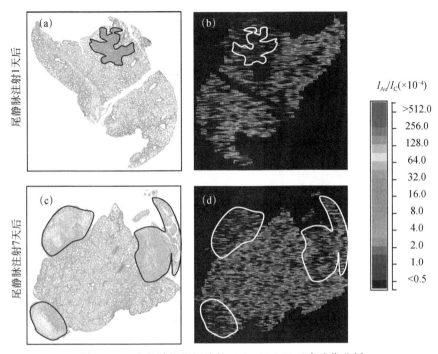

图 21-18 小鼠肺组织切片的 LA-ICP-MS 元素成像分析

小鼠肺癌组织切片的 H&E 染色[（a）、（c）]；Au NPs 标记 hMSCs 在肺癌组织中的分布[（b）、（d）]

21.7 纳米材料原位成像

纳米技术已在生命科学、疾病诊疗、环境健康等领域得到广泛应用。纳米材料具备不同寻常的物理化学性质与生物效应，亟待研究和揭示。因此，近些年利用 LA-ICP-MS 元素成像技术进行纳米材料研究的相关工作越来越多。

2012 年，Drescher 等便利用 LA-ICP-MS 元素成像技术研究了成纤维 3T3 细胞内 Au NPs（25 nm）和纳米银（Ag NPs，50 nm）的分布特征[122]。此外，他们还比较了不同类型细胞对 TiO_2 NPs、Ag NPs、BrightSilica(Ag)和 SiO_2 (29 nm)@Ag 摄入的差异[123]。此外，他们发现经 $HAuCl_4$ 处理后，成纤维细胞 3T3 内会原位生成 Au NPs[124]。除细胞内 NMs，LA-ICP-MS 还可针对体外培养组织内的 NMs 进行成像分析。例如，Arakawa 等将体外培养的成纤维细胞球体（MCS）暴露于 Ag NPs，研究了 Ag NPs 的组织渗透能力[125]。经 24 h 暴露后，Ag NPs 主要分布于 MCS 的边缘，并与 P 和 Zn 的分布具有相关性，可能由于 MCS 外缘丰富的胶原蛋白阻止了 Ag NPs 向 MCS 中心区域的渗透。之后，他们进一步考察了 Ag NPs 粒径、表面修饰及暴露时间对渗透能力的影响[126]。发现，暴露 24 h 后，小粒径 Ag NPs（5 nm）可渗透进入 MCS 内部，而大粒径 Ag NPs（50 nm）仅分布于 MCS 外缘（图 21-19）。相较于聚乙烯吡咯烷酮（PVP）修饰，柠檬酸修饰可增强 Ag NPs 的渗透能力。同时，Ag NPs 还会改变 MCS 的微量元素分布[127]。

利用 LA-ICP-MS 对纳米材料的组织原位成像研究更为广泛，如 Au NPs、Ag NPs、氧化石墨烯（GO）、上转换纳米颗粒（$NaGdF_4$：Yb^{3+}/Er^{3+}@SiO_2、$NaYF_4$：Yb/Tm/Gd、$NaYF_4$：Yb/Er）[128-130]、铱纳米颗粒[131]、硅纳米颗粒（Si NPs）[132]等。NMs 的表面电荷性质会直接影响其体内分布。为揭示表面电荷对 Au NPs 亚器官分布的影响，Vachet 等研究比较了表面正电荷、负电荷及中性 Au NPs 在

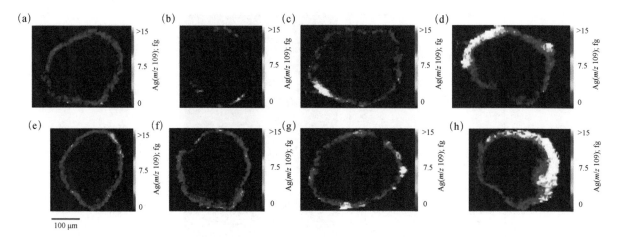

图 21-19 暴露 24 h[（a）～（d）]或 48 h[（e）～（h）]后，MCS 中 Ag NPs 的分布特征[126]

[（a）、（e）]、[（c）、（g）]和[（d）、（h）]分别为 5 nm、20 nm 和 50 nm 柠檬酸修饰的 Ag NPs，[（b）、（f）]为 20 nm PVP 修饰的 Ag NPs

小鼠肝脏、脾脏和肾脏中的亚器官分布特征[133]。元素成像结果表明，肾小球充当着过滤器的作用，对表面正电荷的 Au NPs 具有很强的截留效果，而表面中性和负电荷的 Au NPs 则很少累积于肾小球。在脾脏中，表面正电荷和负电荷的 Au NPs 大量分布于红髓，而表面中性的 Au NPs 主要分布于白髓和边缘区域（图 21-20）。出现上述现象主要是因为，脾脏的红髓负责清除血液中的颗粒物、抗原和死亡血细胞，而脾脏的白髓为免疫系统的一部分，红髓与白髓间的边缘区域则负责物质交换，是参与免疫反应的起始区域。在肝脏中，表面中性的 Au NPs 主要位于库普弗细胞内。Wiemann 与 Karst 等研究了经气管滴注 21 天后，50 nm Ag NPs 在大鼠体内的分布特征[134]。LA-ICP-MS 元素成像结果显示，在纵隔淋巴结中，Ag NPs 主要分布于巨噬细胞和网状纤维内。在肾脏中，Ag NPs 主要分布于近端小管内，而非肾小球。此外，在肝脏和脾脏的免疫细胞类群中也观察到 Ag NPs 的分布。之后，通过构建三维 LA-ICP-MS 图像，他们发现经气管滴注的 Ag NPs（Ag50PVP）主要分布于脾脏的白髓[135]。近期，他们又研究了经气管滴注 Ag NPs 在大鼠肺部的分布特征，发现 Ag NPs 几乎完全分布于肺泡巨噬细胞内，而肺泡隔中的 Ag 含量很低[136]。氧化铁纳米颗粒（ION）可用于 MRI 临床诊断，但机体内源 Fe 会对 ION 产生信号干扰。为区分内源 Fe 与 ION，Faber 等制备了富集同位素 ^{57}Fe 的 ION（^{57}Fe-ION），并利用 LA-ICP-MS 元素成像技术对小鼠体内 ^{57}Fe-ION 标记的循环吞噬细胞进行示踪，成功地定位了小鼠皮下炎症部位的循环吞噬细胞[137]。

环境生物体内的纳米材料的分布特征也是受重点关注的研究方向。例如，Arruda 等利用 LA-ICP-MS 元素成像技术研究了 40 nm Ag NPs 的大豆叶片分布，以及对微量元素 Mn 和 Cu 的影响[138]。他们发现，Ag NPs 暴露后，可在大豆叶片中检测到含量很低的 Ag，并且 Ag NPs 没有显著地影响 Cu 和 Mn 的含量与分布。Reemtsma 与 Kühnel 等比较了斑马鱼（*Danio rerio*）胚胎和大型溞（*Daphnia magna*）对纳米颗粒（Ag NPs、Au NPs 和 Al$_2$O$_3$ NPs）的摄取[139]。LA-ICP-MS 成像结果显示，纳米颗粒主要分布于斑马鱼胚胎绒毛膜表面，难以进入胚胎内部。相反，纳米颗粒会被大型溞摄取，并主要在肠道内累积，也有少量纳米颗粒分布于鳃和眼。之后，他们进一步比较了斑马鱼胚胎中 4 种 NPs（Ag NPs、Au NPs、CuO NPs 和 ZnO NPs）和金属离子（Ag$^+$、Cu^{2+} 和 Zn^{2+}）的分布差异[140]。他们发现，Ag NPs 在胚胎绒毛膜中的累积高达 58%～85%，部分 Ag NPs 可进入玻璃体腔（PVS），只有少量进入胚胎，而 Ag$^+$ 主要位于 PVS 内。Au NPs 的分布与 Ag NPs 类似。CuO NPs 暴露后，可以在绒毛膜和胚胎中观察到 Cu，而 Cu^{2+} 暴露后仅在绒毛膜上检测到 Cu 信号。对于 ZnO NPs 和 Zn^{2+} 暴露，内胚层的 Zn 含量最高，仅有少量 Zn 分布于绒毛膜。

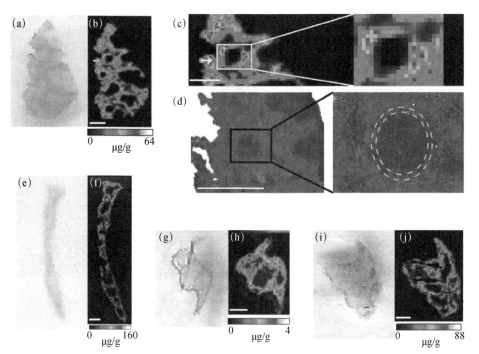

图 21-20　利用 LA-ICP-MS 定位小鼠脾脏内表面正电荷、负电荷以及中性 Au NPs 的分布特征[133]

［(a)、(b)、(e)、(f)］为表面正电荷 Au NPs 的分布；［(g)、(h)］为表面中性 Au NPs 的分布；
［(i)、(j)］为表面负电荷 Au NPs 的分布；(c) 为 (b) 局部区域的放大；(d) 为 H&E 染色的显微图像

在生物体内，LA-ICP-MS 无法直接对碳纳米材料进行有效的检测，但可以利用碳纳米材料的掺杂元素，或元素标记元素成像研究。例如，Liu 等利用石墨烯和氧化石墨烯（GO）的金属杂质（Fe、Co、Cu、Ni、Mn 等）作为元素指纹图谱，对大豆植株中的石墨烯和 GO 进行定位分析[141]。他们发现，石墨烯和 GO 在大豆植株中体现出不同的分布特征。经 200 mg/L 石墨烯或 GO 暴露后，前者的累积次序为根>叶>茎>子叶，而后者的累积次序为根>茎>叶>子叶。通过研究两者的根部分布，发现石墨烯主要分布于表皮，可以进入根瘤，但难以进入韧皮部，而 GO 可以更多地进入韧皮部。Feng 等使用 La/Ce 双元素标记的方法对小鼠体内 GO 的分布和代谢进行了研究[142]。他们发现，经静脉注射 4 h 后，GO 主要分布于肝脏的肝窦周围。在脾脏中，GO 主要分布于红髓和边缘区（MZ）。此外，在肾皮质和髓质中也检测到很强的信号。与上述研究思路不同，Xu 等利用 LA-ICP-MS 元素成像技术评价了 GO 对秀丽隐杆线虫（Caenorhabditis elegans）体内 As(Ⅲ)的影响，发现 GO 可以加速线虫体内 As(Ⅲ)的清除[143]。

在元素分析的基础上，近期多个研究团队成功地发展出了新一代的单颗粒 LA-ICP-MS 成像技术（LA-SP-ICP-MS）。与常规元素成像不同，LA-SP-ICP-MS 可同时获取 NPs 的元素与粒径空间分布信息，展现出更强的 NPs 解析能力[144,145]。例如，为研究 Au NPs 的体内粒径分布情况，Wang 等发展的 LA-SP-ICP-MS 成像技术可实现不同粒径 Au NPs 的组织原位成像[145]。他们发现，经静脉注射 8 h 后，肝脏内粒径为 80 nm 的 Au NPs 并没有发生聚集。Elteren 等则利用 LA-SP-ICP-MS 研究了向日葵根截面中 60 nm Au NPs 的分布，发现 Au NPs 主要分布于根茎和根表面，且粒径并未发生明显变化（图 21-21）[144]。

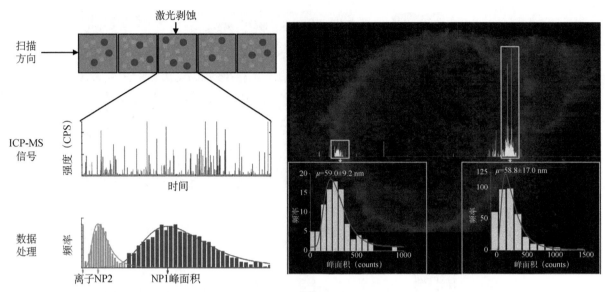

图 21-21　利用 LA-SP-ICP-MS 分析向日葵植株根部横截面中 Au NPs 的分布特征[144]

21.8　展　　望

综上所述，LA-ICP-MS 元素成像技术已被广泛应用于生物组织、细胞及分子的检测分析，为微量元素、重金属、金属药物、纳米材料等领域的研究提供了重要的手段。未来相关研究可以从以下方面突破：

（1）有关 LA-ICP-MS 元素成像的大部分研究工作都采用二维平面成像策略，对三维立体成像的探索还比较有限，这主要受限于样品前处理复杂、分析时间长、后期数据处理难度大等因素，有待于新技术与方法的发展；

（2）在生物分析中，LA-ICP-MS 成像技术主要用于元素的原位解析与检测，针对特定元素的多同位素研究还鲜有报道，如同位素比值等；

（3）因为 LA-ICP-MS 无法对化合物或生物分子进行有效的分离与结构解析，所以有必要与其他分析技术相结合，进一步扩展其适用范围；

（4）通常，LA-ICP-MS 的分辨率达到微米级别，但对生物亚器官组织或细胞的元素成像还远远不够，如何实现亚细胞级别的元素成像还有待于技术突破；

（5）基于元素成像，针对纳米颗粒的 LA-ICP-MS 分析成像技术亟待推进，特别是 LA-SP-ICP-MS 技术的发展，将有助于纳米医学与健康领域的进步。

参 考 文 献

[1] Sylvester P J, Jackson S E. A brief history of laser ablation inductively coupled plasma mass spectrometry (LA-ICP-MS). Elements, 2016, 12(5): 307-310.

[2] Becker J S. Inductively coupled plasma mass spectrometry (ICP-MS) and laser ablation ICP-MS for isotope analysis of

long-lived radionuclides. International Journal of Mass Spectrometry, 2005, 242(2-3): 183-195.

[3] Durrant S F, Ward N I. Recent biological and environmental applications of laser ablation inductively coupled plasma mass spectrometry (LA-ICP-MS). Journal of Analytical Atomic Spectrometry, 2005, 20(9): 821-829.

[4] New E J, Wimmer V C, Hare D J. Promises and pitfalls of metal imaging in biology. Cell Chemical Biology, 2018, 25(1): 7-18.

[5] Susnea I, Weiskirchen R. Trace metal imaging in diagnostic of hepatic metal disease. Mass Spectrometry Reviews, 2016, 35(6): 666-686.

[6] Grijalba N, Legrand A, Holler V, Bouvier-Capely C. A novel calibration strategy based on internal standard-spiked gelatine for quantitative bio-imaging by LA-ICP-MS: Application to renal localization and quantification of uranium. Analytical and Bioanalytical Chemistry, 2020, 412(13): 3113-3122.

[7] Becker J S, Jakubowski N. The synergy of elemental and biomolecular mass spectrometry: New analytical strategies in life sciences. Chemical Society Reviews, 2009, 38(7): 1969-1983.

[8] Sussulini A, Becker J S, Becker J S. Laser ablation ICP-MS: Application in biomedical research. Mass Spectrometry Reviews, 2017, 36(1): 47-57.

[9] Pozebon D, Scheffler G L, Dressler V L. Recent applications of laser ablation inductively coupled plasma mass spectrometry (LA-ICP-MS) for biological sample analysis: A follow-up review. Journal of Analytical Atomic Spectrometry, 2017, 32(5): 890-919.

[10] Sajnog A, Hanc A, Baralkiewicz D. Metrological approach to quantitative analysis of clinical samples by LA-ICP-MS: A critical review of recent studies. Talanta, 2018, 182: 92-110.

[11] Cruz-Alonso M, Lores-Padin A, Valencia E, Gonzalez-Iglesias H, Fernandez B, Pereiro R. Quantitative mapping of specific proteins in biological tissues by laser ablation-ICP-MS using exogenous labels: Aspects to be considered. Analytical and Bioanalytical Chemistry, 2019, 411(3): 549-558.

[12] Galazzi R M, Chacon-Madrid K, Freitas D C, Da Costa L F, Arruda MAZ. Inductively coupled plasma mass spectrometry based platforms for studies involving nanoparticle effects in biological samples. Rapid Communications in Mass Spectrometry, 2020, 34: e8726.

[13] Kavčič A, Mikus K, Debeljak M, Van Elteren J T, Arcon I, Kodre A, Kump P, Karydas A G, Migliori A, Czyzycki M, Vogel-Mikus K. Localization, ligand environment, bioavailability and toxicity of mercury in *Boletus* spp. and *Scutiger pes-caprae* mushrooms. Ecotoxicology and Environmental Safety, 2019, 184: 11.

[14] Santos M C, Wagner M, Wu B, Scheider J, Oehlmann J, Cadore S, Becker J S. Biomonitoring of metal contamination in a marine prosobranch snail (*Nassarius reticulatus*) by imaging laser ablation inductively coupled plasma mass spectrometry (LA-ICP-MS). Talanta, 2009, 80(2): 428-433.

[15] Niehoff A C, Bauer O B, Kroger S, Fingerhut S, Schulz J, Meyer S, Sperling M, Jeibmann A, Schwerdtle T, Karst U. Quantitative bioimaging to investigate the uptake of mercury species in *Drosophila melanogaster*. Analytical Chemistry, 2015, 87(20): 10392-10396.

[16] Pamphlett R, Jew S K, Doble P A, Bishop D P. Elemental imaging shows mercury in cells of the human lateral and medial geniculate nuclei. PloS One, 2020, 15(4): 16.

[17] Pamphlett R, Satgunaseelan L, Jew S K, Doble P A, Bishop D P. Elemental bioimaging shows mercury and other toxic metals in normal breast tissue and in breast cancers. PloS One, 2020, 15(1): 22.

[18] Huang W L, Bai Z Q, Jiao J, Yuan H L, Bao Z A, Chen S N, Ding M H, Liang Z S. Distribution and chemical forms of cadmium in *Coptis chinensis* Franch. determined by laser ablation ICP-MS, cell fractionation, and sequential extraction. Ecotoxicology and Environmental Safety, 2019, 171: 894-903.

[19] Pessoa G D, Lopes C A, Madrid K C, Arruda M A Z. A quantitative approach for Cd, Cu, Fe and Mn through laser ablation imaging for evaluating the translocation and accumulation of metals in sunflower seeds. Talanta, 2017, 167: 317-324.

[20] Basnet P, Amarasiriwardena D, Wu F C, Fu Z Y, Zhang T. Elemental bioimaging of tissue level trace metal distributions in rice seeds (*Oryza sativa* L.) from a mining area in China. Environmental Pollution, 2014, 195: 148-156.

[21] Egger A E, Grabmann G, Gollmann-Tepekoylu C, Pechriggl E J, Artner C, Turkcan A, Hartinger C G, Fritsch H, Keppler B K, Brenner E, Grimm M, Messner B, Bernhard D. Chemical imaging and assessment of cadmium distribution in the human body. Metallomics, 2019, 11(12): 2010-2019.

[22] Van Malderen S J M, Laforce B, Van Acker T, Vincze L, Vanhaecke F. Imaging the 3D trace metal and metalloid distribution in mature wheat and rye grains via laser ablation-ICP-mass spectrometry and micro-X-ray fluorescence spectrometry. Journal of Analytical Atomic Spectrometry, 2017, 32(2): 289-298.

[23] Van Malderen S J M, Laforce B, Van Acker T, Nys C, De Rijcke M, De Rycke R, De Bruyne M, Boone M N, De Schamphelaere K, Borovinskaya O, De Samber B, Vincze L, Vanhaecke F. Three-dimensional reconstruction of the tissue-specific multielemental distribution within *Ceriodaphnia dubia* via multimodal registration using laser ablation ICP-mass spectrometry and X-ray spectroscopic techniques. Analytical Chemistry, 2017, 89(7): 4161-4168.

[24] Marillo-Sialer E, Black J R, Paul B, Kysenius K, Crouch P J, Hergt J M, Woodhead J D, Hare D J. Construction of 3D native elemental maps for large biological specimens using LA-ICP-MS coupled with X-ray tomography. Journal of analytical atomic spectrometry, 2020, 35(4): 671-678.

[25] Wu B, Chen Y X, Becker J S. Study of essential element accumulation in the leaves of a Cu-tolerant plant *Elsholtzia splendens* after Cu treatment by imaging laser ablation inductively coupled plasma mass spectrometry (LA-ICP-MS). Analytica Chimica Acta, 2009, 633(2): 165-172.

[26] Niehoff A C, Schulz J, Soltwisch J, Meyer S, Kettling H, Sperling M, Jeibmann A, Dreisewerd K, Francesconi K A, Schwerdtle T, Karst U. Imaging by elemental and molecular mass spectrometry reveals the uptake of an arsenolipid in the brain of *Drosophila melanogaster*. Analytical Chemistry, 2016, 88(10): 5258-5263.

[27] Togao M, Nakayama S M M, Ikenaka Y, Mizukawa H, Makino Y, Kubota A, Matsukawa T, Yokoyama K, Hirata T, Ishizuka M. Bioimaging of Pb and STIM1 in mice liver, kidney and brain using laser ablation inductively coupled plasma mass spectrometry (LA-ICP-MS) and immunohistochemistry. Chemosphere, 2020, 238: 8.

[28] Huang W L, Jiao J, Ru M, Bai Z Q, Yuan H L, Bao Z A, Liang Z S. Localization and speciation of chromium in *Coptis chinensis* Franch. using synchrotron radiation X-ray technology and laser ablation ICP-MS. Scientific Reports, 2018, 8(1): 14.

[29] Vanderschee C R, Kuter D, Chou H, Jackson B P, Mann K K, Bohle D S. Addressing K/L-edge overlap in elemental analysis from micro-X-ray fluorescence: Bioimaging of tungsten and zinc in bone tissue using synchrotron radiation and laser ablation inductively coupled plasma mass spectrometry. Analytical and Bioanalytical Chemistry, 2019, 412(2): 259-265.

[30] Moreno-Gordaliza E, Giesen C, Lazaro A, Esteban-Fernandez D, Humanes B, Canas B, Panne U, Tejedor A, Jakubowski N, Gómez-Gómez M M, Elemental bioimaging in kidney by LA-ICP-MS as a tool to study nephrotoxicity and renal protective strategies in cisplatin therapies. Analytical Chemistry, 2011, 83(20): 7933-7940.

[31] Van Acker T, Van Malderen S J M, Van Heerden M, Mcduffie J E, Cuyckens F, Vanhaecke F. High-resolution laser ablation-inductively coupled plasma-mass spectrometry imaging of cisplatin-induced nephrotoxic side effects. Analytica Chimica Acta, 2016, 945: 23-30.

[32] Moraleja I, Esteban-Fernandez D, Lazaro A, Humanes B, Neumann B, Tejedor A, Mena M L, Jakubowski N,

Gomez-Gomez M M. Printing metal-spiked inks for LA-ICP-MS bioimaging internal standardization: comparison of the different nephrotoxic behavior of cisplatin, carboplatin, and oxaliplatin. Analytical and Bioanalytical Chemistry, 2016, 408(9): 2309-2318.

[33] Moraleja I, Mena M L, Lazaro A, Neumann B, Tejedor A, Jakubowski N, Gomez-Gomez M M, Esteban-Fernandez D. An approach for quantification of platinum distribution in tissues by LA-ICP-MS imaging using isotope dilution analysis. Talanta, 2018, 178: 166-171.

[34] Hucke A, Rinschen M M, Bauer O B, Sperling M, Karst U, Koppen C, Sommer K, Schroter R, Ceresa C, Chiorazzi A, Canta A, Semperboni S, Marmiroli P, Cavaletti G, Schlatt S, Schlatter E, Pavenstadt H, Heitplatz B, Van Marck V, Sparreboom A, Barz V, Knief A, Deuster D, Zehnhoff-Dinnesen A A, Ciarimboli G. An integrative approach to cisplatin chronic toxicities in mice reveals importance of organic cation-transporter-dependent protein networks for renoprotection. Archives of Toxicology, 2019, 93(10): 2835-2848.

[35] Breglio A M, Rusheen A E, Shide E D, Fernandez K A, Spielbauer K K, McLachlin K M, Hall M D, Amable L, Cunningham L L. Cisplatin is retained in the cochlea indefinitely following chemotherapy. Nature Communications, 2017, 8: 9.

[36] Theiner S, Kornauth C, Varbanov H P, Galanski M, Van Schoonhoven S, Heffeter P, Berger W, Egger A E, Keppler B K. Tumor microenvironment in focus: LA-ICP-MS bioimaging of a preclinical tumor model upon treatment with platinum(Ⅳ)-based anticancer agents. Metallomics, 2015, 7(8): 1256-1264.

[37] Niehoff A C, Grunebaum J, Moosmann A, Mulac D, Sobbing J, Niehaus R, Buchholz R, Kroger S, Wiehe A, Wagner S, Sperling M, Von Briesen H, Langer K, Karst U. Quantitative bioimaging of platinum group elements in tumor spheroids. Analytica Chimica Acta, 2016, 938: 106-113.

[38] Theiner S, Van Malderen S J M, Van Acker T, Legin A, Keppler B K, Vanhaecke F, Koellensperger G. Fast high-resolution laser ablation-inductively coupled plasma mass spectrometry imaging of the distribution of platinum-based anticancer compounds in multicellular tumor spheroids. Analytical Chemistry, 2017, 89(23): 12641-12645.

[39] Holzlechner M, Bonta M, Lohninger H, Limbeck A, Marchetti-Deschmann M. Multisensor Imaging: From sample preparation to integrated multimodal interpretation of LA-ICPMS and MALDI MS Imaging data. Analytical Chemistry, 2018, 90(15): 8831-8837.

[40] Trog S, El-Khatib A H, Beck S, Makowski M R, Jakubowski N, Linscheid M W. Complementarity of molecular and elemental mass spectrometric imaging of Gadovist™ in mouse tissues. Analytical and Bioanalytical Chemistry, 2018, 411(3): 629-637.

[41] Birka M, Wentker K S, Lusmoller E, Arheilger B, Wehe C A, Sperling M, Stadler R, Karst U. Diagnosis of nephrogenic systemic fibrosis by means of elemental bioimaging and speciation analysis. Analytical Chemistry, 2015, 87(6): 3321-3328.

[42] Delfino R, Biasotto M, Candido R, Altissimo M, Stebel M, Salome M, Van Elteren J T, Mikus K V, Zennaro C, Sala M, Addobbati R, Tromba G, Pascolo L. Gadolinium tissue deposition in the periodontal ligament of mice with reduced renal function exposed to Gd-based contrast agents. Toxicology Letters, 2019, 301: 157-167.

[43] Fingerhut S, Niehoff A C, Sperling M, Jeibmann A, Paulus W, Niederstadt T, Allkemper T, Heindel W, Holling M, Karst U. Spatially resolved quantification of gadolinium deposited in the brain of a patient treated with gadolinium-based contrast agents. Journal of Trace Elements in Medicine and Biology, 2018, 45: 125-130.

[44] Clases D, Fingerhut S, Jeibmann A, Sperling M, Doble P, Karst U. LA-ICP-MS/MS improves limits of detection in elemental bioimaging of gadolinium deposition originating from MRI contrast agents in skin and brain tissues. Journal of Trace Elements in Medicine and Biology, 2019, 51: 212-218.

[45] Richter H, Bucker P, Dunker C, Karst U, Kircher P R. Gadolinium deposition in the brain of dogs after multiple intravenous administrations of linear gadolinium based contrast agents. PloS One, 2020, 15(2): 16.

[46] Egger A E, Theiner S, Kornauth C, Heffeter P, Berger W, Keppler B K, Hartinger C G. Quantitative bioimaging by LA-ICP-MS: A methodological study on the distribution of Pt and Ru in viscera originating from cisplatin-and KP1339-treated mice. Metallomics, 2014, 6(9): 1616-1625.

[47] Klose M H M, Theiner S, Kornauth C, Meier-Menches S M, Heffeter P, Berger W, Koellensperger G, Keppler B K. Bioimaging of isosteric osmium and ruthenium anticancer agents by LA-ICP-MS. Metallomics, 2018, 10(3): 388-396.

[48] O'neill E S, Kaur A, Bishop D P, Shishmarev D, Kuchel P W, Grieve S M, Figtree G A, Renfrew A K, Bonnitcha P D, New E J. Hypoxia-responsive cobalt complexes in tumor spheroids: Laser ablation inductively coupled plasma mass spectrometry and magnetic resonance imaging studies. Inorganic Chemistry, 2017, 56(16): 9860-9868.

[49] Zhao H W, Wang S H, Nguyen S N, Elci S G, Kaltashov I A. Evaluation of nonferrous metals as potential *in vivo* tracers of transferrin-based therapeutics. Journal of the American Society for Mass Spectrometry, 2016, 27(2): 211-219.

[50] Reifschneider O, Schutz C L, Brochhausen C, Hampel G, Ross T, Sperling M, Karst U. Quantitative bioimaging of *p*-boronophenylalanine in thin liver tissue sections as a tool for treatment planning in boron neutron capture therapy. Analytical and Bioanalytical Chemistry, 2014, 407(9): 2365-2371.

[51] Becker J S, Zoriy M V, Pickhardt C, Palomero-Gallagher N, Zilles K. Imaging of copper, zinc, and other elements in thin section of human brain samples (hippocampus) by laser ablation inductively coupled plasma mass spectrometry. Analytical Chemistry, 2005, 77(10): 3208-3216.

[52] Paul B, Hare D J, Bishop D P, Paton C, Nguyen V T, Cole N, Niedwiecki M M, Andreozzi E, Vais A, Billings J L, Bray L, Bush A I, McColl G, Roberts B R, Adlard P A, Finkelstein D I, Hellstrom J, Hergt J M, Woodhead J D, Doble P A. Visualising mouse neuroanatomy and function by metal distribution using laser ablation-inductively coupled plasma-mass spectrometry imaging. Chemical Science, 2015, 6(10): 5383-5393.

[53] Portbury S D, Hare D J, Sgambelloni C J, Bishop D P, Finkelstein D I, Doble P A, Adlard P A. Age modulates the injury-induced metallomic profile in the brain. Metallomics, 2017, 9(4): 402-410.

[54] Sela H, Cohen H, Karpas Z, Zeiri Y. Distinctive hippocampal zinc distribution patterns following stress exposure in an animal model of PTSD. Metallomics, 2017, 9(3): 323-333.

[55] Opacic M, Ristic A J, Savic D, Selih V S, Zivin M, Sokic D, Raicevic S, Bascarevic V, Spasojevic I. Metal maps of sclerotic hippocampi of patients with mesial temporal lobe epilepsy. Metallomics, 2017, 9(2): 141-148.

[56] Hutchinson R W, Cox A G, McLeod C W, Marshall P S, Harper A, Dawson E L, Howlett D R. Imaging and spatial distribution of beta-amyloid peptide and metal ions in Alzheimer's plaques by laser ablation-inductively coupled plasma-mass spectrometry. Anal Biochem, 2005, 346(2): 225-233.

[57] Hare D J, Raven E P, Roberts B R, Bogeski M, Portbury S D, McLean C A, Masters C L, Connor J R, Bush A I, Crouch P J, Doble P A. Laser ablation-inductively coupled plasma-mass spectrometry imaging of white and gray matter iron distribution in Alzheimer's disease frontal cortex. Neuroimage, 2016, 137: 124-131.

[58] Feng L, Wang J, Li H, Luo X, Li J. A novel absolute quantitative imaging strategy of iron, copper and zinc in brain tissues by Isotope dilution laser ablation ICP-MS. Analytica Chimica Acta, 2017, 984: 66-75.

[59] McInnes L E, Noor A, Kysenius K, Cullinane C, Roselt P, McLean C A, Chiu F C K, Powell A K, Crouch P J, White J M, Donnelly P S. Potential diagnostic imaging of Alzheimer's disease with Copper-64 complexes that bind to amyloid-beta plaques. Inorg Chemistry, 2019, 58(5): 3382-3395.

[60] Hare D, Reedy B, Grimm R, Wilkins S, Volitakis I, George J L, Cherny R A, Bush A I, Finkelstein D I, Doble P. Quantitative elemental bio-imaging of Mn, Fe, Cu and Zn in 6-hydroxydopamine induced Parkinsonism mouse models.

Metallomics, 2009, 1(1): 53-58.

[61] Hare D J, Lei P, Ayton S, Roberts B R, Grimm R, George J L, Bishop D P, Beavis A D, Donovan S J, Mccoll G, Volitakis I, Masters C L, Adlard P A, Cherny R A, Bush A I, Finkelstein D I, Doble P A. An iron-dopamine index predicts risk of parkinsonian neurodegeneration in the substantia nigra pars compacta. Chemical Science, 2014, 5(6): 2160-2169.

[62] Matusch A, Fenn L S, Depboylu C, Klietz M, Strohmer S, McLean J A, Becker J S. Combined elemental and biomolecular mass spectrometry imaging for probing the inventory of tissue at a micrometer scale. Analytical Chemistry, 2012, 84(7): 3170-3178.

[63] Matusch A, Depboylu C, Palm C, Wu B, Hoglinger G U, Schafer M K, Becker J S. Cerebral bioimaging of Cu, Fe, Zn, and Mn in the MPTP mouse model of Parkinson's disease using laser ablation inductively coupled plasma mass spectrometry (LA-ICP-MS). Journal of the American Society for Mass Spectrometry, 2010, 21(1): 161-171.

[64] Gardner B, Dieriks B V, Cameron S, Mendis L H S, Turner C, Faull R L M, Curtis M A. Metal concentrations and distributions in the human olfactory bulb in Parkinson's disease. Scientific Reports, 2017, 7(1): 10454.

[65] De Vega R G, Fernández-Sánchez M L, Pisonero J, Eiro N, Vizoso F J, Sanz-Medel A. Quantitative bioimaging of Ca, Fe, Cu and Zn in breast cancer tissues by LA-ICP-MS. Journal of Analytical Atomic Spectrometry, 2017, 32(3): 671-677.

[66] De Vega R G, Sanchez M L F, Eiro N, Vizoso F J, Sperling M, Karst U, Medel A S. Multimodal laser ablation/desorption imaging analysis of Zn and MMP-11 in breast tissues. Analytical and Bioanalytical Chemistry, 2018, 410(3): 913-922.

[67] Andersen M K, Krossa S, Hoiem T S, Buchholz R, Claes B S R, Balluff B, Ellis S R, Richardsen E, Bertilsson H, Heeren R M A, Bathen T F, Karst U, Giskeodegard G F, Tessem M B. Simultaneous detection of zinc and its pathway metabolites using MALDI MS imaging of prostate tissue. Analytical Chemistry, 2020, 92(4): 3171-3179.

[68] Ali M H M, Rakib F, Al-Saad K, Al-Saady R, Goormaghtigh E. An innovative platform merging elemental analysis and FTIR imaging for breast tissue analysis. Scientific Reports, 2019, 9(1): 9854.

[69] Doble P A, Miklos G L G. Distributions of manganese in diverse human cancers provide insights into tumour radioresistance. Metallomics, 2018, 10(9): 1191-1210.

[70] Hachmoller O, Aichler M, Schwamborn K, Lutz L, Werner M, Sperling M, Walch A, Karst U. Investigating the influence of standard staining procedures on the copper distribution and concentration in Wilson's disease liver samples by laser ablation-inductively coupled plasma-mass spectrometry. Journal of Trace Elements in Medicine and Biology, 2017, 44: 71-75.

[71] Boaru S G, Merle U, Uerlings R, Zimmermann A, Flechtenmacher C, Willheim C, Eder E, Ferenci P, Stremmel W, Weiskirchen R. Laser ablation inductively coupled plasma mass spectrometry imaging of metals in experimental and clinical Wilson's disease. Journal of Cellular and Molecular Medicine, 2015, 19(4): 806-814.

[72] Hachmoller O, Zibert A, Zischka H, Sperling M, Groba S R, Grunewald I, Wardelmann E, Schmidt H H, Karst U. Spatial investigation of the elemental distribution in Wilson's disease liver after D-penicillamine treatment by LA-ICP-MS. Journal of Trace Elements in Medicine and Biology, 2017, 44: 26-31.

[73] Muller J C, Lichtmannegger J, Zischka H, Sperling M, Karst U. High spatial resolution LA-ICP-MS demonstrates massive liver copper depletion in Wilson disease rats upon methanobactin treatment. Journal of Trace Elements in Medicine and Biology, 2018, 49: 119-127.

[74] Ackerman C M, Weber P K, Xiao T, Thai B, Kuo T J, Zhang E, Pett-Ridge J, Chang C J. Multimodal LA-ICP-MS and nanoSIMS imaging enables copper mapping within photoreceptor megamitochondria in a zebrafish model of Menkes disease. Metallomics, 2018, 10(3): 474-485.

[75] Muller J C, Horstmann M, Traeger L, Steinbicker A U, Sperling M, Karst U. Mu XRF and LA-ICP-TQMS for

quantitative bioimaging of iron in organ samples of a hemochromatosis model. Journal of Trace Elements in Medicine and Biology, 2019, 52: 166-175.

[76] Costas-Rodriguez M, Van Acker T, Hastuti A, Devisscher L, Van Campenhout S, Van Vlierberghe H, Vanhaecke F. Laser ablation-inductively coupled plasma-mass spectrometry for quantitative mapping of the copper distribution in liver tissue sections from mice with liver disease induced by common bile duct ligation. Journal of Analytical Atomic Spectrometry, 2017, 32(9): 1805-1812.

[77] Chery C C, Gunther D, Cornelis R, Vanhaecke F, Moens L. Detection of metals in proteins by means of polyacrylamide gel electrophoresis and laser ablation-inductively coupled plasma-mass spectrometry: Application to selenium. Electrophoresis, 2003, 24(19-20): 3305-3313.

[78] Becker J S, Zoriy M, Krause-Buchholz U, Becker J S, Pickhardt C, Przybylski M, Pompe W, Rodel G. In-gel screening of phosphorus and copper, zinc and iron in proteins of yeast mitochondria by LA-ICP-MS and identification of phosphorylated protein structures by MALDI-FT-ICR-MS after separation with two-dimensional gel electrophoresis. Journal of Analytical Atomic Spectrometry, 2004, 19(9): 1236-1243.

[79] Bandura D R, Ornatsky O, Liao L. Characterization of phosphorus content of biological samples by ICP-DRC-MS: Potential tool for cancer research. Journal of Analytical Atomic Spectrometry, 2004, 19(1): 96-100.

[80] Becker J S, Mounicou S, Zoriy M V, Becker J S, Lobinski R. Analysis of metal-binding proteins separated by non-denaturating gel electrophoresis using matrix-assisted laser desorption/ionization mass spectrometry (MALDI-MS) and laser ablation inductively coupled plasma mass spectrometry (LA-ICP-MS). Talanta, 2008, 76(5): 1183-1188.

[81] Polatajko A, Azzolini M, Feldmann I, Stuezel T, Jakubowski N. Laser ablation-ICP-MS assay development for detecting Cd-and Zn-binding proteins in Cd-exposed *Spinacia oleracea* L. Journal of Analytical Atomic Spectrometry, 2007, 22(8): 878-887.

[82] Xu M, Bijoux H, Gonzalez P, Mounicou S. Investigating the response of cuproproteins from oysters (*Crassostrea gigas*) after waterborne copper exposure by metallomic and proteomic approaches. Metallomics, 2014, 6(2): 338-346.

[83] Becker J S, Lobinski R, Becker J S. Metal imaging in non-denaturating 2D electrophoresis gels by laser ablation inductively coupled plasma mass spectrometry (LA-ICP-MS) for the detection of metalloproteins. Metallomics, 2009, 1(4): 312-316.

[84] Xu M, Frelon S, Simon O, Lobinski R, Mounicou S. Development of a non-denaturing 2D gel electrophoresis protocol for screening *in vivo* uranium-protein targets in *Procambarus clarkii* with laser ablation ICP MS followed by protein identification by HPLC-Orbitrap MS. Talanta, 2014, 128: 187-195.

[85] Ballihaut G, Pecheyran C, Mounicou S, Preud'homme H, Grimaud R, Lobinski R. Multimode detection (LA-ICP-MS, MALDI-MS and nanoHPLC-ESI-MS2) in 1D and 2D gel electrophoresis for selenium-containing proteins. TrAC-Trends in Analytical Chemistry, 2007, 26(3): 183-190.

[86] Ballihaut G, Claverie F, Pecheyran C, Mounicou S, Grimaud R, Lobinski R. Sensitive detection of selenoproteins in gel electrophoresis by high repetition rate femtosecond laser ablation-inductively coupled plasma mass spectrometry. Analytical Chemistry, 2007, 79(17): 6874-6880.

[87] Tastet L, Schaumloffel D, Lobinski R. ICP-MS-assisted proteomics approach to the identification of selenium-containing proteins in selenium-rich yeast. Journal of Analytical Atomic Spectrometry, 2008, 23(3): 309-317.

[88] Bierla K, Bianga J, Ouerdane L, Szpunar J, Yiannikouris A, Lobinski R. A comparative study of the Se/S substitution in methionine and cysteine in Se-enriched yeast using an inductively coupled plasma mass spectrometry (ICP MS)-assisted proteomics approach. Journal of Proteomics, 2013, 87: 26-39.

[89] Romaris-Hortas V, Bianga J, Moreda-Pineiro A, Bermejo-Barrera P, Szpunar J. Speciation of iodine-containing proteins

in *Nori* seaweed by gel electrophoresis laser ablation ICP-MS. Talanta, 2014, 127: 175-180.

［90］He Y, Esteban-Fernandez D, Neumann B, Bergmann U, Bierkandt F, Linscheid M W. Application of MeCAT-Click labeling for protein abundance characterization of *E. coli* after heat shock experiments. Journal of Proteomics, 2016, 136: 68-76.

［91］Hu S H, Zhang S C, Hu Z C, Xing Z, Zhang X R. Detection of multiple proteins on one spot by laser ablation inductively coupled plasma mass spectrometry and application to immuno-microarray with element-tagged antibodies. Analytical Chemistry, 2007, 79(3): 923-929.

［92］Waentig L, Jakubowski N, Hardt S, Scheler C, Roos P H, Linscheid M W. Comparison of different chelates for lanthanide labeling of antibodies and application in a Western blot immunoassay combined with detection by laser ablation (LA-)ICP-MS. Journal of analytical atomic spectrometry, 2012, 27(8). DOI: 10.1039/c2ja30068k.

［93］Roos P H, Venkatachalam A, Manz A, Waentig L, Koehler C U, Jakubowski N. Detection of electrophoretically separated cytochromes P450 by element-labelled monoclonal antibodies via laser ablation inductively coupled plasma mass spectrometry. Analytical and Bioanalytical Chemistry, 2008, 392(6): 1135-1147.

［94］Waentig L, Jakubowski N, Roos P H. Multi-parametric analysis of cytochrome P450 expression in rat liver microsomes by LA-ICP-MS. Journal of Analytical Atomic Spectrometry, 2011, 26(2): 310-319.

［95］Waentig L, Techritz S, Jakubowski N, Roos P H. A multi-parametric microarray for protein profiling: simultaneous analysis of 8 different cytochromes via differentially element tagged antibodies and laser ablation ICP-MS. Analyst, 2013, 138(21): 6309-6315.

［96］Kanje S, Herrmann A J, Hober S, Mueller L, Next generation of labeling reagents for quantitative and multiplexing immunoassays by the use of LA-ICP-MS. Analyst, 2016, 141(23): 6374-6380.

［97］Rodriguez-Menendez S, Fernandez B, Garcia M, Alvarez L, Fernandez M L, Sanz-Medel A, Coca-Prados M, Pereiro R, Gonzalez-Iglesias H. Quantitative study of zinc and metallothioneins in the human retina and RPE cells by mass spectrometry-based methodologies. Talanta, 2018, 178: 222-230.

［98］Giesen C, Mairinger T, Khoury L, Waentig L, Jakubowski N, Panne U. Multiplexed immunohistochemical detection of tumor markers in breast cancer tissue using laser ablation inductively coupled plasma mass spectrometry. Analytical Chemistry, 2011, 83(21): 8177-8183.

［99］Cruz-Alonso M, Fernandez B, Alvarez L, Gonzalez-Iglesias H, Traub H, Jakubowski N, Pereiro R. Bioimaging of metallothioneins in ocular tissue sections by laser ablation-ICP-MS using bioconjugated gold nanoclusters as specific tags. Microchimica Acta, 2017, 185(1): 64.

［100］Cruz-Alonso M, Fernandez B, Garcia M, Gonzalez-Iglesias H, Pereiro R. Quantitative imaging of specific proteins in the human retina by laser ablation ICP MS using bioconjugated metal nanoclusters as labels. Analytical Chemistry, 2018, 90(20): 12145-12151.

［101］Cruz-Alonso M, Fernandez B, Navarro A, Junceda S, Astudillo A, Pereiro R. Laser ablation ICP-MS for simultaneous quantitative imaging of iron and ferroportin in hippocampus of human brain tissues with Alzheimer's disease. Talanta, 2019, 197: 413-421.

［102］De Vega R G, Clases D, Ferndndez-Sdnchez M L, Eiro N, Gonzalez L O, Vizoso F J, Doble P A. Sanz-Medel A. MMP-11 as a biomarker for metastatic breast cancer by immunohistochemical-assisted imaging mass spectrometry. Analytical and Bioanalytical Chemistry, 2019, 411(3): 639-646.

［103］Zhang X, Liu R, Yuan Q, Gao F, Li J, Zhang Y, Zhao Y, Chai Z, Gao L, Gao X. The precise diagnosis of cancer invasion/metastasis via 2D laser ablation mass mapping of metalloproteinase in primary cancer tissue. ACS Nano, 2018, 12(11): 11139-11151.

[104] Wang M, Zheng L N, Wang B, Chen H Q, Zhao Y L, Chai Z F, Reid H J, Sharp B L, Feng W Y. Quantitative analysis of gold nanoparticles in single cells by laser ablation inductively coupled plasma-mass spectrometry. Analytical Chemistry, 2014, 86(20): 10252-10256.

[105] Zhai J, Wang Y L, Xu C, Zheng L N, Wang M, Feng W Y, Gao L, Zhao L N, Liu R, Gao F P, Zhao Y L, Chai Z F, Gao X Y. Facile approach to observe and quantify the α$_{IIb}$β$_3$ integrin on a single-cell. Analytical Chemistry, 2015, 87(5): 2546-2549.

[106] Lohr K, Borovinskaya O, Tourniaire G, Panne U, Jakubowski N. Arraying of single cells for quantitative high throughput laser ablation ICP-TOF-MS. Analytical Chemistry, 2019, 91(18): 11520-11528.

[107] Zheng L N, Sang Y B, Luo R P, Wang B, Yi F T, Wang M, Feng W Y. Determination of silver nanoparticles in single cells by microwell trapping and laser ablation ICP-MS determination. Journal of Analytical Atomic Spectrometry, 2019, 34(5): 915-921.

[108] Zhang X Y, Zheng L N, Wang H L, Shi J W, Feng W Y, Li L, Wang M. Elemental bio-imaging of biological samples by laser ablation-inductively coupled plasma-mass spectrometry. Chinese Journal of Analytical Chemistry, 2016, 44(11): 1646-1651.

[109] Van Malderen S J M, Vergucht E, De Rijcke M, Janssen C, Vincze L, Vanhaecke F. Quantitative determination and subcellular imaging of Cu in single cells via laser ablation-ICP-Mass spectrometry using high-density microarray gelatin standards. Analytical Chemistry, 2016, 88(11): 5783-5789.

[110] Costas-Rodríguez M, Colina-Vegas L, Solovyev N, De Wever O, Vanhaecke F. Cellular and sub-cellular Cu isotope fractionation in the human neuroblastoma SH-SY5Y cell line: Proliferating versus neuron-like cells. Analytical and Bioanalytical Chemistry, 2019, 411(19): 4963-4971.

[111] Rodriguez-Menendez S, Fernández B, Gonzalez-Iglesias H, Garcia M, Alvarez L, Alonso J I G, Pereiro R. Isotopically enriched tracers and inductively coupled plasma mass spectrometry methodologies to study zinc supplementation in single-cells of retinal pigment epithelium *in vitro*. Analytical Chemistry, 2019, 91(7): 4488-4495.

[112] Theiner S, Schweikert A, Van Malderen S J M, Schoeberl A, Neumayer S, Jilma P, Peyrl A, Koellensperger G. Laser ablation-inductively coupled plasma time-of-flight mass spectrometry imaging of trace elements at the single-cell level for clinical practice. Analytical Chemistry, 2019, 91(13): 8207-8212.

[113] Herrmann A J, Techritz S, Jakubowski N, Haase A, Luch A, Panne U, Mueller L. A simple metal staining procedure for identification and visualization of single cells by LA-ICP-MS. Analyst, 2017, 142(10): 1703-1710.

[114] Lohr K, Traub H, Wanka A J, Panne U, Jakubowski N. Quantification of metals in single cells by LA-ICP-MS: Comparison of single spot analysis and imaging. Journal of Analytical Atomic Spectrometry, 2018, 33(9): 1579-1587.

[115] Mueller L, Herrmann A J, Techritz S, Panne U, Jakubowski N. Quantitative characterization of single cells by use of immunocytochemistry combined with multiplex LA-ICP-MS. Analytical and Bioanalytical Chemistry, 2017, 409(14): 3667-3676.

[116] Van Malderen S J M, Van Acker T, Laforce B, De Bruyne M, De Rycke R, Asaoka T, Vincze L, Vanhaecke F. Three-dimensional reconstruction of the distribution of elemental tags in single cells using laser ablation ICP-mass spectrometry via registration approaches. Analytical and Bioanalytical Chemistry, 2019, 411(19): 4849-4859.

[117] Buckle T, Van Der Wal S, Van Malderen S J M, Muller L, Kuil J, Van Unen V, Peters R J B, Van Bemmel M E M, Mcdonnell L A, Velders A H, Koning F, Vanhaecke F, Van Leeuwen F W B. Hybrid imaging labels: Providing the link between mass spectrometry-based molecular pathology and theranostics. Theranostics, 2017, 7(3): 624-633.

[118] Van Acker T, Buckle T, Van Malderen S J M, Van Willigen D M, Van Unen V, Van Leeuwen F W B, Vanhaecke F. High-resolution imaging and single-cell analysis via laser ablation-inductively coupled plasma-mass spectrometry for

the determination of membranous receptor expression levels in breast cancer cell lines using receptor-specific hybrid tracers. Analytica Chimica Acta, 2019, 1074: 43-53.

[119] Managh A J, Edwards S L, Bushell A, Wood K J, Geissler E K, Hutchinson J A, Hutchinson R W, Reid H J, Sharp B L. Single cell tracking of gadolinium labeled $CD4^+T$ cells by laser ablation inductively coupled plasma mass spectrometry. Analytical Chemistry, 2013, 85(22): 10627-10634.

[120] Reifschneider O, Wentker K S, Strobel K, Schmidt R, Masthoff M, Sperling M, Faber C, Karst U. Elemental bioimaging of thulium in mouse tissues by laser ablation-ICPMS as a complementary method to heteronuclear proton magnetic resonance imaging for cell tracking experiments. Analytical Chemistry, 2015, 87(8): 4225-4230.

[121] Nakada N, Kuroki Y. Cell tracking of chromium-labeled mesenchymal stem cells using laser ablation inductively coupled plasma imaging mass spectrometry. Rapid Commun Mass Spectrom, 2019, 33(20): 1565-1570.

[122] Drescher D, Giesen C, Traub H, Panne U, Kneipp J, Jakubowski N. Quantitative imaging of gold and silver nanoparticles in single eukaryotic cells by laser ablation ICP-MS. Analytical Chemistry, 2012, 84(22): 9684-9688.

[123] Drescher D, Zeise I, Traub H, Guttmann P, Seifert S, Buchner T, Jakubowski N, Schneider G, Kneipp J. In situ characterization of SiO_2 nanoparticle biointeractions using brightsilica. Advanced Functional Materials, 2014, 24(24): 3765-3775.

[124] Drescher D, Traub H, Buchner T, Jakubowski N, Kneipp J. Properties of in situ generated gold nanoparticles in the cellular context. Nanoscale, 2017, 9(32): 11647-11656.

[125] Arakawa A, Jakubowski N, Flemig S, Koellensperger G, Rusz M, Iwahata D, Traub H, Hirata T. High-resolution laser ablation inductively coupled plasma mass spectrometry used to study transport of metallic nanoparticles through collagen-rich microstructures in fibroblast multicellular spheroids. Analytical and Bioanalytical Chemistry, 2019, 411(16): 3497-3506.

[126] Arakawa A, Jakubowski N, Koellensperger G, Theiner S, Schweikert A, Flemig S, Iwahata D, Traub H, Hirata T. Imaging of Ag NP transport through collagen-rich microstructures in fibroblast multicellular spheroids by high-resolution laser ablation inductively coupled plasma time-of-flight mass spectrometry. Analyst, 2019, 144(16): 4935-4942.

[127] Arakawa A, Jakubowski N, Koellensperger G, Theiner S, Schweikert A, Flemig S, Iwahata D, Traub H, Hirata T. Quantitative imaging of silver nanoparticles and essential elements in thin sections of fibroblast multicellular spheroids by high resolution laser ablation inductively coupled plasma time-of-flight mass spectrometry. Analytical Chemistry, 2019, 91(15): 10197-10203.

[128] Kostiv U, Rajsiglova L, Luptakova D, Pluhacek T, Vannucci L, Havlicek V, Engstova H, Jirak D, Slouf M, Makovicky P, Sedlacek R, Horak D. Biodistribution of upconversion/magnetic silica-coated $NaGdF_4$：Yb^{3+}/Er^{3+} nanoparticles in mouse models. RSC Advances, 2017, 7(73): 45997-46006.

[129] Li Q, Wang Z, Chen Y, Zhang G. Elemental bio-imaging of PEGylated $NaYF_4$：Yb/Tm/Gd upconversion nanoparticles in mice by laser ablation inductively coupled plasma mass spectrometry to study toxic side effects on the spleen, liver and kidneys. Metallomics, 2017, 9(8): 1150-1156.

[130] Khabir Z, Guller A E, Rozova V S, Liang L E, Lai Y J, Goldys E M, Hu H H, Vickery K, Zvyagin A V. Tracing upconversion nanoparticle penetration in human skin. Colloids and Surfaces B: Biointerfaces, 2019, 184: 11.

[131] Buckley A, Warren J, Hodgson A, Marczylo T, Ignatyev K, Guo C, Smith R. Slow lung clearance and limited translocation of four sizes of inhaled iridium nanoparticles. Particle and Fibre Toxicology, 2017, 14(1): 5.

[132] Ko J A, Furuta N, Lim H B. New approach for mapping and physiological test of silica nanoparticles accumulated in sweet basil (*Ocimum basilicum*) by LA-ICP-MS. Analytica Chimica Acta, 2019, 1069: 28-35.

[133] Elci S G, Jiang Y, Yan B, Kim S T, Saha K, Moyano D F, Tonga G Y, Jackson L C, Rotello V M, Vachet R W. Surface charge controls the suborgan biodistributions of gold nanoparticles. ACS Nano, 2016, 10(5): 5536-5542.

[134] Wiemann M, Vennemann A, Blaske F, Sperling M, Karst U. Silver nanoparticles in the lung: Toxic effects and focal accumulation of silver in remote organs. Nanomaterials, 2017, 7(12): 441.

[135] Bishop D P, Grossgarten M, Dietrich D, Vennemann A, Cole N, Sperling M, Wiemann M, Doble P A, Karst U, Quantitative imaging of translocated silver following nanoparticle exposure by laser ablation-inductively coupled plasma-mass spectrometry. Analytical Methods, 2018, 10(8): 836-840.

[136] Reifschneider O, Vennemann A, Buzanich G, Radtke M, Reinholz U, Riesemeier H, Hogeback J, Koppen C, Grossgarten M, Sperling M, Wiemann M, Karst U. Revealing silver nanoparticle uptake by macrophages using SR-μXRF and LA-ICP-MS. Chemical Research in Toxicology, 2020, 33(5): 1250-1255.

[137] Masthoff M, Buchholz R, Beuker A, Wachsmuth L, Kraupner A, Albers F, Freppon F, Helfen A, Gerwing M, Holtke C, Hansen U, Rehkamper J, Vielhaber T, Heindel W, Eisenblatter M, Karst U, Wildgruber M, Faber C. Introducing specificity to iron oxide nanoparticle imaging by combining ^{57}Fe-based MRI and mass spectrometry. Nano Letters, 2019, 19(11): 7908-7917.

[138] Chacon-Madrid K, Arruda M A Z. Internal standard evaluation for bioimaging soybean leaves through laser ablation inductively coupled plasma mass spectrometry: A plant nanotechnology approach. Journal of Analytical Atomic Spectrometry, 2018, 33(10): 1720-1728.

[139] Bohme S, Stark H J, Kühnel D, Reemtsma T. Exploring LA-ICP-MS as a quantitative imaging technique to study nanoparticle uptake in *Daphnia magna* and zebrafish (*Danio rerio*) embryos. Analytical and Bioanalytical Chemistry, 2015, 407(18): 5477-5485.

[140] Bohme S, Baccaro M, Schmidt M, Potthoff A, Stark H J, Reemtsma T, Kuhnel D. Metal uptake and distribution in the zebrafish (*Danio rerio*) embryo: Differences between nanoparticles and metal ions. Environmental Science: Nano, 2017, 4(5): 1005-1015.

[141] Zhang T, Liu Q, Wang W, Huang X, Wang D, He Y, Liu J, Jiang G. Metallic fingerprints of carbon: Label-free tracking and imaging of graphene in plants. Analytical Chemistry, 2020, 92(2): 1948-1955.

[142] Liang S S, Wang B, Li X, Chu R X, Yu H Y, Zhou S, Wang M, Chen H Q, Zheng L N, Chai Z F, Feng W Y. *In vivo* pharmacokinetics, transfer and clearance study of graphene oxide by La/Ce dual elemental labelling method. Nanoimpact, 2020, 17.

[143] Dai H, Liu Y, Wang J J, Nie Y G, Sun Y X, Wang M D, Wang D Y, Yang Z, Cheng L, Wang J, Weng J, Wang Q Q, Wang F Y, Wu L J, Zhao G P, Xu A. Graphene oxide antagonizes the toxic response to arsenic via activation of protective autophagy and suppression of the arsenic-binding protein LEC-1 in *Caenorhabditis elegans*. Environmental Science: Nano, 2018, 5(7): 1711-1728.

[144] Metarapi D, Sala M, Vogel-Mikus K, Selih V S, Van Elteren J T. Nanoparticle analysis in biomaterials using laser ablation-single particle-inductively coupled plasma mass spectrometry. Analytical Chemistry, 2019, 91(9): 6200-6205.

[145] Li Q, Wang Z, Mo J M, Zhang G X, Chen Y R, Huang C C. Imaging gold nanoparticles in mouse liver by laser ablation inductively coupled plasma mass spectrometry. Scientific Reports, 2017, 7(1): 2965.

第 22 章　原子光谱技术在纳米材料分析中的应用

(谭志强[①]，白庆胜[①]，陈　强[①]，刘岩婉晶[②]，刘景富[①]*，江桂斌[①])

▶ 22.1　不同介质中纳米材料的识别和定量 / 622

▶ 22.2　尺寸表征 / 628

▶ 22.3　表面性质分析 / 632

▶ 22.4　生物体内成像 / 633

▶ 22.5　形态分析 / 634

▶ 22.6　展望 / 638

---

①中国科学院生态环境研究中心，北京；②国科大杭州高等研究院，杭州。*通讯作者联系方式：　jfliu@rcees.ac.cn

本章导读

- 总结原子光谱技术在不同基质样品中纳米材料识别和定量中的应用。
- 阐述原子光谱技术对纳米材料尺寸表征的基本原理。
- 简要介绍原子光谱技术在纳米材料表面荷电、修饰剂等分析中的应用；重点介绍质谱成像技术在生物体内纳米材料分析中的应用；介绍原子光谱技术在金属纳米材料形态分析中的应用，并着重介绍在线联用技术的发展和应用。
- 展望原子光谱技术在纳米材料分析应用的发展方向。

自 20 世纪末，纳米材料逐渐成为科学研究的热点。纳米材料是指在三维空间中至少有一维处于纳米尺寸（1～100 nm）的新型材料，具有优异的光学、力学、电学、磁学、热学等方面性能，已经广泛应用于能源、化学、生物学、物理学、环境学、医学、食品等领域[1]。与此同时，纳米材料在生产、加工、使用和处置过程中，不可避免地进入到水、土壤、大气等环境介质和生物体中并参与整个生态系统的循环[2]，由此引发的环境和健康风险问题已经引起科学界的广泛关注。因此，建立和发展可靠的纳米材料分析表征技术是纳米材料精准合成制备、安全应用以及环境与健康风险评估等研究的必要前提。

原子光谱技术作为传统分析表征技术的重要组成，在绝大部分金属元素和少数非金属元素的准确定性和定量分析中具有举足轻重的地位。随着现代科技的发展，原子光谱技术的分析性能有了大幅度提高，同时其分析对象也逐渐扩展到纳米材料领域。目前，原子光谱技术可为不同介质中纳米材料的识别和定量、尺寸表征、表面性质分析、生物体内成像以及形态分析等前沿研究提供可靠技术支持，极大地推动了相关研究领域的发展。

22.1 不同介质中纳米材料的识别和定量

22.1.1 环境介质

水环境基质相对简单，其中的纳米材料浓度水平也普遍较低，通常在定量分析前需要引入分离、富集等前处理技术。常见的金属纳米颗粒的前处理技术有超滤、离心、萃取等。其中，萃取技术操作简便，无需特殊仪器设备辅助，非常易于在常规实验室推广和普及。与传统的液液萃取技术相比，分散液液微萃取技术（DLLME）使用有机溶剂的量大大降低，且具有操作简单、成本低、富集倍数高、萃取效率高等特点。Liu 等采用表面活性剂 Triton X-114（TX-114）作分散剂，将 DLLME 与电热蒸发-电感耦合等离子体质谱（ETV-ICP-MS）联用，建立了环境水样中痕量金纳米颗粒的定量方法[3]。在萃取过程中，硫代硫酸根作为金离子的掩蔽试剂，有效消除了其对金纳米颗粒萃取的干扰。这种萃取方法对粒径在 17～108 nm 范围内不同修饰剂（如柠檬酸、巯基十一烷基酸、巯基丁二酸、聚乙烯吡咯烷酮）的金纳米颗粒均具有很高的富集倍数。收集的萃取相可直接进样于 ETV-ICP-MS 进行检测，该方法对金纳米颗粒的检出限为 2.2 ng/L，被成功应用于河水、湖水、自来水等实际水样中金纳米颗粒的识别和定量。类似地，Selahle 等使用基于 TX-114 的 DLLME 有效分离和富集了工业废水样品中的二氧化钛和氧化锌两种纳米颗粒，并通过电感耦合等离子体发射光谱（ICP-OES）对富集相中的这两种纳米颗粒进行识别和定量[4]。以上研究表明，DLLME 在富集纳米材料的同时，还可以有效去除基质干扰，是一种适用于不同环境基质水体中纳米材料分离富集的有效手段。

与液液微萃取技术相比，浊点萃取（CPE）不使用任何挥发性的有毒有机试剂，是一种更加绿色的分离技术。该技术主要是利用表面活性剂胶束的增溶作用和浊点现象达到分离富集的目的。Liu等首次发现使用TX-114作萃取剂能够可逆地浓缩/分离纳米材料，而且纳米材料的粒径大小和形貌在相转移和富集后的TX-114相中均不发生变化[5]。在此基础上，该课题组建立了基于CPE从环境水样中分离和富集痕量银纳米颗粒的方法[6]。研究表明，在银纳米颗粒等电点的pH附近可获得较高的萃取效率，而且加入一定浓度的盐（如35 mmol/L 硝酸钠）可有效促进相分离并显著提高萃取效率。此外，通过引入硫代硫酸根与基质中的银离子发生络合反应，所产生的络合物难以进入TX-114富集相，从而有效消除银离子对富集相中银纳米颗粒定量分析的干扰。透射电镜（TEM）结果表明，萃取前聚乙烯吡咯烷酮修饰的银纳米颗粒的粒径范围为9～73 nm[图22-1（a）]，商品银纳米颗粒的粒径分布范围为6～58 nm[图22-1（c）]；萃取后TX-114中聚乙烯吡咯烷酮修饰的银纳米颗粒的粒径范围为9～94 nm[图22-1（b）]，商品银纳米颗粒的粒径分布范围为5～47 nm[图22-1（d）]，表明萃取前后的两种银纳米颗粒粒径分布范围基本一致。TX-114富集相中的银纳米颗粒经微波消解后采用ICP-MS进行定量，方法的检出限为0.006 μg/L，相当于34.3 fmol/L 银纳米颗粒。由于CPE能够保持银纳米颗粒的粒径大小和形貌，该课题组采用该方法对腐殖酸存在下银离子生成的银纳米颗粒进行了萃取和富集，并采用ICP-MS对富集的银纳米颗粒进行识别和定量，初步揭示了银纳米颗粒天然来源的可能途径，为分析和追溯环境中其他纳米材料来源提供了新思路[7]。接着，该课题组选取富同位素 ^{107}Ag 的银纳米颗粒和富同位素 ^{109}Ag 的银离子，采用同位素示踪方法进一步研究了在环境相关条件下银纳米颗粒与银离子之间的相互转化过程[8]。研究发现，在简单的银纳米颗粒水溶液中，在没有还原性物质加入情况下，银纳米颗粒主要发生氧化反应。当有腐殖酸存在时，银离子可被还原生成银纳米颗粒，并同时伴随银纳米颗粒的氧化。因此，在实际水环境中银纳米颗粒的氧化和还原过程共存，二者处于动态平衡状态。

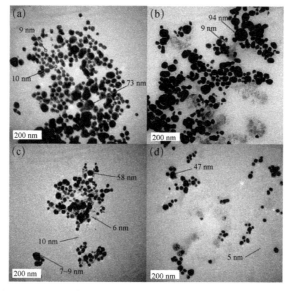

图 22-1　CPE 萃取前后银纳米颗粒粒径分布的 TEM 图[6]

(a) 萃取前聚乙烯吡咯烷酮修饰的银纳米颗粒；(b) 萃取后 TX-114 相中聚乙烯吡咯烷酮修饰的银纳米颗粒；
(c) 萃取前的商品银纳米颗粒；(d) 萃取后 TX-114 相中的商品银纳米颗粒

固相萃取技术包括液相和固相的物理萃取过程。在萃取过程中，固相萃取剂可选择性地吸附目标分析物，最后用较小体积的合适溶剂把分析物洗脱下来，从而达到分离、净化和富集的目的。基于磷酸与钛离子之间的强亲和力，Wang 等制备了可选择性富集二氧化钛纳米颗粒的磷酸功能化超顺磁性氧化铁复合材料固相萃取剂[9]。该课题组利用色散校正密度泛函理论,深入探讨了这种复合材料对二氧化钛纳米颗粒和钛离子吸附能力差异的原因。这种复合材料对二氧化钛纳米颗粒的富集倍数高达400,吸附容量为462 mg/g(25 nm 二氧化钛纳米颗粒)。采用ICP-MS对富集的二氧化钛纳米颗粒进行定量,检出限为17 ng/L。该方法为评估实际水环境中二氧化钛纳米颗粒的浓度水平提供了可靠的定量方法。另外,该课题组还制备了聚丙烯酰胺-乙烯基吡啶-甲叉双丙烯酰胺整体柱用来分离和富集金纳米颗粒,并将其与ICP-MS在线联用实现金纳米颗粒的识别和定量(如图22-2所示)[10]。在该聚合物整体柱中,金纳米颗粒表面修饰剂柠檬酸和柱床表面吡啶/酰胺基团通过静电作用和氢键作用结合,实现了对金纳米颗粒的分离富集。该整体柱具有高选择性和强抗基质干扰能力,且在萃取过程中可以保持金纳米颗粒的原始形貌。对于3 nm柠檬酸修饰的金纳米颗粒,该方法的检出限为1.17 fmol/L,样品通量为6 h^{-1}。本方法具有检出限低、样品消耗体积小、萃取/解吸动力学快、线性范围宽、样品通量大且无需消解等优点,并成功应用于实际环境水样中金纳米粒子的分析(加标回收率:77%～103%)。为了进一步简化固相萃取技术的前处理操作,Zhou等以聚偏二氟乙烯微孔滤膜作为固相萃取圆盘,建立了环境水样中痕量银纳米颗粒的快速、高倍富集分离方法[11]。在本方法中,孔径为0.45 μm的聚偏二氟乙烯微孔滤膜可选择性地萃取银纳米颗粒,且当银离子浓度在银纳米颗粒含量的1～10倍范围内,共存的银离子对银纳米颗粒的萃取无显著影响。采用2%(m/V)FL-70水溶液为洗脱剂和ICP-MS识别和定量,该方法对银纳米颗粒的富集倍数和检出限分别为1000 ng/L 和 0.2 ng/L。萃取前后银纳米颗粒的形貌和粒径分布基本保持不变,因此本方法可望在环境水体中银纳米颗粒的归趋、毒理效应和机制研究中发挥重要作用。

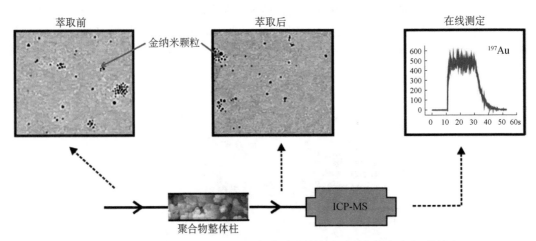

图 22-2　聚合物整体柱与ICP-MS在线联用系统识别和定量金纳米颗粒[10]

近年来,微固相萃取作为一种快速高效的分离富集手段受到分析工作者的关注。萃取材料在微固相萃取中起着主导作用,对萃取效率和选择性均有较大的影响。Zhou等发现采用水热法合成的磁性四氧化三铁颗粒经老化后作为萃取材料,可选择性萃取银、氯化银、硫化银等含银纳米颗粒,而不萃取银离子[12]。对吸附在磁性颗粒表面的含银纳米颗粒,可采用分步洗脱后分别定量测定。先用2%(V/V)醋酸选择性洗脱银纳米颗粒和氯化银纳米颗粒,再用含10 mmol/L 硫脲的2%(V/V)醋酸混合溶液洗脱硫化银纳米颗粒,最后采用ICP-MS分别测定。该方法对含银纳米颗粒的检出限为0.007～0.035 μg/L,为研究银纳米颗粒和银离子的硫化过程提供了可靠的技术支持。

除了金属纳米颗粒外,水环境中还存在低浓度的非金属纳米颗粒,如碳质纳米颗粒。然而,由于环境水体中广泛存在的天然有机质的干扰以及传统原子光谱仪对碳元素的灵敏度普遍较低,使得环境水体中低浓度碳质纳米颗粒(如氧化石墨烯)的识别和定量方法非常匮乏。Tan等基于光诱导与尖端放电-原子发射光谱（PD-OES）在线联用系统（如图22-3所示），构建了水体中低浓度氧化石墨烯的识别和定量分析平台。该分析平台被成功用于实时监控模拟日光下低浓度（如0.2 mg/L C）氧化石墨烯的光化学降解过程，为揭示实际水环境中痕量浓度氧化石墨烯的转化和归趋规律提供了技术支撑[13]。

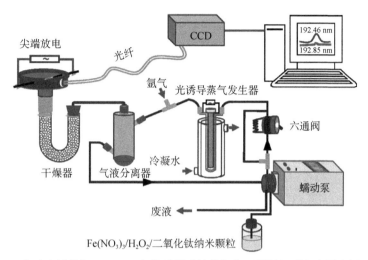

图 22-3　基于光诱导与PD-OES在线联用系统的氧化石墨烯识别和定量分析平台[13]

与环境水体相比，土壤和沉积物中纳米材料的含量相对较高，但其基质比较复杂，且其中的纳米材料容易发生异质团聚，在矿物表面有明显的吸附趋势，因此对这些介质中纳米材料识别和定量前往往需要引入合适的样品前处理技术将其与土壤基质分离。虽然已有研究者尝试使用超声提取、CPE等操作从土壤和沉积物介质中预分离纳米材料，但操作比较烦琐，而且还可能会改变纳米材料的原始形貌。如图22-4所示，Li等使用焦磷酸钠作提取剂，有效实现了银纳米颗粒和土壤胶体的分离，并最大限度地保持了银纳米颗粒的原始形貌[14]。但是，在后续离心和过滤过程中，银纳米颗粒会不可避免地损失，而且由于单颗粒(SP)-ICP-MS定量方法粒径检出限限制的原因，该方法不能提供尺寸小于30 nm的银纳米颗粒的浓度信息。为了深入理解纳米材料在土壤中的吸附过程，Torrent等[15]使用全反射X射线荧光光谱法（TXRF）对银纳米颗粒在土壤中的吸附动力学进行了原位分析。研究发现，聚乙烯吡咯烷酮修饰的银纳米颗粒在6小时内即可完全被土壤吸附。TXRF不仅可对土壤悬浮液中的银纳米颗粒进行识别和直接定量，还可提供土壤中其他多元素组成和含量的基本信息，为深入研究金属纳米颗粒与土壤或沉积物中其他共存物的相互作用提供了可靠分析方法。

沙尘是大气细颗粒的重要组分，其主要成分为二氧化硅纳米颗粒。同时，二氧化硅纳米颗粒也是人类生产和使用量最大的一类纳米材料。因此，天然源和人为源的二氧化硅纳米颗粒在实际环境中共存，而识别不同来源的二氧化硅纳米颗粒将有助于推动大气细颗粒的溯源研究。多接收等离子体质谱仪（MC-ICP-MS）可对各类样品中的金属稳定同位素进行高准确度、高灵敏度和高精密度的分析，被尝试用于识别和定量水环境中天然源和人为源的银纳米颗粒[16]。近期，Yang等利用MC-ICP-MS获取了不同来源二氧化硅纳米颗粒二维同位素指纹图谱，发现它们的硅同位素和氧同位素组成上均表现出一定的差异性[图22-5（a）和（b）]，根据这些同位素指纹信息可判断二氧化硅纳米颗粒的来源[17]。另外，该课题组利用机器学习对样品的同位素指纹信息构建了分类模型[图22-5（c）]，得到了每个样品的不同来源的概率值，总体正确率达到了93.3%，从而实现了二氧化硅纳米颗粒的定量及精准溯源。为识别和定量大气中的四氧化三铁纳米颗粒，该课题组使用碱加热消解对细颗粒石英滤膜样品进行完全消解，并利用3D打印技术设计和制造了一种自动循环磁性分离装置对消解后的样品进行高效地分离提取，随后使用乙酸溶液对收集的磁性组分进行纯化，得到纯净的四氧化三铁纳米颗粒成分[18]。收集的四氧化三铁纳米颗粒经过微波消解后，采用ICP-MS定量铁元素的含量即实

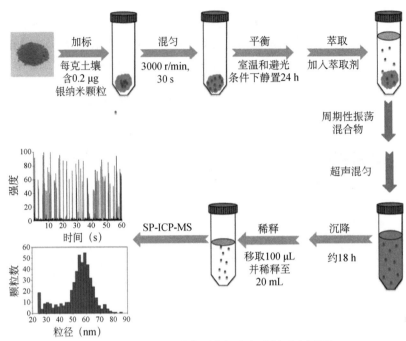

图 22-4 土壤中银纳米颗粒提取和分析示意图[14]

现对四氧化三铁纳米颗粒的准确定量。对四氧化三铁纳米颗粒的公路交通源和非公路交通源贡献的估算结果表明，公路交通源和非公路交通源对空气中四氧化三铁纳米颗粒的贡献相当。最后，根据大气中四氧化三铁纳米颗粒的浓度和粒径分布，结合人体暴露评估模型，他们估算了北京地区成年人通过呼吸作用对四氧化三铁纳米颗粒的日均暴露量。该研究为科学评估空气中四氧化三铁纳米颗粒的健康风险提供了重要科学数据和方法学支撑。

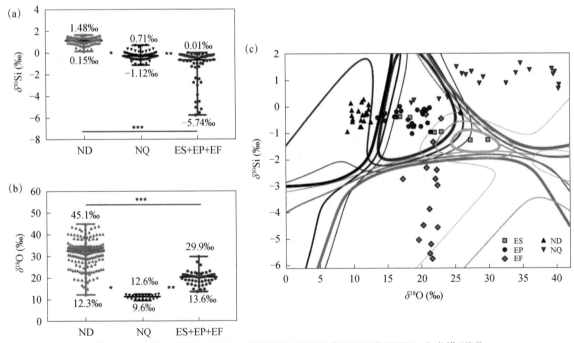

图 22-5 不同来源的二氧化硅纳米颗粒的同位素指纹及机器学习分类模型[17]

（a）硅同位素指纹；（b）氧同位素指纹；（c）硅-氧二维同位素指纹及机器学习分类模型。ND：天然硅藻土；NQ：天然石英；ES：人造溶胶凝胶法二氧化硅纳米颗粒；EP：人造沉淀法二氧化硅纳米颗粒；EF：人造气相法二氧化硅纳米颗粒

22.1.2 生物基质

纳米材料在环境介质中不可避免的释放，大大提高了其进入生物体的可能性。纳米材料的浓度是影响其生物毒性的最主要因素。Gray 等采用四甲基氢氧化铵对大型蚤组织进行了提取，然后采用 SP-ICP-MS 对其中的金和银两种纳米颗粒进行了识别和定量[19]。该方法对鱼组织和水中 100 nm 金和银纳米颗粒的分析结果基本一致（如图 22-6 所示），对生物样品的加标回收率为 83%～121%，表明该方法非常适用于生物基质中纳米材料的定量。近年来，越来越多的毒理学研究选取无脊椎动物（例如秀丽隐杆线虫）做模型生物。Johnson 等[2]采用 SP-ICP-MS 研究了秀丽隐杆线虫对 30 nm 和 60 nm 两种尺寸金纳米颗粒的摄入过程。研究发现，四甲基氢氧化铵处理前后金纳米颗粒的粒径分布没有发生明显变化，但在线虫体内的金纳米颗粒粒径分布发生了明显改变，可能因为金纳米颗粒发生了氧化或溶解。细胞中纳米材料的定量和识别是评价其细胞毒性的必要前提。Yu 等采用基于 TX-114 的 CPE 方法，从 HepG2 细胞中成功分离出银纳米颗粒和银离子。收集的银纳米颗粒和银离子经微波消解后，采用 ICP-MS 进行定量[20]。采用该方法，进一步研究了银纳米颗粒在 HepG2 细胞中的摄入过程。细胞经过 10 mg/L 银纳米颗粒暴露 24 h 后，每 10^4 个细胞中约含有 67.8 ng 的 Ag，其中约 10.3%为银离子。该方法为研究银纳米颗粒在生物介质中的转化及毒性提供了可靠的定量手段。

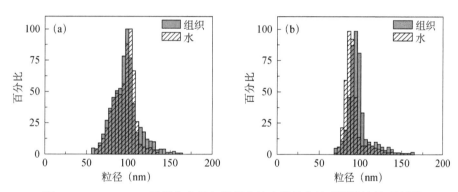

图 22-6　SP-ICP-MS 识别和定量鱼组织和纯水样品中纳米颗粒结果对比[19]

(a) 98 μg/L 金纳米颗粒；(b) 4.8 μg/L 银纳米颗粒

与 ICP-MS 相比，时间飞行质谱（TOF-MS）的分辨率更高，可实现高含量基体中微量和痕量成分的分析，其精密度和检出限都有了很大的改善。Soria 等使用液相色谱-四极杆飞行时间质谱（LC-QTOF-MS）联用系统研究了柠檬酸和聚乙烯吡咯烷酮修饰的两种银纳米颗粒对拟南芥植物生长的影响[21]。由 QTOF-MS 提供的代谢物谱图可知，银纳米颗粒可从根向叶和花芽移动。此外，非靶向和靶向代谢组学结果均表明，处理后的植物代谢组发生了明显变化，而且银处理过的拟南芥中积累了特定的植物鞘氨醇和 N-酰基乙醇胺。综合拟南芥中的银浓度与代谢产物分布变化信息，可以更好地理解纳米材料与生物体之间的相互作用。类似地，为了阐明植物对纳米材料的吸收过程，Wagener 等将小麦、甘蓝型油菜、大麦暴露于经银和二氧化铈纳米颗粒处理后的污泥中，采用飞行时间二次离子质谱（TOF-SIMS）对植物组织中的纳米颗粒进行了识别和定量[22]。他们在一些组织中观察到了大尺寸银和二氧化铈的团聚体，如在韧皮部中出现的这些团聚体颗粒表明纳米颗粒可以向上运输到植物其他部位。

激光剥蚀(LA)-ICP-MS 可提供元素在待测样品中的二维和三维元素分布图，为细胞中纳米材料的分析提供了可靠手段。Wang 等基于 LA-ICP-MS 建立了细胞内金纳米颗粒的识别和定量方法[23]。采用这种方法，得到的单个细胞中金元素的平均含量与消解细胞所得结果一致（如图 22-7 所示）。该方法对细胞中金的检出限为 1.5 fg，为单细胞金属组学研究提供了高灵敏的定量工具，也为评价金属纳米材料的生物效应及安全性提供了可靠分析方法。

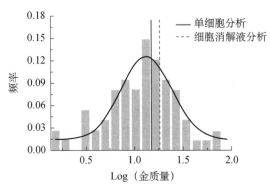

图 22-7　不同方法定量细胞内金纳米颗粒的比较[23]

与 LA-ICP-MS 类似，TXRF 无需复杂的样品前处理过程，可实现多元素同时分析，而且对于复杂基体的样品可单独进行校准和定量分析。Fernandez-Ruiz 等[24]首次将 TXRF 用于评估给药小鼠静脉内不同时间的金纳米棒的生物积累和清除。他们分析了在不同时间、小鼠主要器官（例如肝脏、脾脏、脑和肺等）中的金纳米棒生物蓄积动力学，并同时分析了通过尿液排泄途径清除金纳米棒的动力学。研究发现，金纳米棒在肝脏和脾脏中迅速积累，而在所考察的时间（如 1 个月）内并未在脑和肺中积累。另外，尿液分析结果未显示金纳米棒的生物积累现象。该研究说明，TXRF 是一种功能强大、用途广泛、精确度高的复杂生物基质中纳米材料定量分析技术，有望用于生物体内纳米材料的代谢行为研究。

22.1.3　食品基质

纳米材料的独特性能推动了其在食品领域中的应用，如被用作食品包装材料和食品添加剂。二氧化钛纳米颗粒是许多权威食品管理机构（如欧盟和美国食品药品监督管理局）允许使用的食品添加剂，但欧盟明确要求含纳米材料添加剂的食品必须提供纳米材料信息（如种类、浓度、尺寸等）。Kim 等[25]采用热硫酸和消解催化剂对 88 种添加纳米颗粒的食品样品消解，然后使用 ICP-OES 对消解液中的钛含量进行了定量分析。TEM 和能量色散 X 射线光谱结果表明这些食品中二氧化钛纳米颗粒的尺寸均小于 100 nm。研究发现糖果、口香糖和巧克力中的二氧化钛纳米颗粒的平均浓度分别为 0.36 mg/g、0.04 mg/g 和 0.81 mg/g。目前关于二氧化钛纳米颗粒的生物毒性尚无定论，食品中纳米材料健康风险的评估仍需要更多毒理学方面的数据来支撑。

22.2　尺寸表征

纳米材料的物理化学性能和生物毒性与其尺寸密切相关。另外，纳米材料的尺寸也是决定其团聚、迁移、转化和与其他共存物相互作用的重要因素[26]。

22.2.1 响应信号强度

原子吸收光谱技术是依据吸光度与原子浓度成正比的关系实现待测元素的定量分析，而金属纳米颗粒的原子化过程与对应金属离子存在区别，导致它们吸收光谱峰分布存在差异[27]。Panyabut 等使用电热原子吸收光谱法（ETAAS）研究了 1400～2200℃的不同雾化温度下，尺寸从 5 nm 到 100 nm 的金纳米颗粒和银纳米颗粒随时间变化的原子吸收峰分布[28]。结果表明，金纳米颗粒或银纳米颗粒最大吸光度出现的时间随粒径的增加而线性增加，因此 ETAAS 可用于提供特定尺寸纳米颗粒的粒径信息。为了实现不同尺寸纳米颗粒混合物的尺寸表征，该研究引入了非对称流场流分离技术来预分离金纳米颗粒混合物，然后采用 ETAAS 对收集的分离组分进行尺寸表征。采用这种方法得到的金纳米颗粒的尺寸信息与 TEM 结果非常吻合[29]。由于传统的原子吸收光谱仪每次只能测定一个元素，因此该类方法不适用于多种不同材料纳米颗粒共存或合金纳米颗粒尺寸的快速准确表征。

与原子吸收光谱技术相比，基于 ICP-MS 的分析技术具有灵敏度高、线性范围宽、可同时分析多种元素等特点。除用作识别和定量分析外，SP-ICP-MS 也是一种可靠的尺寸表征技术，且具有灵敏度高、适用粒径范围广、分析速度快等特点。样品引入系统是决定 SP-ICP-MS 粒径检出限的关键部件，Franze 等对三种常用样品引入系统（即带有 Peltier 冷却旋风喷雾室的 PFA 微雾化器、带有加热的旋风喷雾室和三级 Peltier 冷却脱溶剂系统的 PFA 微雾化器及单分散雾化器）的性能进行了优化和评估[30]。通过对比这三种系统对 30 nm、50 nm、60 nm、80 nm 和 100 nm 的银纳米颗粒的表征结果，该课题组发现三级 Peltier 冷却脱溶剂系统的 PFA 微雾化器和单分散雾化器可显著提高样品传输效率并降低粒径检出限。Pace 等较早验证了 SP-ICP-MS 对实际环境水体中银纳米颗粒粒径的表征能力，并指出该方法在研究低浓度银纳米颗粒的团聚、溶解等环境行为研究中的应用潜力[31]。Liu 等采用 SP-ICP-MS 对 20～200 nm 的金纳米颗粒的尺寸进行了表征（如图 22-8 所示）[32]。经研究发现停留时间是影响数据质量的关键因素。在优化后的停留时间下，该研究发展的 SP-ICP-MS 系统的线性动态尺寸范围为 10～70 nm。另外，通过使用低提取电压、碰撞池 KED 模式和高分辨率模式等，可以将该系统的尺寸测量范围扩展到大约 200 nm。通过分析含有 20～200 nm 金纳米颗粒的混合物，证明了该系统对多分散金纳米颗粒的尺寸表征同样具有良好的分辨率。Yang 等使用泊松统计理论讨论了停留时间和颗粒数浓度对 SP-ICP-MS 分析性能的影响，也发现采用短的停留时间和低的颗粒数浓度可明显提高结果的准确性[33]。样品中纳米颗粒的均匀性（特别是团聚体存在）对 SP-ICP-MS 的线性尺寸响应范围有一定的影响。为解决这个问题，Rakcheev 等将水动力学色谱与 SP-ICP-MS 在线联用，实现对金纳米颗粒及其团聚体的分析表征[34]。该课题组根据 SP-ICP-MS 提供的水动力学色谱分离后的颗粒质量和有效直径，计算了团聚体平均相对密度和维数，并用来追踪团聚体在团聚过程中随时间的尺寸变化。类似地，为解决上述问题，Tan 等构建了电喷雾-差分电迁移率分析（ES-DMA）与 SP-ICP-MS 联用系统[35]。在该联用系统中，ES-DMA 可有效实现纳米颗粒与其团聚体的分离。该方法结合了 ES-DMA 高尺寸分辨率和可调性以及 SP-ICP-MS 低检出限的优点，不仅可以分离和表征不同纳米颗粒团聚体尺寸，而且还可以检测环境相关浓度下不同团聚体的表观密度。虽然 SP-ICP-MS 在低浓度纳米材料尺寸表征中表现出独特的优势，但仍然有许多挑战需要克服。例如，虽然 SP-ICP-MS 在已知颗粒的组成和几何形状条件下很容易确定颗粒物尺寸，但是很难准确确定实际环境中未知组成或形状纳米颗粒的尺寸，因此还需要结合其他纳米颗粒表征技术进行佐证[31]。另外，由于 SP-ICP-MS 仪器工作原理的限制，目前该方法尚无法区分金属离子和较小粒径（如小于 10 nm）的金属纳米颗粒。

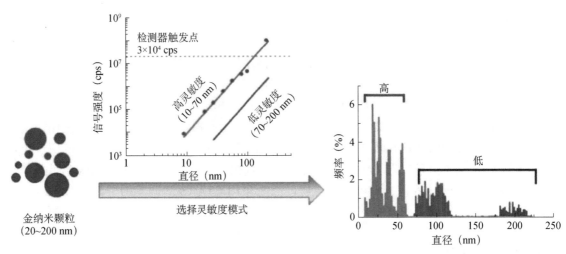

图 22-8　SP-ICP-MS 表征金纳米颗粒尺寸[32]

小角度 X 射线散射法是一种新兴的纳米颗粒尺寸表征手段。该技术通过分析 X 射线照射样品纳米尺度的电子密度不均匀区产生 X 射线小角散射，来提供纳米材料的尺寸分布信息。Geertsen 等围绕检测计数/模拟阈值等参数，提出了一种检测调整方法，扩大了该方法对粒径表征的线性范围，对 30～150 nm 不同尺寸金纳米颗粒的表征结果与 SP-ICP-MS 表征结果非常吻合[36]。

22.2.2　保留行为

纳米材料在色谱、电泳、流场流分离、整体柱等分离技术中的保留行为与其尺寸存在明显相关性，将这些技术与原子光谱技术联用，可为复杂样品中宽粒径范围纳米颗粒尺寸表征提供有效的分析方法。

在尺寸排阻色谱（SEC）中，大尺寸颗粒不能进入多孔材料颗粒内的孔隙，而是在多孔材料颗粒间隙随溶剂迁移，小尺寸颗粒能够进入多孔材料颗粒内孔隙，因此在通道内经历较长时间流出。Zhou 等基于 SEC-ICP-MS 在线联用系统，构建了复杂基质中痕量纳米颗粒的表征和测定平台（如图 22-9 所示）[37]。由于纳米材料的保留时间与其 TEM 测定的粒径呈线性相关，因此可将保留时间换算成粒径，通过仪器校正，得到纳米颗粒的粒径分布。同时，由于不同粒径的纳米颗粒的 ICP-MS 响应与其金属离子相似，因此可以实现不同粒径纳米颗粒和金属离子的同时定量。此外，通过 ICP-MS 的多元素同时检测，还可以获得纳米颗粒的元素组成。该平台具有很高的粒径分辨率（~0.27 nm）和非常低的定量检出限（如银纳米颗粒、金纳米颗粒和二氧化铈纳米颗粒的检出限分别为：0.11 μg/L、0.14 μg/L 和 0.07 μg/L）。另外，该平台不仅可以识别不同组成的痕量金属纳米颗粒（如硫化银纳米颗粒、银包金核壳纳米颗粒，以及金纳米颗粒和银纳米颗粒的混合物），还可区分多种尺寸差异极小的纳米颗粒（如 NIST 不同批次的金纳米颗粒标物，合成方法相同、但批次不同的银纳米颗粒），为分析和追溯复杂基质中纳米颗粒提供了有效途径。近期，Yan 等开发了基于薄层色谱与 LA-ICP-MS 表征金纳米颗粒的方法[38]。在薄层色谱分离过程中，金纳米颗粒的保留时间长短与其尺寸大小呈现线性负相关。该方法为实际环境水样中不同粒径的金纳米颗粒的尺寸表征和定量提供了另一可靠分析方法，但纳米材料与固定相的不可逆吸附可能会影响色谱柱或载板的使用寿命。

与色谱不同，毛细管电泳（CE）和流场流分离的分离通道内没有固定相，可以有效避免纳米材料的吸附问题。由于不同尺寸纳米颗粒的电泳迁移率或扩散系数存在差异，因此它们在 CE 或流场流分离中具有不同的保留时间。Liu 等将 CE 与 ICP-MS 联用，建立了快速表征银纳米颗粒的方法[39]。

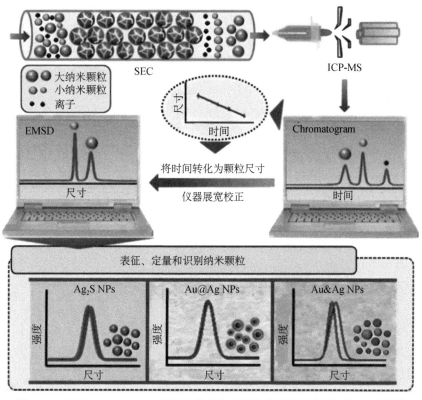

图 22-9　基于 SEC-ICP-MS 的复杂基质中痕量纳米颗粒的表征和测定平台[37]

在该方法中，不同粒径的银纳米颗粒具有不同的表观电泳淌度，而且银纳米颗粒的粒径与通过毛细管的迁移时间具有明显相关性。根据银纳米颗粒表观电泳淌度和保留时间的相关性可拟合出银的粒径，且该拟合值与 TEM 表征结果非常吻合。该系统无须样品制备，可非常方便地用于纳米材料商品和医用品的质量控制，实现环境水样中纳米材料的快速筛查和尺寸表征。Poda 等采用非对称流场流分离与 ICP-MS 联用技术，表征了以 10 nm 为增量的 10～80 nm 范围的银纳米颗粒，并通过 TEM 验证了该方法的准确性[40]。此外，该课题组还采用该联用技术对沉积物和无脊椎动物组织提取液中银纳米颗粒的粒径分布进行了表征，发现银纳米颗粒尺寸表现出明显增大趋势，这表明颗粒表面结构可能发生了变化。目前，基于 CE 和流场流分离的在线联用系统面临的主要问题是如何扩展这两类技术的应用范围以及如何进一步地推广和普及。

作为一种新型样品前处理技术，有机聚合物整体柱集分离、净化、富集于一体，具有制备简单、富集倍数高、应用范围广等特点。Liu 等以半胱胺作为解吸剂，详细考察了不同粒径金纳米颗粒在亲水性（聚丙烯酰胺-4-乙烯基吡啶-乙二醇二甲基丙烯酸酯）整体柱上的吸附和解吸行为，得到了不同粒径金纳米颗粒的临界解吸浓度[41]。通过分析临界解吸剂浓度与金纳米颗粒粒径之间的依赖关系，发现金纳米颗粒临界解吸剂浓度的自然对数与金纳米颗粒的粒径之间呈线性关系，并据此建立了聚合物整体柱与 ICP-MS 在线联用表征金纳米颗粒粒径（3～40 nm）的新方法。如图 22-10 所示，建立的方法对金纳米颗粒的尺寸分离分辨率约为 20 nm，可以用于 3 nm 和 20 nm 金纳米颗粒以及 10 nm 和 30 nm 金纳米颗粒的粒径分离以及定量分析。与基于色谱技术分离金纳米颗粒的方法相比，该研究建立的在线联用系统能同时分离和富集金纳米颗粒，且具有操作简单、成本较低和分析速度快等优点。

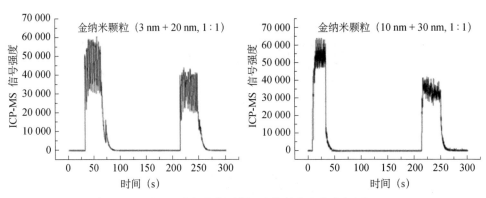

图 22-10 不同尺寸金纳米颗粒混合物的分离和表征[41]

22.3 表面性质分析

为了提高纳米材料的稳定性和生物相容性，通常需要在其表面进行修饰。例如，将蛋白质、多肽、DNA 等生物大分子修饰在金纳米颗粒表面，可大大提高金纳米颗粒在生物介质中的稳定性和特异结合能力，更好地发挥其在生物成像、疫苗接种、药物传输等领域中的作用。

表面荷电是纳米材料的重要性质之一，影响纳米材料的稳定性及其与共存物的相互作用。Qu 等基于 CE-ICP-MS 在线联用系统研究了表面修饰剂对金纳米颗粒在 CE 通道内迁移行为的影响[42]。研究发现，柠檬酸或鞣酸等小分子修饰的金纳米颗粒的电泳迁移率大小与直径大小呈正相关关系，而聚乙烯吡咯烷酮或聚乙二醇等大分子修饰的金纳米颗粒的电泳迁移率与直径呈负相关关系。牛血清白蛋白在纳米颗粒表面形成了蛋白冠结构，因此样品之间的荷电比之差最小，导致这些金纳米颗粒的电泳迁移率波动最小。在研究环境基质影响的研究中，该课题组发现腐殖酸对纳米颗粒的迁移行为影响较小。该研究结果强调了通过使用表面修饰剂和基质匹配的标准物来建立尺寸校准的必要性，有助于提高对实际样品中未知组成纳米材料尺寸表征的准确度。

表面修饰剂配体的性质和密度是纳米材料表面电荷和反应活性的决定因素。激光解析离子化质谱（LDI-MS）主要用于分子质量小于 1000 Da 配体的分析，而基质辅助激光解吸电离（MALDI）的引入可以大大提高 LDI-MS 对高分子质量配体的分析能力。Nicolardi 等构建了一种基于 LDI、MALDI 和超高分辨率傅里叶变换离子回旋共振质谱的检测平台，实现了金纳米颗粒表面分子质量大于 1000 Da 修饰剂配体的分析。该平台对修饰剂配体表现出超高的分辨能力，有望提供纳米材料表面化学配体的指纹图谱信息[43]。Hinterwirth 等根据金与硫浓度比值与金纳米颗粒尺寸之间的线性相关性，从线性回归斜率计算出配体表面覆盖率（如图 22-11 所示）[44]。该方法与纳米颗粒浓度无关，因此样品制备过程中金纳米颗粒的损失也不会影响测定结果。另外，本研究还分析了具有不同链长和亲水性/亲脂性金纳米颗粒的配体，并通过加标回收验证了该方法在金纳米颗粒表面修饰剂表征中的准确性。与碳纳米材料的识别和定量类似，其表面修饰剂的表征也是一个难点。最近，Yang 等利用自行研发的蒸气发生-PD-OES 系统建立了碳纳米材料表面羧基的定量分析方法[45]。在定量之前，先用浓盐酸将多壁碳纳米管、石墨烯、氧化石墨烯等典型碳纳米材料表面上的羧基转化为羧酸，然后引入碳酸氢钠反应生成 CO_2，随后将 CO_2 吹扫进入 PD-OES 测定。由于 PD-OES 对 CO_2 的检测具有很高的灵敏度，该方法对 10 mg 碳纳米材料的羧基检出限为 0.1 μmol/g。该方法采用邻苯二甲酸氢钾用作定量碳纳米材料羧基的校准标准，弥补了缺少基准物质的不足。

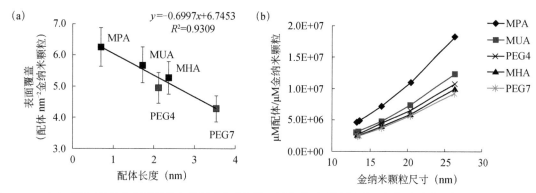

图 22-11 （a）配体长度对表面覆盖率的影响[正方形，黑色为巯基链烷酸，灰色为巯基-聚乙二醇-羧酸)]；
（b）从图（a）的结果和不同类型表面改性的粒径计算得出的每个金纳米颗粒的配体总数[44]

22.4　生物体内成像

近年来随着纳米技术的不断革新，基于纳米材料成像探针的种类和数量呈现快速增长趋势，为医疗诊断和治疗提供了重要工具。常用的生物成像技术包括荧光成像、光热成像、质谱成像等。其中，质谱成像技术，无需复杂样品前处理，也不需要对待测物进行标记，具有灵敏度高、特异性强、高通量等特点。

LDI-MS 可以通过测定纳米材料表面的配体来追踪生物体内的纳米材料。在测定过程中，配体充当"质量条形码"，而纳米材料可选择性地吸收电离配体的激光能量，因此这种方法能够同时检测具有不同尺寸和材质的纳米材料。Yan 等采用 LDI-MS 对小鼠中的金纳米颗粒进行多重成像，直接观察到金纳米颗粒在小鼠肝脏和脾脏内分布情况（如图 22-12 所示）[46]。成像结果表明，金纳米颗粒易于在肝脏和脾脏中累积。LDI-MS 具有灵敏度高、样品需求量少等优点，非常有希望发展成为生物体内纳米材料活检和临床分析的常用方法。Chen 等利用 LDI-MS 无标记质谱成像方法对小鼠体内的碳纳米管、氧化石墨烯和碳点进行了分析[47]。研究选取了本征碳簇指纹信号分析，克服了常规质谱仪对高分子量含碳纳米颗粒分析的难题。成像结果表明，大多数碳纳米管和碳点都存在于肾脏的外部薄壁组织中。其中，在脾的边缘区域碳纳米管的浓度最高。这些发现有助于解释碳纳米材料与生物相互作用的详细机制。该方法具有成像和定量分析的双重功能，非常有潜力发展为生物体内纳米材料成像分析的通用技术。

对不易吸收激光的化合物，激光解析离子化过程往往需要加大辐射能量，这可能会破坏样品分子结构，因此也极大限制了该技术在生物大分子分析中的应用。MALDI 采用常规离子化方法即可较为容易地将大分子解离为分子碎片并得到其质谱信息，可有效解决激光解析离子化过程中遇到的难题。但是，将 MALDI-MS 应用于生物样品分析和成像会受到有机基质晶体的尺寸限制。使用银纳米颗粒代替有机基质可有效解决这个问题。银纳米颗粒不仅可以有效吸收紫外激光辐射，还可将能量转移到分析物上并促进分析物的解吸，构成了适合于分析物阳离子化的银离子源[48]。MALDI-MS 可用于纳米材料的生物成像，但目前尚无法实现对纳米材料的定量分析。Chen 等采用近场解析电离质谱实现了单细胞中金纳米颗粒和银纳米颗粒的成像，其成像分辨率可达 250 nm 像素尺寸，成像结果与激光扫描共聚焦显微镜结果非常吻合[49]。近期，该课题组基于自行构建的微透镜光纤激光解吸电离质谱成像平台，对癌细胞中负载抗癌药物的叶酸修饰四氧化三铁纳米颗粒进行了原位成像研究[50]。

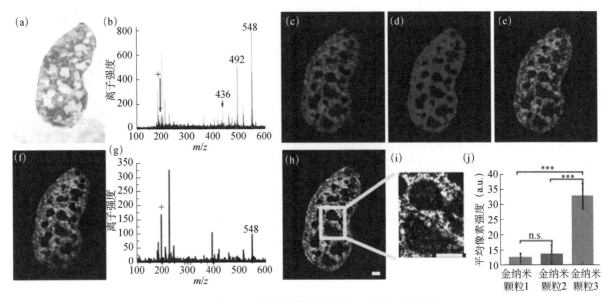

图 22-12　小鼠脾脏内金纳米颗粒的 LDI-MS 图像[46]

（a）小鼠脾组织的光学图像；（b）脾的红髓区域的质谱；（c）～（e）红髓区域金纳米颗粒图像；（f）红髓区域金离子图像；
（g）白髓区域金纳米颗粒图像；（h）金纳米颗粒的重叠图像；（i）金纳米颗粒分布的放大图像；（j）白髓区域中金纳米颗粒的生物分布

通过该平台，可以获得四氧化三铁纳米颗粒、叶酸修饰剂、抗癌药物等在细胞内的分布。该方法在纳米医药、纳米生物学、纳米毒理学等领域中均有非常广阔的应用前景。

LA-ICP-MS 被认为是纳米材料生物成像的另一有效分析方法。该方法具有以下优点：①样品需要量少；②多元素同时分析；③线性范围宽；④检出限低；⑤可通过矩阵匹配的标准实现校准。在 LA-ICP-MS 分析中，样品引入是通过逐行激光烧蚀样品完成的。检测时，对同位素的多行扫描结果被转换为代表目标元素局部分布的二维强度分布。Arakawa 等利用该方法对成纤维细胞多细胞球体薄片中的银纳米颗粒进行了成像和定量分析[51]。他们设计了与矩阵匹配的校准标准品，并使用带有银纳米颗粒的悬浮液和多元素标准品的非接触式压电驱动阵列点样仪进行打印。该方法对银、镁、磷、钾、锰、铁、钴、铜、锌等元素的检出限均在飞克级水平。孵育 48 小时后，银纳米颗粒在多细胞球体的外缘明显富集，但在细胞核中未被检出。Blaske 等结合了无损微 XRF 样品通量高和 LA-ICP-MS 灵敏度高的优点，充分发挥了它们在纳米材料成像中的技术优势[52]。无损微 XRF 可在非常短的时间内提供钙、磷、硫等常量元素的高分辨率图像，而 LA-ICP-MS 可提供银或锆等微量元素的图像，为生物体内纳米材料的成像分析提供了快速准确、高灵敏的技术手段。

22.5　形 态 分 析

金属纳米材料在生成和使用的过程中会不可避免地释放到环境中，除以颗粒的形式存在外，还可能以游离态、络合态或不溶态等形式存在，而且不同形态之间会发生相互转化，从而影响其环境归趋。尽管金属纳米颗粒的毒性机制尚存在一些争议，但已有研究证实金属纳米颗粒的浓度、大小、赋存形态（如团聚体）以及相应的金属离子浓度是影响其生物毒性的重要因素[53-55]。金属纳米颗粒独特的表面结构和反应活性增加了其向金属离子动态转化的可能性（如一般颗粒粒径越小，越容易释放金属离子）。为了科学评估金属纳米颗粒的暴露风险以及阐明金属纳米颗粒的生物毒性机制，亟需建立金属纳米颗粒不同形态的分离分析方法。

22.5.1 离线方式

常用的膜过滤、超滤、离心等离线分离技术虽然操作简便，但往往只能实现纳米颗粒和相应离子的初步分离，而且还存在受基质干扰影响大、管壁吸收、回收率低等问题。因此，开发分离效率高、抗干扰能力强、回收率高的分离方法是纳米颗粒形态分析的重点。

在采用 CPE 定量银纳米颗粒的基础上[6]，Chao 等基于银纳米颗粒与银离子在 TX-114 相分配系数的差异，建立了 CPE 与 ICP-MS 联用测定环境水体和抗菌产品中总银、银纳米颗粒以及总银离子含量的方法[56]。样品中的总银由样品直接消解后采用 ICP-MS 定量，银纳米颗粒经 CPE 分离并经消解后由 ICP-MS 定量，总银离子含量由总银含量扣除银纳米颗粒含量获得。该方法对于抗菌产品中总银、银纳米颗粒和总银离子的定量限分别为 0.2 μg/L、0.4 μg/L 和 3 μg/L。用该方法对六种市售含银纳米颗粒抗菌产品进行了分析，却只在三种产品中检测到银纳米颗粒。本研究表明，由于样品中银存在不同的形态，常用的样品消解后直接测定总银含量来定量产品中银纳米颗粒的方法是不可靠的。

实际水环境中，银纳米颗粒往往与不同形态的银离子（如自由溶解态银离子、总银离子以及吸附态银离子）共存。其中，自由溶解态浓度是银离子在环境中迁移、分配以及在生物体内累积的驱动力，是解释其可给性和生物有效性评价的关键参数。Chao 等将中空纤维支载液膜萃取技术与 ICP-MS 结合，建立了银纳米颗粒分散液和市售含银纳米颗粒产品浸泡液中自由溶解态银离子和总银离子含量的测定方法[57]。如图 22-13 所示，该方法采用微耗损采样模式选择性地萃取富集银纳米颗粒分散液中自由溶解态的银离子，采用耗尽性采样模式富集银纳米颗粒分散液中总银离子，富集完成后分别采用 ICP-MS 直接进行测定。将测定的总银离子含量扣除自由溶解态银离子含量，即可获得银纳米颗粒表面吸附态银离子的含量。该方法成功用于三种合成银纳米颗粒分散液以及市售银纳米颗粒产品浸泡液中自由溶解态银离子、总银离子和吸附态银离子含量的测定。

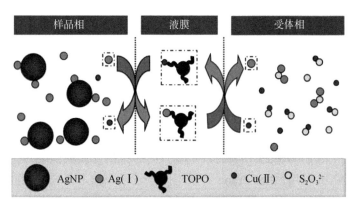

图 22-13　中空纤维支载液膜萃取不同形态银离子原理图[57]

磁固相微萃取技术具有操作简便、相分离快、富集因子高等特点。Su 等将磁固相萃取与 ICP-MS 联用建立了环境水样品中金纳米颗粒和金离子的形态分析方法[58]。负载铝离子的四氧化三铁@二氧化硅@亚氨基二乙酸磁性纳米颗粒可以同时吸附金纳米颗粒和金离子，后续分别采用硫代硫酸钠和氨水对金离子和金纳米颗粒进行选择性顺序洗脱。用这种磁性纳米颗粒作为萃取剂可以有效地分离和预富集 14～140 nm 的金纳米颗粒，且与目标金纳米颗粒表面的修饰剂种类无关。此外，在萃取过程中，金纳米颗粒的尺寸和形状均未发生明显变化，而且洗脱溶液无需消解即可直接引入 ICP-MS 测定。

该方法对常见的共存离子具有良好的抗干扰能力,可以通过在样品溶液中添加 10 mg/L 的铝离子消除浓度高达 30 mg/L 的溶解性有机物的干扰。该方法被成功用于天然水、污水、海水等不同环境水体样品中金纳米颗粒和金离子的形态分析。另外,该课题组还制备了一种高效吸附银离子和银纳米颗粒的富含咪唑基团的磁性超交联聚合物材料[59]。采用硫代硫酸钠和硝酸可实现银离子和银纳米颗粒的顺序洗脱。采用石墨炉原子吸收光谱仪测定收集的银离子和银纳米颗粒,方法检出限分别为 7.3 ng/L 和 8.2 ng/L。该方法操作简单、快速、易于批量操作,适用于宽浓度范围的银离子和宽粒径范围、不同修饰剂的银纳米颗粒的分离和测定。

22.5.2 在线方式

尽管 CPE、中空纤维支载液膜萃取、磁性固相微萃取等技术具有操作简便、运行成本低、富集倍数高等优点,但在复杂基质样品中多种尺寸纳米颗粒和离子共存的形态分析中仍存在应用局限。有效的分离技术与高灵敏的原子光谱检测技术相结合,为复杂基质样品中纳米材料的形态分析提供了可靠手段。

SEC 是一种比较常见的根据目标物的尺寸进行分离的技术。Zhou 等基于尺寸排阻原理,建立了银离子与银纳米颗粒(1～100 nm)基线分离的 LC 分离过程[60]。通过在线联用的 ICP-MS 可直接测定样品中的银离子,而银纳米颗粒的含量通过消解后 ICP-MS 直接测定的总银扣除银离子含量获得(如图 22-14 所示)。该方法可在 5 min 内实现样品中 1～100 nm 范围内的全部银纳米颗粒与银离子的快速液相分离,以及银离子的 ICP-MS 在线联用定量。将该方法与之前建立的 CPE 结合 ICP-MS 的方法进行比较,发现这两种方法测定的抗菌产品中银纳米颗粒含量非常接近,且抗菌产品中银离子为主要的存在形式。考虑到 CPE 中高浓度银离子对银纳米颗粒萃取影响的问题,该 SEC-ICP-MS 联用方法在抗菌产品中银形态分析方面则更具有优势。本研究为银纳米颗粒的生物转化和毒性效应探究提供了关键技术。另外,该联用方法还被用于小粒径氧化镍、氧化钴、氧化锌、氧化铜、二氧化铈等典型金属氧化物纳米颗粒和金属离子的快速基线分离和测定[61],非常有潜力成为金属纳米颗粒形态分析的通用标准方法。银纳米颗粒的高度动态性使不同形态银所诱导的毒性效应存在争议,因此研究银纳米颗粒在细菌体内的形态分布能够为其毒性机制探究提供可靠的数据支持。Dong 等以大肠杆菌为模型,采用 SEC-ICP-MS 联用技术分离和定量大肠杆菌细胞内颗粒态银、络合银离子以及自由态银离子[62]。首先,为了完全排除吸附在细胞壁表面银的干扰,使用溶菌酶裂解并去除细胞壁,再利用四甲基氢氧化铵消解细胞并提取不同形态银。细胞壁表面吸附银的含量可通过总暴露量扣除细胞内及细胞外基质中的银含量求得,进而得到细菌中银的生物分布信息。该方法对于大肠杆菌中不同形态银均具有较低的检出限。在实际暴露实验中,研究发现细胞内和细胞壁表面吸附银含量分别占总暴露量的 5.98%～15.21% 及 25.13%～64.43%,同时银纳米颗粒能够转化形成络合态或自由态银离子。基于 SEC-ICP-MS 联用系统,Yang 等采用同位素示踪技术研究了银纳米颗粒和银离子在水稻中的摄入、转化和迁移等行为[63]。研究发现,银离子主要保留在根表面,而银纳米颗粒存在于根内部,说明水稻可对银纳米颗粒直接摄入或者银离子被还原为银纳米颗粒。秧苗内高浓度的银离子表明表面银纳米颗粒发生了体内氧化。该研究结果证实,银离子和银纳米颗粒在植物内发生了相互转化,且银纳米颗粒可作为银离子的载体或源进入植物体内。为阐明硒化镉/硫化锌量子点的细胞摄入、清除及其对 HepG2 细胞的毒性,Peng 等采用 SEC-ICP-MS 联用技术研究了四种硒化镉/硫化锌量子点在 HepG2 细胞中的存在形态[64]。从 ICP-MS 测定镉的结果可知,硒化镉/硫化锌量子点在细胞中主要以两种形态(QD-1 和 QD-2)存在。通过对比保留时间、TEM 表征和荧光检测等方法分析,证明 QD-1 是一种类似量子点的纳米颗粒。通过反相高效液相色谱和 ICP-MS 及电喷雾四极杆飞行时间串联质谱同时联用方法分析，证明 QD-2 是一类含有 Cd 的金属硫蛋白（Cd-MTs）。不同培育条件（如培育时间、培育浓度、清除时间等）下，量子点在细胞中均主要以 QD-1 和 QD-2 两种形态存在。而且，QD-1 和 QD-2 的含量随着培育浓度和培育时间的增加而增加。因此，该研究为从分子水平上揭示量子点的细胞毒性机理提供了新思路。

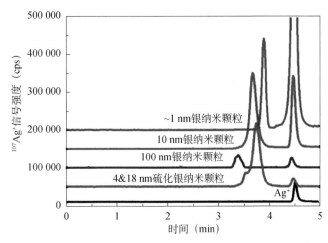

图 22-14　基于 SEC-ICP-MS 的不同尺寸银纳米颗粒和银离子的分离色谱图[60]

除了能够对银纳米颗粒进行粒径表征外，CE-ICP-MS 方法还能区分银纳米颗粒与银离子[39]。采用该方法，在 10～200 μg/L 线性范围内，银离子和三种粒径的银纳米颗粒的线性相关系数均在 0.999 以上，迁移时间和峰面积的相对标准偏差分别在 1.7%～5.4% 和 2.4%～2.8% 范围内，表明该方法具有非常好的重现性和精确性。在市售标注含银纳米颗粒厨房除菌剂和妇科洗液产品中，仅有部分产品检测到银纳米颗粒，其他被测产品中仅检测到银离子。这些结果说明该方法可以应用于含纳米材料产品的快速筛查和质量控制。

无机离子纳米材料中存在阳离子交换现象，即纳米材料中的阳离子能很快地被其他阳离子交换[65]。Huang 等发现银离子能与碲化镉量子点发生阳离子交换现象，而银纳米颗粒则不能发生该现象[66]。基于这个发现，结合阳离子交换的信号放大作用以及化学蒸气发生高的样品引入效率，该研究建立了一种简单灵敏的基于氢化物发生-原子荧光/ICP-MS 测定银离子及银纳米颗粒形态的方法（如图 22-15 所示）。当将银离子加入到碲化镉量子点中时，银能像一颗子弹一样打入到碲化镉中让几十倍的镉离子四散出来。与此相反的是，即使小到 10 nm 的银纳米颗粒加入到反应溶液中也不能将镉离子置换出来。该方法不仅操作简单、运行成本低、灵敏度高，而且适用于单细胞中的银离子和银纳米颗粒的形态分析。

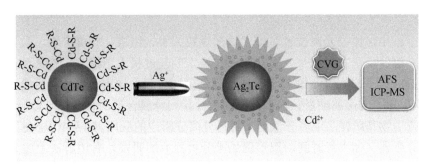

图 22-15　基于阳离子交换反应的银离子和银纳米颗粒形态分析过程示意图[66]

R 表示烷基，CVG 表示化学蒸气发生法场流分离是一种适用于大分子、胶体和颗粒等高效分离的技术[67]。遗憾的是，目前国内对该技术的研究非常少。场流分离中，分离主要是由基于垂直于载流流动方向上的外加场与目标物自身扩散的相互作用完成的。按照外加场的不同，场流分离技术包括流场流分离、沉降场流分离、热场流分离、电场流分离等，其中流场流分离技术理论最为成熟、应用最为广泛[68,69]。Tan等在国内较早开展了一种新型流场流分离技术，即中空纤维流场流分离技术的研究[70]。该课题组基于中空纤维流场流分离与紫外可见吸收检测器、动态光散射仪、ICP-MS在线联用系统，实现了μg/L浓度水平的5种不同粒径（1.4 nm、10 nm、20 nm、40 nm和60 nm）银纳米颗粒以及2种不同形态（游离或弱结合态和强结合态）银离子的在线分离、识别、表征及定量分析（如图22-16所示），为实际水环境中不同形态银的浓度水平调查提供了准确、可靠、高灵敏的分析方法。在此基础上，该课题组深入研究了环境相关浓度水平银纳米颗粒的团聚、银离子的光化学还原以及实际水环境介质（如污水处理厂进水）中银纳米颗粒和银离子的相互转化过程[71]。这些研究证实了中空纤维流场流分离技术在复杂基质样品中低浓度银纳米颗粒和银离子形态分析中的实用性，揭示了环境相关浓度下银纳米颗粒和银离子的环境行为与高浓度情况下的研究结果的显著差异性，强调了纳米材料环境行为的研究从环境相关浓度水平出发才更具实际环境意义。

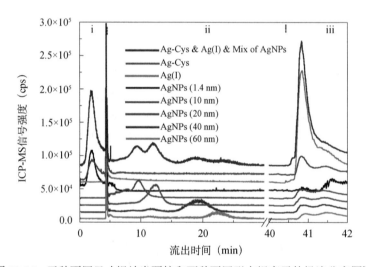

图22-16 五种不同尺寸银纳米颗粒和两种不同形态银离子的场流分离图[71]

22.6 展　　望

在纳米材料分析中，原子光谱技术可提供纳米颗粒大小、组成、粒度分布、数量和质量元素浓度、修饰剂配体密度等信息。近年来，SP-ICP-MS技术的发展极大地推动了对低浓度纳米材料分析的相关研究。然而，由于SP-ICP-MS仪器工作原理的限制，目前该方法尚无法区分离子态金属和较小粒径（如小于10 nm）的金属纳米颗粒以及不同尺寸相同金属含量的合金纳米颗粒，这也限制了该技术在实际环境或生物基质样品中多分散纳米材料分离分析中的应用。随着ICP-MS进样系统的不断升级和新型质量分析器（如飞行时间、扇形磁场等分析器）的研发，有望解决上述SP-ICP-MS的应用瓶颈问题。

发展生物体内纳米材料的成像技术是纳米医学精准诊断和治疗的前提，同时也是科学评估纳米材料环境和健康风险的必然要求，目前这方面的研究刚刚起步，这也推动了新型原子光谱技术（如激光质谱纳米成像技术）的不断涌现和发展。

实际样品基质一般比较复杂，开发合适的样品前处理技术，可拓展原子光谱技术在实际样品中的应用范围。另外，将色谱、电泳、整体柱、场流分离等分离技术与原子光谱技术在线联用，是提高原子光谱技术在实际样品中多分散纳米材料分析实用性的另一有效途径。

随着我国科技投入的不断增加，原子光谱分析方法将不断发展完善，一批自主研发的仪器设备与传统原子光谱技术的结合，催生一系列适用性更广的新方法，必将在纳米材料制备和纳米产品生命周期评价及环境风险评估中扮演更加重要的角色。

参 考 文 献

[1] Garnett E, Mai L Q, Yang P D. Introduction: 1D nanomaterials/nanowires. Chemical Reviews, 2019, 119 (15): 8955-8957.

[2] Johnson M E, Hanna S K, Bustos A R M, Sims C M, Elliott L C C, Lingayat A, Johnston A C, Nikoobakht B, Elliott J T, Holbrook R D, Scoto K C K, Murphy K E, Petersen E J, Yu L L, Nelson B C. Separation, sizing, and quantitation of engineered nanoparticles in an organism model using inductively coupled plasma mass spectrometry and image analysis. ACS Nano, 2017, 11 (1): 526-540.

[3] Liu Y, He M, Chen B B, Hu B. Ultra-trace determination of gold nanoparticles in environmental water by surfactant assisted dispersive liquid liquid microextraction coupled with electrothermal vaporization-inductively coupled plasma-mass spectrometry. Spectrochimica Acta Part B: Atomic Spectroscopy, 2016, 122: 94-102.

[4] Selahle S K, Nomngongo P N. Quantification of TiO_2 and ZnO nanoparticles in wastewater using inductively coupled plasma optical emission spectrometry. Environmental Toxicology & Chemistry, 2019, 101 (3-6): 204-214.

[5] Liu J F, Liu R, Yin Y G, Jiang G B. Triton X-114 based cloud point extraction: A thermoresversible approach for separation/concentration and dispersion of nanomaterials in aqueous phase. Chemical Communications, 2009, 12: 1514-1516.

[6] Liu J F, Chao J B, Liu R, Tan Z Q, Yin Y G, Wu Y, Jiang G B. Cloud point extraction as an advantageous preconcentration approach for analysis of trace silver nanoparticles in environmental waters. Analytical Chemistry, 2009, 81 (15): 6496-6502.

[7] Yin Y G, Liu J F, Jiang, G. B. Sunlight-induced reduction of ionic Ag and Au to metallic nanoparticles by dissolved organic matter. ACS Nano, 2012, 6 (9): 7910-7919.

[8] Yu S J, Yin Y G, Chao J B, Shen M H, Liu J F. Highly dynamic silver nanoparticles in aquatic environments: Chemical and morphology change induced by oxidation of Ag^0 And reduction of Ag^+. Environmental Science & Technology, 2014, 48 (1): 403-411.

[9] Wang Y, Chen B B, Wang B S, He M, Hu B. Phosphoric acid functionalized magnetic sorbents for selective enrichment of TiO_2 nanoparticles in surface water followed by inductively coupled plasma mass spectrometry detection. Science of the Total Environment, 2020, 703: 135464.

[10]. Zhang L, Chen B B, He M, Liu X L, Hu B. Hydrophilic polymer monolithic capillary microextraction online coupled to ICPMS for the determination of carboxyl group-containing gold nanoparticles in environmental waters. Analytical Chemistry, 2015, 87 (3): 1789-1796.

[11] Zhou X X, Lai Y J, Liu R, Li S S, Xu J W, Liu J F. Polyvinylidene fluoride micropore membranes as solid phase

extraction disk for preconcentration of nanoparticulate silver in environmental waters. Environmental Science & Technology, 2017, 51 (23): 13816-13824.

[12]] Zhou X X, Li Y J, Liu J F. Highly efficient removal of silver-containing nanoparticles by aged iron oxide magnetic particles. ACS Sustainable Chemistry & Engineering, 2017, 5 (6): 5468-5476.

[13] Tan Z Q, Wang B W, Yin Y G, Liu Q, Li X, Liu J F. *In situ* tracking photodegradation of trace graphene oxide by the online coupling of photoinduced chemical vapor generation with a point discharge optical emission spectrometer. Analytical Chemistry, 2020, 92 (1): 1549-1556.

[14] Li L, Wang Q, Yang Y, Luo L, Ding R, Yang Z G, Li H P. Extraction method development for quantitative detection of silver nanoparticles in environmental soils and sediments by single particle inductively coupled plasma mass spectrometry. Analytical Chemistry, 2019, 91 (15): 9442-9450.

[15] Torrent L, Iglesias M, Hidalgo M, Margui E. Analytical capabilities of total reflection X-Ray fluorescence spectrometry for silver nanoparticles determination in soil adsorption studies. Spectrochimica Acta Part B: Atomic Spectroscopy, 2016, 126: 71-78.

[16] Lu D W, Liu Q, Zhang T Y, Cai Y, Yin Y G, Jiang G B. Stable silver isotope fractionation in the natural transformation process of silver nanoparticles. Nature Nanotechnology, 2016, 11 (8): 682-686.

[17] Yang X Z, Liu X, Zhang A Q, Lu D W, Li G, Zhang Q H, Liu Q, Jiang G B. Distinguishing the sources of silica nanoparticles by dual isotopic fingerprinting and machine learning. Nature Communications, 2019, 10: 1620.

[18] Zhang Q H, Lu D W, Wang D Y, Yang X Z, Zuo P J, Yang H, Fu Q, Liu Q, Jiang G B. Separation and tracing of anthropogenic magnetite nanoparticles in the urban atmosphere. Environmental Science & Technology, 2020, 54 (15): 9274-9284.

[19] Gray E P, Coleman J G, Bednar A J, Kennedy A J, Ranville J F, Higgins C P. Extraction and analysis of silver and gold nanoparticles from biological tissues using single particle inductively coupled plasma mass spectrometry. Environmental Science & Technology, 2013, 47 (24): 14315-14323.

[20] Yu S J, Chao J B, Sun J, Yin Y G, Liu J F, Jiang G B. Quantification of the uptake of silver nanoparticles and ions to HepG2 cells. Environmental Science & Technology, 2013, 47 (7): 3268-3274.

[21] Soria N G C, Montes A, Bisson M A, Atilla-Gokcumen G E, Aga D S. Mass spectrometry-based metabolomics to assess uptake of silver nanoparticles by arabidopsis thaliana. Environmental Science: Nano, 2017, 4 (10): 1944-1953.

[22] Wagener S, Jungnickel H, Dommershausen N, Fischer T, Laux P, Luch A. Determination of nanoparticle uptake, distribution, and characterization in plant root tissue after realistic long-term exposure to sewage sludge using information from mass spectrometry. Environmental Science & Technology, 2019, 53 (9): 5416-5426.

[23] Wang M, Zheng L N, Wang B, Chen H Q, Zhao Y L, Chai Z F, Reid H J, Sharp B L, Feng W Y. Quantitative analysis of gold nanoparticles in single cells by laser ablation inductively coupled plasma-mass spectrometry. Analytical Chemistry, 2014, 86 (20): 10252-10256.

[24] Fernandez-Ruiz R, Redrejo M J, Friedrich E J, Ramos M, Fernandez T. Evaluation of bioaccumulation kinetics of gold nanorods in vital mammalian organs by means of total reflection X-ray fluorescence spectrometry. Analytical Chemistry, 2014, 86 (15): 7383-7390.

[25] Kim N, Kim C, Jung S, Park Y, Lee Y, Jo J, Hong M, Lee S, Oh Y, Jung K. Determination and identification of titanium dioxide nanoparticles in confectionery foods, marketed in south Korea, using inductively coupled plasma optical emission spectrometry and transmission electron microscopy. Food Additives and Contaminants Part A: Chemistry Analysis Control Exposure & Risk Assessment, 2018, 35 (7): 1238-1246.

[26] Lowry G V, Gregory K B, Apte S C, Lead J R. Transformations of nanomaterials in the environment. Environmental

Science & Technology, 2012, 46 (13): 6893-6899.

[27] Resano M, Mozas E, Crespo C, Briceno J, Menoyo J D, Belarra M A. Solid sampling high-resolution continuum source graphite furnace atomic absorption spectrometry to monitor the biodistribution of gold nanoparticles in mice tissue after intravenous administration. Journal of Analytical Atomic Spectrometry, 2010, 25 (12): 1864-1873.

[28] Panyabut T, Sirirat N, Siripinyanond A. Use of electrothermal atomic absorption spectrometry for size profiling of gold and silver nanoparticles. Analytica Chimica Acta, 2018, 1000: 75-84.

[29] Mekprayoon S, Siripinyanond A. Performance evaluation of flow field-flow fractionation and electrothermal atomic absorption spectrometry for size characterization of gold nanoparticles. Journal of Chromatography A, 2019, 1604: 460493.

[30] Franze B, Strenge I, Engelhard C. Single particle inductively coupled plasma mass spectrometry: Evaluation of three different pneumatic and piezo-based sample introduction systems for the characterization of silver nanoparticles. Journal of Analytical Atomic Spectrometry, 2012, 27 (7): 1074-1083.

[31] Pace H E, Rogers N J, Jarolimek C, Coleman V A, Gray E P, Higgins C P, Ranville J F. Single particle inductively coupled plasma-mass spectrometry: A performance evaluation and method comparison in the determination of nanoparticle size. Environmental Science & Technology, 2012, 46 (22): 12272-12280.

[32] Liu J Y, Murphy K E, MacCuspie R I, Winchester M R. Capabilities of single particle inductively coupled plasma mass spectrometry for the size measurement of nanoparticles: A case study on gold nanoparticles. Analytical Chemistry, 2014, 86 (7): 3405-3414.

[33] Yang Y, Long C L, Li H P, Wang Q, Yang Z G. Analysis of silver and gold nanoparticles in environmental water using single particle-inductively coupled plasma-mass spectrometry. Science of the Total Environment, 2016, 563: 996-1007.

[34] Rakcheev D, Philippe A, Schaumann G E. Hydrodynamic chromatography coupled with single particle-inductively coupled plasma mass spectrometry for investigating nanoparticles agglomerates. Analytical Chemistry, 2013, 85 (22): 10643-10647.

[35] Tan J J, Liu J Y, Li M D, El Hadri H, Hackley V A, Zechariah M R. Electrospray-differential mobility hyphenated with single particle inductively coupled plasma mass spectrometry for characterization of nanoparticles and their aggregates. Analytical Chemistry, 2016, 88 (17): 8548-8555.

[36] Geertsen V, Barruet E, Gobeaux F, Lacour J L, Tache O. Contribution to accurate spherical gold nanoparticle size determination by single-particle inductively coupled mass spectrometry: A comparison with small-angle X-ray scattering. Analytical Chemistry, 2018, 90 (16): 9742-9750.

[37] Zhou X X, Liu J F, Jiang G B. Elemental mass size distribution for characterization, quantification and identification of trace nanoparticles in serum and environmental waters. Environmental Science & Technology, 2017, 51 (7): 3892-3901.

[38] Yan N, Zhu Z L, Jin L L, Guo W, Gan Y Q, Hu S H. Quantitative characterization of gold nanoparticles by coupling thin layer chromatography with laser ablation inductively coupled plasma mass spectrometry. Analytical Chemistry, 2015, 87 (12): 6079-6087.

[39] Liu L H, He B, Liu Q, Yun Z J, Yan X T, Long Y M, Jiang G B. Identification and accurate size characterization of nanoparticles in complex media. Angewandte Chemie-International Edition, 2014, 53 (52): 14476-14479.

[40] Poda A R, Bednar A J, Kennedy A J, Harmon A, Hull M, Mitrano D M, Ranville J F, Steevens J. Characterization of silver nanoparticles using flow-field flow fractionation interfaced to inductively coupled plasma mass spectrometry. Journal of Chromatography A, 2011, 1218 (27): 4219-4225.

[41] Liu X L, Chen B B, Cai Y B, He M, Hu B. Size-based analysis of AuNPs by online monolithic capillary microextraction-ICPMS. Analytical Chemistry, 2017, 89 (1): 560-564.

[42] Qu H, Linder S W, Mudalige T K, Surface coating and matrix effect on the electrophoretic mobility of gold nanoparticles: A capillary electrophoresis-inductively coupled plasma mass spectrometry study. Analytical and Bioanalytical Chemistry, 2017, 409 (4): 979-988.

[43] Nicolardi S, van der Burgt Y E M, Codee J D C, Wuhrer M, Hokke C H, Chiodo F. Structural characterization of biofunctionalized gold nanoparticles by ultrahigh-resolution mass spectrometry. ACS Nano, 2017, 11 (8): 8257-8264.

[44] Hinterwirth H, Kappel S, Waitz T, Prohaska T, Lindner W, Lammerhofer M. Quantifying thiol ligand density of self-assembled monolayers on gold nanoparticles by inductively coupled plasma-mass spectrometry. ACS Nano, 2013, 7 (2): 1129-1136.

[45] Yang R, Lin Y, Liu B Y, Su Y B, Tian Y F, Hou X D, Zheng C B. Simple universal strategy for quantification of carboxyl groups on carbon nanomaterials: Carbon dioxide vapor generation coupled to microplasma for optical emission spectrometric detection. Analytical Chemistry, 2020, 92 (5): 3528-3534.

[46] Yan B, Kim S T, Kim C S, Saha K, Moyano D F, Xing Y Q, Jiang Y, Roberts A L, Alfonso F S, Rotello V M, Vachet R W. Multiplexed imaging of nanoparticles in tissues using laser desorption/ionization mass spectrometry. Journal of the American Chemical Society, 2013, 135 (34): 12564-12567.

[47] Chen S M, Xiong C Q, Liu H H, Wan Q Q, Hou J, He Q, Badu-Tawiah A, Nie Z X. Mass spectrometry imaging reveals the sub-organ distribution of carbon nanomaterials. Nature Nanotechnology, 2015, 10 (2): 176-182.

[48] Sekula J, Niziol J, Rode W, Ruman T. Silver nanostructures in laser desorption/ionization mass spectrometry and mass spectrometry imaging. Analyst, 2015, 140 (18): 6195-6209.

[49] Chen X L, Wang T T, Yin Z B, Hang W. Single-cell imaging of AuNPs and AgNPs by near-field desorption ionization mass spectrometry. Journal of Analytical Atomic Spectrometry, 2020, 35 (5): 927-932.

[50] Meng Y F, Chen X L, Wang T T. Hang W, Li X P, Nie W. Micro-lensed fiber laser desorption mass spectrometry imaging reveals subcellular distribution of drugs within single cells. Angewandte Chemie-International Edition, 2020, 59 (41): 17864-17871.

[51] Arakawa A, Jakubowski N, Koellensperger G, Theiner S, Schweikert A, Flemig S, Iwahata D, Traub H, Hirata T. Quantitative imaging of silver nanoparticles and essential elements in thin sections of fibroblast multicellular spheroids by high resolution laser ablation inductively coupled plasma time-of-flight mass spectrometry. Analytical Chemistry, 2019, 91 (15): 10197-10203.

[52] Blaske F, Reifschneider O, Gosheger G, Wehe C A, Sperling M, Karst U, Hauschild G, Hoell S. Elemental Bioimaging of nanosilver-coated prostheses using X-ray fluorescence spectroscopy and laser ablation-inductively coupled plasma-mass spectrometry. Analytical Chemistry, 2014, 86 (1): 615-620.

[53] Shen M H, Zhou X X, Yang X Y, Chao J B, Liu R, Liu J F. Exposure medium: Key in identifying free Ag^+ as the exclusive species of silver nanoparticles with acute toxicity to *Daphnia magna*. Scientific Reports, 2015, 5: 9674.

[54] Hao F, Liu Q S, Chen X, Zhao X C, Zhou Q F, Liao C Y, Jiang G B. Exploring the heterogeneity of nanoparticles in their interactions with plasma coagulation factor XII. ACS Nano, 2019, 13 (2): 1990-2003.

[55] Liu S J, Xia T. Continued efforts on nanomaterial-environmental health and safety is critical to maintain sustainable growth of nanoindustry. Small, 2020, 16 (21): 2000603.

[56] Chao J B, Liu J F, Yu S J, Feng Y D, Tan Z Q, Liu R, Yin Y G. Speciation analysis of silver nanoparticles and silver ions in antibacterial products and environmental waters via cloud point extraction-based separation. Analytical Chemistry, 2011, 83 (17): 6875-6882.

[57] Chao J B, Zhou X X, Shen M H, Tan Z Q, Liu R, Yu S J, Wang X W, Liu J F. Speciation analysis of labile and total silver(Ⅰ) in nanosilver dispersions and environmental waters by hollow fiber supported liquid membrane extraction.

Environmental Science & Technology, 2015, 49 (24): 14213-14220.

[58] Su S W, Chen B B, He M, Xiao Z W, Hu B. A novel strategy for sequential analysis of gold nanoparticles and gold ions in water samples by combining magnetic solid phase extraction with inductively coupled plasma mass spectrometry. Journal of Analytical Atomic Spectrometry, 2014, 29 (3): 444-453.

[59] Zhao B S, He M, Chen B B, Hu B. Ligand-assisted magnetic solid phase extraction for fast speciation of silver nanoparticles and silver ions in environmental water. Talanta, 2018, 183: 268-275.

[60] Zhou X X, Liu R, Liu J F. Rapid chromatographic separation of dissoluble Ag(Ⅰ) and silver-containing nanoparticles of 1~100 nanometer in antibacterial products and environmental waters. Environmental Science & Technology, 2014, 48 (24): 14516-14524.

[61] Zhou X X, Liu J F, Geng F L. Speciation analysis of metal oxide nanoparticles and their ionic counterparts in water samples by size exclusion chromatography coupled to ICP-MS. NanoImpact, 2016, 1: 13-20.

[62] Dong L J, Lai Y J, Yu S J, Liu J F. Speciation analysis of the uptake and biodistribution of nanoparticulate and ionic silver in *Escherichia coli*. Analytical Chemistry, 2019, 91 (19): 12525-12530.

[63] Yang Q Q, Shan W Y, Hu L G, Zhao Y, Hou Y Z, Yin Y G, Liang Y, Wang, F Y, Cai Y, Liu J F, Jiang G B. Uptake and transformation of silver nanoparticles and ions by rice plants revealed by dual stable isotope tracing. Environmental Science & Technology, 2019, 53 (2): 625-633.

[64] Peng L, He M, Chen B B, Wu Q M, Zhang Z L, Pang D W, Zhu Y, Hu B. Cellular uptake, elimination and toxicity of CdSe/ZnS quantum dots in HepG2 cells. Biomaterials, 2013, 34 (37): 9545-9558.

[65] Son D H, Hughes S M, Yin Y D, Alivisatos A P. Cation exchange reactions-in ionic nanocrystals. Science, 2004, 306 (5698): 1009-1012.

[66] Huang K, Xu K L, Tang J, Yang L, Zhou J R, Hou X D, Zheng C B. Room temperature cation exchange reaction in nanocrystals for ultrasensitive speciation analysis of silver ions and silver nanoparticles. Analytical Chemistry, 2015, 87 (13): 6584-6591.

[67] Giddings J C. Field-flow fractionation: Analysis of macromolecular, colloidal, and particulate materials. Science, 1993, 260 (5113): 1456-1465.

[68] Williams S K R, Runyon J R, Ashames A A. Field-flow fractionation: Addressing the nano challenge. Analytical Chemistry, 2011, 83 (3): 634-642.

[69] Moon M H. Flow field-flow fractionation: Recent applications for lipidomic and proteomic analysis. TrAC-Trends in Analytical Chemistry, 2019, 118: 19-28.

[70] Tan Z Q, Liu J F, Guo X R, Yin Y G, Byeon S K, Moon M H, Jiang G B. Toward full spectrum speciation of silver nanoparticles and ionic silver by on-line coupling of hollow fiber flow field-flow fractionation and minicolumn concentration with multiple detectors. Analytical Chemistry, 2015, 87 (16): 8441-8447.

[71] Tan Z Q, Yin Y G, Guo X R, Amde M, Moon M H, Liu J F, Jiang G B. Tracking the transformation of nanoparticulate and ionic silver at environmentally relevant concentration levels by hollow fiber flow field-flow fractionation coupled to ICPMS. Environmental Science & Technology, 2017, 51 (21): 12369-12376.

第23章 食品分析、卫生检验与临床检验

(刘丽萍[①],陈绍占[①],刘　洋[①],吴永宁[①*])

▶ 23.1　食品分析研究进展　/ 645

▶ 23.2　临床样品分析研究进展　/ 660

▶ 23.3　展望　/ 669

[①]北京市疾病预防控制中心、国家食品安全风险评估中心,北京。*通讯作者联系方式:wuyongning@cfsa.net

第23章 食品分析、卫生检验与临床检验

本章导读

- 食品是人类生存与发展的最基本物质，为人类提供维持生命和身体健康的能量及多种营养与功效成分。化学元素是构成人体的基本成分，对人体的生长发育、健康与疾病起着重要作用。微量元素在生物体内的含量虽少，但在生命活动过程中起着十分重要的作用。
- 食品安全与健康营养一直是社会关注的热点，重金属与微量元素也一直受到广泛关注，原子光谱技术广泛应用于食品、生物样品中重金属和微量元素的分析检测。本章主要介绍原子光谱、色谱-原子光谱联用技术在食品安全、临床检验中的应用。

23.1 食品分析研究进展

为了保证人们的饮食安全，世界各国都制订了微量元素和有害元素在不同食品基质中的限量标准，同时制定了与各类卫生标准相匹配的标准检测方法。我国《食品安全国家标准 食品中污染物限量》（GB 2762—2017）[1]规定了食品中铅、镉、汞、砷、锡、镍、铬的限量；目前原子光谱技术广泛应用于食品安全与营养健康的元素分析检测中。

23.1.1 现行有效的标准方法

食品中重金属、微量元素的含量主要采用原子吸收光谱、原子荧光光谱、电感耦合等离子体发射光谱（ICP-OES）、电感耦合等离子体质谱（ICP-MS）进行检测，食品安全国家标准 GB 5009 系列标准检测方法规定了 K、Na、Ca、Mg、P、Fe、Mn、Zn、Cu、Pb、F、Cr、Cd、Sb、Ge、Ni、Sn、As、Al、Hg 以及植物性食品中稀土元素的分析方法，见表 23-1。

表 23-1 原子光谱技术在食品安全国家标准中的应用

序号	标准名称		测定方法
1	食品中总砷及无机砷的测定[2]（GB 5009.11—2014）	总砷	①电感耦合等离子体质谱法：检出限为 0.003 mg/kg
			②氢化物发生原子荧光光谱法：检出限为 0.010 mg/kg
		无机砷	①液相色谱-原子荧光光谱法：检出限为稻米 0.02 mg/kg；水产动物 0.03 mg/kg；婴儿辅助食品 0.02 mg/kg
			②液相色谱-电感耦合等离子质谱法：检出限为稻米 0.01 mg/kg；水产动物 0.02 mg/kg；婴儿辅助食品 0.01 mg/kg
2	食品中铅的测定（GB 5009.12—2017）		①石墨炉原子吸收光谱法：检出限为 0.02 mg/kg
			②电感耦合等离子体质谱法：检出限为 0.02 mg/kg
			③火焰原子吸收光谱法：检出限为 0.4 mg/kg
3	食品中铜的测定（GB 5009.13—2017）		①石墨炉原子吸收光谱法：检出限为 0.02 mg/kg
			②火焰原子吸收光谱法：检出限为 0.2 mg/kg
			③电感耦合等离子体质谱法：检出限为 0.05 mg/kg
			④电感耦合等离子体发射光谱法：检出限为 0.2 mg/kg
4	食品中锌的测定（GB 5009.14—2017）		①火焰原子吸收光谱法：检出限为 1 mg/kg
			②电感耦合等离子体发射光谱法：检出限为 0.5 mg/kg
			③电感耦合等离子体质谱法：检出限为 0.5 mg/kg

续表

序号	标准名称	测定方法	
5	食品中镉的测定 （GB 5009.15—2014）	石墨炉原子吸收光谱法：检出限为 0.001 mg/kg	
6	食品中总汞及有机汞的测定[3] （GB 5009.17—2021）	总汞	①原子荧光光谱法：检出限为 0.003 mg/kg ②直接进样测汞法：检出限为 0.0002 mg/kg ③电感耦合等离子体质谱法：检出限为 0.001 mg/kg ④冷原子吸收光谱法：检出限为 0.002 mg/kg
		有机汞	①液相色谱-原子荧光光谱联用方法：检出限为 0.008 mg/kg ②液相色谱-电感耦合等离子体质谱联用法：检出限为 0.005 mg/kg
7	食品中铁的测定 （GB 5009.90—2016）	①火焰原子吸收光谱法：检出限为 0.75 mg/kg ②电感耦合等离子体发射光谱法：检出限为 1 mg/kg ③电感耦合等离子体质谱法：检出限为 1 mg/kg	
8	食品中钾、钠的测定 （GB 5009.91—2017）	①火焰原子吸收光谱法：钾的检出限为 0.2 mg/100 g，钠的检出限为 0.8 mg/100 g ②火焰原子发射光谱法：钾的检出限为 0.2 mg/100 g，钠的检出限为 0.8 mg/100 g ③电感耦合等离子体发射光谱法：钾的检出限为 7 mg/kg，钠的检出限为 3 mg/kg ④电感耦合等离子体质谱法：钾的检出限为 1 mg/kg，钠的检出限为 1 mg/kg	
9	食品中钙的测定 （GB 5009.92—2016）	①火焰原子吸收光谱法：检出限为 0.5 mg/kg ②电感耦合等离子体发射光谱法：检出限为 5 mg/kg ③电感耦合等离子体质谱法：检出限为 1 mg/kg	
10	食品中硒的测定 （GB 5009.93—2017）	①氢化物原子荧光光谱法：检出限为 0.002 mg/kg ②荧光分光光度法：检出限为 0.01 mg/kg ③电感耦合等离子体质谱法：检出限为 0.01 mg/kg	
11	植物性食品中稀土元素的测定[4] （GB 5009.94—2012）	电感耦合等离子体质谱法：检出限为（μg/kg）Sc 0.6，Y 0.3，La 0.4，Ce 0.3，Pr 0.2，Nd 0.2，Sm 0.2，Eu 0.06，Gd 0.1，Tb 0.06，Dy 0.08，Ho 0.03，Er 0.06，Tm 0.03，Yb 0.06，Lu 0.03	
12	食品中铬的测定 （GB 5009.123—2014）	石墨炉原子吸收光谱法：检出限为 0.01 mg/kg	
13	食品中锗的测定 （GB/T 5009.151—2003）	①原子荧光光谱法 ②原子吸收分光光度法	
14	食品中锡的测定 （GB 5009.16—2014）	氢化物原子荧光光谱法：定量限为 2.5 mg/kg	
15	食品中镍的测定 （GB 5009.138—2017）	石墨炉原子吸收光谱法：检出限为 0.02 mg/kg	
16	食品中铝的测定 （GB 5009.182—2017）	①电感耦合等离子体质谱法：检出限为 0.5 mg/kg ②电感耦合等离子体发射光谱法：检出限为 0.5 mg/kg ③石墨炉原子吸收光谱法：检出限为 0.3 mg/kg	
17	食品中镁的测定 （GB 5009.241—2017）	①火焰原子吸收光谱法：检出限为 0.6 mg/kg ②电感耦合等离子体发射光谱法：检出限为 5 mg/kg ③电感耦合等离子体质谱法：检出限为 1 mg/kg	
18	食品中锰的测定 （GB 5009.242—2017）	①火焰原子吸收光谱法：检出限为 0.2 mg/kg ②电感耦合等离子体发射光谱法：检出限为 0.1 mg/kg ③电感耦合等离子体质谱法：检出限为 0.1 mg/kg	
19	食品中多元素的测定[5] （GB 5009.268—2016）	①电感耦合等离子体质谱法 ②电感耦合等离子体发射光谱法	

2015 年 9 月 21 日颁布，2016 年 3 月 21 日实施的 GB 5009.11—2014《食品安全国家标准　食品

中总砷及无机砷的测定》[2]，将液相色谱-原子荧光光谱法（LC-AFS）和液相色谱-电感耦合等离子体质谱法（LC-ICP-MS）纳入食品中无机砷的测定，食品中无机砷经稀硝酸热浸提后，以液相色谱进行分离，分离后的目标化合物以原子荧光光谱仪或电感耦合等离子体质谱仪进行测定。

2021年9月7日颁布，2022年3月7日实施的GB 5009.17—2021《食品安全国家标准　食品中总汞及有机汞的测定》[3]中"第一篇　食品中总汞的测定"，增加直接进样测汞法为第二法，引用GB5009.268中的电感耦合等离子体质谱法为第三法，保留原子荧光光谱分析法为第一法、冷原子吸收光谱法为第四法；修订了原子荧光光谱分析法和冷原子吸收光谱法中不合理的内容，提升了原方法的适用性。"第二篇　食品中甲基汞的测定"，保留液相色谱-原子荧光光谱法为第一法，修订了方法线性浓度范围、流动相浓度，增加了无机汞、甲基汞、乙基汞的分离度确认，增加了稻米、食用菌样品基质；增加液相色谱-电感耦合等离子体质谱法为第二法，适用于水产品、食用菌、大米等基质中甲基汞的测定。

2012年5月17日颁布，2012年7月17日实施的《食品安全国家标准　植物性食品中稀土元素的测定》（GB 5009.94—2012）[4]为植物性食物中稀土元素的测定提供了有效的测定方法。该标准适用于谷类粮食、豆类、蔬菜、水果、茶叶等植物性食品中钪（Sc）、钇（Y）、镧（La）、铈（Ce）、镨（Pr）、钕（Nd）、钐（Sm）、铕（Eu）、钆（Gd）、铽（Tb）、镝（Dy）、钬（Ho）、铒（Er）、铥（Tm）、镱（Yb）、镥（Lu）的测定。植物中微量稀土的分析方法也有些报道，其中ICP-MS以灵敏度高、精密度好、线性范围宽、干扰少等优点，在稀土元素痕量分析方面最具优势。

2016年12月23日颁布，2017年6月23日实施的GB 5009.268—2016《食品安全国家标准　食品中多元素的测定》[5]给出了硼、钠、镁、铝、钾、钙、钛、钒、铬、锰、铁、钴、镍、铜、锌、砷、硒、锶、钼、镉、锡、锑、钡、汞、铊、铅、磷共计27种元素的分析方法，其中，第一法电感耦合等离子体质谱法（ICP-MS）可以分析检测除磷以外的26种元素；而第二法电感耦合等离子体发射光谱法（ICP-OES）适用于铝、硼、钡、钙、铜、铁、钾、镁、锰、钠、镍、磷、锶、钛、钒、锌16种元素的测定。此标准方法涵盖了以往食品安全与营养健康中涉及的多个元素，利用ICP-MS及ICP-OES技术的高灵敏度、高精密度、高抗干扰能力、多种元素同时测定的特点，将方法的实用性扩展至整个食品种类，为食品安全与营养健康提供了有效支持。

23.1.2　元素总量分析进展

除了以上成熟的原子光谱技术已纳入GB 5009系列的标准方法，目前还有直接进样测汞和测镉的方法应用在食品安全分析中。

23.1.2.1　直接进样测汞和测镉的方法

常用的原子光谱技术均需要对样品进行前处理，由于汞沸点低，极易挥发，测定食品等固体样品中汞含量时，前处理成为测定方法的瓶颈。直接进样测汞仪将样品前处理装置和原子光谱技术集于一身，对待测样品无需经过传统的酸消解处理，样品经高温灼烧及催化热解后，汞被还原成汞单质，用金汞齐富集或直接通过载气带入检测器，在253.7 nm波长处测量汞的原子吸收信号，或以汞灯激发检测汞的原子荧光信号，外标法定量，能直接测定固体或液体样品中的汞含量，与传统方法相比，具有取样量少、灵敏度高、准确快速的特点。无需进行样品前处理，大大减少了硝酸等试剂的使用。目前对于样品中痕量汞的直接进样法，主要有基于催化热解冷原子吸收、催化热解原子荧光和塞曼效应的原子吸收三种机理的方法。

刘丽萍等[6-11]采用催化热解冷原子吸收法的直接测汞仪对不同类型的食品，如婴幼儿配方乳粉、水产品、茶叶、食用菌、谷类、乳类、鱼类、蔬菜水果等膳食样品进行汞含量的分析测定，无论固体样品还是液体样品均可采用金汞齐富集后，利用汞原子蒸气在 253.7 nm 处共振线的特征吸收进行测定。无需样品前处理，方法灵敏度高、简便，能快速测定各类食品中总汞的含量。

谢科[17]采用电热解-塞曼原子吸收直接测汞仪，基于塞曼效应原子吸收光谱分析原理，利用塞曼效应在光学角度对杂质信号进行背景校正，结合热解技术直接测定固体样品中汞含量，样品在仪器内直接电热解，无需样品消解等前处理过程，可快速检测调味品中总汞含量，有证标准物质测定值均在标识值范围内。测定的调味品中总汞含量与原子荧光法测定结果一致。

杨娟等[13,14]采用催化热解原子荧光原理，结合运用还原催化和金汞齐富集技术，建立了直接测定固体、液体食品中汞含量的方法。测定了海带、紫菜、牡蛎、扇贝、大虾等食品样品，方法无需复杂的样品前处理过程，分析全过程未引入化学试剂，是一种高效、环保的分析方法。

Wang 等[15]介绍了一种新型固体进样汞镉分析仪，建立了采用金丝和钨丝捕集汞和镉后原子荧光光谱连续测定固体样品中痕量汞和镉的直接进样测定方法。方法 Hg 和 Cd 的检出限分别可达到 0.7 pg 和 0.5 pg，具有较好的分析灵敏度和精密度，适用于食品、农产品中汞、镉的现场测定，具备快速、准确的特点。

对于固体进样直接测定方法，为了保证测定的准确性，样品均匀性非常重要。样品需采用粉碎机粉碎到一定的粒径，再准确称取适量的样品，上机测定。样品均需进行平行样测定，以减少随机误差，提高准确度。

目前此技术已纳入 2021 年颁布的食品安全国家标准 GB 5009.17—2021 中。

23.1.2.2 碘的测定

碘是人体必需的微量元素，主要参与甲状腺素的合成，与人体的生长发育密切相关，碘缺乏将严重影响智力发育。碘缺乏病是一种生物地球化学性疾病，食用富含碘的食品可有效补充人体所需要的碘。但摄入过量碘也将对健康造成一定的危害。因此分析检测食品中碘含量具有重要的意义。一般食品中的碘以碘酸盐、碘化物、单质碘以及有机碘的形式存在。目前原子光谱技术测定食品中碘含量的方法主要为原子吸收光谱法和电感耦合等离子体质谱法。

由于原子吸收光谱法无法直接测定位于真空紫外区的碘，通常可以使用原子吸收光谱技术间接测定食品中的碘。陈丽娟[16]建立了间接测定鸡蛋中微量碘的方法，在抗坏血酸作用下，加入样品中的 Cu^{2+} 被还原为 Cu^+，与样品中 I^- 反应生成 CuI 沉淀，加入邻苯二甲酸氢钾缓冲溶液离心后，测定溶液中剩余 Cu^{2+} 含量，可间接得到样品中碘的含量，该方法检出限 1.0 μg/mL。Yebra[17]通过原子吸收光谱法测定银，间接测定牛奶中的碘。样品经氢氧化钾灰化后，在灰分溶解液中加入硝酸银，溶液中的碘化物与银离子形成碘化银沉淀，沉淀经氨水冲洗后溶于硫代硫酸盐中，用火焰原子吸收法测定溶液中银从而间接测定样品中碘含量。

碘的质量数为 127，天然丰度 100%，由于没有质谱干扰，非常适宜采用电感耦合等离子体质谱技术测定。常规的 ICP-MS 技术通常使用酸体系消化处理样品，但碘在酸体系中稳定性不好，将影响测定的准确性及重现性。对于碘的分析测定可通过稀释样品或采用氢氧化钾（KOH）、氢氧化铵（NH_4OH）和四甲基氢氧化铵（TMAH）等碱性溶液处理样品。刘丽萍等[18]采用 TMAH 和过氧化氢（H_2O_2）超声提取样品，碲（128）为内标，ICP-MS 法测定乳制品中碘含量。方法线性范围为 0～400 μg/L，检出限为 0.09 μg/L，相关系数 r 均大于 0.9990，RSD 小于 5%，加标回收率在 89.3%～98.0% 之间。吕超[44]采用类似的方法测定了饮料、果汁和啤酒中的碘含量。张妮娜[20]进一步优化了肉类膳食样品的提取条件，通过冷冻脱脂、甲酸沉淀蛋白后采用 ICP-MS 测定其碘含量。丁玉龙等[21]以硝酸消解乳制品，铕做内标，使用 ICP-MS 测定其中的碘含量，方法检出限为 0.02 μg/kg，线性范围为 0～

100 μg/L。谭玲[22]以 ^{115}In 为内标，采用 ICP-MS 法测定矿泉水中的碘含量，方法检出限为 53.4 ng/L。张慧敏[23]采用 TMAH 溶液于 90℃烘箱或水浴摇床提取样品，采用铼为内标，ICP-MS 测定碘含量，方法检出限为 5.6 μg/kg。

由于 TMAH 试剂难以纯化，空白值较高，使用此类碱性溶液通常会导致溶液中的碳含量较高，采用 ICP-MS 分析前需进一步稀释样品，从而导致检出限较高。近年来，有学者提出了微波诱导燃烧法（microwave-induced combustion，MIC），此方法是在封闭的容器中基于样品燃烧，通过氧化去除有机物，减少待测元素挥发损失。样品燃烧后，气体被吸收到氨水等适宜的溶液中，采用 ICP-MS 测定碘含量。微波诱导燃烧法在样本消解中具有一定的优势，可一定程度上减轻记忆效应。Jussiane[24]采用 MIC 法处理食用面粉样品，25 mmol/L NH$_4$OH 为吸收液，使用 ICP-MS 测定面粉中碘，方法检出限为 0.76 ng/g，加标回收率为 90%～95%，RSD 小于 3%。Rondan[25]采用 MIC 法处理水稻样品，加入 20 bar 氧气，6 mL 50 mmol/L NH$_4$OH 为吸收液，用超纯水稀释定容后采用 ICP-MS 测定碘含量，加标回收率为 94%～97%，检出限 0.0008 mg/kg。Toralles[26]采用 MIC 法处理全蛋及其组分（蛋清和蛋黄），6 mL 50 mmol/L NH$_4$OH 为吸收液，用超纯水稀释定容后采用 ICP-MS 测定碘含量。采用 MIC 方法空白背景较低，可获得低的检出限。

23.1.2.3　多元素检测

国家食品安全风险监测标准操作规程，从 2015 年开始采用多元素同时测定的标准操作规程，目前已建立 46 个元素同时测定的多元素测定方法——2020 年国家食品安全风险监测标准操作规程[27]，规定了采用电感耦合等离子体质谱法测定蔬菜、水果、食用菌、谷类粮食、婴幼儿谷类辅助食品、虾蛄、茶叶、枣及贝类中 46 种元素的多元素测定方法，该规程适用于蔬菜、水果、食用菌、谷类粮食、婴幼儿谷类辅助食品、虾蛄、茶叶、枣及贝类中铝、锰、铜、锌、铬、钡、钼、银、钴、钒、硼、镉、铅、硒、锑、铍、镍、铊、铀、钍、锡、砷、钾、钠、钙、镁、铁、锂、锶、汞及钪、钇、镧、铈、镨、钕、钐、铕、钆、铽、镝、钬、铒、铥、镱、镥的含量测定。称取一定量试样加入一定量的硝酸和过氧化氢，采用微波消解或密闭高压罐消解进行样品消解，采用电感耦合等离子体质谱法测定 46 种元素。

23.1.2.4　三重串联四极杆电感耦合等离子体质谱技术（ICP-MS/MS）

三重串联四极杆电感耦合等离子体质谱（ICP-MS/MS）实现了离子的选择性提取与反应，为硫、磷等难分析元素的准确测定奠定了技术基础。ICP-MS/MS 首先通过第一级四极杆（Q$_1$）实现了指定质荷比离子的筛选，将指定质荷比的离子送入二级碰撞/反应池（ORC），同时过滤掉其他质荷比的干扰离子；根据目标离子性质在 ORC 中通入不同种类碰撞/反应气，消除多原子离子干扰，或生成目标多原子离子以将目标离子与同质荷比干扰离子区分；之后通过第三级四极杆（Q$_2$）实现对 ORC 中生成的目标质荷比离子进行选择分析，实现高灵敏度、低背景干扰检测。目前，ICP-MS/MS 技术已经成功应用于食品中硫、磷及农药的测定。

王丙涛等[28]研究建立了 ICP-MS/MS 检测食品中 P、As 和 Se 含量的检测方法。分析了 P、As 和 Se 在检测过程中常见的质谱干扰和电离效率，比较了标准模式、He 模式、H$_2$ 模式、O$_2$ 模式及双四极杆等各种检测模式，发现采用 O$_2$ 模式双质谱对 P、Se 的检测最为准确，As 测定采用 He 模式可实现精准检测。根据优化后的检测条件，经过 Q$_1$ 时同位素 M 进入碰撞反应池，与 O$_2$ 反应生成的氧化物 MO 进入 Q$_2$ 被采集，通过二级质谱质量转移可以消除干扰。P 在 2.0～500 μg/L 范围内、As 和 Se 在 0.5～200 μg/L 浓度范围内均具有良好的线性，加标回收率在 94.0%～101.5%之间，RSD 小于 2.1%。通过不同模式的考察，实现了复杂基质样品中 P、As 和 Se 的准确测定。王晓伟等[29]针对植物源性中药材中二氧化硫（SO_2）超标问题进行了研究，基于三重串联四极杆电感耦合等离子体质谱（ICP-MS/MS）技术建立了快捷、准确的二氧化硫检测方法。该方法可以有效地降低背景中多原子离子的干扰，提高检测灵敏度，使用氧气为反应气，对 $^{32}S^{16}O_4^{2-}$ 和 $^{34}S^{16}O_4^{2-}$ 的检出限分别为 5.48 μg/L 和 9.76 μg/L，线性范围 0.02~100 mg/L，可实现中草药中 SO_2 浓度的准确测定。Guo 等[30]基于串联电感耦合等离子体四极杆质谱的质量转移策略，建立了食品和茶叶中总氟的测定方法，为解决电感耦合等离子体质谱法测定氟时灵敏度低和干扰大的问题提供了新的途径。

Nelson 等[31]首次应用气相色谱-电感耦合等离子体-质谱/质谱（GC-ICP-MS/MS）法对各种食品基质中含磷农药进行了分析检测，该方法对有机磷农药检测具有灵敏度高和选择性好的优点，表明 GC-ICP-MS/MS 在超痕量磷、硫和含氯化合物的分析中具有潜力。Nelson 等[32]使用 GC-ICP-MS/MS 通过测量农药中的 P 和 S（以及 Cl 和 Br）来测定农药残留，GC-ICP-MS/MS 提供了非常出色的选择性和特异性，灵敏度优于目前成熟的农残测定方法。该研究方法也可应用于其他化合物的测定，如有机磷化学试剂、溴代阻燃剂、聚合物添加剂以及香精和香料中痕量硫化合物。GC-ICP-MS/MS 方法适用于通过测量杂原子含量对有机磷和有机硫农药进行高选择性和高灵敏度的检测，为有机磷农药的测定提供了更灵敏的方法。

23.1.3 元素形态分析

元素形态：某一元素以不同的同位素组成、不同的电子组态或价态以及不同的分子结构等存在的特定形式。元素形态分析：环境和生物样品中与生命有关的特定元素的一个或多个化学、物理形态的定性定量分析过程。

随着社会的发展，形态分析在环境、食品、生物分析中越来越占有重要的地位，由于元素在环境中的迁移转化规律，元素的毒性、生物利用度、有益作用及其在生物体内的代谢行为在相当大的程度上取决于该元素存在的化学形态。从 20 世纪 60 年代日本水俣病-甲基汞中毒事件开始，元素形态分析得到了普遍重视和迅速发展，特别是对汞、砷、铅、硒、锡、碘、铬等元素形态分析的研究。

23.1.3.1 砷形态的研究进展

砷是一种类金属元素，天然存在于食品和水中。各形态的砷毒性依次为砷化氢（H_3As）>亚砷酸盐[As(Ⅲ)]>砷酸盐[As(Ⅴ)]>甲基砷酸（MMA）>二甲基砷酸（DMA）>三甲基砷氧化物（TMAO）>砷胆碱（AsC）>砷甜菜碱（AsB），无机砷的毒性最大，有机胂的毒性较小。砷的危害主要源于无机砷，无机砷暴露主要来源是食品和水。人类长期摄入无机砷含量高的食品可导致皮肤损害、发育毒性、神经毒性及皮肤癌等。AsB 是鱼类和其他海产品中砷的主要砷形态，AsC 被认为是砷甜菜碱的前体，AsB 和 AsC 通常被认为是无毒的。其他砷化合物，如 MMA、DMA 与 TMAO 一起常被发现在海洋生物中，也与砷糖和砷脂一起存在于海带和海藻中。阿散酸（p-ASA）、洛克沙胂（ROX）、4-羟基苯胂酸（4-HPAA）、4-硝基苯胂酸（4-NPAA）等有机胂制剂作为饲料添加剂可以促进畜禽生长，但具有一定的毒性。稻米中检出的含砷化合物一般为 As(Ⅴ)、As(Ⅲ)、MMA 和 DMA。野生食用菌中也常检测出多种含砷化合物。针对食品中存在的不同毒性的含砷化合物，我国《食品安全国家标准 食品中污染物限量》（GB 2762—2017）规定了食品中总砷和无机砷的限量。

色谱分离和原子光谱检测技术联用综合了色谱的高效分离和元素检测器的高选择性优点；色谱和原子光谱联用技术选择性好、灵敏度高，已成为砷形态分析的重要手段。在砷形态分析中，常用的检测技术包括原子吸收光谱（AAS）、原子荧光光谱（AFS）、电感耦合等离子体质谱（ICP-MS）与高效液相色谱联用分析砷形态。毛细管电泳也常与氢化物发生原子荧光光谱（HG-AFS）或电感耦合等离子体质谱（ICP-MS）联用进行砷形态分析测定。

1. 高效液相色谱-原子吸收光谱联用技术（HPLC-AAS）

以往 HPLC-AAS 具有灵敏度高、操作简便的特点，是砷形态分析中常用的方法之一。但采用氢化物发生技术进行砷形态分析时，需注意 AsB、AsC、砷糖等有机胂不能直接用这种方法分析检测，因为它们不能形成挥发性氢化物，需要进一步氧化破坏 C—As 键，把有机胂转化成可挥发性砷化合物。当 HPLC 用于砷形态分析时，AsB 和 As(Ⅲ)通常在许多色谱分离条件下分不开，得到的峰是重叠的。为解决此问题，可以采用两次进样，一次经氧化，另一次不经氧化，通过差值法得到 AsB 的含量。

Villa-Lojo 等[33]采用微波萃取，以 $K_2S_2O_8$ 作为氧化剂，使用阴离子交换柱(Sep-Pak 柱)分离测定鱼组织中的 As(Ⅲ)、As(Ⅴ)、MMA、DMA 和 AsB、AsC。分两次进样，分析 As(Ⅲ)、As(Ⅴ)、MMA 和 DMA 时，微波萃取时只使用水，流动相是磷酸缓冲液（pH=6.0）；分析 AsB、AsC 时，用 $K_2S_2O_8$ 作氧化剂，流动相为水；从而实现六种砷形态的分析测定。结果显示，采用微波萃取对阴离子形态的砷进行分析，As(Ⅲ)的测量值稍有增加，其他砷形态增加值并不明显。Niedzielski 等[34]采用 HPLC-HG-AAS 测定了牛肝菌子实体中的 As(Ⅲ)、As(Ⅴ)和 DMA。

2. 高效液相色谱-原子荧光光谱联用技术（HPLC-AFS）

氢化物发生过程，分析物可以与试样中共存物分离，仅气态氢化物被引入到检测器中，而光谱干扰和化学干扰大大降低，AFS 的检出限得到较大的改善。

刘淑晗等[35]建立了南极磷虾及其制品中 6 种砷形态的高效液相色谱-（紫外）氢化物发生原子荧光光谱[HPLC-(UV)-HG-AFS]分析方法。样品经 1%硝酸浸泡过夜后，于 70℃水浴提取 2 h，冷却至室温，加入 0.5 mL 3%乙酸，混匀，于 4℃冰箱放置 5 min 后，加入正己烷去脂，以 8000 r/min 离心 15 min，弃去上层溶液，重复 1 次，过膜后上机测定。在最佳实验条件下，As(Ⅲ)、As(Ⅴ)、MMA、DMA、AsB 和 AsC 的检出限(LOD, S/N≥3)均为 0.01 mg/kg，定量限(LOQ, S/N≥10)均为 0.03 mg/kg。采用该方法测定了南极磷虾及其制品中 6 种砷形态，无机砷含量均低于水产品及其制品中无机砷的国家限量标准。

Cui 等[36]针对动物源性食品中有机胂制剂的残留问题，建立了一种 HPLC-(UV)-HG-AFS 法测定动物源性食品中有机胂制剂的分析方法，包括对氨基苯胂酸（p-ASA）、硝苯胂酸胂（NIT）和洛克沙胂（ROX）。该方法采用 C_{18} 柱，流动相为 50 mmol/L KH_2PO_4（含 0.1% V/V 三氟乙酸，pH=2.43）。采用加速溶剂提取法（ASE）处理样品，对动物肉中有机胂制剂进行提取。p-ASA、NIT 和 ROX 的检出限和定量限分别为 0.24 ng/mL、0.74 ng/mL、0.41 ng/mL 和 0.72 ng/mL、2.24 ng/mL、1.24 ng/mL。

Farías 等[37]研究了高效液相色谱-氢化物发生-原子荧光光谱联用法（HPLC-HG-AFS）测定水稻中的 As(Ⅲ)、As(Ⅴ)、MMA、DMA。样品经（103±2）℃干燥，采用 0.28 mol/L 硝酸为提取剂，于（95±3）℃提取 90 min。采用 NIST 标准物质"大米粉"进行质量控制，结果显示总的砷形态含量占总砷含量的 99.7%。

3. 高效液相色谱-电感耦合等离子体质谱联用技术（HPLC–ICP–MS）

高效液相色谱与电感耦合等离子体质谱联用技术具有操作简单、灵敏度高、线性范围宽等优点，是当前最主要的砷形态分析技术之一。目前，HPLC-ICP-MS 技术主要用于稻米[38-43]、海产品[44-52]、畜禽肉制品[53-55]、蘑菇[56-59]、乳制品[60-62]等不同基质中砷形态的分析测定。利用 HPLC-ICP-MS 进行砷形态分析时，多采用反相色谱或离子交换色谱进行分离。

吴池莹等[38]选用 0.15 mol/L 硝酸作为提取液，加入过氧化氢，80℃微波辅助萃取 20 min，采用 LC-ICP-MS 对食品中无机砷进行测定。结果表明：过氧化氢可将亚砷酸盐[As(Ⅲ)]完全氧化成砷酸盐[As(Ⅴ)]，从而使出峰时间与有机胂重合的 As(Ⅲ)分离开。Jinadasa 等[40]将超声辅助萃取和涡旋辅助分散微固相萃取相结合，采用离子印迹聚合物作为选择性吸附剂，对水稻提取物中的无机砷[As(Ⅲ)和 As(Ⅴ)]进行分离和预富集，采用 HPLC-ICP-MS 进行测定。在最佳条件下研究了该过程的分析性能：以 1∶1 甲醇/超纯水为萃取剂，50 mg 吸附剂，萃取时 pH 8.0，在 40%振幅下连续超声 1.0 min，1000 r/min 涡旋 1.0 min，用超纯水以 1000 r/min 的涡流洗脱 1.0 min，预浓缩系数为 10。As(Ⅲ)和 As(Ⅴ)的检出限分别为 0.20 μg/kg 和 0.41 μg/kg，灵敏度远低于欧盟制定的大米和含大米产品中无机砷的限量。

吕超等[44]采用 HPLC-ICP-MS 联用技术建立了 As(Ⅲ)、AsB、DMA、MMA、As(Ⅴ) 5 种砷形态的分析方法。以乙酸溶液对水产膳食样品进行超声提取，采用 HPLC-ICP-MS 分析其砷形态，线性范围为 2.5~500 μg/L，检出限为 0.6~0.9 μg/L；测定结果表明水产品及水产类膳食中砷形态的主要存在形式为砷甜菜碱。

三重四极杆电感耦合等离子体质谱（ICP-MS/MS）具有抗干扰强、灵敏度高等优点，近年来逐渐应用于食品中砷形态分析。Schmidt 等[45]采用液相色谱与三重四极杆电感耦合等离子体质谱联用法（LC-ICP-MS/MS）测定经过不同烹饪方法处理后的海鲜中的 AsB、As(Ⅲ)、DMA、MMA 和 As(Ⅴ)。使用 O_2 作为反应气将 ^{75}As 转化为 $^{75}As^{16}O$ 后进行测定。研究了烹饪处理（煮、煎和炒）及添加或不添加香料（盐、柠檬汁和大蒜）对黑鳍鲨和亚洲虎虾中砷形态的影响。结果表明，烹饪处理实际上不影响海鲜中砷形态的稳定性。研究发现水煮后海鲜中的分析物损失很大（从 15%到 45%）。由于不同海产品所含有的砷形态种类和砷形态含量不同，特别是可食用海藻含有多种砷糖，且砷糖含量大大超过其他类型砷形态的总和，因此仅采用阴离子交换 LC-ICP-MS 技术无法实现对海产中砷化合物的完全分离和准确测定。通过使用阴离子交换柱和阳离子交换柱相互补充，可以实现对海产品中复杂砷形态化合物的分离和检测。Zmozinski 等[47]利用阳离子交换 LC-ICP-MS 分析测定了海产品中 As(Ⅲ)、As(Ⅴ)、MMA 和 DMA，同时采用阴离子交换 LC-ICP-MS 分析检测了同一海产品中的 AsB、AsC 和 TMAO，成功避免了 AsB 对 MMA 和 MMA 对 TMAO 的干扰。

有机胂制剂作为饲料添加剂，可以促进畜禽生产、提高饲料转化率和提高产蛋量，但也具有一定的毒副作用，需要对其含量进行检测。陈绍占等[53]建立了畜禽肉制品中 As(Ⅲ)、MMA、DMA、As(Ⅴ)、p-ASA、ROX、4-HPAA 和 4-NPAA 8 种砷形态化合物的 HPLC-ICP-MS 分析方法。以甲醇-水（1∶1，V/V）为提取剂，采用超声水浴进行前处理，选取 Dionex IonPac AS14 阴离子交换柱为分析柱，50 mmol/L $(NH_4)_2HPO_4$（pH=7.3）和水为流动相，梯度洗脱，对畜禽肉制品中的砷形态化合物进行分析测定。吴思霖等[54]建立了有效分离检测鸡肉及鸡肝样品中 ROX、ASA、NPAA、卡巴胂（CBS）、As(Ⅴ)、As(Ⅲ)、MMA、DMA、AsB 和 AsC 共 10 种砷形态化合物的 HPLC-ICP-MS 分析方法。采用 10%甲醇为提取液，碳酸铵溶液为流动相，以阴离子分析柱进行分离，ICP-MS 测定，通过对实际样品进行分析测定，在鸡肝中检出 ASA 和 As(Ⅲ)。Zhao 等[55]采用 HPLC-ICP-MS 联用技术测定了中国市场上鸡肉中的砷形态，评估了从中国 10 个城市采集的鸡肉中砷含量和形态。在 81 个成对的生熟样品中，90%以上的样品中检出 ASA 和 ROX，表明中国有机胂制剂的使用很普遍，首次发现 ASA 比 ROX 对鸡肉中总 As 的贡献更大。鸡肉中砷形态显示出相当大的地域差异，从具有较高含量的 ROX 和 ASA 的城市中检测到较高含量的无机砷（iAs），这表明有机胂制剂的使用增加了鸡肉中 iAs 的水平。

Nearing 等[56]用 ICP-MS 测定总砷，用 HPLC-ICP-MS 和 X 射线吸收光谱（XAS）技术测定了蘑菇中的砷形态。报道了 AsB 是马勃科和伞菌科的主要成分，但在原木蘑菇中普遍不存在，说明微生物群落可能影响蘑菇中砷形态的存在形式。采用 XAS 分析中,首次在蘑菇中发现了 As(Ⅲ)-硫化合物。Komorowicz 等[57]使用 ICP-DRC-MS 和 IC/ICP-DRC-MS 技术对中国云南省采集的三种蘑菇中的总砷和砷形态进行了定量分析,实验结果证明 AsB、DMA、As(Ⅲ)、MMA 和 As(Ⅴ)在这些蘑菇中的浓度范围很宽,主要取决于菌种,可能也与它们生长的地域有关。样品经酶辅助提取后,通过改良的 Unified BARGE 方法模拟人胃肠道中的酶促环境来评估砷的生物可利用性。由于 IC/ICP-DRC-MS 技术仅能测定相对简单的砷化合物;因此采用酶辅助提取后,使用 SEC-UV-Vis/ICP-DRC-MS 进行测量,以评估蛋白结合砷的存在。Chen 等[58]采用 IC-ICP-MS 法对食用菌中的 AsB、AsC、MMA、DMA、As(Ⅲ)和 As(Ⅴ)进行了测定,选用 Dionex IonPac AS19 阴离子交换柱为分析柱,以超纯水和 50 mmol/L $(NH_4)_2CO_3$ (pH=9.7)为流动相进行梯度洗脱,20 min 内可很好的分离测定 6 种砷形态。Zou 等[59]采用 HPLC-ICP-MS/MS 联用法测定食用菌中 As(Ⅲ)、As(Ⅴ)、MMA、DMA、AsB 和 AsC。分析测定了 8 种蘑菇 266 个样品中砷形态,结果表明,大多数野生食用菌都含有有机胂,主要是 AsB 和 AsC。但是,假蜜环菌中的无机砷含量(3.63 mg/kg)和部分栽培的姬松茸中的无机砷含量(最高 4.50 mg/kg)相对较高,对消费者的健康构成潜在风险。

丁宁等[60]采用碳酸铵-甲醇(V/V,99∶1)溶液提取,建立了 HPLC-ICP-MS 同时测定乳制品中 AsC、AsB、As(Ⅲ)、As(Ⅴ)、MMA、DMA 和 p-ASA 7 种砷形态的分析方法。结果表明,8 min 内 7 种砷形态可很好的分离测定。赵建国等[61]建立了酶解析-HPLC-ICP-MS 法测定乳粉与生乳中 AsB、As(Ⅲ)、As(Ⅴ)、MMA、DMA 的方法,采用碱性蛋白酶水解将吸附于蛋白质中的砷解析后,用质量分数 12%柠檬酸溶液进一步去除杂质后于 8000 r/min 离心 10 min,用 PEP-2SPE 小柱净化后分析测定。Stiboller 等[62]报道了通过 HPLC-ICP-MS 测定母乳中痕量砷形态的方法。比较了单四级杆和三重四极杆质量分析仪对牛奶中存在的天然氯化物的干扰消除作用,发现三重四极杆仪器可更有效克服 $ArCl^+$干扰,无需使用梯度洗脱的 HPLC 条件即可消除 $ArCl^+$干扰。该方法使用三氟乙酸沉淀蛋白质后加入二氯甲烷或二溴甲烷除脂,将水相进行阴离子交换和阳离子交换/混合模式色谱分析,分别以碳酸氢铵水溶液和吡啶缓冲液为流动相,分析测定两个挪威母乳样品发现了 AsB、DMA 和一个未知的砷形态,未检出无机砷。

4. 毛细管电泳和电感耦合等离子体质谱联用技术(CE-ICP-MS)

毛细管电泳(capillary electrophoresis,CE)采用电驱动进行分离,具有高效的分离效率、水相运行和多种分离模式等优点,十分适合于砷形态分析。

CE-ICP-MS 联用技术常用于测定藻类和鱼类中 As(Ⅲ)、As(Ⅴ)、MMA、DMA、AsB 和 AsC 等砷形态[63-65]。在检测紫菜、海带和食用海藻等砷含量高、以砷糖为主的样品中显示出突出的优势,只需要改变 CE 运行缓冲溶液的 pH 就可消除砷糖对 DMA 或其他砷形态的干扰[66];而使用 IC-ICP-MS 时,需要分别使用阴离子交换柱和阳离子交换柱,通过 2 次分析才能实现对所有砷形态的准确测定。使用 CE-ICP-MS 联用技术,还可实现对 2 种不同元素形态同时分析。Liu 等[67]采用 CE-ICP-MS 联用技术,同时分析测定了 6 种砷形态和 5 种硒形态,分析时间仅需 10 min,砷和硒的检出限均可达到 ppb 级别。相比于 LC-ICP-MS,CE-ICP-MS 的缺点也十分明显,稳定性和重现性较差、操作条件较苛刻等。目前 CE-ICP-MS 技术主要用于研究,较少应用于实际样品的砷形态分析测定。

23.1.3.2 汞形态的研究进展

汞是影响人类和生态系统健康的剧毒元素之一,也是研究最多的环境污染物之一。在自然界中,汞以金属汞、无机汞和有机汞的形态存在。无机汞有一价及二价的化合物,有机汞有甲基汞、二甲基汞、乙基汞、苯甲基汞等。环境中存在各种各样的汞形态,汞的化学形态决定了它的生物利用度、传输、持久性和对人体的影响，不同形态的汞，毒性不同，有机汞的毒性大于无机汞，其中甲基汞的毒性最强。日本水俣湾有机汞污染导致的水俣病患者有 2248 人（其中死亡 1004 人）。该事件引起了世界范围对甲基汞的关注，各国针对鱼肉中甲基汞含量都制定了严格的限量标准：国际食品法典委员会（CAC）限定鱼类产品中甲基汞的含量为 0.5 mg/kg；我国鱼类中甲基汞限定标准同样为 0.5 mg/kg。欧盟已立法要求不仅要测定总汞的含量，而且要区分不同汞化合物形态的含量，因此对甲基汞和汞的不同形态化合物进行分析研究非常必要。近年来，对甲基汞和不同存在形式汞化合物的分析研究已成为分析化学研究领域的一个热点。

汞形态分析常用的分离方法包括气相色谱法、液相色谱法和毛细管电泳法；常用的检测技术为原子荧光光谱和电感耦合等离子体质谱。

1. 高效液相色谱-原子荧光光谱联用技术（HPLC-AFS）

吕超等[68]建立了盐酸提取-高效液相色谱与原子荧光联用技术（HPLC-AFS）测定水产中无机汞（Hg^{2+}）、甲基汞（MeHg）、乙基汞（EtHg）的分析方法。采用 5 mol/L 盐酸，超声萃取 2 h，加入 10 g/L 半胱氨酸溶液络合无机汞，滴加 6 mol/L 氢氧化钠溶液中和至中性。由测定结果可知，水产品中汞化合物形态主要是以 MeHg 为主，个别样品中含有少量 Hg^{2+}。

Brombach 等[69]建立了固相萃取-高效液相色谱-冷蒸气发生-原子荧光光谱法（SPE-HPLC-CV-AFS）测定大米中的 MeHg。以不同的大米样品为研究对象，研究了该方法在痕量水平测定大米和大米制品中 MeHg 的可靠性。

Grijalba 等[70]以离子液体为流动相，采用反相高效液相色谱-紫外-冷蒸气发生-原子荧光光谱法（RP-HPLC-UV-CV-AFS）分离测定 Hg^{2+}、MeHg 和 EtHg。经过优化选择，以甲醇和 0.4%（V/V）1-甲基-3-辛基氯化咪唑鎓-100 mmol/L 氯化钠-20 mmol/L 柠檬酸盐缓冲液（pH=2.0）梯度洗脱，12 min 可实现汞形态的分离测定。

Liu[71]建立了以光诱导化学蒸气发生（CVG）为接口，在线联用汞-半胱氨酸-IC-AFS 快速测定海产中 MeHg 的方法。以 3%乙腈、1%（w/w）L-半胱氨酸、20 mmol/L 吡啶和 160 mmol/L 甲酸为流动相，（pH 为 2.4），用 Hamilton PRP-X200 为分析柱，在 7 min 内完成了 Hg^{2+} 和 MeHg 的分离，这两种物质在紫外线辐射下被甲酸还原，在线转换为 Hg^0，用原子荧光光谱仪（AFS）测量。

Li 等[72]开发了一种新型的 UV/纳米-ZrO_2/HCOOH 系统，首次将其制成雾化装置，作为 HPLC 和 AFS 之间的在线接口进行汞的形态分析和测定。紫外线在纳米 ZrO_2 的导带上产生的电子将汞形态还原为汞冷蒸气，采用 AFS 测定。使用 HPLC-（UV/nano-ZrO_2/HCOOH）-AFS 测定 Hg^{2+}、MeHg 和 EtHg 的检出限分别为 24 ng/L、13 ng/L 和 16 ng/L。

Ai 等[73]利用 Fe_3O_4 磁性纳米颗粒（MNPs）制备了一种新型、绿色、高效的柱后氧化方法，无需微波/紫外线照射即可在线将氢化物发生/冷蒸气发生（HG/CV）的非活性组分转化为活性组分，并将其应用于 HPLC-HG/CV-AFS 法。选择 Hg^{2+}、MeHg、EtHg 和苯基汞（PhHg）进行分析，Hg^{2+}、MeHg、EtHg 和 PhHg 的检出限（LOD）分别为 0.7 μg/L、1.1 μg/L、0.8 μg/L 和 0.9 μg/L（以 Hg 计）。通过分析肌肉组织标准物质（DORM-2 鱼）和水样标准物质验证方法的准确性。

2. 气相色谱-原子荧光光谱联用技术（GC-AFS）

Jókai 等[74]报道了一种测定匈牙利市售的典型鱼类食品样品中 MeHg 的分析方法，该方法以氢氧化钾和氢氧化钠对样品进行碱消化，然后用四苯基硼酸钠水溶液进行苯基化衍生和固相微萃取（SPME）。采用有证参考物质（CRM）BCR-464、TORT-2 和 BRC 710 对 SPME-GC-pyrolysis-AFS 法进行验证，并分析测定了 16 种市售水产品及消费量大的海鲜样品（鲱鱼、沙丁鱼和鳕鱼），样品中 MeHg 浓度在 0.016～0.137 μg/g 的范围内。

近年来的研究表明，MeHg 可被植物（如水稻）吸收和积累，成为人类通过饮食接触 MeHg 的重要途径之一。Jiménez-Moreno 等[75]建立了一种快速、灵敏同时测定水稻和水生植物中 MeHg 和 Hg^{2+} 的方法，采用微波萃取汞形态，经乙基化衍生，气相色谱-热解原子荧光（GC-pyro-AFS）进行检测分析。采用酸（6 N HNO_3）和碱（四甲基氢氧化铵，TMAH）为萃取剂，在优化好的条件下，40 min 可完成 MeHg 和 Hg^{2+} 的同时测定。对于超低汞含量样品的分析，需要增加氮气蒸发预富集步骤，在酸萃取的情况下，两种汞形态检出限可达到 0.7～1.0 ng/g。

刘沛明[76]建立了萃取-水相乙基化衍生结合 GC-CVAFS 测定大米和鱼样品中 MeHg 的方法。应用 25%氢氧化钾-甲醇溶液碱提取样品，再滴加浓盐酸中和，用有机溶剂 CH_2Cl_2 萃取，之后将样品中的 MeHg 萃取预富集至水相，经乙基化反应后用 GC-CVAFS 分析测定。崔颖等[77]采用 30%硝酸溶液提取海产动物样品，用 2 mol/L 醋酸钠-醋酸缓冲溶液调节 pH 后，用吹扫捕集-气相色谱-冷原子荧光光谱仪测定其中的 MeHg 和 EtHg 含量。使用衍生试剂，将 MeHg 转化为甲基丙基汞，EtHg 转化为乙基丙基汞，吹扫捕集进行富集并进一步消除了基体干扰。

邹学权等[78]采用丙基化衍生-吹扫捕集-原子荧光光谱法定量测定动物源食品中 MeHg 与 EtHg，样品经 25%氢氧化钾-甲醇溶液提取后，加入缓冲溶液，用四丙基硼化钠衍生试剂将 MeHg 与 EtHg 转化为易挥发的衍生产物，在线吹扫捕集富集后采用 AFS 测定 MeHg 与 EtHg。

王维洁等[79]建立了水产品中 MeHg 残留的快速测定方法。样品经过 25%氢氧化钾-甲醇溶液提取，调 pH 为中性，加入醋酸钠缓冲溶液，采用四乙基硼化钠进行衍生后，进入全自动烷基汞分析仪检测 MeHg 含量。

3. 高效液相色谱-电感耦合等离子体质谱联用技术（HPLC-ICP-MS）

刘丽萍等[80]建立了 HPLC-ICP-MS 联用技术测定水产品中汞形态化合物的分析方法。样品加入 5 mol/L 盐酸，经超声萃取离心。取上清液滴加 35%氨水至中性，加入 2%半胱氨酸络合，离心过膜后使用 C_{18} 色谱柱，5%甲醇-0.06 mol/L 乙酸铵-0.1%半胱氨酸为流动相，采用 HPLC-ICP-MS 测定，检出限为 0.5～0.8 μg/L。

冯晓青等[81]建立了基于微探头超声破碎为基础的盐酸浸提和酶水解提取的前处理方法，用于快速分析海鲜样品中 MeHg。分别采用 7 mol/L 盐酸溶液和含 0.75%(W/V)蛋白酶 XIV 及 2.5%(V/V)2-巯基乙醇的提取液从海鲜样品中提取汞，用含有 0.1%(W/V)的 L-半胱氨酸和 0.1%(W/V)L-半胱氨酸 $HCl·H_2O$ 流动相在 C_{18} 色谱柱可在 4 min 内实现汞形态的分离。

Zhu 等[82]通过 $F_3O_4@SiO_2@γ$-巯基丙基三甲氧基硅烷（γ-MPTS）磁性纳米颗粒提取鱼样中 Hg^{2+}、MeHg 和 PhHg，采用高比例的甲醇为流动相，采用磁固相萃取（MSPE）-高效液相色谱（HPLC）-电感耦合等离子体质谱法（ICP-MS）方法在 8 min 内实现三种汞形态的快速分离，Hg^{2+}、MeHg 和 PhHg 的检出限为 0.49～0.74 ng/L。

龚燕等[83]采用（0.1%L-半胱氨酸＋5%甲醇）的提取液提取稻米中 Hg^{2+} 和 MeHg，以 0.1% L-半胱氨酸＋0.06 mol/L 乙酸铵＋5%甲醇为流动相，用 Venusil® MP C_{18} 色谱柱分离，以电感耦合等离子体质谱分析测定 Hg^{2+} 和 MeHg；方法的检出限（LOD）为 0.001 mg/kg。

贾彦博等[84]建立了 HPLC-ICP-MS 法，用于测定东海乌参样品中的 Hg^{2+}、MeHg、EtHg 和 PhHg 等 4 种汞形态的含量。样品经 25% KOH 微波萃取 1 h，采用 Agilent Zorbax Plus C_{18} 色谱柱分离，以 10 mmol/L 醋酸铵+0.12%L-半胱氨酸缓冲盐及 2%甲醇作为流动相进行梯度洗脱，采用 HPLC-ICP-MS 进行分析测定。Hg^{2+}、MeHg、EtHg 和 PhHg 可在 10 min 内完成分离，检出限分别为 0.8 ng/mL、0.2 ng/mL、0.2 ng/mL 和 0.8 ng/mL。

Zou 等[85]报道了野生食用菌可以从周围环境中积累大量的汞，可对人体健康构成威胁。该研究采用 HPLC-ICP-MS 同时测定野生食用菌中 Hg^{2+}、MeHg、EtHg 和 PhHg，采用 C_8 色谱柱快速分离，方法线性范围为 0~50 μg/L，检出限为 0.6~4.5 μg/kg。

4. 气相色谱-电感耦合等离子体质谱联用技术（GC-ICP-MS）

Chung 等[86]建立了 GC-ICP-MS 同时测定 MeHg、EtHg 的可靠、灵敏的方法。先用胰酶提取样品，再用盐酸提取。用四苯基硼酸钠缓冲水溶液对提取液中的 MeHg 和 EtHg 进行衍生，采用 GC-ICP-MS 对样品进行了分析。分别测定了有证参考物质 NIST SRM 1947（苏必利尔湖鱼）、NIST SRM 1566b（牡蛎组织）和 NRC Tort-2（龙虾肝胰腺）中 MeHg 的含量，测定值与标准值相吻合。MeHg 和 EtHg 的方法检出限均为 0.3 μg/kg（以 Hg 计）。

Valdersnes 等[87]与欧洲标准化组织（CEN）合作开发了测定海产品中 MeHg 的同位素稀释-气相色谱-电感耦合等离子体质谱分析方法（ID-GC-ICP-MS）。方法过程包括：在组织样品中加入甲基汞（$Me^{201}Hg$），用四甲基氢氧化铵分解，调节 pH，用四乙基硼化钠衍生后，用正己烷萃取衍生后的 MeHg；之后用 GC-ICP-MS 对样品进行分析，并用同位素稀释方程计算结果。利用有证标准物质 NIST SRM 1566b、NIST SRM 2977、NRCC TORT2、NRCC DORM 3、NRCC DOLT 4 和 ERM CE 464 考察了方法的准确性。

Esteban-Fernández 等[88]采用双同位素稀释-气相色谱-电感耦合等离子体质谱法（GC-ICP-MS）测定了新鲜鱼样和冻干鱼样中的 MeHg。使用 25% TMAH 作为萃取液，在 50℃超声水浴中萃取鱼样。超声处理后，离心，取 1 mL 上清液，加 4 mL 醋酸钠缓冲溶液、0.4 mL 四丙基硼化钠溶液和 1 mL 正己烷。缓慢摇动 10 min 后，收集有机相（必要时离心），直接用 GC-ICP-MS 分析测定。

Dressler 等[89]建立了流动注射冷蒸气发生-电感耦合等离子体质谱法（FI-CVG-ICP-MS）和 GC-ICP-MS 测定红酒中总汞、Hg^{2+} 和 MeHg 的方法。使用 1%（M/V）四苯基硼酸钠溶液进行衍生，汞形态被萃取后通过 GC-ICP-MS 进行分析测定。Hg^{2+} 和 MeHg 的检出限分别为 0.77 μg/L 和 0.80 μg/L。

23.1.3.3 硒的研究进展

硒是人体必需微量元素，具有抗氧化、提高机体免疫力的功能。缺硒会导致人体免疫力下降，导致肿瘤、心血管等疾病，长期缺硒会罹患克山病和大骨节病等疾病。由于在生物体内硒的阈值较窄，硒摄入过量也会导致中毒。硒的生物活性和毒性不仅取决于硒总量，还与硒的化学形态息息相关。由于人体摄入的硒几乎全部来自于膳食，因此对食品中硒形态的分析测定具有重要意义。目前常用的分析检测技术有高效液相色谱-原子吸收光谱联用技术（high performance liquid chromatography-atomic absorption spectrometry，HPLC-AAS）、高效液相色谱-原子荧光光谱联用技术（high performance liquid chromatography-atomiomic fluorescence spectrometry，HPLC-AFS）、高效液相色谱-电感耦合等离子体质谱联用技术（high performance liquid chromatography-inductively coupled plasma mass spectrometry，HPLC-ICP-MS）、毛细管电泳-电感耦合等离子体质谱联用技术（capillary electrophoresis-inductively coupled plasma mass spectrometry，CE-ICP-MS）等。

1. 高效液相色谱-原子吸收光谱联用技术（HPLC-AAS）

Niedzielski 等[90]采用 HPLC-HG-AAS 测定富硒蘑菇中的亚砷酸根[Se(Ⅳ)]和砷酸根[Se(Ⅵ)]，蘑菇样品采用磷酸溶液超声水浴提取，使用 Supelco LC-SAX1 和 Zorbax SAX 阴离子色谱柱，流动相为 100 mmol/L 磷酸氢二钠和 10 mmol/L 磷酸二氢钾混合液进行分离，检测出了 Se(Ⅳ)和 Se(Ⅵ)。

2. 高效液相色谱-原子荧光光谱联用技术（HPLC-AFS）

龚洋等[91]采用 HPLC-HG-AFS 测定富硒稻谷中 Se(Ⅳ)、Se(Ⅵ)、硒代蛋氨酸（SeMet）和硒代胱氨酸（SeCys$_2$），样品采用链霉蛋白酶 E 于 37℃水浴超声提取后，经 Hamilton 阴离子色谱柱，以 40 mmol/L 磷酸氢二铵（pH=6.0）为流动相进行分离测定，硒形态的方法检出限为 0.01～0.06 mg/kg，发现富硒稻谷中主要以 SeMet 为主。

Zhou 等[92]采用 HPLC-AFS 测定不同硒源处理后平菇中的 SeCys$_2$、甲基硒代半胱氨酸（MeSeCys）、Se(Ⅳ)、SeMet、Se(Ⅵ)五种硒形态，平菇样品采用 Tris-HCl 缓冲液（pH=7.5）和蛋白酶 E 溶液于 37℃振荡提取，采用 Hamilton PRP-X100 阴离子交换色谱柱，以 60 mmol/L 磷酸氢二铵（pH=6.0）为流动相进行分析测定，实验表明平菇中的主要硒形态是 SeMet。钟银飞[93]采用 LC-AFS 测定大蒜中不同硒形态，样品用甲醇水（1∶2）溶液超声提取，离心后氮吹浓缩，用 PRP-X100 阴离子交换色谱柱及流动相 60 mmol/L 磷酸氢二铵溶液（pH=6.0）进行分析，SeCys$_2$、Se(Ⅳ)、SeMet 检出限分别为 60 μg/L、10 μg/L、55 μg/L。

尚德荣等[94]用 HPLC-HG-AFS 检测水产品中的硒形态，样品经水 70℃超声提取后，采用 Hamilton PRP-X100 色谱柱，流动相为 40 mmol/L 磷酸氢二铵（pH=6.0）HPLC-HG-AFS 测定 SeCys、Se(Ⅳ)、SeMet、Se(Ⅵ)四种硒形态，检出限分别为 1.2 ng/mL、1.4 ng/mL、0.8 ng/mL、2.1 ng/mL，结果显示扇贝中的硒形态为 SeCys 和 SeMet，海参中的硒形态为 SeCys 和 Se(Ⅳ)。韩婷婷等[95]也采用 HPLC-HG-AFS 检测胶州湾海产品中的硒形态，样品采用 1.5 mol/L 氢氧化钾溶液于 70℃超声提取，采用 AS11-HC 阴离子色谱柱，20 mmol/L 碳酸氢钠溶液/2%乙腈（pH 12～13）为流动相分析测定了 SeCys$_2$、SeMet、Se(Ⅳ)、Se(Ⅵ)，检出限分别为 1.53 μg/L、1.72 μg/L、0.30 μg/L、1.06 μg/L，在样品中主要检出了 SeCys$_2$ 和 SeMet。Xie 等[96]采用 1.5 mol/L 氢氧化钾和甲醇于室温下振荡提取海产品，调 pH 至 6.5 后离心过膜，以 Hamilton PRP-X100 色谱柱，30 mmol/L 磷酸二氢铵和甲醇（39∶1）为流动相，用 IC-HG-AFS 测定海产品中的 SeCys、SeMet、Se(Ⅳ)，检出限分别为 1.66 μg/L、0.99 μg/L、1.10 μg/L，海产品中主要硒形态是 SeCys 和 SeMet，未检出 Se(Ⅳ)。

钟永生等[97]用 HPLC-AFS 检测富硒鸡蛋中的 Se(Ⅳ)、Se(Ⅵ)、MeSeCys、SeCys$_2$、SeMet 五种硒形态，样品用超纯水超声 30 min，加 0.1 g 蛋白酶 K 40℃恒温水浴，酶解提取后，用 Hamilton PRP-X100 阴离子色谱柱，以 40 mmol/L 磷酸二氢钠和 20 mmol/L 氯化钠（pH=6.0）为流动相进行分析，检出限为 0.03～0.08 ng/mL。富硒鸡蛋中主要以 SeMet 为主，还检出少量的 Se(Ⅳ)。

Hu 等[98]采用 HPLC-HG-AFS 测定蛹虫草中的四种硒形态 SeCys$_2$、Se(Ⅳ)、SeMet 和 Se(Ⅵ)，样品用 25 mmol/L 乙酸铵超声提取，采用 Hamilton PRP-X100 阴离子交换色谱柱，以 40 mmol/L 磷酸氢二铵（pH=6.0）为流动相进行分析，蛹虫草中的硒形态以 SeCys$_2$ 和 SeMet 为主，还检出一个未知硒形态。

张妮娜等[99]用 LC-HG-AFS 测定富硒保健品中的硒形态，对以亚硒酸钠为硒源的保健品用 0.6 mol/L 盐酸溶液于 65℃水浴超声提取，以富硒酵母为硒源的保健品采用蛋白酶 K 和水于 37℃水浴提取后，采用 HamiltonPRP-X100 阴离子交换柱，以 40 mmol/L 磷酸氢二铵（pH=6.0）为流动相进行分析，Se(Ⅵ)、Se(Ⅳ)、SeCys$_2$、SeMet 检出限分别为 0.85 μg/L、1.07 μg/L、0.91 μg/L 和 1.73 μg/L，以亚硒酸钠作为硒源的样品检出 Se(Ⅳ)，以富硒酵母为硒源的 2 份样品中检出 Se(Ⅳ)、SeCys$_2$ 和 SeMet。

3. 高效液相色谱-电感耦合等离子体质谱联用技术（HPLC-ICP-MS）

陈绍占等[100]采用 HPLC-ICP-MS 测定谷类食品中硒形态，样品溶于纯水后加入蛋白酶 XIV 37℃水浴超声提取，采用 Hamilton PRP-X100 阴离子色谱柱，以 40 mmol/L 磷酸氢二铵（pH=5.0）和 60 mmol/L 磷酸氢二铵（pH=6.0）为流动相，梯度洗脱，SeCys$_2$、MeSeCys、Se(Ⅳ)、SeMet、Se(Ⅵ)检出限分别为 2.5 μg/L、5.0 μg/L、2.5 μg/L、10.0 μg/L 和 5.0 μg/L，谷类食品中主要以 SeMet 为主。

Gao 等[101]采用 HPLC-ICP-MS 测定水稻中 SeCys、SeMeCys、Se(Ⅳ)、SeMet、Se(Ⅵ)，样品溶于水采用蛋白酶 XIV 和脂肪酶在 37℃提取后，使用 StableBond C_{18} 色谱柱，以 0.5 mmol/L 四丁基氢氧化铵（TBAH）和 10 mmol/L 乙酸铵（pH=5.5）为流动相进行分析，检出限在 0.02～0.12 μg/L。水稻样品中以 SeMet 为主要，检出少量的 SeCys、Se(Ⅳ)或 Se(Ⅵ)。秦冲等[102]采用 HPLC-ICP-MS 检测富硒小麦中的硒形态，样品用蛋白酶 XIV 和超纯水进行微波萃取后，用 Hamiltion PRP-X100 阴离子交换色谱柱，以 6 mmol/L 柠檬酸为流动相（pH=5.0）进行分析，$SeCys_2$、Se(Ⅳ)、SeMet、Se(Ⅵ)检出限分别为 0.23 μg/L、0.15 μg/L、0.30 μg/L、0.16 μg/L。

Zhong 等[103]采用 HPLC-ICP-MS 测定富硒大蒜苗中的硒形态，样品采用 0.1 mol/L HCl 超声提取后，采用 C_{18} 分析柱，以流动相 20 mmol/L 乙酸铵（pH=5.6）/5%甲醇进行分析，发现大蒜样品中含有 Se(Ⅳ)、MeSeCys、SeMet 及两个未知硒形态，其中以 MeSeCys 为主要形态。SeMet 和 MeSeCys 的检出限为 5.0 μg/L 和 1.2 μg/L。姚真真等[104]采用 HPLC-ICP-MS 测定富硒苹果中的五种硒形态，样品用 5 mol/L 柠檬酸超声提取，选用 Hamiltion PRP-X100 阴离子交换色谱柱，流动相为 5 mmol/L 柠檬酸溶液（pH=5.0）进行分析，$SeCys_2$、MeSeCys、Se(Ⅳ)、SeMet、Se(Ⅵ)的检出限分别为 0.6 μg/L、0.7 μg/L、1 μg/L、0.9 μg/L、1 μg/L。Maneetong 等[105]采用 HPLC-ICP-MS 测定富硒甘蓝中的硒形态，样品采用 10%甲醇溶液（含 0.1 mol/L HCl）超声提取后，采用 Inertsil C_{18} 为分析柱，以 8 mmol/L 牛血清白蛋白（BSA）、4 mmol/L 三氟乙酸（TFA）（pH=4.5）为流动相进行分析，在富硒甘蓝中分离出 SeMet、MeSeCys 及两个未知硒形态。吴雅颖等[106]采用 HPLC-ICP-MS 测定竹笋中 Se(Ⅵ)、Se(Ⅳ)、SeMet、$SeCys_2$、MeSeCys 和 SeEt 六种硒形态，样品用纯水溶解后加入链酶蛋白酶 E 于 37℃超声提取，色谱柱为 Hamilton PRP-X100，流动相为 5 mmol/L 柠檬酸溶液（pH=4.7），方法检出限在 0.4～4.6 μg/L，富硒雷竹笋中含有 SeMet、$SeCys_2$、Se(Ⅳ)、Se(Ⅵ)、MeSeCys，其中以 SeMet 为主要的硒形态。熊珺等[107]采用 HPLC-ICP-MS 测定食品大蒜、蘑菇、卷心菜中的六种硒形态 Se(Ⅳ)、Se(Ⅵ)、$SeCys_2$、SeMet、MeSeCys 和 SeEt，样品溶于纯水加入 40 mg 蛋白酶 K 于 37℃恒温水浴提取，采用 Dionex IonPac AS11 为色谱柱，7.5 mmol/L 磷酸氢二铵溶液（pH=11）为流动相进行分析，MeSeCys、SeMet、SeEt、Se(Ⅳ)、$SeCys_2$ 和 Se(Ⅵ)的检出限分别为 0.25 μg/L、0.20 μg/L、0.35 μg/L、0.15 μg/L、0.30 μg/L、0.15 μg/L。在大蒜中检出 MeSeCys，卷心菜中检出 SeMet，蘑菇检出 Se(Ⅳ)、SeMet 和 $SeCys_2$ 三种硒形态。倪张林等[108]用 HPLC-ICP-MS 测定富硒蔬菜中的硒形态，样品用胃蛋白酶和 0.1% HCl 溶液于 37℃下超声萃取，采用 Hamilton PRP-X100 阴离子色谱柱，以 5 mmol/L 的柠檬酸溶液（pH=4.7）为流动相进行分析，Se(Ⅳ)、Se(Ⅵ)、SeMet、$SeCys_2$、MeSeCys 定量限分别为 5 μg/kg、10 μg/kg、10 μg/kg、5 μg/kg、5 μg/kg，富硒蔬菜中以 $SeCys_2$、MeSeCys 和 SeMet 为主。

Zhang 等[109]用 HPLC-ICP-MS 对扇贝组织中的硒形态进行检测，扇贝样品采用木瓜蛋白酶 30～40℃超声后加入混合酶溶液（风味酶∶羧肽酶 Y∶胰蛋白酶为 3∶1∶1）于 45℃水解，用 Dinoex IonPac AS11 阴离子色谱柱，流动相为 10 mmol/L 碳酸氢钠和 2%乙腈（pH=11），MeSeCys、$SeCys_2$、SeMet、Se(Ⅳ)、Se(Ⅵ)五种硒形态的检出限依次为 0.69 μg/L、0.48 μg/L、0.93 μg/L、0.53 μg/L、1.22 μg/L。扇贝组织均含有 MeSeCys、$SeCys_2$、SeMet 和 Se(Ⅳ)。

陈贵宇等[110]采用 HPLC-ICP-MS 检测富硒茶叶中的硒形态，样品采用 10 mg/mL 蛋白酶 K 超声提取后，Hamilton PRP-X100 阴离子色谱柱，流动相 A：10 mmol/L 柠檬酸溶液，流动相 B：超纯水，进行梯度洗脱，$SeCys_2$、MeSeCys、Se(Ⅳ)、SeMet、Se(Ⅵ)5 种硒形态检出限为 0.13～1.09 μg/L。富硒茶叶中检出了 MeSeCys 和 SeMet。

Lipiec 等[111]采用 HPLC-ICP-MS 检测鸡蛋中 SeCys、Se(Ⅳ)和 SeMet 三种硒形态，蛋清采用 30 mmol/L Tris-HCl 缓冲液（pH=7.5）37℃水浴提取，蛋黄采用 7 mol/L 尿素溶液和 30 mmol/L Tris-HCl 缓冲液（pH=7.5）超声提取。色谱柱采用 Alltima C_8，流动相 A：含 0.1%HFBA 的水溶液，流动相 B：

含 0.1%HFBA 的甲醇溶液，梯度洗脱，结果表明：蛋黄中以 SeCys 为主，蛋清中以 SeMet 为主。检出限分别为 0.06 μg/g、0.003 μg/g、0.01 μg/g。

Cao 等[112]使用 HPLC-ICP-MS 测定葛根中的硒形态，葛根样品采用 20 mg/mL 蛋白酶 K 和脂肪酶于 37℃提取后，用 5-磺基水杨酸（SSA）-功能性的二氧化硅涂层磁性纳米颗粒（SMNP）对样品中的硒形态富集，用 Hamilton PRP-X100 阴离子色谱柱，流动相 8 mmol/L 柠檬酸（pH=5.0），Se(Ⅳ)、Se(Ⅵ)、SeMet、SeCys$_2$ 检出限分别是 0.002 ng/mL、0.002 ng/mL、0.004 ng/mL、0.002 ng/mL。样品中检出 SeCys$_2$、Se(Ⅳ)、SeMet、Se(Ⅵ)。曹玉嫔等[113]用 HPLC-ICP-MS 测定牛蒡和三七中 SeCys$_2$、Se(Ⅳ)、SeMet、Se(Ⅵ)四种硒形态的方法，样品采用 20 mg/mL 蛋白酶 K 和脂肪酶在 37℃水浴超声提取后，分析柱 Hamilton PRP-X100，用流动相 8 mmol/L 柠檬酸（pH=5.0）进行分析。牛蒡和三七中均检出 SeCys$_2$、Se(Ⅳ)、SeMet、Se(Ⅵ)，这两种样品中无机硒含量均高于有机硒。

王丙涛等[114]及杨修斌等[115]均采用 HPLC-ICP-MS 测定转基因大豆中的硒形态，将样品溶于纯水加入蛋白酶 37℃水浴振荡提取，用 Hamilton PRP-X100 阴离子色谱柱，分别用流动相 5 mmol/L 及 10 mmol/L 柠檬酸溶液（pH=4.5）进行分离，前者分离了 Se(Ⅵ)、Se(Ⅳ)、SeMet 和 SeCys$_2$ 四种硒形态，后者分离出 Se(Ⅳ)、Se(Ⅵ)、SeMet、SeCys$_2$、SeEt 五种硒形态。大豆中检出 SeMet、Se(Ⅳ)、SeCys$_2$、Se(Ⅵ)，还有少量未知硒形态，检出限为 0.3～3.6 μg/L。

魏琴芳等[116]将样品溶于纯水加入链霉蛋白酶 E 于 37℃超声提取后，采用 C$_{18}$ 反相色谱柱，以 30 mmol/L 磷酸氢二铵、0.5 mmol/L 四丁基溴化铵和 2%甲醇（pH=6）为流动相，采用 LC-ICP-MS 测定大豆样品中的 SeCys$_2$、MeSeCys、Se(Ⅳ)、SeMet 和 Se(Ⅵ)五种硒形态，检出限为 0.5～1.0 μg/L，在大豆样品中检出 SeMet。

4. 毛细管电泳-电感耦合等离子体质谱法联用技术（CE-ICP-MS）

Zhao 等[117]建立了 CE-ICP-MS 测定富硒大米中 4 种硒形态方法，样品溶于纯水加入蛋白酶和脂肪酶于 37℃加热提取后，采用缓冲液 20 mmol/L NaH$_2$PO$_4$-10 mmol/L Na$_2$B$_4$O$_7$-0.2 mmol/L CTAB（pH=8.6）用 CE-ICP-MS 分析 SeCys、SeMet、Se(Ⅳ)、Se(Ⅵ)，方法检出限为 0.1～0.9 ng/mL，在富硒水稻中检出 SeMet。

23.1.3.4 铬的研究进展

铬（Cr）是人体必需微量元素，参与糖代谢调节、促进蛋白质合成和生长发育等。铬缺乏可导致葡萄糖和脂类代谢的改变，与糖尿病、心血管和神经系统疾病有关。铬的毒性与其价态密切相关，铬从-2 到+6 价态都存在，三价铬[Cr(Ⅲ)]和六价铬[Cr(Ⅵ)]是自然界中最常见的稳定价态，而 Cr(Ⅱ)、Cr(Ⅳ)和 Cr(Ⅴ)不稳定，容易转化成 Cr(Ⅲ)或 Cr(Ⅵ)。其中金属铬(0 价)无毒，Cr(Ⅵ)毒性大，具有致畸、致癌和致突变等毒性。国际癌症研究机构把 Cr(Ⅵ)归为"Ⅰ类致癌物"。Cr(Ⅵ)比 Cr(Ⅲ)易吸收 Cr(Ⅲ)是人和动物必需的微量元素，而 Cr(Ⅵ)属于严格控制的有害物质。

杨海锋等[118]采用磷酸缓冲溶液（0.5 mol/L K$_2$HPO$_4$ 和 0.5 mol/L KH$_2$PO$_4$ 混合溶液）及碱性提取液（0.5 mol/L NaOH 和 0.28 mol/L Na$_2$CO$_3$ 混合溶液）提取食用菌中的 Cr(Ⅲ)和 Cr(Ⅵ)，采用 Agilent BIO WAX 色谱柱，流动相为 0.075 mol/L HNO$_3$ 和 0.6 mmol/L EDTA-2Na，采用高效液相色谱电感耦合等离子体质谱联用技术（HPLC-ICP-MS）测定，建立的食用菌中 Cr(Ⅲ)和 Cr(Ⅵ)的 HPLC-ICP-MS 检测方法；线性范围 0.4～20 μg/L，Cr(Ⅲ)和 Cr(Ⅵ)的方法检出限为 0.03 mg/kg。

倪张林等[119]建立了干食用菌中 Cr(Ⅲ)和 Cr(Ⅵ)的 LC-ICP-MS 方法。采用微波灰化技术对食用菌样品进行灰化处理，灰化样品用乙二胺四乙酸二钠盐(EDTA-2Na)稳定其中的 Cr(Ⅲ)，采用阴离子交换柱，60 mmol/L 硝酸(pH=9.3)为流动相，HPLC-ICP-MS 测定。Cr(Ⅲ)和 Cr(Ⅵ)线性范围为 0.5～50 μg/L，定量限均为 0.5 μg/L。

高勇兴[120]通过阴离子色谱柱分析了乳粉中的 Cr(Ⅲ)和 Cr(Ⅵ)，选用 0.5 mL 33%硝酸为提取液，70 mmol/L NH$_4$NO$_3$（pH=7.1±0.1）为淋洗液，采用 HPLC-ICP-MS 分析。Hernandez 等[121]采用微波辅助萃取，乙二胺四乙酸络合、HPLC-ICP-MS 进行乳制品中三价铬的分析，采用 CS5A 离子交换柱，以 2 mol/L 的硝酸为流动相，分析测定 Cr(Ⅲ)和 Cr(Ⅵ)，其中 Cr(Ⅲ)的定量限为 38 μg/kg。

李琴等[122]建立了液相色谱法与 HPLC-ICP-MS 测定牛乳中不同价态铬的方法。使用 IonPac AG11 阴离子交换色谱柱、NH$_4$NO$_3$ 淋洗液，可以很好地分离溶液中的 Cr(Ⅲ)和 Cr(Ⅵ)，当进样量为 100 μL 时，检出限 Cr(Ⅲ)和 Cr(Ⅵ)分别为 0.25 μg/L 与 0.03 μg/L。

樊祥等[123]采用 HPLC-ICP-MS 测定食品中 Cr(Ⅵ)含量，同时间接测定 Cr(Ⅲ)含量。总铬含量采用电感耦合等离子体质谱法进行测定，样品中 Cr(Ⅵ)经 0.1 mol/L 的磷酸氢二钾溶液(pH 7.9～8.1)提取，采用 Hamilton PRP-X100 色谱柱，60 mmol/L 硝酸铵（pH=9.0）流动相，HPLC-ICP-MS 测定 Cr(Ⅵ)，线性范围为 0.4～10.0 μg/L，方法检出限为 0.03 mg/kg，Cr(Ⅲ)含量通过总铬减去 Cr(Ⅵ)得到。

陈东等[124]采用 HPLC-ICP-MS 检测双份饭样品中不同价态铬含量，样品经过冷冻干燥挥干水分，置于微波消解管中，加入 5%的四丁基氢氧化铵溶液，在 120℃条件下微波消解 30 min，离心过滤后，采用 Agilent Bio-WAX 色谱柱，0.075 mol/L NH$_4$NO$_3$ 溶液（pH=7）为流动相，HPLC-ICP-MS 测定 Cr(Ⅲ)和 Cr(Ⅵ)，方法线性范围为 0.5～50 μg/L，Cr(Ⅲ)和 Cr(Ⅵ)的检出限分别为 0.32 μg/kg 和 0.35 μg/kg。

Unceta 等[125]采用阴离子色谱柱，以乙二胺四乙酸为萃取剂，LC-ICP-MS 分析测定了膳食补充剂中 Cr(Ⅲ)和 Cr(Ⅵ)。选用 50 mmol/L 的乙二胺四乙酸溶液在电热板上提取样品中铬形态，将 Cr(Ⅲ)稳定为 Cr(Ⅲ)-乙二胺四乙酸络合物，以 2 mmol/L 乙二胺四乙酸溶液作流动相。

葛晓鸣等[126]建立一种超声辅助提取，HPLC-ICP-MS 测定食品中水溶性铬价态的方法，样品经乙二胺四乙酸二钠-硝酸铵混合溶液提取后，加入铁氰化钾和乙酸锌溶液沉淀蛋白质，过 0.45 μm 滤膜，采用 Dionex IonPac AG 7 色谱柱，以 EDTA-硝酸铵溶液为流动相，HPLC-ICP-MS 分析测定 Cr(Ⅲ)和 Cr(Ⅵ)，线性范围为 0.5～50 ng/mL，方法检出限 10 μg/kg。

23.2 临床样品分析研究进展

微量元素是生物组织的基本组成成分，但当它们的浓度远超过完成其生物学功能所需量时，就可以变为有毒的成分。1996 年，联合国粮食及农业组织（FAO）、国际原子能机构（IAEA）、世界卫生组织（WHO）联合组织的专家委员会将存在于人体中的微量元素分为：人体必需微量元素、人体可能必需微量元素、有潜在毒性但低剂量时可能具有必需功能的微量元素。

微量元素在维持人类健康中发挥着重要作用。微量元素与人体的内分泌、免疫、感染、生长发育以及神经系统结构与功能有关。同时，依靠着生物体内的稳态平衡，将其数量维持在狭小的正常范围之内，当摄入过多超越其调节功能所需要量时，或摄入不足导致缺乏时可引起平衡紊乱甚至发生疾病。目前已知多种疾病的发生、发展都与微量元素有关，如癌症、心血管病、脑病、神经系统病患、风湿性炎症、地方病、营养缺乏症等。研究表明，患病者与健康人的体液或脏器间以及病变组织与正常组织之间，微量元素的含量有差异，微量元素的量变与病变原因或病变结果之间存在联系。

随着现代科学仪器和分析技术的迅速发展和应用，提高了微量元素痕量分析的灵敏度和准确度，为深入研究元素的生物学作用提供了重要条件。电感耦合等离子体发射光谱（ICP-OES）、电感耦合等离子体质谱（ICP-MS）、原子吸收光谱（AAS）和原子荧光光谱（AFS）等原子光谱技术由于选择性强、准确度高、分析速度快和灵敏度高等特点，已广泛应用于生物组织中微量元素的含量、存在形态和分布的研究，为疾病的正确诊断和监测、病理研究等提供重要信息。

生物样品中微量元素的分析研究是一门新的学科领域，与化学、生物学、营养学、环境卫生、毒理学以及临床医学等有着密切联系，与人体健康息息相关。常量元素和微量元素的分析测定将为疾病诊断及新陈代谢的研究提供科学依据。下面简要介绍原子光谱技术在临床样品分析中的新应用。

23.2.1 现行有效的标准方法

以往人们比较关注职业暴露、环境污染和突发事件对人体造成的健康伤害。采用原子光谱技术进行重金属、有害元素的监测，原子吸收、原子荧光技术早已应用于临床样品的分析检测，电感耦合等离子体发射光谱（ICP-OES）、电感耦合等离子体质谱（ICP-MS）以灵敏度高、准确可靠、多元素同时测定也逐步用于生物样品分析检测中，已相继颁布了中华人民共和国国家职业卫生标准和卫生行业相关标准，表23-2列出了原子光谱技术分析血样的标准方法，表23-3列出了原子光谱技术分析尿样的标准方法。以往均是采用原子吸收、原子荧光技术测定生物样品中单一元素，2018年颁布的《尿中多种金属同时测定电感耦合等离子体质谱法》，首次将多元素同时测定纳入卫生行业标准GBZ/T 308—2018中，尿液样品经过处理可以直接测定钒、铬、钴、镉、铊、铅六种元素。

由于砷化合物种类繁多，砷化合物的毒性各不相同，各种不同食品经代谢后排出体外的砷化合物形态也各不相同，总砷含量不能正确反映人体经砷暴露后的代谢情况，2019年实施的卫生行业标准WS/T 635—2018首次将元素形态分析引入砷形态的测定中，对代谢后尿中不同形态的砷化物进行分析测定，有效反映砷中毒和砷暴露相关疾病的情况，为有效分析监测人群近期砷暴露提供支持。

表23-2 原子光谱法分析血液样品的标准方法

序号	标准名称		测定方法
1	血中铅的测定	GBZ/T 316.1—2018	石墨炉原子吸收光谱法，检出限 7 μg/L
		GBZ/T 316.2—2018	电感耦合等离子体质谱法，检出限 0.17 μg/L
		GBZ/T 316.3—2018	原子荧光光谱法，检出限 6.7 μg/L
2	血中铬的测定	GBZ/T 315—2018	石墨炉原子吸收光谱法，检出限 0.47 μg/L
3	血中镉的测定	GBZ/T 317.1—2018	石墨炉原子吸收光谱法，检出限 0.24 μg/L
		GBZ/T 317.2—2018	电感耦合等离子体质谱法，检出限 0.08 μg/L
4	血中镍的测定	GBZ/T 314—2018	石墨炉原子吸收光谱法，检出限 1.13 μg/L

表23-3 原子光谱法分析尿样的标准方法

序号	标准名称		测定方法
1	尿中铅的测定	GBZ/T 303—2018	石墨炉原子吸收法，检出限 1.0 μg/L
2	尿中铬的测定	GBZ/T 306—2018	石墨炉原子吸收光谱法，检出限 0.21 μg/L
3	尿中镉的测定	GBZ/T 307.1—2018	石墨炉原子吸收光谱，检出限 0.17 μg/L
		GBZ/T 307.2—2018	电感耦合等离子体质谱法，检出限 0.08 μg/L
4	尿中镍的测定	WS/T 44—1996	石墨炉原子吸收光谱法，检出限 1.4 μg/L
5	尿中铍的测定	WS/T 46—1996	石墨炉原子吸收光谱法，检出限 0.09 μg/L
6	尿中铜的测定	WS/T 94—1996	石墨炉原子吸收光谱法，检出限 2.0 μg/L
7	尿中锌的测定	WS/T 95—1996	火焰原子吸收光谱法，检出限 0.01 μg/L
8	尿中砷的测定	WS/T 474—2015	氢化物发生原子荧光法，检出限 0.50 μg/L

续表

序号	标准名称		测定方法
9	尿中锑的测定	GBZ/T 302—2018	原子荧光光谱法,检出限 0.06 μg/L
10	尿中铝的测定	GBZ/T 304—2018	石墨炉原子吸收光谱法,检出限 2.50 μg/L
11	尿中锰的测定	GBZ/T 305—2018	石墨炉原子吸收光谱法,检出限 1.5 μg/L
12	尿中多种金属同时测定	GBZ/T 308—2018[127]	电感耦合等离子体质谱法,钒、铬、钴、镉、铊、铅的检出限分别为 0.36 μg/L、0.57 μg/L、0.09 μg/L、0.06 μg/L、0.32 μg/L、1.67 μg/L
13	尿中碘的测定	WS/T 107.1—2016	砷铈催化分光光度法,检出限 2 μg/L
		WS/T 107.2—2016[128]	电感耦合等离子体质谱法,检出限 0.4 μg/L
	尿中砷形态测定	WS/T 635—2018[129]	液相色谱-原子荧光法,亚砷酸盐、一甲基砷、二甲基砷检出限均 0.5 μg/L

23.2.2 微量元素测定

近些年,原子吸收光谱(AAS)和原子荧光光谱(AFS)、电感耦合等离子体发射光谱法(ICP-OES)和电感耦合等离子体质谱(ICP-MS)等原子光谱技术越来越多地应用于生物样品的检测。特别是电感耦合等离子体质谱以其灵敏度高、多元素同时测定广泛应用于血和尿样品的测定中。

Meyer 等[130]应用 ICP-MS/MS 方法分析人血清中的多元素,包括 Mg、Ca、Fe、Cu、Zn、Mo、Se、I、As 和 Cd。取 200 μL 血清样品,加入 250 μL HNO_3、250 μL H_2O_2、250 μL 去离子水、25 μL 含 100 μg/L Rh 和 1000 μg/L ^{77}Se 的内标溶液混合处理,在 65℃下和在 95℃消解 40 min,获得澄清透明溶液。用 1.5 mL 去离子水稀释消解液,得到最终浓度含 10%的 HNO_3 以及 1 μg/L Rh 和 10 μg/L ^{77}Se 的内标物,用于分析测定。由于碘在较高温度的酸性条件下会形成挥发性化合物,因此碘元素采用碱性溶液处理,将 25 μL 人血清用含有 1 μg/L Rh 的 0.1 mol/L NaOH 处理后稀释至 1 mL。该溶液可直接用于 ICP-MS/MS 测量。多元素采用 Agilent ICP-QQQ-MS 8800 系统监测,碰撞池采用 He 模式(Ca、Mg、Fe、Cu、Zn、I 和 Cd)或 O_2/H_2 模式(Se、Mo 和 As)。该方法检出限(LOD):Mg、Ca、Fe、Cu、Zn、Se、Mo、I、Cd 和 As 分别为 2.34 μg/L、40.88 μg/L、2.05 μg/L、0.10 μg/L、0.24 μg/L、0.04 μg/L、0.025 μg/L、0.10 μg/L、0.001 μg/L 和 0.01 μg/L。

Johnson-Davis 等[131]采用 ICP-MS 测量血清中 Cr、Co、Cu、Mn、Ni、Se 和 Zn 七种元素。取 50 μL 血清样品用 950 μL 含 0.05% Triton X-100 和 1% HNO_3 的溶液稀释,涡旋混合。Agilent 7900 ICP-MS 检测,采用 He 碰撞池模式消除多原子干扰,检测质量数:^{52}Cr、^{55}Mn、^{59}Co、^{60}Ni、^{65}Cu、^{66}Zn、^{78}Se 和内标元素 ^{156}Gd、^{193}Ir。该方法 Cr、Co、Mn 的定量限为 1 μg/L,Ni 的定量限为 2 μg/L,Se 的定量限为 12 μg/L,Cu 和 Zn 的定量限为 10 μg/dL。

Jones 等[132]改进了 ICP-MS 测量全血中 Pb、Cd、Hg、Se 和 Mn 的方法,用于监测急性中毒的全血样品。取全血样品以含 0.4% V/V TMAH,1%甲醇,0.01%吡咯烷二硫代氨基甲酸铵,0.05% Triton X-100 的稀释剂混合均匀,使用 PerkinElmer SCIEX ELAN® DRC II ICP-MS,检测同位素 ^{80}Se、^{55}Mn、^{202}Hg、^{114}Cd 和 ^{208}Pb。采用动态反应池消除干扰,甲烷(CH_4)消除 $^{40}Ar_2^+$的背景信号。Pb、Cd、Hg、Mn 和 Se 的检出限分别为 0.07 μg/dL、0.10 μg/dL、0.28 μg/dL、0.99 μg/dL 和 24.5 μg/L。

党瑞等[133]探讨了 ICP-MS 测定人血浆样品中 21 种金属元素的前处理方法。实验中取 0.2 mL 血浆样品,前处理方法分为:①硝酸直接稀释法,用 6% HNO_3 直接稀释血浆样品,并定容至 5 mL;②HNO_3-H_2O_2 水浴消解法,取 0.3 mL HNO_3 和 0.3 mL H_2O_2 加入到血浆样品中,于沸水浴中消解 4 h 至溶液澄清透明,用纯水定容至 5 mL,混匀;③HNO_3-H_2O_2 烘箱高温消解法,130℃烘箱中高温消解 2 h,其余步骤与水浴消解法一致。采用 Agilent 7700x ICP-MS 检测,分别检测 ^{7}Li、^{24}Mg、^{27}Al、

^{42}Ca、^{47}Ti、^{51}V、^{52}Cr、^{55}Mn、^{56}Fe、^{59}Co、^{60}Ni、^{63}Cu、^{66}Zn、^{75}As、^{78}Se、^{88}Sr、^{95}Mo、^{111}Cd、^{137}Ba、^{205}Tl 和 ^{208}Pb。这 21 种元素的检出限介于 0.001~1.52 μg/L,标准曲线线性相关系数大于 0.9991。采用在血浆样品中加入 21 种多元素混合标准溶液的加标回收实验对三种前处理方法进行评估,其中 HNO_3 直接稀释法、HNO_3-H_2O_2 水浴消解法、HNO_3-H_2O_2 烘箱高温消解法的加标回收率范围分别为 77.45%~125.2%、77.45%~128.6%、82.27%~118.8%。其中 HNO_3-H_2O_2 烘箱高温消解处理方法操作简便、干扰小、重现性好、准确度高。

王俊等[134]用 3% HNO_3 稀释尿样 20 倍,以 ^{45}Sc、^{89}Y 和 ^{159}Tb 作为内标,ICP-MS 测定尿液中 19 种元素,使用碰撞反应池模式(反应气为 NH_3 或 O_2)检测 ^{51}V、^{52}Cr、^{55}Mn、^{58}Ni、^{59}Co、^{63}Cu、^{64}Zn、^{120}Sn、^{121}Sb、^{130}Te、^{75}As 和 ^{80}Se,用标准模式检测 ^{9}Be、^{85}Rb、^{88}Sr、^{114}Cd、^{208}Pb、^{133}Cs 和 ^{205}Tl。该方法测量的 ^{9}Be、^{51}V、^{52}Cr、^{55}Mn、^{58}Ni、^{59}Co、^{63}Cu、^{64}Zn、^{75}As、^{80}Se、^{85}Rb、^{88}Sr、^{114}Cd、^{120}Sn、^{121}Sb、^{130}Te、^{133}Cs、^{205}Tl 和 ^{208}Pb 19 种元素的检出限分别为 2.7 ng/L、2.0 ng/L、2.7 ng/L、1.5 ng/L、3.9 ng/L、0.8 ng/L、5.8 ng/L、35.3 ng/L、32.4 ng/L、21.1 ng/L、1.4 ng/L、4.2 ng/L、1.1 ng/L、9.1 ng/L、1.1 ng/L、8.6 ng/L、0.1 ng/L、0.2 ng/L 和 1.0 ng/L。

刘柳等[135]采用碰撞/反应池技术的 ICP-MS 检测人尿样品中 21 种痕量元素。选取均匀的健康人尿液用超纯水 1:1 稀释制备本底尿液,再用 4 倍体积的 5% HNO_3 和 0.02% Triton X-100 溶液稀释,配制成尿基质溶液。人尿样品处理:吸取 1.0 mL 解冻的尿液样品再用尿基质溶液定容至 5 mL,混匀,60℃水浴超声 1 h,冷却摇匀待测,标准系列采用尿基质溶液配制。采用 PerkinElmer NexION 300 ICP-MS 检测,选用 He 碰撞气体 KED 模式,检测元素的质量数:^{9}Be、^{27}Al、^{51}V、^{53}Cr、^{55}Mn、^{57}Fe、^{59}Co、^{60}Ni、^{65}Cu、^{68}Zn、^{69}Ga、^{75}As、^{82}Se、^{88}Sr、^{109}Ag、^{114}Cd、^{133}Cs、^{138}Ba、^{205}Tl、^{206}Pb、^{207}Pb、^{208}Pb 和 ^{238}U。21 种元素的检出限介于 0.004~12.08 μg/L 之间。测量了北京市某地区 21 件人体尿液样品,结果显示 Cr 和 Ga 两种元素未检出,实际尿样中各元素间浓度差异较大。

汤昌海等[136]报道了采用 ICP-MS 测定人尿液中 Cr、Mn、Ni、Cd 和 Pb 五种金属元素的方法。收集的尿液加入 1% HNO_3 稀释 4 倍,使用 PerkinElmer NexION300X ICP-MS 进行检测,其中 Cr、Mn 和 Ni 采用碰撞(He)模式去除干扰,检测元素的质量数:^{52}Cr、^{55}Mn、^{60}Ni、^{111}Cd、^{208}Pb、^{59}Co,内标为 ^{115}In 和 ^{209}Bi。Cr、Mn、Ni、Cd 和 Pb 的方法检出限分别为 0.233 μg/L、0.074 μg/L、0.197 μg/L、0.017 μg/L 和 0.054 μg/L。

Markiewicz 等[137]采用 ICP-MS 分析羊水样品中 18 种常规和有害元素。羊水样品来自怀孕的健康志愿者,分娩前用无菌针头和注射器经腹穿刺取羊水标本 5 mL。样品 4℃下 3000 r/min 离心 10 min,在-80℃下冷冻保存。取 1 mL 样品加入 0.5 mL 65% HNO_3 和 0.5 mL 30% H_2O_2 进行微波消解,消解液用纯水稀释定容至 10 mL,采用 PerkinElmer SCIEX Elan DRC II ICP-MS 进行检测,检测元素的质量数:^{24}Mg、^{27}Al、^{44}Ca、^{51}V、^{52}Cr、^{55}Mn、^{59}Co、^{60}Ni、^{63}Cu、^{66}Zn、^{78}Se、^{88}Sr、^{91}AsO、^{111}Cd、^{121}Sb、^{138}Ba、^{208}Pb 和 ^{238}U,内标同位素为 ^{45}Sc、^{74}Ge、^{103}Rh 和 ^{159}Tb。其中 Ba、Cd、Co、Cu、Mg、Ni、Pb、Sr、U、Zn 采用标准模式检测,As、Sb、Se 碰撞池通入 O_2,Al、Ca、Cr、Mn、V 反应气选用 NH_3。其中 Al、As、Ba、Ca、Cd、Co、Cr、Cu、Mg、Mn、Ni、Pb、Sb、Se、Sr、U、V 和 Zn 的方法检出限分别为 2.6 μg/L、0.043 μg/L、0.38 μg/L、56 μg/L、0.013 μg/L、0.0061 μg/L、0.052 μg/L、0.072 μg/L、0.88 μg/L、0.025 μg/L、0.51 μg/L、0.16 μg/L、0.028 μg/L、0.094 μg/L、0.037 μg/L、0.000 45 μg/L、0.0027 μg/L、2.0 μg/L。测定结果表明,在羊水样品中,Ca 浓度最高、U 浓度最低,总体上羊水中的元素浓度低于其母体血清浓度。

Chantada-Vázquez 等[138]报道了通过激光剥蚀 ICP-MS 法测定干血清斑中多元素的方法。血清样品直接点样在 Whatman 滤纸上,在恒温恒湿条件下密封保存。分析时用穿孔器在干血清斑的中心切割直径为 3 mm 的圆片,用 Nd:YAG 213 nm 紫外 New wave UP213 激光烧蚀系统进样,采用 Thermo

Finnigan ELEMENT XR ICP-MS 分析，检测元素的质量数：^{27}Al、^{44}Ca、^{63}Cu、^{56}Fe、^{39}K、^{6}Li、^{24}Mg、^{55}Mn、^{98}Mo、^{23}Na、^{31}P、^{85}Rb、^{82}Se、^{51}V 和 ^{66}Zn。采用该方法进行多元素的分析定量限为 21 μg/L～221 mg/L，用有证标准物质进行质量控制。采用此方法对成年人血清样品进行的检测结果与使用微波消解进行样品前处理方法的测定结果一致。

人体代谢所需碘主要来源于食物，进入人体的碘在胃和小肠被迅速吸收入，血碘被甲状腺摄取生成甲状腺素，血清碘能较准确反映机体近期碘营养状况和甲状腺功能。当人体摄入足量碘时，大部分从尿液排出，尿液中碘含量水平在一定程度上反映了人体碘营养水平。

电感耦合等离子体质谱法以其独特的接口技术将 ICP 离子源与质谱快速扫描、高灵敏度及干扰少的优点相结合，已广泛应用于医疗、生物等各个领域。2017 年国家卫生行业标准《尿中碘的测定 第 2 部分：电感耦合等离子体质谱法》（WS/T 107.2—2016）颁布，采用标准加入 ICP-MS 法对尿中碘含量进行测量，检出限为 0.4 μg/L。温权等[139]也采用直接稀释-电感耦合等离子体质谱法测尿中碘含量，样品经 0.25%的四甲基氢氧化铵稀释 20 倍后直接进样分析，采用耐高盐接口、基体匹配和内标校正降低尿样中有机成分和高盐组分的非质谱干扰，方法检出限为 0.12 ng/mL。黄淑英等[140]以 2.5 g/L 盐酸肼-1.0 g/L 氯化铵-0.5%盐酸-2.0%乙醇溶液为稀释剂，按稀释剂和样品（19∶1）稀释后采用电感耦合等离子体质谱法测定尿液中碘含量，检出限为 0.4 μg/L。禹松林等[141]采用氨水∶异丙醇∶水作为稀释剂，10 倍稀释，铼作为内标，采用 ICP-MS 法测定稀释混匀后的尿液和血清中的碘含量。检出限为 0.87 μg/L。张亚平[143]采用铼作为内标，以抗坏血酸（2.0 g/L）、氯化铵（1.0 g/L）、乙醇胺（0.1%）和乙醇溶液（1.0%）为稀释剂，ICP-MS 法测定 20 倍稀释后的血清中碘含量。Yu 等[144]样品经 1.5%异丙醇和 7 mmol 一水合氨稀释，使用 ICP-MS 测定人尿和血清中的碘，方法的检出限为 0.87 μg/L。

23.2.3 元素形态分析

23.2.3.1 砷形态的研究进展

含砷化合物的形态分析研究很普遍，因为不同的砷形态在生物和环境中的毒性差异很大。砷作为类金属元素，同时具有金属和非金属元素的性质，在环境中以 As(Ⅲ)和 As(Ⅴ)的无机形式存在，As(Ⅲ)比 As(Ⅴ)毒性大，主要由三价有机代谢物引起毒性。除了游离无机的砷形态，含砷有机物也是临床基质中砷形态研究的重要对象，包括一甲基砷酸（MMAV）、二甲基砷酸（DMAV）、三甲基氧化砷（TMAO）、砷甜菜碱（AsB）、砷胆碱（AsC）和砷糖（AsS）。分析临床体液样品中更具毒性的无机砷和不同砷形态含量比总砷含量对健康评估的意义更大。砷中毒可导致死亡，根据不同的暴露剂量、途径和时间，砷的摄入可引起胃肠道刺激，如胃痛、恶心、呕吐和腹泻。其他毒理学效应包括血液学损伤，如红细胞和白细胞生成减少，导致疲劳、心律失常和血管损伤，皮肤形态的改变。增加罹患膀胱癌、肝癌、肺癌和皮肤癌的风险。由于砷在环境中持续存在，人类可以通过包括空气和饮用水的途径暴露，食物是人体砷中毒最常见的接触途径之一，因此通过检测血液、尿液中砷形态的种类，以监测摄入不同砷形态的生物转化和代谢途径具有非常重要的意义。

Contreras-Acuña 等[145]对食用海鲜后的志愿者提供的血清和尿样品中的砷代谢物形态进行检测。尿样用超纯水 1∶5 稀释，离心，过膜后分析；血清样品用 2∶1 甲醇/水沉淀蛋白，混均离心后取上清液氮吹至近干，用流动相溶解上机分析。采用 Hamilton PRP X-100 阴离子交换柱进行分离，流动相选择 50 mmol/L（pH 8.5）碳酸铵；或采用 Supelcosil LC-SCX 阳离子交换柱，流动相为 20 mmol/L（pH 2.5）吡啶，采用 HPLC-ICP-MS 检测。阴离子交换柱法可以分离 As(Ⅲ)、As(Ⅴ)、DMA^V、MMA^V 和 AsC 的砷形态，由于 AsB、四甲基砷（TETRA）和 TMAO 的共洗脱，需要以 20 mmol/L 吡啶为流动相采用阳离子交换柱分离测定。方法中 As(Ⅲ) 和 As(Ⅴ) 的检出限分别为 3.5 ng/g 和 2.4 ng/g。研究发现食用海产品后，人血清和尿液中砷化合物含量增加，观察到人体体液中砷形态的比例、转化和排泄方式，砷酸盐被还原成亚砷酸盐，随后被甲基化为 MMA。一甲基砷酸通常被还原为三价形式 ($MMA^{Ⅲ}$)，然而由于其不稳定性，在体液中甲基化转化为二甲基砷酸 ($DMA^{Ⅴ}$)。虽然来自职业暴露的无机砷可以以完整的以无机砷、MMA 和 DMA 的形式排出。AsB 在海产品中含量较高，在体内只有少量被代谢。其他有机砷化合物，如砷糖，几乎完全代谢成 DMA。

Nguyen 等[146]通过分析血清和尿液样品评估含砷地表水的暴露情况。血清样品经过 25%三氯乙酸、乙腈和去离子水混合处理，离心后取上清液分析；尿液样品用 1/9 (V/V) 甲醇/水溶液稀释 10 倍，过膜后分析。采用 Hamilton PRP-X100 阴离子交换柱，流动相为含 0.05% Na_2EDTA、5%甲醇的 5 mmol/L 和 50 mmol/L $(NH_4)_2CO_3$ 溶液，梯度洗脱，用 ICP-MS 检测五种砷形态。AsB、As(Ⅲ)、DMA、MMA 和 As(Ⅴ) 的检出限分别为 0.38 ng/mL、1.5 ng/mL、0.35 ng/mL、0.40 ng/mL 和 0.27 ng/mL。研究发现在志愿者的血清和尿液样品中检出五种砷形态，尿液样品以 AsB 和 DMA 形式存在，血清中主要以 DMA、As(Ⅲ) 和 AsB 形式存在。

林琳等[147]采用 HPLC-ICP-MS 测定了砷中毒司法鉴定和砷剂治疗患者血液和尿液中 AsC、AsB、As(Ⅲ)、DMA、MMA 和 As(Ⅴ) 六种砷形态，血液或尿液样品用含 0.1%曲拉通的 50 mmol/L Na_2EDTA 溶液等体积混合，经超声 40 min、3500 r/min 离心后，取上清液再与等体积乙腈混合以沉淀蛋白，再次离心后取上清液经 Hamilton PRP-X100 阴离子分析柱，以 100 mmol/L (pH 9.5) $(NH_4)_2CO_3$ 溶液为流动相，梯度洗脱，HPLC-ICP-MS 分析测定。血液中砷形态的检出限为 1.66~10 ng/mL，尿液中砷形态的检出限为 0.5~10 ng/mL。分别对食用和未食用海鲜的志愿者尿液样品进行砷形态分析，食用海鲜组尿液中检出 AsB，平均含量是未食用海鲜组的 2.38 倍，DMA 平均含量是未食用海鲜组的 1.89 倍，食用海鲜组尿液中检出 MMA。

有多名研究者报道了尿液样品中砷形态的分析结果。Morton 等[148]分析了半导体行业工人尿液样品中的五种砷形态，尿样经 2.5 mmol/L $(NH_4)_2CO_3$ 溶液稀释 10 倍后，采用 Thermo HyperREZXP SAX 阴离子交换柱，2.5 mmol/L 和 50 mmol/L $(NH_4)_2CO_3$ 流动相梯度洗脱，LC-ICP-MS 分析测定内分离 AsB、As(Ⅲ)、DMA、MMA 和 As(Ⅴ)，检出限在 0.03~0.09 µg/L 之间。结果显示，65 名工人与对照人群相比，尿液中的砷含量相近且浓度很低，AsB 是尿液中的主要砷形态，这可能与膳食相关，未见职业砷暴露发生。

Sen 等[149]采用含 2%甲醇的 10.0 mmol/L (pH 5.8) KH_2PO_4 溶液将尿样品稀释 10 倍，用含 2%甲醇的 (pH 9.0) 5 mmol/L $(NH_4)_2HPO_4$ 和 5 mmol/L NH_4NO_3 为流动相，采用 Hamilton PRP-X100 阴离子色谱柱，HPLC-ICP-MS 分析检测尿中砷形态。AsC、AsB、As(Ⅲ)、DMA、MMA 和 As(Ⅴ) 的检出限在 0.04~0.16 µg/L 范围内。

Serrano 等[150]调查了西班牙人群尿液样本中的总砷和五种砷形态，选用 5 mmol/L $(NH_4)_2HPO_4$ (pH 9.2) 和 5 mmol/L NH_4NO_3 溶液为流动相，尿样经流动相 10 倍稀释，Hamilton PRP-X100 阴离子交换柱分离，HPLC-ICP-MS 分析检测。AsB、As(Ⅲ)、DMA、MMA 和 As(Ⅴ) 的检出限分别为 1.0 µg/L、1.4 µg/L、1.9 µg/L、1.8 µg/L 和 1.4 µg/L。

Chen 等[151]报道分析了经口服 As_2O_3 治疗的急性早幼粒细胞白血病患者唾液中砷形态，患者唾液样品用纯水稀释 3 倍，过滤膜后分析，经 Phenomenex ODS-3 C18 反相柱分离，流动相为 5 mmol/L 四丁基氢氧化铵+5%甲醇和 3 mmol/L 丙二酸 (pH 5.65)，HPLC-ICP-MS 检测分析，发现患者唾液中的砷形态以 As(Ⅲ) 为主，检测浓度范围为 0.3~181.5 ng/mL，占总砷的 71.8%。

牛黄解毒片为常用中药，具有解热泻火、抗炎镇痛等作用。牛黄解毒片中的雄黄主要成分为二硫化二砷 (As_2S_2)，本无毒性，但超量、长期摄入有砷中毒风险。陈绍占等[152]研究了经牛黄解毒片暴露后大鼠血清中砷形态，取经牛黄解毒片暴露 30 天后的大鼠血清，用乙腈 60℃水浴超声提取后离心过 0.45 μm 滤膜，采用 Dionex IonPac A19 分析柱，以含 3%甲醇的 20 mmol/L（pH 9.7）$(NH_4)_2CO_3$ 溶液为流动相，Agilent 7700x ICP-MS 测定。AsB、DMA、As(Ⅲ)、MMA 和 As(Ⅴ)的检出限范围为 0.05～0.10 μg/L，在经牛黄解毒片暴露后大鼠血清中检出的砷形态主要为 DMA 和一未知物砷形态。

陈绍占等[153]还报道了雄黄在大鼠脏器中的代谢情况，取雄黄染毒大鼠的肝脏和肾脏组织，均质后采用乙酸/水超声提取，离心过滤后用 Dionex IonPac AS19 阴离子交换柱，20 mmol/L (pH 9.7) $(NH_4)_2CO_3$ 为流动相，采用 HPLC-ICP-MS 分析五种砷形态。AsB、DMA、As(Ⅲ)、MMA 和 As(Ⅴ)检出限范围为 0.3～0.5 μg/L，雄黄代谢后在大鼠肝脏内主要以 DMA 形式存在，在肾脏中的砷形态主要以 DMA、MMA、As(Ⅲ)和一未知的砷化合物。

23.2.3.2 汞形态的研究进展

汞以无机物和有机物的形式存在于环境和生物基质中，在临床样品的形态分析中，常见的汞形态包括二价无机汞，以及甲基汞（MeHg）、乙基汞（EtHg）等有机汞化合物。金属汞（Hg^0）和无机汞由于受采矿活动、工厂释放和牙科汞合金等影响而存在于空气中从而产生接触暴露。Hg^0 在空气中氧化生成无机形态的汞，使汞变得可溶，促进在环境中的沉积和积累。在血液中，汞被氧化成 Hg^{2+} 的形态，并可能被甲基化。无机汞主要积聚在肾脏和肝脏，急性暴露可导致肾衰竭。

汞的所有形态都是有毒的，其中烷基汞，尤其 MeHg 是微生物对 Hg^{2+} 的甲基化产物，毒性最大。它们具有高溶解度、通过肺部和胃肠道快速吸收，从而具有很高的生物利用度，蓄积在肾脏、肝脏和大脑等器官中。汞可在食物链中富集，生物放大效应增加了摄入海鲜导致食物中毒的风险。MeHg 是一种高脂溶性有机化合物，具有神经毒性。由于汞毒理学的重要性，目前已采用多种方法测定头发、指甲、血液、尿液和组织中的总汞含量，并通过 HPLC-ICP-MS 联用进行形态分析。

一些研究报道了使用 HPLC-ICP-MS 成功地测定了如血液成分、头发和尿液等临床样品中的汞形态。在研究中主要应用 C_8、C_{18} 反相色谱柱，流动相的选择略有不同，包括使用不同浓度的甲醇、2-巯基乙醇、甲酸、L-半胱氨酸和醋酸铵等。

Rodrigues 等[154]报道了采用 HPLC-ICP-MS 检测血液中的 MeHg 和 Hg^{2+} 的方法。血液采用含 0.10%（V/V）HCl、0.05%（m/V）L-半胱氨酸、0.10%（V/V）2-巯基乙醇稀释液稀释后，选用 C_{18} 反相色谱柱，流动相为 0.05%（V/V）巯基乙醇、0.4%（m/V）L-半胱氨酸、0.06 mol/L 乙酸铵和 5%（V/V）甲醇，HPLC-ICP-MS 分析测定，MeHg 和 Hg^{2+} 的检出限分别为 0.1 μg/L 和 0.25 μg/L。

de Souza 等[155]采用 HPLC 分离、汞冷蒸气发生和 ICP-MS 联用技术进一步提高了测定血清中汞形态灵敏度的方法。血浆样品经含 0.10%（V/V）HCl、0.05%（m/V）L-半胱氨酸、0.10%（V/V）2-巯基乙醇的溶液提取，采用 C_8 反相色谱柱，3%甲醇和含 0.5% 2-巯基乙醇、0.05%甲酸的流动相分离，在色谱柱和雾化器之间通入 0.06%（w/V）$NaBH_4$ 还原剂使汞形成蒸气发生。再采用 ICP-MS 分析，Hg^{2+}、EtHg、MeHg 的检测限分别为 12 ng/L、5 ng/L、4 ng/L。

Sogame 等[156]用 HPLC-ICP-MS 法检测血液样本中 MeHg 和 Hg^{2+}。人全血样品中加入 0.25 mL 含有 1.5%（w/V）L-半胱氨酸和 7%（V/V）HCl 溶液，混合、超声提取。离心后，用 0.015 mL 10%（w/V）三氯乙酸溶液去除蛋白，离心过滤，上清液进样分析。ZORBAX SB-C_{18} 色谱柱，流动相为 5%（V/V）甲醇、0.1%（V/V）2-巯基乙醇和 0.018%（V/V）HCl。MeHg 和 Hg^{2+} 检测的线性范围分别为 0.08～60 ng/mL 和 0.05～2.5 ng/mL。

Tsoi 等[157]采用固相微萃取法（SPME）提取，HPLC-ICP-MS 分析尿液样品中汞形态，采用的 Supelco 聚二甲基硅氧烷/二乙烯基苯（PDMS/DVB）固相微萃取纤维提取含有（5 mol/L）NaCl 的尿液样品，75℃水浴萃取，纤维在小瓶顶部提取 45 min。将萃取纤维取出插入充满流动相的解离瓶

5 min，使吸附物溶于溶液，将分析物通过流动相进入 C_{18} 色谱柱分离，流动相为 0.4%（w/V）L-半胱氨酸、0.05%（V/V）2-巯基乙醇、5%（V/V）甲醇的水溶液，ICP-MS 检测。方法 MeHg 和 EtHg 的检出限均为 0.06 μg/L。

彭国俊等[158]采用 HPLC-ICP-MS 测定头发和生物样品中 4 种汞形态。取人发标样用 5 mol/L HCl 溶液室温超声萃取二次，离心上清液过 0.45 μm 滤膜用 C_{18} 色谱柱分离，流动相为 2 mmol/L（pH 6.8）乙酸铵、0.1% L-半胱氨酸，8%～70%甲醇梯度洗脱，3 min 内分离了 Hg^{2+}、MeHg、EtHg 和苯基汞（PhHg），检出限范围为 0.05～0.08 μg/L。

张兰等[159]用类似的方法也测定了人发标样（GBW 07601）中的 4 种汞形态。采用 5 mol/L HCl 溶液和 3% L-半胱氨酸体系在 80℃下微波辅助萃取头发样品，采用 C_{18} 色谱柱，流动相 A 相为 10 mmol/L（pH 7.5）乙酸铵和 0.12% L-半胱氨酸，B 相为甲醇，梯度洗脱，测定结果表明人发标准物质含有 Hg^{2+} 和 MeHg。

23.2.3.3 硒形态的研究进展

硒是人体和生物必需的微量元素，主要通过硒蛋白参与许多重要的生理过程，硒直接或间接发挥抗氧化功能，增加细胞对氧化应激的防御。但过量会导致硒中毒，硒的毒性同样与化学形态密切相关，亚硒酸根[Se(Ⅳ)]比硒酸根[Se(Ⅵ)]毒性更大，而硒化氢（H_2Se）具有最高毒性。临床样品基质中通常分析的硒形态有 Se(Ⅳ)、Se(Ⅵ)、硒代蛋氨酸（SeMet）、甲基硒代半胱氨酸（MeSeCys）、硒代胱氨酸（$SeCys_2$）、谷胱甘肽过氧化物酶结合硒（SeGpX）、硫氧还蛋白还原酶结合硒（SeTrxR）、硒蛋白 P（SePP）、三甲基硒离子（$TMSe^+$）和硒糖。

da Silva 等[160]使用阴离子交换柱法测量了人尿和生物样品中硒形态。取尿液样品用 25 mmol/L 乙酸铵缓冲液稀释 2 倍后，过 0.45 μm 滤膜上机测定。用 Hamilton PRP X100 阴离子交换柱，流动相为 25 mmol/L 和 250 mmol/L（pH 5.17）的乙酸铵缓冲液，梯度洗脱，采用 ICP-MS 测定其中的 Se(Ⅳ)、Se(Ⅵ)、SeMet 和 $SeCys_2$，检出限范围为 0.2～0.4 μg/L。

刘源等[161]采用阴离子交换柱，HPLC-ICP-MS 测定人尿中 $SeCys_2$、MeSeCys、SeMet、Se(Ⅳ)和 Se(Ⅵ) 5 种硒形态。人尿样品用纯水稀释 3 倍，离心过 0.22 μm 滤膜，采用 Hamilton PRP-X100 色谱柱分离，以含 1%甲醇的 40 mmol/L（pH 5）$(NH_4)_2HPO_4$ 为流动相，等度洗脱，HPLC ICP-MS 检测，检出限为 0.2～0.5 μg/L。测定结果显示人尿中硒形态主要以 $SeCys_2$ 为主，同时含有少量 MeSeCys、SeMet、无机硒及未知含硒化合物。

为了提取、分析生物样品中微量的硒代氨基酸和硒代寡肽，Mao 等[162]研发了磺化聚苯乙烯-二氧化钛（PSP-TiO_2）杂合包被材料用于搅拌棒吸附萃取（stir bar sorptive extraction，SBSE）对尿液样品进行前处理，采用 HPLC-ICP-MS 联用技术分析 $SeCys_2$、MeSeCys、SeMet、硒代乙硫氨酸（SeEt）和硒代寡肽（γ-谷氨酰基-硒-甲基硒代半胱氨酸和硒代谷胱甘肽）。尿液经稀释、过滤，用搅拌棒吸附萃取提取硒形态，采用 C_{18} MG 色谱柱分离，以含有 0.1% 三氟乙酸、0.3%甲醇的 25 mmol/L（pH 3.1）的乙酸/乙酸铵为流动相，HPLC-ICP-MS 分离测定，$SeCys_2$、MeSeCys、SeMet、γ-GluMeSeCys、SeEt 和 GS-Se-SG 这 6 种硒形态的检出限在 545.9～158.8 ng/L（^{82}Se）之间。

据文献报道，海鲜和肉类是人类摄入含硒营养物质的主要来源，而硒通过硒糖等生物代谢途径排泄，为了监测尿液样品中的硒形态，Terol 等[163]建立了高温液相色谱（HTLC）与 ICP-MS 的联用技术，快速分离了人尿样品中硒糖 1[甲基 2-乙酰氨基-2-脱氧-1-硒代-β-D-半乳糖苷(methyl 2-acetamido-2-deoxy-1-seleno-β-D-galactopyranoside)]、硒糖 2[甲基 2-乙酰氨基-2-脱氧-1-硒代-β-D-吡喃葡萄糖苷(methyl 2-acetamido-2-deoxy-1-seleno-β-D-glucopyranoside)]和三甲基硒离子（$TMSe^+$）三种硒形态。尿液样品当天采集，混匀后过 0.45 μm 滤膜，采用 HTLC 色谱柱，以含 2%甲醇的纯水做流动相，经 ICP-MS 分析。可在 7 min 内分离硒糖 1、硒糖 2 和 TMSe⁺ 三种硒形态，而不受尿液基质中 Se(Ⅳ)、Se(Ⅵ)、SeMet 和 SeCys₂ 等其他常见硒形态的干扰。TMSe⁺、硒糖 1 和硒糖 2 的检出限分别为 0.5 ng/mL、0.3 ng/mL 和 0.4 ng/mL。

Liang 等[164]监测了长期汞暴露人群服用富硒酵母后的血清硒形态。采集的血清样品用 0.05 mol/L 的乙酸铵（pH 7）缓冲液稀释 4 倍，离心过 0.22 μm 滤膜。样品用 Hamilton PRP-X100 阴离子交换柱分离，用含 2%甲醇的 0.5 mmol/L 柠檬酸铵（pH 3.7）的 A 相和 2%甲醇 100 mmol/L 柠檬酸铵（pH 8.0）的 B 相梯度洗脱。用 He 和 H₂ 的碰撞池模式 ICP-MS 分析，SeCys、Se(Ⅳ)、SeMet 和 Se(Ⅵ)检出限分别为 0.34~1.38 μg/L。研究表明，补充硒后血清中总硒和 Se(Ⅳ)浓度均升高，但血清中存在的大量硒以未知硒蛋白形态存在。

Kokarnig 等[165]为了研究食用富硒膳食后志愿者血清和尿液中含硒代谢产物，分别优化了不同色谱条件分离多种硒形态，用 HPLC-ICP-MS 检测了已知硒形态，并用 ESI-MS 确证。收集的血清通过与等体积乙腈混合，离心取上清液，蒸干后用流动相溶解；尿液过膜后采用 ICP-MS 测定。实验中采用的色谱条件为：①Hamilton PRP-X200 阳离子交换柱，以 20 mmol/L 甲酸铵（pH 3.0）流动相分离 TMSe、MeSeCys 或硒糖 3；②Chrompack Ionospher 5C 阳离子交换柱，流动相 10 mmol/L 吡啶（pH 2.5），可分离硒糖 3；③Hamilton PRP-X100 阴离子交换柱 20 mmol/L 丙二酸（pH 9.5），分离 Se(Ⅳ)和 Se(Ⅵ)；④Waters Atlantis d C₁₈ 反相色谱柱，60 mmol/L 乙酸铵（pH 3.0）流动相，分离硒糖 1 和 SeMet。通过 HPLC-ICP-MS 测得的硒形态，除了 TMSe，检出限均为 0.25 ng/mL。本研究初步揭示了 Se(Ⅳ)、Se(Ⅵ)、SeMet、MeSeCys 和硒化酵母等富硒膳食来源的代谢途径，提示除 Se(Ⅳ)外，富硒饮食可增加血液、尿液中硒含量，且硒糖 1 是普遍的代谢产物。

23.2.3.4 铬形态的研究进展

铬是一种过渡金属，主要以三价铬 Cr(Ⅲ)和六价铬 Cr(Ⅵ)的形式存在。Cr(Ⅲ)是人体必需的微量元素，作用于葡萄糖耐量因子，维持人体代谢酶类的活性、促进胆固醇和脂肪酸的生成、增强胰岛素的敏感性，维持正常耐糖量。然而 Cr(Ⅵ)却有强烈的毒性，它以 CrO_4^{2-} 等阴离子形式存在，氧化性强，溶解性高，极易对人体和动物产生危害。Cr(Ⅵ)体积小，与 Cr(Ⅲ)相比更容易通过细胞膜的碳酸盐、硫酸盐和磷酸盐通道，被转运入细胞。随后被还原为 Cr(Ⅲ)，产生大量活性氧自由基中间体，损伤细胞结构和 DNA，造成癌变。由于不同氧化态的铬的营养价值、毒性差异很大，通过临床样本具体分析 Cr(Ⅲ)和 Cr(Ⅵ)的含量至关重要。

Wang 等[166]建立了离子对反相色谱法对铬暴露工人的尿液样品的形态分析方法。尿液用含 0.5 mmol/L EDTA 和 2 mmol/L TEA 处理液稀释 4 倍并络合 Cr(Ⅲ)，分离采用 Hamilton PRP-1 色谱柱，2.0 mmol/L（pH 3.5）TEA 为流动相，Thermo X7 ICP-MS 检测 $^{52}Cr^+$，Cr(Ⅲ)和 Cr(Ⅵ)的检出限为 0.03 μg/L。

金属植入物在骨科手术中有很大应用。然而，其在人体生理环境内释放出的金属离子受到广泛关注。铬是金属植入物的重要成分，需要检测如全血、关节积液和组织等临床样本中铬的含量，评估金属中毒风险。Pechancováa 等[167]用离子对色谱联用 ICP-MS 技术对全髋或膝关节置换术患者的血液和关节积液样品进行了总铬定量和铬形态分析。样品超滤后用四丁基碘化铵（TBAI）和 EDTA（1∶9）稀释，形成[Cr(Ⅲ)-EDTA]和[Cr(Ⅵ)-TBA]络合物，采用 ZORBAX Eclipse XDB-C8 反相柱分离，流动相为含 0.6 mmol/L EDTA 二钠盐的 5 mmol/L TBAI（pH 7）缓冲液，HPLC-ICP-MS 检测 ^{52}Cr。检出限分别为 0.13 μg/L 和 0.14 μg/L，研究表明，金属植入物患者的关节积液样本中铬的含量显著高于对照组，但是在血液中总铬和两种铬形态没有相关的差异，大多数 Cr(Ⅲ)与蛋白质结合存在。

23.2.3.5 铂形态的研究进展

铂类化合物是高效的抗癌药物，在目前的肿瘤化疗组合方案中被广泛使用。已上市的铂类化疗药物有三代，常用的铂类药物主要为顺铂、卡铂、奥沙利铂等，其抗肿瘤的机制是铂与肿瘤细胞的 DNA 形成络合物，抑制肿瘤细胞复制，产生强烈的细胞毒性，发挥抗肿瘤的作用。但是，含铂药物具有很多毒性和副作用，如肾毒性、耳毒性、消化系统毒性和骨髓抑制等，并作为重金属在体内蓄积。因此，通过尿液、血液等临床样品进行含铂化合物的治疗药物监测，对于指导临床用药剂量和减轻毒性具有重要意义。

刘德晔等[168]建立了 HPLC-ICP-MS 快速测定血浆中顺铂、卡铂、奥沙利铂的方法。取正常人的血浆样品加标混合后，离心后取上清液过 0.45 μm 的滤膜。分析采用 Dionex Acclaim PA2 C_{18} 柱，100%纯水为流动相，ICP-MS 检测丰度最高的 ^{195}Pt 同位素，该方法依次洗脱顺铂、卡铂和奥沙利铂。方法检出限以铂计为 0.04 ng/mL，标准曲线范围在 0.5～100 ng/mL。三种含铂化合物的线性相关系数 r 均大于 0.9995，加标回收率在 84%～102%之间，相对标准偏差在 0.9%～6.4%之间。

23.3 展 望

原子光谱技术在食品安全、食品营养分析检测中发挥了巨大的作用，为重金属与微量元素的检测提供了技术保障。随着技术的发展、研究的深入，元素形态分析目前还不能满足工作的需要，不同基质样品中砷形态、硒形态、铬形态、锡形态、铅形态的研究还有漫长的道路需要我们去探索。

原子光谱技术在临床样品中的应用虽然有很长的历史，但主要停留在有害元素的分析检测方面，随着社会对营养关注度的提升，也随着检测技术的发展，微量元素与疾病健康的关系会有更深入的研究，原子光谱技术在元素总量和元素形态分析方面的标准方法、研究将会有更大的发展和进步。

参 考 文 献

[1] 食品安全国家标准, 食品中污染物限量. GB 2762—2017. 中华人民共和国国家卫生和计划生育委员会, 2017.
[2] 食品安全国家标准, 食品中总砷及无机砷的测定. GB 5009.11—2014. 中华人民共和国国家卫生和计划生育委员会, 2014.
[3] 食品安全国家标准, 食品中总汞及有机汞的测定. GB 5009.17—2021. 中华人民共和国国家卫生健康委员会　国家市场监督管理总局, 2021.
[4] 食品安全国家标准, 植物性食品中稀土元素的测定. GB 5009.94—2012. 中华人民共和国卫生部, 2012.
[5] 食品安全国家标准, 食品中多元素的测定. GB 5009.268—2016. 中华人民共和国国家卫生和计划生育委员会, 2016.
[6] 刘丽萍, 张妮娜, 李筱薇, 吕超, 毛红. 直接测汞仪测定食品中的总汞. 中国食品卫生杂志, 2010, 22(1): 19-23.
[7] 张妮娜, 王小艳, 刘丽萍, 毛红, 王颖. 直接测汞仪法快速测定婴幼儿配方乳粉中汞. 卫生研究, 2015, 44(1): 129-131.
[8] 王维平, 刘佳琪, 薛阿喜. 水产品中痕量汞的快速直接检测研究. 中国水运: 下半月, 2017, 17(12): 243-244.
[9] 董娇, 岑琴, 张君. 直接测汞仪测定茶叶中的汞. 食品安全导刊, 2018, (12): 56-57.

[10] 周宇, 贾宏新, 郑凤娥, 黄旭. 测汞仪直接测定食品中总汞. 中国无机分析化学, 2015, 5(3): 5-7.

[11] 周春艳, 胡黎黎, 顾万江, 覃梅, 王正虹, 何健. 直接测汞仪测定野生食用菌中总汞含量及暴露风险评估. 中国卫生检验杂志, 2017, 27(17): 2448-2451.

[12] 谢科, 韩枫, 宫智勇, 吴永宁. 电热解-塞曼原子吸收光谱法测定调味品中的总汞. 武汉工业学院学报, 2013, 32(2): 77-81.

[13] 杨娟, 陈建忠, 杨清华, 平文卉. 全自动测汞仪直接测定食品中总汞. 中国卫生检验杂志, 2016, (6): 782-783.

[14] 徐君辉, 冯礼, 沈飚, 晁铎源. 固体进样测汞装置与原子荧光联用测定海产品中的汞. 分析仪器, 2016, (3): 46-49.

[15] Wang B, Feng L, Mao X, Liu J, Yu C, Ding L, Qian Y. Direct determination of trace mercury and cadmium in food by sequential electrothermal vaporization atomic fluorescence spectrometry using tungsten and gold coil traps. Journal of Analytical Atomic Spectrometry, 2018, 33(7): 1209-1216.

[16] 陈丽娟, 冯玲, 罗荻, 柳莎, 姚程炜, 黄宝美. 间接原子吸收光谱法测定鸡蛋中碘含量. 食品科学, 2012, 33(12): 247-249.

[17] Yebra M C , Bollaín M H. A simple indirect automatic method to determine total iodine in milk products by flame atomic absorption spectrometry. Talanta, 2010, 82(2): 828-833.

[18] 刘丽萍, 吕超, 谭玲, 李筱薇. 电感耦合等离子体质谱法测定乳制品中碘含量的方法研究. 质谱学报, 2010(3): 138-142.

[19] 吕超, 刘丽萍, 谭玲. 电感耦合等离子体质谱法测定饮料、啤酒及果汁中的碘. 中国食品卫生杂志, 2010, 22(4): 347-350.

[20] 张妮娜, 刘丽萍, 吴可欣, 王正. 电感耦合等离子体-质谱法测定居民肉类膳食中碘的方法研究.中国食品卫生杂志, 2014, 26(2): 153-155.

[21] 丁玉龙, 葛宇, 徐红斌, 段文峰, 于学雷, 田芳, 宋畅. 微波消解-电感耦合等离子体质谱法测定乳制品中总碘. 分析测试技术与仪器, 2016, 22(3): 184-188.

[22] 谭玲, 刘丽萍, 吕超, 王京宇. 电感耦合等离子体质谱法(ICP-MS)测定矿泉水的碘. 中国卫生检验杂志, 2009, (12): 2750-2751.

[23] 张慧敏, 黎雪慧, 刘桂华, 黄铁军. 电感耦合等离子体质谱技术测定痕量碘方法研究.中国卫生检验杂志, 2014, 24(4): 493-495.

[24] Jussiane S S, Lisarb O D, Angelica C F, Vanize C C, Marcia F M, Fabio A D, Erico M M. Flore S. Determination of bromine and iodine in edible flours by inductively coupled plasma mass spectrometry after microwave-induced combustion. Microchemical Journal, 2017, 133：246-250.

[25] Rondan F S, Crizel M G, Hartwig C A, Novo D L R, Moraes D P, Sandra M, Paola A M, Mareia F. Ultra-trace determination of bromine and iodine in rice by ICP-MS after microwave-induced combustion. Journal of Food Composition and Analysis, 2018, 66: 199-204.

[26] Toralles I G, Coelho G S Jr, Costa V C, Cruz S M, Flores E M M, Mesko M F. A fast and feasible method for Br and I determination in whole egg powder and its fractions by ICP-MS. Food Chemistry, 2017, 221: 877-883.

[27] 刘丽萍, 陈绍占, 周天慧. 食品安全风险监测标准操作规程 食品中多元素测定. 国家食品安全风险评估中心发布. 2020.

[28] 王丙涛, 赵旭, 涂小珂, 罗洁, 林燕奎, 葛丽雅, 颜治, 严冬.ICP-MS/MS 检测食品中磷、硒、砷的含量. 现代食品科技, 2017, 33(7): 295-300.

[29] 王晓伟, 刘景富, 关红, 王小艳, 邵兵, 张晶, 刘丽萍, 张妮娜. 三重串联四极杆电感耦合等离子体质谱法测定植源性中药材中总硫含量. 光谱学与光谱分析, 2016, 36(2): 527-531.

[30] Guo W, Jin L, Hu S, Guo Q. Method development for the determination of total fluorine in foods by tandem inductively

coupled plasma mass spectrometry with a mass-shift strategy. Journal of Agricultural and Food Chemistry 2017;65(16): 3406-3412.

[31] Nelson J, Hopfer H, Silva F, Wilbur S, Chen J M, Shiota O K, Wylie P L. Evaluation of GC-ICP-MS/MS as a new strategy for specific heteroatom detection of phosphorus, sulfur, and chlorine determination in foods. Journal of agricultural and food chemistry, 2015, 63(18): 4478-4483.

[32] Nelson J, Hopfer H, Silva F, Wilbur S, Chen J, Ozawa K S, Wylie P L.利用基于 GC-ICP-MS/MS 的磷和硫检测方法测定食品中的农药. 安捷伦科技（中国）有限公司应用简报, 2015-10-30.

[33] Villa-Lojo M C, Alonso-Rodríguez E, López-Mahía P, Muniategui-Lorenzo S, Prada-Rodríguez D. Coupled high performance liquid chromatography-microwave digestion-hydride generation-atomic absorption spectrometry for inorganic and organic arsenic speciation in fish tissue. Talanta, 2002, 57: 741-750.

[34] Niedzielski P, Mleczek M, Magdziak Z, Siwulski M, Kozak L. Selected arsenic species: As(Ⅲ), As(Ⅴ) and dimethylarsenic acid (DMAA) in Xerocomus badius fruiting bodies. Food Chemistry, 2013, 141: 3571-3577.

[35] 刘淑晗, 张海燕, 娄晓祎, 孔聪, 汪宇, 史永富, 黄宣运, 沈晓盛. 高效液相色谱-(紫外)氢化物发生原子荧光光谱法测定南极磷虾及其制品中 6 种砷形态. 分析测试学报, 2019, 38(9): 1085-1090.

[36] Cui J, Xiao Y B, Dai L, Zhao X H, Wang Y. Speciation of organoarsenic species in food of animal origin using accelerated solvent extraction (ASE) with determination by HPLC-hydride generation-atomic fluorescence spectrometry (HG-AFS). Food Analytical Methods, 2013, 6(2): 370-379.

[37] Farías S S, Londonio A, Quintero C, Befani R, Soro M, Smichowski P. On-line speciation and quantification of four arsenical species in rice samples collected in Argentina using a HPLC-HG-AFS coupling. Microchemical Journal, 2015, 120: 34-39.

[38] 吴池莹, 樊祥, 张润何, 霍忆慧, 陆滕翀, 孙诗倩. 微波辅助萃取结合 LC-ICP/MS 法测定食品中无机砷. 分析试验室, 2019, 38(3): 295-299.

[39] 苏祖俭, 胡曙光, 蔡文华, 杨杏芬, 王晶, 范建彬, 黄泓耀, 黄伟雄. 建立无机砷形态的检测方法及在大米基质检测中的应用. 中华预防医学杂志, 2018, 52(10): 994-1002.

[40] Jinadasa K K, Peña-Vázquez E, Bermejo-Barrera P, Moreda-Piñeiro A. Ionic imprinted polymer-vortex-assisted dispersive micro-solid phase extraction for inorganic arsenic speciation in rice by HPLC-ICP-MS. Talanta, 2020, 121418.

[41] Narukawa T, Matsumoto E, Nishimura T, Hioki A. Reversed phase column HPLC-ICP-MS conditions for arsenic speciation analysis of rice flour. Analytical Sciences, 2015, 31(6): 521-527.

[42] Fang Y, Pan Y, Li P, Xue M, Pei F, Yang W, Ma N, Hu Q. Simultaneous determination of arsenic and mercury species in rice by ion-pairing reversed phase chromatography with inductively coupled plasma mass spectrometry. Food Chemistry, 2016, 213: 609-615.

[43] Narukawa T, Chiba K, Sinaviwat S, Feldmann J. A rapid monitoring method for inorganic arsenic in rice flour using reversed phase-high performance liquid chromatography-inductively coupled plasma mass spectrometry. Journal of Chromatography A, 2017, 1479: 129-136.

[44] 吕超, 刘丽萍, 董慧茹, 李筱薇. 高效液相色谱-电感耦合等离子体质谱联用技术测定水产类膳食中 5 种砷形态的方法研究. 分析测试学报, 2010, 29(5): 465-468.

[45] Schmidt L, Landero J A, Santos R F, Mesko M F, Mello P A, Flores E M M, Caruso J A. Arsenic speciation in seafood by LC-ICP-MS/MS: Method development and influence of culinary treatment. Journal of Analytical Atomic Spectrometry, 2017, 32(8): 1490-1499.

[46] Matsumoto-Tanibuchi E, Sugimoto T, Kawaguchi T, Sakakibara N, Yamashita M. Determination of inorganic arsenic in

seaweed and seafood by LC-ICP-MS: Method validation. Journal of AOAC International, 2019, 102(2): 612-618.

[47] Zmozinski A V, Llorente-Mirandes T, Lopez-Sanchez J F, Da Silva M M. Establishment of a method for determination of arsenic species in seafood by LC-ICP-MS. Food Chemistry, 2015, 173: 1073-1082.

[48] Moreda-Piñeiro J, Alonso-Rodríguez E, Romarís-Hortas V, Moreda-Piñeiro A, López-Mahía P, Muniategui-Lorenzo S, Prada-Rodríguez D, Bermejo-Barrera P. Assessment of the bioavailability of toxic and non-toxic arsenic species in seafood samples. Food Chemistry, 2012, 130: 552-560.

[49] Whaley-Martin K J, Koch I, Moriarty M, Reimer K J. Arsenic speciation in blue mussels (*Mytilus edulis*) along a highly contaminated arsenic gradient. Environmental Science and Technology, 2012, 46: 3110-3118.

[50] Moreda-Piñeiro A, Moreda-Piñeiro J, Herbello-Hermelo P, Bermejo-Barrera P, Muniategui-Lorenzo S, López-Mahía P, Prada-Rodríguez D. Application of fast ultrasound water-bath assisted enzymatic hydrolysis-high performance liquid chromatography-inductively coupled plasma-mass spectrometry procedures for arsenic speciation in seafood materials. Journal of Chromatography A, 2011, 1218(39): 6970-6980.

[51] Jia Y, Wang L, Ma L, Yang Z. Speciation analysis of six arsenic species in marketed shellfish: Extraction optimization and health risk assessment. Food Chemistry, 2018, 244: 311-316.

[52] Schmidt L, Landero J A, Novo D L R, Duarte F A, Mesko M F, Caruso J A, Flores E M M. A feasible method for as speciation in several types of seafood by LC-ICP-MS/MS. Food Chemistry, 2018, 255: 340-347.

[53] 陈绍占, 刘丽萍, 杜振霞, 马辉. 高效液相色谱-电感耦合等离子体质谱联用技术测定畜禽肉制品中八种砷形态化合物. 质谱学报, 2015, 36(1): 33-39.

[54] 吴思霖, 王欣美, 于建, 潘晨, 王柯. 高效液相色谱-电感耦合等离子体质谱联用测定鸡肉及鸡肝中10种砷形态化合物. 分析测试学报, 2018, 37(4): 113-117.

[55] Zhao D, Wang J, Yin D, Li M, Ma L Q. Arsanilic acid contributes more to total arsenic than roxarsone in chicken meat from Chinese markets. Journal of Hazardous Materials, 2020, 383: 121178.

[56] Nearing M M, Koch I, Reimer K J. Arsenic speciation in edible mushrooms. Environmental Science and Technology, 2014, 48(24): 14203-14210.

[57] Komorowicz I, Hanć A, Lorenc W, Barałkiewicz D, Falandysz J, Wang Y. Arsenic speciation in mushrooms using dimensional chromatography coupled to ICP-MS detector. Chemosphere, 2019, 233: 223-233.

[58] Chen S, Guo Q, Liu L. Determination of arsenic species in edible mushrooms by high-performance liquid chromatography coupled to inductively coupled plasma mass spectrometry. Food Analytical Methods, 2017, 10(3): 740-748.

[59] Zou H, Zhou C, Li Y, Yang X, Wen J, Song S, Li C, Sun C. Speciation analysis of arsenic in edible mushrooms by high-performance liquid chromatography hyphenated to inductively coupled plasma mass spectrometry. Food Chemistry, 2020, 327: 127303.

[60] 丁宁, 蒋小良, 江礼华, 周宇, 王国胜, 苏淑坛, 钟月香. 超声辅助提取HPLC-ICP-MS同时测定乳制品中7种砷形态. 中国乳品工业, 2018, 46(5): 46-48.

[61] 赵建国, 李琴, 陶大利, 姜金斗. 酶解析-HPLC-ICP-MS同时测定生乳及乳粉中砷形态及质量浓度. 中国乳品工业, 2015, 43(2): 50-53.

[62] Stiboller M, Raber G, Gjengedal E L F, Eggesbo M, Francesconi K A. Quantifying inorganic arsenic and other water-soluble arsenic species in human milk by HPLC/ICPMS. Analytical Chemistry, 2017, 89(11): 6265-6271.

[63] 赵云强, 郑进平, 杨明伟, 付凤富. 毛细管电泳-电感耦合等离子体质谱法测定藻类中6种不同形态的砷化合物. 色谱, 2011, 29: 111-114.

[64] 陈发荣, 郑立, 王志广, 孙杰, 韩力挥, 王小如. 毛细管电泳-电感耦合等离子体质谱测定蓝点马鲛中砷化学形态.

光谱学与光谱分析, 2014, 34: 1675-1678.

[65] 陈发荣, 郑立, 韩力挥, 孙杰, 尹晓斐, 刘峰, 王小如. 毛细管电泳-电感耦合等离子体质谱(CE-ICP-MS)联用测定干海产品中的六种砷形态化合物. 食品工业科技, 2014, 35(19): 304-307.

[66] Yang G, Zheng J, Chen L, Lin Q, Zhao Y, Wu Y, Fu F. Speciation analysis and characterisation of arsenic in lavers collected from coastal waters of Fujian, south-eastern China. Food Chemistry, 2012, 132: 1480-1485.

[67] Liu L, Yun Z, He B, Jiang G. Efficient interface for online coupling of capillary electrophoresis with inductively coupled plasma-mass spectrometry and its application in simultaneous speciation analysis of arsenic and selenium. Analytical Chemistry, 2014, 86: 8167-8175.

[68] 吕超, 刘丽萍, 董慧茹, 李筱薇. 盐酸提取-液相色谱-原子荧光联用技术检测水产品中甲基汞等汞化合物. 分析试验室, 2010, (2): 70-74.

[69] Brombach C C, Ezzeldin M F, Chen B, Corns W T, Krupp E. Quick and robust method for trace determination of MeHg in rice and rice products without derivatisation. Analytical methods, 2015, 7(20): 8584-8589.

[70] Grijalba A C, Quintas P Y, Fiorentini E F, Wuilloud R G. Usefulness of ionic liquids as mobile phase modifiers in HPLC-CV-AFS for mercury speciation analysis in food. Journal of Analytical Atomic Spectrometry, 2018, 33(5): 822-834.

[71] Liu Q Y. Determination of mercury and methylmercury in seafood by ion chromatography using photo-induced chemical vapor generation atomic fluorescence spectrometric detection. Microchemical Journal, 2010, 95(2): 255-258.

[72] Li H, Xu Z, Yang L, Wang Q. Determination and speciation of Hg using HPLC-AFS by atomization of this metal on a UV/nano-ZrO$_2$/HCOOH photocatalytic reduction unit. Journal of Analytical Atomic Spectrometry, 2015, 30(4): 916-921.

[73] Ai X, Wang Y, Hou X, Yang L, Zheng C, Wu L. Advanced oxidation using Fe$_3$O$_4$ magnetic nanoparticles and its application in mercury speciation analysis by high performance liquid chromatography-cold vapor generation atomic fluorescence spectrometry. Analyst, 2013, 138(12): 3494-3501.

[74] Jókai Z, Abrankó L, Fodor P. SPME-GC-Pyrolysis-AFS determination of methylmercury in marine fish products by alkaline sample preparation and aqueous phase phenylation derivatization. Journal of Agricultural and Food Chemistry, 2005, 53(14): 5499-5505.

[75] Jiménez-Moreno M, Lominchar M Á, Sierra M J, Millán R, Martín-Doimeadios R C R. Fast method for the simultaneous determination of monomethylmercury and inorganic mercury in rice and aquatic plant. Talanta, 2018, 176: 102-107.

[76] 刘沛明. 萃取-水相乙基化衍生结合 GC-CVAFS 法测定大米和鱼样品中的甲基汞. 中国测试, 2015, 41(6): 56-59.

[77] 崔颖, 肖亚兵, 王禹, 高健会. P&T-GC-CVAFS 法测定海产动物中的甲基汞和乙基汞. 食品研究与开发, 2016, 37(1): 145-148.

[78] 邹学权, 张文华, 陈孜. 丙基化衍生-吹扫捕集-原子荧光光谱法同时测定动物源食品中的甲基汞和乙基汞. 浙江化工, 2018, 49(9): 54-58.

[79] 王维洁, 陈朋, 王萍亚, 罗海军, 戴意飞, 赵巧灵. 全自动烷基汞分析仪测定水产品中甲基汞的方法研究. 食品工业, 2018, 39(4): 326-328.

[80] 刘丽萍, 吕超, 王颖. 液相色谱-电感耦合等离子质谱联用技术测定水产品中汞化合物形态分析方法探讨. 分析测试学报, 2010, 29(8): 767-771.

[81] 冯晓青, 徐瑞, 汪怡, 王芹, 王露, 宋鑫, 时玉军, 杭学宇. 微探头超声破碎辅助提取-HPLC-ICP/MS 快速测定海产品中甲基汞. 分析仪器, 2018, 219(4): 122-127.

[82] Zhu S, Chen B, He M, Huang T, Hu B. Speciation of mercury in water and fish samples by HPLC-ICP-MS after

magnetic solid phase extraction. Talanta, 2017, (171): 213-219.

[83] 龚燕, 尚晓虹, 赵馨. HPLC-ICP-MS 联用测定稻米中的无机汞和甲基汞. 化学试剂, 2018, 40(3): 257-260.

[84] 贾彦博, 陆吉琛, 朱蓓, 林舒忆, 洪春来. 高效液相色谱-电感耦合等离子体质谱联用法测定东海乌参样品中的二价汞、甲基汞、乙基汞和苯基汞. 食品安全质量检测学报, 2016, 7(11): 4609-4613.

[85] Zou H, Zhou C, Li Y, Yang X, Sun C. Speciation analysis of mercury in wild edible mushrooms by high-performance liquid chromatography hyphenated to inductively coupled plasma mass spectrometry. Analytical and Bioanalytical Chemistry, 2020, 412: 2829-2840.

[86] Chung W C, Chan T P. A reliable method to determine methylmercury and ethylmercury simultaneously in foods by gas chromatography with inductively coupled plasma mass spectrometry after enzymatic and acid digestion. Journal of Chromatography A, 2011, 1218(9): 1260-1265.

[87] Valdersnes S, Maage A, Fliegel D, Julshamn K. A method for the routine determination of methylmercury in marine tissue by GC isotope dilution-ICP-MS. Journal of AOAC International, 2012, 95(4): 1189-1194.

[88] Esteban-Fernández D, Mirat M, de la Hinojosa M I M, Alonso J I G. Double spike isotope dilution GC-ICP-MS for evaluation of mercury species transformation in real fish samples using ultrasound-assisted extraction. Journal of Agricultural and Food Chemistry, 2012, 60(34): 8333-8339.

[89] Dressler V L, Santos C M M, Antes F G, Bentlin F R S, Pozebon D, Flores E M M. Total mercury, inorganic mercury and methyl mercury determination in red wine. Food Analytical Methods, 2012, 5(3): 505-511.

[90] Niedzielski P, Mleczek M, Siwulski M, Rzymski P, Gąsecka M, Kozak L. Supplementation of cultivated mushroom species with selenium: bioaccumulation and speciation study. European Food Research and Technology, 2015, 241(3): 419-426.

[91] 龚洋, 雷正达, 李莉, 何瑞, 李进春. 液相色谱——氢化物发生原子荧光光谱法测定富硒稻谷中硒的形态. 食品与发酵科技, 2020, 56(2): 105-108.

[92] Zhou F, Dinh Q T, Yang W, Wang M, Xue M, Bañuelos G S, Liang D. Assessment of speciation and *in vitro* bioaccessibility of selenium in Se-enriched *Pleurotus ostreatus* and potential health risks. Ecotoxicology and Environmental Safety, 2019, 185: 109-675.

[93] 钟银飞. 液相-原子荧光(LC-AFS)联用技术测定大蒜中不同形态硒化合物含量研究. 食品安全导刊, 2019, (33): 52-53.

[94] 尚德荣, 秦德元, 赵艳芳, 翟毓秀, 宁劲松, 丁海燕. 高压液相色谱-氢化物发生-原子荧光光谱(HPLC-HG-AFS)联用技术检测水产品中硒的赋存形态. 食品安全质量检测学报, 2013, (6): 1847-1852.

[95] 韩婷婷, 崔鹤, 宋田, 姬泓巍, 李慧新, 朱倩林, 蔡峰, 张涛. 离子色谱-氢化物发生-原子荧光光谱（IC-HG-AFS）联用技术检测胶州湾海产品中硒的赋存形态. 食品工业科技, 2016, 37(18): 81-84.

[96] Xie X, Feng C, Ye M, Wang C. Speciation determination of selenium in seafood by high-performance ion-exchange chromatography-yydride generation-atomic fluorescence spectrometry. Food Analytical Methods, 2015, 8(7): 1739-1745.

[97] 钟永生, 万承波, 林黛琴. 富硒鸡蛋中微量元素硒的形态分析. 江西化工, 2019, (4): 30-32.

[98] Hu T, Liu L, Chen S, Wu W, Xiang C, Guo Y. Determination of selenium species in Cordyceps militaris by High-performance liquid chromatography coupled to hydride generation atomic fluorescence spectrometry. Analytical Letters, 2018, 51(14): 2316-2330.

[99] 张妮娜, 任武洁, 刘丽萍, 陈绍占, 王小艳, 王晓伟. 阴离子交换色谱-氢化物发生-原子荧光光谱联用测定片剂类富硒保健品中 4 种形态硒. 中国食品卫生杂志, 2017, 29(6): 702-707.

[100] 陈绍占, 唐德剑, 李晓玉, 夏曾润, 刘丽萍. 谷类食品中硒形态超声酶提取-高效液相色谱-电感耦合等离子体质

谱法测定. 中国公共卫生, 2017, 36(1): 130-136.

[101] Gao H, Chen M, Hu X, Chai S, Qin M, Cao Z. Separation of selenium species and their sensitive determination in rice samples by ion-pairing reversed-phase liquid chromatography with inductively coupled plasma tandem mass spectrometry. Journal of Separation Science, 2017, 41(2): 432-439.

[102] 秦冲, 施畅, 万秋月, 王磊, 刘爱琴, 安彩秀. HPLC-ICP-MS法测定富硒小麦中硒的形态. 食品研究与开发, 2019, 40(02): 140-144.

[103] Zhong N, Zhong L, Hao L, Luan C, Li X. Speciation of selenium in enriched garlic sprouts by high-performance liquid chromatography coupled with inductively coupled plasma-mass spectrometry. Analytical Letters, 2015, 48(1): 180-187.

[104] 姚真真, 哈雪姣, 马智宏, 王北洪, 刘静, 李冰茹. 高效液相色谱-电感耦合等离子体质谱法检测富硒苹果中5种硒形态. 食品安全质量检测学报, 2018, 9(3): 475-480.

[105] Maneetong S, Chookhampaeng S, Chantiratikul A, Chinrasri O, Thosaikham W, Sittipout R, Chantiratikul P. Hydroponic cultivation of selenium-enriched kale (*Brassica oleracea* var. *alboglabra* L.) seedling and speciation of selenium with HPLC-ICP-MS. Microchemical Journal, 2013, 108: 87-91.

[106] 吴雅颖, 桂仁意, 汤鋆, 程建中, 杨萍. HPLC-ICP-MS联用技术测定竹笋中六种硒形态. 营养学报, 2014, 36(5): 494-498.

[107] 熊珺, 覃毅磊, 龚亮, 赖毅东. HPLC-ICP-MS在线联用分析食品中无机硒和硒氨基酸的形态. 食品工业科技, 2017, 38(4): 67-72.

[108] 倪张林, 汤富彬, 张玮, 屈明华. HPLC-DRC-ICP-MS测定富硒蔬菜中的硒形态. 分析试验室, 2013, 32(2): 39-43.

[109] Zhang Q, Yang G. Selenium speciation in bay scallops by high performance liquid chromatography separation and inductively coupled plasma mass spectrometry detection after complete enzymatic extraction. Journal of Chromatography A, 2014, 1325: 83-91.

[110] 陈贵宇, 潘煜辰, 李清清, 冷桃花, 李君绩, 葛宇. 高效液相色谱-电感耦合等离子质谱法分析富硒茶叶中硒的形态. 食品科学, 2018, 39(8): 155-159.

[111] Lipiec E, Siara G, Bierla K, Ouerdane L, Szpunar J. Determination of selenomethionine, selenocysteine, and inorganic selenium in eggs by HPLC-inductively coupled plasma mass spectrometry. Analytical and Bioanalytical Chemistry, 2010, 397(2): 731-741.

[112] Cao Y, Yan L, Huang H, Deng B. Selenium speciation in radix puerariae using ultrasonic assisted extraction combined with reversed phase high performance liquid chromatography-inductively coupled plasma-mass spectrometry after magnetic solid-phase extraction with 5-sulfosalicylic acid functionalized magnetic nanoparticles. Spectrochimica Acta Part B: Atomic Spectroscopy, 2016, 122: 172-177.

[113] 曹玉嬫, 闫丽珍, 黄红丽, 邓必阳. 超声辅助提取结合高效液相色谱-电感耦合等离子体质谱联用技术测定牛蒡和三七中硒形态. 分析化学, 2015, 43(9): 1329-1334.

[114] 王丙涛, 王清龙, 赵琼晖, 颜治, 杨修斌. 转基因大豆对硒的富集作用和形态分布研究. 食品安全质量检测学报, 2014, 5(11): 3476-3481.

[115] 杨修斌, 王丙涛, 俞坤, 颜治, 林起辉, 赵琼晖, 林燕奎, 王超. HPLC-ICP-MS联用检测转基因大豆中的硒形态. 现代食品科技, 2015, 31(2): 280-284.

[116] 魏琴芳, 贾彦博, 胡文彬, 洪春来, 韦燕燕. 液相色谱-电感耦合等离子质谱联用法测定大豆中的5种硒形态. 食品安全质量检测学报, 2020, 11(8): 2456-2461.

[117] Zhao Y, Zheng J, Yang M, Yang G, Wu Y, Fu F. Speciation analysis of selenium in rice samples by using capillary electrophoresis-inductively coupled plasma mass spectrometry. Talanta, 2011, 84(3): 983-988.

[118] 杨海锋, 陈磊, 宋卫国, 叶少丹, 梁立韵, 黄南, 姚春霞. HPLC-ICP-MS联用测定食用菌中的Cr(Ⅲ)和Cr(Ⅵ). 食

用菌学报, 2018, 25(2): 126-131.

[119] 倪张林, 汤富彬, 屈明华, 莫润宏. 微波灰化-液相色谱-电感耦合等离子体质谱联用测定干食用菌中的三价铬和六价铬. 色谱, 2014, 32(02): 174-178.

[120] 高勇兴. HPLC-ICP-MS 法分析乳粉中铬的形态. 广州化工, 2017, 45(19): 90-92.

[121] Hernandez F, Jitaru P, Cormant F, Noël L, Guérin T. Development and application of a method for Cr(Ⅲ) determination in dairy products by HPLC-ICP-MS. Food Chemistry, 2018, 240.

[122] 李琴, 周红, 陶大利, 姜金斗. 牛乳中铬形态研究与风险分析. 中国乳品工业, 2016, 44(3): 51-53, 56.

[123] 樊祥, 程甲, 张润何, 吴池莹, 陆滕翀, 霍忆慧, 孙诗倩, 许权辉. 高效液相色谱-电感耦合等离子体质谱法测定食品中的六价铬含量. 食品安全质量检测学报, 2018, 9(21): 5704-5708.

[124] 陈东, 王晓伟, 赵榕, 邵兵, 王正, 吴国华, 薛颖. 高效液相色谱-电感耦合等离子体质谱法同时测定双份饭样品中 Cr(Ⅲ)和 Cr(Ⅵ). 分析化学, 2015, 43(12): 1901-1905.

[125] Unceta N, Astorkia M, Abrego Z, Gómez-Caballero A, Goicolea M A, Barrio R J. A novel strategy for Cr(Ⅲ) and Cr(Ⅵ) analysis in dietary supplements by speciated isotope dilution mass spectrometry. Talanta. 2016, 154: 255-262.

[126] 葛晓鸣, 冯睿, 孙骥, 叶海雷, 金婉芳, 陈先锋, 马明, 彭锦峰. 超声辅助提取-高效液相色谱-电感耦合等离子体质谱法分析食品中水溶性铬价态. 食品安全质量检测学报, 2019, 10(3): 670-675.

[127] GBZ/T 308—2018.尿中多种金属同时的测定 电感耦合等离子体质谱法. 中华人民共和国国家卫生健康委员会, 2018.

[128] WS/T 107.2—2016.尿中碘的测定 电感耦合等离子体质谱法. 中华人民共和国国家卫生健康委员会, 2016.

[129] WS/T 635—2018. 尿中砷形态测定 液相色谱-原子荧光法. 中华人民共和国国家卫生健康委员会, 2018.

[130] Meyer S, Markova M, Pohl G, Marschall T A, Pivovarova O, Pfeiffer A F H, Schwerdtle T. Development, validation and application of an ICP-MS/MS method to quantify minerals and (ultra-)trace elements in human serum. Journal of Trace Elements in Medicine and Biology, 2018, 49: 157-163.

[131] Johnson-Davis KL, Farnsworth C, Law C, Parker R. Method validation for a multi-element panel in serum by inductively coupled plasma mass spectrometry (ICP-MS). Clinical Biochemistry, 2020, 82: 90-98.

[132] Jones D R, Jarrett J M, Tevis D S, Franklin M, Mullinix N J, Wallon K L, Quarles C D, Caldwell K L, Jones R L. Analysis of whole human blood for Pb, Cd, Hg, Se, and Mn by ICP-DRC-MS for biomonitoring and acute exposures. Talanta, 2017, 162：114-122.

[133] 党瑞, 唐思思, 李俊伟, 朱金峰, 王德才, 管春梅. 电感耦合等离子体质谱法测定血浆中 21 种金属元素的样品前处理方法比较. 现代预防医学, 2015, 42(4): 678-681.

[134] 王俊, 谭洪兴, 刘惠坚, 廖琴琴, 徐健, 刘小立. 动态反应池电感耦合等离子体质谱法快速监测人体尿液中 19 种元素的含量. 卫生研究, 2012, 41(4): 654-657.

[135] 刘柳, 刘喆, 杨一兵, 王秦, 徐东群. KED-ICP-MS 法测定尿样中 21 种无机元素. 光谱学与光谱分析, 2019, 39(4): 1262-1266.

[136] 汤昌海, 李争云, 马微. ICP-MS 法测定尿液中铅、镉、锰、镍和铬. 工业卫生与职业病, 2016, 42(5): 383-385.

[137] Markiewicz B, Sajnóg A, Lorenc W, Hanć A, Komorowicz I, Suliburska J, Kocyłowski R, Barałkiewicz D. Multielemental analysis of 18 essential and toxic elements in amniotic fluid samples by ICP-MS: Full procedure validation and estimation of measurement uncertainty. Talanta, 2017, 174: 122-130.

[138] Chantada-Vázquez M P, Moreda-Piñeiro J, Cantarero-Roldán A, Bermejo-Barrera P, Moreda-Piñeiro A. Development of dried serum spot sampling techniques for the assessment of trace elements in serum samples by LA-ICP-MS. Talanta, 2018, 186: 169-175.

[139] 温权, 曾胜波, 梁肇海, 蔡红霓.直接稀释-电感耦合等离子体质谱法测尿中碘. 中国地方病防治杂志, 2020,

35(2): 166-168.

[140] 张慧敏, 黎雪慧, 刘桂华, 黄铁军. 电感耦合等离子体质谱技术测定痕量碘方法研究. 中国卫生检验杂志, 2014, 24(4): 493-495.

[141] 黄淑英, 张亚平, 张淑琼, 李呐, 黄嫣红. 尿中碘的无需基尿匹配电感耦合等离子体质谱测定方法. 中华地方病学杂志, 2017, 36(12): 920-926.

[142] 禹松林, 程倩, 韩建华, 周伟燕, 程歆琦, 侯立安, 高冉, 苏薇, 李志, 邱玲. 电感耦合等离子体质谱检测人尿液和血清中碘临床方法的建立. 中华检验医学杂志, 2016, 39(12): 917-921.

[143] 张亚平, 李卫东, 黄淑英, 李秀维, 徐署东, 王海燕, 李呐, 刘婷婷. 电感耦合等离子体质谱法测定血清中碘的标准化方法研究. 中华地方病学杂志, 2020, 39(3): 215-220.

[144] Yu S L, Yin Y C, Cheng Q, Han J H, Cheng X Q, Guo Y, Sun D D, Xie S W, Qiu L. Validation of a simple inductively coupled plasma mass spectrometry method for detecting urine and serum iodine and evaluation of iodine status of pregnant women in Beijing. Scandinavian Journal of Clinical and Laboratory Investigation, 2018, 78(6): 501-507.

[145] Contreras-Acuña M, García-Barrera T, García-Sevillano M A, Ariza G. Arsenic metabolites in human serum and urine after seafood (*Anemonia sulcata*) consumption and bioaccessibility assessment using liquid chromatography coupled to inorganic and organic mass spectrometry. Microchemical Journal, 2014, 112: 56-64.

[146] Nguyen M H, Pham T D, Nguyen T L, Vu H A, Ta T T, Tu M B, Nguyen T H Y, Chu D B. Speciation analysis of arsenic compounds by HPLC-ICP-MS: Application for human serum and urine. Journal of Analytical Methods in Chemistry, 2018, (5576): 1-8.

[147] 林琳, 张素静, 徐渭聪, 骆如欣, 马栋, 沈敏. 血液和尿液中砷形态化合物的 HPLC-ICP-MS 分析. 法医学杂志, 2018, 34(1): 37-43.

[148] Morton J & Leese E. Arsenic speciation in clinical samples: Urine analysis using fast micro-liquid chromatography ICP-MS. Analytical and Bioanalytical Chemistry, 2011, 399: 1781-1788.

[149] Sen I, Zou W, Alvaran J, Nguyen L, Ryszard G, She J W. Development and validation of a simple and robust method for arsenic speciation in human urine using HPLC/ICP-MS. Journal of AOAC International, 2015, 98(2): 517-523.

[150] Serrano I N, Ballesteros M T L, Pacheco S S F, Álvarez S I, Colón J L L. Total and speciated urinary arsenic levels in the Spanish population. Science of the Total Environment, 2016, 571: 164-171.

[151] Chen B W, Cao F L, Yuan C G, Lu X F, Shen S W, Zhou J, Le X C. Arsenic speciation in saliva of acute promyelocytic leukemia patients undergoing arsenic trioxide treatment. Analytical and Bioanalytical Chemistry, 2013, 405: 1903-1911.

[152] 陈绍占, 刘丽萍, 杜宏举, 金鹏飞, 周天慧. 高效液相色谱-电感耦合等离子体质谱法分析经牛黄解毒片暴露后大鼠血清中砷形态. 质谱学报, 2017, 38(2): 177-186.

[153] 陈绍占, 杜振霞, 刘丽萍, 姜泓. 高效液相色谱-电感耦合等离子质谱法分析雄黄在大鼠脏器中代谢的砷形态. 分析化学, 2014, 42(3): 349-354.

[154] Rodrigues J LSouza S S, Souza V O, Barbosa F. Methylmercury and inorganic mercury determination in blood by using liquid chromatography with inductively coupled plasma mass spectrometry and a fast sample preparation procedure. Talanta, 2010, 80: 1158-1163.

[155] de Souza S S, Campiglia A D, Barbosa F. A simple method for methylmercury, inorganic mercury and ethylmercury determination in plasma samples by high performance liquid chromatography-cold-vapor-inductively coupled plasma mass spectrometry. Analytica Chimica Acta, 2013, 761: 11-17.

[156] Sogame Y & Tsukagoshi A. Development of a liquid chromatography-inductively coupled plasma mass spectrometry method for the simultaneous determination of methylmercury and inorganic mercury in human blood. Journal of

Chromatography B, 2020, 1136: 121855.

[157] Tsoi Y K, Tam S, Leung K S. Rapid speciation of methylated and ethylated mercury in urine using headspace solid phase microextraction coupled to LC-ICP-MS. Journal of analytical atomic spectrometry, 2010, 25: 1758-1762.

[158] 彭国俊, 朱晓艳, 陈建国, 金献忠, 陈少鸿, 冯睿, 魏丹毅. 高效液相色谱-电感耦合等离子体质谱联用快速测定水样和生物样品中的 4 种汞形态. 分析实验室, 2015, 34(1): 44-48.

[159] 张兰, 陈玉红, 王英峰, 施燕支. 微波辅助萃取-高效液相色谱-电感耦合等离子体质谱法联用测定生物样品中的汞形态. 环境化学, 2013, 32(11): 2219-2222.

[160] de Silva E G, Mataveli L R V, Arruda M A Z. Speciation analysis of selenium in plankton, Brazil nut and human urine samples by HPLC-ICP-MS. Talanta, 2013, 110: 53-57.

[161] 刘源, 陈绍占, 陈镇, 姚晓慧, 周天慧, 刘丽萍. 高效液相色谱-电感耦合等离子体质谱法测定人尿中硒形态. 分析测试学报, 2020, 39(2): 273-277.

[162] Mao X J, Hu B, He M, Chen B B. High polar organic-inorganic hybrid coating stir bar sorptive extraction combined with high performance liquid chromatography-inductively coupled plasma mass spectrometry for the speciation of seleno-amino acids and seleno-oligopeptides in biological samples. Journal of Chromatography A, 2012, 1256: 32-39.

[163] Terol A, Ardini F, Basso A, Grotti M. Determination of selenium urinary metabolites by high temperature liquid chromatography-inductively coupled plasma mass spectrometry. Journal of Chromatography A, 2015, 1380: 112-119.

[164] Hu L, Dong Z Q, Huang X H, Li Y F, Li B, Qu L Y, Wang G P, Gao Y X, Chen C Y. Analysis of small molecular selenium species in serum samples from mercury-exposed people supplemented with selenium-enriched yeast by anion exchange-inductively coupled plasma mass spectrometry. Chinese Journal of Analytical Chemistry, 2011, 39(4): 466-470.

[165] Kokarnig S, Tsirigotaki A, Wiesenhofer T, Lackner V, Francesconi K A, Pergantis S A, Kuehnelt D. Concurrent quantitative HPLC-mass spectrometry profiling of small selenium species in human serum and urine after in gestion of selenium supplements. Journal of Trace Elements in Medicine and Biology, 2015, 29: 83-90.

[166] Wang H J, Du X M, Wang M, Wang T C, Ou-Yang H, Wang B, Zhu M T, Wang Y, Jia G, Feng W Y. Using ion-pair reversed-phase HPLC ICP-MS to simultaneously determine Cr(Ⅲ) and Cr(Ⅵ) in urine of chromate workers. Talanta, 2010, 81: 1856-1860.

[167] Pechancová R, Pluháček T, Gallo J, Milde D. Study of chromium species release from metal implants in blood and joint effusion: Utilization of HPLC-ICP-MS. Talanta, 2018, 185: 370-377.

[168] 刘德晔, 朱醇, 马永建, 吉文亮, 刘华良. 高效液相色谱-电感耦合等离子体质谱测定血浆中铂类抗癌药物. 分析试验室. 2012, 31(7): 75-79.

第24章　地质样品中微量及痕量元素分析

（黄小文[①]，漆　亮[①*]）

- 24.1　微量元素分析 / 680
- 24.2　稀土元素分析 / 683
- 24.3　稀有金属元素分析 / 684
- 24.4　稀散金属元素分析 / 684
- 24.5　贵金属元素分析 / 685
- 24.6　卤族元素分析 / 686
- 24.7　展望 / 688

①中国科学院地球化学研究所，贵阳。*通讯作者联系方式：qiliang@mail.gyig.ac.cn

本章导读

- 酸溶法+电感耦合等离子体质谱（ICP-MS）测定是地质样品中微量元素（包括稀土元素）分析的主要方法，不同酸介质的选择可以解决部分元素回收率低的问题。
- 基体匹配的标样对于硫化物和铂族矿物等的原位微量元素分析仍然重要。
- 传统的样品化学前处理或者制备方法与激光剥蚀（LA）-ICP-MS的结合是原位元素分析的重要发展方向之一。
- 三稀金属矿石类型多样、成分复杂，需要考虑元素的赋存状态、挥发性质和基体干扰等因素而选择合适的分析方法。
- 卤族元素的湿化学分析法需要考虑挥发性丢失的问题，而原位分析标样的选择至关重要。

地质样品的化学成分承载着成岩、成矿以及环境污染程度等重要信息，准确分析地质样品的化学组成对于正确理解各种地质和环境过程至关重要。根据不同的研究目的，需要测定地质样品的微量元素、稀土元素、贵金属以及稀有、稀散金属含量等。由于不同元素在不同的地质样品中的含量存在差异，且赋存形式不一致，因此需要采取不同的分析测试方法以达到准确分析的目的。随着近年来战略性关键金属矿产资源的研究不断深入[1,2]，对地质样品的分析测试提出了新的要求[3]。本章追踪了近十年来（2010～2020年）地质样品中微量及痕量元素分析的最新进展，并按照元素的分类对地质样品化学成分分析的化学前处理和仪器测试进行了详细的阐述，旨在为地质与环境领域的科研工作者提供分析方法方面的参考。

24.1 微量元素分析

24.1.1 湿化学法

地质样品的微量元素分析是化学组成分析的重要内容，由于基体组成的差异，微量元素含量差别较大，微量元素赋存形式也有所不同，如何有效地将这些微量元素全部提取出来成了关键。Hu 和 Qi[4]系统总结了地质样品的前处理方法，包括敞口酸溶、密闭酸溶、微波消解、部分溶解、干法灰化、碱熔、火试金、卡洛斯管、高压灰化以及高温熔融等。总的来讲，不同的样品溶解方法具有不同的优势，且都存在局限性，一个化学前处理流程无法将所有元素都有效提取出来。化学消解法是经典的样品前处理方法，适用的样品类型丰富，且因其获得的溶液具有非常好的均一性，仍然是电感耦合等离子体质谱（inductively coupled plasma mass spectrometry，ICP-MS）、电感耦合等离子体原子发射光谱（inductively coupled plasma-optical emission spectrometry/atomic emission spectrometry，ICP-OES/AES）等主流元素分析仪器的重要进样方式。

在过去的十年，研究者对传统的酸溶法和碱熔法进行了详细的评估，并提出了一些新的样品消解方案。Cotta 和 Enzweiler[5]对经典的 HF-HNO$_3$ 密闭溶样法进行了改进并使用高温消解仪，在溶样之前加入 Mg（50 mg 样品需加入 6～7 mg MgO）并设置合适的温度（220℃）以阻止难溶 AlF$_3$ 的形成。作者同时考察了不同的碱熔方法，通过增加碱性烧结液的蒸干步骤，有效提高了 13 种微量元素（Ni、Sc、Co、Sr、Ba、Zr、Nb、Cd、Sn、Sb、Hf、Pb 和 Th）的回收率。Zhang 等[6]重新评估了经典的 HF-HNO₃ 密闭溶样法，发现 HNO₃ 的加入抑制了 HF 分解难溶硅酸盐矿物的能力，提出只使用 HF 进行第一步酸溶，在样品粉末颗粒足够细小（比如<30 μm）和样品量足够少（<100 mg）时，可避免出现氟化物沉淀。Chen 等[7]认为难溶氟化物沉淀可以通过加入 1 mL 逆王水和 0.5 mL HF，在 190℃条件下密闭消解 2 h，再次溶解。封闭压力酸溶（HF-HNO₃）可以分解大部分地质样品，且大部分元素能够获得理想的回收率，但 Zr、Hf、REEs、Rb、Th、U 等元素在硝酸复溶过程不易提取[8]，导致了这些元素的回收率明显偏低，张保科等[9]提出用盐酸复溶代替硝酸复溶，且增加称样量（25 mg 增至 100 mg）和延长溶样时间（12 h 延长到 48 h）可以明显提高这些元素的回收率。针对三酸（HCl-HNO₃-HF）或四酸（HCl-HNO₃-HF-HClO₄）敞口溶样难溶元素回收率偏低和易挥发性元素回收率不稳定的问题，王君玉等[10]提出用五酸（HCl-HNO₃-HF-HClO₄-H₂SO₄）溶样，改善了难溶元素和易挥发元素的回收率。

针对长英质岩石难溶矿物如锆石等分解不完全问题，Zhang 等[11]采用氟化氢铵代替敞口式消解法中的氢氟酸，提高了消解温度，有效分解了难溶矿物，提高了 Zr、Hf 和重稀土元素（HREE）等的回收率，且大大缩短了溶样时间；采用常压消解环境，避免了难溶氟化物 AlF₃ 的形成，特别适合富铝样品如铝土矿的微量元素分析[12]。张彦辉等[13]自行设计加工了电加热增压消解装置，在同一消解罐中先微波消解（180℃，60 min）再增压消解（200～220℃，40～50 min），实现了难溶矿物（含铌钽铀矿物、锆石等）岩石样品的完全溶解。Wang 等[14]对比了不同的溶样装置对微量元素含量测定的影响，发现密闭溶样杯消解能准确测定挥发性元素 S、Se 和 Te，但是 W 的空白较高，而高压消解仪较密闭溶样杯 Ba、Bi、Tl 的空白要高。

24.1.2 原位分析方法

近年来，在矿物原位微量元素分析方面，也取得了重要进展，并应用于矿床成因和矿产勘查研究中[15]。国内外学者已成功建立了碳酸盐矿物[16]、含水硅酸盐矿物（如角闪石、绿帘石、电气石和透闪石）[17]、石英[18-20]、磷灰石[21-23]、磁铁矿[24-29]、黄铜矿[30]、闪锌矿[31]等激光剥蚀电感耦合等离子体质谱（laser ablation inductively coupled plasma mass spectrometry，LA-ICP-MS）微量元素分析方法。硫化物中一些微量级别的铂族元素也实现了原位的 LA-ICP-MS 分析[32,33]。其中碳酸盐矿物、含水硅酸盐矿物和磁铁矿微量元素的分析[16,17,24,28]，多采用 Liu 等[34]提出的无内标-多外标校正方法，即假设所有元素的氧化物含量之和假设为 100%。无内标-多外标校正方法的优点是无需知道内标元素的准确含量，便可直接利用 LA-ICP-MS 分析矿物中所有主量和微量元素的含量，省去了电子探针（electron probe micro-analyzer，EPMA）测试内标元素含量的步骤，节约了时间和成本，但为了使分析结果更加准确，需分析尽可能多的元素。另外一种校正方法是单内标-（单或多）外标校正方法[35]，外部标样多为人工合成硅酸盐玻璃或者硫化物，需要预先用 EPMA 准确测定内标元素的含量，且需考虑基体匹配问题。以磁铁矿微量元素分析为例，当以 Fe 为内标元素时，需要较大束斑（>25 μm）和 Fe 含量相对较高的硅酸盐玻璃（如 GSE-1G）作为外部校正样才能获得准确的结果[25,27]；而采用铁含量较高的玄武质玻璃 BCR-2G、BIR-1G、BHVO-2G 和 GSE-1G 作为外标，以 Fe 元素强度进行内部标准化，可以很好地消除基体效应，获得准确的结果[24]。Pisiak 等[36]对比了三种微量元素含量计算的三种标准化方法：①采用纯磁铁矿化学计量学的 Fe 含量（72 wt%）作为内标元素含量；②采用电子探针测定的实际 Fe 含量作为内标元素含量；③采用无内标-多外标校正方法。结果表明，无内标-多外标校正方法能够获得与电子探针分析更加接近的结果，证明了该校正方法的优势。

矿物微区面扫描技术的发展也大大提高了矿物微量元素及其同位素研究的空间分辨率[18,37-40]。Zhu 等[39]利用 LA-ICP-MS 对含磁铁矿、黄铁矿和菱铁矿的样品进行了面扫描，发现用 Fe 作为内标元素比用 S 能够获得更加准确的结果，且磁铁矿和菱铁矿之间的基体效应可以忽略。汪方跃等[40]建立了 LA-ICP-MS 面扫描分析技术并利用 Matlab 软件开发了一套数据处理程序，利用该技术可以在 2 小时内分析 3 mm × 3 mm 区域，并同时给出元素或比值在二维平面的分布特征。除了各类型矿物，近些年也实现了透明和半透明矿物单个流体包裹体的 LA-ICP-MS 成分分析[20,41,42]。Hammerli 等[42]利用 LA-ICP-MS 分析已知 Cl 和 Br 含量的方柱石，评估了激光束斑大小、等离子体能量、多原子干扰对 Cl 和 Br 含量分析的影响，建立了方柱石及其流体包裹体（16 μm）中的 Cl 和 Br 含量分析方法，该方法也可以用于磷灰石、黑云母、角闪石和闪锌矿的流体包裹体成分分析。当分析闪锌矿的流体包裹体（25 μm）时，Br 的检测限约为 8 μg/g，Cl 的检测限高于 500 μg/g。蓝廷广等[20]通过人工合成石英 NaCl-H$_2$O-Rb-Cs 和 NaCl-KCl-CaCl$_2$-H$_2$O-Rb-Cs 流体包裹体，使用显微测温 NaCl 等效盐度（电价平衡方法）为内标，NIST610 为外标，建立了流体包裹体成分 LA-ICP-MS 分析方法。

基体匹配的标样对于一些矿物的微量元素分析非常重要，尤其是挥发性和亲铁亲铜元素，由于自然样品的不均一性，很少能直接作为标样[21]，而更多标样需要人工合成[43]。Danyushevsky 等[44]研制了硫化物标样 STDGL2b2，该标样由 25%锌精矿和 75%的磁黄铁矿烧制而成，并添加了微量元素，该标样的使用提高了硫化物原位微量元素分析的准确性。赵令浩等[45]采用铳镍试金法制备了含铂族元素的硫化物标样，可用于硫化物微区原位痕量铂族元素的分析。飞秒激光的引入，有效克服了纳秒激光原位分析基体匹配、元素分馏问题，并提供了更高的空间分辨率[46,47]。

24.1.3 传统样品制备方法与原位分析的结合

传统的样品前处理方法与 LA-ICP-MS 分析相结合已成为全岩样品微量元素分析的一个重要手段。全岩粉末熔融玻璃与 LA-ICP-MS 的结合可以用于测定不同岩石样品的主量和微量元素含量[48-50]，但是由于熔融制样过程中挥发性元素 Pb 和 Zn 等容易发生丢失，针对这一问题，朱律运等[51]改进并设计了一种双铱带高温炉，用该装置熔样有效抑制了挥发性元素的丢失，但造成了不同程度的 Cr、Ni 和 Cu 损失。Loewen 和 Kent[52]考察了 LA-ICP-MS 测定硅酸盐玻璃中度挥发性元素（Cd、Sn、Pb、Zn、Cu、Mo)的元素分馏行为，认为其主要与 He 气流的稳定性和流速有关，单间剥蚀室普遍存在这个问题，而双间剥蚀室由于能提供稳定的气流可以消除挥发性元素的分馏效应。Zhu 等[53]设计了一种新的样品粉末熔融装置，由碳化硼干锅和钼加热条构成，无需加入助熔剂便可以合成非常均一的硅酸岩玻璃。He 等[54]采用钼纸包裹硅酸岩粉末样品置于石墨管中进行高温熔融，实现了无助熔剂的硅酸岩玻璃制备方法。岩石或矿物粉末压饼方法，也常用于制作标样，直接用于 LA-ICP-MS 的主量和微量元素分析，但是压饼常存在样品不均一的问题。Garbe-Schönberg 和 Müller[55]借助高速行星球磨机和玛瑙工具，在水悬浮液中采用湿磨方案获得了闪长岩的纳米级别颗粒（<1.5 μm），这些极细的颗粒保证了粉末压饼的均一性，提高了 LA-ICP-MS 分析的准确性。Mukherjee 等[56]考察了压饼过程中粉末颗粒大小、压力大小、黏合剂类型对样品均一性影响，并提出为了满足 LA-ICP-MS 分析时的均一性，一般样品颗粒直径均值须小于 10 μm，而对于长英质岩石比如花岗岩，粉末颗粒直径均值最好小于 5 μm。Zhang 等[57]通过对岩石粉末用氟化氢铵进行预溶解，并在 400℃条件下蒸干，得到非常均一的岩石粉末（<8.5 μm），实现了硅酸盐样品 LA-ICP-MS 元素含量的准确测定。Peters 和 Pettke[58]采用湿磨法获得纳米级岩石粉末颗粒，使用微晶纤维素作为黏合剂进行压饼，获得的硅酸岩样品用于 LA-ICP-MS 分析，实现了助熔剂元素 Li 和 B，亲铜元素 As、Sb、Tl、In 和 Bi 的准确测定。最近，Liao 等[59]使用 LA-ICP-MS 对硅酸岩样品溶液直接进行分析，该方法改变了溶液样品传统雾化器进样的方式，有效地将水有关的质谱干扰降低了 1~2 个数量级。而且溶液的基体效应不明显，不同的稀释倍数（80~2000 倍）和不同浓度（2%~30% HNO$_3$）酸介质均获得相似的结果。与固体

LA-ICP-MS 分析相比，元素的灵敏度提高了 70～250 倍，大多数元素的检测限降低了 2 个数量级，且时间分辨有关的元素分馏可以忽略。与传统雾化器进样相比，在化学前处理过程酸和超纯水的用量减少了 20～100 倍，节约了分析成本，并减少了环境污染。由于该方法很好地解决了 ICP-MS 测试时的基体效应问题，可用于直接分析具有复杂基体的样品溶液，比如海水、高盐度卤水及饮料等[60]。

24.2 稀土元素分析

对于大多数岩石样品而言，稀土元素和其他微量元素通常采用同一分析流程进行分离富集和质谱测定。在最近的一些综述文章中，研究者系统地总结了稀土元素分析的化学前处理方法[61]和仪器测试技术[62,63]，为地质样品中稀土元素的分析提供了很好的参考。稀土元素分析通常采用酸溶或者碱熔+酸提取的样品消解方法，酸溶需要使用一系列的强酸（包括 HF、HNO_3、HCl、$HClO_4$），以保证样品中难溶组分如锆石和黏土等的完全分解；碱熔方法能快速有效地消解样品，但是由于使用大量的 Na_2O_2 或者 NaOH 熔剂，酸提取后的溶液中可溶解的固体物质较多，如果不加以分离将影响 ICP-MS 的测定[61]。对于地质样品，稀土元素分析常用的仪器包括 ICP-MS、ICP-AES/OES、中子活化分析仪（neutron activation analyzer，NAA）和 X 射线荧光光谱仪（X-ray fluorescence spectrometer，XRF）[62,63]。激光剥蚀和电热蒸发与 ICP-MS 或 ICP-AES/OES 的联用，避免了烦琐的化学前处理流程，可以直接用于固体样品或者悬浮液的分析[63]。激光诱导击穿光谱（LIBS）在原位测定稀土元素方面也显示出一定的潜力[64]。

对于一些特殊的样品，由于非常低的稀土含量或者不易消解的问题，需要采用特定的分析流程。超基性岩中稀土含量非常低（<1 μg/g），且富镁基体在用 $HF-HNO_3$ 溶解时将产生大量氟化物沉淀，针对这些问题，Sun 等[65]提出用 H_3BO_3 溶解氟化物沉淀，用阴离子交换和共沉淀方法将稀土与主要基体元素 Fe、Mg、Ba、Ca 和 Cr 分离，实现了超基性岩样品低空白、低检测限和高回收率的稀土元素分析。条带状铁建造中磁铁矿的稀土含量非常低（一般为 ng/g 级），且基体 Fe 含量太高，常规的分析方法很难准确测定这些样品中稀土元素含量，Li 等[66]提出分别采用阴离子交换树脂（AG1-X8，200～400 目）和阳离子交换树脂（AG50-X12，200～400 目）进行基体元素与稀土的分离，并用高分辨 ICP-MS 进行稀土元素测定。该方法具有低的空白值，阴离子交换法为 0.029 ng/g La 和 0.018 ng/g Lu；阳离子交换法为 0.151 ng/g La 和 0.031 ng/g Lu；阴离子和阳离子交换法的方法检测限分别为 0.014～0.054 ng/g 和 0.030～0.120 ng/g，因此阴离子交换法在测定磁铁矿中超低含量稀土更具优势[66]。Liu 等[67,68]提出采用泡沫聚氨酯（polyurethane foam，PUF）可以有效去除 Fe 基体，并富集低含量的稀土元素，适合用于铁含量较高的矿物中稀土元素分析，比如黄铁矿、磁铁矿、铁橄榄石等。

独居石是重要的稀土赋存矿物，但是不易完全分解，Padmasubashini 和 Satyanarayana[69]对比了硫酸溶解和 KHF_2-NaF 熔融的样品分解效果，发现两种方法获得的稀土元素含量相似，但是锆只能在氟化物熔融条件下定量回收。金伯利岩中通常含有大量的难溶矿物，如金红石、钛铁矿等，Kumar 等[70]采用 Na_2O_2 碱熔方法，彻底分解难溶矿物，实现了稀土元素 ICP-MS 的准确测定，方法检测限为 0.12～1.54 μg/g。在一些伟晶岩矿床中，稀土元素通常赋存在富 U、Nb、Ta 的难溶矿物如铌铁矿、钽铁矿及铌钇矿中，如果采用常规酸溶方法，Nb、Ta 很容易发生水解生成氢氧化物沉淀，影响分析结果，一些学者提出采用无水 $KHSO_4$[71]或者正磷酸二氢钠和焦磷酸四钠[72]熔融样品，并用草酸浸取使 REE、Y 和 Th 转为草酸盐沉淀而与干扰元素 U 和易水解元素 Nb 和 Ta 等分离。海相沉积物中含有难溶矿物如锆石等，酸溶法很难使样品完全溶解，Saha 等[73]采用 $LiBO_2$ 和 $Li_2B_4O_7$ 进行样品熔融，实现了样品的完全分解，ICP-MS 测定结果和 NAA 相似，除了 La 含量有一定差别。

除了考虑不同的化学前处理方法对于稀土元素的回收率影响，质谱测试时的干扰校正也非常关键。王冠等[74]对高分辨 ICP-MS 不同分析模式下元素干扰情况进行了评估，结果表明在低、中分辨率模式下，轻稀土元素 Ce、Pr、Nd、Sm 的氧化物和 Ba 的氧化物干扰明显，Gd 的测定值明显偏离真实值；在高分辨模式下，只有 ^{157}Gd 受 ^{141}Pr^{16}O 的干扰严重，当样品的 Pr/Gd 比值大于 100 时，Gd 的含量必须进行数学校正。

24.3 稀有金属元素分析

由于三稀（稀土、稀有、稀散）矿石样品的复杂性，需要测定的元素较多，一种方法可能无法测定全部元素，需要多个分析方法的结合。目前有很多综述文章总结了三稀金属的分析方法[3,75]。屈文俊等[3]总结了十种常用的三稀元素分析测试方法，根据三稀金属矿石的类型和所需测定的元素可选择一种或多种分析方法。李刚等[76]总结了铌和钽的分析测试方法。本节将重点介绍在稀有金属元素测定方面的一些新进展。

Tomascak 等[77]系统介绍了不同类型样品 Li 含量分析的化学流程。De La Rosa 等[78]建立了核反应分析法测定电气石和其他含 Li 矿物中 Li 含量，即利用加速质子诱导 ^{7}Li 和 ^{4}He 反应，该方法实现了 µg/g 级别的 Li 分析。

赵学沛[79]采用 HCl-HNO$_3$-HClO$_4$-HF 混合酸溶样，采用酒石酸-稀盐酸-H$_2$O$_2$ 定容，用 ICP-AES 测定了稀有金属矿中的锂、铍、铌、钽、锡等元素。李志伟等[80]采用 HCl-HNO$_3$-HClO$_4$-HF-H$_2$SO$_4$ 溶样，以 3～4 滴氢氟酸、5%硫酸、5%过氧化氢提取体系替代常规的有机酸（酒石酸等）提取体系，实现了 ICP-OES 同时测定稀有金属矿选矿试验各阶段产品中不同含量的铌、钽、锂、铍等元素。程祎等[81]利用高压密闭消解-电感耦合等离子体质谱法测定地质样品中铌、钽、锆、铪和 16 种稀土元素。针对钼、铌、钽分析中出现的样品溶解不完全、元素易水解问题，高会艳[82]建立了不同铌、钽含量地质样品分析方法，对于铌、钽含量较低且易于分解的样品，采用硝酸-氢氟酸-硫酸混合酸恒温电热板消解 ICP-MS 方法测定，对于铌、钽含量高且难溶的样品，采用过氧化钠高温熔融 ICP-AES 方法测定。为了增强 ICP-AES 测试铌和钽时的灵敏度，姚玉玲等[83]选择乙醇作增敏剂，提高了雾化效率，当乙醇浓度为 6%时，原子线 Nb 292.781 nm、Ta 240.063 nm 的灵敏度分别增强了 180.5%和 265.5%，检出限也降低了一半。刘江斌等[84]介绍了 XRF 同时测定地质样品中铌、钽、锆、铪、铈、镓、钪、铀等稀有元素的含量。陈静等[85]指出在利用 XRF 测定地质样品中低含量铌、钽时，标准曲线要根据实际情况合理回归，将测试后的样品与化学方法对照。

在利用密闭 HF-HNO$_3$ 溶解花岗闪长岩样品时，锆石、金红石等难溶矿物中的锆很难完全回收。针对这一问题，黎卫亮等[86]增加了赶硅过程，通过加大氢氟酸用量，使锆的溶出率提高到 95%左右，实现了花岗闪长岩中微量锆的准确测定。

24.4 稀散金属元素分析

已有综述文章总结了不同的稀散元素分析方法，包括碲[87]、锗[88]、铼[89-91]。程秀花等[92]系统总结了稀有金属元素镓、铟、铊、锗、硒和碲等的样品消解、分离富集和仪器测试方法。本节将选择性地介绍在稀散金属元素测定方面的最新进展。

Kaya 和 Volkan[93]设计了一种新型的 $GeCl_4$ 生成装置，即在普通的氧化二氮-乙炔火焰原子吸收光谱仪中安装了电热炉、蒸气发生和气-液分离装置，可以用于从复杂基体中分离和富集 Ge，降低了 Ge 的检测限。Audétat 等[94]标定了一个自然石英样品的微量元素组成，可用于原位测定石英中微量元素 Ti、Al、Li、Fe、Mn、Ga 和 Ge 的标样。陈波等[95]提出了同一流程 ICP-MS 直接测定 Ge、Se 和 Te 的方法，即采用硝酸-氢氟酸-硫酸消解样品，50%硝酸提取，3%乙醇定容，未使用盐酸则避免了 Ge 和 Te 的挥发性丢失，硝酸-乙醇介质明显提高了 Se 和 Te 的灵敏度。

高贺凤等[96]针对封闭酸溶和敞开酸溶消解样品的缺点提出半密闭式酸分解法（氢氟酸、盐酸、硝酸、高氯酸混合酸），ICP-MS 测定 Ga，且使用经验系数 0.005 扣除了 $^{55}Mn^{16}O$ 对 ^{71}Ga 的影响，方法检出限为 0.06 μg/g。多个研究系统考察了 ICP-MS 测定 Ga、Ge、In、Cd 和 Tl 时的元素干扰及相应的消除方法[97-99]。Blokhin 等[99]系统考察了四极杆和高分辨 ICP-MS 测定铁锰结壳中 Ga 含量的质谱干扰，研究发现对于四极杆 ICP-MS 来说，无论是否使用碰撞池，由于 $^{138}Ba^{2+}$ 对 ^{69}Ga 干扰，无法准确测定该同位素；而 $^{55}Mn^{16}O$ 对 ^{71}Ga 的干扰可以通过 He 碰撞池消除，相反 $^{142}Ce^{2+}$ 和 $^{142}Nd^{2+}$ 的干扰却增强了。使用高分辨 ICP-MS 可以同时消除基体元素对 ^{69}Ga 和 ^{71}Ga 的干扰。

黎卫亮等[100]考察了不同浓度甲醇、乙醇、丙酮对 ICP-MS 分析镓、铟、铊、锗、碲的信号强度影响，发现 4%乙醇能够明显增强质谱信号，元素的检测限为 0.01～0.02 μg/g。

吴峥等[101]改进了石墨炉升温程序，省去了灰化步骤，使石墨炉升温时间由原来的 90～100 s 降至 24 s，提高了石墨炉原子吸收光谱法测定 Cd 的工作效率。利用 ICP-MS 测定地质样品中的 Cd 时，通常容易受到 Zr 或 Mo 的干扰，逆王水溶样可以消除大部分 Zr 有关的氢氧化物干扰，在气溶胶进入等离子体之前加入 Ar 气，可以有效消除残余的 Zr 氢氧化物和 Mo 的氧化物或氢氧化物干扰[102]，优化条件后的 Zr 和 Mo 的氧化物或氢氧化物产率降低了 90%，MoOH/Mo 或 MoO/Mo 的比值为 0.005%，ZrOH/Zr 的比值为 0.007%。

贺攀红等[103]采用硝酸-盐酸-氢氟酸-高氯酸溶样，直接加入浓盐酸煮沸将 Se^{6+} 还原为 Se^{4+}，在试液中加入铁盐溶液或在硼氢化钠还原剂中加入铁氰化钾抑制了 Cu^{2+} 的干扰，建立了氢化物发生器与电感耦合等离子体发射光谱仪联用测定铀矿地质样品中的痕量硒方法。

24.5 贵金属元素分析

24.5.1 微量贵金属元素分析

贵金属元素，包括 Au、Ag 和铂族元素（platinum-group elements，PGE）等，它们在地质样品中可以以微量元素形式存在，但在多数情况下为痕量元素。由于含量的差别，分析方法也不同。地质和环境样品中的微量-痕量 Au、Pd 和 Pt 的分离和预富集方法包括化学改性吸附剂、含硫吸附剂萃取剂、离子印迹聚合物、微波消解、流体注射等，常用的分析仪器包括 ICP-MS、ICP-OES、电热原子吸收光谱（electrothermal atomic absorption spectrometry，ETAAS）、NAA，较少使用的为 XRF 和火焰原子吸收光谱（flame atomic absorption spectrometer，FAAS）[104]。例如，Mladenova 等[105]利用固相萃取剂半胱氨酸改性硅胶分离 Au、Pd 和 Pt，然后采用 ETAAS 和 ICP-OES 测定，获得的检测限为 0.002～0.06 μg/L。Balaram 等[106]利用二异丁基酮萃取地质样品中的 Au，采用 FAAS 和石墨炉原子吸收光谱（graphite furnace atomic absorption spectrometer, GFAAS）进行 Au 含量分析，检测限分别为 20 ng/g 和 0.1 ng/g。液-液萃取常用来预富集水样品中的 Au 和 Ag，然后采用 ETAAS 或者 FAAS 测定[107,108]。刘军等[109]建立了一次密闭酸溶，组合富集剂分离富集（活性炭、改性活性炭、717 阴离子交换树脂），ICP-MS 测定 Au、Ag、Pt、Pd 的方法，方法检出限分别为 Au 0.097 ng/g、Pt 0.14 ng/g、Pd 0.17 ng/g、Ag 0.082 ng/g。赵延庆[110]采用聚氨酯泡沫塑料吸附，ICP-MS 测定地质化探样品中的 Au，方法检测限为 0.12 ng/g。

24.5.2 痕量贵金属元素分析

如果 Re、Au 和 PGE 等以痕量-超痕量元素形式存在，那么它们的分析不同于常规的微量分析方法，通常采用同位素稀释法，且需要更加复杂的样品前处理流程进行分离富集[111]。对于 PGE 的分析，针对低含量地质样品，Qi 等[112]设计了大容量（120 mL）的特氟龙溶样罐，并改进了溶样方法，即先用 HF 进行去硅，蒸干后的残余物用 HF 和 HNO_3 在高压密闭的特氟龙溶样罐中 190℃下溶解 48 小时。该方法简化了分析流程，增加了样品取样量，克服了块金效应，可以有效地溶解超基性到酸性岩石以及硫化物矿石，不足之处是无法分析挥发性的 Os。传统的卡洛斯管一次性使用，无法清洗，实验成本较高，Qi 等[113]设计了一种新型的、可重复使用的大容量（200 mL）卡洛斯管，并配备了可拆卸的玻璃+特氟龙塞子和不锈钢密封套件，通过设定合适的溶样条件，该卡洛斯管可以用于低含量地质样品的 PGE 和 Re-Os 同位素分析。Li 等[114]建立了同位素稀释法 ICP-MS 和负热电离质谱仪（negative thermal ionization mass spectrometry, N-TIMS）同时测定 PGE、Re 和 Os 同位素的分析流程，样品逆王水溶解后，CCl_4 萃取 Os，残余溶液经阳离子树脂（AG50W-X8），获得含 Re、PGE 和干扰元素 Mo、Zr、Hf 的溶液，再经 N-苯甲酰基苯基羟胺（N-benzoyl-N-phenylhydroxylamine, BPHA）树脂将 PGE-Re 与 Mo、Zr、Hf 等基体元素分开。Ishikawa 等[115]对高温逆王水卡洛斯管溶样，同位素稀释法 ICP-MS 测定高度亲铁元素（Re、Au、PGE）含量进行了评估，考察了 HF 去硅过程对不同类型样品溶解效果的影响，研究表明最优的溶样方案为：1~2 g 样品在卡洛斯管用逆王水溶解，240℃保持 72 h，CCl_4 萃取 Os，离心之后残余固体用 HF 再次消解，最后是 Re 和其他 PGE 的分离。HF 去硅过程可以有效提高基性岩中 Ru 的回收率，而对超基性岩和沉积岩的亲铁元素回收率无明显影响。Wang 和 Becker[116]实现了同一化学分析流程同时测定全岩样品中硫属元素（S、Se、Te）和高度亲铁元素的含量，并指出这些元素在岩石粉末中存在不同尺度的异质性。

24.6 卤族元素分析

卤族元素，简称卤素（氟、氯、溴、碘），存在于地质与环境样品中，卤素的准确含量测定对于岩石、矿产资源的成因与勘查研究以及环境资源的评价具有重要意义。对于环境样品中的卤素，通常采用电化学分析方法[117]和高温热水解离子色谱法[118]，而地质样品则采用 XRF、EPMA、NAA、ICP-MS、电热蒸发电感耦合等离子体质谱（electrothermal vaporization-inductively coupled plasma mass spectrometry, EV-ICP-MS）和 LA-ICP-MS 较多些[119]。

24.6.1 湿化学法

ICP-MS 是测定全岩样品卤素含量较常用的方法[120,121]。Oliveira 等[121]总结了 ICP-MS 或

ICP-OES 测定各种类型样品中碘含量的方法，并指出碱性介质是样品溶解的最佳选择。He 等[122]建立了氟化氢铵敞口溶样扇形磁场电感耦合等离子体质谱（sector field inductively coupled plasma mass spectrometry，SF-ICP-MS）测定岩石、土壤和沉积物中 Cl、Br 和 I 的方法。该方法使用 100 mg 样品，加入 400 mg NH_4HF_2 在 220℃条件下敞开溶解 2 小时，可以实现 Cl、Br 和 I 的定量回收，由于样品溶解过程中始终保持碱性环境，有效抑制了卤素的挥发性丢失。

24.6.2 原位分析方法

XRF 是一种无损分析方法，可以直接用于粉末压饼的元素分析，在进行卤素分析时非常关键的是选择合适的标样和基体校正策略。Krishna 等[123]利用波长色散 X 射线荧光光谱仪（wavelength dispersive X-ray fluorescence spectrometry，WDXRF）对各类岩浆岩、土壤和硫化物矿石标样进行了卤素含量的分析，发现玄武岩比正长岩标样和流纹岩标样在测定各类岩石中卤素含量时，更加适合用于含量校正，且测定的不确定度主要来自于这些卤素在样品中的不均一性（取样量为 2 g，颗粒直径为 74 μm）。Li 等[124]利用相似的方法测定了海洋地质样品中卤素的含量，通过标准添加法设定了一系列标样，经验系数用来校正 F 和 Cl 的基体效应，K α 和 K β 峰分别用于校正 Br 和 I 的基体效应，F、Cl、Br 和 I 的方法检测限（积分时间为 100 s）分别为 100 μg/g、5 μg/g、0.5 μg/g 和 10 μg/g。

EV-ICP-MS 方法可以直接固体进样，省去了繁杂的化学前处理过程，但是需克服高温裂解和电热蒸发过程中卤素挥发性丢失的问题[125,126]。de Gois 等[126]用该方法测定了煤中 Br 和 Cl 的含量，考察了改良剂 Pd、Pd+Al、Pd+Ca 对卤素热稳定性的影响，发现 Pd 和 Ca 的混合物在抑制挥发性丢失方面效果最佳，高温裂解和电热蒸发温度可以分别达到 700℃和 1900℃，Br 和 Cl 的定量限分别为 0.03 μg/g 和 7 μg/g。Cui 等[125]采用该方法测定了岩石、土壤和沉积物中低含量碘。为了防止碘的挥发性丢失，装样前预先在样品室加入 4 μg $Pd(NO_3)_2$ 和 4 μg 抗坏血酸改良剂以提高碘的热稳定性，然后在电热蒸发高温分解过程中加入 200 μg 柠檬酸钠以提高碘的运移和离子化效率。利用该方法大约 12～15 min 可以完成一个样品的分析，岩石、土壤和沉积物的检测限分别为 11 ng/g、9 ng/g 和 8 ng/g。

EPMA 和 LA-ICP-MS 是常用的原位分析矿物卤素含量的方法。例如，磷灰石中的 F 和 Cl 通常用 EPMA 测定[127-129]。但是 EPMA 在测定 Br 含量时容易受到 Al 的 X 射线信号干扰，Zhang 等[120]定量评估了矿物和硅酸岩玻璃中 Al_2O_3 含量与 Br 信号之间的关系，用于准确扣除 Al 的干扰，建立了含 Al 和 Br 矿物如方钠石和方柱石的 Br 含量测定方法。Marks 等[119]对比了多种分析方法测定岩浆磷灰石中卤素及部分微量元素的含量，方法包括 EPMA、LA-ICP-MS、二次离子质谱（secondary ion mass spectrometry，SIMS）、高温水解结合离子色谱、傅里叶变换红外光谱、NAA 和全反射 X 射线荧光光谱（total reflection X-ray fluorescence spectrometry，TXRF）。该研究强调了地学中不太常用的 TXRF 可以同时测定磷灰石中 Br 和 Cl 以及一些微量元素（如 Sr、Ce、Fe、Mn、As）含量，该方法具有低检测限（<1 μg/g）、小样品量（mg 级别）及快速、经济等特点。利用 Durango 磷灰石作为内部参考标样，SIMS 的 Br 和 Cl 分析结果与 EPMA、NAA 和 TXRF 结果吻合。该研究同时强调了卤素在同一样品磷灰石颗粒内部和颗粒之间不均匀分布的特点。Ansberque 等[130]采用扫描电镜的能谱仪（scanning electron microscope-energy dispersive X-ray spectrometer，SEM-EDS）分析了磷灰石中的 F，LA-ICP-MS 分析了磷灰石中的 Cl 和微量元素 Sr、Y、REE 的含量。SEM-EDS 和 LA-ICP-MS 均具有分析速度快的特点，前者每小时最多可进行约 60 次分析，后者每小时最多可进行约 120 次分析。而且与 EPMA 相比，SEM-EDS 在低电流情况下具有更高的计数，卤素的挥发性迁移明显减弱。SEM-EDS 和 LA-ICP-MS 的结合克服了两种仪器的缺点，即前者无法检测低 Cl 磷灰石中的 Cl 含量（检测限约为 0.1wt%），由于 F 具有较高的电离能，后者无法测定 F 含量。流体包裹体中的 Cl、Br 和

I 通常采用 LA-ICP-MS 测定[131-134]。在这些研究中，通常使用方柱石标样（Sca17）作为外部校正标样，利用已知成分的合成流体包裹体进行方法可靠性验证。

与其他方法相比，中子活化分析有两个优点，一是可以更好定量化样品前处理过程中卤素的损失，二是可以减少空白或者污染的影响[135,136]。Sekimoto 和 Ebihara[135]利用该方法测定了 9 件沉积岩和 3 件流纹岩样品的 Cl、Br 和 I 含量，降低了 γ 射线光谱的背景和优化了前处理流程。通过对比 ICP-MS 测定得到的 Br 和 I 含量发现，NAA 分析结果要相对高些，指示 ICP-MS 分析前处理高温水解过程并不能有效地富集 Br 和 I。在随后的研究中，作者补充测定了美国地调局各类地球化学参考物质，包括安山岩、玄武岩、碳酸岩、煤、浸染状金矿石等，并与其他分析方法进行了对比[136]。

除了以上一些常用的分析方法，也有学者提出采用稀有气体分析方法测定岩石和矿物中的 Cl、Br 和 I[137-139]。该方法类似于稀有气体质谱 K-Ar 或 Ar-Ar 定年方法，样品需要预先进行照射，稀有气体提取，然后进行质谱测定，前处理过程相对复杂。Kendrick 等[139]使用离子探针（sensitive high resolution ion microprobe，SHRIMP）测定了硼酸锂助熔剂制成的岩石粉末玻璃的 F 含量，发现 1080℃ 熔融制样时并不会丢失 F，但是该方法制成的玻璃 F 含量均一性较差。

24.7 展　　望

地质样品化学成分复杂，同一元素在不同的样品中含量存在差异，单一分析方法无法完成所有元素的分析，需要根据测定的元素选择最佳测试方法。对于微量元素的分析，酸溶法仍然是最常用的方法，不同酸介质的选择可以解决部分元素回收率低的问题；原位微量元素的分析实现了矿物或者合成玻璃等化学成分微区、无损、快速的分析，但基体匹配标样的制备对于硫化物和铂族矿物等的微量元素分析仍然是亟待解决的问题；传统的样品化学前处理或者制备方法与 LA-ICP-MS 的结合充分利用了两种方法的优点，是一个重要发展趋势；对于稀土元素含量较低的样品比如超基性岩或者含有难溶矿物的样品如金伯利岩和伟晶岩，需要采用特殊的化学溶样和预富集流程，以达到更好的稀土元素回收率；三稀金属矿石类型多样、赋存的元素不同，需要根据所测的元素和基体类型选择合适的分析方法，需要考虑元素的赋存状态、挥发性质和基体干扰等因素；贵金属元素 Au 和铂族元素等的分析需要考虑其含量级别，若以微量元素形式存在，则可采用微量元素的常用一些分析方法，若以痕量元素形式存在，则需采用同位素稀释法；卤族元素的分析既可以采用化学法也可以采用原位分析方法，化学法需要考虑挥发性丢失的问题，而原位分析方法标样的选择至关重要。总之，简单、快速、准确、低检测限及低成本是地质样品化学成分分析的重要发展趋势。

参 考 文 献

[1] 毛景文, 杨宗喜, 谢桂青, 袁顺达, 周振华. 关键矿产——国际动向与思考. 矿床地质, 2019, 38(4): 689-698.

[2] 翟明国, 吴福元, 胡瑞忠, 蒋少涌, 李文昌, 王汝成, 王登红, 齐涛, 秦克章, 温汉捷. 战略性关键金属矿产资源: 现状与问题. 中国科学基金, 2019, 33(2): 106-111.

[3] 屈文俊, 王登红, 朱云, 樊兴涛, 李超, 温宏利. 稀土稀有稀散元素现代仪器测试全新方法的建立. 地质学报, 2019, 93(6): 1514-1522.

[4] Hu Z, Qi L. Sample digestion methods // Holland H D, Turekian K K, Editors. Treatise on Geochemistry. 2nd Edition. Elsevier, 2014: 87-109.

[5] Cotta A J B, Enzweiler J. Classical and new procedures of whole rock dissolution for trace element determination by ICP-MS. Geostandards Geoanalytical Research, 2012, 36(1): 27-50.

[6] Zhang W, Hu Z, Liu Y, Chen L, Chen H, Li M, Zhao L, Hu S, Gao S. Reassessment of HF/HNO$_3$ decomposition capability in the high-pressure digestion of felsic rocks for multi-element determination by ICP-MS. Geostandards Geoanalytical Research, 2012, 36(3): 271-289.

[7] Chen S, Wang X, Niu Y, Sun P, Duan M, Xiao Y, Guo P, Gong H, Wang G, Xue Q. Simple and cost-effective methods for precise analysis of trace element abundances in geological materials with ICP-MS. Science Bulletin, 2017, 62(4): 277-289.

[8] 何红蓼, 李冰, 韩丽荣, 孙德忠, 王淑贤, 李松. 封闭压力酸溶-ICP-MS 法分析地质样品中 47 个元素的评价. 分析试验室, 2002, 21(5): 8-12.

[9] 张保科, 温宏利, 王蕾, 马生凤, 巩爱华. 封闭压力酸溶-盐酸提取-电感耦合等离子体质谱法测定地质样品中的多元素. 岩矿测试, 2011, 30(6): 737-744.

[10] 王君玉, 吴葆存, 李志伟, 韩敏, 钟莅湘. 敞口酸溶-电感耦合等离子体质谱法同时测定地质样品中 45 个元素. 岩矿测试, 2011, 30(4): 440-445.

[11] Zhang W, Hu Z, Liu Y, Chen H, Gao S, Gaschnig R M. Total rock dissolution using ammonium bifluoride (NH$_4$HF$_2$) in screw-top Teflon vials: A new development in open-vessel digestion. Analytical Chemistry, 2012, 84(24): 10686-10693.

[12] 张文, 胡兆初, 漆亮, 刘勇胜, 高山. 氟化氢铵消解-电感耦合等离子体质谱法测定铝土矿 37 个微量元素. 北京: 中国地球科学联合学术年会论文集, 2015.

[13] 张彦辉, 张良圣, 常阳, 范增伟, 郭冬发. 增压-微波消解电感耦合等离子体质谱法测定含难溶矿物岩石样品中的微量元素. 铀矿地质, 2018, 34(2): 105-111.

[14] Wang Z, Becker H, Wombacher F. Mass fractions of S, Cu, Se, Mo, Ag, Cd, In, Te, Ba, Sm, W, Tl and Bi in geological reference materials and selected carbonaceous chondrites determined by isotope dilution ICP-MS. Geostandards Geoanalytical Research, 2015, 39(2): 185-208.

[15] 张乐骏, 周涛发. 矿物原位 LA-ICPMS 微量元素分析及其在矿床成因和预测研究中的应用进展. 岩石学报, 2017, 33(11): 3437-3452.

[16] Chen L, Liu Y, Hu Z, Gao S, Zong K, Chen H. Accurate determinations of fifty-four major and trace elements in carbonate by LA-ICP-MS using normalization strategy of bulk components as 100%. Chemical Geology, 2011, 284(3-4): 283-295.

[17] 陈春飞, 刘先国, 胡兆初, 宗克清, 刘勇胜. LA-ICP-MS 微区原位准确分析含水硅酸盐矿物主量和微量元素. 地球科学——中国地质大学学报, 2014, 39(5): 525-536.

[18] Rusk B, Koenig A, Lowers H. Visualizing trace element distribution in quartz using cathodoluminescence, electron microprobe, and laser ablation-inductively coupled plasma-mass spectrometry. American Mineralogist, 2011, 96(5-6): 703-708.

[19] Flem B, Müller A. *In situ* analysis of trace elements in quartz using laser ablation inductively coupled plasma mass spectrometry // Quartz: Deposits, Mineralogy and Analytics. Berlin: Springer, 2012: 219-236.

[20] 蓝廷广, 胡瑞忠, 范宏瑞, 毕献武, 唐燕文, 周丽, 毛伟, 陈应华. 流体包裹体及石英 LA-ICP-MS 分析方法的建立及其在矿床学中的应用. 岩石学报, 2017, 33(10): 3239-3262.

[21] Chew D M, Babechuk M G, Cogné N, Mark C, O'Sullivan G J, Henrichs I A, Doepke D, McKenna C A. (LA, Q)-ICPMS trace-element analyses of Durango and McClure Mountain apatite and implications for making natural

LA-ICPMS mineral standards. Chemical Geology, 2016, 435: 35-48.

[22] She Y-W, Song X-Y, Yu S-Y, Chen L-M, Zheng W-Q. Apatite geochemistry of the Taihe layered intrusion, SW China: Implications for the magmatic differentiation and the origin of apatite-rich Fe-Ti oxide ores. Ore Geology Reviews, 2016, 78: 151-165.

[23] Mao M, Rukhlov A S, Rowins S M, Spence J, Coogan L A. Apatite trace element compositions: A robust new tool for mineral exploration. Economic Geology, 2016, 111(5): 1187-1222.

[24] 孟郁苗, 黄小文, 高剑峰, 戴智慧, 漆亮. 无内标-多外标校正激光剥蚀等离子体质谱法测定磁铁矿微量元素组成. 岩矿测试, 2016, 35(6): 585-594.

[25] 张德贤, 戴塔根, 胡毅. 磁铁矿中微量元素的激光剥蚀-电感耦合等离子体质谱分析方法探讨. 岩矿测试, 2012, 31(1): 120-126.

[26] Savard D, Barnes S J, Dare S, Beaudoin G. Improved calibration technique for magnetite analysis by LA-ICP-MS. Mineralogical Magazine, 2012, 76(6): 2329.

[27] Nadoll P, Koenig A E. LA-ICP-MS of magnetite: Methods and reference materials. Journal of Analytical Atomic Spectrometry, 2011, 26(9): 1872-1877.

[28] Huang X-W, Zhou M-F, Qi L, Gao J-F, Wang Y-W. Re-Os isotopic ages of pyrite and chemical composition of magnetite from the Cihai magmatic-hydrothermal Fe deposit, NW China. Mineralium Deposita, 2013, 48(8): 925-946.

[29] Gao J-F, Zhou M-F, LightFoot P C, Wang C Y, Qi L, Sun M. Sulfide saturation and magma emplacement in the formation of the Permian Huangshandong Ni-Cu sulfide deposit, Xinjiang, Northwestern China. Economic Geology, 2013, 108: 1833-1848.

[30] Cook N J, Ciobanu C L, Danyushevsky L V, Gilbert S. Minor and trace elements in bornite and associated Cu-(Fe)-sulfides: A LA-ICP-MS study. Geochimica et Cosmochimica Acta, 2011, 75(21): 6473-6496.

[31] Cook N J, Ciobanu C L, Pring A, Skinner W, Shimizu M, Danyushevsky L, Saini-Eidukat B, Melcher F. Trace and minor elements in sphalerite: A LA-ICPMS study. Geochimica et Cosmochimica Acta, 2009, 73(16): 4761-4791.

[32] Gilbert S, Danyushevsky L, Robinson P, Wohlgemuth-Ueberwasser C, Pearson N, Savard D, Norman M, Hanley J. A comparative study of five reference materials and the Lombard meteorite for the determination of the platinum-group elements and gold by LA-ICP-MS. Geostandards Geoanalytical Research, 2013, 37(1): 51-64.

[33] Chen L-M, Song X-Y, Danyushevsky L V, Wang Y-S, Tian Y-L, Xiao J-F. A laser ablation ICP-MS study of platinum-group and chalcophile elements in base metal sulfide minerals of the Jinchuan Ni-Cu sulfide deposit, NW China. Ore Geology Reviews, 2015, 65: 955-967.

[34] Liu Y, Hu Z, Gao S, Günther D, Xu J, Gao C, Chen H. *In situ* analysis of major and trace elements of anhydrous minerals by LA-ICP-MS without applying an internal standard. Chemical Geology, 2008, 257(1-2): 34-43.

[35] Norman M D, Pearson N J, Sharma A, Griffin W L. Quantitative analysis of trace elements in geological materials by laser ablation ICPMS: Instrumental operating conditions and calibration values of NIST glasses. Geostandards and Geoanalytical Research, 1996, 20(2): 247-261.

[36] Pisiak L K, Canil D, Lacourse T, Plouffe A, Ferbey T. Magnetite as an indicator mineral in the exploration of porphyry deposits: A case study in till near the Mount Polley Cu-Au deposit, British Columbia, Canada. Economic Geology, 2017, 112(4): 919-940.

[37] 范宏瑞, 李兴辉, 左亚彬, 陈蕾, 刘尚, 胡芳芳, 冯凯. LA-(MC)-ICPMS 和(Nano)SIMS 硫化物微量元素和硫同位素原位分析与矿床形成的精细过程. 岩石学报, 2018, 34(12): 3479-3496.

[38] 周伶俐, 曾庆栋, 孙国涛, 段晓侠. LA-ICPMS 原位微区面扫描分析技术及其矿床学应用实例. 岩石学报, 2019, 35(7): 1964-1978.

[39] Zhu Z-Y, Cook N J, Yang T, Ciobanu C L, Zhao K-D, Jiang S-Y. Mapping of sulfur isotopes and trace elements in sulfides by LA-(MC)-ICP-MS: Potential analytical problems, improvements and implications. Minerals, 2016, 6(4): 110.

[40] 汪方跃, 葛粲, 宁思远, 聂利青, 钟国雄. 一个新的矿物面扫描分析方法开发和地质学应用. 岩石学报, 2017, 33(11): 3422-3436.

[41] Pettke T, Oberli F, Audétat A, Guillong M, Simon A C, Hanley J J, Klemm L M. Recent developments in element concentration and isotope ratio analysis of individual fluid inclusions by laser ablation single and multiple collector ICP-MS. Ore Geology Reviews, 2012, 44: 10-38.

[42] Hammerli J, Rusk B, Spandler C, Emsbo P, Oliver N H S. In situ quantification of Br and Cl in minerals and fluid inclusions by LA-ICP-MS: A powerful tool to identify fluid sources. Chemical Geology, 2013, 337: 75-87.

[43] 吴石头, 王亚平, 许春雪. 激光剥蚀电感耦合等离子体质谱元素微区分析标准物质研究进展. 岩矿测试, 2015, 34(5): 503-511.

[44] Danyushevsky L, Robinson P, Gilbert S, Norman M, Large R, McGoldrick P, Shelley M. Routine quantitative multi-element analysis of sulphide minerals by laser ablation ICP-MS: Standard development and consideration of matrix effects. Geochemistry: Exploration, Environment, Analysis, 2011, 11(1): 51-60.

[45] 赵令浩, 詹秀春, 胡明月, 孙冬阳, 范晨子, 袁继海, 蒯丽君, 屈文俊. 铳镍试金技术制备含铂族元素硫化物微区分析标准样品的可行性. 岩矿测试, 2013, 32(5): 694-701.

[46] Jochum K P, Stoll B, Weis U, Jacob D E, Mertz-Kraus R, Andreae M O. Non-matrix-matched calibration for the multi-element analysis of geological and environmental samples using 200 nm femtosecond LA-ICP-MS: A comparison with nanosecond lasers. Geostandards Geoanalytical Research, 2014, 38(3): 265-292.

[47] Li Z, Hu Z, Liu Y, Gao S, Li M, Zong K, Chen H, Hu S. Accurate determination of elements in silicate glass by nanosecond and femtosecond laser ablation ICP-MS at high spatial resolution. Chemical Geology, 2015, 400: 11-23.

[48] Kon Y, Murakami H, Takagi T, Watanabe Y. The development of whole rock analysis of major and trace elements in XRF glass beads by fsLA-ICPMS in GSJ geochemical reference samples. Geochemical Journal, 2011, 45(5): 387-416.

[49] Leite T D F, Escalfoni Jr R, da Fonseca T C O, Miekeley N. Determination of major, minor and trace elements in rock samples by laser ablation inductively coupled plasma mass spectrometry: Progress in the utilization of borate glasses as targets. Spectrochimica Acta Part B: Atomic Spectroscopy, 2011, 66(5): 314-320.

[50] Jenner F E, O'Neill H S C. Major and trace analysis of basaltic glasses by laser-ablation ICP-MS. Geochemistry Geophysics Geosystems, 2012, 13(3): 1-17.

[51] 朱律运, 刘勇胜, 胡兆初, 高山, 王晓红, 田滔. 玄武岩全岩元素含量快速、准确分析新技术: 双铱带高温炉与LA-ICP-MS 联用法. 地球化学, 2011, 40(5): 407-417.

[52] Loewen M W, Kent A J R. Sources of elemental fractionation and uncertainty during the analysis of semi-volatile metals in silicate glasses using LA-ICP-MS. Journal of Analytical Atomic Spectrometry, 2012, 27(9): 1502-1508.

[53] Zhu L, Liu Y, Hu Z, Hu Q, Tong X, Zong K, Chen H, Gao S. Simultaneous determination of major and trace elements in fused volcanic rock powders using a hermetic vessel heater and LA-ICP-MS. Geostandards Geoanalytical Research, 2013, 37(2): 207-229.

[54] He Z, Huang F, Yu H, Xiao Y, Wang F, Li Q, Xia Y, Zhang X. A flux-free fusion technique for rapid determination of major and trace elements in silicate rocks by LA-ICP-MS. Geostandards Geoanalytical Research, 2016, 40(1): 5-21.

[55] Garbe-Schönberg D, Müller S. Nano-particulate pressed powder tablets for LA-ICP-MS. Journal of Analytical Atomic Spectrometry, 2014, 29(6): 990-1000.

[56] Mukherjee P K, Khanna P P, Saini N K. Rapid determination of trace and ultra trace level elements in diverse silicate

rocks in pressed powder pellet targets by LA-ICP-MS using a matrix-independent protocol. Geostandards Geoanalytical Research, 2014, 38(3): 363-379.

[57] Zhang W, Hu Z, Liu Y, Yang W, Chen H, Hu S, Xiao H. Quantitative analysis of major and trace elements in NH_4HF_2-modified silicate rock powders by laser ablation-inductively coupled plasma mass spectrometry. Analytica Chimica Acta, 2017, 983: 149-159.

[58] Peters D, Pettke T. Evaluation of major to ultra trace element bulk rock chemical analysis of nanoparticulate pressed powder pellets by LA-ICP-MS. Geostandards Geoanalytical Research, 2017, 41(1): 5-28.

[59] Liao X, Hu Z, Luo T, Zhang W, Liu Y, Zong K, Zhou L, Zhang J. Determination of major and trace elements in geological samples by laser ablation solution sampling-inductively coupled plasma mass spectrometry. Journal of Analytical Atomic Spectrometry, 2019, 34(6): 1126-1134.

[60] Liao X, Luo T, Zhang S, Zhang W, Zong K, Liu Y, Hu Z. Direct and rapid multi-element analysis of wine samples in their natural liquid state by laser ablation ICPMS. Journal of Analytical Atomic Spectrometry, 2020, 35(6): 1071-1079.

[61] Pinto F G, Junior R E, Saint'Pierre T D. Sample preparation for determination of rare earth elements in geological samples by ICP-MS: A critical review. Analytical Letters, 2012, 45(12): 1537-1556.

[62] Gorbatenko A A, Revina E I. A review of instrumental methods for determination of rare earth elements. Inorganic Materials, 2015, 51(14): 1375-1388.

[63] Zawisza B, Pytlakowska K, Feist B, Polowniak M, Kita A, Sitko R. Determination of rare earth elements by spectroscopic techniques: A review. Journal of Analytical Atomic Spectrometry, 2011, 26(12): 2373-2390.

[64] Bhatt C R, Jain J C, Goueguel C L, McIntyre D L, Singh J P. Determination of rare earth elements in geological samples using laser-induced breakdown spectroscopy (LIBS). Applied spectroscopy, 2018, 72(1): 114-121.

[65] Sun Y, Sun S, Wang C Y, Xu P. Determination of rare earth elements and thorium at nanogram levels in ultramafic samples by inductively coupled plasma-mass spectrometry combined with chemical separation and pre-concentration. Geostandards Geoanalytical Research, 2013, 37(1): 65-76.

[66] Li W, Jin X, Gao B, Wang C, Zhang L. Analysis of ultra-low level rare earth elements in magnetite samples from banded iron formations using HR-ICP-MS after chemical separation. Analytical Methods, 2014, 6(15): 6125-6132.

[67] Liu Y, Xue D, Li W, Li C, Wan B. A simple method for the precise determination of multi-elements in pyrite and magnetite by ICP-MS and ICP-OES with matrix removal. Microchemical Journal, 2020, 158: 105221.

[68] Liu Y, Xue D-s, Li W, Li C. Determination of ultra-trace rare earth elements in iron minerals using HR-ICP-MS after chemical purification by polyurethane foam. Journal of Analytical Atomic Spectrometry, 2020, 35(10): 2156-2164.

[69] Padmasubashini V, Satyanarayana K. Determination of rare earth elements, yttrium, thorium, and other trace elements in monazite samples by inductively coupled plasma mass spectrometry. Atomic Spectroscopy, 2013, 34(1): 6-14.

[70] Kumar S, Pandey S, Kumar S. Determination of rare earth elements in Indian kimberlite using inductively coupled plasma mass spectrometer (ICP-MS). Journal of Radioanalytical Nuclear Chemistry, 2012, 294(3): 419-424.

[71] Krishnakumara M, Satyanarayanab K, Mukkantic K. Synergistic separation of rare earth elements (REEs, La-Lu), Y and Th From U-, Nb-, and Ta-rich refractory minerals for determination by ICP-AES. Atomic Spectroscopy, 2015, 36(2): 74-81.

[72] Khorge C R, Patwardhan A A. Separation and determination of REEs and Y in columbite-tantalite minerals by ICP-OES: A rapid approach. Atomic Spectroscopy, 2018, 39(2): 75-80.

[73] Saha M C, Alteb R S, Roya N K. Quantitative determination of rare earth elements in marine sediment samples by ICP-MS and NAA technique. Atomic Spectroscopy, 2015, 36(3): 109-115.

[74] 王冠, 李华玲, 任静, 杨波, 胡志中. 高分辨电感耦合等离子体质谱法测定地质样品中稀土元素的氧化物干扰研

究. 岩矿测试, 2013, 32(4): 561-567.

[75] 元艳, 方金东, 董学林. 地质样品中三稀金属元素分析方法进展. 资源环境与工程, 2014, 28(1): 89-93.

[76] 李刚, 姚玉玲, 李婧祎, 赵朝辉, 罗涛, 李崇瑛. 铌钽元素分析技术新进展. 岩矿测试, 2018, 37(1): 1-14.

[77] Tomascak P B, Magna T, Dohmen R. Methodology of lithium analytical chemistry and isotopic measurements // Advances in Lithium Isotope Geochemistry. Cham: Springer, 2016: 5-18.

[78] De La Rosa N, Kristiansson P, Nilsson E J C, Ros L, Pallon J, Skogby H. Quantification of lithium at ppm level in geological samples using nuclear reaction analysis. Journal of Radioanalytical and Nuclear Chemistry, 2018, 317(1): 253-259.

[79] 赵学沛. 多种酸溶矿 ICP-AES 测定稀有金属矿中锂铍铌钽锡. 化学研究与应用, 2017, 29(11): 1714-1718.

[80] 李志伟, 赵晓亮, 李珍, 王烨, 王君玉. 敞口酸熔-电感耦合等离子体发射光谱法测定稀有多金属矿选矿样品中的铌钽和伴生元素. 岩矿测试, 2017, 36(6): 594-600.

[81] 程祎, 李志伟, 于亚辉, 刘军, 韩志轩, 孙勇, 吴林海. 高压密闭消解-电感耦合等离子体质谱法测定地质样品中铌、钽、锆、铪和16种稀土元素. 理化检验-化学分册, 2020, 56(7): 782-787.

[82] 高会艳. ICP-MS 和 ICP-AES 测定地球化学勘查样品及稀土矿石中铌钽方法体系的建立. 岩矿测试, 2014, 33(3): 312-320.

[83] 姚玉玲, 吴丽琨, 刘卫, 李刚. 乙醇增敏-电感耦合等离子体发射光谱法测定矿石及选冶样品中的铌钽. 岩矿测试, 2015, 34(2): 224-228.

[84] 刘江斌, 赵峰, 余宇, 党亮, 张旺强, 陈月源. X 射线荧光光谱法同时测定地质样品中铌钽锆铪铈镓钪铀等稀有元素. 岩矿测试, 2010, 29(1): 74-76.

[85] 陈静, 高志军, 陈冲科, 刘延霞, 张明炜. X 射线荧光光谱法分析地质样品的应用技巧. 岩矿测试, 2015, 34(1): 91-98.

[86] 黎卫亮, 程秀花, 余娟, 刘欢. 高压密闭酸溶-电感耦合等离子体质谱法测定花岗闪长岩中的微量锆. 岩矿测试, 2016, 35(1): 32-36.

[87] 葛小莹, 王艳龙, 焦圣兵, 罗善霞, 戚洪友, 侯慧敏. 地质样品中碲的测定方法研究. 化学工程师, 2011, 25(8): 46-48.

[88] 周建, 朱利亚, 赵青. 锗分析方法的研究进展. 理化检验-化学分册, 2012, 48(1): 122-128.

[89] 张琦, 蔡玉曼. 电感耦合等离子质谱测定矿石中铼的研究进展. 分析化学进展, 2018, 8(1): 11-17.

[90] 罗善霞, 焦圣兵. 地质样品中铼的分离和测定方法研究进展. 冶金分析, 2013, 33(2): 22-27.

[91] 黄小文, 漆亮, 高剑峰. 铼-锇同位素分析样品预处理研究进展. 岩矿测试, 2011, 30(1): 90-103.

[92] 程秀花, 唐南安, 张明祖, 黎卫亮, 王鹏, 陈陆洋. 稀有分散元素分析方法的研究进展. 理化检验: 化学分册, 2013, 49(6): 757-764.

[93] Kaya M, Volkan M. Germanium determination by flame atomic absorption spectrometry: An increased vapor pressure-chloride generation system. Talanta, 2011, 84(1): 122-126.

[94] Audétat A, Garbe-Schönberg D, Kronz A, Pettke T, Rusk B, Donovan J J, Lowers H A. Characterisation of a natural quartz crystal as a reference material for microanalytical determination of Ti, Al, Li, Fe, Mn, Ga and Ge. Geostandards Geoanalytical Research, 2015, 39(2): 171-184.

[95] 陈波, 刘洪青, 邢应香. 电感耦合等离子体质谱法同时测定地质样品中锗硒碲. 岩矿测试, 2014, 33(2): 192-196.

[96] 高贺凤, 王超, 张立纲. 电感耦合等离子体质谱法精确测定地质样品中的微量元素镓. 岩矿测试, 2013, 32(5): 709-714.

[97] 熊英, 吴赫, 王龙山. 电感耦合等离子体质谱法同时测定铜铅锌矿石中微量元素镓铟铊钨钼的干扰消除. 岩矿测试, 2011, 30(1): 7-11.

[98] 孙朝阳, 董利明, 贺颖婷, 杨利华, 郑存江. 电感耦合等离子体质谱法测定地质样品中钪镓锗铟镉铊时的干扰及其消除方法. 理化检验-化学分册, 2016, 52(9): 1026-1030.

[99] Blokhin M G, Zarubina N V, Mikhailyk P E. Inductively coupled plasma mass spectrometric measurement of gallium in ferromanganese crusts from the Sea of Japan. Journal of Analytical Chemistry, 2014, 69(13): 1237-1244.

[100] 黎卫亮, 程秀花, 张明祖, 陈陆洋. 乙醇增强-电感耦合等离子体质谱法测定地质样品中镓铟铊锗碲. 冶金分析, 2014, 34(3): 13-18.

[101] 吴峥, 牟乃仓, 王龙山, 张飞鸽, 高登峰, 王光照. 快速程序升温-石墨炉原子吸收光谱法测定地质样品中痕量镉. 岩矿测试, 2011, 30(2): 186-189.

[102] Xu Q, Guo W, Jin L, Guo Q, Hu S. Determination of cadmium in geological samples by aerosol dilution ICP-MS after inverse aqua regia extraction. Journal of Analytical Atomic Spectrometry, 2015, 30(9): 2010-2016.

[103] 贺攀红, 杨珍, 荣耀, 龚书浩, 龚治湘. 氢化物发生-电感耦合等离子体发射光谱法测定铀矿地质样品中痕量硒. 岩矿测试, 2016, 35(2): 139-144.

[104] Mladenova E, Karadjova I, Tsalev D L. Solid-phase extraction in the determination of gold, palladium, and platinum. Journal of Separation Science, 2012, 35(10-11): 1249-1265.

[105] Mladenova E, Dakova I, Karadjova I, Karadjov M. Column solid phase extraction and determination of ultra-trace Au, Pd and Pt in environmental and geological samples. Microchemical Journal, 2012, 101: 59-64.

[106] Balaram V, Mathur R, Satyanarayanan M, Sawant S, Roy P, Subramanyam K S V, Kamala C T, Anjaiah K V, Ramesh S L, Dasaram B. A rapid method for the determination of gold in rocks, ores and other geological materials by F-AAS and GF-AAS after separation and preconcentration by DIBK extraction for prospecting studies. Mapan, 2012, 27(2): 87-95.

[107] Ashkenani H, Taher M A. Use of ionic liquid in simultaneous microextraction procedure for determination of gold and silver by ETAAS. Microchemical Journal, 2012, 103: 185-190.

[108] Soylak M, Yilmaz E. Ionic liquid-based method for microextraction/enrichment of gold from real samples and determination by flame atomic absorption spectrometry. Atomic Spectroscopy, 2013, 34(1): 15-19.

[109] 刘军, 闫红岭, 连文莉, 陈浩凤, 王琳, 于亚辉. 封闭溶矿-电感耦合等离子体质谱法测定地质样品中金银铂钯. 冶金分析, 2016, 36(7): 25-33.

[110] 赵延庆. 聚氨酯泡沫塑料吸附-电感耦合等离子体质谱法测定地质化探样品中金. 冶金分析, 2016, 36(7): 34-38.

[111] 漆亮, 黄小文. 地质样品铂族元素及Re-Os同位素分析进展. 矿物岩石地球化学通报, 2013, 32(2): 171-189.

[112] Qi L, Gao J, Huang X, Hu J, Zhou M-F, Zhong H. An improved digestion technique for determination of platinum group elements in geological samples. Journal of Analytical Atomic Spectrometry, 2011, 26(9): 1900-1904.

[113] Qi L, Gao J F, Zhou M F, Hu J. The design of re-usable Carius tubes for the determination of rhenium, osmium and platinum-group elements in geological samples. Geostandards and Geoanalytical Research, 2013, 37(3): 345-351.

[114] Li J, Jiang X Y, Xu J F, Zhong L F, Wang X C, Wang G-Q, Zhao P-P. Determination of platinum-group elements and Re-Os isotopes using ID-ICP-MS and N-TIMS from a single digestion after two-stage column separation. Geostandards and Geoanalytical Research, 2014, 38(1): 37-50.

[115] Ishikawa A, Senda R, Suzuki K, Dale C W, Meisel T. Re-evaluating digestion methods for highly siderophile element and [187]Os isotope analysis: Evidence from geological reference materials. Chemical Geology, 2014, 384: 27-46.

[116] Wang Z, Becker H. Abundances of sulfur, selenium, tellurium, rhenium and platinum-group elements in eighteen reference materials by isotope dilution sector-field ICP-MS and negative TIMS. Geostandards Geoanalytical Research, 2014, 38(2): 189-209.

[117] 王瑞侠, 周享春, 陆光汉. 电化学分析方法测定卤素的研究进展. 理化检验-化学分册, 2012, 48(9): 1123-1128.

[118] 彭炳先, 吴代赦. 高温热水解离子色谱法快速同时测定粘土中的卤素. 分析化学, 2013, 41(10): 1499-1504.

[119] Marks M A W, Wenzel T, Whitehouse M J, Loose M, Zack T, Barth M, Worgard L, Krasz V, Eby G N, Stosnach H. The volatile inventory (F, Cl, Br, S, C) of magmatic apatite: An integrated analytical approach. Chemical Geology, 2012, 291: 241-255.

[120] Zhang C, Lin J, Pan Y, Feng R, Almeev R R, Holtz F. Electron probe microanalysis of bromine in minerals and glasses with correction for spectral interference from aluminium, and comparison with microbeam synchrotron X-ray fluorescence spectrometry. Geostandards Geoanalytical Research, 2017, 41(3): 449-457.

[121] Oliveira A A, Trevizan L C, Nobrega J A. Iodine determination by inductively coupled plasma spectrometry. Applied Spectroscopy Reviews, 2010, 45(6): 447-473.

[122] He T, Hu Z, Zhang W, Chen H, Liu Y, Wang Z, Hu S. Determination of Cl, Br, and I in geological materials by sector field inductively coupled plasma mass spectrometry. Analytical Chemistry, 2019, 91(13): 8109-8114.

[123] Krishna A K, Mohan K R, Satyanarayanan M. Non-destructive sampling method for environmental and geochemical exploration studies and XRF determination of Cl, F, and Br. Atomic Spectroscopy, 2012, 33(3): 100-108.

[124] Li X, Wang Y, Zhang Q. Determination of halogen levels in marine geological samples. Spectroscopy Letters, 2016, 49(3): 151-154.

[125] Cui Y, Jin L, Li H, Hu S, Lian Y. Direct determination of trace Iodine in geological samples by solid sampling electrothermal vaporization-inductively coupled plasma mass spectrometry. Atomic Spectroscopy, 2020, 41(2): 87-92.

[126] de Gois J S, Pereira É R, Welz B, Borges D L G. Simultaneous determination of bromine and chlorine in coal using electrothermal vaporization inductively coupled plasma mass spectrometry and direct solid sample analysis. Analytica Chimica Acta, 2014, 852: 82-87.

[127] Pan L-C, Hu R-Z, Wang X-S, Bi X-W, Zhu J-J, Li C. Apatite trace element and halogen compositions as petrogenetic-metallogenic indicators: Examples from four granite plutons in the Sanjiang region, SW China. Lithos, 2016, 254: 118-130.

[128] Stock M J, Humphreys M C S, Smith V C, Johnson R D, Pyle D M, EIMF. New constraints on electron-beam induced halogen migration in apatite. American Mineralogist, 2015, 100(1): 281-293.

[129] Teiber H, Scharrer M, Marks M A W, Arzamastsev A A, Wenzel T, Markl G. Equilibrium partitioning and subsequent re-distribution of halogens among apatite-biotite-amphibole assemblages from mantle-derived plutonic rocks: Complexities revealed. Lithos, 2015, 220: 221-237.

[130] Ansberque C, Mark C, Caulfield J T, Chew D M. Combined in-situ determination of halogen (F, Cl) content in igneous and detrital apatite by SEM-EDS and LA-Q-ICPMS: A potential new provenance tool. Chemical Geology, 2019, 524: 406-420.

[131] Fusswinkel T, Giehl C, Beermann O, Fredriksson J R, Garbe-Schönberg D, Scholten L, Wagner T. Combined LA-ICP-MS microanalysis of iodine, bromine and chlorine in fluid inclusions. Journal of Analytical Atomic Spectrometry, 2018, 33(5): 768-783.

[132] Hammerli J, Rusk B, Spandler C, Emsbo P, Oliver N H. *In situ* quantification of Br and Cl in minerals and fluid inclusions by LA-ICP-MS: A powerful tool to identify fluid sources. Chemical Geology, 2013, 337: 75-87.

[133] Leisen M, Boiron M-C, Richard A, Dubessy J. Determination of Cl and Br concentrations in individual fluid inclusions by combining microthermometry and LA-ICPMS analysis: Implications for the origin of salinity in crustal fluids. Chemical Geology, 2012, 330: 197-206.

[134] Seo J H, Guillong M, Aerts M, Zajacz Z, Heinrich C A. Microanalysis of S, Cl, and Br in fluid inclusions by LA-ICP-MS. Chemical Geology, 2011, 284(1-2): 35-44.

[135] Sekimoto S, Ebihara M. Accurate determination of chlorine, bromine, and iodine in sedimentary rock reference samples by radiochemical neutron activation analysis and a detailed comparison with inductively coupled plasma mass spectrometry literature data. Analytical Chemistry, 2013, 85(13): 6336-6341.

[136] Sekimoto S, Ebihara M. Accurate determination of chlorine, bromine and iodine in US Geological Survey geochemical reference materials by radiochemical neutron activation analysis. Geostandards Geoanalytical Research, 2017, 41(2): 213-219.

[137] Ruzié-Hamilton L, Clay P L, Burgess R, Joachim B, Ballentine C J, Turner G. Determination of halogen abundances in terrestrial and extraterrestrial samples by the analysis of noble gases produced by neutron irradiation. Chemical Geology, 2016, 437: 77-87.

[138] Kendrick M A. High precision Cl, Br and I determinations in mineral standards using the noble gas method. Chemical Geology, 2012, 292: 116-126.

[139] Kendrick M A, D'Andres J, Holden P, Ireland T. Halogens (F, Cl, Br, I) in thirteen USGS, GSJ and NIST international rock and glass reference materials. Geostandards Geoanalytical Research, 2018, 42(4): 499-511.

第 25 章 标准与标准物质

(冯流星[①],张见营[①],王 松[①],巢静波[①],逯 海[①],任同祥[①],唐一川[①],李红梅[①*])

▶ 25.1 概述 / 698

▶ 25.2 标准物质技术发展 / 702

▶ 25.3 标准物质在典型领域的应用 / 731

▶ 25.4 展望 / 736

[①]中国计量科学研究院,北京。*通讯作者联系方式:lihm@nim.ac.cn

本章导读

- 简要介绍标准物质的定义、作用，与分析技术、标准方法的关系，主要准物质数据库。
- 重点介绍各个领域标准物质定值技术的发展，包括同位素、高纯物质、生命科学、纳米颗粒、微区分析以及统计理论等。
- 最后介绍标准物质在典型领域的应用，包括仪器性能评价、食品安全、环境保护、临床诊断等领域。

25.1 概　　述

25.1.1 标准物质概述

标准物质（reference material，RM）定义：具有足够均匀和稳定的特定特性的物质，其特性被证实适用于测量中或标称特性检查中的预期用途。有证标准物质（certified reference material，CRM）定义：附有由权威机构发布的文件，提供使用有效程序获得的具有不确定度和溯源性的一个或多个特性量值的标准物质。标准物质特性量测量的总平均值即为该特性量的标准值。标准值的总不确定度由三个部分组成：标准物质定值过程中产生的不确定度、物质的不均匀性和随时间变化引起的变动性（不稳定性）所引入的不确定度。

标准物质的主要特性包括均匀性、稳定性、准确性和溯源性。标准物质的均匀性定义为：与物质的一种或几种特性相关的具有相同结构或组成的状态。当对该物质进行检验时，不论样品是否取自同一最小包装单元，检验具有规定大小的样品，若被测量的特性值在规定的不确定度范围内，则该标准物质对这一特性来说是均匀的。标准物质的稳定性定义为：在规定的时间和空间环境下，标准物质特性量值保持在规定范围内的性质（或能力）。由此可见标准物质的稳定性表示该标准物质的特性量值随时间变化的特性，时间间隔的长短是由这一特性决定的，通常把这个时间间隔称为标准物质的有效期。有效期越长，表明标准物质的稳定性越好。准确性是指标准物质具有严格定义的和准确计量的标准值（也称保证值或鉴定值）特性，是被鉴定特性量之真值的最佳估计，标准值与真值的偏离不超过计量不确定度。确定标准物质特性量值的过程就是标准物质的定值，定值是基于定性基础上，定量获取标准物质特性量值的过程，包括定值方法选择，测量方法的确认与控制，测量仪器的计量校准、测量溯源性研究、测量数据的统计学处理及评估测量不确定度。溯源性是计量（或测量）学的属性，定义为：通过一条具有确定不确定度的连续的比较链，使测量结果、测量标准值能够与规定的参考基、标准，通常是与国家测量基、标准或国际测量基、标准联系起来的特性。很显然，一切具有共同溯源目标，合理赋予不确定度的不间断的比较测量将具有溯源性。标准物质可通过以下公认的基本方式实现其量值溯源：①溯源至 SI 单位或其导出单位，最常用的就是使用基准方法（primary method）。基准方法有坚实的理论基础和严格的数学表达的方法，测量过程可以完全清晰地被描述，精密度、准确度、测量范围和稳定性已经过严谨的研究与验证，具有最高测量水平。目前国际公认的基准方法有库仑法、同位素稀释质谱法、重量法、凝固点下降法。②溯源至其他公认的计量标准，比较常见的是使用基准物质、有证标准物质进行校准，或通过国家计量实验室、指定研究机构进行校准来实现溯源。③溯源至国际公认、准确的定义，实现某一特定单位的复现，如传统标度 pH 标准物质的定值、浊度标准物质的定值。

25.1.2 分析技术、标准方法与标准物质的关系

标准物质的作用犹如一把尺子，它所衡量的对象涉及化学、生物、工程、物理等众多特性或成分。标准物质可用于分析方法评价、分析仪器评价、待测样品测试、分析环境评价、实验人员与检测实验室能力的评价等。使用标准物质对于改进分析工作质量、提高分析准确度、保证分析结果的有效性具有十分重要的意义，为科技进步与创新、重大决策以及经济和社会发展中所涉及的公平贸易、标准制定、实施和验证、民生保障等提供了坚实的支撑。近些年来，国内外十分重视能够出具公正分析数据的实验室质量体系建设，标准物质作为确保数据准确性与公正性的必不可少的重要工具，其应用需求迅猛增长。

众所周知，原子光谱类仪器大多是依据电磁物理理论设计，综合了机械、电磁、真空和计算机等学科的技术，以测量吸光度、原子离子、分子离子为目的。仪器原理和结构决定了：原子谱线、不同质量的粒子、离子在其中运动的行为不同，极易产生系统误差，更何况来自外部电、磁和热辐射的干扰。而且，在样品制备、样品引入、样品电离（即原子/离子转化）和离子检测（包括模/数转换）过程中也都有可能引进误差。使得测量值与"真实值"之间难免出现偏移。如何由测量值求得样品"真实值"，始终是广大原子光谱分析工作者所追求的目的。经过长时间的实践证明，借助标准物质校准原子光谱仪器，检验、评价测量方法是保证测量结果有效性、可靠性和测量捷径的最好方法。

同时，原子光谱类分析技术在标准物质研制过程中也正在发挥着越来越重要的作用。标准物质是一种技术含量较高的特殊"产品"。标准物质的均匀性、稳定性和特性量值是反映标准物质特点的主要参数和指标，技术含量高，成为标准物质研制重点。在国内外不同等级标准物质研制过程中，原子光谱技术，包括原子吸收（AAS）、电感耦合等离子体发射光谱（ICP-OES）、电感耦合等离子体质谱（ICP-MS）等是相关重要标准物质均匀性、稳定性检验和特性量值确定的首选和普遍使用的手段。

文本技术标准是对标准化领域中需要协调统一的技术事项所制订的标准。它是根据不同时期的科学技术水平和实践经验，针对具有普遍性和重复出现的技术问题，提出的最佳解决方案，是从事科研、设计、工艺、检验等技术工作中共同遵守的技术依据。我国目前原子光谱领域的技术标准主要是国家标准、行业标准等，涉及环境、食品等多个分析领域，如环境领域中原子吸收分光光度法相关的国家及行业技术标准就有70余项，原子荧光相关的技术标准上百项，此外还包括电感耦合等离子体发射光谱、电感耦合等离子体质谱等相关的技术标准。技术标准与标准物质有着密切的关系，首先在技术标准的制修订中需要标准物质对标准中的技术方法进行验证，其次在标准的执行过程中，也需要相关标准物质为分析过程提供质量保证。可以说，标准物质是技术标准制定和实施的基础。

25.1.3 主要标准物质数据库

对于原子光谱分析类标准物质，国内外都没有专门的标准物质资源库或专属类别。下面列举国际上最常见的标准物质库有关信息，用户可根据领域和用途在一些标准物质资源库中获取。

25.1.3.1 国际标准物质数据库

国际标准物质数据库（COMAR）是一个由志愿合作国际组织建立、维护，以互联网为基础的有证标准物质信息服务系统。COMAR的使命就是传播可利用的有证标准物质信息，帮助测试、试验人员找到他们所需的有证标准物质。COMAR数据库收录了25个国家220个研制机构制备的上万种有证标准物质，是目前国际上收录有证标准物质品种信息最多的数据库。COMAR是一个非商业的国际组织网络，数据库可免费使用。在COMAR数据库中的有证标准物质，按照应用领域分为8个主领

域及 10 个以上的分领域：①黑色金属（ferrous metals）；②有色金属（nonferrous metals）；③工业材料（industrial materials）；④物理特性（physical properties）；⑤有机物（organics）；⑥无机物（inorganics）；⑦生活质量（quality of life）；⑧生物和临床（biological & clinical）。

为了获得相关信息，使用者可直接上网查询 COMAR 网站（网址：http://www.comar.bam.de），也可向国家编码中心咨询。

25.1.3.2 美国国家标准与技术研究院（NIST）标准物质

NIST 的前身为美国国家标准局，成立于 1901 年，是世界上开展标准物质研制最早的机构，是设在美国商务部科技管理局下的非管理机构，负责国家层面上的计量基础研究。NIST 在研究权威化学及物理测量技术的同时，还提供以 NIST 为商标的各类标准参考物质（standard reference material，SRM），该商标以使用权威测量方法和技术而闻名。目前，NIST 共提供 1500 多种标准参考物质，形成了世界领先的、较为完善的标准物质体系，在保证美国国内化学测量量值的有效溯源及分析测量结果的可靠性方面发挥了重要的支撑作用。总体来讲，美国 NIST 与其他美国商业标准物质生产者的关系是：NIST 作为国家计量院将其研究重点放在高端标准物质及测量标准的研究上，建立国家的高端量值溯源体系，商业标准物质将其量值通过 NTRM、校准等各种形式溯源至 NIST，以改善标准物质的供应状况，满足不同层次用户对标准物质日益增长的需求。NIST 在国际标准化组织（ISO）制定的国际计量学通用基本术语有证标准物质（CRM）、标准物质（RM）的定义外，又将其标准物质分为 SRM（标准参考物质）、RM（标准物质）及 NTRM（可溯源至 NIST 的标准物质）。美国 NIST 标准物质的查询可通过 NIST 官方网站（网址：https://www.nist.gov/srm）进行查询。

25.1.3.3 欧洲标准物质

欧洲标准物质（ERM）是由欧洲三个主要标准物质研制机构，即比利时的欧盟联合研究中心标准物质和测量研究所、德国材料与测试技术研究院和英国政府化学家实验室通过密切协作的方式创作的标准物质品牌。

ERM 分为六个主类别，每个主类别含有若干个次类别，共计三十六个次类别。每个主类别以一个字母代表，用另外一个字母代表主类中的次类别。表 25-1 中收录了各个主类和次类别。

表 25-1 欧洲标准物质分类表/编码

主类别	字母	次类别	字母	编码
纯度、浓度、活性的非基体标准物质	A	固体或液体无机化合物和元素（纯的和溶液）	A	AAXXX
		气体（纯的和混合物）	B	ABXXX
		固体或液体有机分子（纯的和溶液）	C	ACXXX
		有机大分子	D	ADXXX
		同位素标记物	E	AEXXX
		其他	Z	AZXXX
食品、农产品和相关基体标准物质	B	饮用水和饮料	A	BAXXX
		动物物质	B	BBXXX
		植物物质	C	BCXXX
		上述未包括的加工食品和食物	D	BDXXX
		动物饲料	E	BEXXX
		转基因物质	F	BFXXX
		其他	Z	BZXXX

续表

主类别	字母	次类别	字母	编码
认定成分的环境和相关基体标准物质	C	水（河水、海水、地下水）	A	CAXXX
		废弃物、污物和沥出物	B	CBXXX
		土壤、沉积物、淤泥	C	CCXXX
		植物植被物质	D	CDXXX
		动物生物指示剂物质	E	CEXXX
		飞灰、燃料灰、焚化灰	F	CFXXX
		其他	Z	CZXXX
认定成分的健康相关基体标准物质	D	人体液（血清、尿）	A	DAXXX
		人组织（头发、骨头、牙齿等）	B	DBXXX
		其他	Z	DZXXX
认定成分的工业和工程标准物质	E	铁合金	A	EAXXX
		有色金	B	EBXXX
		聚合物、塑料	C	ECXXX
		玻璃、陶瓷	D	EDXXX
		矿物、矿石、岩石、黏土	E	EEXXX
		燃料（煤、柴油）	F	EFXXX
		半导体	G	EGXXX
		其他	Z	EZXXX
物理特性标准物质	F	机械特性（例如硬度、冲击韧性、黏度）	A	FAXXX
		光学特性（例如波长、吸光材料）	B	FBXXX
		热物质（例如导热率、热值）	C	FCXXX
		形态特性（例如粒度尺寸、表面积）	D	FDXXX
		其他	Z	FZXXX

欧洲标准物质的研制过程中，在基础材料的选择、候选物的加工、均匀性检查、稳定性检验、定值、储存等相关技术问题完全按照国际公认指南执行，特性量值大都经过国际比对，在严格的审批之后才可出售、使用。使用欧洲标准物质数据库搜索，用户可以找到关于 ERM 有证标准物质的所有信息。欧洲标准物质网站信息为 http://www.erm-crm.org/html/homepage.hrm。

25.1.3.4 中国国家标准物质资源共享平台

中国国家标准物质资源共享平台是目前国内最为权威的标准物质网络服务平台，由中国计量科学院负责建立。该平台是以实现标准物质资源共享为目标的公益性信息服务系统，2005 年开始对外服务，收录一级、二级有证标准物质十三大类。表 25-2 给出了各类标准物质的编号和名称。

表 25-2 我国十三类标准物质的编号和名称

编号	名称
01	钢铁成分分析标准物质
02	有色金属及金属中气体成分分析标准物质
03	建材成分分析标准物质
04	核材料成分分析及放射性测量标准物质
05	高分子材料特性测量标准物质
06	化工产品成分分析标准物质

续表

编号	名称
07	地质矿产成分分析标准物质
08	环境化学分析标准物质
09	临床化学分析与药品成分分析标准物质
10	食品成分分析标准物质
11	煤炭石油成分分析和物理特性测量标准物质
12	工程技术特性测量标准物质
13	物理特性及物理化学特性测量标准物质

标准物质资源共享平台提供下列服务：标准物质信息分类浏览；标准物质信息多渠道查询；标准物质技术咨询；标准物质网上订购；标准物质业界新闻、管理法规和技术规范的浏览与检索；标准物质查新服务；标准物质相关国际比对信息；标准物质相关会议信息发布、参会注册等。平台网站查询 http://www.ncrm.org.cn（国家标准物质资源共享平台）。

25.2 标准物质技术发展

25.2.1 同位素标准物质

25.2.1.1 同位素标准物质研制的必要性

同位素丰度比是国内外公认的痕量、超痕量成分的权威测量方法——同位素稀释质谱法的核心。同位素基标准物质作为化学计量的基础和量值传递与溯源的一种重要手段，主要用于快速、方便、准确地校准质谱仪器，为获得准确可靠的元素同位素丰度比测量结果提供保障。

随着同位素稀释质谱法在越来越多的领域得到应用，如营养元素[1]和有机分子化合物同位素示踪[2]、元素形态分析[3]、生物医学研究等，各行业对同位素丰度基准的需求越来越强。而质谱仪器的快速更新，质谱测量技术的不断发展以及与各种在线测量、分离技术的联用（如超声雾化、流动注射、气相色谱、液相色谱，氢化物发生器、电热蒸发、激光烧蚀等），为同位素技术提供了更广阔的发展空间和巨大的应用潜力，进而同位素技术的广泛应用也为更多的学科和领域提供了更丰富和更有价值的信息。同时，质谱分析技术的广泛应用和仪器数量迅猛增长，又促使对同位素标准物质的需求在种类和数量上不断增加。

近年来，国际上对原子量测量、同位素基标准物质和同位素丰度测量数据的关注度不断提高，每两年国际纯粹与应用化学联合会（IUPAC）同位素丰度与原子量委员会讨论的同位素测量和原子量测量的报告数量均在不断增长[4]。国际计量大会（CGPM）、国际计量委员会（CIPM）、国际计量局（BIPM）和 IUPAC 都设有专门机构负责组织协调研究工作，交流化学计量研究成果和标准物质的信息，使其向着更高的层次发展，扩大应用范围，为同位素丰度基准研究创造了良好氛围。随着科学技术的发展和国内整体分析测量水平的提高，对测量结果的溯源性和准确性的要求越来越强烈。一般的分析测试技术和方法已不能满足许多研究领域或行业的更为广阔和深入发展的要求。

另外，随着社会的发展，在生产活动的质量控制、各行业和学科的技术发展以及事关人民生活物质和生活环境的质量监督等方面，对广泛应用基标准体系和基标准物质也提出了更高的要求，既需要不断扩充基标准物质的种类和数量，又需要不断提高基标准物质的品质。

目前国内已有大型质谱仪器如热电离同位素质谱仪、多接收电感耦合等离子体质谱仪、高分辨质谱等上百台，其他类型质谱仪器不计其数，在仪器校准等方面对同位素丰度标准物质已经形成很大的市场需求，相应的标准物质品种和数量亟待扩充，同位素基标准体系的建立和完善，同位素基标准物质的研制已经成为亟待解决的重要课题。

25.2.1.2 同位素标准物质定值方法

绝对质谱法（也称为校正质谱法）是国际公认的同位素测量中的权威方法（或基准方法）[5]。方法原理是选择某元素的两种或三种高纯、高浓缩同位素试剂，准确分析这些试剂的化学纯度和同位素丰度，然后用高精密度天平称重配制一系列具有准确配制值的校正样品，用以测量该质谱仪的质量偏移校正系数。然后在相同的实验条件下分析标准物质和校正样品中硅的同位素丰度比，利用校正样品测得的质量偏移校正系数对标准物质的同位素丰度比进行校正，从而获得可溯源至国际基本单位的同位素丰度比标定值。具体技术路线图见图 25-1。

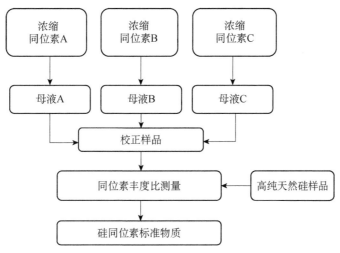

图 25-1 同位素标准物质定值技术路线图

25.2.1.3 同位素标准物质研究现状

目前国际上开展同位素标准物质研制的研究机构如表 25-3 所列[4]，这也是同位素丰度与原子量委员会（CIAAW）认可的标准物质研制单位。美国国家标准与技术研究院（NIST）、美国地质调查局（USGS）、欧盟联合研究中心理事会-健康、消费者和标准物质（EU JRC Directorate F-Health, Consumers and Reference Materials）和国际原子能机构（IAEA）研制的标准物质在相关领域得到了较为广泛的应用。中国计量科学研究院（NIM）作为被 CIAAW 认可的十二家研制单位之一，在过去二十多年里，采用校正质谱法先后研制了 13 种元素的 60 种同位素标准物质（见表 25-4），包括天然丰度组成的同位素标准物质和浓缩同位素标准物质，以满足同位素丰度组成分析以及利用同位素稀释法开展的元素含量分析研究。

表 25-3 同位素标准物质的资源信息列表

研究机构	网址
美国国家标准与技术研究院（NIST，早期的 NBS）	http://www.nist.gov/srm/
欧洲标准物质与测量技术研究院（IRMM）	http://irmm.jrc.ec.europa.eu/
国际原子能机构（IAEA）	http://nucleus.iaea.org/rpst/
美国地质调查局（USGS）	http://isotopes.usgs.gov/

续表

研究机构	网址
国际海洋物理科学协会（IAPSO），OSIL	http://www.osil.co.uk/
德国联邦材料研究与测试中心（BAM）	http://www.bam.de
欧洲标准物质（ERM）	http://www.erm-crm.org/
加拿大国家研究委员会（NRC）	http://www.nrc-cnrc.gc.ca/
国际地球分析师协会（IAGeo）	http://www.geoanalyst.org/ http://9zdip.w4yserver.at/products_iageo.html
美国新布伦士威克实验室（NBL）	http://www.nbl.doe.gov
法国原子能和替代能源委员会（CEA）	http://www.eurisotop.com/
中国计量科学研究院（NIM）	http://en.nim.ac.cn/new-measurement-standards http://www.ncrm.org.cn/English/Home/Index.aspx

更多资源可访问：http://www.rminfo.nite.go.jp/english/link/link2.html。

表 25-4　NIM 研制的同位素标准物质汇总表

标准物质编号	标准物质名称	标准物质特性量
GBW 04601~04611	钐同位素溶液标准物质	
GBW 04447~04457	硒同位素溶液标准物质	
GBW 04465~04475	锌同位素丰度比溶液标准物质	
GBW 04612~04622	镉同位素溶液标准物质	
GBW 04623	铱同位素溶液标准物质	同位素丰度比
GBW 04504	钼同位素溶液标准物质	
GBW 04505	^{97}Mo 浓缩同位素溶液标准物质	
GBW 04506	^{100}Mo 浓缩同位素溶液标准物质	
GBW 04503	天然丰度硅同位素标准物质	
NIM-RM 2707	铜同位素溶液标准物质	
GBW 04440	钕同位素丰度标准物质	
GBW 04444	铁同位素丰度标准物质	同位素丰度
GBW 04445	铁同位素丰度标准物质	
GBW 04446	铁同位素丰度标准物质	
GBW 04441	^{111}Cd 浓缩同位素稀释剂标准物质	
GBW 04442	^{207}Pb 浓缩同位素稀释剂标准物质	
GBW 04443	^{202}Hg 浓缩同位素稀释剂标准物质	同位素丰度及浓度
GBW 04462	^{54}Fe 浓缩同位素稀释剂标准物质	
GBW 04463	^{65}Cu 浓缩同位素稀释剂标准物质	
GBW 04464	^{67}Zn 浓缩同位素稀释剂标准物质	

表 25-4 所列同位素标准物质均为具有绝对量值的同位素标准物质，借助以上标准物质可以获得能溯源至国际基本单位制单位的同位素丰度比或者同位素丰度。但是在实际的应用过程中，对于有些同位素分析研究而言，同位素丰度比的相对变化值（δ）更有现实意义，也因此 CIAAW 推荐了部分元素的同位素参考标准物质（$\delta=0$），如表 25-5 所示。

表 25-5　部分 $\delta=0$ 参考标准物质列表

原子序数	元素	$\delta=0$ 参考标准物质	原子序数	元素	$\delta=0$ 参考标准物质
1	氢	VSMOW, SLAP	14	硅	NBS28
3	锂	LSVEC	16	硫	VCDT, IAEA-S-1
5	硼	NIST SRM 951a	17	氯	SMOC, NIST SRM 975
6	碳	VPDB, LSVEC	20	钙	NIST SRM 915a
7	氮	Air N_2, USGS32	24	铬	NIST SRM 979
8	氧	VPDB, VSMOW, SLAP	26	铁	IRMM-014
12	镁	DSM3	28	镍	NIST SRM 986

续表

原子序数	元素	δ=0 参考标准物质	原子序数	元素	δ=0 参考标准物质
29	铜	NIST SRM 976	47	银	NIST SRM 978a
30	锌	IRMM-3702	48	镉	NIST SRM 3108, BAM-1012
31	镓	NIST SRM 994	75	铼	NIST SRM 989
32	锗	NIST SRM 994	76	锇	IAG-CRM-4
34	硒	NIST SRM 3149	78	铂	IRMM-010
35	溴	SMOB	80	汞	NRC NIMS-1
37	铷	NIST SRM 984	81	铊	NIST SRM 997
38	锶	NIST SRM 987	82	铅	NIST SRM 981, ERMM-3800
42	钼	NIST SRM 3134	92	铀	NIST SRM 950a

25.2.2 高纯基准物质

在化学测量中，高纯物质（基准物质）作为化学成分量溯源的基础原料，其化学纯度直接影响着标准物质特性量值的准确度和不确定度。高纯基准物质作为化学计量溯源体系的源头，具有最高的溯源层级，是量值可以准确溯源到 SI 国际单位制（kg 与 mol）的重要一环[6]。以 Cu 元素含量测量为例，其化学成分分析的完整量值溯源链以及各个环节的目标不确定度如图 25-2 所示，Cu 溶液标准物质是采用纯物质（BAM-Y001）通过重量法制备而成，逐级往下量传。若要获得不确定度在 0.03% 的溶液标准物质，纯度测量的目标不确定度需在 0.01% 以内[7]。高纯物质的纯度一般以主成分的百分含量（%）或者 kg/kg 表示，在工业领域也常以 N（nigh）表示，例如 5N 表示纯度为 99.999%，5N7 表示纯度为 99.9997%。

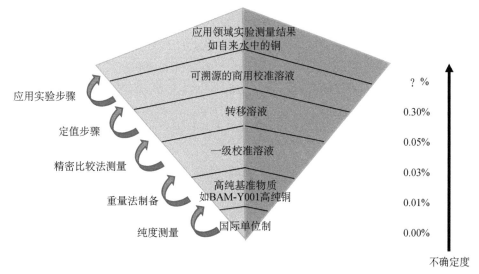

图 25-2　Cu 元素化学成分分析量值溯源链及不确定度[7]

化学计量基准作为国家化学测量量值溯源体系的源头，代表国家的最高测量水平。鉴于高纯基准物质在化学量值溯源链中的特殊作用，国际计量组织十分重视高纯基准物质纯度的测量，在国际化学计量发展规划中，高纯基准物质纯度高准确度测量是着重发展的方向之一。由于纯物质的缺乏，导致元素分析的溯源源头缺失，严重影响标准溶液量值的准确性。法国国家计量院（LNE）比较了多种校准溶液，发现 30% 与其标准值不一致[8]，这种不一致会造成使用不同标准溶液校准的测量结果缺乏一致性，因此研制纯度测量方法以及高纯基准物质，建立不同元素的高纯溯源基准迫在眉睫。

25.2.2.1 高纯基准物质的制备与提纯

在高纯基准物质研制中，为了满足纯度测量不确定度要求（<0.01%），一般需要对现有的候选材料进行提纯，降低杂质的含量水平。这是因为材料越纯，采用杂质扣除法测量纯度时，纯度的测量不确定度越容易达到目标要求（<0.01%）。不同金属的提纯工艺以及提纯前后纯度如表25-6所示。

表25-6 不同金属的提纯结果

材料	提纯前纯度*	提纯后纯度*	扣除杂质*	提纯工艺	年份	参考文献
Zn	4N5	6N5	16	真空蒸馏	2010	[9]
Ga	5N2	7N2	24	区域熔融	2009	[10]
Mg	1N5	4N	4	电子提纯	2016	[11]
Hf	2N7	5N	18	氢等离子体熔炼	2011	[12]

*由于纯度计算只是扣除部分杂质，此处纯度并不是其总纯度（total purity），只用于展示提纯效果。

25.2.2.2 高纯基准物质纯度定值方法

1. 纯度定值方法概述

高纯基准物质的纯度测量方法主要有主含量分析法与杂质扣除法。主含量分析主要有滴定法（库仑滴定测量法、EDTA容量法）、沉淀重量法等。对于主含量测量方法，由于其测量精度较差，从而使得纯度测量的不确定度在$10^{-3} \sim 10^{-4}$量级（恒电流库仑滴定法测量化合物的不确定可以到10^{-5}量级，但不适用于高纯金属），因此难以满足99.99%以上的高纯物质的纯度测量需求。根据质量平衡法，高纯物质的纯度还可以通过准确分析杂质的含量，采用杂质扣除法间接计算获得，公式如式（25-1）所示。

$$P = \left[1 - \sum(m_1 + \cdots + m_n)\right] \times 100\% \tag{25-1}$$

式中，P为纯度，m为杂质元素的质量分数，n为总的杂质元素种类。

根据属性及其测量手段的不同，可以将元素周期表中从H到U共92种元素分为4类，如图25-3元素属性以及测量手段的分类所示。目前，德国联邦材料研究与测试研究院（BAM）[13]与中国计量科学研究院（NIM）[14]在研制纯度标准物质时，均根据该分类方法对特定的杂质选用特定的方法进行测量。

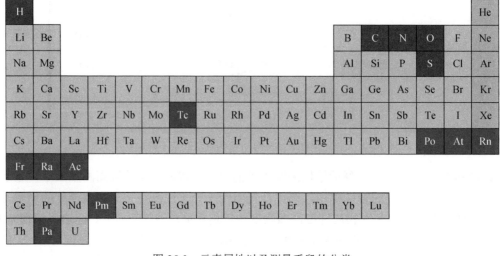

图25-3 元素属性以及测量手段的分类

1) 金属元素与类金属元素（73种）

金属与类金属元素是数量最多的元素，主要采用 GD-MS、HR-ICP-MS、Q-ICP-MS、FI-ICP-MS、ICP-OES、ET-AAS 等，并配合采用 LA-ICP-MS 对难溶元素（铂族元素）的消解情况进行核验。

2) 非金属元素（5种）、惰性气体元素（5种）与放射性元素（9种）

非金属元素（气体元素）包括 C、H、O、N、S，其检测主要采用载气热抽取的方法进行检测。惰性气体元素包括 He、Ne、Ar、Kr、Xe，由于其含量极少，也可以根据理论进行估算。放射性元素包括 Tc、Po、At、Rn、Fr、Ra、Ac、Pm、Pa，常依据放射性元素的来源对其含量进行合理评估。

为了加强纯物质纯度测量方法研究，2005年国际计量委员会（CIPM）物质的量咨询委员会（CCQM）组织了高纯 Ni 的纯度的实验室间比对（CCQM-P62），为了方便起见，选了6种易测量元素（Ag、Al、Cu、Fe、Pb、Zn）含量总和作为评估其纯度的依据，各个元素的含量在亚 mg/kg 量级。比对结果表明[15]，7家实验室的测量结果（6种元素总和）在 3.1~25.4 mg/kg 之间，在8倍范围内一致，而要实现纯度的结果等效一致，需要使得杂质测量结果在 30% 范围内一致，可见本次比对结果远远不能满足要求，纯度的准确测量任重而道远。2007~2008年，CCQM 又组织了新一轮的实验室间比对（CCQM-P107），测量高纯 Zn 中 6 种元素（Ag、Bi、Cd、Cr、Ni、Tl）的总和，测量结果[16]在 30% 不确定度范围内等效一致。2012年，CCQM 组织了高纯 Zn 中 6 种（Ag、Al、Cd、Cr、Ni、Tl）元素总和的国际比对（CCQM-K72），测量结果[17]在 30% 目标不确定度范围内获得等效一致，国际比对 CCQM-K72 结果[17]如图 25-4 所示。2014年，CCQM 组织了高纯 Zn 纯度的国际比对（CCQM-P149）考察各个国家实验室在全元素测量的能力与水平[18]，这也是第一次国际组织的总纯度测量比对，结果表明虽然有个别元素（特别是非金属）一致性较差，但对于纯度而言，在 0.011% 不确定度范围内，纯度的测量结果等效一致。

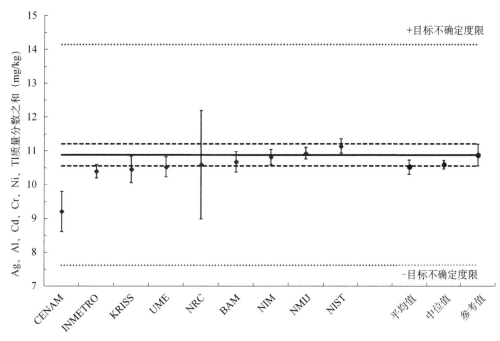

图 25-4　国际比对 CCQM-K72 结果[17]

2. 痕量杂质的原子光谱测量方法

为了满足纯度的测量不确定度要求（<0.01%），除了对候选材料进行提纯外，还需要降低杂质的测量不确定度。目前，占据大多数的金属与类金属杂质元素（73种）的测量主要采用原子光谱的方

法，一是基于溶液进样的 ICP-MS 等方法，二是基于固体直接进样的 GD-MS 等方法。

ICP-MS 相对于 ICP-OES 等光谱技术具有更高的灵敏度，因此更加适用于高纯材料的痕量杂质分析。由于水、酸空白的影响，大部分元素的 HR-ICP-MS 的检出限在仍在 $10^{-9}\sim10^{-5}$ 量级[19]，采用流动注射（FI）、基体分离的方式可以降低检出限，但仍不及固体直接方法。

GD-MS 由于灵敏度高、取样代表性好（溅射束斑一般在 5～8 mm），所以是目前高纯材料痕量杂质测量最有效的工具。作为固体直接分析手段，由于省去了样品稀释的过程，避免了元素的污染，从而使得 GD-MS 的检出限比 HR-ICP-MS 低 1～3 个数量级，如图 25-5 所示，大大提升了高纯物质的纯度测量水平。

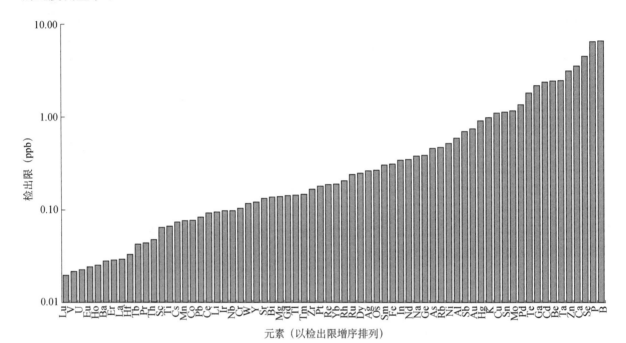

图 25-5　GD-MS 测量各元素的检出限

对于 GD-MS 而言，由于缺乏基体匹配的标准物质，准确的定量分析困难，测量结果溯源不清晰，杂质测量的不确定度较大（200%[14]），与其他方法的测量结果可比性差。值得庆幸的是，在高纯材料纯度分析中，当杂质含量在 μg/kg 量级时，即使杂质分析的不确定度较大（200%），对纯度的准确测量影响也不大，满足纯度分析要求。然而当杂质含量较高时（0.1 mg/kg<x<1 mg/kg），若要实现纯度的准确测量（不确定度<0.01%），杂质的测量不确定度需要控制在 30%以内[15-16]。

近年来，NIM 采用高纯 Cu 粉末为基体开发了粉末掺杂法与高温熔融法制备校正样品[20]，对不同元素的相对灵敏度因子（RSF）进行了校正。粉末掺杂法是往高纯金属粉末中掺杂标准溶液，经过烘干、摇匀、压片后制备得到校正样品，然后校正 GD-MS 仪器，获得各个元素在特定基体中的 RSF。熔融法是在对掺杂的粉末进行熔融，可以获得均匀性更好的校正样品，但由于挥发以及偏析，仅仅适用于部分元素的校正。采用基体匹配的 RSF 校正，杂质测量的不确定度可以降低到 10%以内。

25.2.2.3　高纯基准物质研究现状

国际上现有的高纯金属纯度标准物质见表 25-7。早期，由于技术的限制以及认识的不足，计量院等单位在研制高纯基准物质时，只是分析少量金属杂质，而且多是半定量分析。实际上，由于在高纯材料中，非金属杂质往往占据主要成分，因此这种局部纯度（partial purity）并不能直接替代满足计量学要求、考虑所有杂质元素的总纯度（total purity）。例如，高纯Cu[21]（BAM-B-primary-Cu-1），商家提供的纯度为99.9999%，经过全杂质扣除计算其总纯度为99.944%±0.017%，显然两者具有很大的差异。杂质元素种类测量不完全使得其纯度不满足计量学意义上的纯度定义，导致纯度量值不准确或不确定度较大，从而通过量值传递过程降低了相关标准物质的准确度。2004年，BAM首次采用全杂质扣除法研制了高纯金属纯度基准物质高纯Cu（BAM-Y001），并于同年相继发布了另外9种（Fe、Si、Pb、Sn、W、Bi、Ga、NaCl、KCl）纯度基准物质[6]，纯度的不确定度在10^{-5}水平，比我国20世纪90年代采用主成分测量的不确定度（10^{-4}）低一个数量级。在"十三五"期间，NIM先后研制了高纯Cu、Si、Au、Ag、Pt、In等纯度标准物质，纯度>99.999%，纯度的不确定度在10^{-6}，为目前国际最高水平。

表25-7 国际上现有的高纯金属纯度标准物质

信息来源	产品编号	名称	纯度值	杂质种类	测量方法	年份
IRMM	IRMM-521B	Ni (0.5 mm wire)	>99.99%	19	NAA，ICP，AAS	1988
	IRMM-521R	Ni (0.1 mm foil)	>99.99%	75	HR-ICP-MS，NAA	1988
	IRMM-522A	Cu (0.1 mm foil)	>99.995%	30	SSMS，NAA	1993
	IRMM-522B	Cu (1.0 mm foil)	>99.995%	30	SSMS，NAA	1993
	IRMM-522C	Cu (0.5 mm wire)	>99.995%	30	SSMS，NAA	1993
	IRMM-522D	Cu (1.0 mm wire)	>99.995%	30	SSMS，NAA	1993
	IRMM-523A	Al (0.1 mm foil)	>99.999 5%	21	SSMS，NAA	1990
	IRMM-523B	Al (1.0 mm foil)	>99.999 5%	21	SSMS，NAA	1990
	IRMM-523C	Al (1.0 mm wire)	>99.999 5%	21	SSMS，NAA	1990
	IRMM-524A	Fe (0.1 mm foil)	>99.996%	11	SSMS，NAA	1993
	IRMM-524B	Fe (0.5 mm wire)	>99.996%	11	SSMS，NAA	1993
	IRMM-525A	Nb (0.02 mm foil)	>99.98%	17	SSMS，NAA，ICP-MS	1990
	IRMM-525C	Nb (0.5mm wire)	>99.98%	17	SSMS，NAA，ICP-MS	1990
	IRMM-526A	Nb (0.02 mm foil)	>99.994%	14	SSMS，NAA，ICP-MS	1990
	IRMM-526B	Nb (0.1 mm foil)	>99.994%	14	SSMS，NAA，ICP-MS	1990
	IRMM-526C	Nb (0.5mm wire)	>99.994%	14	SSMS，NAA，ICP-MS	1990
	IRMM-529	Rh (0.05 mm foil)	>99.97%	65	SSMS，NAA，ICP-MS	1999
	IRMM-531A	Ti (0.1 mm foil)	>99.94%	19	GD-MS，ICP-OES，ICP-MS，INAA	1999
	IRMM-531B	Ti (0.5 mm foil)	>99.94%	19	GD-MS，ICP-OES，ICP-MS，INAA	1999
BAM	BAM-Y001	High purity copper	0.999 97±0.000 01（kg/kg）	91	ICP-MS，ET-AAS，CGHE，PAA，等	2004
	BAM-Y002	High purity iron	0.999 86±0.000 04（kg/kg）	91	ICP-MS，ET-AAS，CGHE，PAA，等	2004
	BAM-Y003	High purity silicon	0.999 91±0.000 07（kg/kg）	91	ICP-MS，ET-AAS，CGHE，PAA，等	2004
	BAM-Y004	High purity lead	0.999 92±0.000 06（kg/kg）	91	ICP-MS，ET-AAS，CGHE，PAA，等	2004
	BAM-Y005	High purity tin	0.999 91±0.000 06（kg/kg）	91	ICP-MS，ET-AAS，CGHE，PAA，等	2004
	BAM-Y006	High purity tungsten	0.999 81±0.000 10（kg/kg）	91	ICP-MS，ET-AAS，CGHE，PAA，等	2004
	BAM-Y007	High purity bismuth	0.999 90±0.000 07（kg/kg）	91	ICP-MS，ET-AAS，CGHE，PAA，等	2004
	BAM-Y008	High purity gallium	0.999 92±0.000 07（kg/kg）	91	ICP-MS，ET-AAS，CGHE，PAA，等	2004
沈阳冶炼厂	GBW02751	金纯度标准物质	99.994%±0.037%	主含量测定	恒电位库仑滴定法	1994
	GBW02752	银纯度标准物质	99.994%±0.044%	主含量测定	恒电位库仑滴定法	1994
	GBW02753	铜纯度标准物质	99.99%±0.031%	主含量测定	恒电位库仑滴定法	1994
NIM	GBW02142	高纯铜纯度标准物质	99.999 6%±0.000 2%	91	GD-MS，ICP-MS，CGHE	2016
	GBW02143	高纯硅纯度标准物质	99.999 5%±0.000 3%	94	GD-MS，ICP-MS，CGHE	2016

续表

信息来源	产品编号	名称	纯度值	杂质种类	测量方法	年份
NIM	GBW02793	高纯金纯度标准物质	99.999 5%±0.000 2%	94	GD-MS，ICP-MS，CGHE	2019
	GBW02794	高纯银纯度标准物质	99.999 7%±0.000 2%	94	GD-MS，ICP-MS，CGHE	2019

AAS：原子吸收法；CGHE：载气热抽取-红外/热导法；ET-AAS：电热蒸发-原子吸收光谱；GD-MS：辉光放电质谱；HR-ICP-MS：高分辨电感耦合等离子体质谱；ICP-MS：电感耦合等离子体质谱；ICP-OES：电感耦合等离子体原子发射光谱；NAA：中子活化分析；PAA：光子活化分析。

25.2.2.4 高纯基准物质技术发展趋势

在高纯基准物质研究中，随着人们认识的发展以及测量手段的进步，高纯基准物质呈现出如下特点：①纯度分析由直接分析（主成分分析）逐步发展为间接分析（杂质扣除）；②杂质测量由部分测量逐步发展为全杂质测量；③测量方法越来越多样化，并且检出限越来越低。原子光谱技术在高纯材料纯度分析以及纯度基准物质研制中占据着不可替代的地位。随着技术的进步，检出限会越来越低，测量不确定度会越来越低，更好地服务于高纯基准物质的研制；而反过来，高纯基准物质将会保证溶液标准物质的准确性，会更好地保证原子光谱的准确测量。

25.2.3 基于原子光谱技术的生命分析用标准物质

25.2.3.1 原子光谱技术与生命分析

原子光谱的主要分析对象是金属元素。金属元素是生命存在的重要物质基础之一。金属生物分子（metallobiomolecules）如金属蛋白（metalloproteins）、金属酶（metalloenzymes）以及其他含有金属的生物分子以及各种游离的金属离子如K^+、Na^+、Ca^{2+}、Mg^{2+}等在人类生命活动中具有重要作用。过去许多与生物体系有关的问题是由元素的总量来说明的，现在人们越来越认识到生物体中元素化学形态的重要性，如氧化态、配体性质以及分子结构。发展生物体中金属及其化合物的新形态分析技术和方法，以及金属元素的形态与生物活性分子相互作用研究的方法学至关重要[22]。由于具有较强的分离能力和灵敏特效的检测能力，色谱-光谱联用技术已成为金属组学研究的主要方法。由于生物体中的金属和类金属与生物配体通常形成难挥发的共价化合物，故 HPLC 和电泳是常用的分离技术。HPLC 适用于小分子化合物和基体较简单的样品的分离，对于较复杂的样品可用多维色谱分离，而排阻色谱、电泳、毛细管电泳、毛细管电色谱和聚丙烯酰胺凝胶电泳适用于大分子化合物和复杂样品的分离。ICP-MS 由于具有高的灵敏度且分析快速，是一种理想的检测手段，而 ES-MS 则被用于表征分子结构。

大多数元素溶液基体标准物质以及复杂基体标准物质的主要定值技术，都是通过原子光谱类方法如 AAS、ICP-OES、ICP-MS 等完成。除了对金属元素标准物质进行定值外，近年来，基于原子光谱（如 ICP-MS）的生物大分子定量技术正在成为生命科学中大分子测量的重要手段。高分辨电感耦合等离子质谱（HR-ICP-SF-MS），具有较高的灵敏度和分辨率，可显著降低等离子体中形成的多原子离子干扰，使超痕量分析降低到 fg/g，被认为有希望成为蛋白质定量的新工具。与生物质谱不同的是，ICP-MS 的离子源工作在约 7000 K 高温条件，进入 ICP 蛋白质分子的化学键很容易被打开而成为原子，之后原子被电离为一价离子进入质谱仪。所以，在 ICP-MS 不但可以分析元素周期表中的几乎所有元素，而且这种元素的特征响应（element-specific response）与其原子所处的分子环境无关。这意味着在 ICP-MS 中，如果定量蛋白质中某一元素，可以使用这种元素的任何形态作为标准溶液。在 ICP-MS 可以测量的所有元素中，硫和磷是最适合作为蛋白质定量分析的内标元素。这是由于：一方面，硫磷是蛋白质中常见的元素，多数蛋白质都含有半胱氨酸或甲硫氨酸这样的含硫氨基酸；另一方面，硫原子多以共价键稳定地存在于蛋白质分子中。如果某种蛋白质已经由生物质谱鉴定，或者这种蛋白质分子的氨基酸序列和其中含有的硫原子数已知，那么就可以通过 HPLC/ICP-MS 联用系统分离蛋白质，并通过在线测量其中的硫而对蛋白质绝对定量。将灵敏、准确的同位素稀释技术应用于蛋白质定量分析，无疑将为解决定量蛋白质组学研究提供一条理想的途径。基于硫元素的定量方法和基于肽段酶解法具有以下几方面优势。一方面，可以避免上述肽段酶解法中酶切条件优化困难、寻找目标肽段复杂，以及目标肽段合成和浓度认证存在误差的不足。另一方面，基于硫元素的分析方法比肽段法更直接追踪到 SI 单位，前者能够通过多肽中的硫含量溯源至 SI，而后者中用作定量的肽段首先要进行合成和浓度认证。因为硫元素标准物质比实验中表征的肽段的不确定性更小，因此，基于硫元素的分析方法可以使不确定度最小化[23-24]。

25.2.3.2 原子光谱技术在国际比对及标准物质研制中的应用

由于 ICP-MS 技术在生物大分子测量方面具有独特的优势，该方法在国际计量领域也是获得了广泛的关注。欧洲计量合作组织（EURAMET）近年来组织了多个金属组学领域相关的欧盟计量项目，包括欧盟计量联合研究计划（EMRP）（Metrology for metalloproteins），以及美国国家标准与技术研究院（NIST）项目"基于质谱的硒蛋白鉴定与表征"（Selenium protein identification and profiling by mass spectrometry）等。同时，国际上许多先进国家计量院包括英国政府化学家实验室（LGC）、德国联邦物理技术研究院（PTB）、韩国标准科学研究院（KRISS）等也都正在开展基于同位素稀释的 ICP-MS 蛋白质定量研究。此外，国际计量领域在相关方法研究的基础上，还将该系列从元素到蛋白质的定量技术应用于国际比对中，其中在国际计量局组织的 CCQM-P191 胰岛素纯度国际比对（图 25-6）[25-26]以及 CCQM-K 115.b/P55.2 缩宫素纯度国际比对（图 25-7）中，中国计量科学研究院采用基于硫元素测量的 ID-ICP-MS 方法实现了胰岛素和缩宫素纯度的准确测量，比对结果均位于十几个国际计量院（包括 NIST、PTB、LGC 等）的中间位置，达到国际等效一致，表明 NIM 在采用 ID-ICP-MS 测量生物大分子方面的国际先进水平，可以保证该技术应用于标准物质定值的准确性和可靠性。

在定值技术充分研究的基础上，中国计量科学研究院针对阿尔茨海默病（AD）重要生物标志物 β 淀粉样多肽无法溯源的现状，采用氨基酸水解同位素稀释质谱法和基于硫元素的同位素稀释质谱法作为两种独立参考方法[27]，研制了 $A\beta_{40}$ 和 $A\beta_{42}$ 两种标准物质（NIM-RM 5303～5304），可用于 AD 症诊断中 Aβ 检测的溯源标准，还可用于该标志物检测方法的验证和实验室质量控制等领域。

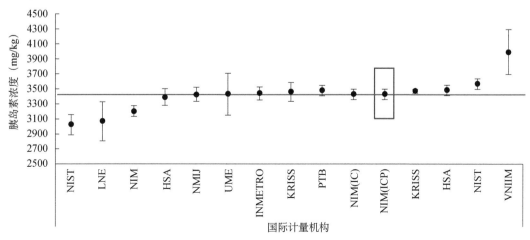

图 25-6 胰岛素纯度测量国际比对（CCQM-P191）[24]

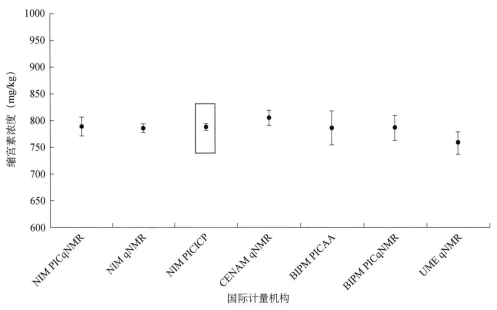

图 25-7　缩宫素纯度测量国际比对（CCQM-K115.b/P55.2）[28]

25.2.4　纳米颗粒标准物质

25.2.4.1　纳米颗粒标准物质的作用

纳米材料一般指一维或多维尺度在 1~100 nm 范围内的材料，可分为零维、一维和二维纳米材料。纳米颗粒为零维纳米材料，即三维受限于纳米尺度的材料。根据来源可分为天然纳米颗粒和人工纳米颗粒。在过去几十年，因纳米尺度对物质产生的特殊性质，开发出大量新产品、新材料，导致越来越多的人工纳米颗粒进入消费品、食品及工业产品中。目前，针对纳米颗粒在生物体中作用机理尚不明确，其经过食品、环境进入生物体内的迁移转化是否安全尚需进一步研究，主要技术难点是缺乏有效的食品、环境及生物体中痕量纳米颗粒的准确定量和表征技术。因此，发展稳健的复杂基体中纳米颗粒的定量、追溯技术和研制标准物质成为亟需解决的任务。

纳米颗粒标准物质在定量和表征方面的作用如下[29,30]：①校准仪器，以建立被测信号与待测量之间的关系。要求标准物质是近理想化的分散液或易于分散的粉末、纳米颗粒是球形且尽可能是单分散的。②建立校准标准，如在单颗粒-电感耦合等离子体质谱法（sp-ICP-MS）检测时，以标准物质尺寸为标准建立粒径与响应信号之间的关系，可消除离子态与纳米颗粒之间离子化效率差异造成的影响。③方法验证，以评定检测技术性能、保证测量方法的稳健性，要求不同类型的纳米颗粒需采用相近尺寸的标准物质进行验证。④风险评估，在无纳米级标准物质的情况下，合适的标准物质可作为体外和体内毒性试验的对照以研究纳米材料导致的毒性机制。⑤计量溯源，根据有证标准物质确定检测结果的可靠程度并保证测量结果的溯源性。

25.2.4.2　纳米颗粒标准物质研究现状

Warheit 等[31]按照相对优先的顺序，以纳米颗粒性质对毒性的影响进行排序：粒径及粒径分布和表面积>结晶度>聚集态>组成和表面涂层>表面反应性>合成方法>纯度。纳米颗粒作为一种"新"物质，标准物质的研制需可靠的分析方法来评估均匀性、稳定性及通过计量有效的方法获得可靠值，检测技术和方法的可靠性则需经过有证标准物质验证。因此，需以逐步迭代方式进行标准物质的研制和方法验证[32]，即以"不完美"标准物质作为第一步，用于相关检测方法的开发和验证，从而更精准地检测和表征有证标准物质候选物。

美国 NIST 根据医疗和生物领域对于金纳米颗粒标记的生物分析发展需求，于 2008 年成功研制了世界上第一批纳米金标准物质（RM 8011、8012 和 8013），粒径分别为 10 nm、30 nm 和 60 nm。分别采用原子力显微镜（AFM）、扫描电镜（SEM）、透射电镜（TEM）、微分迁移率分析（DMA）、动力学光散射（DLS）、小角度 X 射线散射（SAXS）六种技术测定了颗粒粒径，分别列出了不同测定方法获得的粒径结果作为标准值，这也反映了纳米尺度下 AFM、SEM 和 TEM 等方法的计量挑战。2015 年，美国 NIST 采用包衣和冷冻干燥技术，将 PVP-纳米银粉饼在真空下密封，克服了纳米银在氧气或高湿环境中易氧化释放银离子的问题，保证了纳米银标准物质（RM 8017）的稳定性。该标准物质提供了 DLS 测量的平均粒径和 ICP-OES 获得的质量浓度参考信息而非标准值，虽具有相关的不确定度，但并不包括所有不确定度来源，也从另一侧面反映了多种方法之间尚缺乏足够的一致性。为制定有效的风险评估和管理计划，美国国家纳米技术计划倡议纳米环境、健康及安全研究战略，NIST 作为牵头机构开发了硅纳米颗粒标准物质（RM 8027），用于评估纳米材料生物反应的体外分析及实验室间比对。

2008 年，欧盟联合研究中心理事会-健康、消费者和标准物质（EU JRC Directorate F-Health, Consumers and Reference Materials）研制了纳米二氧化硅标准物质，用于验证 DLS 和离心液体沉降（CLS）两种粒度测量方法以及候选二氧化硅胶体的均匀性、稳定性考察。定值方法为 DLS、CLS、TEM/SEM，且每一种检测方法都经实验室间比对，以确保遵循指定检测程序可用于不同实验室获得重复性的测量结果。经 30 个具有能力的实验室参与，以获得的等效球形粒径对颗粒跟踪分析（PTA）和非对称流场流分级（AF4）进行方法研究。除金属及金属氧化物纳米颗粒标准物质外，有机纳米颗粒中聚苯乙烯球最为普遍，中国计量科学研究院（NIM）、日本国家计量研究所（NIMJ）、NIST 均有生产。

目前尚无复杂基体纳米颗粒标准物质，原因在于直接将纳米颗粒加入到基体中可能改变颗粒特性，这也是基体标准物质的研制难点。JRC 目前正在研究复杂基体中纳米颗粒标准物质的可行性，如开展液体食品中基于电子显微镜技术和 FFF-ICP-MS 的二氧化硅粒径和浓度测定方法、中子活化分析和 TEM、sp-ICP-MS 对鸡肉中总银浓度和纳米银的定量和表征方法研究[33]，但在标准物质稳定性、开发足够精确的均匀性和稳定性评估方法方面仍然存在挑战。

纳米颗粒标准物质研制中最主要的困难是维持其稳定性和均匀性，稳定性不仅指颗粒的尺寸，也包括颗粒数量浓度，即在储存过程中纳米颗粒不发生团聚、溶解等情况，且分散均匀。随着世界各国及区域组织对纳米材料潜在风险评价中技术要求的进一步提高，常规的粒径和化学浓度（元素总量）检测已不能满足相关要求，还必须对纳米材料的颗粒数量浓度、粒径、粒径分布、存在状态等进行检测和表征，因此对标准物质多特性量值的需求进一步增加。

25.2.4.3 纳米颗粒定量和表征中的原子光谱技术

单颗粒电感耦合等离子体质谱技术是近年出现的纳米颗粒粒径、数量和质量浓度、聚集状态的分析方法，其结合了形态鉴别和定量分析双重优势，具有灵敏度高、直观性强的特点。自 Degueldre 等[34]提出单颗粒分析理论以来，已用于多种纳米颗粒的测定和表征。sp-ICP-MS 中最常使用的是四极杆质谱，该技术相比其他纳米分析技术具有极低的检出限和较高灵敏度。新一代 ICP-MS 的驻留时间和采集频率分别达到 10 μs 和 10^5 Hz，能够在高离子浓度下更有效进行基线分离，降低检出限。三重四极杆电感耦合等离子体质谱因具有双质量选择能力，有助于解决基质干扰问题。扇形场电感耦合等离子体质谱的应用，则解决了多原子离子、同质异位素的干扰问题，有利于环境样品中纳米颗粒的测定。多接收电感耦合等离子体质谱和飞行时间电感耦合等离子体质谱能够进行多元素分析，已用于对核壳结构纳米材料的分析，根据纳米颗粒的特殊指纹信息可确定其来源。

当纳米颗粒存在多分散尺寸分布或存在高浓度离子、有机质等复杂基体时，常采用分离技术与ICP-MS联用相结合的方式[35]，如场流分级分离（FFF）、流体动力学色谱（HDC）、高效液相色谱（HPLC）、毛细管电泳（CE）等与ICP-MS联用。如FFF-ICP-MS已用于抗菌产品中Ag、防晒霜中TiO_2、咖啡奶霜中SiO_2以及环境水体中的纳米颗粒的检测。HDC对团聚体的结构破坏较小、保留因子几乎不受颗粒性质（尤其是表面涂层）的影响。HPLC、CE与ICP-MS联用可用于分析纳米颗粒在生物体内的转化过程与毒性机制。CE-ICP-MS能够根据纳米颗粒包覆材料或表面电荷对其进行区分。AF4-sp-ICP-MS可同时检测纳米颗粒质量和流体动力学直径，从而区分不同特性纳米颗粒，例如涂层厚度和聚集状态[36]。CE-sp-ICP-MS则能够对不同粒径包覆材料进行区分[37]。

为考察sp-ICP-MS对于纳米颗粒粒径和颗粒数量浓度测定的方法性能，JRC的Linsinger等[38]组织对不同实验室sp-ICP-MS测定纳米银粒径、质量和颗粒数量浓度的结果进行评估和考察，虽然实验室间再现性较差，但证明该方法适用于纳米颗粒的检测和定量。2015年，NIST的Montoro Bustos等[39]以纳米金标准物质为样品组织了实验室间比对，以评估sp-ICP-MS方法对粒径检测的重现性、精密度、准确度等参数，纳米金粒径的检测结果一致，但颗粒数量浓度的测定仍然存在较大差异。2017年，JRC的Weigel等[40]组织国际实验室间比对，证实了sp-ICP-MS对于鸡肉中纳米银的中值粒径和颗粒数量浓度检测的能力。为进一步评估sp-ICP-MS计量溯源和测定能力，2018年，国际计量局（BIPM）无机工作组（CCQM IAWG）组织了纳米金颗粒数量浓度的CCQM-P194研究型国际比对，其中50%的实验室采用sp-ICP-MS进行分析（图25-8）。中国计量科学研究院通过对传输效率计算方式、驻留时间、稀释试剂、进样方式等参数的优化，建立了基于sp-ICP-MS技术的金纳米颗粒数量浓度和粒径的准确测定方法[41]，颗粒数量浓度结果与比对参考值偏差<5%，测定结果等效一致，扩展不确定度<15%（$k=2$）。该比对首次采用化学手段进行纳米金颗粒数量浓度的国际比对，形成的检测和表征能力可为其他金属纳米颗粒的准确测定和表征、标准物质定值提供参考。

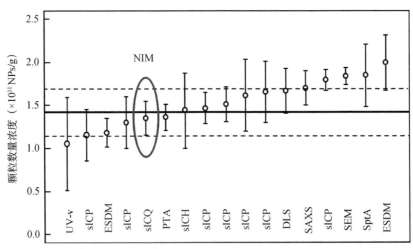

图25-8　CCQM-P194纳米金颗粒数量浓度比对结果

25.2.5　微区分析用标准物质

25.2.5.1　微区分析用标准物质的制备方法

1. 筛选组成均匀的天然矿物

该方法是通过筛选自然界中组成均匀、目标元素含量适中的天然矿物（锆石、橄榄石、金红石以及富含某些微量元素的石英等），通过均匀性检验和定值后，切割分发用作行业内标准物质，主要在地质研究领域使用。以锆石为例，理想中的标样应该满足：①同位素组成分布均匀；②具有较高的 U 元素含量（大于几十 μg/g 水平）；③初始 Pb 含量可忽略不计；④具有足够的样品量以供全球分发使用[42]。目前，研究中使用比较多的锆石标样有 91500[43]、Plesovice[44]、Mud Tank[45]、GJ-1[46]、M257[47]、TEMORA[48]、清湖[49]、蓬莱[50]等。天然标样的筛选制备困难，同时样品量是有限的，难以满足相关研究领域持续且日益增长的需求。

2. 样品熔融玻璃法

样品熔融玻璃法是目前在微区标物制备中应用最广的方法，其方法原理如图 25-9 所示：将样品在高温下变成液体，然后倒模冷却固化加工成型。该方法中熔融过程的存在使得添加元素更容易分布均匀，同时一次制备批量大，能满足标物大量制备的需求，制得的标物中元素含量的 RSD 一般在 5%~10%。但是，高温处理过程中使得该方法不适用于易挥发元素和还原性元素的标物制备；同时，温度的剧烈变化会导致同位素分馏，制得的样品组成和预期组成会有较大的偏差；另外，对于一些难熔矿物，尤其是花岗岩，采用该方法制备时难以完全溶解，导致相应的 Zr、Hf 等元素的分布不均匀[51]。

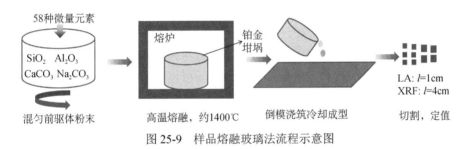

图 25-9　样品熔融玻璃法流程示意图

文献中对于该方法的实验优化主要有三个方向：①提高制得样品的均匀性，代表是吴石头等[52]通过二次粉碎熔融的制备过程，显著提高了制得玻璃片中 Zr、Hf 等难熔元素的均匀性，RSD%优于 10%。②降低样品制备的温度，减少挥发性、还原性元素的损失。Monsels 等[53]通过使用硼酸锂碱溶（样品：溶剂=1∶4）后熔融的办法，在 1100℃条件下制得了铝土矿玻璃标物，实现了矿物中 30 种元素的测量。③改善熔融方法，中国地质大学（武汉）胡兆初等[54]利用高能激光作为能源熔融硅酸盐矿物，在几十 ms 时间内实现了样品的熔融冷却，避免了化学助剂和熔融容器的使用。

3. 样品粉末压片法

样品粉末压片法是制备 XRF 和 LA-ICP-MS 微区分析标物常用的方法，具体过程如图 25-10 所示：先将样品研磨成微米级或纳米级的粉末，然后通过高压压片机将固体粉末压制成型。该方法操作简便，便于添加微量元素或稀释剂，制得的标物最大限度地保留了基体样品的组成，克服了高温熔融导致的样品损失与同位素分馏。该方法的不足体现在：①压制成型的标物其表面结构强度和天然矿物基体有很大差别，激光剥蚀效果会有很大差别；②掺杂元素的均匀性如何保证是一个难题；③在样品粉碎及压制成型过程中如何有效避免污染也是一个难题[51]。

粉末压片法的关键有两个：一个是样品粉末的制备，另一个是压制成型。粉末颗粒的粒径是影响压制效果、激光剥蚀行为、元素/同位素分布均匀性唯一和最重要的影响因素，10 μm 以下的样品颗粒大小是保证微区分析均匀性的前提[55]。样品压制成型效果会影响制得标物的分析性能，根据矿物的三相图，在高温高压成矿条件下对样品粉末进行压制成型，从而尽可能地保证制得的标物与天然矿物的基体匹配，这应该是未来粉末压片法制备微区标物的发展方向。

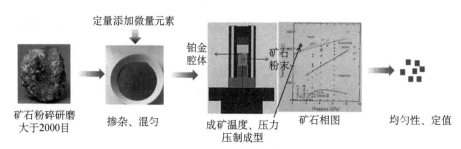

图 25-10　样品粉末压片法流程示意图

25.2.5.2　微区分析用标准物质均匀性检验与定值方法

1. 微区分析标准物质均匀性检验

研制微区分析标准物质时，首先得保证所制备标准物质的均匀性，这里的均匀性应包含两个方面：一方面是样品与样品之间的均匀性，即不同片之间的均匀性；另一方面是微区均匀性，即单片样品上微米级区域内的均匀性。关于微区分析标物的均匀性判定标准，目前相近的只有岩石粉末均匀性判定的国家标准，即 JJF 1343—2012《标准物质定值的通用原则及统计学原理》，根据微区标物研制的特点制定相适应的样品微区均匀性判定的国家标准是目前行业发展亟需的[56]。

2. 微区标准物质定值方法

经均匀性验证后的微区标物候选物的定值方法主要有两种：一种是固体消解、分离后以溶液方法测整体含量，根据待测元素的种类、含量和同位素组成可以选择合适的仪器进行测量。该方法的优点是基体简单、测量精度高、重复性好，但测得的是一个平均值，不能反映样品内部的元素分布，因此只能用在经均匀性检测后的候选物定值。另一种标物定值方法是直接固体分析，代表有激光剥蚀-电感耦合等离子体质谱（LA-ICP-MS）、二次离子质谱（SIMS）、电子探针分析（EPMA）等。该方法能反映元素的空间分布，但受到的干扰较多，部分元素的检出限与测量重复性不能保证，测量结果的溯源性难以保证。

25.2.5.3　微区分析用标准物质现状

1. 国际上微区分析用标准物质现状

1）美国 NIST 系列玻璃片标准物质

NIST 61x 系列玻璃片标准物质是目前微区分析中使用最广泛的标准物质，它是由美国国家标准与技术研究院（NIST）研制，主要包含 NIST 610~616 四种不同浓度梯度的熔融玻璃片[57]。四种玻璃片的主量元素一致，均为 SiO_2（~70%）、Na_2O（~13.5%）、CaO（~12%）和 Al_2O_3（~2%），添加的微量元素有 66 种，浓度从 NIST 610 到 NIST 616 依次递减，分别约为 400 mg/kg、40 mg/kg、1 mg/kg 和 0 mg/kg。其中，NIST 610 和 NIST 612 中元素含量较高、元素种类较全、分布均匀，是目前最常用的微区分析标准物质。

2）美国 USGS 系列玻璃片标准物质

USGS 系列玻璃片标准物质是由美国地质调查局（USGS）研制[58]，包含 BCR-2G、BHVO-2G、BIR-1G、NKT-1G、TB-1G、GSC-1G、GSD-1G 和 GSE-1G 等。这些玻璃片标准物质除了 NTK-1G 是霞石岩基体的，其余均是玄武岩基体的。GSC-1G、GSD-1G 和 GSE-1G 三个是系列合成玄武岩玻璃标准物质，添加微量元素种类为 52 种，含量分别约为 3 mg/kg、30 mg/kg 和 300 mg/kg。其余的标准物质都是由 USGS 相应的基体成分标准物质经高温熔融冷却制得。USGS 的系列玻璃标准物质在地质领域广泛使用。

3）欧盟玻璃片标准物质

欧盟主要有两种玻璃片标准物质，MPI-DING 系列玻璃片标准物质由德国马普化学研究所的 Dingwell 博士研制[59]，包含 6 种岩石基体，分别为玄武岩（KL2-G、ML3B-G）、安山岩（StHs6/80-G）、科马提岩（GOR128-G、GOR132-G）、流纹岩（ATHO-G）、石英闪长岩（T1-G）和橄榄岩（BM90/21-G）；BAM 系列包含 BAM-S005-A 和 BAM-S005-B 是由德国联邦材料研究与测试研究所（BAM）研制，是碱石灰基质玻璃，主要用作 X 射线荧光（XRF）的校准[60]。

4）美国 USGS 粉末压片标准物质

USGS 粉末压片标准物质包含碳酸岩、硫化物和磷酸岩三种基体，因为这三种基体在自然界难以找到含量和均匀性满足要求的天然矿物，同时因易挥发分解或还原性强而不能使用熔融玻璃法制备。其中，碳酸岩基体标准物质为 MACS-3，其定值元素为 47 种，包含 3 种主量元素和 44 种微量元素，微量元素中稀土元素含量约为 10 mg/kg，其他元素约为 50 mg/kg；硫化物基体标准物质为 MASS-1，其定值元素为 24 种，包含 4 种主量元素和 20 种微量元素，微量元素含量约为 50 mg/kg；磷酸岩基体标准物质有 MAPS-4/5 两种，其中 MAPS-4 中的微量元素大约是 MAPS-5 的 10 倍。

2. 国内微区标准物质研制现状

国内微区分析标准物质研制还处于起步阶段。国家地质实验测试中心开展微区分析标准物质研制的时间较早，代表是 Hu 等[61]在 2011 年报道的 CSGS 系列熔融玻璃基体标准物质，包含四种基体：西藏碱性玄武岩、霓霞正长岩、北京土壤和安山岩。中国科学院地质与地球物理研究所在基体同位素标准物质研制方面经验丰富，先后研制了方解石氧碳同位素（GBW04481）、锆石氧铪同位素（GBW04482）、文石锶同位素（GBW04704）和锆石铀铅年龄（GBW04705）四个国家标准物质。吴石头等[62]于 2019 年研制了 ARM 系列安山岩含量微区分析标准物质，添加 54 种微量元素，浓度分别约为 500 mg/kg、50 mg/kg 和 5 mg/kg。中国地质大学（武汉）在微区分析方面具有丰富的经验，Feng 等[63]于 2018 年报道了四种硫化物基体标准物质，并对其中两种的 Pb 同位素进行分析；Ke 等[64]于 2019 年报道了使用稀土元素掺杂的钨单晶作为实验室标准物质；Huang 等[65]于 2020 年报道了一种新的锆石标准物质 SA01，用于 U-Pb 年代学研究和 Hf、O 地球化学研究。西北大学在微区分析标物制备方面集中于用粉末压片法制备硫化物实验室标准。中国科学院广州地球化学研究所于 2017 年与英国政府化学家实验室（LGC）合作研制了一种 U-Th/He 定年的锆石标样 LGC-1[66]。中国计量科学研究院目前已建立了微区分析平台，并开展了相关微区元素含量和同位素组成标物研制工作。

25.2.5.4 微区分析用标准物质技术发展趋势

随着微区分析方法的不断开发以及应用范围的不断扩展，对元素含量分析的标物的需求量会越来越大，对标物基体种类与元素含量不确定度水平的要求会越来越高；而随着地质、环境、考古等领域对同位素原位分析需求的日益增长，未来除了使用来自于天然矿物的同位素标准，成矿条件下高温高压压片制得的基体匹配的同位素标物也将得到快速发展，最终走出实验室标样水平，与元素含量标物并驾齐驱；此外，随着微区分析技术的发展，亟需建立相关的行业标准、计量标准。

25.2.6 统计学理论

25.2.6.1 正态分布检验

正态性检验主要有三类方法：一是计算综合统计量，如动差法、夏皮罗-威尔克法、达戈斯提诺法、Shapiro-Francia 法；二是正态分布的拟合优度检验，如偏态系数和峰态系数法、χ^2 检验、对数似然比检验、柯尔莫可洛夫-斯米洛夫检验；三是图示法（正态概率图 normal probability plot），如分位数图、百分位图和稳定化概率图等。本节主要介绍偏态系数和峰态系数法、夏皮罗-威尔克法和达戈斯提诺法。

使用偏态系数和峰态系数法检验的条件为：总体具有在偏度和峰度方向上偏离正态的先验信息。而夏皮罗-威尔克法及达戈斯提诺法是在分布与正态偏离的形式没有任何事先了解的情况下进行的，此两种检验称为公用型检验，前者适用于测定次数较少时（$3 \leqslant n \leqslant 50$），后者适用于测定次数较多时（$50 \leqslant n \leqslant 1000$）。

1. 偏态系数和峰态系数法

设对某特性量进行测定，得到一组独立测量结果，将数据按由小到大顺序排列，表示如下：

$$x_1、x_2、\cdots、x_n$$
$$(x_1 \leqslant x_2 \leqslant \cdots \leqslant x_n)$$
$$\bar{x} = \sum_{i=1}^{n} x_i / n$$

记

$$m_2 = \sum_{i=1}^{n} (x_i - \bar{x})^2 / n$$

$$m_3 = \sum_{i=1}^{n} (x_i - \bar{x})^3 / n$$

$$m_4 = \sum_{i=1}^{n} (x_i - \bar{x})^4 / n$$

称 $A = |m_3| / \sqrt{(m_2)^3}$ 为偏态系数，它用于检验不对称性；$B = m_4 / (m_2)^2$ 为峰态系数，它用于检验峰态。

若该组测量数据服从正态分布，则其偏态系数 A 和峰态系数 B 应分别小于相应的临界值 A_1 和落入区间 $B_1 - B'_1$ 中，A_1 和 $B_1 - B'_1$ 的值与要求的置信概率 P 和测量次数 n 有关，其值分别见表 25-8、表 25-9。

表 25-10 中列出某样品中镱（Yb）含量的测定结果。

表 25-8　不对称性检验的临界值 A_1

n	P=0.95	P=0.99	n	P=0.95	P=0.99	n	P=0.95	P=0.99
8	0.99	1.42	100	0.39	0.57	800	0.14	0.20
9	0.97	1.41	125	0.35	0.51	850	0.14	0.20
10	0.95	1.39	150	0.32	0.46	900	0.13	0.19
12	0.91	1.34	175	0.30	0.43	950	0.13	0.18
15	0.85	1.26	200	0.28	0.40	1000	0.13	0.18
20	0.77	1.15	250	0.25	0.36	1200	0.12	0.16
25	0.71	1.06	300	0.23	0.33	1400	0.11	0.15
30	0.66	0.98	350	0.21	0.30	1600	0.10	0.14
35	0.62	0.92	400	0.20	0.28	1800	0.10	0.13
40	0.59	0.87	450	0.19	0.27	2000	0.09	0.13
45	0.56	0.82	500	0.18	0.26	2500	0.08	0.11
50	0.53	0.79	550	0.17	0.24	3000	0.07	0.10
60	0.49	0.72	600	0.16	0.23	3500	0.07	0.10
70	0.46	0.67	650	0.16	0.22	4000	0.06	0.09
80	0.43	0.63	700	0.15	0.22	4500	0.06	0.08
90	0.41	0.60	750	0.15	0.21	5000	0.06	0.08

表 25-9　峰态检验的临界值 B_1-B'_1

n	P 0.95	0.99	n	P 0.95	0.99
7	1.41~3.55	1.25~4.23	200	2.51~3.57	2.37~3.98
8	1.46~3.70	1.31~4.53	250	2.55~3.52	2.42~3.87
9	1.53~3.86	1.35~4.82	300	2.59~3.47	2.46~3.79
10	1.56~3.95	1.39~5.00	350	2.62~3.44	2.50~3.72
12	1.64~4.05	1.46~5.20	400	2.64~3.41	2.52~3.67
15	1.72~4.13	1.55~5.30	450	2.66~3.49	2.55~3.63
20	1.82~4.17	1.65~5.36	500	2.67~3.37	2.57~3.60
25	1.91~4.16	1.72~5.30	550	2.69~3.35	2.58~3.57
30	1.98~4.11	1.79~5.21	600	2.70~3.34	2.60~3.54
35	2.03~4.10	1.84~5.13	650	2.71~3.33	2.61~3.52
40	2.07~4.06	1.89~5.04	700	2.72~3.31	2.62~3.50
45	2.11~4.00	1.93~4.94	750	2.73~3.30	2.64~3.48
50	2.15~3.99	1.95~4.88	800	2.74~3.29	2.65~3.46
75	2.27~3.87	2.08~4.59	850	2.74~3.28	2.66~3.45
100	2.35~3.77	2.18~4.39	900	2.75~3.28	2.66~3.43
125	2.40~3.71	2.24~4.24	950	2.76~3.27	2.67~3.42
150	2.45~3.65	2.29~4.13	1000	2.76~3.26	2.68~3.41

表 25-10　某物质中镱的质量分数测定原始数据

K	x_K	K	x_K	K	x_K	K	x_K
1	2.40	11	2.92	21	3.08	31	3.20
2	2.40	12	2.92	22	3.09	32	3.20
3	2.50	13	3.00	23	3.10	33	3.21
4	2.70	14	3.00	24	3.10	34	3.27
5	2.70	15	3.01	25	3.10	35	3.37
6	2.70	16	3.03	26	3.12	36	3.37
7	2.83	17	3.04	27	3.12	37	3.38
8	2.86	18	3.04	28	3.16	38	3.43
9	2.86	19	3.08	29	3.19	39	3.65
10	2.90	20	3.08	30	3.19	40	3.68

使用表 25-10 中的数据计算偏态系数 A 及峰态系数 B 如下：

$$\bar{x} = (2.40+2.40+2.50+\cdots+3.65+3.68)/40 = 3.050$$

$$m_2 = [(2.40-3.050)^2+(2.40-3.050)^2+(2.50-3.050)^2+$$
$$\cdots+(3.65-3.050)^2+(3.68-3.050)^2]/40 = 0.078$$

$$m_3 = [(2.40-3.050)^3+(2.40-3.050)^3+(2.50-3.050)^3+$$
$$\cdots+(3.65-3.050)^3+(3.68-3.050)^3]/40 = -0.0055$$

$$m_4 = [(2.40-3.050)^4+(2.40-3.050)^4+(2.50-3.050)^4+$$
$$\cdots+(3.65-3.050)^4+(3.68-3.050)^4]/40 = 0.021$$

$$A = \frac{|m_3|}{\sqrt{(m_2)^3}} = 0.0055/\sqrt{(0.078)^3} = 0.254$$

$$B = \frac{m_4}{(m_2)^2} = \frac{0.021}{(0.078)^2} = 3.452$$

对 $P=0.95$，$n=40$，查表 25-8 得 $A_1=0.59$；对 $P=0.95$，$n=40$，查表 25-8 得区间 2.07~4.06。由于 0.254<0.59，2.07<3.503<4.06，故可接受此组数据为正态分布。

2. 夏皮罗-威尔克（Shapiro-Wilk）法

将一组测量数据按由小到大的顺序排列。

夏皮罗-威尔克法检验的统计量是：

$$W = \left\{\sum a_K \left[X_{n+1-K} - X_K\right]\right\}^2 / \sum_{K=1}^{n}\left(X_K - \bar{X}\right)^2$$

式中，分子下标的 K 值，当测量次数 n 是偶数时为 $1\sim n/2$，当测量次数是奇数时则为 $1\sim(n-1)/2$；系数 a_K 是与 n 及 K 有关的特定值，见表 25-11。

该统计量 W 的判据是，当 $W>W(n, P)$ 时，则接受测定数据为正态分布。$W(n, P)$ 是与测量次数 n 及置信概率 P 有关的数值，其值见表 25-12。

表 25-11 系数 a_K 的值

K \ n	2	3	4	5	6	7	8	9	10
1	0.7071	0.7071	0.6872	0.6646	0.6431	0.6233	0.6052	0.5888	0.5739
2	—	—	0.1677	0.2413	0.2806	0.3031	0.3164	0.3244	0.3291
3	—	—	—	—	0.0875	0.1401	0.1743	0.1976	0.2141
4	—	—	—	—	—	—	0.0561	0.0947	0.1224
5	—	—	—	—	—	—	—	—	0.0399

K \ n	11	12	13	14	15	16	17	18	19	20
1	0.5601	0.5475	0.5359	0.5251	0.5150	0.5056	0.4968	0.4886	0.4808	0.4734
2	0.3315	0.3325	0.3325	0.3318	0.3306	0.3290	0.3273	0.3253	0.3232	0.3211
3	0.2260	0.2347	0.2412	0.2460	0.2495	0.2521	0.2540	0.2553	0.2561	0.2565
4	0.1429	0.1586	0.1707	0.1802	0.1878	0.1939	0.1988	0.2027	0.2059	0.2085
5	0.0695	0.0922	0.1099	0.1240	0.1353	0.1447	0.1524	0.1587	0.1641	0.1686
6	—	0.0303	0.0539	0.0727	0.0880	0.1005	0.1109	0.1197	0.1271	0.1334
7	—	—	—	0.0240	0.0433	0.0593	0.0725	0.0837	0.0932	0.1013
8	—	—	—	—	—	0.0196	0.0359	0.0496	0.0612	0.0711
9	—	—	—	—	—	—	—	0.0163	0.0303	0.0422
10	—	—	—	—	—	—	—	—	—	0.0140

K \ n	21	22	23	24	25	26	27	28	29	30
1	0.4643	0.4590	0.4542	0.4493	0.4450	0.4407	0.4366	0.4328	0.4291	0.4254
2	0.3185	0.3156	0.3126	0.3098	0.3069	0.3043	0.3018	0.2992	0.2968	0.2944
3	0.2578	0.2571	0.2563	0.2554	0.2543	0.2533	0.2522	0.2510	0.2499	0.2487
4	0.2119	0.2131	0.2139	0.2145	0.2148	0.2151	0.2152	0.2151	0.2150	0.2148
5	0.1736	0.1764	0.1787	0.1807	0.1822	0.1836	0.1848	0.1857	0.1864	0.1870
6	0.1399	0.1443	0.1480	0.1512	0.1539	0.1563	0.1584	0.1601	0.1616	0.1630
7	0.1092	0.1150	0.1201	0.1245	0.1283	0.1316	0.1346	0.1372	0.1395	0.1415
8	0.0804	0.0878	0.0941	0.0997	0.1046	0.1089	0.1128	0.1162	0.1192	0.1219
9	0.0530	0.0618	0.0696	0.0764	0.0823	0.0876	0.0923	0.0965	0.1002	0.1036
10	0.0263	0.0368	0.0459	0.0539	0.0610	0.0672	0.0728	0.0778	0.0822	0.0862
11	—	0.0122	0.0228	0.0321	0.0403	0.0476	0.0540	0.0598	0.0650	0.0697
12	—	—	—	0.0107	0.0200	0.0284	0.0358	0.0424	0.0483	0.0537
13	—	—	—	—	—	0.0094	0.0178	0.0253	0.0320	0.0381
14	—	—	—	—	—	—	—	0.0084	0.0159	0.0227
15	—	—	—	—	—	—	—	—	—	0.0076

K \ n	31	32	33	34	35	36	37	38	39	40
1	0.4220	0.4188	0.4156	0.4127	0.4096	0.4068	0.4040	0.4015	0.3989	0.3964
2	0.2921	0.2898	0.2876	0.2854	0.2834	0.2813	0.2794	0.2774	0.2755	0.2737
3	0.2475	0.2463	0.2451	0.2439	0.2427	0.2415	0.2403	0.2391	0.2380	0.2368
4	0.2145	0.2141	0.2137	0.2132	0.2127	0.2121	0.2116	0.2110	0.2104	0.2098
5	0.1874	0.1878	0.1880	0.1882	0.1883	0.1883	0.1883	0.1881	0.1880	0.1878
6	0.1641	0.1651	0.1660	0.1667	0.1673	0.1678	0.1683	0.1686	0.1689	0.1691

续表

K\n	31	32	33	34	35	36	37	38	39	40
7	0.1433	0.1449	0.1463	0.1475	0.1487	0.1496	0.1505	0.1513	0.1520	0.1526
8	0.1243	0.1265	0.1284	0.1301	0.1317	0.1331	0.1344	0.1356	0.1366	0.1376
9	0.1066	0.1093	0.1118	0.1140	0.1160	0.1179	0.1196	0.1211	0.1225	0.1237
10	0.0899	0.0931	0.0961	0.0988	0.1013	0.1036	0.1056	0.1075	0.1092	0.1108
11	0.0739	0.0777	0.0812	0.0844	0.0873	0.0900	0.0924	0.0947	0.0967	0.0986
12	0.0585	0.0629	0.0669	0.0706	0.0739	0.0770	0.0798	0.0824	0.0848	0.0870
13	0.0435	0.0485	0.0530	0.0572	0.0610	0.0645	0.0677	0.0706	0.0733	0.0759
14	0.0289	0.0344	0.0395	0.0441	0.0484	0.0523	0.0559	0.0592	0.0622	0.0651
15	0.0144	0.0206	0.0262	0.0314	0.0361	0.0404	0.0444	0.0481	0.0515	0.0546
16	—	0.0068	0.0131	0.0187	0.0239	0.0287	0.0331	0.0372	0.0409	0.0444
17	—	—	—	0.0062	0.0119	0.0172	0.0220	0.0264	0.0305	0.0343
18	—	—	—	—	—	0.0057	0.0110	0.0158	0.0203	0.0244
19	—	—	—	—	—	—	—	0.0053	0.0101	0.0146
20	—	—	—	—	—	—	—	—	—	0.0049

K\n	41	42	43	44	45	46	47	48	49	50
1	0.3940	0.3917	0.3894	0.3872	0.3850	0.3830	0.3808	0.3789	0.3770	0.0351
2	0.2719	0.2701	0.2684	0.2667	0.2651	0.2635	0.2620	0.2604	0.2589	0.2574
3	0.2357	0.2345	0.2334	0.2323	0.2313	0.2302	0.2291	0.2281	0.2271	0.2260
4	0.2091	0.2085	0.2078	0.2072	0.2065	0.2058	0.2052	0.2045	0.2038	0.2032
5	0.1876	0.1874	0.1871	0.1868	0.1865	0.1862	0.1859	0.1855	0.1851	0.1847
6	0.1693	0.1694	0.1695	0.1695	0.1695	0.1695	0.1695	0.1693	0.1692	0.1691
7	0.1531	0.1535	0.1539	0.1542	0.1545	0.1548	0.1550	0.1551	0.1553	0.1554
8	0.1384	0.1392	0.1398	0.1405	0.1410	0.1415	0.1420	0.1423	0.1427	0.1430
9	0.1249	0.1259	0.1269	0.1278	0.1286	0.1293	0.1300	0.1306	0.1312	0.1317
10	0.1123	0.1136	0.1149	0.1160	0.1170	0.1180	0.1189	0.1197	0.1205	0.1212
11	0.1004	0.1020	0.1035	0.1049	0.1062	0.1073	0.1085	0.1095	0.1105	0.1113
12	0.1891	0.0909	0.0927	0.0943	0.0959	0.0972	0.0986	0.0998	0.1010	0.1020
13	0.0782	0.0804	0.0824	0.0842	0.0860	0.0876	0.0892	0.0906	0.0919	0.0932
14	0.0677	0.0701	0.0724	0.0745	0.0765	0.0783	0.0801	0.0817	0.0832	0.0846
15	0.0575	0.0602	0.0628	0.0651	0.0673	0.0694	0.0713	0.0713	0.0731	0.0764
16	0.0476	0.0506	0.0534	0.0560	0.0584	0.0607	0.0628	0.0648	0.0667	0.0685
17	0.0379	0.0411	0.0442	0.0471	0.0497	0.0522	0.0546	0.0568	0.0588	0.0608
18	0.0283	0.0318	0.0352	0.0383	0.0412	0.0439	0.0465	0.0489	0.0511	0.0532
19	0.0188	0.0227	0.0263	0.0296	0.0328	0.0357	0.0385	0.0411	0.0436	0.0459
20	0.0094	0.0136	0.0175	0.0211	0.0245	0.0277	0.0307	0.0335	0.0361	0.0386
21	—	0.0045	0.0087	0.0126	0.0163	0.0197	0.0229	0.0259	0.0288	0.0314
22	—	—	—	0.0042	0.0081	0.0118	0.0153	0.0185	0.0215	0.0211
23	—	—	—	—	—	0.0039	0.0076	0.0111	0.0143	0.0174
24	—	—	—	—	—	—	—	0.0073	0.0071	0.0101
25	—	—	—	—	—	—	—	—	—	0.0035

表 25-12　$W(n, P)$ 的值

n	P=0.99	P=0.95	n	P=0.99	P=0.95	n	P=0.99	P=0.95	n	P=0.99	P=0.95
3	0.753	0.767	15	0.835	0.881	27	0.894	0.923	39	0.917	0.939
4	0.687	0.748	16	0.844	0.887	28	0.896	0.924	40	0.919	0.940
5	0.686	0.762	17	0.851	0.892	29	0.898	0.926	41	0.920	0.941
6	0.713	0.788	18	0.858	0.897	30	0.900	0.927	42	0.922	0.942
7	0.730	0.803	19	0.863	0.901	31	0.902	0.929	43	0.923	0.943
8	0.749	0.818	20	0.868	0.905	32	0.904	0.930	44	0.924	0.944
9	0.764	0.829	21	0.873	0.908	33	0.906	0.931	45	0.926	0.945
10	0.781	0.842	22	0.878	0.911	34	0.908	0.933	46	0.927	0.945
11	0.792	0.850	23	0.881	0.914	35	0.910	0.934	47	0.928	0.946
12	0.805	0.859	24	0.884	0.916	36	0.912	0.935	48	0.929	0.947
13	0.814	0.866	25	0.888	0.918	37	0.914	0.936	49	0.929	0.947
14	0.825	0.874	26	0.891	0.920	38	0.916	0.938	50	0.930	0.947

对表 25-12 所列数据，按夏皮罗-威尔克法进行检验，见表 25-13。a_K 值由表 25-11 中查出（其中 $n=40$，$P=0.45$）。

表 25-13　夏皮罗-威尔克法检验计算

K	X_K	X_{n+1-K}	$X_{n+1-K}-X_K$	a_K	K	X_K	X_{n+1-K}	$X_{n+1-K}-X_K$	a_K
1	2.40	3.68	1.28	0.3964	11	2.92	3.19	0.27	0.0986
2	2.40	3.65	1.25	0.2737	12	2.92	3.19	0.27	0.0870
3	2.50	3.43	0.93	0.2368	13	3.00	3.16	0.16	0.0759
4	2.70	3.38	0.68	0.2098	14	3.00	3.12	0.12	0.0651
5	2.70	3.37	0.67	0.1878	15	3.01	3.12	0.11	0.0546
6	2.70	3.37	0.67	0.1691	16	3.03	3.10	0.07	0.0444
7	2.83	3.27	0.44	0.1526	17	3.04	3.10	0.06	0.0343
8	2.86	3.21	0.35	0.1376	18	3.04	3.10	0.06	0.0244
9	2.86	3.20	0.34	0.1237	19	3.08	3.09	0.01	0.0146
10	2.90	3.20	0.30	0.1108	20	3.08	3.08	0	0.0049

经计算，得：$\bar{X}=3.050$ $\sum(X_K-\bar{X})^2=3.11$，$W=1.72\times1.72/3.11=0.957$；查表 25-12，$W(n,P)=0.940$（其中 $n=40$，$P=0.95$）。

由于 0.957>0.940，故接受该组数据为正态分布。

3. 达戈斯提诺（D'Agostino）法

将数据按由小到大顺序排列。

检验的统计量为

$$Y=\sqrt{n}\left[\frac{\sum\left[\left(\frac{n+1}{2}-K\right)(X_{n+1-K}-X_K)\right]}{n^2\sqrt{m_2}}-0.282\,094\,79\right]/0.029\,985\,98$$

$$m_2=\sum_{i=1}^{n}(x_i-\bar{x})^2/n$$

式中，n 为测定次数；下标 K 的值，当 n 是偶数时为 $1\sim n/2$；当 n 是奇数时为 $1\sim(n-1)/2$。

该统计量的判据是：当置信概率为 95% 时，Y 值应落入区间 a–a' 范围内。当置信概率为 99% 时，Y 值应落入区间 b–b' 范围内。上述区间值见表 25-14。

表 25-14　达戈斯提诺法检验临界区间

n	区间		n	区间	
	a–a' ($p=0.95$)	b–b' ($p=0.99$)		a–a' ($p=0.95$)	b–b' ($p=0.99$)
50	−2.74~1.06	−3.91~1.24	450	−2.25~1.65	−3.06~2.09
60	−2.68~1.13	−3.81~1.34	500	−2.24~1.67	−3.04~2.11
70	−2.64~1.19	−3.73~1.42	550	−2.23~1.68	−3.02~2.14
80	−2.60~1.24	−3.67~1.48	600	−2.22~1.69	−3.00~2.15
90	−2.57~1.28	−3.61~1.54	650	−2.21~1.70	−2.98~2.17
100	−2.54~1.31	−3.57~1.59	700	−2.20~1.71	−2.97~2.18
150	−2.45~1.42	−3.41~1.75	750	−2.19~1.72	−2.96~2.20
200	−2.39~1.50	−3.30~1.85	800	−2.18~1.73	−2.94~2.21
250	−2.35~1.54	−3.23~1.93	850	−2.18~1.74	−2.93~2.22
300	−2.32~1.53	−3.17~1.98	900	−2.17~1.74	−2.92~2.23
350	−2.29~1.61	−3.13~2.03	950	−2.16~1.75	−2.91~2.24
400	−2.27~1.63	−3.09~2.06	1000	−2.16~1.75	−2.91~2.25

以某物质中钴（Co）质量分数（μg/g）为例（共 67 个数据），按达戈斯提诺法检验其数据正态性。数据按由小到大顺序排列，见表 25-15。

表 25-15　某物质中钴的质量分数测定原始数据

K	X_K	X_{n+1-K}	$X_{n+1-K}-X_K$	$\frac{n+1}{2}-K$	K	X_K	X_{n+1-K}	$X_{n+1-K}-X_K$	$\frac{n+1}{2}-K$
1	11.7	14.5	2.8	33	18	12.8	13.6	0.8	16
2	11.9	14.5	2.6	32	19	12.8	13.6	0.8	15
3	11.9	14.3	2.4	31	20	12.8	13.6	0.8	14
4	12.1	14.2	2.1	30	21	12.8	13.6	0.8	13
5	12.2	14.1	1.9	29	22	12.8	13.5	0.7	12
6	12.2	14.1	1.9	28	23	12.8	13.5	0.7	11
7	12.4	14.1	1.7	27	24	12.9	13.5	0.6	10
8	12.4	14.0	1.6	26	25	13.0	13.4	0.4	9
9	12.6	13.8	1.2	25	26	13.0	13.4	0.4	8
10	12.6	13.8	1.2	24	27	13.1	13.4	0.3	7
11	12.6	13.8	1.2	23	28	13.1	13.4	0.3	6
12	12.6	13.8	1.2	22	29	13.1	13.4	0.3	5
13	12.6	13.7	1.1	21	30	13.1	13.3	0.2	4
14	12.7	13.7	1.0	20	31	13.1	13.3	0.2	3
15	12.7	13.7	1.0	19	32	13.1	13.3	0.2	2
16	12.7	13.7	1.0	18	33	13.1	13.3	0.2	1
17	12.7	13.7	1.0	17	34	13.2			

因测定数据数目为奇数，故对 K 求和为 $1\sim(67-1)/2=33$。由表 25-15 数据得

$$\sum\left[\left(\frac{n+1}{2}-K\right)\left(X_{n+1-K}-X_K\right)\right]=801.4$$

$$\sqrt{m_2}=0.630\ 264\ 42$$

$$Y=\frac{\sqrt{67}\left(\dfrac{801.4}{67^2\times 0.630\ 264\ 42}-0.282\ 094\ 79\right)}{0.029\ 985\ 98}=0.32$$

由表 25-14 可见 Y 值介于相应的区间 a–a' 之间，故接受该组数据为正态分布。

25.2.6.2　离群值判断

在一组平行测定中，若有个别数据与平均值差别较大，则把此数据视为可疑值，也称离群值。如果统计学上认为应该舍弃的数据留用了，势必会影响其平均值的可靠性。相反，本应该留用的数据被舍弃，虽然精密度提高，但却夸大了平均值的可靠性。

离群值检验方法较为常用的有：格拉布斯法、拉依达法、$4d$ 检验法、肖维勒法、狄克逊法、t 检验法、Q 检验法、偏度-峰度检验法和柯克伦法。本节主要介绍格拉布斯法、狄克逊法和柯克伦法。

当离群值数量仅为 1 时，格拉布斯法综合犯错的可能性最低，推荐使用格拉布斯法。当限定检出离群值的个数大于 1 时，格拉布斯法检验的结果不是最优的，一般采用狄克逊法。

1. 格拉布斯（Grubbs）法

在一组测定值中，如某测定值 x_i，有残差 $\upsilon_i=x_i-\bar{x}$。当 $|\upsilon_i|>\lambda(\alpha,n)S$ 时，则 x_i 应被剔除。$\lambda(\alpha,n)$ 是与测量次数及给定的显著性水平 α 有关的数值。可查表 25-16 得到 $\lambda(\alpha,n)$ 值。

2. 狄克逊（Dixon）法

将测定数据按由小到大的顺序排列：

$$X_{(1)}\leqslant X_{(2)}\leqslant\cdots\leqslant X_{(n-1)}\leqslant X_{(n)}$$

表 25-16　$\lambda_{(\alpha,n)}$ 数值表

n	1%	5%	n	1%	5%	n	1%	5%	n	1%	5%
3	1.155	1.155	28	3.199	2.876	53	3.507	3.151	78	3.663	3.297
4	1.496	1.481	29	3.218	2.893	54	3.516	3.158	79	3.669	3.301
5	1.764	1.715	30	3.236	2.908	55	3.524	3.166	80	3.673	3.305
6	1.973	1.887	31	3.253	2.924	56	3.531	3.172	81	3.677	3.309
7	2.139	2.020	32	3.270	2.938	57	3.539	3.180	82	3.682	3.315
8	2.274	2.126	33	3.286	2.953	58	3.546	3.186	83	3.687	3.319
9	2.387	2.215	34	3.301	2.965	59	3.553	3.193	84	3.691	3.323
10	2.482	2.290	35	3.316	2.979	60	3.560	3.199	85	3.695	3.327
11	2.564	2.355	36	3.330	2.991	61	3.566	3.205	86	3.699	3.331
12	2.636	2.412	37	3.343	3.003	62	3.573	3.212	87	3.704	3.335
13	2.699	2.462	38	3.356	3.014	63	3.579	3.218	88	3.708	3.339
14	2.755	2.507	39	3.369	3.025	64	3.586	3.224	89	3.712	3.343
15	2.806	2.549	40	3.381	3.036	65	3.592	3.230	90	3.716	3.347
16	2.852	2.585	41	3.393	3.046	66	3.598	3.235	91	3.720	3.350
17	2.894	2.620	42	3.404	3.057	67	3.605	3.241	92	3.725	3.355
18	2.932	2.651	43	3.415	3.067	68	3.610	3.246	93	3.728	3.358
19	2.968	2.681	44	3.425	3.075	69	3.617	3.252	94	3.732	3.362
20	3.001	2.709	45	3.435	3.085	70	3.622	3.257	95	3.736	3.365
21	3.031	2.733	46	3.445	3.094	71	3.627	3.262	96	3.739	3.369
22	3.060	2.758	47	3.455	3.103	72	3.633	3.267	97	3.744	3.372
23	3.087	2.781	48	3.464	3.111	73	3.638	3.272	98	3.747	3.377
24	3.112	2.802	49	3.474	3.120	74	3.643	3.278	99	3.750	3.380
25	3.135	2.822	50	3.483	3.128	75	3.648	3.282	100	3.754	3.383
26	3.157	2.841	51	3.491	3.136	76	3.654	3.287			
27	3.178	2.859	52	3.500	3.143	77	3.658	3.291			

按表 25-17 计算 r_1 值和 r_n 值。若 $r_1 > r_n$，且 $r_1 > f_{(\alpha,n)}$，则判定 $X_{(1)}$ 为异常值；若 $r_n > r_1$，且 $r_n > f_{(\alpha,n)}$，则判定 $X_{(n)}$ 为异常值；若 r_1 及 r_n 值均小于 $f_{(\alpha,n)}$ 值，则所有数据保留。

表 25-17　$f_{(\alpha,n)}$ 值、r_1 及 r_n 值计算公式表

n	统计量	$f_{(\alpha,n)}$ $\alpha=1\%$	$f_{(\alpha,n)}$ $\alpha=5\%$	n	统计量	$f_{(\alpha,n)}$ $\alpha=1\%$	$f_{(\alpha,n)}$ $\alpha=5\%$
3		0.994	0.970	11	$r_1 = \dfrac{X_{(3)} - X_{(1)}}{X_{(n-1)} - X_{(1)}}$ 和	0.709	0.619
4	$r_1 = \dfrac{X_{(2)} - X_{(1)}}{X_{(n)} - X_{(1)}}$ 和	0.926	0.829	12		0.660	0.583
5	$r_n = \dfrac{X_{(n)} - X_{(n-1)}}{X_{(n)} - X_{(1)}}$	0.821	0.710	13	$r_n = \dfrac{X_{(n)} - X_{(n-2)}}{X_{(n)} - X_{(2)}}$ 中的较大者	0.638	0.557
6	中的较大者	0.740	0.628	14	$r_1 = \dfrac{X_{(3)} - X_{(1)}}{X_{(n-2)} - X_{(1)}}$ 和	0.670	0.586
7		0.680	0.569	15		0.647	0.565
8	$r_1 = \dfrac{X_{(2)} - X_{(1)}}{X_{(n-1)} - X_{(1)}}$ 和	0.717	0.608	16	$r_n = \dfrac{X_{(n)} - X_{(n-2)}}{X_{(n)} - X_{(3)}}$ 中的较大者	0.627	0.546
9	$r_n = \dfrac{X_{(n)} - X_{(n-1)}}{X_{(n)} - X_{(2)}}$	0.672	0.564	17	$r_1 = \dfrac{X_{(3)} - X_{(1)}}{X_{(n-2)} - X_{(1)}}$ 和	0.610	0.529
10	中的较大者	0.635	0.530	18	$r_n = \dfrac{X_{(n)} - X_{(n-2)}}{X_{(n)} - X_{(3)}}$ 中的较大者	0.594	0.514

续表

n	统计量	$f_{(\alpha,n)}$ $\alpha=1\%$	$f_{(\alpha,n)}$ $\alpha=5\%$	n	统计量	$f_{(\alpha,n)}$ $\alpha=1\%$	$f_{(\alpha,n)}$ $\alpha=5\%$
19		0.580	0.501	25	$\eta = \dfrac{X_{(3)} - X_{(1)}}{X_{(n-2)} - X_{(1)}}$ 和 $r_n = \dfrac{X_{(n)} - X_{(n-2)}}{X_{(n)} - X_{(3)}}$ 中的较大者	0.517	0.443
20		0.567	0.489	26		0.510	0.436
21	$\eta = \dfrac{X_{(3)} - X_{(1)}}{X_{(n-2)} - X_{(1)}}$ 和 $r_n = \dfrac{X_{(n)} - X_{(n-2)}}{X_{(n)} - X_{(3)}}$ 中的较大者	0.555	0.478	27		0.502	0.429
22		0.544	0.468	28	$\eta = \dfrac{X_{(3)} - X_{(1)}}{X_{(n-2)} - X_{(1)}}$ 和 $r_n = \dfrac{X_{(n)} - X_{(n-2)}}{X_{(n)} - X_{(3)}}$ 中的较大者	0.495	0.423
23		0.535	0.459	29		0.489	0.417
24		0.526	0.451	30		0.483	0.412

例如：某低合金钢中钼的含量，在重复性条件下进行 14 次独立分析，测量值按从小到大排列，分别为 0.354%、0.357%、0.358%、0.359%、0.359%、0.361%、0.363%、0.363%、0.364%、0.367%、0.368%、0.369%、0.372%、0.390%，试判断离群值。

方法 1：格拉布斯法

$$\upsilon = x_{14} - \bar{x} = 0.390\% - 0.3646\% = 0.0254\%$$
$$\lambda(0.05, 14) = 2.507$$
$$\lambda(0.05, 14) * s = 2.507 \times 0.0089\% = 0.0223\%$$
$$\upsilon > \lambda(0.05, 14) * s$$

0.390% 为离群值。

方法 2：狄克逊法

计算得 $\bar{x} = 0.3636\%$，$s = 0.0089\%$。测量数据的样本量 $n=14$，计算狄克逊检验统计量：

$$D_{14} = r_{22} = \frac{x_n - x_{n-2}}{x_n - x_3} = \frac{0.390\% - 0.369\%}{0.390\% - 0.358\%} = 0.656$$

$$D'_{14} = r'_{22} = \frac{x_3 - x_1}{x_{n-2} - x_1} = \frac{0.358\% - 0.354\%}{0.369\% - 0.354\%} = 0.267$$

$$f(0.05, 14) = 0.587$$
$$D_{14} > f(0.05, 14)$$

0.390% 为离群值。

3. 柯克伦（Cochran）法

给定 p 个由相同的 n 次重复测试结果计算的标准偏差 s_i。柯克伦检验统计量 C 定义为

$$C = \frac{s_{\max}^2}{\sum_{i=1}^{p} s_i^2}$$

式中，$s_{\max}$ 是这组标准差中的最大值。

如果统计量 C 大于临界值（见表 25-18），则认为该组数据的标准差与其他组数据间存在显著性差异。

表 25-18 柯克伦检验的临界值

p	n=2		n=3		n=4		n=5		n=6	
	1%	5%	1%	5%	1%	5%	1%	5%	1%	5%
2	—	—	0.995	0.975	0.979	0.939	0.959	0.906	0.937	0.877
3	0.993	0.967	0.942	0.871	0.883	0.798	0.834	0.746	0.793	0.707
4	0.968	0.906	0.864	0.768	0.781	0.684	0.721	0.629	0.676	0.590
5	0.928	0.841	0.788	0.684	0.696	0.598	0.633	0.544	0.588	0.506
6	0.883	0.781	0.722	0.616	0.626	0.532	0.564	0.480	0.520	0.445
7	0.838	0.727	0.664	0.561	0.568	0.480	0.508	0.431	0.466	0.397
8	0.794	0.680	0.615	0.516	0.521	0.438	0.463	0.391	0.423	0.360
9	0.754	0.638	0.573	0.478	0.481	0.403	0.425	0.358	0.387	0.329
10	0.718	0.602	0.536	0.445	0.447	0.373	0.393	0.331	0.357	0.303
11	0.684	0.570	0.504	0.417	0.418	0.348	0.366	0.308	0.332	0.281
12	0.653	0.541	0.475	0.392	0.392	0.326	0.343	0.288	0.310	0.262
13	0.624	0.515	0.450	0.371	0.369	0.307	0.322	0.271	0.291	0.243
14	0.599	0.492	0.427	0.352	0.349	0.291	0.304	0.255	0.274	0.232
15	0.575	0.471	0.407	0.335	0.332	0.276	0.288	0.242	0.259	0.220
16	0.553	0.452	0.388	0.319	0.316	0.262	0.274	0.230	0.246	0.208
17	0.532	0.434	0.372	0.305	0.301	0.250	0.261	0.219	0.234	0.198
18	0.514	0.418	0.356	0.293	0.288	0.240	0.249	0.209	0.223	0.189
19	0.496	0.403	0.343	0.281	0.276	0.230	0.238	0.200	0.214	0.181
20	0.480	0.389	0.330	0.270	0.265	0.220	0.229	0.192	0.205	0.174
21	0.465	0.377	0.318	0.261	0.255	0.212	0.220	0.185	0.197	0.167
22	0.450	0.365	0.307	0.252	0.246	0.204	0.212	0.178	0.189	0.160
23	0.437	0.354	0.297	0.243	0.238	0.197	0.204	0.172	0.182	0.155
24	0.425	0.343	0.287	0.235	0.230	0.191	0.197	0.166	0.176	0.149
25	0.413	0.334	0.278	0.228	0.222	0.185	0.190	0.160	0.170	0.144
26	0.402	0.325	0.270	0.221	0.215	0.179	0.184	0.155	0.164	0.140
27	0.391	0.316	0.262	0.215	0.209	0.173	0.179	0.150	0.159	0.135
28	0.382	0.308	0.255	0.209	0.202	0.168	0.173	0.146	0.154	0.131
29	0.372	0.300	0.248	0.203	0.196	0.164	0.168	0.142	0.150	0.127
30	0.363	0.203	0.241	0.198	0.191	0.159	0.164	0.138	0.145	0.124
31	0.355	0.286	0.235	0.193	0.186	0.155	0.159	0.134	0.141	0.120
32	0.347	0.280	0.229	0.188	0.181	0.151	0.155	0.131	0.138	0.117
33	0.339	0.273	0.224	0.184	0.177	0.147	0.151	0.127	0.134	0.114
34	0.332	0.267	0.218	0.179	0.172	0.144	0.147	0.124	0.131	0.111
35	0.325	0.262	0.213	0.175	0.168	0.140	0.144	0.121	0.127	0.108
36	0.318	0.256	0.208	0.172	0.165	0.137	0.140	0.118	0.124	0.106
37	0.312	0.251	0.204	0.168	0.161	0.134	0.137	0.116	0.121	0.103
38	0.306	0.246	0.200	0.164	0.157	0.131	0.134	0.113	0.119	0.101
39	0.300	0.242	0.196	0.161	0.154	0.129	0.131	0.111	0.116	0.099
40	0.294	0.237	0.192	0.158	0.151	0.126	0.128	0.108	0.114	0.097

注：p=给定水平下的实验室数；n=每个单元的测试重复数。

25.2.6.3 单因素方差分析法在标准物质均匀性评估中的应用

1. 单因素方差分析法

假定有 a 个标准物质单元，对单元 i 进行了 n_i 次测量（$i=1,\cdots,a$），对于简单的标准物质瓶间均匀性评估，所有单元的重复测量次数 n_j 都相同，即 $n_1=n_2=\cdots=n_a=n_0$。可采用单因素方差分析法进行分析。统计模型如下：

$$x_{ij}=\mu+\delta_i+\varepsilon_{ij}$$

式中，x_{ij} 为第 i 个单元的第 j 个观测值；μ 为所有可能结果总体（观测结果 x_{ij} 假定从中产生）的（真）均值；δ_i 为单元 i 对结果的影响效应，即单元 i 与 μ 的（真）偏差；ε_{ij} 为第 i 个单元的第 j 个观测值的随机误差，又称残差项。

$$MS_{\text{between}} = \frac{\sum_{i=1}^{a} n_i (\bar{x}_i - \bar{\bar{x}})^2}{a-1}, \text{自由度} \nu = a-1$$

$$MS_{\text{within}} = \frac{\sum_{i=1}^{a} \sum_{j=1}^{n_i} (x_{ij} - \bar{x}_i)^2}{\sum_{i=1}^{a} n_i - a}, \text{自由度} \nu = \sum_{i=1}^{a} n_i - a$$

其中：

$$\bar{x}_i = \frac{\sum_{j=1}^{n_i} x_{ij}}{n_i}$$

$$\bar{\bar{x}} = \frac{1}{\sum_{i=1}^{a} n_i} \sum_{i=1}^{a} \sum_{j=1}^{n_i} x_{ij}$$

可由组间均方 MS_{between}、组内均方 MS_{within} 以及 n_0 计算单元间标准偏差 s_{bb}，并作为单元间不均匀性引入的不确定度分量 u_{bb}：

$$s_{bb}^2 = \max\left(\frac{MS_{\text{between}} - MS_{\text{within}}}{n_0}, 0\right)$$

其中：

$$n_0 = \frac{1}{a-1}\left[\sum_{i=1}^{a} n_i - \frac{\sum_{i=1}^{a} n_i^2}{\sum_{i=1}^{a} n_i}\right]$$

例：研究对象为土壤中铬标准物质。单元（瓶）间均匀性研究数据见表 25-19。

表 25-19　土壤中铬的单元（瓶）间均匀性评估测量数据　　　　（单位：mg/kg）

瓶号	结果 No.1	结果 No.2	结果 No.3
1	121.30	128.74	119.91
2	120.87	121.32	119.24
3	122.44	122.96	123.45
4	117.60	119.66	118.96
5	110.65	112.34	110.29
6	117.29	120.79	121.42
7	115.27	121.45	117.48
8	118.96	123.78	123.29
9	118.67	116.67	114.58
10	126.24	123.51	126.20
11	128.65	122.02	121.93
12	126.84	124.72	123.14
13	122.61	128.48	126.20
14	118.95	123.82	118.11
15	118.74	118.23	117.38
16	119.74	121.78	121.01
17	121.21	123.28	116.38

续表

瓶号	结果 No.1	结果 No.2	结果 No.3
18	129.30	124.10	122.02
19	136.81	129.80	128.47
20	127.81	117.66	122.90

这些数据也可以用组平均值、标准偏差及组内样本数来代表。表 25-20 是根据表 25-19 计算得到的每单元（瓶）测量结果平均值、标准偏差和测量次数。表 25-21 为土壤中铬单元（瓶）间均匀性研究的方差分析（ANOVA）表。

表 25-20 每瓶测量结果平均值、方差和测量次数

瓶号	平均值（mg/kg）	方差	测量次数
1	123.32	22.54	3
2	120.48	1.20	3
3	122.95	0.26	3
4	118.74	1.10	3
5	111.09	1.20	3
6	119.83	4.95	3
7	118.07	9.81	3
8	122.01	7.04	3
9	116.64	4.18	3
10	125.32	2.45	3
11	124.20	14.85	3
12	124.90	3.45	3
13	125.76	8.76	3
14	120.29	9.50	3
15	118.12	0.47	3
16	120.84	1.06	3
17	120.29	12.54	3
18	125.14	14.06	3
19	131.69	20.08	3
20	122.79	25.76	3

表 25-21 土壤中铬单元（瓶）间均匀性研究的方差分析（ANOVA）表

变差来源	平方和（sum of squares）	自由度（Df）	均方 M（mean square）
单元间	1037.1	19	54.59
单元内	330.5	40	8.26
总和	1367.6	59	

单元间方差可用下式计算：

$$s_{\text{between}}^2 = \frac{MS_{\text{between}} - MS_{\text{within}}}{n_0} = \frac{54.59 - 8.26}{3} = 15.44 \text{ mg}^2/\text{kg}^2$$

单元间标准偏差是该方差的平方根：

$$s_{\text{bb}} = \sqrt{15.44} = 3.93 \text{ mg/kg}$$

重复性标准偏差可由 MS_{within} 计算得到

$$s_r = \sqrt{MS_{\text{within}}} = \sqrt{8.26} = 2.87 \text{ mg/kg}$$

2. 测量方法重复性欠佳时的单元间均匀性研究

研究对象为猪肾脏组织中 γ-谷氨酰转移酶标准物质。采用氯化钠溶液稀释猪肾组织并匀浆。反

复沉淀后，先后采用 DEAE-Trisacryl 层析柱和羟磷灰石色谱柱进行纯化。基体选用不会改变部分纯化酶的催化活性的牛血清白蛋白（浓度为 60 g/L）。分装过程通过定期采集安瓿瓶（共 101 瓶）并称重的方法进行核查，未发现分装过程存在趋势。最后将样品冻干。

通过检查样品中的污染酶、纯度、含水率及残留氧含量，证明其适合作为标准物质原料。

为了量化单元间变动性，对 20 个安瓿样品各进行 6 次测量，数据按安瓿分组，进行方差（ANOVA）分析，用于计算单元内标准偏差（s_r）和单元间标准偏差（s_{bb}）。表 25-22 给出了包括均方、标准偏差估计值在内的方差分析结果。

表 25-22 猪肾脏组织分析结果

平均值	67.78 IU/L
$MS_{between}$	1.76 IU²/L²
MS_{within}	1.63 IU²/L²
s_r	1.28 IU/L（1.88%）
s_{bb}	0.147 IU/L（0.22%）
u'_{bb}	0.196 IU/L（0.29%）

单元间标准偏差用下式计算：

$$s_{bb}^2 = \max\left(\frac{MS_{between} - MS_{within}}{n_0}, 0\right) = \frac{1.76 - 1.63}{6} = 0.0217 \text{ IU}^2/\text{L}^2$$

则单元间标准偏差 s_{bb} 的估计值为 0.147 IU/L，相对标准偏差为 0.22%。组内标准偏差的自由度为 100。重复性标准偏差可由 MS_{within} 计算：

$$s_r = \sqrt{MS_{within}} = \sqrt{1.63} = 1.28 \text{ IU/L}$$

该标准偏差过大，无法为低至 0.22% 的单元间相对不均匀性提供充足证据。因此，采用更为保守的方式计算由单元间不均匀性引入的不确定度：

$$u'_{bb} = \sqrt{\frac{MS_{within}}{n_0}} \cdot \sqrt[4]{\frac{2}{v_{MS_{within}}}} = \sqrt{\frac{1.63}{6}} \cdot \sqrt[4]{\frac{2}{100}} = 0.196 \text{ IU/L}$$

该表达式可看作单元内标准偏差的标准不确定度的估计值。建议采用 s_{bb} 和 u'_{bb} 中的较大值，即 0.196 IU/L 作为与单元间变动性相关的不确定度。

25.2.6.4 一元线性回归在标准物质稳定性检验中的应用

常采用的线性拟合模型可由下式表示：

$$Y = \beta_0 + \beta_1 X + \varepsilon$$

式中，β_0 为截距；β_1 为斜率；ε 为随机误差项，通常假定为平均值为 0 的正态分布。

假定有 n 对 (X,Y) 的观测值 (x_i, y_i)，x_i 代表观测时间，y_i 代表所关注特性的观测值，每对 (x_i, y_i) 通过下式相关：

$$y_i = \beta_0 + \beta_1 x_i + \varepsilon_i$$

可通过线性最小二乘法计算得到截距 β_0 和斜率 β_1 真值的估计值 b_0 和 b_1，以及相应的标准偏差 $s(b_0)$ 和 $s(b_1)$。

b_1 和 b_0 计算公式如下：

$$b_1 = \frac{\sum_{i=1}^{n}(x_i - \bar{x})(y_i - \bar{y})}{\sum_{i=1}^{n}(x_i - \bar{x})^2} \quad b_0 = \bar{y} - b_1 \bar{x}$$

式中，$\bar{x}$ 为所有时间点的平均值，$\bar{y}$ 为所有观测值的平均值。

b_1 的标准偏差 $s(b_1)$ 由下式给出：

$$s(b_1) = \frac{S}{\sqrt{\sum_{i=1}^{n}(x_i - \bar{x})^2}}$$

式中，s 为直线上每点的标准偏差，计算如下：

$$s^2 = \frac{\sum_{i=1}^{n}(y_i - b_0 - b_1 x_i)^2}{n-2}$$

继而可得到 b_0 的标准偏差 $s(b_0)$：

$$s(b_0) = s(b_1)\sqrt{\frac{\sum_{i=1}^{n} x_i}{n}}$$

典型的对线性模型假设的评估包括：

（1）计算残差 $[r_i = y_i - (b_0 + b_1 x_i)]$，对时间绘图，然后检查弯曲程度、轮次间效应、不同时间点在分散性上的显著差异或异常值；

（2）绘制残差的正态图（或 Q-Q 图），检查是否存在非正态性；

（3）作为选项，可对残差的组间显著性差异、不同时间点在分散性上的显著差异或非正态性进行统计学检验。

在通过以上评估确认满足基本回归假设的有效性后，基于斜率 b_1 的标准偏差 $s(b_1)$，可用 t 检验法判断 b_1 与 0 的显著差异。采用下式计算 t 统计值：

$$t_{b_1} = \frac{b_1}{s(b_1)}$$

将其与 95% 置信概率下、自由度为 $n-2$ 的双尾学生 t 分布临界值比较，若 t_{b_1} 大于该临界值，则表明斜率 b_1 与 0 有显著性差异，观察到了不稳定性趋势。也可用 F 检验来进行显著性判断。

例：对土壤中铬标准物质进行稳定性研究。实验数据列于表 25-23。

表 25-23　土壤中铬的稳定性数据

时间（月）	铬的含量测试结果平均值（mg/kg）
0	97.76
12	101.23
24	102.14
36	97.72

由于没有物理或化学模型能够真实地描述土壤中铬的降解机理，故可选用线性模型作为该候选标准物质的经验模型。理想的预期是直线的截距（在不确定度范围内）等于该标准物质的定值结果；同时，直线的斜率趋近于零。

该例采用每个时间点铬含量的测试结果平均值进行回归。计算斜率和截距的估计值 b_1、b_0：

$$b_1 = \frac{\sum_{i=1}^{n}(x_i - \bar{x})(y_i - \bar{y})}{\sum_{i=1}^{n}(x_i - \bar{x})^2} = \frac{4.74}{720} = 0.006\,583$$

$$b_0 = \bar{y} - b_1 \bar{x} = 99.7125 - (0.006\,583 \times 18) = 99.594$$

直线上每点的标准偏差可用下式计算：

$$s^2 = \frac{\sum_{i=1}^{n}(y_i - b_0 - b_1 x_i)^2}{n-2} = \frac{15.94}{2} = 7.973$$

取平方根，得 $s=2.8237$ mg/kg，与斜率相关的不确定度为

$$s(b_1) = \frac{S}{\sqrt{\sum_{i=1}^{n}(x_i - \overline{x})^2}} = \frac{2.8237}{\sqrt{720}} = 0.105\,233$$

在自由度为 $n-2=2$、置信水平 $p=0.95$（95%的显著水平）条件下，t 检验的临界值等于 4.30。由于：

$$|\beta_1| < t_{0.95, n-2} \cdot s(\beta_1)$$

故斜率并不显著，因此，没有观察到不稳定性。

根据标准物质的计划寿命设置证书有效期为 36 个月，长期稳定性引起的不确定度贡献为

$$u_{lts} = s(b_1) * t_{cert} = 0.105\,233 \times 36 = 3.78 \text{ mg/kg}$$

本例受到稳定性数据质量较差因素的制约。可以预计，完全由土壤中铬的不稳定性产生的影响要小得多。

25.3　标准物质在典型领域的应用

25.3.1　标准物质在仪器性能评价及检定、校准中的应用

25.3.1.1　标准物质在仪器性能评价中的应用

仪器性能评价的对象是新仪器，所谓"新"仪器可以在两个层级定义，一种是完全的"新"，原理、构造都具有独创性，市场上没有同类产品的"新"仪器，这类仪器可以简称为"新研发"；另一种是对于生产厂家本身而言的"新"仪器，市场上有同类产品，但对研发企业而言属于"新产品"。两种情况在进行仪器性能评价时既有区别也有相似，相似在于一种新的分析仪器研发出来，在投入市场之前对其功用、性能等方面需要做出全面的判断和检验，这种需求是相似的。区别在于"新研发"如何进行性能评价，需要进行大量测量实验，数据采集，制定出合理的评价方法，而"新产品"则有更多可以借鉴参考的评价方法，例如：型式评价大纲、仪器国标等。

针对化学分析类设备而言，仪器性能评价的最主要的目的是准确地评价分析仪器的测量能力。根据测量对象的不同、测量方法的差异，在具体的评价指标、评价项目和评价方法上需要有针对性地制定评价细则。参考 RB/T160—2017《分析化学仪器设备验证与综合评价指南》内容，对化学分析类仪器性能评价主要的评价内容包括：稳定性、基线噪声、检出限、线性范围、测量准确度、测量重复性、测量复现性等几个方面。"新研发"仪器性能评价的主要评价项目也必然涵盖在内，这些指标在实际测量过程中大部分都需要采用标准物质来进行测量，与普通样品相比，标准物质的可溯源性无疑对评价结果的可溯源提供有力的技术支撑。在已经颁布的光谱类仪器型式评价大纲中，标准物质均被明确作为计量器具规范使用。由此可见标准物质在仪器性能评价中具有重要作用，是对仪器性能评价结果可比、可溯源的有效保证。

25.3.1.2　检定校准用标准物质现状（原子光谱类）

开展化学计量的检定和校准，保证量值准确、一致，是化学计量工作的重要任务之一，其技术依据是计量检定规程或计量校准规范，由于化学计量本身的特殊性，标准物质在标准器组成中通常都作

为主标准器使用。原子光谱类仪器作为无机元素分析的主要检测手段在各类检测实验室中均占据重要地位，常见的原子光谱类仪器大部分都有对应的检定规程或校准规范，同时也有按规程或规范要求所研制的配套标准物质。表 25-24 即列出了几种典型原子光谱类设备及其对应的规程、规范及标准物质。

表 25-24　设备对应规程、规范及标准物质

设备名称	规程名称及编号	标准物质名称及编号
原子吸收分光光度计	原子吸收分光光度计检定规程 JJG 694—2009	原子吸收检定用标准物质 GBW(E) 130079
火焰光度计	火焰光度计检定规程 JJG 630—2007	火焰光度计检定用标准物质 GBW(E) 130109
原子荧光光度计	原子荧光光度计检定规程 JJG 939—2009	原子荧光光度计用砷、锑混合溶液标准物质 GBW(E) 130537～130540
四极杆电感耦合等离子体质谱仪	四极杆电感耦合等离子体质谱仪校准规范 JJF 1159—2006	ICP-MS 仪器校准用标准物质 GBW(E) 130242～130244
发射光谱仪	发射光谱仪检定规程 JJG 768—2005	ICP 光谱仪检定用标准物质 GBW(E) 130286～130289
液相色谱-原子荧光联用仪	液相色谱-原子荧光联用仪检定规程 JJG 1151—2018	砷形态混合溶液标准物质 GBW(E)082204

从表 25-24 可以看出，当前的化学计量已经越来越规范化、标准化，逐渐形成了由新设备开始市场化→被广泛使用→计量机构开展对新设备仪器性能的研究→对新设备的计量要求做出规范→制定出针对该类设备的规程（规范）；与此同时研制出与新规程（规范）配套的新标准物质，从而实现该标准物质最终对该设备的计量检定（校准）这样一个完整的计量溯源链，见图 25-11。

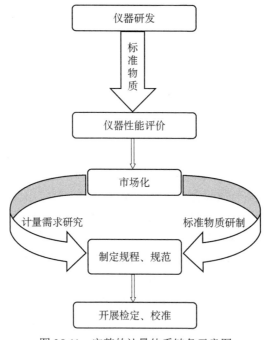

图 25-11　完整的计量体系链条示意图

25.3.2　标准物质在食品安全及环境保护中的应用

25.3.2.1　标准物质在食品安全中的应用

标准物质和食品安全质量控制之间有着紧密的联系。标准物质的正确选择与使用，能为产品的检测提供良好的标准，在实际的操作过程中，能有效地进行分析仪器的检定、校准，能对测量方法进行有效性评价等。

标准物质主要应用于检测仪器的校准、评价测量方法、评价测量结果的准确度、监控分析结果（质量控制）、产品评价与仲裁，同时还能对工作人员的技术水平进行检验与考核。

1. 标准物质在食品安全测量方法评价中的应用

食品基体作为较为复杂的检测对象，其前处理过程是否有效、检测方法是否准确均对检测的结果造成直接影响。因此在对新基体开展新目标物检测或新方法建立时，必须使用标准物质对其准确性进行评价。通过分析实验室检测结果与标准物质证书值间差异大小，可以判定新方法较标物定值时方法是否足够准确。

2. 标准物质在食品安全测量结果准确度评价中的应用

测量审核作为能力验证计划的有效补充，是衡量实验室测量结果准确度评价的重要技术依据。实验室通过对被测样品进行实际测试，将测试结果与参考值进行比较，从而获得测量审核的判定结果。使用标准物质作为考核样品，不但可以确保被测样品的均匀性及稳定性，还能提供更为准确的参考值以及不确定度，从而有效确保测量审核判定结果的准确性。

3. 标准物质在结果仲裁中的应用

在日常食品安全检测过程中，当两家检测实验室的检测结果出现较大差异，且各方意见无法统一时，一般会由国家级实验室进行试验，并把试验结果作为最终结果，即仲裁结果。由于仲裁试验责任重大，因此对检测方法、检测准确性的要求格外严格。使用相似基体的标准物质，首先可以对检测方法是否可行进行评价，其次借由标准物质测量过程中出现的问题，为实际样品的定值规避风险及不确定因素，从而得到较高的试验精度。

25.3.2.2　标准物质在环境保护中的应用

中国计量科学研究院多年来针对水、土壤/沉积物、植物、海洋生物等环境要素，运用质谱、光谱、色谱以及联用技术开展元素及形态的准确测定方法研究和标准物质研制，典型元素和形态标准物质见图 25-12；自 2005 年起参加和组织了 20 余项元素和形态分析的国际或区域比对，参加的典型元素国际比对见图 25-13，均取得等效一致；在此基础上，构建多种技术手段结合的核心测量能力和高层级量值溯源体系，结合不确定度评定等方法参数评估，建立了水、沉积物、海洋生物、燃油、植物、RoHS 检测相关产品中多种重金属和汞、砷、锡和硒元素形态校准与检测能力（CMC）共 67 项，并在国际关键比对数据库（KCDB）中公布。元素和形态溶液、基体标准物质为环境和食品领域形态分析、相关标准制修订中方法验证以及复杂基体中元素和形态检测过程中的方法确认提供了有力的量值溯源保证。

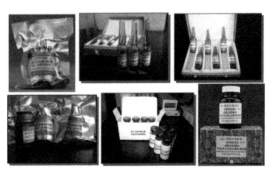

图 25-12　元素与形态标准物质

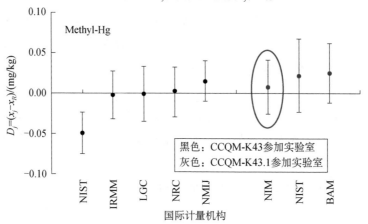
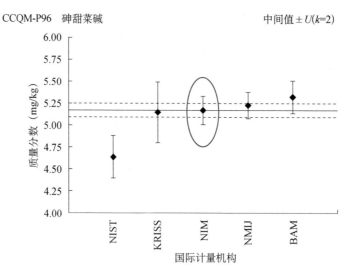
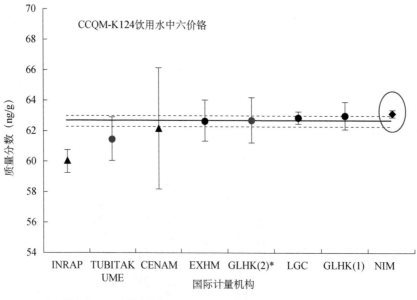

图 25-13　参加的元素形态国际比对

为更好地支持环境、食品相关标准的制定和实施，中国计量科学研究院在原国家质检总局食品安全专项（ASPAQ0807等）、科技计划项目（AKY1324等）等项目支持下，自2003年起即开展了元素形态分析溶液标准物质和复杂环境、食品等基体中元素和形态标准物质的研制。基于液相色谱、气相色谱与原子光谱联用技术，建立了汞、砷、锡、硒、铅、铬、溴等多种元素形态分析权威测量方法；形成了同位素稀释质谱、库仑分析准确测定元素总量与HPLC-ICP-MS、GC-ICP-MS测定形态纯度的有机金属化合物纯度测定方法，缩短了纯度分析溯源链，研制了5个系列20种元素形态溶液标准物质，解决了典型有机汞、砷、硒、锡等典型元素形态的量值溯源问题；研制了鱼肉中总汞和甲基汞、牡蛎中元素和有机锡形态、沉积物中元素和痕量甲基汞形态、冻干人尿中砷形态等基体标准物质。以沉积物中元素和痕量甲基汞为例，通过对ICP-MS测定过程中基体效应、光谱和氧化物干扰消除、流程空白严格控制等建立了基于同位素稀释质谱法的Pb、Hg、Cd、Cr、Cu、Zn等元素高准确度测定方法，将标准物质中上述元素的不确定度降低至2%。参加了"CCQM-K127土壤中污染元素和其他元素测定"国际比对，测定结果等效一致，沉积物中13种元素进入校准与测量能力国际关键比对数据库；将高灵敏度的吹扫捕集-气相色谱-电感耦合等离子体质谱（PT-GC-ICP-MS）与高准确度的同位素稀释质谱法结合，通过优化吹扫捕集条件、稀释剂与样品平衡时间等条件，实现了沉积物中超痕量甲基汞（~5 ng/g），低至6%测量扩展不确定度（$k=2$）的准确测定，标准物质中甲基汞特性量值扩展不确定度（$k=2$）小于14%，成为目前世界上甲基汞含量最低的基体标准物质，并参加了土壤中甲基汞相关标准制定过程中的方法验证。

多年来研制的标准物质为环境监测提供量值溯源标准的同时，形成的校准与检测能力有力支持了环境领域检测、能力验证、相关标准的制修订、环境监测专业技术人员大比武等对于能力评价与考核工作。自2012年以来，为相关环境监测及疾控系统提供水质、土壤监测能力验证和技术人员大比武样品和定值服务15次，参加实验室累计5800余家，为第三方检测实验室提供测量审核服务20次，为水、土壤污染调查和环境监测数据质量提升提供了有力支持。

25.3.3 β淀粉样多肽（Aβ）含量标准物质在临床阿尔茨海默病诊断领域的应用

阿尔茨海默病（Alzheimer's disease，AD）是不可逆的神经退行性疾病，随着人口的老龄化，AD的发病率越来越高，AD病的致病机理和临床治疗已引起了广泛关注。众多临床研究表明，血液、脑脊液和脑组织内的Aβ水平异常与AD的病程进展密切相关，Aβ已成为目前研究AD的重要生物标志物之一。因此，Aβ的高准确度测量方法建立和相关标准物质的研制对于AD症的诊断、治疗和药物研发等方面至关重要。

当前，文献中报道的临床检验中针对重组β淀粉样多肽的定量方法主要包括酶联免疫法、电化学方法、LC-MS/MS法、毛细管电泳结合紫外等。然而，这些方法大都基于免疫学原理，不同厂家甚至不同批次的试剂盒均会对测量结果产生偏差，导致检测结果的不一致性。为了保证临床检验中AD病标志物检测结果的准确性和可溯源性，在ISO 17025中，参考测量程序中使用的已有认证标准物质的质量保证是至关重要的先决条件。欧洲体外诊断（IVD）指令要求市场上所有用于诊断的标准物质均可追溯到现有的高阶参考材料或程序。在肽或蛋白质定量中，由于可通过同位素稀释ICP-MS对硫的准确测量从而实现目标蛋白质的绝对定量。该方法相对于目前基于氨基酸水解的同位素稀释法更为准确可靠，且可直接溯源到SI单位，其在蛋白质标准物质，尤其是大分子纯度标准物质定值中具有重要的应用前景。在目前Aβ的标准物质研制中，虽然文献中

报道了人脑脊液中 $A\beta_{42}$ 有证标准物质 ERM®-DA480（浓度 0.45 μg/L）、ERM®-DA481（浓度 0.72 μg/L）和 ERM®-DA482（浓度 1.22 μg/L），但还未曾有过对 Aβ 纯品标准物质的报道，Aβ 纯品标准物质位于溯源链的顶端，是基体标准物质的溯源源头。因此，亟需建立具有可溯源到 SI 单位、高准确度和高精密度纯 β 淀粉样多肽的绝对定量参考测量方法和标准物质，以满足开展高端溯源研究的需求。

中国计量科学研究院在国家重点研发计划项目课题"高准确度原位微区定量与成像计量溯源技术研究与临床应用"（2017YFF0205402）中。采用氨基酸水解同位素稀释质谱法和基于硫元素的同位素稀释质谱法作为两种独立参考方法对两种阿尔茨海默病标志物 $A\beta_{40}$ 和 $A\beta_{42}$ 进行定值。在氨基酸水解法中，选取的氨基酸以苯丙氨酸（Phe）、丙氨酸（Ala）、亮氨酸（Leu）为目标氨基酸，通过相应的同位素标记（^{13}C 和 ^{15}N 标记）的氨基酸对样品中的 Phe、Ala 和 Leu 进行定量，并根据样品中的氨基酸序列实现对 $A\beta_{40}$ 和 $A\beta_{42}$ 含量的定值（图 25-14）。在基于硫元素的同位素稀释法中，通过 HPLC-HR-ICP-MS 联用技术，采用柱后在线连续添加已知的浓度 ^{34}S 同位素稀释剂，通过在线连续监测 $^{32}S/^{34}S$ 的信号强度随时间变化，积分同位素信号强度与时间坐标轴形成的峰面积，结合同位素稀释公式，计算出硫元素的含量，进一步结合蛋白质的分子量及所含硫元素的含量，最终实现 $A\beta_{40}$ 和 $A\beta_{42}$ 标准物质浓度的定值。两种独立的定值方法结果等效一致，采用两种方法结果的平均值作为标准物质量值。该标准物质的量值和扩展不确定度分别为：$A\beta_{40}$：(7.58 ± 0.30) μg/g 和 $A\beta_{42}$：(7.62 ± 0.30) μg/g。该标准物质已经过迈克生物股份有限公司、吉林大学等单位应用，量值准确可靠，不确定度评定合理，可用于 Aβ 检测的溯源标准，还可用于该标志物检测方法的验证和实验室质量控制等领域。

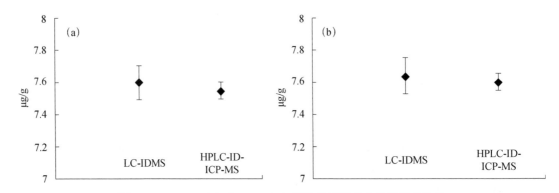

图 25-14　$A\beta_{40}$（a）和 $A\beta_{42}$（b）标准物质两种定值结果的比较

25.4　展　　望

如上所述，原子光谱技术不仅在食品、环境等传统领域的标准物质研究中发挥了重要作用，而且在生物大分子测量、原位微区分析等新型学科的标准物质定值技术中展现出新的活力，为保证传统和新兴领域中的准确测量和质量控制提供了重要手段。未来原子光谱技术在化学计量和标准物质技术中的相关研究可以从以下方面突破：

（1）目前的同位素标准物质大都为溶液基体，针对复杂基体中特定同位素分析的标准物质依然较少；

（2）针对大数据计量与人工智能，开发适用于原子光谱分析的多组学、多成分量标准物质；

（3）原子光谱技术大都仅限于元素或同位素分析，对于生命科学领域复杂生物基体中的大分子测量方面的计量技术还很不足。因此有必要与其他分离分析技术相结合，实现多种复杂目标物的准确测量，研制相关的标准物质；

（4）针对原子光谱原位成像技术，通常LA-ICP-MS的分辨率只能达到微米级别，但对生物亚器官组织或细胞的元素成像还远远不够，如何实现亚细胞级别的元素成像还有待于技术突破；

（5）基于元素成像，针对纳米颗粒的LA-ICP-MS分析成像技术亟待推进，特别是LA-SP-ICP-MS技术的发展，将有助于纳米医学与健康领域的进步。

参 考 文 献

[1] Cai J, Ren T X, Lu J X, Wu J H, Mao D Q, Li W D, Zhang Y, Li M, Piao J H, Yang L C, Ma Y X, Wang J, Yang X G. Physiologic requirement for iron in pregnant women, assessed using the stable isotope tracer technique. Nutrition & Metabolism, 2020, 17 (1): 33-42.

[2] Putz M, Piper T, Thevis M. Identification of trenbolone metabolites using hydrogen isotope ratio mass spectrometry and liquid chromatography/high accuracy/high resolution mass spectrometry for doping control analysis. Frontiers in Chemistry, 2020, 8: 1-15.

[3] Bacon J R, Butler O T, Cairns W R L, Cook J M, Davidson C M, Cavoura O, Mertz-Kraus R. Atomic spectrometry update: A review of advances in environmental analysis. Journal of Analytical Atomic Spectrometry, 2020, 35 (1): 9-53.

[4] Brand W A, Coplen T B, Vogl J, Rosner M, Prohaska T. Assessment of international reference materials for isotope-ratio analysis (IUPAC Technical Report). Pure and Applied Chemistry, 2014, 86 (3): 425-467.

[5] 中国计量科学研究院. JJF1508—2015. 同位素丰度测量基准方法. 中华人民共和国国家计量技术规范: 2015.

[6] Kipphardt H, Matschat R, Rienitz O, Schiel D, Gernand W, Oeter D. Traceability system for elemental analysis. Accreditation and Quality Assurance, 2006, 10 (11): 633-639.

[7] Kipphardt H, Matschat R, Panne U. Metrology in chemistry: A rocky road. Mikrochimica Acta, 2008, 162: 35-41.

[8] Rivier C, Labarraque G, Marschal A. Proceedings of the 9th International Metrology Congress. Bordeaux, France, 1999.

[9] Gopala A, Kipphardt H, Matschat R, Panne U. Process methodology for the small scale production of m6N5 purity zinc using a resistance heated vacuum distillation system. Materials Chemistry and Physics, 2010, 122 (1): 151-155.

[10] Ghosh K, Mani V N, Dhar S. Numerical study and experimental investigation of zone refining in ultra-high purification of gallium and its use in the growth of GaAs epitaxial layers. Journal of Crystal Growth, 2009, 311 (6): 1521-1528.

[11] Park J, Jung Y, Kusumah P, Dilasari B, Ku H, Kim H, Kwon K, Lee C K. Room temperature magnesium electrorefining by using non-aqueous electrolyte. Metals and Materials International, 2016, 22 (5): 907-914.

[12] Mimura K, Matsumoto K, Isshiki M. Purification of hafnium by hydrogen plasma arc melting. Materials Transactions, 2011, 52 (2): 159-165.

[13] Matschat R, Heinrich H, Czerwensky M, Kuxenko S, Kipphardt H. Multielement trace determination in high purity advanced ceramics and high purity metals. Bulletin of Materials Science, 2005, 28 (4): 361-366.

[14] Zhou T, Richter S, Matschat R, Kipphardt H. Determination of the total purity of a high-purity silver material to be used as a primary standard for element determination. Accreditation and Quality Assurance, 2013, 18 (4): 341-349.

[15] Kipphardt H, Matschat R. Purity assessment for providing primary standards for elemental determination: A snap shot of

international comparability. Mikrochimica Acta, 2008, 162 (1): 269-275.

[16] Kipphardt H, Matschat R, Vogl J, Gusarova T, Czerwensky M, Heinrich H, Hioki A, Konopelko L A, Methven B, Miura T. Purity determination as needed for the realisation of primary standards for elemental determination: Status of international comparability. Accreditation and Quality Assurance, 2010, 15 (1): 29-37.

[17] Vogl J, Kipphardt H, Del Rocio Arvizu Torres M, Manzano J V L, Rodrigues J M, De Sena R C, Yim Y, Heo S W, Zhou T, Turk G C. Final report of the key comparison CCQM-K72: Purity of zinc with respect to six defined metallic analytes. Metrologia, 2014, 51: 08008.

[18] Vogl J, Kipphardt H, Richter S, Bremser W, Del Rocio Arvizu Torres M, Manzano J V L, Buzoianu M, Hill S, Petrov P, Goenagainfante H. Establishing comparability and compatibility in the purity assessment of high purity zinc as demonstrated by the CCQM-P149 intercomparison. Metrologia, 2018, 55 (2): 211-221.

[19] Pattberg S, Matschat R. Determination of trace impurities in high purity copper using sector-field ICP-MS: Continuous nebulization, flow injection analysis and laser ablation. Fresenius' Journal of Analytical Chemistry, 1999, 364 (5): 410-416.

[20] Zhang J Y, Zhou T, Tang Y C, Cui Y J, Li J Y. Determination of relative sensitivity factors of elements in high purity copper by doping-melting and doping-pressed methods using glow discharge mass spectrometry. Journal of Analytical Atomic Spectrometry, 2016, 31 (11): 2182-2191.

[21] Kipphardt H, Matschat R, Panne U. Development of SI traceable standards for element determination. Chimia, 2009, 63 (10): 637-639.

[22] Wang M, Feng W Y, Zhao Y L. ICP-MS-based strategies for protein quantification. Mass Spectrometry Reviews, 2010, 29: 326-348.

[23] Feng L X, Zhang D, Wang J, Li H M. Simultaneous quantification of proteins in human serum via sulfur and iron using HPLC coupled to post-column isotope dilution mass spectrometry. Analytical Methods, 2014, 6 (19): 7655-7662.

[24] Feng L X, Zhang D, Wang J, Shen D R, Li H M. A novel quantification strategy of transferrin and albumin in human serum by species-unspecific isotope dilution laser ablation inductively coupled plasma mass spectrometry (ICP-MS). Analytica Chimica Acta, 2015, 884: 19-25.

[25] Lee H S, Kim S H, Jeong J S, Lee Y M, Yim Y H. Sulfur-based absolute quantification of proteins using isotope dilution inductively coupled plasma mass spectrometry. Metrologia, 2015, 52 (5): 619-627.

[26] BIPM, Consultative Committee for Amount of Substance: Metrology in chemistry and biology (CCQM). https://www.bipm.org/utils/common/pdf/ CC/CCQM/CCQM24: 2018.

[27] 霍中, 冯流星, 李红梅, 熊金平. 基于硫同位素稀释质谱法的β淀粉样多肽绝对定量研究. 分析化学, 2020, 47: 1931-1937.

[28] Josephs R D, Daireaux A, Li M, Choteau T, Martos G, Westwood S, Wielgosz R I, Li H, Wang S, Feng L, Huang T, Pan M, Zhang T, Gonzalez-Rojano N, Balderas-Escamilla M, Perez-Castorena A, Perez-Urquiza M, Ün I, Bilsel M. Pilot study on peptide purity-synthetic oxytocin. Metrologia, 2020, 57: Number 1A.

[29] Linsinger T P J, Roebben G, Solans C, Ramsch R. Reference materials for measuring the size of nanoparticles. Trends in Analytical Chemistry, 2011, 30 (1): 18-27.

[30] Stefaniak A B, Hackley V A, Roebben G, Ehara K, Hankin S, Postek M T, Lynch I, Fu W E, Linsinger T P, Thunemann A F. Nanoscale reference materials for environmental, health and safety measurements: Needs, gaps and opportunities. Nanotoxicology, 2013, 7 (8): 1325-1337.

[31] Warheit D B, Borm P J, Hennes C, Lademann J. Testing strategies to establish the safety of nanomaterials: Conclusions of an ECETOC workshop. Inhalation Toxicology, 2007, 19 (8): 631-643.

[32] Grombe R, Charoud-Got J, Emteborg H, Linsinger T P, Seghers J, Wagner S, von der Kammer F, Hofmann T, Dudkiewicz A, Llinas M, Solans C, Lehner A, Allmaier G. Production of reference materials for the detection and size determination of silica nanoparticles in tomato soup. Analytical and Bioanalytical Chemistry, 2014, 406 (16): 3895-3907.

[33] Grombe R, Allmaier G, Charoud-Got J, Dudkiewicz A, Emteborg H, Hofmann T, Larsen E H, Lehner A, Llinàs M, Loeschner K, Mølhave K, Peters R J, Seghers J, Solans C, von der Kammer F, Wagner S, Weigel S. Linsinger T P J. Feasibility of the development of reference materials for the detection of Ag nanoparticles in food: neat dispersions and spiked chicken meat. Accreditation and Quality Assurance, 2015, 20 (1): 3-16.

[34] Degueldre C, Favarger P Y. Colloid analysis by single particle inductively coupled plasma-mass spectroscopy: A feasibility study. Colloids and Surfaces A: Physicochemical and Engineering Aspects, 2003, 217 (1-3): 137-142.

[35] Meermann B, Nischwitz V. ICP-MS for the analysis at the nanoscale: A tutorial review. Journal of Analytical Atomic Spectrometry, 2018, 33 (9): 1432-1468.

[36] Huynh K A, Siska E, Heithmar E, Tadjiki S, Pergantis S A. Detection and quantification of silver nanoparticles at environmentally relevant concentrations using asymmetric flow field-flow fractionation online with single particle inductively coupled plasma mass spectrometry. Analytical Chemistry, 2016, 88 (9): 4909-4916.

[37] Mozhayeva D, Engelhard C. Separation of silver nanoparticles with different coatings by capillary electrophoresis coupled to ICP-MS in single particle mode. Analytical Chemistry, 2017, 89 (18): 9767-9774.

[38] Linsinger T P, Peters R, Weigel S. International interlaboratory study for sizing and quantification of Ag nanoparticles in food simulants by single-particle ICPMS. Analytical and Bioanalytical Chemistry, 2014, 406 (16): 3835-3843.

[39] Montoro Bustos A R, Petersen E J, Possolo A, Winchester M R. Post hoc interlaboratory comparison of single particle ICP-MS size measurements of NIST gold nanoparticle reference materials. Analytical Chemistry, 2015, 87 (17): 8809-8817.

[40] Weigel S, Peters R, Loeschner K, Grombe R, Linsinger T P J. Results of an interlaboratory method performance study for the size determination and quantification of silver nanoparticles in chicken meat by single-particle inductively coupled plasma mass spectrometry (sp-ICP-MS). Analytical and Bioanalytical Chemistry, 2017, 409 (20): 4839-4848.

[41] Chao J B, Wang J R, Zhang J R. Accurate determination and characterization of gold nanoparticles based on single particle inductively coupled plasma mass spectrometry. Chinese Journal of Analytical Chemistry, 2020, 48 (7): 946-954.

[42] Bracciali L. Coupled zircon-rutile U-Pb chronology: LA ICP-MS dating, geological significance and applications to sediment provenance in the Eastern Himalayan-Indo-Burman Region. Geosciences, 2019, 9 (11): 467.

[43] Wedenbeck M, Alle P. Three natural zircon standards for U-Th-Pb Lu-Hf trace element and REE analysis. Geostandards Newsletter, 1995, 19 (1): 1-23.

[44] Sláma J, Košler J, Condon D J, Crowley J L, Gerdes A, Hanchar J M, Horstwood M S A, Morris G A, Nasdala L, Norberg N, Schaltegger U, Schoene B, Tubrett M N, Whitehouse M J. Plešovice zircon: A new natural reference material for U-Pb and Hf isotopic microanalysis. Chemical Geology, 2008, 249 (1-2): 1-35.

[45] Gain S E M, Gréau Y, Henry H, Belousova E, Dainis I, Griffin W L, O'Reilly S Y. Mud tank zircon: Long-term evaluation of a reference material for U-Pb dating, Hf‐isotope analysis and trace element analysis. Geostandards and Geoanalytical Research, 2019, 43 (3): 339-354.

[46] Jackson S E, Pearson N J, Griffin W L, Belousova E A. The application of laser ablation-inductively coupled plasma-mass spectrometry to *in situ* U-Pb zircon geochronology. Chemical Geology, 2004, 211: 47-69.

[47] Nasdala L, Hofmeister W, Norberg N, Mattinson J M, Corfu F, Dörr W, Kamo S L, Kennedy A K, Kronz A, Reiners P W, Frei D, Kosler J, Wan Y S, Götze J, Häger T, Kröner A, Valley J W. Zircon M257: A homogeneous natural reference material for the ion microprobe U-Pb analysis of zircon. Geostandards and Geoanalytical Research, 2008, 32: 247-265.

[48] Black L P, Kamo S L, Allen C M, Davis D W, Aleinikoff J N, Valley J W, Mundil R, Campbell I H, Korsch R J, Williams I S, Foudoulis C. Improved $^{206}Pb/^{238}U$ microprobe geochronology by the monitoring of a trace-element-related matrix effect; SHRIMP, ID-TIMS, ELA-ICP-MS and oxygen isotope documentation for a series of zircon standards. Chemical Geology, 2004, 205: 115-140.

[49] 唐国强, 龚冰, 刘宇, 侯可军, 胡兆初, 李献华, 杨岳衡, 李武显, 李秋立. Qinghu(清湖)锆石: 一个新的 U-Pb 年龄和 O, Hf 同位素微区分析工作标样. 科学通报, 2013, 58 (20): 1954-1961.

[50] Li X H, Long W G, Li Q L, Liu Y, Zheng Y F, Yang Y H, Chamberlain K R, Wan D F, Guo C H, Wang X C, Tao H. Penglai zircon megacrysts: A potential new working reference material for microbeam determination of Hf-O isotopes and U-Pb age. Geostands and Geoanalytical Research, 2010, 34 (2): 117-134.

[51] Zhang W, Hu Z C. Recent advances in sample preparation methods for elemental and isotopic analysis of geological samples. Spectrochimica Acta Part B: Atomic Spectroscopy, 2019, 160: 105690.

[52] Wu S T, Karius V, Schmidt B C, Simon K, Wörner G. Comparison of ultrafine powder pellet and flux-free fusion glass for bulk analysis of granitoids by laser ablation-inductively coupled plasma-mass spectrometry. Geostandards and Geoanalytical Research, 2018, 42: 575-591.

[53] Monsels D A, Van Bergen M J, Mason P R D. Determination of trace elements in bauxite using laser ablation inductively coupled plasma-pass spectrometry on lithium borate glass beads. Geostandards and Geoanalytical Research, 2018, 42: 239-251.

[54] Zhang C X, Hu Z C, Zhang W, Liu Y S, Zong K Q, Li M, Chen H H, Hu S H. Green and fast laser fusion technique for bulk silicate rock analysis by laser ablation-inductively coupled plasma mass spectrometry. Analytical Chemistry, 2016, 88 (20): 10088-10094.

[55] Mukherjee P K, Khanna P P, Saini N K. Rapid determination of trace and ultra trace level elements in diverse silicate rocks in pressed powder pellet targets by LA-ICP-MS using a matrix-independent protocol. Geostandards and Geoanalytical Research, 2014, 38: 363-379.

[56] 吴石头, 王亚平, 许春雪. 激光剥蚀电感耦合等离子体质谱元素微区分析标准物质研究进展. 岩矿测试, 2015, 34 (5): 503-511.

[57] National Institute of Standards and Technology. Trace Elements in Glass Standard Reference Material 610 612 614 616. 2012.

[58] Guillong M, Hametner K, Reusser E, Wilson S A, Günther D. Preliminary characterisation of new glass reference materials (GSA-1G, GSC-1G, GSD-1G and GSE-1G) by laser ablation-inductively coupled plasma-mass spectrometry using 193 nm, 213 nm and 266 nm wavelengths. Geostandards and Geoanalytical Research, 2005, 29 (3): 285-302.

[59] Dingwell D, Bagdassarov N, Bussod N. Magma Rheology. Mineralogical Association of Canada Short Course on Experiments at High Pressure and Applications to the Earth's Mantle, 1993: 131-196.

[60] Matschat R, Dette A, Guadagnino E. The certification of the mass fraction of arsenic(Ⅲ) oxide, barium oxide, cadmium oxide, cerium(Ⅳ) oxide, chloride, cobalt oxide, chromium(Ⅲ) oxide, copper (Ⅰ/) oxide, iron (Ⅲ) oxide, manganese(Ⅱ) oxide, molybdenum(Ⅵ) oxide, nickel(Ⅱ) oxide, lead(Ⅱ) oxide, antimony (Ⅲ) oxide, selenium, tin(Ⅳ) oxide, sulfur trioxide, strontium oxide, titanium(Ⅳ) oxide, vanadium(Ⅴ) oxide, zinc oxide, and zirconium(Ⅳ) oxide in soda lime glass, BAM—S005-A and BAM—S005-B, 2005.

[61] Hu M Y, Fan X T, Stoll B, Kuzmin D, Liu Y, Liu Y S, Sun W D, Wang G, Zhan X C, Jochum K P. Preliminary characterisation of new reference materials for microanalysis: Chinese geological standard glasses CGSG-1, CGSG-2, CGSG-4 and CGSG-5. Geostandards and Geoanalytical Research, 2011, 35 (2): 235-251.

[62] Wu S T, Wörner G, Jochum K P, Stoll B, Simon K, Kronz A. The preparation and preliminary characterisation of three

synthetic Andesite reference glass materials (ARM-1, ARM-2, ARM-3) for *in situ* microanalysis. Geostandards and Geoanalytical Research, 2019, 43 (4): 567-584.

[63] Feng Y T, Zhang W, Hu Z C, Liu Y S, Chen K, Fu J L, Xie J Y, Shi Q H. Development of sulfide reference materials for *in situ* platinum group elements and S-Pb isotope analyses by LA-(MC)-ICP-MS. Journal of Analytical Atomic Spectrometry, 2018, 33 (12): 2172-2183.

[64] Ke Y Q, Zhou J Z, Yi X Q, Sun Y J, Shao J F, You S Y, Wang W, Tang Y Z, Tu C Y. Development of REE-doped $CaWO_4$ single crystals as reference materials for *in situ* microanalysis of scheelite via LA-ICP-MS. Journal of Analytical Atomic Spectrometry, 2020, 35 (5): 886-895.

[65] Huang C, Wang H, Jiang J H, Ramezani J, Yang C, Zhang S B, Yang Y H, Xia X P, Feng L J, Lin J, Wang T T, Ma Q, He H Y, Xie L W, Wu S T. SA01-A proposed zircon reference material for microbeam U-Pb age and Hf-O isotopic determination. Geostandards and Geoanalytical Research, 2019, 44 (1): 103-123.

[66] Tian Y T, Vermeesch P, Danišík M, Condon D J, Chen W, Kohn B, Schwanethal J, Rittner M. LGC-1: A zircon reference material for in-situ (U-Th)/He dating. Chemical Geology, 2017, 454: 80-92.